Environmental Science

EARTH AS A LIVING PLANET

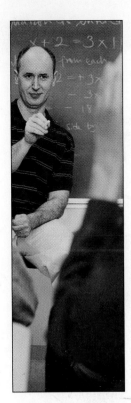

The next great environmental leader…

… may be sitting right in your classroom!
Every one of your students has the potential
to make a difference. And realizing that
potential starts right here, in your course.

When students succeed in your course—
when they stay on-task and make the breakthrough that turns confusion into
confidence—they are empowered to realize
the possibilities for greatness that lie within each of them. We know
your goal is to create an environment where students reach their full
potential and experience the exhilaration of academic success that will
last them a lifetime. *WileyPLUS* can help you reach that goal.

Wiley**PLUS** is an online suite of resources—including
the complete text—that will help your students:

- come to class better prepared for your lectures
- get immediate feedback and context-sensitive help on assignments and quizzes
- track their progress throughout the course

"I just wanted to say how much this program helped me
in studying… I was able to actually see my mistakes and
correct them. … I really think that other students should
have the chance to use *WileyPLUS*."

Ashlee Krisko, *Oakland University*

www.wiley.com/college/wileyplus

80% of students surveyed said it improved their understanding of the material.

FOR INSTRUCTORS

WileyPLUS is built around the activities you perform in your class each day. With WileyPLUS you can:

Prepare and Present
Create outstanding class presentations using a wealth of resources such as PowerPoint™ slides, image galleries, interactive simulations, and more. You can even add materials you have created yourself.

Create Assignments
Automate the assigning and grading of homework or quizzes by using the provided question banks, or by writing your own.

Track Student Progress
Keep track of your students' progress and analyze individual and overall class results.

Now Available with WebCT and Blackboard!

"It has been a great help, and I believe it has helped me to achieve a better grade."

Michael Morris,
Columbia Basin College

FOR STUDENTS

You have the potential to make a difference!
WileyPLUS is a powerful online system packed with features to help you make the most of your potential and get the best grade you can!

With WileyPLUS you get:

- A complete online version of your text and other study resources.

- Problem-solving help, instant grading, and feedback on your homework and quizzes.

- The ability to track your progress and grades throughout the term.

For more information on what *WileyPLUS* can do to help you and your students reach their potential, please visit www.wiley.com/college/*wileyplus*.

76% of students surveyed said it made them better prepared for tests. *

*Based on a survey of 972 student users of *WileyPLUS*

THE WILEY BICENTENNIAL—KNOWLEDGE FOR GENERATIONS

*E*ach generation has its unique needs and aspirations. When Charles Wiley first opened his small printing shop in lower Manhattan in 1807, it was a generation of boundless potential searching for an identity. And we were there, helping to define a new American literary tradition. Over half a century later, in the midst of the Second Industrial Revolution, it was a generation focused on building the future. Once again, we were there, supplying the critical scientific, technical, and engineering knowledge that helped frame the world. Throughout the 20th Century, and into the new millennium, nations began to reach out beyond their own borders and a new international community was born. Wiley was there, expanding its operations around the world to enable a global exchange of ideas, opinions, and know-how.

For 200 years, Wiley has been an integral part of each generation's journey, enabling the flow of information and understanding necessary to meet their needs and fulfill their aspirations. Today, bold new technologies are changing the way we live and learn. Wiley will be there, providing you the must-have knowledge you need to imagine new worlds, new possibilities, and new opportunities.

Generations come and go, but you can always count on Wiley to provide you the knowledge you need, when and where you need it!

WILLIAM J. PESCE
PRESIDENT AND CHIEF EXECUTIVE OFFICER

PETER BOOTH WILEY
CHAIRMAN OF THE BOARD

Environmental Science

EARTH AS A LIVING PLANET

SIXTH EDITION

DANIEL B. BOTKIN

Professor Emeritus
Department of Ecology, Evolution, and Marine Biology
University of California, Santa Barbara

President
The Center for the Study of the Environment
Santa Barbara, California

EDWARD A. KELLER

Professor of Environmental Studies and Earth Science
University of California, Santa Barbara

With assistance from:
Dorothy B. Rosenthal
Mel S. Manalis
Valery A. Rivera
Diane Perez-Botkin

BICENTENNIAL
1807
WILEY
2007
BICENTENNIAL

JOHN WILEY & SONS, INC.

About the Cover:
The forest canopy has been called "the last biotic frontier." Recent innovations in canopy access has opened this part of our living planet to exploration and scientific discovery. Professor Nalini Nadkarni, a forest ecologist at The Evergreen State College, and one of the pioneers of forest canopy research, uses modified mountain climbing techniques to safely ascend into tall tropical and temperate treetops. Her work has focused on understanding the ecological roles and importance of canopy dwelling plants and animals in rainforest ecosystems. Here, she climbs a Big leaf Maple tree (Acer macrophyllum) to study the ways in which branch dwelling mosses intercept nitrogen from rainfall and participate in whole ecosystem nutrient cycles.

Associate Editor Merillat Staat
Senior Production Editor Patricia McFadden
Senior Marketing Manager Clay Stone
Marketing Manager Ashaki Charles
Editorial Assistant Stephen Reiss
Senior Designer Kevin Murphy
Creative Director Harry Nolan
Senior Illustration Editor Sandra Rigby
Photo Editor Ellinor Wagner
Senior Media Editor Linda Muriello
Production Management Preparé, Inc.

This book was set in 10/12 Galliard by Prepare, Inc. and printed and bound by Von Hoffmann Corporation. The cover was printed by Von Hoffmann Corporation.

This book is printed on acid-free paper. ∞

ISBN 9780470049907

Printed in the United States of America

10 9 8 7 6 5 4 3 2 1

DEDICATIONS

For My Sister, Dorothy B. Rosenthal
who has been a source of inspiration, support, ideas, and
books to read, and is one of my harshest and best critics.
Dan Botkin

and

For Valery Rivera
who contributed so much to this book and
is a fountain of inspiration in our work and lives.
Ed Keller

Daniel B. Botkin is President of The Center for the Study of Environment, and Professor Emeritus of Ecology, Evolution, and Marine Biology, University of California, Santa Barbara, where he has been on the faculty since 1978, serving as Chairman of the Environmental Studies Program from 1978 to 1985. For more than three decades, Professor Botkin has been active in the application of ecological science to environmental management. He is the winner of the Mitchell International Prize for Sustainable Development and the Fernow Prize for International Forestry, and he has been elected to the California Environmental Hall of Fame.

Trained in physics and biology, Professor Botkin is a leader in the application of advanced technology to the study of the environment. The originator of widely used forest gap-models, he has conducted research on endangered species, characteristics of natural wilderness areas, the biosphere, and global environmental problems. During his career, Professor Botkin has advised the World Bank about tropical forests, biological diversity, and sustainability; the Rockefeller Foundation about global environmental issues; the government of Taiwan about approaches to solving environmental problems; and the state of California on the environmental effects of water diversion on Mono Lake. He served as the primary advisor to the National Geographic Society for its centennial edition map on "The Endangered Earth." He directed a study for the states of Oregon and California concerning salmon and their forested habitats.

He has published many articles and books about environmental issues. His latest books are *Beyond the Stoney Mountains: Nature in the American West from Lewis and Clark to Today* (Oxford University Press), *Strange Encounters: Adventures of a Renegade Naturalist* (Penguin/Tarcher), *The Blue Planet* (Wiley), *Our Natural History: The Lessons of Lewis and Clark* (Oxford University Press), *Discordant Harmonies: A New Ecology for the 21st Century* (Oxford University Press), and *Forest Dynamics: An Ecological Model* (Oxford University Press).

Professor Botkin was on the faculty of the Yale School of Forestry and Environmental Studies (1968–1974) and was a member of the staff of the Ecosystems Center at the Marine Biological Laboratory, Woods Hole, MA (1975–1977). He received a B.A. from the University of Rochester, an M.A. from the University of Wisconsin, and a Ph.D. from Rutgers University.

Edward A. Keller was chair of the Environmental Studies and Hydrologic Sciences Programs from 1993 to 1997 and is Professor of Earth Science at the University of California, Santa Barbara, where he teaches earth surface processes, environmental geology, environmental science, river processes, and engineering geology. Prior to joining the faculty at Santa Barbara, he taught geomorphology, environmental studies, and earth science at the University of North Carolina, Charlotte. He was the 1982–1983 Hartley Visiting Professor at the University of Southampton, a Visiting Fellow in 2000 at Emmanuel College of Cambridge University, England, and receipent of the Easterbrook Distinguished Scientist award from the Geological Society of America in 2004.

Professor Keller has focused his research efforts into three areas: studies of Quaternary stratigraphy and tectonics as they relate to earthquakes, active folding, and mountain building processes; hydrologic process and wildfire in the chaparral environment of Southern California; and physical habitat requirements for the endangered Southern California steelhead trout. He is the recipient of various Water Resources Research Center grants to study fluvial processes and U.S. Geological Survey and Southern California Earthquake Center grants to study earthquake hazards.

Professor Keller has published numerous papers and is the author of the textbooks *Environmental Geology*, *Introduction to Environmental Geology* and (with Nicholas Pinter) *Active Tectonics* (Prentice-Hall). He holds bachelor's degrees in both geology and mathematics from California State University, Fresno; an M.S. in geology from the University of California; and a Ph.D. in geology from Purdue University.

PREFACE

What Is Environmental Science?

Environmental science is a group of sciences that attempt to explain how life on the Earth is sustained, what leads to environmental problems, and how these problems can be solved.

Why Is This Study Important?

- We depend on our environment. People can live only in an environment with certain kinds of characteristics and within certain ranges of availability of resources. Because modern science and technology give us the power to affect the environment, we have to understand how the environment works, so that we can live within its constraints.

- People have always been fascinated with nature, which is, in its broadest view, our environment. As long as people have written, they have asked three questions about ourselves and nature:

    What is the nature like when it is undisturbed by people?

    What are the effects of people on nature?

    What are the effects of nature on people?

    Environmental science is our modern way of seeking answers to these questions.

- We enjoy our environment. To keep it enjoyable, we must understand it from a scientific viewpoint.

- Our environment improves the quality of our lives. A healthy environment can help us live longer and more fulfilling lives.

- It's just fascinating.

What Is the "Science" in Environmental Science?

Many sciences are important to environmental science. These include biology (especially ecology, that part of biology that deals with the relationships among living things and their environment), geology, hydrology, climatology, meteorology, oceanography, and soil science.

How Is Environmental Science Different from other Sciences?

- It involves many sciences.

- It includes sciences, but also involves related non-scientific fields that have to do with how we value the environment, from environmental philosophy to environmental economics.

- It deals with many topics that have great emotional effects on people, and therefore are subject to political debate and to strong feelings that often ignore scientific information.

What Is Your Role as a Student and as a Citizen?

Your role is to understand how to think through environmental issues so that you can arrive at your own decisions.

What Are the Professions That Grow out of Environmental Science?

Many professions have grown out of the modern concern with environment, or have been extended and augmented by modern environmental sciences. These include park, wildlife, and wilderness management; urban planning and design; landscape planning and design; conservation and sustainable use of our natural resources.

Goals of This Book

Environmental Science: Earth as a Living Planet provides an up-to-date introduction to the study of the environment. Information is presented in an interdisciplinary perspective necessary to deal successfully with environmental problems. The goal is to teach you, the student, how to think through environmental issues.

Critical Thinking

We must do more than simply identify and discuss environmental problems and solutions. To be effective, we must know what science is and is not. Then, we need to develop critical thinking skills. Critical thinking is so important that we have made it the focus of its own chapter, Chapter 2. With this in mind, we have also developed *Environmental Science* to present the material in a factual and unbiased format. Our goal is to help you think through the issues, not tell you what to think. To this purpose, at the end of each chapter, we present "Critical Thinking Issues." Critical thinking is further emphasized throughout the text in analytical discussions of topics, evaluation of perspectives, and integration of important themes, which are described in detail later.

Interdisciplinary Approach

The approach of *Environmental Science* is interdisciplinary in nature. Environmental science integrates many disciplines, including the natural sciences, in addition to fields such as anthropology, economics, history, sociology, and philosophy of the environment. Not only do we need the best ideas and information to deal successfully with our environmental problems, but we also must be aware of the cultural and historical contexts in which we make decisions about the environment. Thus, the field of environmental science also integrates the natural sciences with environmental law, and environmental impact, and environmental planning.

■ Themes

Our book is based on the philosophy that six threads of inquiry are of particular importance to environmental science. These key themes, called threads of inquiry, are woven throughout the book.

These six key themes are discussed in more detail in Chapter 1. They are also revisited at the end of each chapter and are emphasized in the Closer Look boxes, each of which is highlighted by an icon suggesting the major underlying theme of the discussion. In many cases, more than one theme is relevant.

Human Population

Underlying nearly all environmental problems is the rapidly increasing human population. Ultimately, we cannot expect to solve environmental problems unless the total number of people on Earth is an amount the environment can sustain. We believe that education is important to solving the population problem. As people become more educated, and as the rate of literacy increases, population growth tends to decrease.

Sustainability

Sustainability is a term that has gained popularity recently. Speaking generally, it means that a resource is used in such a way that it continues to be available. However, the term is used vaguely, and it is something experts are struggling to clarify. Some would define it as ensuring that future generations have equal opportunities to access the resources that our planet offers. Others would argue that sustainability refers to types of developments that are economically viable, do not harm the environment, and are socially just. We all agree that we must learn how to sustain our environmental resources so that they continue to provide benefits for people and other living things on our planet.

A Global Perspective

Until recently it was common to believe that human activity caused only local, or at most regional, environmental change. We now know that human activities can affect the environment globally. An emerging science known as Earth System Science seeks a basic understanding of how our planet's environment works as a global system. This understanding can then be applied to help solve global environmental problems. The emergence of Earth System Science has opened up a new area of inquiry for faculty and students.

The Urban World

An ever-growing number of people are living in urban areas. Unfortunately, our urban centers have long been neglected, and the quality of the urban environment has suffered. It is here that we experience the worst of air pollution, waste disposal problems, and other stresses on the environment. In the past we have centered our studies of the environment more on wilderness than the urban environment. In the future we must place greater focus on towns and cities as livable environments.

People and Nature

People seem to be always interested—amazed, fascinated, pleased, curious—in our environment. Why is it suitable for us? How can we keep it that way? We know that people and our civilizations are having major effects on the environment, from local ones (the street where you live) to the entire planet (we have created a hole in the Earth's ozone layer, which can affect us and many forms of life).

Science and Values

Finding solutions to environmental problems involves more than simply gathering facts and understanding the scientific issues of a particular problem. It also has much to do with our systems of values and issues of social justice. To solve our environmental problems, we must understand what our values are and which potential solutions are socially just. Then we can apply scientific knowledge about specific problems and find acceptable solutions.

Organization

Our text is divided into eight parts. Part I provides a broad overview of the key themes in *Environmental Science*, the scientific method, and thinking critically about the environment. Part II presents the study of Earth as a system, emphasizing how systems work and the basic biochemical cycles of our planet. Part III focuses on life and the environment and includes subjects such as human population, ecosystems, biological diversity, biological productivity and energy flow, and restoration and recovery of ecosystem response to disturbance. Part IV presents living resources from a sustainability viewpoint, and topics covered include world food supply, agriculture and environment, plentiful and endangered species, forest ecology, conserving and managing life in the oceans, environmental health and toxicology and natural hazards. Part V introduces and discusses basic principles of energy, fossil fuels and environment, alternative energy, and nuclear energy. Part VI discusses water supply, use, and management, and water pollution treatment. Part VII concerns the atmosphere, from global issues such as climate, global warming, and stratospheric ozone depletion to regional issues such as acid rain to local issues including urban air pollution and indoor air pollution. Part VIII concerns relationships between environment and society. Topics include environmental economics, the urban environment, integrated waste management, minerals and the environment, environmental impact and planning, and how we might achieve sustainability.

Special Features

In writing *Environmental Science* we have designed a text that incorporates a number of special features that we believe will help teachers to teach and students to learn. These include the following:

■ A **Case Study** introduces each chapter. The purpose is to interest students in the chapter's subject and to raise important questions on the subject matter. For example, in Chapter 15, which deals with Environmental Health and Toxicology, the Case Study introduces the problem of sex reversal in frogs and asks the question, "Are we participating in an unplanned experiment on how herbicides might transform the bodies of people?"

■ **Learning Objectives** are introduced at the beginning of each chapter to help students focus on what is important in the chapter and what they should achieve after reading and studying the chapter.

■ **A Closer Look** is the name of special learning modules that present more detailed information concerning a particular concept or issue. For example, A Closer Look 5.1 (Matter and Energy) discusses some basic

physics and chemistry. Many of these special features contain figures and data to enrich the reader's understanding, and relate back to the book themes.

■ Near the end of each chapter, a **Critical Thinking Issue** is presented to encourage critical thinking about the environment and to help students understand how the issue may be studied and evaluated. For example, Chapter 21 presents the environmental issue of how polluted waters can be restored. The issue in Chapter 18 examines the important environmental question of whether or not we should raise the gasoline tax.

■ Following the Summary, a special section, **Reexamining Themes and Issues**, reinforces the six major themes of the textbook.

■ **Study Questions** for each chapter provide a study aid, emphasizing critical thinking.

■ **Further Readings** are provided with each chapter so that students may expand their knowledge by reading additional sources of information (both print and electronic) on the environment.

■ **References** cited in the text are provided at the end of the book as notes for each chapter. These are numbered according to their citation in the text. We believe it's important that introductory textbooks carefully cite sources of information used in the writing. These are provided to help students recognize those scholars whose work we depend on, and so that students may draw upon these references as needed for additional reading and research.

Changes in the Sixth Edition

Environmental science is a rapidly developing set of fields. The scientific understanding of environment changes rapidly. Even the kinds of science, and the kinds of connections between science and our ways of life, change. Also, the environment itself is changing rapidly: Populations grow; species become threatened or released from near-extinction; our actions change. To remain contemporary, a textbook in environmental science requires frequent updating.

Other changes and special features in the sixth edition include:

■ A reorganized Chapter 1, extensively covering sustainability.

■ A new chapter on natural disasters.

■ Major modifications of the content of many chapters.

■ Improvements to our companion Web site (www.wiley.com/college/botkin) provides activities for students and resources for instructors, including online quizzing, and a variety of news video clips and animations.

- Many new photos and illustrations.
- New and updated Case Studies, Closer Look boxes, and Critical Thinking Issues.

Augmentation of Web Site References

Valid information is becoming increasingly available over the Web, and easy access to these data is of great value. Government data that used to take weeks of library search are available almost instantly over the Web. For this reason, we have greatly augmented the number of Web site references and have gathered them all on the book's companion Web site.

New Case Studies

Each chapter begins with a case study that helps the student learn about the chapter's topic through a specific example. A major improvement in the sixth edition is the replacement of some older case studies with new ones that discuss current issues and are more closely integrated into the chapter. New case studies are listed following the Table of Contents.

New Critical Thinking Issues

Each chapter ends with a discussion of an environmental issue, with critical thinking questions for the students. This is one of the ways that the text is designed to help students learn to think for themselves about the analysis of environmental issues. In the sixth edition, some older environmental issues have been replaced with new ones, and these have been more closely integrated into the text. These are listed following the Table of Contents.

■ Supplementary Materials

Environmental Science, Sixth Edition, features a full line of teaching and learning resources developed to help professors create a more dynamic and innovative learning environment. For students, we offer tools to build their ability to think clearly and critically. For the convenience of both the professors and students, we provide teaching and learning tools on the Web and on a CD-ROM.

For Students

Student Web Site (www.wiley.com/college/botkin)

A completely new, redesigned, content-rich Web site has been created to provide enrichment activities and resources for students. These features include review of Learning Objectives, online quizzing, Virtual Field Trips, interactive Environmental Debates, a map of regional case studies, critical thinking readings, glossary and flash-cards, Web links to important data and research in the field of environmental studies, and video and animations covering a wide array of selected topics.

For Instructors

Instructor's Resource CD-ROM

The Instructor's Resource CD (IRCD) is a multi-platform CD-ROM. The Instructor's Resource CD includes the Instructor's Resource Guide, the Test Bank, the Computerized Test Bank, jpeg files for all the illustrations in the text, and lecture PowerPoint™ presentations. The IRCD is designed to address the needs of all teaching levels. These assets are also available to professors at the Botkin/Keller text Web site (www.wiley.com/college/ botkin).

Instructor's Resource Guide

The Instructor's Resurce Guide, prepared by James Morris of the University of South Carolina, is available on both the IRCD and the Botkin/Keller Web site (www.wiley.com/college/botkin). The IRG provides useful tools to highlight key concepts from each chapter. Each chapter includes the following topics: Lecture Launchers that incorporate technology and opening thought questions; Discussion of Selected Sections from the text, which highlight specific definitions, equations, and examples; and Critical Thinking Activities to encourage class discussion.

Test Bank

The Test Bank, prepared by Nicholas Pinter of Southern Illinois University, is available on both the Instructor's Resource CD and the Botkin/Keller Web site (www.wiley. com/college/botkin). The Test Bank includes approximately 2,000 questions, in multiple-choice, short-answer, and essay formats. The Test Bank is provided in a word.doc format for your convenience to use and edit for your individual needs. For this edition, the author has created many new questions and has labeled the boxed applications according to the six themes and issues set forth in the text. In addition, the author has created questions for the theme boxes and emphasized the themes in many of the questions throughout the test bank.

Computerized Test Bank

The Computerized Test Bank (CTB) is a multi-platform CD-ROM (Diploma/Exam) that is also available on the Wiley Botkin/Keller Web site (www.wiley.com/ college/botkin). The CTB includes all the files from the Test Bank, but puts them into a dynamic computerized format. The easy-to-use test-generating program fully supports graphics, print tests, student answer sheets, and answer keys quickly and easily. The software's advanced features allow you to create an exam to your exact specifications, with an easy-to-use interface.

PowerPoint™ Presentations

Prepared by Diane Ramirez, these presentations are tailored to the text's topical coverage and are designed to convey key concepts, illustrated by embedded text art.

Advanced Placement® Guide for Environmental Science

Prepared by Brian Kaestner of Saint Mary's Hall, these are available on the Instructor's Resource Web site (www.wiley.com/college/botkin). The Advanced Placement Guide provides a useful tool for high school instructors who are teaching the AP® Environmental Science course. This Guide will help teachers to focus on the key concepts of every chapter to prepare students for the Advanced Placement® Exam. Each chapter includes a Chapter Overview that incorporates critical thinking questions, Key Topics important to the exam, a correlation list of Multiple Choice and Free Response questions from the 1998 and 2003 AP® exams, and Web links to Laboratories and Activities that reinforce key topics.

Instructor's Web Site

The resources provided on the Instructor's Resource CD will also be available on the instructor section of the Wiley Botkin/Keller Web site (www.wiley.com/college/botkin).

Wiley PLUS

Wiley PLUS provides an integrated suite of teaching and learning resources, including an online version of the text, in one easy-to-use Web site. Organized around the essential activities you perform in class, *Wiley PLUS* helps you:

- **Prepare and Present.** Create class presentations using a wealth of Wiley-provided resources, including an oline version of the textbook, PowerPoint slides, animations, and more—making your preparation time more efficient. You may easily adapt, customize, and add to this content to meet the needs of your course.

- **Create Assignments.** Automate the assigning and grading of homework or quizzes by using Wiley-provided question banks or by writing your own. Student results will be automatically graded and recorded in your gradebook. *Wiley PLUS* can link homework problems to the relevant section of the online text, providing students with context-sensitive help.

- **Track Student Progress.** Keep track of your students' progress via an instructor's gradebook, which allows you to analyze individual and overall class results to determine the student's progress and level of understanding.

- **Administer Your Course.** *Wiley PLUS* can easily be integrated with other course management systems, gradebooks, or other resources you are using in your class, providing you with the flexibility to build your course in your own way.

Daniel B. Botkin
Edward A. Keller

ACKNOWLEDGMENTS

Completion of this book was only possible due to the cooperation and work of many people. To all those who so freely offered their advice and encouragement in this endeavor, we offer our most sincere appreciation. We are indebted to our colleagues who made contributions: Dorothy A. Rosenthal for writing Chapter 2 and the environmental issues at the end of each chapter; Mel S. Manalis for assistance in developing the energy chapters, global warming, and stratospheric ozone depletion discussions; Marc J. McGinnis for assistance in helping develop discussions concerning environmental law; and Harold Morowitz for helpful suggestions on how to introduce the basic energy concepts. Diane Perez reviewed and edited the manuscript to improve its clarity.

We greatly appreciate the work of our editor Merillat Staat at John Wiley & Sons, for support, encouragement, assistance, and professional work. We extend thanks to our production editor Trish McFadden, who did a great job and made important contributions in many areas; to Kevin Murphy for a beautiful interior design and striking cover; Ellinor Wagner for photo research; Sandra Rigby for the illustration program; and to Diana Perez for excellent editing. Thanks go out also for editorial assistance from Stephen Reiss. The extensive supplements package was enhanced through the efforts of Linda Muriello.

Of particular importance to the development of the book were the individuals who read the book chapter by chapter and provided valuable comments and constructive criticism. This was a particularly difficult job given the wide variety of topics covered in the text, and we believe that the book could not have been successfully completed without their assistance. These reviewers are offered our special gratitude:

Reviewers of Previous Editions

John All, Western Kentucky University
William B.N. Berry, University of California, Berkeley
Christopher P. Bloch, Texas Tech University
Jennifer M. Rhode, Georgia College and State University
Jason E. Box, Ohio State University
Rupali Datta, University of Texas at San Antonio
Walter Illman, The University of Iowa
Don Lotter, Imperial Valley College
Heidi Marcum, Baylor University
Mark A. McGinley, Monroe Community College
Maren L. Reiner, University of Richmond
Randall, Repic, University of Michigan, Flint

Bradley R. Reynolds, University of Tennessee at Chattanooga
Jennifer M. Rhode, Georgia College and State University
Michael Toscano, Delta College
Robert J. Andres, University of North Dakota
James W. Bartolome, University of California, Berkeley
Anthony Benoit, Three Rivers Technical Community College
William Berry, University of California, Berkeley
Renée E. Bishop, Penn State Worthington Scranton
Kelly D. Cain, University of Wisconsin
Richard Clements, Chattanooga State Technical Community College
Terence H. Cooper, University of Minnesota
Nate Currit, Pennsylvania State University
William Davin, Berry College
David S. Duncan, University of South Florida
Jean Dupon, Menlo College
David J. Eisenhour, Morehead State University
Brian D. Fath, Towson University
Richard S. Feldman, Marist College
John P. Harley, Eastern Kentucky University
Syed E. Hasan, University of Missouri
Joseph Hobbs, University of Missouri
Dan F. Ippolito, Anderson University
Frances Kennedy, State University of West Georgia
Eric Keys, Arizona State University
John Kinworthy, Concordia University
Allen H. Koop, Grand Valley State University
Janet Kotash, Moraine Valley Community College
Ernesto Lasso de la Vega, International College
Tim Lyon, Ball State University
John S. Mackiewicz, University at Albany, State University of New York
Eric F. Maurer, University of Cincinnati
Deborah L. McKean, University of Cincinnati
James Morris, University of Southern Carolina
Kathleen A. Nolan, St. Francis College
William D. Pearson, University of Louisville
Julie Phillips, De Anza College
John Pratte, Kennesaw State University
Donald C. Rizzo, Marygrove College
Carlton Rockett, Bowling Green State University
Angel Rodriguez, Broward Community College
John Rueter, Portland State University
Robert M. Sanford, University of Southern Maine
Christian Shorey, University of Iowa
Roger Sedjo, Resources for the Future, Washington, D.C.
James H. Speer, Indiana State University
Richard Waldren, University of Nebraska, Lincoln
Carole L. Ziegler, University of San Diego
Diana Anderson, Northern Arizona University
Susan Beatty, University of Colorado, Boulder

Brian Beeder, Morehead State University
Alan Bjorkman, North Park University
Charles Bomar, University of Wisconsin—Stout
Rosanna Cappellato, Emory University
W.B. Clapham, Jr., Cleveland State University
Jerry Delsol, Modesto Junior College
Deborah Freile, Berry College
Nancy Goodyear, Bainbridge College
Paul Grogger, University of Colorado
Herbert Grossman, Pennsylvania State University
Gian Gupta, University of Maryland
Lonnie Guralnick, Western Oregon University
Raymond Hames, University of Nebraska
Alan Holyoak, Manchester College
Donald Humphreys, Temple University
James Jensen, SUNY, Buffalo
David Johnson, Michigan State University
John Kinworthy, Concordia University
Thomas Klee, Hillsborough Community College
Mark Knauss, Shorter College
Ned Knight, Linfield College
Steven Kolmes, University of Portland
Matthew Laposata, Kennesaw State University
Timothy McCay, Colgate University
Michele Morek, Brescia University
Nancy Ostiguy, Pennsylvania State University
John Pratte, Kennesaw State University
Jeffrey Schneider, SUNY, Oswego
Jill Scheiderman, SUNY, Dutchess Community College
Peter Schwartzman, Knox College
Daniel Sivek, University of Wisconsin
Meg Stewart, Vassar College
Richard Stringer, Harrisburg Area Community College
Janice Swab, Meredith College
Jeffrey Tepper, Valdosta State University
Richard Vance, UCLA
Sarah Warren, North Carolina State University
William Winner, Oregon State
Bruce Wyman, McNeese State University
Richard Zingmark, University of South Carolina
Marc Abrams, Pennsylvania State University
Michele Barker-Bridges, Pembroke State University (NC)
David Beckett, University of Southern Mississippi
Mark Belk, Brigham Young University
Kristen Bender, California State University, Long Beach
Gary Booth, Brigham Young University
Grace Brush, Johns Hopkins University
John Campbell, Northwest Community College (WY)
Ann Causey, Prescott College (AZ)

Simon Chung, Northeastern Illinois State
Thomas B. Cobb, Bowling Green State University
Jim Dunn, University of Northern Iowa
Robert Feller, University of South Carolina
Andrew Friedland, Dartmouth College
Douglas Green, Arizona State University
James H. Grosklags, Northern Illinois University
Bruce Hayden, University of Virginia
David Hilbert, San Diego State University
Peter Kolb, University of Idaho
Henry Levin, Kansas City Community College
Hugo Lociago, University of California, Santa Barbara
Tom Lowe, Ball State University
Mel Manalis, University of California, Santa Barbara
Earnie Montgomery, Tulsa Junior College, Metro Campus
Walter Oechel, San Diego State University
C. W. O'Rear, East Carolina University
Stephen Overmann, Southeast Missouri State University
Martin Pasqualetti, Arizona State University
David Pimental, Cornell University
Joseph Simon, University of South Florida
Lloyd Stark, Pennsylvania State University
Laura Tamber, Nassau Community College (NY)
Ann Zimmerman, University of Toronto

Reviewers of This Edition

Marvin Baker, University of Oklahoma
Mary Benbow, University of Manitoba
Grady Blount, Texas A&M University, Corpus Christi
John Bounds, Sam Houston State University
Vincent Breslin, SUNY, Stony Brook
Bonnie Brown, Virginia Commonwealth University
Annina Carter, Adirondack Community College
Peter Colverson, Mohawk Valley Community College
Harry Corwin, University of Pittsburgh
Craig Davis, Ohio State University
Craig Davis, University of Colorado
David Johnson, Michigan State University
S. B. Joshi, York University
Stephen Malcolm, Western Michigan University
James Melville, Mercy College
Chris Migliaccio, Miami-Dade Community College-Wolfson
Clayton Penniman, Central Connecticut State University
Jeffrey Schneider, SUNY, Oswego

Daniel B. Botkin
Edward A. Keller

BRIEF CONTENTS

CONTENTS

4
The Human Population and the Environment 55

5
The Biogeochemical Cycles 75

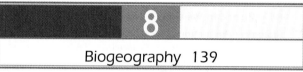

14

Wildlife, Fisheries, and Endangered Species 261

15

Environmental Health, Pollution, and Toxicology 291

16
Natural Disasters and Catastrophes 315

17
Energy: Some Basics 345

18
Fossil Fuels and the Environment 364

Environmental
Science

EARTH AS A LIVING PLANET

SIXTH EDITION

DANIEL B. BOTKIN

Professor Emeritus
Department of Ecology, Evolution, and Marine Biology
University of California, Santa Barbara

President
The Center for the Study of the Environment
Santa Barbara, California

EDWARD A. KELLER

Professor of Environmental Studies and Earth Science
University of California, Santa Barbara

With assistance from:
Dorothy B. Rosenthal
Mel S. Manalis
Valery A. Rivera
Diane Perez-Botkin

JOHN WILEY & SONS, INC.

CHAPTER 1

Key Themes in Environmental Sciences

Learning Objectives

Certain themes are basic to environmental science. After reading this chapter, you should understand:

■ That people and nature are intimately connected.

■ Why rapid human population growth is the fundamental environmental issue.

■ What sustainability is and why we must learn to sustain our environmental resources.

■ How human beings affect the environment of the entire planet.

■ Why urban environmentals need attention.

■ Why solutions to environmental problems involve making value judgments based on scientific knowledge.

■ What the precautionary principle is and why it is important.

The rich diversity of life on and around coral reefs is illustrated in this photo from the Red Sea, Egypt.

CASE STUDY

Shrimp, Mangroves, and Pickup Trucks: Local and Global Connections Reveal Major Environmental Concerns

Maitri Visetak owns a small plot of land along the coast of southern Thailand; he wanted to improve life for his family, and he succeeded. A growing demand for shrimp as a luxury food and overfishing of wild shrimp had fueled growth of the world market for farmed shrimp from a $1.5 billion industry 30 years ago to an $8 billion business today. In the early 1990s, Mr. Visetak began farming shrimp in two small ponds (0.2 hectare, or 0.5 acre; Figure 1.1a). Within two years, he had accumulated enough capital to purchase two pickup trucks—in Thailand, a clear indication of financial success. By then, though, his ponds were contaminated with shrimp waste, antibiotics, fertilizers, and pesticides. Shrimp could no longer live in the ponds. And there was an even more widespread effect: Pollutants escaping from the ponds threatened survival of the area's mangrove trees (Figure 1.1b). Like thousands of other shrimp farmers in Southeast Asia, India, Africa, and Latin America, Mr. Visetak considered abandoning these ponds and moving on to others.

Maitri Visetak is trying to feed his family in the best way he knows how, but along with thousands of other shrimp farmers in the world, he is unwittingly contributing to destruction of coastal mangroves, one of the world's valuable ecosystems. Half of the world's mangrove forests have been destroyed and with them a major source of food for local human populations and breeding grounds for much of the tropical world's sea life. The United Nations Environment Program has estimated that one-fourth of the destruction of mangroves can be traced to shrimp farming. Environmentalists have become alarmed, and in many areas local people have staged protests against shrimp farming. With the world's population expected to increase from 6.2 billion to 9 billion by the middle of the twenty-first century, concern over the world's mangrove forests is growing.[1–6]

(a)

(b)

Figure 1.1 ■ Sustainability. (a) Shrimp farms such as this one threaten the survival of mangrove forests. (b) Mangroves on the banks of Indian River, Isle of Dominica, West Indies. Mangrove trees grow in coastal wetlands. Their specialized roots can survive immersion in ocean water at high tide and exposure to the drying sun at low tide. Swamps formed by mangroves provide habitat for many kinds of ocean life and are important for commercial fisheries in many parts of the world.

Maitri Visetak's story illustrates the major themes of environmental science. First, people and nature are intimately connected, and changes in one lead to changes in the other. Second, human population increase is a major contributor to environmental problems. Third, industrial development and urbanization have serious environmental consequences. Fourth, unsustainable use of resources must be replaced with sustainable practices. Fifth, local changes can have global effects. Sixth, environmental issues involve values and attitudes as well as scientific understanding. Maitri Visetak's story also illustrates important questions that we all must face: Which individual actions contribute to environmental degradation? What actions can people, both as individuals and as groups, take to limit environmental damage?

1.1 Major Themes of Environmental Science

At the beginning of the modern era—in A.D. 1—the number of people in the world was probably about 100 million, one-third of the present population of the United States. In 1960 the world contained 3 billion people. Our population has more than doubled in the last 40 years, to 6.6 billion people today. In the U.S. population increase is often apparent when we travel. Urban traffic snarls, long lines to enter state national parks, and hampered attempts to get tickets to popular attractions are all symptoms of a growing population. If recent human population growth rates continue, our numbers could reach 9 billion before 2040. The problem is that the Earth has not grown any larger and the abundance of its resources has not increased. How, then, can the Earth sustain all these people? And what is the maximum number of people that could live on the Earth—not just for a short time, but *sustained* over a long time?

Estimates of how many people the planet can support range from 2.5 billion to 40 billion. Why do the estimates vary so widely? Because the answer depends on what quality of life people are willing to accept. The poorer that quality, the greater the number of people that can be squeezed onto the Earth's surface. How many people the Earth can sustain depends on *science and values* and is also a question about *people and nature*. The more people packed onto the Earth, the less room and resources for wild animals and plants, wilderness, areas for recreation, and other aspects of nature—and the faster Earth's resources will be used. The answer also depends on how the people are distributed on the Earth—whether they are concentrated mostly in cities or spread evenly across the land.

Although the environment is complex and environmental issues seem sometimes to cover an unmanageable number of topics, the science of the environment comes down to the central topics just mentioned: the human population, urbanization, and sustainability within a global perspective. These issues have to be evaluated in light of the interrelations between people and nature. And the answers ultimately depend on both science and nature.

Therefore this book approaches environmental science through six interrelated themes:

- *Human population growth* (the environmental problem).
- *Sustainability* (the environmental goal).
- *A global perspective* (solving many environmental problems requires a global solution).
- *An urbanizing world* (most of us live and work in urban areas).
- *People and nature* (we share a common history with nature).
- *Science and values* (science provides solutions. Which one is chosen is in part a value judgement).

You may ask, "If this is all there is to it, what is in the rest of this book?" (See A Closer Look 1.1). The answer lies with the old saying: "The devil is in the details." The solution to specific environmental problems requires specific knowledge. The six themes help us see the big picture and provide a valuable background. The opening case study illustrates linkages among the themes and the importance of details. The shrimp farmer, Maitri Visetak, would not cause a major environmental problem if

A CLOSER LOOK 1.1

A Little Environmental History

A brief historical explanation will help clarify what we seek to accomplish. Before 1960, few people had ever heard the word *ecology*, and the term *environment* meant little as a political or social issue. Then came the publication of Rachel Carson's landmark book, *Silent Spring* (Houghton Mifflin, Boston, 1960, 1962). At about the same time, several major environmental events occurred, such as oil spills along the coasts of southern California and Massachusetts and highly publicized threats of extinction to many species, including whales, elephants, and songbirds. The environment became a popular issue.

As with any new social or political issue, at first relatively few people recognized its importance. Those who did found it necessary to stress the problems—to emphasize the negative—in order to bring public attention to environmental concerns. Adding to the limitations of the early approach to environmental issues was a lack of scientific knowledge and practical know-how. Environmental sciences were in their infancy. Some people even saw science as part of the problem.

The early days of modern environmentalism were dominated by confrontations between those labeled environmentalists and those labeled anti-environmentalists. Stated in the simplest terms, environmentalists believed that the world was in peril. To them, economic and social development meant the destruction of the environment and ultimately the end of civilization, the extinction of many species, and perhaps the extinction of human beings. Their solution was a new worldview that depended only secondarily on facts, understanding, and science. In contrast, once again stated in the simplest term, the anti-environmentalists believed that social and economic health and progress were necessary—whatever the environmental effects—if people and civilization were to prosper. From their perspective, environ-

mentalists represented a dangerous and extreme view, with a focus on the environment to the detriment of people—a focus they thought would destroy the very basis of civilization and lead to the ruin of our modern way of life.

Today, the situation has changed considerably. Public opinion polls repeatedly show that people around the world rank the environment among the most important social and political issues. There is no longer a need to prove that environmental problems are serious.

Significant progress has been made in many areas of environmental science (although our scientific understanding of the environment still lags behind our need to know). Advances have also been made in the creation of legal frameworks for the management of the environment, thus providing a new basis for addressing environmental issues. The time is now ripe to seek truly lasting, more rational solutions to environmental problems.

he were the only one farming shrimp. It is the huge number of people who want to eat shrimp and who need ways to make a living that makes this local problem a worldwide one.

In this chapter we introduce the six themes with brief examples, showing the linkages among them and touching on the importance of specific knowledge that will be the concern of the rest of the book. We start with human population growth.

1.2 Human Population Growth

The John Eli Miller Family

John Eli Miller was an ordinary American except for one thing—when he died in Middlefield, Ohio, in the mid–twentieth century, he was the head of the largest family in the United States (Figure 1.2). He was survived by 5 children, 61 grandchildren, 338 great-grandchildren, and 6 great-great-grandchildren. Within his lifetime, John

Miller witnessed a family population explosion. What was perhaps even more remarkable was that the explosion started with a family of just 7 children—not all that unusual for the nineteenth-century United States.[7]

During most of John Miller's life, his family was not unusually large. It is just that he lived long enough to find out what simple multiplication can do, and he lived in a time when the death rate among infants, children, and young adults was very low compared with typical death rates during the history of most human populations. Of 7 children born to John Miller, 5 survived him; of 63 grandchildren, 61 survived him; and of 341 great-grandchildren (born to 55 married grandchildren—an average of slightly more than 6 children per parent)—338 survived him.

John Miller's family emphasizes a major factor in our modern population explosion. Modern technology, modern medicine, and the supply of food, clothing, and shelter have decreased death rates and increased the net rate of growth. As a result, the human population has increased greatly, threatening the environment.

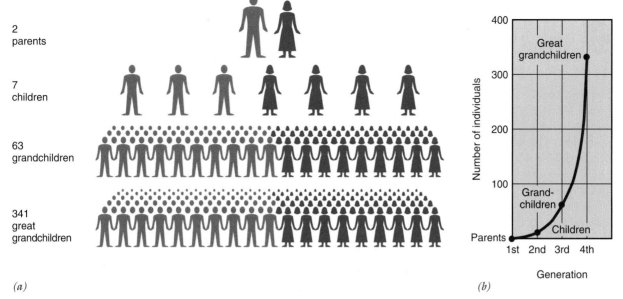

2 parents		
7 children		
63 grandchildren		
341 great grandchildren		

(a)

(b)

Figure 1.2 ■ The population bomb starts with little sparks. (*a*) A simplified family tree of four generations of the John Eli Miller family. (*b*) The population explosion of the John Eli Miller family shown in graphic form.

Our Rapid Population Growth

The most dramatic increase in the history of the human population occurred in the last part of the twentieth century and continues today into the early 21st century. As mentioned earlier, in merely the past 40 years, the human population of the world more than doubled, increasing

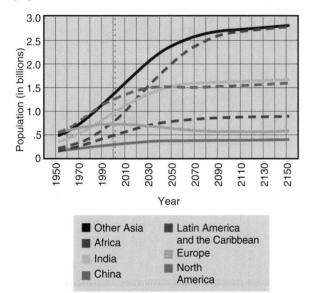

Figure 1.3 ■ Population change since 1950 projected to the year 2150 for major areas of the world, medium fertility scenario. The population of Africa will nearly quadruple. The only major area whose population is projected to drop over time is Europe—from 728 million to 595 million, a decline of 18% over 155 years. [*Source:* Population Division, Department of Economic and Social Affairs, United Nations Secretariat, *World Population Projections to 2150* (New York: United Nations, 1998).]

from 2.5 billion to over 6.6 billion. Figure 1.3 illustrates the rapid explosion of the human population, sometimes referred to as the population bomb.[8] The figure breaks down the increases by region.

Human population growth is, in some important ways, *the* underlying issue of the environment. Much current environmental damage is directly or indirectly the result of the very large number of people on the Earth and our rate of increase. As you will see in Chapter 4, where we consider the human population in more detail, for most of human history the total population was small and the average long-term rate of increase was low relative to today's population growth rate.[9, 10]

Although it is customary to think of the population as increasing continuously without declines or fluctuations, the growth of the human population has not been a steady march. For example, great declines occurred during the time of the Black Death. (See A Closer Look 1.2.)

African Famines

Famine is one of the things that happen when a human population exceeds its environmental resources. Famines have occurred in recent decades in Africa. In the mid-1970s, following a drought in the Sahel region, 500,000 Africans starved to death and several million more were permanently affected by malnutrition.[12] Starvation in African nations gained worldwide attention some 10 years later, in the 1980s.[13, 14] In one year during that period, as many as 22 African nations suffered catastrophic food shortages and 150 million Africans faced starvation. Although there have not been such spectacularly acute

The Black Death

The epidemic disease bubonic plague, commonly known as the Black Death, spread throughout Europe during the fourteenth century. The most severe episodes occurred between 1347 and 1351, but there were many recurrences throughout the century (Figure 1.4).[11] The disease is caused by the bacteria *Yersinia pestis* and is spread by fleas that live on rodents. It was first recorded in Western history as a major human problem in the seventh century in the Roman Empire and in North Africa. The plague probably first appeared in India in the seventh century and spread rapidly north and west. Not until the fourteenth century did another major epidemic occur. The plague again spread rapidly, reaching Italy in 1348 and Spain, France, Scandinavia, and central Europe within two years. In England, one-fourth to one-third of the population died within a single decade, although mortality varied widely by region. Entire towns were abandoned, and the production of food for the remaining population was jeopardized.

The Black Death had many environmental and economic consequences. For example, the great reduction in the labor force led to an increase in wages and is believed to have been a contributing factor to a subsequent increase in the standard of living. Much agricultural land was abandoned because no one was available to work it.

As this example illustrates, human populations have not always increased continuously but have suffered setbacks and declines. The bubonic plague is one of the best-recorded and best-known setbacks in human history. It is likely, though, that other such episodes, resulting from changes in climate, declines in the food supply, or environmental catastrophes, have happened many times in human history.

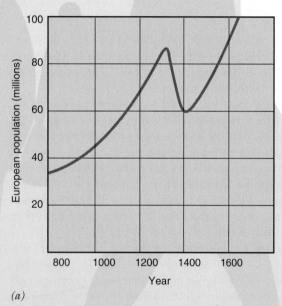

(a)

(b)

Figure 1.4 ■ (*a*) The change in the population of Europe during the time of the Black Death. (*b*) During the fourteenth century, the Black Death killed many people in Europe. The reduction in the human population had economic and environmental effects. This fourteenth-century illustration of two plague victims is from a miniature from the Toggenburg Bible. Note the swellings over the bodies, characteristic of the plague.

famines in the last decade as occurred in the 1970s and 1980s, there is a continuing food crisis in southern Africa, particularly in Malawi, Zambia, and Zimbabwe.[15]

Famine in Africa has had multiple interrelated causes. One, as suggested, is drought. Although drought is not new to Africa, the size of the population affected by drought is new. In addition, deserts in Africa appear to be spreading, in part because of changing climate but also because of human activities. Poor farming practices have increased erosion, and deforestation may be helping to make the environment drier. The control and destruction of food has sometimes been used as a weapon in political disruptions (Figure 1.5).

Famines in Africa illustrate another key theme: people and nature. People affect the environment, and the environment also affects people. The environment affects agriculture, and agriculture also affects the environment. Human population growth in Africa has severely stretched

Figure 1.5 ■ Values and science. Social conditions affect the environment, and the environment affects social conditions. Political disruption in Somalia (illustrated by a Somalian boy with a gun, left photo) interrupted farming and food distribution, leading to starvation. Overpopulation, climate change, and poor farming methods also lead to starvation, which in turn promotes social disruption. Famine has been common in parts of Africa since the 1980s, as illustrated by gifts of food from aid agencies in southern Sudan (right photo).

the capacity of the land to provide sufficient food and has threatened its future productivity.

This situation involves yet another key theme: science and values. Scientific knowledge has led to increases in agricultural production and to a better understanding of population growth and what is required to conserve natural resources. With this knowledge, we are forced to confront a choice: Which is more important, the survival of people alive today or conservation of the environment, on which future food production and human life depend?[16] Answering this question demands *value judgments* and the information and knowledge with which to make such judgments. For example, we must determine whether we can continue to increase agricultural production without destroying the very environment on which agriculture and, indeed, the persistence of life on Earth depend. Put another way, a technical, scientific investigation provides a basis for a value judgment.

The human population is doubling every few decades, but human effects on the environment are growing even faster.[17] Human beings cannot escape the laws of population growth, which are discussed in several chapters. The broad science and value question is: What will we do about the increase in our own species and its impact on our planet and our future?

1.3 Sustainability and Carrying Capacity

The story of the Eli Miller family and the recent famines in Africa bring up one of the central environmental questions of our time: What is the maximum number of people the Earth can sustain? That is, what is the sustainable human carrying capacity of the Earth? Much of this book will deal with knowledge that helps answer this question. However, there is little doubt that we are using our renewable environmental resources faster than they can be replenished—that is, we are using these resources *unsustainably*. In general, we are using forests and fish faster than they can regrow and eliminating habitats of endangered species and wildlife faster than they can be replenished. We are extracting minerals, oil, and groundwater without sufficient concern for their limits or the need to recycle them. As a result, there is a present shortage of some resources and an expectation of more shortages in the future. Clearly, we must learn how to sustain our environmental resources so that they continue to provide benefits for people and other living things on our planet.

Sustainability: The Environmental Objective

The environmental statement of the 1990s was "saving our planet." Is Earth's very survival really in danger? In the long view of planetary evolution, it is certain that planet Earth will survive us. Our sun is likely to last another several billion years, and if all humans became extinct in the next few years, life would flourish on our planet. The changes we have made in the landscape, the atmosphere, and the waters would last for a few hundreds or thousands of years but would (in a modest period of time) be cleansed by natural processes. What we are concerned with, as environmentalists, is the quality of the *human* environment on Earth for us today and for our children.

Environmentalist agree that sustainability must be achieved, but we are unclear at present how to achieve it, in part because the word is used to mean different things, often leading to confusion and to people taking cross-purposes. **Sustainability** refers to resources and their environment. In this book, sustainability has two scientific definitions: (1) **sustainable resource harvest**, such as a sustainable supply of timber, means that the same quantity of that resource can be harvested each year (or other harvest interval) for an unlimited or specified amount of time without decreasing the ability of the resources to produce the same harvest level. (2) A **sustainable ecosystem** is an ecosystem from which we are harvesting a resource that is still able to maintain its essential functions and properties.

At the level of our society, we can define sustainability as ensuring that future generations have equal opportunity to the resources that our planet offers, or (at a minimum) that future generations inherit an environment with human-induced environmental damage no greater than that of today. Others would argue that sustainability refers to types of development that are economically viable, do not harm the environment, and are socially just (that is, the development is fair to all people). Two points are particularly significant to understanding sustainability:[18]

- Sustainability means for an unspecified long period of time.
- Sustainable growth is an oxymoron as any steady growth (fixed percentage growth per year) produces large numbers in modest periods of time. (See Exponential Growth in Chapter 3.) Other types of growth are possible—for example, economic growth, growth of wind or solar power, or recovery of an endangered species.

One of the environmental paradigms of the twenty-first century will be sustainability, but how will it be obtained? We have begun to consider what is known as the *sustainable global economy*. By economy, environmentalists mean the careful management and wise use of the planet and its resources, analogous to the management of money and goods considered by economists. By focusing on the concept of a sustainable global economy we are assuming that under present conditions the global economy is *not* sustainable. Increasing numbers of people have resulted in pollution of the land, air, and water to an extent that the ecosystems upon which people depend are in danger of collapse. What then are attributes of a sustainable economy in the information age?[18]

- Populations of humans and other organisms living in harmony with the natural support systems such as air, water, and land (including ecosystems).
- An energy policy that does not pollute the atmosphere, cause climatic change such as global warming, or present unacceptable risk (a political or social decision).

- A plan for renewable resources such as water, forests, grasslands, agricultural lands, and fisheries that will not deplete the resources or damage ecosystems.
- A plan for nonrenewable resources that does not damage the regional to global environment while ensuring a share of our nonrenewable resources is left to future generations.
- A social, legal, and political system that is dedicated to sustainability with a democratic mandate to produce such an economy.

Recognizing the fact that population is the environmental problem, we should keep in mind that a sustainable global economy will not be constructed around a completely stable global population. Rather, such an economy requires a realization that the size of the human population will fluctuate within some stable range, necessary to maintain healthy relationships with other components of the environment.

To achieve a sustainable global economy, it is necessary that we[18]

- Develop an effective population-control strategy. This will, at least, require more education of people since literacy and population growth are inversely related.
- Completely restructure our energy programs. A sustainable global economy is probably impossible if it is based on use of fossil fuels. New energy plans will utilize the concept of an integrated energy policy, with more emphasis on renewable energy sources (such as solar and wind). Finally, conservation of energy must have a central place in our energy management plans.
- Institute economic planning, including development of a tax structure that will encourage population control and wise use of resources. Financial aid for developing countries is absolutely necessary to narrow the gap between rich and poor nations.
- Implement social, legal, political, and educational changes that provide for maintenance of a quality local, regional, and global environment. This must be a serious commitment that all the people of the world will cooperate to achieve.

Moving Toward Sustainability: Some Criteria

Stating that we wish to develop a sustainable future acknowledges that our present practices are not sustainable. Indeed, continuing in our present paths of overpopulation, resource consumption, and pollution will not lead to sustainability. What is needed is development of new concepts that will mold industrial, social, and environmental interests into an integrated, harmonious system. In other words, we need to develop a new paradigm—a new pattern—as an alternative to our present model for running society and creating wealth.[19] The new paradigm might be described in the following ways.[20]

Carrying Capacity of the Chinook Salmon

One approach to determining the carrying capacity and the sustainable harvest of a resource is to examine historical records and see if previous catch levels were maintained. One problem of determining carrying capacity using this method is illustrated by the history of commercial fishing of chinook salmon in the Columbia River of the Pacific Northwest. Figure 1.6 graphs the annual catch of salmon from 1866 (just after the Civil War, when commercial fishing began in earnest on the river) to 1966.[21] The catch increased rapidly from 1860 to 1880, then declined somewhat and fluctuated between 16 and 36 million pounds until 1920, when it declined again in a highly varying pattern.

From this history, what, if any, levels of chinook salmon are sustainable? The high catches between 1890 and 1920 are about the same, so a person in charge of managing the salmon fishery in 1920 might reasonably have concluded that the salmon had a sustainable catch of between 16 and 36 million pounds. However, a person who had only the information after 1950, when the catch represents the tail of a declining curve, would conclude that the approximate level of catch in the early 1960s was sustainable. This graph illustrates the difficulty in determining a truly sustainable harvest simply from historical records.

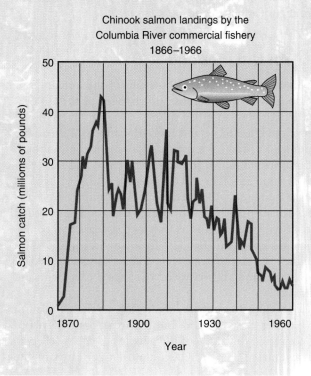

Chinook salmon landings by the Columbia River commercial fishery 1866–1966

Figure 1.6 ■ The commercial catch of chinook salmon on the Columbia River between 1866 and 1966. [*Source:* J. M. Van Hyning, "Factors Affecting the Abundance of Fall Chinook Salmon in the Columbia River," *Research Reports of the Fish Commission of Oregon* 4(1), Figure 14, p. 38.]

- Evolutionary rather than revolutionary. Development of a sustainable future will require an evolution in our values that involves our lifestyles as well as social, economic, and environmental justice.

- Inclusive, not exclusive. All peoples of Earth must be included. This means bringing the people of the world to a higher standard of living in a sustainable way that will not compromise our environment.

- Proactive, not reactive. We must plan for change and for events such as human population problems, resource shortages, and natural hazards rather than waiting for them to surprise us and then reacting. This may involve application of the precautionary principle, which is discussed with science and values (Section 1.7).

- Attracting, not attacking. People must be attracted to the new paradigm because it is right and just. Those who speak for our environment should not take a hostile stand but should attract people to the path of sustainability through sound scientific argument and appropriate values.

- Assisting the disadvantaged, not taking advantage. This involves issues of environmental justice. All people have the right to live and work in a safe, clean environment. Working people around the globe need to receive a living wage sufficient to support their families. Exploitation of workers to reduce costs of manufacturing goods or growing food diminishes us all.

Carrying Capacity of the Earth

Carrying capacity is a concept related to sustainability. It is usually defined as the maximum number of individuals of a species that can be sustained by an environment without decreasing the capacity of the environment to sustain that same amount in the future. When we ask "What is the maximum number of people that Earth can sustain?" we are asking about Earth's carrying capacity—and we are also asking about sustainability.

As mentioned before, the desirable human carrying capacity depends in part on our values. Do we want those who follow us to live short lives in crowded surroundings without a chance to enjoy Earth's scenery and diversity of life? Or do we hope that our descendants will have a life of high quality and good health? Once we choose a goal for the quality of life, we can use scientific information to understand what the carrying capacity might be and how we might achieve it. (See A Closer Look 1.3.)

1.4 A Global Perspective

One solution to famines, such as those that have occurred in recent decades in Africa, is better food distribution throughout the world. Although the total amount of food produced worldwide each year is still sufficient to feed all the world's people, famines occur because local food production is insufficient for the human population in some places and because worldwide transportation of food is inadequate. In this way, local famines are a global environmental problem.

The recognition that, worldwide, civilization can change the environment at a global level is relatively recent. As discussed in detail in later chapters, scientists now believe that emissions of modern chemicals are changing the ozone layer high in the atmosphere. Scientists also believe that burning fossil fuels increases the concentration of greenhouse gases in the atmosphere, which may change Earth's climate. These atmospheric changes suggest that the actions of many groups of people at many locations affect the environment of the entire world.[22, 23] Another new idea explored in later chapters is that nonhuman life also affects the environment of our whole planet and has changed it over the course of several billion years. These two new ideas have profoundly affected our approach to environmental issues.

Awareness of the global interactions between life and the environment has led to the development of the **Gaia hypothesis**, originated by British chemist James Lovelock and American biologist Lynn Margulis.[22] The Gaia hypothesis (discussed in Chapter 3) proposes that the environment at a global level has been profoundly changed by life over the history of life on Earth and that these changes have tended to improve the chances for the continuation of life. Because life affects the environment at a global level, the environment of our planet is different from that of a lifeless one.

1.5 An Urban World

In part because of the rapid growth of the human population and in part because of changes in technology, we are becoming an urban species, and our effects on the environment are more and more the effects of urban life (Figure 1.7). With economic development comes urbanization; people move from farms to cities and then perhaps to suburbs. Cities and towns increase in size. Because cities are commonly located near rivers and along coastlines, urban sprawl often overtakes the good agricultural land of river floodplains as well as the coastal wetlands, which are important habitats for many rare and endangered species. As urban areas expand, wetlands are filled in, forests cut, and soils covered over with pavement and buildings.

In developed countries, about 75% of the population live in urban areas and 25% in rural areas; but in developing countries, only 40% of the people are city dwellers.[17] It is estimated that by 2025 almost two-thirds of the

Figure 1.7 ■ An urban world and a global perspective. When the United States is viewed at night from space, the urban areas show up as bright lights. The number of urban areas reflects the urbanization of our nation.

Figure 1.8 ■ A photo of Los Angeles taken from the air shows the large extent of a megacity.

1.6 People and Nature

Today we stand at the threshold of a major change in our approach to environmental issues. Two paths lie before us. One path is to assume that environmental problems are the result of human actions and that the solution is simply to stop these actions based on the notion, popularized some 40 years ago, that people are separate from nature. This path has produced many advances but also many failures. It has emphasized confrontation and emotionalism. It has been characterized by a lack of understanding of basic facts about the environment and how natural ecological systems function, as well as a willingness to base solutions on political ideologies and on ancient myths about nature.

The second path is to begin with a scientific analysis of an environmental controversy and to move from confrontation to cooperative problem solving. It accepts the connection between people and nature. It offers the potential for long-lasting, successful solutions to environmental problems. One purpose of this book is to take the student down the second pathway.

People and nature are intimately integrated. Each affects the other. We depend on nature in many ways. We depend on nature directly for many material resources such as wood, water, and oxygen in the air. We depend on nature indirectly through what are called "public service functions." For examples, soil is necessary for plants and therefore for us (Figure 1.9); the atmosphere provides a climate in which we can live; the ozone layer high in the atmosphere protects us from damaging ultraviolet radiation; trees absorb some air pollutants; wetlands can cleanse water. And we depend on nature for beauty and recreation—for the needs of our inner selves—as people always have.

Simultaneously, we affect nature. For as long as people have had tools, including fire, they have changed nature, often in ways that we like, prefer, and have considered "natural." It can be argued that it is natural for organisms to change their environment. Elephants topple trees,

population—5 billion people—will live in cities. Only a few urban areas had populations over 4 million in 1950. In 1999, Tokyo, Japan, was the world's largest city. In 2015, Tokyo will still be the world's largest city, with a projected population of 28.9 million. The number of **megacities**—urban areas with at least 8 million inhabitants—increased from 2 (New York City and London) in 1950 to 23 in 1995 (including Los Angeles and New York City) (Figure 1.8). Most megacities—17—are in the developing world. It is estimated that by 2015 the world will have 36 megacities, 23 of them in Asia.[24, 25]

In the past, environmental organizations often focused on nonurban issues—wilderness, endangered species, and natural resources, including forests, fisheries, and wildlife. Although these will remain important issues, in the future we must place more emphasis on urban environments and on the effects of urban environments on the rest of the planet.

Figure 1.9 ■ (*a*) Cross section of a soil; (*b*) earthworms are among the many soil animals important to maintain soil fertility and structure.

(*a*)

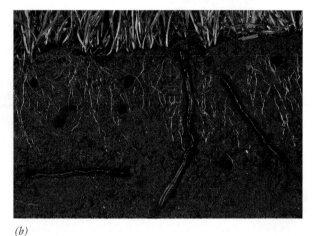

(*b*)

Figure 1.10 ■ Land cleared by African elephants, Travo National Park, Kenya. [*Source:* D. B. Botkin.]

changing forests to grasslands, and people cut down trees and plant crops (Figure 1.10). Who is to say which is more natural? In fact, few organisms do *not* change their environment.

People have known this for a long time, but the idea that people might change nature to their advantage was unpopular in the last decades of the twentieth century. At that time the word *environment* suggested something separate—"out there"—so that people were perceived apart from nature. Today, environmental sciences are showing us how people and nature connect—and in what ways this is beneficial to both.

As the environment becomes more and more recognized as important, we become more Earth-centered. We seek to spend more time in nature for recreation and spiritual activities. We accept that we have evolved on and with the Earth and are not separate from it. And we understand the need to celebrate our union with nature as we work toward sustainability.

Most people recognize that we must seek sustainability not only of the environment but also of our economic activities so that humanity and the environment can persist together. The dichotomy of the twentieth century is giving way to a new unity: the idea that a sustainable environment and a sustainable economy may be compatible—people and nature are intertwined, and a success for one involves a success for the other.

1.7 Science and Values

Deciding what to do about an environmental problem involves both values and science, as we have already seen. We must choose what we want the environment to be. To make this choice, we must first know what is possible. That requires knowing and understanding the implications of the scientific data. Once we know our options, we can select from among them. What we choose is determined by our values. An example of a value judgment regarding the world's human environmental problem is the choice between the desire of an individual to have many children and the need to find a way to limit the human population worldwide.

Once we have chosen a goal based on knowledge and values, we have to find a way to attain that goal. This step also requires knowledge. The more technologically advanced and powerful our civilization, the more knowledge is required. For example, current fishing methods make it possible for us to harvest very large numbers of chinook salmon from the Columbia River, and demand for salmon encourages us to harvest as many as possible. To determine whether chinook salmon are sustainable, we must know how many there are now and how many there have been in the past. We must also understand the processes of birth and growth for this fish, its food requirements, its habitat, its life cycle, and so forth—all the factors that ultimately determine the abundance of salmon in the Columbia River.

Consider, in contrast, the situation almost two centuries ago. When Lewis and Clark first made an expedition to the Columbia, they found many small villages of Native Americans who depended in large part on the fish in the river for food (Figure 1.11). The human population was

Figure 1.11 ■ Native Americans fishing for salmon on the Columbia River.

small, and the methods of fishing were simple. The maximum number of fish the people could catch probably posed no threat to the salmon. These people could fish without scientific understanding of numbers and processes. (This example does not suggest that prescientific societies lacked an appreciation for the idea of sustainability. On the contrary, many so-called primitive societies held strong beliefs about the limits of harvests.)

Precautionary Principle

Science and values come to the forefront when we think about what action to take about a perceived environmental problem for which the science is only partially known. This is often the case because all science is preliminary and subject to analysis of new data, ideas, and tests of hypotheses. Even with careful scientific research,

Figure 1.12 ■ The city of San Francisco, with its scenic bay-side environment, has adopted the Precautionary Principle.

it can be difficult, even impossible, to prove with absolute certainty how relationships between human activities and other physical and biological processes lead to local and global environmental problems, such as global warming, depletion of ozone in the upper atmosphere, loss of biodiversity, species extinction, and declining resources. For this reason, in 1992 the Rio Earth Summit on Sustainable Development listed as one of its principles what we now define as the **Precautionary Principle**. Basically, it says that when there is a threat of serious, perhaps even irreversible, environmental damage, we should not wait for scientific proof before taking precautionary steps to prevent potential harm to the environment.

The Precautionary Principle requires critical thinking about a variety of environmental concerns, such as the manufacture and use of chemicals including pesticides, herbicides, and drugs; the use of fossil fuels and nuclear energy; the conversion of land from one use to another (for example, from rural to urban); and the management of wildlife, fisheries, and forests.[26]

One important question in applying the Precautionary Principle is how much scientific evidence we should have before taking action on a particular environmental problem. The principle recognizes the need to evaluate all the scientific evidence we have and draw provisional conclusions while continuing our scientific investigation, which may provide additional or more reliable data. For example, when considering environmental health issues related to the use of a pesticide, we may have a lot of scientific data, but with gaps, inconsistencies, and other scientific uncertainties. Those in favor of continuing to use that pesticide may argue that there isn't enough proof to ban it. Others may argue that absolute proof of safety is necessary before a new pesticide is used. Those advocating the Precautionary Principle would argue that we should

continue to investigate but, to be on the safe side, should not wait to take cost-effective precautionary measures to prevent environmental damage or health problems.

What constitutes a cost-effective measure? Certainly we would need to examine the benefits and costs of taking a particular action versus taking no action. Other economic analyses may also be appropriate.[26, 27]

The Precautionary Principle is emerging as a new tool for environmental management and has been adopted by the city of San Francisco (Figure 1.12) and the European Union. There will always be arguments over what constitutes sufficient scientific knowledge for decision making. Nevertheless, the Precautionary Principle, even though it may be difficult to apply, is becoming a common part of environmental analysis with respect to environmental protection and environmental health issues. It requires us to apply the principle of environmental unity and predict potential consequences before they occur. The Precautionary Principle is a *proactive*, rather than a *reactive*, tool. That is, we can use it when we see real trouble coming, rather than reacting to big trouble that has already arisen.

Placing a Value on the Environment

How do we place a value on any aspect of our environment? How do we choose between two different concerns? The value of the environment is based on eight justifications: utilitarian (materialistic), ecological, aesthetic, recreational, inspirational, creative, moral, and cultural.

The **utilitarian justification** sees some aspect of the environment as valuable because it benefits individuals economically or is directly necessary to human survival. For example, mangrove swamps provide shrimp that are the basis of the livelihood of the fisherman in the opening case study. The **ecological justification** is that an

How Can We Preserve the World's Coral Reefs?

Coral reefs are among the largest, oldest, most diverse, and most beautiful communities of plants and animals. Today, many coral reefs have been seriously damaged or are at risk. Scientists estimate that approximately 10% have already been destroyed, while another 30% are threatened. The major threats to reefs are direct or indirect results of human activities. Almost 60% of the world's reefs are threatened by human activities, including coastal development, destructive fishing practices, overexploitation of resources, and marine pollution.

The pieces of coral that most people know from souvenir and jewelry shops are limestone skeletons secreted by colonies of animals related to sea anemones and jellyfish. Like their relatives, these small individual coral animals, or polyps, use tentacles equipped with stinging cells to capture food. In addition, polyps obtain nourishment from photosynthetic algae that live in their cells. When polyps die, their skeletons remain while the next generation of individuals secrete new material. Thus reefs grow slowly by accretion. Coral reefs that exist today are 5,000 to 10,000 years old. By taking the brunt of the force of waves, coral reefs protect coastlines from erosion, a function that has been estimated to have a value of $50,000 a year per square foot. In addition, reefs can provide humans with living resources (fish) and services (tourism, coastal protection) worth $375 billion per year.

Coral reefs provide homes for a vast variety of plants and animals. Approximately 25% of all marine organisms, about 1 million species, are associated with coral reefs. Reef organisms are the source of many useful chemicals and medicines, and scientists are currently searching for others. The plant and animal species found around coral reefs are linked in intricate ways, so that removing only one or two key elements may cause a catastrophic collapse. For example, overfishing in the waters off the Cook Islands in the South Pacific in the 1980s removed most of the reef's parrot fish and sea urchins, both of which feed on algae. Soon algae overgrew the reef, and the entire community of reef life collapsed.

Coral reefs have long been the main source of protein for tropical people, who today number approximately 1 billion. Because of modern transportation and preservation methods, fish and other food organisms from coral reefs are now eaten by many other people as well. In fact, reef fish constitute about 15% of the entire worldwide catch. Unfortunately, because many of the world's reefs are being overfished, some species are now rare and endangered.

Recently some consumers, especially in Asian countries, have been placing a high premium on eating fish that are alive when they reach restaurants. The demand for fish for aquariums also fuels the demand for live fish. To obtain live fish, many fishermen use dynamite to stun fish or cyanide to poison them temporarily. Both methods can kill or damage other organisms, and dynamite can destroy the reef material itself. When fish are lodged in crevices in the reef, fishermen may use crowbars to pry apart the coral so they can reach the fish. Fishing is not the only threat to coral reefs, however. The limestone material that forms the bulk of a reef is sometimes mined for use as a construction material. Millions of tourists from around the world who flock to reef areas to fish, swim, dive, and enjoy their beauty pose an additional threat. But perhaps the greatest threat to coral reefs comes from increasing population in the tropics. Population densities greater than that of the New Jersey coast (about 500 people per square kilometer) are found in parts of tropical Asia and the Caribbean. Half a billion people live within one kilometer of a coral reef. In many areas, raw or inadequately treated sewage is released near shore. Runoff from land development and deforestation adds to the burden of sediment and pollution. Degradation of the coral reef in Kaneohe Bay in Hawaii due to sewage and other runoff was dramatically reversed in the 1980s when the sewage was diverted to the open ocean. But with increasing urbanization and population growth around the bay in the 1990s, recovery slowed and perhaps even reversed.

Critical Thinking Questions

1. How does the current state of the world's coral reefs illustrate each of the six key themes of this book?
2. What are the utilitarian, ecological, aesthetic, and moral justifications for preserving coral reefs?
3. If Maitri Visetak were making his living from fishing rather than farming shrimp, how might he view the preservation of coral reefs? What arrangements could be made to meet his needs but at the same time preserve coral reefs in his area?
4. What things can you do in your everyday life to contribute to the preservation of coral reefs?

ecosystem is necessary for the survival of some species of interest to us, or that the system itself provides some benefit. For example, mangrove swamps provide habitat for marine fish, and although we do not eat mangrove trees, we may eat the fish that depend on them. Therefore conservation of the mangrove is important ecologically. And the mangroves are habitat for many noncommercial species, some endangered. As another example, burning coal and oil adds greenhouse gases to the atmosphere, which may lead to a change in climate that could affect the entire Earth. Such ecological reasons form a basis for the conservation of nature that is essentially enlightened self-interest.

Aesthetic justification has to do with our appreciation of the beauty of nature. For example, many people find wilderness scenery beautiful and would rather live in a world with wilderness than without it. One way we enjoy nature's beauty is to seek recreation in the outdoors. The aesthetic and recreational justifications are gaining a legal basis. The state of Alaska acknowledges that sea otters have an important role related to recreation: People observe and photograph the otters and enjoy viewing them in a wilderness setting (*recreational justification*). Many examples illustrate the importance of the aesthetic values of the environment. When people grieve following the death of a loved one, they typically seek out places with grass, trees, and flowers, and thus we decorate our graveyards. Conservation of nature can be based on its benefits to the human spirit (*inspirational justification*)—to benefit what is sometimes called our inner selves. Nature is an aid to human creativity (the *creative justification*). Artists and poets, among others, find a source of their creativity in their contact with nature. This is a widespread reason that people like nature, but it is rarely used in formal environmental arguments. Although popular discussions of environmental issues might make aesthetic, recreational, and inspirational elements seem superficial as justifications for the conservation of nature, in fact, beauty in their surroundings is of profound importance to people. Frederick Law Olmsted, the great American landscape planner, argued that plantings of vegetation provide medical, psychological, and social benefits and are essential to city life.[23]

Moral justification has to do with the belief that various aspects of the environment have a right to exist and that it is our moral obligation to allow them to continue or help them to persist. Moral arguments have been extended to many nonhuman organisms, to entire ecosystems, and even to inanimate objects. For example, the historian Roderick Nash has written an article entitled "Do Rocks Have Rights?" that discusses such moral justification.[28] And the United Nations General Assembly World Charter for Nature, signed in 1982, states that species have a moral right to exist.

The analysis of environmental values is the focus of a new discipline known as environmental ethics. Another concern of environmental ethics is our obligation to future generations: Do we have a moral obligation to leave the environment in good condition for our descendants, or are we at liberty to use environmental resources to the point of depletion within our own lifetimes?

Summary

◼ Six threads, or themes, run through this text: the urgency of the population issue, the importance of urban environments, the need for sustainability of resources, the importance of a global perspective, people and nature, and the role of science and values in the decisions we face.

◼ People and nature are intertwined. Each affects the other.

◼ The human population grew at a rate unprecedented in history in the twentieth century. Population growth is the underlying environmental problem.

◼ Sustainability, the environmental goal, is a long-term process to maintain a quality environment for future generations. Sustainability is becoming an important environmental paradigm for the twenty-first century.

◼ When the impact of technology is combined with the impact of population, the impact on the environment is multiplied.

◼ In an increasingly urban world, we must focus much of our attention on the environments of cities and on the effects of cities on the rest of the environment.

◼ Determining the Earth's carrying capacity for people and levels of sustainable harvests of resources is difficult but crucial if we are to plan effectively to meet our needs in the future. Estimates of the carrying capacity of the Earth for people range from 2.5 to 40 billion. The difference has to do with the quality of life projected for people—the poorer the quality of life, the more people can be packed onto the Earth.

◼ Awareness of how people at a local level affect the environment globally gives credence to the Gaia hypothesis. Future generations will need a global perspective on environmental issues.

◼ Placing a value on various aspects of the environment requires knowledge and understanding of the science but also depends on our judgments concerning the uses and aesthetics of the environment and on our moral commitments to other living things and to future generations.

◼ The Precautionary Principle is emerging as a powerful new tool for environmental management.

REEXAMINING THEMES AND ISSUES

These six key themes lead to many questions, including the following:

Human Population

What is more important: the quality of life of a person alive today or the quality of life of future generations?

Sustainability

What is more important: abundant resources today—as much as we want and can obtain—or the persistence of these resources for future generations?

Global Perspective

What is more important: the quality of your local environment or the quality of the global environment—the environment of the entire planet?

Urban World

What is more important: human creativity and innovation, including arts, humanities, and science, or the persistence of certain endangered species? Must this always be a tradeoff, or are there ways to have both?

People and Nature

If people have altered the environment for much of the time our species has been on Earth, what then is "natural"?

Science and Values

Does nature know best, so that we never have to ask what environmental goal we should seek, or do we need knowledge about our environment, so that we can make the best judgments given available information?

Key Terms

aesthetic justification **15**
carrying capacity **9**
ecological justification **13**
Gaia hypothesis **10**

megacities **11**
moral justification **15**
Precautionary
   Principle **13**

sustainability **8**
sustainable ecosystem **8**
sustainable resource
   harvest **8**

utilitarian
   justification **13**

Study Questions

1. Refer to Figure 1.6. How would you respond to the statement "the catch of salmon in the Columbia River is sustainable at 1960s levels"?

2. In what ways do the effects on the environment of a resident of a large city differ from the effects of someone living on a farm? In what ways are the effects similar?

3. Programs have been established to supply food from Western nations to starving people in Africa. Some people argue that such food programs, which may have short-term benefits, actually increase the threat of starvation in the future. What are the pros and cons of international food relief programs?

4. What are the values involved in deciding whether to create an international food relief program? What are five kinds of information required to determine the long-term effects of such programs?

5. Which of the following are global environmental problems? Why?
   a. The growth of the human population.
   b. The furbish lousewort, a small flowering plant found in the state of Maine. It is so rare that it has been seen by few people and is considered endangered.
   c. The blue whale, listed as an endangered species under the U.S. Marine Mammal Protection Act.
   d. A car that has air-conditioning.

e. Seriously polluted harbors and coastlines in major ocean ports.

6. How could you determine (a) the carrying capacity and (b) the sustainable yield of salmon in the Columbia River? (Refer to Figure 1.6.)

7. Is it possible that all the land on Earth will sometime in the future become one big city? If not, why not? To what extent does the answer depend on the following:
a. Global environmental considerations.
b. Scientific information.
c. Values.

Further Reading

Botkin, D. B. 2000. *No Man's Garden: Thoreau and a New Vision for Civilization and Nature.* Washington, D.C.: Island Press. Discusses many of the central themes of this textbook, with special emphasis on values and science and on an urban world. Henry David Thoreau's life and works illustrate approaches that can be beneficial in our dealing with modern environmental issues.

Botkin, D. B. 1990. *Discordant Harmonies: A New Ecology for the 21st Century.* New York: Oxford University Press. An analysis of the myths that lie behind attempts to solve environmental issues.

Kent, M. M. 1990. *World Population: Fundamentals of Growth.* Washington, D.C.: Population Reference Bureau. Facts and data about the growing human population.

Kessler, E., ed. 1992. "Population, Natural Resources and Development," *AMBIO* 21(1). A special issue of the journal *AMBIO* addressing many problems concerning human population growth and its economic and environmental implications.

Leopold, A. 1949. *A Sand County Almanac.* New York: Oxford University Press. Perhaps, along with Rachel Carson's *Silent Spring*, one of the most influential books of the post–World War II and pre–Vietnam War era about the value of the environment. Leopold defines and explains the land ethic and writes poetically about the aesthetics of nature.

Lutz, W. 1994. *The Future of World Population.* Washington, D.C.: Population Reference Bureau. Summary of current information on population trends and future scenarios of fertility, mortality, and migration.

Nash, R. F. 1988. *The Rights of Nature: A History of Environmental Ethics.* Madison: University of Wisconsin Press. An introduction to environmental ethics.

CHAPTER 2

Science as a Way of Knowing: Critical Thinking about the Environment

Forms of life seem so incredible and so well fitted for their environment that we wonder how they have come about. This question leads us to seek to understand different ways of knowing.

Learning Objectives

Science is a process of refining our understanding of nature through continual questioning and active investigation. It is more than a collection of facts to be memorized. After reading this chapter, you should understand that:

■ Thinking about environmental issues requires thinking scientifically.

■ Scientific knowledge is acquired through observations of the natural world that can be tested through additional observations and experiments and can be disproved.

■ Scientific understanding is not fixed, but changes over time as new data, observations, theories, and tests become available.

■ Deductive and inductive reasoning are different, and both are used in scientific thinking.

■ Every measurement involves some degree of approximation—that is, uncertainty—and that a measurement without a statement about its degree of uncertainty is meaningless.

■ Technology, the application of scientific knowledge, is not science, but science and technology interact, stimulating growth in each other.

■ Decision-making about environmental issues involves society, politics, culture, economics, and values, as well as scientific information.

Birds at Mono Lake:
Applying Science to Solve
an Environmental Problem

Mono Lake is a large salt lake in California, just east of the Sierra Nevada and across these mountains from Yosemite National Park (Figure 2.1). More than a million birds use the lake; some feed and nest there, some stop on their migrations to feed. Within the lake, brine shrimp and brine fly larvae grow in great abundance, providing food for the birds. The shrimp and fly larvae, in turn, feed on algae and bacteria that grow in the lake (Figure 2.2).

The lake persisted for thousands of years in a desert climate because streams from the Sierra Nevada—fed by mountain snow and rain—flowed into it. But in the 1940s, the city of Los Angeles diverted all stream water—beautifully clear water—to provide 17% of the water supply for the city. The lake began to dry out. It covered 60,000 acres in the 1940s, but only 40,000 by the 1980s.

Environmental groups expressed concern that the lake would soon become so salty and alkaline that all the brine shrimp and flies would die, the birds would no longer be able to nest or feed there, and the beautiful lake would become a hideous eyesore—much like what happened to the Aral Sea in Asia. The Los Angeles Department of Water and Power argued that everything would be all right, because rain falling directly on the lake and water flowing underground would provide ample water for the lake. "Save Mono Lake" became a popular bumper sticker in California, and the argument about the future of the lake raged for more than a decade.

Scientific information was needed to answer the key questions: Without stream input, how small would the lake become? Would it become too salty and alkaline for the shrimp, fly larvae, algae, and bacteria? If so, when?

The state of California set up a scientific panel to study of the future of Mono Lake. The panel discovered that two crucial pieces of knowledge needed to answer these questions had not been

studied: the size and shape of the basin of the lake (so one could determine the lake's volume, and from this how its salinity and alkalinity would change) and the rate at which water evaporated from the lake (to determine whether and how fast the lake would become too dry to sustain life within it). New research was commissioned that answered these questions. The answers: Yes, the lake would become too salty for the shrimp, fly larvae, algae, and bacteria, and would do so by about 2003.[1] With this scientific information in hand, the courts decided that Los Angeles would have to stop all removal of the water that flowed into Mono Lake. By 2006, the lake still had not recovered to the level required by the counts, indicating that diversion of water had been undesirable for the lake and its ecosystem.

Scientific information told Californians what would happen and when. *What to do* about the lake was a question of values. In the end, decisions based on values and scientific knowledge were made by the courts, which stopped the city from diverting any of the stream waters that flowed into the lake. The birds, scenery, brine shrimp, and brine flies were saved.

Figure 2.1 ■ Mono Lake.

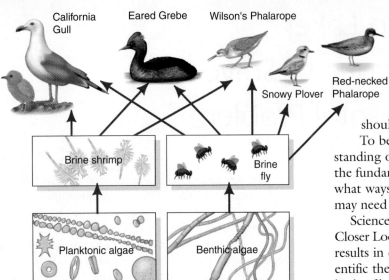

Figure 2.2 ■ The Mono Lake food chain. The arrows show who feeds on whom. Just five species of birds are the top predators. This lake is one of the simpler ecosystems.

Science provided knowledge about what would happen and what management approaches were possible. This knowledge was combined with values about people and nature to choose policies and actions.[2]

2.1 Understanding What Science Is (and What It Isn't)

Modern civilization depends on science and its application—from iPods to atomic weapons. Therefore, each of us needs to understand what science is and what it is not. "What is science?" became a headline topic in January 2005 when the school board of Dover, Pennsylvania, proposed that a theory called "intelligent design" be taught along with the scientific theory of evolution. In December 2005, a federal judge ruled that this was unconstitutional. The judge stated that intelligent design was a religious viewpoint and therefore not science, and that a religious belief could not be taught in biology courses as an alternative to evolution.*

So what is science, and how does it differ from other ways of knowing? And what is *environmental science*?

Science as a Way of Knowing

The complexity of environmental sciences raises several questions: How can we understand ecological phenom-

ena? How can we take a scientific approach to the great, ancient questions about nature around us, the source of great wonder at the complexity of life and the marvelous adaptations of living things? How can we seek answers to practical questions about the effects of people on nature and the actions people should take in solving environmental problems?

To begin to tackle the problem of a scientific understanding of life and its environment, it is useful to review the fundamentals of the scientific method and consider in what ways ecology can fit that mold and in what ways it may need to cast new molds.

Science has a long history in Western civilization. (See A Closer Look 2.1.) Science is a process, a way of knowing. It results in conclusions, generalizations, and sometimes scientific theories and even scientific laws. These comprise a body of beliefs. Often people confuse the process of science with a fixed set of beliefs—the results. But science does not lead so much to a fixed set of beliefs as to a set of beliefs that, at the present time, allow us to explain all known observations about a kind of phenomenon and make predictions about that kind of phenomenon.

Science is a process of discovery—a continuing process whose essence is change in ideas. Often, the fact that scientific ideas change seems frustrating. Why can't scientists agree on what is the best diet for people? Why is a chemical considered dangerous in the environment for a while and then determined not to be? Why do scientists in one decade believe that fire in nature is an undesirable disturbance and in a later decade decide that it is important and natural? Is there going to be global warming or not? Can't scientists just find out the truth for each of these questions, once and for all, and agree on it?

Rather than looking to science for answers to such questions, it is more accurate to think of science as a continuing adventure of making increasingly better approximations about how the world works. Sometimes changes in ideas are small, and the major context remains the same. Sometimes a science undergoes a fundamental revolution in ideas.

Science is one way of looking at the world. It begins with observations about the natural world, such as: How many birds nest at Mono Lake? What species of algae live in the lake? Under what conditions do the algae live? From these observations, scientists formulate hypotheses that can be tested. For example, brine shrimp die when the salinity of the water reaches three times that of seawater. Modern science does not deal with things that cannot be tested by

*Laurie Goodstein and Kenneth Chang contributed reporting from New York for this article, "Issuing Rebuke, Judge Rejects Teaching of Intelligent Design," *New York Times*, December 21, 2005.

A CLOSER LOOK 2.1

A Brief History of Science

Thinking scientifically about the environment is as old as science itself. Science had its beginnings in the ancient civilizations of Babylonia and Egypt. There, observations of the environment were carried out primarily for practical reasons, such as planting crops, or for religious reasons, such as using the positions of the planets and stars to predict events. These ancient practices differed from modern science in that they did not distinguish between science and technology or between science and religion. As we will see in this chapter, science is a way of knowing. Technology is the means people use to obtain physical needs and benefits. In our scientific age, technology has become the application of scientific knowledge to achieve some practical goal.

Because of their interest in ideas, the ancient Greeks developed a more theoretical approach to science. Knowledge for its own sake, rather than for practical purposes, became a primary goal. At the same time, the Greeks' philosophical approach began to move science away from religion and toward philosophy.

Modern science is usually considered to have its roots in the end of the sixteenth and the beginning of the seventeenth centuries, with the development of the scientific method by Gilbert (magnets), Galileo (physics of motion), and Harvey (circulation of blood). Unlike earlier classical scientists—who asked "Why?" in the sense of "For what purpose?"—they made important discoveries by asking "How?" in the sense of "How does it work?" Galileo also pioneered the use of numerical observations and mathematical models. The scientific method, which quickly proved

very successful in advancing knowledge, was first described explicitly by Francis Bacon in 1620. Although not a practical scientist himself, Bacon recognized the importance of the scientific method, and his writings did much to promote scientific research.[3]

Our cultural heritage, then, gives us several ways of thinking about the environment, including the kind of thinking we do in everyday life and the kind of thinking scientists do (Table 2.1). There are many similarities between thinking in modern everyday life and in science. This means that thinking scientifically is something all of us can do. At the same time, there are crucial differences. Ignoring these differences can lead to invalid conclusions, in general, and to serious errors in decisions about the environment.

Table 2.1 • Knowledge in Everyday Life Compared with Knowledge in Science

Factor in	Everyday Life and in	Science
Goal	To lead a satisfying life (implicit)	To know, predict, and explain (explicit)
Requirements	Context-specific knowledge; no complex series of inferences; can tolerate ambiguities and lack of precision	General knowledge; complex, logical sequences of inferences; must be precise and unambiguous
Resolution of questions	Through discussion, compromise, consensus	Through observation, experimentation, logic
Understanding	Acquired spontaneously through interacting with world and people; criteria not well defined	Pursued deliberately; criteria clearly specified
Validity	Assumed, no strong need to check; based on observations, common sense, tradition, authorities, experts, social mores, faith	Must be checked; based on replications, converging evidence, formal proofs, statistics, logic
Organization of knowledge	Network of concepts acquired through experience; local, not integrated	Organized, coherent, hierarchical, logical; global, integrated
Acquisition of knowledge	Perception, patterns, qualitative; subjective	Plus formal rules, procedures, symbols, statistics, mental models; objective
Quality control	Informal correction of errors	Strict requirements for eliminating errors and making sources of error explicit

Source: Based on F. Reif and J. H. Larkin, "Cognition in Scientific and Everyday Domains: Comparison and Learning Implications," *Journal of Research in Science Teaching* 28(9), pp. 733–760. Copyright © 1991 by National Association for Research in Science Teaching. Reprinted by permission of John Wiley & Sons.

observation, such as the ultimate purpose of life or the existence of a supernatural being. Science also does not deal with questions that involve values, such as standards of beauty or issues of good and evil—for example, whether

the scenery at Mono Lake is beautiful. Both science and values are important, as the case study of Mono Lake illustrates; that is why this connection is one of the key themes of this book. The criterion by which we decide whether a statement is in the realm of science is:

Whether it is possible, at least in principle, to disprove the statement.

Disprovability

It is generally agreed today that the essence of the scientific method is **disprovability** (see Figure 2.3; a diagram that will be helpful throughout this chapter). A statement can be said to be scientific if someone can state a method by which it could be disproved. Thus, if you can think of a test that could disprove a statement, then that statement can be said to be scientific. If you cannot think of such a test, then the statement is said to be nonscientific. For example, consider the crop circles discussed in A Closer Look 2.2. One Web site states that some people believe the crop circles are a "spiritual nudge" which is "designed to awaken us to our larger context and milieu, which is none other than our collective earth soul." Whether or not this is true, it does not seem to be a statement open to disproof. The statement that "the scenery at Mono Lake is pretty" also is not disprovable. On the other hand, the statement that "more than 50% of the people who visit Mono Lake find the scenery beautiful" could be tested by public opinion surveys and therefore can be treated as a scientific statement—actually as a hypothesis (discussed later).

There are many ways of looking at the world, such as the religious, aesthetic, and moral views; they are not science, however, because their assertions are not open to disproof in the scientific sense. They are based ultimately on faith, beliefs, and cultural and personal choices. The distinction between a scientific statement and a nonscientific statement is not a value judgment—the distinction is not meant to imply that science is the only "good" kind of knowledge. The distinction is simply a philosophical one about kinds of knowledge and logic. To say that these other ways of looking at the world are not science is not to denigrate them. Each way of viewing the world gives us a different way of perceiving and of making sense of our world, and each is valuable to us.

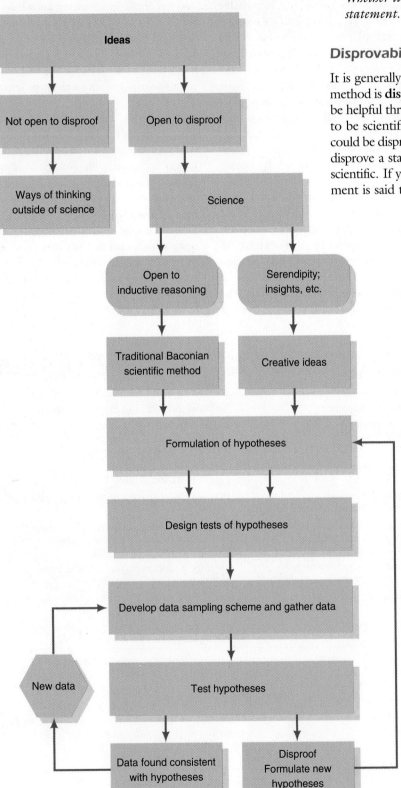

Schematic diagram of the scientific method

Figure 2.3 ■ Schematic diagram of the scientific method. This diagram shows the steps in the scientific method, both traditional and nontraditional, as explained in the text.

The Case of the Mysterious Crop Circles

For 13 years, circular patterns appeared "mysteriously" in grain fields in southern England (Figure 2.4). Proposed explanations included aliens, electromagnetic forces, whirlwinds, and pranksters. The mystery generated a journal and a research organization headed by a scientist, as well as a number of books, magazines, and clubs devoted solely to crop circles. Scientists from Great Britain and Japan brought in scientific equipment to study the strange patterns. Then, in September 1991, two men confessed to having created the circles by entering the fields along paths made by tractors (to disguise their footprints) and dragging planks through the fields. When they made their confession, they demonstrated their technique to reporters and some crop-circle experts.[4, 5]

In spite of their confession, some people continue to believe that the crop circles have some alternative causes. Crop-circle organizations continue to exist and now have Web sites. For example, a report published on the World Wide Web in 2003 stated that "strange orange lightning" was seen one evening and that crop circles appeared the next day.[4, 6]

How is it that so many people, including some scientists, took, and still take, crop circles seen in England seriously? The answer is that they misunderstood the scientific method and engaged in fallacious reasoning—and that at some level some people want to believe in a mysterious cause for the circles and choose to ignore standard scientific analyses and methods. The failure of some people to think critically about crop circles causes no harm, but the same type of thinking applied to other, more serious environmental issues, can have serious consequences. There are two kinds of failures regarding the scientific method illustrated by crop circles: first, that the scientific method is used incorrectly; and second, that scientific information is rejected by choice. Both failures occur in environmental problems.

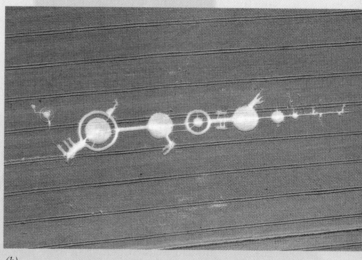

Figure 2.4 ■ (*a*) A crop circles close-up at the Vale Pwesey in southern England in July 1990. (*b*) Crop circles seen from the air make distinctive patterns.

Assumptions of Science

Science makes certain assumptions about the natural world. To understand what science is, it is important to be aware of these assumptions.

■ Events in the natural world follow patterns that can be understood through careful observation and scientific analysis, which we will describe later.

■ These basic patterns and the rules that describe them are the same throughout the universe.

■ Science is based on a type of reasoning known as *induction*; it begins with specific observations about the natural world and extends to generalizations.

■ Generalizations can be subjected to tests that may disprove them. If such a test cannot be devised, then a generalization cannot be treated as a scientific statement.

■ Although new evidence can disprove existing scientific theories, science can never provide absolute proof of the truth of its theories.

The Nature of Scientific Proof

One source of serious misunderstanding about science is the use of the word *proof*, which most students encounter in mathematics, particularly in geometry. Proof in mathematics and logic involves reasoning from initial definitions and assumptions. If a conclusion follows logically from these **premises**, the conclusion is said to be proven. This process is known as **deductive reasoning**.

An example of deductive reasoning is the following syllogism, or series of logically connected statements.

> *Premise: A straight line is the shortest distance between two points.*
> *Premise: The line from A to B is the shortest distance between points A and B.*
> *Conclusion: Therefore, the line from A to B is a straight line.*

Note that the conclusion in this syllogism follows directly from the premises.

Deductive proof does not require that the premises be true, only that the reasoning be foolproof.

Logically valid but untrue statements can result from false premises, as in the following example:

> *Premise: Humans are the only toolmaking organisms.*
> *Premise: The woodpecker finch uses tools.*
> *Conclusion: Therefore, the woodpecker finch is a human being.*

In this case, the concluding statement must be true if both the preceding statements are true. However, we know that the conclusion is not only false but ridiculous. If the second statement is true (which it is), then the first cannot be true. The conclusion that a bird, the woodpecker finch, which uses cactus spines to remove insects from tree limbs (Figure 2.5), is human follows logically in the syllogism but defies common sense.

Because science is based on observations, its conclusions are only as true as the premises from which they are deduced. *The rules of deductive logic govern only the process of moving from premises to conclusion.* Science, in contrast, requires not only logical reasoning but also correct premises. Returning to the example of the woodpecker finch, to be scientific the three statements should be expressed conditionally (with reservation):

> *If humans are the only toolmaking organisms*
>   *and*
> *the woodpecker finch is a toolmaker,*
>   *then*
> *the woodpecker finch is a human being.*

Figure 2.5 ■ A woodpecker finch in the Galápagos Islands uses a twig to remove insects from a hole in a tree, demonstrating tool use by nonhuman animals.

When we formulate generalizations based on a number of observations, we are engaging in **inductive reasoning**. To illustrate, one of the birds that feeds at Mono Lake is the eared grebe. The "ears" are a fan of golden feathers that occur behind the eyes of males during the breeding season. Let us define birds with these golden feather fans as eared grebes (Figure 2.6). If we always observe that the breeding male grebes have this feather fan, we may make the inductive statement "All male eared grebes have golden feathers during the breeding season." What we really mean is, "All of the male eared grebes we have seen in the breeding season have golden feathers." We never know

Figure 2.6 ■ Male eared grebe.

when our very next observation will turn up a bird that is like a male eared greed in all ways except that it lacks these feathers in the breeding season. This is not impossible, somewhere, due to a mutation.

Proof in inductive reasoning is therefore very different from proof in deductive reasoning. When we say something is proven in induction, what we really mean is that it has a very high degree of probability. **Probability** is a way of expressing our certainty (or uncertainty)—our estimation of how good our observations are, how confident we are of our predictions.

When we have a fairly high degree of confidence in our conclusions in science, we often forget to state the degree of certainty or uncertainty. Instead of saying, "There is a 99.9% probability that . . . ," we say, "It has been proved that. . . ." Unfortunately, many people interpret this as a deductive statement, meaning the conclusion is absolutely true, which has led to much misunderstanding about science. Although science begins with observations and therefore inductive reasoning, deductive reasoning is useful in helping scientists analyze whether conclusions based on inductions are logically valid. *Scientific reasoning combines induction and deduction*—different but complementary ways of thinking.[7]

What we have just described is the classic scientific method. Scientific advances, however, often begin with instances of insight—leaps of imagination, which are then subjected to the stepwise inductive process. Some scientists have made major advances by being in the right place at the right time and knowing the right things at the right time. For example, penicillin was discovered "by accident" in 1928 when Sir Alexander Fleming was studying the pus-producing bacteria *Staphylococcus aureus*. A culture of these bacteria was accidentally contaminated by the green fungus *Penicillium notatum*. Fleming noticed that the bacteria did not grow in those areas of a culture where the fungus grew. He isolated the mold, grew it in a fluid medium, and found that it produced a substance that killed many of the bacteria that caused diseases. Eventually, this discovery led other scientists to develop an injectable agent to treat diseases. *Penicillium notatum* is a common mold found on stale bread. No doubt many others had seen the mold, perhaps even noticing that other strange growths on bread did not overlap with *Penicillium notatum*. It took Fleming's knowledge and observational ability for this piece of "luck" to occur.

2.2 Measurements and Uncertainty

A Word about Numbers in Science

We communicate scientific information in several ways. One is the written word for conveying synthesis, analysis, and conclusions. When we add numbers to our analysis, we obtain another dimension of understanding that goes beyond qualitative understanding and synthesis of a problem. Using numbers and statistical analysis allows us to visualize relationships in graphs and make predictions. It also allows us to analyze the strength of a relationship and in some cases discover a new relationship.

People in general put more faith in the accuracy of measurements than do scientists. Scientists realize that every measurement is only an approximation. Measurements are limited; they depend on the instruments used and the people who use the instruments. Measurement uncertainties are inevitable; they can be reduced, but they can never be completely eliminated. Because all measurements have some degree of uncertainty, a measurement is meaningless unless it is accompanied by an estimate of its uncertainty.

Consider the loss of the *Challenger* space shuttle in 1986, the first major space shuttle accident, which appeared to be the result of the failure of rubber O-rings that were supposed to hold sections of rockets together. Imagine a simplified scenario in which an engineer is given a rubber O-ring used to seal fuel gases in a space shuttle. She is asked to determine the flexibility of the O-rings under different temperature conditions to help answer the questions "At what temperature do the O-rings become brittle and subject to failure?" and "At what temperature(s) is it unsafe to fire the shuttle?"

After doing some tests, the engineer says that the rubber becomes brittle at −1°C (30°F). Is it safe to launch the shuttle at 0°C (32°F)? At this point, you do not have enough information to answer the question. You assume that the temperature data may have some degree of uncertainty, but you have no idea how large it is. Is the uncertainty ±5°C, ±2°C, or ±0.5°C? To make a reasonably safe and economically sound decision about whether to launch the shuttle, you must know the uncertainty of the measurement.

Dealing with Uncertainties

There are two sources of uncertainty. One is the real variability of nature. The other is the fact that every measurement has some error. Measurement uncertainties and other errors that occur in experiments are called *experimental errors*. Errors that occur consistently, such as those resulting from incorrectly calibrated instruments, are *systematic errors*.

Scientists traditionally include a discussion of experimental errors when they report results. Often, error analysis leads to greater understanding and sometimes even to important discoveries. For example, the existence of the eighth planet in our solar system, Neptune, was discovered when scientists investigated apparent inconsistencies—observed "errors"—in the orbit of the seventh planet, Uranus. Measurement uncertainties can be reduced by improving the instruments used and by requiring a standard procedure in making measurements. They can also be reduced by using carefully designed experiments and appropriate statistical procedures. However, uncertainties are inherent in every measurement and can never be completely eliminated.

As difficult as it is for us to live with uncertainty, that is the nature of nature, as well as the nature of measurement and of science. We need to use our understanding of the uncertainties in measurements to read reports of scientific studies critically, whether they appear in science journals or in popular magazines and newspapers. (See A Closer Look 2.3.)

Accuracy and Precision

Certain measurements have been made very carefully by many people over a long period, and accepted values have been determined. *Accuracy* is the extent to which a measurement agrees with the accepted value. *Precision* is the degree of exactness with which a quantity is measured. A very precise measurement may not be accurate. If the accepted value for the temperature of boiling water at sea level is $100.000°C$ ($212.000°F$) and you measure it as $99.875°C$ ($211.775°F$), your measurement is just as precise as the accepted value—that is, it is measured to the nearest $1/1000$ degree—but it is still slightly inaccurate. The accuracy of observations is checked by comparison with observations made by scientists.

Although it is important to make measurements as precisely as possible, it is equally important not to report measurements with more precision than they warrant. Doing so conveys a misleading sense of both precision and accuracy. A measurement based on a ruler divided into tenths of a centimeter should be precise to the nearest hundredth of a centimeter, but no more. A small bone that appears to end exactly at the 2.1-cm mark should be reported as 2.10 cm long, not as 2.100 cm. The latter would imply that the right-hand zero was read directly, when in fact it was estimated.

2.3 Observations, Facts, Inferences, and Hypotheses

It is important to distinguish between observations and inferences, which are ideas based on observations. **Observations**, the basis of science, may be made through any of the five senses or by instruments that measure beyond what we can sense. An **inference** is a generalization that arises from a set of observations. When what is observed about a particular thing is agreed on by all, or almost all, it is often called a **fact**.

We might observe that a substance is a white, crystalline material with a sweet taste. We might *infer* from these observations alone that the substance is sugar. Before this inference can be accepted as fact, however, it must be subjected to further tests. Confusing observations with inferences and accepting untested inferences as facts are kinds of sloppy thinking often described by the phrase "thinking makes it so."

When scientists wish to test an inference, they convert it into a statement that can be disproved. This type of statement is known as a **hypothesis**. A hypothesis contin-

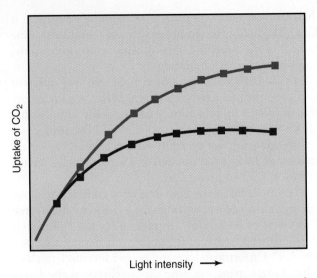

Figure 2.7 ■ *Dependent and Independent variables.* Photosynthesis as affected by light. In this diagram, photosynthesis is represented by carbon dioxide (CO_2) uptake. Light is the independent variable (CO^2), uptake is the dependent variable. The blue and red lines represent two plants with different responses to light.

ues to be accepted until it is disproved. If a hypothesis that has not been disproved, it still has not been proved true in the deductive sense; it has only been found to be probably true until and unless evidence to the contrary is found.

Typically, hypotheses take the form of if–then statements. For example, a scientist is trying to understand how growth of a plant will change with the amount of light it receives. He measures the rate of photosynthesis at a variety of light intensities (Figure 2.7). The rate of photosynthesis is called the **dependent variable** because it is affected by, and in this sense depends on, the amount of light, which is called the **independent variable**. The independent variable is also sometimes called a **manipulated variable**, because the scientist deliberately changes, or manipulates, it. The dependent variable is then referred to as a **responding variable**—one that responds to changes in the manipulated variable.

As trees grow, many **variables** exist. Some, like the position of the North Star, can be assumed to be irrelevant. Others, like the duration of daylight, are potentially relevant. In testing a hypothesis, a scientist tries to keep all relevant variables constant, except for the independent and dependent variables. This practice is known as *controlling variables*. In a **controlled experiment**, the experiment is compared to a standard, or control—an exact duplicate of the experiment except for the condition of the one variable being tested (the independent variable). Any difference in outcome (dependent variable) between the experiment and the control can be attributed to the effect of the independent variable.

An important aspect of science, but one often overlooked in descriptions of the scientific method, is the need to define or describe variables in exact terms that can be understood by all scientists. The least ambiguous way to define or describe a variable is in terms of what one would

A CLOSER LOOK 2.3

Measurement of Carbon Stored in Vegetation

A number of people have suggested that a partial solution to global warming might be a massive worldwide program of tree planting. Trees take carbon dioxide (an important greenhouse gas) out of the air in the process of photosynthesis. Because they live a long time, trees can store carbon for decades, even centuries. But how much carbon can be stored? Many books and reports published during the past 20 years contained numbers representing the total stored carbon in the vegetation of Earth, but all were presented without any estimate of error (Table 2.2). Without an estimate of that uncertainty, the figures are meaningless, yet important environmental decisions have been based on them.

Recent studies have reduced the error by replacing guesses and extrapolations with scientific sampling techniques, similar to the procedures used to predict the outcomes of elections. Even these improved data would be meaningless, however, without an estimate of error. The new figures show that the earlier estimates were three to four times too large, grossly overestimating the storage of carbon in vegetation and, therefore, the contribution tree planting could make in offsetting global warming.

Table 2.2 • Estimates of Above-Ground Biomass in the North American Boreal Forest				
Source	Biomass[a] (kg/m^2)	Carbon[b] (kg/m^2)	Total Biomass[c] (10^9 metric tons)	Total Carbon[c] (10^9 metric tons)
This study[d]	4.2 ± 1.0	1.9 ± 0.4	22 ± 5	9.7 ± 2
Previous estimates[e]				
1	17.5	7.9	90	40
2	15.4	6.9	79	35
3	14.8	6.7	76	34
4	12.4	5.6	64	29
5	5.9	2.7	30	13.8

Source: D. B. Botkin and L. Simpson, "The First Statistically Valid Estimate of Biomass for a Large Region," *Biogeochemistry* 9 (1990): 161–274. Reprinted by permission of Kluwer Academic, Dordrecht, The Netherlands.

[a]Values in this column are for total above-ground biomass. Data from previous studies giving total biomass have been adjusted using the assumption that 23% of the total biomass is in below-ground roots. Most references use this percentage; Leith and Whittaker use 17%. We have chosen to use the larger value to give a more conservative comparison.

[b]Carbon is assumed to be 45% of total biomass following R. H. Whittaker, *Communities and Ecosystems* (New York: Macmillan, 1974).

[c]Assuming our estimate of the geographic extent of the North American boreal forest: 5,126,427 km^2 (324,166 mi^2).

[d]Based on a statistically valid survey; above-ground woodplants only.

[e]Lacking estimates of error: Sources of previous estimates by number (1) G. J. Ajtay, P. Ketner, and P. Duvigneaud, "Terrestrial Primary Production and Phytomass," in B. Bolin, E. T. Degens, S. Kempe, and P. Ketner, eds., *The Global Carbon Cycle* (New York: Wiley, 1979), pp. 129–182. (2) R. H. Whittaker and G. E. Likens, "Carbon in the Biota," in G. M. Woodwell and E. V. Pecam, eds., *Carbon and the Biosphere* (Springfield, Va.: National Technical Information Center, 1973), pp. 281–300. (3) J. S. Olson, H. A. Pfuderer, and Y. H. Chan, *Changes in the Global Carbon Cycle and the Biosphere*, ORNL/EIS-109 (Oak Ridge, Tenn.: Oak Ridge National Laboratory, 1978). (4) J. S. Olson, I. A. Watts, and L. I. Allison, *Carbon in Live Vegetation of Major World Ecosystems*, ORNL-5862 (Oak Ridge, Tenn.: Oak Ridge National Laboratory, 1983). (5) G. M. Bonnor, *Inventory of Forest Biomass in Canada* (Petawawa, Ontario: Canadian Forest Service, Petawawa National Forest Institute, 1985).

have to do to duplicate the variable's measurement. Such definitions are called **operational definitions**. Before carrying out an experiment, both the independent and dependent variables must be defined operationally. Operational definitions allow other scientists to repeat experiments exactly and to check on the results reported.

While performing an experiment, a scientist must record the values of the input (independent variable) and the output (dependent variable). These values are referred to as *data* (singular: *datum*). They may be numerical, **quantitative data** or nonnumerical, **qualitative data**. In our example, qualitative data would be the species of a tree; quantitative data would be the mass in grams or the diameter in centimeters.

Knowledge in an area of science grows as more hypotheses are supported. Because hypotheses in science are continually being tested and evaluated by other scientists, science has a built-in self-correcting feedback system.

Scientists use accumulated knowledge to develop **explanations** that are consistent with currently accepted hypotheses. Sometimes an explanation is presented as a model. A **model** is "a deliberately simplified construct of nature."[8] It may be a physical working model, a pictorial model, a set of mathematical equations, or a computer simulation. For example, the U.S. Army Corps of Engineers has a physical model of San Francisco Bay, open to the public to view. It is a miniature in a large aquarium, with the topography of the bay reproduced to scale and into which water flows following tidal patterns. Elsewhere, the Army Corps develops mathematical equations and computer simulations that attempt to explain some aspects of such water flow.

As new knowledge accumulates, models may have to be revised or replaced, with the goal of finding models more consistent with nature.[8] Computer simulation of the atmosphere has become important in scientific analysis of the possibility of global warming. Computer simulation is becoming important for biological systems as well, such as simulations of forest growth (Figure 2.8). *Models that offer broad, fundamental explanations of many observations are called* **theories**.

The ideas discussed in this section are usually referred to as the **scientific method** and can be presented as a series of steps:

1. Make observations and develop a question about the observations.
2. Develop a tentative answer to the question—a hypothesis.
3. Design a controlled experiment to test the hypothesis (implies identifying and defining independent and dependent variables).
4. Collect data in an organized form, such as a table.
5. Interpret the data visually through graphs, quantitatively using statistical analysis, or other means.
6. Draw a conclusion from the data.
7. Compare the conclusion with the hypothesis and determine whether the results support or disprove the hypothesis.
8. If the hypothesis is shown to be consistent with observations in some limited experiments, conduct additional experiments to test it further. If the hypothesis is rejected, make additional observations and construct a new hypothesis (Figure 2.9).

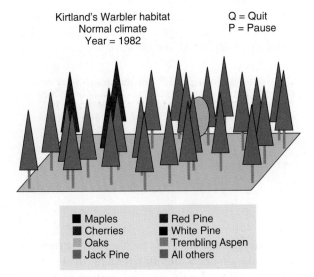

Kirtland's Warbler habitat
Normal climate
Year = 1982

Q = Quit
P = Pause

- ■ Maples
- ■ Cherries
- ■ Oaks
- ■ Jack Pine
- ■ Red Pine
- ■ White Pine
- ■ Trembling Aspen
- ■ All others

Figure 2.8 ■ A computer simulation of forest growth. Shown here is a screen display of individual trees, whose growth is forecast year by year, depending on environmental conditions. In this computer run, only three types of trees are present. This kind of model is becoming increasingly important in environmental sciences. [*Source:* From *JABOWA-II* by D. B. Botkin. Copyright © 1993 by D. B. Botkin. Used by permission of Oxford University Press, Inc.]

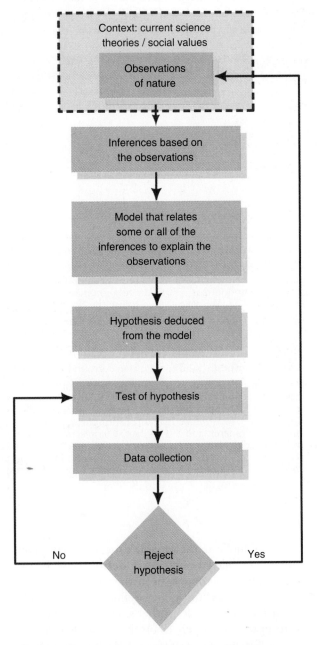

Figure 2.9 ■ Scientific investigation as a feedback process. [*Source:* Modified from C. M. Pease and J. J. Bull, *Bioscience* 42 (April 1992): 293–298.]

2.4 A Word about Creativity and Critical Thinking

Creativity in science, as in other areas of knowledge, has to do with original thoughts (those never stated before). Many people have the ability to be creative, but in science it helps to be intensely curious about how something in nature works. Sometimes creativity comes as an inspiration or sudden idea—such as the possible answer to an old question. Suppose you are trying to find out why a particular species of tree is dying in a particular region. You may have a creative hunch that it's due to human-caused acid rain (rainfall that is acidic due to air pollution; see Chapter 23). This is a hypothesis that requires testing, and this is where critical thinking in applying the scientific method enters the picture. To think critically, you think of ways to test a hypothesis and use your creative skills to form alternative hypotheses that will also be tested. You synthesize what is known about the question you are asking, then gather and analyze data in an attempt to disprove your hypotheses. You examine all sides of the problem you are studying, evaluate each in terms of the available evidence—physical, chemical, and biological data—and reach a tentative conclusion.

You may also think creatively and critically about a particular environmental problem without gathering your own data and testing hypotheses. In this case, you obtain and synthesize available data, analyze the data in terms of its uncertainties, examine all sides of the problem to the extent possible, and draw a tentative conclusion. Some scientists have made important creative discoveries by re-examining available data gathered for other purposes. For example, gathering data on the abundance of tiny single-celled marine organisms (*Foraminifera*) and the sulfur particles they produce and release into the atmosphere led to the discovery that these small particles of sulfur compounds serve as nuclei around which water vapor condenses and droplets form. They are important in making clouds that produce precipitation in the ocean and on land when the clouds move over land. This discovery was important to our understanding about the global linkages among ocean water, organisms, chemical compounds, clouds, and rain.

2.5 Misunderstandings about Science

The scientific method as we have described it so far does not take into account differences among the various disciplines of science. The logic of research is not the same in physics, for example, as in biology. In fact, within biology itself, research on evolution, for example, differs in significant ways from research on ecology. It is much more realistic to speak of the *methods of science* than of the scientific method.

Theory in Science and Language

A common misunderstanding about science arises from confusion between use of the word *theory* in science and in everyday language. A **scientific theory** is a grand scheme that relates and explains many observations and is supported by a great deal of evidence. In contrast, in everyday usage, a theory can be a guess, a hypothesis, a prediction, a notion, a belief. We often hear the phrase "It's just a theory." That may make sense in everyday language but not in the language of science. In fact, theories have tremendous prestige and are considered the greatest achievements of science.[9]

Further misunderstanding arises when scientists use the word *theory* in several different senses.[7] For example, we may encounter references to a currently accepted, widely supported theory, such as the theory of evolution by natural selection; a discarded theory, such as the theory of inheritance of acquired characteristics; a new theory, such as the theory of evolution of multicellular organisms by symbiosis; and a model dealing with a specific or narrow area of science, such as the theory of enzyme action.[9]

One of the most important misunderstandings about the scientific method pertains to the relationship between research and theory. Although theory is usually presented as growing out of research, in fact theories also guide research. The very observations a scientist makes occur in the context of existing theories. At times, discrepancies between observations and accepted theories become so great that a scientific revolution occurs; old theories are discarded and replaced with new or significantly revised theories.[10]

Science and Technology

Another misunderstanding about science occurs when science is confused with technology. As noted earlier, science is a search for understanding of the natural world, whereas technology is the application of scientific knowledge in an attempt to benefit human beings. Science often leads to technological developments, just as new technologies lead to scientific discoveries. The telescope began as a technological device, such as an aid to sailors, but when Galileo used it to study the heavens, it became a source of new scientific knowledge. That knowledge stimulated the technology of telescope making, leading to the production of better telescopes, which in turn led to further advances in astronomy.

Science is limited by the technology available. Before the invention of the electron microscope, scientists were limited to magnifications of 1,000 times and to studying objects about the size of one-tenth of a micrometer. (A micrometer is 1/1,000,000 of a meter, or 1/1,000 of a millimeter.) The electron microscope enabled scientists to view objects far smaller by magnifying more than 100,000 times. The electron microscope, a basis for new science, was also the product of science. Without prior scientific knowledge about electron beams and how to focus them, the electron microscope could not have been developed.

Most of us do not come in direct contact with science in our daily lives; instead, we come in contact with the products of science—technological devices such as computers, iPods, and microwave ovens. Thus, people tend to confuse the products of science with science itself. As you study science, it will help if you keep in mind the distinction between science and technology.

Science and Objectivity

One myth about science is the myth of objectivity, or value-free science—the idea that scientists are capable of complete objectivity independent of their personal values and the culture in which they live and that science deals only with objective facts. Objectivity is certainly a goal of scientists, but it is unrealistic to think that they can be totally free of influence from their social environments and personal values. A more realistic view is to admit that scientists do have biases and to try to identify these biases rather than ignore them. In some ways, this situation is similar to that of measurement error. It is inescapable, and we can best deal with it by recognizing it and estimating its effects.

To find examples of how personal and social values affect science, we have only to look at recent controversies about environmental issues, such as whether or not to adopt more stringent automobile emission standards. Genetic engineering, nuclear power, and the preservation of threatened or endangered species involve conflicts among science, technology, and society.

The idea that science is not entirely value-free should not be taken to mean that fuzzy thinking is acceptable in science. It is still important to think critically and logically about science and related social issues. Without the high standards of evidence held up as the norm for science, we run the danger of accepting unfounded ideas about the world. When we confuse what we would like to believe with what we have the evidence to believe, we have a weak basis for making critical environmental decisions—decisions that could have far-reaching and serious consequences.

Science, Pseudoscience, and Frontier Science

Some ideas presented as scientific are in fact not scientific, because they are inherently untestable, lack empirical support, or are based on faulty reasoning or poor scientific methodology. Such ideas are referred to as **pseudoscientific** (*pseudo* means "false").

Pseudoscientific ideas arise from various sources, as shown in Figure 2.10. With more research, however, some of these frontier models may move into the realm of accepted science, and new ideas will take their place at the advancing frontier.[11] Research may not support other hypotheses at the frontier, and these will be discarded by scientists.

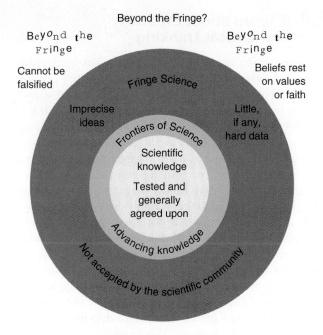

Figure 2.10 ■ Beyond the fringe? A diagrammatic view of different kinds of knowledge and ideas.

Some people continue to believe in discarded scientific ideas, or pseudoscience. For example, although scientists long ago discarded the notion that the movements of the heavenly bodies could affect the personalities and fates of humans, a substantial number of people believe in or are not sure about astrology (in one study, 25% believed and 22% were not sure).[12] Interestingly, many of the things that astrologers learned about the movements of the stars and planets were so accurate that they became the basis for astronomy, the scientific study of the heavens. Parts of astrology moved into accepted science, parts into pseudoscience.

Accepted science may merge into frontier science, which in turn may merge into more far-out ideas, or fringe science. Really wild ideas may be considered beyond the fringe.

Although scientists have no trouble distinguishing between accepted science and pseudoscience, they do have trouble identifying ideas at the frontier that will become accepted and those that will be relegated to pseudoscience. That trouble arises because science is a process of continual investigation. This ambiguity at the frontiers of science leads many people to accept some frontier science before it has been completely verified and to confuse pseudoscience with frontier science. (See the discussion of the Gaia hypothesis in Chapter 3.)

2.6 Environmental Questions and the Scientific Method

Environmental sciences deal with especially complex systems and include a relatively new set of sciences. Therefore, the process of scientific study has not always neatly followed the formal scientific method discussed earlier in this chapter. Often, observations are not used to develop formal hypotheses. Controlled laboratory experiments have been the exception rather than the rule. Much environmental research has been limited to field observations of processes and events that have been difficult to subject to controlled experiments.

Environmental research presents several obstacles to following the classic scientific method. The long time frame of many ecological processes relative to human lifetimes, professional lifetimes, and lengths of research grants poses problems for establishing statements that can in practice be subject to disproof. What do we do if a theoretical disproof through direct observation would take a century or more? Other obstacles include difficulties in setting up adequate **experimental controls** in field studies, in developing laboratory experiments of sufficient complexity, and in developing theory and models for complex systems. Throughout this text, we present differences between the "standard" scientific method and the actual approach that has been used in environmental sciences.

An Example: The California Condor

An environmental question that has been difficult to investigate using the traditional scientific method is the problem of the California condor. In the 1970s, the total population of this species, America's largest bird in terms of wingspan, had declined to 22. The bird had never been abundant since European settlement of North America, and the numbers had declined over the years. Several suggestions were made in the early 1980s about what to do to save this species from extinction.

One suggestion was to remove all the condors from the wild and attempt to breed them in captivity in zoos. Then, once the numbers had increased, some of the birds could be released back into the wild. Another suggestion was to improve the habitat of the condors, especially by providing more food and returning more of their native range to grasslands through controlled burning. Grasslands had been more prevalent in the condors' habitat prior to the European settlement. Over much of the condors' natural range, fire had been suppressed during the twentieth century, and areas that had been comparatively open grasslands had reverted to dense shrublands called chaparral. It is difficult for the large condors to land and take off in dense shrubland and hard for them to see the carcasses of dead animals, on which they feed.

In theory, scientists could have taken half of the 22 remaining condors out of the wild and tried captive breeding and left the other half in the wild and modified their habitat. But a total of 22 individuals was considered too few to take any chances with, and an experimental approach attempting two different kinds of treatment of the condors seemed too risky. The result was that all the condors were captured and a program of captive breeding begun (Figure 2.11).

By the 1990s, enough condors had been bred that it was believed safe to try to reintroduce some of them into the wild. Today there are more than 280 condors, 139 in the wild.[13] At first, when the condors were reintroduced, they had trouble finding food and using that food, and depended on supplementary feeding, and they did not begin to breed in the wild. There was some concern whether the zoo-raised animals would be able to care for themselves. But in 2003 the first wild condor chick was successfully fledged, and to date five chicks have hatched and fledged. Also the condors have begun to find food on their own, although the restora-

(a)

(b)

Figure 2.11 ■ (a) Condors in a captive breeding program. (b) Released condors in their natural habitat.

tion projects continue to provide supplementary food. So the restoration of the California condor appears to becoming a success, and it exemplifies how we can help endangered species, including what kinds of work and scientific understanding is required to be successful.

To summarize this discussion, you will find that environmental sciences—young sciences dealing with complex phenomena—are filled with unanswered questions, and that many of these questions do not seem open to the traditional scientific method. They require new approaches that retain the essential key of the scientific method: the ability to disprove a statement. The large number of unanswered questions and the difficulty in answering them may seem discouraging. But they can also be viewed as an adventure in which doubt and uncertainty are a wonderful challenge. The famous Nobel Prize–winning physicist Richard Feynman expressed this idea when he said, "You cannot understand science and its relation to anything else unless you understand and appreciate [science as] the great adventure of our time. You do not live in your time unless you understand that this is a tremendous adventure and a wild and exciting thing."[14] We can approach environmental science in this spirit and seek a set of scientific methods that will allow us to answer its fascinating questions.

Some Alternatives to Direct Experimentation

So how have environmental scientists tried to answer these difficult questions? Several approaches have been taken, including use of historical records and observations of modern catastrophes and disturbances.

Historical Evidence

Ecologists have made use of both human and ecological historical records. A classic example is a study of the history of fire in the Boundary Waters Canoe Area (BWCA) of Minnesota, 1 million acres of boreal forests, streams, and lakes well known for recreational canoeing.

Murray ("Bud") Heinselman had lived near the BWCA for much of his life and was instrumental in having it declared a wilderness area. A forest ecological scientist, Heinselman set out to determine the past patterns of fires in this wilderness. Those patterns are important to the maintenance of the wilderness. If the wilderness has been characterized by fires of a specific frequency, then it can be argued that this frequency is required to maintain the area in its most "natural" state.

Heinselman used three kinds of historical data: written records, tree-ring records, and buried records (fossil and prefossil organic deposits). Trees of the boreal forests, like most trees that are conifers or angiosperms (flowering plants), produce annual growth rings. If a fire burns through the bark of a tree, it leaves a scar, just as a serious burn leaves a scar on human skin. The tree grows over the scar, depositing a new growth ring for each year. (Figure 2.12 shows fire scars and

Figure 2.12 ■ Cross section of a tree showing fire scars and tree rings. Together these allow scientists to date fires and to average time between fires.

tree rings on a cross section of a tree.) By examining cross sections of trees, it is possible to determine the date of each fire and the number of years between fires. From written and tree-ring records, Heinselman found that since the seventeenth century, the BWCA forests had burned an average of once per century. However, the frequency of fires varied over time. Furthermore, buried charcoal dated using carbon-14 revealed that fires could be traced back more than 30,000 years.[15]

The three kinds of historical records provided important evidence about fire in the history of the BWCA. At the time Heinselman did his study, the standard hypothesis was that fires were bad for forests and should be suppressed. The historical evidence provided a disproof of this hypothesis. It showed fires to be a natural and an integral part of the forest and demonstrated that the forest had persisted with fire for a very long time. Thus, the use of historical information meets the primary requirement of the scientific method—the ability to disprove a statement. Historical evidence is a major source of data that can be used to test scientific hypotheses in ecology.

Modern Catastrophes and Disturbances as Experiments

Sometimes a large-scale catastrophe provides a kind of modern ecological experiment. The volcanic eruption of Mount St. Helens in 1980 supplied such an experiment, destroying vegetation and wildlife over a wide area. The recovery of plants, animals, and ecosystems following this explosion gave scientists insights into the dynamics of ecological systems and provided some surprises. The main surprise was the rapidity with which vegetation recovered and wildlife returned to parts of the mountain. In other ways, the recovery followed expected patterns in ecological succession (see Chapter 9).

(a)   (b)

Figure 2.13 ■ Forested land in Yellowstone National Park (*a*) during and (*b*) after the 1988 fire. The fire was a catastrophic event, but a natural one.

A more recent example of a catastrophe that has become the subject of ecological research was the 1988 wildfire in Yellowstone National Park (Figure 2.13).

It is important to point out that the greater the quantity and the better the quality of ecological data prior to such a catastrophe, the more that can be learned from the response of ecological systems to the event. This calls for careful monitoring of the environment.

2.7 Science and Decision Making

Like the scientific method, the process of making decisions is sometimes presented as a series of steps:

1. Formulate a clear statement of the issue to be decided.
2. Gather the scientific information related to the issue.
3. List all alternative courses of action.
4. Predict the positive and negative consequences of each course of action and the probability that each consequence will occur.
5. Weigh the alternatives and choose the best solution.

Such a procedure is a good guide to rational decision making, but it assumes a simplicity not often found in real-world issues. It is difficult to anticipate all the consequences of a course of action, and unintended consequences are at the root of many environmental problems. Often the scientific information is incomplete and even controversial. For example, the insecticide DDT causes eggshells of birds that feed on insects to be so thin that unhatched birds die. When DDT first came into use, this consequence was not predicted. Only when populations of species such as the brown pelican became seriously endangered did people become aware of it.

In the face of incomplete information, scientific controversies, conflicting interests, and emotionalism, how can environmental decisions be made? We need to begin with the scientific evidence from all relevant sources and with estimates of the uncertainties in each. Where scientists disagree about the interpretation of data, it may be possible to develop a consensus or a series of predictions based on the different interpretations. The impacts of the scenarios need to be identified and the risks associated with each analyzed in comparison with the benefits. Avoiding emotionalism and resisting slogans and propaganda are essential to developing sound approaches to environmental issues. Ultimately, however, environmental decisions are policy decisions negotiated through the political process. Policymakers are rarely professional scientists; generally, they are political leaders and ordinary citizens. Therefore, the scientific education of those in government and business, as well as of all citizens, is crucial.

2.8 Learning about Science

Science is an open-ended process of finding out about the natural world. In contrast, science lectures and texts are usually summaries of the answers arrived at through this process, and science homework and tests are exercises in finding the right answer.

Therefore, students often perceive science as a body of facts to be memorized and view lectures and texts as authoritative sources of absolute truths about the world. In contrast, scientists view scientific knowledge as currently accepted truth, always subject to change as new observations and interpretations are made. Students tend not to question the material in textbooks and lectures, whereas the essence of science is questioning and critically examining accepted truths.

Evaluating Media Coverage

The following questions will help you evaluate a report on science:[14]

1. Where is the report? Is it in a scientific journal, a serious newspaper, a popular science magazine, or a tabloid?

2. Is the report based on observations of actual occurrences? Were those observations made by more than one person? more than a few?

3. Are the sources of the report identified specifically? Are the sources specific, named scientists, scientific journals, or scientific organizations?

4. Does the report provide evidence that the claims are supported by other members of the scientific community?

5. Does the evidence for the claims seem sufficient? Is there contradictory evidence that might offset the evidence given for them?

6. Do the claims follow logically from the evidence? Do the claims violate reason (for example, a 99-year-old woman gives birth to baby)? Is there a simpler explanation for the observations?

7. Is there a clear basis for suspecting bias on the part of the source or the writer of the report?

8. Can you describe the line of reasoning that leads from the evidence to the claims?

Viewing science as a collection of facts may lead students to see science as difficult to understand and remember. In contrast, scientists emphasize the use of facts to form coherent pictures of the world that have the power to explain many phenomena. These pictures (models, theories) are so powerful that scientists find them easy to remember and expect that students will also.

Learning about science requires solving problems. Here, too, the attitudes of students and scientists differ. Whereas students may look to formulas, type problems, and algorithms to help solve problems, scientists look to general principles, critical thinking, and creativity.

We encourage you to learn environmental science in an active mode, to be critical of what you hear, what you see, and what you read—including this textbook. We encourage you to try to understand what science is—its assumptions, methods, and limitations—and to apply this understanding to your study of environmental science. Most of all, we hope that you will see not isolated facts to be learned by rote but the connections among facts that reflect interdependence and unity in the environment.

2.9 Science and Media Coverage

Most media reports on scientific issues deal with new discoveries, fringe science, and pseudoscience. It is important to analyze the claims in such reports to decide in which category they should be placed. Critical reading and listening require thinking carefully about whether a statement is based on observations and data, an objective interpretation of data, an interpretation based on the experience of an expert in the area, or a subjective opinion.

The need for critical examination applies to statements made by scientists as well as by other sources. On the one hand, scientists' training in analyzing data may make them more qualified than the general public to decide certain complex issues; on the other, as discussed before, scientists' own values, interests, and cultural backgrounds may influence their interpretations of data and bias their pronouncements. The situation is complicated by the fact that some scientists become "media scientists," whose opinions are sought on a wide variety of topics, some of which they have not studied as scientists. The opinions of experts do need to be heard, but appealing to scientists as authorities outside their area of expertise is contrary to the anti-authoritarian nature of science.

Summary

■ Science is one path to critical thinking about the natural world. Its goal is to gain an understanding of how nature works. Decisions on environmental issues must begin with an examination of the relevant scientific evidence. However, environmental decisions also require careful analysis of economic, social, and political consequences. Solutions will reflect religious, aesthetic, and ethical values as well.

■ Science begins with careful observations of the natural world, from which scientists formulate hypotheses. Whenever possible, scientists test hypotheses with controlled experiments.

■ Although the scientific method is often taught as a prescribed series of steps, it is better to think of it as a general guide to scientific thinking, with many variations.

■ Scientific knowledge is acquired through inductive reasoning, in which general conclusions are based on specific observations. Conclusions arrived at through induction can never be proved with certainty. Because of the inductive nature of science, it is possible to disprove hypotheses, but it is not possible to prove them with 100% certainty.

How Do We Decide What to Believe about Environmental Issues?

When you read about an environmental issue in a newspaper or magazine, how do you decide whether to accept the claims made in the article? Are they based on scientific evidence, and are they logical?

Scientific evidence is based on observations, but media accounts often rely mainly on inferences (interpretations) rather than evidence. Distinguishing inferences from evidence is an important first step in evaluating articles critically.

Second, it is important to consider the source of a statement. Is the source a reputable scientific organization or publication? Does the source have a vested interest that might bias the claims? When sources are not named, it is impossible to judge the reliability of claims.

If a claim is based on scientific evidence presented logically from a reliable, unbiased source, it is appropriate to accept the claim tentatively, pending further information. Practice your critical evaluation skills by reading the article in the box and answering the critical thinking questions.

Critical Thinking Questions

1. What is the major claim made in the article?
2. What evidence does the author present to support the claim?
3. Is the evidence based on observations, and is the source of the evidence reputable and unbiased?
4. Is the argument for the claim, whether or not based on evidence, logical?
5. Would you accept or reject the claim?
6. Even if the claim were well supported by evidence based on good authority, why would your acceptance be only tentative?

CLUE FOUND IN DEFORMED FROG MYSTERY

BY MICHAEL CONLON
Reuters News Agency (as printed in the *Toronto Star*)
November 6, 1996

A chemical used for mosquito control could be linked to deformities showing up in frogs across parts of North America, though the source of the phenomenon remains a mystery. "We're still at the point where we've got a lot of leads that we're trying to follow but no smoking gun," says Michael Lannoo of Ball State University in Muncie, Ind. "There are an enormous number of chemicals that are being applied to the environment and we don't understand what the breakdown products of these chemicals are," says Lannoo, who heads the U.S. section of the worldwide Declining Amphibian Population Task Force.

He says one suspect chemical was Methoprene, which produces a breakdown product resembling retinoic acid, a substance important in development. "Retinoic acid can produce in the laboratory all or a majority of the limb deformities that we're seeing in nature," he says. "That's not to say that's what's going on. But it is the best guess as to what's happening." Methoprene is used for mosquito control, among other things, Lannoo says.

Both the decline in amphibian populations and the deformities are of concern because frogs and related creatures are considered "sentinel" species that can provide early warnings of human risk. The skin of amphibians is permeable and puts them at particular risk to agents in the water.

Lannoo says limb deformities in frogs had been reported as far back as 1750, but the rate of deformities showing up today was unprecedented in some species. Some were showing abnormalities that affected more than half of the population of a species living in certain areas, he adds. He says he doubted that a parasite believed to have been the cause of some deformities in frogs in California was to blame for similar problems in Minnesota and nearby states. Deformed frogs have been reported in Minnesota, Wisconsin, Iowa, South Dakota, Missouri, California, Texas, Vermont, and Quebec. The deformities reported have included misshapen legs, extra limbs, and missing or misplaced eyes.

■ Measurements are approximations that may be more or less exact, depending on the measuring instruments and the people who use the instruments. A measurement is meaningful when it is accompanied by an estimate of the degree of uncertainty, or error.

■ Accuracy in measurement is the extent to which the measurement agrees with an accepted value. Precision is the degree of exactness with which a measurement is made. A precise measurement may not be accurate. The estimate of uncertainty provides information on the precision of a measurement.

■ A general statement that relates and explains a great many hypotheses is called a theory. Theories are the greatest achievements of science.

■ Critical thinking can help us distinguish science from pseudoscience. It can also help us recognize possible bias on the part of scientists and in the media. Critical thinking involves questioning and synthesizing what is learned in order to achieve knowledge, rather than merely acquire information.

REEXAMINING THEMES AND ISSUES

This chapter summarizes the scientific method, which is essential to the analysis and solution of environmental problems. The scientific method is critical to understanding the **human population** problem and to developing sound approaches to **sustainability**. The **global perspective** on environment arises out of new findings in environmental science. The importance of our increasingly **urbanized world** is best understood with the assistance of scientific investigation.

Solutions to environmental problems require both **values and knowledge**. Understanding the scientific method is especially important if we are going to understand the connection between **values and knowledge** and the relationship between **people and nature**. Ultimately, environmental decisions are policy decisions, negotiated through the political process. Policymakers often lack sufficient understanding of the scientific method, leading to false conclusions.

Uncertainty is part of the nature of measurement and science. We must learn to accept uncertainty as part of our attempt to conserve and use our natural resources.

Key Terms

controlled experiment **26**
deductive reasoning **24**
dependent variable **26**
disprovability **22**
experimental controls **31**
explanation **28**
fact **26**

hypothesis **26**
independent variable **26**
inductive reasoning **24**
inference **26**
manipulated variable **26**
model **28**
observations **26**

operational definitions **27**
premises **24**
probability **25**
pseudoscientific **30**
qualitative data **27**
quantitative data **27**
responding variable **26**

scientific method **28**
scientific theory **29**
theories **28**
variable **26**

Study Questions

1. Which of the following are scientific statements and which are not? What is the basis for your decision in each case?
   a. The amount of carbon dioxide in the atmosphere is increasing.
   b. Condors are ugly.
   c. Condors are endangered.
   d. Today there are 280 condors.
   e. Crop circles are a sign from Earth to us that we should act better.
   f. Crop circles can be made by people.
   g. The fate of Mono Lake is the same as the fate of the Aral Sea.

2. What is the logical conclusion of each of the following syllogisms? Which conclusions correspond to observed reality?
   a. All men are mortal.
      Socrates is a man.
      Therefore_____
   b. All sheep are black.
      Mary's lamb is white.
      Therefore_____

   c. All Amazons are women.
      No men are Amazons.
      Therefore_____
   d. All elephants are animals.
      All animals are living beings.
      Therefore_____

3. Which of the following statements are supported by deductive reasoning and which by inductive reasoning?
   a. The sun will rise tomorrow.
   b. The square of the hypotenuse of a right triangle is equal to the sum of the squares of the other two sides.
   c. Only male deer have antlers.
   d. If $A = B$ and $B = C$, then $A = C$.
   e. The net force acting on a body equals its mass times its acceleration.

4. The accepted value for the number of inches in a centimeter is 0.3937. Two students mark off a centimeter on a piece of paper and then measure the distance using a ruler (in inches). Student A finds the distance equal to 0.3827 in., and student B finds it equal to 0.39 in. Which measurement is more accurate? Which

is more precise? If student B measured the distance as 0.3900 in., what would be your answer?

5. a. A teacher gives five students each a metal bar and asks them to measure the length. The measurements obtained are 5.03, 4.99, 5.02, 4.96, and 5.00 cm. How can you explain the variability in the measurements? Are these systematic or random errors?

   b. The next day, the teacher gives the students the same bars but tells them that the bars have contracted because they have been in the refrigerator. In fact, the temperature difference would be too small to have any measurable effect on the length of the bars. The students' measurements, in the same order as in part (a), are 5.01, 4.95, 5.00, 4.90, and 4.95 cm. Why are the students' measurements different from those of the day before? What does this illustrate about science?

6. Identify the independent and dependent variables in each of the following:
   a. Change in the rate of breathing in response to exercise.
   b. The effect of study time on grades.
   c. The likelihood that people exposed to smoke from other people's cigarettes will contract lung cancer.

7. a. Identify a technological advance that resulted from a scientific discovery.
   b. Identify a scientific discovery that resulted from a technological advance.
   c. Identify a technological device you have used today. What scientific discoveries were necessary before the device could be developed?

8. What is fallacious about each of the following conclusions?
   a. A fortune cookie contains the statement "A happy event will occur in your life." Four months later, you find a $100 bill. You conclude that the fortune was correct.
   b. A person claims that aliens visited Earth in prehistoric times and influenced the cultural development of humans. As evidence, the person points to ideas among many groups of people about beings who came from the sky and performed amazing feats.
   c. A person observes that light-colored animals almost always live on light-colored surfaces, whereas dark forms of the same species live on dark surfaces. The person concludes that the light surface causes the light color of the animals.
   d. A person knows three people who have had fewer colds since they began taking vitamin C on a regular basis. The person concludes that vitamin C prevents colds.

9. Find a newspaper article on a controversial topic. Identify some loaded words in the article—that is, words that convey an emotional reaction or a value judgment.

10. Identify some social, economic, aesthetic, and ethical issues involved in a current environmental controversy.

Further Reading

American Association for the Advancement of Science (AAAS). 1989. *Science for All Americans*. Washington, D.C.: AAAS. This report focuses on the knowledge, skills, and attitudes a student needs to be scientifically literate.

Botkin, D. B. 2001. *No Man's Garden: Thoreau and a New Vision for Civilization and Nature*. Washington, D.C.: Island Press. The author discusses how science can be applied to the study of nature and to problems associated with people and nature. He also discusses science and values.

Grinnell, F. 1992. *The Scientific Attitude*. New York: Guilford. Examples from biomedical research are used to illustrate the processes of science (observing, hypothesizing, experimenting) and how scientists interact with each other and with society.

Kuhn, Thomas S. 1996. *The Structure of Scientific Revolutions*. Chicago: University of Chicago Press. This is a modern classic in the discussion of the scientific method, especially regarding major transitions in new sciences like environmental sciences.

McCain, G., and E. M. Segal. 1982. *The Game of Science*. Monterey, Calif.: Brooks/Cole. The authors present a lively look into the subculture of science.

Sagan, C. 1995. *The Demon-Haunted World*. New York: Random House. The author argues that irrational thinking and superstition threaten democratic institutions and discusses the importance of scientific thinking to our global civilization.

CHAPTER 3

The Big Picture: Systems of Change

Amboseli National Reserve, Kenya, Africa. Elephants are an important species for eco-tourism activities. The mountain in the background is Mount Kilimanjaro.

Learning Objectives

Changes in systems may occur naturally or may be induced by humans. Many complex and far-reaching interactions can result. After reading this chapter, you should understand:

■ Why solutions to many environmental problems involve the study of systems and rates of change.

■ How positive and negative feedback operate in a system.

■ What the implications of exponential growth and doubling time are.

■ That natural disturbances and changes in systems such as forests, rivers, and coral reefs are important to their continued existence.

■ What an ecosystem is and why sustained life on Earth is a characteristic of ecosystems.

■ What the Gaia hypothesis is and how life on Earth has affected Earth itself.

■ What the principle of uniformitarianism is and how it can be used to anticipate future changes.

■ Why the principle of environmental unity is important in studying environmental problems.

■ Why environmental problems are difficult to solve.

■ How human activities amplify the effects of natural disasters.

CASE STUDY

Amboseli National Reserve

Environmental change is often caused by a complex web of interactions, and the most obvious answer to the question of what caused a particular change may not be the right answer. Amboseli National Reserve, formally Amboseli National Park, is a case in point. In the short span of a few decades, this reserve, located in southern Kenya at the foot of Mount Kilimanjaro, underwent a significant environmental change. An understanding of physical, biological, and human-use factors—and how these factors are linked—was needed to explain what happened.

Figure 3.1 shows the boundary of the reserve and the major geologic units. The park is centered on an ancient lake bed, remnants of which include the seasonally flooded Lake Amboseli and some swampland. Mount Kilimanjaro is a well-known volcano, composed of alternating layers of volcanic rock and ash deposits. Rainfall that reaches the slopes of Mount Kilimanjaro infiltrates the volcanic material (becomes groundwater) and moves slowly down the slopes to saturate the ancient lake bed, eventually emerging at springs in the swampy, seasonally flooded land. The groundwater becomes very saline (salty) as it percolates through the lake bed, since salt stored in the lake bed sediments dissolves easily when the sediments are wet.

Before the mid-1950s, the dominant vegetation in the park was fever tree woodlands, mostly acacia trees, which provided habitat for mammalian species such as kudu, baboons, vervet monkeys, leopards, and impalas. Starting in the 1950s, and accelerating in the 1960s, these woodlands disappeared and were replaced by short grass and brush, which provided habitat for typical plains animals such as zebras and wildebeest. During these decades, the annual amounts of rainfall increased.[1, 2]

Amboseli is a semiarid environment, and since the mid-1970s has remained a grassland with scattered brush and few trees. During the past three decades, the mean daily temperature has increased dramatically. During the same time, rainfall has continued to vary from year to year by a factor of four, but with no regular pattern of change through time. Some of the increase in temperature may be due to global warming, but the warming has been much too great to be the result of only global warming. The slopes of Mount Kilimanjaro above Amboseli have less forest cover than they did 25 years ago because a lot of land has been transformed to agricultural uses. Loss of trees exposes dark soils that absorb solar energy, and this could cause local warming. In addition, there has been a significant decrease in snow and ice cover on the high slopes and summit of the mountain. Snow and ice reflect sunlight. As snow and ice decrease and dark rock is exposed, more solar energy is absorbed at the surface, warming it. Therefore, decrease in snow and ice will cause some local warming. Melting ice also feeds runoff to groundwater that feeds springs at seasonal Lake Amboseli, keeping groundwater levels close to the surface.[3]

Loss of the woodland habitat was initially blamed on overgrazing of cattle by the Maasai people (Figure 3.2) and damage to the trees from elephants (Figure 3.3). Environmental scientists eventually rejected these hypotheses as the main causes of the environmental

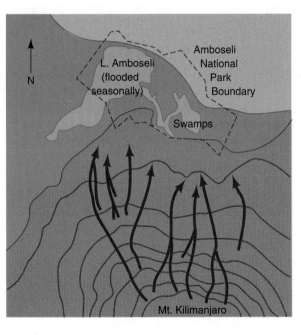

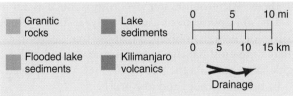

Figure 3.1 ■ Generalized geologic and landform map of Amboseli National Reserve, southern Kenya, Africa, and Mount Kilimanjaro. [*Source:* After T. Dunn and L. B. Leopold, *Water in Environmental Planning* (San Francisco: Freeman, 1978).]

Figure 3.2 ■ Maasai people grazing cattle in Amboseli National Reserve, Kenya. Grazing activities were prematurely blamed for loss of fever tree woodlands.

Figure 3.3 ■ Elephant feeding on a yellow-bark acacia tree. Elephant damage to trees is considered a factor in loss of woodland habitat in Amboseli National Reserve. However, elephants probably play a relatively minor role compared with oscillations in climate and groundwater conditions.

change. Their careful work showed that changes in rainfall and soils were the primary culprits, rather than people or elephants.[1, 2]

Let us look at the evidence more closely to see why the scientists rejected the overgrazing and elephant-damage hypotheses. Investigators were surprised to note that most dead trees were in an area that had been free of cattle since 1961, which was before the major decline in the woodland environment. Furthermore, some of the woodlands that suffered the least decline had the highest density of people and cattle. These observations suggested that overgrazing by cattle was not responsible for loss of the trees. Elephant damage was thought to be a major factor because elephants had stripped bark from more than 83% of the trees in some areas and had pushed over some younger, smaller trees. However, researchers concluded that elephants played only a secondary role in changing the habitat. As the density of fever trees and other woodland plants decreased, the incidence of damage caused by elephants increased. In other words, elephant damage interacted with some other, primary factor in changing the habitat.

Research on rainfall, groundwater history, and soils suggested that the Amboseli National Reserve area is very sensitive to changing amounts of rainfall. During dry periods, the salty groundwater sinks lower in the earth, and soil near the surface has a relatively low salt content. The fever trees grow well in the nonsalty soil. During wet periods, the groundwater rises close to the surface, bringing with it salt, which invades the root zones of trees and kills them. The groundwater level rose as much as 3.5 m (11.4 ft) in response to unusually wet years in the 1960s. Analysis of the soils confirmed that the tree stands that suffered the most damage were those associated with highly saline soils. As the trees died, they were replaced by salt-tolerant grasses and low brush.[1,2]

Evaluation of the historic record from early European explorers, accounts from Maasai herders, and fluctuating lake levels in other East African lakes suggested that before 1890, there had been another period of above-normal rainfall and loss of woodland environment. Thus, the scientists concluded that cycles of greater and lesser rainfall change hydrology and soil conditions, which in turn change the plant and the animal life of the area.[1] Cycles of wet and dry periods can be expected to continue, and associated with these will be changes in the soils, distribution of plants, and abundance and types of animals present.[1]

Ecotourism is a multimillion dollar industry at Amboseli, as tourists flock to view wildlife, especially elephants. In late 2005, the status of Amboseli was changed by the Kenya Wildlife Service to a national reserve, managed by the Maasai People. Local management will hopefully provide economic incentive to maintain and enhance conservation. Environmental justice argues that local people have the right to manage and gain from the resources on their historic land. We hope the Maasai will seize the moment and manage the reserve wisely.

The Amboseli story illustrates how environmental science attempts to work out sequences of events that follow a particular change. At Amboseli, rainfall cycles change hydrology and soil conditions, which in turn change the vegetation and animals of the area. In this chapter, we examine environmental systems and changes that occur as a result of natural and human-induced processes.

3.1 Systems and Feedback

In environmental science at every level, we must deal with a variety of systems that range from simple to complex. We must therefore understand systems and how different parts of systems interact with one another. We begin our discussion of systems by defining what a system is and how it can operate.

System Defined

A **system** is a set of components or parts that function together to act as a whole. A single organism (such as your body) is a system, as are a sewage treatment plant, a city (Figure 3.4), a river (Figure 3.5), and your dorm room. On a much different scale, the entire Earth is a system.

Systems may be open or closed. An **open system** is not generally contained within boundaries, and some energy or material (solid, liquid, or gas) moves into or out of the system. The ocean is an open system with regard to water, because water moves into the ocean from the atmosphere and out of the ocean to the atmosphere. Conversely, in a **closed system**, no such movements take place. Earth is a closed system (for all practical purposes) with regard to material.

Systems respond to inputs and have outputs. Your body, for example, is a complex system. If you are hiking and see a grizzly bear, the sight of the bear is an input. Your body reacts to that input: The adrenaline level in your blood goes up, your heart rate increases, and the hair on your head and arms may rise. Your response—perhaps moving slowly away from the bear—is an output. We call these responses feedback.

Figure 3.4 ■ Lake Michigan, Lincoln Park, and the city of Chicago make up an example of a large, complex urban system that includes air, water, and land resources. Urban systems are particularly important in environmental science because more and more people are living in urban areas.

Figure 3.5 ■ The Owens River, on the east side of the Sierra Nevada in California, is a system that includes water, sediment, vegetation, and animals such as fish and insects that function together as a whole.

Feedback

Feedback occurs when the output of the system also serves as an input and leads to further changes in the system. A good example of feedback is human temperature regulation. If you go out in the sun and get hot, the increase in temperature affects your sensory perceptions (input). If you stay in the sun, your body responds physiologically: Your skin pores open, and you are cooled by evaporating water (you sweat). The cooling is output, and it is also input to your sensory perceptions. You may also respond behaviorally: Because you feel hot (input), you walk into the shade and your temperature returns to normal.

In our example, an increase in temperature is followed by a response that leads to a decrease in temperature. This is an example of **negative feedback**, in which an increase in output leads to a later decrease. Negative feedback is self-regulating, or stabilizing; it usually keeps a system in a relatively constant condition.

Positive feedback occurs when an increase in output leads to a further increase in the output. A fire starting in a forest provides an example of positive feedback. The wood may be slightly damp at the beginning and so may not burn well. Once a fire starts, wood near the flame dries out and begins to burn, which in turn dries out a greater quantity of wood and leads to a larger fire. The larger the fire, the more wood is dried and the more rapidly the fire increases. Positive feedback, sometimes called "a vicious cycle," is destabilizing.

Environmental damage can occur when human use of the environment leads to positive feedback. Off-road vehicle use—including bicycles (Figure 3.6), for example— may cause positive feedback with respect to soil erosion. These vehicles' churning tires, designed to grip the earth, erode the soil and uproot plants, increasing the rate of erosion. As more soil

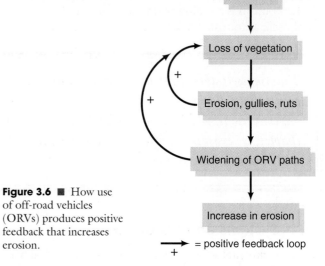

Figure 3.6 ■ How use of off-road vehicles (ORVs) produces positive feedback that increases erosion.

is exposed to erosion by running water, ruts and gullies are carved. Drivers then avoid the ruts and gullies, driving on adjacent sections that are not as eroded, thus widening paths and further increasing erosion. The gullies themselves cause an increase in erosion because they concentrate runoff and have steep side slopes. Once formed, gullies tend to grow in length, width, and depth, causing additional erosion (Figure 3.7). Eventually, an area of intensive off-road vehicle use may become a wasteland of eroded paths and gullies. Positive feedback has made the situation get worse and worse.

Some situations can be examined in terms of both positive and negative feedback loops. Changes in human population in large cities present an example, as illustrated in Figure 3.8. Positive feedback, which increases the population in cities, may occur when people perceive greater opportunities in cities and hope for a higher standard of living. Negative feedback may result from air and water pollution, disease, crime, and discomfort, if these factors encourage some people to migrate from the cities to rural areas.

Practicing your critical thinking skills, you may ask, "Is negative feedback generally desirable, and is positive feedback generally undesirable?" Reflecting on this question, we can see that although negative feedback is self-regulating, it may in some instances not be desirable. The period of time over which the positive or negative feedback occurs is the important factor.

Figure 3.7 ■ Off-road vehicle damage of rare plants living on coastal dunes near San Luis Obispo, California. Note the tire tracks extending into the dune field.

For example, suppose we are interested in restoring the ecology of Yellowstone National Park through the reintroduction of wolves. We will expect positive feedback in the population for a time as the number of wolves grows. In this case, positive feedback for a period of time is desirable because it produces a change we want. We might also envision a system in a state that is stable but undesirable. An example is a polluted stream in an urban environment. Urban runoff and its associated pollutants, such as oil and other chemicals from streets, entering the stream's system may, through negative feedback mechanisms, reach a stable state between the water and the pollutants in the stream. However, most people would consider this system to be undesirable. A channel restoration project might be implemented to control pollutants by collecting and treating

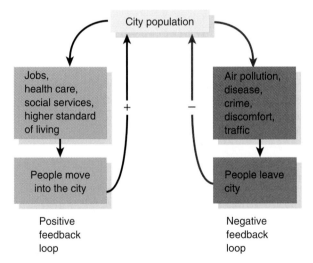

Figure 3.8 ■ Potential positive and negative feedback loops for changes of human population in large cities. The left side of the figure shows that as jobs, health care, and a higher standard of living increase, so do migration and city population. Conversely, the right side of the figure shows that increases in air pollution, disease, crime, discomforts, or traffic tend to reduce the city pollution. [*Source:* Modified from M. Maruyama. "The Second Cybernetics: Deviation-Amplifying Mutual Causal Processes," *American Scientist* 51 (1963):164–670. Reprinted by permission of *American Scientist*, magazine of Sigma Xi, The Scientific Research Society.)

them before they enter the stream. As a result, the stream might reach a new state that is ecologically more desirable.

We can see that whether we view positive or negative feedback as desirable depends on the system and potential changes. Nevertheless, some of the major environmental problems we face today result from positive feedback mechanisms that are out of control. These include resource use and growth of human population, among others.

Interestingly, throughout most of human history, strong negative feedback cycles resulted in a very low growth in human population. Disease and limited capacity to produce food kept growth low. However, in the past hundred years, modern medicine, sanitation, and agricultural practices turned negative feedback into positive feedback, and rapid increase in human population occurred.

3.2 Exponential Growth

A particularly important example of positive feedback occurs with **exponential growth**. Simply stated, growth is exponential when it occurs at a constant *rate* per time period (rather than a constant *amount*). For instance, suppose you have $1,000 in the bank, and it grows at 10% per year. The first year, $100 in interest is added to your account. The second year, you earn more because you earn 10% on the new total amount, $1,100. The greater the amount, the greater the interest earned, so the money (or the population, or some other quantity) increases by larger and larger amounts. When we plot data in which exponential growth is occurring, the curve we obtain is said to be "J" shaped. It looks like a skateboard ramp, starting out nearly flat and then rising steeply. (The actual shape depends on the scale of the units of the curve.) Figure 3.9 shows two typical exponential growth curves.

Calculating exponential growth involves two related factors: the rate of growth measured as a percentage and the doubling time in years. The **doubling time** is the time necessary for the quantity being measured to double. A useful rule is that the doubling time is approximately equal to 70 divided by the annual percentage growth rate. Working It Out 3.1 describes exponential growth calculations and explains why 70 divided by the annual growth rate is the doubling time. If you understand algebra and natural logarithms, you should find the example interesting. However, it is the general principles that are most important.

One significant principle of exponential growth is that it is incompatible with the concept of sustainability, which is a long-term process (decades to hundreds of years or more). In fact, the term *sustainable growth* is an oxymoron—a self-contradiction. Even at modest growth rates, the number of whatever is growing will eventually reach extraordinary levels that are impossible to maintain.[4,5]

Exponential growth has interesting (and sometimes alarming) consequences, as illustrated in a fictional story by Albert Bartlett.[4] Imagine a hypothetical strain of bacteria in which each bacterium divides into two every 60 seconds

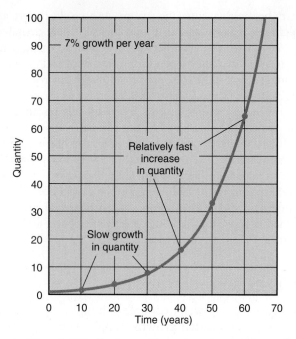

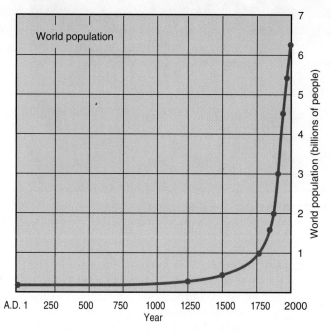

Figure 3.9 ■ (*a*) Idealized curve illustrating exponential growth. Growth rate is constant at 7%, and time necessary to double the quantity is constant at 10 years. Notice that growth is slow at first and much faster after several doubling times. For example, the quantity changes from 2 to 4 (absolute increase of 2) for the dou- bling from 10 to 20 years. It increases from 32 to 64 (absolute increase of 32) during the doubling from 50 to 60 years. (*b*) Human population increase for the last 2000 years. (*Source*: Data from U.S. Department of State.)

Exponential Growth

If the quantity of something (say, the number of humans on Earth) increases or decreases at a fixed fraction per unit of time, whose symbol is k (for example, $k = + 0.02$ per year), then the quantity is changing exponentially. With positive k, we have exponential growth. With negative k, we have exponential decay.

Growth rate R is defined as the percent change per unit of time—that is, $k = R/100$. Thus, if $R = 2\%$ per year, then $k = +0.02$ per year.

The equation to describe exponential growth is

$$N = N_0\, e^{kt}$$

where N is the future value of whatever is being evaluated; N_0 is present value; e, the base of natural logarithms, is a constant 2.71828; k is as defined above; and t is the number of years over which the growth is to be calculated. This equation can be solved using a simple hand calculator, and a number of interesting environmental questions can be answered. For example, assume that we want to know what the world population is going to be in the year 2020, given that the population in 2003 is 6.3 billion and the population is growing at a constant rate of 1.36% per year ($k = 0.0136$). We can estimate N, the world population for the year 2020, by applying the preceding equation:

$$N = (6.3 \times 10^9) \times e^{(0.0136 \times 17)}$$
$$= 6.3 \times 10^9 \times e^{0.2312}$$
$$= 6.3 \times 10^9 \times 2.71828^{0.2312}$$
$$= 7.94 \times 10^9, \text{ or } 7.94 \text{ billion persons}$$

The doubling time for a quantity undergoing exponential growth (i.e., increasing by 100%) can be calculated by the following equation:

$$2\, N_0 = N_0{}^{ekT_d}$$

where T_d is the doubling time.

Take the natural logarithm of both sides

$$\ln 2 = kT_d \text{ and } T_d = \ln 2/k$$

Then remembering that $k = R/100$

$$T_d = 0.693/(R/100)$$
$$= 100\,(0.693)/R$$
$$= 69.3/R, \text{ or about } 70/R$$

This result is our general rule—that the doubling time is approximately 70 divided by the growth rate. For example, if $R = 10\%$ per year, then $T_d = 7$ years.

(the doubling time is 1 minute). Assume that one bacterium is put in a bottle at 11:00 A.M. The bottle (its world) is full at 12:00 noon. When was the bottle half full? The answer is 11:59 A.M. If you were a bacterium in the bottle, at what time would you realize that you were running out of space? There is no single answer to this question. Consider, though, that at 11:58 the bottle was 75% empty, and at 11:57 it was 88% empty. Now assume that at 11:58 some farsighted bacteria realized that the population was running out of space and started looking around for new bottles. Let's suppose that they were able to find and move into three more bottles. How much time did they buy? Two additional minutes. They will run out of space at 12:02 P.M. If they had found 16 additional bottles, how much more time would they have?

The story of bacteria in bottles, though obviously hypothetical, illustrates the power of exponential growth. We look at exponential growth and doubling time again in Chapter 4, when we consider the growth of the human population. Here, we simply note that many systems in nature display exponential growth for some period of time, so it is important that we be able to recognize it. In particular, it is important to recognize exponential growth in a positive feedback cycle, as accompanying changes may be very difficult to control or stop.

3.3 Environmental Unity

Our discussion of positive and negative feedback sets the stage for another fundamental concept in environmental science: **environmental unity**. Simply stated, environmental unity means that it is impossible to change only one thing; everything affects everything else. Of course, this is something of an overstatement. The concept is not absolutely true; the extinction of a species of snails in North America, for instance, is hardly likely to change the flow characteristics of the Amazon River. However, many aspects of the natural environment are closely linked. Changes in one part of a system often have secondary and tertiary effects within the system and effects on adjacent systems. Earth and its ecosystems are complex entities in which any action may have several or many effects. The case study of Amboseli National Reserve is a good example of the principle of environmental unity. We can find many other examples in both human-made and natural settings.

An Urban Example

Consider the changes occurring in midwestern U.S. cities such as Chicago and Indianapolis during a major shift in land use from forests or agricultural land to urban development. Clearing the land for urban use increases runoff and the amount of sediment eroded from the land (soil erosion). Increased runoff from streets and sediment eroded from bare ground during construction affect the form and shape of the river channel. The river carries more sediment, and some of it is deposited on the bottom of the channel, reducing channel depth and increasing flood hazard.

Eventually, as more land is paved, the amount of sediment eroded from the land decreases, runoff increases further, and the streams readjust to a lower sediment load (the amount of sediment carried by the stream) and more runoff. The readjustment is a form of negative feedback inherent in streams and rivers.

Urbanization is also likely to pollute the streams or otherwise change water quality. The increased fine sediment makes the water muddy, and chemicals from street and yard runoff pollute the stream. These changes affect the biological systems in the stream and adjacent banks. Thus, land-use conversion can set off a series of changes in the environment, and each change is likely to trigger additional changes.

A Forest Example

The interaction among forests, streams, and fish in the Pacific Northwest provides another example of environmental unity. In the redwood forests of northern California and southern Oregon, large pieces of woody debris, such as tree trunks and roots, are necessary to form and maintain nearly all the pool environments in small streams (Figure 3.10). Large pieces of redwood fall

Figure 3.10 ■ Stream processes are significantly modified by fallen redwood trees. The large tree trunk in the central portion of the photograph produced a small pool, which is a good fish habitat. The site is Prairie Creek, Redwood State Park, California.

naturally into streams and partially block their flow, producing pools of deeper water. These pools provide much of the rearing habitat for young salmon, which spend part of their lives in the streams before migrating to the ocean.

It was formerly common practice to remove woody debris from streams because it was thought to block the migration of adult salmon attempting to return to spawning beds in the streams. We now know that this practice degrades the fish habitat. Stream restoration projects now often place large woody debris into channels to improve the fish habitat. The role of large woody debris in stream processes and salmon habitat illustrates the value of studying relations between physical and biological systems to help provide for sustainable fish populations. Such studies are at the heart of environmental science.

3.4 Uniformitarianism

Earth and its life-forms have changed many times, but the processes necessary to sustain life and the environment for life have occurred throughout much of Earth's history. The principle that physical and biological processes presently forming and modifying Earth can help explain the geologic and evolutionary history of Earth is known as **uniformitarianism**. The principle may be more simply stated as "the present is the key to the past." For example, if a deposit of gravel and sand found at the top of a mountain is similar to stream gravels found today in an adjacent valley, we may infer by uniformitarianism that a stream once flowed in a valley where the mountaintop is now.

Uniformitarianism was first suggested in 1785 by the Scottish scientist James Hutton, known as the father of geology. Charles Darwin was impressed by the concept of uniformitarianism, and the concept pervades his ideas on biological evolution. Today, uniformitarianism is considered one of the fundamental principles of the biological and Earth sciences.

Uniformitarianism does not demand or even suggest that the magnitude and frequency of natural processes remain constant. Obviously, some processes do not extend back through all of geologic time. For example, the early Earth atmosphere did not contain free oxygen. However, for the past several billion years, the continents, oceans, and atmosphere have been similar to those of today. We assume that physical and biological processes that form and modify the Earth's surface have not changed significantly over this period. To be useful from an environmental standpoint, the principle of uniformitarianism has to be more than a key to the past. We must turn it around and say that a study of past and present processes is the key to the future. That is, we can assume that in the future the same physical and biological processes will operate, although the rates will vary as the environment is influenced by natural change and human activity. Geologically short-lived landforms such as beaches (Figure 3.11) and lakes will continue to appear and disappear in response to storms, fires, volcanic eruptions, and earthquakes. Extinctions of animals and plants will continue in spite of, as well as because of, human activity. We want to improve our ability to predict what the future may bring, and uniformitarianism can assist in this task.

Figure 3.11 ■ This beach on the island of Bora Bora, French Polynesia, is an example of a geologically short-lived landform, vulnerable to rapid change from storms and other natural processes.

3.5 Changes and Equilibrium in Systems

Uniformitarianism, then, suggests that changes in natural systems may be predictable. We turn next to an examination of how systems may change. This will involve looking at the relation of system inputs and outputs.

Where the input into a system is equal to the output (Figure 3.12*a*), there is no net change in the size of the reservoir (the amount of whatever is being measured), and the system is said to be in a **steady state**. The steady state is a dynamic equilibrium, because material or energy is entering and leaving the system in equal amounts. The opposing processes occur at equal rates. An approximate steady state may occur on a global scale, such as in the balance between incoming solar radiation and outgoing radiation from Earth, or on the smaller scale of a university, where new freshmen begin their studies and seniors graduate at about the same rate.

When the input into the system is less than the output (Figure 3.12*b*), the size of the reservoir declines. For example, if a resource, such as groundwater, is consumed faster than it can be naturally or humanly replaced, it may be used up. Conversely, in a system where input exceeds output (Figure 3.12*c*), the reservoir will increase. Examples are the buildup of heavy metals in lakes and the pollution of groundwater.

By using rates of change or input–output analysis of systems, we can derive an average residence time for objects or material moving through a system. The **average residence time** is the time it takes for a given part of the total reservoir of a particular material to be cycled through the system. To compute the average residence time when the size of the reservoir and the rate of transfer are constant, we divide the total size of the reservoir by the average rate of transfer through that reservoir. For example, suppose the university mentioned above has 10,000 students. Each year 2,500 freshmen start and 2,500 seniors graduate. The average residence time for students is 10,000 divided by 2,500, or 4 years.

Average residence time has important implications for environmental systems. A system such as a small lake with an inlet and an outlet and a high transfer rate of water has a short residence time for water. (See Working It Out 3.2.) On the one hand, that makes the lake especially vulnerable to change if, for example, a pollutant is introduced. On the other hand, the pollutant soon leaves the lake. Large systems with a slow rate of transfer of water, such as oceans, have a long residence time and are much less vulnerable to quick change. However, once polluted, large systems with slow transfer rates are difficult to clean up.

Let us look more closely at system inputs and outputs. Inputs to systems may be thought of as causes and outputs or responses as effects. For example, we may add a nitrogen fertilizer to an orchard of orange trees. Adding the fertilizer is an input (or cause), and the output (or effect) is the number of oranges the tree produces.

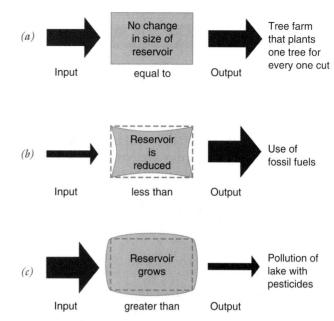

Figure 3.12 ■ Major ways in which a reservoir, or stock, of some material can change. (*Source:* Modified from P. R. Ehrlich, A. H. Ehrlich, and J. P. Holvren, *Ecoscience: Population, Resources, Environment,* 3rd ed. [San Francisco: W. H. Freeman, 1977].) Row (*a*) represents steady-state conditions, rows (*b*) and (*c*) are examples of negative and positive changes in storage.

If the relationship between a cause (input) and effect (output) is strictly proportional for all values, then we call this relationship *linear.* Input and output coupled with feedback in a system may result in relationships between cause and effect that are *nonlinear,* and there may be *delays* in the response.[5] Some relationships in systems are linear over a particular range of input and then become nonlinear. For example, if you apply 0.25 kg of fertilizer per orange tree and the yield increases by 5%, and you then apply 0.50 kg per tree and the yield increases by 10%, and you then apply 0.75 kg per tree and the yield increases by 15%, the relationship is linear over these values of input of fertilizer. But what if you apply 50 kg per tree in the hope of increasing the yield by 1,000%? You would probably damage or kill the tree and the yield would be zero!

Over the entire range from 0.25 to 50 kg, then, the relationship between cause and effect changes. You might also note delays in response. When you add fertilizer, for example, it takes time for it to enter the soil and be used by the tree. Many responses to environmental inputs (including human population change; pollution of the land, water, and air; and use of resources) are nonlinear and may involve delays that must be recognized if we are to understand and solve environmental problems.

With an understanding of input and output, we have a framework for interpreting some of the changes that may affect systems. An idea that has been used and defended in the study of our natural environment is that natural systems that have not been affected by human activity tend toward some sort of steady state, or dynamic equilibrium.

Average Residence Time (ART)

The *average residence time* (ART) for a chemical element or compound is an important concept in evaluating many environmental problems. ART is defined as the ratio of the size of a reservoir or pool of some material—say, the amount of water in a reservoir—to the rate of transfer through the reservoir. The equation is

$$ART = S/F$$

where S is the size of the reservoir and F is the rate of transfer.

Knowing the ART for a particular chemical in the environment—as, for example, a pollutant in the air, water, or soil—allows for a more quantitative understanding of that pollutant. We can better evaluate the nature and extent of the pollutant in time and space, assisting development of strategies to reduce or eliminate the pollutant.

Let's look at a simple example. Figure 3.13 shows a map of Big Lake, a reservoir of water impounded by a dam. The lake has three rivers that feed a combined 10 m^3/sec of water into the lake, and the outlet structure releases an equal 10 m^3/sec. We assume evaporation of water from the lake is negligible in this simplified example. A water pollutant, MTBE (methyl tertiary – butyl ether), is also present in the lake. MTBE is added to gasoline to help reduce emissions of carbon monoxide. It is toxic; in small concentrations of 20–40 µg/l (thousandths of grams per liter) in water, it smells like turpentine and is nauseating to some people. MTBE readily dissolves in water and so travels with it. Sources of MTBE in Big Lake are urban runoff from Bear City gasoline stations, gasoline spills on land or in the lake, and boats in the lake with gasoline-burning engines. Concern over MTBE in California has led to the decision to stop adding it to gasoline. We can ask several questions concerning the water and MTBE in Big Lake.

1. What is the ART of water in the lake?
2. What is the amount of MTBE in the lake, the rate (amount per time) of MTBE being put into the lake, and the ART of MTBE in the lake? Because the water and MTBE move together, their ARTs should be the same—we can verify this.

ART of Water in Big Lake

For these calculations, use multiplication factors and conversions in Appendices B and C at the end of this book.

$$ART_{water} = \frac{S}{F} = ART_{water} = \frac{1,000,000,000 \text{ m}^3}{10 \text{ m}^3/\text{sec}}$$

$$\text{or } \frac{10^9 \text{ m}^3}{10 \text{ m}^3/\text{sec}}$$

The units m^3 cancel out and

$$ART = 100,000,000 \text{ sec or } 10^8 \text{ sec}$$

Convert 10^8 sec to years:

$$\frac{\text{seconds}}{\text{year}} = \frac{60 \text{ sec}}{1 \text{ minute}} \times \frac{60 \text{ minute}}{1 \text{ hour}} \times \frac{24 \text{ hour}}{1 \text{ day}} \times \frac{365 \text{ day}}{1 \text{ year}}$$

Canceling units and multiplying, there are 31,536,000 sec/year, which is

$$3.1536 \times 10^7 \text{ sec/year}$$

Then the ART for Big Lake is

$$\frac{100,000,000 \text{ sec}}{31,536,000 \text{ sec/yr}} \text{ or } \frac{10^8 \text{ sec}}{3.1536 \times 10^7 \text{ sec/yr}}$$

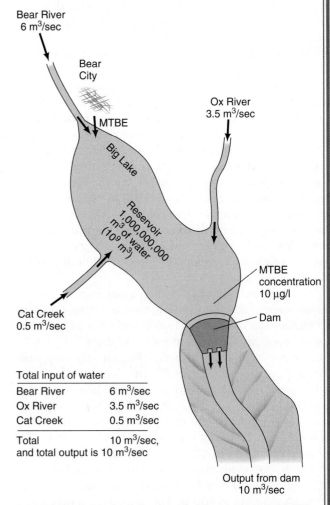

Bear River
6 m³/sec

Bear City

MTBE

Ox River
3.5 m³/sec

Big Lake

Reservoir 1,000,000,000 m³ of water (10⁹ m³)

MTBE concentration 10 µg/l

Dam

Cat Creek
0.5 m³/sec

Output from dam
10 m³/sec

Total input of water	
Bear River	6 m³/sec
Ox River	3.5 m³/sec
Cat Creek	0.5 m³/sec
Total	10 m³/sec, and total output is 10 m³/sec

Figure 3.13 ■ Idealized diagram of a lake system with MTBE contamination.

The ART for water in Big Lake is 3.17/year.

ART for MTBE in Big Lake

Concentration of MTBE in water near the dam is measured as 10 µg/l. Then the total amount of MTBE in the lake (size of reservoir or pool of MTBE) is the product of volume of water in the lake and concentration of MTBE:

$$10^9 \text{ m}^3 \times \frac{10^3 \text{ l}}{\text{m}^3} \times \frac{10 \text{ µg}}{\text{l}} = 10^{13} \text{ µg or } 10^7 \text{g}$$

which is 10^4 kg, or 10 metric tons, of MTBE.

The output of water from Big Lake is 10 m^3/sec, and this contains 10 µg/l of MTBE; the transfer rate of MTBE (g/sec) is

$$\text{MTBE/sec} = \frac{10 \text{ m}^3}{\text{sec}} \times \frac{10^3 \text{ l}}{\text{m}^3} \times \frac{10 \text{ µg}}{\text{l}} \times \frac{10^{-6} \text{ g}}{\text{µg}} = 0.1 \text{ g/sec}$$

Because we assume that input and output of MTBE are equal, the input is also 0.1 g/sec.

$$\text{ART}_{\text{MTBE}} = \frac{S}{F} = \frac{10^7 \text{ g}}{0.1 \text{ g/sec}} = 10^8 \text{ sec, or 3.17 years}$$

Thus, as we suspected, the ARTs of the water and MTBE are the same. This results because MTBE is dissolved in the water. If it attached to the sediment in the lake, the ART_{MTBE} would be much longer. Chemicals with large reservoirs or small rates of transfer tend to have long ARTs. In this exercise we have calculated the ARTs of water in Big Lake as well as the input, total amount, and ART of MTBE.

Sometimes this is called the *balance of nature*. Certainly, negative feedback operates in many natural systems and may tend to hold a system at equilibrium. Nevertheless, we need to ask how often the equilibrium model really applies.

If we examine natural systems in detail and perform our evaluation over a variety of time frames, it is evident that a steady state, or dynamic equilibrium, is seldom obtained or maintained for very long. Rather, systems are characterized not only by human-induced disturbances but also by natural disturbances (sometimes called natural disasters, such as floods and wildfires). Thus, changes over time can be expected. In fact, studies of such diverse systems as forests, rivers, and coral reefs suggest that disturbances due to natural events such as storms, floods, and fires are necessary for the maintenance of those systems.

The environmental lesson is that systems change naturally. If we are going to manage systems for the betterment of the environment, we need to gain a better understanding of the following:[6,7]

- Types of disturbances and changes that are likely to occur
- Time periods over which changes occur
- The importance of each change to the long-term productivity of the system

These concepts are at the heart of understanding the principles of environmental unity and sustainability.

3.6 Earth and Life

We turn next from a general discussion of systems to a more direct focus on Earth as a living planet. Earth formed approximately 4.6 billion years ago when a cloud of interstellar gas known as a solar nebula collapsed, creating protostars and planetary systems. Life on Earth began approximately 1.6 billion years later (3 billion years ago) and since that time has profoundly affected the planet. Since the emergence of life, many kinds of organisms have evolved, flourished, and died, leaving only their fossils to record their place in history.

Several million years ago, there were evolutionary beginnings for the eventual dominance of humans on Earth. Eventually, the fossil record tells us, we, too, will disappear. The brief moment of humanity in Earth history may not be particularly significant. However, to us living now and to the human generations still to come, how we affect our environment is important.

Human activities increase and decrease the magnitude and frequency of some natural Earth processes. For example, rivers may rise and flood the surrounding countryside regardless of human activities, but the magnitude and frequency of flooding may be greatly increased or decreased by human activity. In order to predict the long-range effects of such processes as flooding, we must be able to determine how our future activities will change the rates of physical processes.

From a biological and geological point of view, we know that the ultimate fate of every species is extinction. However, humans have accelerated this fate for many species. As the human population has increased, a parallel increase in the extinction of species has occurred. These extinctions are closely related to land-use change—to agricultural and urban uses that change the ecological conditions of an area. Some species are domesticated or cultivated, and their numbers grow; others are removed as pests.

3.7 Earth as a Living System

Earth as a planet has been profoundly altered by the life that inhabits it. Earth's air, oceans, soils, and sedimentary rocks are very different from what they would be on a lifeless planet. In many ways, life helps control the makeup of the air, oceans, and sediments.

Life interacts with its environment on many levels. A single bacterium in the soil interacts with the air, water, and particles of soil around it within the space of a fraction of a cubic centimeter. A forest extending hundreds of square kilometers interacts with large volumes of air, water, and soil. All of the oceans, all of the lower atmosphere, and all of the near-surface part of the solid Earth are affected by life.

A general term, **biota**, is used to refer to all living things (animals and plants, including microorganisms) within a given area—from an aquarium to a continent to Earth as a whole. The region of Earth where life exists is known as the **biosphere**. It extends from the depths of the oceans to the summits of mountains. The biosphere includes all life as well as the lower atmosphere and the oceans, rivers, lakes, soils, and solid sediments that actively interchange materials with life. All living things require energy and materials. In the biosphere, energy is received from the sun and the interior of Earth and is used and given off as materials are recycled.

To understand what is required to sustain life, consider the following question: How small a part of the biosphere could be isolated from the rest and still sustain life? Suppose you put parts of the biosphere into a glass container and sealed it. What minimum set of contents would sustain life? If you placed a single green plant in the container along with air, water, and some soil, the plant could make sugars from water and from carbon dioxide in the air. It could also make many organic compounds, including proteins and woody tissue, from sugars and from inorganic compounds in the soil. But no green plant can decompose its own products and recycle the materials. Eventually, your green plant would die.

We know of no single organism, population, or species that both produces all its own food and completely recycles all its own metabolic products. For life to persist, there must be several species within an environment that includes fluid media—air and water—to transport materials and energy. Such an environment is an ecosystem, our next important topic of discussion.

3.8 Ecosystems

An **ecosystem** is a community of organisms and its local nonliving environment in which matter (chemical elements) cycles and energy flows. It is a fundamental principle that *sustained life on Earth is a characteristic of ecosystems*, not of individual organisms or populations or single species.

The Nature of Ecosystems

The term *ecosystem* is applied to areas of all sizes, from the smallest puddle of water to a large forest or the entire global biosphere. Ecosystems differ greatly in composition—that is, in the number and kinds of species, the kinds and relative proportions of nonbiological constituents, and the degree of variation in time and space. Sometimes the borders of an ecosystem are well defined, as in the transition from the ocean to a rocky coast or from a pond to the surrounding woods. Sometimes the borders are vague, as in the subtle gradation of forest to prairie in Minnesota and the Dakotas or from grasslands to savannas or forests in East Africa. What is common to all ecosystems is not physical structure—size, shape, variations of borders—but the existence of the processes we have mentioned—the flow of energy and the cycling of chemical elements.

Ecosystems can be natural or artificial. A pond constructed as part of a waste treatment plant is an artificial ecosystem. Ecosystems can be natural or managed, and the management can vary over a wide range of actions. Agriculture can be thought of as partial management of certain kinds of ecosystems.

Natural ecosystems carry out many public service functions for us. Wastewater from houses and industries is often converted to drinkable water by passage through natural ecosystems such as soils. Pollutants, such as those in the smoke from industrial plants or in the exhaust from automobiles, are often trapped on leaves or converted to harmless compounds by forests.

The Gaia Hypothesis

Our discussion of Earth as a system—life in its environment, the biosphere, and ecosystems—leads us to the question of how much life on Earth has affected our planet. In recent years, the **Gaia hypothesis**—named for Gaia, the Greek goddess Mother Earth—has become a hotly debated subject.[8] The hypothesis states that life manipulates the environment for the maintenance of life. For example, some scientists believe that algae floating near the surface of the ocean influence rainfall at sea and the carbon dioxide content of the atmosphere, thereby significantly affecting the global climate. It follows, then, that the planet Earth is capable of physiological self-regulation.

According to James Lovelock, a British scientist who has been developing the Gaia hypothesis since the early 1970s, the idea of a living Earth is probably as old as humanity.[8] James Hutton, whose theory of uniformitarianism was discussed earlier, stated in 1785 that he believed Earth to be a superorganism and compared the cycling of nutrients from soils and rocks in streams and rivers to the circulation of blood in an animal.[8] In this metaphor, the rivers are the arteries and veins, the forests are the lungs, and the oceans are the heart of Earth.

The Gaia hypothesis is really a series of hypotheses. The first is that life, since its inception, has greatly affected the planetary environment. Few scientists would disagree. The second hypothesis asserts that life has altered Earth's environment in ways that have allowed life to persist. Certainly, there is some evidence that life has had such an effect on Earth's climate. A popularized extension of the Gaia hypothesis is that life *deliberately* (consciously) controls the global environment. Few scientists accept this idea.

The extended Gaia hypothesis may have merit in the future, however. Humans have become conscious of our effects on the planet, some of which influence future changes in the global environment. Thus, the concept that humans can consciously make a difference in the future of our planet is not as extreme a view as would once have been thought. The future status of the human environment may depend, in part, on actions we take now and in coming years. This aspect of the Gaia hypothesis exemplifies the key theme of thinking globally, which was introduced in Chapter 1.

The decisions we make in managing our global environment depend on our values as well as our understanding of how Earth works (another key theme from Chapter 1). With this in mind, we explore in greater depth how human processes are linked to environmental change.

3.9 Why Solving Environmental Problems Is Often Difficult

Global environmental systems—whether the entire planet or the hydrosphere, lithosphere, biosphere, or human population—are open systems characterized by poorly defined boundaries and the transfer of material and energy. They are inherently difficult to work with. Global environmental problems are particularly difficult because of the following three characteristics:[5] (1) exponential growth and the positive feedback that accompanies growth, (2) lag times between stimuli and responses of systems that can be decades or longer, and (3) consequences of events that have the potential to result in irreversible changes on the human time scale. Let's look at each of these in turn.

1. *Exponential growth:* Systems undergoing exponential growth, such as human population increase, or use of resources tend to experience exponential growth related to a rate of growth and doubling time. As we saw in this chapter and will see again in Chapter 4, the consequences of exponential growth and its accompanying positive feedback can be dramatic, leading to incredible increases of what is being evaluated or measured.

2. *Lag time:* **Lag time** is the time between a stimulus and the response of a system. If the lag time is very short, consequences are rather easier to identify. For example, release of a toxic gas from a chemical plant may have near-immediate consequences to people living near the plant. If there is a long delay between stimulus and response, then the resulting changes are much more difficult to recognize. When we are dealing with biological resources such as fish or people, the result of long lag times or delays from exponential growth may lead to what is sometimes called **overshoot and collapse**.[5,9] As

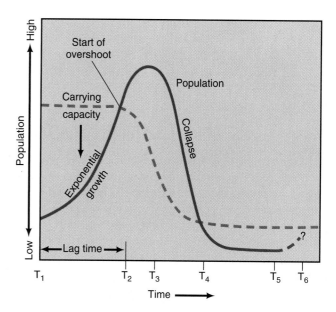

Figure 3.14 • The concept of overshoot, illustrating the influence of exponential growth, lag time, and collapse on carrying capacity. Carrying capacity starts out relatively high, but as exponential growth increases the population beyond the carrying capacity, overshoot occurs and population collapses. If environmental damage occurs as a result of overuse and damage to resources on which the carrying capacity depends, then the carrying capacity also crashes, as shown here. [*Source*: Modified after D.H. Meadows and others 1992.]

an example, Figure 3.14 shows the relationship between carrying capacity and the human population. The carrying capacity starts out being much higher than the human population, but as exponential growth occurs, eventually the population exceeds the carrying capacity and overshoot occurs. This eventually results in the collapse of population to some level below the new carrying capacity, which has also been reduced. The lag time is the time of exponential growth of population before it exceeds the carrying capacity. A similar scenario may be posited for harvesting of species of fish or trees.

3. *Irreversible consequences.* Adverse consequences of environmental change do not necessarily lead to irreversible consequences. Some do, however, and these lead to particular problems. When we talk about irreversible consequences, we mean consequences that may not be easily rectified on a human scale of decades or a few hundred years. A good example of this is soil erosion or harvesting of old-growth forest. With respect to soil erosion, the consequences to productivity of crops may not be reached until the crops no longer have their roots in the active soil, with its nutrients necessary for producing a crop. There may be a long lag time of soil erosion until this occurs, but once the soil is finally eroded, it may take hundreds or thousands of years for a new soil to form—and so the consequences are irreversible.[5] Similarly, with logging of old-growth forest, when these forests are harvested it may take hundreds

Is the Gaia Hypothesis Science?

According to the Gaia hypothesis, Earth and all living things form a single system, with interdependent parts, communication among these parts, and the ability to self-regulate. Are the Gaia hypothesis and its component hypotheses science, fringe science, or pseudoscience? Is the Gaia hypothesis anything more than an attractive metaphor? Does it have religious overtones? Answering these questions is more difficult than answering similar questions about, say, crop circles, as described in Chapter 2. Analyzing the Gaia hypothesis forces us to deal with some of our most fundamental ideas about science and life.

Critical Thinking Questions

1. What are the main hypotheses included in the Gaia hypothesis?
2. What kind of evidence would support each hypothesis?
3. Which of the hypotheses can be tested?
4. Is each hypothesis a science, fringe science, or pseudoscience?
5. Some scientists have criticized James E. Lovelock, who formulated the Gaia hypothesis, for using the term *Gaia*. Lovelock responds that it is better than referring to a "biological cybernetic system with homeostatic tendencies." What does this phrase mean?
6. What are the strengths and weaknesses of the Gaia hypothesis?

of years for them to be restored. Lag times may be even longer if the soils have been damaged or eroded as the consequences of timber harvesting.

In summary, we see that exponential growth, long lag time, and the possibility of irreversible consequences have special implications for environmental problems and finding solutions to those problems. We now recognize the potential dangers of exponential growth and realize that when these are coupled to long lag times and irreversible consequences, we must pay special attention to finding solutions. Thus, again we see the importance of the principle of environmental unity, which states that one activity or change often leads to a sequence of changes, some of which may be difficult to recognize. Recognition of lag time and the irreversible consequences associated with exponential growth is paramount in addressing environmental problems.

Most changes brought on by human activity involve rather slow processes with cumulative effects. For example, our present global warming began with the Industrial Revolution, when people started burning massive amounts of fossil fuels. Levels of carbon dioxide in the atmosphere added by human activities have nearly doubled since the Industrial Revolution and are causing human-enhanced global warming. Certainly, we are aware that climate may change naturally and that such change may be rapid (see Chapter 23). However, present global warming is significant and results in part from human activity.

The decline of fisheries is another process that generally takes place over a relatively long time. In such cases, recognizing that irreversible damage may have been done often involves crossing a threshold that is difficult to identify in time to avoid problems. More commonly, we recognize that a threshold has been exceeded after we begin to identify

consequences of a particular activity. For example, we have painfully come to the conclusion that the crash of the salmon population in the Pacific Northwest is the combined result of fishing, building of dams, and land-use practices such as timber harvesting. We did not recognize the crash until it occurred and we reflected on past activities. This example further supports the idea that environmental science problems are often complex, with linkages among various parts of ocean, river, land, and forest systems.

Change can also be chaotic. This occurs when some small change is amplified, resulting in complex and perhaps periodic activity or behavior. Chaos theory is a mathematical description of change that has been used to describe the behavior of a variety of systems, including fluctuations of populations and changes in circulation and patterns of air current in the atmosphere. An often-cited hypothetical example is a scenario in which the beating of the wings of a hummingbird in Brazil causes, through a series of amplified activities, a hurricane in Miami. Although this example seems unlikely to occur in nature, many surprises confront us when we consider natural Earth systems. For example, changes in the temperature of the Pacific Ocean cause far-ranging changes in storms, floods, and other natural hazards on a global basis (see Chapter 23).

As stated, one of our goals in understanding the role of human processes in environmental change is to help manage our global environment. To accomplish this, we need to be able to predict changes before they occur. But as the examples above demonstrate, prediction presents great challenges. Although some changes are anticipated, others come as a surprise. As we learn to apply the principles of environmental unity and uniformitarianism more skillfully, we will be better able to anticipate changes that would otherwise have been surprises.

Summary

- A system is a set of components or parts that function together as a whole. Environmental studies deal with complex systems at every level, and solutions to environmental problems often involve understanding systems and rates of change.

- Systems respond to inputs and have outputs. Feedback is a special kind of system response. Positive feedback is destabilizing, whereas negative feedback tends to stabilize or encourage more constant conditions in a system.

- Relationships between the input (cause) and output (effect) of systems may be nonlinear and may involve delays.

- The principle of environmental unity, simply stated, holds that everything affects everything else. It emphasizes linkages among parts of systems.

- The principle of uniformitarianism can help predict future environmental conditions on the basis of the past and the present.

- A particularly important aspect of positive feedback is exponential growth, in which the increase per time period is a constant fraction or percentage of the current amount. Exponential growth involves two factors: the rate of growth and the doubling time.

- Changes in systems can be studied through input–output analysis. The average residence time is the average time it takes for the total reservoir of a particular material to be cycled through the system.

- Life on Earth began about 3 billion years ago and since that time has profoundly changed our planet. Sustained life on Earth is a characteristic not of individual organisms or populations but of ecosystems—local communities of interacting populations and their nonbiological environments.

- The general term *biota* refers to all living things, and the region of Earth where life exists is known as the *biosphere*.

- The Gaia hypothesis states that life on Earth, through a complex system of positive and negative feedback, regulates the planetary environment to help sustain life.

- Exponential growth, long lag times, and the possibility of irreversible change combine to make solving environmental problems difficult.

REEXAMINING THEMES AND ISSUES

Human Population

The human population of Earth is experiencing a variety of positive feedback mechanisms leading to an increasing population. Of particular concern are local or regional increases in population density (number of people per unit area), which strain resources and lead to human suffering and economic damage.

Sustainability

Negative feedback is stabilizing. If we are to have a sustainable human population and use our resources sustainably, then we need to set in place a series of negative feedbacks within our agricultural, urban, and industrial systems.

Global Perspective

This chapter introduced Earth as a system. One of the most fruitful areas for environmental research remains the investigation of relationships between physical and biological processes on a global scale. More of these relationships must be discovered if we are to solve environmental problems related to such issues as potential global warming, ozone depletion, and disposal of toxic waste.

Urban World

The concepts of environmental unity and uniformitarianism are particularly appropriate in urban environments, where land-use changes result in a variety of changes that affect physical and biochemical processes.

People and Nature

People and nature are linked in complex ways in systems that are constantly changing. Some changes are not related to human activity, but many are—and human-caused changes from local to global in scale are accelerating.

Science and Values

Our discussion of the Gaia hypothesis reminds us that we still know very little about how our planet works and how physical, biological, and chemical systems are linked. What we do know is that we need more scientific understanding. This understanding will be driven in part by the value that we place on our environment and on the well-being of other living things.

Key Terms

average residence time **47**
biosphere **50**
biota **50**
closed system **41**
doubling time **43**

ecosystem **50**
environmental unity **45**
exponential growth **43**
feedback **42**
Gaia hypothesis **50**

lag time **51**
negative feedback **42**
open system **41**
overshoot and collapse **51**
positive feedback **42**

steady state **47**
system **41**
uniformitarianism **46**

Study Questions

1. How does the Amboseli National Reserve case history exemplify the principle of environmental unity?

2. What is the difference between positive and negative feedback in systems? Provide an example of each.

3. What is the main point concerning exponential growth? Is exponential growth good or bad?

4. Why is the idea of equilibrium in systems somewhat misleading in regard to environmental questions? Is the establishment of a balance of nature ever possible?

5. Why is the concept of the ecosystem so important in the study of environmental science? Should we be worried about disturbing ecosystems? Under what circumstances should we worry or not worry?

6. Is the Gaia hypothesis a true statement of how nature works, or is it simply a metaphor? Explain.

7. How might you use the principle of uniformitarianism to help evaluate environmental problems? Is it possible to use this principle to help evaluate the potential consequences of too many people on Earth?

8. Why does overshoot occur, and what could be done to anticipate and avoid it?

Further Reading

Abelson, P. H. 1990. "Global Change." *Science* 249:1085. This editorial outlines what the United States is doing to address the challenges of global change and what it can do in the future.

Bunyard, P., ed. 1996. *Gaia in Action: Science of the Living Earth.* Edinburgh: Floris Books. This book presents investigations into implications of the Gaia hypothesis.

Ehrlich, P. R., A. H. Ehrlich, and J. P. Holdren. 1970. *Ecoscience.* San Francisco: Freeman. Although this is an older book, Chapter 2 provides a good overview of the physical world and how systems and changes may affect our environment.

Lovelock, J. 1995. *The Ages of Gaia: A Biography of Our Living Earth.* New York: Norton. This small book explains the Gaia hypothesis, presenting the case that life very much affects our planet and in fact may regulate it for the benefit of life.

CHAPTER 4

The Human Population and the Environment

A veterinary worker in western France in 2006 trying to vaccinate ibises against a strain of the bird flu virus.

Learning Objectives

In 2005 hurricanes, earthquakes, and tsunamis caused many human deaths and, along with the new emerging bird flu and other possible diseases, seemed to threaten even greater human catastrophes in the future. But the human population has been growing rapidly and seemingly steadily for decades. What is happening? How do we account for both the growth of our population and possible threats to it? After reading this chapter, you should understand:

■ Ultimately, there can be no long-term solutions to environmental

problems unless the human population stops increasing.

■ Two major questions about the human population involve what controls its rate of growth and how many people Earth can sustain.

■ The rapid increase in the human population has occurred with little or no change in the maximum lifetime of an individual.

■ Modern medical practices and improvements in sanitation, control of disease-spreading organisms, and supplies of human necessities have decreased death rates and acceler-

ated the net rate of human population growth.

■ Countries with a high standard of living have moved more quickly to a lower birth rate than have countries with a low standard of living.

■ Although we cannot predict with absolute certainty what the future human carrying capacity of Earth will be, understanding of human population dynamics can help us make useful forecasts.

CASE STUDY

Death in Indonesia from the Great Tsunami of 2004

On December 26, 2004, an earthquake in the ocean off the western coast of Sumatra, one of the main islands of Indonesia, created a tsunami, a tidal wave that destroyed many coastal areas and killed an estimated 230,000 people in Indonesia, Sri Lanka, South India, Thailand, and other countries, some as far away as South Africa (Figure 4.1). The following October, another major earthquake struck Pakistan, killing an estimated 73,000. News of both disasters spread rapidly, and billions of dollars of aid flowed to help the victims; the world felt the suffering of the many who survived and lost family and friends.

But this disaster was made even more tragic by the fact that the human population, growing at just 1.2% per year, replaced the number lost in those catastrophes in just a few days. There are so many of us—6.48 billion at the latest estimate—that even at a low rate of increase the number of people added to our population is huge— 84 million more of us every year, an average of 230,000 more people every day. The growth curve of the popula-

tion showed barely a ripple. Even the population of Indonesia, 210 million, would recover the total number lost in the tsunami in no more than a month. In the United States, the population reached 300 million in October, 2006, and is expected to reach 400 million in just 40 years!

The human population has been increasing rapidly for many centuries except for short periods of catastrophes, and it has become commonplace to believe that this increase will continue indefinitely. But the disasters of the last two years are starting to make people wonder if perhaps our population has turned a corner. In 2005, concern grew about a possible pandemic arising from an Asian bird flu, a viral disease of birds first found in domestic poultry but that is rapidly adapting to human beings. What is going on with our population? And more important, how can we forecast what will happen to our numbers in the future? Seeking that understanding is the purpose of this chapter, and it is fundamental for any student of Environmental Science.

Figure 4.1 ■ The great Tsunami of December 26, 2004 killed many people and destroyed many houses near the coast, as shown here for the city of Banda Aceh, Indonesia, in a photograph taken three weeks later, on Saturday, Jan. 22, 2005.

> *If the human population continues to grow rapidly, it will ultimately overwhelm the environment. That is why human population growth is a major theme of this textbook.*

4.1 How Populations Change Over Time: Basic Concepts of Population Dynamics

Basic Concepts

A **population** is a group of individuals of the same species living in the same area or interbreeding and sharing genetic information. A **species** is all individuals that are capable of interbreeding. A species is made up of populations. Five key properties of any population are: abundance, which is the size of a population—now, in the past, and in the future—**birth rates, death rates, growth rates**, and **age structure**.

Populations change over time and over space. The general study of population changes is called **population dynamics**. How rapidly a population changes depends on the growth rate, which is the difference between the birth rate and the death rate. (See Table 4.1 for other useful terms and Working It Out 4.1.)

Age Structure

An important factor in population growth is the population age structure, the proportion of the population of each age. The age structure of a population affects current and future birth rates, death rates, and growth rates; has an impact on the environment; and has implications for current and future social and economic conditions.

We can picture an age structure as if it were a pile of blocks, one for each age group, where the size of each block represents the number of people in that group (Figure 4.2). Although age structures can take many shapes, four general types are most important to our discussion: a pyramid, a column, an inverted pyramid (top-heavy), and a column with a bulge. The pyramid age structure occurs in a population with many young people and a high death rate at each age—and therefore a short average lifetime. A column shape occurs where the birth rate and death rate are low and a high percentage of the population is elderly. A bulge occurs if some event in the past caused a high birth rate or death rate for some age group but not others.

Table 4.1   Human Population Terms
Crude birth rate: number of births per 1,000 individuals per year; "crude" because population age structure is not taken into account
Crude death rate: number of deaths per 1,000 individuals per year
Crude growth rate: net number added per 1,000 individuals per year; also equal to crude birth rate minus crude death rate
Fertility: pregnancy or the capacity to become pregnant or to have children
General fertility rate: number of live births expected in a year per 1,000 women aged 15 to 49 years, considered the childbearing years
Age-specific birth rate: number of births expected per year among a fertility-specific age group of women in a population
Total fertility rate (TFR): average number of children expected to be born to a woman throughout her childbearing years
Cause-specific death rate: number of deaths from one cause per 100,000 total deaths
Incidence rate: number of people contracting a disease during a time period, usually measured per 100 people
Prevalence rate: number of people afflicted by a disease at a particular time
Case fatality rate: percentage of people who die once they contract a disease
Morbidity: general term meaning the occurrence of disease and illness in a population
Rate of natural increase (RNI): birth rate minus death rate, implying annual rate of population growth not including migration
Doubling time: number of years it takes for a population to double, assuming a constant rate of natural increase
Infant mortality rate: annual number of deaths of infants under age 1 per 1,000 live births
Life expectancy at birth: average number of years a newborn infant can expect to live under current mortality levels
GNP per capita: gross national product (GNP), which includes the value of all domestic and foreign output, per person

Source: C. Haub and D. Cornelius, *World Population Data Sheet* (Washington, D.C.: Population Reference Bureau, 1998).

Forecasting Population Change

Populations change in size through births, deaths, immigration (arrivals from elsewhere), and emigration (departures to go elsewhere). We can write a formula to represent population change:

$$P_2 = P_1 + (B - D) + (I - E)$$

where P_1 is the number of individuals in a population at time 1, P_2 is the number of individuals in that population at some later time 2, B is the number of births in the period from time 1 to time 2, D is the number of deaths from time 1 to time 2, I is the number entering as immigrants, and E is the number leaving as emigrants.

Ignoring for the moment immigration and emigration, how rapidly a population changes depends on the growth rate, which is the difference between the birth rate and the death rate (see Table 4.1 for other useful terms). The human population growth rate is usually expressed in the rate per 1,000, called the crude rate, rather than the more familiar percentage, which is the rate per 100. For example, in 1999 the crude death rate in the United States was 9, meaning that 9 of every 1,000 people died. (The same information expressed as a percentage is a rate of 0.9%.) In 1999 the crude birth rate in the United States was 15.[1] The crude growth rate is the net change—the birth rate minus the death rate. Thus the crude growth rate in 1999 in the United States was 6. For every 1,000 people at the beginning of 1999, there were 1,006 at the end of the year.

Continuing for the moment to ignore immigration and emigration, we can state that how rapidly a population grows depends on the difference between the birth rate and the death rate. The birth rate (usually denoted by b) is the fraction or percentage of the population born in a unit of time. The death rate (d) is the fraction or percentage of the population that dies in a unit of time. The growth rate of a population is then

$$g = b - d$$

Note that in all these cases, the units are numbers per unit of time.

Recall from Chapter 3 that doubling time—the time it takes a population to reach twice its present size—can be estimated by the formula

$$T = 70/\text{annual growth rate}$$

where T is the doubling time and the annual growth rate is expressed as a percentage. For example, a population growing 2% per year would double in approximately 35 years.

Birth rates, death rates, and growth rates can be calculated from the number of births and deaths during some time period and the total population at some specific time during that period. Letting N equal the total number of individuals in the population, the birth rate (b) is the number of births per unit time (B) divided by the total population (N), or

$$b = B/N$$

The death rate is the number of deaths per unit time (D) divided by the total population (N), or

$$d = D/N$$

The growth rate (g) is the result of the number of births minus the number of deaths per unit time divided by the total number in the population (N), or

$$g = (B - D)/N \quad \text{or} \quad g = G/N$$

It is important to be consistent in using the population at the beginning, middle, or end of the period. Usually, the number at the beginning or the middle is used.

Consider an example: There were 19,700,000 people in Australia in mid-2002, and there were 394,000 births from 2002 to 2003. The birth rate, b, calculated against the mid-2002 population was 394,000/19,700,000, or 2%. During the same period, there were 137,900 deaths; the death rate, d, was 137,900/19,700,000, or 0.7%. The growth rate, g, was (394,000 − 137,900)/19,700,000, or 1.3%.[1]

Age structure varies considerably by nation (Figure 4.2). Kenya's pyramid-shaped age structure illustrates a rapidly growing population heavily weighted toward youth. In developing countries today, about 34% of the populations are under 15 years of age. Such an age structure indicates that the population will grow very rapidly in the future, when the young reach marriage and reproductive ages. It suggests that the future for such a nation requires more jobs for the young, and it has many other social implications that go beyond the scope of this book.

In contrast, the age structure of the United States is more like a column, showing a population with slow growth, and Italy's slightly top-heavy pyramid shows a nation with declining growth. Elderly people make up a small percentage (3%) of Kenya's population but a much larger percentage of the populations of the United States and Italy (13% and 17%, respectively).[1]

Age structure provides insight into a population's history, its current status, and its likely future. For example, the baby boom that occurred after World War II in the United States

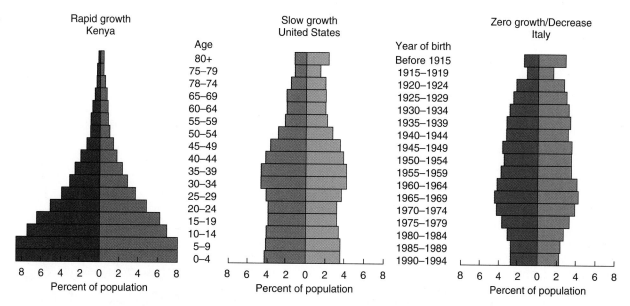

Figure 4.2 ■ Age structure of Kenya, the United States, and Italy, 1995. The bars to the left are males, to the right are females. (*Sources:* U.S. Bureau of the Census, "U.S. Population Estimates by Age, Sex, and Race: 1990 to 1995," PPL-41, February 14, 1996; Council of Europe, *Recent Demographic Developments in Europe 1997*, Table 1–1; United Nations, *The Sex and Age Distribution of the World Populations—The 1996 Revision*, 500–1.)

(a great increase in births from 1946 through 1964) forms a pulse in the population that can be seen as a bulge in the age structure, especially of those aged 40 to 50 in 2000 (see Figure 4.2). A secondary, smaller bulge results from offspring of the first baby boom, which can be seen as a slight increase in 5- to 15-year-olds. This second peak shows that the baby-boom pulse is moving through the age structure. Each baby boom increases demand for social and economic resources; for example, schools were crowded when the baby boomers were of primary and secondary school age.

One economic implication of age structure involves care of the elderly. In preindustrial and nonindustrial societies, average lifetimes are short, children care for their parents, and therefore it benefits parents to have many children. In modern technological societies, family size is smaller, and care for older people is distributed throughout the society through taxes, so that those who work provide funds to care for those who cannot. Parents tend to benefit when their children are well educated and have high-paying jobs. Rather than relying on a large family in which each child has fewer resources, parents tend to have fewer children and invest more in each. This makes *zero population growth* possible. However, a shift from a youthful age structure (like Kenya's) to an elderly age structure (like Italy's) means that a smaller percentage of the population works— and therefore fewer tax monies are available for elder care.

A population heavily weighted toward the elderly poses problems for a nation. The easiest way to increase tax income is to increase the percentage of young people and thereby promote rapid population growth. Thus, short-term economic pressures at a national level can lead to political policies supporting rapid population growth, which is not in the long-term interest of the nation.

4.2 Kinds of Population Growth

Exponential Growth

As discussed in Chapter 3, a population experiencing exponential growth increases by a constant percentage per unit of time. The growth rate of the human population increased but varied during the first part of the twentieth century, peaking in 1965–1970 at 2.1% because of improvements in health care, medicine, and food production. Thus, the human population actually increased at a rate faster than the rate of exponential growth. This increase in the human growth rate has stopped, however, and the growth rate is declining globally. As mentioned, it is now approximately 1.2%.[1]

A Brief History of Human Population Growth

The history of the human population (see A Closer Look 4.1) can be viewed as consisting of four major periods:

1. In the early period of hunters and gatherers, the world's total human population was probably less than a few million.

2. A second period, beginning with the rise of agriculture, allowed a much greater density of people and the first major increase in the human population.

3. The Industrial Revolution, with improvements in health care and the supply of food, led to a rapid increase in the human population.

4. Today, the rate of population growth has slowed in wealthy, industrialized nations but continues to increase rapidly in many poorer, less developed nations (Figures 4.3 and 4.4).

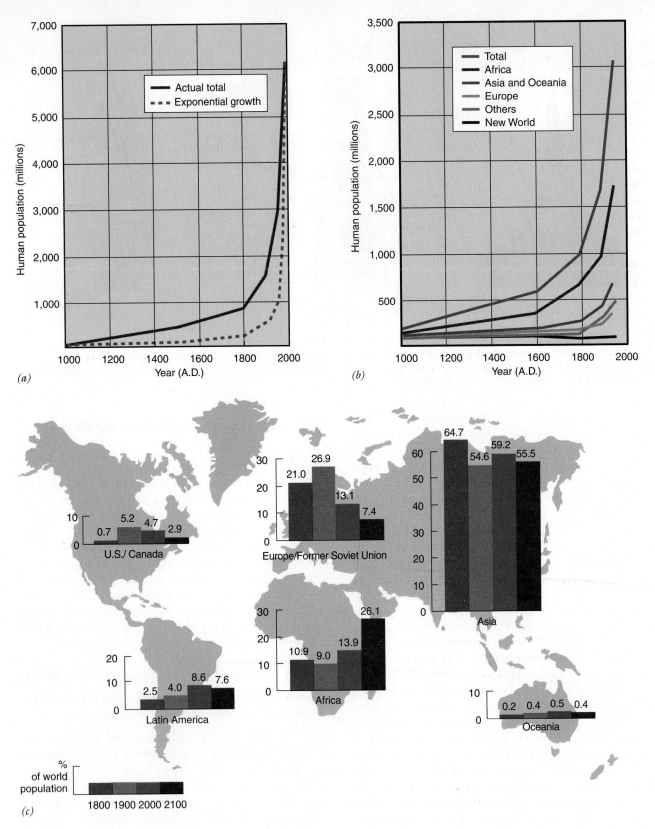

Figure 4.3 ■ World human population since 1000 A.D. (*a*) Actual increase compared with an exponential curve. Note that the rate of increase of the human population has exceeded an exponential rate. With an exponential curve, the net rate of increase is constant. In reality, the net rate of increase grew as a result of several factors—most important for our discussion, a decline in the death rate. (*b*) Human population from 1000 A.D. to 2000 A.D. by major geographic region. (*c*) Past and forecasted human population growth rate by geographic region and major nations. The main feature of this graph is that most growth in the 21st century is occuring in the poorer, developing nations. Less developed nations now account for 99% of the world population growth.[1] (*Sources:* M. M. Kent and L. A. Crews, *World Populations: Fundamentals of Growth* [Washington, D.C.: Population Reference Bureau, 1990] and *2005 Population Data Sheet*, 2005, Population Reference Bureau, Washington, D.C.)

Growth of the Human Population

Stage 1. Hunters and Gatherers

From the first evolution of humans to the beginning of agriculture.[2]

Population density: About 1 person per 130–260 km² in the most habitable areas.

Total human population: As low as one-quarter million, less than the population of modern small cities like Hartford, Connecticut, and certainly less than a few million, which is fewer people than now live in many of our largest cities.

Average rate of growth: The average annual rate of increase over the entire history of human population is less than 0.00011% per year.

Stage 2. Early, Preindustrial Agriculture

Beginning sometime between 9000 B.C. and 6000 B.C. and lasting approximately until the sixteenth century.

Population density: With the domestication of plants and animals and the rise of settled villages, human population density increased greatly, to about 1 or 2 people/km2 or more, beginning a second period in human population history. (Even today, primitive people who practice agriculture have population densities greatly exceeding those of hunters and gatherers.)

Total human population: About 100 million by A.D. 1 and 500 million by A.D. 1600 (see Figure 4.3.).

Average rate of growth: Perhaps about 0.03%, which was large enough to increase the human population from 5 million in 10,000 B.C. to about 100 million in A.D. 1. The Roman Empire accounted for about 54 million. From A.D. 1 to A.D. 1000, the population increased to 200–300 million.

Stage 3. The Machine Age

Some experts say that this period marked the transition from agricultural to literate societies, when better medical care and sanitation were factors in reducing the death rate.

Total human population: About 900 million in 1800, almost doubling in the next century and doubling again (to 3 billion) by 1960.

Average rate of growth: By 1600, about 0.1% per year, with rate increases of about 0.1% every 50 years until 1950. This rapid increase occurred because of the discovery of causes of diseases, invention of vaccines, improvements in sanitation, other advances in medicine and health, and advances in agriculture that led to a great increase in the production of food, shelter, and clothing.

Stage 4. The Modern Era

Total human population: Reaching and exceeding 6.6 billion.

Average rate of growth: The growth rate of the human population reached 2% in the middle of the twentieth century and has declined to 1.2%.[1]

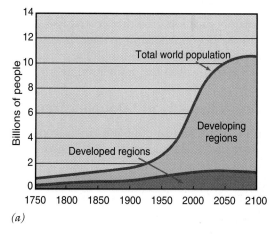

(a)

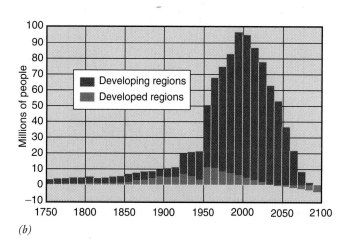

(b)

Figure 4.4 ■ Logistic growth curve. World population is shown as total numbers (*a*) and growth per decade (*b*), by development status, 1750–2100. For example, during the decade from 1980 to 1990, 82 million people were added each year, for a total addition of 820 million people.

How Many People Have Lived on Earth?

How many people have lived on Earth? Of course, before written history, there were no censuses. The first estimates of population in Western civilization were attempted in the Roman era. During the Middle Ages and the Renaissance, scholars occasionally estimated the number of people. The first modern census was taken in 1655 in the Canadian colonies by the French and the British.[3] The first series of regular censuses taken by a country began in Sweden in 1750, and the United States has taken a census every decade since 1790. Most countries began much later. The first Russian census, for example, was taken in 1870. Even today, many countries do not take censuses or do not do so regularly. The population of China has only recently begun to be known with any accuracy. By studying modern primitive peoples and applying principles of ecology, however, we can gain a rough idea of the total number of people who may have lived on Earth. Summing all the values, including those since the beginning of written history, about 50 billion people are estimated to have lived on Earth.[4] If this is so, then, surprisingly, the more than 6.6 billion people alive today represent more than 10% of all of the people who have ever lived.

It is also interesting to look at the total cumulative population over the history of humans, which is explored in A Closer Look 4.2.

4.3 Present Human Population Rates of Growth

At present, the world population numbers considerably more than 6.6 billion people, with an annual growth rate of approximately 1.2% (see Figures 4.3 and 4.4). At this rate, 84 million people are added to Earth's population in a single year, a number greater than the 2005 population of Germany and more than two and a half times the population of Canada.[1] Human population trends vary greatly among the major regions of the world (see Figure 4.3) and among countries. In India, the current population growth rate is 1.7%, in Northern Europe, it averages 0.2%.[1] The growth rate of the population of the United States has declined and is now 0.6% (Figure 4.5).

Thus, there seems to be a correlation between poverty and population growth. The poorer a nation, the more likely the population growth rate will be high; the wealthier a nation—and the higher the average per-capita income—the lower the population growth rate. Since a high growth rate works against an increase in per-capita income, poor nations are in danger of a positive feedback (see Chapter 3): the more people, the greater the growth rate; the greater the growth rate, the more people.

4.4 Projecting Future Population Growth

With human population growth a central issue, it is important that we develop methods to forecast what will happen to our population in the future. One of the simplest approaches is to calculate the doubling time.

Exponential Growth and Doubling Time

Recall from Chapter 3 that doubling time, a concept used frequently in discussing human population growth, is the time required for a population to double in size (see Working It Out 4.1). The standard way to calculate doubling time is to assume that the population is growing exponentially (has a constant growth rate). We can then estimate doubling time by dividing 70 by the annual growth rate stated as a percentage.

The doubling time based on exponential growth is very sensitive to growth rate; that is, it changes quickly as the growth rate changes (Figure 4.6). A few examples demonstrate this sensitivity. With a current population growth of

Figure 4.5 ■ United States population 1790 to 2000. Note that the rate of growth of the United States population is slowing.

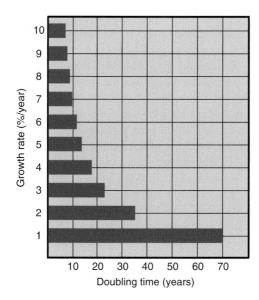

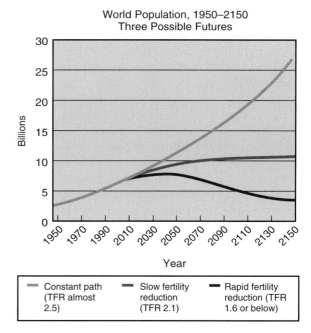

World Population, 1950–2150
Three Possible Futures

- Constant path (TFR almost 2.5)
- Slow fertility reduction (TFR 2.1)
- Rapid fertility reduction (TFR 1.6 or below)

Figure 4.6 ■ Doubling time changes rapidly with the population growth rate. Because the world's population is increasing at a rate between 1% and 2%, we expect it to double within the next 35 to 70 years.

Figure 4.7 ■ Exponential and logistic growth curves. Three possible paths of future world population growth, as projected by the United Nations using the logistic curve, for three different Total Fertility Rates (the expected number of children a woman will have during her life). The constant path assumes that the 1998 growth rate will continue unchanged, resulting in an exponential increase. The slow fertility reduction path assumes that the world's fertility will decline to reach replacement level by the year 2050 and that the world's population will stabilize at about 11 billion by the 22nd century. The rapid fertility reduction path assumes that the world's fertility will go into decline in the 21st century, peaking at 7.7 billion in 2050 and dropping to 3.6 billion by 2150. These are theoretical curves. Total Fertility Rate has remained high and is now 2.7. [*Source:* United Nations Population Division, 1998.]

0.6%, the United States has a doubling time of 70 divided by 0.6, or 117 years. In contrast, the current growth rate of Nicaragua is 2.7%, giving that nation a doubling time of 26 years. Northern Europe, with an annual rate of about 0.2%, has a doubling time of 350 years. The world's most populous country, China, has the same growth rate as the United States, and therefore the same 117-year doubling time.[1]

No population can sustain an exponential rate of growth indefinitely. Eventually the population will run out of food and space and become increasingly vulnerable to catastrophes, as we are beginning to observe. A population of 100 increasing at 5% per year, for example, would grow to 1 billion in less than 325 years. If the human population had increased at this rate since the beginning of recorded history, it would now exceed all the known matter in the universe.

The Logistic Growth Curve

An exponentially growing population theoretically increases forever, but on Earth, which is limited in size, this is not possible. If a population cannot increase forever, what changes in the population can we expect over time? One of the first suggestions made about the human population is that it would follow a smooth S-shaped curve known as the **logistic growth curve**. The population would increase exponentially only temporarily. After that, the rate of growth would gradually decline (i.e., the population would increase more slowly) until an upper population limit, called the **logistic carrying capacity**, was reached (Figure 4.4*a* and Figure 4.7). Once the logistic carrying capacity had been reached, the population would remain at that number.

The logistic growth curve was first suggested in 1838 by a European scientist, P. F. Verhulst, as a theory for the growth of animal populations. It has been applied widely to the growth of many animal populations, including those important in wildlife management, endangered species, and those in fisheries (see Chapter 13), as well as the human population. Unfortunately, there is little evidence that human populations—or any animal populations, for that matter—actually follow this growth curve.

The logistic curve involves assumptions that are unrealistic for humans and for other mammals. These assumptions include a constant environment, a constant carrying capacity, and a homogeneous population (one in which all individuals are identical in their effects on each other). The logistic curve is especially unlikely if death rates continue to decrease from improvements in health care, medicine, and food supplies. Once a human population has benefited from these improvements, it must pass through what has become known as the demographic transition to achieve **zero population growth**, which could then lead to a stabilized population. The demographic transition is discussed later in this chapter.

Forecasting Human Population Growth Using the Logistic Curve

Though unrealistic, the logistic curve has been the method used for most long-term forecasts of the size of human populations in specific nations. This S-shaped curve first rises steeply upward and then changes slope, curving toward the horizontal carrying capacity (Figures 4.4a). The point at which the curve changes is the inflection point. Until a population has reached the inflection point, we cannot project the final logistic size. Unfortunately for those who want to make this calculation, the human population has not yet made the bend around the inflection point. Typically, forecasters have dealt with this problem by assuming that today's population is just reaching the inflection point. This standard practice inevitably leads to a great underestimate of the maximum population. For example, one of the first projections of the upper limit of the U.S. population, made in the 1930s, assumed that the inflection point had occurred then. That assumption resulted in an estimate that the final population of the United States would be approximately 200 million. That number has long since been exceeded; indeed, the U.S. population has passed 300 million.[5]

The United Nations has made a series of projections based on current birth rates and death rates and assumptions about how these rates will change. These projections form the basis for the curves presented in Figure 4.7. The logistic projections assume that (1) mortality will fall everywhere and level off when female life expectancy reaches 82 years; (2) fertility will reach replacement levels everywhere between 2005 and 2060; (3) there will be no major worldwide catastrophe. This approach projects an equilibrium world population of 10.1–12.5 billion.[6] Developed countries would experience population growth from 1.2 billion today to 1.9 billion, but populations in developing countries would increase from 4.5 billion to 9.6 billion. Bangladesh (an area the size of Wisconsin) would reach 257 million; Nigeria, 453 million; and India, 1.86 billion. In these projections, the developing countries contribute 95% of the increase.[4, 6]

4.5 The Demographic Transition

The **demographic transition** is a three-stage pattern of change in birth rates and death rates that has occurred during the process of industrial and economic development of Western nations. It leads to a decline in population growth.

A decline in the death rate is the first stage of the demographic transition (Figure 4.8).[2] In a nonindustrial country, birth rates and death rates are high, and the growth rate is low.[6] With industrialization, health and sanitation improve, and the death rate drops rapidly. The birth rate remains high, however, and the population enters stage II, a period with a high growth rate. Most European nations passed through this period in the eighteenth and nineteenth centuries.

As education and the standard of living increase and as family-planning methods become more widely used, the population reaches stage III. The birth rate drops toward the death rate, and the growth rate therefore decreases, eventually to a low or zero growth rate. However, the birth rate declines only if families believe there is a direct connection between future economic well-being and funds spent on the education and care of their young. Such families have few children and put all their resources toward the education and well-being of those few.

Historically, parents have preferred to have large families. Without other means of support, parents can depend on children for a kind of "social security" in their old age, and children help with many kinds of hunting, gathering, and low-technology farming. Unless a change in attitude occurs among parents—unless they see more benefits from a few well-educated children than from many poorer children—nations face a problem in making the transition from stage II to stage III (see Figure 4.8c).

Some developed countries are approaching stage III, but it is an open question whether other developing nations will make the transition before a serious population crash occurs. *The key point here is that the demographic transition will take place only if parents come to believe that having a small family is to their benefit. Here we see again the connection between science and values.* Scientific analysis can show the value of small families, but this knowledge must become part of cultural values to have an effect.

4.6 Population and Technology

The danger that the human population poses to the environment is the result of two factors: the number of people and the impact of each person on the environment. When there were few people on Earth and technology was limited, human impact was local. In that situation, overuse of a local resource had few or no large or long-lasting effects. The fundamental problem now is that there are so many people and our technologies are so powerful that our effects on the environment are global and important.

The simplest way to characterize the total impact of the human population on the environment is as follows. The total environmental effect is the average impact of an individual multiplied by the total number of individuals,[7] or

$$T = P \times I$$

where P is the population size—the number of people—and I is the average environmental impact per person. The impact per person varies widely. The average impact of a person who

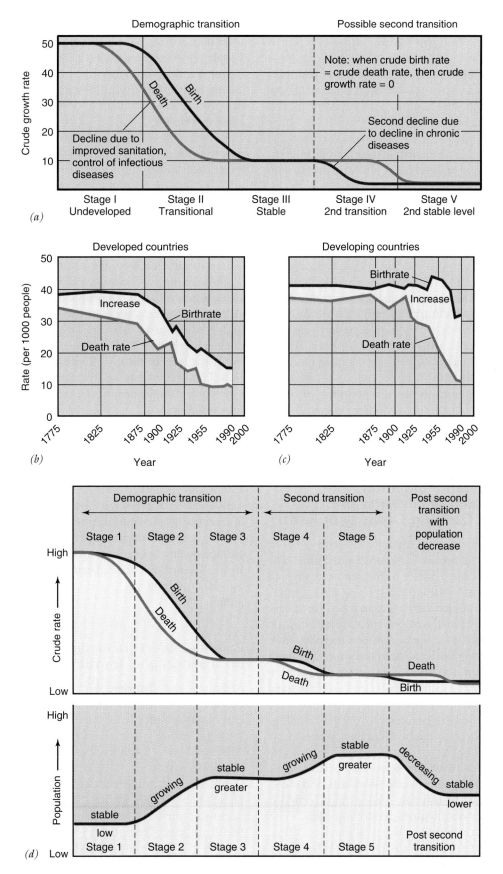

Figure 4.8 ■ The demographic transition: (*a*) theoretical, including possible fourth and fifth stages that might take place in the future; (*b*) as has taken place for developed countries since 1775; (*c*) as may be occurring in developing nations since 1775; and (*d*) the growth changes from negative to positive and back again during this transition. (*Source:* M. M. Kent and K. A. Crews, *World Population: Fundamentals of Growth* [Washington, D.C.: Population Reference Bureau, 1990]. Copyright 1990 by the Population Reference Bureau, Inc. Reprinted by permission.)

lives in the United States is much greater than the impact of a person who lives in a low-technology society. But even in a poor low-technology nation like Bangladesh, the sheer number of people leads to large-scale environmental effects. Almost 200 years ago, Thomas Malthus foresaw the human population problem (see A Closer Look, 4.3).

Modern technology increases the use of resources and also enables us to affect the environment in many new ways, compared with hunters and gatherers or people who farmed with simple wooden and stone tools. For example, before the invention of chlorofluorocarbons (CFCs), used as propellants in spray cans and as coolants in refrigerators and air conditioners, we were not causing depletion of the ozone layer in the upper atmosphere. Similarly, before we started driving automobiles, there was much less demand for steel, little demand for oil, and much less air pollution. These linkages between people and the global environment illustrate the global theme and the people and nature theme of this book.

The population-times-technology equation reveals a great irony involving two standard goals of international aid: improving the standard of living and slowing the overall human population growth. Improving the standard of living increases the total environmental impact, countering the environmental benefits of a decline in population growth.

4.7 The Human Population, the Quality of Life, and the Human Carrying Capacity of Earth

What is the **human carrying capacity** of Earth—that is, how many people can live on Earth at the same time? The answer depends on what quality of life people desire and are willing to accept.

Estimates of the human carrying capacity of Earth have typically involved two methods (See the Critical Thinking Issue in this chapter). One method is to extrapolate from past growth. This approach, as discussed earlier, assumes that the population will follow an S-shaped logistic growth curve, so that it will gradually level off (see Figures 4.4 and 4.7).

The second method can be referred to as the "packing-problem" approach. This method simply considers how many people might be packed onto Earth, not taking into sufficient account the need for lands and oceans to provide food, water, energy, construction materials, and scenic beauty and the need to maintain biological diversity. It could be called the "standing-room-only approach." This has led to very high estimates of the total number of people that might occupy Earth—as many as 50 billion.

More recently, a philosophical movement has developed at the other extreme, known as "deep ecology." This philosophy makes sustaining the biosphere the primary moral imperative for people. Its proponents argue that the whole Earth is necessary to sustain life. Therefore, everything else must be sacrificed to the goal of sustaining the biosphere. People are considered active agents of destruction of the biosphere, and therefore the total number of people should be greatly reduced. Estimates based on this rationale for the *desirable* number of people vary greatly, from a few million up.

Between the packing-problem approach and deep-ecology approach are a number of options. It is possible to set goals in between these extremes, but each of these goals is a value judgment, again reminding us of the theme of *science and values*. What constitutes a desirable quality of life is a value judgment. What kind of life is possible is affected by technology, which in turn is affected by science. And scientific understanding tells us what is required to meet each quality-of-life level.

The options vary according to the quality of life for the average person. If all the people of the world were to live at the same level as those of the United States, with our high resource use, then the carrying capacity would be comparatively low. If all the people of the world were to live at the level of those in Bangladesh, with all of its risks as well as its poverty and its heavy drain on biological diversity and scenic beauty, the carrying capacity would be much higher.

In summary, the acceptable carrying capacity is not simply a scientific issue; it is an issue combining science and values, one of the themes of this book. Science plays two roles. First, by leading to new knowledge that leads to new technology, it makes possible both a greater impact per individual on Earth's resources and a higher density of human beings. Second, scientific methods can be used to forecast a probable carrying capacity once a goal for the average quality of life, in terms of human values, is chosen. In this second use, science can tell us the implications of our value judgments, but it cannot provide those value judgments.

Potential Effects of Medical Advances on Demographic Transition

Although the demographic transition is traditionally defined as consisting of three stages, advances in treating chronic health problems such as heart disease can lead a stage III country to a second decline in the death rate. This could bring about a second transitional phase of population growth (stage IV), in which the birth rate would remain the same while the death rate fell. A second stable phase of low or zero growth (stage V) would be achieved only when the birth rate declined to match the decline in the death rate. Thus, there is danger of a new spurt of growth even in industrialized nations that have passed through the standard demographic transition.

Recent medical advances in the understanding of aging and the potential of new biotechnology to increase both the average longevity and the maximum lifetime of human beings have major implications for the growth of the human

The Prophecy of Malthus

Almost 200 years ago, the English economist Thomas Malthus eloquently stated the human population problem. He based his argument on three simple premises:[8]

- Food is necessary for people to survive.
- "Passion between the sexes is necessary and will remain nearly in its present state"—so children will continue to be born.
- The power of population growth is "indefinitely greater than the power of Earth to produce subsistence."

Malthus reasoned that it would be impossible to maintain a rapidly multiplying human population on a finite resource base. His projections of the ultimate fate of humankind were dire, as dismal a picture as that painted by the most extreme pessimists of today. The power of population growth is so great, he wrote, that "premature death must in some shape or other visit the human race. The vices of mankind are active and able ministers of depopulation, but should they fail, sickly seasons, epidemics, pestilence and plague, advance in terrific array, and sweep off their thousands and ten thousands." Malthus recognized the possibility of potential disease threats like the bird flu that concerns us today. Should these fail, "gigantic famine stalks in the rear, and with one mighty blow, levels the population with the food of the world."

Critics of Malthus continue to point out that his predictions have yet to come true. Whenever things have looked bleak, technology has provided a way out, allowing us to live at greater densities. These critics have argued that our tech-nologies will continue to save us from a Malthusian fate and that therefore we need not worry about the growth of the human population.

Who is correct? Ultimately, in a finite world, Malthus must be correct about the final outcome of unchecked growth. He may have been wrong about the timing; he did not anticipate the capability of technological changes to delay the inevitable. But although some people believe that Earth can support many more people than it does now, in the long run there must be an upper limit. The basic issue that confronts us is this: How can we achieve a constant world population, or at least halt the increase in the population, in a way that is most beneficial to most people? This is undoubtedly one of the most important questions that has ever faced humanity.

population. As these medical advances take place, the death rate will drop and the growth rate will increase even more. Thus, a prospect that is positive from each individual's point of view—a longer, healthier, and more active life—could have negative effects on the environment. We will therefore ultimately need to decide among the following choices: Stop medical research dealing with chronic diseases of old age and attempts to increase people's maximum lifetime; reduce the birth rate; or do neither and wait for Malthus's projections to come true—for famine, environmental catastrophes, and epidemic diseases to cause large and sporadic episodes of human death. The first choice seems inhumane, but the second is highly controversial, so doing nothing and waiting for Malthus's projections may be what actually happens, a future that nobody wants. For the people of the world this is one of the most important issues concerning science and values and people and nature.

Human Death Rates and the Rise of Industrial Societies

We return now to further consideration of the first stage in the demographic transition. We can get an idea of the first stage by comparing a modern industrialized country, such as Switzerland, which has a crude death rate of 8 per 1,000, with a developing nation, such as Sierra Leone, which has a crude death rate of 24.[1] Modern medicine has greatly reduced death rates from disease in countries such as Switzerland, particularly with respect to death from acute or epidemic diseases.

An *acute* or *epidemic disease* appears rapidly in the population, affects a comparatively large percentage of it, and then declines or almost disappears for a while, only to reappear later. Epidemic diseases typically are rare but have occasional outbreaks during which a large proportion of the population is infected. Influenza, plague, measles, mumps, and cholera are examples of epidemic diseases. A *chronic disease*, in contrast, is always present in a population, typically occurring in a relatively small but relatively constant percentage of the population. Heart disease, cancer, and stroke are examples.

The great decrease in the percentage of deaths due to acute or epidemic diseases can be seen in a comparison of causes of deaths in Ecuador in 1987 and the United States in 1900, 1987, and 1998 (Figure 4.9).[9] In Ecuador, a developing nation, acute diseases and those listed as "all others" accounted for about 60% of mortality in 1987. In the United States in 1987, these accounted for only 20% of mortality. Chronic diseases account for about 70% of mortality in the modern United States. In contrast, these accounted for

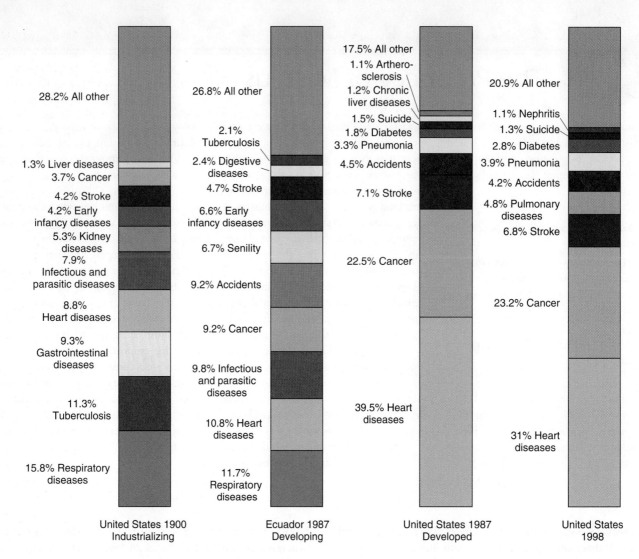

Figure 4.9 ■ Causes of mortality in industrializing, developing, and industrialized nations. (*Sources:* U. S. 1900, Ecuador 1987, and U.S. 1987 data from M. M. Kent and K. A. Crews, *World Population: Fundamentals of Growth* [Washington, D.C.: Population Reference Bureau, 1990]. Copyright 1990 by the Population Reference Bureau, Inc. Reprinted by permission. *National Vital Statistics Report* 48 [11], July 24, 2000.)

less than 20% of the deaths in the United States in 1900 and about 33% in Ecuador in 1987. Ecuador in 1987, then, resembled the United States of 1900 more than it resembled the United States of either 1987 or 1998.

Although outbreaks of the well-known traditional epidemic diseases have declined greatly during the past century in industrialized nations, there is now concern that the incidence of these diseases may increase due to several factors. One is that as the human population grows, people live in new habitats, where previously unknown diseases occur. Another is that strains of disease organisms have developed resistance to antibiotics and other modern methods of control.

A broader view of why diseases are likely to increase comes from an ecological and evolutionary perspective (which will be explained in later chapters). Stated simply, the more than 6.6 billion people on Earth constitute a great resource and opportunity for other species: We offer

a huge and easily accessible host. It is naive to think that other species will not take advantage of the opportunity we present. From this perspective, the future promises more diseases, rather than fewer. This is a new perspective. In the mid–twentieth century, it was easy to believe that modern medicine would eventually cure all diseases and that most people would live the maximum human life span.

The sudden occurrence of a new disease in February 2003, severe acute respiratory syndrome (SARS), demonstrated that modern transportation and the world's huge human population could lead to the rapid spread of epidemic diseases. Jet airliners daily carry vast numbers of people and goods around the world. The disease began in China, perhaps spread from some wild animal to human beings. In part this is because China has become much more open to foreign travelers, with more than 90 million people visiting this country in a recent year.[10] By late spring 2003, SARS had spread to two dozen countries. Quick action led by the World Health Organization (WHO) contained the disease, which as of this writing appears well under control.[11]

West Nile virus is another example of how rapidly and widely diseases spread today. Before 1999, West Nile virus occurred in Africa, West Asia, and the Middle East, but not in the New World.

Related to encephalitis, West Nile virus is spread by mosquitoes, which bite and thus infect birds, which then bite people. It reached the Western Hemisphere through infected birds. It has now been found in more than 25 species of birds native to the United States, including crows, the bald eagle, and the black-capped chickadee—a bird that is a common visitor to bird feeders in the U.S. Northeast. Fortunately, in human beings this disease appears to last only a few days and rarely causes severe symptoms.[12] The 1918 world flu virus is estimated to have killed as many as 50 million people in one year, probably more deaths than any other single epidemic in human history. It spread around the world in the autumn, striking otherwise healthy young adults in particular. Many died within hours! By the spring of 1919, the virus had virtually disappeared.[11] In 2006 Dr. Julie L. Gerberding, director of the U.S. Centers for Disease Control and Prevention, said that bird flu might be on the brink of creating an epidemic as deadly and infectious as the 1918 flu epidemic.[12]

Longevity and Its Effect on Population Growth

The **maximum lifetime** is the genetically determined maximum possible age to which an individual of a species can live. **Life expectancy** is the average number of years an individual can expect to live given the individual's present age. (The term is often applied, without qualification, to mean the life expectancy of a newborn.) Life expectancy is much higher in developed, more prosperous nations. The highest life expectancies today are in Japan (82 years), Iceland (81), Sweden (81), and Australia, Canada, France, Italy, Norway, Spain, and Switzerland (each with 80). The United States, the richest country in the world, is not among the top 10 in life expectancy.[1] The lowest life expectancy is currently 35 years in Botswana, Lesotho, and Swaziland. The 10 national shortest life expectancies are all in Africa. Technically, life expectancy is an age-specific number: Each age class within a population has its own life expectancy. For general comparison, however, the life expectancy at birth is used.

A surprising aspect of the second and third periods in the history of human population is that population growth occurred with little or no change in the maximum lifetime. What changed were birth rates, death rates, population growth rates, age structure, and average life expectancy. In fact, studies of the age at death carved on tombstones tell us that the chances of a person 75 years old living to be age 90 were greater in ancient Rome than they are today in England (Figure 4.10). These studies also suggest that death rates were much higher among young people in Rome than in twentieth-century England. In ancient Rome, the life expectancy of a one-year-old was about 22 years, whereas in twentieth-century England it was about 50 years. Life expectancy in twentieth-century England was greater than in ancient Rome for all ages until about age 55, after which the life expectancy appears to have been higher for ancient Romans than for a twentieth-century Briton. This would suggest that many hazards of modern life may be concentrated more on the aged. Pollution-induced diseases are one factor in this change.

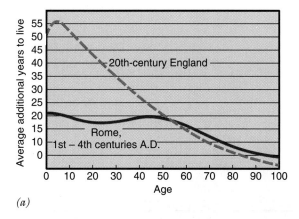

(a)

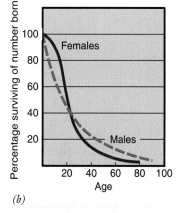

(b)

Figure 4.10 ■ (a) Life expectancy in ancient Rome and 20th-century England. This graph shows the average additional number of years one could expect to live having reached a given age. For example, a 10-year-old in England could expect to live about 55 more years; a 10-year-old in Rome could expect to live about 20 more years. Among the young, life expectancy was greater in 20th-century England than in ancient Rome. However, the graphs cross at about age 60. An 80-year-old Roman could expect to live longer than an 80-year-old Briton. The graph for Romans is recon- structed from ages given on tombstones. (b) Approximate survivorship curve for Rome for the first four centuries A.D. The percent surviving decreases rapidly in the early years, reflecting the high mortality rates for children in ancient Rome. Females had a slightly higher survivorship rate until age 20, after which males had a slightly higher rate. (*Source:* Modified from G. E. Hutchinson, *An Introduction to Population Ecology* [New Haven, Conn.: Yale University Press, 1978]. Copyright 1978 by Yale University Press. Used by permission.)

4.8 Limiting Factors

Basic Concepts

On our finite planet, human populations will eventually be limited by some factor or combination of factors. We can group limiting factors into those that affect a population during the year in which they become limiting (short-term factors), those whose effects are apparent after 1 year but before 10 years (intermediate-term factors), and those whose effects are not apparent for 10 years (long-term factors). Some factors fit into more than one category, having, say, both short-term and intermediate-term effects.

An important *short-term* factor is the disruption of food distribution in a country, commonly caused by drought, or a shortage of energy for transporting food (see Chapters 11 and 12).

Intermediate-term factors include desertification (see Chapter 11); dispersal of certain pollutants, such as toxic metals, into waters and fisheries; disruption in the supply of non-renewable resources, such as rare metals used in making steel alloys for transportation machinery; and a decrease in the supply of firewood or other fuels for heating and cooking.

Long-term factors include soil erosion, a decline in groundwater supplies, and climate change. Changes in resources available per person suggest that we may have already exceeded the long-term human carrying capacity of Earth. For example, wood production peaked at 0.67 m^3/person (0.88 yd^3/person) in 1967, fish production at 5.5 kg/person (12.1 lb/person) in 1970, beef at 11.81 kg/person (26.0 lb/person) in 1977, mutton at 1.92 kg/person (4.21 lb/person) in 1972, wool at 0.86 kg/person (1.9 lb/person) in 1960, and cereal crops at 342 kg/person (754.1 lb/person) in 1977.[15] Before these peaks were reached, per-capita production of each resource had grown rapidly.

4.9 How Can We Achieve Zero Population Growth?

We have surveyed several aspects of population dynamics. We now return to our earlier question: How can we achieve zero population growth (ZPG)? This would be a condition where the human population, on average, neither increases nor decreases. Much of environmental concern has focused on how to lower the human birth rate and decrease our population growth.

Age of First Childbearing

The simplest and one of the most effective means of slowing population growth is to delay the age of first childbearing by women.[18] As more women enter the workforce and as education levels and standards of living increase, this delay occurs naturally. Social pressures that lead to deferred marriage and childbearing can also be effective (Figure 4.11).

Countries with high growth rates have early marriages. In South Asia and in sub-Saharan Africa, about 50% of women marry between the ages of 15 and 19. In Bangladesh, women marry at age 16 on average, whereas in Sri Lanka the average age for marriage is 25. The World Bank estimates that if Bangladesh adopted Sri Lanka's marriage pattern, families could average 2.2 fewer children.[17] Increases in the marriage age could account for 40 to 50% of the drop in fertility required to achieve zero population growth for many countries.

Figure 4.11 ■ As more and more women enter the workforce and establish professional carrers, the average age of first child-bearing tends to increase. The combination of an active life-style that includes children is illustrated here by the young mother jogging with her child in Perth, Australia.

How Many People Can Earth Support?

The human population is considerably more than 6.6 billion. Estimates of how many people the planet can support range from 2.5 billion to 40 billion. Why do the estimates vary so widely?

An estimate of 2.5 billion assumes that we maintain current levels of food production and that everyone eats as well as Americans do now—that is, 30 to 40% more calories than needed. The estimate of 40 billion assumes that all the remaining flat land of the world can be used to produce food, although in fact most of it is too cold or too dry to farm. What is a realistic carrying capacity? What factors need to be considered to answer this question?

Food Supply

World grain production has apparently leveled off since reaching its highest levels in the mid-1980s. The production of grain from 1984 to 1994 remained at approximately 1.7 billion tons, up from 631 million tons in 1950. The remarkable increase in productivity after 1950 resulted from the development of high-yielding varieties, use of chemical fertilizers, application of pesticides, and doubling of cropland acreage. If the present harvest were distributed evenly and everyone ate a vegetarian diet, it could support 6 billion people. As the world population has continued to grow, the per-capita allotment of grain has been falling since 1984, when it stood at 346 kg per capita. By 1994 it had fallen to 311 kg per capita.

Land and Soil Resources

Almost all the usable agricultural land, approximately 1.5 billion hectares (3.7 billion acres), is already being cultivated. An increase of 13% in agricultural lands is possible but would be costly. Land area devoted to raising crops has dropped since 1950 to 1.7 hectares (4.2 acres) per capita and will likely continue to drop, to approximately 1 hectare (2.5 acres) per capita by 2025 if present population projections hold. More soil is lost each year to erosion (about 26 billion metric tons) than is formed.

Water Resources

Water suitable for drinking and irrigation amounts to only a small proportion (less than 3%) of the water on Earth. Underground reservoirs are being depleted on the order of feet per year but are being replaced in inches or even fractions of inches per year. Per-capita water consumption varies from 350 – 1,000 L (371 – 1,060 qt) a day in the developed countries to 2 – 5 L (2.1 – 5.3 qt) a day in rural areas, where people may obtain water directly from streams or primitive wells.

Net Primary Production

Human and domestic animals use about 4% of the net primary production of the world's land area and 2% of that of the oceans. (Primary production is discussed in Chapter 8.)

Population Density

Population density varies greatly, from 3,076 people/km² on the tiny island of Malta to 66 people/km² in Africa as a whole. Bangladesh has 2,261 people/km²; the Netherlands, 1,002/km²; and Japan, 869/km².

Technology

Carrying capacity is not merely a matter of numbers of people. It also involves the impact they have on the world's resources—most critically on energy resources. Multiplying population by per-capita energy consumption by the population gives a relative measure of the impact people have on the environment. By that measure, each American has the impact of 35 people in India or 140 people in Bangladesh.

Critical Thinking Questions

1. Some people claim that agricultural technology can improve food production so that it continues to outpace, or at least keep pace with, population growth, thereby increasing Earth's carrying capacity above 6 billion. What evidence supports this point of view? What evidence refutes it?

2. The land area of the world is approximately 150 million km² (57.9 million mi²). What would the world population be if the average density were 400 people/km²? Is this a sound basis for determining carrying capacity? Explain.

3. In recent decades, some scientists have debated whether the basic limitation on carrying capacity is determined by population or by technology. What is your position, and how do you support it?

4. What factors, in addition to the six listed in this box, do you think should be considered in determining the planet's carrying capacity?

Birth Control: Biological and Societal

Another simple means of decreasing birth rates is breast-feeding, which can delay resumption of ovulation.[20] This is used consciously as a birth-control method by women in a number of countries; in the mid-1970s the practice of breastfeeding provided more protection against conception in developing countries than did family-planning programs, according to the World Bank.[6]

Nevertheless, much emphasis is placed on the need for family planning.[19] Traditional methods range from abstinence to induction of sterility with natural agents. Modern methods include the birth-control pill, which prevents ovulation through control of hormone levels; surgical techniques for permanent sterility; and mechanical devices. Contraceptive devices are used widely in many parts of the world, especially in East Asia, where data show that 78% of women use them. In Africa, only 18% of women use them; in Central and South America, the numbers are 53% and 62%, respectively.[1]

Abortion is also widespread. Although now medically safe in most cases, abortion is one of the most controversial methods from a moral perspective. Ironically, it is one of the most important birth-control methods in terms of its effects on birth rates—approximately 46 million abortions are performed each year.[20]

National Programs to Reduce Birth Rates

Reducing birth rates requires a change in attitude, knowledge of the means of birth control, and the ability to afford these means. As we have seen, a change in attitude can occur simply with an increase in the standard of living. In many countries, however, it has been necessary to provide formal family-planning programs to explain the problems arising from rapid population growth and describe the benefits to individuals of reduced population growth. These programs also provide information about birth-control methods and provide access to these methods.[23] The choice of population-control methods is an issue that involves social, moral, and religious beliefs, which vary from country to country.

The first country to adopt an official population policy was India in 1952. Few developing countries had official family-planning programs before 1965. Since 1965, many such programs have been introduced, and the World Bank has lent $4.2 billion to more than 80 countries to support "reproductive" health projects.[20, 24] Although most countries now have some kind of family-planning program, effectiveness varies greatly. A wide range of approaches have been used, from simply providing more information to promoting and providing means for birth control, to offering rewards and imposing penalties. Penalties usually take the form of taxes. Ghana, Malaysia, Pakistan, Singapore, and the Philippines have used a combination of methods, including limits on tax allowances for children and maternity benefits. Tanzania has restricted paid maternity leave for women to a frequency of once in three years. Singapore does not take family size into account in allocating government-built housing, so larger families are more crowded. Singapore also gives higher priority in school admission to children from smaller families.[6] Some countries, including Bangladesh, India, and Sri Lanka, have paid people to be voluntarily sterilized. In Sri Lanka, this practice has applied only to families with two children, and a voluntary statement of consent is signed.

China has one of the oldest and most effective family-planning programs. In 1978, China adopted an official policy to reduce its human population growth from 1.2% in that year to zero by the year 2000. Although in 2000 China's growth rate had slowed to 1.0% a year rather than to zero, the program has done much to curb the country's rapid population growth. The Chinese program placed emphasis on single-child families. The government used education, a network of family planning that provides information and means for birth control, and a system of rewards and penalties.

Summary

▧ The human population is the underlying environmental issue, because most current environmental damage results from the very high number of people on Earth and their great power to change the environment.

▧ Throughout most of our history, the human population and its average growth rate were small. The growth of the human population can be divided into four major phases. Although the population has increased in each phase, the current situation is unprecedented.

▧ Countries whose birth rates have declined have experienced a demographic transition marked by a decrease in death rates followed by a decrease in birth rates. Many developing nations have experienced a great decrease in their death rates but still have a very high birth rate. It remains an open question whether some of these nations will be able to achieve a lower birth rate before reaching disastrously high population levels.

▧ The maximum population Earth can sustain and how large a population will ultimately be attained by human beings are controversial questions. Standard estimates suggest that the human population will reach 10 to 16 billion before stabilizing.

▧ How the human population might stabilize or be stabilized raises questions concerning science and values, and people and nature.

▧ Considerable lags in the responses of human populations to changes in birth rates and death rates occur of the age structure of the population. A population that achieves replacement-level fertility will continue to grow for several generations.

▧ One of the most effective ways to lower a population's growth rate is to decrease the age of first childbearing. This also involves relatively few societal and value issues.

REEXAMINING THEMES AND ISSUES

Human Population

Our discussion in this chapter reemphasizes the point that there can be no long-term solution to our environmental problems unless the human population stops growing at its present rate. This makes the problem of human population a top priority.

Sustainability

As long as the human population continues to grow, it is doubtful that our other environmental resources can be made sustainable.

Global Perspective

Although the growth rate of the human population varies from nation to nation, the overall environmental effects of the rapidly growing human population are global. For example, the increase in use of fossil fuels in Western nations since the beginning of the Industrial Revolution has affected the entire world. The growing demand for fossil fuels and their increasing use in developing nations are also having a global effect.

Urban World

One of the major patterns in the growth of the human population is the increasing urbanization of the world. Cities are not self-contained but are linked to the surrounding environment, depending on it for resources and affecting environments elsewhere. Urban development often leads to encroachment on highly valued natural areas, especially because cities are typically located where water, transportation, and material resources are readily available.

People and Nature

As with any species, the growth rate of the human population is governed by fundamental laws of population dynamics. We cannot escape these basic rules of nature. People greatly affect the environment, and the idea that human population growth is *the* underlying environmental issue illustrates the deep connection between people and nature.

Science and Values

The problem of human population exemplifies the connection between values and knowledge. Scientific and technological knowledge has helped us cure diseases, decrease death rates, and thereby increase the growth of the human population. Our ability today to forecast human population growth provides much useful knowledge, but what we do with this knowledge is subject to intense debate around the world, because values are so important in relation to birth control and family size. We can solve the problem of human population only by confronting the connection between the way we value human life and our scientific knowledge about human population and its environmental effects.

Key Terms

Study Questions

1. What are the principal reasons that the human population has grown so rapidly in the twentieth century?

2. Why is it important to consider the age structure of a human population?

3. Three characteristics of a population are birth rate, growth rate, and death rate. How has each been affected by (a) modern medicine, (b) modern agriculture, and (c) modern industry?

4. What is meant by the statement "What is good for an individual is not always good for a population"?

5. Strictly from a biological point of view, why is it difficult for a human population to achieve a constant size?

6. What environmental factors are likely to increase the chances of an outbreak of an epidemic disease?

7. Why is it so difficult to predict the growth of Earth's human population?

8. Before the beginning of the Industrial Revolution and major scientific advancements, what factors tended to decrease the size of the human population?

9. To which of the following can we attribute the great increase in human population since the beginning of the Industrial Revolution: changes in human (a) birth rates, (b) death rates, (c) longevity, or (d) death rates among the very old? Explain.

10. What is the demographic transition? When would one expect replacement-level fertility to be achieved—before, during, or after the demographic transition?

11. Based on the history of human populations in various countries, how would you expect the following to change as per-capita income increased: (a) birth rates, (b) death rates, (c) average family size, and (d) age structure of the population? Explain.

Further Reading

Brown, L. R., G. Gardner, and B. Halueil. 1999. *Beyond Malthus: Nineteen Dimensions of the Population Challenge*, New York: W. W. Norton. A discussion of recent changes in human population trends and their implications.

Cohen, J. E. 1995. *How Many People Can the Earth Support?* New York: Norton. A detailed discussion of world population growth, Earth's human carrying capacity, and factors affecting both.

Ehrlich, P. R., and A. H. Ehrlich. 2004. *One with Nineveh: Politics, Consumption, and the Human Future*. Washington, D.C.: Island Press. An extended discussion of the effects of the human population on the world's resources, and on the question of the carrying capacity of Earth for our species.

Kent, M. M., and K. A. Crews. 1990. *World Population: Fundamentals of Growth*. Washington, D.C.: Population Reference Bureau. A clear and concise introduction to population concepts and historical and present trends in the human population.

Kessler, E., ed. 1992. "Population, Natural Resources and Development," *AMBIO* 21(1). A special issue of the journal *AMBIO* addressing many problems concerning human population growth and its economic and environmental implications. It includes a description of several case studies.

McKee, J. K. 2003. *Sparing Nature: The Conflict Between Human Population Growth and Earth's Biodiversity*, New Brunswick, N.J.: Rutgers University Press. One of the few recent books about human populations.

CHAPTER 5

The Biogeochemical Cycles

The city of Seattle, Washington, is located between Puget Sound (right) and Lake Washington (left). View is to the south.

Learning Objectives

Life is composed of many chemical elements, which exist in the right amounts, the right concentrations, and the right ratios to one another. If these conditions are not met, then life is limited. The study of chemical availability and biogeochemical cycles is important to the solution of many environmental problems. After reading this chapter, you should understand:

■ What the major biogeochemical cycles are.

■ What the major factors and processes that control biogeochemical cycles are.

■ Why some chemical elements cycle quickly and some slowly.

■ How each major component of Earth's global system (the atmosphere, waters, solid surfaces, and life) are involved and linked with biogeochemical cycles.

■ How the biogeochemical cycles most important to life, especially the carbon cycle, generally operate.

■ How humans affect biogeochemical cycles.

Lake Washington

Humans may unknowingly alter the natural cycling of chemicals to the detriment of the environment. Science can provide potential solutions, but our value systems dictate whether we are willing to pursue these solutions. The following story concerning Lake Washington shows that adverse effects of chemical cycling can be addressed and reduced.

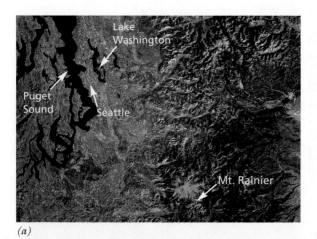

(a)

(b)

Figure 5.1 ■ (*a*) Thematic image of the Seattle, Washington, region (about 24,000 km²). Bodies of water (e.g., Puget Sound and Lake Washington) are dark blue/black; urban areas (e.g., Seattle) are pink/light-blue/gray; forested lands are dark green; clear-cut (logged) lands are pink/yellow; and snow and ice on Mount Rainier and other areas are baby blue. (*b*) Lake Washington in the summer of 1998 at a popular recreational beach on the eastern shoreline.

The city of Seattle, Washington, lies between two major bodies of water—saltwater Puget Sound to the west and freshwater Lake Washington to the east (Figure 5.1*a*). Beginning in the 1930s, the freshwater lake was used for disposal of sewage. By 1959, 11 sewage treatment plants had been constructed along the lake. The plants removed disease-causing organisms and much of the organic matter that had formerly entered the lake. Each day, 76,000 m³ (about 20 million gallons) of treated wastewater flowed from the plants into the lake. Smaller streams that fed the lake brought in small amounts of untreated sewage. The lake's response to the introduction of treated wastewater was a major bloom of undesirable algae and photosynthetic bacteria, which reduced water clarity and the general aesthetics of the lake.[1]

Seattle's mayor appointed an advisory committee to determine what might be done. Scientific research showed that phosphorus in the treated wastewater had stimulated the growth of algae and bacteria. Although the treatment process had eliminated disease-causing organisms, it had not removed phosphorus, which came from common laundry detergent and other urban sources. The phosphorus, when it entered the lake, acted as a potent fertilizer, posing a chemical problem for the lake's ecosystem. Puget Sound is a large body of ocean water with a rapid rate of exchange with the Pacific Ocean. Because of this rapid exchange, the phosphorus would be quickly diluted to a very low concentration.

To solve the problem, the committee advised the city to discharge wastewater from the sewage treatment plants into Puget Sound rather than into the lake.

The change was made by 1968, and the lake improved rapidly. The diversion of the phosphorus-rich wastewater greatly reduced the amount of phosphorus cycling through the lake's ecosystem. In terms of input–output analysis, introduced in Chapter 3, the input of phosphorus was greatly reduced, and so the concentration of the nutrient in the lake water was also reduced. By the time a year had passed, the unpleasant algae had decreased, and surface waters had become much clearer than they had been five years before. Oxygen concentrations in deep water increased immediately to levels greater than those observed in the 1930s, favoring an increase in fish.[1]

Urbanization around Lake Washington (Figure 5.1*b*) continues to be a concern. Urban runoff is a potential source of water pollution that may degrade the quality of streams or lakes it enters.

Lake Washington is a large urban lake in a highly urbanized area. The lake has excellent water quality for an urban lake. However, good water quality is not a given considering lake history and potential for urban growth. Spring algae blooms still occur from human-introduced nutrients in the lake. Herbicides enter the lake from surface runoff, as do other pollutants.

Two main rivers flow into Lake Washington. In the past, those rivers supported large runs of king salmon, but due to loss of habitat and fishing pressure, the fish have been listed as endangered. Restoring the king salmon will involve the rivers where the salmon spawn and are reared as well as the lake they move through to and from the ocean. The lake ecosystem also includes sockeye salmon that spawn in the rivers but are reared in the lake. Survival of young sockeye in the lake has declined in recent years, and studies are ongoing to understand why.

Lake Washington for all its positive improvement remains vulnerable to further change. The future of the water resources and salmon is in the hands of the people of Seattle. They are ultimately responsible to protect and enhance lake ecosystems for future generations. Given past history and current activity and interest, they seem up to the task.

The story of Lake Washington illustrates the importance of understanding how chemicals cycle in ecosystems. The story confirms several key themes from Chapter 1. As the population in Seattle increased, so did its sewage. The scientific study of the lake and the discovery of the role of phosphorus reflected the value the people of Seattle placed on Lake Washington. Recognizing the need for a high-quality environment in an urban area, they developed a plan to sustain the water quality of their urban lake. They "acted locally" to solve a water pollution problem and provided a positive example of pollution abatement.

5.1 How Chemicals Cycle

Earth is a particularly good planet for life from a chemical point of view. The atmosphere contains plenty of water and oxygen, which we and other animals need to breathe. In many places, soils are fertile, containing the chemical elements necessary for plants to grow; and Earth's bedrock contains valuable metals and fuels. Of course, some parts of the Earth's surface are not perfect for life—deserts with little water, chemical deserts (such as the middle regions of the oceans) where nutrients necessary for life are not abundant, and certain soils in which some of the chemical elements required for life are lacking or others toxic to life are present.[2] Scientific questions reflecting our values for a quality environment are: What kinds of chemical processes benefit or harm the environment, ourselves, and other life-forms? How can we manage chemicals in the environment to improve and sustain ecosystems from the local to the global level? To answer these questions, we need to know how chemical elements cycle, and this is our starting point.

Biogeochemical Cycles

The term *chemical* refers here to an element such as carbon (C) or phosphorus (P), or to a compound, such as water (H_2O). (See A Closer Look 5.1 for a discussion of matter and energy.) A **biogeochemical cycle** is the complete path a chemical takes through the four major components, or reservoirs, of Earth's system: atmosphere, hydrosphere (oceans, rivers, lakes, groundwaters, and glaciers), lithosphere (rocks and soils), and biosphere (plants and animals). A biogeochemical cycle is *chemical* because it is chemicals that are cycled, *bio-* because the cycle involves life, and *geo-* because a cycle may include atmosphere, water, rocks, and soils.

Consider as an example an atom of carbon (C) in carbon dioxide (CO_2) emitted from burning coal (which is

Matter and Energy

The universe as we know it consists of two entities: matter and energy. Matter is the material that makes up our physical and biological environments (you are composed of matter). Energy is the ability to do work. The first law of thermodynamics—also known as the law of conservation of energy or the first energy law—states that energy cannot be created or destroyed but can change from one form to another. This law stipulates that the total amount of energy in the universe does not change.

Our sun produces energy through nuclear reactions at high temperatures and pressures that change mass (a measure of the amount of matter) into energy. At first glance, this may seem to violate the law of conservation of energy. However, this is not the case, because energy and matter are interchangeable. Albert Einstein first described the equivalence of energy and mass in his famous equation $E = mc^2$, where E is energy, m is mass, and c is the velocity of light in a vacuum, such as outer space (approximately 300,000 km/s or 186,000 mi/s). Because the velocity of light squared is a very large number, even a small amount of change in mass produces very large amounts of energy.[3]

Energy, then, may be thought of as an abstract mathematical quantity that is always conserved. This means that it is impossible to get something for nothing when dealing with energy; it is impossible to extract more energy from any system than the amount of energy that originally entered the system. In fact, the second law of thermodynamics states that you cannot break even. When energy is changed from one form to another, it always moves from a more useful form to a less useful one. Thus, as energy moves through a real system and is changed from one form to another, energy is conserved, but it becomes less useful. We return to the two energy laws when we discuss energy in ecosystems (Chapter 9) and in modern society (Chapter 17).

We turn next to a brief introduction to the basic chemistry of matter, which will help you in understanding biogeochemical cycles. An atom is the smallest part of a chemical element that can take part in a chemical reaction with another atom. An element is a chemical substance composed of identical atoms that cannot be separated by ordinary chemical processes into different substances. Each element is

given a symbol. For example, the symbol for the element carbon is C, and that for phosphorus is P.

A model of an atom (Figure 5.2) shows three subatomic particles: neutrons, protons, and electrons. The atom is visualized as a central nucleus composed of neutrons with no electrical charge and protons with positive charge. A cloud of electrons, each with a negative charge, revolves about the nucleus. The number of protons in the nucleus is unique for each element and is the atomic number for that element. For example, hydrogen, H, has one proton in its nucleus, and its atomic number is 1. Uranium has 92 protons in its nucleus, and its atomic number is 92. A list of known elements with their atomic numbers, called the Periodic Table, is shown in Figure 5.5.

Electrons in our model of the atom are arranged in shells (representing energy levels), and the electrons closest to the nucleus are bound tighter to the atom than those in the outer shells. Electrons have negligible mass compared with neutrons or protons; therefore, nearly the entire mass of an atom is in the nucleus.

The sum of the number of neutrons and protons in the nucleus of an atom is

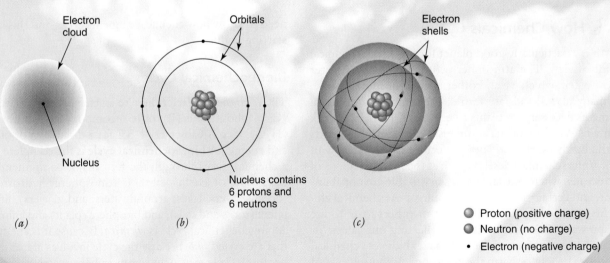

Figure 5.2 ■ (*a*) Idealized diagram showing the basic structure of an atom as a nucleus surrounded by a cloud of electrons. (*b*) Conceptualized model of an atom of carbon with six protons and six neutrons in the nucleus and six orbiting electrons in two

energy shells. (*c*) Three-dimensional view of (*b*). Size of nucleus relative to size of electron shells is greatly exaggerated. [*Source:* After F. Press, and R. Siever *Understanding Earth* (New York: Freeman, 1994).]

known as the atomic weight. Atoms of the same element always have the same atomic number (the same number of protons in the nucleus), but they can have different numbers of neutrons and therefore different atomic weights. Two atoms of the same element with different numbers of neutrons in their nuclei and different atomic weights are known as isotopes of that element. For example, two isotopes of oxygen are ^{16}O and ^{18}O, where 16 and 18 are the atomic weights. Both isotopes have an atomic number of 8, but ^{18}O has two more neutrons than ^{16}O. Such study is proving very useful in learning how Earth works. For example, the study of oxygen isotopes has resulted in a better understanding of how the global climate has changed. This topic is beyond the scope of our present discussion but can be found in many basic textbooks on oceanography.

An atom is chemically balanced in terms of electric charge when the number of protons in the nucleus is equal to the number of electrons. However, an atom may lose or gain electrons, changing the balance in the electrical charge. An atom that has lost or gained electrons is called an ion. An atom that has lost one or more electrons has a net positive charge and is called a cation. For example, the potassium ion K^+ has lost one electron, and the calcium ion Ca^{2+} has lost two electrons. An atom that has gained electrons has a net negative charge and is called an anion. For example, O^{2-} is an anion of oxygen that has gained two electrons.

A compound is a chemical substance composed of two or more atoms of the same or different elements. The smallest unit of a compound is a molecule. For example, each molecule of water, H_2O, contains two atoms of hydrogen and one atom of oxygen held together by chemical bonds. Minerals that form rocks are compounds, as are most chemical substances found in a solid, liquid, or gaseous state in the environment.

The atoms that constitute a compound are held together by *chemical bonding*. The four main types of chemical bonds are covalent, ionic, Van der Waals, and metallic. It is important to recognize that when talking about chemical bonding of compounds, we are dealing with a complex subject. Although some compounds have a particular type of bond, many other compounds have more than one type of bond. With this caveat in mind, let's define each type of bond. *Covalent bonds* result when atoms share electrons. This sharing takes place in the region between the atoms, and the strength of the bond is related to the number of pairs of electrons that are shared. Some important environmental compounds are held together solely by covalent bonds. These include carbon dioxide (CO_2) and water (H_2O). Covalent bonds are stronger than *ionic bonds*, which form as a result of attraction between cations and anions. An example of an environmentally important compound with ionic bonds is table salt (mineral halite), or sodium chloride ($NaCl$). Compounds with

ionic bonds such as sodium chloride tend to be soluble in water and thus dissolve easily, as salt does. Van der Waals bonds are weak forces of attraction between molecules that are not bound to each other. Such bonding is much weaker than either covalent or ionic bonding. For example, the mineral graphite (which is the "lead" in pencils) is black and consists of sheets of carbon atoms that easily part or break from one another because the bonds are of the weak Van der Waals type. Finally, metallic bonds are those in which electrons are shared, as with covalent bonds. However, they differ because in metallic bonding the electrons are shared by all atoms of the solid rather than by specific atoms. As a result, the electrons can flow. For example, the mineral and element gold is an excellent conductor of electricity and pounds easily into thin sheets because the electrons have the freedom of movement that is characteristic of metallic bonding.

In summary, in the study of the environment, we are concerned with matter (chemicals) and energy that moves in and between the major components of the Earth system. An example is the element carbon, which moves through the atmosphere, hydrosphere, lithosphere, and biosphere in a large variety of compounds. These include carbon dioxide (CO_2) and methane (CH_4), which are gases in the atmosphere; sugar ($C_6H_{12}O_6$) in plants and animals; and complex hydrocarbons (compounds of hydrogen and carbon) in coal and oil deposits.

made up of fossilized plants hundreds of millions of years old). The carbon atom is released into the atmosphere and then taken up by a plant and incorporated into a seed. The seed is eaten by a mouse. The mouse is eaten by a coyote, and the carbon atom is expelled following digestion as scat onto the soil. Decomposition of the scat allows our carbon atom to enter the atmosphere again. It may also enter another organism, such as an insect, which utilizes the scat as a resource.

Chemical Reactions

It is important to acknowledge in our discussion of how chemical cycles work that the emphasis is on chemistry. Many chemical reactions occur within and between the living and nonliving portions of ecosystems. A **chemical reaction** is a process in which new chemicals are formed

from elements and compounds that undergo a chemical change. For example, a simple reaction between rainwater (H_2O) and carbon dioxide (CO_2) in the atmosphere produces weak carbonic acid (H_2CO_3):

$$H_2O + CO_2 \rightarrow H_2CO_3$$

This weak acid reacts with earth materials, such as rock and soil, to release chemicals into the environment. The released chemicals include calcium, sodium, magnesium, and sulfur, with smaller amounts of heavy metals such as lead, mercury, and arsenic. The chemicals appear in various forms, such as compounds and ions in solution.

Many other chemical reactions determine whether chemicals are available to life. For example, photosynthesis is a series of chemical reactions by which living green plants, with sunlight as an energy source, convert carbon dioxide (CO_2)

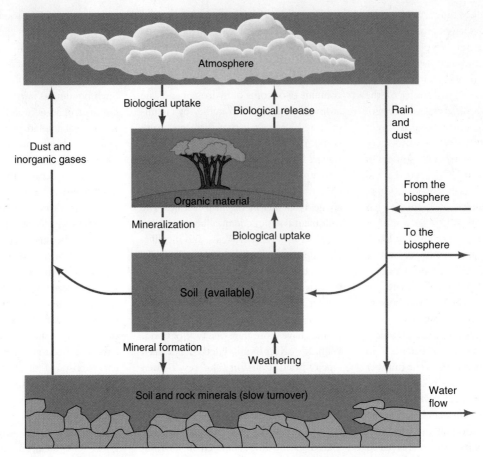

Figure 5.3 ■ A generalized cycling of a chemical in an ecosystem. Chemical elements cycle within an ecosystem or exchange between an ecosystem and the biosphere. Organisms exchange elements with the nonliving environment; some elements are taken up from and released to the atmosphere, and others are exchanged with water and soil or sediments. The parts of an ecosystem can be thought of as storage compartments for chemicals. The chemicals move among these compartments at different rates and remain within them for different average lengths of time. For example, the soil in a forest has an active part, which rapidly exchanges elements with living organisms, and an inactive part, which exchanges elements slowly (as shown in the lower part of the diagram). Generally, life benefits if needed chemicals are kept within the ecosystem and are not lost through geologic processes, such as erosion, that remove them from the ecosystem.

and water (H_2O) to sugar ($C_6H_{12}O_6$) and oxygen (O_2). The general chemical reaction for photosynthesis is

$$6CO_2 + 6H_2O \xrightarrow{\text{sunlight}} C_6H_{12}O_6 + 6O_2$$

Photosynthesis produces oxygen as a by-product, and that is why we have free oxygen in our atmosphere. We return to the topic of photosynthesis later in this chapter (A Closer Look 5.3).

After considering the two chemical reactions and applying critical thinking, you may recognize that both reactions combine water and carbon dioxide, but the products are very different: carbonic acid in one combination and a sugar in the other. How can this be so? The answer lies in an important difference between the simple reaction in the atmosphere to produce carbonic acid and the production of sugar and oxygen in the series of reactions of photosynthesis. Green plants use the energy from the sun, which they absorbed through the chemical chlorophyll. Thus, active solar energy is converted to a stored chemical energy in sugar.

Perhaps the simplest way to think of a biogeochemical cycle is as a "box-and-arrow" diagram, which shows where a chemical is stored and the pathways along which it is transferred from one storage place to another. (See Figure 5.3 and A Closer Look 5.2.) We can consider biogeochemical cycles on any spatial scale of interest to us, from a single ecosystem to the whole Earth. It is often useful to consider such a cycle from a global perspective. The problem of potential global warming, for example, calls for an understanding of the cycling of carbon transferred into and out of Earth's atmosphere. Sometimes it is useful to consider a cycle at a more local level, as in the case study of Lake Washington. The key that unifies all these cycles is the involvement of the four principal components of the Earth system: lithosphere, atmosphere, hydrosphere, and biosphere. By their nature, chemicals in these four major components have different average times of storage. In general, the average residence time of chemicals is long in rocks, short in the atmosphere, and intermediate in the hydrosphere and biosphere.

A Biogeochemical Cycle

The simplest way to visualize a biogeochemical cycle is as a box-and-arrow diagram, where the boxes represent places where a chemical is stored (called storage compartments) and the arrows represent pathways of transfer (Figure 5.4*a*). The rate of transfer or flux is the amount per unit time of a chemical that enters or leaves a storage compartment. When a chemical enters a storage compartment from another compartment,

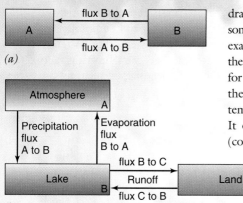

(a)

(b)

Figure 5.4 ■ Basic parts of a biochemical cycle. (*a*) A and B are storage compartments. Chemicals flow from one compartment to another. (*b*) Some components of the hydrologic cycle.

we say that the receiving compartment is a sink. For example, the forests of the world (which are a storage compartment for carbon) may function as a sink for carbon from the atmosphere, sequestering it in wood, leaves, and roots. The amount of carbon transferred from the atmosphere to the forest on a global basis is the flux, which can be measured in units, such as billions of metric tons of carbon per year.

A biogeochemical cycle is generally drawn for a single chemical element, but sometimes it is drawn for a compound—for example, water (H_2O). Figure 5.4*b* shows the basic elements of a biogeochemical cycle for water, which represents three parts of the hydrological cycle. Water is stored temporarily in a lake (compartment B). It enters the lake from the atmosphere (compartment A) as precipitation and from the land around the lake as runoff (compartment C). It leaves the lake through evaporation to the atmosphere or runoff to a surface stream or subsurface flows.

In each compartment, we can identify an average length of time that an atom is stored before it is transferred. This is called the residence time.

As an example, consider a salt lake with no transfer out except by evaporation. Assume that the lake contains 3,000,000 m^3 (106 million ft^3) of water and the evaporation is 3,000 m^3/day (106,000 ft^3/day). Surface runoff into the lake is also 3,000 m^3/day, so the volume of water in the lake remains constant. We can calculate the average residence time of the water in the lake as the volume of the lake divided by the evaporation rate, or 3,000,000 m^3 divided by 3,000 m^3/day, which is 1,000 days (or 2.7 years).

Another crucial aspect of a biogeochemical cycle is the set of factors or processes that control the flow from one compartment to another. To understand a biogeochemical cycle, these factors and processes should be quantified and understood. For example, understanding how air temperature and wind velocity vary across a lake is important in understanding the rate of evaporation of water from the lake.

5.2 Environmental Questions and Biogeochemical Cycles

With a general idea of how chemicals cycle, we next consider some of the environmental questions that the science of biogeochemical cycles can help answer. These questions include the following:

Biological Questions

■ What factors, including chemicals necessary for life, place limits on the abundance and growth of organisms and their ecosystems?

■ What toxic chemicals might be present that adversely affect the abundance and growth of organisms and their ecosystems?

■ How can people improve the production of a desired biological resource?

■ What are the sources of chemicals required for life, and how might we make these more readily available?

■ What problems occur when a chemical is too abundant, as was the case with phosphorus in Lake Washington?

Geologic Questions

■ What physical and chemical processes control the movement and storage of chemical elements in the environment?

■ How are chemical elements transferred from solid earth to the water, atmosphere, and life-forms?

■ How does the long-term storage of chemical elements (for thousands of years or longer) in rocks and soils affect ecosystems on local to global scales?

Atmospheric Questions

■ What determines the concentrations of elements and compounds in the atmosphere?

■ Where the atmosphere is polluted as the result of human activities, how might we alter a biogeochemical cycle to reduce the pollution?

Hydrologic Questions

■ What determines whether a body of water will be biologically productive?

■ When a body of water becomes polluted, how can we alter biogeochemical cycles to reduce the pollution and its effects?

5.3 Biogeochemical Cycles and Life: Limiting Factors

The first question on the preceding list concerns chemicals necessary for life and limits imposed by these chemicals. We consider this issue next.

All living things are made up of chemical elements, but of the 103 known elements, only 24 are required for life processes (Figure 5.5). These 24 are divided into the **macronutrients**, elements required in large amounts by all life, and **micronutrients**, elements required either in small amounts by all life or in moderate amounts by some forms of life and not at all by others.

The macronutrients, in turn, include the "big six," the elements that form the fundamental building blocks of life. These are carbon, hydrogen, nitrogen, oxygen, phosphorus, and sulfur. Each element plays a special role in organisms. Carbon is the basic building block of organic compounds. Along with oxygen and hydrogen, carbon forms carbohydrates. Nitrogen, along with these other three, makes proteins. Phosphorus is the "energy element"; it occurs in the compounds called ATP and ADP, important in the transfer and use of energy within cells.

In addition to the "big six," other macronutrients also play important roles. Calcium, for example, is the structure element, occurring in bones of vertebrates, shells of shellfish, and wood-forming cell walls of vegetation. Sodium and potassium are important to nerve signal transmission. Many of the metals required by living things are necessary for specific enzymes. (An enzyme is a complex organic compound that acts as a catalyst—it causes or speeds up chemical reactions such as digestion.)

For any form of life to persist, chemical elements must be available at the right times, in the right amounts, and in the right concentrations relative to each other. When this does not happen, a chemical can become a **limiting factor**, preventing the growth of an individual, population, or species,

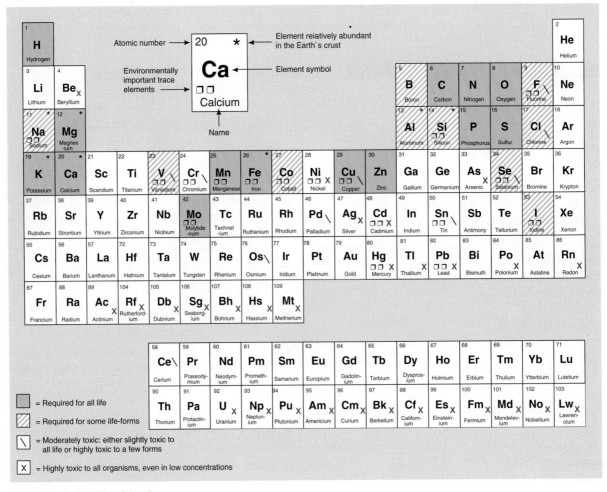

Figure 5.5 ■ Periodic table of the elements.

or even causing its local extinction. Limiting factors were discussed in Chapter 4 and will be discussed further in Chapter 11. Chemical elements may also be toxic to some life-forms and ecosystems. Mercury, for example, is toxic even in low concentrations. Copper and some other elements are required in low concentrations for life processes but are toxic when present in high concentrations.

5.4 General Concepts Central to Biogeochemical Cycles

Although there are as many biogeochemical cycles as there are chemicals, certain general concepts hold true for these cycles.

- Some chemical elements cycle quickly and are readily regenerated for biological activity. Oxygen and nitrogen are among these. Typically, these elements have a gas phase and are present in Earth's atmosphere, and/or they are easily dissolved in water and are carried by the hydrologic cycle.

- Other chemical elements are easily tied up in relatively immobile forms and are returned slowly, by geological processes, to where they can be reused by life. Typically, these elements lack a gas phase and are not found in significant concentrations in the atmosphere. They also are relatively insoluble in water. Phosphorus is an example of this kind of chemical.

- Chemicals whose biogeochemical cycles include a gas phase and that are stored in the atmosphere tend to cycle rapidly. Those without an atmospheric phase are likely to end up as deep-ocean sediment and recycle slowly.

- Since life evolved, it has greatly altered biogeochemical cycles, and this alteration has changed our planet in many ways, such as in the development of the fertile soils on which agriculture depends.

- The continuation of processes that control biogeochemical cycles is essential to the long-term maintenance of life on Earth.

- Through modern technology, we have begun to transfer chemical elements among air, water, and soil at rates comparable to natural processes. These transfers can benefit society, as when they improve crop production, but they can also pose environmental dangers, as illustrated by the opening case study. To live wisely with our environment, we must recognize the positive and negative consequences of altering biogeochemical cycles. Then we must attempt to accentuate the positive and minimize the negative.

Discussion of biogeochemical cycles beyond the general concepts listed above requires an understanding of geologic and hydrologic cycles. Of particular importance are geologic processes linked to cycling of chemicals in the biosphere.

5.5 The Geologic Cycle

Throughout the 4.6 billion years of Earth's history, rocks and soils have been continuously created, maintained, changed, and destroyed by physical, chemical, and biological processes. Collectively, the processes responsible for formation and change of Earth materials are referred to as the **geologic cycle** (Figure 5.6). The geologic cycle is best described as a group of cycles: tectonic, hydrologic, rock, and biogeochemical.

The Tectonic Cycle

The **tectonic cycle** involves creation and destruction of the solid outer layer of Earth, the lithosphere. The lithosphere is about 100 km (60 mi) thick and is broken into several large segments called plates, which are moving relative to one another (Figure 5.7). The slow movement of these large segments of Earth's outermost rock shell is referred to as **plate tectonics**. The plates "float" on denser material and move at rates of 2 to 15 cm/year (0.8 to 5.9 in./year), or about as fast as your fingernails grow. The tectonic cycle is driven by forces originating deep within the Earth. Closer to the surface, rocks are deformed by spreading plates, which produce ocean basins, and collisions of plates, which produce mountain ranges.

Plate tectonics has important environmental effects. Moving plates change the location and size of continents, altering atmospheric and ocean circulation and thereby altering climate. Plate movement has also created ecological islands by breaking up continental areas. When this happens, closely related life-forms are isolated from one another for millions of years, leading to the evolution of new species. Finally, boundaries between plates are geologically active areas, and most volcanic activity and earthquakes occur there. Earthquakes occur when the brittle upper lithosphere fractures along faults. Movement of several meters between plates can occur within a few seconds or minutes, in contrast to the slow, deeper plate movement described above.

Three types of plate boundaries occur: divergent, convergent, and transform faults.

- A *divergent plate* boundary occurs at a spreading ocean ridge, where plates are moving away from one another and new lithosphere is produced. This process, known as seafloor spreading, produces ocean basins.

- A *convergent plate boundary* occurs when plates collide. When a plate composed of relatively heavy ocean basin rocks dives, or subducts, beneath the leading edge of a plate composed of lighter continental rocks, a subduction zone is present. Such a convergence may produce linear coastal mountain ranges, such as the Andes in South America. When two plates, each composed of lighter continental rocks, collide, a continental mountain range such as the Himalayas in Asia may form.[4, 5]

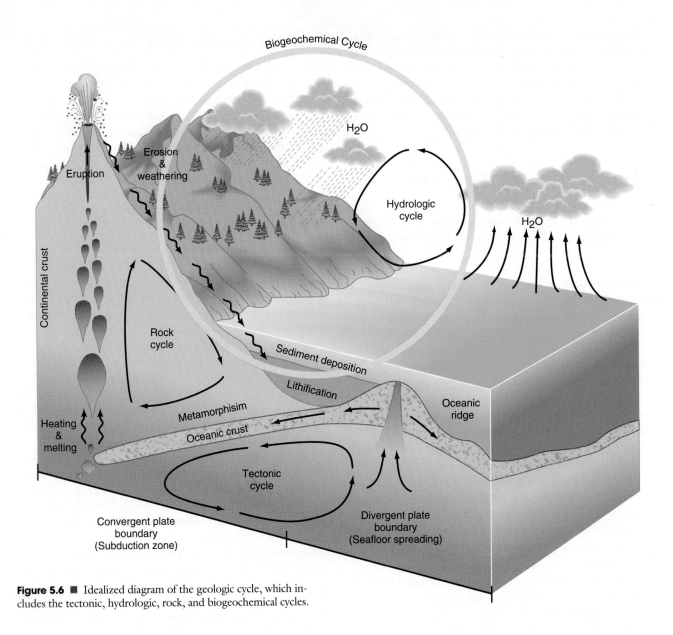

Figure 5.6 ■ Idealized diagram of the geologic cycle, which includes the tectonic, hydrologic, rock, and biogeochemical cycles.

■ *A transform fault boundary* occurs where one plate slides past another. An example is the San Andreas Fault in California, which is the boundary between the North American and Pacific plates. The Pacific plate is moving north relative to the North American plate at about 5 cm/year (2 in./year). As a result, Los Angeles is moving slowly toward San Francisco, about 500 km (300 mi) north. If this motion continues, in about 10 million years, San Francisco will be a suburb of Los Angeles.

The Hydrologic Cycle

The **hydrologic cycle** (Figure 5.8) is the transfer of water from the oceans to the atmosphere to the land and back to the oceans. The processes involved include evaporation of water from the oceans; precipitation on land; evaporation from land; and runoff from streams, rivers, and subsurface groundwater. The hydrologic cycle is driven by solar energy, which evaporates water from oceans, freshwater bodies, soils, and vegetation. Of the total 1.3 billion km^3 of water on Earth, about 97% is in oceans and about 2% is in glaciers and ice caps. The rest is in freshwater on land and in the atmosphere. Although it represents only a small fraction of the water on Earth, the water on land is important in moving chemicals, sculpting landscape, weathering rocks, transporting sediments, and providing our water resources. The water in the atmosphere—only 0.001% of the total on Earth—cycles quickly to produce rain and runoff for our water resources.

Annual rates of transfer from the storage compartments in the hydrologic cycle are shown in Figure 5.8. These rates of transfer define a global water budget. If we sum

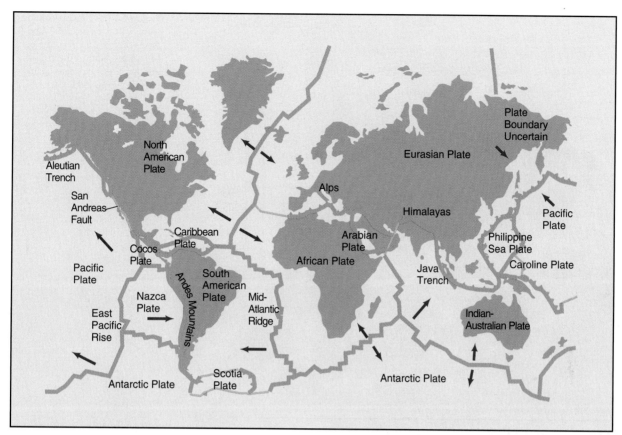

Figure 5.7 ■ Generalized map of Earth's lithospheric plates. Divergent plate boundaries are shown as heavy lines (for example, the Mid-Atlantic ridge). Convergent boundaries are shown as barbed lines (for example, the Aleutian trench). Transform fault boundaries are shown as yellow thinner lines (for example, the San Andreas Fault). Arrows indicate directions of relative plate motions. [*Source:* Modified from B. C. Birchfiel, R. J. Foster, E. A. Keller, W. N. Melhorn, D. G. Brookins, L. W. Mintz, and H. V. Thurman, *Physical Geology: The Structures and Processes of the Earth* (Columbus, Ohio: Merrill, 1982).]

the arrows going up in the figure, and then sum the arrows going down, we find that the two sums are the same, 577,000 km^3/year. Similarly, precipitation on land (119,000 km^3/year) is balanced by evaporation from land plus surface and subsurface runoff.

Especially important from an environmental perspective is that rates of transfer on land are small relative to what's happening in the ocean. For example, most of the water that evaporates from the ocean falls again as precipitation into the ocean. On land, most of the water that falls as precipitation comes from evaporation of water from land. This means that regional land-use changes, such as the building of large dams and reservoirs, can change the amount of water evaporated into the atmosphere and change the location and amount of precipitation on land—water we depend on to raise our crops and supply water for our urban environments. Furthermore, as we pave over large areas of land in cities, storm water runs off more quickly and in greater volume, thereby increasing flood hazards. Bringing water into semiarid cities by pumping groundwater or transporting water from distant mountains through aqueducts may increase evaporation, thereby increasing humidity and precipitation in a region.

We can see from Figure 5.8 that approximately 60% of water that falls by precipitation on land each year evaporates to the atmosphere. A smaller component (about 40%) returns to the ocean by surface and subsurface runoff. This small annual transfer of water supplies resources for rivers and urban and agricultural lands. Unfortunately, distribution of water on land is far from uniform. As a result, water shortages occur in some areas. As human population increases, water shortages will become more frequent in arid and semiarid regions, where water is naturally nonabundant.

At the regional and local level, the fundamental hydrologic unit of the landscape is the drainage basin (also called a watershed or catchment). A **drainage basin** is the area that contributes surface runoff to a particular stream or river. The term *drainage basin* is usually used in evaluating the hydrology of an area, such as the stream flow or runoff from hill slopes. Drainage basins vary greatly in size, from less than a hectare (2.5 acres) to millions of square kilometers. A drainage basin is usually named for its main stream or river, such as the Mississippi River drainage basin.

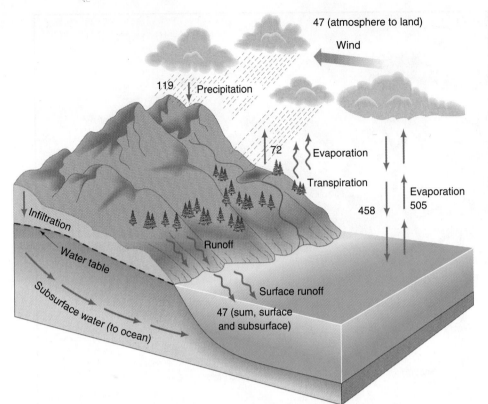

47 (atmosphere to land)

Wind

119 ↓ Precipitation

72 Evaporation

Transpiration

458 Evaporation 505

Infiltration

Water table

Subsurface water (to ocean)

Runoff

Surface runoff

47 (sum, surface and subsurface)

Figure 5.8 ■ The hydrologic cycle, showing the transfer of water (thousands of km³/yr) from the oceans to the atmosphere to the continents and back to the oceans again. [*Source:* From P. H. Gleick, *Water in Crisis* (New York: Oxford University Press, 1993).]

Storage Compartments of Water

Compartment	Vol. (thousands of km³)	Percentage of Total Water
Ocean	1,338,000	96.5
Glaciers and ice caps	24,064	1.74
Shallow groundwater	10,530	0.76
Lakes	176.4	0.013
Soil moisture	16.5	0.001
Atmosphere	12.9	0.001
Rivers	2.12	0.0002

The Rock Cycle

The **rock cycle** consists of numerous processes that produce rocks and soils. The rock cycle depends on the tectonic cycle for energy and on the hydrologic cycle for water. As shown in Figure 5.9, rock is classified as igneous, sedimentary, or metamorphic. These types of rock are involved in a worldwide recycling process. Internal heat from the tectonic cycle produces igneous rocks from molten material near the surface, such as lava from volcanoes. These new rocks weather when exposed at the surface. The freezing of water in cracks of rocks creates a physical weathering process. The water expands when it freezes, breaking the rocks apart. Physical weathering makes smaller particles of rocks from bigger ones, producing sediment such as gravel, sand, and silt. Chemical weathering occurs when the weak acids in water dissolve chemicals from rocks. The sediments and dissolved chemicals are then transported by water, wind, or ice (glaciers).

Weathered materials accumulate in depositional basins, such as the oceans. Sediments in depositional basins are compacted by overlying sediments and converted to sedimentary rocks by compaction and the cementing together of particles. After sedimentary rocks are buried to sufficient depths (usually tens to hundreds of kilometers), they may be altered by heat, pressure, or chemically active fluids and transformed to metamorphic rocks. Deeply buried rocks may be transported to the surface by an uplift associated with plate tectonics and then subjected to weathering.

You can see in Figure 5.9 that life processes play an important role in the rock cycle through the addition of organic carbon to rocks. Through the addition of organic carbon, rocks such as limestone, which is mostly calcium carbonate (the material of seashells and bones), as well as fossil fuel resources such as coal, are produced.

Plate tectonic processes of uplift and subsidence of rocks, along with erosion, produce Earth's varied topography. The spectacular Grand Canyon of the Colorado River in Arizona (Figure 5.10), sculpted from mostly sedimentary rocks, is one example. Another is the beautiful Tower Karst in China (Figure 5.11); these resistant blocks of limestone have survived chemical weathering and erosion that removed the surrounding rocks.

Our discussion of geologic cycles has emphasized tectonic, hydrologic, and rock-forming processes. We can now begin to integrate biogeochemical processes into the picture.

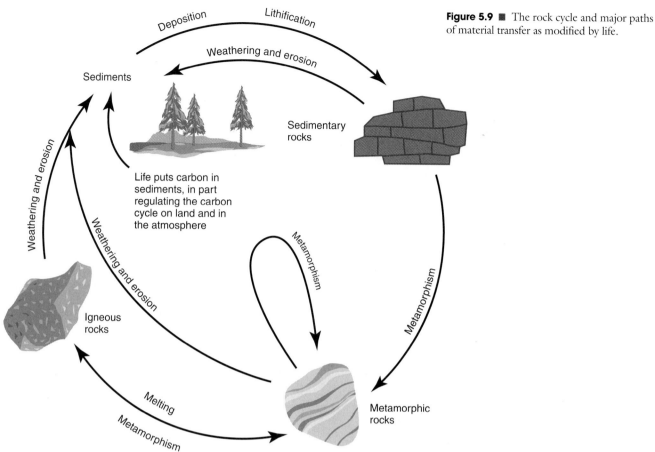

Deposition   Lithification

Weathering and erosion

Sediments

Sedimentary rocks

Weathering and erosion

Weathering and erosion

Life puts carbon in sediments, in part regulating the carbon cycle on land and in the atmosphere

Metamorphism

Metamorphism

Igneous rocks

Melting

Metamorphism

Metamorphic rocks

Figure 5.9 ■ The rock cycle and major paths of material transfer as modified by life.

Figure 5.10 ■ In response to slow tectonic uplift of the region, the Colorado River has eroded through the sedimentary rocks of the Colorado plateau to produce the spectacular scenery of the Grand Canyon. The river in recent years has been greatly modified by dams and reservoirs above and below the Grand Canyon. Sediment once carried to the Gulf of California is now deposited in reservoirs. As a result of the Glen Canyon Dam, the water flowing through the Grand Canyon is clearer and colder, resulting in changes in the abundance of sandbars and species of fish. The presence of this upstream dam has changed the hydrology and environment of the Colorado River in the Grand Canyon.

Figure 5.11 ■ This landscape in the People's Republic of China features Tower Karst, steep-sided hills or pinnacles composed of limestone. The rock has been slowly dissolving through chemical weathering. The pinnacles and hills are remnants of the weathering and erosion processes.

5.6 Biogeochemical Cycling in Ecosystems

When we ask questions about what chemicals might limit the abundance of a specific individual organism, population, or species, we look for the answer first at the ecosystem level. As explained in Chapter 3, an ecosystem is a community of different species and their nonliving environment in which energy flows and chemicals cycle. The boundaries that we choose for this investigation may be somewhat arbitrary, selected for convenience of measurement and analysis. On land, we often evaluate biogeochemical cycles for a fundamental element of the landscape, usually a drainage basin. Freshwater bodies—lakes, ponds, and bogs—are also convenient landforms for analysis of ecosystems and biogeochemical cycling.

Chemical cycling in an ecosystem begins with inputs from outside. On land, chemical inputs to an ecosystem come from the atmosphere via rain, wind-transported dust (called dry fallout), and volcanic ash from eruptions and from adjoining land via stream flow from flooding and groundwater from springs. Ocean and freshwater ecosystems have the same atmospheric and land inputs (including large, nearshore submarine springs). Ocean ecosystems, in addition, have inputs from ocean currents and submarine vents (hot springs) at divergent plate boundaries.

Chemicals cycle internally within an ecosystem through air, water, rocks, soil, and food chains by way of physical transport and chemical reactions. With the death of individual organisms, decomposition through chemical reactions returns chemical elements to other parts of the ecosystem. In addition, living organisms release some chemical elements directly into an ecosystem. Defecation by animals and dropping of fruit by plants are examples.

An ecosystem can lose chemical elements to other ecosystems. For example, rivers transport chemicals from the land to the sea. An ecosystem that has little loss of chemical elements can function in its current condition for longer periods than a "leaky" ecosystem that loses chemical elements rapidly. All ecosystems, however, lose chemicals to some extent. Therefore, all ecosystems require some external inputs of chemicals.

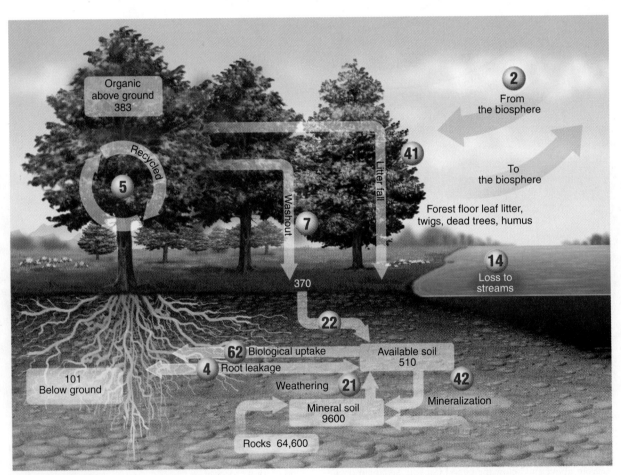

Figure 5.12 ■ Annual calcium cycle in a forest ecosystem. The circled numbers are flux rates in kilograms per hectare per year. The other numbers are the amounts stored in kilograms per hectare. Unlike sulfur, calcium does not have a gaseous phase, although it does occur in compounds as part of dust particles. Calcium is highly soluble in water in its inorganic form and is readily lost from land ecosystems in water transport. The information in this diagram was obtained from Hubbard Brook Ecosystem. [*Source:* G. E. Likens, F. H. Bormann, R. S. Pierce, J. S. Eaton, and N. M. Johnson, *The Biogeochemistry of a Forested Ecosystem*, 2nd ed. (New York: Springer-Verlag, 1995).]

Ecosystem Cycles of a Metal and a Nonmetal

Within an ecosystem, different chemical elements may have very different pathways, as illustrated in Figure 5.12 for calcium and in Figure 5.13 for sulfur. The calcium cycle is typical of a metallic element, and the sulfur cycle is typical of a nonmetallic element.

An important difference between these cycles is that calcium, like most metals, does not form a gas on Earth's surface. Therefore, calcium is not present as a gas in the atmosphere. In contrast, sulfur forms several gases, including sulfur dioxide (a major air pollutant and component of acid rain; see Chapter 24) and hydrogen sulfide (swamp or rotten egg gas, usually produced biologically).

Because sulfur has gas forms, it can be returned to an ecosystem more rapidly than can calcium. Annual input of sulfur from the atmosphere to a forest ecosystem has been measured to be 10 times that of calcium. For this reason, calcium and other elements without a gas phase are more likely to become limiting factors.

Chemical Cycling and the Balance of Nature

For life to be sustained indefinitely within an ecosystem, energy must be continuously added, and the store of essential chemicals must not decline. We have already discussed the common belief that, without human interference, life would be sustained indefinitely in a steady state, or "balance of nature." We have also discussed the belief that life tends to function to preserve an environment beneficial to itself. Both beliefs presume that chemical elements necessary for life exist in a dynamic steady state within an ecosystem.[6] These beliefs can be rephrased as a scientific hypothesis: Without human disturbance, the net storage of chemical elements within an ecosystem will remain constant over time. However, as noted, inevitably, some fraction of the chemical elements stored in an ecosystem is lost and must be replaced. Are ecosystems, then, ever in a dynamic steady state in regard to chemical elements? Studies indicate that they are not, because rates of input and output of chemicals do not balance, and concentrations of some chemicals decrease over time.

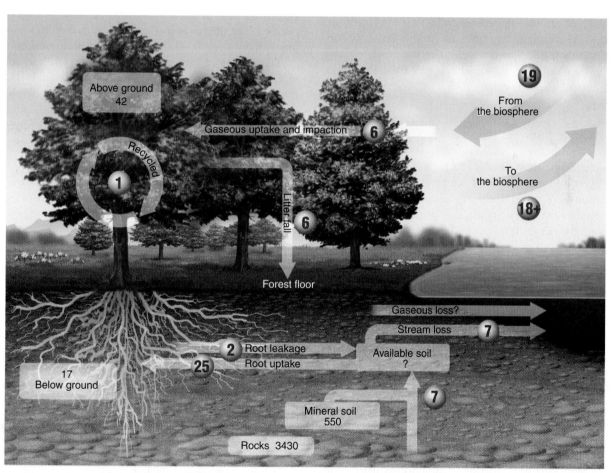

Figure 5.13 ■ Annual sulfur cycle in a forest ecosystem. The circled numbers are the flux rates in kilograms per hectare per year. The other numbers are the amounts stored in kilograms per hectare. Sulfur has a gaseous phase as H_2S and SO_2. The information in this diagram was obtained from Hubbard Brook Ecosystem. [*Source:* G. E. Likens, F. H. Bormann, R. S. Pierce, J. S. Eaton, and N. M. Johnson, *The Biogeochemistry of a Forested Ecosystem*, 2nd ed. (New York: Springer-Verlag, 1995).]

5.7 Some Major Global Chemical Cycles

Earlier in this chapter, we asked what chemical elements limit the abundance of life. We pointed out that the chemical elements required by life are divided into two major groups: macronutrients, which are required by all forms of life in large amounts, and micronutrients, which are either required by all forms of life in small amounts or required by only certain life-forms. In this section, we consider the global cycles of three macronutrients—carbon, nitrogen, and phosphorus. We focus on these in part because they are three of the "big six"—the elements that are the basic building blocks of life. Each is also related to important environmental problems that have attracted attention in the past and will continue to do so in the future.

The Carbon Cycle

Carbon is the element that anchors all organic substances, from coal and oil to DNA (deoxyribonucleic acid), the compound that carries genetic information. Although of central importance to life, carbon is not one of the most abundant elements in Earth's crust. It contributes only 0.032% of the weight of the crust, ranking far behind oxygen (45.2%), silicon (29.5%), aluminum (8.0%), iron (5.8%), calcium (5.1%), and magnesium (2.8%.).[7,8]

The major pathways and storage reservoirs of the **carbon cycle** are shown in Figure 5.14. Notice that carbon has a gaseous phase as part of its cycle. It occurs in Earth's atmosphere as carbon dioxide (CO_2) and methane (CH_4), both greenhouse gases (see Chapter 23). Carbon enters the atmosphere through the respiration of living things, through fires that burn organic compounds, and by diffusion from the ocean. It is removed from the atmosphere by photosynthesis of green plants, algae, and photosynthetic bacteria. (See A Closer Look 5.3.) Over the past 3 billion years of Earth's history, the rate at which biological processes have removed carbon dioxide from the atmosphere has exceeded the rate of addition. Therefore, Earth's atmosphere has much less carbon dioxide than would occur on a lifeless Earth.

Carbon occurs in the ocean in several inorganic forms, including dissolved carbon dioxide as well as carbonate (CO_3^{2-}) and bicarbonate (HCO_3^-). It also occurs in organic compounds of marine organisms and their products, such as seashells ($CaCO_3$). Carbon enters the ocean from the atmosphere by simple diffusion of carbon dioxide. The carbon dioxide then dissolves and is converted to carbonate and bicarbonate. Marine algae and photosynthetic bacteria obtain the carbon dioxide they use from the water in one of these three forms. Carbon is transferred from the land to the ocean in rivers and streams as dissolved carbon, including organic compounds, and as organic particulates (fine particles of organic matter). Winds also transport small organic particulates from the land to the ocean. Transfer via rivers and streams makes up a relatively small fraction of the total global carbon flux to the oceans. However, on the local and regional scale, input of carbon from rivers to nearshore areas such as deltas and salt marshes, which are often highly biologically productive, is important.

Carbon enters the biota through photosynthesis and is returned to the atmosphere or waters by respiration or by wildfire. When an organism dies, most of its organic material decomposes to inorganic compounds, including carbon dioxide. Some carbon may be buried where there is not sufficient oxygen to make this conversion possible or where the temperatures are too cold for decomposition. In these locations, organic matter is stored. Over years, decades, and centuries, storage of carbon occurs in wetlands, including parts of floodplains, lake basins, bogs, swamps, deep-sea sediments, and near-polar regions. Over longer periods (thousands to several million years), some carbon may be buried with sediments that become sedimentary rocks. This carbon is transformed into fossil fuels such as natural gas, oil, and coal. Nearly all of the carbon stored in the lithosphere exists as sedimentary rocks. Most of this is in the form of carbonates such as limestone, much of which has a direct biological origin.

The cycling of carbon dioxide between land organisms and the atmosphere is a large flux. Approximately 15% of the total carbon in the atmosphere is taken up by photosynthesis and released by respiration on land annually. Thus, as noted, life has a large effect on the chemistry of the atmosphere.

The Missing (Unidentified) Carbon Sink

Because carbon forms two of the most important greenhouse gases—carbon dioxide and methane—much research has been devoted to understanding the carbon cycle. However, at a global level, some key issues remain unanswered. For example, monitoring of atmospheric carbon dioxide levels over the past several decades suggests that of the approximately 8.5 units released per year by human activities into the atmosphere, approximately 3.2 units remain there. It is estimated that about 2.4 units diffuse into the ocean. This leaves 2.9 units unaccounted for.[9,10] Several hundred or more million tons of carbon are burned each year from fossil fuel and end up somewhere unknown to science. Inorganic processes do not account for the fate of this **"missing carbon sink."** Either marine or land photosynthesis, or both, must provide the additional flux. At this time, however, scientists do not agree on which processes dominate or in what regions of Earth this missing flux occurs.

Why can't scientists agree on these basic issues? Unfortunately, there are many uncertainties. For example, when uncertainties in measuring carbon are considered, the missing carbon sink has a flux of 2.9 ± 1.1 billion metric tons/yr. The uncertainty is about 40% of

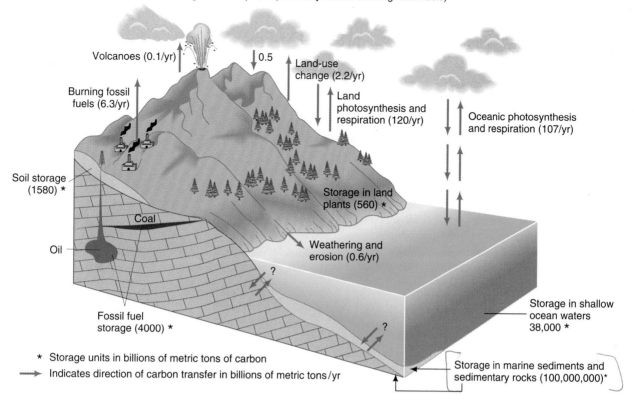

Storage in atmosphere (750 + 3/yr due to burning fossil fuels) *

Volcanoes (0.1/yr)   0.5

Burning fossil fuels (6.3/yr)

Land-use change (2.2/yr)

Land photosynthesis and respiration (120/yr)

Oceanic photosynthesis and respiration (107/yr)

Soil storage (1580) *

Coal

Oil

Storage in land plants (560) *

Weathering and erosion (0.6/yr)

?

Fossil fuel storage (4000) *

?

Storage in shallow ocean waters 38,000 *

* Storage units in billions of metric tons of carbon

→ Indicates direction of carbon transfer in billions of metric tons/yr

Storage in marine sediments and sedimentary rocks (100,000,000)*

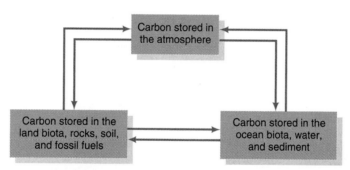

Carbon stored in the atmosphere

Carbon stored in the land biota, rocks, soil, and fossil fuels

Carbon stored in the ocean biota, water, and sediment

Figure 5.14 ■ (*a*) Generalized global carbon cycle. (*b*) Parts of the carbon cycle are simplified to illustrate the cyclic nature of the movement of carbon. [*Source:* Modified from G. Lambert, *La Recherche* 18 (1987): 782–783, with some data from R. Houghton, *Bulletin of the Ecological Society of America* 74, no. 4 (1993): 355–356, and R. Houghton, *Tellus* 55B, no. 2 (2003): 378–390.]

the suspected flux. Including estimated uncertainties yields the following global budget of carbon for the mid-1990s, in units of billions of metric tons of carbon (GtC) per year.[11–13]

Atmospheric increase		Emissions from fossil fuels		Net emissions from changes in land use		Oceanic uptake		Missing carbon sink
3.2 ± 0.2	=	6.3 ± .4	+	2.2 ± .8	−	2.4 ± .7	−	2.9 ± .1.1

Since the beginning of the Industrial Revolution in the mid-nineteenth century, the unidentified sink has apparently steadily increased in size (Figure 5.17). It is believed that a possible sink to account for part of the missing carbon (about 0.7 ± 0.8 billion metric tons/yr) can be found in

the terrestrial ecosystems, including forests and, to a lesser extent soils. Of particular importance are those forests in the Northern Hemisphere that are recovering from timber harvesting during the past two centuries and are growing fast today. Fast-growing forests remove carbon from the atmosphere at an increasing rate as biomass is added. However, fires and melting of frozen ground in the boreal forests also release carbon, and it's not certain whether the forests comprise a net sink or a net source of carbon to the atmosphere.[14] This is reflected by the large uncertainty in the terrestrial flux of 0.7 ± 0.8 GtC per year. The uncertainty of ±0.8 units is greater than the estimated flux of 0.7 units! Two major uncertainties in estimating carbon fluxes are rates of land-use change, especially those cleared for

Photosynthesis and Respiration

Carbon dioxide enters biological cycles through photosynthesis, the process whereby the cells of living organisms (such as plants) convert energy from sunlight in a series of chemical reactions to chemical energy. In the process, carbon dioxide and water are combined to form organic compounds such as simple sugars and starch, with oxygen as a by-product (Figure 5.15). Carbon leaves living biota through respiration, the process by which organic compounds are broken down to release gaseous carbon dioxide. For example, animals (including people) take in air, which has a relatively high concentration of oxygen. Oxygen is absorbed by blood in the lungs. Through respiration, carbon dioxide is released into the atmosphere. Figure 5.16 illustrates the role of photosynthesis and respiration, along with other processes, in the carbon cycle of a lake.

Energy from sun

CO_2

O_2

H_2O

At the cell level: chlorophyll (green plant absorbs sunlight)

General Photosynthesis: chemical reaction

$$6CO_2 + 6H_2O \xrightarrow{\text{sunlight}} C_6H_{12}O_6 + 6O_2$$

$$\text{carbon} + \text{water} \xrightarrow{\text{sunlight}} \text{sugar (glucose)} + \text{oxygen}$$

Figure 5.15 ■ Idealized diagram illustrating photosynthesis for a green plant (tree) and generalized reaction.

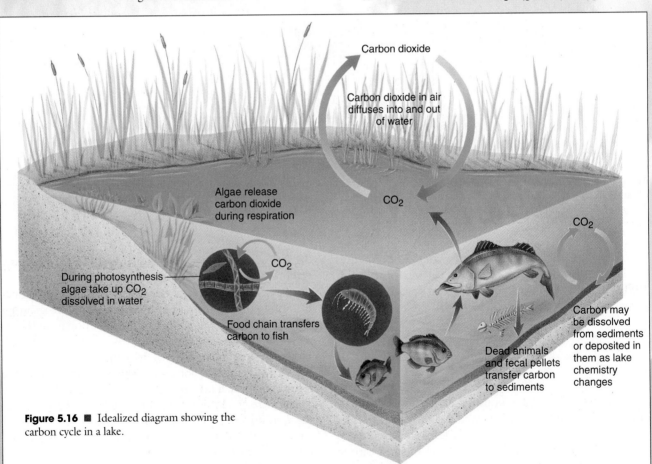

Carbon dioxide

Carbon dioxide in air diffuses into and out of water

CO_2

CO_2

Algae release carbon dioxide during respiration

During photosynthesis algae take up CO_2 dissolved in water

CO_2

Food chain transfers carbon to fish

Dead animals and fecal pellets transfer carbon to sediments

Carbon may be dissolved from sediments or deposited in them as lake chemistry changes

Figure 5.16 ■ Idealized diagram showing the carbon cycle in a lake.

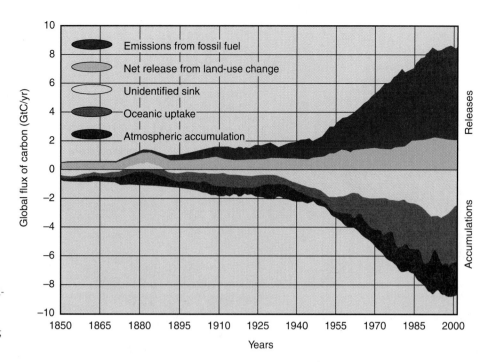

Figure 5.17 ■ Global flux of carbon, 1850–2000. The missing (unidentified) sink has varied considerably in recent decades. Modified after Woods Hole Research Center, Global Carbon Cycle, "The Missing Carbon Sink," 2004

agriculture, and the amount of carbon in ecosystem storage compartments affected by human use, especially those cleared for agriculture, harvested, or burned.[12, 13] Uncertainties will be reduced in the future if we are more successful in measuring and monitoring land-use change (deforestation, burning, and clearing) and estimating the flux of carbon in the ecosystems and the atmosphere. Because of changes in land-use and the burning of fossil fuels, the missing carbon sink is not static but changes on annual and decadal time scales. It was about 1 GtC in 1940, 2 GtC in the 1980s, 3 GtC in the mid-1990s, and back to 2 GtC in 2000 (see Figure 5.17).[12,13] The missing carbon problem illustrates the complexity of biogeochemical cycles, especially ones in which the biota play an important role.

The carbon cycle will continue to be an important area of research because of its significance to global climate investigations, especially to global warming.[15,16] To better understand this issue, we turn next to the interactions between the carbon and tectonic cycles through the carbon-silicate cycle.

The Carbon–Silicate Cycle

Carbon cycles rapidly among the atmosphere, oceans, and life. However, over geologically long periods of time, the cycling of carbon becomes intimately involved with the cycling of silicon. The combined **carbon–silicate cycle** is therefore of geologic importance to the long-term stability of the biosphere over periods that exceed one-half billion years.[17]

The carbon–silicate cycle begins when carbon dioxide in the atmosphere dissolves in the water to form weak carbonic acid (H_2CO_3) that falls as rain (Figure 5.18). As the mildly acidic water migrates through the ground, it weathers rocks

and facilitates the erosion of Earth's abundant silicate-rich rocks. Among other products, weathering and erosion release calcium ions (Ca^{2+}) and bicarbonate (HCO_3^-) ions. These ions enter the ground and surface waters and eventually are transported to the ocean. Calcium and bicarbonate ions make up a major portion of the chemical load that rivers deliver to the oceans.

Tiny floating marine organisms in the ocean use the calcium and bicarbonate to construct their shells. When these organisms die, the shells sink to the bottom of the ocean, where they accumulate as carbonate-rich sediments. Eventually, carried by moving tectonic plates, they enter a subduction zone, where they are subjected to increased heat, pressure, and partial melting. The resulting magma releases carbon dioxide, which rises in volcanoes and is released into the atmosphere. This process provides a lithosphere-to-atmosphere flux of carbon.

The long-term carbon–silicate cycle (Figure 5.18) and the short-term carbon cycle (Figure 5.14) interact to affect levels of CO_2 and O_2 in the atmosphere. For example, the burial of organic material in an oxygen-poor environment amounts to a net increase of photosynthesis (which produces O_2) over respiration (which produces CO_2). Thus, if burial of organic carbon in oxygen-poor environments increases, the concentration of atmospheric oxygen will increase. Conversely, if more organic carbon escapes burial and is oxidized to produce CO_2, then the CO_2 concentration in the atmosphere will increase.[18] Long-term changes in CO_2 and O_2 have been profoundly important in global change processes over geologic time, far beyond what we normally consider in environmental science.

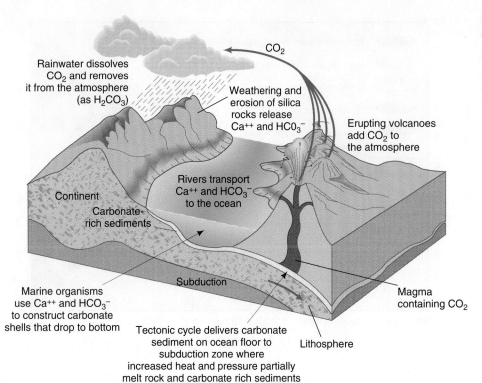

Figure 5.18 ■ Idealized diagram showing the carbon–silicate cycle. [*Source:* Modified from J. E. Kasting, O. B. Toon, and J. B. Pollack, "How Climate Evolved on the Terrestrial Planets," *Scientific American* 258 (1988): 2.]

The Nitrogen Cycle

Nitrogen is essential to life because it is necessary for the manufacture of proteins and DNA. Free nitrogen (N_2 uncombined with any other element) makes up approximately 80% of Earth's atmosphere. However, many organisms cannot use this nitrogen directly. Some, such as animals, require nitrogen in an organic compound. Others, including plants, algae, and bacteria, can take up nitrogen either as the nitrate ion (NO_3^-) or the ammonium ion (NH_4^+). Because nitrogen is a relatively unreactive element, few processes convert molecular nitrogen to one of these compounds. Lightning oxidizes nitrogen, producing nitric oxide. In nature, essentially all other conversions of molecular nitrogen to biologically useful forms are conducted by bacteria.

The **nitrogen cycle** is one of the most important and most complex of the global cycles (Figure 5.19). The process of converting inorganic, molecular nitrogen in the atmosphere to ammonia or nitrate is called **nitrogen fixation**. Once in these forms, nitrogen can be used on land by plants and in the oceans by algae. Through chemical reactions, bacteria, plants, and algae then convert these inorganic nitrogen compounds into organic ones, and the nitrogen becomes available to ecological food chains. When organisms die, other bacteria convert the organic compounds containing nitrogen back to ammonia, nitrate, or molecular nitrogen, which enters the atmosphere. The process of releasing fixed nitrogen back to molecular nitrogen is called **denitrification**.

Nearly all organisms depend on nitrogen-converting bacteria. Some organisms have evolved symbiotic relationships with these bacteria. For example, the roots of the pea family have nodules that provide a habitat for the bacteria. The bacteria obtain organic compounds for food from the plants, and the plants obtain usable nitrogen. Such plants can grow in otherwise nitrogen-poor environments. When these plants die, they contribute nitrogen-rich organic matter to the soil, thereby improving the soil's fertility. Alder trees, too, have nitrogen-fixing bacteria as symbionts in their roots. (Symbionts are organisms in symbiotic relationships.) These trees grow along streams, and their nitrogen-rich leaves fall into the streams and increase the supply of the element in a biologically usable form to freshwater organisms.

Nitrogen-fixing bacteria are also symbionts in the stomachs of some animals, particularly the cud-chewing animals. These animals, which include buffalo, cows, deer, moose, and giraffes, have a specialized four-chambered stomach. The bacteria provide as much as half of the total nitrogen needed by the animals, with the rest provided by protein in the green plants the animals eat.

In terms of availability for life, nitrogen falls somewhere between carbon and phosphorus. Like carbon, nitrogen has a gaseous phase and is a major component of Earth's atmosphere. Unlike carbon, however, it is not very reactive, and its conversion depends heavily on biological activity. Thus, the nitrogen cycle is not only essential to life but also is primarily driven by life.

In the early part of the twentieth century, scientists discovered that industrial processes could convert molecular nitrogen into compounds usable by plants. This greatly increased the availability of nitrogen in fertilizers. Today,

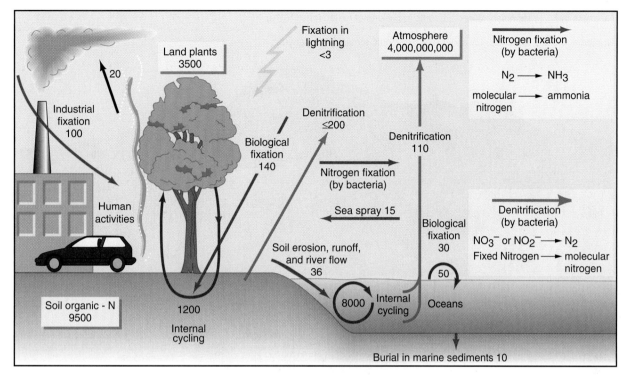

Figure 5.19 ■ The global nitrogen cycle. Numbers in boxes indicate amounts stored, and numbers with arrows indicate annual flux, in 10^{12} g N_2. Note that the industrial fixation of nitrogen is nearly equal to the global biological fixation. [*Source:* Data from R. Söderlund and T. Rosswall, in *The Handbook of Environmental Chemistry*, Vol. 1, Pt. B, ed. O. Hutzinger (New York: Springer-Verlag, 1982), and W. H. Schlosinger, *Biogeochemistry: An Analysis of Global Change* (San Diego: Academic Press, 1997), p. 386.]

industrial fixation of nitrogen is a major source of commercial nitrogen fertilizer. The amount of industrial fixed nitrogen is about 50% of the amount fixed in the biosphere. Nitrogen in agricultural runoff is a potential source of water pollution.

Nitrogen combines with oxygen in high-temperature atmospheres. As a result, many modern industrial combustion processes produce oxides of nitrogen. These processes include the burning of fossil fuels in gasoline and diesel engines. Thus, oxides of nitrogen, which are air pollutants, are indirect results of modern industrial activity and modern technology. Nitrogen oxides play a significant role in urban smog (see Chapter 24).

In summary, nitrogen compounds are a bane and a boon for society and for the environment. Nitrogen is required for all life, and its compounds are used in many technological processes and in modern agriculture. But nitrogen is also a source of air and water pollution.

The Phosphorus Cycle

Phosphorus, one of the "big six" elements required in large quantities by all forms of life, is often a limiting nutrient for plant and algal growth. However, if phosphorus is too abundant, it can cause environmental problems, as illustrated by the story of Lake Washington that opened this chapter.

Unlike carbon and nitrogen, phosphorus does not have a gaseous phase on Earth (Figure 5.20). Thus, the **phospho-**

rus cycle is significantly different from the carbon and nitrogen cycles. The rate of transfer of phosphorus in Earth's system is slow compared with that of carbon or nitrogen. Phosphorus exists in the atmosphere only in small particles of dust. In addition, phosphorus tends to form compounds that are relatively insoluble in water. Consequently, phosphorus is not readily weathered chemically. It does occur commonly in an oxidized state as phosphate, which combines with calcium, potassium, magnesium, or iron to form minerals.

Phosphorus enters the biota through uptake as phosphate by plants, algae, and some bacteria. In a relatively stable ecosystem, much of the phosphorus that is taken up by vegetation is returned to the soil when the plants die. Nevertheless, some phosphorus is inevitably lost to ecosystems. It is transported by rivers to the oceans, either in a water-soluble form or as suspended particles.

An important way in which phosphorus returns from the ocean to the land involves ocean-feeding birds, such as the Chilean pelican. These birds feed on small fish, especially anchovies, which in turn feed on tiny ocean plankton. Plankton thrive where nutrients such as phosphorus are present. Areas of rising oceanic currents known as upwellings are such places. Upwellings occur near continents where the prevailing winds blow offshore, pushing surface waters away from the land and allowing deeper waters to rise and replace those moved offshore. Upwellings carry nutrients, including phosphorus, from the depths of the oceans to the surface.

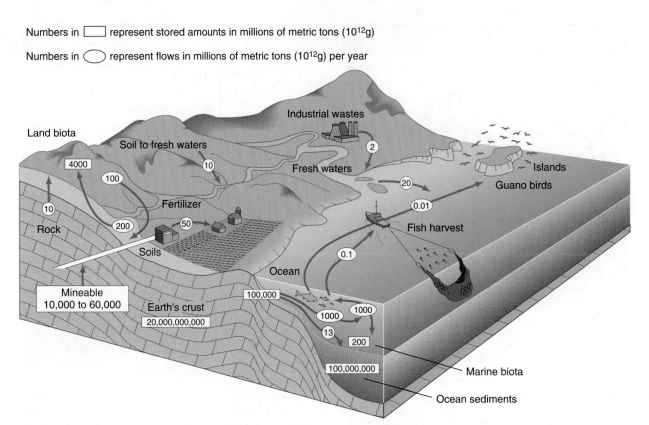

Numbers in ☐ represent stored amounts in millions of metric tons (10^{12} g)

Numbers in ◯ represent flows in millions of metric tons (10^{12} g) per year

Figure 5.20 ■ The global phosphorus cycle. Phosphorus is recycled to soil and land biota through geologic processes that uplift the land and erode rocks, by birds that produce guano, and by human beings. Although Earth's crust contains a very large amount of phosphorus, only a small fraction of it can be mined by conventional techniques. Therefore, phosphorus is an expensive resource to produce. Values of the amount of phosphorus stored or in flux are compiled from various sources. Estimates are approximate to the order of magnitude. [*Sources:* Based primarily on C. C. Delwiche and G. E. Likens, "Biological Response to Fossil Fuel Combustion Products," in *Global Chemical Cycles and Their Alterations by Man*, ed. W. Stumm (Berlin: Abakon Verlagsgesellschaft, 1977), pp. 73–88; and U. Pierrou, "The Global Phosphorus Cycle," in *Nitrogen, Phosphorus and Sulfur—Global Cycles*, eds. B. H. Svensson and R. Soderlund (Stockholm: *Ecological Bulletin*, 1976, pp. 75–88).]

The fish-eating birds nest on offshore islands, where they are protected from predators. Over time, their nesting sites become covered with their phosphorus-laden excrement, called guano. The birds nest by the thousands, and deposits of guano accumulate over centuries. In relatively dry climates, guano hardens into a rocklike mass that may be up to 40 m (130 ft) thick. The guano results from a combination of biological and nonbiological processes. Without the plankton, fish, and birds, the phosphorus would have remained in the ocean. Without the upwellings, the phosphorus would not have been available.

Guano deposits were once major sources of phosphorus for fertilizers. In the mid-1800s, as much as 9 million metric tons per year of guano deposits were shipped to London from islands near Peru. Today, most phosphorus fertilizers come from mining of phosphate-rich sedimentary rocks containing fossils of marine animals. The richest phosphate mine in the world is Bone Valley, 40 km east of Tampa, Florida. Between 10 and 15 million years ago, Bone Valley was the bottom of a shallow sea where marine invertebrates lived and died.[19] Through tectonic processes, Bone Valley was slowly uplifted, and in the 1880s and 1890s phosphate ore was discovered there. Today, Bone Valley provides more than one-third of the world's phosphate production and three-fourths of U.S. production.

Total U.S. reserves of phosphorus are estimated at 2.2 billion metric tons, a sufficient quantity to supply our needs for several decades. However, if the price of phosphorus increases as high-grade deposits are exhausted, phosphorus from lower-grade deposits can be mined at a profit. Florida is thought to have 8.1 billion metric tons of phosphorus that can be recovered with existing mining methods if the price is right.[19] Mining, of course, may have negative effects on the land and ecosystems. For example, in some phosphorus mines, huge pits and waste ponds have scarred the landscape, damaging biologic and hydrologic resources. Balancing the need for phosphorus with the adverse environmental impacts of mining is a major environmental issue. Following phosphate extraction, land disrupted by open-pit phosphate mining, shown in Figure 5.21, is reclaimed to pasture land as mandated by law.

Figure 5.21 ■ Large open-pit phosphate mine in Florida (similar to Bone Valley), with piles of waste material. The land in the upper part of the photograph has been reclaimed and is being used for pasture.

Summary

■ Biogeochemical cycles are the major way in which elements important to Earth processes and life are moved through the atmosphere, hydrosphere, lithosphere, and biosphere.

■ Biogeochemical cycles can be described as a series of reservoirs, or storage compartments, and pathways, or fluxes, between reservoirs.

■ In general, some chemical elements cycle quickly and are readily regenerated for biological activity. Elements whose biogeochemical cycles include a gaseous phase in the atmosphere tend to cycle more rapidly.

■ Our modern technology has begun to alter and transfer chemical elements in biogeochemical cycles at rates comparable to those of natural processes. Some of these activities are beneficial to society, but others pose dangers.

■ To be better prepared to manage our environment, we must recognize both positive and negative consequences of activities that transfer chemical elements and deal with them appropriately.

■ Biogeochemical cycles tend to be complex, and Earth's biota has greatly altered the cycling of chemicals through the air, water, and soil. Continuation of these processes is essential to the long-term maintenance of life on Earth.

■ Every living thing, plant or animal, requires a number of chemical elements. These chemicals must be available at the appropriate time and in the appropriate form and amount.

■ Chemicals can be reused and recycled, but in any real ecosystem some elements are lost over time and must be replenished if life in the ecosystem is to persist.

■ Change and disturbance of natural ecosystems are the norm. A steady state, in which the net storage of chemicals in an ecosystem does not change with time, cannot be maintained.

■ There are many uncertainties in measuring either the amount of a chemical in storage or the rate of transfer between reservoirs. For example, the global carbon cycle includes a large sink that science has not yet been able to locate.

How Are Human Activities Affecting the Nitrogen Cycle?

Scientists estimate that nitrogen deposition to Earth's surface will double in the next 25 years. What is causing this increase, and how will it affect the nitrogen cycle?

The natural rate of nitrogen fixation on land is estimated to be 140 teragrams (Tg) of nitrogen a year (1 teragram = 1 million metric tons). Human activities, such as use of fertilizers, draining of wetlands, clearing of land for agriculture, and burning of fossil fuels, are causing additional nitrogen to enter the environment. Currently, human activities are responsible for more than half of the fixed nitrogen that is deposited on land. Before the twentieth century, fixed nitrogen was recycled by bacteria with no net accumulation. Since 1900, however, the use of commercial fertilizers has increased exponentially (see the graph). Nitrates and ammonia from burning fossil fuels have increased about 20% in the last decade or so. These inputs have overwhelmed the denitrifying part of the nitrogen cycle and the ability of plants to use fixed nitrogen.

Nitrate ions in the presence of soil or water may form nitric acid. With other acids in the soil, nitric acid can leach out chemicals important to plant growth, such as magnesium and potassium. When these chemicals are depleted, more toxic ones, such as aluminum, may be released, damaging tree roots. Acidification of soil by nitrate ions is also harmful to organisms. When toxic chemicals wash into streams, they can kill fish. Excess nitrates in rivers and along coasts can cause algae to overgrow, with effects similar to those described earlier for Lake Washington. High levels of nitrates in drinking water from streams or groundwater contaminated by fertilizers are a health hazard.

The nitrogen and carbon cycles are linked because nitrogen is a component of chlorophyll, the molecule plants use in photosynthesis. Because nitrogen is a limiting factor on land, it has been predicted that increasing levels of global nitrogen may increase plant growth.

Recent studies have suggested that a beneficial effect from increased nitrogen would be short-lived, however. As plants use additional nitrogen, some other factor will become limiting.

When that occurs, plant growth will slow, and so will the uptake of carbon dioxide. More research is needed to understand the interactions between carbon and nitrogen cycles and to predict long-term effects of human activities.

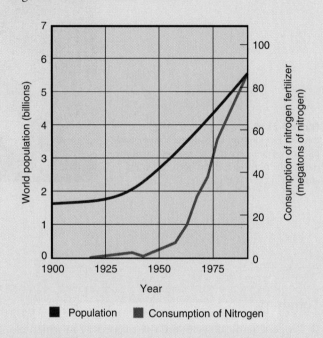

Critical Thinking Questions

1. Compare the rate of human contributions to nitrogen fixation with the natural rate.

2. How does the change in fertilizer use relate to the change in world population? Why?

3. Develop a diagram to illustrate the links between the nitrogen and carbon cycles.

4. Make a list of ways in which human activities could be modified to reduce human contributions to the nitrogen cycle.

REEXAMINING THEMES AND ISSUES

Human Population

Through modern technology, we are transferring some chemical elements through the air, water, soil, and biosphere at rates comparable to those of natural processes. As our population increases, so does our utilization of resources and so do these rates of transfer. This is a potential problem because eventually the rate of transfer for a particular chemical may become so large that pollution of the environment results.

Sustainability

If we are to sustain a high-quality environment, the major biogeochemical cycles must transfer and store the chemicals necessary to maintain healthy ecosystems. That is one reason why understanding biogeochemical cycles is so important. For example, the release of sulfur into the atmosphere is degrading air quality at local to global levels. As a result, the United States is striving to control these emissions.

Global Perspective

The major biogeochemical cycles discussed in this chapter are presented from a global perspective. Through ongoing research, scientists are trying to better understand how major biogeochemical cycles work. For example, the carbon cycle and its relationship to the burning of fossil fuels and the storage of carbon in the biosphere and oceans are being intensely investigated. Results of these studies are assisting in the development of strategies to reduce carbon emissions. These strategies are implemented at the local level, at power plants, and in cars and trucks that burn fossil fuels.

Urban World

Our society has concentrated the use of resources in urban regions. As a result, the release of various chemicals into the biosphere, soil, water, and atmosphere is often greater in urban centers, resulting in biogeochemical cycles that cause pollution problems.

People and Nature

Humans, like other animals, are linked to natural processes and nature in complex ways. We change ecosystems through land-use changes and the burning of fossil fuels, both of which change biogeochemical cycles, especially the carbon cycle that anchors life and affects Earth's climate.

Science and Values

Our understanding of biogeochemical cycles is far from complete. There are large uncertainties in the measurement of fluxes of chemical elements such as nitrogen, carbon, phosphorus, and others. We are studying biogeochemical cycles because we believe that understanding them will allow us to solve environmental problems. Which problems we address first will reflect the values of our society.

Key Terms

biogeochemical cycle **77**
carbon–silicate cycle **93**
carbon cycle **90**
chemical reaction **79**
denitrification **94**

drainage basin **85**
geologic cycle **83**
hydrologic cycle **84**
limiting factor **82**
macronutrients **82**

micronutrients **82**
missing carbon sink **90**
nitrogen cycle **94**
nitrogen fixation **94**
phosphorus cycle **95**

plate tectonics **83**
rock cycle **86**
tectonic cycle **83**

Study Questions

1. Why is an understanding of biogeochemical cycles important in environmental science? Explain your answer with two examples.

2. What are some of the general rules that govern biogeochemical cycles, especially the transfer of material?

3. Identify the major aspects of the carbon cycle and the environmental concerns associated with it.

4. What are differences in the geochemical cycles for phosphorus and nitrogen, and why are the differences important in environmental science?

5. How can the carbon–silicate cycle provide a negative-feedback mechanism to control the temperature of Earth's atmosphere?

Further Reading

Berner, R. A., and E. K. Berner. 1996. *Global Environment: Water, Air, and Geochemical Cycles.* Upper Saddle River, N.J.: Prentice-Hall. This is a good discussion of environmental geochemical cycles, focusing on Earth's air and water systems.

Kasting, J. F., O. B. Toon, and J. B. Pollack. 1988. "How Climate Evolved on the Terrestrial Planets," *Scientific American* 258(2):90–97. This paper provides a good discussion of the carbonate–silicate cycle and why it is important in environmental science.

Lerman, A. 1990. "Weathering and Erosional Controls of Geologic Cycles," *Chemical Geology* 84:13–14. Natural transfer of elements from the continents to the oceans is largely accomplished by erosion of the land and transport of dissolved material in rivers.

Post, W. M., T. Peng, W. R. Emanual, A. W. King, V. H. Dale, and D. L. DeAngelis. 1990. "The Global Carbon Cycle," *American Scientist* 78:310–326. The authors describe the natural balance of carbon dioxide in the atmosphere and review why the global climate hangs in the balance.

Schlesinger, W. H. 1997. *Biogeochemistry: An Analysis of Global Change,* 2nd ed. San Diego: Academic Press. This book provides a comprehensive and up-to-date overview of the chemical reactions on land, in the oceans, and in the atmosphere of Earth.

CHAPTER 6

Ecosystems and Ecosystem Management

Silver Lake on the Croton River, NY; a typical view of the Eastern Deciduous Forest of North America, home to the whitetailed deer.

Learning Objectives

Life on Earth is sustained by ecosystems, which vary greatly but have certain attributes in common. After reading this chapter, you should understand:

■ Why the ecosystem is the basic system that supports life and allows it to persist.

■ What food chains, food webs, and trophic levels are.

■ What the concept of ecosystem management involves.

■ How conservation and management of the environment might be improved through ecosystem management.

■ How ecosystems perform "public service."

The Acorn Connection

As young children, most people learn that acorns grow into oak trees. In fact, most acorns do *not* grow into trees but become food for mice, chipmunks, squirrels, and deer.

In the woodlands of the northeastern United States, where oak trees abound, large crops of acorns (Figure 6.1*f*) are produced every three to four years. Acorns are rich in proteins and fats and are an excellent source of nutrition. A steady supply of acorns would be an excellent food base for the woodland animals.

The production of acorns is affected by the amount of light and rain, the temperature patterns over the year, and the quality of the soil. Scientists reason that if oaks produced the same number of acorns each year, the populations of animals that feed on them would grow so large that very few acorns would survive.

In reality, the number of acorns produced varies from year to year, with "mast" years—years of high production—occurring occasionally. In the years between bumper crops of acorns, mice populations decline. With the next bumper crop, there are more acorns than can be eaten by the consumers of acorns, so many acorns survive to become oaks. Also, because of the abundance of food, the mice populations increase.

White-footed mice (Figure 6.1*b*), feeders on acorns, also carry tick larvae (Figure 6.1*a*). When the ticks feed on the blood of the mice, they inject the microorganisms responsible for Lyme disease into the mice. Mice populations are highest during the summer following a bumper crop of acorns, and so are tick larvae.

(a)   (b)   (c)

(d)   (e)   (f)

Figure 6.1 ■ The tick that carries Lyme disease (*a*) feeds on both the white-footed mouse (*b*) and the white-tailed deer (*c*). Oak leaves (*d*) are an important food for the deer (*c*) and for gypsy moths (*e*), while oak acorns (*f*) are important food for the mouse. But the mouse also eats the moths. The more mice, the fewer gypsy moths, but the more ticks.

In later stages of their life cycle, ticks attach to other animals, including deer (Figure 6.1c). As deer brush against plants, ticks are deposited. The ticks can be picked up by people brushing against the plants as they walk past. If an infected tick bites a person, the person may contract Lyme disease. As second-growth forest area has increased and deer populations have soared, Lyme disease has become the most common tick-borne disease in the United States.[1, 2]

Why has forest area increased? Beginning in colonial times, the forests of the northeastern United States were cleared to make space for farming and settlements, to provide fuel, and to provide timber for commercial uses. As coal, oil, and gas replaced wood as a primary fuel, and as farming moved westward to the more fertile Great Plains, fields that had been cleared were abandoned. In many areas, the maximum clearing occurred around 1900. Since then, forests have grown back.

But let's return to the mice. In addition to feeding on acorns (and other grains), mice feed on insects, including larvae of the gypsy moth. Gypsy moth larvae (Figure 6.1e) feed on leaves of trees and are particularly fond of oak leaves. Studies suggest that in years when mice populations are low—the years between bumper crops of acorns—gypsy moth populations can increase

dramatically. During these periodic outbreaks, gypsy moth larvae can virtually denude an area, stripping the leaves from the trees. Oaks that have lost most or all of their leaves may not produce bumper crops of acorns.

Once the leaves are off the trees, more light reaches the ground, and seedlings of many plants that could not do well in deep forest shade begin to grow. As a result, other species of trees may gain a foothold in the forest and change its species profile. Of course, the next generation of gypsy moth larvae find little to eat, and the population of gypsy moths begins to decline again.

Abundant acorns draw deer into the woods, where they browse on small plants and tree seedlings. Ticks drop off the deer and lay eggs in the leaf litter. When the eggs hatch, the larvae attach to mice, and the cycle of Lyme disease continues. Deer do not eat ferns, however, and in areas where deer populations are dense, many ferns but few wildflowers and tree seedlings are found.

Predators are also affected by the periodic nature of acorn crops. For example, birds that feed on gypsy moth larvae lose a food source when moth populations are low. When moth populations are high, however, bird nests are more exposed to predation because trees lose so many leaves.[2]

The acorn connection illustrates many of the basic characteristics of ecosystems and ecological communities. First, all of the living parts of the oak forest community depend on the nonliving parts of the ecosystem for their survival: water, soil, air, and the light that provides energy for photosynthesis. Second, the members of the ecological community affect the nonliving parts of the ecosystem. When gypsy moths denude an area, for example, more sunlight can reach the forest floor. Third, the living organisms in the ecosystem are connected in complex relationships that make it difficult to change one thing without changing many others. Fourth, the relationships among the members of the ecological community are dynamic and constantly changing. Many species are adapted to and benefit from a changing environment, as shown by the advantages provided oaks by varying acorn production. Fifth, the implication for human management of ecosystems is that any management practice involves trade-offs. In this case, managing the forest to protect people against Lyme disease only results in more potential for gypsy moth damage.[1]

Figure 6.2 ■ The eastern deciduous forest of the United States is a major recreational resource, as illustrated here by these hikers looking out from a patch of this kind of forest to the Hudson River, Croton Point, NY. This makes the acorn connection all the more of an environmental problem.

6.1 The Ecosystem: Sustaining Life on Earth

We tend to associate life with individual organisms, for the obvious reason that it is individuals that are alive. But sustaining life on Earth requires more than individuals or even single populations or species. Life is sustained by the interactions of many organisms functioning together, in ecosystems, interacting through their physical and chemical environments. Sustained life on Earth, then, is a characteristic of ecosystems, not of inividual organisms or populations.[3] To understand important environmental issues, such as conserving endangered species, sustaining renewable resources, and minimizing the effects of toxic substances, we must understand certain basic principles about ecosystems.

Basic Characteristics of Ecosystems

Ecosystems have several fundamental characteristics.

- **Structure.** An ecosystem is made up of two major parts: nonliving and living. The nonliving part is the physical-chemical environment, including the local atmosphere, water, and mineral soil (on land) or other substrate (in water). The living part, called the **ecological community**, is the set of species interacting within the ecosystem.
- **Processes.** Two basic kinds of processes must occur in an ecosystem: a cycling of chemical elements and a flow of energy.
- **Change.** An ecosystem changes over time and can undergo development through a process called **succession**, which is discussed in Chapter 9.

The processes that occur in an ecosystem are necessary for the life of the ecological community, but no member of that community can carry out these processes alone. That is why we have said that sustained life on Earth, rather than individuals or populations, is a characteristic of ecosystems.

We can see this by looking at cycling in an ecosystem. As mentioned in Chapter 5, chemical cycling is complex. Each chemical element required for growth and reproduction must be made available to each organism at the right time, in the right amount, and in the right ratio relative to other elements. These chemical elements must also be recycled—converted to a reusable form. Wastes are converted into food, which is converted into wastes, which must be converted once again into food, with the cycling going on indefinitely, if the ecosystem is to remain viable.

For complete recycling of chemical elements to take place, several species must interact. In the presence of light, green plants, algae, and photosynthetic bacteria produce sugar from carbon dioxide and water. From sugar and inorganic compounds, they make many other organic compounds, including proteins and woody tissue. But no green plant can decompose woody tissue back to its original inorganic compounds. Other forms of life—primarily bacteria and fungi—can decompose organic matter; but they cannot produce their own food. Instead, they obtain energy and chemical nutrition from the dead tissues on which they feed.

Theoretically, at its simplest, the ecological community in an ecosystem consists of at least one species that produces its own food from inorganic compounds in its environment and another species that decomposes the wastes of the first species, plus a fluid medium (air, water, or both).

We turn next to a more detailed discussion of ecological communities—in particular, food chains in ecological communities.

Ecological Communities and Food Chains

We have identified an ecological community as the set of interacting species that makes up the living part of an ecosystem. In practice, the term *ecological community* is defined by ecologists in two ways. One method is to define the community as a set of *interacting* species found in the same place and functioning together to make possible the persistence of life. That is essentially the definition we used earlier. A problem with this definition is that it is often difficult in practice to know the entire set of interacting species. Ecologists therefore may use a pragmatic or operational definition, in which the community consists of all the species found in an area, whether or not they are known to interact. Animals in different cages in a zoo could be called a community according to this definition.

One way in which individuals in a community interact is by feeding on one another. Energy, chemical elements, and some compounds are transferred from creature to creature along **food chains**, the linkage of who feeds on whom. In more complex cases, these linkages are called **food webs**.

Ecologists group the organisms in a food web into trophic levels. A **trophic level** consists of all those organisms in a food web that are the same number of feeding levels away from the original source of energy. The original source of energy in most ecosystems is the sun. In other cases, it is the energy in certain inorganic compounds.

Green plants, algae, and certain bacteria produce sugars through the process of photosynthesis, using only the energy of the sun and carbon dioxide (CO_2) from the air, so they are grouped into the first trophic level. Organisms in the first trophic level, which make their own food and inorganic chemicals and a source of energy, are called **autotrophs**. *Herbivores*, organisms that feed on plants, algae, or photosynthetic bacteria, are members of the second trophic level. *Carnivores*, or meat-eaters, that feed directly on herbivores make up the third trophic level. Carnivores that feed on third-level carnivores are in the fourth trophic level, and so on.

Food chains and food webs are often quite complicated and thus difficult to analyze. A detailed look at one of the simplest food chains is provided in A Closer Look 6.1. Next, we look briefly at several more-complicated food chains.

A Terrestrial Food Chain

An example of terrestrial food chains and trophic levels is shown in Figure 6.5. This north temperate woodland food web existed in North America before European settlement and includes human beings. The first trophic level, autotrophs, includes grasses, herbs, and trees. The second trophic level, herbivores, includes mice, an insect

Hot Spring Ecosystems in Yellowstone National Park

Perhaps the simplest natural ecosystem is a hot spring such as those found in geyser basins in Yellowstone National Park, Wyoming.[4] Few organisms can live in these hot springs because the environment is so severe. Water in parts of the springs is close to the boiling point. In addition, some springs are very acidic and others are very alkaline; either extreme makes a harsh environment. Some of the organisms that can live in hot springs are brightly colored and give these pools the striking appearance for which they are famous (Figure 6.3).

Typically, the springs have a wide range of water temperatures, from almost boiling near the source to much cooler near the edges, especially in the winter, when there may be snow on the ground next to a spring. In a typical alkaline hot spring, the hottest waters, between 70° and 80°C (158°–176°F), are colored bright yellow-green by photosynthetic blue-green bacteria, one of the few kinds of photosynthetic organisms that can survive in hot springs. In slightly cooler waters, 50° to 60°C (122–140°F), thick mats of bacteria and algae accumulate, some becoming 5 cm thick. These mats are formed by long strings of photosynthetic bacteria and algae. As the flowing springwater passes over the mats, the long strings of cells trap and hold single-cell algae.

First trophic level. Photosynthetic bacteria and algae make up the spring's first trophic level, which is composed of autotrophs—organisms that make their own food from inorganic chemicals and a source of energy. In the hot springs, as in most communities, the source of energy is sunlight (Figure 6.4).

Second trophic level. Some flies, called ephydrid flies, live in the cooler areas of the springs. One species, *Ephydra bruesi*, lays bright orange-pink egg masses on stones and twigs that project above the mat. The fly larvae feed on the bacteria and algae. Since these flies eat only plants, they are

Figure 6.3 ■ One of the many hot springs in Yellowstone National Park. The bright greenish color comes from photosynthetic bacteria, one of the few kinds of organisms that can survive in the hot temperatures and chemical conditions of the springs.

herbivores. These form the second trophic level.

Third trophic level. Another fly, called the dolichopodid fly, is carnivorous and feeds on the eggs and larvae of the herbivorous flies. Dragonflies, wasps, spiders, tiger beetles, and one species of bird, the killdeer, also feed on the herbivorous flies. The herbivorous ephydrid flies also have a parasite, a red mite, that feeds on fly eggs and travels by attaching

itself to the bodies of the adult flies. Another parasite, a small wasp, lays its eggs within the fly larvae. All these form the third trophic level.

Fourth trophic level. Wastes and dead organisms of all trophic levels are fed on by decomposers, which in the hot springs are primarily bacteria. These form the fourth trophic level.

The entire hot springs community of organisms—photosynthetic bacteria and

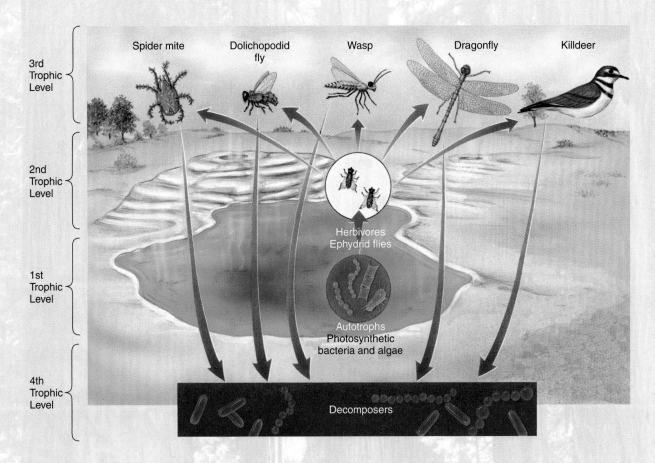

Figure 6.4 ■ Food web of a Yellowstone National Park hot spring.

algae, herbivorous flies, carnivores, and decomposers—is maintained by two factors: (1) sunlight, which provides an input of usable energy for the organisms; and (2) a constant flow of hot water, which provides a continual new supply of chemical elements required for life and a habitat in which the bacteria and algae can persist.

Even though this is one of the simplest ecological communities in terms of the numbers of species, a fair number of species are found. About 20 species in all are important in this ecosystem. The eco-

logical community they form has been sustained for long periods in these unusual habitats.

Another interesting aspect of the hot springs ecosystem is species dominance. Dominant species are those that are most abundant or otherwise most important in the community. (We discuss this in connection with diversity in Chapter 7.) As noted earlier, in the hot springs community, the species of photosynthetic bacteria or algae that is dominant changes with the temperature; one species dominates the hotter springs and hottest regions

within a spring, and another species dominates cooler waters.

Because the algae are so brightly colored, this spatial patterning in dominance is readily apparent to visitors. It was striking to one of the earliest explorers of Yellowstone, a trapper named Osborne Russell, who visited the springs in the 1830s and 1840s. He wrote that one boiling spring about 100 m (328 ft) across had three distinct colors: "From the west side for one-third of the diameter it was white, in the middle it was pale red, and the remaining third on the east light sky blue."[5]

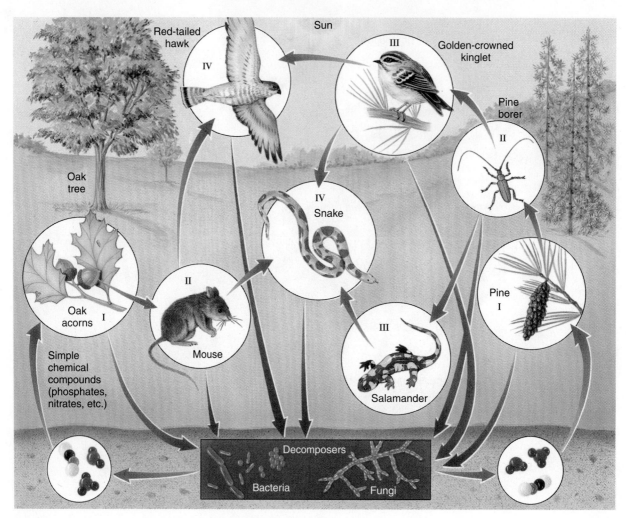

Figure 6.5 ■ A typical terrestrial food web. Roman numerals identify trophic levels.

called the pine borer, and other animals (such as deer) not shown here. The third trophic level, carnivores, includes foxes and wolves, hawks and other predatory birds, spiders, and predatory insects. People are *omnivores* (eaters of both plants and animals) and feed on several trophic levels. In Figure 6.5, people would be included in the fourth trophic level, the highest level in which they would take part. **Decomposers**, such as bacteria and fungi, feed on wastes and dead organisms of all trophic levels. Decomposers are also shown here on the fourth level.

An Oceanic Food Chain

In oceans, food webs involve more species and tend to have more trophic levels than they do in the hot springs described in A Closer Look 6.1 or the terrestrial ecosystem just considered. In a typical oceanic ecosystem (Figure 6.6), microscopic single-cell planktonic algae and planktonic photosynthetic bacteria are in the first trophic level. Small invertebrates called *zooplankton* and some fish feed on the algae and photosynthetic bacteria, forming the second trophic level. Other fish and invertebrates feed on these herbivores and form the third trophic level. The great

baleen whales filter seawater for food, feeding primarily on small herbivorous zooplankton (mostly crustaceans), and thus the baleen whales are also in the third level. Some fish and marine mammals, such as killer whales, feed on the predatory fish and form higher trophic levels.

The Food Web of the Harp Seal

In the abstract, a diagram of a food web and its trophic levels seems simple and neat; but in reality, food webs are complex, because most creatures feed on several trophic levels. For example, consider the food web of the harp seal (Figure 6.7). The harp seal is shown at the fifth level.[6] It feeds on flatfish (fourth level), which feed on sand lances (third level), which feed on euphausiids (second level), which feed on phytoplankton (level 1). But the harp seal actually feeds at several trophic levels, from the second through the fourth, and it feeds on predators of some of its prey and thus is a competitor with some of its own food.[6] A species that feeds on several trophic levels typically is classified as belonging to the trophic level above the highest level from which it feeds. Thus, we place the harp seal on the fifth trophic level.

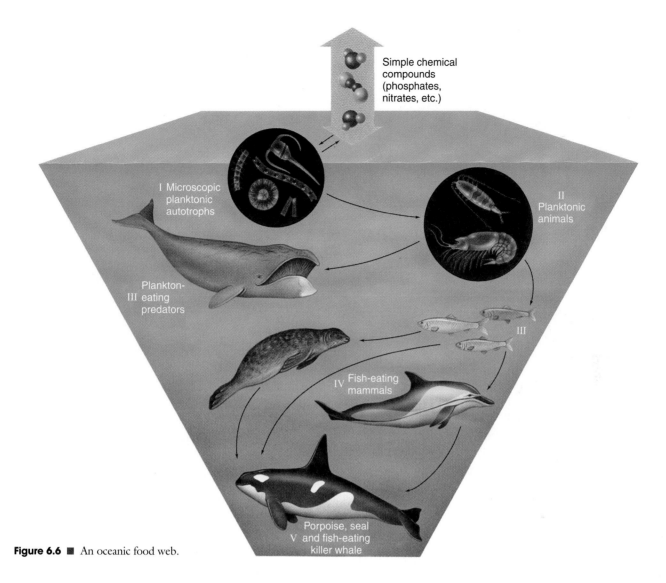

Simple chemical
compounds
(phosphates,
nitrates, etc.)

I Microscopic
planktonic
autotrophs

II Planktonic
animals

Plankton-
III eating
predators

III

IV Fish-eating
mammals

Porpoise, seal
V and fish-eating
killer whale

Figure 6.6 ■ An oceanic food web.

6.2 The Community Effect

Species can interact directly through food chains, as we have just seen. They also interact directly through symbiosis and competition, discussed in the next chapter. But a species can also affect other species *indirectly*, by affecting a third, a fourth, or many other species that, in turn, affect the second species. In addition, a species can affect the nonliving environment, which then affects a group of species in the community. Changes in that group affect another group. Such indirect and more complicated interactions are referred to as **community-level interactions**.

Interactions at the community level are illustrated by the sea otters of the Pacific Ocean. In fact, the community-level interactions of the sea otter are at the heart of some arguments in favor of conservation of this species.

Sea otters feed on shellfish, including sea urchins and abalone (Figure 6.8*a*). Sea otters originally occurred throughout a large area of the Pacific Ocean coasts, from northern Japan northeastward along the Russian and

Alaskan coasts, and southward along the coast of North America to Morro Hermoso in Baja California, to Mexico.[7] The otters were brought almost to extinction by commercial hunting for their fur during the eighteenth and nineteenth centuries; they have one of the finest furs in the world. By the end of the nineteenth century, there were too few otters left to sustain commercial exploitation, and there was concern that the species would become extinct.

A small population survived and has increased since then, so that today there are approximately 111,500 sea otters. According to the Marine Mammal Center, approximately 2,000 sea otters live along the coast of California, a few hundred in Washington and British Columbia, and 100,000 along the Aleutian Islands of Alaska. The sea otter population in Russian waters is about 9,000.[8] Legal protection of the sea otter by the U.S. government began in 1911 and continues under the U.S. Marine Mammal Protection Act of 1972 and the Endangered Species Act of 1973.

The sea otter has been a focus of controversy and research. On the one hand, fishermen argue that the sea ot-

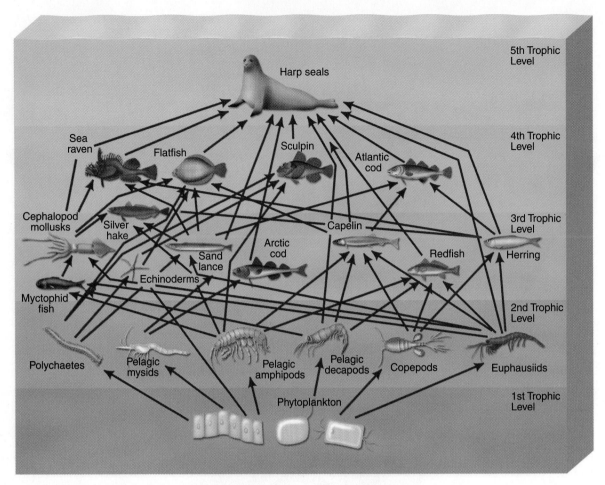

Figure 6.7 ■ Food web of the harp seal showing how complex a real food web can be.

ter population has recovered—in fact, recovered too much. In this view, there are too many sea otters today, and they interfere with commercial fishing because they take large amounts of abalone.[9] On the other hand, conservationists argue that sea otters have an important community-level role, necessary for the persistence of many oceanic species. They say that there are still too few sea otters for this role to be maintained at a satisfactory level.

What is this important role? It is made up of many effects on the community that result from sea otters' feeding on sea urchins. Sea urchins are a preferred food of sea otters. Sea urchins, in turn, feed on kelp, large brown algae that form undersea "forests" and provide important habitat for many species. Sea urchins graze along the bottoms of the beds, feeding on the base of kelp, called *holdfasts*, which attach the kelp to the bottom. When holdfasts are eaten through, the kelp floats free and dies.

Where sea otters are abundant, as on Amchitka Island in the Aleutian Islands, kelp beds are abundant and there are few sea urchins (Figure 6.8*b*). At nearby Shemya Island, which lacks sea otters, sea urchins are abundant and there is little kelp (Figure 6.8*c*).[7] Experimental removal of sea urchins has led to an increase in kelp.[10]

Otters, then, affect the abundance of kelp, but the influence is indirect. Sea otters neither feed on kelp nor protect individual kelp plants from attack by sea urchins. Sea otters reduce the number of sea urchins. With fewer sea urchins, less kelp is destroyed. With more kelp, there is more habitat for many other species; so sea otters indirectly increase the diversity of species.[9,11] Thus, sea otters have a community-level effect. This example shows that such effects can occur through food chains and can alter the distribution and abundance of individual species.

A species such as the sea otter that has a large effect on its community or ecosystem is called a **keystone species**, or a key species.[12] Its removal or a change in its role within the ecosystem changes the basic nature of the community.

Community-level effects demonstrate the reality behind the concept of an ecological community; they show us that certain processes can take place only because of a set of species interacting together. These effects also suggest that an ecological community is more than the sum of its parts—a perception called the *holistic view*. (Refer back to the discussion of environmental unity in Chapter 3.)

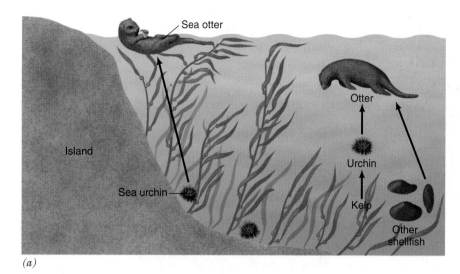

(a)

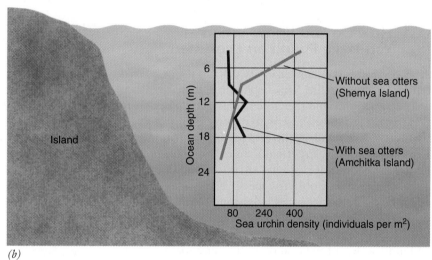

(b)

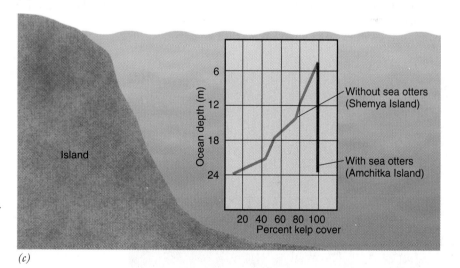

(c)

Figure 6.8 ■ The effect of sea otters on kelp. (*a*) Sea otters feed on shellfish, including sea urchins. The sea urchins feed on kelp. When there are sea otters, there are few urchins (*b*) but abundant kelp (*c*).

In spite of their importance, community-level interactions are often difficult to recognize. One difficulty is knowing when and how species interact. Community-level interactions are not always as clear as those involving the sea otter. Even there, considerable scientific research was required to under-stand the interactions. Adding to the complexity, the set of species that make up an ecological community is not fixed completely but varies within the same kind of ecosystem from time to time and place to place. This brings us to the question of how to identify ecosystems.

Figure 6.9 ■ Sometimes the transition from one ecosystem to another is sharp and distinct, as in the transition from lake to forest at Lake Moraine in Banff National Park, Alberta, Canada.

6.3 How Do You Know When You Have Found an Ecosystem?

An ecosystem is the minimal entity that has the properties required to sustain life. This implies that an ecosystem is real and important and therefore that we should be able to find one easily. However, ecosystems vary greatly in structural complexity and in the clarity of their boundaries. Ecosystems differ in size, from the smallest puddle of water to a large forest. Ecosystems and their communities differ in composition, from a few species in the small space of a hot spring to many species interacting over a large area of the ocean. Furthermore, ecosystems differ in the kinds and relative proportions of their nonbiological constituents and in their degree of variation in time and space.

Sometimes the borders of the ecosystem are well defined, such as the border between a lake and the surrounding countryside (Figure 6.9). But sometimes the transition from one ecosystem to another is gradual, as in the transition from desert to forest on the slopes of the San Francisco Mountains in Arizona and in the subtle gradations from grasslands to savannas in East Africa and from boreal forest to tundra in the far north, where the trees thin out gradually.

A commonly used practical delineation of the boundary of an ecosystem on land is the **watershed**. Within a watershed, any drop of rain that reaches the ground flows out in the same stream. Topography (the lay of the land) determines the watershed. When a watershed is used to define the boundaries of an ecosystem, the ecosystem is unified in terms of chemical cycling. Some classic experimental studies of ecosystems have been conducted on forested watersheds in U.S. Forest Service experimental areas, including the Hubbard Brook experimental forest in New Hampshire (Figure 6.10) and the Andrews experimental forest in Oregon.

What all ecosystems have in common is not a particular physical size or shape but the processes we have mentioned: the flow of energy and the cycling of chemical elements. Ecological communities change over time, and it is the interactions among the species—a dynamic set of processes—that are the key to the community concept.

6.4 Ecosystem Management

Ecosystems can be natural or artificial or a combination of both. An artificial pond that is a part of a waste-treatment plant is an example of an artificial ecosystem.

Figure 6.10 ■ The V-shaped logged area in this picture is the famous Hubbard Brook ecosystem study. Here, a watershed defines the ecosystem, and the V shape is an entire watershed cut as part of the experiment.

How Are the Borders of an Ecosystem Defined?

The borders between ecosystems may be well defined or gradual. Those considered well defined include freshwater streams. Such ecosystems are often studied separately from surrounding ecosystems by researchers with different training and using different methods. Research on streams in southeast Alaska in which salmon spawn has raised questions about the practice of studying aquatic and terrestrial ecosystems separately.

Salmon are anadromous fish—fish that come from the ocean to spawn in freshwater streams. In southeast Alaska, enormous numbers of salmon spawn in over 5,000 streams. Although salmon are born in freshwater, they migrate to the ocean, where most of their growth occurs. After they return to their home streams, they spawn and die. In one sense, therefore, salmon are a means of transporting resources from the ocean to freshwater. Because of their large numbers, salmon have the potential to make significant contributions to organic and mineral content of streams.

Salmon have a high lipid content compared with many other fish and are thus a good energy source for the animals that prey on them. In addition, their decay adds nitrogen, phosphorus, carbon, and other inorganic elements to freshwater. In one lake in western Alaska, for example, 24 million fish add 170 tons of phosphorus to the lake each year—an amount equal to or greater than recommended rates for applying fertilizers to trees. When the fish die, their carcasses decay and provide nourishment for algae, fungi, and bacteria. Invertebrates feed on these and on decaying bits of fish. Other fish feed on the invertebrates. Finally, bears and other carnivores eat salmon, both live and dead, during their upstream migration. In that way, nutrients derived from salmon pass into the soil and vegetation surrounding the streams.

Spawning fish have higher proportions of heavy isotopes of nitrogen and carbon (^{15}N and ^{13}C). These can be used to trace the relative contributions of anadromous fish to the nitrogen and carbon content of organisms in the food web. One such study showed that spawning salmon contributed 10.9% of the nitrogen found in invertebrate predators and 17.5% in the foliage of riparian plants. While it is not surprising to find aquatic invertebrates, which feed on salmon eggs and juveniles, with large amounts of nitrogen derived from salmon, researchers were surprised at the high levels in streamside vegetation. When terrestrial mammals and birds feed on salmon, their feces and any uneaten salmon carcasses decay and add nutrients to the soil, where they can be taken up through the roots of plants. In southeast Alaska, over 40 species of mammals and birds feed on salmon. Salmon migrations attract large numbers of predators to streams and lakes. Salmon and other anadromous fish thus appear to link the ocean, freshwater, and land to an extent that is only beginning to be appreciated.

Critical Thinking Questions

1. Given the intricate connections between the aquatic and terrestrial ecosystems along salmon streams, how would you define the boundaries of the ecosystems?

2. When more adult salmon reach the spawning grounds than are needed to maintain the population, some are considered excess. How might the research described here affect that view?

3. Some biologists have called salmon a keystone species. Given what you know about keystone species, how would you argue for or against this designation?

4. In recent years, the numbers of anadromous fish along the Pacific Coast of North America have declined precipitously because of overfishing and habitat destruction. What effects would you predict this might have on the ecology of freshwater streams and their adjoining land areas?

5. What types of management decisions about fish, wildlife, and forests would follow from recognizing the connection between aquatic and terrestrial ecosystems?

Ecosystems can also be managed, and the management can include a large range of actions. Agriculture can be thought of as partial management of certain kinds of ecosystems (see Chapters 11 and 12), as can forests managed for timber production (see Chapter 13). Wildlife preserves are examples of partially managed ecosystems (see Chapters 13 and 14).

Sometimes, when we manage or domesticate individuals or populations, we separate them from their ecosystems. We also do this to ourselves (see Chapter 3). When we do this, we must replace the ecosystem functions of energy flow and chemical cycling with our own actions. This is what happens in a zoo, where we must provide food and remove the wastes for individuals separated from their natural environments.

The ecosystem concept, then, lies at the heart of the management of natural resources. When we try to conserve species or manage natural resources so that they are sustainable, we must focus on their ecosystem and make sure that it continues to function. If it doesn't, we must replace or supplement ecosystem functions with our own actions. Ecosystem management, however, involves more than compensating for changes we make in ecosystems. It means managing and conserving life on Earth by considering chemical cycling, energy flow, community-level interactions, and the natural changes that take place within ecosystems.

Summary

■ An ecosystem is the simplest entity that can sustain life. At its most basic, an ecosystem consists of several species and a fluid medium (air, water, or both). The ecosystem must sustain two processes—the cycling of chemical elements and the flow of energy.

■ The living part of an ecosystem is the ecological community, a set of species connected by food webs and trophic levels. A food web or chain describes who feeds on whom. A trophic level consists of all the organisms that are the same number of feeding steps from the initial source of energy.

■ Community-level effects result from indirect interactions among species, such as those that occur when sea otters influence the abundance of sea urchins.

■ Ecosystems are real and important, but it is often difficult to define the limits of a system or to pinpoint all the interactions that take place.

■ Ecosystem management is considered key to the successful conservation of life on Earth.

REEXAMINING THEMES AND ISSUES

Human Population

The human population depends on many ecosystems that are widely dispersed around the globe. Our modern technology may appear to make us independent of these natural systems. In fact, though, the more connections we establish through modern transportation and communication, the more kinds of ecosystems we depend on. Therefore, the ecosystem concept is one of the most important we will learn about in this book.

Sustainability

The ecosystem concept is at the heart of managing for sustainability. When we try to conserve species or manage living resources so that they are sustainable, we must focus on their ecosystem and make sure that it continues to function.

Global Perspective

Our planet has sustained life for approximately 3.5 billion years. To understand how Earth as a whole has sustained life for such a long time, we must understand the ecosystem concept, because the environment at a global level must meet the same basic requirements as any local ecosystem.

Urban World

Cities are embedded in larger ecosystems. But like any life-supporting system, a city must meet basic ecosystem needs. This is accomplished through connections between cities and surrounding environments. Together, these function as ecosystems or sets of ecosystems. To understand how we can create pleasant and sustainable cities, we must understand the ecosystem concept.

People and Nature

The feelings we get when we hike through a park or near a beautiful lake are as much a response to an ecosystem as to individual species. This illustrates the deep connection between people and ecosystems. Also, many effects we have on nature are at the level of an ecosystem, not just on an individual species.

Science and Values

The introductory case study concerning acorns, mice, deer, and Lyme disease illustrates the interactions between values and scientific knowledge about ecosystems. Science can tell us how organisms like deer and mice interact. This knowledge confronts us with choices. Do we want to have many deer and mice and to live with Lyme disease? Do we want to invest in ecological and medical resources to find a better way to control that disease? The choice we make depends on our values.

Key Terms

autotrophs **105**
community-level
  interactions **109**

decomposers **108**
ecological community **105**
food chains **105**

food webs **105**
keystone species **110**
succession **105**

trophic level **105**
watershed **112**

Study Questions

1. What is the difference between an ecosystem and an ecological community?

2. In what ways would an increase in the number of sea otters and a change in their geographic distribution benefit fishermen? In what ways would these changes be a problem for fishermen?

3. Based on the discussion in this chapter, would you expect a highly polluted ecosystem to have many species or few species?

4. Is our species a keystone species? Explain.

5. Which of the following are ecosystems? Which are ecological communities? Which are neither?
   a. Chicago
   b. A 1,000-hectares farm in Illinois
   c. A sewage-treatment plant
   d. The Illinois River
   e. Lake Michigan

Further Reading

Bormann, F. H., and G. E. Likens. 1979. *Pattern and Process in a Forested Ecosystem*. New York: Springer-Verlag. A synthetic view of the northern hardwood ecosystem, including its structure, function, development, and relationship to disturbance.

Molles, M. C. *Ecology: Concepts and Applications*. New York: McGraw-Hill. Currently, this is one of the most popular introductory ecology textbooks.

Odum, Eugene, G. W. Barrett 2004. *Fundamentals of Ecology*. Duxbury, Brooks/Cole. Odum's original textbook was a classic, especially in providing one of the first serious introductions to ecosystem ecology. This is the latest update of the late authority's work, done with his protégé.

Rockwood, L.L. 2006. *Introduction to Population Ecology*. Oxford, England: Blackwell Publishing Professional. This is a new, up-to-date introduction to a part of ecology crucial to management of natural resources.

CHAPTER 7

Biological Diversity

Emperor penguins survive the Antarctic winter with amazing adaptions. How they survive these winters fascinated many people who made the movie "March of the Penguins", one of the most popular documentary films.

Learning Objectives

People have long wondered how the amazing diversity of living things on Earth came to be. This diversity has developed through biological evolution and is affected by interactions among species and by the environment. After reading this chapter, you should understand:

■ How mutation, natural selection, migration, and genetic drift lead to evolution of new species.

■ Why people value biological diversity.

■ How people can affect biological diversity.

■ How biological diversity may affect biological production, energy flow, chemical cycling, and other ecosystem processes.

■ What major environmental problems are associated with biological diversity.

■ Why so many species have been able to evolve and persist.

■ The concepts of the ecological niche and habitat.

Grizzly Bears and Emperor Penguins

As long as people have written, they have written about the amazing diversty of life. Two movies released in 2005, *March of the Penguins* and *Grizzly Man,* reinforce this eternal fascination people have with the diversity of life on Earth, the wonderous adaptation of species to their environments, and how life's diversity makes us feel connected to all that is around us. This great diversity of life has prompted three recurring questions: How did it come about? Is it important—that is, does the continuation of any life on Earth depend on the great diversity of life? And what is necessary to sustain that diversity?

Take the emperor penguins as an example of a species' seemingly incredible adaptation to a challenging environment (Figure 7.1). They breed in the middle of Antarctica, in one of the coldest and harshest environments on Earth. The males and females take turns keeping the eggs warm while their mate travels up to 80 km (50 miles) to the ocean to feed and then return. To protect themselves from the worst cold of the storms, the penguins huddle in large groups and take turns moving from the cold outer edge of the group to the warmer, protected middle.[1]

Or think about grizzly bears (Figure 7.1). They are large animals, the males weigh up to 360 kg (800 lb), and are among the most dangerous bears. They were the only animal that, unprovoked, attacked the members of the Lewis and Clark expedition during their travels in 1805 in what is now Montana. Grizzly bears are omnivorous and are famous for catching salmon as the salmon are attempting to swim upriver to spawn. The salmon have their own incredibly complicated life cycle, and the grizzlies hitch a ride, so to speak, on them, taking advantage of the fact that the salmon have fed and fattened up in the ocean, and then carry—in their body muscles and fat—chemically rich organic compounds, which provide a highly nutritious food for the bears, who otherwise live in a comparatively chemically barren environment. How did each of these species come about, and how did their interactions develop?

How interdependent are species? Would the grizzly become extinct *without* salmon? Could prey such as salmon *require* predators in some sense, perhaps regulating their abundance? (That predators might regulate the abundance

Figure 7.1 ■ People are generally fascinated by the North American grizzly bear, one of the continent's most dangerous animals, and one of the few that made unprovoked attacks on Lewis and Clark during their expedition west in 1804 to 1806

of a prey is an ancient theory and one that recurs in conservation science.) These questions are not only ancient, but they are today the subject of much scientific interest. In this chapter we will explore the basic principles of biodiversity.

7.1 What Is Biological Diversity?

Biological diversity refers to the variety of life-forms, commonly expressed as the number of species in an area, or the number of genetic types in an area. However, discussions about biological diversity are complicated by the fact that people mean various things when they talk about it. They may mean conservation of a single rare species, of a variety of habitats, of the number of genetic varieties, of the number of species, or of the relative abundance of species. These concepts are interrelated, but each has a distinct meaning.

Newspapers and television frequently cover the problem of disappearing species around the world and the need to conserve these species. Before we can intelligently discuss the issues involved in conserving the diversity of life, we must understand how this diversity came to be. This chapter first addresses the principles of biological evolution are then turns to biological diversity itself: its various meanings, how interactions among species increase or decrease diversity, and how the environment affects diversity.

7.2 Biological Evolution

The big question about biological diversity is: How did it all come about? This is a question people have asked as long as they have written. Before modern science, the diversity of life and the adaptations of living things to their environment seemed too amazing to have come about by chance. The only possible explanation seemed to be that this diversity was created by God (or gods). People were fascinated by this diversity, and were familiar with it, as illustrated by the famous medieval tapestry, The Hunting of the Unicorn. In the example shown here (Figure 7.2), a great variety of plants and animals (including frogs and insects) are represented accurately and with great detail. Except for the imaginary unicorn in the center, the tapestry's drawing are familiar to naturalists today. The great Roman philosopher and writer Cicero put it succinctly: "Who cannot wonder at this harmony of things, at this symphony of nature which seems to will the well-being of the world?" He concluded that "everything in the world is marvelously ordered by divine providence and wisdom for the safety and protection of us all".[2]

With the rise of modern science, however, other explanations became possible. In the nineteenth century, Charles Darwin found an explanation that became known as **biological evolution.** Biological evolution refers to the change in inherited characteristics of a population from generation to generation. It can result in new species—populations that can no longer reproduce with members of the original species. Along with self-reproduction, biological evolution is one of the features that distinguish life from everything else in the universe.

According to the theory of biological evolution, new species arise as a result of competition for resources and the difference among individuals in their adaptations to envi-

Figure 7.2 ■ People have long loved the diversity of life. Here a Dutch Medieval Tapestry, *The Lady with the Unicorn* (late 15th century), celebrates the great diversity of life. Except for the mythological unicorn, all the plants and animals shown are real and drawn with great accuracy.

ronmental conditions. Since the environment continually changes, which individuals are best adapted changes too. As Charles Darwin wrote, "can it be doubted, from the struggle each individual has to obtain subsistence, that any minute variation in structure, habits, or instincts, adapting that individual better to the new [environmental] conditions, would tell upon its vigor and health? In the struggle it would have a better *chance* of surviving; and those of its offspring that inherited the variation, be it ever so slight, would also have a better *chance.*" Sounds plausible, but how does this evolution occur? Four processes lead to evolution: mutation, natural selection, migration, and genetic drift.

Mutation

Genes, contained in chromosomes within cells, are inherited—passed from one generation to the next. A *genotype* is the genetic makeup of an individual or group. Genes are made up of a complex chemical compound called deoxyribonucleic acid (DNA). DNA in turn is made up of chemical building blocks that form a code, a

(a)

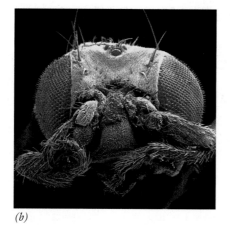
(b)

(c)

Figure 7.3 ■ A normal fruit fly (*a*), and a fruit fly with an antennae mutation (*b*). Trandescantia is a small flowering plant used in the study of effects of mutagens (*c*). The color of stamen hairs in the flower (pink versus clear) is the result of a single gene, and changes when that gene is mutated by radiation or certain chemicals, such as ethylene chloride.

kind of alphabet of information. The DNA alphabet consists of four letters (specific nitrogen-containing compounds, called bases), which are combined in pairs: (A) adenine, (C) cytosine, (G), guanine, and (T) thymine. How these letters are combined in long strands determines the "message" interpreted by a cell to produce specific compounds.

Sets of the four base pairs form a **gene**, which is a single piece of genetic information. The number of base pairs that make up a gene varies. To make matters more complex, some base pairs found in DNA are nonfunctional—they are not active and do not determine any chemicals that are produced by the cell. Furthermore, some genes affect the activity of others, turning those other genes on or off. And creatures such as ourselves have genes that limit the number of times a cell can divide—and therefore determine maximum longevity.

When a cell divides, the DNA is reproduced and each new cell gets a copy. Sometimes an error in reproduction changes the DNA and therefore changes inherited characteristics. Sometimes an external agent comes in contact with DNA and alters it. Radiation, such as X rays and gamma rays, can break the DNA apart or change its chemical structure. Certain chemicals also can change DNA. So can viruses. When DNA changes in any of these ways, then it is said to have undergone **mutation**.

In some cases, a cell or offspring with a mutation cannot survive. In other cases, the mutation simply adds variability to the inherited characteristics. But in still other cases, individuals with mutations are so different from their parents that they cannot reproduce with normal offspring of their species, so a new species has been created (Figure 7.3).

Natural Selection

When there is variation within a species, some individuals may be better suited to the environment than others. (Change is not always for the better. Mutation can result in a new species whether or not that species is better adapted than its parental species to the environment.) Organisms whose biological characteristics make them better able to survive and reproduce in their environment leave more offspring than others. Their descendants form a larger proportion of the next generation. In this way, these individuals are more "fit" for the environment; this process of increasing the proportion of offspring is called **natural selection**. Which inherited characteristics lead to more offspring depends on the specific characteristics of an environment, and as the environment changes over time, the characteristics' "fit" will also change. In summary, natural selection has four primary characteristics:

1. Inheritance of traits from one generation to the next and some variation in these traits—that is, genetic variability.
2. Environmental variability.
3. Differential reproduction that varies with the environment.
4. Influence of the environment on survival and reproduction.

Natural selection is illustrated in A Closer Look 7.1, which describes how the mosquitoes that carry malaria develop resistance to DDT and how the microorganism that causes malaria develops a resistance to quinine, a treatment for the disease.

When natural selection takes place over a long time, a number of characteristics can change. The accumulation of these changes may be so great that the present generation can no longer reproduce with individuals that have the original DNA structure, resulting in a new species. A **species** is a group of individuals that can (and at least occasionally do) reproduce with each other.

Geographic Isolation and Migration

Sometimes two populations of the same species become geographically isolated from each other for a long time. During that time, the two populations may change so

Natural Selection: Mosquitoes and the Malaria Parasite

Malaria poses a great threat to 2.4 billion people—over one-third of the world's population—living in more than 90 countries, most of them located in the tropics. In the United States, Florida has recently experienced a small but serious malaria outbreak. Worldwide, an estimated 300 to 400 million people are infected each year, 1.1 million of whom die.[3] In Africa alone, more than 3,000 children die daily from this disease.[4]

Once thought to be caused by filth or bad air (hence the name *malaria*, from the Latin for "bad air"), malaria is actually caused by parasitic microbes (four species of the protozoa *Plasmodium*). These microbes affect and are carried by *Anopheles* mosquitoes, which then transfer the protozoa to people.

One solution to the malaria problem, then, would be the eradication of *Anopheles* mosquitoes. By the end of World War II, scientists had discovered that the pesticide DDT was extremely effective against *Anopheles* mosquitoes. They had also found chloroquine highly effective in killing *Plasmodium* parasites. (Chloroquine is an artificial derivative of quinine, a chemical from the bark of the quinine tree that was an early treatment for malaria.)

In 1957 the World Health Organization (WHO) began a $6 billion campaign to rid the world of malaria using a combination of DDT and chloroquine. At first, the strategy seemed successful. By the mid-1960s, malaria was nearly gone or had been eliminated from 80% of the target areas.

However, success was short-lived. The mosquitoes began to develop a resistance to DDT, and the protozoa became resistant to chloroquine. In many tropical areas, the incidence of malaria worsened. For example, as a result of the WHO program, the number of cases in Sri Lanka had dropped from 1 million to only 17 by 1963. But by 1975, 600,000 cases had been reported, and the actual number is believed to be four times higher. And now there are 500 million cases of malaria resulting in 1 million deaths a year. Resistance among the mosquitoes to DDT became widespread, and resistance of the protozoa to chloroquine was found in 80% of the 92 countries where malaria was a major killer.[3,5]

The mosquitoes and the protozoa developed this resistance through natural selection. When they were exposed to DDT and chloroquine, the susceptible individuals died. The most resistant organisms survived and passed their resistant genes to their offspring. Since the susceptible individuals died, they left few or no offspring, and any offspring they left were susceptible.

Thus, a change in the environment—the human introduction of DDT and chloroquine—caused a particular genotype to become dominant in the populations. A practical lesson from this experience is that if we set out to eliminate a disease-causing species, we must attack it completely at the outset and destroy all the individuals before natural selection leads to resistance. But sometimes this may be an impossible task, in part because of the natural genetic variation in the target species.

Since the drug chloroquine is generally ineffective now, new drugs have been developed to prevent malaria. However, these second- and third-line drugs will eventually become unsuccessful, too, as a result of the same process of biological evolution by natural selection. This process is speeded up by the ability of the *Plasmodium* to rapidly mutate. In South Africa, for example, the protozoa became resistant to mefloquine immediately after the drug became available as a treatment.

An alternative is to develop a vaccine against the *Plasmodium* protozoa. Biotechnology has made it possible to map the structure of these malaria-causing organisms. Scientists are currently mapping the genetic structure of *P. falciparum*, the most deadly of the protozoa, and expect to finish within several years. With this information, they expect to create a vaccine containing a variety of the species that is benign in human beings but that produces an immune reaction.[6]

In addition, scientists are mapping the genetic structure of *Anopheles gambiae*, the carrier mosquito. This project could provide insight about genes that could prevent development of the malaria parasite within the mosquito. In addition, it could identify genes associated with insecticide resistance and provide clues to developing a new pesticide.[6]

The development of resistance to DDT by mosquitoes and to chloroquine by *Plasmodium* is an example of biological evolution in action today. With the aid of biotechnology, scientists are working to understand the specific chemical structure of the inheritance of characteristics.

much that they can no longer reproduce together even when they are brought back into contact. In this case, two new species have evolved from the original species. This can happen even if the genetic changes are not more fit but simply different enough to prevent reproduction.

Ironically, the loss of geographic isolation can also lead to a new species. This can happen when one population of a species migrates into a habitat previously occupied by another population of that species, thereby changing gene frequency in that habitat. For example, this change in gene frequency can result from the migration of seeds of flowering plants blown by wind or carried in the fur of mammals—if the seed lands in a new habitat, the environment may be different enough to favor genotypes not as favored by natural selection in the parents' habitat. Natural selection, in combination with geographic isolation and subsequent migration, can thus lead to new dominant genotypes and eventually to new species.

Migration has been an important evolutionary process over geologic time (a period long enough for geologic changes to take place). For example, during intervals between recent ice ages and at the end of the last ice age, Alaska and Siberia were connected by a land bridge that permitted the migration of plants and animals. When the land bridge was closed off by rising seawater, populations that had migrated to the New World were cut off, and new species evolved. The same thing happened much longer ago when marsupials reached Australia and were then cut off from other continents.

The word *evolution* in the term *biological evolution* has a special meaning. Outside biology, *evolution* is used broadly to mean the history and development of something. For example, book reviewers talk about the evolution of the plot of a novel, meaning how the story unfolds. Geologists talk about the evolution of Earth, which simply means Earth's history and the geologic changes that have occurred over that history. Within biology, however, the term has a more specialized meaning. Biological evolution is a one-way process. Once a species is extinct, it is gone forever. You can run a machine, such as a mechanical grandfather clock, forward and backward. But when a new species evolves, it cannot evolve backward into its parents.

Genetic Drift

Another process that can lead to evolution is **genetic drift**—changes in the frequency of a gene in a population due not to mutation, selection, or migration but simply to chance. Chance may determine which individuals become isolated in a small group from a larger population and thus which genetic characteristics are most common in that isolated population. The individuals may not be better adapted to the environment; in fact, they may be more poorly adapted or neutrally adapted. Genetic drift can occur in any small population and may also present problems when a small group is by chance isolated from the main population. For example, bighorn sheep live in the mountains of the southwestern deserts of the United States. In the summer, these sheep feed high up in the mountains, where it is cooler and wetter and there is more vegetation. Before high-density European settlement of the region, the sheep could move freely and sometimes migrated from one mountain to another by descending into the valleys and crossing them in the winter. In this way, large numbers of sheep interbred.

With the development of cattle ranches and other human activities, many populations of bighorn sheep could no longer migrate among the mountains by crossing the valleys. These sheep became isolated in very small groups—commonly, a dozen or so—so chance may play a large role in what inherited characteristics remain in the population.

Genetic drift can improve the adaptation of these populations. Suppose, for example, that you begin a study of mosquitoes that carry the malaria parasite (see A Closer Look 7.1). You want to study their resistance to DDT, so you go out and collect a sample of the mosquitoes. After you grow the population and obtain a number of generations of offspring, you begin your research by exposing them to DDT—and none of them die. By chance, the ones you collected were all resistant to DDT, as are their offspring. Your experiment ends up with a population completely resistant to DDT, unlike the wild population, which has a few individuals who are resistant.

But genetic drift is generally considered a serious problem when populations become very small, as happens with endangered species. It can be a problem for a rare or endangered species for two reasons: (1) Characteristics that are less adapted to existing environmental conditions may dominate, making survival of the species less likely; and (2) the small size of the population reduces genetic variability and hence the ability of the population to adapt to future changes in the environment.

Evolution as a Game

Biological evolution is so different from other processes that it is worthwhile to spend some extra time exploring this idea. There are no simple rules that species must follow to win or to stay in the game of life. Sometimes when we try to manage species, we assume that evolution will follow simple rules. But species play tricks on us; they adapt or fail to adapt over time in ways that we did not anticipate. Such unexpected outcomes result from our failure to understand fully how species have evolved in relation to their ecological situations. Nevertheless we continue to hope and plan as if life and its environment will follow simple rules. This is true even for the most recent work in genetic engineering.

Complexity is a feature of evolution. Species have evolved many intricate and amazing adaptations that have allowed them to persist. It is essential to realize that these adaptations have evolved not in isolation but in the context of relationships to other organisms and to the environment. The environment sets up a situation within which evolution, by natural selection, takes place. The great ecologist G. E. Hutchinson referred to this interaction in the title of one of his books, *The Ecological Theater and the Evolutionary Play.* Here, the ecological situation—the condition of the environment and other species—is the scenery and theater within which natural selection occurs, and natural selection results in a story of evolution played out in that theater—over the history of life on Earth.[7]

7.3 Basic Concepts of Biological Diversity

Now that we have explored biological evolution, we can turn to biological diversity. To develop workable policies for conserving biological diversity, we must be clear about the meaning of the term. Biological diversity involves the following concepts:

- *Genetic diversity:* the total number of genetic characteristics of a specific species, subspecies, or group of species. In terms of genetic engineering and our new understanding of DNA, this could mean the total base-pair sequences in DNA; the total number of genes, active or not; or the total number of active genes.
- *Habitat diversity:* the different kinds of habitats in a given unit area.
- *Species diversity,* which, in turn, has three qualities: *species richness*—the total number of species; *species evenness*—the relative abundance of species; and *species dominance*—the most abundant species.

To understand the differences among species richness, species evenness, and species dominance, imagine two ecological communities, each with 10 species and 100 individuals, as illustrated in Figure 7.4. In the first community (Figure 7.4a), 82 individuals belong to a single species, and the remaining nine species are represented by two individuals each. In the second community (Figure 7.4b), all the species are equally abundant; each therefore has 10 individuals. Which community is more diverse?

At first, one might think that the two communities have the same species diversity because they have the same number of species. However, if you walked through both communities, the second would appear more diverse. In the first community, most of the time you would see individuals only of the dominant species (in the case shown in Figure 7.4a, elephants); you probably would not see many of the other species at all. In the second community, even a casual visitor would see many of the species in a short time. The first community would appear to have relatively little diversity until it was subjected to careful study. You can test the probability of encountering a new species in either community by laying a ruler down in any direction on Figures 7.4a and 7.4b and counting the number of species that it touches.

As this example suggests, merely counting the number of species is not enough to describe biological diversity. Species diversity has to do with the relative chance of seeing species as much as it has to do with the actual number present. Ecologists refer to the total number of species in an area as *species richness*, the relative abundance of species as *species evenness*, and the most abundant species as *dominant.*

7.4 The Evolution of Life on Earth

For the mosquitoes and their malaria parasite (see A Closer Look 7.1), evolution occurred rapidly. In contrast, during most of Earth's history, evolution seems to have proceeded on average much more slowly. How do we know about the history of evolution? In part from the study of fossils. The earliest known fossils, 3.5 billion years old, are microorganisms that appear to be ancestral forms of bacteria and what some microbiologists now call Archaea.[8] For the next 2 billion years, only such microbial forms lived on Earth.

Amazingly, these organisms greatly changed the global environment, especially by altering the chemistry of the atmosphere. A major way this change came about was from photosynthesis, a capability that evolved during those 2 billion years. As with all photosynthetic organisms, these early ones removed carbon dioxide from the atmosphere and released large amounts of oxygen into it (illustrating our ongoing assertion that life has always changed the environment on a global scale). This led to a high concentration of oxygen in the atmosphere (familiar to us today), setting the ecological stage for the evolution of new forms of life. That free oxygen allowed the evolution of respiration, which paved the way for oxygen-breathing organisms, including, eventually, humans.

The earliest fossils of multicellular organisms appear in approximately 600-million-year-old rocks in southern Australia. These had shells, gills, filters, efficient guts, and a circulatory system, and in these ways they were relatively advanced—they must have had ancestors that do not appear in known fossils, but in which these organs and systems evolved. Among these were jellyfish-like animals, trilobites, mollusks (clams), echinoderms (such as sea urchins), and sea snails.

During this first major period of multicellular life, called the *Cambrian period*, which lasted until about 500 millions years ago, living things remained in the oceans. Almost 100 million years later during the Silurian period, plants evolved to live on land.

For larger, multicellular organisms, life on land required some major innovations, including the following:

- Structural support, needed because, while aquatic organisms are buoyed up by the water, on land gravity becomes a real force with which to contend.
- An internal aquatic environment, with a plumbing system giving it access to all parts of the organism and devices for conserving the water against losses to the surrounding atmosphere.
- Means for exchanging gases with air instead of with water.
- A moist environment for the reproductive system, essential for all sexually reproducing organisms.

The first fish to venture onto land, an obscure group called the crossopterygians, did so in the Devonian period (about 400 million years ago). These gave rise to the amphibians.

Figure 7.4 ■ Diagram illustrating the difference between species evenness, which is the relative abundance of each species, and species richness, which is the total number of species. Figures (*a*) and (*b*) have the same number of species but different relative abundances. Lay a ruler across each diagram and count the number of species the edge crosses. Do this several times, and determine how many species are in each diagram (*a*) and (*b*). See text for explanation of results.

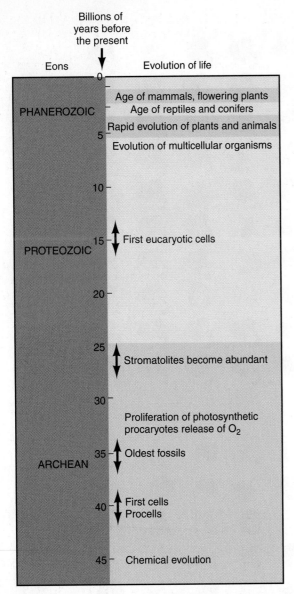

Figure 7.5 ■ The evolution of life on Earth from 4.6 billion years ago to the present. The rates at which new organisms appear and of biological diversity both increase with time.

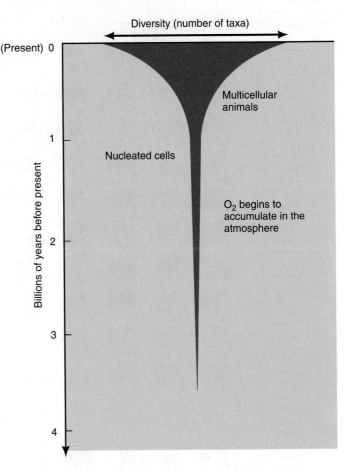

Figure 7.6 ■ A simplified representation of global diversity through geologic time.

The crossopterygians had several features that served to make the transition possible. Their lobelike fins, for example, were preadapted as limbs, complete with small bones to form the limb. They also had internal nostrils characteristic of air-breathing animals. Being fish, the crossopterygians already had a serviceable blood system that was adequate for making a start on land. Fossils of amphibians occur even later, in the *Devonian period* about 360 million years ago. Water conservation, however, never became a strong point with amphibians: they retain permeable skins to this day, which is one reason they have never become independent of the aquatic environment.

The earliest land plants were seedless, and the earliest of these could reproduce only in water, so they were limited to wet habitats. These plants reached their peak in dominance of the land in the Carboniferous period (see the ap-

pendices at the end of the book for dates for these periods). Seed plants evolved during the Devonian, starting with conifers with naked seeds (plants called *gymnosperms*, which means "naked seed"). The last frontiers for plants—so far, at least—were dry steppes, savannas, and prairies. These were not colonized until the *Tertiary period*, about 55 million years ago, when grasses evolved (Figures 7.5 and 7.6).

Despite their limitations, the amphibians ruled the land for many millions of years during the Devonian period. They had one difficulty that limited their expansion into many niches: They never met the reproductive requirement for life on land. In most species, the female amphibian lays her eggs in the water, the male fertilizes them there after a courtship ritual, and the young hatch as fish (tadpoles). Like the seedless plants, the amphibians—with one foot on the land, so to speak—have remained tied to the water for breeding. Some became quite large (2–3 m, or 2–3 yd, long). One branch evolved to become reptiles; the rest that survive are frogs, toads, newts, salamanders, and limbless water "snakes" that seem to have decided that, after all, they prefer a fish's life.

The reptiles freed themselves from the water by evolving an egg that could be incubated outside of the water and by getting themselves a watertight skin. These two "inventions" gave them the versatility to occupy terrestrial niches that the amphibians had missed because of

(a)   (b)

Figure 7.7 ■ (a) T. Rex skeleton (b) Drawing of newly discovered transitional early mammal living during to the age of dinosaurs.

sters). This resulted in the production of the two orders of dinosaurs (the largest quadrupeds ever to walk the Earth) and gave rise to two new vertebrate classes—mammals and birds (Figure 7.7).

Mammals are in many ways better equipped to occupy terrestrial niches than were the great reptiles. It is difficult to pick out a single mammalian "invention" comparable to the jaws of fish or the reptilian egg, for mammals, which are fine-tuned quadrupeds, are adapted to a faster and versatile life than reptiles. The mammalian "invention" is perhaps just that: a set of interdependent improvements managed by a more capable brain and supported by a faster metabolism. The placental uterus is sometimes regarded as the key to mammalian success; but it's really only a piece of equipment mandated by the delicate intricacy of the fetus that lives in it, especially its brain.

Thus life evolved on Earth, bringing us, in the most general way, to the present, where we confront problems about the great diversity of life. The mechanism of biological evolution, the rate at which species evolved and became extinct, and the kinds of environments in which species evolve provide essential background to understanding today's biological diversity issues.

During the history of life on Earth, evolution generally proceeded comparatively slowly, as did the extinction of species. But major catastrophes, including the crashing of asteroids onto Earth, rapidly changed the environment at a global level, extinguished many species in a comparatively short time, and opened up niches to which new species then evolved (Figures 7.7 and 7.8).

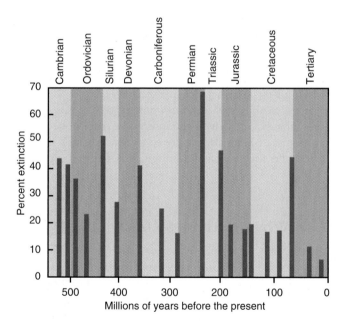

Figure 7.8 ■ A surprising number great extinction events have occurred during the last 500 million years. The percentage of extinction was determined from the disappearance of genera of well-skeletonized animals.

their bondage to the water. The egg in a hard shell did for reptilian diversity what jaws did for diversity in fishes. Originating in the Carboniferous coal swamps (about 375 million years ago), by the Jurassic period (some 185 million years later) the reptiles had moved onto the land, up into the air, and back to the water (as veritable sea mon-

Environmental change at many scales of time and space is a characteristic of our planet. Species have evolved within this environment and adapted to it. As a result, many species require certain kinds and rates of change. When we slow down or speed up environmental change, we impose novel risks on species.

7.5 The Number of Species on Earth

Many species have come and gone on Earth. But how many exist today? No one knows the exact number because new species are discovered all the time, especially in little-explored areas such as tropical rain forests. For example, since 1992, five new mammals have been discovered in Laos and two more have been "rediscovered." The rediscovered animals were known from a few specimens but had not been

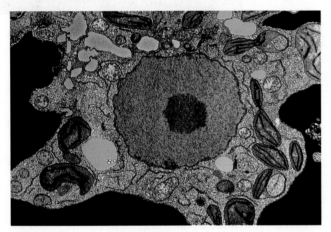

(a)

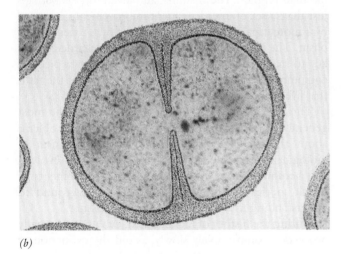

(b)

Figure 7.9 ■ Photomicrograph of (*a*) a eukaryote cell and (*b*) a bacterial (prokaryote) cell. From these images you can see that the eukaryotic cell has a much more complex structure, including many organelles.

seen for a long time. The new species include (1) the spindle-horned oryx (which is not only a new species but also represents a previously unknown genus); (2) the small black muntjak; (3) the giant muntjak (the muntjak, also known as "barking deer," is a small deer; the giant muntjak is so called because it has large antlers); (4) the striped hare (whose nearest relative lives in Sumatra); and (5) a new species of civet cat. The rediscovered species are a wild pig, previously known only from a skull found by a hunter in the 1890s, and Roosevelt's muntjak, previously known from one specimen obtained in the 1920s.

That such a small country with a long history of human occupancy would have so many new mammal species suggests how little we still know about the total biological diversity on Earth. Some 1.5 million species have been named, but biologists estimate that the total number is probably considerably higher, from 3 million to as many as 100 million (Table 7.1).

Many people often think in terms of two major kinds of life: animals and plants. Scientists, however, group living things on the basis of evolutionary relationships—a biological genealogy. In the recent past, scientists classified life into five kingdoms: animals, plants, fungi, protists, and bacteria. New evidence from the fossil record and studies in molecular biology suggest that it may be more appropriate to describe life as existing in three major domains, one called Eukaryota or Eukarya, which includes animals, plants, fungi, and protists (mostly single-celled organisms); Bacteria; and Archaea.[8, 9] Eukarya have a nucleus and other organelles; Bacteria and Archaea do not. Archaea used to be classified among Bacteria, but they have substantial molecular differences that suggest ancient divergence in heritage (Figure 7.9).

Insects and plants make up most of the known species (see Table 7.1). Many of the insects are tropical beetles in rain forests. Mammals, the group to which people belong, include a comparatively small number of species, only about 4,400.

7.6 Why Are There So Many Species?

Since species compete with one another for resources, wouldn't the losers drop out, leaving only a few winners? How did so many different species survive? Part of the answer lies in the different ways in which organisms interact, and part of the answer lies with the idea of the ecological niche.

Interactions between Species

Fundamentally, species interact in three ways: competition, in which the outcome is negative for both groups; symbiosis, which benefits both participants; and predation–parasitism, in which the outcome benefits one and is detrimental to the other. Each type of interaction affects evolution, the persistence of species, and the overall diversity of life.

Table 7.1 • Number of Species by Major Form of Life

Taxonomic Category	Common Name or Example	Major Group Subtotal Minimum	Major Group Subtotal Maximum	Number of Species Minimum	Number of Species Maximum
Virus	Virus	1,000	1,000	1,000	1,000
Monera/Bacteria	Bacteria	4,800	10,000	4,800	10,000
Archaebacteria				-	-
Cyanobacteria	Blue-green algae			-	-
Fungi		71,760	116,260		
Chytridiomycota				-	-
Zygomycota	Zygote fungi			600	1,100
Ascomycota	Cup or sac fungi			30,000	60,000
Basidiomycota	Club fungi			16,000	30,000
Deuteromycota	Imperfect fungi			25,000	25,000
Microsporidia				-	-
Glomeromycota				160	160
Lichens	Old man's beard	13,500	13,500	13,500	13,500
Protista/Protoctist		80,710	194,760		
Ciliophora	Ciliates			8,000	10,000
Myxomycota	Slime molds			560	560
Cryptophyta				-	-
Dinophyta	Dinoflagellates			4,000	4,000
Oomycota	Egg molds			-	-
Euglenophyta	Euglenoids			800	1,000
Dinoflagellata				1,000	1,100
Sarcodina	Amoeba			40,000	40,000
Sporozoa/ Apicomplexa				3,600	3,600
Chrysophyta	Golden algae			850	11,000
Bacillariophyta	Diatoms			10,000	100,000
Phaeophyta	Brown algae			900	1,500
Rhodophyta	Red algae			4,000	6,000
Chlorophyta	Green algae			7,000	16,000
Plantae		478,365	529,705	-	-
Bryophyta	Mosses			10,000	15,000
Hepatophyta	Liverworts			6,000	10,000
Anthocerophyta	Hornworts			100	100
Psilophyta—now in Pterophyta	Whisk ferns			-	-
Lycophyta	Club mosses			1,000	1,200
Sphenophyta (sp)	Horsetails			15	15
Filicinophyta	See Pterophyta			-	-
Pterophyta	Fems, horsetails, whisk fems			10,000	12,000
Coniferophyta	Conifers			550	600
Cycadophyta	Cycads			100	185
Ginkgophyta	Ginkgo			1	1
Gnetophyta	Ephedra			70	75
Anthophyta	Flowering plants			230,000	250,000
Monocotyledons	Lily			50,000	70,000
Dicotyledons	Maple tree			170,000	170,000
Gymnosperms	Pine tree			529	529
Animalia		873,084	1,870,019		
Porifera	Sponges			5,000	9,000
Placozoa	Trichoplaxes			1	1
Ectoprocta/ Bryozoans				4,000	4,500

Taxonomic Category	Common Name or Example	Major Group Subtotal Minimum	Major Group Subtotal Maximum	Number of Species	
				Minimum	Maximum
Cnidaria	Corals, jellies, hydras			9,000	10,000
Platyhelminthes	Flatworms			12,200	20,000
Kinorhyncha				150	150
Rotifera	Rotifers			1,800	2,000
Phoronida	Marine worms			12	20
Brachiopoda	Lamp shells			300	335
Acanthocephala	Thorny-headed worms			1,100	1,100
Loricifera				10	100
Nematoda	Roundworms			12,000	500,000
Mollusca	Molluscs			50,000	110,000
Ctenophora	Comb jellies			80	100
Annelida	Segmented worms			12,000	16,500
Priapula				15	17
Cycliophora				1	1
Onychophora	Velvet worms			80	110
Arthropoda	*Arthropods*				
Tardigrada	Water bears			400	800
Cheliceriformes	Horseshoe crabs, spiders			57,000	75,000
Uniramia				-	-
Insecta	*Insects*				
Anoplura				2,400	2,400
Blattodea	Cockroaches			4,000	4,000
Coleoptera	Beetles			290,000	500,000
Dermaptera	Earwigs			1,000	1,200
Diptera	Flies, mosquitos, etc.			80,000	151,000
Hemiptera	"True bugs"			55,000	85,000
Hymenoptera	Ants, bees, wasps, etc.			90,000	125,000
Isoptera	Termits			2,000	2,000
Lepidoptera	Butterflies, moths			112,000	140,000
Odonata	Dragonflies damselflies			5,000	5,000
Orthoptera	Grasshoppers, Crickets, etc.			13,000	30,000
Phasmida	Stick insects			2,600	2,600
Phthiraptera	Sucking lice			2,400	2,400
Siphonaptera	Fleas			1,200	2,400
Thysanura	Silverfish			450	450
Trichoptera	Caddisflies			7,000	7,100
Chordata					
Hemichordata	Acorn worms			85	85
Urochordates	Tunicate, sea squirts			1,250	3,000
Vertebrata		38,550	56,650		
Chondrichthyes	Sharks, rays etc			750	850
Osteichthyes	Bony fish			20,000	30,000
Amphibia	Amphibians			200	4,800
Reptilia	Reptiles			5,000	7,000
Aves	Birds			8,600	9,000
Mammalia	Mammals			4,000	5,000
TOTAL ALL LIFE		**1,523,219**	**2,735,244**		

References:

Wilson, E. O. *The Diversity of Life*. Cambridge, Mass: Harvard University Press, Belknap Press

Margulis, L. & Schwartz, K.V. *Five Kingdoms* (3rd ed.). New York: W. H. Freeman & Co.

Campbell, N. A. & Reece, J. B. *Biology* (7th ed.) Pearson/Benjamin Cummings

Mader, S.S. *Biology* (6th ed.) WCB McGraw-Hill

The Competitive Exclusion Principle

The **competitive exclusion principle** states that two species that have exactly the same requirements cannot coexist in exactly the same habitat. Garrett Hardin expressed the idea most succinctly: "Complete competitors cannot coexist."[10] This principle would seem to result in only a few species surviving on Earth.

The American gray squirrel and the British red squirrel illustrate the competitive exclusion principle. The American gray squirrel was introduced into Great Britain because some people thought it was attractive and would be a pleasant addition to the landscape. Thus, its introduction was not accidental but intentional. In fact, about a dozen attempts were made, the first perhaps as early as 1830. By the 1920s, the American gray squirrel was well established in Great Britain, and in the 1940s and 1950s its numbers expanded greatly.

Today, the American gray squirrel is a problem; it competes with the native red squirrel and is winning. The two species have almost exactly the same habitat requirements. At present, there are 2.5 million grey squirrels in Great Britain, and only 160,000 red squirrels. Although red squirrels used to be found in deciduous woodlands throughout the lowlands of central and southern Britain, they now are common only in Cambria, Northumberland, and Scotland, with scattered populations in East Anglia, in Wales, on the Isle of Wight, and on islands in Poole Harbor, Dorset.[11] If present trends continue, the red squirrel may disappear from the British mainland in the next 20 years.

One reason for the shift in the balance of these species may be that the main source of food during winter for red squirrels is hazelnuts, while gray squirrels prefer acorns. Thus, red squirrels have a competitive advantage in areas with hazelnuts, and gray squirrels have the advantage in oak forests. When gray squirrels were introduced, oaks were the dominant mature trees in Great Britain; about 40% of the trees planted were oaks.

The introduction of the gray squirrel into Great Britain illustrates one of the major causes of modern threats to biological diversity: the introduction by people of species into new habitats. Introduced competitors frequently threaten native species. We will look at this issue again in Chapter 8.

So, according to the competitive exclusion principle, complete competitors cannot coexist; one will always exclude the other. Therefore, we might expect that over a very long period this would reduce the number of species over a large area. Taking this idea to its logical extreme, we could imagine Earth with very few species—perhaps one green plant on the land, one herbivore to eat it, one carnivore, and one decomposer. If we added four species for the ocean and four for freshwater, we would have only 12 species on our planet.

Being a little more realistic, we could take into account adaptations to major differences in climate and other environmental aspects. Perhaps we could specify 100 environmental categories: cold and dry, cold and wet, warm and dry, warm and wet, and so forth. Even so, we would expect that within each environmental category, competitive exclusion would result in the survival of only a few species. Allowing four species per major environmental category would result in only 400 species. Yet about 1.4 million species have been named, and scientists speculate that many more millions may exist—so many that we do not have even a good estimate. How can they all coexist?

7.7 Niches: How Species Coexist

The niche concept explains how so many species can coexist, and this concept is introduced most easily by experiments done with small, common insects—flour beetles (*Tribolium*), which, as their name suggests, live on wheat flour. Flour beetles make good experimental subjects because they require only small containers of wheat flour to live and are easy to grow (in fact, too easy; if you don't store your flour at home properly, you will find these little beetles happily eating in it).

The flour beetle experiments work like this: A specified number of beetles of two species are placed in small containers of flour—each container with the same number of beetles of each species. The containers are then maintained at various temperature and moisture levels—some are cool and wet, others warm and dry. Periodically, the beetles in each container are counted. This is very easy. The experimenter just puts the flour through a sieve that lets the flour through but not the beetles. Then he counts the number of beetles of each species and puts the beetles back in their container to eat, grow, and reproduce for another interval. Eventually, one species always wins—some of its individuals continue to live in the container while the other species goes extinct. So far, it would seem that there should only be one species of *Tribolium*. But which species survives depends on temperature and moisture. One species does better when it is cold and wet, the other when it is warm and dry (Figure 7.10). Curiously, when conditions are in between, sometimes one species wins and sometimes the other, seemingly randomly; but invariably one persists while the second becomes extinct. So the competitive exclusion principle holds for these beetles, but both species can survive in a complex environment—one that has cold and wet habitats as well as warm and dry habitats. In no location, however, do the species coexist.

The little beetles provide us with the key to the coexistence of many species. Species that require the same resources can coexist by utilizing those resources under different environmental conditions. So it is habitat complexity that allows complete competitors—and not-so-complete competitors—to coexist[12] because they avoid competing with each other.

Professions and Places: The Ecological Niche and the Habitat

The flour beetles are said to have the same ecologically functional niche, which means they have the same *profession*—eating flour. But they have different *habitats*. Where a species lives is its **habitat**, but what it does for a living (its profession) is its **ecological niche**.[13] Suppose you have a

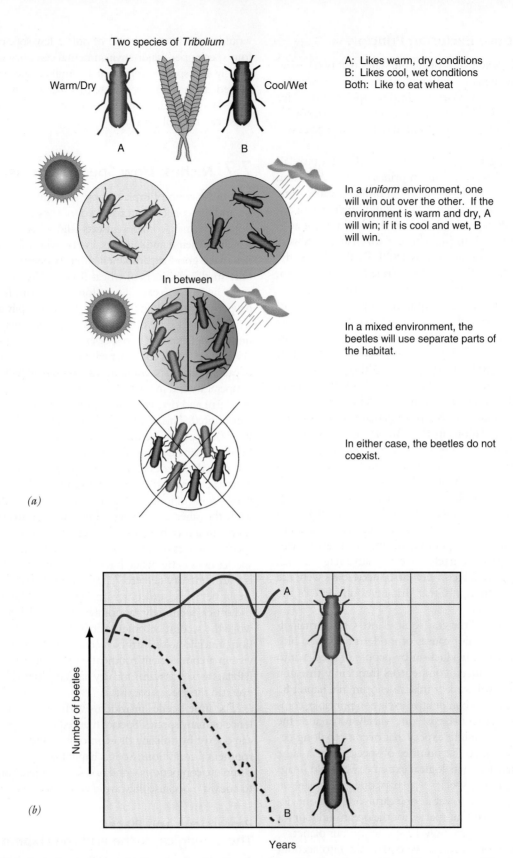

Figure 7.10 ■ A classical experiment with flour beetles. Two species of flour beetles are placed in small containers of flour. Each container is kept at a specified temperature and humidity. Periodically, the flour is sifted and the beetles counted and then returned to their containers. Which species persists is observed and recorded. (*a*) The general process illustrating competitive exclusion in these species; (*b*) Results of a specific, typical experiment under warm, dry conditions.

neighbor who is a bus driver. Where your neighbor lives and works is your town—that's his habitat. What your neighbor does is drive a bus—that's his niche. Similarly, if someone says, "Here comes a wolf," you think not only of a creature that inhabits the northern forests (its habitat) but also of a predator that feeds on large mammals (its niche).

Understanding the niche of a species is useful in assessing the impact of development or of changes in land use. Will the change remove an essential requirement for some species' niche? A new highway that makes car travel easier might eliminate your neighbor's bus route (an essential part of his habitat) and thereby eliminate his profession (his niche). Other things could also eliminate his niche. Suppose a new school were built so that all the children could walk to school. Then a bus driver would not be needed; his niche no longer existed in your town. In the same way, cutting a forest may drive away prey and eliminate the niche of the wolf.

Measuring Niches

The ecological niche is a useful idea, but, as scientists, we want to be able to measure it—to make it quantitative. How can we do that? How do you measure a profession? One answer is to describe the niche as the set of all environmental conditions under which a species can persist and carry out its life functions. This measured niche is known as the Hutchinsonian niche, after G. E. Hutchinson, who first suggested it.[14] It is illustrated by the distribution of two species of flatworm, a tiny worm that lives on the bottom of freshwater streams. A study was made of two species of these small worms in Great Britain, where it was found that some streams contained one species, some the other, and still others both.[15]

The stream waters are cold at their source in the mountains and become progressively warmer as they flow downstream. Each species of flatworm occurs within a specific range of water temperatures. In streams where species A occurs alone, it is found from 6° to 17°C (42.8°–62.6°F) (Figure 7.11a). Where species B occurs alone, it is found from 6° to 23°C (42.8°–73.4°F) (Figure 7.11b). When they occur in the same stream, their temperature ranges are much narrower. Species A lives in the upstream sections, where the temperature ranges from 6° to 14°C (42.8°–57.2°F), and species B lives in the warmer downstream areas, where temperatures range from 14° to 23°C (57.2°–73.4°F) (Figure 7.11c).

The temperature range in which species A occurs when it has no competition from B is called its *fundamental temperature niche*. The set of conditions under which it persists in the presence of B is called its *realized temperature niche*. The flatworms show that species divide up their habitats so that they use resources from different parts of it.

Of course, temperature is only one aspect of the environment. Flatworms also have requirements in terms of the acidity of the water and other factors. We could create graphs for each of these factors, showing the range within which A and B occurred. The collection of all those graphs would constitute the complete Hutchinsonian description of the niche of a species.

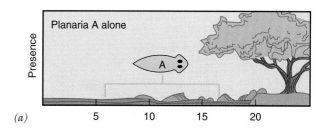

(a)

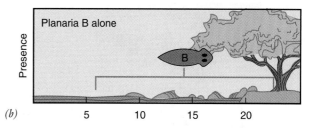

(b)

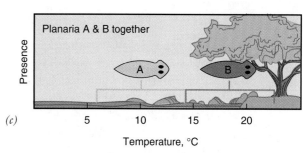

(c)

Temperature, °C

Figure 7.11 ■ The occurrence of freshwater flatworms in cold mountain streams in Great Britain. (*a*) The presence of species A in relation to temperature in streams where it occurs alone. (*b*) The presence of species B in relation to temperature in streams where it occurs alone. (*c*) The temperature range of both species in streams where they occur together. Inspect the three graphs; what is the effect of each species on the other?

A Practical Implication: From the discussion of the competitive exclusion principle and the ecological niche, we learn something important about the conservation of species. If we want to conserve a species in its native habitat, we must make sure that all the requirements of its niche are present. Conservation of endangered species is more than a matter of putting many individuals of that species into an area; all the life requirements for that species must also be present—we have to conserve not only a population, but its habitat and its niche.

Symbiosis

Our discussion up to this point might leave the impression that species interact mainly through competition—by interfering with one another. But **symbiosis** is also important. This term, derived from a Greek word meaning "living together," describes a relationship between two organisms that is beneficial to both and enhances each organism's chances of persisting. Each partner in symbiosis is called a **symbiont**.

Symbiosis is widespread and common; most animals and plants have symbiotic relationships with other species. We humans have symbionts—microbiologists tell us that about

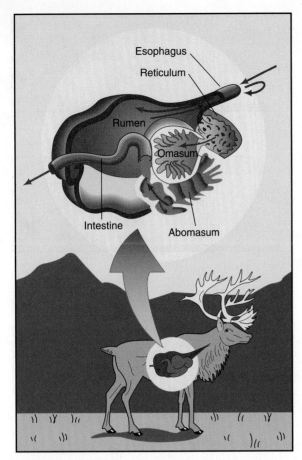

Figure 7.12 ■ The stomach of a reindeer illustrates complex symbiotic relationships. For example, in the rumen, bacteria digest woody tissue the reindeer could not otherwise digest. The result is food for the reindeer and food and a home for the bacteria, which could not survive in the local environment outside.

Esophagus
Reticulum
Rumen
Omasum
Intestine
Abomasum

10% of a person's body weight is actually the weight of symbiotic microorganisms in the intestines. The resident bacteria help our digestion; we provide a habitat that supplies all their needs; both we and they benefit. We become aware of this intestinal community when it changes—for example, when we travel to a foreign country and ingest new strains of bacteria. Then we suffer a well-known traveler's malady, gastrointestinal upset.

Another important kind of symbiotic interaction occurs between certain mammals and bacteria. A reindeer on the northern tundra may appear to be alone but carries with it many companions. Like domestic cattle, the reindeer is a ruminant, with a four-chambered stomach (Figure 7.12) teeming with microbes (a billion per cubic centimeter). In this partially closed environment, the respiration of microorganisms uses up the oxygen ingested by the reindeer while eating. Other microorganisms digest cellulose, take nitrogen from the air in the stomach, and make proteins. The bacterial species that digest the parts of the vegetation that the reindeer cannot digest itself (in particular, the cellulose and lignins of cell walls in woody tissue) require a peculiar environment: They can survive only in an environment without oxygen. One of the few places on Earth's surface

where such an environment exists is the inside of a ruminant's stomach.[16] The bacteria and the reindeer are symbionts, each providing what the other needs; and neither could survive without the other. They are therefore called **obligate symbionts**.

A Practical Implication: We can see that symbiosis promotes biological diversity and that if we want to save a species from extinction, we must save not only its habitat and niche but also its symbionts. This suggests another important point that will become more and more evident in later chapters: The attempt to save a single species almost invariably leads us to conserve a group of species, not just a single species or a particular physical habitat.

Predation and Parasitism

Predation–parasitism is the third way in which species interact. *Predation* occurs when an organism (a predator) feeds on other live organisms (prey), usually of another species. *Parasitism* occurs when one organism (the parasite) lives on, in, or within another (the host) and depends on it for existence but makes no useful contribution to it and may harm it.

Predation can increase the diversity of prey species. Think again about the competitive exclusion principle. Suppose two species are competing in the same habitat and have the same requirements. One will win out. But if a predator feeds on the more abundant species, it can keep that prey species from overwhelming the other. Both might persist whereas, without the predator, only one would. For example, some studies have shown that a moderately grazed pasture has more species of plants than an ungrazed one. The same seems to be true for natural grasslands and savannas. Without grazers and browsers, then, African grasslands and savannas might have fewer species of plants.

7.8 Environmental Factors That Influence Diversity

Species are not uniformly distributed over Earth's surface; diversity varies greatly from place to place. For instance, suppose you were to go outside and count all the species in a field or any open space near where you are reading this book (that would be a good way to begin to learn for yourself about biodiversity). The number of species you found would depend on where you are. If you live in northern Alaska or Canada, Scandinavia, or Siberia, you would probably find a significantly smaller number of species than if you live in the tropical areas of Brazil, Indonesia, or central Africa. Variation in diversity is partially a question of latitude—in general, greater diversity occurs at lower latitudes. Diversity also varies within local areas. If you count species in the relatively sparse environment of an abandoned city lot, for example, you will find quite a different number than if you counted species in an old, long-undisturbed forest.

The large-scale geographic pattern in the distribution of species, called *biogeography*, is the topic of the next chap-

Table 7.2 • Some Major Factors That Increase and Decrease Biological Diversity

A. Factors that tend to increase diversity

1. A physically diverse habitat.
2. Moderate amounts of disturbance (such as fire or storm in a forest or a sudden flow of water from a storm into a pond).
3. A small variation in environmental conditions (temperature, precipitation, nutrient supply, etc.).
4. High diversity at one trophic level increases the diversity at another trophic level. (Many kinds of trees provide habitats for many kinds of birds and insects.)
5. An environment highly modified by life (e.g., a rich organic soil).
6. Middle stages of succession.
7. Evolution.

B. Factors that tend to decrease diversity

1. Environmental stress.
2. Extreme environments (conditions near the limit of what living things can withstand).
3. A severe limitation in the supply of an essential resource.
4. Extreme amounts of disturbance.
5. Recent introduction of exotic species (species from other areas).
6. Geographic isolation (being on a real or ecological island).

ter. For now, we will look at some of the environmental factors that influence diversity locally. Table 7.2 summarizes several of these factors.

The species and ecosystems that occur on the land change with soil type and topography: slope, aspect (the direction the slope faces), elevation, and nearness to a drainage basin. These factors influence the number and kinds of plants. The kinds of plants, in turn, influence the number and kinds of animals. Some of the possible interrelationships are illustrated in Figure 7.13.[17]

Such a change in species can be seen with changes in elevation in mountainous areas like these at the Grand

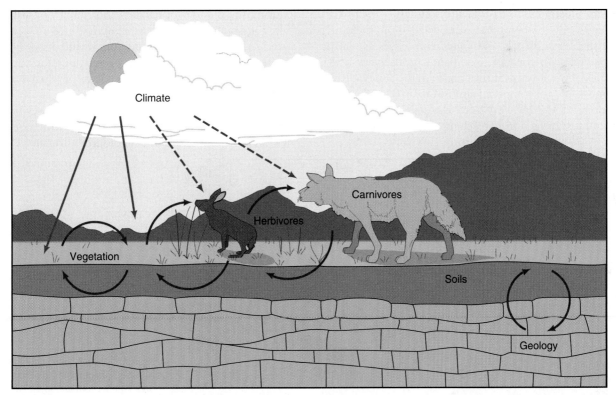

Figure 7.13 ■ Interrelationships among climate, geology, soil, vegetation, and animals. What lives where depends on many factors. Climate, geologic features (bedrock type, topographic features), and soils influence vegetation. Vegetation in turn influences soils and the kinds of animals that will be present. Animals affect the vegetation. Arrows represent a causal relationship; the direction is from cause to effect. A dashed arrow indicates a relatively weak influence, and a solid arrow a relatively strong influence.

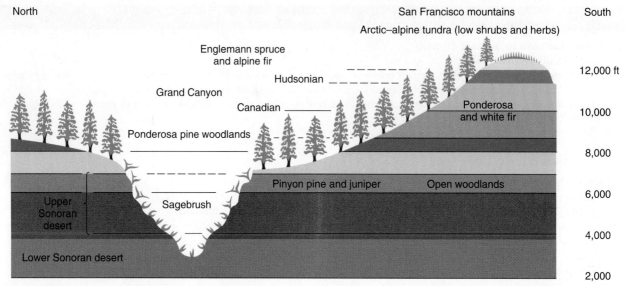

Figure 7.14 ■ Change in the relative abundance of a species over an area or a distance is referred to as an *ecological gradient*. Such a change can be seen with changes in elevation in mountainous areas. The altitudinal zones of vegetation in the Grand Canyon of Arizona and the nearby San Francisco Mountains are shown. (*Source*: From C. B. Hunt, *Natural Regions of the United States and Canada* [San Francisco: W. H. Freeman, 1974] Copyright 1974 by W. H. Freeman.)

Canyon and the nearby San Francisco Mountains of Arizona (Figure 7.14). Although such patterns are most easily seen in vegetation, they occur for all organisms. See, for example, the pattern of distribution of African mammals on Mount Kilimanjaro (Figure 7.15).

Some habitats harbor few species because they are stressful to life, as a comparison of vegetation in two areas of Africa illustrates. In eastern and southern Africa, well-drained, sandy soils support diverse vegetation, including many species of *Acacia* and *Combretum* trees as well as many grasses. In contrast, woodlands on the very heavy clay soils of wet areas near rivers, such as the Sengwa River in Zimbabwe, are composed almost exclusively of a single species called *Mopane*. Very heavy clay soils store water and prevent most oxygen from reaching roots. As a result, only tree species with very shallow roots survive.

Moderate environmental disturbance can also increase diversity. For example, fire is a common disturbance in many forests and grasslands. Occasional light fires produce a mosaic of recently burned and unburned areas. These patches favor different kinds of species and increase overall diversity.

Of course, people also affect diversity. In general, urbanization, industrialization, and agriculture decrease diversity, reducing the number of habitats and simplifying habitats. (See, for example, the effects of agriculture on habitats, discussed in Chapter 11.) In addition, we intentionally favor specific species and manipulate populations for our own purposes, as when a person plants a lawn or when a farmer plants a single crop over a large area.

Most people don't think of cities as having any beneficial effects on biological diversity. Indeed, the development of cities tends to reduce biological diversity. This is,

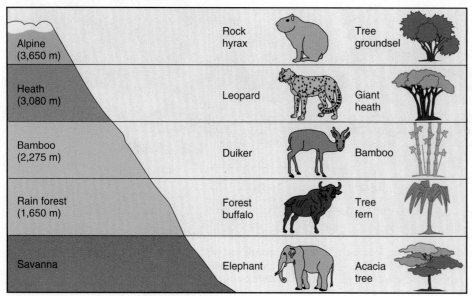

Figure 7.15 ■ Changes in the distribution of animals with elevation on a typical mountain in Kenya. (*Source:* From C. B. Cox, I. N. Healey, and P. D. Moore, *Biogeography* [New York: Halsted, 1973].)

Why Preserve Biodiversity?

Preserving the diversity of life on Earth has become an accepted goal for many people. But when that goal comes into conflict with other goals, such as economic development, the question becomes *How much diversity and at what cost?* In 1980, the International Union for Conservation of Nature and Natural Resources (IUCN) summarized the reasons for conserving diversity. Read the statements reproduced below and answer the questions that follow.

An Ethical Basis for Preserving Biodiversity

1. The world is an interdependent whole made up of natural and human communities. The well-being and health of any one part depends on the well-being and health of the other parts.

2. Humanity is part of nature, and humans are subject to the same immutable ecological laws as are all other species on the planet.

3. All life depends on the uninterrupted functioning of natural systems that ensures the supply of energy and nutrients, so ecological responsibility among all people is necessary for the survival, security, equity, and dignity of the world's communities.

4. Human culture must be built on a profound respect for nature, a sense of being at one with nature, and a recognition that human affairs must proceed in harmony and in balance with nature.

5. The ecological limits within which we must work are not limits to human endeavor; instead, they give direction and guidance as to how human affairs can sustain environmental stability and diversity.

6. All species have an inherent right to exist. The ecological processes that support the integrity of the biosphere and its diverse species, landscapes, and habitats are to be maintained. Similarly, the full range of human cultural adaptations to local environments is to be enabled to prosper.

7. Sustainability is the basic principle of all social and economic development.

8. Personal and social values should be chosen to accentuate the richness of flora, fauna, and human experience. This moral foundation will enable the many utilitarian values of nature—for food, health, science, technology, industry, and recreation—to be equitably distributed and sustained for future generations.

9. The well-being of future generations is a social responsibility of the present generation. Therefore, the present generation should limit its consumption of nonrenewable resources to the level that is necessary to meet the basic needs of society and ensure that renewable resources are nurtured for their sustainable productivity.

10. All persons must be empowered to exercise responsibility for their own lives and for the life of Earth. They must therefore have full access to educational opportunities, political enfranchisement, and sustaining livelihoods.

11. Diversity in ethical and cultural outlooks toward nature and human life is to be encouraged by promoting relationships that respect and enhance the diversity of life, regardless of the political, economic, or religious ideology in a society.

Critical Thinking Questions

1. Which of the preceding statements are scientific—that is, which are based on repeatable observations, can be tested, and are supported by evidence?

2. Using the table that follows, identify which statement from the argument for preserving biodiversity is implied by each type of value. Write the number of the statement in the right-hand column. A statement may fit into more than one category.

Type of Value	Source of Value of Living Organisms	Statement Number
Ethical	The fact that they are alive	
Aesthetic	Their beauty and the rewards humans derive from their beauty	
Economic	The direct and indirect ways in which they benefit humans	
Ecological	Their contributions to the health of the ecosystem	
Intellectual	Their contributions to knowledge	
Emotive	The sense of awe and wonder they inspire in humans	
Religious	Having been created by a supernatural being or force	
Recreational	Sport, tourism, and other recreations	

3. How would you rank the types of values given?

4. Which type or types of values could be used to support preservation of smallpox viruses? all 1,200 (or more) species of beetles in the tropical rain forest? the northern spotted owl? dolphins? sharks?

in part, because cities have typically been located at good sites for travel, such as along rivers or near oceans, where biological diversity is often high. However, in recent years we have begun to realize that cities can contribute in important ways to the conservation of biological diversity. We discuss this topic in Chapter 28.

The number of species per unit area varies over time as well as space. The changes occur over many time periods, from very short (a year) to moderate (years, decades), to even longer (centuries or more in a forest), and finally to very long geologic periods. As an example of the latter, diversity of animals in the early Paleozoic era (beginning about 500 million years ago) was much lower than in the late Mesozoic (beginning about 130 million years ago) and the Cenozoic (beginning about 75 million years ago) eras.[15]

7.9 Genetic Engineering and Some New Issues about Biological Diversity

Our understanding of evolution today owes a lot to the modern science of molecular biology and the practice of genetic engineering, which are creating a revolution in how we think about and deal with species. At present, scientists have essentially the complete DNA code for five species: the fruit fly (*Drosophila*), a nematode worm called *C. elegans* (a very small worm that lives in water); yeast; a small weed plant, thale cress (*Arabidopsis thaliana*); and ourselves. Scientists focused on these species either because they are of great interest to us (as with ourselves) or because they are relatively easy to study—having either few base pairs (the nematode worm) or having genetic characteristics that were well known (the fruit fly).

Environmental Issues as Information Issues

The amount of information contained in DNA is enormous. Thale cress's DNA consists of 125 million base pairs, and this is comparatively few. Important crop plants have even larger numbers of base pairs. Rice has 430 million and wheat more than 16 billion! The base pairs in thale cress appear to make up approximately 25,000 genes, but many are duplicates; so there are probably about 15,000 unique genes that determine what thale cress will be like. This is about the same number of genes in the nematode worm and the fruit fly. In contrast, the number of human genes is estimated to be between 30,000 and 130,000.

A Practical Implication. Scientists can now manipulate DNA and can therefore manipulate inherited characteristics of crops, bacteria, and other organisms, giving them new combinations of characteristics not found before and therefore demonstrating that characteristics are inherited and can be altered, as predicted by evolutionary theory. These new capabilities pose novel problems and hold new promise for biological diversity. On the one hand, we may be able to help rare, endangered species by increasing their genetic variability or by overcoming some of their less adaptive genetic characteristics that result from genetic drift. On the other hand, we may inadvertently create superpests, predators, or competitors of endangered species. Such new organisms may be to our benefit, but we must be careful not to release into the environment new strains that can reproduce rapidly and become unexpected pests. Genetic engineering poses new challenges for the environment, as we will discuss in later chapters. (We discuss some environmental implications of genetic engineering in Chapters 11 and 13.)

Summary

■ Biological evolution—the change in inherited characteristics of a population from generation to generation—is responsible for the development of the many species of life on Earth. Four processes that lead to evolution are mutation, natural selection, migration, and genetic drift.

■ Biological diversity involves three concepts: genetic diversity (the total number of genetic characteristics), habitat diversity (the diversity of habitats in a given unit area), and species diversity. Species diversity, in turn, involves three ideas: species richness (the total number of species), species evenness (the relative abundance of species), and species dominance (the most abundant species).

■ About 1.4 million species have been identified and named. Insects and plants make up most of these species. With further explorations, especially in tropical areas, the number of identified species, especially of invertebrates and plants, will increase.

■ Species engage in three basic kinds of interactions: competition, symbiosis, and predation–parasitism. Each type of interaction affects evolution, the persistence of species, and the overall diversity of life. It is important to understand that organisms have evolved together so that predator, parasite, prey, competitor, and symbiont have adjusted to one another. Human interventions frequently upset these adjustments.

■ The competitive exclusion principle states that two species that have exactly the same requirements cannot coexist in exactly the same habitat; one must win. The reason that more species do not die out from competition is that they have developed a particular niche and thus avoid competition.

■ The number of species in a given habitat is affected by many factors, including latitude, elevation, topography, the severity of the environment, and the diversity of the habitat. Predation and moderate disturbances, such as fire, can actually increase the diversity of species. The number of species also varies over time. Of course, people affect diversity as well.

Human Population

The growth of human populations has decreased biological diversity. If the human population continues to grow, pressures will continue on endangered species, and maintaining existing biological diversity will be an ever-greater challenge.

Sustainability

Sustainability involves more than just having many individuals of a species. For a species to persist, its life requirements must be present and its habitat must be in good condition. A diversity of habitats enables more species to persist.

Global Perspective

For several billion years, life has affected the environment on a global scale. These global effects have in turn affected biological diversity. Life added oxygen to the atmosphere and removed carbon dioxide, thereby making animal life possible.

Urban World

People have rarely thought about cities as having any beneficial effects on biological diversity. However, in recent years, there has been a growing realization that cities can contribute in important ways to the conservation of biological diversity. This topic will be discussed in Chapter 28.

People and Nature

People have always treasured the diversity of life, but we have been one of the main causes of the loss in diversity.

Science and Values

Perhaps no environmental issue causes more debate, is more central to arguments over values, or has greater emotional importance to people than biological diversity. Concern with specific endangered species has been at the heart of many political controversies. The path to resolving these conflicts and debates involves a clear understanding of the values at issue as well as knowledge about species and their habitat requirements and the role of biological diversity in life's history on Earth.

Key Terms

biological diversity **118**
biological evolution **118**
competitive exclusion
    principle **129**

ecological niche **129**
gene **119**
genetic drift **121**
habitat **129**

migration **121**
mutation **119**
obligate symbionts **132**
natural selection **119**

species **119**
symbiosis **131**
symbiont **131**

Study Questions

1. Why do introduced species often become pests?
2. On which of the following planets would you expect a greater diversity of species?
   a. A planet with intense tectonic activities
   b. A tectonically dead planet (Remember that tectonics refers to the geologic processes that involve the movement of tectonic plates and continents, processes that lead to mountain building and so forth.)
3. You are going to conduct a survey of national parks. What relationship would you expect to find between the number of species of trees and the size of the parks?
4. A city park manager has run out of money to buy new plants. How can the park labor force alone be used to increase the diversity of (a) trees and (b) birds in the parks?
5. A plague of locusts visits a farm field. Soon after, many kinds of birds arrive to feed on the locusts. What changes occur in animal dominance and diversity? Begin with the time before the locusts arrive and end with the time after the birds have been present for several days.
6. What will happen to total biodiversity if (a) the emperor penguin becomes extinct? (b) the grizzly bear becomes extinct?
7. What is the difference between habitat and niche?
8. There are more than 600 species of trees in Costa Rica, most of which are in the tropical rain forests. What might account for the coexistence of so many species with similar resource needs?
9. Which of the following can lead to populations that are less adapted to the environment than were their ancestors?
   a. Natural selection
   b. Migration
   c. Mutation
   d. Genetic drift

Further Reading

Leveque, C., and J. Mounolou. 2003. *Biodiversity.* New York: John Wiley.

Charlesworth, B., and C. Charlesworth. 2003. *Evolution: A Very Short Introduction.* Oxford: Oxford University Press.

Darwin, C. A. 1859. *The Origin of Species by Means of Natural Selection, or the Preservation of Proved Races in the Struggle for Life.* London: Murray. Reprinted variously. A book that marked a revolution in the study and understanding of biotic existence.

Margulis, Lynn, Michael Dolan, and Kathryn Delisle (Illustrator), and Christie Lyons. *Diversity of Life: The Illustrated Guide to the Five Kingdoms,* 2nd Spiral edition. Sudbury: Jones & Barlett Publishers.

Novacek M. J. (ed.). 2001 *The Biodiversity Crisis: Losing What Counts.* New York: An American Museum of Natural History Book. New Press.

CHAPTER 8

Biogeography

Burmese pythons, introduced and living in Florida's Everglades National Park, grow to 20 feet and pose an environmental problem.

Learning Objectives

If we are to conserve biological diversity, we must understand the large-scale, global patterns of biogeography. After reading this chapter, you should understand:

■ How climate, bedrock, and soils affect the geography of life.

■ What biotic provinces and biomes are, and how they differ.

■ How plate tectonics affects biogeography.

■ What island biogeography is, and what it implies for the general geography of life.

■ What are the geographic patterns of the Earth's 17 major biomes.

■ How people affect the geography of life.

■ How introducing exotic species typically affects habitats.

Be Careful Where You Put Unwanted Pets: Pythons in the Everglades

The Burmese python is an endangered species in its homeland, an area from northeastern India to southern China and some islands of the East Indies. It has been overharvested because its skins are used in the international fashion industry, and its blood and gall are used in Asian folk medicine.

Some consider this a beautiful animal because of its colorful skin (chapter opening photo). In recent years, more than 14,000 baby pythons have been imported into the United States and sold as pets. But the Burmese python is one of the world's six largest snakes, reaching almost 7 m (20 ft) long and weighing 90 kg (200 lb)! What do you do when the cute, colorful, and exotic pet becomes huge? You try to sell it, and if that doesn't work, you may just let it go. The *National Geographic News* reported in 2004 that the Everglades National Park was "overrun" with them and that park rangers had captured 68 of them.[1] Fortunately, the native Everglades alligator is a successful predator on the python and appears to be helping to control its abundance (Figure 8.1*a*).

So what started out as an apparently good-willed gesture, having a pet and possibly helping to save an endangered species, turned out to be, in a new habitat for the animal, a serious problem.[2] The story of the Burmese python illustrates the kinds of problems that arise when we fail to understand the natural geography of living things and to respect that geography. How can we help an endangered species like the Burmese python without destroying habitats that are not native to it (Figure 8.1*b*)? In this chapter we will consider this problem and others related to biogeography. As you read this chapter, remember that the introduction of exotic species in new habitats is one of the major ways that human actions threaten biodiversity.[3]

(a)

(b)

Figure 8.1 ■ (*a*) Florida alligators have become predators of Burmese pythons introduced into the Everglades National Park. (*b*) Both the alligator and the Burmese python live in mangrove swamps, one of the park's important kinds of habitats.

8.1 Why Were Introductions of New Species into Europe So Popular Long Ago?

Today, human introduction of species into new habitats is one of the most troublesome problems with life on Earth. The irony is that some introductions, such as crops and decorative plants, have been of great benefit. In 1749 Linneaus, one of the first scientific botanists and the father of the modern taxonomy of plants, sent a colleague, Peter Kalm, to North America to collect plants to decorate the gardens of Europe. Western Europe's climate was similar to the climate in parts of China and eastern North America, but those two areas had a much greater variety of plant species. For this reason, people who wanted beautiful gardens in Europe during the eighteenth century sought to augment native vegetation with flowering trees, shrubs, and herbs from the New World. But why were there fewer species in Europe? And if introductions of exotic species are such a problem today, why were these introductions into Europe from North America not a problem then? This puzzle is explained by theories of **biogeography**—large-scale, global patterns—beginning with the concepts of the biotic province and the biome.

8.2 Wallace's Realms: Biotic Provinces

In 1876, the great British biologist Alfred Russell Wallace (co-discoverer of the theory of biological evolution with Charles Darwin), suggested that the world could be divided into six biogeographic regions on the basis of fundamental features of the animals found in those areas.[4] Wallace referred to these regions as **realms** and named them Nearctic (North America), Neotropical (Central and South America), Palaearctic (Europe, northern Asia, and northern Africa), Ethiopian (central and southern Africa), Oriental (the Indian subcontinent and Malaysia), and Australian. These have become known as *Wallace's realms* (Figure 8.2). The recognition of these worldwide patterns in animal species was the first step in understanding biogeography.

Let us consider in more depth the idea of the relationships among living things—a kind of genealogical tree. All living organisms are classified into groups called **taxa**, usually on the basis of their evolutionary relationships or similarity of characteristics. (Linnaeus, whom we mentioned before, played a crucial role in working all this out.) The hierarchy of these groups (from largest and most inclusive to smallest and least inclusive) begins with a domain or kingdom. The plant kingdom is made up of divisions. The animal kingdom is made up of phyla (singular: *phylum*). A phylum or division is, in turn, made up of classes, which are made up of orders, which are made up of families, which are made up of genera (singular: *genus*), which are made up of species.

In each major biogeographic area (Wallace's realm), certain families of animals are dominant, and animals of these families fill the ecological niches (see Chapter 7). Animals filling a particular ecological niche in one realm are of different genetic stock from those filling the same niche in the other realms. For example, bison and pronghorn antelope are among the large mammalian herbivores in North America. Rodents such as the capybara fill the same niches in South America, and kangaroos fill them in Australia. In central and southern Africa, many species, including giraffes and antelopes, fill these niches.

This is the basic concept of Wallace's realms, and it is still considered valid and has been extended to all

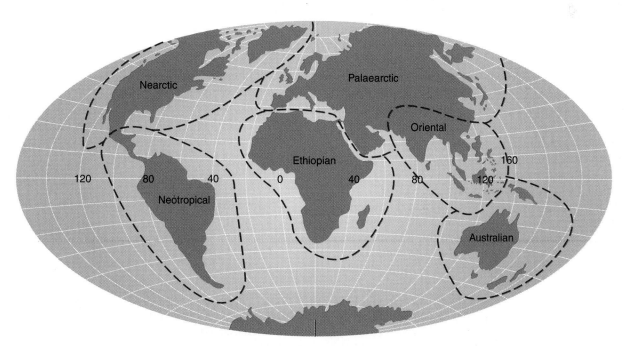

Figure 8.2 ■ The main biogeographic realms for animals are based on genetic factors. Within each realm, the vertebrates are in general more closely related to each other than to vertebrates filling similar niches in other realms.

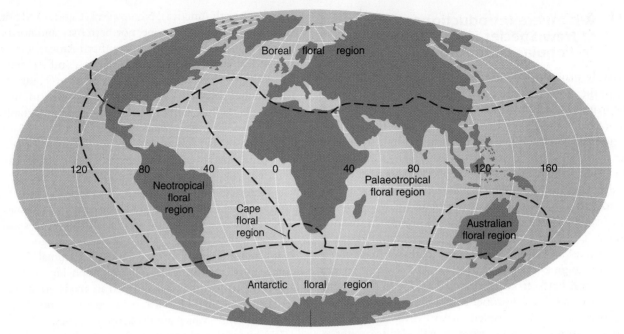

Figure 8.3 ■ The major vegetation realms are also based on genetic factors. Thus, plants within a realm are more closely related to each other than to plants of other realms.

life-forms,[5] including plants (Figure 8.3)[6, 7] and invertebrates. These realms are now referred to as **biotic provinces**.[8] A biotic province is a region inhabited by a characteristic set of taxa (species, families, orders), bounded by barriers that prevent the spread of those distinctive kinds of life to other regions and the immigration of foreign species.[9] So in a biotic province, organisms share a common genetic heritage but may live in a variety of environments as long as they are genetically isolated from other regions.

This was Wallace's observation. But how did this come about? Wallace did not have the benefit of our modern understanding of geologic processes to explain how distinct biological groups could have evolved in isolation from one another. The modern explanation is that continental drift, caused by plate tectonics, has periodically joined and separated the continents (see the discussion in Chapter 5).[10, 11] Unification of continents allowed genetic mixing; separation imposed geographic isolation. Continental unification and land bridges enabled organisms to enter new habitats. Continental separation led to genetic isolation and the evolution of new species.

Aha! This at least partially explains why introductions of exotics from one part of Earth to another can cause problems. Within a realm, species are more likely to be related and to have evolved and adapted in the same place for a long time. But when people bring home a species from far away, they are likely to be introducing a species that is unrelated, or only distantly related, to native species. And this new and unrelated species has not evolved and adapted in the presence of the home species, so ecological and evolutionary adjustments are yet to take place. Sometimes the introduction brings in a superior competitor. That is what is happening with the Burmese python in the Everglades.

8.3 Biomes

The biome is another major biogeographic pattern. It is a kind of ecosystem, such as a desert, a tropical rain forest, a grassland. Here's how it works: Similar environments provide similar opportunities for life and similar constraints. As a result, similar environments lead to the evolution of organisms similar in form and function (but not necessarily in genetic heritage or internal makeup) and similar ecosystems. This is known as the *rule of climatic similarity* and leads to the concept of the **biome**. The close relationship between environment and kinds of life-forms is shown in Figure 8.4.

Plants that grow in deserts of North America and East Africa illustrate the idea of a biome (see Figure 8.5). The desert euphorbia of South Africa looks similar to North American desert plants, but is not closely related to them. They belong to different biological families. Geographically isolated for 180 million years, they have been subjected to similar climates, which imposed similar stresses and opened up similar ecological opportunities. On both continents, desert plants evolved to adapt to these stresses and potentials, and have come to look alike and prevail in like habitats.

The ancestral differences between these look-alike plants can be found in their flowers, fruits, and seeds, which change the least over time and thus provide the best clues to the genetic history of a species. The Joshua tree and saguaro cactus of North America and the giant Euphorbia of East and Southern Africa are tall, have succulent green stems that replace the leaves as the major sites of photosynthesis, and have spiny projections, but these plants are not closely related. The Joshua tree is a member of the agave family, the saguaro is a member of the cactus family, and the Euphorbia is a member of the spurge family.

Their similar shapes result from evolution in similar desert climates, a process known as **convergent evolution**. The Euphorbia and the Joshua tree are in the same biome but in different biotic provinces. They function similarly and have the same niche, but they are not closely related.

So here is the difference between a biotic province and a biome: A biotic province is based on who is related to whom. A biome is based on niches and habitats. Species within a biotic province, in general, are more closely related to each other than to species in other provinces. In two different biotic provinces, the same ecological niche will be filled with species that perform a specific function and may look very similar to each other but have quite different genetic ancestries. In this way, a biotic province is an evolutionary unit.

The strong relationship between climate and life suggests that if we know the climate of an area, we can make a pretty good prediction about what biome will be found there, what its approximate biomass (amount of living matter) and production will be, and what the shapes and forms of dominant organisms will be.[12, 13]

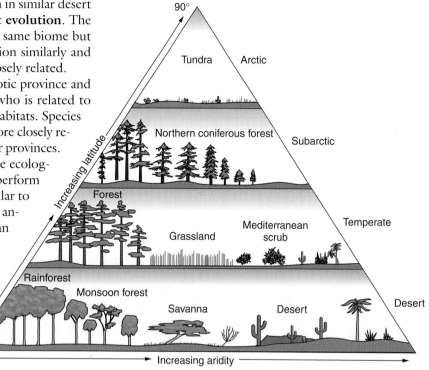

Figure 8.4 ■ Simplified diagram of the relationship between precipitation and latitude and Earth's major land biomes. Here, latitude serves as an index of average temperature, so latitude can be replaced by average temperature in this diagram. [*Source:* Figure 27-4, p. 293, from *Physical Geography of the Global Environment* by Harm de Blij, Peter O. Muller, and Richard S. Williams, edited by Harm de Blij, copyright 2004 by Oxford University Press, Inc. Used by permission of Oxford University Press, Inc.]

(a)   (b)   (c)

Figure 8.5 ■ Convergent evolution. Given sufficient time and similar climates in different areas, species similar in shape and form will tend to occur. The Joshua tree (*a*) and saguaro cactus (*b*) of North America look similar to the giant Euphorbia (*c*) of East Africa. All three are tall, have green succulent stems that replace the leaves as the major sites of photosynthesis, and have spiny projections. But these plants are not closely related; the Joshua tree is a member of the agave family, the saguaro is a member of the cactus family, and the Euphorbia is a member of the spurge family. Their similar shapes result from evolution under similar desert climates, a process known as convergent evolution.

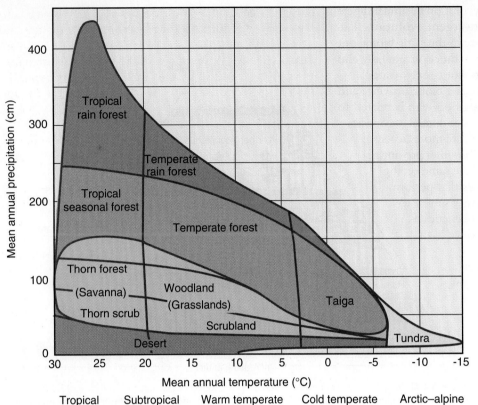

Figure 8.6 ■ Another view of the relationship between climate and vegetation (compare to Figure 8.4). A pattern of types of vegetation in relation to precipitation and temperature. Boundaries between types are approximate. Note that deserts occur over a wide range of annual temperatures, from 30°–5°C, as long as rainfall is less than about 50 cm/year. The warmer the climate, the more rainfall is required to change from desert to another biome. [*Source:* Adapted from R. H. Whittaker, *Communities and Ecosystems*, 2nd ed. (New York: Macmillan, 1975).]

The general relationship between biome type and the two most important climatic factors—rainfall and temperature—is diagrammed in Figure 8.6.

Another important process that influences life's geography is **divergent evolution**. In this process, a population is divided, usually by geographic barriers. Once separated into two populations, each evolves separately, but the two groups retain some characteristics in com-

mon. It is now believed that the ostrich (native to Africa), the rhea (native to South America), and the emu (native to Australia) have a common ancestor but evolved separately (Figure 8.7). In open savannas and grasslands, a large bird that can run quickly but feed efficiently on small seeds and insects has certain advantages over other organisms seeking the same food. Thus, these species maintained the same characteristics in widely separated areas.

(a)

(b)

(c)

Figure 8.7 ■ Divergent evolution. These three large, flightless birds evolved from a common ancestor but are now found in widely separated regions: (*a*) the ostrich in Africa, (*b*) the rhea in South America, and (*c*) the emu in Australia.

Both convergent and divergent evolution increase biological diversity.

People make use of convergent evolution when they move decorative and useful plants around the world. Cities that lie in similar climates in different parts of the world now share many of the same decorative plants. Bougainvillea, a spectacularly bright flowering shrub originally native to Southeast Asia, decorates cities as distant from each other as Los Angeles and the capital of Zimbabwe. In New York City and its outlying suburbs, Norway maple from Europe and the tree of heaven and gingko tree from China grow along with such native species as sweet gum, sugar maple, and pin oak. People intentionally introduced the Asian and European trees.

8.4 Geographic Patterns of Life within a Continent

Our discussion so far has focused on continental similarities and differences among species and on biological diversity. The same concepts—convergent evolution, divergent evolution, common ancestry, and similar environments—led to geographic patterns within a continent. The theory of continental drift provides us with a moving picture show of huge landmasses moving ponderously over Earth's surface, periodically isolating and remixing groups of organisms and leading to an increase in the diversity of species. If each continent were a uniform plot of land in a uniform climate, there would be fewer potential ecological niches (see Chapter 7), and therefore biological diversity within a continent would be low.

Plate tectonics results in continents with complex topography, including mountain ranges and alterations in drainage patterns and the paths of rivers. These can be barriers to the migration of species, leading to geographic isolation within a continent. When the Rocky Mountains began to form 90 million years ago, they created a barrier to various kinds of nonmountain life-forms. As a result of this—as well as climatic changes, some of which stemmed partially from the mountain building—California's vegetation is quite distinct from that found at similar elevations and latitudes east of the Rocky Mountains.

Patterns of life on a continent are also affected by the proximity of a habitat to an ocean or other large body of water, by the ocean currents near shore, and by the location relative to mountain ranges, latitude, and longitude. Figure 8.8 shows the pattern of life from west to east across North America (see A Closer Look 8.1).

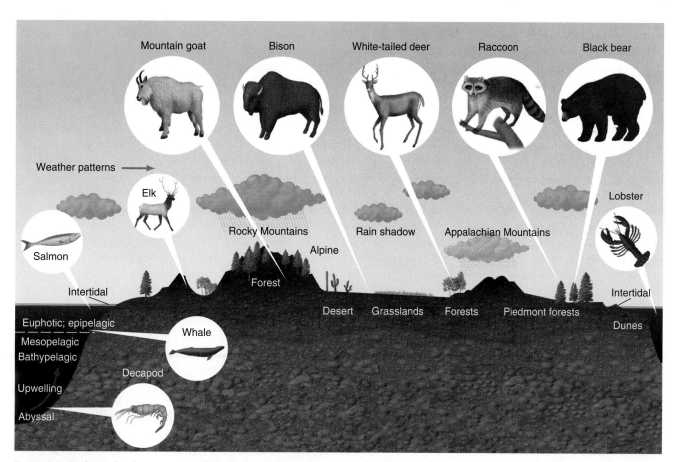

Figure 8.8 ■ Generalized cross section of North America showing weather, landforms, and the geography of life. Characteristic species of each region are shown. The weather patterns move from west to east.

A Biogeographical Cross Section of North America

A generalized cross section of the United States shows the relationships among weather patterns, topography, and biota (Figure 8.8). Off the West Coast of the United States in the Pacific basin occur the pelagic ecosystems, where sufficient light for photosynthesis penetrates the waters. This region is populated by small, mainly single-cell algae. Other oceanic zones with too little light for photosynthesis are populated by animals that feed on dead organisms that sink from above. Near the shore, particularly in areas of upwelling, such as along the California coast, are abundant algae, fish, birds, shellfish, and marine mammals. Where the tides and waves alternately cover and uncover the shore, a long, thin line of intertidal ecosystems is found, dominated by kelp and other large algae that are attached to the ocean bottom; by shellfish, such as mussels, barnacles, abalone, crabs, and other invertebrates; and by shorebirds, such as sandpipers.

Weather systems move generally from west to east in the Northern Hemisphere. As air masses are forced over the coastal mountains and Rocky Mountains, they are cooled, and the moisture in them condenses to form clouds and rain. The West Coast is an area of moderate temperature because water has a high capacity to store heat and the Pacific Ocean modulates the land's temperature. The annual precipitation increases with elevation on the western slopes of the mountains. In the south, moving east from the southern California coast, rainfall remains low until the mountains force the air high enough to condense much of its moisture. In general, the colder, wetter heights of the mountains support coniferous forests.

Along the coasts of Washington and Oregon, cool temperatures year-round lead to heavy rains near the shore, producing an unusual temperate-climate rain forest. The best-known example occurs in the Olympia National Forest on the northwestern edge of the state of Washington.

The eastern slopes of the coastal ranges form a so-called rain shadow. First, the air that passes over these eastern slopes gives up most of its moisture to the mountains; as a result, it is dry as it passes to the east. In addition, the air sinks to lower elevations, is warmed, and can hold more moisture. This dry air tends to take up moisture from the ground, producing the deserts of Utah, California, Arizona, and New Mexico. Whereas annual rainfall on the Olympic Peninsula of Washington reaches 375 cm/year (150 in./year), east of the Cascade Mountains it falls to 20 cm/year (8 in./year).

The same effect occurs in the Rockies. Less than 160 km (100 mi) west of Denver, in the Rocky Mountains, the annual rainfall is 100 cm (40 in.). In the Great Plains, 160 km east of Denver, the annual rainfall is only 30–40 cm (12–16 in.). Average annual rainfall increases steadily eastward: 50 cm (20 in.) at Dodge City, Kansas; 70 cm (27.5 in.) near Lincoln, Nebraska; and 90 cm (35.5 in.) near Kansas City, Missouri.[14]

The biomes reflect these changes in rainfall. Just east of Denver are shortgrass prairies, which become mixed-grass prairies (a mixture of shortgrass and tall-grass prairies) and then tall-grass prairies as we continue eastward. Rainfall reaches levels sufficient to support forests farther east, near the South Dakota–Minnesota border in the North, where the annual rainfall reaches 50–64 cm (20–25 in.). From there to the East Coast, the eastern deciduous forest (dominated by trees that lose their leaves during the winter) and the boreal forest of eastern North America predominate.

These ecological borders, such as those between the shortgrass and tall-grass prairies, are sometimes subtle, but they were quite striking to some of the early travelers in the West. One such traveler, Josiah Gregg, wrote in his journal in 1831 that to the west of Council Grove,

Kansas, at the border of the tall-grass and shortgrass prairies, the "vegetation of every kind is more stinted—the gay flowers more scarce, and the scant timber of a very inferior quality," whereas to the east he found the prairies to have "a fine and productive appearance truly rich and beautiful."[14]

The patterns described for the United States occur worldwide. One sees changes with elevation from warm, dry-adapted woodlands to moist, cool-adapted woodlands in Spain, where beech and birch, characteristic of middle and northern Europe (Germany, Scandinavia), are found at high elevations and alpine tundra is found at the summits. Similar patterns occur in Venezuela, where a change in elevation from sea level to 5,000 m (16,404 ft) at the summits of the Andes is equivalent to a latitudinal change from the Amazon basin to the southern tip of the South American continent. The seasonality of rainfall, as well as the total amount, often determines which ecosystems occur in an area.

Two other general concepts of biogeography, illustrated by both the latitudinal patterns from the Arctic to the tropics and the altitudinal patterns from mountaintops to valleys, are that (1) the number of species declines as the environment becomes more stressful and (2) on land, the height of vegetation decreases as the environment becomes more stressful (Figure 8.9). These concepts apply to most stresses, including those that human beings impose by adding pollutants to the environment, decreasing the fertility of soils or otherwise impoverishing habitats, and increasing the rate of environmental disturbance. From these concepts, we can predict that highly polluted and disturbed landscapes and seascapes will have few species and that, on the land, the dominant species of plants will have small stature.

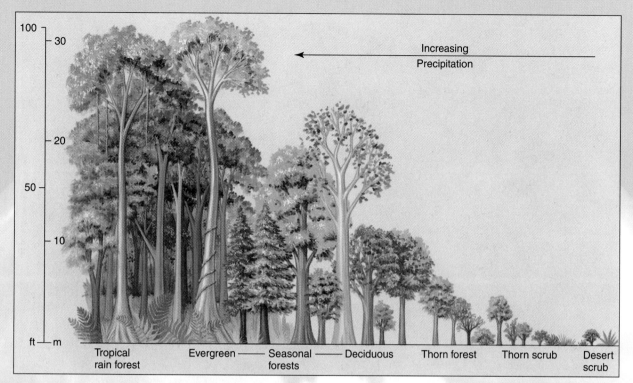

100 — 30
20
50
10
ft — m

Increasing
Precipitation

Tropical
rain forest

Evergreen — Seasonal
forests

Deciduous

Thorn forest

Thorn scrub

Desert
scrub

Figure 8.9 ■ Environmental stress and biogeography. Certain general patterns can be found as an environment becomes more stressful. This diagram shows the effects of increasing water stress. Where rainfall is plentiful, vegetation is abundant, and there are forests of tall trees of many species. As rainfall lessens, the size of the plants decreases to small trees, then shrubs and grasses, then scattered plants. The total biomass decreases and, in general, the number of species decreases. Similar changes accompany increases in other kinds of stress, including the stress of certain pollutants. [*Source:* Adapted from R. H. Whittaker, *Communities and Ecosystems*, 2nd ed. (New York: Macmillan, 1975).]

8.5 Island Biogeography

The many jokes and stories about people becoming castaways on an island have a basis in facts about the biogeography of islands. Islands have fewer species than continents, and the smaller the island, the fewer the species, on average. Also, the farther away an island is from a continent, the fewer species it will have. These two observations form the basis for the *theory of island biogeography*.

Darwin's visit to the Galápagos Islands gave him his most powerful insight into biological evolution.[15] There he found many species of finches that were related to a single species found elsewhere. On the Galápagos, each species was adapted to a different niche.[16] Darwin suggested that finches isolated from other species on the continents eventually separated into a number of groups, each adapted to a more specialized role. This process is called **adaptive radiation**. On the Hawaiian Islands, a finchlike ancestor evolved into several species, including fruit and seed eaters, insect eaters, and nectar eaters, each with a beak adapted for its specific food (Figure 8.10).[17] We can make several generalizations about species diversity on islands:

■ The two sources of new species on an island are migration from the mainland and evolution of new species in place (as with Darwin's finches in the Galápagos).

■ Islands have fewer species than continents.

■ The smaller the island, the fewer the species, as can be seen in the number of reptiles and amphibians in various West Indian islands.

■ The farther the island from a mainland (continent), the fewer the species (Figure 8.11).[18]

Why are these generalizations so? Small islands tend to have fewer habitat types. And some habitats on a small island may be too small to support a population large enough to have a good chance of surviving for a long time. A small population might be easily extinguished by a storm, flood, or other catastrophe or disturbance. Every species is subject to risk of extinction by predation, disease (parasitism), competition, climatic change, or habitat alteration. Generally, the smaller the population, the greater its risk of extinction. And the smaller the island, the smaller the population of a particular species that can be supported.

The farther the island from the mainland, the harder it will be for an organism to travel the distance. In addition,

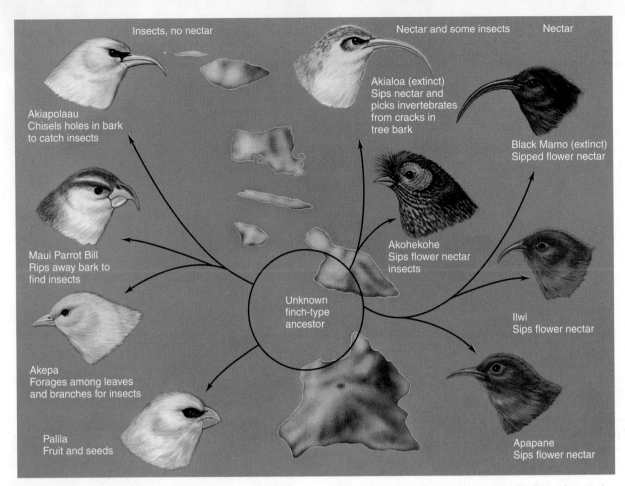

Figure 8.10 ■ Evolutionary divergence among honeycreepers in Hawaii. Sixteen species of birds, each with a beak specialized for its food, evolved from a single ancestor. Eight of the species are shown here. The species evolved to fit ecological niches that, on the North American continent, had previously been filled by other species not closely related to the ancestor. [*Source:* From C. B. Cox, I. N. Healey, and P. D. Moore, *Biogeography* (New York: Halsted, 1973).]

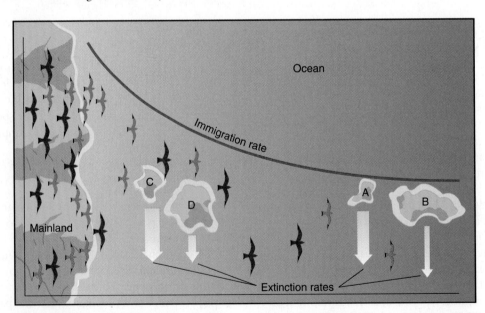

Figure 8.11 ■ Idealized relation of an island's size, distance from the mainland, and number of species. The nearer an island is to the mainland, the more likely it is to be found by an individual, and thus higher the rate of immigration. The larger the island, the larger the population it can support, and the higher the chance of persistence of a species—small islands have a higher rate of extinction. The average number of species therefore depends on the rate of immigration and the rate of extinction. Thus, a small island near the mainland may have the same number of species as a large island far from the mainland. The thickness of the arrow represents the magnitude of the rate. [*Source:* Modified from R. H. MacArthur and E. O. Wilson, *The Theory of Island Biogeography* (Princeton, NJ: Princeton University Press, 1967).]

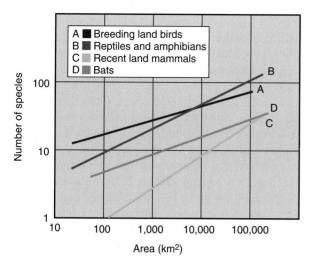

Figure 8.12 ■ Islands have fewer species than do mainlands. The larger the island, the greater the number of species. This general rule is shown by a graph of the number of species of birds, reptiles and amphibians, land mammals, and bats for islands in the Caribbean. [*Source:* Modified from Wilcox, B., ed., *IUCN Red List of Threatened Animals* (Gland, Switzerland: IUCN, 1988).]

a smaller island is a smaller "target," less likely to be found by individuals of any species.

A final generalization about island biogeography is that over a long time, an island tends to maintain a rather constant number of species, which is the result of the rate at which species are added minus the rate at which they become extinct. These numbers follow the curves shown in Figure 8.12. For any island, the number of species of a particular life-form can be predicted from the island's size and distance from the mainland.

The concepts of island biogeography apply not just to real islands in an ocean but also to ecological islands. An **ecological island** is a comparatively small habitat separated from a major habitat of the same kind. For example, a pond in the Michigan woods is an ecological island relative to the Great Lakes that border Michigan. A small stand of trees within a prairie is a forest island. A city park is also an ecological island. Is a city park large enough to support a population of a particular species? To know whether it is, we can apply the concepts of island biogeography.

8.6 Biogeography and People

We have seen that biogeography affects biological diversity. Changes in biological diversity in turn affect people and the living resources that we depend on. These effects extend from individuals to civilizations. For example, the last ice ages had dramatic effects on plants and animals and thus on human beings. Europe and Great Britain have fewer native species of trees than other temperate regions of the world. Only 30 tree species are native to Great Britain (that is, they were present prior to human settlement), although hundreds of species grow there today.

Why are there so few native species in Europe and Great Britain? Because of the combined effects of climatic change and the geography of European mountain ranges. In Europe, major mountain ranges run east–west, whereas in North America and Asia, the major ranges run north–south. During the past 2 million years, Earth has experienced several episodes of continental glaciation, when glaciers several kilometers thick expanded from the Arctic over the landscape. At the same time, glaciers formed in the mountains and expanded downward. Trees in Europe were caught between the ice from the north and the ice from the mountains and had few refuges; many species became extinct. In contrast, in North America and Asia, as the ice advanced, tree seeds could spread southward, where they became established and produced new plants. Thus, the tree species "migrated" southward and survived each episode of glaciation.[19]

Since the rise of modern civilization, these ancient events have had many practical consequences. As we mentioned before, soon after Europeans discovered North America, they began to import **exotic species** (a species introduced into a new geographic area) of trees and shrubs to Europe and Great Britain, where these were used to decorate gardens, homes, and parks and formed the basis of much of the commercial forestry in the region. For example, in the famous gardens of the Alhambra in Granada, Spain, Monterey cypress from North America are grown as hedges and cut in elaborate shapes. Douglas fir and Monterey pine are important commercial timber trees in Great Britain and Europe. These are only two examples of how knowledge of biogeography—being able to predict what will grow where based on climatic similarity—has been used for both aesthetic and economic benefits.

As mentioned in Chapter 7, people have altered biodiversity mainly by (1) direct hunting, which can cause extinction or serious decline in a species; (2) directly disrupting habitats; and (3) introducing exotic species into new habitats. The last is particularly relevant to this chapter.

Human introduction of exotics has had mixed results. On the one hand, the major foods of the world come from only a few species, and these species have been introduced widely by people. Without such introductions, people could not live in great numbers in most locations (see Chapter 12). The use of exotic species has also beautified landscapes. In addition, many pets that are important to people, such as cats and dogs, are routinely introduced into new habitats by people.

On the other hand, as this chapter's opening case study and Critical Thinking Issue show, introductions of exotics into new habitats have often had disastrous ecological consequences. This leads to some general rules:

■ Unless there is a very good reason to introduce a species into a new habitat, don't do it.

■ If you are going to introduce a species into a new habitat, do it very carefully. Check first about the natural pests and parasites of the species you plan to introduce. Are some of them essential to keep the species from reaching undesirable abundances? Are some of them likely to cause problems themselves?

As the technology of travel increases, inadvertent introductions of exotics become more likely. We must therefore increase our awareness of the problems associated with these introductions, be on our guard to know when introductions happen, and prevent those that are clearly undesirable or whose desirability is unknown.

8.7 Earth's Biomes

Earth has 17 major biomes: tundra, taiga (boreal forests), temperate deciduous forests, temperate rain forests, temperate woodlands, temperate shrublands, temperate grasslands, tropical rain forests, tropical seasonal forests and savannas, deserts, wetlands, freshwaters, intertidal areas, open oceans, benthos, upwellings, and hydrothermal vents. The first rule of moving species around the planet is: It's less likely to be harmful if you move a species within its biotic province. The second rule is: Moving a species into the same biome from a different biotic province is likely to be harmful. The third rule is: Local moves are less likely to be harmful than global moves (from one continent to another). This does not mean that we should stop all introductions of species, but it does mean that such introductions have to be done very carefully, especially introductions from one part of a biome to another, across continents.

To know when such moves are likely to cause problems, we need to know about Earth's biomes. The next part of this chapter describes Earth's major biomes and their locations.

Figure 8.13 shows the major land biomes. Subtypes occur within these major types. Biomes are usually named for the dominant vegetation (for example, coniferous forests, grasslands); for the dominant shape and form, or physiognomy, of the dominant organisms (forest, shrubland); or for the dominant climatic conditions (cold desert, warm desert).

In case you think the biome concept is abstract, take a look at the satellite image of Earth from space (Figure 8.14). The biomes show up and give Earth's surface a distinct pattern, a pattern that corresponds with a world map of average summer temperature (Figure 8.15)—showing that the importance of climate to biomes, discussed earlier, is also a reality. For example, where July average temperatures are above 30°C, you will see deserts (Figures 8.14 and 8.15) in the Americas and in Africa. Boreal forests occur where July average temperatures are below 20°C. Precipitation patterns, combined with temperature patterns, lead to an even closer correspondence between climate and biomes, as revealed by vegetation. Vegetation is the form of life most visible from space, but other forms of life have similar geographic relationships.

Climate also correlates with biological *productivity*. Warm, wet climates favor high vegetation production—as long as other factors, such as availability of chemical elements that plants require, are not limiting. Cold or dry climates do not generally allow such high rates of vegetation growth.

Biological diversity varies among biomes. A geographic pattern that has long intrigued and puzzled biologists is the general decline in biological diversity with increasing latitude. Mexico has 23,000 plant species. Brazil, Colombia, Peru, Madagascar, Mexico, India, China, Indonesia, and Australia contain about 60% of the world's named plant species.[20] In contrast, boreal forests have relatively low

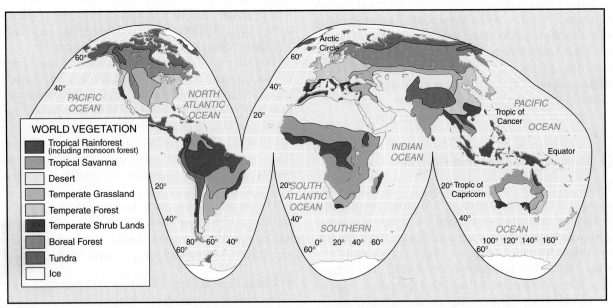

Figure 8.13 ■ Global distribution of the major land biomes. [*Source:* Figure 27-1, p. 290, from *Physical Geography of the Global Environment* by Harm de Blij, Peter O. Muller, and Richard S. Williams. Edited by Harm de Blij, copyright 2004 by Oxford University Press, Inc. Used by permission of Oxford University Press, Inc.]

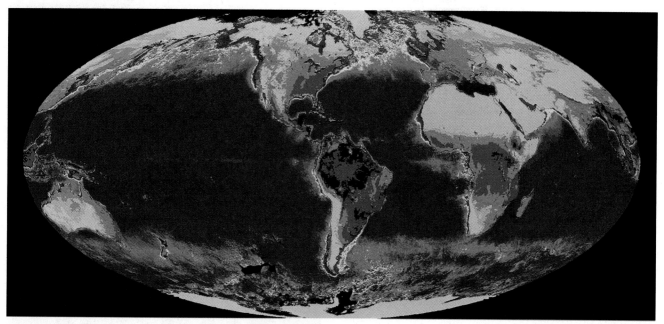

Figure 8.14 ■ NASA Landsat imagery of the vegetation of Earth from space.

biological diversity—while hundreds of tree species occur in a tropical rain forest, a typical boreal forest has fewer than ten and often no more than five tree species. The Serengeti Plains of east Africa have dozens of species of large mammals, while in the boreal forest you find fewer than ten—moose, deer, bear (one or a few species), mountain lions.

What explains this pattern? Ecologists continue to debate the cause of latitudinal patterns in biological diversity. Our

discussion so far suggests one theory—simply that the more favorable the temperature and precipitation for life, the more diverse it will be. But the same pattern seems to hold in the oceans, so high diversity cannot simply be linked to rainfall. Another theory is that the greater the *variability* of climate, the lower the diversity. According to this theory, temperature and rainfall in tropical rain forests not only remain within boundaries good for life, but also remain relatively constant

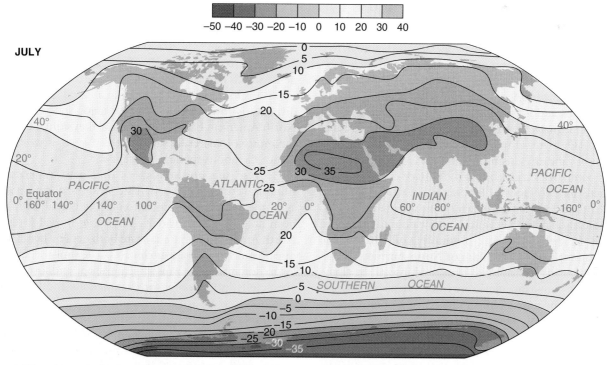

Figure 8.15 ■ Mean sea-level air temperatures (°C) for July. [*Source:* Figure 8-11, p. 93, from *Physical Geography of the Global Environment* by Harm de Blij, Peter O. Muller, and Richard S. Williams. Edited by Harm de Blij, copyright 2004 by Oxford University Press, Inc. Used by permission of Oxford University Press, Inc.]

throughout the year. Meanwhile, high latitudes, such as around Fairbanks, Alaska, have warm summers—the daily temperatures can be about 20°C—but in the winter the temperatures are very low. The basis of this theory lies with the niche concept (see Chapter 7). Part of the explanation seems to be that where environment varies greatly, each species must be a generalist—adapted to a wide range of environmental conditions—so there are fewer available niches. Where environment is relatively constant, species can become specialists, with a narrow range of tolerance, and therefore can divide the environment into many niches (see the discussion of the Hutchinsonian niche in Chapter 7).

But most likely the pattern is the result of several factors, as there are exceptions to the general rule that relatively constant climates have relatively high diversity. For example, in Israel, areas long disturbed by the grazing of goats and sheep and by human settlements have high vegetation diversity. The causes of geographic patterns in biological diversity is a topic that has fascinated people for thousands of years, and a complete discussion goes beyond what is possible in this book. The important thing to remember is that here is an ancient question still not completely answered, and ready for more scientific insights in the future.

8.8 The Geography of Life on the Planet Earth

In the remainder of this chapter, we take a closer look at the 17 major biomes of Earth.

Tundra

Tundras are treeless plains that occur in the harsh climates of low rainfall and low average temperatures (Figure 8.16). The dominant vegetation includes grasses and their relatives (sedges), mosses, lichens, flowering dwarf shrubs, and mat-forming plants. As the environment becomes harsher, dwarf shrubs and grasslike plants give way to mosses and lichens and finally to bare rock surfaces with occasional lichens. The extreme tundra occurs in Antarctica, where the major land organism in some areas is a lichen that grows within rocks, just below the surface.

There are two kinds of tundra: arctic, which occurs at high latitudes, and alpine, which occurs at high elevations. The vegetation of both is similar, but the dominant kinds of animals are different. Arctic tundras typically have some large mammals, such as caribou of North America, as well as important small mammals, birds, and insects. In alpine tundras, the dominant animals are small rodents and insects. This is partly because alpine tundras occupy comparatively small, isolated areas, whereas arctic tundras cover the large territories required for populations of large mammals.

Parts of tundra have *permafrost*—permanently frozen ground—which is extremely fragile. When disturbed by such activities as road development, permafrost areas may be permanently changed or take a very long time to recover.

Figure 8.16 ■ Tundra biome. Shown here is arctic tundra in Denali National Park, Alaska, with mountain avens in bloom.

Taiga, or Boreal Forests

The **taiga**, or boreal forest, biome includes the forests of the cold climates of high latitudes and high altitudes. Taiga forests are dominated by conifers, especially spruces, firs, larches, and some pines. Aspens and birches are important flowering trees (Figure 8.17). Typically, boreal forests form dense stands of relatively small trees, usually less than 30 m tall. Boreal forests cover very large areas. Their biological diversity is low—only about 20 major species occur in North American boreal forests—but they contain some of the most commercially valuable trees, such as white pine, spruce of various species, and cedar.

Because the northern areas of North America and Eurasia were connected by land bridges during past ice ages, the animals and vegetation of the boreal forest have been able to spread widely. Thus, the boreal forests of North America and Eurasia share both genetic heritage (biotic provinces) and similar climates. The similar climate has led to dominant life-forms similar in shape and form (biome characteristics). Moose, for example, are found on both continents, as are small flowering plants called *Saxifraga flagellaris*. Note that with the boreal forest, biotic province and biome overlap. The dominant animals of boreal forests include a few large mammals (moose, deer, wolves, and bears), small carnivores (foxes), small rodents (squirrels and rabbits), many insects, and migratory birds, especially waterfowl and carnivorous land birds, such as owls and eagles.

Figure 8.17 ■ The boreal forest and a moose, a characteristic mammal of this biome.

Disturbances—particularly fires, storms, and outbreaks of insects—are common in the boreal forests. For example, the entire million acres of the Boundary Waters Canoe Area of Minnesota burns over (through numerous small fires) an average of once each century, and individual forest stands are rarely more than 90 years old.

Boreal forests contain some of Earth's largest remaining wilderness areas and are appreciated for conservation of wilderness and for various kinds of recreation. The fur-bearing animals of boreal and tundra biomes have long been of commercial value, although that value has declined greatly in Western nations because of concerns about the animals' welfare.

Temperate Deciduous Forests

Temperate forests occur in climates somewhat warmer than those of the boreal forest. These forests grow throughout North America, Eurasia, and Japan and have many genera in common but different species. Dominant vegetation includes tall deciduous trees; common species are maples, beeches, oaks, hickories, and chestnuts, typically taller than trees of the boreal forest. These forests are important economically for their hardwood trees, which are used for furniture, among other things. Temperate deciduous forests are among the biomes most changed by human beings because they occur in regions long dominated by civilization, including much of China, Japan, Western Europe, the United States, and the urbanized parts of Canada.

Large mammals in this biome depend on young forests. Because the deep shade of temperate forests allows little vegetation growth near the ground, there is little vegetation that ground-dwelling animals can reach. Dominant animals therefore tend to be small mammals that live in trees (such as squirrels) and those (such as mice) that feed on soil organisms and small plants. Birds and insects are abundant.

There are few remaining uncut or unplowed stands of temperate forest, and some of them are important nature preserves. As with the boreal forest biome, fire is a natural, recurring feature, although the frequency of fire in many temperate forests is typically lower than in the boreal forests. In this biome, hunting-and-gathering cultures appear to have contributed to the frequency of fire in many regions.

Temperate Rain Forests

Temperate rain forests occur where temperatures are moderate and precipitation exceeds 250 cm/year. Such rain forests are rare but spectacular. The dominant trees are evergreen conifers, in contrast to the temperate deciduous forests, where flowering deciduous trees dominate. An intriguing question is why evergreens dominate the temperate rain forests but not the deciduous temperate forests. The best explanation appears to be that winters are wet and relatively mild in the temperate rain forests, so evergreen trees have an advantage—they can carry out photosynthesis and grow when the temperature warms above freezing in the winter, while the deciduous plants cannot. In contrast, in the temperate deciduous forests, winter temperatures remain below freezing, and the conifers are at a disadvantage—they must pay the metabolic costs of maintaining green needles without receiving a metabolic benefit.[21]

The temperate rain forests are the giant forests (Figure 8.18). In the Northern Hemisphere, they include the redwood forests of California and Oregon, where the tallest trees in the world exist, as well as forests in the state

Figure 8.18 ■ Temperate rain forest biome. Moss covers Sitka spruce in Olympic National Park, a famous example of the temperate rain forest of the Pacific Northwest of North America.

of Washington and adjacent Canada, dominated by such large trees as Douglas firs and western cedars. The trees grow taller than 70 m and are long-lived; Douglas firs live more than 400 years, redwoods several thousand.

Temperate rain forests also occur in the Southern Hemisphere. The best known of these are the forests of western New Zealand. The trees here are typically smaller than those in North America.

Temperate rain forests have comparatively low diversity of plants and animals. Why? In part because the abundant growth of the dominant vegetation produces very deep shade in which few other plants can grow and so provide food for herbivores.

This biome is important economically—redwoods, Douglas firs, and western cedars are major North American timber crops. It is also important culturally—the home of Native American tribes who want to continue their traditional ways of using forest resources. In addition, it is a focus of concern for biological conservation of the magnificent ancient forests, the many streams, and the associated species, such as salmon, the spotted owl, and the marbled murrelet, an oceanic bird that nests in these forests.[22]

Temperate Woodlands

Temperate woodlands occur where the temperature patterns are like those of deciduous forests but the climate is slightly drier. In North America, such areas occur from New England south to Georgia and the Caribbean islands. Temperate woodlands are dominated by small trees, such as piñon pines and evergreen oaks. The stands tend to be open, with wide spaces between trees, allowing considerable light to reach the ground. These generally pleasant areas are often used for recreation.

Fire is a common disturbance, and many species are adapted to it and require it. Because the pines typical of these areas are often fast-growing and produce wood good for timber, pulp, and paper, temperate woodlands are often economically valuable. Forest plantations occur in this biome. The pleasant combination of trees and grasses, along with the abundance of some large mammals such as mule deer in Ponderosa pine, has also made these a focus for biological conservation, once again bringing science and values to the fore.

Temperate Shrublands

In still drier climates, temperate shrublands called **chaparral** occur. These miniature woodlands are dominated by dense stands of shrubs that rarely exceed a few meters in height. Chaparral occurs in Mediterranean climates—climates with low rainfall that is concentrated in the cool season. It is found along the coast of California and in Chile, South Africa, and the Mediterranean region. While only about 5% of Earth's land area is in this biome, it is among the most attractive to people because of its mod-

erate, sunny climate. Therefore, it has been highly modified around the world by human action, and relatively few examples of native chaparral remain.

Typically, chaparral vegetation is distinctively aromatic; an example is sage. Some scientists believe the aromatic compounds produced by the plants are a kind of chemical warfare—that these compounds are toxic to competing plants and give the producers of them a competitive advantage. Although this explanation seems plausible, establishing that it is true has been difficult, and its proof still eludes scientists. There are few large mammals; reptiles and small mammals are characteristic. The animals and plants of this biome have little economic value at present, but the biome is important for watersheds and erosion control.

The vegetation is adapted to fire; many species regenerate rapidly, and some actually promote fire by producing abundant fuel in the form of dead twigs and branches. As a result, stands are rarely more than 50 years old. When intense precipitation follows fire, erosion can be exceptionally severe until renewed vegetation again protects the slopes. Temperate shrublands tend to be favorite sites for human settlement, as illustrated by the location of the ancient Greek and Roman civilizations in the Mediterranean area. Managing fires in chaparral is especially important in temperate shrublands, as was made clear almost every year when wildfires in both chaparral and conifer forests burned many homes in the Los Angeles basin.

Vegetation from the world's temperate woodlands is used to decorate public and private gardens, streets, and other public areas in cities and towns in this biome throughout the world.

Temperate Grasslands

Temperate grasslands occur in regions too dry for forests and too moist for deserts. Dominant plant species are grasses and other flowering plants, many of which are perennials with extensively developed roots. Temperate grasslands cover large areas of Earth—or did before many were converted to agriculture. They include the great North American prairies (Figure 8.19), which originally covered more land than any other biome in the United States; the steppes of Eurasia; the plains of eastern and southern Africa; and the pampas of South America.

The soils often have a deep organic layer, formed by the decaying roots of the grasses and the decaying stems and leaves of the prairie plants. The result is some of the best soils in the world for agriculture. By volume, most of the food of the world comes from this biome—all of the small grains and most of the large hoofed herbivores that provide meat for people, including cattle and American bison.

In North America, unplowed prairie is rare. People have made a considerable effort in recent years to restore prairies, a labor-intensive task. Prior to European settlement, the North American prairie extended from central Minnesota west to the Rocky Mountains and from Texas and

Figure 8.19 ■ Temperate grassland biome. The great American prairie was one of the world's largest areas of temperate grasslands before European settlement of North America.

Oklahoma northward into central Saskatchewan, Canada. Other grasslands occurred in the high plains of northern California and in eastern Oregon and Washington.

Grasslands are home to the highest abundance and greatest diversity of large mammals: the wild horses, asses, and antelopes of Eurasia; the once-huge herds of bison that roamed the prairies of the American West, along with pronghorn; the kangaroos of Australia; and the antelope and other large herbivores of Africa. Fossil evidence suggests that grasslands and grazing mammals evolved together, beginning about 60 million years ago. Grassland plants are adapted to certain kinds of grazing, and animals are important in the transport of seeds. The animals, of course, require edible grasses and forbs (flowering plants that are not grasses, trees, or shrubs). Thus, there is a kind of symbiosis among the animals and plants of this biome, which is essential for its continuation.

Fire is a natural, recurring feature, and in most locations both fire and grazing are necessary for the persistence of grasses and forbs. If fire and grazing are eliminated, the land tends to become desert shrubland in the drier regions and open woodlands in the wetter areas. In many grasslands, hunting-and-gathering cultures appear to have contributed to the fire frequency and in this way increased the area and persistence of this biome.

Tropical Rain Forests

Tropical rain forests occur where the average temperature is high and relatively constant throughout the year and where rainfall is high and relatively frequent throughout the year. Such conditions occur in northern South America, Central America, western Africa, northeastern Australia, Indonesia, the Philippines, Borneo, Hawaii, and parts of Malaysia. Tropical rain forests have long been home to hunting-and-gathering cultures, but relatively few civilizations have been able to persist in this biome.

Tropical rain forests are famous for their diversity of vegetation. Hundreds of species of trees may be found within a few square kilometers. Typically, some trees are very tall, and some, such as palm trees, remain relatively small. Some plants, like bromeliads and some ferns, grow on trees (Figure 8.20).

Figure 8.20 ■ Tropical rain forest biome. The vegetation along the Segama River in Borneo illustrates lowland tropical rain forests.

Figure 8.21 ■ Desert biome. These desert lands are at White Sands National Monument, New Mexico. Deserts vary greatly in the amount of vegetation they contain; this desert is relatively sparse, and some have no vegetation.

Approximately two-thirds of the 300,000 known species of flowering plants occur in the tropics, mainly in tropical rain forests.[20] Many species of animals occur as well. The mammals tend to live in trees, but some are ground dwellers. Insects and other invertebrates are abundant and show a high diversity. Rain forests occur in some of the most remote regions of Earth and remain poorly known; many undiscovered species are believed to exist there.

Except for dead organic matter at the surface, soils in this biome tend to be very low in nutrients. Most chemical elements (nutrients) are held in the living vegetation, which has evolved to survive in this environment; otherwise, rainfall would rapidly remove many chemical elements necessary for life.

Tropical Seasonal Forests and Savannas

Tropical seasonal forests occur at low latitudes, where the average temperature is high and relatively constant throughout the year and rainfall is abundant but very seasonal. Such forests are found in India and Southeast Asia, Africa, and South and Central America. In areas of even lower rainfall, tropical savannas—grasslands with scattered trees—are found. These include the savannas of Africa, which, along with the grasslands, have the greatest abundance of large mammals remaining anywhere in the world. The number of plant species is high as well.

Disturbances, including fires and the impact of herbivory on the vegetation, are common but may be necessary to maintain these areas as savannas; otherwise, they would revert to woodlands in wetter areas or to shrublands in drier areas. In still drier climates, savannas are replaced by shrublands, characterized by small shrubs, a generally low abundance of vegetation, and a low density of vertebrate animals.

Deserts

Deserts occur in the driest regions where vegetation can survive, typically where the rainfall is less than 50 cm/year.

Most deserts—such as the Sahara of North Africa and the deserts of the southwestern United States (Figure 8.21), Mexico, and Australia—occur at low latitudes. However, cold deserts occur in the basin and range areas of Utah and Nevada and in parts of western Asia.

Most deserts have a considerable amount of specialized vegetation, as well as specialized vertebrate and invertebrate animals. Soils often have little or no organic matter but abundant nutrients and need only water to become very productive. Disturbances are common in the form of occasional fires; occasional cold weather; and sudden, infrequent, and intense rains that cause flooding.

Relatively few large mammals live in deserts. The dominant animals of warm deserts are nonmammalian vertebrates (snakes and reptiles). Mammals are usually small, like the kangaroo mice of North American deserts.

Wetlands

Wetlands include freshwater swamps, marshes, and bogs and saltwater marshes. All have standing water: The water table is at the surface, and the ground is saturated with water (Figure 8.22). Standing water creates a special soil environment with little oxygen, so decay takes place slowly, and only plants with specialized roots can survive. Bogs—wetlands with a stream input but no surface water outlet—are characterized by floating mats of vegetation. Swamps and marshes are wetlands with surface inlets and outlets.

Dominant plants are small, ranging from small trees—such as the mangroves of warm coastal wetlands and the black spruces and larches of the North—to shrubs, sedges, and mosses. Small changes in elevation make a great difference. On slight rises, roots can obtain oxygen, and small trees can grow; in lower areas are patches of open water with algae and mosses.

Although wetlands occupy a relatively small portion of Earth's land area, they are important in the biosphere. In the oxygenless soils, bacteria survive that cannot live in high-oxygen atmospheres. These bacteria carry out chemical processes, such as the production of methane and

Figure 8.22 ■ Wetland biome. Wetlands include areas of open standing water and areas of herbs, grasses, shrubs, and trees that can withstand persistent or frequent flooding.

hydrogen sulfide, that have important effects in the biosphere. In addition, over geologic time, wetland environments produced the vegetation that today is coal.

Saltwater marshes are important breeding areas for many oceanic animals and are home to many invertebrates. Dominant animals include crabs and other shellfish, such as clams. Saltwater marshes are therefore an important economic resource. Dominant animals in freshwater wetlands include many species of insects, birds, and amphibians; few mammals are exclusive inhabitants of this biome. The larger swamps of warm regions are famous for large reptiles and snakes, as well as for a relatively high diversity of mammals where topographic variation includes small upland areas.

Although, because of the high water table, people tend not to live within wetlands, this biome can produce many edible plants; plants useful for making things such as baskets and other containers; and animals, including fish, that provide food. Wetlands are often used for recreation and have been a favorite biome for many naturalists and conservationists. Henry David Thoreau liked wetlands the best of all kinds of biomes.[23]

Freshwaters

Freshwater lakes, ponds, rivers, and streams make up a very small portion of Earth's surface but are critical to our water supply for homes, industry, recreation, and agriculture and play important ecological roles (Figure 8.23). Rivers and streams are also important in the biosphere as major transporters of materials from land to ocean.

Floating algae are common dominants, referred to as *phytoplankton*. Along shores and in shallow areas are rooted flowering plants, such as water lilies. Animal life is often abundant. Open waters have numerous small invertebrate animals (collectively called *zooplankton*), both herbivores and carnivores, and many species of finfish and shellfish.

Estuaries—areas at the mouths of rivers, where river water mixes with ocean water—are rich in nutrients. They usually support an abundance of fish and are important breeding sites for many commercially significant fish. Furthermore, many species that spend a greater portion of their life cycle in other biomes depend on freshwaters for reproduction or for food, as well as for drinking water. Freshwaters, then, are among the most important biomes for life's diversity.

Freshwaters are also among the areas most altered by human activities, especially by modern technology. Throughout much of the history of civilization, people have recognized the importance of freshwaters, but at the same time rivers and streams were heavily used as a way to get rid of wastes. Waterpower was one of the first sources of nonbiological energy, and rivers have long provided major transportation routes for people. Impoundment of streams and rivers (for example, by building dams and reservoirs), along with channelization to make transportation on them easier, has led to major changes in many freshwaters.

In recent years, the importance of freshwaters and their surrounding riparian and wetland areas has been recognized, and a major change has occurred in Western civilization's attitude toward rivers and streams. A generation ago, waterpower was considered one of the "cleanest" and most environmentally friendly forms of energy production. Today, there is growing interest in protecting free-flowing rivers and streams, whose complex channels, backwaters, meanders, floodplains, and seasonal ponds are important to fish, waterbirds, and

Figure 8.23 ■ Freshwater biome.

other wildlife. Much conservation effort is now being expended to restore freshwaters. For example, during the 2004 to 2006 bicentennial of the Lewis and Clark expedition, restoration of the Missouri River became a major focus for river restoration. On the Missouri River is a new kind of national wildlife refuge, the Big Muddy National Wildlife Refuge, which is a series of meanders along the river, like beads on a watery string. Engineers and ecologists are cooperating to re-create the complex backwater habitats on these meanders, providing habitats for fish, birds, and mammals.

Meanwhile, however, in developing nations, waterpower remains one of the least expensive sources of energy, and the development of major dams continues, as with the Three Gorges Dam in China, recently completed. We can expect freshwaters to be a major focus of environmental conflicts in the next decades.[24]

Intertidal Areas

The intertidal biome is made up of areas exposed alternately to air during low tide and ocean waters during high tide (Figure 8.24). Constant movement of waters transports nutrients into and out of these areas, which are usually rich in life and important to people as a direct source of food and as a spawning and breeding ground for many important foods. As a result, they are major economic resources. Large algae are found here, from giant kelp of temperate and cold waters to algae of coral reefs in the tropics. Birds and attached shellfish are usually abundant and are economically important. Near-shore areas are often important breeding grounds for many species of fish and shellfish, often also of economic significance.

The near-shore part of the oceanic environment is most susceptible to pollution from land sources. It is often heavily polluted by human activities, because major cities and civilizations tend to develop at the mouths of major rivers and along productive intertidal coastlines. In addition, as a major recreational area, it is subject to considerable alteration by people. Some of the oldest environmental laws concern the rights to use resources of this biome, and today major legal conflicts continue about access to intertidal areas and harvesting of biological resources.

Disturbances are common in the intertidal biome. Indeed, some of the most extreme variations in environmental conditions occur in this biome, among which are daily changes in sea level with the tides, seasonal changes in tidal minima and maxima, and ocean storms. Adaptation to these disturbances is essential to survival within the intertidal area. Consider, for example, a barnacle or mussel that twice daily experiences a change from a cool or cold saline water environment to direct exposure to bright sunlight and the highly oxygenated atmosphere.

Figure 8.24 ■ Intertidal biome. A rocky intertidal biome on the Atlantic coast of North America.

Open Ocean

Called the *pelagic region*, the open ocean biome includes open waters in all of the oceans. These vast areas tend to be low in nitrogen and phosphorus—chemical deserts with low productivity and low diversity of algae. Many species of large animals occur but at low density.

Benthos

The bottom portion of oceans is called the *benthos* (deeps). The primary input of food is dead organic matter that falls from above. The waters are too dark for photosynthesis, so no plants grow there.

Upwellings

Deep-ocean waters are cold and dark, and life is scarce. However, these waters are rich in nutrients because of numerous creatures that die in surface waters and sink. (See Chapter 9 for a discussion of energy flow in ocean ecosystems.) Upward flows, or upwellings, of deep-ocean waters bring nutrients to the surface, allowing abundant growth of algae, and animals that depend on algae. Upwellings occur off the west coast of North America, South America, West Africa, and near the Arctic and Antarctic ice sheets. In some areas, deeper waters are brought to the surface by winds that push coastal waters away from shore. These fertile upwelling zones are among the most important regions for the production of commercial fish.

Hydrothermal Vents

Hydrothermal vents, a recently discovered biome, occur in the deep ocean, where plate tectonic processes create vents of hot water with a high concentration of sulfur compounds. These sulfur compounds provide an energy basis for chemosynthetic bacteria, which support giant clams, worms, and other unusual life-forms. Water pressure is high, and temperatures range from the boiling point in waters of vents to the frigid (about 4°C) waters of the deep ocean.

Summary

■ To conserve biological diversity, we must understand the large-scale global patterns of life. This is known as biogeography.

■ Geographic isolation leads to the evolution of new species. Wallace's realms, or biotic provinces, are major geographic divisions (generally continents) based on fundamental features of the species found in them. Species filling specific niches within one realm are of different stock from those filling the same niches in other realms.

■ The rule of climatic similarity holds that similar environments lead to the evolution of biota and biological communities similar in external form and function but not in genetic heritage or internal makeup. Areas of climatic similarity with similar biota are known as biomes. A biome is a kind of ecosystem; examples are desert, grasslands, and rain forest.

■ Convergent evolution occurs when two genetically dissimilar species that inhabit geographically separate parts of a biome develop along similar lines and have similar external form and function. Divergent evolution occurs when several species evolve from a common ancestral species but develop separately because of geographic isolation.

■ The study of island life has led to a theory of island biogeography that includes several important concepts. One is that islands have fewer species than mainlands because of their smaller size and distance from the mainland. Another is that the smaller an island and the farther it is from the mainland, the fewer species the island will contain.

■ Ecological islands—habitats separated from the main part of a biome—show the same diversity characteristics as physical islands. The smaller the ecological island and the greater its distance from its "mainland," the fewer species it can support.

■ Earth has 17 major biomes, each with its own characteristic dominant shapes and forms of life. Biomes vary in their importance to people, and some are of great importance. Most biomes have been heavily altered by human actions. Understanding the major characteristics of these biomes is important to the conservation and sustainable use of their resources.

■ People have long introduced exotic species into new habitats, sometimes creating benefits, often causing new problems. From the study of biogeography, certain general rules can be set down concerning the introduction of exotic species. The primary rule is this: Unless there is a very clear and good reason to introduce an exotic species into a new habitat, don't do it; and take precautions to prevent such introductions from occurring inadvertently.

Escape of an Exotic Species

About 20 years ago, a delicate seaweed named *Caulerpa taxifolia* (Figure 8.25) was brought from its native habitat in the Pacific Ocean to a zoo in Germany, where it was cultivated and used to embellish saltwater aquarium exhibits, a seemingly harmless action. The seaweed was such a success that samples were sent to other institutions, including the Oceanographic Museum in Monaco.

Within about five years of its introduction there, an unfortunate accident took place. The seaweed was inadvertently flushed into the Mediterranean when exhibit tanks were cleaned. This accident might seem innocuous, but considering it so would ignore the tremendous power of species to act as biological invaders.

Once freed in the Mediterranean, *Caulerpa* quickly changed its growth pattern and adapted to its new habitat along the

southern coast of France. This may have occurred through a mutation, through hybridization with native seaweeds, or because its genetic code contained information that allowed for considerable plasticity. Whatever the exact genetic explanation, today *Caulerpa* grows about six times larger in the Mediterranean than it does in its native Pacific Ocean. It also tolerates colder temperatures, surviving in waters as chilly as 10°C, compared with a limit of 21°C in its home waters.

Over the past two or three years, *Caulerpa* has spread to the Adriatic, and it now appears to threaten the entire Mediterranean with its ability to choke out competing algae. It grows on rocks, sand, and mud, unlike most seaweeds, which do best on only one kind of substrate. It grows so profusely that it blankets competing native seaweeds, excluding them; and it appears to be toxic to local animals that feed on seaweeds, such as sea urchins, which will not touch it.

Controlling this invader has proved very difficult. When mechanically removed, it soon recovers to form even denser stands. Scientists are seeking a biological control method and are currently studying an exotic snail for release.

Thus, a seemingly harmless algae that gracefully decorated indoor aquariums has become an invading monster, affecting algae and animal life in the Mediterranean Sea, with consequences for commerce, recreation, and aesthetics. The problem caused by *Caulerpa taxifolia* is the result of the geography of life and the alteration of this geography by human action. The question that faces us today is: How can we control *Caulerpa taxifolia*?

Critical Thinking Questions

1. Suppose you were in charge of a program to control or eliminate this pest from the Mediterranean. Would you try to eliminate *Caulerpa* completely or just reduce its abundance?
2. Develop a plan for the control of this algae, based on material in this and previous chapters. What kinds of biogeographic exploration might help you develop a plan to control this species?
3. If nothing is done, what would be a likely long-term outcome?

Figure 8.25 ■ *Caulerpa taxifolia*. This delicate seaweed brought from the Pacific Ocean has become a pest in the Mediterranean.

Human Population

The spread of human beings around the world has been a major factor in the past several thousand years in altering the biogeography of Earth. As long as people have traveled and migrated to new areas, they have brought with them animals and plants, and their fungal and bacterial companions. The introduction of these species into new areas has had major effects on native species, including the extinction of species. Human alteration of the landscape is increasing because of our advancing technology and growing population, and this also alters biogeography. Trade in animal products such as rhinoceros horns and elephant ivory has led to the threat of extinction for some species. Today, our species is one of the greatest causes of alterations of biogeography.

Sustainability

How we alter the biogeography of species affects the sustainability of human societies, of commercial production of living resources, and the conservation of biological diversity. When we move species from one area to another, we can have major effects on the sustainability of native species. The guide to the future, with respect to biogeography, is caution. Do not introduce new species unless they seem essential, and then do so very carefully, testing the introduction to make sure that the species does not become a pest.

Global Perspective

Biogeography is one of the fundamentally global features of life on Earth. The abilities to invade and persist in new environments are essential properties of life. But while the ability to invade is a positive biological force, it can also cause problems. Because human beings are now spread around the world, we are a global factor with a great influence on biogeography.

Urban World

In our urbanized world we transport species, especially flowering plants, to decorate our cities and make them more livable. When done carefully, so that the introduced species do not become invasive pests, this can greatly improve urban life. Linnaeus understood this and led the early importation of plants from the New World to decorate gardens in the cities of Europe. Frederick Law Olmsted, the father of landscape planning in America, viewed the world's vegetation as a "palette" with which we could decorate our cities. Cities can also be a positive factor in the conservation of biological diversity. Some species, such as the peregrine falcon, have been successfully introduced into cities, and this has helped the recovery of such endangered species. Urban gardens, even the small backyards of residents, can host plants that are rare and endangered and that in turn can provide food for threatened and endangered birds.

People and Nature

This theme pervades the entire discussion of biogeography. Everywhere we turn, we see our effects on the geographic distribution of life, from the summits of mountains to deep-sea vents. People have introduced species into new habitats, with desirable and undesirable results. To forecast whether an introduction will be desirable, we need to understand the concepts of biotic provinces, biomes, and island biogeography. Biogeography is a key to understanding how people affect life on Earth.

Science and Values

The effects human beings have on biogeography are revealed by scientific research, as are the ways in which these effects are caused. But whether we choose to introduce a species into a new habitat or allow the extinction of a native species where we live depends on our values. We need science to tell us what can happen and warn of future potential effects on the distribution of species. With this knowledge, we can make decisions based on our values.

Key Terms

adaptive radiation **147**
biogeography **141**
biome **142**
biotic province **142**

chaparral **154**
convergent
    evolution **143**
divergent evolution **144**

ecological island **149**
exotic species **149**
realms **141**
taiga **152**

taxa **141**
tundra **152**

Study Questions

1. Design a program to conserve the endangered Burmese python. Your program can include a nature preserve, international agreements about trade, U.S. Laws, and preserves for the species in new habitats. The goals are to save the species from extincion and prevent it from becoming a pest anywhere in the world.

2. You are assigned the job of finding a new major food crop, a plant that will provide food but will not become a pest where it is introduced. How would you search for this crop, and where would you search?

3. Paul Martin, a well-known anthropologist, has suggested that the African elephant be reintroduced into North America to replace the mammoth and mastodon that once lived here, but went extinct at the end of the last ice age. He believes that people killed off these animals, and it is our moral imperative to reintroduce the elephant. Debate this issue, presenting the major arguments for and against it, in terms of biogeography.

4. In Jules Verne's classic novel *The Mysterious Island*, a group of Americans find themselves on an isolated volcanic island inhabited by kangaroos and large rodents closely related to the agoutis of South America. Why is this situation unrealistic? What would make this co-occurrence possible?

5. What do we learn from biogeography that helps us understand why there are so many species on Earth? Make a list of the major factors that lead to high biodiversity.

6. What are three ways in which people have altered the distribution of living things?

7. From the perspective of biogeography, why do people attach so much importance to the conservation of tropical rain forests?

8. What ideas from the theory of island biogeography might explain why there are large mammals in arctic tundra but not in alpine tundra?

9. If you were to travel to Mars, which is dry and has wide daily variations in temperature, and were to search for life, what kind of biome would you search for first?

10. Suppose you were going to build a spacecraft for long-term voyages and you planned to use an ecological life-support system—that is, use ecosystems to provide food; to recycle oxygen, carbon dioxide, and water; and to treat wastes. What biomes do you believe would be most important to take along? Would you create a "new" biome?

Further Reading

Elton, C. S. 2000. *The Ecology of Invasions by Animals and Plants.* New York: Oxford University Press. (Reprinted with new introduction by Daniel Simberloff.) A classic work by one of the major ecologists of the twentieth century.

Lomolino, Mark V. and James H. Brown. 2005. *Biogeography*, Third Edition Sinauer Associates. A broad multidisciplinary view of the geography of life.

Quammen, D. 1997. *The Song of the Dodo: Island Biogeography in an Age of Extinction.* New York: Scribner; Reprint edition. A not too technical, well-written, book about biogeography.

Rosenzweig, M. L. 2003. *Species Diversity in Space and Time.* New York: Cambridge University Press. A modern scientific treatment of species diversity.

CHAPTER 9

Biological Productivity and Energy Flow

Poor Farmers Plough Field with Oxen Farmers Walter and Tinashe Hondomoto plough a field in Zimbawe. African nations have become major food importers, since farmers like these are unable to produce enough to feed all the people of the region

Learning Objectives

To conserve and manage our biological resources successfully, we must understand the basic concepts of energy, energy flow in ecosystems, and biological production. After reading this chapter, you should understand that:

- Energy flow determines the upper limit of the production of biological resources.
- The first and second laws of thermodynamics tell us the limits of energy production and efficiency.

- Energy flows one way through an ecosystem.
- A basic quality of life is its ability to create order from energy on a local scale.

Can the World Produce Enough Food for Africa?

Agriculture in Africa is in crisis; 200 million people are malnourished (Figures 9.1 and 9.2). There are many causes: droughts, wars, and social strife, and well as poor farming practices that have degraded soils widely. In the past 50 years, African nations have, in total, suffered an astonishing 186 coups and 26 major wars. During the same time, too little investment has been made in rural areas, so that crop productivity has not kept pace with the human population. Agricultural soils in sub-Saharan Africa is degraded in about 72% of arable land and 31% of pasture land. In some countries—Ethiopia and Eritrea, for example—some of the farmland has degraded so badly that it may not be productive in any reasonable future.[1]

Human populations have grown rapidly in Africa; sub-Saharan population increased from 335 million in 1975 to 751 million in 2005—more than twofold. Ironically, the increased incidence of AIDS has reduced the number of healthy farmers, so that there are more people to feed but fewer who are feeding them.

The result is that a continent that was more than self-sufficient in food at independence in the mid-twentieth century is now a massive food importer. In 1966–1970, net exports averaged 1.3 million metric tons of food a year. But by the end of the 1970s, Africa was importing 4.4 million metric tons of staple foods a year; by the mid-1980s this total had increased to 10 million metric tons. By 2004, wheat imports alone to subsahara Africa total 9.6 million metric tons.

Starvation and malnutrition in Africa are not a surprise; in his book *The Population Bomb*, published in 1969, Paul Ehrlich wrote that this was likely to be the case. Since this crisis was predicted, it should have been avoided.[2]

Production of food in Africa raises the basic biological question: Just how productive can any life-form be? What are the limits of the growth of crops, or any kind of life, in a year? What factors limit growth? An important place for us to start is with an understanding of the basic ecological principles of biological growth, production, and energy flow, and these are the topics of this chapter.

Figure 9.1 ■ A starving Sudanese family wait in line at a feeding center run by medical charity Medecins Sans Frontieres in the village of Paliang, about 160 km (99 miles) northwest of the southern town of Rumbek, May 25, 2005. They are among the many Sudanese who need important food to avert south Sudan's worst food crisis since a 1998 famine, in which at least 60,000 people died.

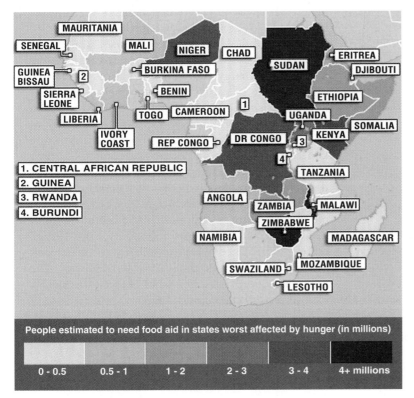

People estimated to need food aid in states worst affected by hunger (in millions)

| 0 - 0.5 | 0.5 - 1 | 1 - 2 | 2 - 3 | 3 - 4 | 4+ millions |

Figure 9.2 ■ African nations most affected by food shortages.
Source: BBC News. (http://news.bbc.co.uk/2/shared/spl/hi/africa/05/crisis_map/html/1.stm) May 4, 2006.

9.1 How Much Can We Grow?

Determining how much organic matter can be produced in any time period is important to many environmental topics, especially those that concern biological resources. How many bushels of wheat can a farmer produce in a field in a year? What is the upper limit of food that can be produced for all the people on Earth? What is the limit on the number of whales in the ocean? What is the maximum production that we can expect of forests?

Many factors can limit the growth of living things, but the ultimate limit on production of organic matter is energy flow. To estimate the actual production and the maximum possible production of organic matter of any kind, it is necessary to understand basic concepts of energy, and energy flow in ecosystems.

9.2 Biological Production

The total amount of organic matter on Earth or in any ecosystem or area is called its **biomass**. Biomass is usually measured as the amount per unit surface area of Earth (for example, as grams per square meter [g/m^2] or metric tons per hectare [MT/ha]).

Biomass is increased through biological production (growth). Change in biomass over a given period is called *net production*. **Biological production** is the capture of usable energy from the environment to produce organic compounds in which that energy is stored. In photosynthesis, the environmental energy is from visible light. This light is transferred to energy in the chemical bonds of the organic compounds. This capture is often referred to as energy "fixation," and it is often said that the organism has "fixed" energy. Three measures are used for biological production: biomass, energy stored, and carbon stored. We can think of these measures as the currencies of production. (General relationships for calculating production are given in Working It Out 9.1 and 9.2. Common units of measure of production are given in the appendix.)

Two Kinds of Biological Production

There are two kinds of biological production. Some organisms make their own organic matter from a source of energy and inorganic compounds. These organisms,

Equations for Production, Biomass, and Energy Flow

We can write a general relation between biomass (B) and net production (NP):

$$B_2 = B_1 + NP$$

where B_2 is the biomass at the end of the time period, B_1 is the amount of biomass at the beginning of the time period, and NP is the change in biomass during the time period (Figure 9.3). Thus,

$$NP = B_2 - B_1$$

General production equations are given as

$$GP = NP + R$$
$$NP = GP - R$$

where GP is gross production, NP is net production, and R is respiration.

The three currencies of energy flow are biomass, energy content, and carbon content. The average of the energy in vegetation is approximately 21 kilojoules per gram (kJ/g). Energy content of organic matter varies. Ignoring bone and shells, woody tissue contains the least energy per gram, about 17 kJ/g; fat contains the most, about 38 kJ/g; and muscle contains approximately 21–25 kJ/g. Leaves and shoots of green plants have 21–23 kJ/g; roots have about 19 kJ/g.[3]

The kilojoule (1 kJ = 1,000 J = 0.24 kcal) is the International System (SI) unit preferred in scientific notation for energy and work. It replaces the calorie or kilocalorie of earlier studies of energy flow. The kilocalorie is the amount of energy required to heat a kilogram of water 1 degree Celsius (from 15.5° to 16.5°C). (The calorie, which is one-thousandth of a kilocalorie, is the amount of energy required to heat a gram of water the same 1 degree Celsius.) Note that the calorie referred to in diet books is actually the kilocalorie, though for convenience it is referred to in popular literature as a calorie. To keep all this straight, just remember that almost nobody uses the "little" calorie, regardless of what they call it. To compare, an average apple contains about 419 kJ or 100 kcal. The calorie is typically used in studies of diets; the joule is used in physics and engineering.

(a)

(b)

Figure 9.3 ■ Net production: (*a*) A Field of corn at the end of the growing season; (*b*) the same field at the beginning of the growing season. Refer to Working It Out 9.1. You can think of the mature crop shown in (*a*) as B_2 and the field at the beginning of the growing season (*b*) as B_1. The difference between (*a*) and (*b*) illustrates net primary production (*NPP*).

introduced in the discussion of trophic levels in Chapter 6, are called **autotrophs** (meaning self-nourishing). The autotrophs include green plants (those containing chlorophyll), such as herbs, shrubs, and trees; algae, which are usually found in water but occasionally grow on land; and certain kinds of bacteria that grow in water.

The production carried out by autotrophs is called **primary production**. Most autotrophs make sugar from sunlight, carbon dioxide, and water in a process called **photosynthesis**, which releases free oxygen (see Working It Out 9.1 and Working It Out 9.2). Some autotrophic bacteria can derive energy from inorganic sulfur compounds; these bacteria are referred to as **chemoautotrophs**. Such bacteria have been discovered in deep-ocean vents, where they provide the basis for a strange ecological community. Chemoautotrophs are also found in muds of marshes, where there is no free oxygen.

Other kinds of life cannot make their own organic compounds from inorganic ones and must feed on other living things. These are called **heterotrophs**. All animals, including human beings, are heterotrophs, as are fungi, many kinds of bacteria, and many other forms of life. Production by het-

Energy Equalities

For Those Who Make Their Own Food (autotrophs)

Photosynthesis—the process by which autotrophs make sugar from sunlight, carbon dioxide, and water—is defined as

$$6CO_2 + 6H_2O = C_6H_{12}O_6 + 6O_2$$

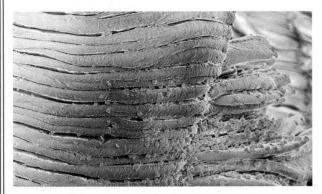

Figure 9.4 ■ Deep-sea vent chemosynthetic bacteria.

Chemosynthesis takes place in certain environments. In chemosynthesis, the energy in hydrogen sulfide (H_2S) is used by certain bacteria to make simple organic compounds. The reactions differ among species and depend on characteristics of the environment (Figure 9.4).

Net production for autotrophs is given as

$$NPP = GPP - R_a$$

where NPP is net primary production, GPP is gross primary production, and R_a is the respiration of autotrophs.

For Those Who Do Not Make Their Own Food (heterotrophs)

Secondary production of a population is given as

$$NSP = B_2 - B_1$$

where NSP is net secondary production, B_2 is the biomass at time 2, and B_1 is the biomass at time 1. The change in biomass is the result of the addition of weight of living individuals, the addition of newborns and immigrants, and loss through death and emigration. The biological use of energy occurs through respiration, most simply expressed as

$$C_6H_{12}O_6 + 6O_2 = 6CO_2 + 6H_2O + Energy$$

erotrophs is called **secondary production** because it depends on the production of autotrophic organisms. This dependence is the basis of the food web described in Chapter 6. (The associated energy flow is diagrammed in Figure 9.5.)

Once an organism has obtained new organic matter, it can use the energy in that organic matter to do things: to move, to make new kinds of compounds, to grow, to reproduce, or store it for future uses. The use of energy from organic matter by most heterotrophic and autotrophic organisms is accomplished through **respiration**. In respiration, an organic compound is combined with oxygen to release energy and produce carbon dioxide and water (see Working It Out 9.2). The process is similar to the burning of organic compounds but takes place within cells at much lower temperatures through enzyme-mediated reactions. *Respiration is the use of biomass to release energy that can be used to do work.* Respiration returns to the environment the carbon dioxide that had been removed by photosynthesis.

Gross and Net Production

The production of biomass and its use as a source of energy by autotrophs include three steps:

1. An organism produces organic matter within its body.
2. It uses some of this new organic matter as a fuel in respiration.
3. It stores some of the newly produced organic matter for future use.

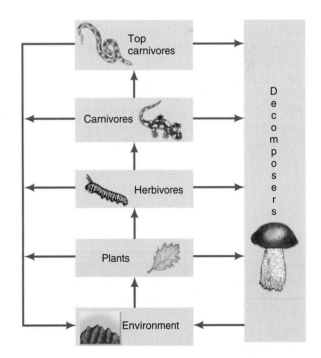

Figure 9.5 ■ Energy pathways through an ecosystem. Usable energy flows from the external environment (the sun) to the plants, then to the herbivores, carnivores, and top carnivores. Death at each level transfers energy to decomposers. Energy lost as heat is returned to the external environment.

The first step, production of organic matter before use, is called **gross production**. The amount left after utilization is called **net production**.

Net production = Gross production – Respiration.

The difference between gross and net production is like the difference between a person's gross and net income. Your gross income is the amount you are paid. Your net income is what you have left after taxes and other fixed costs. Respiration is like the necessary expenses that are required in order for you to do your work.

The gross production of a tree—or any other plant—is the total amount of sugar it produces by photosynthesis before any is used. Within living cells in a green plant, some of the sugar is oxidized in respiration. Energy is used to convert sugars to other carbohydrates, those carbohydrates to amino acids, amino acids to proteins and new leaf tissue. Energy is also used to transport material within the plant to roots, stems, flowers, and fruits. Some energy is lost as heat in the transfer. Some energy is used to make other organic compounds in other parts of the plant: cell walls, proteins, and so forth. Some is stored in these other parts of the plant for later use. For woody plants like trees, it includes new wood laid down in the trunk, new buds that will develop into leaves and flowers the next year, and new roots.

9.3 Energy Flow

Energy is a difficult and abstract concept. When we buy electricity, what are we buying? We cannot see it or feel it, even if we have to pay for it.[4]

At first glance, energy flow seems simple enough: We take energy in and use it, just like a machine. But if we dig a little deeper into this subject, we discover a philosophical importance: We learn what distinguishes life and life-containing systems from the rest of the universe.

Although most of the time energy is invisible to us, with infrared film we can see the differences between warm and cold objects, and we can see some factors about energy flow that affect life. With infrared film, warm objects appear red, and cool objects blue. Figure 9.6 shows birch trees in a New Hampshire forest, both as we see them, using standard film, and with infrared film, which shows tree leaves bright red, indicating that they have been warmed by the sun and are absorbing and reflecting energy, whereas the white birch bark remains cooler. The ability of tree leaves to absorb energy is essential; it is this source of energy that ultimately supports all life in a forest. Energy flows through life, and *energy flow* is a key concept.

All life requires energy. Energy is the ability to do work, to move matter. As anyone who has dieted knows, our weight is a delicate balance between the energy we take in through our food and the energy we use. What we do not use and do not pass on, we store. Our use of energy, and whether we gain or lose weight, follows the laws of physics. This is not only true for people, it is also true of all populations of living things, of all ecological communities and ecosystems, and of the entire biosphere.

Ecosystem energy flow is the movement of energy through an ecosystem from the external environment through a series of organisms and back to the external environment. It is one of the fundamental processes common to all ecosystems.

Energy enters an ecosystem by two pathways. One is the pathway already discussed: Energy fixed by organisms. In the second pathway, heat energy is transferred by the air or water currents or by convection through soils and sediments and warms living things. For instance, when a warm air mass passes over a forest, heat energy is transferred from the air to the land and to the organisms.

What about production within an ecosystem? We saw earlier that net production equals gross production minus respiration. As stated, this relationship is true not only for an individual organism but also for an ecological community or an ecosystem (see Working It Out 9.3). In an ecosystem, the gross production is simply the gross production of all the autotrophs. The ecosystem's net production is the amount of biomass added in some time period after all utilization, including the respiration of autotrophs and heterotrophs, has taken place.

WORKING IT OUT 9.3

Ecosystem Equalities

For an ecosystem:

$$GEP = GPP$$

where *GEP* is gross ecosystem production and *GPP* is gross primary production.

$$R_e = R_a + R_h$$

where R_e is net ecosystem respiration, R_a is respiration of autotrophs, and R_h is respiration of heterotrophs.

$$NEP = GEP - R_e$$

where *NEP* is net ecosystem production, *GEP* is gross ecosystem production, and R_e is net ecosystem respiration.

Figure 9.6 ■ Making energy visible. Top: A birch forest in New Hampshire as we see it, using normal photographic film (*a*) and the same forest photographed with infrared film (*b*). Red color means warmer temperatures; the leaves are warmer than the surroundings because they are heated by sunlight. Bottom: A nearby rocky outcrop as we see it, using normal photographic film (*c*) and the same rocky outcrop photographed with infrared film (*d*). Blue means that a surface is cool. The rocks appear deep blue, indicating that they are much cooler than the surrounding trees.

(a)　(b)

(c)　(d)

9.4 The Ultimate Limit on the Abundance of Life

What ultimately limits the amount of organic matter that can be produced? What limits the maximum rate of that production? How closely do ecosystems, species, populations, and individuals approach this limit? Are any of these near to being as productive as possible? The answer lies in the laws of thermodynamics.

The Laws of Thermodynamics

The *law of conservation of energy* states that in any physical or chemical change, energy is neither created nor destroyed but merely changed from one form to another. The law of conservation of energy is also called the *first law of thermodynamics* (discussed in A Closer Look 5.1).

If the total amount of energy is always conserved—if it remains constant—then why can't we just recycle energy inside our bodies? Similarly, why can't energy be recycled in ecosystems and in the biosphere?

Let us imagine how that might work, say, with frogs and mosquitoes. Frogs eat insects, including mosquitoes. Mosquitoes suck blood from vertebrates, including frogs. Consider an imaginary closed ecosystem consisting of water, air, a rock for frogs to sit on, frogs, and mosquitoes. In this system, the frogs get their energy from eating the mosquitoes, and the mosquitoes get their energy from biting the frogs (Figure 9.7). Such a closed system would be a biological perpetual-motion machine. It could continue indefinitely without an input of any new material or energy. This sounds nice, but, unfortunately, it is impossible. Why? The general answer is found in the *second law of thermodynamics*, which addresses how energy changes in form.

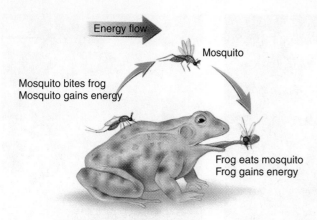

Figure 9.7 ■ An impossible ecosystem.

Energy always changes from a more useful, more highly organized form to a less useful, disorganized form. That is, energy cannot be completely recycled to its original state of organized, high-quality usefulness. For this reason, the mosquito–frog system will eventually stop when not enough useful energy is left. (There is also a more mundane reason: Only female mosquitoes require blood and then only in order to reproduce. Mosquitoes are otherwise herbivorous.)

From the discussion presented in A Closer Look 9.1, we reach a new understanding of a basic quality of life.[3] It is the ability to create order on a local scale that distinguishes life from its nonliving environment. This ability requires obtaining energy in a usable form, and that is why we eat. This principle is true for every ecological level: individual, population, community, ecosystem, and biosphere. Energy must continually be added to an ecological system in a usable form. Energy is inevitably degraded into heat, and this heat must be released from the system. If it is not released, the temperature of the system will increase indefinitely. The net flow of energy through an ecosystem, then, is a one-way flow.

Based on what we have said about the energy flow through an ecosystem, we can see that an ecosystem must lie between a source of usable energy and a sink for degraded (heat) energy. The ecosystem is said to be an *intermediate system* between the energy source and the energy sink. The energy source, ecosystem, and energy sink together form a **thermodynamic system**. The ecosystem can undergo an increase in order, called a *local increase*, as long as the entire system undergoes a decrease in order, called a *global decrease*. To put all this simply, creating local order involves the production of organic matter. Producing organic matter requires energy; organic matter stores energy.

Energy Efficiency and Transfer Efficiency

How efficiently do living things use energy? This is an important question for the management and conservation of all biological resources. We would like biological resources to be efficient in their use of energy—to produce much biomass from a given amount of energy.

No system can be 100% efficient. As energy flows through a food web, it is degraded, and less and less is usable. Generally, the more energy an organism gets, the more it has for its own use. However, organisms differ in how efficiently they use the energy they obtain. A more efficient organism has an advantage over a less efficient one.

Efficiency can be defined for both artificial and natural systems: machines, individual organisms, populations, trophic levels, ecosystems, and the biosphere.[5] *Energy efficiency* is defined as the ratio of output to input, and it is usually further defined as the amount of useful work obtained from some amount of available energy. Efficiency has different meanings to different users. From the point of view of a farmer, an efficient corn crop is one that converts a great deal of solar energy to sugar and uses little of that sugar to produce stems, roots, and leaves. In other words, the most efficient crop is the one that has the most harvestable energy left at the end of the season. A truck driver views an efficient truck just the opposite. For him, an efficient truck uses as much energy as possible from its fuel and stores as little energy as possible (in its exhaust). When we view organisms as food, we define efficiency as the farmer does, in terms of energy storage (net production from available energy). When we are energy users, we define efficiency as the truck driver does, in terms of how much useful work we accomplish with the available energy.

A common ecological measure of energy efficiency is called food-chain efficiency, or **trophic-level efficiency**, which is the ratio of production of one trophic level to the production of the next lower trophic level. (See Chapter 6.) This efficiency is never very high. Green plants convert only 1–3% of the energy received from the sun during the year to new plant tissue. The efficiency with which herbivores convert the potentially available plant energy into herbivorous energy is usually less than 1%, as is the efficiency with which carnivores convert herbivores into carnivorous energy. It is frequently written in popular literature that the transfer is 10%—for example, that 10% of the energy in corn can be converted into energy in a cow. However, this is a managed ecological efficiency rather than the natural, trophic-level efficiency. In natural ecosystems, the organisms in one trophic level tend to take in much less energy than the potential maximum amount available to them, and they use more energy than they store for the next trophic level.

Consider an example. At Isle Royale National Park, an island in Lake Superior, wolves feed on moose in a natural wilderness. A pack of 18 wolves kill an average of one moose approximately every 2.5 days,[6] resulting in a trophic-level efficiency of wolves of about 0.01%. Wolves use most of the energy they take in from eating moose, especially in the search for prey.[7] From the wolves' point of view, wolves are efficient, but from the point of view of someone who wants to feed on wolves, they appear inefficient.

The rule of thumb for ecological trophic energy efficiency is that more than 90% (usually much more) of all energy transferred between trophic levels is lost as heat.

The Second Law of Thermodynamics

To better understand why we cannot recycle energy, imagine a closed system (a system that receives no input after the initial input) containing a pile of coal, a tank of water, air, a steam engine, and an engineer (Figure 9.8). Suppose the engine runs a lathe that makes furniture. The engineer lights a fire to boil the water, creating steam to run the engine. As the engine runs, the heat from the fire gradually warms the entire system.

When all the coal is completely burned, the engineer will not be able to boil any more water, and the engine will stop. The average temperature of the system is now higher than the starting temperature. The energy that was in the coal is dispersed throughout the entire system, much of it as heat in the air.

Why can't the engineer recover all that energy, recompact it, put it under the boiler, and run the engine? The answer is found in the second law of thermodynamics. Physicists have discovered that no real use of energy can ever be 100% efficient. Whenever useful work is done, some energy is inevitably converted to heat. Collecting all the energy dispersed in this closed system would require more energy than could be recovered.

Our imaginary system begins in a highly organized state, with energy compacted in the coal. It ends in a less organized state, with the energy dispersed throughout the system as heat. The energy has been degraded, and the system is said to have undergone a decrease in order. The measure of the decrease in order (the disorganization of energy) is called **entropy**. The engineer did produce some furniture, converting a pile of lumber into nicely ordered tables and chairs. The system had a local increase of order (the furniture) at the cost of a general increase in disorder (the state of the entire system). All energy of all systems tends to flow toward states of increasing entropy.

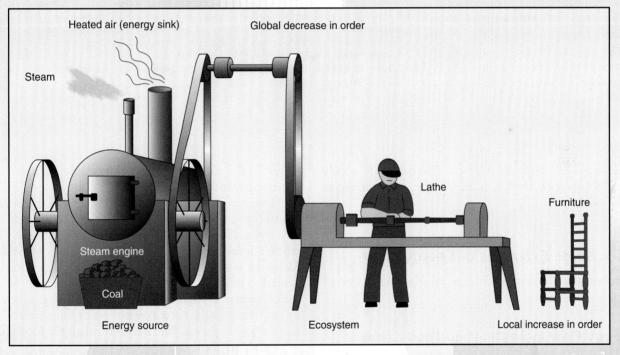

Figure 9.8 ■ A system closed to the flow of energy.

Less than 10% (approximately 1% in natural ecosystems) is fixed as new tissue. In highly managed ecosystems, such as ranches, the efficiency may be greater. But even in such systems, it takes an average of 3.2 kg (7 lb) of vegetable matter to produce 0.45 kg (1 lb) of edible meat. Cattle are among the least efficient producers, requiring around 7.2 kg (16 lb) of vegetable matter to produce 0.45 kg (1 lb) of edible meat. Chickens are much more efficient, using approximately 1.4 kg (3 lb) of vegetable matter to produce 0.45 kg (1 lb) of eggs or meat. Much attention has been paid to the idea that humans should eat at a lower trophic level in order to use resources more efficiently. (See Critical Thinking Issue, "Should People Eat Lower on the Food Chain?")

Ecological Efficiencies

Table 9.1 gives some values for growth efficiency, or gross production efficiency (*P*/*C*), which is the ratio of the material produced (*P* is the net production) by an organism or population to the material consumed (*C*). The amount consumed is normally much less than the maximum amount available. Estimates show, for example, that less than 1% to about 20% of the leaves available in forests and woodlands are consumed annually by leaf-eating insects.[9]

Table 9.1 also gives examples of net growth efficiency, or net production efficiency (*P*/*A*). This is the ratio of the material produced (*P*) to the material assimilated (*A*), which is less than the material consumed because some food taken in is discharged as waste and never used by an organism.

Table 9.1 • Ecological Efficiencies for Animal Populations

	Ecological Efficiencies (%)	
	---	---
Trophic Types	Net Production Efficiency (*P*/*A*)	Gross Production Efficiency (*P*/*C*)
Terrestrial Animals		
Microorganisms[a]	~40	
Invertebrates		
Herbivores	20–40	8–27
Carnivores	10–37	~34
Saprophages	17–40	5–8
Vertebrates		
Herbivores[a]	2–10	
Carnivores[a]	2–10	
Aquatic Animals		
Fishes[a,b]	—	1–7

Sources: T. Penczak, *Comp. Biochem. Physiology* 101 (1992): 791–798; D. E. Reichle, "The Role of Soil Invertebrates in Nutrient Cycling," in U. Lohm and T. Persson, eds., *Ecol. Bull.* (Stockholm) 25 (1977): 145–156; and M. Schaefer, "Secondary Production and Decomposition," in E. Rohrig and B. Ulrich, eds., *Temperate Deciduous Forests*, vol. 7 of *Ecosystems of the World* (Amsterdam: Elsevier, 1991).
[a] Data are based on characteristic values for trophic levels and populations.
[b] Populations in a tropical river.

Many other kinds of energetic efficiencies are widely used in ecological studies. Some of these are described in A Closer Look 9.2.[8]

9.5 Some Examples of Energy Flow

We conclude the chapter with several examples of energy flow in ecosystems.

Energy Flow in an Old-Field Food Chain

In an old field in Michigan, meadow mice feed on grasses and herbs, and least weasels feed on mice (this is just one of many food chains in this old field).[10] The first step in the flow of energy is the fixation of light energy by photosynthesis in the leaves of grasses, herbs, and shrubs. In this step, the energy in light is transferred to and stored in sugar or carbohydrates in the plants. Recall that this energy—the energy that is stored by autotrophs before any is used—is called gross primary production. Some of the energy is used immediately by the leaves to keep their own life processes going. As explained earlier, the amount

stored by the autotrophs after using what is needed is net primary production.

As we have seen, only a small fraction of the energy available to each trophic level is used for net production of new tissue. A large fraction of the energy available to each trophic level is used in respiration. In the old field in Michigan, about 15% of the vegetation's gross production is used in respiration; 68% of the energy taken up by the mice is used in respiration; and 93% of the energy taken up by the least weasel is used in respiration. Only a part of the energy flow in the old field moves through the food chain of vegetation–mice–weasels. One reason is that mice eat only the seeds of the plant. Most of the energy remains in the vegetation until it is transferred from dead vegetation to other animals, fungi, and bacteria by what is called the decomposer food chain.

Energy Flow in a Stream or River

In most ecosystems, the original fixing of energy occurs within the ecosystem; however, some freshwater streams are an exception. The amount of organic matter produced

Should People Eat Lower on the Food Chain?

The energy content of a food chain is often represented by an *energy pyramid*, such as the one shown here in Figure 9.9*a* for a hypothetical, idealized food chain. In an energy pyramid, each level of the food chain is represented by a rectangle whose area is more or less proportional to the energy content for that level. For the sake of simplicity, the food chain shown here assumes that each link in the chain has one and only one source of food.

Assume that if a 75-kg (165-lb) person ate frogs (and some people do!), he would need 10 a day, or 3,000 a year (approximately 300 kg or 660 lb). If each frog ate 10 grasshoppers a day, the 3,000 frogs would require 9,000,000 grasshoppers a year to supply their energy needs, or approximately 9,000 kg (19,800 lb) of grasshoppers. A horde of grasshoppers of that size would require 333,000 kg (732,600 lb) of wheat to sustain them for a year.

As the pyramid illustrates, energy content decreases at each higher level of the food chain. The result is that the amount of energy at the top of a pyramid is related to the number of layers the pyramid has. For example, if people fed on grasshoppers rather than frogs, each person could probably get by on 100 grasshoppers a day. The 9,000,000 grasshoppers could support 300 people for a year, rather than only one. If, instead of grasshoppers, people ate wheat, then 333,000 kg of wheat could support 666 people for a year.

This argument is often extended to suggest that people should become herbivores (vegetarians, in lay parlance) and eat directly from the lowest level of all food chains, the autotrophs. Consider, however, that humans can eat only parts of some plants. Herbivores can eat some parts of plants that humans cannot eat and some plants that humans cannot eat at all. When people eat

these herbivores, more of the energy stored in plants becomes available for human consumption. The most dramatic example of this is in aquatic food chains. Because people cannot digest most kinds of algae, which are the base of most aquatic food chains, they depend on eating fish that eat algae and fish that eat other fish. So if people were to become entirely herbivorous, they would be excluded from many food chains. In addition, there are major areas of Earth where crop production damages the land but grazing by herbivores does not. In those cases, conservation of soil and biological diversity lead to arguments that support the use of grazing animals for human food. This creates an environmental issue: How low on the food chain should people eat?

Critical Thinking Questions

1. Why does the energy content decrease at each higher level of a food chain? What happens to the energy that is lost at each level?

2. The pyramid diagram uses mass as an indirect measure of the energy value for each level of the pyramid. Why is it appropriate to use mass to represent energy content?

3. Using the average of 21 kJ of energy to equal 1 g of completely dried vegetation (see Working It Out 9.3) and assuming that wheat is 80% water, what is the energy content of the 333,000 kg of wheat shown in the pyramid?

4. Make a list of the environmental arguments for and against an entirely vegetarian diet for people. What might be the consequences for U.S. agriculture if everyone in the country began to eat lower on the food chain?

5. How low do you eat on the food chain? Would you be willing to eat lower? Explain.

(a)

1 person

3,000 frogs

9,000,000 grasshoppers

333,000 kg of wheat

(b)

(c)

Figure 9.9 ■ (*a*) Energy pyramid (*b*) Grasshoppers (*c*) Frogs that eat Grasshoppers.

by algae living in a stream is small relative to the amount of organic matter that falls into the stream from dead leaves and twigs of vegetation on the land.[11] *Detritivores* (organisms that feed on dead organic material) are common in streams and feed mainly on this deposited vegetation. Some of the animals are shredders that tear up leaves; others feed on the smaller pieces.

Other grazing animals move along the surfaces of rocks and scrape off attached algae. Many stream predators are larvae of land-dwelling insects, such as dragonflies. Some animals capture prey from the land or air, as in the case of trout that catch flying insects.

An extreme case of a food chain based on external food input occurs in the floodplain of the Amazon River basin, in which fish feed on fruits and nuts carried into the streams during the rainy season. Here, the production of herbivorous fish exceeds what would be possible from aquatic primary production alone, yielding an abundant food supply for the human populations of the region.[11]

Energy Flow in Ocean Ecosystems

Several ocean food chains start with phytoplankton that live near the ocean surface where sunlight penetrates and oxygen levels are comparatively high. One of these food chains continues near the ocean surface where a variety of animals feed on those algae—these animals include tiny floating invertebrates and some of Earth's huge whales, the baleen whales that sweep up the plankton as they swim. In turn, these animals are fed on by other animals that live near the surface or spend much of their time in the upper ocean.

A second, curious food chain exists primarily deep in the ocean. It begins at the surface with the production of phytoplankton and includes surface feeding on those algae by small invertebrate animals. Wastes released by these animals, including fecal material and dead individuals, sink and descend to the ocean depths. There, deep-living animals feed on this "rain" of organic material, produced far away, high above them where sunlight shines. It is for us a curious and comparatively little known kind of life, so different from our land habitat. It is a subject of much research and discovery.

Chemosynthetic Energy Flow in the Ocean

Earlier, we mentioned organisms that make their own food from energy in sulfur compounds. This process creates a curious class of food chains in the depths of the oceans. These food chains support previously unknown life-forms. The basis of the food chains is *chemosynthesis*, in which the source of energy is not sunlight but hot, inorganic sulfur compounds emitted from vents in the ocean floor.

Sulfur-laden water is emitted from hot-water vents at depths of 2,500–2,700 m (8,200–8,900 ft) in areas where

Figure 9.10 ■ Among Earth's most curious creatures are giant worms that live in deep-sea vents and feed on chemosynthetic organisms.

flowing lava causes seafloor spreading. A rich biological community exists in and around the vents, including large white clams up to 20 cm (7.8 in) in diameter, brown mussels, and white crabs. Clams and mussels filter chemoautotrophic bacteria and particles of dead organic matter from the water. Some vent communities contain limpets, pink fish, tube worms, and octopuses. Among the most curious creatures found in vents are giant worms, some 3.9 m (12.8 ft) long (Figure 9.10).[12]

Large areas of ocean have low productivity; combined, however, oceans account for a major portion of total energy fixed. Highly productive areas of the oceans occur in upwelling zones, which occur when deep-ocean waters, rich in nutrients from dead organic material, flow upward, allowing abundant growth of algae and photosynthetic bacteria. Herbivores and carnivores move organic nutrients through the food chain. Although only 1/1,000 of the oceans' surface has natural upwellings, these zones account for more than 44% of the fish eaten by the world's human population.[13]

Summary

■ The study of energy flow is important in determining limits on food supply and on the production of all biological resources, such as wood and fiber.

■ In every ecosystem, energy flow provides a foundation for life and thus imposes a limit on the abundance and richness of life. The amount of energy available to each trophic level in a food chain depends not only on the strength of the energy source but also on the efficiency with which the energy is transferred along the food chain.

■ Energy is fixed by autotrophs—organisms that make their own food from energy and small inorganic compounds. The initial energy comes from two sources: light (mainly sunlight) and small sulfur compounds. Plants, algae, and some bacteria are autotrophs.

■ Only autotrophs can make their own food; all other organisms are heterotrophs, which must feed on other organisms.

■ Biological production is the production of new organic matter, which we measure as change in biomass, change in stored energy, or change in stored carbon. Another way to think about biological production is that it is the change in biomass over time.

■ Gross production is production measured before any utilization. Net production is the amount stored (not used) at the end of some time period. Respiration uses stored energy, so net production equals gross production minus respiration.

■ The laws of thermodynamics connect life to order in the universe. The second law of thermodynamics tells us that order always decreases when any real process occurs in the universe. However, life is more ordered than its environment. The ability to create order is the essence of what we get from our food.

■ Energy efficiency is the ratio of output to input, or the amount of useful work obtained from some amount of available energy. Trophic-level efficiency is the ratio of production of one trophic level to the production of the next lower trophic level. This efficiency is never very high, often only about 1%.

REEXAMINING THEMES AND ISSUES

Human Population

The ultimate limit on the human population and its use of resources is set by available energy, although many other factors can set limits well below the maximum.

Sustainability

Ecological energy flow sets an upper limit to sustainable biological production.

Global Perspective

From a cosmic perspective, Earth is a small planet where life uses the one-way flow of energy from the sun to create life's order: organisms, species, ecosystems, and the environment of the whole planet.

Urban World

One of the fundamental characteristics of life is that it uses energy to create order on a local scale. From an ecological perspective, cities are an extreme example of this process.

People and Nature

Efficiency of energy storage is one of people's primary practical concerns about life. The abundant production of life is one of the qualities people enjoy most about nature.

Science and Values

Until we know the upper limit on biological production, we cannot know the practical limits of our harvest and use of living resources. Understanding energy flow is fundamental to understanding the limits of our actions and what we can consider a good—that is, sustainable—set of activities.

Key Terms

autotrophs **166**
biological
  production **165**
biomass **165**
chemoautotrophs **166**

ecosystem energy
  flow **168**
entropy **171**
gross production **168**
heterotrophs **166**

net production **168**
photosynthesis **166**
primary
  production **166**
respiration **167**

secondary
  production **167**
thermodynamic
  system **170**
trophic-level
  efficiency **170**

Study Questions

1. What is the difference between gross production and net production? Primary and secondary production?

2. What's the meaning of the statement "Any living or life-containing system is always more ordered than its nonliving environment"?

3. Keep track of the food you eat during one day and make a food chain linking yourself with the sources of those foods. Determine the biomass (grams) and energy (kilocalories) you have eaten. Using an average of 5 kcal/g, then using the information on food packaging or assuming that your net production is 10% efficient in terms of the energy intake, how much additional energy might you have stored during the day? What is your weight gain from the food you have eaten?

4. Referring to question 3, what amount of vegetation did you eat during one day? If vegetation was 1% efficient in converting sunlight to organic matter stored as net production, how much sunlight was required to provide the vegetation you took in during the day?

Further Reading

Relative to many other topics discussed in this book, energy flow and productivity have a long history, and the most useful and interesting references tend to be classics published several decades ago. Although in most cases we try to provide the most up-to-date references, with this subject we believe some of the easiest-to-read and most important references are among the classical earlier works.

Blum, H. F. 1962. *Time's Arrow and Evolution.* New York: Harper & Row. A very readable book discussing how life is connected to the laws of thermodynamics and why this matters.

Gates, D. M. 1980. *Biophysical Ecology.* New York: Springer-Verlag. A discussion about how energy in the environment affects life.

Morowitz, H. J. 1979. *Energy Flow in Biology.* Woodbridge, Conn.: Oxbow. The most thorough and complete discussion available about the connection between energy and life, at all levels, from cells to ecosystems to the biosphere.

Morowitz, H. J. 1981. "The Six Million Dollar Man." In *The Wine of Life and Other Essays on Societies, Energy, and Living Things.* New York: Bantam. A fun essay about the second law of thermodynamics and life.

Peterson, R. O. 1995. *The Wolves of Isle Royale: A Broken Balance.* Minocqua, Wis.: Willow Creek Press. A firsthand account of Rolf Peterson's 25-year association with a long-running study of the wild wolves of Isle Royale National Park and their primary prey, the moose.

Schrödinger, E. (ed. Roger Penrose) 1992. *What Is Life?: With Mind and Matter and Autobiographical Sketches (Canto).* Cambridge: Cambridge University Press. The original statement about how the use of energy differentiates life from other phenomena in the universe. Easy to read, and a classic.

Sherman, K. 1990. *Large Marine Ecosystems: Patterns, Processes, and Yields.* Portland, Ore.: Book News. A book based on an AAAS symposium dealing with the possible impacts of global change on ocean productivity. It discusses managing large marine ecosystems as multinational units in order to sustain biomass yields of major coastal regions.

CHAPTER 10

Ecological Restoration

Sunset in Yosemite, a nineteenth-century painting by Albert Bierstadt, illustrates the idea of the balance of nature. Nature is portrayed as still (in a single state), beautiful, and without people.

Learning Objectives

Restoration ecology is a new field. In this chapter, we explore the concepts of restoration ecology, with a special emphasis on how ecosystems restore themselves through the process of ecological succession. After reading this chapter, you should understand:

- What ecological restoration means.
- What kinds of goals are possible for ecological restoration.
- What basic approaches, methods, and limits apply to restoration.
- How an ecosystem restores itself through ecological succession after a disturbance.

- What role disturbances play in the persistence of ecosystems.
- How physical forces and biological processes affect the land.
- Why ecosystems do not remain in a steady state.

CASE STUDY

The Hands That Rock the Cradle of Civilization: Demise and Possible Restoration of the Tigris–Euphrates Marshlands

The famous cradle of civilization, the land between the Tigris and Euphrates rivers, is so called because the waters from these rivers and the wetlands they formed made possible one of the earliest sites of agriculture, and from this the beginnings of Western civilization. This well-watered land in the midst of a major desert was also one of the most biologically productive areas in the world, used by many species of wildlife, including millions of migratory birds.

Ironically, the huge and famous wetlands between these two rivers, land that today is within Iraq, have been greatly diminished by the very civilization that they helped create. "We can see from the satellite images that by 2000, all of the marshes were pretty much drained, except for 7 percent on the Iranian border," said Dr. Curtis Richardson, director of the Duke University Wetland Center.[1]

A number of events of the modern age led to the marsh's destruction. Beginning in the 1960s, Turkey and Syria began to build dams upriver in the Tigris and Euphrates to provide irrigation and electricity, and now these number more than 30. Then in the 1980s Saddam Hussein had dikes and levees built to divert water from the marshes so that oil fields under the marshes could be drilled.

For at least 5,000 years, the Ma'adan people—the Marsh Arabs—lived in these marshes. But the Iran–Iraq War (1980–1988) killed many of them and also added to the destruction of the wetlands (Figure 10.1*a* and *b*).

Today efforts are under way to restore the wetlands. According to the United Nations Environment Program, since the early 1970s the area of the wetlands has increased by 58%.[2] But some scientists believe that there has been little improvement, and the question remains here, as in many places around the world: Can ecosystems be restored once people have seriously changed them?

The purpose of this chapter is to provide an understanding of what is possible and what can be done to restore damaged ecosystems.

In recent years, a new field called **restoration ecology** has developed within the science of ecology. Its goal is to return damaged ecosystems to some set of conditions considered functional, sustainable, and "natural." Whether restoration can always be successful is still an open question. For some ecosystems and species, success appears achievable, but often success has required great effort.

10.1 Restore to What?

The idea of ecological restoration raises a curious question: Restore to what?

The Balance of Nature

Until the second half of the twentieth century, the predominant belief in Western civilization was that any natural area—a forest, a prairie, an intertidal zone—left undisturbed by people achieved a single condition that would persist indefinitely. This condition, as mentioned in Chapter 3, has been known as the **balance of nature**. The major tenets of a belief in the balance of nature are as follows:

1. Nature undisturbed achieves a permanency of form and structure that persists indefinitely.
2. If it is disturbed and the disturbing force is removed, nature returns to exactly the same permanent state.
3. In this permanent state of nature, there is a "great chain of being" with a place for each creature (a habitat and a niche) and each creature in its appropriate place.

These ideas had their roots in Greek and Roman philosophies about nature, but they have played an important role

(a)

(b)

Figure 10.1 ■ (*a*) A Marsh Arab village in the famous wetlands of Iraq, said to be one of the places where Western civilization originated. The people in this picture are among an estimated 100,000 Ma'adan people who now live in their traditional marsh villages, many having returned recently. These marshes are among the most biologically productive areas on Earth. (*b*) Map of the Fertile Crescent, where the Marsh Arabs live, and is called the Cradle of Civilization. It is the land between the Tigris (the eastern river) and Euphrates (the western river) in what is now Iraq. Famous cities of history, such as Nineveh, developed here, made possible by the abundant water and good soil. The gray area shows the known original extent of the marshes, the bright green their present area.

in modern environmentalism as well. In the early twentieth century, ecologists formalized the belief in the balance of nature. They said that succession proceeded to a fixed, classic condition, which they called a **climax state** and defined as a steady-state stage that would persist indefinitely and have maximum organic matter, maximum storage of chemical elements, and maximum biological diversity. At that time, people thought that wildfires were always detrimental to wildlife, vegetation, and natural ecosystems. *Bambi*, a Walt Disney movie of the 1930s, expressed this belief, depicting a fire that brought death to friendly animals. In the United States, Smokey Bear is a well-known symbol employed for many decades by the U.S. Forest Service to warn visitors to national forests to be careful with fire and avoid setting wildfires. The message is that wildfires are always harmful to wildlife and ecosystems.

All of this suggests a belief that the balance of nature does in fact exist. But if that were true, the answer to the question "Restore to what?" would be simple: Restore to the original, natural, permanent condition. The method of restoration would be simple, too: Get out of the way and let nature take its course. Since the second half of the twentieth century, though, ecologists have learned that nature is not constant and that forests, prairies—all ecosystems—undergo change. Moreover, since change has been a part of natural ecological systems for millions of years, many species have adapted to change. Indeed, many require specific kinds of change in order to survive.

Dealing with change—natural and human-induced— poses questions of human values as well as science. This is illustrated by wildfires in forests, grasslands, and shrublands, which can be extremely destructive to human life

and property. Scientific understanding tells us that fires are natural and that some species require them. But whether we choose to allow fires to burn, or light fires ourselves, is a matter of values. In 1991 a wildfire that started in chaparral shrubland in Santa Barbara, California, burned just 83 minutes but did $500 million worth of damage. A few years later, a wildfire in Oakland, California, did $1 billion worth of damage. Restoration ecology depends on science to discover what used to be, what is possible, and how different goals can be achieved. The selection of goals for restoration is a matter of human values.

The Boundary Waters Canoe Area Wilderness: An Example of the Naturalness of Change

One of the best-documented examples of natural disturbance is the role of fire in the northern woods of North America. The Boundary Waters Canoe Area—an area of more than 400,000 hectares (a million acres) in northern Minnesota designated as wilderness under the U.S. Wilderness Act— exemplifies nature relatively undisturbed by people. The area is no longer open to logging or other direct disturbances by people. In the early days of European exploration and settlement of North America, French voyagers traveled through this region hunting and trading for furs. In some places, logging and farming were common in the nineteenth and early twentieth centuries, but for the most part the land has been relatively untouched. Despite the lack of human influence, the forests show a persistent history of fire. Fires occur somewhere in this forest almost every year, and on average the entire area burns once in a century. Fires cover areas large enough to be visible by satellite remote sensing (Figure 10.2).

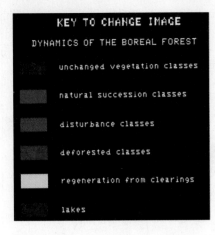

Figure 10.2 ■ Forest fires can be natural or caused by people. This figure shows the change in a large area of the Superior National Forest in Minnesota between 1973 and 1983, as observed by the Landsat satellite. The black boundaries show a central corridor where logging is permitted, surrounded by the Boundary Waters Canoe Area at the top and bottom. This is an area that is protected from all uses except certain kinds of recreation. The bright yellow shows areas that were clear of trees in 1973 but had regenerated to young forest by 1983. Most of this change is due to regrowth following a large fire that burned both inside and outside the wilderness. Red areas were forested in 1973 but cleared in 1983. Most of these are outside the wilderness, and some of these are due to logging (red) and some to fire or storms. Greens show areas that were forested both years.[3]

When fires occur in the forests of the Boundary Waters Canoe Area at natural rates and natural intensities, they have some beneficial effects. For example, trees in unburned forests appear more susceptible to insect outbreaks and disease. Thus, recent ecological research suggests that wilderness depends on change and that succession and disturbance are continual processes. The landscape is dynamic.[3]

Goals of Restoration: What Is "Natural"?

With the examples of the wetlands in Iraq and forests of the Boundary Waters Canoe Area, we can now return to the question "Restore to what?" If an ecosystem passes naturally through many different states, and all of them are "natural," and if change itself—including certain kinds of wildfire—is natural, then what can it mean to "restore" nature? And how can restoration that involves such things as wildfires occur without undue damage to human life and property?

Can we restore an ecological system to any one of its past states and claim that this is natural and successful restoration? A frequently accepted answer is that restoration means restoring an ecosystem to its historical range of variation and to an ability to sustain itself and its crucial functions, including the cycling of chemical elements (see Chapter 5), the flow of energy (Chapter 9), and the maintenance of the biological diversity that existed previously (Chapters 7 and 8). According to this interpretation, ecological restoration means restoring processes and a set of conditions that are known to have existed for that ecosystem. From this point of view, we examine populations that have declined and ecosystems that have been damaged and try to learn what is lacking. Then we go about trying to restore what is lacking.

But a wide variety of answers have been put forward to answer the question: What is restoration? At the extreme are those who argue that all human impacts on nature are "unnatural" and therefore undesirable and that the only true goal of restoration is to bring nature back to a condition that existed prior to human influence. The anthropologist Paul S. Martin takes this position. He proposes that the only truly "natural" time is before any significant human influence occurs. Specifically, he suggests that restoration to conditions of 10,000 B.C.—before farming—should be our goal. He even suggests introducing the African elephant into North America to replace the mastodon, whose extinction, he argues, was the result of hunting by Indians.[4]

But, in general, new ways of thinking about restoration leave open choice, which is a matter of science and values. Science tells us what nature has been and what it can be; our values determine what we want nature to be. There is no single perfect condition. However, for some of the

Table 10.1 • Some Possible Restoration Goals	
Goal	**Approach**
1. Pre-Industrial	Maintain ecosystems as they were in A.D. 1500
2. Presettlement (e.g., of North America)	Maintain ecosystems as they were about A.D. 1492
3. Preagriculture	Maintain ecosystems as they were about 5,000 B.C.
4. Before any significant impact of human beings	Maintain ecosystems as they were about 10,000 B.C.
5. Maximum production	Independent of a specific time
6. Maximum diversity	Independent of a specific time
7. Maximum biomass	Independent of old growth
8. Preserve a specific endangered species	Whatever stage it is adapted to
9. Historical range of variation	Create the future like the known past

goals in Table 10.1, specific conditions are especially desirable. It is possible to restore an ecosystem so that most of the time it supports conditions that people desire for one reason or another.

10.2 What Needs to Be Restored?

Ecosystems of all types have undergone degradation and need restoration. However, certain kinds of ecosystems have undergone especially widespread loss and degradation and are therefore a focus of current attention. In addition to forests and wetlands, attention has focused on grasslands, especially the North American prairie; streams and rivers and the riparian zones alongside them; lakes; and habitats of threatened and endangered species. Also included are areas that people desire to restore for aesthetic and moral reasons, showing once again that restoration involves values. In this section, restoration of wetlands, rivers, streams, and prairies is discussed briefly.

Wetlands, Rivers, and Streams

In North America, large areas of both freshwater and coastal wetlands have been greatly altered during the past 200 years (Figure 10.3). It is estimated that California, for example, has lost more than 90% of its wetlands, both freshwater and coastal, and that the total wetland loss for the United States is about 50% (see Chapter 21). It is not only the cradle of civilization that has suffered; wetlands around the world are affected.

One of the largest and most expensive restoration projects in the United States is the restoration of the Kissimmee River in Florida. This river was channelized, or straightened, by the U.S. Army Corps of Engineers to provide ship passage through Florida. However, although the river and its adjacent ecosystems were greatly altered, shipping never developed, and now several hundred million dollars must be spent to put the river back as it was before. The task will include restoring the meandering flow of the river channel and replacing the soil layers in the order in which they had lain on the bottom of the river prior to channelization (see Chapter 21).

Everglades National Park is also the focus of a variety of restoration efforts, and a large effort is under way to restore bottomland—that is, wetland—forests of the lower Mississippi Valley (see discussion in Chapter 14).

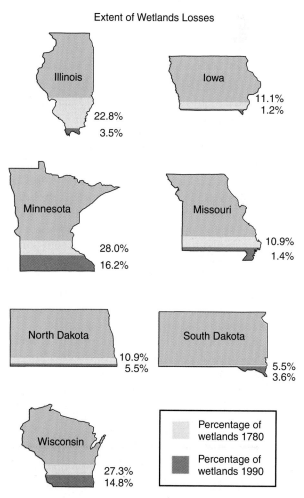

Figure 10.3 ■ Loss of wetlands in the midwestern United States. These maps illustrate the extensiveness of wetland losses. In some states, the losses are even greater. For example, it is estimated that about 90% of the wetlands in California have been lost. [*Source:* Based on Scientific Assessment and Strategy Team, *Science for Floodplain Management into the 21st Century* (Faber, 1996), p. 84.]

(a)

(b)

Figure 10.4 ■ Primary succession. (*a*) Forests developing on new lava flows in Hawaii and (*b*) at the edge of a retreating glacier.

Prairie Restoration

As noted in Chapter 8, prairies once occupied more land in the United States than any other kind of ecosystem. Today, only a few small remnants of prairie remain.

Prairie restoration is of two kinds. In a few places, original prairie exists that has never been plowed. Here, the soil structure is intact, and restoration is simpler. One of the best known of these areas is the Kanza Prairie near Manhattan, Kansas. In other places, where the land has been plowed, restoration is more complicated. Nevertheless, the restoration of prairies has gained considerable attention in recent decades, and restoration of prairie on previously plowed and farmed land is occurring in many midwestern states. The Allwine Prairie, which is within the city limits of Omaha, Nebraska, has been undergoing restoration from farm to prairie for many years. Prairie restoration has also been taking place near Chicago.

A peculiarity about the history of prairies is that although most prairie land was converted to agriculture, this was not done along roads and railroads, so that long, narrow strips of unplowed native prairie remain on these rights-of-way. In Iowa, for example, prairie once covered more than 80% of the state—11 million hectares (28 million acres). More than 99.9% of the prairie land has been converted to other uses, primarily agriculture, but along roadsides there are 242,000 hectares (600,000 acres) of prairie—more than in all of Iowa's county, state, and federal parks. These roadside and railway stretches of prairie provide some of the last habitats for native plants, and restoration of prairies elsewhere in Iowa is making use of these habitats as seed sources.[5]

10.3 When Nature Restores Itself: The Process of Ecological Succession

We have been discussing people's attempts to restore damaged ecosystems. Often, the damage has been caused by people, but natural areas are subject to natural disturbances as well. Storms and fires, for example, have always been a part of the environment.[4] Recovery of disturbed ecosystems can also occur naturally, through a process called **ecological succession**. This natural recovery can occur if the damage is not too great. Sometimes, though, the recovery takes longer than people would like.

We can classify ecological succession as either primary or secondary. **Primary succession** is the initial establishment and development of an ecosystem where one did not exist previously. **Secondary succession** is reestablishment of an ecosystem following disturbances. In secondary succession, there are remnants of a previous biological community, including such things as organic matter and seeds. Forests that develop on new lava flows (Figure 10.4*a*) and at the edges of retreating glaciers (Figure 10.4*b*) are examples of primary succession. Forests that develop on abandoned pastures or following hurricanes, floods, or fires are examples of secondary succession (see A Closer Look 10.1).

Succession is one of the most important ecological processes, and the patterns of succession have many management implications. We see examples of succession all around us. When a house lot is abandoned in a city, weeds begin to grow. After a few years, shrubs and trees can be found; secondary succession is taking place. A farmer weeding a crop and a homeowner weeding a lawn are both fighting against the natural processes of secondary succession.

An Example of Forest Secondary Succession

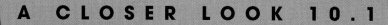

Within a few years after a field is abandoned, seeds of many kinds sprout, some of short-lived weedy plants and some of trees (Figure 10.5a). After a few years, certain species, generally referred to as pioneer species, become established. In the Poconos of Pennsylvania, red cedar is such a species (Figure 10.5b). In New England, white pine, pin cherry, white birch, and yellow birch are especially abundant. These trees are fast-growing in bright light and have widely distributed seeds. For example, seeds of red cedar are eaten by birds and dispersed; birches have very light seeds that are widely dispersed by the wind. After several decades, forests of the pioneer species are well established, forming a dense stand of trees (Figure 10.5c).

Once the initial forest is established, other species begin to grow and become important. Typical dominants in the northeastern United States are sugar maple and beech. These later-successional trees are slower-growing than the ones that came into the forest first, but they have other characteristics that make them well adapted to the later stages of succession. These species are what foresters call shade-tolerant: They grow relatively well in the deep shade of the redeveloping forest.

After three or four decades, most of the short-lived species have matured, borne fruit, and died. Since they cannot grow in the shade of a forest that has been reestablished, they do not regenerate. For example, after five or six decades a New England forest is a rich mixture of birches, maples, beeches, and other species. The trees vary in size, but the trees that dominate now are generally taller than those dominating at earlier stages. After one or two centuries, such a forest will be composed mainly of shade-tolerant species.

(a)

(b)

(c)

Figure 10.5 ■ A series of photographs of succession on an abandoned pasture in the Poconos: (a) a second-year field; (b) young red cedar trees, grasses, and other plants some years after a field was abandoned; (c) mature forests developed along old stone farm walls.

Patterns in Succession

Succession occurs in most kinds of ecosystems, and when it occurs, it follows certain general patterns. We will now consider succession in three classic cases involving forests: (1) on dry sand dunes along the shores of the Great Lakes in North America, (2) in a northern freshwater bog, and (3) in an abandoned farm field.

Dune Succession

Sand dunes are continually being formed along sandy shores and then breached and destroyed by storms. In Indiana, on the shores of Lake Michigan, soon after a dune is formed, dune grass invades. This grass has special adaptations to the unstable dune. Just under the surface, it puts out runners with sharp ends (if you step on one, it will hurt). The dune grass rapidly forms a complex network of underground runners, crisscrossing almost like a coarsely woven mat. Above the ground, the green stems carry out photosynthesis, and the grasses grow.

Once the dune grass is established, its runners stabilize the sand, and seeds of other plants are less easily buried too deep or blown away. The seeds germinate and grow, and an ecological community of many species begins to develop. The plants of this early stage tend to be small, grow well in bright light, and withstand the harshness of the environment—high temperatures in the summer, low temperatures in the winter, and intense storms.

Slowly, larger plants, such as eastern red cedar and eastern white pine, are able to grow on the dunes. Eventually, a forest develops, which may include species such as beech and maple. Such a forest can persist for many years, but at some time a severe storm breaches even these heavily vegetated dunes, and the process begins again (Figure 10.6).

Figure 10.6 ■ Dune succession on the shores of Lake Michigan. Dune grass shoots appear scattered on the slope, where they emerge from underground runners.

Bog Succession

A bog is an open body of water with surface inlets—usually small streams—but no surface outlet. As a result, the waters of a bog are quiet, flowing slowly if at all. Many bogs that exist today originated as lakes that filled depressions in the land, which in turn were created by glaciers during the Pleistocene ice age.

Succession in a northern bog, such as Livingston Bog in Michigan (Figure 10.7), begins when a sedge (grasslike herbs) puts out floating runners (Figure 10.8*a, b*). These runners form a complex matlike network, similar to that formed by dune grass. The stems of the sedge grow on the runners and carry out photosynthesis. Wind blows particles onto the mat, and soil, of a kind, develops. Seeds of other plants land on the mat and do not sink into the water. They can germinate. The floating mat becomes thicker, and small shrubs and trees, adapted to wet environments, grow. In the North, these include species of the blueberry family.

Figure 10.7 ■ Livingston Bog, a famous bog in the northern part of Michigan lower peninsula.

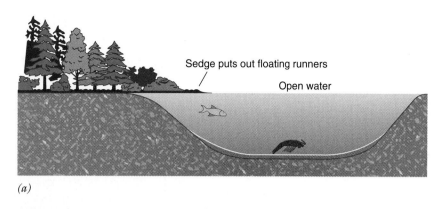

(a)

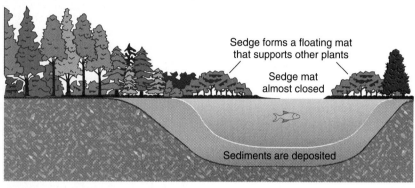

(b)

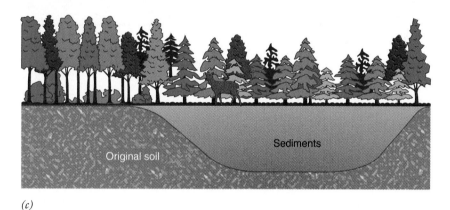

(c)

Figure 10.8 ■ Diagram of bog succession. Open water (*a*) is transformed through formation of a floating mat of sedge and deposition of sediments (*b*) into wetland forest (*c*).

The bog also fills in from the bottom as streams carry fine particles of clay into it (Figure 10.8*b*, *c*). At the shoreward end, the floating mat and the bottom sediments meet, forming a solid surface. But farther out, a quaking bog occurs. You can walk on this quaking bog mat; and if you jump up and down, all the plants around you bounce and shake. The mat is really floating. Eventually, as the bog fills in from the top and the bottom, trees that can withstand wetter conditions—such as northern cedar, black spruce, and balsam fir—grow. The formerly open water bog becomes a wetland forest.

Old-Field Succession

In the eastern United States, such as the states of New England, a great deal of land was cleared and farmed in the eighteenth and nineteenth centuries.

Today, much of this land has been abandoned for farming and allowed to grow back to forest (see Figure 10.5). The first plants to enter the abandoned farmlands are small plants adapted to the harsh and highly variable conditions of a clearing—a wide range of temperatures and precipitation. As these plants become established, other, larger plants enter. Eventually, large trees grow, such as sugar maple, beech, yellow birch, and white pine, forming a dense forest.

General Patterns of Succession

Reviewing these three examples of ecological succession involving forests, you can see common elements in them, even though the environments are very different. Common elements of succession in such cases include the following:

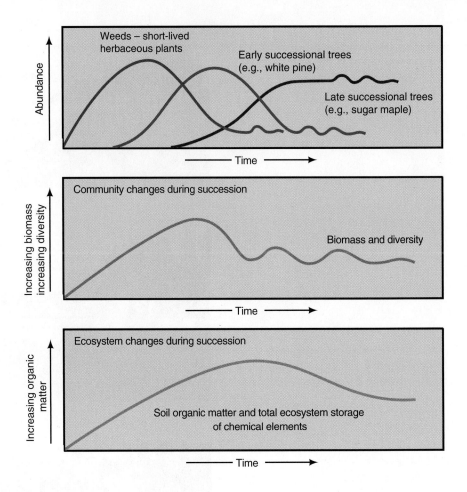

Figure 10.9 ■ Graphs showing changes in biomass and diversity with succession.

1. An initial kind of vegetation specially adapted to the unstable conditions. These plants are typically small in stature, with adaptations that help stabilize the physical environment.
2. A second stage with plants still of small stature, rapidly growing, with seeds that spread rapidly.
3. A third stage in which larger plants, including trees, enter and begin to dominate the site.
4. A fourth stage in which mature forest develops.

Although we list four stages, it is common practice to combine the first two and to speak of early, middle, and late **successional stages**. These general patterns of succession can be found in most ecosystems, although the species differ. The stages of succession are described here in terms of vegetation, but similarly adapted animals and other life-forms are associated with each stage. We discuss other general properties of the process of succession later in this chapter.

Species characteristic of the early stages of succession are called pioneers, or **early-successional species**. They have evolved and are adapted to the environmental conditions in early stages of succession. Plant species that dominate late stages of succession, called **late-successional species**, tend to be slower-growing and longer-lived. These species have evolved and are adapted to environmental

conditions in the late stages. For example, they grow well in shade and have seeds that, while not as widely dispersing, can persist a rather long time.

In early stages of succession, biomass and biological diversity increase (Figure 10.9). In middle stages of succession, we find trees of many species and many sizes. Gross and net production (see Chapter 9) change during succession as well: Gross production increases, and net production decreases. Chemical cycling also changes: The organic material in the soil increases, as does the amount of chemical elements stored in the soils and trees.[6] We look more closely at chemical cycling changes in the next section.

10.4 Succession and Chemical Cycling

One of the important effects of succession is a change in the storage of chemical elements necessary for life. On land, the storage of chemical elements (including nitrogen, phosphorus, potassium, and calcium, essential for plant growth and function) generally increases during the progression from the earliest stages of succession to middle succession. There are two reasons for this.

First, organic matter stores chemical elements; as long as there is an increase in organic matter within the ecosystem, there will be an increase in the storage of chemical elements. This is true for live and dead organic matter. In addition, many plants have root nodules containing bacteria that can assimilate atmospheric nitrogen, which is then used by the plant in a process known as *nitrogen fixation.*

The second reason is indirect: The presence of live and dead organic matter helps retard erosion. Both organic and inorganic soil can be lost to erosion because of the effects of wind and water. Vegetation tends to prevent such losses and therefore causes an increase in total stored material.

Organic matter in soil contributes to the storage of chemical elements in two ways. First, it contains chemical elements itself. Second, dead organic matter functions as an ion-exchange column that holds on to metallic ions that would otherwise be transported in the groundwater as dissolved ions and lost to the ecosystem. As a general rule, the greater the volume of soil and the greater the percentage of organic matter in the soil, the more chemical elements will be retained.

However, the amount of chemical elements stored in a soil depends not only on the total volume of soil but also on its storage capacity for each element. The chemical storage capacity of soils varies with the average size of the soil particles. Soils composed mainly of large, coarse particles, like sand, have a smaller total surface area and can store a smaller quantity of chemical elements. Clay, which is made up of the smallest particles, stores the greatest quantity of chemical elements.

Soils contain greater quantities of chemical elements than do live organisms. However, much of what is stored in a soil may be relatively unavailable, or may only become available slowly because the elements are tied up in complex compounds that decay slowly. In contrast, the elements stored in living tissues are readily available to other organisms through food chains.

Rates of cycling and average storage times are system characteristics (as discussed in Chapter 3). Soils store more elements than does live tissue but cycle them at a slower rate. The increase in chemical elements that occurs in the early and middle stages of succession does not continue indefinitely. If an ecosystem persists for a very long period with no disturbance, it will experience a slow but definite loss of stored chemical elements. Thus, the ecosystem will slowly run downhill and become depauperate—literally, impoverished—and thus less able to support rapid growth, high biomass density, and high biological diversity (Figure 10.10).[7] The changes in chemical cycling during disturbance and successional recovery are discussed further in A Closer Look 10.2.

10.5 Species Change in Succession: Do Early-Successional Species Prepare the Way for Later Ones?

To restore an ecosystem, it is important to understand what causes one species to replace another during the succession process. If we understand these causes and effects,

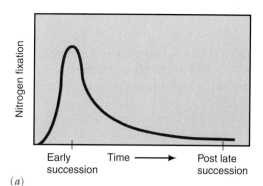

(a)

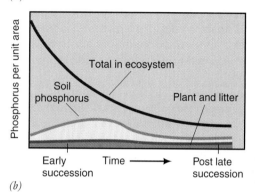

(b)

Figure 10.10 ■ (*a*) Hypothesized changes in soil nitrogen during the course of soil development. (*b*) Change in total soil phosphorus over time with soil development. [*Source:* P. M. Vitousek and P. S. White, "Process Studies in Forest Succession," in D. C. West, H. H. Shugart, and D. B. Botkin, eds., *Forest Succession: Concepts and Applications* (New York: Springer-Verlag, 1981), Figure 17.1, p. 269.]

we can use them to better restore ecosystems. Earlier and later species in succession may interact in three ways: through (1) facilitation, (2) interference, or (3) life history differences. If they do not interact, the result is termed *chronic patchiness* (Table 10.2).[8,9]

Table 10.2 • Patterns of Interaction among Earlier and Later Species in Succession

1. *Facilitation.* One species can prepare the way for the next (and may even be necessary for the occurrence of the next).

2. *Interference.* Early-successional species can, for a time, prevent the entrance of later-successional species.

3. *Life history differences.* One species may not affect the time of entrance of another; two species may appear at different times during succession because of differences in transport, germination, growth, and longevity of seeds.

4. *Chronic patchiness.* Succession never occurs, and the species that enters first remains until the next disturbance.

Source: J. H. Connell and R. O. Slatyer, "Mechanisms of Succession in Natural Communities and Their Role in Community Stability and Organization," *American Naturalist* III (1977): 1119–1144; S. T. A. Pickett, S. L. Collins, and J. J. Armesto, "Models, Mechanisms, and Pathways of Succession," *Botanical Review* 53 (1987): 335–371.

Facilitation

In dune and bog succession, the first plant species—dune grass and floating sedge—prepare the way for other species to grow. This is called **facilitation** (early-successional species facilitate the establishment of later-successional species; Figure 10.11a).

Facilitation has been found to take place in tropical rain forests.[10] Early-successional species speed the reappearance of the microclimatic conditions that occur in a mature forest. Because of the rapid growth of early-successional plants, the temperature, relative humidity, and light intensity at the soil surface in tropical forests can reach levels similar to those of a mature rain forest after only 14 years.[6] Once these conditions are established, species that are adapted to deep forest shade can germinate and persist.

Knowing the role of facilitation can be useful in the restoration of damaged areas. Plants that facilitate the presence of others should be planted first. On sandy areas, for example, dune grasses can help hold the soil before attempts are made to plant larger shrubs or trees.

Interference

Facilitation does not always occur. Sometimes, instead, certain early-successional species interfere with the entrance of other species (Figure 10.11b). For example, in the old fields of mid-Atlantic states, such as the states from Connecticut to Virginia, among the early-successional species are grasses that form dense cover, including the prairie grass little bluestem. The living and dead stems of this grass form a mat so dense that seeds of other plants that fall onto it cannot reach the ground and therefore do not germinate. The grass interferes with the entrance of other plant species—a process called, naturally enough, **interference**.

Interference does not last forever, however. Eventually, some breaks occur in the grass mat—perhaps from surface water erosion, the death of a patch of grass from disease, or removal by fire. Breaks in the grass mat allow seeds of trees such as red cedar to reach the ground. Red cedar is also adapted to early succession because its seeds are spread rapidly and widely by birds who feed on them and because this species can grow well in the bright light and otherwise harsh conditions of early succession. Once started, red cedar soon grows taller than the grasses, shading them so much that they cannot grow. More ground is open, and the grasses are eventually replaced.

In parts of Asia, interference occurs where bamboo grows, and, in tropical areas, where another, smaller grass, *Imperata*, grows. Like little bluestem in the United States, these grasses form stands so dense that seeds of other, later-successional species cannot reach the ground, germinate, or obtain enough light, water, and nutrients to sur-

vive. *Imperata* either replaces itself or is replaced by bamboo, which then replaces itself.[10] Once established, *Imperata* and bamboo appear able to persist for a long time. Once again, when and if breaks occur in the cover of these grasses, other species can germinate and grow, and a forest eventually develops.

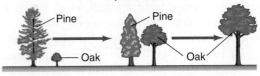

Facilitation—pine provides shade that helps oaks

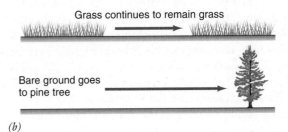

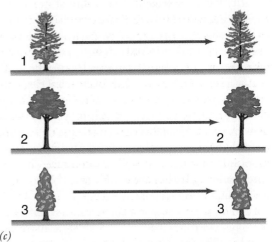

Figure 10.11 ■ Two of the three patterns of interaction among species in ecological succession. (*a*) Facilitation. As Henry David Thoreau observed in Massachusetts more than 100 years ago, pines provide shade and act as "nurse trees" for oaks. Pines do well in openings. If there were no pines, few or no oaks would survive. Thus the pines facilitate the entrance of oaks. (*b*) Interference. Some grasses that grow in open areas form dense mats that prevent seeds of trees from reaching the soil and germinating. The grasses interfere with the addition of trees. (*c*) Chronic patchiness. Earlier-entering species neither help nor interfere with other species; instead, as in a desert, the physical environment dominates.

Changes in Chemical Cycling During a Disturbance

When an ecosystem is disturbed by fire, storms, or human activities, changes occur in chemical cycling. For example, when a forest is burned, complex organic compounds, such as wood, are converted to smaller inorganic compounds, including carbon dioxide, nitrogen oxides, and sulfur oxides. Some of the inorganic compounds from the wood are lost to the ecosystem during the fire as vapors that escape into the atmosphere and are distributed widely or as particles of ash that are blown away, but some of the ash falls directly onto the soil. These compounds are highly soluble in water and readily available for vegetation uptake. Therefore, immediately after a fire, the availability of chemical elements increases. This is true even if the ecosystem as a whole has undergone a net loss in total stored chemical elements.

If sufficient live vegetation remains after a fire, then the sudden, temporary increase, or *pulse*, of newly available elements is taken up rapidly, especially if it is followed by a moderate amount of rainfall (enough for good vegetation growth but not so much as to cause excessive erosion). The pulse of inorganic nutrients can then lead to a pulse in the growth of vegetation and an increase in the amount of stored chemical elements in the vegetation. This in turn boosts the supply of nutritious food for herbivores, which can subsequently undergo a population increase. The pulse in chemical elements in the soil can therefore have effects that extend throughout the food chain.

Other disturbances on the land produce effects similar to those of fire. For example, severe storms, such as hurricanes and tornadoes, knock down and kill vegetation. The vegetation decays, increasing the concentration of chemical elements in the soil, which are then available for vegetation growth. Storms also have another effect in forests: When trees are uprooted, chemical elements that were near the bottom of the root zone are brought to the surface, where they are more readily available.

Knowledge of the changes in chemical cycling and the availability of chemical elements from the soil during succession can be useful to us in restoring damaged lands. We know that nutrients must be available within the rooting depth of the vegetation. The soil must have sufficient organic matter to hold on to nutrients. Restoration will be more difficult where the soil has lost its organic matter and has been leached. A leached soil has lost nutrients as water drains through it, especially acidic water, dissolving and carrying away chemical elements. Heavily leached soils, such as those subjected to acid rain and acid mine drainage, pose special challenges to the process of land restoration.

Life History Differences

In other cases, species do not so much affect one another. Rather, differences in the life histories of the species allow some to arrive first and grow quickly, while others arrive later and grow more slowly. An example of such a **life history difference** is seed disbursal. The seeds of early-successional species are readily transported by wind or animals and so reach a clearing sooner and grow faster than seeds of late-successional species. In many forested areas of eastern North America, for example, birds eat the fruit of cherries and red cedar, and their droppings contain the seeds, which are spread widely.

In contrast, late-successional species can have life histories that bring them in much later. Although sugar maple, for example, can grow in open areas, its seeds take longer to travel and its seedlings can tolerate shade. Beech produces large nuts that store a lot of food for a newly germinated seedling. This helps the seedling establish itself in the deep shade of a forest until it is able to feed itself through photosynthesis. But these seeds are heavy and are moved relatively short distances by seed-eating animals.

Chronic Patchiness

A fourth possibility is that species just do not interact and that succession, as it has been described, does not take place. This is called **chronic patchiness** (Figure 10.11*c*), and it is the case in some deserts. For example, in the warm deserts of California, Arizona, and Mexico, the major shrub species grow in patches, which often consist of mature individuals with few seedlings. These patches tend to persist for long periods until there is a disturbance.[11] Similarly, in highly polluted environments, a sequence of species replacement may not occur.

What kinds of changes occur during succession depends on the complex interplay between life and its environment. Life tends to build up, or aggrade, whereas nonbiological processes in the environment tend to erode and degrade. In harsh environments, where energy or chemical elements required for life are limited and disturbances are frequent, the physical, degrading environment dominates, and succession does not occur.

How Can We Evaluate Constructed Ecosystems?

What happens when restoring damaged ecosystems is not an option? In such cases, those responsible for the damage may be required to establish alternative ecosystems to replace the damaged ones.

An example involved some saltwater wetlands on the coast of San Diego County, California. In 1984, construction of a flood-control channel and two projects to improve interstate freeways damaged an area of saltwater marsh. The projects were of concern because California had lost 91% of its wetland area since 1943 and the few remaining coastal wetlands were badly fragmented. In addition, the damaged area provided habitat for three endangered species: the California least tern, the light-footed clapper rail, and a plant called the salt-marsh bird's beak.

The California Department of Transportation, with funding from the Army Corps of Engineers and the Federal Highway Administration, was required to compensate for the damage by constructing new areas of marsh in the Sweetwater Marsh National Wildlife Refuge. To meet these requirements, eight islands, known as the Connector Marsh, with a total area of 4.9 hectares, were constructed in 1984. An additional 7-hectares area, known as Marisma de Nacion, was established in 1990.

Goals for the constructed marsh, which were established by the U.S. Fish and Wildlife Service, included the following:

1. Establishment of tide channels with sufficient fish to provide food for the California least tern.

2. Establishment of a stable or increasing population of salt-marsh bird's beak for three years.

3. Selection of the Pacific Estuarine Research Laboratory (PERL) at San Diego State University to monitor progress on the goals and conduct research on the constructed marsh. In 1997 PERL reported that goals for the least tern and bird's beak had been met but that attempts to establish a habitat suitable for the rail had been only partially successful.

During the past decade, PERL scientists have conducted extensive research on the constructed marsh to determine the reasons for its limited success. They found that rails live, forage, and nest in cordgrass more than 60 cm tall. Nests are built of dead cordgrass attached to stems of living cordgrass so that the nests can remain above the water as it rises and falls. If the cordgrass is too short, the nests are not high enough to avoid being washed out during high tides. Researchers suggested that the coarse soil used to construct the marsh did not retain the amount of nitrogen needed for cordgrass to grow tall. Adding nitrogen-rich fertilizer to the soil resulted in taller plants in the constructed marsh, but only if the fertilizer was added on a continuing basis.

Another problem is that the diversity and numbers of large invertebrates, which are the major food source of the rails, are lower in the constructed marsh than in natural marshes. PERL researchers suspect that this, too, is linked to low nitrogen levels. Because nitrogen stimulates the growth of algae and plants,

Species	Mitigation Goals	Progress in Meeting Requirements	Status as of 2006
California least tern	Tidal channels with 75% of the fish species and 75% of the number of fish found in natural channels	Met standards	FWS recommended change from endangered to threatened
Salt-marsh bird's beak	Through reintroduction, at least 5 patches (20 plants each) that remain stable or increase for 3 years	Did not succeed on constructed islands but an introduced population on natural Sweetwater Marsh thrived for 3 years (reached 140,000 plants); continue to monitor because plant is prone to dramatic fluctuations in population	Still listed as endangered
Light-footed clapper rail	Seven home ranges (82 ha), each having tidal channels with: a. Forage species equal to 75% of the invertebrate species and 75% of the number of invertebrates in natural areas b. High marsh areas for rails to find refuge during high tides c. Low marsh for nesting with 50% coverage by tall cordgrass d. Population of tall cordgrass that is self-sustaining for 3 years	Constructed Met standards Sufficient in 1996 but two home ranges fell short in 1997 All home ranges met low marsh acreage requirement and all but one met cordgrass requirement and six lacked sufficient tall cordgrass Plant height can be increased with continual use of fertilizer but tall cordgrass is not self-sustaining	Still listed as endangered; in 2005, eight captive-raised birds were released

Note: FWS stands for US Fish and Wildlife Service

which provide food for small invertebrates, and these in turn provide food for larger invertebrates, low nitrogen can affect the entire food chain.

Critical Thinking Questions

1. Make a diagram of the food web in the marsh showing how the clapper rail, cordgrass, invertebrates, and nitrogen are related.

2. The headline of an article about the Sweetwater Marsh project in the April 17, 1998, issue of *Science* declared, "Restored Wetlands Flunk Real-World Test." Based on the information you have about the project, would you agree or disagree with this judgment? Explain your answer.

3. How do you think one can decide whether a constructed ecosystem is an adequate replacement for a natural ecosystem?

4. The term *adaptive management* refers to the use of scientific research in ecosystem management. In what ways has adaptive management been used in the Sweetwater Marsh project? What lessons from the project could be used to improve similar projects in the future?

10.6 Applying Ecological Knowledge to Restore Heavily Damaged Lands and Ecosystems

An example of how ecological succession can aid in the restoration of heavily damaged lands is the effort being made to undo mining damage in Great Britain, where mining has caused widespread destruction to land. In Great Britain, where some mines have been used since medieval times, approximately 55,000 hectares (136,000 acres) have been damaged by mining. Recently, programs have been initiated to remove toxic pollutants from the mines and mine tailings, to restore these damaged lands to useful biological production, and to restore the visual attractiveness of the landscape.[12]

One area damaged by a long history of mining lies within the British Peak District National Park, where lead has been mined since the Middle Ages and waste tailings are as much as 5 m (16.4 ft) deep. The first attempts to restore this area used a modern agricultural approach: heavy application of fertilizers and planting of fast-growing agricultural grasses to revegetate the site rapidly. These grasses quickly green on the good soil of a level farm field, and it was hoped that, with fertilizer, they would do the same in this situation. But after a short period of growth, the grasses died. On the poor soil, leached of its nutrients and lacking organic matter, erosion continued, and the fertilizers that had been added were soon leached away by water runoff. As a result, the areas were shortly barren again.

When the agricultural approach failed, an ecological approach was tried, using knowledge about ecological succession. Instead of planting fast-growing but vulnerable agricultural grasses, ecologists planted slow-growing native grasses known to be adapted to minerally deficient soils and the harsh conditions that exist in cleared areas. In choosing these plants, the ecologists relied on their observations of what vegetation first appeared in areas of Great Britain that had undergone succession naturally.[12] The result of the ecological approach has been successful restoration of once-damaged lands (Figure 10.12).

Figure 10.12 ■ An old lead-mining area in Great Britain, now undergoing restoration. Restoration involves planting of early-successional native grasses adapted to low-nutrient soils with little physical structure.

Heavily damaged landscapes are to be found in many places. Restoration similar to that in Great Britain is done in the United States to reclaim lands damaged by strip mining. In such cases, restoration sometimes begins during the mining process rather than afterward. Similar methods could also be used to restore areas once occupied by buildings in cities.

Summary

- Restoration of damaged ecosystems is a major new emphasis in environmental sciences and is developing into a new field. Restoration involves a combination of human activities and natural processes of ecological succession.

- Disturbance, change, and variation in the environment are natural, and ecological systems and species have evolved in response to these changes.

- When ecosystems are disturbed, they undergo a process of recovery known as ecological succession, the establishment and development of an ecosystem. Knowledge of succession is important in the restoration of damaged lands.

■ During succession there is usually a clear, repeatable pattern of changes in species. Some species, called early-successional species, are adapted to the first stages, when the environment is harsh and variable but when necessary resources may be available in abundance. This contrasts with late stages in succession, when biological effects have modified the environment and reduced some of the variability but also have tied up some resources. Typically, early-successional species are fast-growing, whereas late-successional species are slow-growing and long-lived.

■ Biomass, production, diversity, and chemical cycling change during succession. Biomass and diversity peak in mid-succession, increasing at first to a maximum, then declining and varying over time.

■ Changes in the kinds of species found during succession can be due to facilitation, interference, or simply life history differences. In facilitation, one species prepares the way for others. In interference, an early-successional species prevents the entrance of later-successional ones. Life history characteristics of late-successional species sometimes slow their entrance into an area.

REEXAMINING THEMES AND ISSUES

Human Population

If we degrade ecosystems to the point where their recovery from disturbance is slowed or they cannot recover at all, then we reduce the local carrying capacity of those areas for human beings. For this reason, an understanding of the factors that determine ecosystem restoration is important to developing a sustainable human population.

Sustainability

Life tends to build up; nonbiological forces in the environment tend to tear down. By helping ecosystems to develop, we promote sustainability. Heavily degraded land, such as land damaged by pollution or overgrazing, loses the capacity to recover—to undergo ecological succession. Knowledge of the causes of succession can be useful in restoring ecosystems and thereby achieving sustainability.

Global Perspective

Each degradation of land takes place locally, but such degradation has been happening around the world since the beginnings of civilization. Ecosystem degradation is therefore now a global issue.

Urban World

In cities, we generally eliminate or damage the processes of succession and the ability of ecosystems to recover. As our world becomes more and more urban, we must learn to maintain these processes within cities as well as in the countryside. Ecological restoration is an important way to improve city life.

People and Nature

Restoration is one of the most important ways that people can compensate for their undesirable effects on nature.

Science and Values

Because ecological systems naturally undergo changes and exist in a variety of conditions, there is no single "natural" state for an ecosystem. Rather, there is the process of succession, with all of its stages. In addition, there are major changes in the species composition of ecosystems over time. While science can tell us what conditions are possible and have existed in the past, which ones we choose to promote in any location is a question of values. Values and science are intimately integrated in ecological restoration.

Key Terms

balance of nature **178**
chronic
   patchiness **189**
climax state **179**

early-successional
   species **186**
ecological succession **182**
facilitation **188**

interference **188**
late-successional
   species **186**
life history difference **189**

primary succession **182**
restoration ecology **178**
secondary succession **182**
successional stages **186**

Study Questions

1. Farming has been described as managing land to keep it in an early stage of succession. What does this mean, and how is it achieved?

2. Redwood trees reproduce successfully only after disturbances (including fire and floods), yet individual redwood trees may live more than 1,000 years. Is redwood an early- or late-successional species?

3. Why could it be said that succession does not take place in a desert shrubland (an area where rainfall is very low and the only plants are certain drought-adapted shrubs)?

4. Develop a plan to restore an abandoned field in your town to natural vegetation for use as a park. The following materials are available: bales of hay; artificial fertilizer; and seeds of annual flowers, grasses, shrubs, and trees.

5. Oil has leaked for many years from the gasoline tanks of a gas station. Some of the oil has oozed to the surface. As a result, the gas station has been abandoned and revegetation has begun to occur. What effects would you expect this oil to have on the process of succession?

6. Refer to the Iraqi marshes of the opening case study. Assume that there is no hope of changing water diversion from the many dams upstream on the Tigris and Euphrates rivers. Develop a plan to restore the marshes, considering in particular the decrease in the area of the marshes.

7. Early in the twentieth century, a large meteorite collided with the earth in Siberia and destroyed boreal forests over a large area. In what ways might restoration and succession following this large-scale disturbance differ from restoration and succession after a fire burned a few hectares in the same forest? (See Chapter 8 for information about boreal forests.)

Further Reading

Botkin, D. B. 1992. *Discordant Harmonies: A New Ecology for the 21st Century.* New York: Oxford University Press.

Botkin, D. B. 2001. *No Man's Garden: Thoreau and a New Vision for Civilization and Nature.* Washington, D.C.: Island Press.

Falk, Donald A., and Joy B. Zedler. 2005. *Foundations of Restoration Ecology.* Washington, D.C.: Island Press. A new and important book, written by two of the world's experts on ecological restoration.

Higgs, E. 2003. *Nature by Design: People, Natural Process, and Ecological Restoration.* Cambridge, Mass.: MIT Press. A book that discusses the broader perspective on ecological restoration, including philosophical aspects.

CHAPTER 11

Producing Enough Food for the World: How Agriculture Depends on Environment

A modern organic farm. According to the USDA, organic farming is one of the fastest-growing sectors in U.S. agriculture.

Learning Objectives

The big question about farming and the environment is: Can we produce enough food to feed Earth's growing human population, and do this sustainably? The major agricultural challenges facing us today are to increase the productivity of the land, acre by acre, hectare by hectare; to distribute food adequately around the world; to decrease the negative environmental effects of agriculture; and to avoid creating new kinds of environmental problems as agriculture advances. After reading this chapter, you should understand:

■ What it means to take an ecological perspective on agriculture.

■ How agroecosystems differ from natural ecosystems.

■ How the food supply depends on the environment.

■ What role limiting factors play in determining crop yield.

■ How the concept of sustainability applies to agriculture.

■ How the growing human population, the loss of fertile soils, and the lack of water for irrigation can affect future food shortages worldwide.

■ The relative importance of food distribution and food production.

■ The potential benefits and environmental effects of genetic engineering of crops.

Food for China

China, with 1.3 billion people, one-fifth of the world's population, is the most populous country on Earth. Between 1980 and 1995, China's population grew by 200 million people—about two-thirds of the present population of the United States—to reach 1.2 billion. Although its growth rate is expected to slow somewhat in the coming decades, population experts predict that there will be about 1.5 billion Chinese by 2025(Figure 11.1a).[1] Farming is, of course, an ancient practice in China. China's agriculture is not keeping up with its population growth. However, its population continues to increase, up 16% since 1995.[2] But the area harvested for rice has decreased since 1995, down 6% from 31.1 million hectares (ha) to 29.3.[3] Can China's food production increase to keep pace with its growing population? If not, China may need to import more grain than other countries can spare. China's increased demand for world grain supplies could raise food prices dramatically and precipitate famines in other areas of the world (Figure 11.1b).

To gain some idea of China's potential impact on the world's food supply, consider this: To supply two more beers a year to each person in China would require all the grain produced in Norway. If the average Chinese ate as much fish as the average citizen of Japan, China would consume the entire world's fish catch. All the chickens necessary to reach China's goal of 200 eggs per person per year would require all the grain exported by Canada—the world's second-largest grain exporter.[4]

In just two years, between 1959 and 1961, 30 million Chinese starved to death during a famine brought on by a state policy of modernization that forced millions of farmers to work on large construction projects. In the wake of famine, China attempted to slow population growth through a "one couple, one child" policy (see Chapter 4) and to increase food production through extensive agricultural reform. Between 1980 and 1995, the total fertility rate in China fell from 4.8 children per woman to fewer than 2 children. In the same period, China achieved a remarkable increase in annual grain production from the 200 kg per person needed to maintain a minimal level of physical activity to 300 kg.

What would be done so that China could produce enough food for its people? The country has relatively little cropland—approximately 100 million hectares (ha), or 0.08 hectares per capita. By comparison, India has 0.19 hectares per capita—170 million hectares of land with which to feed its population of more than 1 billion.[1] Industrialization, including the vast, almost completed hydroelectric Three Gorges Dam, is eliminating some farmland. China has been losing agricultural land to roads, railroads, and manufacturing plants at the rate of 1.6% per year.[5]

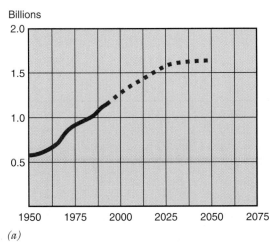

(a)

(b)

Figure 11.1 ■ (a) Population of China, 1950–1994, with projections to 2050. [*Source:* L. R. Brown, *Who Will Feed China? Wakeup Call for a Small Planet* (New York: Norton, 1995), p. 36.] (b) Traditional farming in China.

At the same time, with half of its cropland under irrigation, China is facing a severe water crisis because of seasonal variations in rainfall, periodic floods, and diversion of water to nonagricultural uses. And with increased prosperity, Chinese are eating higher up the food chain—consuming more meat—which requires more grain production per person. Whether China's agricultural resources can sustain its population in coming decades will be a critical test of whether the world can find a sustainable balance among food production, consumer demand, and population growth.[6, 7, 8, 9, 10, 11]

> *The needs of China, then, illustrate several of the themes of this book. We see once again that the human population is a fundamental, underlying problem that needs to be viewed from a global perspective. The desires and demands of the people of China, including an improvement in the material quality of their lives, will put pressure on world food production. This leads to questions about the sustainability of agriculture, both within a nation and worldwide, and ultimately leads to decisions that involve values and science.*

11.1 Can We Feed the World?

Can we produce enough food to feed Earth's growing human population? Can we grow crops sustainably, so that both crop production and agricultural ecosystems remain viable? Can we produce this food without seriously damaging other ecosystems that receive the wastes of agriculture? These are basic environmental questions about agriculture. To answer them, we first must understand how crops grow and how productive they can be—the topic of this chapter. Then we have to consider the environmental effects of agriculture—the subject of the following chapter.

Of all human activities, agriculture has arguably been proved the most sustainable, simply because people have farmed the Nile Valley, the fertile crescent of the Middle East, rice fields in China, and elsewhere for thousands of years. Few, if any, other human activities have been maintained in the same place for equally long times.

Even so, great concern remains about the sustainability of most agriculture. Where crops have been sustainably produced, farming has changed local ecosystems. Perhaps the most notable exception is the Nile Valley: Before the building of the Aswan High Dam, annual floods deposited new soil every year, and farming did not displace this

	Total Land Area (sq km)	Human Population (Millions)	People per Area	Crop Area (sq km)	Crop Area Per Person (sq km)	Crop Land as % of Total Land
Location						
Asia	30,988,970	3,823	123.37	16,813,750	0.044	54%
Africa	29,626,570	850	28.69	11,460,700	0.135	39%
N. and C. America	21,311,580	507	23.79	6,189,030	0.122	29%
S. America	17,532,370	936	53.39	5,842,850	0.062	33%
Europe	22,093,160	362	16.39	4,836,410	0.134	22%
Australia	7,682,300	19	2.47	4,395,000	2.313	57%
World	130,043,970	6,301	48.45	49,734,060	0.079	38%

Table 11.1 • Land, People, and Agriculture, 2006

Source: FAO Statistics 2006 http://faostat.fao.org/faostat/.

Note: Data are available for crops until 2003, hence, some population values in this table will differ from those elsewhere in the chapter, which are for 2005.

process. In most places, however, farming degrades soil; fertilizers and pesticides affect soil, water, and downstream ecosystems; and many other changes occur that are described in this chapter and the next.

The history of agriculture is a series of human attempts to overcome environmental limitations and problems. Each new solution has created new environmental problems, which in turn have required their own solutions. Thus, in seeking to improve agricultural systems, we should expect some undesirable side effects and be ready to cope with them.

There are threats to the existing land in agricultural production, including human developmental pressures to build cities and suburbs and to flood land to create new hydroelectric power sources. In 2005, the great Asian tsunami reminded the world that natural catastrophes affect agricultural land (Figure 11.2).

A surprisingly large percentage of the world's land area is in agriculture: approximately 11% of the total land area of the world excluding Antarctica—an area about the size of South and North America combined—enough to make agriculture a human-induced biome (Table 11.1 and Figure 11.3).

The percentage of land in agriculture varies considerably among continents, from 30% of the land in Europe to 6% in Australia. In the United States, the current situation is still good, so good that considerable farmland is "banked"— farmers are paid not to grow crops, because at present there is more than enough production. As more land is needed, this banked farmland will go back into production.

Here's one of the big problems: In the future, when the human population doubles, the production of agriculture

Figure 11.2 ■ Tsunami aftermath. Destroyed rice fields in Indonesia.

must double just to meet the present level of per-capita food consumption, and for some people the food supply right now isn't adequate. If there are no increases in production per unit area, an additional area equal to the entire New World will have to be put into farmland. Where will we find it? Think about biogeography (Chapter 8). Which biomes should we alter greatly? Land best suited for agriculture has already been put to this use. And existing good farmland is already under pressure for conversion to cities, towns, and suburbs. As the human population increases, cities and suburbs will continue to grow (see Chapter 28) and more of the

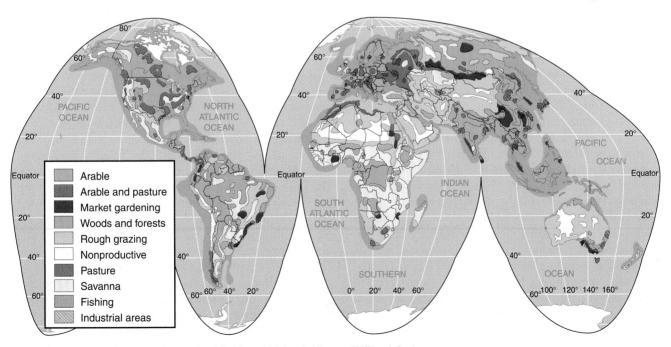

Figure 11.3 ■ World land use showing arable (farmable) land. (*Source:* Phillips Atlas.)

Figure 11.4 ■ Irrigated farmland in Saudi Arabia as seen from space. Although productive where such agriculture was previously impossible, such irrigation in desert areas increases world demand for freshwater.

best agricultural land will be converted to other uses. This suggests that future generations will depend not on better farmland and better farming conditions but on trying to make do on increasingly marginal land. It is a huge challenge.

The world food supply is also greatly influenced by social disruptions and social attitudes, which affect the environment and in turn affect agriculture. In Africa, social disruptions since 1960 have included 12 wars, 70 coups, and 13 assassinations. Such social instability makes sustained agricultural yields difficult.[12] So does variation in weather, the traditional bane of farmers.[13, 14]

So the key to food production in the future appears to be increased production per unit area, probably on worse and worse land, and with the risk of increasing environmental damage. Can this be done? Some agricultural scientists and agricultural corporations believe that production per unit area will continue to increase, partially through advances in genetically modified crops (GMCs). This new methodology, however, raises some important potential environmental problems, discussed in Chapter 12. Furthermore, increased production in the past has depended on increased use of water and fertilizers (Figure 11.4). Water is a limiting factor in many parts of

the world and will become a limiting factor in more areas in the future (Chapters 21 and 22).

11.2 How We Starve

People "starve" in two ways: undernourishment and malnourishment. Undernourishment results from a lack of sufficient calories in available food, so that one has little or no ability to move or work and eventually dies from the lack of energy. Malnourishment results from a lack of specific chemical components of food, such as proteins, vitamins, or other essential chemical elements. Both are global problems.

Widespread undernourishment manifests itself as famines that are obvious, dramatic, and fast-acting. Malnourishment is long-term and insidious. Although people may not die outright, they are less productive than normal and can suffer permanent impairment and even brain damage. Among the major problems of undernourishment are marasmus, progressive emaciation caused by a lack of protein and calories; kwashiorkor, a lack of sufficient protein in the diet, which in infants leads to a failure of neural development and therefore to learning disabilities (Figure 11.6); and chronic hunger, which occurs when people have enough food to stay alive but not enough to lead satisfactory and productive lives. This means that world food production must provide adequate nutritional quality, not just total quantity.

The supply of protein has been the major nutritional-quality problem. Animals provide the easiest protein food source for people, but depending on animals for protein raises several questions of values, including ecological ones (Is it better to eat lower on the food chain?), environmental ones (Do domestic animals erode soil faster than crops?), and ethical ones (Is it morally right to eat animals?). How people answer these questions affects approaches to agriculture and thereby the environmental effects of agriculture. Once again, the theme of science and values arises.

Since the end of World War II, rarely has a year passed without a famine somewhere in the world.[15] Food emergencies affected 34 countries worldwide at the end of the twentieth century. Varying weather patterns in Africa, Latin America, and Asia, as well as an inadequate international trade in food, contributed to these emergencies.[19] Examples include famines in Ethiopia (1984–1985), Somalia (1991–1993), and the 1998 crisis in Sudan (see also Chapter 9). Africa remains the continent with the most acute food shortages, due to adverse weather and civil strife.[20] The distribution problem is clearly illustrated by these recent famines. Food distribution fails because poor people cannot buy food and pay for its delivery, because transportation is lacking or too expensive, or because food is withheld for political or military reasons. Although there is considerable international trade in food, most of the trade is among rich nations.

Another common solution is food aid among nations, where one nation provides food to another or gives or lends

Daily calories per capita

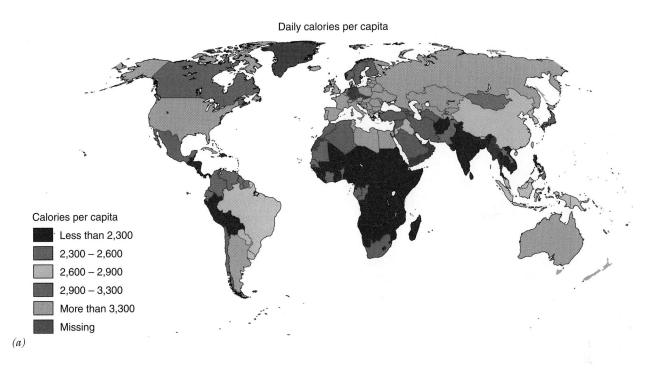

Calories per capita

- ■ Less than 2,300
- ■ 2,300 – 2,600
- ■ 2,600 – 2,900
- ■ 2,900 – 3,300
- ■ More than 3,300
- ■ Missing

(a)

Percentage of population undernourished (1997–1999)

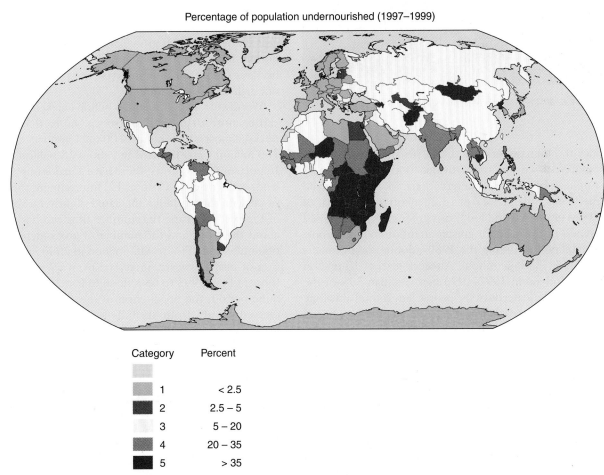

Category	Percent
1	< 2.5
2	2.5 – 5
3	5 – 20
4	20 – 35
5	> 35

(b)

Figure 11.5 ■ (*a*) Daily intake of calories worldwide. (*b*) Where people are undernourished. The percentage is the share of the country's total population that is undernourished. [*Source*: World Resources Institute Web site: http://www.wri.org/.]

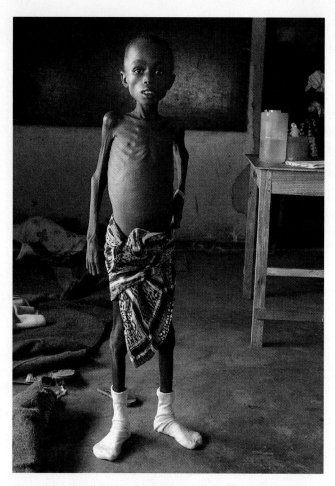

Figure 11.6 ■ Photograph of a child suffering from kwashiorkor.

money to purchase food. In the 1950s and 1960s, only a few industrialized countries provided food aid, using stocks of surplus food. A peak in international food aid occurred in the 1960s, when a total of 13.2 million tons per year of food was given. A world food crisis in the early 1970s raised awareness of the need for greater attention to food supply and stability. But during the 1980s, donor commitments totaled only 7.5 million tons. A record level of 15 million tons of food aid in 1992–1993 met less than 50% of the minimum caloric needs of the people fed. If food aid alone is to bring the world's malnourished people to a desired nutritional status, an estimated 55 million tons will be required by the year 2010—more than six times the amount available in 1995.[16]

When a group of people starve, the world feels sorrow for them. Humanitarian gestures are important, but in themselves such efforts cannot solve the world's food problem. Food aid is a short-term answer. In the long run, when food distribution is the primary problem, the best solution is to increase local production. Ironically, food aid can work against increased availability of locally grown food. Free food undercuts local farmers; they cannot compete with it. Availability of food grown locally also avoids

disruptions in distribution and the need to transport food over long distances. The only complete solution to famine is to develop long-term sustainable agriculture locally. The old saying "Give a man a fish and feed him for a day; teach him to fish and feed him for life" is true. We must develop and teach agricultural techniques that can be maintained over long periods without using up resources.

11.3 What We Eat and What We Grow

Crops

Of Earth's half-million plant species, only about 3,000 have been used as agricultural crops and only 150 species have been cultivated on a large scale. In the United States, 200 species are grown as crops. Most of the world's food is provided by only 14 crop species. In approximate order of importance, these are wheat, rice, maize, potatoes, sweet potatoes, manioc, sugarcane, sugar beet, common beans, soybeans, barley, sorghum, coconuts, and bananas (Figures 11.7 and 11.8). Of these, six provide more than 80% of the total calories consumed by human beings either directly or indirectly.[17]

Some crops, called forage, are grown as food for domestic animals. These include alfalfa, sorghum, and various species of grasses grown as hay. Alfalfa is the most important forage crop in the United States, where 14 million hectares (about 30 million acres) are planted in alfalfa—one-half the world's total.

Worldwide, people keep 14 billion chickens, 1.3 billion cattle, more than 1 billion sheep, more than a billion ducks, almost a billion pigs, 700 million goats, more than 160 million water buffalo, and about 18 million camels.[18] Important food sources, these have a major impact on the land, as discussed in Chapter 12. Interestingly, the number of cattle in the world has increased slightly, by about 0.2% in the past 10 years; the number of sheep has remained about the same; and the number of goats increased from 660 million in 1995 to 807 million in 2005. The production of beef, however, increased from 57 million metric tons (MT) in 1995 to 63 million MT in 2005. During the same period, the production of meat from chickens increased greatly, from 46 million MT to 70 million MT, and meat from pigs increased from 80 million MT to more than 100 million MT in the same period.[2]

Most cattle live on rangeland or pasture. **Rangeland** provides food for grazing and browsing animals without plowing and planting; **pasture** is plowed, planted, and harvested to provide forage for animals. More than 34 million km^2 are in permanent pasture worldwide—an area larger than the combined sizes of Canada, the United States, Mexico, Brazil, Argentina, and Chile.[2]

There is a large world trade in small grains. Only the United States, Canada, Australia, and New Zealand are major exporters; the rest of the world's nations are net

(a)   (b)   (c)

Figure 11.7 ■ Some of the world's major crops, including (*a*) wheat, (*b*) rice, and (*c*) soybeans. See text for a discussion of the relative importance of these three crops.

importers. In 2005, world small-grain production was 2.2 billion tons, a record crop.[2] World small-grain production was 0.8 billion metric tons in 1961, reached 1 billion in 1966, then doubled to 2 billion in 1996, a remarkable increase in 30 years. But production has re-mained relatively flat since then. The question we must ask, and cannot answer at this time, is whether this means that the world's carrying capacity for small grains has been reached or simply that the demand is not growing (Figure 11.9).

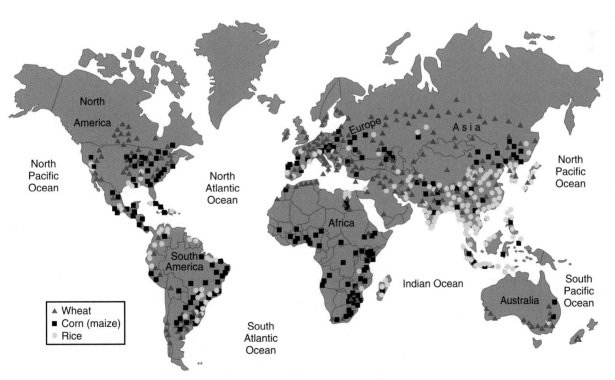

Figure 11.8 ■ Geographic distribution of world production of a few major small-grain crops.

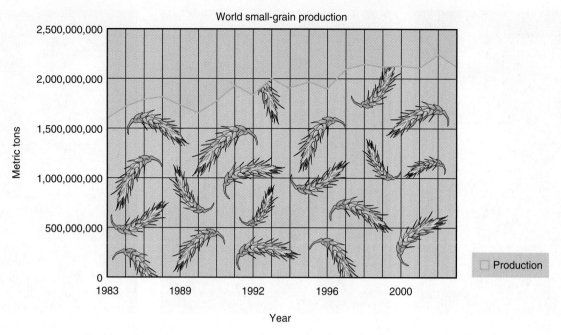

Figure 11.9 ■ World small-grain production since 1983. (*Source*: FAO statistics FAOSTATS Web site.)

Aquaculture

In contrast to food obtained on the land, most marine and freshwater food is still obtained by hunting. Environment and the hunt for fish in Earth's fisheries are discussed in Chapter 14. Hunting wild fish has not been sustainable (see Chapter 14), and **aquaculture**, the farming of food in aquatic habitats—both marine and freshwater—is an important source of protein, is growing rapidly, and could be one of the major solutions to providing nutritional quality. Popular aquacultural products include carp, tilapia, oysters, and shrimp, but in many nations other species are farm-raised and culturally important, such as yellowtail (important in Japan); crayfish (United States); eels and minnows (China); catfish (southern and midwestern United States); salmon (Norway and the United States); trout (United States); plaice, sole, and the Southeast Asian milkfish (Great Britain); mussels (France, Spain, and Southeast Asian countries); and sturgeon (Ukraine). A few species—trout and carp—have been subject to genetic breeding programs.[21]

Although relatively new in the United States, aquaculture has a long history elsewhere, especially in China, where at least 50 species are grown, including finfish, shrimp, crab, other shellfish, sea turtles, and sea cucumbers (a marine animal).[22] In the Szechuan area of China, fish are farmed in more than 100,000 hectares (about 250,000 acres) of flooded rice fields. This is an ancient practice that can be traced back to a treatise on fish culture written by Fan Li in 475 B.C.[21]

Aquaculture can be extremely productive on a per-area basis, especially because flowing water brings food into the pond or enclosure from outside. Although the area of Earth that can support freshwater aquaculture is small, we can expect this kind of aquaculture to increase in the future and become a more important source of protein. In China and other Asian countries, farmers often grow several species of fish in the same pond, exploiting their different ecological niches. Ponds developed mainly for carp, a bottom-feeding fish, also contain minnows, which feed at the surface on leaves added to the pond.

Sometimes fishponds use otherwise wasted resources, such as fertilized water from treated sewage; some fishponds exist in natural hot springs (Idaho) and contain warmed water used in cooling electric power plants (Long Island, New York; Great Britain).[21]

Mariculture, the farming of ocean fish, although producing a small part of the total marine fish catch, has grown rapidly in the last decades and will likely continue to do so. Mariculture of abalone and oysters, whose natural production is limited, is increasing. In the United States and Canada, for example, researchers are working to learn how to attract the young, swimming stages of these shellfish to areas where they can be conveniently grown and harvested.

Oysters and mussels are grown on rafts that are lowered into the ocean, a common practice in the Atlantic Ocean in Portugal and in the Mediterranean in such nations as France. As filter feeders, these animals obtain

Figure 11.10 ■ An oyster farm in Poulsbo, Washington. Oysters are grow on artificial pilings in the intertidal zone.

food from water that moves past them in currents. Because a small raft is exposed to a large volume of water, and thus a large volume of food, rafts can be extremely productive. Mussels grown on rafts in bays of Galicia, Spain, produce 300 metric tons per hectare, whereas public harvesting grounds of wild shellfish in the United States yield only about 10 kg/ha.[21] Oysters and mussels are also grown on artificial pilings in the intertidal zone in the state of Washington (Figure 11.10).

11.4 An Ecological Perspective on Agriculture

Farming creates novel ecological conditions (Figure 11.11). These **agroecosystems** differ from natural ecosystems in six ways.

■ In farming we try to stop ecological succession and keep the agroecosystem in an early-successional state (see Chapter 10). Most crops are early-successional species, which means that they do best when sunlight, water, and chemical nutrients in the soil are abundant. Under natural conditions, crop species would eventually be replaced by later-successional plants. Any attempt to prevent natural successional processes from occurring requires time and effort on our part. Most crops are planted on cleared land, which is then kept clear of other vegetation. In contrast, when a clearing is created by a natural disturbance, such as a fire or a storm, the vegetation returns—first early-successional species and then later ones.

■ Another way that most agroecosystems differ from natural ecosystems is **monoculture**—large areas planted with a single species or even a single strain or subspecies, such as a single hybrid of corn. The downside of monoculture is that it makes the entire crop vulnerable to attack by a single disease or a single change in environmental conditions. Repeated planting of a single species can reduce the soil content of certain essential elements, thereby reducing overall soil fertility. This can be counteracted to a certain extent by artificial fertilizers and by the ancient practice of **crop rotation**. In crop rotation, different crops are planted in turn in the same field, with the field occasionally left fallow. A fallow field is allowed to grow a cover crop (sometimes planted, sometimes whatever germinates) that is not harvested for at least one season. Often the vegetation that grows on the fallow field is plowed under to add to soil fertility.

■ Crops are planted in neat rows, which makes life easy for pests because the crop plants have no place to hide. In natural ecosystems, many species of plants grow mixed together in complex patterns, so it is harder for pests to find their favorite victims.

■ Farming greatly simplifies biological diversity and food chains. Most pest-control methods reduce the abundance and diversity of natural predators and therefore make the agroecosystem more susceptible to undesirable changes.

■ Plowing is unlike any natural soil disturbance—nothing in nature repeatedly and regularly turns over the soil to a specific depth. Plowing exposes the soil to erosion and damages its physical structure, leading to a decline in organic matter and a loss of chemical elements to erosion. This is illustrated by studies in Santa Barbara County, California, of previously plowed land that has been abandoned for agriculture. Instead of returning to its original California oak woodlands, the plowed land either becomes shrub–herb vegetation or undergoes succession to oak woodlands much more slowly than in nontilled land.

■ The newest difference is genetic modification of crops—a novel situation.

Figure 11.11 ■ How farming changes an ecosystem. It converts complex ecosystems of high species diversity and structural diversity to a monoculture of uniform structure. The soil is greatly modified. See text for additional information about agricultural effects on ecosystems.

Pre-agricultural ecosystem

Agroecosystem

11.5 Limiting Factors

High-quality agricultural soil has all the chemical elements required for plant growth and a physical structure that lets both air and water move freely through the soil, yet it retains water well. The best agricultural soils have a high organic content and a mixture of sediment particle sizes. Small particles, especially fine clays, help to retain moisture and chemical elements; larger particles, including sand and pebbles, help the flow of water. But different crops require different soils. Lowland rice grows in flooded ponds and requires a heavy, water-saturated soil, while watermelons grow best in very sandy soil.

Soils rarely have everything a crop needs. The question for a farmer is: What needs to be added or done to make a soil more productive for a crop? The traditional answer is

that, at any time, just one factor is limiting. If that factor can be improved, the soil will be more productive; if that single factor is not improved, nothing else will make a difference. The idea that some single factor determines the growth and therefore the presence of a species is known as Liebig's law of the minimum, after Justus von Liebig, a nineteenth-century agriculturalist who is credited with first stating this idea. He knew that crops required a number of nutrients in the soil and that crop yields could be increased by adding these as fertilizers. However, the factor that caused an increase varied from time to time and place to place. A general statement of Liebig's law is: The growth of a plant is affected by one **limiting factor** at a time—the one whose availability is the least in comparison to the needs of a plant.

But the reality can be much more complicated. Crops require about 20 chemical elements. These must be available in the right amounts, at the right times, and in the right proportions to each other. It is customary to divide these life-important chemical elements into two groups, **macronutrients** and **micronutrients**. A macronutrient is a chemical element required by all living things in relatively large amounts. Macronutrients are sulfur, phosphorus, magnesium, calcium, potassium, nitrogen, oxygen, carbon, and hydrogen. A micronutrient is a chemical element required in small amounts—either in extremely small amounts by all forms of life or in moderate to small amounts for some forms of life. Micronutrients are often rarer metals, such as molybdenum, copper, zinc, manganese, and iron. (Macronutrients and micronutrients are also discussed in Chapter 4.)

If Liebig were always right, then environmental factors would always act one by one to limit the distribution of living things. But there can be exceptions to this rule. For example, nitrogen is a necessary part of every protein, and proteins are essential building blocks of cells. Enzymes, which make many cell reactions possible, contain nitrogen. A plant given little nitrogen and phosphorus might not make enough of the enzymes involved in taking up and using phosphorus. Increasing nitrogen to the plant might therefore increase the plant's uptake and use of phosphorus. If this were so, the two elements would have a **synergistic effect**. In a synergistic effect, a change in availability of one resource affects the response of an organism to some other resource.

So far we have discussed effects of chemical elements when they are in short supply. But it is also possible to have too much of a good thing—most chemical elements become toxic when they are present in concentrations that are too high. As a simple example, plants die when they have too little water but also when they are flooded, unless they have specific adaptations to living in water. So it is with chemical elements required for life. In this section, we have discussed the problems of too little; we discuss the problems of too much in Chapter 15.

The older a soil, the more likely it is to lack trace elements, because as a soil ages its chemical elements tend to be leached by water from the upper layers to deeper layers (see Chapter 10). When crucial chemical elements are leached below the reach of crop roots, the soil becomes infertile. Striking cases of soil-nutrient limitations have been found in Australia, which has some of the oldest soils in the world—on land that has been above sea level for many millions of years, during which time severe leaching has taken place. Sometimes trace elements are required in extremely small amounts. For example, it is estimated that in certain Australian soils adding an ounce of molybdenum to a field increases the yield of grass by 1 ton/year.

The idea of a limiting growth factor, originally used in reference to crop plants, has been extended by ecologists to include all life requirements for all species in all habitats.

11.6 The Future of Agriculture

We can identify three major technological approaches to agriculture. One is modern mechanized agriculture, where production is based on highly mechanized technology that has a high demand for resources—including land, water, and fuel—and makes little use of biologically based technologies. Another approach is resource-based agriculture, which is based on biological technology and conservation of land, water, and energy. An offshoot of the second is organic food production, where crops are grown without artificial chemicals (including pesticides), where genetic engineering of crops is not used, and where ecological control methods are employed. The third is bioengineering.

In mechanized agriculture, production is determined by economic demand and limited by that demand, not by resources. In resource-based agriculture, production is limited by environmental sustainability and the availability of resources, and economic demand usually exceeds production (Figure 11.12).

The history of agriculture can be summarized most simply as consisting of four stages:

1. Resource-based agriculture and what we now call organic agriculture were introduced about 10,000 years ago.
2. A shift to mechanized, demand-based agriculture occurred during the Industrial Revolution of the eighteenth and nineteenth centuries.
3. A return to resource-based agriculture began in the twentieth century, using new technologies.
4. Today there is a growing interest in organic agriculture as well as a potentially large-scale use of genetically engineered crops. (See A Closer Look 11.1 and 11.2.)

What can be done to help crop production keep pace with human population growth? Since there are so many plant species, perhaps some yet unused ones could provide new sources of food and grow in environments little used for agriculture. Those interested in the conservation of biological diversity urge a search for such new crops on the grounds that this is one utilitarian justification for the conservation of species. It is also suggested that some of these new crops may be easier on the environment and therefore more likely to allow sustainable agriculture. But it may be that over the long history of human existence those species that are edible have been found, and the number is small. Research is under way to seek new crops or plants that have been eaten locally but whose potential for widespread, intense cultivation has not been tested. Among current candidates are guayule, crambe, pigeon peas, and grain amaranth.[22]

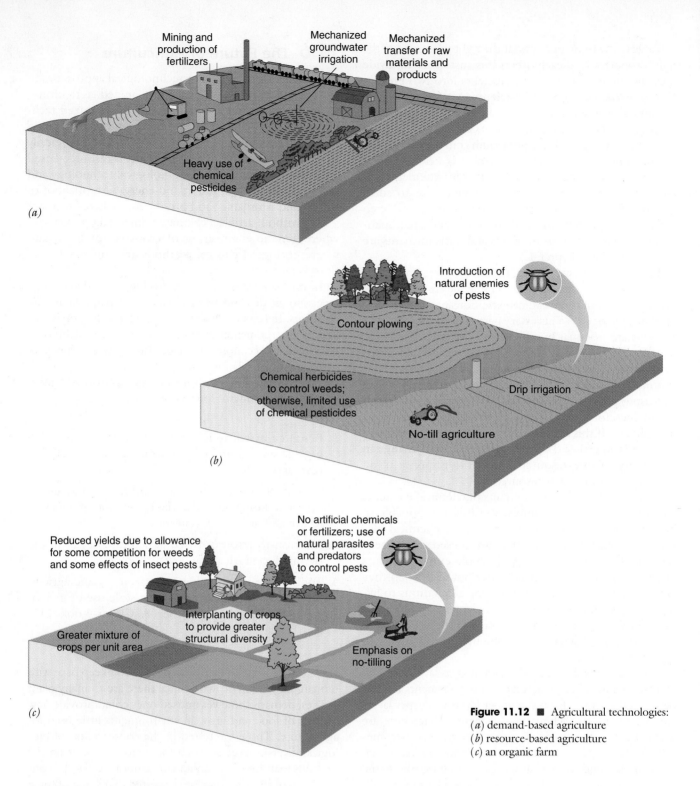

Figure 11.12 ■ Agricultural technologies:
(*a*) demand-based agriculture
(*b*) resource-based agriculture
(*c*) an organic farm

11.7 Increasing the Yield per Acre

Future increases in agricultural production will likely come mainly from the development of high-yield strains of crops. During the twentieth century, rapid strides were made in increasing the production of crops per unit area. Some of these advances, which involved the development of new hybrids, became known as the green revolution. Several other methods of increasing the food supply may hold some promise as well, although always with limitations.

The Green Revolution

The **green revolution** is the name attached to post–World War II programs that have led to the development of new strains of crops with higher yields, better resistance to disease, or better ability to grow under poor conditions. One advancement of the green revolution was the development of superstrains of rice at the International Rice Research Institute in the Philippines (see Figure 11.14). Although hybridization

Traditional Farming Methods

In industrialized countries of temperate zones, there is a long history of plowing to clear land for agriculture; but in less industrialized, tropical areas, there is a history of agricultural methods that depend on clearing the vegetation without plowing. Where the loss of nutrients from the soil occurs rapidly following clearing, as in some tropical rain forests, the traditional practice is to cut the forest in small patches but not cut it completely (Figure 11.13). Some shrubs and herbaceous plants are left. Several crops are planted together among existing vegetation. Crops are harvested for a few years. Then the land is allowed to grow back to a forest. The natural process of secondary succession—the redevelopment of the ecosystem—is allowed to occur. In fact, these farming practices promote this redevelopment and increase the conservation of chemical elements in the ecosystem. After the forest has grown back, the process is repeated.

This kind of agriculture has many names. It is sometimes called cultivation with forest or bush fallow. In Latin America, it is called milpa agriculture; in Great Britain, swidden agriculture; in western Africa, fang agriculture. In this type of agriculture, a mixture of crops is utilized, including root, stem, and fruit crops. For example, in western Africa, fang agriculture includes yams (a root crop) plus maize; in Southeast Asia, root crops are grown with rice and millet or with rice and maize.[21]

In theory, this method could be sustainable if human population density remained low. Erosional losses are minimized, and the soil eventually recovers its fertility. Uncut vegetation provides future seed sources. When human population pressures are low, a longer time occurs between the periods of use of any site. This is known as a long rotation period. Under high population pressures, such as occur in many places today, the rotation period is much shorter, and the land may not be able to recover sufficiently from previous use. In such cases, production is not sustainable.

For many years, agricultural experts from developed nations viewed this method of agriculture as a poor process with low, short-term productivity, used only by primitive peoples. Now it is understood that this kind of agriculture is well suited to high-rainfall lands where soils readily become impoverished when the land is completely cleared. The mixture of crops allows different species to contribute to soil fertility in different ways. Some perennial plants slow physical erosion; native legumes add nitrogen to the soil; and so forth.

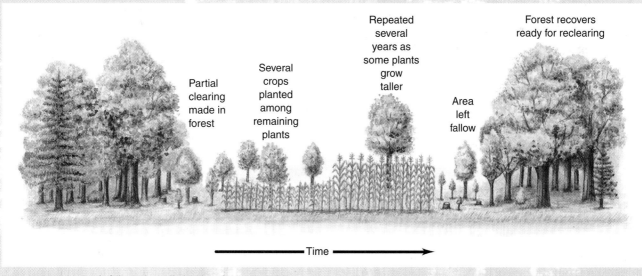

Partial clearing made in forest

Several crops planted among remaining plants

Repeated several years as some plants grow taller

Area left fallow

Forest recovers ready for reclearing

Time

Figure 11.13 ■ Bush fallow, also called milpa, fang, or swidden agriculture. Over time, secondary succession will take place on land partially cleared by slash and burn.

of rice vastly increased rice production per acre, the new strains required greater use of fertilizers and as much as four to seven times the water. And in some cases they produced a rice that was not considered desirable to eat.

Another development in the green revolution was strains of maize with improved disease resistance at the International Maize and Wheat Improvement Center in Mexico.

Improved Irrigation

Better irrigation techniques could improve crop yield and reduce overall water use. Drip irrigation—from tubes that drip water slowly—greatly reduces the loss of water from evaporation and increases yield. However, it is expensive and thus most likely to be used in developed nations or nations with a large surplus of hard currency—in other words, in few of the countries where hunger is most severe.

Some people suggest that in the future we will rely increasingly on artificial agriculture, such as *hydroponics*, which is the growing of plants in a fertilized water solution on a completely artificial substrate in an artificial environment, such as a greenhouse. This approach is extremely expensive and unlikely to be effective where hunger is the greatest.

11.8 Organic Farming

Future farming will have to be easier on the environment than past farming if agriculture is to be widely sustainable. Organic farming is often suggested as one of the solutions. **Organic farming** is typically considered to have three qualities: It is more like natural ecosystems than monocultures; it minimizes negative environmental impacts; and the food that results from it does not contain artificial compounds. According to the U.S. Department of Agriculture (USDA), organic farming has been one of the fastest-growing sectors in U.S. agriculture, although it still occupies a small fraction of U.S. farmland and contributes only a small amount of agriculture income. By the end of the twentieth century it amounted to about $6 billion—much less than the agricultural production of California.

There are about 12,000 organic farmers in the United States, and the number is growing 12% per year.[23] The USDA began certification of organic farming in 2002; certified organic cropland more than doubled, and the number of farmers certifying their products grew by 40%. All organic farmers had to be USDA-certified by 2002.[24] In the United States, more than 1.3 million acres are certified as organic. In the 1990s, organic milk cows increased from 2,300 to 12,900, and organic layer hens increased from 44,000 to more than 500,000. In the United States only 0.01 % of the land planted in corn and soybeans was grown under certified organic farming systems in the mid-1990s; about 1% of dry peas and tomatoes were grown organically, and about 2% of apples, grapes, lettuce, and carrots were organically grown. On the high end, nearly one-third of U.S. buckwheat, herb, and mixed vegetable crops were grown under organic farming conditions.[25]

11.9 Alternatives to Monoculture

An important trade-off is implicit in the choice to plant a single hybrid. Each year, seed companies make use of climate forecasts for the growing season and knowledge of the most likely strains of diseases and insect pests in the area to be planted. They then develop hybrids resistant to those strains and well adapted to the forecast weather. If the predictions are correct, crop production can be very high. If the predictions are incorrect, crop production in the entire area can be very low.

An alternative is to plant a mixture of crops and/or a broad range of genotypes at a particular time and place. This approach is typical of the preindustrial agriculture still found in many developing countries, and it is promoted today by organic farmers. It gives lower average yearly production but reduces the risk of very low production years. Monoculture trades off long-term stability for the opportunity to have very high production this year. Which approach to choose is a question of values and therefore part of the theme of science and values.

Regions of Earth differ greatly in their capacity for crop production. Factors that influence which crops are produced in which areas include tradition, access to technology and supplies, and local politics. In the United States alone, states differ greatly in their crop production (Figure 11.15). California has the greatest value of crop production in the United States, totaling $12.4 billion in 1998 and producing 36.4 million tons of fruits, nuts, and vegetables in that year. This accounts for about half of all U.S. production.[28] Much of California would be too dry to support this food production without irrigation.

In the high plains east of the Rocky Mountains (where short-grass prairie originally grew), rangeland and irrigated farmland are common. There, irrigated cropland is mainly planted in small grains (corn, wheat, etc.). Farther east in Nebraska are areas where winter wheat is most important; spring wheat is important in the Dakotas. The Midwest is the corn belt. In northern states from Minnesota to Maine, dairy products and hay for cattle are important. In the Southeast, major crops include cotton, tobacco, vegetables, and fruits.

11.10 Eating Lower on the Food Chain

Some people believe that it is ecologically unsound to use domestic animals as food on the grounds that eating each step farther up a food chain leaves much less food to eat per acre. This argument is as follows (you will remember this from the Critical Thinking Issue in Chapter 9): No organism is 100% efficient. Only a fraction of the energy in food taken in is converted to new organic matter. Crop plants may convert 1–10% of sunlight to edible food, and cows may convert only 1–10% of hay and grain

A CLOSER LOOK 11.2

Potential Future Advances in Agriculture

New Genetic Strains and Hybrids

From the beginning, farming has affected the genetics of domesticated animals and plants, as people selected strains that were easy to grow and harvest (Figure 11.14). The act of farming makes certain species abundant where they were rare; thus, farming selectively favors certain genotypes (populations with certain genetic characteristics). Features that make a species or a genotype a weaker competitor under natural conditions sometimes make it more desirable as a crop. For example, wild relatives of wheat lose their seeds when the seeds are ripened and gently shaken by wind or animals. If you try to cut wild wheat stems and take them home, most of the seeds will be shaken off and few will be left when you arrive. This is an adaptation that helps spread the seeds.

Some individuals of wild wheat have a mutation that makes the seeds remain on the stalk. In the wild, these mutants leave fewer offspring and do not persist in the population; in the wild, the mutants are less fit genetically (see Chapter 7). Early farmers selected these mutants because they were easier to collect, transport, and use. In this way, people changed selective pressures on wheat and hastened the evolution of wheat strains that are useful to us but that could not survive on their own. One might also say that the result has been a symbiotic relationship between people and these otherwise less competitive forms of wheat.

People also domesticated wheat by moving it to habitats to which it was not originally adapted, then developing new strains that could persist in these environments and also changing the environmental conditions of the new habitat to better suit the wheat.[26] Corn (maize) went through a similar process of domestication. Domestic animals have been bred to make them more docile and bet-

Figure 11.14 ■ Experimental rice plots at the International Rice Research Institute, Philippines, showing visual crop variation based on the use of fertilizers.

ter producers of the meat and dairy products that we prefer. Modern agriculture has carried this process a step further with the frequent intentional development of hybrids of different genotypes, bred to overcome new strains of disease and changes in climate.

New Crops

Developing new crops by domesticating species that are currently wild offers considerable potential. Although new crops are unlikely to replace current crop species as major food sources in this century, there is great interest in new crops to increase production in marginal areas and to increase the production of nonfood products, such as oils. The development of new crops has been a continuing process in the history of agriculture. As people spread around the world, new crops were discovered and transported from one area to another. The process of introduction and increase in the production of crops continues.

Among the likely candidates for new crops are amaranth for seeds and leaves;

Leucaena, a legume useful for animal feed; and triticale, a synthetic hybrid of wheat and rye. A promising source of new crops is the desert; none of the 14 major crops are plants of arid or semiarid regions, yet there are vast areas of desert and semidesert. The United States has 200,000 million hectares (about 500,000 million acres) of arid and semiarid rangeland. In Africa, Australia, and South America, the areas are even greater. Several species of plants can be grown commercially under arid conditions, allowing us to use a biome for agriculture that has been little used in this way in the past. Examples are guayule (a source of rubber), jojoba (for oil), bladderpod (for oil from seeds), and gumweed (for resin). Jojoba, a native shrub of the American Sonoran Desert, produces an extremely fine oil, remarkably resistant to bacterial degradation, which is useful in cosmetics and as a fine lubricant. Jojoba is now grown commercially in Australia, Egypt, Ghana, Iran, Israel, Jordan, Mexico, Saudi Arabia, and the United States.[27]

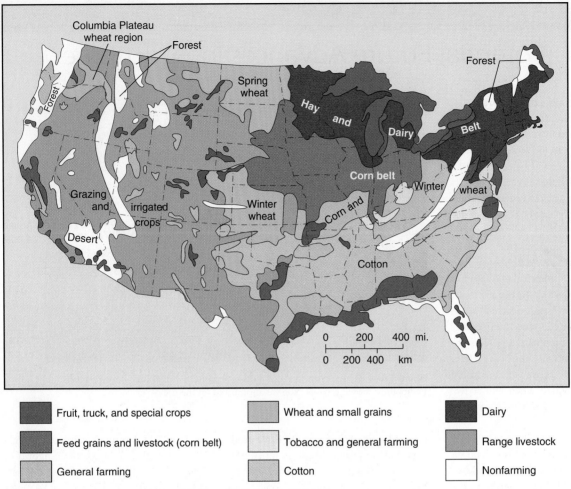

Figure 11.15 ■ Major types of agricultural production in the United States.

Legend:

- Fruit, truck, and special crops
- Wheat and small grains
- Dairy
- Feed grains and livestock (corn belt)
- Tobacco and general farming
- Range livestock
- General farming
- Cotton
- Nonfarming

to meat. Thus, the same area could produce 10–100 times more vegetation than meat per year (see the Critical Thinking Issue in Chapter 9). This holds true for the best agricultural lands, which have deep, fertile soils on level ground.

As with so many issues, however, a simple generalization does not apply to all cases. Land too poor for crops that people can eat can be excellent rangeland, with grasses and woody plants that domestic livestock can eat (Figure 11.16 and Figure 11.17). These lands occur on steeper

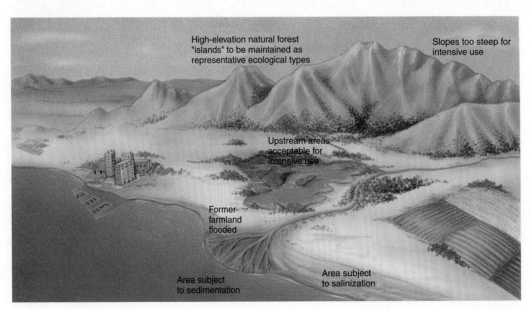

Figure 11.16 ■ Physical and ecological considerations in watershed development—such as slope, elevation, floodplain, and river delta location—limit land available for agriculture.

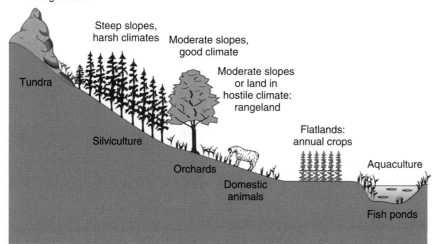

slopes, with thinner soils or with less rainfall. Thus, from the point of view of sustainable agriculture, there is value in rangeland or pasture. The wisest approach to sustainable agriculture involves a combination of different kinds of land use: using the best agricultural lands for crops, using poorer lands for pastures and rangelands, and avoiding the use of the best lands for grain production for animal feed.

Another problem with the argument that we should eat lower on the food chain is that food is more than just calories, and animals are important sources of proteins and minerals. Animals provide the major source of protein in human diets—56 million metric tons of edible protein per year worldwide. In the United States, 75% of the protein, 33% of the energy, and most of the calcium and phosphorus in human nutrition come from animal products.

A third factor to keep in mind is that domestic animals are often used for other purposes, such as plowing, carrying loads, and transportation; and they are sources of wool and leather as well as food. In addition, their excrement is an important fertilizer and, in some areas of the world, an important—sometimes the only—fuel for fire. In these ways, the use of animals as part of food production represents an increase in efficiency.

Some people maintain vegetarian diets because of specific dietary problems. Others do not eat meat for moral, ethical, or religious reasons. Thus, the decision about when and whether to use animals as part of food production is an issue of both science and values.

11.11 Genetically Modified Food: Biotechnology, Farming, and Environment

The discovery that DNA is the universal carrier of genetic information has led to the development of a completely new technology known as genetic engineering. Now that the chemistry of inheritance is understood, scientists have been able to develop methods to transfer specific genetic characteristics from one individual to another, from one population to another, and from one species to another. This has led to the genetic modification of crops, with major implications for agriculture. The development and use of **genetically modified crops (GMCs)** has given rise to new environmental controversies as well as a promise of increased agricultural production.

Genetic engineering in agriculture involves several different practices, which we can group as follows: faster and more efficient ways to develop new hybrids; the introduction of the "terminator gene" (discussed in the following chapter); and the transfer of genetic properties from widely divergent kinds of life. These three practices have quite different potentials and problems. Of the three, hybridization is the least novel—crop hybridization has become a standard methodology in modern agriculture; biotechnology adds a new way to create hybrids, but it is not an entirely new process. In contrast, the second and third practices were never previously possible.

There is considerable interest in the potential for genetic engineering to develop strains of crops with entirely new characteristics. One focus of this research is the development of new crops that have the same symbiotic relationship found in legumes (members of the pea family) so that they "fix" nitrogen (convert atmospheric gaseous nitrogen to a form that can be used by green plants). Recall that bacteria grow in the root nodules of legumes. The bacteria live on substances produced by the legumes; in turn, the bacteria fix nitrogen. Legumes are often rotated with other crops so that the soil is enriched in nitrogen. It may be possible to develop new strains of corn and other crops that, along with new strains of bacteria, can form a symbiotic nitrogen-fixing relationship. Such an accomplishment would increase the production of these crops and reduce the need for fertilizers.

Another goal of agricultural genetic engineering is the development of strains with improved tolerance of drought, cold, heat, and toxic chemical elements. For example, one effort is aimed at developing wheat that is resistant to high levels of aluminum, an element that has negative effects on many plants.[29] Yet another goal is to create crops that produce their own pesticides. This is discussed in the next chapter. Although genetic modifications have proved to have great benefit, they involve both limitations and environmental concerns. These, too, are discussed in the next chapter.

The land area planted with GMCs has grown rapidly from the first plantings in 1996 to 145 million acres in 2002 (Figure 11.18). Today more than 58 million hectares (145 million acres) are planted with genetically modified crops, over two-thirds of which are in the United States. Among the major genetically modified crops are corn, cotton, soybeans, canola, squash, and papaya.[30] It is difficult for the average consumer to avoid food and nonfood agricultural products from GMCs, given that three-quarters of the soybeans, almost a third of the corn, and more than half of the plants that produce canola oil grown in the United States are genetically modified;[30] in addition, at present it is not possible to separate GMC products from non-GMC products when they reach the retail market.

The jury is out as to whether the benefits will outweigh undesirable effects discussed in the next chapter. What is important for environmental science is that the use of genetically modified crops is under way before the environmental effects are well understood. As with many new technologies of the industrial age, application has preceded environmental investigation and understanding. The challenge for environmental science is to gain an understanding of environmental effects of GMCs quickly.

11.12 Climate Change and Agriculture

Climate change is more likely to decrease yield than to increase it, because areas with the best soils in the world also happen to have climates well suited to agriculture. Thus, most climatic changes are likely to make things worse. If global warming takes place as forecast by models of global climate, major disruptions to agriculture will result.[31] The best climates for agriculture may shift from their present locations to locations farther north. For example, the climate of the midwestern corn belt of the United States, which has been so good for crop production, could move northward, *favoring* Canada. This could have a real impact on the *amount* of grain produced because soils in Canada are generally not as suitable as U.S. midwestern soils for grain production. Global warming could also lead to an increase in evapotranspiration (loss of water from the soil both by evaporation and by transpiration from plants) in many midlatitude areas.[32] Supplying water for irrigation would become an even greater problem than it is today. Those concerned about the future of agriculture must take the possibility of global warming into account in their planning.

Plants require water, and much modern agriculture involves irrigation (Figure 11.20). Sources of irrigation wa-

Figure 11.18 ■ Genetically modified crops have created some major controversies, as indicted by this demonstration in a corn field in Villanueva de Gallego in north-eastern Spain in 2003. The slogan on the left banner reads "Stop, genetic pollution". Spanish Agriculture, Food and Fishing Minister Miguel Arias Canete is portrayed close to a huge corn ear.

Will There Be Enough Water to Produce Food for a Growing Population?

Between 2000 and 2025, scientists estimate, the world population will increase from 6.6 billion to 7.8 billion,[34] approximately double what it was in 1974. To keep pace with the growing population, the United Nations Food and Agriculture Organization predicts that food production will have to double by 2025, and so will the amount of water consumed by food crops. Will the supply of freshwater be able to meet this increased demand, or will water supply limit global food production?

Growing crops consume water through transpiration (loss of water from leaves as part of the photosynthetic process) and evaporation from plant and soil surfaces. The volume of water consumed by crops worldwide—including rainwater and irrigated water—is estimated at 3,200 billion m^3 per year. An almost equal amount of water is used by other plants in and near agricultural fields; thus, it takes 7,500 billion m^3 per year of water to supply crop ecosystems around the world (see Table 11.2). Grazing and pasture land account for another 5,800 billion m^3 and evaporation from irrigated water another 500 billion m^3, for a total of 13,800 billion m^3 of water per year for food production, or 20% of the water evaporated and transpired worldwide. By 2025, therefore, humans will be appropriating almost half of all the water available to life on land for growing food for their own use. Where will the additional water come from?

Although the amount of rainwater cannot be increased, it can be more efficiently used through farming methods such as terracing, mulching, and contouring. Forty percent of the global food harvest now comes from irrigated land, and some scientists estimate that the volume of irrigated water available to crops will have to triple by 2025—a volume equal to that of 24 Nile Rivers or 110 Colorado Rivers.[29] A significant saving of water can therefore come from more efficient irrigation methods, such as improved sprinkler systems, drip irrigation, night irrigation, and surge flow.

Additional water could be diverted from other uses to irrigation. But this may not be as easy as it sounds because of competing needs for water. For example, if water were provided to the 1 billion people in the world who currently lack drinking and household water, less would be available for growing crops. And the new billions of people to be added to the world population in the next decades will also need water. Already, humans use 54% of the world's runoff. Increasing this to more than 70%, as will be required to feed the growing population, may result in loss of freshwater ecosystems, decline in world fisheries, and extinction of aquatic species.

Table 11.2 • Estimated Water Requirements of Food and Forage Crops	
Crop	*Liters/kg*
Potatoes	500
Wheat	900
Alfalfa	900
Sorghum	1,110
Corn	1,400
Rice	1,912
Soybeans	2,000
Broiler chicken	3,500*
Beef	100,000*

* Includes water used to raise feed and forage.

Source: D. Pimentel et al., "Water Resources: Agriculture, the Environment, and Society," *Bioscience* 4, no. 2 (February 1997): 100.

In many places, groundwater and aquifers are being used faster than they are being replaced—a process that is unsustainable in the long run. Many rivers are already so heavily used that they release little or no water to the ocean. These include the Ganges and most other rivers in India, the Huang He (Yellow River) in China, the Chao Phraya in Thailand, the Amu Dar'ya and Syr Dar'ya in the Aral Sea basin, and the Nile and Colorado rivers.

Two hundred years ago, Thomas Malthus put forth the proposition that population grows more rapidly than the ability of the soil to grow food and that at some time the human population will outstrip the food supply (see A Closer Look 5.1). Malthus might be surprised to know that by applying science and technology to agriculture, food production has so far kept pace with population growth. For example, between 1950 and 1995, the world population increased 122% while grain productivity increased 141%. Since 1995, however, grain production has slowed down (see Figure 11.19), and the question remains as to whether Malthus will be proved right in the twenty-first century. Will science and technology be able to solve the problem of water supply for growing food for people, or will water prove a limiting factor in agricultural production?

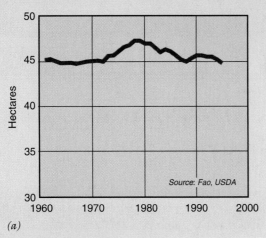

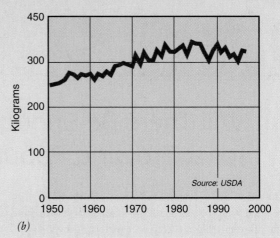

Figure 11.19 ■ (*a*) World irrigated area per thousand people, 1961–1995. [*Source:* L. R. Brown, M. Renner, and C. Flavin, *Vital Signs: 1998* (New York: Norton, 1998), p. 47.] (*b*) World grain production per person, 1950–1977. [*Source:* L. R. Brown, M. Renner, and C. Flavin, *Vital Signs: 1998* (New York: Norton, 1998), p. 29.]

Critical Thinking Questions

1. How might changes in diet in developed countries affect water availability?

2. How might global warming affect estimates of the amount of water needed to grow crops in the twenty-first century?

3. Withdrawing water from aquifers faster than the replacement rate is sometimes referred to as "mining water." Why do you think this term is used?

4. Many countries in warm areas of the world are unable to raise enough food, such as wheat, to supply their populations. Consequently, they import wheat and other grains. How is this equivalent to importing water?

5. Malthusians are those who believe that sooner or later, unless population growth is checked, there will not be enough food for the world's people. Anti-Malthusians believe that technology will save the human race from a Malthusian fate. Analyze the issue of water supply for agriculture from both points of view.

ter include groundwater; water diversion from nearby streams, rivers, and natural lakes; and artificial reservoirs. But large-scale irrigation projects cause environmental problems (discussed in Chapter 12). Construction of reservoirs changes the local environment. Some habitats disappear. Stream patterns change, and erosion rates increase in the watershed of the reservoir (see Chapter 21).

Figure 11.20 ■ In 2005 the annual monsoon—the rainy season—was delayed in northern Indian, and the resulting drought destroyed crops like those shown in this photograph.

Summary

- The basic environmental questions about agriculture are: Can we produce enough food to feed Earth's growing human population? Can we grow crops sustainably, so that both crop production and agricultural ecosystems maintain their viability? Can we produce this food without seriously damaging other ecosystems that receive the wastes of agriculture?
- Agriculture changes the environment; the more intense the agriculture, the greater the changes.
- From an ecological perspective, agriculture is an attempt to keep an ecosystem in an early-successional stage.
- The history of agriculture can be viewed as a series of attempts to overcome environmental limitations and problems. Each new solution has created new environmental problems, which have in turn required their own solutions.
- Farming greatly simplifies ecosystems, creating short and simple food chains, resulting in large areas planted in a single species or genetic strain grown in regular rows, reducing species diversity, and reducing the organic content and overall fertility of soils.
- These simplifications open farmed land to pests, both predators and parasites.
- Biotechnology makes genetic manipulation possible, creating entirely new ways to produce new crops and new strains within a crop. This holds potential for in-

creased production and for the production of *new* products from crops and domesticated animals, *but* it also poses novel environmental threats and raises many issues regarding science and values.

- At present, world food production is adequate in quantity and quality (nutritional value) to feed the current human population. Our modern food problem is the result of two factors: the great increase in human populations, which outstrips local food production in many areas, and inadequate world food distribution.

- Today, inadequate distribution is the most important cause of starvation. However, in the future, if the human population continues to increase, there will be a limit on the ability of Earth to produce sufficient food.

- If the human population continues to grow at something like its present rate, it will double within the next hundred years, meaning that world food production will have to double just to provide the same amount and quality of food per person available today.

- As living standards rise in nations such as China, the demand for higher-quality food will increase. As a result, per-capita food demand will grow faster than the human population.

- In the past, we have increased world food production by (1) increasing the area devoted to food production and (2) increasing the production per unit area. The first method dominated most of human history; but during the twentieth century, major increases were due largely to increasing yields per unit area.

- Some of the best farmland in the world is being withdrawn from food production and put to other uses, including urbanization and suburbanization.

- Future global environmental problems could decrease world food production. These problems include possible effects of global warming (Chapter 23) and possible transportation of pests from one part of the world to another (Chapter 8).

- Many agricultural experts believe that production per unit area can increase through the use of genetically engineered crops. However, there are concerns that these crops may cause major, novel environmental problems. The jury is out as to whether such crops will be a net benefit or a net disbenefit.

REEXAMINING THEMES AND ISSUES

Human Population

Our modern food problem is the result of the great increase in human population growth, which outstrips local food production in many areas, and an inadequate food distribution system for this growing population. Human population growth will eventually be limited by total food production.

Sustainability

A major goal of agriculture must be to achieve sustainable food production in any location. This requires the development of farming methods that do not damage soils, eliminate water supplies, cause extinction of wild relatives of crops or of potential new food species, or lead to permanent pollution downstream. The world as a whole is moving from demand-based to resource-based agriculture. The latter is more consistent with a sustainable approach to agriculture.

From an ecological perspective, agriculture is an attempt to keep an ecosystem in a specific stage, usually an early, highly productive successional stage. In this way, agriculture works against natural mechanisms of sustainability, and we must compensate for this.

Global Perspective

Most of the world's food is provided by only 14 crop species. A challenge for the future is to search the world's diversity of life for additional food species. Food is a global resource in that it is traded internationally, and the availability of food in any area is the result of both local production and global agricultural markets.

Urban World

The artificiality of urban environments leads many city dwellers to believe that they are independent of the environment. But when urban people in Brazil starved, they broke into food-storage facilities and caused social disruption, thus illustrating the intimate connection between urban life and agricultural production.

People and Nature

Agriculture is one of the major ways in which people have changed nature. Through agriculture, people have directly changed more than 10% of the Earth's land surface and indirectly affected a larger area. The development of agriculture allowed the human population to increase greatly, and modern agriculture is necessary to sustain the huge number of people on Earth.

Science and Values

The fundamental ethical questions we face concerning our food supply are: Shall we continue to attempt to feed more and more people? Should we attempt this at the risk of sacrificing habitats of noncrop species, natural ecosystems, and landscapes; reducing water supply for other purposes; and increasing the erosion of landscape? Or shall we attempt to limit our numbers and also limit total food production?

Key Terms

agroecosystem **203**
aquaculture **202**
crop rotation **203**
genetically modified
   crops **211**

green revolution **206**
limiting factor **204**
macronutrient **205**
mariculture **202**

micronutrient **205**
monoculture **203**
organic farming **208**
pasture **200**

rangeland **200**
synergistic effect **205**

Study Questions

1. What will be the best way to feed the world in the next 10 years? the next 100 years?

2. A city garbage dump is filled; it is suggested that the area be turned into a farm. What factors in the dump might make it a good area to farm, and what might make it a poor area to farm?

3. How might we use our knowledge of succession to make agriculture sustainable?

4. Ranching wild animals—that is, keeping them fenced but never tamed—has been suggested as a way to increase food production in Africa, where wildlife is abundant. Based on this chapter, what are the environmental advantages and disadvantages of such game ranching?

5. Explain what is meant by the following statement: The world food problem is one of distribution, not of production. What are the major solutions to this world food problem?

6. You are sent into the Amazon rain forest to look for new crop species. In what kinds of habitats would you look? What kinds of plants would you look for?

7. How does agriculture simplify an ecosystem? In what ways is this simplification beneficial to people? In what ways does it pose problems for a sustainable food supply?

8. A vegetable garden is planted in a vacant lot in a city. Peas and beans grow well, but tomatoes and lettuce do poorly. What is the likely problem? How could it be corrected?

9. A second vegetable garden is planted in another vacant lot. Nothing grows well. Outside the city, in otherwise similar environments, vegetables grow vigorously. What might explain the difference?

10. Should organic farming be allowed to include genetically modified organisms? Why or why not?

Further Reading

Borgstrom, G. 1965. *The Hungry Planet: The Modern World at the Edge of Famine*. New York: Macmillan. A classic book by one of the leaders of agricultural change.

Cunfer, G. 2005. *On the Great Plains: Agriculture and Environment*. College Station: Texas A&M University Press. Uses the history of European agriculture applied to the American Great Plains as a way to discuss the interaction between nature and farming.

Manning, R. 2004. *Against the Grain: How Agriculture Has Hijacked Civilization*. New York: North Point Press. An important iconoclastic book in which the author attributes many of civilization's ills—from war to the spread of disease—to the development and use of agriculture. In doing so, he discusses many of the major modern agricultural issues.

Seymour, John, and Deirdre Headon (eds.). 2003. *The Self-sufficient Life and How to Live It*. Cambridge: DK ADULT. Ever think about becoming a farmer and leading an independent life? This book tells you to do it. It is an interesting, alternative way to learn about agriculture. The book is written for a British climate, but the messages can be applied generally.

CHAPTER 12

Effects of Agriculture on the Environment

Cattle graze along the upper Missouri River, polluting it with their manure, and increasing erosion as they trample the ground near the river. This is one way that aquiculture affects the environment.

Learning Objectives

Agriculture changes the environment in many ways, both locally and globally. After reading this chapter, you should understand:

■ How agriculture can lead to soil erosion, how severe the problem is, what methods are available to minimize erosion, and how these methods have reduced soil erosion in the United States.

■ How farming can deplete soil fertility and why agriculture in

most cases requires the use of fertilizers.

■ Why some lands are best used for grazing and how overgrazing can damage land.

■ What causes desertification.

■ How farming creates conditions that tend to promote pest species, the importance of controlling pests (including weeds), and the problems associated with chemical pesticides.

■ How alternative agricultural methods—including integrated pest management, no-till agriculture, mixed cropping, and other methods of soil conservation—can provide major environmental benefits.

■ That genetic modification of crops could improve food production and benefit the environment but perhaps also could create new environmental problems.

Clean-Water Farms

Steve Burr, a farmer near Salina, Kansas, has 300 acres of crops and 400 acres of grassland on which he raises cattle (Figure 12.1). In 1994, Steve became concerned about the amount of erosion on his cropland and the effects of fertilizers, pesticides, and livestock wastes on water quality. He decided to make two major changes in the management of his farm. By converting some of the crop acreage to grass, he reduced erosion and chemical pollution in the runoff from his land. At the same time, he divided his grazing land into sections, called paddocks, and rotated his animals through them so that they had a constant supply of fresh forage. The cows now dropped their wastes evenly as they moved through the series of paddocks, and the animal wastes no longer piled up in feedlots, where they posed a major disposal problem. The rotation system also prevented overgrazing. In addition, Steve spent less money on dry feed for his cows, allowing him to increase the size of his herd and generate more income. This method of raising livestock is known as *intensive rotational grazing* or management-intensive grazing.

Steve Burr is not alone in his efforts to adopt farm-management practices that reduce pollution and improve water quality while improving his balance sheet. The Burr farm is one of 36 participants in the Kansas Rural Center's Clean Water Farms Project, which began in 1995. The goal of the project is to farm in a manner beneficial both to the environment and to the economics of farming.[1] More and more farmers in many other states are using intensive rotational grazing systems, and American farmers are not the only ones adopting this more environmentally benign and economically advantageous approach to raising livestock. Nor is it only in Kansas that these programs are working. A recent study of 280 sustainable farming projects in 57 of the world's poorest nations shows that such sustainable farming practices on average increased crop production by 79%. At the same time, these projects are making soils more sustainable and aiding biodiversity.[2, 3]

Figure 12.1 ■ Steve Burr rotates his cattle on this grassland to prevent overgrazing on his farmland.

Intensive rotational grazing is just one of many approaches to farming that are sustainable. Others include crop rotation, use of cover crops to reduce fertilizer needs and erosion, composting livestock wastes, no-till farming, integrated pest and weed management, and redesigning livestock waste management and watering systems.

This case study shows that practices that are environmentally benign can be economically advantageous. The environmental advantages of such alternative approaches to agriculture and the environmental effects of various forms of agriculture are the subjects of this chapter.

12.1 How Agriculture Changes the Environment

Agriculture is both one of humanity's and civilization's greatest triumphs and the source of some of its greatest environmental problems. Agriculture has an ancient lineage, going back thousands of years, and it has always changed the local environment. Nothing in nature resembles a plow or does what a plow does, so as soon as plowing was invented, the environmental effects of agriculture increased.

Environmental effects of agriculture expanded greatly with the scientific-industrial revolution. Major environmental problems that result from agriculture include soil erosion; sediment transport and deposition downstream; on-site pollution from overuse and secondary effects of fertilizers and pesticides; off-site pollution of other ecosystems, of soils, water, and air; deforestation; desertification; degradation of aquifers; salinization; accumulation of toxic metals; accumulation of toxic organic compounds; and loss of biodiversity.

12.2 The Plow Puzzle

Here is a curious puzzle about agriculture and the plow: There are big differences between the soils of an unplowed forest and the soils of previously forested land that has been plowed and used for crops for several thousand years. In Italy, for example, iron plows were pulled by oxen many centuries ago.[4] These differences were observed and written about by one of the originators of the modern study of the environment, George Perkins Marsh. Born in Vermont in the nineteenth century, Marsh became the American ambassador to Italy and Egypt. While in Italy, he was so struck by the differences in the soils of the forests of his native Vermont and the soils that had been farmed for thousands of years on the Italian peninsula that he made this a major theme in his landmark book *Man and Nature*. The farmland he observed in Italy had once been forests. But while the soil in Vermont was rich in organic matter and had definite layers, the soil of Italian farmland had little organic matter and lacked definite layers (Figure 12.2*a, b*).

One would expect that farming in such soil would eventually become unsustainable, but much of the farmland in Italy and France has been in continuous use since pre-Roman times and is still highly productive. How can this be? And what has been the long-term effect of such agriculture on the environment? The answers lie within this chapter.

12.3 Our Eroding Soils

The American Dust Bowl of the 1930s increases the puzzle about the plow. Soils are keys to sustainable farming. Farming easily damages soils (see A Closer Look 12.1). When land is cleared of its natural vegetation, such as for-est or grassland, the soil begins to lose its fertility. Some of this occurs by physical erosion. The good news is that, due to improved farming practices, soil erosion rates have decreased in the United States. In 2001, the most recent year for which government data are available, 42 million hectares (104 million acres) had serious, excessive erosion. But this was down 37% from the 69 million hectares (170 million acres) with such erosion in 1982.[2] The bad news is that, in general, the decreased rate still is greater than the rate of regeneration of new soil.[3]

Soil erosion became a national issue in the United States in the 1930s, when intense plowing, combined with a major drought, loosened the soil over large areas. The soil blew away, creating dust storms that buried automobiles and houses, destroyed many farms, impoverished many people, and led to a large migration of farmers from Oklahoma and other western and midwestern states to California. The human tragedies of the Dust Bowl were made famous by John Steinbeck's novel *The Grapes of Wrath*, later a popular movie starring Henry Fonda (Figure 12.3).

The land that became the Dust Bowl had been part of America's great prairie, where grasses rooted deep, creating a heavily organic soil a meter or more down, protecting the soil from water and wind. When the plow turned over those roots, the soil was exposed directly to sun, rain, and wind, which further loosened the soil. It was a great tragedy of that time and a lesson people thought would be remembered forever. But soil continues to erode.

The introduction of heavy earthmoving machinery after World War II added to the problem by further damaging the soil structure so important for crop production.

As the Dust Bowl made clear, when a forest or prairie is cleared for agriculture, the soil changes. The original soil developed over a long period; it is typically rich in organic matter and therefore rich in chemical nutrients, and it also provides a physical structure conducive to plant growth. When the original vegetation is cleared and the land is planted in crops, most of whose organic matter is harvested and removed, there is less input of dead organic matter to the soil, and the soil is exposed to sunlight, which warms it and speeds the rate of decomposition of its organic matter. For these reasons, the amount of organic matter declines, and the soil's physical structure becomes less conducive to plant growth.

The rate of loss of fertility is sometimes measured as the time required for the soil to lose one-half of its original store of the chemical elements necessary for crops. How long this takes depends partly on the climate. It happens much faster in warmer and wetter areas, such as tropical rain forests, than in colder or drier areas, such as those where the natural vegetation is a temperate-zone grassland or forest.

Traditionally, farmers combated the decline in soil fertility by using organic fertilizers, such as animal manure. These have the advantage of improving both chemical and physical characteristics of soil. But organic fertilizers can

Figure 12.2 ■ (a) A heavily plowed soil. The soil appears as a uniform color, the result of being turned over frequently by a plow, so that what were once distinct layers are mixed together. (b) Unplowed forest soil. Distinct layers, called horizons, are visible: a black organic layer at the top; then a whitish layer bleached of most life-requiring chemical elements; and third an orangish layer where water has deposited the chemical elements eroded from above.
(c) Idealized diagram of a soil, showing soil horizons.

(a)   (b)

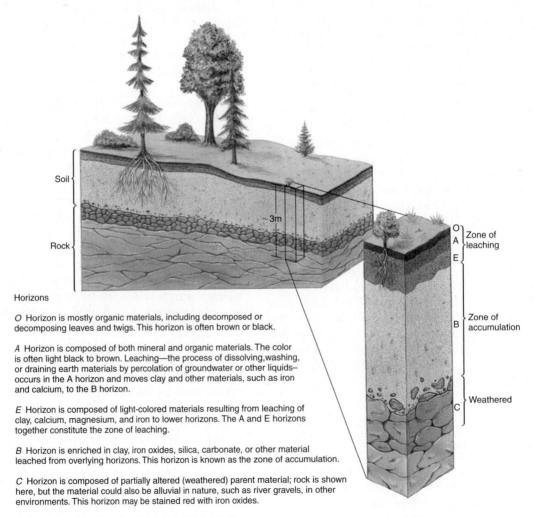

Horizons

O Horizon is mostly organic materials, including decomposed or decomposing leaves and twigs. This horizon is often brown or black.

A Horizon is composed of both mineral and organic materials. The color is often light black to brown. Leaching—the process of dissolving, washing, or draining earth materials by percolation of groundwater or other liquids— occurs in the A horizon and moves clay and other materials, such as iron and calcium, to the B horizon.

E Horizon is composed of light-colored materials resulting from leaching of clay, calcium, magnesium, and iron to lower horizons. The A and E horizons together constitute the zone of leaching.

B Horizon is enriched in clay, iron oxides, silica, carbonate, or other material leached from overlying horizons. This horizon is known as the zone of accumulation.

C Horizon is composed of partially altered (weathered) parent material; rock is shown here, but the material could also be alluvial in nature, such as river gravels, in other environments. This horizon may be stained red with iron oxides.

(c)   *R* Unweathered (unaltered) parent material. (Not shown)

(a)

(b)

Figure 12.3 ■ The Dust Bowl. Poor agricultural practices and a major drought created the Dust Bowl, which lasted about 10 years during the 1930s. Heavily plowed lands lacking vegetative cover blew away easily in the dry winds, creating dust storms (*a*) and burying houses and trucks (*b*).

have drawbacks, especially under intense agriculture on poor soils. In such situations, they do not provide enough of the chemical elements needed to replace what is lost. The development of industrially produced fertilizers, commonly called "chemical" or "artificial" fertilizers, was a major factor in the great increases in crop production in the twentieth century. Among the most important advances were industrial processes to convert molecular nitrogen gas in the atmosphere to nitrate that can be used directly by plants. Phosphorus is mined, usually from a fossil source that was biological in origin, such as deposits of bird guano on islands used for nesting (Figure 12.4). Nitrogen, phosphorus, and other elements are combined in proportions that are appropriate for specific crops in specific locations.

Figure 12.4 ■ A phosphorus mine on a guano island. The material being mined is thousands of years of bird droppings. The birds feed on ocean fish and nest on islands. In dry climates, their droppings accumulate, and have been a major source of phosphorus for agriculture for centuries.

Since the end of World War II, human food production activities have seriously damaged more than 1 billion hectares (2.47 billion acres) of land (about 10.5% of the world's best soil), equal in area to China and India. Overgrazing, deforestation, and destructive crop practices have damaged approximately 9 million hectares (22 million acres) to the point that recovery will be difficult; restoration of the rest will require serious actions.[3]

In the United States, about one-third of the country's topsoil has been lost, resulting in 80 million hectares (198 million acres) either totally ruined by soil erosion or made only marginally productive.[4] Although soil loss in the United States continues, measurements suggest that rates have been considerably reduced as a result of improvements in plowing and the use of no-till agriculture (discussed later in this chapter). For example, the drainage area of Coon Creek, Wisconsin, an area of 360 km2, has been heavily farmed. This stream watershed was the subject of a detailed study in the 1930s by the United States Soil Conservation Service. Then the area was restudied in the 1970s and 1990s. Measurements at these three times indicate that present soil erosion is only 6% of what occurred in the 1930s.[5,6]

Although soil scientists seem to agree in general that soil erosion has decreased in the United States since the 1930s, there is a lack of consensus about the rate of decline and percentage of decline in soil erosion for the United States as a whole. Estimates suggest that, on average, soil erosion in the United States has declined from 17 metric tons per hectare per year (t/ha/yr) to about 13 t/ha/yr.[7]

12.4 Where Eroded Soil Goes: Sediments Also Cause Environmental Problems

Soil eroded from one location has to go somewhere else. A lot of it travels down streams and rivers and is deposited at their mouths. U.S. rivers carry about 3.6 billion metric

Soils

To most of us, soils are just what we step on; we don't think much about them—they're just "dirt." But soils are a key to life on the land, affecting life and affected by it. If you look at them closely, soils are quite remarkable. You won't find anything like Earth soil on Mars or Venus or the moon. Why not? Because water and life have greatly altered the land surface. Geologically, soils are earth materials modified over time by physical, chemical, and biological processes into a series of layers called *soil horizons*. Each kind of soil has its own chemical composition. Soils develop over time—sometimes a very long time, perhaps thousands of years. If you dig carefully into a soil so that you leave a nice, clean vertical side, you will see the soil's layers—that is, if the soil has not been disturbed by a plow or other human activity and if the soil developed in a climate that promoted certain processes. In a northern forest, a soil is dark at the top, then has a white powdery layer, pale as ash, then a brightly colored layer, which is usually much deeper than the white one and is typically orangish. Below that is a soil whose color is close to that of the bedrock (which geologists call "the parent material," for obvious reasons). We call the layers *horizon* (Figure 12.2*c*).

Overall, water flows down through the soil. Rainwater is naturally slightly acid because it has some carbon dioxide from the air dissolved in it, and this forms carbonic acid, a mild acid. Rainwater has a pH of about 5.5. It doesn't taste sour or acid, but it is acid enough to leach metals from the soil. As a result, minerals such as iron, calcium, and magnesium are leached from the upper horizons (A and E) and may be deposited in a lower horizon (B) (Figure 12.2*c*). The upper horizons are usually full of life and are viewed by ecologists as complex ecosystems, or ecosystem

units (horizons O and A). Decomposition is the name of the game as fungi, bacteria, and small animals live on what plants and animals on the surface produce and deposit. Actual chemical decomposition of organic compounds from the surface is done by bacteria and fungi, the great chemical factories of the biosphere. Soil animals, such as earthworms, eat leaves, twigs, and other remains and break them into smaller pieces that are easier for the fungi and bacteria to process. The animals affect the rate of chemical reactions in the soil. There are also predators on soil animals, so there is a soil ecological food chain.

What a soil is like is determined by climate, parent material (bedrock), topography, biological activity, and time. The soil horizons shown in Figure 12.2*c* are not necessarily all present in any one soil. Very young soils may have only an upper A horizon over a C horizon, whereas mature soils may have nearly all the horizons shown.

Soil fertility is the capacity of a soil to supply nutrients necessary for plant growth. Soils that have formed on geologically young materials—for example, glacial deposits in northern Indiana and Illinois that form the famous midwestern corn belt—are nutrient-rich. Soils in semiarid regions such as the Great Central Valley of California are often nutrient-rich and need only water to become very productive for agriculture; soils in humid areas and tropics may be heavily leached and relatively nutrient-poor due to the high rainfall. In such soils, nutrients may be cycled through the organic-rich upper horizons; and if forest cover is removed, reforestation may be very difficult (see Chapter 13). Soils that accumulate certain clay minerals in semiarid regions may swell when

they get wet and shrink as they dry out, cracking roads, walls, buildings, and other structures. Expansion and contraction of soils in the United States cause billions of dollars' worth of property damage each year.

Soils with small clay particles (less than 0.004 mm in diameter) retain water well and retard the movement of water because the spaces between the particles are very small. Soils with coarser grains (greater than 0.06 mm in diameter), such as sand or gravel, have relatively large spaces between grains, so water moves quickly through them. Soils with a mixture of clay and sand can retain water well enough for plant growth but also drain well. Soils with a high percentage of organic matter also retain water and chemical nutrients for plant growth. It is an advantage to have good drainage, so a coarse-grained soil is a good place to build your house. If you are going to farm, you'll do best in a loam soil that has a mixture of particle sizes. Thus, the type of soil particles present is important in determining where to build a house and where to farm and in siting facilities such as landfills, where retention of pollutants on site is an objective (see Chapter 29).

Coarse-grained soils, especially those composed primarily of sand (0.06–2.0 mm in diameter), are particularly susceptible to erosion by water and wind. Soils composed of coarser (heavier) particles or finer particles that are usually more cohesive (held together by clay minerals) are more resistant to erosion.

It is difficult to think of a human use of the near-surface land environment that does not involve consideration of the soils present. As a result, the study of soils continues to be an important part of environmental sciences.

tons per year (4 billion U.S. tons/year) of sediment, 75% of it from agricultural lands. That's more than 25,000 pounds of sediment for each person in the United States.

Of this total, 2.7 billion metric tons/year (3 billion U.S. tons/year) are deposited in reservoirs, rivers, and lakes. Eventually, these sediments fill in otherwise productive waters, destroying some fisheries. In tropical waters, sediments entering the ocean can destroy coral reefs near a shore.

Sedimentation has chemical environmental effects as well. Nitrates, ammonia, and other fertilizers carried by sediments enrich the waters downstream. This enrichment, called *eutrophication*, promotes the growth of algae. It's a straightforward process: Fertilizers that were meant to increase the growth of crops have the same effect on algae in the water. But people generally do not want water enriched with algae, because the dead algae are decomposed by bacteria that, in turn, remove oxygen from the water. As a result, fish can no longer live in the water. The water becomes thick with a greenish-brown mat, unpleasant for recreation and a poor base for drinking water. Sediments also can transport toxic chemical pesticides. Since the 1930s, agriculture-induced sedimentation has decreased with the decrease in the rate of soil erosion. Even so, taking into account the costs of dredging and the decline in the useful life of reservoirs, sediment damage costs the United States about $500 million a year.

Making Soils Sustainable

Soil forms continuously. In ideal farming, the amount of soil lost would never exceed the amount of new soil produced. Production is slow—on good lands, a layer of soil 1 mm deep, thinner than a piece of paper, forms at a rate ranging from one per decade to one in 40 years. Sustainability of soils can be aided by fall plowing, multiculture (planting several crops intermixed in the same field), terracing, crop rotation, contour plowing, and no-till agriculture. At this point in our discussion, we have reached a partial answer to the question "How could farming be sustained for thousands of years, while the soil has been degraded?" We must recognize a distinction between the sustainability of a product (in this case crops) and the sustainability of the ecosystem. *In agriculture, crop production can be sustained, but the ecosystem may not be.* And if the ecosystem is not sustained, then people must provide additional input of energy and chemical elements to replace what is lost.

Contour Plowing

Plowed furrows make paths for water to flow, and if the furrows go downhill, then the water moves rapidly along them, increasing the erosion rate. In **contour plowing**, the land is plowed perpendicular to the slopes and as horizontally as possible.

Contour plowing has been the single most effective way to reduce soil erosion. This was demonstrated by an experiment on sloping land planted in potatoes. Part of the land was plowed in uphill and downhill rows, and part was contour-plowed. The uphill and downhill section lost 32 metric tons/ha (14.4 tons/acre) of topsoil; the contour-plowed section lost only 0.22 metric ton/ha (0.1 ton/acre) (Figure 12.5a). It would take almost 150 years for the contour-plowed land to erode as much as the tra-

(a)

(b)

Figure 12.5 ■ Alternative agricultural plowing and tilling methods: (*a*) contour strip crops in the midwestern United States; (*b*) no-till soybean crop planted in wheat stubble on a Kansas farm. Both methods reduce erosion. No-till, however, requires extensive use of pesticides.

ditionally plowed land eroded in a single year! In addition to drastically reducing soil erosion, contour plowing uses less fuel and time. Even so, today contour plowing is used on only a small fraction of the land in the United States. For example, of Minnesota's 4 million hectares (10 million acres) of cropland, only 530,000 hectares (1.3 million acres) are contour-plowed.

No-Till Agriculture

An even more efficient way to slow erosion is to avoid plowing altogether. **No-till agriculture** (also called *conservation tillage*) involves not plowing the land, using herbicides and integrated pest management (discussed later in this chapter) to keep down weeds, and allowing some weeds to grow. Stems and roots that are not part of the commercial crop are left in the fields and allowed to decay in place (Figure 12.5b). In contrast to standard modern approaches,

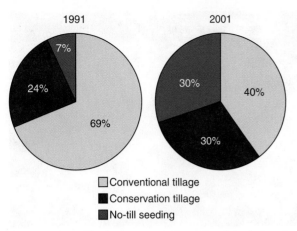

Figure 12.6 ■ To plow or not to plow: Canada's choices. In Canada, the percentage of no-till agriculture increased greatly in a decade. (*Source:* Statistics Canada's Internet Site, http://www.statcan.ca, 2003.)

the goal in no-till agriculture is to suppress and control weeds but not to eliminate them at the expense of soil conservation. Worldwide, no-till agriculture is increasing but is practiced on only 5% of the world's farmland.[4] Paraguay leads the world with 55% of its farmland in no-till. The United States, with 17.5% in no-till, lags behind many other nations. Argentina has 45%, Brazil 39%, and Canada reached 30% in 2001, up from 24% a decade earlier (Figure 12.6). An additional benefit of no-till agriculture is that it reduces the release of carbon dioxide from the soil, which occurs when the soil is plowed. Thus no-till agriculture is one way to reduce the effects of global warming.[4]

12.5 Controlling Pests

From an ecological point of view, *pests* are undesirable competitors, parasites, or predators. The major agricultural pests are insects that feed mainly on the live parts of plants, especially leaves and stems; nematodes (small worms), which live mainly in the soil and feed on roots and other plant tissues; bacterial and viral diseases; weeds (plants that compete with the crops); and vertebrates (mainly rodents and birds) that feed on grain or fruit. Even today, with modern technology, the total losses from all pests are huge; in the United States, pests account for an estimated loss of one-third of the potential harvest and about one-tenth of the harvested crop. Preharvest losses are due to competition from weeds, diseases, and herbivores; postharvest losses are largely due to herbivores.[4]

We tend to think that the major agricultural pests are insects, but in fact weeds are the major problem. Farming produces special environmental and ecological conditions that tend to promote weeds. Remember that the process of farming is an attempt to (1) hold back the natural processes of ecological succession, (2) prevent migrating organisms from entering an area, and (3) prevent natural

interactions (including competition, predation, and parasitism) between populations of different species.

Because a farm is maintained in a very early stage of ecological succession and is enriched by fertilizers and water, it is a good place not only for crops but also for other early-successional plants. These noncrop and therefore undesirable plants are what we call *weeds*. A weed is just a plant in a place we do not want it to be. Recall that early-successional plants tend to be fast-growing and to have seeds that are easily blown by the wind or spread by animals. These plants spread and grow rapidly in the inviting habitat of open, early-successional croplands.

There are about 30,000 species of weeds, and in any year a typical farm field is infested with between 10 and 50 of them. Weeds compete with crops for all resources: light, water, nutrients, and just plain space to grow. The more weeds, the less crop. Some weeds can have a devastating effect on crops. For example, the production of soybeans is reduced by 60% if a weed called cocklebur grows three individuals per meter (one individual per foot).[5]

12.6 The History of Pesticides

Before the Industrial Revolution, farmers could do little to prevent pests except remove them when they appeared or use farming methods that tended to decrease their density. For example, slash-and-burn agriculture (also known as swidden agriculture) allows succession to take place. The greater diversity of plants and the long time between the use of each plot reduces the density of pests (see Chapter 11). Preindustrial farmers also planted aromatic herbs and other vegetation that repel insects.

With the beginning of modern science-based agriculture, people began to search for chemicals that would reduce the abundances of pests. More important, they searched for a "magic bullet"—a chemical (referred to as a *narrow-spectrum pesticide*) that would have a single target, just one pest, and not affect anything else. But this proved elusive. We have already seen that living things have many chemical reactions in common (see Chapters 4 and 6), so a chemical that is toxic to one species is likely to be toxic to another. The story of the scientific search for pesticides is the search for a better and better magic bullet. The earliest pesticides were simple inorganic compounds that were widely toxic. One of the earliest was arsenic, a chemical element toxic to all life, including people. It was certainly effective in killing pests, but it killed beneficial organisms as well and was very dangerous to use.

A second stage in the development of pesticides began in the 1930s and involved petroleum-based sprays and natural plant chemicals. Many plants produce chemicals as a defense against disease and herbivores, and these chemicals are effective pesticides. Nicotine, from the tobacco plant, is the primary agent in some insecticides still in wide use today. However, although natural plant pesticides are comparatively safe, they were not as effective as desired.

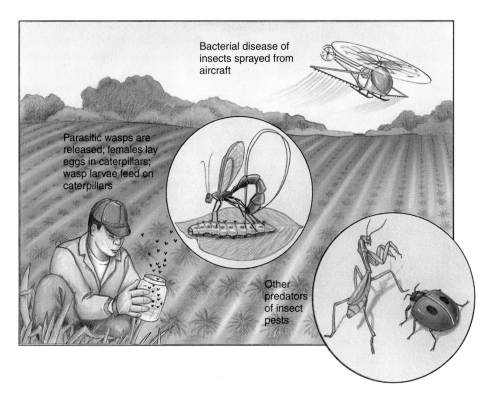

Bacterial disease of insects sprayed from aircraft

Parasitic wasps are released; females lay eggs in caterpillars; wasp larvae feed on caterpillars

Other predators of insect pests

The third stage in the development of pesticides was the development of artificial organic compounds. Some, like DDT, are broad-spectrum, but more effective than natural plant chemicals. These chemicals have been important to agriculture, but unexpected environmental effects keep cropping up, and the magic bullet has remained elusive. For example, aldrin and dieldrin have been widely used to control termites as well as pests on corn, potatoes, and fruits. Dieldrin is about 50 times as toxic to people as DDT. These chemicals are designed to remain in the soil and typically do so for years, but they are readily drained from tropical rain-forest soils. As a result, they have spread widely and are found in organisms in arctic waters. The chemicals accumulate in people. A study of breast-fed babies in Australia showed that every day 88% took in an amount that exceeded the World Health Organization standard for a daily intake.[6]

As a result, a fourth stage in the development of pesticides began, which returned to biological and ecological knowledge. This was the beginning of modern **biological control**, the use of biological predators and parasites to control pests. One of the most effective of these is a bacterium named *Bacillus thuringiensis*, known as BT, which causes a disease that affects caterpillars and the larvae of other insect pests. Spores of BT are sold commercially (you can buy them at your local garden store and use this method for your home garden). BT has been one of the most important ways to control epidemics of gypsy moths, an introduced moth whose larvae periodically strip most of the leaves from large areas of forests in the eastern United States. BT has proved safe and effective—safe because it causes disease only in specific insects and is harmless to people and other mammals,

and because, as a natural biological "product," its presence and its decay are nonpolluting.

Another group of effective biological control agents are small wasps that are parasites of caterpillars. The wasps lay their eggs on the caterpillars; the larval wasps then feed on the caterpillars, killing them. These wasps tend to have very specific relationships (one species of wasp will be a parasite of one species of pest), and so they are both effective and narrow-spectrum (Figure 12.7).

And in the list of biological control species we cannot forget ladybugs, which are predators of many pests. You can buy these, too, at many garden stores and release them in your garden.

Another technique to control insects involves the use of sex pheromones, chemicals released by most species of adult insects (usually the female) to attract members of the opposite sex. In some species, pheromones have been shown to be effective up to 4.3 km (2.7 mi) away. These chemicals have been identified, synthesized, and used as bait in insect traps, in insect surveys, or simply to confuse the mating patterns of the insects involved.

12.7 Integrated Pest Management

While biological control works well, it has not solved all problems with agricultural pests. As a result, a fifth stage developed, known as **integrated pest management (IPM)**. IPM uses a combination of methods, including biological control, certain chemical pesticides, and some methods of planting crops. A key idea underlying IPM is that the goal can be control rather than complete elimination of a pest. This is justified for several reasons.

Economically, it becomes more and more expensive to eliminate a greater and greater percentage of a pest, while the value of that elimination, in terms of crops to sell, becomes less and less. This suggests that it makes economic sense to leave some of the pests and eliminate only enough to provide benefit. In addition, allowing some of a pest population to remain, small but controlled, does less damage to ecosystems, soils, water, and air. Some like to think of IPM as an ecosystem approach to pest management, because it makes use of the characteristics of ecological communities and ecosystems as discussed in Chapters 4, 6, and 10.

Another characteristic of IPM is the attempt to move away from monoculture of a single strain growing in perfectly regular rows. Studies have shown that just the physical complexity of a habitat can slow the spread of parasites. In effect, a pest like a caterpillar or mite is trying to find its way through a maze. If the maze consists of regular rows of nothing but what the pest likes to eat, the maze problem is easily solved by the dumbest of animals. But if there are several species, even two or three, arranged in a more complex pattern, the pests have a hard time finding their prey.

No-till or low-till agriculture is another feature of IPM, because this helps natural enemies of some pests to build up in the soil (plowing destroys the habitats of these pest enemies).

Control of the oriental fruit moth, which attacks a number of fruit crops, is an example of IPM biological control (Figure 12.8). The moth was found to be a prey of a species of wasp, *Macrocentrus ancylivorus*,[7] and introducing the wasp into fields helped control the moth. Interestingly, in peach fields the wasp was more effective when strawberry fields were nearby. The strawberry fields provided an alternative habitat for the wasp, especially important for overwintering.[5] As this example shows, spatial complexity and biological diversity also become part of the IPM strategy.

Although artificial pesticides are used, they are used along with the other techniques, so the application of these pesticides can be sparing and specific.[8] This would also greatly reduce the costs to farmers for pest control.[4]

Current agricultural practices in the United States involve a combination of approaches, but in most cases they are more restricted than an IPM strategy. Biological control methods are used to a comparatively small extent. They are the primary tactics for controlling vertebrate pests (mice, voles, and birds) that feed on lettuce, tomatoes, and strawberries in California but are not major techniques for grains, cotton, potatoes, apples, or melons. Chemicals are the principal control methods for insect pests. For weeds, the principal controls are methods of land culture. The use of genetically resistant stock is important for disease control in wheat, corn, cotton, and some vegetable crops, such as lettuce and tomatoes.

Monitoring Pesticides in the Environment

World use of pesticides exceeds 2.5 billion kg (5 billion pounds), and in the United States it exceeds 680 million kg (1,200 million pounds) (Figure 12.9). The total amount paid for these pesticides is $32 billion worldwide and $11 billion in the United States.[9]

Once applied, these chemicals may decompose in place or may be blown by the wind or transported by surface and subsurface waters, meanwhile continuing to decompose. Sometimes the initial breakdown products (the first, still complex chemicals produced from the original pesticides) are toxic, as is the case with DDT (see A Closer Look 12.2). Eventually, the toxic compounds are decomposed to their original inorganic or simple, nontoxic organic compounds. However, for some chemicals, this can take a very long time.

Herbicides account for about 60% of pesticides found in the nation's waters. Surprisingly little is known about the history of the concentrations of pesticides in the major rivers of America. For example, there is no well-established program to monitor the changes in the concentration of pesticides in

(a)

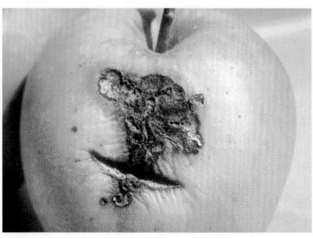

(b)

Figure 12.8 ■ The oriental fruit moth larvae, a pest of fruit crops, is controlled by a parasite wasp that attacks the larvae. (*a*) The larvae. (*b*) Apples damaged by the moth.

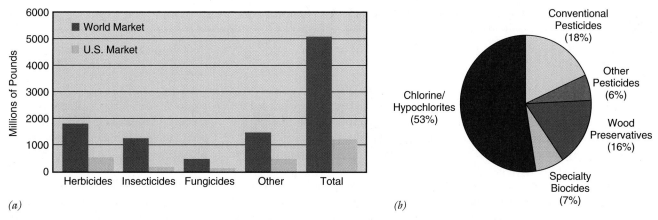

(a) *(b)*

Figure 12.9 ■ World use of pesticides. (*a*) total ammonts, (*b*) percentage by main type [*Sources:* EPA (2006). "2000–2001 Pesticide Market Esimates: Usage," EPA.]

the Missouri River, one of the longest rivers in the world, although recently there have been spot measures of this concentration.[10] The Missouri drains one-sixth of the United States, much of it from the major agricultural states.

Where do all these pesticides go? How long do they last in the environment, both on the site where they were deposited and downstream and downwind? What is the concentration of these in our waters? To establish useful standards for pesticide levels in the environment, and to understand the environmental effects of pesticides, it is necessary to monitor the concentrations. Public health standards and environmental-effects standards have been established for some but not all of these compounds. The United States Geological Survey has established a network for monitoring 60 sample watersheds throughout the nation. These are medium-size watersheds, not the entire flow from the nation's major rivers. One such watershed is that of the Platte River, a major tributary of the Missouri River.

The most common herbicides used for growing corn, sorghum, and soybeans along the Platte River were alachlor, atrazine, cyanazine, and metolachlor, all organonitrogen herbicides. Monitoring of the Platte near Lincoln, Nebraska, suggested that during heavy spring runoff, concentrations of some herbicides might be reaching or exceeding established public health standards. But this research is just beginning, and it is difficult to reach definitive conclusions as to whether present concentrations

are causing harm in public water supplies or to wildlife, fish, algae in freshwater, or vegetation. Advances in knowledge give us much more information, on a more regular basis, about how much of many artificial compounds are in our waters, but we are still unclear about their environmental effects. A wider and better program to monitor pesticides in water and soil is important to provide a sound scientific basis for dealing with pesticides.

12.8 Genetically Modified Crops

Remember from Chapter 11 that the genetic modification of organisms currently uses three methods: (1) faster and more efficient development of new hybrids, (2) introduction of the "terminator gene," and (3) transfer of genetic properties from widely divergent kinds of life (Figure 12.10).

Each of these poses different potential environmental problems. Here, we need to keep in mind a general rule of environmental actions: If actions we take are similar in kind and frequency to natural changes, then the effects on the environment are likely to be benign. This is because species have had a long time to evolve and adapt to these changes. In contrast, changes that are novel—that do not occur in nature—are more likely to have negative or undesirable environmental effects, both direct and indirect. We can apply this rule to the three categories of genetically engineered crops.

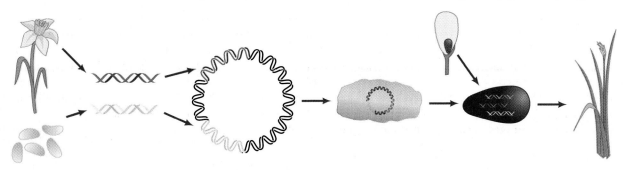

Figure 12.10 ■ An example of how crops are genetically modified. [See text for explanation.]

DDT

In 2006, the United Nations World Health Organization endorsed wider use of DDT in Africa to combat malaria. This controversial decision is the latest event in the long history of the use and misuse of this famous pesticide. The real revolution in chemical pesticides—the development of more sophisticated pesticides—began with the end of World War II and the discovery of DDT and other chlorinated hydrocarbons, including aldrin and dieldrin. When DDT was first developed in the 1940s, it seemed to be the long-sought magic bullet, with no short-term effects on people and deadly only to insects. At the time, scientists believed that a chemical could not be readily transported from its original site of application unless it was water-soluble. DDT was not very soluble in water and therefore did not appear to pose an environmental hazard. DDT was used very widely until three things were discovered.

- It has long-term effects on desirable species. Most spectacularly, it decreased the thickness of eggshells as they developed within birds.

- It is stored in oils and fats and is transferred up food chains as one animal eats another. As it is passed up food chains, it becomes concentrated so that the higher an organism is on a food chain, the higher its concentration of DDT. This process is known as food-chain concentration or biomagnification (discussed in detail in Chapter 15).

- The storage of DDT in fats and oils allows the chemical to be transferred biologically even though it is not very soluble in water.

In birds, DDT and the products of its chemical breakdown (known as DDD and DDE) thinned eggshells so that they broke easily, reducing the success of reproduction. This was especially severe in birds that are high on the food chain—predators that feed on other predators, such as the bald eagle, the osprey, and the pelican, which feed on fish that may be predators of other fish.

As a result, DDT was banned in most developed nations—banned in the United States in 1971. Since then, a dramatic recovery has occurred in the populations of affected birds. The brown pelican of the Florida and California coasts, which had become rare and endangered and whose reproduction had been restricted to offshore islands where DDT had not been used, became common again. The bald eagle became abundant again in the north woods, where it can be seen in Voyageurs National Park and the Boundary Waters Canoe Area of northern Minnesota. However, DDT is still being produced in the United States for use in the developing and less developed nations, especially as a control for malaria-spreading mosquitoes.

The use of DDT has had some benefits. It was primarily responsible for eliminating malaria and yellow fever as major diseases, reducing the incidence of malaria in the United States from an average of 250,000 cases a year prior to the spraying program to fewer than 10 per year in 1950. Even for these uses, however, DDT's effectiveness has declined over the years because many species of insects have developed a resistance to it. Nevertheless, DDT continues to be used because it is cheap and sufficiently effective and because people have become accustomed to using it. The U.S. Centers for Disease Control estimates that between 200,000 and 2.7 million people die each year from malaria, 75% of them children in Africa.[11] About 35,000 metric tons of DDT are produced annually in at least five countries, and it is legally imported and used in dozens, including Mexico.

Although people in developed nations believe they are free from the effects of DDT, in fact this chemical is transported back to industrial nations in agricultural products from nations still using the chemical. Also, migrating birds that spend part of the year in malarial regions are still subject to DDT. Thus, despite being banned in the developed nations, DDT remains an important world issue in pest control. (The problem of developing nations' use of pesticides banned in other nations is an issue not only for DDT but also for other chemicals.)

With the banning of DDT in developed nations, other chemicals became more prominent, chemicals that were less persistent in the environment. Among the next generation of insecticides were organophosphates—phosphorus-containing chemicals that affect the nervous system. These chemicals are more specific and decay rapidly in the soil. Therefore, they do not have the same persistence as DDT. But they are toxic to people and must be handled very carefully by those who apply them.

Chemical pesticides have created a revolution in agriculture. However, in addition to the negative environmental effects of chemicals such as DDT, they have other major drawbacks. One problem is secondary pest outbreaks, which occur after extended use (and possibly because of extended use) of a pesticide. Secondary pest outbreaks can come about in two ways: (1) Reducing one target species reduces competition with a second species, which then flourishes and becomes a pest, or (2) the pest develops resistance to the pesticides through evolution and natural selection, which favor those who have a greater immunity to the chemical.[12] Resistance has developed with many pesticides. For example, Dasanit (fensulfothion), an organophosphate, first introduced in 1970 to control maggots that attack onions in Michigan, was originally successful but is now so ineffective that it is no longer used for that crop.

New Hybrids

The development of hybrids within a species is a natural phenomenon (see Chapter 7), and the development of hybrids of major crops, especially of small grains, has been a major factor in the great increase in productivity of twentieth-century agriculture. So, strictly from an environmental perspective, genetic engineering to develop hybrids within a species is likely to be as benign as the development of agricultural hybrids has been with conventional methods.

There is an important caveat, however. Some people are concerned that the great efficiency of genetic modification methods may produce "superhybrids" that are so productive they can grow where they are not wanted and become pests. There is also concern that some of the new hybrid characteristics could be transferred by interbreeding with closely related weeds (Figure 12.11). This could inadvertently create a "superweed" whose growth, persistence, and resistance to pesticides would make it difficult to control. Another environmental concern is that new hybrids might be developed that could grow on more and more marginal lands. The development of crops on such marginal lands might increase erosion and sedimentation and lead to decreased biological diversity in specific biomes. Still another potential problem is that "superhybrids" might require

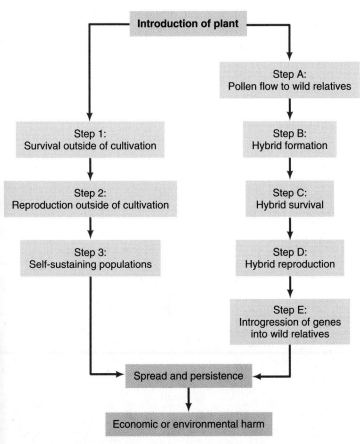

Figure 12.11 ■ Ways in which the genetic characteristics of a modified crop might spread.

much more fertilizer, pesticide, and water. This could lead to greater pollution and the need for more irrigation.

On the other hand, genetic engineering could lead to hybrids that require less fertilizer, pesticide, and water. For example, at present, only legumes (peas and their relatives) have symbiotic relationships with bacteria and fungi that allow them to fix nitrogen. Attempts are under way to transfer this capability to other crops, so that more kinds of crops would enrich the soil with nitrogen and require much less external application of nitrogen fertilizer.

The Terminator Gene

The **terminator gene** makes seeds from a crop sterile. This is done for environmental and economic reasons. In theory, it prevents a genetically modified crop from spreading. It also protects the market for the corporation that developed it: Farmers cannot get around purchasing seeds by using some of their crops' hybrid seeds the next year. But this poses social and political problems. Farmers in less-developed nations and governments of nations that lack genetic-engineering capabilities are concerned that the terminator gene will allow the United States, and a few of its major corporations, to control the world food supply. Concerned observers believe that farmers in poor nations must be able to grow next year's crops from their own seeds because they cannot afford to buy new seeds every year. This is not directly an environmental problem, but it can become an environmental problem indirectly by affecting total world food production, which then affects the human population and how land is used in areas that have been in agriculture.

Transfer of Genes from One Major Form of Life to Another

Most environmental concerns have to do with the third kind of genetic modification of crops: the transfer of genes from one major kind of life to another. This is a novel effect and therefore more likely to have negative and undesirable impacts. In several cases, this type of genetic modification has led to unforeseen and undesirable environmental effects. Perhaps the best known involves potatoes and corn, caterpillars that eat these crops, a disease of caterpillars that controls these pests, and an endangered species, monarch butterflies. Here is what happened.

As discussed earlier, the bacterium *Bacillus thuringiensis* is a successful pesticide, causing a disease in many caterpillars. With the development of biotechnology, agricultural scientists studied the bacteria and discovered the toxic chemical and the gene that caused its production within the bacteria. This gene was then transferred to potatoes and corn so that the biologically engineered plants produced their own pesticide. At first, this was believed to be a constructive step in pest control, because it was no longer necessary to spray a pesticide. However, the genetically engineered potatoes and corn produced the toxic BT

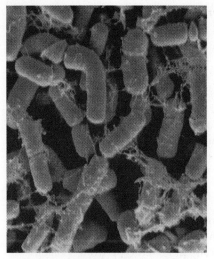  

(*a*) Bacillus thuringiensis *bacteria (a natural pesticide). The gene that caused the pesticide (BT) was placed in corn through genetic engineering.*

(*b*) *BT corn contains its own pesticide in every cell of the plant.*

(*c*) *Pollen from the BT corn is also toxic and when it lands on milkweed, monarch butterflies that eat the milkweed may die.*

Figure 12.12 ■ The flow of the BT toxin from bacteria (*a*) to corn through genetic engineering (*b*) and the possible ecological transfer of toxic substances to monarch butterflies (*c*).

substance in every cell—not just in the leaves that the caterpillars ate, but also in the potatoes and corn sold as food, in the flowers, and in the pollen. This has a potential, not yet demonstrated, to create problems for species that are not intended targets of the BT (Figure 12.12).

A strain of rice has been developed that produces beta-carotene, important in human nutrition. The rice thus has added nutritional benefits, that are particularly valuable for the poor of the world who depend on rice as a primary food. The gene that enables rice to make beta-carotene comes from daffodils, but the modification actually required the introduction of four specific genes and would likely be impossible without genetic-engineering techniques. That is, genes were transferred between plants that would not exchange genes in nature. Once again, the rule of natural change suggests that we should monitor such actions carefully.

Although the genetically modified rice appears to have beneficial effects, the government of India has refused to allow it to be grown in that country.[13] There is much concern worldwide about the political, social, and environmental effects of genetic modification of crops. This is a story in process, one that will change rapidly in the next few years. You can check on these fast-moving events on the textbook's Web site.

12.9 Grazing on Rangelands: An Environment Benefit or Problem?

Almost half of Earth's land area is used as rangeland, and about 30% of Earth's land is *arid* rangeland, land that is easily damaged by grazing, especially during drought

(Figure 12.13). In the United States, more than 99% of rangeland is west of the Mississippi River. Much of the world's rangeland is considered to be in poor condition from overgrazing. In the United States, rangeland conditions have improved since the 1930s, especially in upland areas. However, land near streams and the streams themselves continue to be heavily affected by grazing.

Grazing cattle trample stream banks and release their waste into stream water. Maintaining a high-quality stream environment requires that cattle be fenced behind a buffer zone.

Figure 12.13 ■ Traditional sheep grazing, a practice that has occurred for thousands of years and affects almost half of Earth's land.

The upper Missouri River is famous for its beautiful "white cliffs," but private lands along the river that are used to graze cattle take away from the scenic splendor. Cattle come down to the Missouri River to drink in numbers sufficient to damage the land along the river. The river itself runs heavy with manure. These effects extend to an area near a federally designated wild and scenic portion of the upper Missouri River, and tourists traveling on the Missouri have complained. In recent years, fencing along the upper Missouri River has increased, with small openings to allow cattle to drink, but otherwise restricting what they can do to the shoreline.[14]

Traditional and Industrialized Use of Grazing and Rangelands

Traditional herding practices and industrialized production of domestic animals have different effects on the environment. In modern industrialized agriculture, cattle are initially raised on open range and then transported to feedlots, where they are fattened for market. Feedlots have become widely known in recent years as sources of local pollution. The penned cattle are often crowded and are fed grain or forage that is transported to the feedlot. Manure builds up in large mounds. When it rains, the manure pollutes local streams. Feedlots are popular with meat producers because they are economical for rapid production of good-quality meat. However, large feedlots require intense use of resources and have negative environmental effects.

Traditional herding practices, by comparison, chiefly affect the environment through overgrazing. Goats are especially damaging to vegetation, but all domestic herbivores can destroy rangeland. The effect of domestic herbivores on the land varies greatly with their density relative to rainfall and soil fertility. At low to moderate densities, the animals may actually aid growth of aboveground vegetation by fertilizing soil with their manure and stimulating plant growth by clipping off plant ends in grazing, just as pruning stimulates plant growth. But at high densities, the vegetation is eaten faster than it can grow; some species are lost, and the growth of others is greatly reduced.

The Biogeography of Agricultural Animals

People have distributed cattle, sheep, goats, and horses, as well as other domestic animals, around the world and then promoted the growth of these animals to densities that have changed the landscape. Preindustrial people made such introductions. For example, Polynesian settlers brought pigs and other domesticated animals to Hawaii and other Pacific islands. Since the age of exploration by Western civilization, starting in the fifteenth century, domestic animals have been introduced into Australia, New Zealand, and the Americas. Horses, cows, sheep, and goats were brought to North America after the sixteenth century. The spread of cattle brought new animal diseases and new weeds, which arrived on the animals' hooves and in their manure. Introductions of domestic animals into new habitats have many environmental effects. Two important effects are that (1) native vegetation, not adapted to the introduced grazers, may be greatly reduced and threatened with extinction; and (2) introduced animals may compete with native herbivores, reducing their numbers to a point at which they, too, may be threatened with extinction.

A recent important issue in cattle production is the opening up of tropical forest areas and their conversion to rangeland—for example, in the Brazilian Amazon basin. In a typical situation, the forest is cleared by burning and crops are planted for about four years. After that time, the soil has lost so much fertility that crops can no longer be grown economically. Ranchers then purchase the land, already cleared, and run cattle bred to survive in the hot, humid conditions. After about another four years, the land can no longer support even grazing and is abandoned. In such areas, grazing has greatly impaired the land's capability for many uses, including forest growth.[15] Clearly, this is an unsustainable approach to agriculture and therefore undesirable.

The spread of domestic herbivores around the world is one of the major ways we have changed the environment through agriculture. As the human population increases, and as income and expectations rise, the demand for meat increases. As a result, we can expect greater demand for rangeland and pastureland in the next decades. A major challenge in agriculture will be to develop ways to make the production of domestic animals sustainable.

Carrying Capacity of Grazing Lands

Carrying capacity is the maximum number of a species per unit area that can persist without decreasing the ability of that population or its ecosystem to maintain that density in the future. The carrying capacity of land for cattle varies with rainfall, topography, soil type, and soil fertility.

When the carrying capacity is exceeded, the land is overgrazed. **Overgrazing** slows the growth of the vegetation, reduces the diversity of plant species, leads to dominance by plant species that are relatively undesirable to the cattle, hastens the loss of soil by erosion as the plant cover is reduced, and subjects the land to further damage from the cattle's trampling on it (Figure 12.14). The damaged land can no longer support the same density of cattle.

In areas with moderate to high rainfall evenly distributed throughout the year, cattle can be maintained at high densities; but in arid and semiarid regions, the density drops greatly. In the United States, the carrying capacity for cows drops from 190/km^2 (500/mi^2) in the East to 40/km^2 (100/mi^2) in the region that was prairie and to 4/km^2 (10/mi^2) or less in semiarid and desert regions (Figure 12.15). In Arizona, for example, rainfall is low, and cattle can be maintained only at low densities—one head of cattle for 7–10 hectares (17–25 acres). Near Paso Robles, California, in an area where rainfall is about

Figure 12.14 ■ Soil erosion in northern Natal caused by overgrazing and other land-use practices.

25 cm/year (9.8 in./year)—that is, in desert to semiarid conditions—a ranch where cattle are grazed without artificial irrigation or fertilization supports about one head for 6 hectares (15 acres).

12.10 Desertification: Regional Effects and Global Impact

Deserts occur naturally where there is too little water for substantial plant growth. Because the plants that do grow are too sparsely spaced and unproductive to create a soil rich in organic matter, desert soils are mainly inorganic, coarse, and typically sandy (see the discussion of succession and soils in Chapter 10). When rain does come, it is often heavy, and erosion is severe. The principal climatic condition that leads to desert is low or undependable precipitation. The warmer the climate, the greater the rainfall required to convert an area from desert to nondesert, such as grassland. But even in cooler climates or at higher altitudes, deserts may form if precipitation is too low to support more than sparse plant life. The crucial factor is the amount of water in the soil available for plants to use. Factors that destroy the ability of a soil to store water can create a desert.

Earth has five natural warm desert regions, all of which lie primarily between latitudes 15° and 30° north and south of the equator. These include the deserts of the southwestern United States and Mexico; Pacific Coast deserts of Chile and southern Ecuador; the Kalahari Desert of southern Africa; the Australian deserts that cover most of that continent; and the greatest desert region of all—the desert that extends from the Atlantic Coast of North Africa (the Sahara) eastward to the deserts of Arabia, Iran, Russia, Pakistan, India, and China.[16] Only Europe lacks a major warm desert; Europe lies north of the desert latitudinal band.

Based on climate, about one-third of Earth's land area should be desert, but estimates suggest that 43% of the land is desert. This additional desert area is believed to be a result of human activities.[13] **Desertification** is the deterioration of land in arid, semiarid, and dry subhumid areas due to changes in climate and human activities.[14]

Desertification is a serious global problem. It affects one-sixth of the world's population (about 1 billion people) and threatens 4 billion hectare, one-third of Earth's land area.[5] In the United States it threatens 30% of the land. Land degradation caused by people has altered 73% of drier rangelands (3.3 billion ha) and the soil fertility and structure of 47% of dryland areas with marginal rainfall for crops. Land degradation also affects 30% of dryland areas with high population density and agricultural potential.

A large part of desertification occurs in the poorest countries. These regions include Asia, Africa, and South America. Worldwide, 6 million hectares of land per year are lost to this process, with an estimated economic loss of $40 billion. And the cost of recovery of these lands could reach $10 billion per year.[15]

What Causes Deserts?

Some areas of Earth are marginal lands; even light grazing and crop production can turn them into deserts. In semiarid regions, rainfall is just barely enough to enable the land to produce more vegetation than a desert, and even light grazing is a problem. The leading human causes of desertification are bad farming practices, such as failure to use contour plowing or *simply too much farming* (Figure 12.16); overgrazing (Figure 12.15); the conversion of rangelands to croplands in marginal areas where rainfall is not sufficient to support crops over the long term; and poor forestry practices, including cutting all the trees in an area marginal for tree growth.

In northern China, areas that were once grasslands were overgrazed, and then some of these rangelands were converted to croplands. Both practices led to the conversion of the land to desert. Between 1949 and 1980, some 65,000 km² (25,100 mi², an area larger than Denmark) became desert, and an additional 160,000 km² (61,760 mi²)

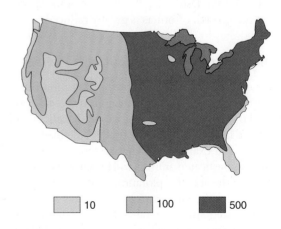

Figure 12.15 ■ Carrying capacity of pasture and rangeland in the United States, in average number of cows per square kilometer. [*Source:* U.S. Department of Agriculture statistics.]

(a)

(b)

Figure 12.16 ◼ (a) Gully soil erosion on cleared and plowed farmland in South Australia. (b) Agricultural runoff carrying heavy sediment load.

these salts may have been in very low concentrations in the irrigation water, over time the salts can build up in the soil to the point at which they become toxic. This effect can sometimes be reversed if irrigation is increased greatly; the larger volume of water then redissolves the salts and carries them with it as it percolates down into the water table.[21]

Preventing Desertification

The first step in preventing desertification is detection of initial symptoms. The major symptoms of desertification are the following:

◼ Lowering of the water table (wells have to be dug deeper and deeper).

◼ Increase in the salt content of the soil.

◼ Reduced surface water (streams and ponds dry up).

◼ Increased soil erosion (the dry soil, losing its organic matter, begins to be blown and washed away in heavy rains).

◼ Loss of native vegetation (not having adapted to desert conditions, native vegetation can no longer survive).[21]

Preventing desertification begins with monitoring these factors. Monitoring aquifers and soils is important in marginal agricultural lands. When we observe undesirable changes, we can try to control the activities producing these changes. Proper methods of soil conservation, forest management, and irrigation can help prevent the spread of deserts. (See Chapters 11 and 21 for a background discussion of soil and farming and irrigation practices.) In addition to the practices discussed earlier, good soil conservation includes the use of windbreaks (narrow lines of trees that help slow the wind) to prevent wind erosion of the soil. A landscape with trees is a landscape with a good chance of avoiding desertification. Practices that lead to deforestation in marginal areas should be avoided. Reforestation, including the planting of windbreaks, should be encouraged.

12.11 Does Farming Change the Biosphere?

People have long recognized local and regional impacts of agriculture, but it is a recent idea that farming might affect Earth's entire life-support system. This possibility came to people's attention in the twentieth century, first with events like the American Dust Bowl, discussed earlier in this chapter, which led some to speculate that such disasters could become worldwide.[22] The idea gained supporters in the late twentieth century as satellites and astronauts gave us views of Earth from space and the idea of a global ecology began to develop. How might farming change the biosphere? First, agriculture changes land cover, resulting in changes in the reflection of light by the land surface, the evaporation of water, the roughness of the surface, and the

are in danger of becoming desert. As a result of desertification, the frequency of sandstorms increased from about three days per year in the early 1950s to an average of 17 days/year in the next decade and to more than 25 days/year by the early 1980s.[17]

Desertlike areas can be created anywhere by poisoning of the soil. Poisoning can result from the application of persistent pesticides or other toxic organic chemicals from industrial processes that lead to improper disposal of toxic chemicals, and from airborne pollutant acidification, excessive manuring in feedlots, and oil or chemical spills. All of these can poison soil, forcing abandonment or reduced agricultural use of lands. Worldwide, chemicals account for about 12% of all soil degradation. Ironically, irrigation in arid areas can also lead to desertification. When irrigation water evaporates, a residue of salts is left behind. Although

Should Rice Be Grown in a Dry Climate?

Water is a precious resource, especially in California, where the average rainfall is low (38–51 cm/year, or 15–20 in./year), the population—33 million—is large and growing, and agricultural water use is high. (Farmers use 46% of the state's water to irrigate 3.5 million hectares, (8.6 million acres), more than in any other state.)[18, 19]

With cities and industries, not to mention fish and other wildlife, in need of water, growing crops that require a lot of water has come under heavy criticism, especially because much of the water that farmers receive is subsidized by the government. Some farmers have responded by reducing the acreage of water-intensive crops and switching to crops that require less water, such as fruits, vegetables, and nuts.

California produces 20% of the nation's rice, making it the second-largest rice-growing state in the United States. The rice produced has a market value of about $215 million and it uses enough water to supply one-fourth of the state's population. Although rice growers are not the biggest water users in the state, they have been particular targets of attack because the flooded fields required for rice are a visible reminder of the amount of water used by agriculture. In addition to high water use, rice growing has had other adverse environmental effects: Its high pesticide and herbicide use contaminates rivers and drinking supplies, and the burning of stubble left after harvesting contributes to air pollution in the valley.

Rice growers have responded to pressure to clean up their act in a number of ways. In the 1990s they decreased water use by 32% and pesticide use by 98%. They have switched to biodegradable pesticides; and they have decreased the burning of stubble by plowing it under, harvesting it, or flooding the fields in winter and allowing the organic matter to rot. Experts are also trying to find ways to protect young salmon, which run in the rivers of the Sacramento Valley, from being pumped into the channels leading to the rice fields. Although drawing water out of the rivers might have a negative impact on salmon, the release of water at the end of the winter, when the rivers are low, could help the spring run of salmon.

And the fields do provide a wetland habitat for many migrating birds and other species, so that winter flooding benefits an even wider diversity and greater number of species. Waterfowl are of particular interest because their numbers declined from 10–12 million in 1967 to 4–5 million in 1990, a period in which the state lost 90% of its wetlands to development and agriculture. About 79% of the yearly destruction of wetlands is attributed to agricultural practices. The drainage of wetlands adds to the loss by about 117,000 acres per year, so the net loss to agricultural lands is 2.38 million acres per year.[11]

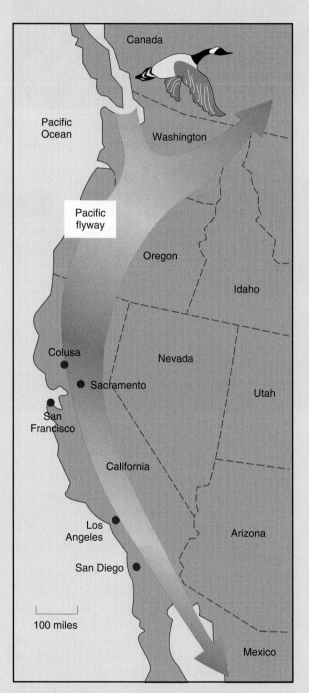

Figure 12.17 ■ The Pacific flyway, used by many birds that stop at agricultural wetlands in California. [*Source:* California Rice Commission 2003.]

Figure 12.18 ■ Dropping water levels in California reservois, 1986–1990: capacity, 38 million acre/feet; average, 22.3 million (1 acre/foot = 325 gallons).

The Sacramento Valley lies along the Pacific flyway (Figures. 12.17 and 12.18), a major migration route for waterfowl. The valley is host to 20% of the ducks of the United States and 50% of all waterfowl in the flyway. Twenty-one wildlife species with special status (endangered, threatened, candidate species, or species of special concern) use the rice fields, attracted by the 114–136 kg (250–300 lb) of rice grains per acre left behind after harvesting and the 273–318 kg (600–700 lb) of invertebrates per acre that grow in the waters.

Critical Thinking Questions

1. Most of the areas where rice is grown have alkaline, hardpan soil, unsuited to other crops. If rice were not grown on the land, it probably would be developed for housing. Each acre of rice requires five acre-feet of water. Less than one acre-foot could supply water for a family of four for a year. If housing lots were one-eighth acre and all housed families of four, how many acre-feet of water per acre of housing would be used in a year? Which uses more, an acre of rice or an acre of people? How would real estate development affect wildlife habitat?

2. Farmers find that the presence of waterfowl in flooded fields speeds up the rotting of stubble. Can you think of at least two reasons for this?

3. Although birds can feed on rice grains in dry fields, they get a more balanced diet by feeding in flooded fields. Why would this be so?

4. Two of the unknowns in this system are the long-term effects of flooding on the ability of the soil to support rice growing and the long-term effects on dryland species, such as rattlesnakes and rats. What is a scientific way of investigating one of these questions?

5. The new rice-growing practices are referred to as "win–win." What is meant by the term in general, and how does this situation illustrate the term?

rate of exchange of chemical compounds (such as carbon dioxide) produced and removed by living things. Each of these changes can have regional and global climatic effects.

Second, modern agriculture increases carbon dioxide (CO_2) in two ways. As a major user of fossil fuels, it increases carbon dioxide in the atmosphere, adding to the buildup of greenhouse gases (discussed in detail in Chapter 23). In addition, clearing land for agriculture speeds the decomposition of organic matter in the soil, transferring the carbon stored in the organic matter into carbon dioxide, which also increases the CO_2 concentration in the atmosphere.

Agriculture can also affect climate through fire. Fires associated with clearing land for agriculture, especially in tropical countries, may have significant effects on the climate because they add small particulates to the atmosphere. Another global effect of agriculture results from the artificial production of nitrogen compounds for use in fertilizer, which may be leading to significant changes in global biogeochemical cycles (see Chapter 3).

Finally, agriculture affects species diversity. The loss of competing ecosystems (because of agricultural land use) reduces biological diversity and increases the number of endangered species.

Summary

■ The Industrial Revolution and the rise of agricultural sciences have led to a revolution in agriculture, with many benefits and some serious drawbacks. These drawbacks have included an increase in soil loss, erosion, and resulting downstream sedimentation, as well as the pollution of soil and water with pesticides, fertilizers, and heavy metals that are concentrated as a result of irrigation.

■ Modern fertilizers have greatly increased the yield per unit area. Modern chemistry has also led to the development of a wide variety of pesticides that have reduced, but not eliminated, the loss of crops to weeds, diseases, and herbivores.

■ Most twentieth-century agriculture has relied on machinery and the use of abundant energy, with relatively little attention paid to the loss of soils, the limits of groundwater, and the negative effects of chemical pesticides.

■ Overgrazing has caused severe damage to lands. It is important to properly manage livestock, including using appropriate lands for grazing and keeping livestock at a sustainable density.

- Desertification is a serious problem that can be caused by poor farming practices and by the conversion of marginal grazing lands to croplands. Additional desertification can be avoided by improving farming practices, planting trees as windbreaks, and monitoring land for symptoms of desertification.
- Two revolutions are occurring in agriculture, one ecological and the other genetic. In the ecological approach to agriculture, pest control will be dominated by integrated pest management. Agriculture will be approached in terms of ecosystems and biomes, taking into account the complexity of these systems. Soil conservation through no-till agriculture and contour plowing will be emphasized, along with water conservation, through methods discussed in Chapter 21. The genetic revolution is already the subject of controversy, offering both benefits and environmental dangers. Dangers will result if genetic modification is used without considering the ecosystem, landscape, biome, and global context in which it is done.

REEXAMINING THEMES AND ISSUES

Human Population

Agriculture is the world's oldest and largest industry; more than one-half of all the people in the world still live on farms. Because the production, processing, and distribution of food alter the environment, and because of the size of the industry, large effects on the environment are unavoidable.

Sustainability

Alternative agricultural methods appear to offer the greatest hope of sustaining agricultural ecosystems and habitats over the long term, but more tests and better methods are needed. As the experience with European agriculture shows, crops can be produced on the same lands for thousands of years as long as sufficient fertilizers and water are available; however, the soils and other aspects of the original ecosystem are greatly changed—these are not sustained. *In agriculture, production can be sustained, but the ecosystem may not be.*

Global Perspective

Agriculture has numerous global effects. It changes land cover, affecting climate at regional and global levels, increasing carbon dioxide in the atmosphere and adding to the buildup of greenhouse gases, which in turn affects climate (discussed in detail in Chapter 23). Fires associated with clearing land for agriculture may have significant effects on the climate because of the small particulates they add to the atmosphere. Genetic modification is a new global issue having not only environmental but also political and social effects.

Urban World

The agricultural revolution enables fewer and fewer people to produce more and more food and leads to greater productivity per acre. Freed from dependence on farming, people flood to cities. This leads to increased urban effects on the land. Thus, agricultural effects on the environment indirectly extend to the cities.

People and Nature

Farming is one of the most direct and large-scale ways that people affect nature. Our own sustainability, as well as the quality of our lives, depends heavily on how we farm.

Science and Values

Human activities have seriously damaged one-fourth of the world's total land area, impacting one-sixth of the world's population (about 1 billion people). Six million hectares of land per year are lost to desertification. A large part of desertification occurs in the poorest countries. Overgrazing, deforestation, and destructive crop practices have caused so much damage that recovery in some areas will be difficult; restoration of the rest will require serious actions. A major value judgment we must make in the future is whether our societies will allocate funds to restore these damaged lands. Restoration requires scientific knowledge, both about present conditions and about actions required for restoration. Will we seek this knowledge and pay the costs for it?

Key Terms

biological control **225**

contour plowing **223**

desertification **232**

integrated pest
management **225**

no-till agriculture **223**

overgrazing **231**

terminator gene **229**

Study Questions

1. Design an integrated pest-management scheme for use in a small vegetable garden in a city lot behind a house. How would this scheme differ from IPM used on a large farm? What aspects of IPM could not be employed? How might the artificial structures of a city be put to use to benefit IPM?

2. You are given $10 billion to reduce the number of deaths worldwide from malaria. (a) You have one year to act. (b) You have ten years to act. In each case, make a plan of action, being specific about the use of pesticides.

3. Under what conditions might grazing cattle be sustainable when growing wheat is not? Under what conditions might a herd of bison provide a sustainable supply of meat when cows might not?

4. Pick one of the nations in Africa that has a major food shortage. Design a program to increase its food production. Discuss how reliable that program might be given the uncertainties that nation faces.

5. How can we avoid another Dust Bowl in the United States?

6. Should genetically modified crops be considered acceptable for "organic" farming?

7. A leading expert proposes a major, expensive program to increase urban gardens around the world. He claims this is one way to solve the world food gap. Decide if urban gardens could be a major source of food. As much as possible, make use of scientific data presented in this chapter, and make necessary calculations to determine the possible increases in world food production from urban gardens.

8. You are about to buy your mother a bouquet of 12 roses for Mother's Day, but you discover that the roses were genetically modified to give them a more brilliant color and to produce a natural pesticide through genetic energy. Do you buy the flowers? Explain and justify your answer based on the material presented in this chapter.

Further Reading

Committee on the Future Role of Pesticides in U.S. Agriculture, Board on Agriculture, and Natural Resources, Board on Environmental Studies and Toxicology. 2000. *The Future Role of Pesticides in U.S. Agriculture*. Washington, D.C.: National Research Council.

McNeely, J. A., and S. J. Scherr. 2003, *Ecoagriculture*. Washington, D.C.: Island Press.

Smil, V. 2000. *Feeding the World*. Cambridge, Mass.: MIT Press.

J. Toy, Terrence, George R. Foster, and Kenneth G. Renard. 2002. *Soil Erosion: Processes, Prediction, Measurement, and Control*. New York: John Wiley.

CHAPTER 13

Forests, Parks, and Landscapes

Conservation and management of forests, parks, and other landscapes can require great effort. Wildfires are often a major environmental concern, as illustrated here by the 2003 Simi Valley fire in Southern California, which burned 180,000 acres. In this picture, a spotter plane flies low over the fire line directing aircraft and helicopter drops, trying to bring the fire under control.

Learning Objectives

Forests and parks are among our most valued resources. Their conservation and management require that we understand landscapes—groups of ecosystems connected together. This is a larger view that includes populations, species, and ecosystems. After reading this chapter, you should understand:

■ What ecological services are provided by landscapes of various kinds.

■ The basic principles of park management.

■ The basic principles of forest management, including its historical context.

■ The roles that parks and nature preserves play in the conservation of wilderness.

Wildfires Raise Questions about How to Manage Parks and Preserves in the Twenty-First Century

Wildfires are common, often large, and frequently cause loss of life and major damage to property. In 2005 there were 66,552 wildfires in the United States (Figure 13.1).[1] These fires burned more than 8.6 million acres, an area larger than the state of Massachusetts. Wildfires are nothing new, anywhere in the world, and certainly not new in American history. The explorers Lewis and Clark recorded a wildfire in what is now North Dakota in 1804 that killed two and injured three Native Americans. In 1825, fires in New Brunswick, Canada, and in Maine, known respectively as the Miramichi and Maine fires, burned 3 million acres and killed 160 people. In October 1871 the Peshtigo fire burned 3.8 million acres in Michigan and Wisconsin and killed 1,500 people. A February 1898 fire burned 3 million acres in South Carolina, killed at least 14, and destroyed many buildings.

As cities and towns have built up in America, the number of structures destroyed by wildfires has increased greatly. In November 1980, the Panorama fire in California destroyed 325 buildings. The famous Oakland, California, fire of 1991 destroyed 2,500 buildings, even though it burned just 1,500 acres. Two years later, the Laguna Hills, California, fire destroyed 366 buildings in just six hours.

The federal government lists "59 historically significant wildland fires" between 1825 and 2005, a period of 180 years; thus, there has been a very large fire in the United States on average every three years.

On January 3, 2006, firefighters fought not one but many fires on the plains of Texas and Oklahoma; on that day one fire about 120 miles west of Dallas burned 20,000 acres and was still out of control. Another near the Texas–Oklahoma border burned 32 homes in one

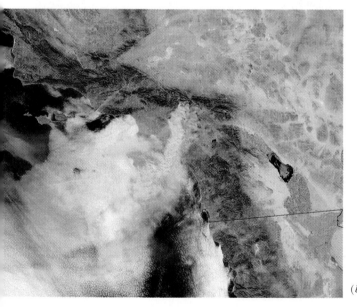

(b)

Figure 13.1 ■ (*a*) In October 2003, wildfires in southern California burned hundreds of thousands of acres, killed 22 people, and destroyed more than 3,400 homes. Some were so large that their smoke could be seen in satellite images. Here the smoke covers parts of the California coastline and the adjacent ocean.

(*b*) A wildfire in the tall grass in the Texas Panhandle was one of many in the Great Plains of the United States, from eastern Colorado to Oklahoma and Texas. These burned a record 700,000 hectares (1.8 million acres).

small town (Figure 13.1*b*). Beginning on Christmas, 2005, 278 homes were destroyed in Texas, and 3,200 people were forced to evacuate. Several fires that burned near Oklahoma City in 2006 were fought on the ground and in the air, with heavy aircraft dropping water and fire-retardant chemicals.[2]

Our society is of two minds about wildland fires. Historically, we are part of the Smokey Bear tradition. "Only you can prevent forest fires," Smokey told us when we were children, and ever since then we have considered it our duty to put out all fires wherever possible (Figure 13.2). The idea behind this message was that all wildfires are bad and are always or mostly caused by people. But today we know that fires are often natural, and that many kinds of life on the land evolved with, are adapted to, and depend on fire. This is true of the prairies of the United States (Figure 13.1), most of the forests of North America, the savannas and plains of East and Southern Africa, where the great wild animals still roam, and in many other ecosystems.

Managing and controlling wildfires, as well as managing wildlands so that wildfires cause the least damage to human life and property, are of major importance (see Figure 13.3). Ironically, the suppression of fires in the past allowed forests and grasslands that used to burn naturally to build up fuel. As a result, when fires break out now, they can be much more destructive than they were in the past. In forests with this heavy fuel load, sometimes the organic matter in the soil burns away and even old, seed-bearing trees that used to survive the fires can be killed. When this occurs, recovery of the forest or grassland takes much longer than before.

One solution—to remove the excess fuel in the forests and grasslands—is controversial. Some claim that timber corporations have proposed this solution so they can harvest trees in national parks and forests that would otherwise be protected. Others oppose any human activity on the grounds that everything we do is artificial and bad. Where cities and suburbs abut wildlands, the problem is even more difficult. Nobody wants a controlled burn—a fire lit on purpose during weather that will supposedly keep it under control—near their homes. In the late twentieth century, the government of California and the U.S. Forest Service joined to carry out some controlled burns. One of these fires got out of control and burned homes, leading the owners to sue the government. The courts ruled that the homeowners cannot sue the federal government because the government is us, and we cannot sue ourselves; but they could sue the state of California.

These recent experiences with wildfires on wildlands point out the problems we face in conserving and managing forests and prairies and other parks and preserves in a modern, heavily urbanized, industrialized, and motorized society. They also illustrate the difficulties we face as we gain new knowledge about the naturalness of environmental change and of wildland fires. These problems, and their possible solutions, are the subject of this chapter.

Figure 13.2 ■ Smokey Bear, the long-standing symbol of fire suppression in the United States, still reigns, as shown in this illustration from his Website.

Figure 13.3 ■ Smoke rises from burned homes in San Diego Oct. 27, 2003, the result of a fast-moving wildfire that destroyed more than 300 homes in a community of 30,000 people.

13.1 Modern Conflicts over Forestland and Forest Resources

In recent decades, conservation of forests has become an international cause célèbre, especially conservation of remaining old-growth forests—notably the giant trees of the rain forests in the North American Pacific Northwest and all tropical rain forests (Figure 13.4). Forestry, however, has a long history as a profession. The professional growing of trees is called **silviculture** (from *silvus*, Latin for "forest," and *cultura*, for "cultivate").

People have long practiced silviculture, much as they have grown crops, but forestry developed into a science-based activity and into what we today consider a profession in the late nineteenth and early twentieth centuries. The first modern U.S. professional forestry school was established at Yale University around the turn of the twentieth century, spurred by growing concerns about the depletion of America's living resources.

Modern conflicts about forests center on the following questions:

■ Should a forest be used only as a resource to provide materials for people and civilization, or should a

forest be used only to conserve natural ecosystems and biological diversity, including specific endangered species?

Figure 13.4 ■ Temperate rain forest on Vancouver Island.

Table 13.1 • Major Forestry Issues

- *Sustainability:* How can we achieve sustainable forestry? (This is the fundamental question.)
- *Clear-cutting:* Is clear-cutting ever good? Should it ever be allowed?
- *Old-growth forests:* Should all old-growth forests be preserved, or should some logging be allowed in some old-growth stands?
- *Plantations:* Are plantations intrinsically bad because they involve intentional manipulation of land to grow trees, or are they one of the keys to achieving the goals of biological conservation of forests?
- *Stream-protection zones:* Should all streams everywhere have a wooded buffer zone in which no logging and no other harvesting or destructive activities are permitted? If so, how wide should that buffer zone be?

- *National forests:* Is the purpose of national forests to provide the nation with a reliable source of timber, or is it to conserve living resources? What is the role of recreation in national forests.
- *Forest fires:* Are forest fires almost all bad, are they occasionally beneficial to a forest, or are they frequently important and essential to forests?
- *Certification:* Should our society certify forestry practices as sustainable? If so, how should this be done and who should do it?
- *Scale of management:* What is the appropriate spatial and temporal scale for management of forests?
- *People's role:* Should people play any active role in the management of forests? If so, what kinds of activities should they undertake, how often, and where?

- Can a forest serve some of both of these functions at the same time and in the same place?
- Can a forest be managed sustainably for either use? If so, how?
- Can a forest be managed for either resources or conservation and also provide recreation and scenic beauty as well as meet the spiritual needs of people? And does this matter anyway?
- What role do forests play in our global environment, such as their effects on climate?

Table 13.1 lists more specific issues in forestry.

Forests have always been important to people; indeed, forests and civilization have always been closely linked. Since the earliest civilizations—in fact, since some of the earliest human cultures—wood has been one of the major building materials and the main source of fuel. Forests provided materials for the first boats and the first wagons. Even today, nearly half the people in the world depend on wood for cooking, and in developing nations wood remains the primary heating fuel.[3]

At the same time, people have appreciated forests for spiritual and aesthetic reasons. There is a long history of sacred forest groves. When Julius Caesar was trying to conquer the Gauls in what is now southern France, he found the enemy difficult to defeat on the battlefield, so he burned the society's sacred groves to demoralize them. Forests held spiritual importance for the Gauls, so Caesar's action served as an early example of psychological warfare. In the Pacific Northwest, the great forests of Douglas fir provided many necessities of life to the Indians, from housing to boats, but were also important to them spiritually.

Forests benefit people and the environment indirectly through what we call **public service functions**. Forests retard erosion and moderate the availability of water (Figure 13.5), improving the water supply from major watersheds to cities. Forests are habitats for endangered species and other wildlife. They are important for recreation, including hiking, hunting, and bird and wildlife viewing. At regional and global levels, forests may be significant factors affecting the climate.

In the early days of the twentieth century, the goal of silviculture was generally to maximize the yield in the harvest of a single resource. The ecosystem was a minor concern, as were nontarget, noncommercial species and associated wildlife. Today, a much broader view dominates, considering the sustainability of timber harvest and of the ecosystem, and considering the entire range of goals for managing forests.

13.2 The Life of a Tree

To solve the big issues about forestry, we need to understand how a tree grows, how an ecosystem works, and how foresters have managed forestland. We begin with a brief summary of the way a tree grows (Figure 13.6).

How a Tree Grows

Leaves of a tree take up carbon dioxide from the air and absorb sunlight. These, in combination with water that is transported up from the roots, provide the energy and chemical elements for leaves to carry out *photosynthesis.* Through photosynthesis, the leaves convert carbon dioxide and water into a simple sugar and molecular oxygen.

The first product of photosynthesis is sugar. Leaves and other parts of the tree combine this sugar (made up of carbon and hydrogen) with other chemical elements to form all the organic compounds necessary for life (see Figure 13.6). For example, the addition of nitrogen in the form of nitrate (an oxide of nitrogen) or ammonia (nitrogen and hydrogen) results in the production of proteins,

Figure 13.5 ■ A forested watershed, showing the effects of trees in evaporating water, retarding erosion, and providing wildlife habitat.

Labels in figure:
Trees intercept water, reducing erosion impact

Water evaporates

Roots anchor the soil

Riparian zone

which in turn make possible enzymes, DNA, and many other large compounds necessary to life. Enzymes are chemicals that allow chemical reactions to take place at much lower temperatures and pressures than would otherwise be required.

Root hairs take up water, along with chemical elements dissolved in the water and small inorganic compounds, such as the nitrate or ammonia necessary to make proteins. Often the process of extracting minerals and compounds from the soil is aided by symbiotic relationships between the tree roots and fungi. Tree roots release sugars and other compounds that are food for the fungi, and the fungi benefit the tree as well. Without these organisms, the tree roots either would not be able to take up elements and compounds or would take them up much more slowly.

Leaves and roots are connected by two transportation systems. Phloem, on the inside of the living part of the bark, transports sugars and other organic compounds down to stems and roots. Xylem, farther inside (Figure 13.6), transports water and inorganic molecules upward to the leaves. Water is transported upward by a sun-powered pump that works as follows. Sunlight heats a leaf, and water from the leaf (primarily from small pores in the leaf called *stomata*) is lost to the atmosphere through the process of transpiration and to a lesser extent through evaporation (together called *evapotranspiration*). Water rises in the tree to replace losses (Figures 13.5 and 13.6).

Tree Niches

Each species of tree has its own niche (see Chapter 6) and is therefore adapted to specific environmental conditions. For example, in boreal forests (see Chapter 8), one of the determinants of a tree niche is the water content of the soil. White birch grows well in dry soils; balsam fir grows well in well-watered sites; and northern white cedar grows well in bogs (Figure 13.7).

Another way in which niches are determined is through tolerance of shade. Some trees can grow only in the bright sun of open areas and are therefore found in clearings. Other species can grow in deep shade; seedlings and saplings of these species can be found within the deep shade of an old forest. Sugar maple and beech are typical of these shade-tolerant trees. Birch and cherry are examples of trees that require bright sunlight and are called "shade-intolerant."

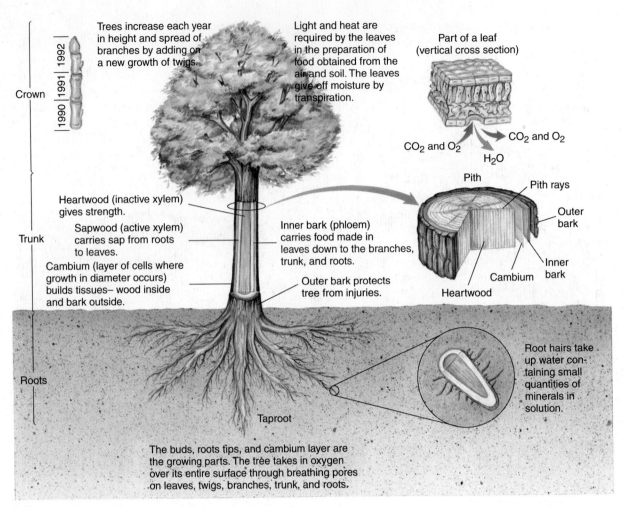

Trees increase each year in height and spread of branches by adding on a new growth of twigs.

Light and heat are required by the leaves in the preparation of food obtained from the air and soil. The leaves give off moisture by transpiration.

Part of a leaf (vertical cross section)

CO_2 and O_2

CO_2 and O_2

H_2O

Crown

1990 | 1991 | 1992

Pith

Pith rays

Outer bark

Heartwood (inactive xylem) gives strength.

Sapwood (active xylem) carries sap from roots to leaves.

Cambium (layer of cells where growth in diameter occurs) builds tissues— wood inside and bark outside.

Inner bark (phloem) carries food made in leaves down to the branches, trunk, and roots.

Outer bark protects tree from injuries.

Inner bark

Cambium

Heartwood

Trunk

Roots

Root hairs take up water containing small quantities of minerals in solution.

Taproot

The buds, roots tips, and cambium layer are the growing parts. The tree takes in oxygen over its entire surface through breathing pores on leaves, twigs, branches, trunk, and roots.

Figure 13.6 ■ How a tree grows. [*Source:* C. H. Stoddard, *Essentials of Forestry Practice*, 3rd ed. (New York: Wiley, 1978).]

Most of the big trees of the western United States require open, bright conditions and certain kinds of disturbances in order to germinate and survive the early stages of their lives. These include coastal redwood, which wins in competition with other species only if both fires and floods occasionally occur; Douglas fir, which begins its growth in openings; and the giant sequoia, whose seeds will germinate only on bare, mineral soil—where there is a thick layer of organic mulch, the sequoia's seeds cannot reach the surface, and they die before they can germinate. We talked about these requirements in somewhat different terms in Chapter 10. Some trees are adapted to early stages in succession, where sites are open and there is bright sunlight. Others are adapted to later stages in succession, where there is a high density of trees.

Understanding the specific requirements of individual tree species helps us to determine where they will grow, where we might best plant them as a commercial crop, and where they might best contribute to biological conservation or to the beauty of a landscape. Clearly, there is no single best set of conditions for a forest, and many kinds of forests adapt to many kinds of conditions. The same is true for all landscapes—grasslands, deserts, rivers, and lakes.

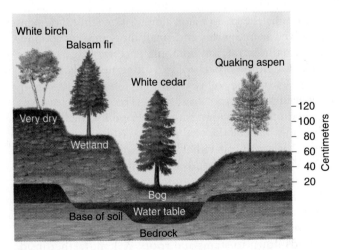

White birch

Balsam fir

White cedar

Quaking aspen

Very dry

Wetland

Bog

Water table

Base of soil

Bedrock

— 120
— 100
— 80
— 60
— 40
— 20

Centimeters

Figure 13.7 ■ Some characteristics of tree niches. Tree species have evolved to be adapted to different kinds of environments. In northern boreal forests, white birch grows on dry sites (and early-successional sites); balsam fir grows in wetter soils, up to wetlands; and white cedar grows in even the wetter sites of northern bogs.

13.3 A Forester's View of a Forest

Traditionally, foresters have managed trees locally in stands. A **stand** is an informal term that foresters use to refer to a group of trees, usually of the same species or group of species and often at the same successional stage. Stands can be small (half a hectare) to medium-size (several hundred hectares). Foresters classify stands on the basis of tree composition. The two major kinds of commercial stands are *even-aged stands*, where all live trees began growth from seeds and roots germinating the same year, and *uneven-aged stands*, which have at least three distinct age classes. In even-aged stands, trees are approximately the same height but differ in girth and vigor.

A forest that has never been cut is called a *virgin forest* or sometimes an **old-growth forest.** A forest that has been cut and has regrown is called a **second-growth forest.** Although the term *old-growth forest* has gained popularity in several well-publicized disputes about forests, it is not a scientific term and does not yet have an agreed-on, precise meaning. Another important management term is **rotation time,** the time between cuts of a stand.

Foresters and forest ecologists group the trees in a forest into the **dominants** (tallest, most common, and most vigorous), **codominants** (fairly common, sharing the canopy or top part of the forest), **intermediate** (forming a layer of growth below dominants), and **suppressed** (growing in the understory).

Productivity of a forest varies according to soil fertility, water supply, and local climate. Foresters classify sites by **site quality,** which is the maximum timber crop the site can produce in a given time. Site quality can decrease with poor management. Traditionally, foresters developed site indexes for the different types of forestlands and derived yield tables to estimate future production. Today, forecasts of forest production rely more and more on computer simulation.

Although forests are complex and difficult to manage, one advantage they have over many other ecosystems is that trees provide easily obtained information that can be a great help to us. For example, the age and growth rate of trees can be measured from tree rings. In temperate and boreal forests, trees produce one growth ring per year.

13.4 Approaches to Forest Management

Managing forests can involve removing poorly formed and unproductive trees (or selected other trees) to permit larger trees to grow more rapidly, planting genetically controlled seedlings, and fertilizing the soil. Forest geneticists breed new strains of trees just as agricultural geneticists breed new strains of corn, wheat, tomatoes, and other crop plants. New "supertrees" are allegedly able to maintain a high rate of growth and increase the total production of forests.

Another aspect of silviculture is the control of diseases and pests. There has been relatively little success in controlling diseases in forests. Tree diseases are primarily fungal.

Often, an insect spreads the fungus from tree to tree. Insect outbreaks tend to occur infrequently, but when they do occur, they can have devastating results. Some insect problems are due to introductions of exotic species. For example, the gypsy moth, intentionally introduced into New England around the turn of the twentieth century as a source of silk, escaped and spread through many eastern states, where it has become a persistent, periodic pest, defoliating trees over large areas when its population explodes. Other insect outbreaks have recurred naturally for a long time. For example, a nineteenth-century New Hampshire gazetteer referred to a "plague of loathesome [*sic*] worms" that removed all the leaves from large areas of forest.[4]

Insects affect trees by defoliating them; by eating the buds at the tops of the trees and destroying the main trunk, causing forked growth; by eating fruits; and by serving as carriers of diseases. Insecticides are sometimes used to combat these pests.

Clear-cutting

Clear-cutting (Figure 13.8) is the cutting of all trees in a stand at the same time. Alternatives to clear-cutting are selective cutting, strip-cutting, shelterwood-cutting, and seed-tree cutting. **Shelterwood-cutting** is the practice of cutting dead and less desirable trees first and later cutting mature trees. As a result, there are always young trees left in the forest. **Seed-tree cutting** removes all but a few seed trees (mature trees with good genetic characteristics and high seed production) to promote regeneration of the forest. In **selective cutting,** individual trees are marked and cut. Sometimes smaller, poorly formed trees are selectively removed; this practice is called **thinning.** At other times, trees of specific species and sizes are removed. For example, some forestry companies in Costa Rica cut only some of the largest mahogany trees, leaving other, less valuable trees to help maintain the ecosystem and permitting some of the large mahogany trees to continue to provide seeds for future generations.

Figure 13.8 ■ A clear-cut forest in western Washington.

In **strip-cutting**, narrow rows of forest are cut, leaving wooded corridors. Strip-cutting offers several advantages. The uncut strips protect regenerating trees from wind and direct sunlight, and these remaining trees provide seeds. In addition, strip-cutting can minimize the negative aesthetic effects of logging by leaving buffer zones and allowing the corridors of forest that remain to be used for recreation and as wildlife habitats.

Experimental Tests of Clear-Cutting

Scientists have tested the effects of clear-cutting.[5, 6] For example, in the U.S. Forest Service Hubbard Brook experimental forest in New Hampshire, an entire watershed was clear-cut and herbicides were applied to prevent regrowth for two years.[6] The results were dramatic. Erosion increased, and the pattern of water runoff changed substantially. The exposed soil decayed more rapidly, and the concentrations of nitrates in the stream water exceeded public health standards.

In another experiment, at the U.S. Forest Service H. J. Andrews experimental forest in Oregon, clear-cutting greatly increased the frequency of landslides, as did the construction of logging roads. In this forest, rainfall is high (about 240 cm, or 94 in., annually) and the trees (mainly Douglas fir, western hemlock, and Pacific silver fir) grow very tall and live a long time.[7]

Clear-cutting also changes chemical cycling in forests and causes the soil to lose chemical elements necessary for life. When a forest is clear-cut, trees are no longer available to take up nutrients. Open to the sun and rain, the ground becomes warmer. This accelerates the process of decay, with chemical elements, such as nitrogen, converted more rapidly to forms that are water-soluble and thus readily lost in runoff during rains (Figure 13.9).[8]

The Forest Service experiments show that clear-cutting can be a poor practice on steep slopes in areas of moderate to heavy rainfall. The worst effects of clear-cutting resulted from the logging of vast areas of North America during the nineteenth and early twentieth centuries. Clear-cutting on such a large scale is neither necessary nor desirable for the best timber production. However, where the ground is level or slightly sloped, where rainfall is moderate, and where the desirable species require open areas for growth, clear-cutting on an appropriate spatial scale may be a useful way to regenerate desirable species. The key here is that clear-cutting is neither all good nor all bad for timber production or forest ecosystems. The use of clear-cutting must be evaluated on a case-by-case basis, taking into account the size of cuts, the environment, and the available species of trees.

Plantation Forestry

Sometimes foresters grow trees in a **plantation**, which is a stand of a single species typically planted in straight rows (Figure 13.10). Usually plantations are fertilized, some-

times by helicopter, and modern machines make harvesting rapid—some remove the entire tree, root and all. Plantation forestry is thus much like modern agriculture. Intensive management like this is common in Europe and parts of the northwestern United States.

Other forests, such as those of New England, are often managed less actively. In these forests, seeds from existing trees are self-sowing. Ecological succession follows without

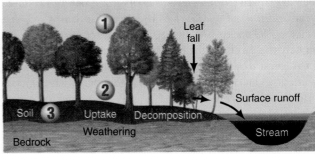

(1) Trees shade ground.

(2) In cool shade, decay is slow.

(3) Trees take up nutrients from soil.

(a)

(1) Branches and so on decay rapidly in open, warm areas.

(2) Soil is more easily eroded without tree roots.

(3) Runoff is greater without evaporation by trees.

(b)

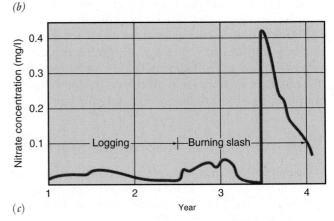

(c)

Figure 13.9 ■ Effects of clear-cutting on forest chemical cycling. Chemical cycling (*a*) in an old-growth forest and (*b*) after clear-cutting. (*c*) Increase in nitrate concentration in streams following logging and the burning of slash (leaves, branches, and other tree debris). [*Sources:* (*a*) and (*b*) adapted from R. L. Fredriksen, "Comparative Chemical Water Quality—Natural and Disturbed Streams Following Logging and Slash Burning," in *Forest Land Use and Stream Environment* (Corvallis: Oregon State University, 1971), pp. 125–137.]

Figure 13.10 ■ A modern forest plantation in the southeastern United States.

management (see Chapter 10). Which approach is best depends on the type of forest, the environment, and the characteristics of the commercially valuable species. The question is: Which is better for trees, ecosystems, and people—plantations, or forests allowed to grow without human interference until it is time to harvest, or something in between? Which is sustainable? Which is better for biodiversity, for landscape beauty, and so forth?

Forest plantations offer an important alternative solution to the pressure on natural forests. If plantations were used where forest production is high, then a comparatively small percentage of the world's forestland could provide all the world's timber. For example, high-yield forests produce 15 to 20 m^3/ha/yr. According to one estimate, if plantations were put on timberland that could produce at least 10 m^3/ha/yr, then 10% of the world's forestland could provide enough timber for the world's timber trade.[9] This could reduce pressure on old-growth forests, on forests important for biological conservation, and on forestlands important for recreation.

13.5 Sustainable Forestry

A major goal today is to have **sustainable forests**. Stated in the most general terms, a sustainable forest is one from which a resource can be harvested at a rate that does not decrease the ability of the forest ecosystem to continue to provide that same rate of harvest indefinitely. In reality, the situation is more complicated.

What Is Forest Sustainability?

There are two basic kinds of ecological sustainability: sustainability of the harvest of a specific resource that grows within an ecosystem; and sustainability of the entire ecosystem—and therefore of many species, habitats, and environmental conditions. For forests, this translates into sustainability of the harvest of timber and sustainability of the forest as an ecosystem. Although sustainability has long been discussed in forestry, we lack adequate scientific data to demonstrate that sustainability of either kind has ever been achieved in forests, except in a few unusual cases.[10]

Certification of Forest Practices

If the data do not indicate whether a particular set of practices has led to sustainable forestry, how can we achieve it? The general approach today is to compare the actual practices of specific corporations or government agencies with practices that we believe to be consistent with sustainability. This has become a formal process called **certification of forestry**, and there are organizations whose main function is to certify forest practices. The catch here is that nobody actually knows whether the beliefs are correct and therefore whether the practices will turn out to be sustainable. Since trees take a long time to grow, and a series of harvests is necessary to prove sustainability, the proof lies in the future. Despite this limitation, certification of forestry is becoming common. As practiced today, it is as much an art or a craft as it is a science.

Worldwide concern about the need for forest sustainability has led to international attempts to ban imports of wood produced from purportedly unsustainable forest practices and to the development of international programs for certification of forest practices. Some European nations have banned the import of certain tropical woods, and some environmental organizations have led demonstrations in support of such bans.

There is a gradual movement away from calling certified forest practices "sustainable," instead referring to "well-managed forests" or "improved management."[11, 12] A small industry has developed consisting of companies that review the management of forests and provide certification. However, uniform criteria have not been established. The search for acceptable uniform criteria has led to an ongoing series of international meetings, including major ones in Montreal and Helsinki.

Certification programs began because environmental organizations were concerned that forests were being cut in an unsustainable fashion. However, the certification process has had an unintended consequence: In some cases it has evolved into a means for producers of wood and wood products (private corporations and nations that depend on wood exports) to reach new markets, keep existing ones, or increase the value of the product (because people will pay more for a "certified" wood product).

Some scientists have begun to call for a new forestry that includes a variety of practices that they believe increase the likelihood of sustainability. Most basic is accepting the dynamic characteristics of forests—that to remain sustainable over the long term, a forest may have to change in the short term. Some of the broader, science-based concerns are spoken of as a group—the need for ecosystem management and a landscape context. Scientists point out that any application of a certification program creates an experiment and should be treated accordingly. Therefore, any new programs that claim to provide sustainable practices must include, for comparison, control areas where no cutting is done and must also include adequate scientific monitoring of the status of the forest ecosystem.

13.6 A Global Perspective on Forests

The functioning of an individual tree provides a basic understanding of how forests can affect the entire biosphere. Vegetation of any kind can affect the atmosphere in four ways (Figure 13.11):

1. By changing the color of the surface and therefore the amount of sunlight reflected and absorbed.
2. By increasing the amount of water transpired and evaporated from the surface to the atmosphere.

3. By changing the rate at which greenhouse gases are released from Earth's surface into the atmosphere.
4. By changing "surface roughness," which affects wind speed at the surface.

In general, vegetation makes the surface darker, so it absorbs more sunlight and reflects less, warming the Earth. The contrast is especially strong between the dark needles of conifers and winter snow in northern forests and between the dark green of shrublands and the yellowish soils of many semiarid climates.

Vegetation in general and forests in particular tend to evaporate more water than bare surfaces. This is because the total surface area of the many leaves is many times larger than the area of the soil surface.

Is this increased evaporation good or bad? That depends on one's goals. Increasing evaporation means that less water runs off the surface. This reduces erosion. Although increased evaporation also means that less water is available for our own water supply and for streams, in most situations the ecological and environmental benefits of increased evaporation outweigh the disadvantages.

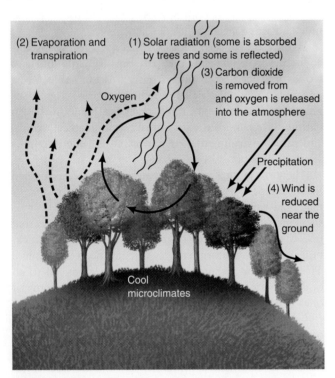

Figure 13.11 ■ Four ways that a forest can affect the atmosphere: (1) some solar radiation is absorbed by trees and some is reflected, changing the local energy budget, compared to a non-forest environment; (2) evaporation and transpiration from trees, together called evapotranspiration, transfers water to the atmosphere; (3) carbon dioxide is removed from the air and oxygen is released to the atmosphere by photosynthesis from trees (carbon dioxide is a greenhouse gas associated with climate change, and reducing the gas cools the temperature of the atmosphere, see Chapter 23); and (4) near-surface wind is reduced because the trees produce a roughness near the ground that slows the wind.

World Forest Area, Global Production, and Consumption of Forest Resources

At the beginning of the twenty-first century, the world contained approximately 3.87 billion hectares (14.7 million square miles) of forested area covering approximately 26.6% of Earth's surface (Figure 13.12).[13] This works out to about 0.6 hectares (about 1 acre) a person. The forest area is up from 3.45 billion hectares (13.1 million square miles) estimated in 1990, but down from 4 billion hectares (15.2 million square miles) in 1980. Countries differ greatly in their forest resources, depending on the potential of their land and climate for tree growth and on their history of land use and deforestation. Ten nations have two-thirds of the world's forests. In descending order, these are the Russian Federation, Brazil, Canada, the United States, China, Australia, the Democratic Republic of the Congo, Indonesia, Angola, and Peru (Figure 13.13).

Developed countries account for 70% of the world's total production and consumption of industrial wood products; developing countries produce and consume about 90% of wood used as firewood. Timber for construction, pulp, and paper makes up approximately 90% of the world timber trade (the rest consists of hardwoods used for furniture, such as teak, mahogany, oak, and maple). North America is the world's dominant supplier.[3] Total global production/consumption is about 1.5 billion m³ annually. (To think of this in terms easier to relate to, a cubic meter of timber is a block of wood 1 m thick on all sides. A million cubic meters would be a block of wood 1 m thick in a square 1 km on a side—or about 1 yard thick covering the area of 5,000 football fields.)

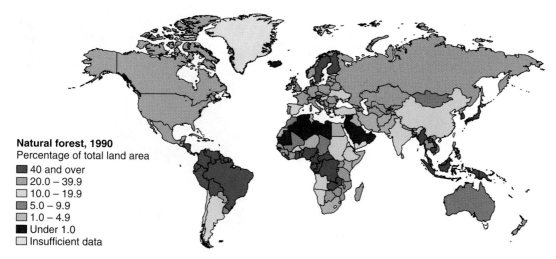

Natural forest, 1990
Percentage of total land area
- 40 and over
- 20.0 – 39.9
- 10.0 – 19.9
- 5.0 – 9.9
- 1.0 – 4.9
- Under 1.0
- Insufficient data

Figure 13.12 ■ The world's forested area. This map shows the percentage of forestland area, by nation. [*Source: State of the World's Forest 2001* (Rome: U.N. Food and Agriculture Organization), available at http://www.fao.org/docrep/U8480E56.jpg.]

The United States has approximately 212 million hectares (524 million acres) of commercial-grade forest, which is defined as forest capable of producing at least 1.4 m³/ha (20 ft³/acre) of wood per year. Commercial timberland occurs in many parts of the United States. Nearly 75% is in the eastern half of the country (about equally divided between the North and South); the rest is in the West (Oregon, Washington, California, Montana, Idaho, Colorado, and other Rocky Mountain states) and in Alaska.

In the United States, 70% of forestland is privately owned, 15% of forests are on U.S. Forest Service lands, and 15% are on other federal lands.[14] Publicly owned forests are primarily in the Rocky Mountain and Pacific Coast states on sites of poor quality and high elevation. In addition to these forests, about 300 million hectares (741 million acres) in the United States are 10% stocked with trees, and another 312 million hectares (770 million acres) are rangeland, including natural grasslands, shrublands, deserts, tundra, coastal marshes, and meadows (Figure 13.14).[15]

In recent years, the world trade in timber has not grown substantially. Thus, the amount traded annually (about 1.5 billion m³, as mentioned earlier) is a reasonable estimate of the total present world demand for the 6.6 billion people on Earth, at their present standards of living. The fundamental questions are whether and how Earth's forests can continue to produce at least this amount of timber for an indefinite period, and whether and how they can produce even more as the world's human population continues to grow and as standards of living rise world-

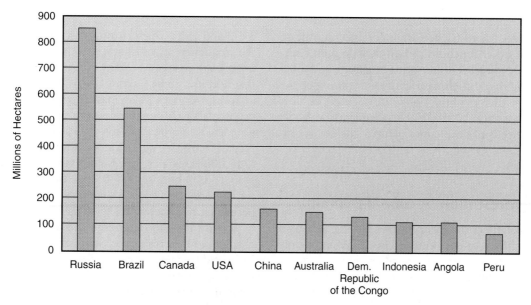

Figure 13.13 ■ Countries with the largest forest areas. [*Source* http://www.mapsofworld.com/world-top-ten/countries-with-most-largest-area-of-forest.html, April 24, 2006.]

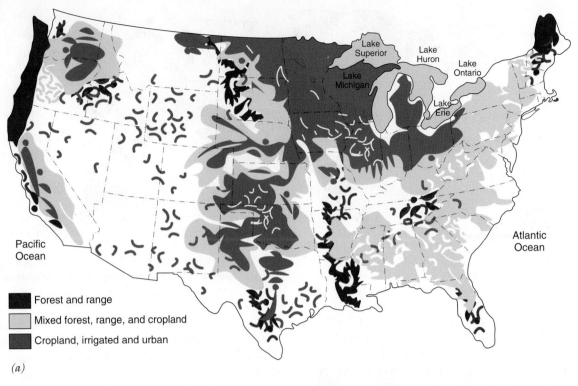

(a)

Forest and range
Mixed forest, range, and cropland
Cropland, irrigated and urban

Figure 13.14 ■ Forest and rangeland in the contiguous United States (lower 48 states): (*a*) land use by area and (*b*) estimated volume of harvested timber (known as *sawtimber volume*). In the late twentieth century, three-fourths of the commercial forests in the lower 48 were in the East, but 70% of the sawtimber volume was in the West. Today, sawtimber volume is shifting eastward because of plantation forestry in the Southeast. [*Sources:* (*a*) U.S. Forest Service, 1980. (*b*) U.S. Forest Service and C. H. Stoddard, *Essentials of Forestry Practice*, 3rd ed. (New York: Wiley, 1978).]

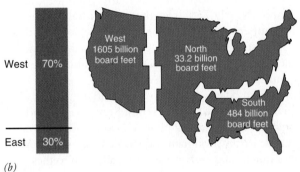

West 70%

East 30%

West 1605 billion board feet

North 33.2 billion board feet

South 484 billion board feet

(b)

wide. All of this has to happen while forests continue to perform their other functions, which include public service functions, biological conservation functions, and functions involving the aesthetic and spiritual needs of people. In terms of the themes of this book, the question is: How can forest production be sustainable, even increase, while meeting the needs of people *and* nature? The answer involves science and values.

13.7 Deforestation: A Global Dilemma

Another global effect of forestry is that cutting forests in one country affects other countries. For example, deforestation is estimated to have increased erosion and caused the loss of 562 million hectares (1.4 billion acres) of soil worldwide, and the estimated annual loss is 5–6 million hectares.[16] Nepal, one of the most mountainous countries in the world, lost more than half its forest cover between 1950 and 1980. Such cutting destabilizes soil, increasing the frequency of landslides, amount of runoff, and sediment load in streams. Many Nepalese streams feed rivers that flow into India (Figure 13.15). Recent heavy flooding in India's Ganges Valley has caused $1 billion a year in property damage and is blamed on the loss of large forested watersheds in other countries.[17] The loss of forest cover in Nepal continues at a rate of about 100,000 hectares (247,000 acres) per year. Reforestation efforts replace less than 15,000 hectares (37,050 acres) per year. If present trends continue, little forestland will remain in Nepal, thus permanently exacerbating India's flooding problems.[18]

It is difficult to determine the worldwide net rate of change in forest resources. Some experts argue that there is a worldwide net increase in forests because large areas in the temperate zone, such as the eastern and midwestern United States, were cleared in the nineteenth and early twentieth centuries and are now regenerating. Most experts, however, disagree with this assessment. Because few

(a)

(b)

Figure 13.15 ■ (*a*) Planting pine trees on the steep slopes in Nepal to replace forests that have been cut. The dark green in the background is yet uncut forest, and the contrast between foreground and background suggests the intensity of clearing that is taking place. (*b*) The Indus River in northern India carries a heavy load of sediment, as shown by the sediments deposited within and along the flowing water and by the color of the water itself. This scene, near the headwaters, shows that erosion takes place at the higher reaches of the river.

forests are successfully managed to achieve sustainability, it seems likely that the world's forests are undergoing a net decline, perhaps rapid. But the fact is, information is lacking on which to base an accurate evaluation. Because forests cover large, often remote areas that are little visited or studied, it is difficult to assess the total amount of forest area. Only recently have programs begun to obtain accurate estimates of the distribution and abundance of forests, and these suggest that past methods overestimated forest biomass by 100–400%.[19]

Accepting these limitations, we find the best estimates suggesting that the rate of deforestation in the twenty-first century is 7.3 million hectares a year—or the loss of an area equal to the size of Panama each year. The good news is that this is 18% less than the average annual loss of 8.9 million hectares in the 1990s.[20]

History of Deforestation

Forests were cut in the Near East, Greece, and the Roman Empire before the modern era. Removal of forests continued northward in Europe as civilization advanced. Fossil records suggest that prehistoric farmers in Denmark cleared forests so extensively that early-successional weeds occupied large areas. In medieval times, Great Britain's forests were cut and many forested areas were eliminated. With colonization of the New World, much of North America was cleared.[21]

The greatest losses in the present century have taken place in South America, where 4.3 million acres have been lost on average per year since 2000 (Figure 13.16*b*).[20] Many of these forests are in the tropics, mountain regions, or high latitudes, places difficult to exploit before the advent of modern transportation and machines.[23] The problem is especially severe in the tropics because of rapid human population growth. Satellite images provide a new way to detect deforestation (Figure 13.16*a*).

Causes of Deforestation

Historically, the two most common reasons people cut forests are to clear land for agriculture and settlement and to use or sell timber for lumber, paper products, or fuel. Logging by large timber companies and local cutting by villagers are both major causes of deforestation. Agriculture is a principal cause of deforestation in Nepal and Brazil, and it was one of the major reasons for clearing forests in New England during the first settlement by Europeans.

The World Firewood Shortage

In many parts of the world, wood is a major energy source. Some 63% of all wood produced in the world, or 2.1 million m^3, is used for firewood. Firewood provides 5% of the world's total energy use,[24] 2% of total commercial energy in developed countries, but 15% of the energy in developing countries, and is the major source of energy for most countries of sub-Saharan Africa, Central America, and continental Southeast Asia.[25]

As the human population grows, use of firewood increases. In this situation, management is essential, including management of woodland stands to improve growth. However, well-planned management of firewood stands has been the exception rather than the rule. Some successful community-based projects are discussed in A Closer Look 13.1.

(a)

Indirect Deforestation

A more subtle cause of the loss of forests is indirect defor-estation—the death of trees from pollution or disease. Acid rain and other pollutants may be killing trees in many areas in and near industrial countries. In Germany, there is talk of *Waldsterben* ("forest death"). The German government has estimated that one-third of the country's forests have suffered damage: death of standing trees, yellowing of nee-dles, or poorly formed shoots. The causes are unclear but appear to involve a number of factors, including acid rain, ozone, and other air pollutants that tend to weaken trees and increase their susceptibility to disease. This problem extends throughout Central Europe and is especially acute in Poland, the Czech Republic, and Slovakia. In the New England area of the United States, similar, curious damage is affecting red spruce.

If global warming occurs as projected by global climate models, indirect forest damage could occur over vast re-gions, with major die-offs in many areas and major shifts in the areas of potential growth for each species of trees.[26] The combination of temperature and rainfall required for various tree species could be altered by global warming, and some species may not continue to grow in their current

(b)

Figure 13.16 ■ *(a)* A satellite image showing clearings in the tropical rain forests in the Amazon in Brazil. The image is in false infrared. Rivers are black, and the bright red is the leaves of the liv-ing rain forest. The straight lines of other colors, mostly light blue to gray, are clearing sites of deforestation by people extending from roads. Much of the clearing is for agriculture. The distance across the image is about 100 km (63 mi). *(b)* An intact South American rain forest with its lush vegetation of many species and a complex vertical structure. This one is in Peru.

Community Forestry

In many parts of the world, people cut nearby forests to meet the needs of small communities. This is particularly true in developing nations, where firewood remains a major fuel and constitutes an important part of energy used. In the past, most government forestry departments concentrated their efforts on government-owned forestland or merely policed their countries' forests. Now, many realize that this approach must change.

Some countries have placed new emphasis on community forestry, in which professional foresters help villagers develop woodlots with the goal of achieving some kind of sustainable local harvest to meet local needs. The United Nations Food and Agriculture Organization (FAO) and the World Bank are supporting these programs. For example, in Malawi, Africa, the World Bank and FAO sponsor a reforestation project in which almost 40% of the households have planted trees. In South Korea, villagers have been reforesting the country at the rate of 40,000 hectares (98,840 acres) a year.

In community forestry, good management practices include limiting access; cutting the slower-growing and poorer-burning species first to promote the growth of better firewood species;

making use of plantations; and supplementing firewood with more easily renewable fuels. However, some of these practices conflict with traditional local activities or are difficult to implement for other reasons.

Such community efforts are impressive, but in total they have only a small effect on the worldwide shortage of firewood. It is not clear whether developing nations can implement successful management policies in time to prevent serious damage to their forests and the land. If alternative fuels for developing nations cannot be found, the effects will be severe, not only for the land but also for the people.

---

locations. Such changes would dwarf damage from other causes, affecting the habitat for many endangered species.[27]

Some suggest that global warming would merely change the location of forests, not their total area or production. However, even if a climate conducive to forest growth were to move to new locations, trees would have to reach these areas. This would take a long time because changes in the geographic distribution of trees depend primarily on seeds blown by the wind or carried by animals. In addition, for production to remain as high as it is now, climates that meet the needs of forest trees would have to occur where the soils also meet these needs. This combination of climate and soils occurs widely now but might become scarcer with large-scale climate change.

13.8 Parks, Nature Preserves, and Wilderness

Landscapes are often protected from harvest and other potentially destructive uses by establishing parks, nature preserves, and legally designated wilderness areas. In addition, some privately held lands have been set aside for biological conservation. Organizations such as the Nature Conservancy, the Southwest Florida Nature Conservancy, and the Land Trust of California purchase lands privately and maintain them as nature preserves. Whether government or private conservation areas succeed better in reaching the goals listed in Table 13.2 is a matter of considerable controversy.

Parks, natural areas, and wilderness provide benefits within their boundaries and can also serve as migratory

Table 13.2 • Goals of Parks, Nature Preserves, and Wilderness Areas

Parks are as old as civilization. The goals of park and nature-preserve management can be summarized as follows:

1. Preservation of unique geological and scenic wonders of nature, such as Niagara Falls and the Grand Canyon
2. Preservation of nature without human interference (preserving wilderness for its own sake)
3. Preservation of nature in a condition thought to be representative of some prior time (e.g., the United States prior to European settlement)
4. Wildlife conservation, including conservation of the required habitat and ecosystem of the wildlife
5. Conservation of specific endangered species and habitats
6. Conservation of the total biological diversity of a region
7. Maintenance of wildlife for hunting
8. Maintenance of uniquely or unusually beautiful landscapes for aesthetic reasons
9. Maintenance of representative natural areas for an entire country
10. Maintenance for outdoor recreation, including a range of activities from viewing scenery to wilderness recreation (hiking, cross-country skiing, rock climbing) and tourism (car and bus tours, swimming, downhill skiing, camping)
11. Maintenance of areas set aside for scientific research, both as a basis for park management and for the pursuit of answers to fundamental scientific questions
12. Provision of corridors and connections between separated natural areas

corridors between other natural areas. Originally, parks were established for specific purposes related to the land within the park boundaries (see A Closer Look 13.2). In the future, the design of large landscapes to serve a combination of land uses—including parks, preserves, and wilderness—needs to become more important and a greater focus for discussion.

Parks and Preserves as Islands

A park is an area set aside for use by people, whereas a nature preserve, although it may be used by people, has as its primary purpose the conservation of some resource, typically a biological one. Every park or preserve is an ecological island of one kind of landscape surrounded by a different kind, or several different kinds, of landscape.

Ecological and physical islands have special ecological qualities (discussed in detail in Chapter 8), and concepts of island biogeography are used in the design and management of parks. Specifically, park and preserve planners know that the size of the park and the diversity of habitats determine the number of species that can be maintained there. Also, the farther the park is from other parks or sources of species, the fewer species are found. Even the shape of a park can determine what species can survive within it.

One of the important differences between a park and a truly natural wilderness area is that a park has definite boundaries. These boundaries are usually arbitrary from an ecological viewpoint and have been established for political, economic, or historical reasons unrelated to the natural ecosystem. In fact, many parks have been developed on what are otherwise considered wastelands, useless for any other purpose.

Even where parks or preserves have been set aside for the conservation of some species, the boundaries are usually arbitrary, and such arbitrariness has caused problems. For example, Lake Manyara National Park in Tanzania, famous for its elephants, was originally established with boundaries incorrect for elephant habits. Before this park was established, elephants would spend part of the year feeding along a steep incline above the lake. At other times of the year, they would migrate down to the valley floor, depending on the availability of food and water. These annual migrations were necessary for the elephants to obtain food of sufficient nutritional quality throughout the year.

When the park was established, farms were laid out along its northern border. These farms crossed the traditional pathways of the elephants, creating two negative effects. First, elephants came into direct conflict with farmers. Elephants crashed through farm fences, eating corn and other crops and causing general disruption. Second, whenever the farmers succeeded in keeping elephants out, the animals were cut off from reaching their feeding ground near the lake. When it became clear that the park boundaries were arbitrary and inappropriate, adjustments were made to extend the boundaries to include

Figure 13.17 ■ The famous main valley of Yosemite National Park.

A Brief History of Parks

The French word *parc* once referred to an enclosed area for keeping wildlife to be hunted. Such areas were set aside for the nobility, excluding the public. An example is the area now known as Depone National Park on the southern coast of Spain. Originally a country home of nobles, today it is one of the most important natural areas of Europe, used by 80% of birds migrating between Europe and Africa.

The first major public park of the modern era was Victoria Park in Great Britain, authorized in 1842. The concept of a national park, whose purposes include protection of nature as well as public access, originated in North America in the nineteenth century. In the twentieth century the purpose of national parks was broadened to emphasize biological conservation, and this idea was applied worldwide.[23]

The world's first designated national park was Yosemite National Park in California, which was made a park by an act signed by President Lincoln in 1864 (Figure 13.17). The term *national park*, however, was first used with the establishment of Yellowstone in 1872.

The purpose of the earliest national parks in the United States was to preserve the unique, awesome landscapes of the country, a purpose the historian Alfred Runte refers to as "monumentalism." In the nineteenth century, Americans considered their national parks a contribution to civilization equivalent to the architectural treasures of the Old World and sought to preserve them as a matter of national pride.[23]

In the second half of the twentieth century, the emphasis of park management became more ecological, with parks established both for scientific research and to maintain examples of representative natural areas. For instance, Zimbabwe established Sengwa National Park (now called Matusadona National Park) solely for scientific research; there are no tourist areas, and tourists are not generally allowed. Its purpose is the study of natural ecosystems with as little human interference as possible so that the principles of wildlife and wilderness management can be better formulated

and understood. Other national parks in the countries of eastern and southern Africa—including those of Kenya, Uganda, Tanzania, Zimbabwe, and South Africa—have been established primarily for viewing wildlife and for biological conservation.

In recent years, the number of national parks throughout the world has increased rapidly. The law establishing national parks in France was first enacted in 1960. Taiwan had no national parks prior to 1980 but now has six. In the United States, the area in national and state parks has increased from less than 12 million hectares (30 million acres) in 1950 to nearly 83.6 million acres today, with much of the increase due to the establishment of parks in Alaska.[24]

The conservation of representative natural areas of a country is an increasingly common goal of national parks. For example, the goal of New Zealand's national park planning is to include at least one area representative of each major ecosystem of the nation, from seacoast to mountain peak.

---

the traditional migratory routes. This eased the conflicts between elephants and farmers.

Because parks isolate populations genetically, they may provide too small a habitat for the maintenance of a minimum safe population size. If parks are to function as biological preserves, they must be adequate in size and habitat diversity to maintain a population large enough to avoid the serious genetic difficulties that can develop in small populations. An alternative, if necessary, is for a manager to move individuals of one species—say, lions in African preserves—from one park to another to maintain genetic diversity.

Conflicts in Managing Parks

The idea of a national, state, county, or city park is well accepted in North America, but conflicts arise over what kinds of activities and what intensity of activities should be allowed in parks. As a recent example, travel into Yellowstone National Park by snowmobile in the winter has become popular, but this has led to air and noise pollution and marred the experience of the park's beauty for many visi-

tors. In 2003, a federal court determined that snowmobile use should be phased out in this park.

Alfred Runte, a historian of America's national parks, explained the heart of the conflict. "This struggle was not against Americans who like their snowmobiles, but rather against the notion that anything goes in the national parks," he said. "The courts have reminded us that we have a different, higher standard for our national parks. Our history proves that no one loses when beauty wins. We will find room for snowmobiles, but just as important room without them, which is the enduring greatness of the national parks."[25]

Many of the recent conflicts relating to national parks have concerned the use of motor vehicles. Voyageurs National Park in northern Minnesota, established in 1974—fairly recently compared with many other national parks—occupies land that was once used by a variety of recreational vehicles and provided livelihoods for hunting and fishing guides and other tourism companies. These people felt that restricting motor-vehicle use would destroy their livelihoods. Voyageurs National Park has

100 miles of snowmobile trails and is open to a greater variety of motor-vehicle recreation than Yellowstone.[26]

Interactions between people and wildlife can also be a problem. While many people like to visit parks to see wildlife, some wildlife, such as grizzly bears in Yellowstone National Park, can be dangerous. There has been conflict in the past between conserving the grizzly and making the park as open as possible for recreation.

How Much Land Should Be in Parks?

Another important controversy in managing parks is what percentage of a landscape should be in parks or nature preserves, especially with regard to the goals of biological diversity. One effort in the United States, called the Wildlands Project, argues that large areas are necessary to conserve ecosystems, so even America's large parks, such as Yellowstone, need to be connected by conservation corridors. The rationale is, in part, the landscape concept discussed at the beginning of this chapter. At the same time, however, the growing human population and the desire for a higher standard of living put pressure on existing parks in many parts of the world.

Nations differ widely in the percentage of their total area set aside as national parks. Costa Rica, a small country with high biological diversity, has more than 12% of its land in national parks.[27] Kenya, a larger nation that also has numerous biological resources, has 7.6% of its land in national parks.[31] In France, an industrialized nation in which civilization has altered the landscape for several thousand years, only 0.7% of the land is in the nation's six national parks. However, France has 38 regional parks that encompass 11% (5.9 million ha) of the nation's area.[28]

The total amount of protected natural area in the United States is more than 104 million hectares (about 240 million acres), approximately 11.2% of the total U.S. land area.[33] However, the percentage of land in parks, preserves, and other conservation areas differs greatly among the states. The West, for example, has vast parks, whereas the six Great Lakes states (Michigan, Minnesota, Illinois, Indiana, Ohio, and Wisconsin), covering an area approaching that of France and Germany combined, allocate less than 0.5% of their land to parks and less than 1% to designated wilderness.[29]

Conserving Wilderness

As a modern legal concept, **wilderness** is an area undisturbed by people. The only people in a wilderness are visitors, who do not remain. The conservation of such an area is a new idea introduced in the second half of the twentieth century and one that is likely to become more important as the human population increases and the effects of civilization become more pervasive throughout the world (Figure 13.18).

The U.S. Wilderness Act of 1964 was a landmark piece of legislation, marking the first time anywhere that wilderness was recognized by national law as a national treasure to be preserved. Under this law, wilderness includes "an area of undeveloped Federal land retaining its primeval character and influence, without permanent improvements or human habitation, which is protected and managed so as to preserve its natural conditions." Such lands are those in which (1) the imprint of human work is unnoticeable, (2) there are opportunities for solitude and for primitive and unconfined recreation, and (3) there are at least 5,000 acres.

The law also recognizes that these areas are valuable for ecological processes, geology, education, scenery, and history. The Wilderness Act required certain maps and descriptions of wilderness areas, resulting in the U.S. Forest Service's Roadless Area Review and Evaluation (RARE I and RARE II), which evaluated lands for inclusion as legally designated wilderness.

Countries with a significant amount of wilderness include New Zealand, Canada, Sweden, Norway, Finland, Russia, and Australia; some countries of eastern and southern Africa; many countries of South America, including parts of the Brazilian and Peruvian Amazon basin; the mountainous high-altitude areas of Chile and Argentina; some of the remaining interior tropical forests of Southeast Asia; and the Pacific Rim countries (parts of Borneo, the Philippines, Papua New Guinea, and Indonesia). In addition, wilderness can be found in the polar regions, including Antarctica, Greenland, and Iceland.

Many countries have no wilderness left to preserve. In the Danish language, even the word for *wilderness* has disappeared, although that word was important in the ancestral languages of the Danes.[30] Switzerland is a country in which wilderness is not a part of preservation. For example, a national park in Switzerland lies in view of the

Figure 13.18 ■ A feeling of wilderness. Wrangell–St. Elias Wilderness Area, Alaska, designated in 1980 and now covering 9,078,675 acres. As the photograph suggests, this vast area gives a visitor a sense of wilderness as a place where a person is only a visitor and human beings seem to have no impact.

Alps—scenery that inspired the English romantic poets of the early nineteenth century to praise what they saw as wilderness and to attach the adjective *awesome*—meaning that it inspired awe in the viewer—to what they saw. But the park is in an area that has been heavily exploited for such activities as mining and foundries since the Middle Ages. All the forests are planted.[30]

In another and perhaps deeper sense, wilderness is an idea and an ideal that can be experienced in many places, such as in Japanese gardens. Henry David Thoreau distinguished between "wilderness" and "wildness." He thought of wilderness as a physical place and wildness as a state of mind. During his travels through the Maine woods in the 1840s, he concluded that wilderness was an interesting place to visit but not to live in. He preferred long walks through the woods and near swamps around his home in Concord, Massachusetts, where he was able to experience a *feeling* of wildness. Thus, Thoreau raised a fundamental question: Can one experience true wildness only in a huge area set aside as a wilderness and untouched by human actions, or can wildness be experienced in small, heavily modified, and naturalistic landscapes, such as those around Concord in the nineteenth century?

As Thoreau suggests, small, local, naturalistic parks may have more value as places of solitude and beauty than some more traditional wilderness areas. In Japan, for instance, there are roadless recreation areas, but they are filled with people. One two-day hiking circuit leads to a high-altitude marsh where people can stay in small cabins. Trash is removed from the area by helicopter. People taking this hike experience a sense of wildness.

In some ways, the answer to the question raised by Thoreau is highly personal. We must discover for ourselves what kind of natural or naturalistic area meets our spiritual, aesthetic, and emotional needs. This is yet another area in which one of our key themes, science and values, is evident.

Conflicts in Managing Wilderness

The legal definition of *wilderness* has given rise to several controversies. The wilderness system in the United States began in 1964 with 3.7 million hectares (9.2 million acres) under U.S. Forest Service control. Today, the United States has 633 legally designated wilderness areas, covering 44 million hectares (106 million acres)—more than 4% of the country. Another 200 million acres meet the legal requirements and could be protected by the Wilderness Act; half of this area is in Alaska, including the largest single area, Wrangel–St. Elias, covering 3.7 million hectares (9 million acres).[31, 32]

Those interested in developing the natural resources of an area, including mineral ores and timber, have argued that the rules are unnecessarily stringent, protecting too much land from exploitation when, they say, there is plenty of wilderness elsewhere. Those who wish to conserve additional wild areas have argued that the interpretation of the U.S. Wilderness Act is too lenient and that mining and logging are inconsistent with the wording of the act. These disagreements are illustrated by the argument over drilling in the Arctic National Wildlife Refuge, discussed in Chapter 18.

The notion of managing wilderness may seem a paradox—a true wilderness would need no management. In fact, though, with the great numbers of people in the world today, even wilderness must be defined, legally set aside, and controlled. We can view the goal of managing wilderness in two ways: in terms of the wilderness itself and in terms of people. In the first instance, the goal is to preserve nature undisturbed by people. In the second, the purpose is to provide people with a wilderness experience.

Legally designated wilderness can be seen as one extreme in a spectrum of environments to manage. The spectrum includes preserves in which some human activities are allowed to be visible—parks designed for outdoor recreation, forests for timber production and various kinds of recreation, hunting preserves, and urban parks—and, at the other extreme, central cities as they are now conceived, open-pit mines, and the like. You can think of many stages in between these on this spectrum.[33] Wilderness management should involve as little direct action as possible, so as to minimize human influence. Ironically, one of the necessities is to control human access so that a visitor has little, if any, sense that other people are present. Access to wilderness by automobiles, light aircraft and helicopters, snowmobiles, and all-terrain off-road vehicles must also be restricted.

Consider, for example, the Desolation Wilderness Area in California, consisting of more than 24,200 hectares (60,000 acres), which in one year had more than 250,000 visitors. Could each visitor really have a wilderness experience there, or was the human carrying capacity exceeded? This is a subjective judgment. If, on one hand, all visitors saw only their own companions and believed they were alone, then the actual number of visitors did not matter for each visitor's wilderness experience. On the other hand, if every visitor found the solitude ruined by strangers, then the management failed, no matter how few people visited the area.

Wilderness designation and management must also take into account adjacent land uses. A wilderness next to a garbage dump or a smoke-emitting power plant is a contradiction in terms. Whether a wilderness can be adjacent to a high-intensity campground or near a city is a more subtle question that must be resolved by citizens.

Today, those involved in wilderness management recognize that wild areas change over time and that these changes should be allowed to occur as long as they are natural. This is different from earlier views that nature undisturbed was unchanging and should be managed so that it did not change. In addition, it is generally argued now that in choosing what activities can be allowed in a wilderness, emphasis should be placed on activities that depend on wilderness

Can Tropical Forests Survive in Bits and Pieces?

Although tropical rain forests occupy only about 7% of the world's land area, they provide habitat for at least half of the world's species of plants and animals (Figure 13.16). Approximately 100 million people live in rain forests or depend on them for their livelihood. Tropical plants provide products such as chocolate, nuts, fruits, gums, coffee, wood, rubber, pesticides, fibers, and dyes. Drugs used to treat high blood pressure, Hodgkin's disease, leukemia, multiple sclerosis, and Parkinson's disease have been made from tropical plants, and medical scientists believe many more are yet to be discovered.

Most of the interest in tropical rain forests has focused on Brazil, whose forests are believed to have more species than any other geographic area. Estimates of destruction in the Brazilian rain forest range from 6% to 12%, but numerous studies have shown that deforested area alone does not adequately measure habitat destruction, because surrounding habitats are also affected (Figure 13.16a). For example, the more fragmented a forest is, the more edges there are, and the greater the impact on the living organisms. Such edge effects vary depending on the species, the characteristics of the land surrounding the forest fragment, and the distance between fragments. For example, a forest surrounded by farmland is more deeply affected than one surrounded by abandoned land in which secondary growth presents a more gradual transition between forest and deforested areas. Some insects, small mammals, and many birds find only

80 m (262.5 ft) a barrier to movement from one fragment to another, whereas one small marsupial has been found to cross distances of 250 m (820.2 ft). Corridors between forested areas also help to offset the negative effects of deforestation on plants and animals of the forest.

Critical Thinking Questions

1. Assuming an edge effect of 1 km, what is the approximate area affected by deforestation of 100 km² (38.6 mi²) in the form of a square, 10 km (6.2 mi) on each side? If the 100-km² area were in the form of 10 rectangles, each 10 km long and 1 km wide, separated from each other by a distance of 5 km (3.1 mi.), how large an area would be affected?

2. What environmental factors at the edge of a fragment would differ from those in the center? How might the differences affect plants and animals at the edge?

3. Why is a simple rule of thumb, such as assuming an edge effect of 1 km, too simplistic as a model of the effects of deforestation? Given that it is too simplistic, what advantages are there in using the rule?

4. Forest fragments are sometimes compared with islands. What are some ways in which this is an appropriate comparison? some ways in which it is not?

(the experience of solitude or the observation of shy and elusive wildlife) rather than on activities that can be carried out elsewhere (such as downhill skiing).

A source of conflict is that wilderness areas frequently contain economically important resources, including timber, mineral ores, and sources of energy. There has been heated debate about whether wilderness areas should be open to the extraction of oil and mineral ores.

Another controversy involves the need to study wilderness versus the desire to leave the wilderness undisturbed. Those in favor of scientific research in the wilderness argue that it is necessary for the conservation of wilderness. Those opposed argue that scientific research contradicts the purpose of a designated wilderness as an area undisturbed by people. One solution is to establish separate research preserves.

Summary

■ In the past, land management for the harvest of resources and the conservation of nature was mostly local, with each parcel of land considered independently. Today, a landscape perspective has developed, and lands used for harvesting resources are seen as part of a matrix that includes lands set aside for the conservation of biological diversity and for landscape beauty.

■ Forests are among civilization's most important renewable resources. Forest management seeks a sustainable harvest and sustainable ecosystems.

■ There are few examples of successful sustainable forestry. As a result, a practice has developed called "certification of sustainable forestry." Certification involves determining which methods appear most consistent with sustainability and then comparing the management of a specific forest with those standards.

■ The continued use of firewood as an important fuel in developing nations is a major threat to forests, given the rapid population growth of those areas. It is doubtful that developing nations can implement successful management programs in time to prevent serious damage to their forests and severe effects on their people.

■ Clear-cutting is a major source of controversy in forestry. Some tree species require clearing to reproduce and grow, but the scope and method of cutting must be examined carefully in terms of the needs of the species and the type of forest ecosystem.

■ Properly managed plantations can relieve pressure on forests.

■ Managing parks for biological conservation is a relatively new idea that began in the nineteenth century. The manager of a park must be concerned with its shape and size. Parks that are too small or the wrong shape may have too small a population of the species for which the park was established and thus may not be able to sustain the species.

■ A special extreme in conservation of natural areas is the management of wilderness. In the United States, the 1964 Wilderness Act provided a legal basis for such conservation. Managing wilderness seems to be a contradiction—trying to preserve an area undisturbed by people requires interference to limit user access and to maintain the wilderness in a natural state, so the area that is not supposed to be influenced by people actually is.

■ Parks, nature preserves, wilderness areas, and actively harvested forests affect one another. The geographic pattern of these areas on a landscape, including corridors and connections among different types, is part of the modern approach to biological conservation and the harvest of forest resources.

REEXAMINING THEMES AND ISSUES

 Human Population

Forests provide essential resources for civilization. As the human population grows, there will be greater and greater demand for these resources. Because forest plantations can be highly productive, we are likely to place increasing emphasis on forest plantations as a source of timber. This would free more forestland for other uses.

 Sustainability

Sustainability is the key to conservation and management of wild living resources. However, sustainable harvests have rarely been achieved for timber production, and sustained ecosystems in harvested forests are even rarer. Sustainability must be the central focus for forest resources in the future.

 Global Perspective

Forests are global resources. A decline in the availability of forest products in one region affects the rate of harvest and economic value of these products in other regions. Biological diversity is also a global resource. As the human population grows, the conservation of biological diversity is likely to depend more and more on legally established parks, nature preserves, and wilderness areas.

 Urban World

We tend to think of cities as separated from living resources, but urban parks are important in making cities pleasant and livable; if properly designed, they can also help to conserve wild living resources.

 People and Nature

Forests have provided essential resources, and often people have viewed them as perhaps sacred but also dark and scary. Today, we value wilderness and forests, but we rarely harvest forests sustainably. Thus, the challenge for the future is to reconcile our dual and somewhat opposing views so that we can enjoy both the deep meaningfulness of forests and their important resources.

 Science and Values

Many conflicts over parks, nature preserves, and legally designated wilderness areas also involve science and values. Science tells us what is possible and what is required to conserve both a specific species and total biological diversity. But what society desires for such areas is, in the end, a matter of values and experience, influenced by scientific knowledge.

Key Terms

certification of
forestry **247**
clear-cutting **245**
codominants **245**
dominants **245**
intermediate **245**

old-growth forest **245**
plantation **246**
public service
functions **242**
rotation time **245**
second-growth forest **245**

seed-tree cutting **245**
selective cutting **245**
shelterwood-cutting **245**
silviculture **241**
site quality **245**
stand **245**

strip-cutting **246**
suppressed **245**
sustainable forests **247**
thinning **245**
wilderness **256**

Study Questions

1. What environmental conflicts might arise when a forest is managed for the multiple uses of (a) commercial timber, (b) wildlife conservation, and (c) a watershed for a reservoir? In what ways could management for one use benefit another?

2. What arguments could you offer for and against the statement "Clear-cutting is natural and necessary for forest management"?

3. Can a wilderness park be managed to supply water to a city? Explain your answer.

4. A park is being planned in rugged mountains with high rainfall. What are the environmental considerations if the purpose of the park is to preserve a rare species of deer? If the purpose is recreation, including hiking and hunting?

5. What are the environmental effects of decreasing the rotation time in forests from an average of 60 years to 10 years? Compare these effects for (a) a woodland in a dry climate on a sandy soil and (b) a rain forest.

6. In a small but heavily forested nation, two plans are put forward for forest harvests. In Plan A, all the forests to be harvested are in the eastern part of the nation, while all the forests of the west are set aside as wilderness areas, parks, and nature preserves. In Plan B, small areas of forests to be harvested are distributed throughout the country, in many cases adjacent to parks, preserves, and wilderness areas. Which plan would you choose? Note that in Plan B, wilderness areas would be smaller than in Plan A.

7. The smallest legally designated wilderness in the United States is Pelican Island, Florida (Figure 13.19), covering 5 acres. Do you think this can meet the meaning of *wilderness* and the intent of the Wilderness Act?

Figure 13.19 ■ Pelican Island Wilderness, Florida. The United States' smallest legally designated wilderness covers 5 acres.

Further Reading

Hendee, J. C. 2002. *Wilderness Management: Stewardship and Protection of Resources and Values*. Golden, Colo.: Fulcrum Publishing. Considered the classic work on this subject.

Kimmins, J. P. 2003. *Forest Ecology*, 3rd ed. Upper Saddle River, N.J.: Prentice Hall. A textbook that applies recent developments in ecology to the practical problems of managing forests.

Runte, A. 1997. *National Parks: The American Experience*. Lincoln: Bison Books of the University of Nebraska. The classic book about the history of national parks in America and the reasons for their development.

CHAPTER 14

Wildlife, Fisheries, and Endangered Species

Baked Blue Hake with Tomato Chutney

Ingredients

4 Blue Hake loins
2 toes finely chopped garlic
1 tsp. salt
1/2 tsp. pepper
1/4 cup margarine
1 small yellow onion, chopped
6 ripe roma tomatoes
1/2 cup sugar
1 cup malt vinegar
2 tsp paprika
1/4 tsp cayenne pepper

Procedure

Tomato Chutney
Wash and chop the tomatoes,
Peel and chop the onions.
Cook the tomatoes and onions gently in a sauce pan,
simmering for 20 - 30 minutes until tender.
Add the salt, paprika, cayenne and half of the vinegar.
Simmer for 45 minutes or until it begins to thicken.
Add the sugar and remaining vinegar, stirring until fully dissolved.
Continue simmering, until the mixture becomes thick, stirring occasionally.
Allow the Chutney to cool to room temperature.

Baked Hake
Heat oven to 400 degrees F.
Oil a 13" x 9" baking pan or use nonstick spray.
In a separate bowl mix fresh garlic, 1/2 tsp. salt, and pepper.
Place the Blue Hake in the pan and spread with the
garlic, salt, and pepper mixture.
Cut the margarine into small pieces and place evenly over the fish.
Bake in a preheated oven for 25 minutes.

Spread the chutney over a plate and place the baked Hake on top.
Serve with couscous or your favorite bread.
Garnish with parsley and fresh lemon if desired.

(a)

(c)

(b)

(a) Fried hake with vegetable couscous.
(b) Blue hake in its natural habitat

Learning Objectives

Wildlife, fish, and endangered species are among the most popular environmental issues today. People love to see wildlife; many people enjoy fishing, make a living from it, or rely on fish as an important part of their diet; and since the nineteenth century, the fate of endangered species has drawn public attention. You would think that by now we would be doing a good job of conserving and managing these kinds of life. This chapter tells you how we are doing. After reading this chapter, you should understand:

■ Why people want to conserve wildlife and endangered species.

■ The importance of the habitat, the ecosystems, and landscape in the conservation of endangered species.

■ Current causes of extinction.

■ Steps we can take to achieve sustainability of wildlife, fisheries, and endangered species.

■ The concepts of species persistence, maximum sustainable yield, the logistic growth curve, carrying capacity, optimum sustainable yield, and minimum viable populations.

Threats to Major World Fisheries

Five major ocean-bottom fish species are on the brink of extinction, declining steadily during the past 17 years. Today they are more than 90% less abundant, according to a 2006 study.[1] These species, which include the Atlantic hake, are important as food.

The twentieth century marked the beginning of scientific-based management of fisheries around the world, and this was supposed to lead to sustainable fisheries. In fact, the long-stated goal of this scientific-based management of fish has been to achieve a "maximum sustainable yield." So what has happened? Here we are in the twenty-first century, and that scientific management is supposed to be continuing. Why then are these major fisheries in trouble? Why, after a century of trying, can we still not have sustainable fish harvests and sustainable populations of fish? And if fish have been one of the longest targets for sustainability, what hope do we have for sustaining any of our living resources?

The five species of fish in danger of extinction include two taken commercially, the roundnose grenadier and the blue hake. The others—onion-eye grenadier, spiny eel, and spinytail skate—are caught as "by-catch"; that is, they are brought on board fishing vessels because they are caught during the process of harvesting fish that can be sold. The by-catch usually dies.

Interestingly, all of these fish are long-lived, up to 60 years, and, like people, they do not mature until their teens. Populations of this kind tend to reproduce slowly and therefore recover slowly from overharvesting.

The fish are caught by bottom trawlers that drag the "trawls" across the ocean floor, where they often destroy everything on the bottom, all the fish and their habitats (Figures 14.1 and 14.2).

These five species were not sought commercially until the 1970s, when they came into vogue as replacements for other species that had been fished out, such as the

How bottom trawling works

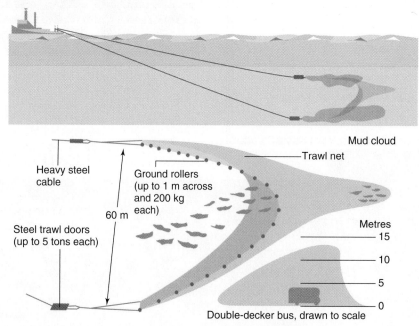

Figure 14.1 ■ How bottom-trawling works. (*Source*: BBC News. http://news.bbc.co.uk/2/hi/science/nature/4581428. stm#bottomtrawl, April 19, 2006)

(a)

(b)

Figure 14.2 ■ Effects of trawling on coral reefs, as shown for a coral reef in the Florida Keys National Marine Sanctuary (*a*) before and (*b*) after trawling. Essentially nothing is left after the trawl has scrapped the reef.

Atlantic cod. Thus, their decline corresponds to the timing of the major increase in their fishing.

This brief discussion suggests that fishing methods that destroy habitats are part of the problem, as is just plain overharvesting. However, to understand why fisheries around the world have been declining for over a century, we have to understand the present basis for the conservation and management of our wild living resources, which is the topic of this chapter.

14.1 Introduction

Wildlife, fisheries, and endangered species are considered together in this chapter because they have a common history of exploitation, management, and conservation, and because modern attempts to manage and conserve them follow the same approaches. Although any form of life, from bacteria and fungi to flowering plants and animals, can become endangered, concern about endangered species has tended to focus on wildlife. We will maintain that focus, but we ask you to be sure to remember that the general principles apply to all forms of life.

When we say that we want to save a species, what is it that we really want to save? There are four possible answers:

1. A wild creature in a wild habitat, as a symbol to us of wilderness.
2. A wild creature in a managed habitat, so the species can feed and reproduce with little interference and so we can see it in a naturalistic habitat.

3. A population in a zoo, so the genetic characteristics are maintained in live individuals.
4. Genetic material only—frozen cells containing DNA from a species for future scientific research.

Which of these goals we choose involves not only science but also values. People have different reasons for wishing to save endangered species—utilitarian, ecological, cultural, recreational, spiritual, inspirational, aesthetic, and moral (see A Closer Look 14.1). Policies and actions differ widely depending on which goal is chosen.

14.2 Traditional Single-Species Wildlife Management

Attempts to apply science to the conservation and management of wildlife and fisheries, and therefore to endangered species, began around the turn of the twentieth century and viewed each species as a single population in isolation. Assumptions included the following:

- The population could be represented by a single number, its total size.
- Undisturbed by human activities, a population would grow to a fixed size, called the "carrying capacity."
- Environment, except for human-induced changes, is constant.

This perception of wildlife and fisheries was formalized in the S-shaped logistic growth equation (Figure 14.3), which we discussed in Chapter 4. Two goals resulted

Reasons for the Conservation of Endangered Species (and of All Life on Earth)

Some important reasons for conserving endangered species can be classified as utilitarian, ecological, aesthetic, moral, and cultural.

Utilitarian Justification

Utilitarian justification is based on the consideration that many wild species might be useful to us and that it is therefore unwise to destroy them before we have a chance to test their uses. Many of the arguments for conserving endangered species, and for biological diversity in general, have focused on the utilitarian justification.

One utilitarian justification is the need to conserve wild strains of grains and other crops. Disease organisms that attack crops evolve continually, and as new disease strains develop, crops become vulnerable. Crops such as wheat and corn depend on the continued introduction of fresh genetic characteristics from wild strains to create new, disease-resistant genetic hybrids.

Related to this justification is the possibility of finding new crops among the many species of plants. Many horticultural crops and products have come from tropical rain forests, and hopes are high that new products will be found.[2] For example, of 275 species found in 1 hectare (0.4 acre) in a Peruvian tropical forest, 72 yielded products with direct economic value.

Another utilitarian justification for biological conservation is that many important chemical compounds come from wild organisms. Digitalis, an important drug in treating certain heart ailments, comes from purple foxglove. Aspirin is a derivative of willow bark. A recent example was the discovery of a cancer-fighting chemical named taxol in the Pacific yew tree (genus name *Taxus*; hence the name of the chemical). Well-known medicines derived from tropical forests include anti-

cancer drugs from rosy periwinkles, steroids from Mexican yams, antihypertensive drugs from serpent wood, and antibiotics from tropical fungi.[25] Some 25% of prescriptions dispensed in the United States today contain ingredients extracted from vascular plants.[26] And these represent only a small fraction of the estimated 500,000 existing plant species. Other plants and organisms may produce useful medical compounds that are as yet unknown. Scientists are testing marine organisms for use in pharmaceutical drugs. Coral reefs offer a promising area of study for such compounds because many coral-reef species produce toxins to defend themselves.

Here are a few examples: Some medicines to treat HIV and herpes come from coral reef sponges. Other chemicals from coral reef organisms are in clinical trials to treat breast and liver cancer and leukemia. A venom from a coral reef snail is in tests as a painkiller, potentially less addictive than morphine.[4, 5]

Some species are also used directly in medical research. For example, the armadillo, one of only two animal species known to contract leprosy, is used to study cures for that disease. Other animals, such as horseshoe crabs and barnacles, are important because of physiologically active compounds they make. Still others may have similar uses as yet unknown to us.

Another utilitarian justification is that many species help to control pollution. Plants, fungi, and bacteria remove toxic substances from air, water, and soils. Carbon dioxide and sulfur dioxide are removed by vegetation, carbon monoxide is reduced and oxidized by soil fungi and bacteria, and nitric oxide is incorporated into the biological nitrogen cycle.[17] Because species vary in their capabilities, diversity of species provides the best range of pollution control.

Tourism provides yet another utilitarian justification. Ecotourism is a growing source of income for many developing countries. Ecotourists value nature, including its endangered species, for aesthetic or spiritual reasons, but the result can be utilitarian.

Ecological Justification

When we reason that organisms are necessary to maintain the functions of ecosystems and the biosphere, we are using an ecological justification for the conservation of these organisms. Individual species, entire ecosystems, and the biosphere provide public service functions essential or important to the persistence of life, and as such they are indirectly necessary for our survival. When bees pollinate flowers, for example, they provide a benefit to us that would be costly to replace with human labor. Trees remove certain pollutants from the air; and some soil bacteria fix nitrogen, converting it from molecular nitrogen in the atmosphere to nitrate and ammonia that can be taken up by other living things. That some such functions involve the entire biosphere reminds us of the global perspective on conserving nature and specific species.

Aesthetic Justification

An aesthetic justification asserts that biological diversity enhances the quality of our lives, providing some of the most beautiful and appealing aspects of our existence. Biological diversity is an important quality of landscape beauty. Many organisms—birds, large land mammals, and flowering plants, as well as many insects and ocean animals—are appreciated for their beauty. This appreciation of nature is ancient. Whatever other reasons Pleistocene people had for creating paintings in caves in France and Spain, their paintings of wildlife, done about 14,000 years ago, are beautiful. The paintings include species that have since become extinct, such as mastodons. Poetry, novels, plays, paintings, and sculpture often celebrate the beauty of nature. It is a very human quality to appreciate nature's beauty and is a strong reason for the conservation of endangered species.

Moral Justification

Moral justification is based on the belief that species have a moral right to exist, independent of our need for them; consequently, in our role as global stewards, we are obligated to promote the continued existence of species and to conserve biological diversity. This right to exist was stated in the UN General Assembly World Charter for Nature, 1982. The U.S. Endangered Species Act also includes statements concerning the rights of organisms to exist. Thus, a moral justification for the conservation of endangered species is part of the intent of the law.

Moral justification has deep roots within human culture, religion, and society. Those who focus on cost–benefit analyses tend to downplay moral justification, but although it may not seem to have economic ramifications, in fact it does. As more and more citizens of the world assert the validity of moral justification, more actions that have economic effects are taken to defend a moral position.

The moral justification has grown in popularity in recent decades, as indicated by increasing interest in the deep-ecology movement, mentioned in an earlier chapter. Arne Næss, one of its principal philosophers, explains: "The right of all the forms [of life] to live is a universal right which cannot be quantified. No single species of living being has more of this particular right to live and unfold than any other species."[6]

Cultural Justification

Certain species, some threatened or endangered, are of great importance to many indigenous peoples, who rely on these species of vegetation and wildlife for food, shelter, tools, fuel, materials for clothing, and medicine. Reduced biological diversity can severely increase the poverty of these people. For poor indigenous people who depend on forests, there may be no reasonable replacement except continual outside assistance, which development projects are supposed to eliminate. Urban residents, too, share in the benefits of biological diversity, even if these benefits may not be apparent or may become apparent too late (see Chapter 28).

from these ideas and the logistic equation: For a species that we intend to harvest, the goal was maximum sustainable yield (MSY); for a species that we wish to conserve, the goal was to have that species reach, and remain at, its carrying capacity. **Maximum sustainable yield** was defined as the population size that yielded maximum production (measured either as a net increase in the number of individuals in the population or as a net change in biomass) that would allow this population to be sustained indefinitely without decreasing its ability to provide the same level of production. More simply, the population was viewed as a factory that could keep churning out exactly the same quantity of a product year after year.

This approach failed. None of the assumptions are true, as you have been learning in this book. Populations cannot be represented only by a single number—their total abundance—and do not remain at a fixed carrying capacity. Nor is the environment constant.

Today a broader view is developing. It acknowledges that a population exists in a changing environment (including human-induced changes), that populations

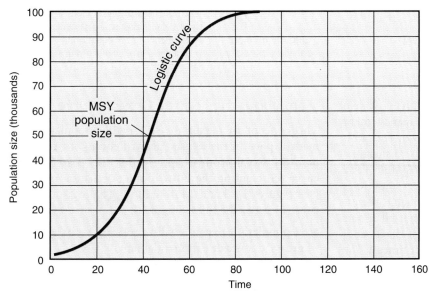

Figure 14.3 ■ The logistic growth curve, which has been the basis for describing and forecasting the growth of animal populations. Although still widely used, it is rarely supported, and often contradicted, by data. MSY is the point of maximum sustainable yield when a logistic population grows the fastest.

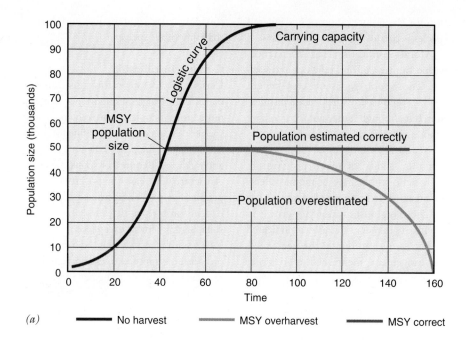

(a)
— No harvest   — MSY overharvest   — MSY correct

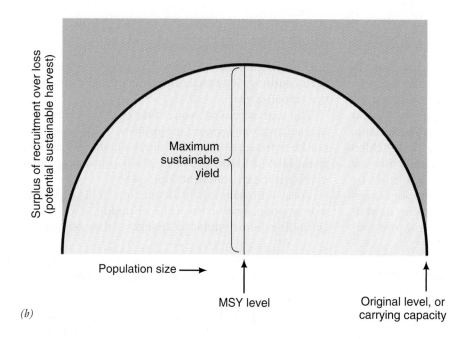

(b)

Figure 14.4 ■ (*a*) The logistic growth curve, showing the carrying capacity and the maximum sustainable yield (MSY) population (where the population size is one-half the carrying capacity). The figure shows what happens to a population when we assume it is at MSY and it is not. Suppose a population grows according to the logistic curve from a small number to a carrying capacity of 100,000 with an annual growth rate of 5%. The correct maximum sustainable yield would be 50,000. When the population reaches exactly that amount and we harvest at exactly the calculated maximum sustainable yield, the population continues to be constant. But if we make a mistake in estimating the size of the population (for example, if we believe that it is 60,000 when it is only 50,000), then the harvest will always be too large, and we will drive the population to extinction. (*b*) Another view of a logistic population. Growth in the population here is graphed against population size. Growth peaks when the population is exactly at one-half the carrying capacity. This is a mathematical consequence of the equation for the curve. It is rarely, if at all, observed in nature.

interact, and that it is necessary to include an ecosystem and landscape context for conservation and management (as we learned in Chapter 13). With this new understanding, the goal for a species to be harvested is to sustain a harvestable population in a sustainable ecosystem. The goal for a threatened or endangered species is sometimes stated as a **minimum viable population**, which is the estimated smallest population that can maintain itself and its genetic variability indefinitely. Other times the goal is stated as the *carrying capacity*, and still other times as the *optimum sustainable population*. (These terms will be explained later in this chapter.)

More about the Logistic Growth Curve

The concept of maximum sustainable population was made explicit in the logistic growth curve, first proposed in 1838. Recall from Chapter 4 that this is an S-shaped curve representing the growth of a population over time (see Figures 14.3 and 14.4*a* and *b*). The logistic growth curve includes the following ideas:

■ A population that is small in relation to its resources grows at a nearly exponential rate.

■ Competition among individuals in the population slows the growth rate.

- The greater the number of individuals, the greater the competition and the slower the rate of growth.
- Eventually, a point is reached, called the "logistic carrying capacity," at which the number of individuals is just sufficient for the available resources.
- At this level, the number of births in a unit time equals the number of deaths, and the population is constant.
- A population can be described simply by its total number.
- Therefore, all individuals are equal.
- The environment is assumed to be constant.

An important result of all of this is that a *logistic population is stable in terms of its carrying capacity*—it will return to that number after a disturbance. If the population grows beyond its carrying capacity, deaths exceed births, and the population declines back to the carrying capacity. If the population falls below the carrying capacity, births exceed deaths, and the population increases. Only if the population is exactly at the carrying capacity do births exactly equal deaths, and then the population does not change.

Carrying capacity is an important term in wildlife management. It has three definitions. The first is the carrying capacity as defined by the logistic growth curve. We will refer to this as the **logistic carrying capacity**. The second definition contains the same idea but is not dependent on that specific equation. It states that the carrying capacity is an abundance at which a population can sustain itself without any detrimental effects that would decrease the ability of that species to maintain that abundance. The third, more recent definition is usually referred to as the **optimum sustainable population** and is the maximum population that can be sustained indefinitely without decreasing the ability of that species *or its habitat or ecosystem* to sustain that population level for some specified time.

Another key concept from the logistic growth curve is the population size that provides the maximum sustainable yield. In the logistic curve, the greatest production occurs when the population is exactly one-half of the carrying capacity (see Figure 14.4). This is nifty, because it makes everything seem simple—all you have to do is figure out the carrying capacity and keep the population at one-half of it. But what seems simple can easily become troublesome. Even if the basic assumptions of the logistic curve were true, which they are not, the slightest overestimate of carrying capacity, and therefore MSY, would lead to overharvesting, a decline in production, and a decline in the abundance of the species. If a population is harvested as if it were actually at one-half its carrying capacity, then unless a logistic population is actually maintained at exactly that number, its growth will decline. Since it is almost impossible to maintain a wild population at some exact number, the approach is doomed from the start.

Despite its limitations, the logistic growth curve was used for all wildlife, especially fisheries and including endangered species, throughout much of the twentieth century.

An Example of Problems with the Logistic Curve

Suppose you are in charge of managing a deer herd for recreational hunting in one of the 50 U.S. states. Your goal is to maintain the population at its MSY level, which, as you can see from Figure 14.4, occurs at exactly one-half of the carrying capacity. At this abundance, the population increases by the greatest number during any time period.

To accomplish this goal, first you have to determine the logistic carrying capacity. You are immediately in trouble because, first, only in a few cases has the carrying capacity ever been determined by legitimate scientific methods (see Chapter 2) and, second, we now know that the carrying capacity varies with changes in the environment. The procedure in the past has been to estimate the carrying capacity by *nonscientific* means and then attempt to maintain the population at one-half that level. This method requires accurate counts each year. It also requires that the environment not vary or, if it does, that it vary in a way that does not affect the population. Since these conditions cannot be met, the logistic curve has to fail as a basis for managing the deer herd.

An interesting example of the staying power of the logistic growth curve can be found in the United States Marine Mammal Protection Act of 1972. This act states that its primary goal is to conserve "the health and stability of marine ecosystems," which is part of the modern approach, and so the act seems to be off to a good start. But then the act states that the secondary goal is to maintain an "optimum sustainable population" of marine mammals. What is this? The wording of the act allows two interpretations. One is the logistic carrying capacity, and the other is the MSY population level of the logistic growth curve. So the act takes us back to square one, the logistic curve.

14.3 Stories Told by the Grizzly Bear and the Bison: Wildlife Management Questions That Require New Approaches

Chapter 2 pointed out that studies in the environmental sciences sometimes do not seem suited to use of the standard scientific method. This is also true of some aspects of wildlife management and conservation. Several examples illustrate the needs and problems.

The Grizzly Bear

A classic example of wildlife management is the North American grizzly bear. An endangered species, the grizzly has been the subject of efforts by the U.S. Fish and Wildlife Service to meet the requirements of the U.S. Endangered Species Act, which include restoring the population of the grizzly.

The grizzly became endangered as a result of hunting and habitat destruction. It is arguably the most dangerous

Figure 14.5 ■ Grizzly bear. Records of bear sightings by Lewis and Clark have been used to estimate their population at the beginning of the nineteenth century.

North American mammal, famous for its unprovoked attacks on people, and has been eliminated from much of its range for that reason. Males weigh as much as 270 kg (600 pounds), females as much as 160 kg (350 pounds). When they rear up on their hind legs, they are almost 3 m (8 ft) tall. No wonder they are frightening. (See Figure 14.5.) Despite this, or perhaps because of it, grizzlies intrigue people. Therefore, watching grizzlies from a safe distance has become a popular recreation.

At first glance, the goal of restoring the grizzly seems simple enough. But then the question arises: Restore to what? One answer is to its abundance at the time of the European discovery and settlement of North America. But it turns out that there is very little historical information about the abundance of the grizzly at that time, so it is not easy to determine the number of grizzlies (or a density of grizzlies per unit area) to select as a "restored" population.

Adding to the difficulty is the lack of a good estimate of the grizzly's present abundance. Unless we know the present abundance, we don't know how far we will have to take the species to "restore" it to some hypothetical past abundance. But the grizzly is difficult to study. It is large and dangerous and tends to be reclusive. The U.S. Fish and Wildlife Service attempted to count the grizzlies in Yellowstone National Park by installing automatic flash cameras that were set off when the grizzlies took a bait. This seemed a good idea, but the grizzlies did not like the flash cameras and destroyed them.[7] At this time we do not have a good scientific estimate of the present number of grizzlies. The National Wildlife Federation lists 1,200 in the contiguous states, 32,000 in Alaska, and about 25,000 in Canada. But these are crude estimates.[7]

How do we arrive at an estimate of a population that existed at a time when nobody thought of counting its members? Where possible, we make use of historical records, as discussed in Chapter 2. We can obtain a crude estimate of the grizzly's abundance at the beginning of the nineteenth century from the journals of the Lewis and Clark expedition. Lewis and Clark did not list the numbers of most wildlife they saw; they simply wrote that they saw "many" bison, elk, and so forth. But the grizzlies were especially dangerous and tended to travel alone, so Lewis and Clark wrote down each encounter, stating the exact number they met. On that expedition, they saw 37 grizzly bears over a distance of approximately 1,000 miles (their records were in miles).[8]

Lewis and Clark saw grizzlies from near what is now Pierre, South Dakota, to what is today Missoula, Montana. A northern and southern geographic limit to the grizzly's range can be obtained from other explorers. Assuming Lewis and Clark could see a half-mile to each side of their line of travel on average, the density of the bears was approximately 3.7 per 100 square miles. If we estimate that the bear's geographic range was an area of 320,000 square miles in the mountain and western plains states, we arrive at a total population of $320,000 \times 0.37$, or about 12,000 bears.

Suppose we phrase this as a hypothesis: "The number of grizzly bears in 1805 in what is now the United States was 12,000." Is this a statement open to disproof? Not without time travel. Therefore, it is not a scientific statement; it can only be taken as an educated guess or, more formally, an assumption or premise. Still, it has some basis in historical documents, and it is better than no information, since we have few alternatives to determine what used to be. We can use this assumption to create a plan to restore the grizzly to that abundance. But is this the best approach?

Another approach is to ask what is the minimum viable population of grizzly bears—forget completely about what might have been the situation in the past and make use of modern knowledge of population dynamics and genetics, along with food requirements and potential production of that food. Studies of existing populations of brown and grizzly bears suggest that only populations larger than 450 individuals respond to protection with rapid growth.[9] Using this approach, we could estimate how many bears appear to be a "safe" number—that is, a number that carries small risk of extinction and loss of genetic diversity. More precisely, we could phrase this statement as "How many bears are necessary so that the probability that the grizzly will become extinct in the next ten years [or some other period that we consider reasonable for planning] is less than 1% [or some other percentage that we would like]?" With appropriate studies, this approach could have a scientific basis. Consider a statement of this kind phrased as a hypothesis: "A population of 450 bears [or some other number] results in a 99% chance that at least one mature male and one mature female will be alive ten years from today." We can disprove this statement, but only by waiting for ten years to go by. Although it is a scientific statement, it is a difficult one to deal with in planning for the present.

The American Bison

Another classic case of wildlife management, or mismanagement, is the demise of the American bison, or buffalo (Figure 14.6*a*). The bison was brought close to extinction in the nineteenth century for two reasons. Bison were hunted because coats made of bison hides had become fashionable in Europe. Bison were also killed as part of warfare against the Plains peoples (Figure 14.6*b*). Colonel R. I. Dodge was quoted in 1867 as saying, "Kill every buffalo you can. Every buffalo dead is an Indian gone."[14]

(a)

(b)

Figure 14.6 ■ (*a*) A bison ranch in the United States. In recent years, interest in growing bison on ranches has increased greatly. In part, the goal is to restore bison to a reasonable percentage of its numbers before the Civil War. In part, bison are ranched because people like them. In addition, there is a growing market for bison meat and other products, including cloth made from bison hair. (*b*) Painting of a buffalo hunt by George Catlin in 1832–1833 at the mouth of the Yellowstone River.

Unlike the grizzly bear, the bison has recovered, in large part because ranchers have begun to find them a profitable animal to raise and sell for meat and other products. Informal estimates, including herds on private and public ranges, suggest there are between 200,000 and 300,000, and bison are said to occur in every state in the United States, including Hawaii—a habitat quite different from their original Great Plains home range.[10] About 20,000 roam wild on public lands in the United States and Canada.[11]

How many bison were there before European settlement of the American West? And how low did their numbers drop? Historical records provide insight. In 1865 a General Mitchell, in response to Indian attacks in the fall of 1864, set fires to drive away the Indians and the buffalo, killing vast numbers of animals.[12] The speed with which bison were almost eliminated was surprising—even to many of those involved in hunting them.

Many early writers tell of immense herds of bison, but few counted them. One exception was General Isaac I. Stevens, who, on July 10, 1853, was surveying for the transcontinental railway in North Dakota. He and his men climbed a high hill and saw "for a great distance ahead every square mile" having "a herd of buffalo upon it." He wrote that "their number was variously estimated by the members of the party—some as high as half a million. I do not think it any exaggeration to set it down at 200,000."[8] He suggested that just one herd had about the same number of bison as exist in total today!

One of the better attempts to estimate the number of buffalo in a herd was made by Colonel R. I. Dodge, who took a wagon from Fort Zarah to Fort Larned on the Arkansas River in May 1871, a distance of 34 miles. For at least 25 of those miles, he found himself in a "dark blanket" of buffalo. Dodge estimated that the mass of animals he saw in one day totaled 480,000. At one point, he and his men traveled to the top of a hill from which he estimated that he could see 6 to 10 miles, and from that high point there appeared to be a single solid mass of buffalo extending over 25 miles. At 10 animals per acre, not a particularly high density, the herd would have numbered 2.7–8.0 million animals.[12]

In the fall of 1868, "a train traveled 120 miles between Ellsworth and Sheridan through a continuous, browsing herd, packed so thick that the engineer had to stop several times, mostly because the buffaloes would scarcely get off the tracks for the whistle and the belching smoke."[12] That spring, a train had been delayed for eight hours while a single herd passed "in one steady, unending stream." We can use accounts of such experiences to set bounds on the possible number of animals seen. At the highest extreme, we can assume that the train bisected a circular herd with a diameter of 120 miles. Such a herd would cover 11,310 square miles, or more than 7 million acres. If we suppose that people exaggerated the density of the buffalo, and there were only 10 per acre—a moderate density for a herd—this single herd would have numbered 70 million animals!

Some might say that this estimate is probably too high, because the herd would more likely have formed a broad, meandering, migrating line, rather than a circle. The impression remains the same—there were huge numbers of buffalo in the American West even as late as 1868, numbering in the tens of millions and probably 50 million or more. Ominously, that same year, the Kansas Pacific Railroad advertised a "Grand Railway Excursion and Buffalo Hunt."[12] Some say that many hunters believed the buffalo could never be brought to extinction because there were so many. This belief was common about all of America's living resources throughout the nineteenth century.

We tend to think that environmentalism is a social and political movement of the twentieth century; to the contrary, after the Civil War there were said to have been angry protests in every legislature over the slaughter of the buffalo. In 1871 the U.S. Biological Survey sent George Grinnell to survey the herds along the Platte River. He estimated that only 500,000 buffalo remained there and that at the then-current rate of killing, the animals would not last long. As late as the spring of 1883, a herd of an estimated 75,000 crossed the Yellowstone River near Miles City, Montana, but fewer than 5,000 reached the Canadian border.[12] By the end of that year—only 15 years after the Kansas Pacific train was delayed for eight hours by a huge herd of buffalo—only a thousand or so buffalo could be found, 256 in captivity and about 835 roaming the plains. A short time later, there were only 50 buffalo wild on the plains.

Today, more and more ranchers are finding ways to maintain bison, and the market for bison meat and other bison products is growing, along with an increasing interest in reestablishing bison herds for aesthetic, spiritual, and moral reasons.

The history of the bison once again raises the question of what we mean by "restore" a population. Even with our crude estimates of original abundances, the numbers would have varied from year to year. So we would have to "restore" bison not to a single number independent of the ability of its habitat to support the population, but to some range of abundances. How do we approach that problem and obtain an estimate of the range?

14.4 Improved Approaches to Wildlife Management

The U.S. Council on Environmental Quality (an office within the executive branch of the federal government), the World Wildlife Fund United States, the Ecological Society of America, the Smithsonian Institution, and the International Union for the Conservation of Nature (IUCN) have proposed four principles of wildlife conservation:

- A safety factor in terms of population size, to allow for limitations of knowledge and the imperfections of procedures. An interest in harvesting a population should not allow the population to be depleted to some theoretical minimum size.

- Concern with the entire community of organisms and all the renewable resources, so that policies developed for one species are not wasteful of other resources.

- Maintenance of the ecosystem of which the wildlife are a part, minimizing risk of irreversible change and long-term adverse effects as a result of use.

- Continual monitoring, analysis, and assessment. The application of science and the pursuit of knowledge about the wildlife of interest and its ecosystem should be maintained and the results made available to the public.

These principles broaden the scope of wildlife management from a narrow focus on a single species to inclusion of the ecological community and ecosystem. They call for a safety net in terms of population size, meaning that no population should be held at exactly the MSY level or reduced to some theoretical minimum abundance. These new principles provide a starting point for an improved approach to wildlife management.

Time Series and Historical Range of Variation

As the history of the buffalo illustrates, we would like to have an estimate of population over a number of years. This set of estimates is called a **time series** and could provide us with a measure of the **historical range of variation**—the known range of abundances of a population or species over some past time interval. Such records exist for few species. One is the American whooping crane (Figure 14.7), America's tallest bird, standing about 1.6 m (5 ft) tall. Because this species became so rare and because it migrated as a single flock, people began counting the total population in the late 1930s. At that time, they saw only 14 whooping cranes. They counted not only the total number but also the number born that year. The difference between these two numbers gives the number dying each year as well. And from this time series, we can estimate the probability of extinction.

The first estimate of the probability of extinction based on the historical range of variation, made in the early 1970s, was a surprise. Although the birds were few, the probability of extinction was less than one in a billion. How could this number be so low? Use of the historical range of variation carries with it the assumption that causes of variation in the future will be only those that occurred during the historical period. For the whooping cranes, one catastrophe—such as a long, unprecedented drought on the wintering grounds—could cause a population decline not observed in the past.

Even with this limitation, this method provides invaluable information. Unfortunately, at present, mathematical estimates of the probability of extinction have been done for just a handful of species. The good news is that the wild whooping cranes on the main flyway have continued to increase, and in 2006 numbered 214.[13] In addition,

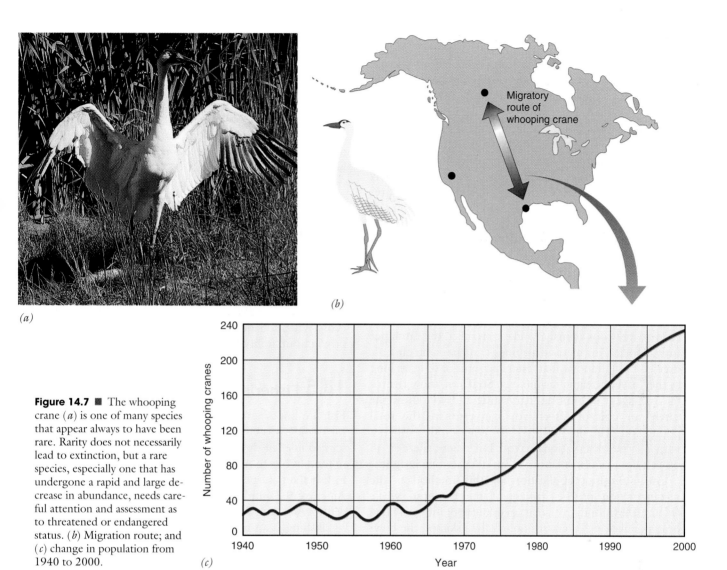

Figure 14.7 ■ The whooping crane (*a*) is one of many species that appear always to have been rare. Rarity does not necessarily lead to extinction, but a rare species, especially one that has undergone a rapid and large decrease in abundance, needs careful attention and assessment as to threatened or endangered status. (*b*) Migration route; and (*c*) change in population from 1940 to 2000.

with help from a variety of governmental and nongovernmental organizations, eastern populations of whooping cranes have been established through breeding programs. The numbers reached 77 nonmigrating birds in Florida and 36 birds migrating between Florida and Wisconsin, so the total number of wild whooping cranes in the world is now more than 300. Moreover, the U.S. Fish and Wildlife Service breeds whooping cranes in Patuxent, Maryland, and has a total of 49, and the International Crane Foundation does the same in Wisconsin, where 29 cranes live, so more than 400 whooping cranes are alive, including wild and captive animals.[14]

Age Structure as Useful Information

An additional key to successful wildlife management is monitoring of the population's age structure (see Chapter 4), which can provide many different kinds of information.

For example, the age structures of the catch of salmon from the Columbia River in Washington for two different periods, 1941–1943 and 1961–1963, were quite different. In the first period, most of the catch (60%) consisted of four-year-olds; three-year-olds and five-year-olds each made up about 15% of the population. Twenty years later, in 1961 and 1962, half the catch consisted of three-year-olds; the number of five-year-olds had declined to about 8%. The total catch had declined considerably. During the period 1941–1943, 1.9 million fish were caught. During the second period, 1961–1963, the total catch dropped to 849,000, just 49% of the total caught in the earlier period. The shift in catch toward younger ages, along with an overall decline in catch, suggests that the fish were being exploited to a point at which they were not reaching older ages. Such a shift in the age structure of a harvested population is an early sign of overexploitation and of a need to alter allowable catches.

Harvests as an Estimate of Numbers

Another method of estimating animal populations is to use the number harvested. Records of the number of buffalo killed were neither organized nor all that well kept, but they were sufficient to give us some idea of the number taken. In 1870, about 2 million buffalo were killed. In 1872, one company in Dodge City, Kansas, handled 200,000 hides. Estimates based on the sum of reports from such companies, together with guesses at how many animals were likely taken by small operators and not reported, suggest that about 1.5 million hides were shipped in 1872 and again in 1873.[8] In those years, buffalo hunting was the main economic activity in Kansas. The Indians were also killing large numbers of buffalo for their own use and for trade. Estimates range to 3.5 million buffalo killed per year, nationwide, during the 1870s.[8] The bison numbered at least in the low millions.

Another way in which harvest counts are used to estimate previous animal abundance is called the **catch per unit effort**. This method assumes that the same effort is exerted by all hunters/harvesters per unit of time, as long as they have the same technology. So if you know the total time spent in hunting/harvesting and you know the catch per unit of effort, you can estimate the total population. The method leads to rather crude estimates with a large observational error; but where there is no other source of information, it can offer unique insights.

An interesting application of this method is the reconstruction of the harvest of the bowhead whale (Figure 14.8*a*) and, from that, an estimate of the total bowhead population. Taken traditionally by Eskimos, the bowhead was the object of "Yankee," or American, whaling (Figure 14.8*b*) from 1820 until the beginning of World War I. (See A Closer Look 14.2 for a general discussion of marine mammals.) Every ship's voyage was recorded, so we know essentially 100% of all ships that went out to catch bowheads. In addition, on each ship a daily log was kept, with records including the number of whales caught, their size in terms of barrels of oil, the sea conditions, the visibility, and the ice conditions. Of these logbooks, 20% still exist, and their entries have been computerized. Using some crude statistical techniques, it was possible to estimate the abundance of the bowhead in 1820, which was 20,000 plus or minus 10,000. Indeed, it was possible to estimate the total catch of whales and the catch for each year—and therefore the entire history of the hunting of this species.

In summary, new approaches to wildlife conservation and management include (1) historical range of abundance; (2) estimation of the probability of extinction based on historical range of abundance; (3) use of age-structure information; and (4) better use of harvests as sources of information. These, along with an understanding of the ecosystem and landscape context for populations, are improving our ability to conserve wildlife.

14.5 Fisheries

Fish are important to our diets, providing about 16% of the world's protein, and they are especially important protein sources in developing countries.[5] Fish provide 6.6% of food in North America (where people are less interested in fish than are people in most other areas), 8% in Latin America, 9.7% in Western Europe, 21% in Africa, 22% in central Asia, and 28% in the Far East.

Fishing is an international trade, but a few countries dominate. Japan, China, Russia, Chile, and the United States are among the major fisheries nations. And commercial fisheries are concentrated in relatively few areas of

Figure 14.8 ■ (*a*) A bowhead whale; (*b*) a nineteenth-century whaling ship. (*a*) (*b*)

Conservation of Whales and Other Marine Mammals

Fossil records show that all marine mammals were originally inhabitants of the land. During the last 80 million years, several separate groups of mammals returned to the oceans and underwent adaptations to marine life. Each group of marine mammals shows a different degree of transition to ocean life. Understandably, the adaptation is greatest for those that began the transition longest ago. Some marine mammals—such as dolphins, porpoises, and great whales—complete their entire life cycle in the oceans and have organs and limbs that are highly adapted to life in the water. They cannot move on the land. Others, such as seals and sea lions, spend part of their time on shore.

Whales

Whales fit into two major categories: baleen and toothed (Figure 14.9a and b). The sperm whale is the only great whale that is toothed; the rest of the toothed group are smaller whales, dolphins, and porpoises. The other great whales, in the baleen group, have highly modified teeth that look like giant combs and act as water filters. Baleen whales feed by filtering ocean plankton.

Drawings of whales have been dated as early as 2200 B.C.[15] Eskimos used whales for food and clothing as long ago as 1500 B.C. In the ninth century, whaling by Norwegians was reported by travelers whose accounts were written down in the court of the English king Alfred.

The earliest whale hunters killed these huge mammals from the shore or from small boats near shore, but gradually whale hunters ventured farther out. In the eleventh and twelfth centuries, Basques hunted the Atlantic right whale from open boats in the Bay of Biscay, off the western coast of France. Whales were brought ashore for processing, and the boats returned to land once the search for a whale was finished.

Eventually, whaling became pelagic: Whalers took to the open ocean and searched for whales from ships that remained at sea for long periods. The whales were brought on board and processed there. This was made possible by the invention of furnaces and boilers for extracting whale oil at sea. Thus, pelagic whaling was a product of the Industrial Revolution. With this invention, whaling grew as an industry. American fleets developed in the eighteenth century in New England; by the nineteenth century, the United States dominated the industry, providing most of the ships and even more of the crews for whaling.[16, 17]

Whales provided many nineteenth-century products. Whale oil was used for cooking, lubrication, and lamps. Whales provided the main ingredients for the base of perfumes. The elongated teeth (whalebone, or baleen) that enable baleen whales to strain the ocean waters for food are flexible and springy and were used for corset stays and other products before the invention of inexpensive steel springs.

Although the nineteenth-century whaling ships are more famous, made popular by such novels as *Moby Dick*, more whales were killed in the twentieth century than in the nineteenth. The resulting worldwide decline in most species of whales made this a global environmental issue.

Conservation of whales has been a concern among conservationists for many years. Attempts to control whaling began with the League of Nations in 1924. The first agreement, the Convention for the Regulation of Whaling, was signed by 21 countries in 1931. In 1946, a conference in Washington, D.C., initiated the International Whaling Commission (IWC), and in 1982 the IWC established a moratorium on commercial whaling.

(a)

(b)

Figure 14.9 ■ (*a*) A sperm whale; (*b*) a blue whale.

Currently, 12 of approximately 80 species of whales are protected.[32]

The IWC has played a major role in reducing (almost eliminating) the commercial harvest of whales. Since the formation of the IWC, no species has become extinct, the total take of whales has decreased, and harvesting of species considered endangered has ceased. Endangered species protected from hunting have had a mixed history (see Table 14.1). Blue whales appear to have recovered somewhat but remain rare and endangered. Gray whales are now relatively abundant, numbering about 26,000.[33] However, global climate change, pollution, and ozone depletion now pose greater risks to whale populations than does whaling.[34]

The establishment of the IWC was a major landmark in wildlife conservation. It was one of the first major attempts by a group of nations to agree on a reasonable harvest of a biological resource. The annual meeting of the IWC has become a forum for discussing international conservation, working out basic concepts of maximum and optimum sustainable yields, and formulating a scientific basis for commercial harvesting. The IWC demonstrates that even an informal commission whose decisions are accepted voluntarily by nations can function as a powerful force for conservation.

In the past, each marine mammal population was treated as if it were isolated, had a constant supply of food, and was subject only to the effects of human harvesting. That is, it was assumed that its growth followed the logistic curve. We now realize that management policies for marine mammals must be expanded to include ecosystem concepts and the understanding that populations interact in complex ways.

The goal of marine mammal management is to prevent extinction and maintain large population sizes rather than to maximize production. For this reason, the Marine Mammal Protection Act, enacted by the United States in 1972, has as its goal an optimum sustainable population (OSP) rather than a maximum or optimum sustainable yield. An OSP means the largest population that can be sustained indefinitely without deleterious effects on the ability of the population or its ecosystem to continue to support that same level.

Some of the great whales remain rare. The largest, the blue whale, is one that continues to be rare, numbering somewhere between 400 and 1,000.[18]

Dolphins and Other Small Cetaceans

Among the many species of small "whales," or cetaceans, are dolphins and porpoises, more than 40 species of which have been hunted commercially or have been killed inadvertently by other fishing efforts.[11] A classic case is the inadvertent catch of the spinner, spotted, and common dolphins of the eastern Pacific. Because these carnivorous, fish-eating mammals often feed with yellowfin tuna, a major commercial fish, more than 7 million dolphins have been netted and killed inadvertently in the past 40 years.[19]

The U.S. Marine Mammal Commission and commercial fishermen have cooperated in seeking methods to reduce dolphin mortality. Research into dolphin behavior helped in the design of new netting procedures that trapped far fewer dolphins. The attempt to reduce dolphin mortality illustrates cooperation among fishermen, conservationists, and government agencies and indicates the role of scientific research in the management of renewable resources.

Table 14.1 • Estimates of the Number of Whales. Whales are difficult to count, and the wide range of estimates indicates this difficulty. Of the baleen whales, the most numerous are the smallest—the minke and the pilot. The only toothed great whale, the sperm whale, is thought to be relatively numerous, but actual counts of this species have been made only in comparatively small areas of the ocean.

Species	Range of Estimates	
	Minimum	Maximum
Blue	400	1,400
Bowhead	6,900	9,200
Fin	27,700	82,000
Gray	21,900	32,400
Humpback	5,900	16,800
Minke	510,000	1,140,000
Pilot	440,000	1,370,000
Sperm	200,000	1,500,000

Source: International Whaling Commission, August 29, 2006.
http://www.iwcoffice.org/conservation/estimate.htm
Except that the estimate for sperm whales is from U.S. NOAA
http://www.nmfs.noaa.gov/pr/species/mammals/cetaceans/spermwhale.htm

Figure 14.10 ■ The world's major fisheries. Red areas are major fisheries; the darker the red, the greater the harvest and the more important the fishery. Most major fisheries occur in areas of ocean upwellings—locations where currents rise, bringing nutrient-rich waters from the depths of the ocean. Upwellings tend to occur near continents.

the world's oceans (Figure 14.10). Continental shelves, which make up only 10% of the oceans, provide more than 90% of the fishery harvest. Fish are abundant where their food is abundant and, ultimately, where there is high production of algae at the base of the food chain. Algae are most abundant in areas with relatively high concentrations of the chemical elements necessary for life, particularly nitrogen and phosphorus. These areas occur most commonly along the continental shelf, particularly in regions of wind-induced upwellings and sometimes quite close to shore.

The world's total fish harvest has increased greatly since the middle of the twentieth century. The total harvest was 35 million metric tons (MT) in 1960. It more than doubled in just 20 years (an annual increase of about 3.6%) to 72 million MT in 1980 and since then has increased to 132,000 MT and seems to be leveling off.[20] The total global fish harvest continues to climb because of increases in the number of boats, improvements in technology, and especially increases in aquaculture production, which also more than doubled between 1992 and 2001, from about 15 million MT to more than 37 million MT. Aquaculture presently provides more than one-fifth of all fish harvested, up from 15% in 1992.[21]

Scientists estimate that there are 27,000 species of fish and shellfish in the oceans. People catch many of these species for food, but only a few kinds provide most of the food—anchovies, herrings, and sardines provide almost 20% (Table 14.2).

Table 14.2 • World Fisheries Catch			
Kind	Harvest (Millions of metric tons)	Percent	Accumulated Percentage
Herring, sardines, and anchovies	25	19.23%	19.23%
Carp and relatives	15	11.54%	30.77%
Cod, hake, and haddock	8.6	6.62%	37.38%
Tuna and their relatives	6	4.62%	42.00%
Oysters	4.2	3.23%	45.23%
Shrimp	4	3.08%	48.31%
Squid and octopus	3.7	2.85%	51.15%
Other mollusks	3.7	2.85%	54.00%
Clams and relatives	3	2.31%	56.31%
Tilapia	2.3	1.77%	58.08%
Scallops	1.8	1.38%	59.46%
Mussels and relatives	1.6	1.23%	60.69%
Subtotal	78.9	60.69%	
TOTAL ALL SPECIES	130	100%	

Source: National Oceanic & Atmospheric Administration World Fisheries.

(a)

(b)

(c)

Figure 14.11 ■ Some modern commercial fishing methods. (a) Trawling using large nets; (b) long-lines have caught a sword-fish; (c) workers on a factory ship.

Although the total marine fisheries catch has increased during the past half-century, the effort required to catch a fish has increased as well. More fishing boats with better and better gear sail the oceans (Figure 14.11). That is why the total catch can increase while the total population of a fish species declines.

The Decline of Fish Populations

Evidence that the fish populations were declining came from the catch per unit effort. A unit of effort varies with the kind of fish sought. For marine fish caught with lines and hooks, the catch rate generally fell from 6–12 fish caught per 100 hooks—the success typical of a previously unexploited fish population—to 0.5–2.0 fish per 100 hooks just 10 years later (Figure 14.11b). These observations suggest that fishing depletes fish quickly—about an 80% decline in 15 years (Figure 14.12). Many of the fish that people eat are predators, and on fishing grounds the biomass of large predatory fish appears to be only about

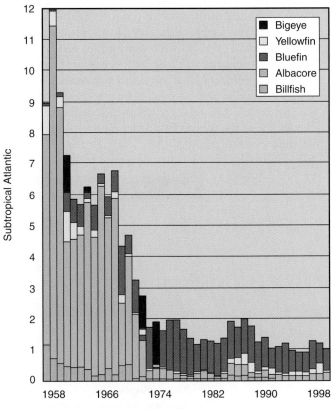

Figure 14.12 ■ Tuna catch decline. The catch per unit of effort—represented here as the number of fish caught per 100 hooks—for tuna and their relatives in the subtropical Atlantic Ocean. The vertical axis shows the number of fish caught per 100 hooks. The catch per unit of effort was 12 in 1958, when heavy modern industrial fishing for tuna began, and had declined rapidly by 1974, to about 2. This pattern occurred worldwide in all the major fishing grounds for these species. [*Source:* Ransom A. Meyers and Boris Worm, "Rapid Worldwide Depletion of Predatory Fish Communities," *Nature* (May 15, 2003).]

Table 14.3 • Problems of Some Major Fisheries

Anchovy: Reached peak in 1970 (10 million metric tons), then declined.

Atlantic herring: Exploitation so high that recruitment was decreased.

Arcto-Norwegian cod: High fishing level followed by four years of poor recruitment.

Downs' stock of herring in the North Sea: Managers failed to grasp stock and recruitment problems.

North Atlantic haddock: Catch averaged 50,000 metric tons for many years; increased to 155,000 in 1965 and 127,000 in 1966, then fell to 12,000 in 1971–1974. In 1973, the International Commission for the Northwest Atlantic Fisheries (ICNAF) established a quota of 6,000 metric tons. Apparently, haddock could sustain a 50,000-metric-ton catch; but when this was tripled, the population declined to a point where only a smaller catch could be sustained.

Atlantic menhaden: Catch peaked at 712,000 metric tons in 1956 but declined to 161,400 by 1969. Fisheries experts believe the drop was due to overfishing.

Salmon: Loss has occurred worldwide wherever salmon and their relatives once used fresh waters to spawn and rear. This has been the result of construction of dams on rivers, channelization of rivers, pollution, overharvesting, and habitat alteration of many kinds.

Pacific sardine: Declined catastrophically from the 1950s through the 1970s.

Source: D. Cushing, *Fisheries Resources of the Sea and Their Management* (London: Oxford University Press, 1975).

10% of preindustrial levels. These changes indicate that the biomass of most major commercial fish has declined greatly, so that we are mining, not sustaining, these living resources.

Species suffering these declines include codfish, flatfishes, tuna, swordfish, sharks, skates, and rays (Table 14.3).[7,8] The North Atlantic, whose George's Bank and Grand Banks have for centuries provided some of the world's largest fish harvests, is suffering. The Atlantic cod-fish catch was 3.7 million MT in 1957, peaked at 7.1 million MT in 1974, then declined to 4.3 million MT in 2000, climbing slightly to 4.7 in 2001.[9] European scientists recently called for a total ban on cod fishing in the North Atlantic, and the European Union came close to accepting this, stopping just short of a total ban and instead establishing a 65% cut in the allowed catch for North Sea cod for 2004 and 2005.[10]

Scallops in the western Pacific show a typical harvest pattern, starting very low in 1964 at 200 MT, increasing rapidly to 5,887 MT in 1975, declining to 1,489 in 1974, increasing to about 7,670 MT in 1993, and then declining to 2,964 in 2002.[21] Catch of tuna and their relatives peaked in the early 1990s at about 730,000 MT and fell to 680,000 MT in 2000, a decline of 14% (Figure 14.12).

Chesapeake Bay, America's largest estuary, was another of the world's great fisheries, famous for oysters and crabs and as the breeding and spawning ground for bluefish, sea bass, and many other commercially valuable species (Figure 14.13). The bay, 200 miles long and 30 miles wide, drains an area of more than 165,000 km^2 from New York State to Maryland and is fed by 48 large rivers and 100 small ones.

Food webs are also complex, adding to the difficulty of managing the Chesapeake Bay fisheries. Typical of marine food webs, the food chain of the bluefish that spawns and breeds in Chesapeake Bay shows links to a number of other species, each requiring its own habitat within the space and depending on processes that have a variety of scales of space and time (Figure 14.14). In addition to this complexity, Chesapeake Bay is influenced by many factors on the surrounding lands in its watershed—runoff from farms, including chicken and turkey farms, that is highly polluted with fertilizers and pesticides; introductions of exotic species; and direct alteration of habitats from fishing and the development of shoreline homes. Add to these the varied salinity of the waters of the bay—freshwater inlets from rivers and streams, seawater from the Atlantic, and brackish water resulting from the mixture of these.

Just determining which of these factors, if any, are responsible for a major change in the abundance of any fish species is difficult enough, let alone finding a solution that is economically feasible and maintains traditional levels of fisheries employment in the bay. The Chesapeake Bay's fisheries resources are at the limit of what environmental sciences can deal with at this time. Scientific theory remains inadequate, as do observations, especially of fish abundance.

Ironically, this crisis has arisen for one of the living resources most subjected to science-based management. How could this have happened? First, management has been based, by and large, on the logistic growth curve, whose problems we have discussed. Second, fisheries are an open resource, subject to the problems of "the tragedy of the commons," a phrase coined by Garrett Hardin. In an open resource, often in international waters, the numbers of fish that may be harvested can be limited only by international treaties, which are not tightly binding. Open resources offer ample opportunity for unregulated or illegal harvest, or harvest contrary to agreements.

Exploitation of a new fishery usually occurs before scientific assessment, so the fish are depleted by the time any reliable information about them is available. Furthermore, some fishing gear is destructive to the habitat. Ground-trawling equipment destroys the ocean floor, ruining habitat for both the target fish and its food. Long-line fishing kills sea turtles and other nontarget surface animals. Large tuna nets have killed many dolphins that are hunting the tuna.

In addition to highlighting the need for better management methods, the harvest of large predators raises

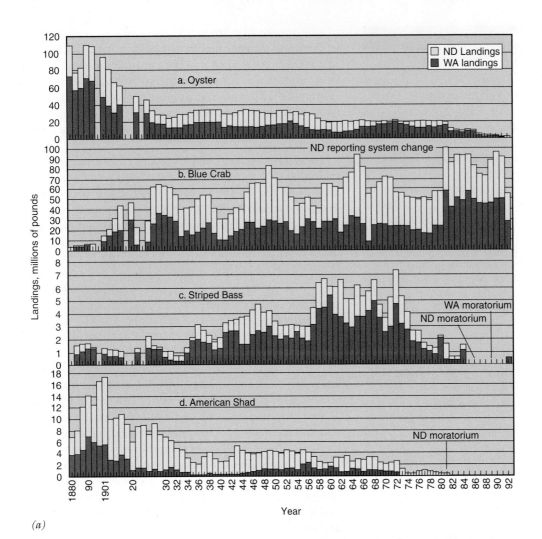

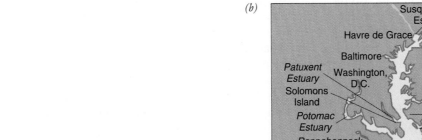

Figure 14.13 ■ (*a*) Fish catch in the Chesapeake Bay. Oysters have declined dramatically. [*Source:* The Chesapeake Bay Foundation.] (*b*) Map of Chesapeake Bay Estuary. [*Source:* U.S. Geological Survey, "The Chesapeake Bay: Geologic Product of Rising Sea Level," 1998.]

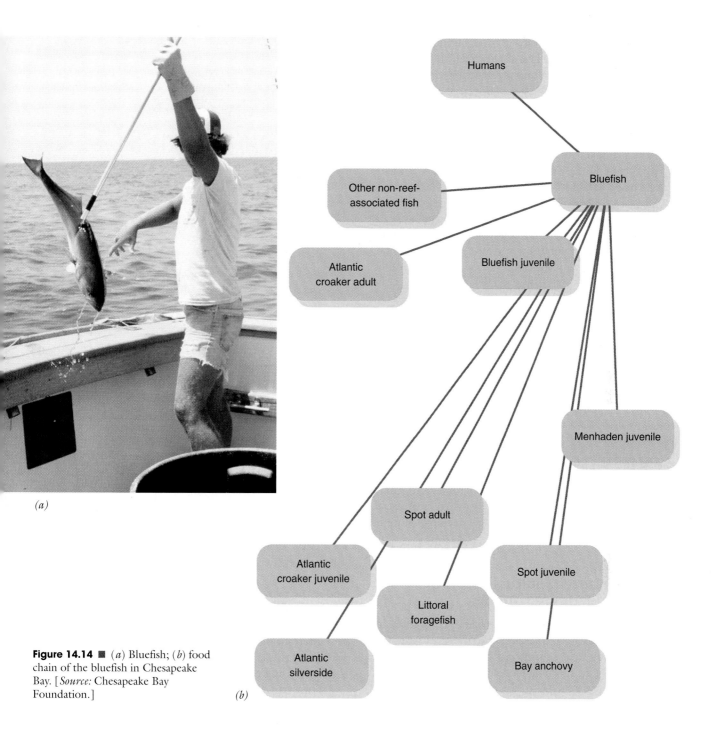

Figure 14.14 ■ (*a*) Bluefish; (*b*) food chain of the bluefish in Chesapeake Bay. [*Source:* Chesapeake Bay Foundation.]

(*a*)

(*b*)

questions about ocean ecological communities, especially whether these large predators play an important role in controlling the abundance of other species.

Human beings began as hunter-gatherers, and some hunter-gatherer cultures still exist. Wildlife on the land used to be a major source of food for hunter-gatherers. It is now a minor source of food for people of developed nations, although still a major food source for some indigenous peoples, such as the Eskimo. In contrast, even developed nations are still primarily hunter-gatherers in the harvesting of fish (see the discussion of aquaculture in Chapter 11).

Can Fishing Ever Be Sustainable?

Suppose you went into fishing as a business and expected a reasonable increase in that business in the first 20 years. The world's ocean fish catch increased from 39 million MT in 1964 to 68 million MT in 2003, a total increase of 77%.[21] From a business point of view, even assuming all fish caught are sold, that is not a rapid improvement in sales. But it is a heavy burden on a living resource. There is a general lesson to be learned here: Few wild biological resources can sustain a harvest at a level that meets even low requirements for a growing business. Thus, most wild biological resources just aren't a good business over the long run. We learned this lesson also from the demise of the bison, discussed earlier, and it is true for whales as well (see A Closer Look 14.2). There have been a few exceptions, such as the several hundred years of fur trading by the Hudson's Bay Company in northern Canada. However, past experience suggests that economically beneficial sustainability is unlikely for most wild populations.

With that in mind, we can turn to farming fish—aquaculture, discussed in Chapters 11 and 12. This has been an important source of food in China for centuries and is an increasingly important food source worldwide. But aquaculture can create its own environmental problems. One of the best-known environmental issues involves Atlantic salmon, the major species of salmon grown as a crop. In 2000, Atlantic salmon were put on the Endangered Species list. One of the explanations proposed for the problems of this species is extensive mariculture of Atlantic salmon in such places as the coast of Maine. Observers have identified two concerns. First, water pollution from salmon excrement and excess food provided for the salmon damage the habitat of wild salmon. Second, breeding between native salmon and farmed salmon not native to a locale may lead to genetic strains that are less fit for a particular stream.

In summary, fish are an important food and world harvests of fish are large, but the fish populations on which the harvests depend are generally declining, easily exploited, and difficult to restore. We desperately need new approaches to forecasting acceptable harvests and establishing workable international agreements to limit catch. This is a major environmental challenge, needing solutions within the next decade.

14.6 The Current Status of Endangered Species

We have managed to convert large populations of wildlife, including fish, to endangered species. With the expanding public interest in rare and endangered species, especially large mammals and birds, it is time for us to turn our attention to these species. First some facts. The number of species of animals listed as threatened or endangered increased from about 1,700 in 1988 to 3,800 in 1996 and 5,188 in 2004 (Table 14.4).[2] The International Union for the Conservation of Nature (IUCN) maintains a list of threatened and endangered animals in a publication known as the *Red List*. The IUCN *Red List of Threatened Species* reports that about 20% of all known species of mammals are at risk of extinction, as well as 12% of known birds, 4% of known reptiles, 31% of amphibians, and 3% of fish, primarily freshwater fish.[23]

The IUCN *Red List of Threatened Species* estimates that 33,798 species of vascular plants (the familiar kind of plants—trees, grasses, shrubs, flowering herbs), or 12.5% of those known, have recently become extinct or endangered.[3] IUCN lists more than 8,000, or approximately 3%, of plants that are threatened (see Table 14.4).[4]

Table 14.4 • Number of Threatened Species

Life-Form	Number Threatened	Percent of Species Known
Vertebrates	5,188	9
Mammals	1,101	20
Birds	1,213	12
Reptiles	304	4
Amphibians	1,770	31
Fish	800	3
Invertebrates	1,992	0.17
Insects	559	0.06
Molluscs	974	1
Crustaceans	429	1
Others	30	0.02
Plants	8,321	2.89
Mosses	80	0.5
Ferns and "Allies"	140	1
Gymnosperms	305	31
Dicots	7,025	4
Monocots	771	1
Total Animals and Plants	**31,002**	**2%**

Source: IUCN Red List
www.iucnredlist.org/info/tables/table1 (2004)

What does it mean to call a species "threatened" or "endangered"? The terms can have strictly biological meanings, or they can have legal meanings. The words *endangered* and *threatened* are defined in the U.S. Endangered Species Act of 1973. The Act says, "The term *endangered species* means any species which is in danger of extinction throughout all or a significant portion of its range other than a species of the Class Insecta determined by the Secretary to constitute a pest whose protection under the provisions of this Act would present an overwhelming and overriding risk to man." In other words, if certain insect species are pests, we want to be rid of them. It is interesting that insect pests can be excluded from protection by this legal definition, but there is no mention of disease-causing bacteria or other microorganisms.

"The term *threatened species*," according to the Act, "means any species which is likely to become an endangered species within the foreseeable future throughout all or a significant portion of its range."

14.7 How a Species Becomes Endangered and Extinct

Extinction is the rule of nature (see the discussion of biological evolution in Chapter 7). **Local extinction** occurs when a species disappears from a part of its range but persists elsewhere. **Global extinction** means a species can no longer be found anywhere. Although extinction is the ultimate fate of all species, the rate of extinctions has varied greatly over geologic time and has accelerated since the Industrial Revolution. From 580 million years ago until the beginning of the Industrial Revolution, about one species per year, on average, became extinct. Over much of the history of life on Earth, the rate of evolution of new species equaled or slightly exceeded the rate of extinction. The average longevity of a species has been about 10 million years.[23] However, as discussed in Chapter 7, the fossil record suggests that there have been periods of catastrophic losses of species and other periods of rapid evolution of new species (see Figures 7.9 and 14.15a–c), which some refer to as "punctuated extinctions." About 250 million years ago, a mass extinction occurred in which approximately 53% of marine animal species disappeared; about 65 million years ago, most of the dinosaurs became extinct. Interspersed with the episodes of mass extinctions, there seem to have

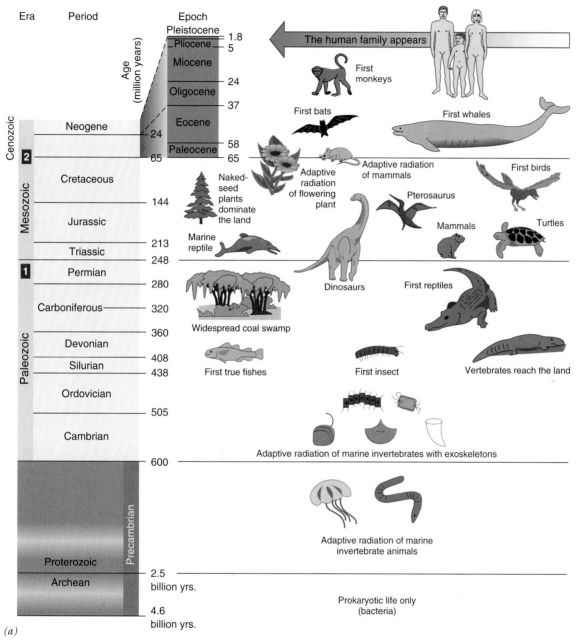

(a)

Figure 14.15 ■ (a) Brief diagrammatic history of evolution and extinction of life on Earth. There have been periods of rapid evolution of new species and episodes of catastrophic losses of species. Two major catastrophes were the Permian loss, which included 52% of marine animals, as well as land plants and animals, and the Cretaceous loss of dinosaurs. (b) Graph of the number of families of marine animals in the fossil records, showing long periods of overall increase in the number of families punctuated by brief periods of major declines. (c) Extinct vertebrate species and subspecies, 1760–1979. The number of species becoming extinct increases rapidly after 1860. Note that most of the increase is due to the extinction of birds. (Sources: [a] D. M. Raup, "Diversity Crisis in the Geological Past," in E. O. Wilson, ed., *Biodiversity* [Washington, D.C.: National Academy Press, 1988], p. 53; derived from S. M. Stanley, *Earth and Life through Time* [New York: W. H. Freeman, 1986]. Reprinted with permission. [b] D. M. Raup and J. J. Sepkoski, Jr., "Mass Extinctions in the Marine Fossil Record," *Science* 215 [1982]:1501–1502. [c] Council on Environmental Quality; additional data from B. Groombridge, England: IUCN, 1993].)

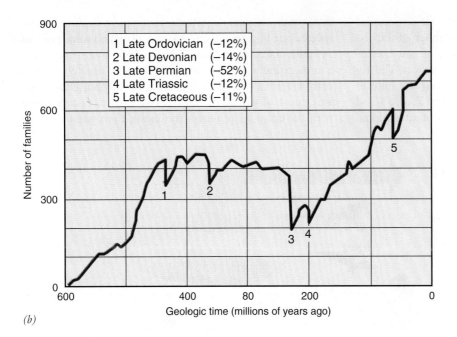

(b)

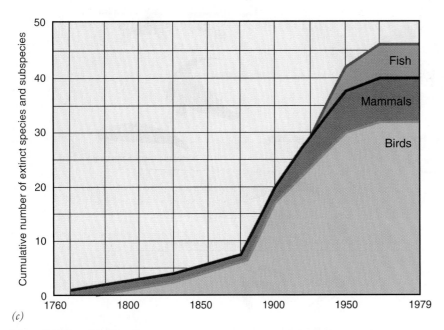

(c)

Figure 14.15 ■ *(continued)*

been periods of hundreds of thousands of years with comparatively low rates of extinction.

An intriguing example of punctuated extinctions occurred about 10,000 years ago, at the end of the last great continental glaciation. At that time, massive extinctions of large birds and mammals occurred: 33 genera of large mammals—those weighing 50 kg (110 lb) or more—became extinct, whereas only 13 genera had become extinct in the preceding 1 or 2 million years (Figure 14.16). Smaller mammals were not so affected,

nor were marine mammals. As early as 1876, Alfred Wallace, an English biological geographer, noted that "we live in a zoologically impoverished world, from which all of the hugest, and fiercest, and strangest forms have recently disappeared." It has been suggested that these sudden extinctions coincided with the arrival, on different continents at different times, of Stone Age people and therefore may have been caused by hunting.[24] Causes of extinction are summarized in A Closer Look 14.3.

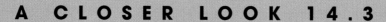

Causes of Extinction

Causes of extinction are usually grouped into four risk categories: population risk, environmental risk, natural catastrophe, and genetic risk. *Risk* here means the chance that a species or population will become extinct owing to one of these causes.

Population Risk

Random variations in population rates (in birth rates and death rates) can cause a species in low abundance to become extinct. This is termed *population risk*. For example, blue whales swim over vast areas of ocean. Because whaling once reduced their total population to only several hundred individuals, the success of individual blue whales in finding mates probably varied from year to year. If in one year most whales were unsuccessful in finding mates, then births could be dangerously low. Such random variation in populations, typical among many species, can occur without any change in the environment. It is a risk especially to species that consist of only a single population in one habitat. Mathematical models of population growth can help calculate the population risk and determine the minimum viable population size.[24]

Environmental Risk

Population size can be affected by changes in the environment that occur from day to day, month to month, and year to year, even though the changes are not severe enough to be considered environmental catastrophes. Environmental risk involves variation in the physical or biological environment, including variations in predator, prey, symbiotic, or competitor species. In some cases, species are so rare and isolated that such normal variations can lead to their extinction.

For example, Paul and Anne Ehrlich described the local extinction of a population of butterflies in the Colorado mountains.[35] These butterflies lay their eggs in the unopened buds of a single species of lupine (a member of the legume family), and the hatched caterpillars feed on the flowers. One year, however, a very late snow and freeze killed all the lupine buds, leaving the caterpillars without food and causing local extinction of the butterflies. Had this been the only population of that butterfly, the entire species would have become extinct.

Natural Catastrophe

A sudden change in the environment that is not the result of human action is a natural catastrophe. Fires, major storms, earthquakes, and floods are natural catastrophes on land; changes in currents and upwellings are ocean catastrophes. For ex-

ample, the explosion of a volcano on the island of Krakatoa in Indonesia in 1883 caused one of recent history's worst natural catastrophes. Most of the island was blown to bits, bringing about local extinction of most life-forms there.

Genetic Risk

Detrimental change in genetic characteristics, not caused by external environmental changes, is called *genetic risk*.[36] Genetic changes can occur in small populations from reduced genetic variation, genetic drift, and mutation (see Chapter 7). In a small population, only some of the possible inherited characteristics will be found. The species is vulnerable to extinction because it lacks variety or because a mutation can become fixed in the population.

Consider the last 20 condors in the wild in California. It stands to reason that this small number was likely to have less genetic variability than the much larger population that existed several centuries ago. This increased the condors' vulnerability. Suppose that the last 20 condors, by chance, had inherited characteristics that made them less able to withstand lack of water. If left in the wild, these condors would have been more vulnerable to extinction than a larger, more genetically variable population.

Figure 14.16 ■ Artist's restoration (drawing) of an extinct saber-toothed cat with prey. The cat is an example of one of the many large mammals that became extinct about 10,000 years ago.

14.8 How People Cause Extinctions and Affect Biological Diversity

People have become an important cause of extinction and an important factor in causing species to become threatened and endangered. Among the ways we cause extinction are:

- By intentional hunting or harvesting (for commercial purposes, for sport, or to control a species that is considered a pest).
- By disrupting or eliminating habitats.
- By introducing exotic species, including new parasites, predators, or competitors of a native species.
- By creating pollution.

People have caused extinctions over a long time, not just in recent years. The earliest people probably caused extinctions through hunting. This practice continues, especially for specific animal products considered valuable, such as elephant ivory and rhinoceros horns. When people learned to use fire, they began to change habitats over large areas. The development of agriculture and the rise of civilization led to rapid deforestation and other habitat changes. Later, as people explored new areas, the introduction of exotic species became a greater cause of extinction (see Chapter 8), especially after Columbus's voyage to the New World, Magellan's circumnavigation of the globe, and the resulting spread of European civilization and technology. The introduction of thousands of novel chemicals into the environment made pollution an increasing cause of extinction in the twentieth century, and pollution control has proved to be a successful way to help species.

The IUCN estimates that 75% of the extinctions of birds and mammals since 1600 have been caused by human beings. Hunting is estimated to have caused 42% of the extinctions of birds and 33% of the extinctions of mammals. The current extinction rate among most groups of mammals is estimated to be 1,000 times greater than the extinction rate at the end of the Pleistocene epoch.[22]

The Good News: Species Whose Status Has Improved

There is some good news about endangered species as a result of human activities. A number of previously endangered species, such as the Aleutian goose of the opening case study, have recovered. Other recovered species include the following:

- The elephant seal, which had dwindled to about a dozen animals around 1900 and now numbers in the hundreds of thousands.
- The sea otter, reduced in the nineteenth century to several hundred and now numbering approximately 10,000.

- Many species of birds endangered because the insecticide DDT caused thinning of eggshells and failure of reproduction. With the elimination of DDT in the United States, many bird species recovered, including the bald eagle, brown pelican, white pelican, osprey, and peregrine falcon.
- The blue whale, thought to have been reduced to about 400 when whaling was still actively pursued by a number of nations. Today, 400 blue whales are sighted annually in the Santa Barbara channel along the California coast, a sizable fraction of the total population.
- The gray whale, which was hunted to near-extinction but has recovered and is abundant along the California coast and in its annual migration to Alaska.

Since the U.S. Endangered Species Act became law in 1973, 40 species have recovered sufficiently to be either reclassified from "endangered" to "threatened" or removed completely from the list. In addition, the U.S. Fish and Wildlife Service—which, along with the National Marine Fisheries Services, administers the Endangered Species Act—lists 33 species that have the potential for reclassification to an improved status.[25]

Can a Species Be Too Abundant? If So, What Do We Do?

Sometimes we succeed too well in increasing the numbers of a species. Case in point: All marine mammals are protected in the United States by the Federal Marine Mammal Protection Act of 1972, which has led to improvement in the status of many marine mammals. Sea lions now number more than 50,000 and have become so abundant as to be local problems.[26] For example, in San Francisco Harbor and in Santa Barbara Harbor, sea lions haul out and sun themselves on boats and pollute the water with their excrement near shore. In one case, so many sea lions hauled out on a sailboat in Santa Barbara Harbor that they sank the boat, and some of the animals were trapped and drowned. (For more information on the history and status of marine mammals, see A Closer Look 14.2.)

Mountain lions have become locally overabundant. In the 1990s, California voters passed an initiative that protected the endangered mountain lion but contained no provisions for management of the lion should it become overabundant, except in cases where it threatened human life and property. Few people thought the mountain lion could ever recover enough to become a problem, but in several cases in recent years mountain lions have attacked and even killed people. Current estimates suggest there may be as many as 4,000 to 6,000 in California.[27] These attacks become more frequent as the mountain lion population grows and as the human population grows and people build houses in what was mountain lion habitat.

(a)

(b)

Figure 14.17 ■ (*a*) A Kirtland's warbler and (*b*) its jack-pine habitat.

14.9 The Kirtland's Warbler and Environmental Change

Many endangered species are adapted to natural environmental change and require it. When human actions eliminate that change, a species can become threatened with extinction. This happened with the Kirtland's warbler, which nests in jack-pine forests in Michigan (Figure 14.17). In 1951 the Kirtland's warbler became the first songbird in the United States to be subject to a complete census, and about 400 nesting males were found. Concern about the species grew in the 1960s and increased when only 201 nesting males were found in the third census, in 1971.[28] Conservationists and scientists tried to understand what was causing the decline, which threatened the species with extinction.

Kirtland's warblers are known to nest only in jack-pine woodlands that are between 6 and 21 years old. At these ages, the trees, 5–20 feet tall, retain dead branches at ground level. Jack pine is a "fire species," which persists only where there are periodic forest fires. Cones of the jack pine open only after they have been heated by fire. The trees are intolerant of shade, able to grow only when their leaves can reach into full sunlight; so even if seeds were to germinate under mature trees, the seedlings could not grow in the shade and would die. Jack pine produces an abundance of dead branches, which some people view as an evolutionary adaptation to promote fires, essential to the survival of the species.

Kirtland's warblers thus require change at rather short intervals—forest fires approximately every 20 to 30 years, which was about the frequency of fires in jack-pine woods in presettlement times.[19] At the time of the first European settlement of North America, jack pine may have covered a large area in what is now Michigan. Even as recently as the

1950s, the pine was estimated to cover nearly 500,000 acres in that state. Small, and poorly formed, jack pine was considered a trash species by commercial loggers and was left alone; but many large fires followed the logging operations when large amounts of slash—branches, twigs, and other economically undesirable parts of the trees—were left in the woods. Elsewhere, fires were set to clear jack-pine areas and promote the growth of blueberries. Some experts think that the population of Kirtland's warblers peaked in the late nineteenth century as a result of these fires. After 1927, fire suppression became the practice, and people were encouraged to replace jack pine with economically more useful species. One result was that the areas conducive to the nesting of the warbler shrank.[29]

Although it may seem obvious today that the warbler requires forest fires, this was not always understood. In 1926, one expert wrote that "fire might be the worst enemy of the bird."[21] Only with the introduction of controlled burning after vigorous advocacy by conservationists and ornithologists was habitat for the warbler maintained. The Kirtland's Warbler Recovery Plan, published by the Department of the Interior and the Fish and Wildlife Service in 1976 and updated in 1985, calls for the creation of 38,000 acres of new habitat for the warbler for which "prescribed fire will be the primary tool used to regenerate nonmerchantable jack-pine stands on poor sites."[29] Today there are an estimated 1,000 singing males, suggesting a total population of 2,000 or more Kirtland's warblers.[30]

14.10 Ecological Islands and Endangered Species

The history of the Kirtland's warbler illustrates that a species may inhabit "ecological islands," which the isolated jack-pine stands of the right age range are for that bird.

(a)

(b)

(c)

Figure 14.18 ■ Ecological islands: (*a*) Central Park in New York City; (*b*) a mountaintop in Arizona where there are bighorn sheep; (*c*) an African wildlife park.

Recall from our discussion in Chapter 8 that an ecological island is an area that is biologically isolated, so a species living there cannot mix (or only rarely mixes) with any other population of the same species (Figure 14.18). Mountaintops and isolated ponds are ecological islands. Real geographic islands are also ecological islands. Insights gained from studies of the biogeography of islands have important implications for the conservation of endangered species and for the design of parks and preserves for biological conservation.

Almost every park is a biological island for some species. A small city park between buildings may be an island for trees and squirrels. At the other extreme, even a large national park is an ecological island. For example, the Masai Mara Game Reserve in the Serengeti Plain, which stretches from Tanzania to Kenya in East Africa, and other great wildlife parks of eastern and southern Africa are becoming islands of natural landscape surrounded by human settlements. Lions and other great cats exist in these parks as isolated populations, no longer able to roam completely freely and to mix over large areas. Other examples are islands of uncut forests left by logging operations and oceanic islands, where intense fishing has isolated parts of fish populations.

How large must an ecological island be to ensure the survival of a species? The size varies with the species but can be estimated. Some islands that seem large to us are too small for species we wish to preserve. For example, a preserve was set aside in India in an attempt to reintroduce the Indian lion into an area where it had been eliminated by hunting and changing patterns of land use. In 1957, a male and two females were introduced into a 95 km^2 (36 mi^2) preserve in the Chakia forest, known as the Chandraprabha Sanctuary. The introduction was carried out carefully and the population was counted annually. There were four lions in 1958, five in 1960, seven in 1962, and eleven in 1965, after which they disappeared and were never seen again.

Why did they go? Although 95 km^2 seems large to us, male Indian lions have territories of 130 km^2 (50 mi^2). Within such a territory, females and young also live. A population that could persist for a long time would need a number of such territories, so an adequate preserve would require 640 to 1,300 km^2 (247 to 500 mi^2). Various other reasons were suggested for the disappearance of the lions, including poisoning and shooting by villagers; but regardless of the immediate cause, a much larger area than was set aside was required for long-term persistence of the lions.

14.11 Using Spatial Relationships to Conserve Endangered Species

The red-cockaded woodpecker (Figure 14.19) is an endangered species of the American Southeast, numbering approximately 15,000.[25] The woodpecker makes its nests

(a)

(b)

Figure 14.19 ■ (*a*) Endangered red-cockaded woodpecker, and (*b*) the bark beetle, food for the woodpecker.

in old dead or dying pines, and one of its foods is the pine bark beetle. To conserve this species of woodpecker, these pines must be preserved. But the old pines are home to the beetles, which are pests to the trees and damage them for commercial logging. This presents an intriguing problem: How can we maintain the woodpecker and its food (the bark beetle) and also maintain productive forests?

The classic twentieth-century way to view the relationship among the pines, the bark beetle, and the woodpecker would be to show a food chain (see Chapter 6). But this alone does not solve the problem for us. A newer approach is to consider the habitat requirements of the pine bark beetle and the woodpecker. Their requirements are somewhat different. But if we overlay a map of one's habitat requirements over a map of the other's, the co-occurrence of habitats can be compared. Beginning with such maps, it becomes possible to design a landscape that would allow the maintenance of all three—pines, beetles, and birds.

Summary

- Modern approaches to management and conservation of wildlife use a broad perspective that considers interactions among species as well as the ecosystem and landscape contexts.

- To successfully manage wildlife for harvest, conservation, and protection from extinction, we need certain information. Measures of total abundance and of births and deaths over a long period will be helpful. We also need to know the habitat, viewed spatially and in quantitative terms. Age structure and other characteristics of a population can help in management, forecasting, and conservation. However, it is often difficult to obtain these data, especially for historical information.

- A common goal of wildlife conservation today is to "restore" the abundance of a species to some previous number, usually a number thought to have existed prior to the influence of modern technological civilization. Information about long-ago abundances is rarely available. Sometimes numbers can be estimated indirectly—for example, by using the Lewis and Clark journals to reconstruct the 1805 population of grizzly bears or using logbooks from ships hunting the bowhead whale. Adding to the complexity, the abundance of wildlife changes over time even in natural systems uninfluenced by modern technological civilization. And historical information often cannot be subjected to formal tests of disproof and therefore does not by itself qualify as scientific. Adequate information exists for relatively few species.

- Another approach is to seek a minimum viable population, a carrying capacity, or an optimal sustainable population or harvest based on data that can be obtained and tested today. This approach abandons the goal of restoring a species to some hypothetical past abundance.

- The good news is that many species once endangered have been successfully restored to an abundance that suggests they are unlikely to become extinct. Success is achieved when the habitat is restored to conditions required by a species. The conservation and management of wildlife presents great challenges but also offers great rewards of long-standing and deep meaning to people.

Should Wolves Be Reestablished in the Adirondack Park?

With an area slightly over 24,000 km², the Adirondack Park in northern New York is the largest park in the lower 48 states. Unlike most parks, however, it is a mixture of private (60%) and public (40%) land and is home to 130,000 people. When European colonists first came to the area, it was, like much of the rest of North America, inhabited by gray wolves. By 1960, wolves had been exterminated in all of the lower 48 states except for northern Minnesota. The last official sighting of a wolf in the Adirondacks was in the 1890s.

Although the gray wolf was not endangered globally—there were more than 60,000 in Canada and Alaska—it was one of the first animals listed as endangered under the 1973 Endangered Species Act. As required, the U.S. Fish and Wildlife Service developed a plan for recovery that included protection of the existing population and reintroduction of wolves to wilderness areas. The recovery plan would be considered a success if survival of the Minnesota wolf population was assured and at least one other population of more than 200 wolves had been established at least 320 km from the Minnesota population.

Under the plan, Minnesota's wolf population increased, and some wolves from that population, as well as others from southern Canada, dispersed into northern Michigan and Wisconsin, each of which had populations of approximately 100 wolves in 1998. Also, 31 wolves from Canada were introduced into Yellowstone National Park in 1995, and that population grew to over 100. By the end of 1998, it seemed fairly certain that the criteria for removing the wolf from the Endangered Species list would soon be met.

In 1992, when the results of the recovery plan were still uncertain, the Fish and Wildlife Service proposed to investigate the possibility of reintroducing wolves to northern Maine and the Adirondack Park. A survey of New York State residents in 1996 funded by Defenders of Wildlife found that 76% of people living in the park supported reintroduction. However, many residents and organizations within the park vigorously opposed reintroduction and questioned the validity of the survey. Concerns focused primarily on the potential dangers to humans, livestock, and pets and the possible impact on the deer population. In response to the public outcry, Defenders of Wildlife established a citizens' advisory committee that initiated two studies by outside experts, one on the social and economic aspects of reintroduction and another on whether there were sufficient prey and suitable habitat for wolves.

Wolves prey primarily on moose, deer, and beaver. Moose have been returning to the Adirondacks in recent years and now number about 40, but this is far less than the moose population in areas where wolves are successfully reestablishing. Beaver are abundant in the Adirondacks, with an estimated population of over 50,000. Because wolves feed on beaver primarily in the spring and the moose population is small, the main food source for Adirondack wolves would be deer.

Deer thrive in areas of early-successional forest and edge habitats, both of which have declined in the Adirondacks as logging has decreased on private forestland and been eliminated altogether on public lands. Furthermore, the Adirondacks are at the northern limit of the range for white-tailed deer, where harsh winters can result in significant mortality. Deer density in the Adirondacks, which is estimated at 3.25/km², is less than that found in wolf habitat in Minnesota, which also has 8,500 moose. If deer were the only prey available, wolves would kill between 2.5 and 6.5% of the deer population, while hunters take approximately 13% each year. Determining whether there is a sufficient prey base to support a population of wolves is complicated by the fact that coyotes have moved into the Adirondacks and occupy the niche once filled by wolves. Whether wolves would add to the deer kill or replace coyotes, with no net impact on the deer population, is difficult to predict.

An area of 14,000 km² in various parts of the Adirondack Park meets criteria established for suitable wolf habitat, but this is about half of the area required to maintain a wolf population for the long term. Based on the average deer density and weight, as well as the food requirements of wolves, biologists estimate that this habitat could support about 155 wolves. However, human communities are scattered throughout much of the park, and many residents are concerned that wolves would not remain on public land and would threaten local residents as well as backcountry hikers and hunters. Also, private lands around the edges of the park, with their greater density of deer, dairy cows, and people, could attract wolves.

Critical Thinking Questions

1. Who should make decisions about wildlife management, such as returning wolves to the Adirondacks—scientists, government officials, or the public?

2. Some people advocate leaving the decision to the wolves—that is, waiting for them to disperse from southern Canada and Maine into the Adirondacks. Study a map of the northeastern United States and southeastern Canada. What do you think is the likelihood of natural recolonization of the Adirondacks by wolves?

3. Do you think that wolves should be reintroduced to the Adirondack Park? If you lived in the park, would that affect your opinion? How would removal of the wolf from the Endangered Species list affect your opinion?

4. Some biologists recently concluded that wolves in Yellowstone and the Great Lakes region belong to a different subspecies, the Rocky Mountain timber wolf, from those that formerly lived in the northeastern United States, the eastern timber wolf. This means that the eastern timber wolf is still extinct in the lower 48 states. Would this affect your opinion about reintroducing wolves into the Adirondacks?

REEXAMINING THEMES AND ISSUES

Human Population

Today, human beings are a primary cause of the extinction of species. People also contributed to extinctions in the past. Nonindustrial societies have caused extinction by such activities as hunting and the introduction of exotic species into new habitats. With the age of exploration in the Renaissance and with the Industrial Revolution, the rate of extinctions accelerated. People altered habitats more rapidly and over greater areas. Hunting efficiency increased, as did the introduction of exotic species into new habitats. As the human population grows, conflicts over habitat between people and wildlife increase. Once again, we find that the human population problem underlies this environmental issue.

Sustainability

At the heart of issues concerning wild living resources is the question of sustainability of species and the ecosystems of which they are a part. One of the key questions is whether we can sustain these resources at a constant abundance. In general, it has been assumed that fish and other wildlife hunted for recreation, such as deer, could be maintained at some constant, highly productive level. Constant production is desirable economically because it would provide a reliable, easily forecast income each year. But despite attempts at direct management, few wild living resources have remained at constant levels. New ideas about the intrinsic variability of ecosystems and populations lead us to question the assumption that such resources can or should be maintained at constant levels.

Global Perspective

Although the final extinction of a species takes place in one locale, the problem of biological diversity and the extinction of species is global because of the worldwide increase in the rate of extinction and because of the growth of the human population and its effects on wild living resources.

Urban World

We have tended to think of wild living resources as existing outside of cities, but there is a growing recognition that urban environments will be more and more important in the conservation of biological diversity. This is partly because cities now occupy many sensitive habitats around the world, such as coastal and inland wetlands. It is also because appropriately designed parks and backyard plantings can provide habitats for some endangered species. As the world becomes increasingly urbanized, this function of cities will become more important.

People and Nature

Wildlife, fish, and endangered species are popular issues. We seem to have a deep feeling of connectedness to many wild animals—we like to watch them, and we like to know that they still exist even when we cannot see them. Wild animals have always been important symbols to people, sometimes sacred ones. The conservation of wildlife of all kinds is therefore valuable to our sense of ourselves, both as individuals and as members of a civilization.

Science and Values

The reasons that people want to save endangered species begin with human values, including values placed on the continuation of life and on the public service functions of ecosystems. Among the most controversial environmental issues in terms of values are the conservation of biological diversity and the protection of endangered species. Science tells us what is possible with respect to conserving species, which species are likely to persist, and which are not. Ultimately, therefore, our decisions about where to focus our efforts in sustaining wild living resources depend on our values.

Key Terms

catch per unit
 effort **272**
global
 extinction **280**

historical range of
 variation **270**
local extinction **280**
logistic carrying
 capacity **267**

maximum sustain-
 able yield **265**
minimum viable
 population **266**

optimum sustainable
 population **267**
time series **270**

Study Questions

1. Why are we so unsuccessful in making rats an endangered species?

2. What have been the major causes of extinction (a) in recent times and (b) before people existed on Earth?

3. As mentioned in the text, the U.S. Fish and Wildlife Service suggested three key indicators of the status of the grizzly bear: (1) sufficient reproduction to offset the existing levels of human-caused mortality, (2) adequate distribution of breeding animals throughout the area, and (3) a limit on total human-caused mortality. Are these indicators sufficient to assure the recovery of this species? What would you suggest instead?

4. Create a plan for the sustainable production of blue hake, the fish discussed in the opening case study.

5. This chapter discussed five justifications for preserving endangered species. Which of them apply to the following? (You can decide that none apply.)
   a. The black rhinoceros of Africa
   b. The Furbish lousewort, a rare small flowering plant of New England, seen by few people
   c. An unnamed and newly discovered beetle from the Amazon rain forest
   d. Smallpox

   e. Wild strains of potatoes in Peru
   f. The North American bald eagle

6. Locate an ecological island close to where you live and visit it. Which species are most vulnerable to local extinction?

7. Oysters were once plentiful in the waters around New York City. Create a plan to restore them to numbers that could be the basis for a commercial harvest.

8. Using information available in libraries, determine the minimum area required for a minimum viable population of the following:
   a. Domestic cats
   b. Cheetahs
   c. The American alligator
   d. Swallowtail butterflies

9. Both a ranch and a preserve will be established for the North American bison. The goal of the ranch owner is to show that bison can be a better source of meat than introduced cattle and at the same time have a less detrimental effect on the land. The goal of the preserve is to maximize the abundance of the bison. How will the plans for the ranch and preserve differ, and how will they be similar?

Further Reading

Botkin, D. B. 2001. *No Man's Garden: Thoreau and a New Vision for Civilization and Nature*. Washington, D.C.: Island Press. A work that discusses deep ecology and its implications for biological conservation, as well as reasons for the conservation of nature, both scientific and beyond science.

Caughley, G., and A. R. E. Sinclair. 1994. *Wildlife Ecology and Management*. London: Blackwell Scientific. A valuable textbook based on new ideas of wildlife management.

MacKay, R. 2002. *The Penguin Atlas of Endangered Species: A Worldwide Guide to Plants and Animals*. New York: Penguin. A geographic guide to endangered species.

Pauly, D., J. Maclean, and J. L. Maclean. 2002. *The State of Fisheries and Ecosystems in the North Atlantic Ocean*. Washington, D. C.: Island Press.

Schaller, George B., and Lu Zhi (photographers.) 2002. *Pandas in the Wild: Saving an Endangered Species*. New York: Aperture Publisher. A photographic essay about scientists attempting to save an endangered species.

CHAPTER 15

Environmental Health, Pollution, and Toxicology

Wild leopard frogs in America have been affected by human-made chemicals in the environment.

Learning Objectives

Serious environmental health problems and diseases may arise from toxic elements in water, air, soil, and even the rocks on which we build our homes. After reading this chapter, you should understand:

■ How the terms *toxic, pollution, contamination, carcinogen, synergism, and biomagnification* are used in environmental health.

■ What the classification and characteristics are of major groups

of pollutants in environmental toxicology.

■ Why there is controversy and concern about synthetic organic compounds such as dioxin.

■ Whether we should be concerned about exposure to human-produced electromagnetic fields.

■ What the dose–response concept is and how it relates to LD-50, TD-50, ED-50, ecological gradients, and tolerance.

■ How the process of biomagnification works and why it is important in toxicology.

■ Why the threshold effects of environmental toxins are important.

■ What the process of risk assessment in toxicology is and why such processes are often difficult and controversial.

C A S E S T U D Y

Demasculinization and Feminization of Frogs in the Environment

The story of wild leopard frogs (see opening photo) from a variety of areas in the midwestern United States sounds something like a science fiction horror story. In affected areas, between 10% and 92% of male frogs exhibit gonadal abnormalities, including retarded development and hermaphroditism, meaning they have both male and female reproductive organs. Other frogs have vocal sacks with retarded growth. Since vocal sacks are used to attract female frogs, these frogs are less likely to mate. What is apparently causing some of the changes in the male frogs is exposure to atrazine, which is the most widely used herbicide in the United States today. The chemical is a weed killer, used primarily in agricultural areas. The region of the United States with the highest frequency of sex reversal (92% of male frogs) is in Wyoming along the North Platte River. The region is not near any large agricultural activity, and the use of atrazine there is not particularly significant. Hermaphrodite frogs are common there because the North Platte River flows from areas in Colorado where atrazine is commonly used. The amount of atrazine released into the environment is estimated to be approximately 7.3 million kg (16 million lb) per year. The chemical degrades in the environment, but the degradation process is longer than the application cycle. Because of its continual application every year, the waters of the Mississippi River basin, which drains about 40% of the lower United States, discharges approximately 0.5 million kg (1.2 million lbs) of atrazine per year to the Gulf of Mexico. Atrazine becomes easily attached to dust particles and has been found in rain, fog, and snow. As a result, atrazine has been found to contaminate groundwater and surface water in regions where it isn't used. The Environmental Protection Agency (EPA) states that up to 3 parts per billion (ppb) of atrazine in drinking water is acceptable, but at this concentration it definitely affects frogs that swim in the water. Other studies around the world have confirmed this. For example, in Switzerland, where atrazine is banned, it commonly occurs with a concentration of about 1 ppb, and that is sufficient to change some male frogs into females. In fact, atrazine can apparently cause sex change in frogs at concentration in the water of as low as one-thirteenth of the level set by the EPA for drinking water.

Of particular interest and importance is the process that causes the changes in the leopard frogs. We begin the discussion with the endocrine system, which is composed of glands that internally secrete hormones directly into the bloodstream. Endocrine hormones, such as testosterone and estrogen, are carried by the blood to parts of the body where they regulate and control functions of growth and sexual development. Testosterone in male frogs is in part responsible for development of male characteristics. The atrazine is believed to switch on a gene that turns testosterone into estrogen (a female sex hormone). It's the hormones, not the genes, that actually regulate the development and structure of reproductive organisms. Frogs are particularly vulnerable during their early development, before and as they metamorphose from tadpoles to adult frogs. This change occurs in the spring, when atrazine levels are often at a maximum level in surface water. Apparently, a single exposure to the chemical may affect the frog's development. Thus, the herbicide is known as a *hormone disrupter*. In a more general sense, substances that interact with the hormone systems of an organism, whether or not they are linked to disease or abnormalities, are known as *hormonally active agents* (HAAs). These HAAs have the ability to trick the organism's body (in this case, the frog's) into believing that the chemicals have a role to play in its functional development. An analogy you might be more familiar with is a computer virus that fools the computer into accepting it as part of the system by which the computer works. Similar to computer viruses, the HAAs interact with an organism and the mechanisms for regulating growth and development, thus disrupting normal growth functions. What happens when HAAs—in particular, hormone disrupters (such as pesticides and herbicides)—are introduced into the system is shown in Figure 15.1. Natural hormones produced by the body send chemical messages to cells, where receptors for the hormone molecules are found on the outside and inside of cells. These natural hormones then transmit instructions to

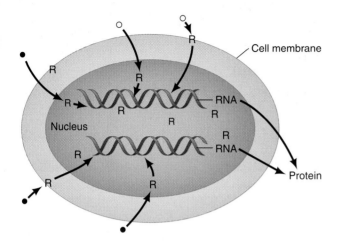

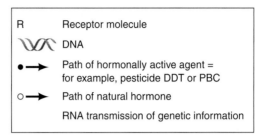

R	Receptor molecule
$\wedge\wedge$	DNA
•→	Path of hormonally active agent = for example, pesticide DDT or PBC
○→	Path of natural hormone
	RNA transmission of genetic information

Figure 15.1 ■ Idealized diagram of hormonally active agents (HAAs) binding to receptors on the surface of and inside a cell. When HAAs, along with natural hormones, transmit information to the cells' DNA, then HAAs may obstruct the role of the natural hormones that produce proteins that in turn regulate the growth and development of an organism.

the cell's DNA, eventually directing development and growth. We now know that chemicals such as some pesticides and herbicides can also bind to the receptor

molecules and either mimic or obstruct the role of the natural hormones. Thus, hormonal disrupters may also be known as HAAs.[1–4]

The story of wild leopard frogs in America dramatizes the importance of carefully evaluating the role of human-made chemicals in the environment. Frogs and other amphibians are on the decline on a global basis, and much research has been directed toward understanding the reduction of populations of frogs and other amphibians. Studies to evaluate past or impending extinction of organisms often centers on global processes such as climate change, but the story of leopard frogs leads us down another path, one associated with human use of the natural environment. It also raises a number of more disturbing questions: Are we participating in an unplanned experiment on how human-made chemicals such as herbicides and pesticides might transform the bodies of living beings, perhaps even people? Are these changes in organisms as a result of exposure to chemicals limited only to certain plants or animals, or are they a forerunner of what we might expect in the future on a much broader scale? Perhaps we will look back on this moment of understanding as a new beginning in meaningful studies that will answer some of these important questions.

15.1 Some Basics

Disease is often due to an imbalance resulting from poor adjustment between the individual and the environment. Disease occurs on a continuum from a state of health to a state of disease. Between these two states is the *gray zone* of suboptimal health, which is a state of unbalance. In the gray zone a person may not be diagnosed with a specific disease but may not be healthy.[5] There are many

gray zones in environmental health, such as the many possible states of suboptimal health from exposure to human-made chemicals including pesticides; additives in processed foods, such as coloring and preservatives; chemical alteration of the structure of food, such as adding artificial saturated fat; exposure to tobacco smoke; exposure to air pollutants such as ozone; exposure to chemicals in gasoline and many household cleaners; and exposure to heavy metals such as mercury or lead. As a result of exposure to chemicals in the environment from human activity, we may be in the midst of an epidemic of chronic disease that is unprecedented in human history.[5]

Disease seldom has a one-cause–one-effect relationship with the environment. Rather, the incidence of a disease depends on several factors, including physical environment, biological environment, and lifestyle. Linkages between these factors are often related to other factors, such as local customs and the level of industrialization. More primitive societies that live directly off the local environment are usually plagued by different environmental health problems than in an urban society. For examples, industrial societies have nearly eliminated such diseases as cholera, dysentery, and typhoid.

People are often surprised to learn that the water we drink, the air we breathe, the soil in which we grow crops, and the rocks on which we build our homes and places of work may affect our chances of experiencing serious environmental health problems and diseases (although, as suggested, direct causative relationships between the environment and disease are difficult to establish). At the same time, the environmental factors that contribute to disease—soil, rocks, water, and air—can also influence our chances of living a longer, more productive life.

Many people believe that soil, water, or air in a so-called natural state must be good and that if human activities have changed or modified them, they have become contaminated, polluted, and therefore bad.[6] This is by no means the entire story; many natural processes, including dust storms, floods, and volcanic processes, can introduce materials harmful to humans and other living things into the soil, water, and air.

A tragic example occurred on the night of August 21, 1986, when there was a massive natural release of carbon dioxide (CO_2) gas from Lake Nyos in Cameroon, Africa. The carbon dioxide was probably initially released from volcanic vents at the bottom of the lake and accumulated there with time. Pressure of the overlying lake water normally kept the dissolved gas at the bottom of the lake. However, the water was evidently agitated by a slide or small earthquake, and the bottom water moved upward. When the CO_2 gas reached the surface of the lake, it was released quickly into the atmosphere. The CO_2 gas, which is heavier than air, flowed downhill from the lake and settled in nearby villages, killing many animals and more than 1,800 people by asphyxiation (Figure 15.2). It was estimated that a similar event could recur within about 20 years, assuming that carbon dioxide continued to be released at the bottom of the lake.[7] Fortunately, a project funded by the U.S. Office of Foreign Disaster Assistance Hazards Reduction Project (scheduled to be completed early in the twenty-first century) is inserting pipes into the bottom of Lake Nyos. The gas-rich water is pumped to the surface where it is safely discharged into the atmosphere. In 2001 a warning system was installed, and one degassing pipe released a little more CO_2 than was seep-

(a)

(b)

Figure 15.2 ■ (a) In 1986, Lake Nyos in Cameroon, Africa, released carbon dioxide that moved down the slopes of the hills to settle in low places, asphyxiating animals and people. (b) Animals asphyxiated by carbon dioxide.

ing naturally into the lake. Recent data suggest that the single pipe now there barely keeps ahead of CO_2 that continues to enter the bottom, so the lake's 500,000 tons of built-up gas have only dropped 6%. At this rate it could take 30 to 50 years to make Lake Nyos safe. In the meantime, there could be another eruption.[8]

Terminology

What do we mean when we use the terms *pollution, contamination, toxic,* and *carcinogen*? A polluted environment is one that is impure, dirty, or otherwise unclean. The term **pollution** refers to the occurrence of an unwanted change in the environment caused by the introduction of harmful materials or the production of harmful conditions (heat, cold, sound). **Contamination** has a meaning similar to

Figure 15.3 ■ This southern California urban stream flows into the Pacific Ocean at a coastal park. The stream water often carries high counts of fecal coliform bacteria. As a result, the stream is a point source of pollution for the beach, which is sometimes closed to swimming following runoff events.

that of *pollution* and implies making something unfit for a particular use through the introduction of undesirable materials—for example, the contamination of water by hazardous waste. The term **toxic** refers to materials (pollutants) that are poisonous to people and other living things. **Toxicology** is the science that studies chemicals that are known to be or could be toxic, and toxicologists are scientists in this field. A **carcinogen** is a particular kind of toxin that increases the risk of cancer. Carcinogens are among the most feared and regulated toxins in our society.

An important concept in considering pollution problems is **synergism**, the interaction of different substances resulting in a total effect greater than the sum of the effects of the separate substances. For example, both sulfur dioxide and coal dust particulates are air pollutants. Either one taken separately may cause adverse health effects, but when they combine, as when sulfur dioxide (SO_2) adheres to the coal dust, the dust with SO_2 is inhaled deeper than sulfur dioxide alone, causing greater damage to lungs. Another aspect of synergistic effects is that the body may be more sensitive to a toxin if it is simultaneously subjected to other toxins.

Pollutants are commonly introduced into the environment by way of **point sources**, such as smokestacks (see A Closer Look 15.1), pipes discharging into waterways, a small stream entering the ocean (Figure 15.3), or accidental spills. **Area sources**, also called *nonpoint sources*, are more diffused over the land and include urban runoff and **mobile sources** such as automobile exhaust. Area sources are difficult to isolate and correct because the problem is often widely dispersed over a region, as in agricultural runoff that contains pesticides (see Chapter 22).

Measuring the Amount of Pollution

How the amount or concentration of a particular pollutant or toxin present in the environment is reported varies widely. The amount of treated wastewater entering Santa Monica Bay in the Los Angeles area is a big number reported in millions of gallons per day. Emission of nitrogen and sulfur oxides into the air is also a big number reported in millions of tons per year. Small amounts of pollutants or toxins in the environment, such as pesticides, are reported in units as parts per million (ppm) or parts per billion (ppb). It is important to keep in mind that the concentration in ppm or ppb may be by volume, mass, or weight. In some toxicology studies, the units used are milligrams of toxin per kilogram of body mass (1 mg/kg is equal to 1 ppm). Concentration may also be recorded as a percent. For example, 100 ppm (100 mg/kg) is equal to 0.01%. (How many ppm is equal to 1%?)

When dealing with water pollution, units of concentration for a pollutant may be milligrams per liter (mg/L) or micrograms per liter (μg/L). A milligram is one-thousandth of a gram, and a microgram is one-millionth of a gram. For water pollutants that do not cause significant change in density of water (1 g/cm^3), a concentration of pollution of 1 mg/L is approximately equivalent to 1 ppm. Air pollutants are commonly measured in units such as micrograms of pollutant per cubic meter of air (μg/m^3).

Units such as ppm, ppb, or μg/m^3 reflect very small concentrations. For example, if you were to use 3 g (one-tenth of an ounce) of salt to season popcorn in order to have salt at a concentration of 1 ppm by weight of the popcorn, you would have to pop approximately 3 metric tons of kernels!

15.2 Categories of Pollutants

A partial classification of pollutants by arbitrary categories is presented below. Examples of other pollutants are discussed in other parts of the book.

Infectious Agents

Infectious diseases, spread from the interactions between individuals and food, water, air, or soil, constitute some of the oldest health problems that humans face. Today, infectious diseases have the potential to pose rapid local to global threats by spreading in hours through airplane travelers. Terrorist activity may also spread diseases. Inhalation anthrax, caused by a bacterium, sent in a powdered form in envelopes through the mail in 2001, killed several people. New diseases are emerging, and previous ones may emerge again. Although we have cured many diseases, we have no known reliable vaccines for others, such as HIV, hantavirus, and dengue fever.

Sudbury Smelters: A Point Source

A famous example of a point source of pollution is provided by the smelters that refine nickel and copper ores at Sudbury, Ontario. Sudbury contains one of the world's major nickel and copper ore deposits. A number of mines, smelters, and refineries lie within a small area. The smelter stacks at one time released large amounts of particulates containing nickel, copper, and other toxic metals into the atmosphere. In addition, because the areas contain a high percentage of sulfur, the emissions included large amounts of sulfur dioxide (SO_2). During its peak output in the 1960s, this complex was the largest single source of sulfur dioxide emissions in North America.

In 1969, regulations were mandated to improve local air quality, forcing a reduction in emissions. Concentrations of sulfur dioxide were reduced locally by more than 50% after 1972. However, attempts to minimize the pollution problem in the immediate vicinity of the smelting operation by increasing smokestack height spread the problem as wind carried the pollutants greater distances. In order to better control emissions from Sudbury, the Ontario government set standards to reduce emissions to less than 365,000 tons per year by 1994 (about 14% of earlier emissions of 2,560,000 tons per year). The goal was achieved by reducing production from the smelters and by treating the emissions to reduce pollution.[9]

As a result of years of pollution, nickel has been found to contaminate soils 50 km (about 31 mi) from the stacks. The forests that once surrounded Sudbury were devastated by decades of acid rain (produced from SO_2 emissions) and the deposition of particulates containing heavy metals. An area of approximately 250 km² (96 mi²) was nearly devoid of vegetation, and damage to forests in the region has been visible over an area of approximately 3,500 km² (1,350 mi²; see Figure 15.4). Secondary effects, in addition to loss of vegetation, include soil erosion and drastic changes in soil chemistry resulting from the influx of the heavy metals.

Reductions in emissions from Sudbury have allowed surrounding areas to slowly begin to recover from these effects. Species of trees once eradicated from some areas have begun to grow again. Recent restoration efforts have included planting of over 7 million trees, 75 species of herbs, moss, and lichens—all of which have contributed to the increase of biodiversity. Lakes that were damaged due to acid precipitation in the area are rebounding and now support populations of plankton and fish.[9] The case of the Sudbury smelters thus provides a positive example of pollution reduction, emphasizing the key theme of thinking globally but acting locally to reduce air pollution. It also illustrates the theme of science and values: Scientists and engineers can design pollution abatement equipment, but spending the money to purchase the equipment reflects what value we place on clean air.

(a)   (b)

Figure 15.4 ■ Lake St. Charles, Sudbury, Ontario prior to restoration. Note high stacks (smelters) in the background and lack of vegetation in the foreground resulting from air pollution (acid and heavy-metal deposition). (b) Recent photo showing regrowth and restoration.

Diseases that can be controlled by manipulating the environment, such as by improving sanitation or treating water, are classified as environmental health concerns. Although there is great concern about the toxins and carcinogens produced in industrial society today, the greatest mortality in developing countries is caused by environmentally transmitted infectious disease. In the United States, thousands of cases of waterborne illness and food poisoning occur each year. These diseases can be spread by people; mosquitoes or fleas; or contact with contaminated food, water, or soil. They can also be transmitted through ventilation systems in buildings.

Some examples of environmentally transmitted infectious diseases are:

- Legionellosis, or Legionnaires' disease, which often occurs where air-conditioning systems have been contaminated by disease-causing organisms.
- Giardiasis, a protozoan infection of the small intestine spread via food, water, or person-to-person contact.
- Salmonella, a food-poisoning bacterial infection spread via water or food.
- Malaria, a protozoan infection transmitted by mosquitoes.
- Lyme Borreliosis, or Lyme disease, transmitted by ticks.
- Cryptosporidosis, a protozoan infection transmitted via water or person-to-person contact (see Chapter 22).[10]
- Anthrax, spread by terrorist activity.

We sometimes hear about epidemics in developing nations. An example is the highly contagious Ebola virus in Africa, which causes external and internal bleeding resulting in death of 80% of those infected. We may tend to think of such epidemics as problems only for developing nations. This belief may give us a false sense of security! Monkeys and bats spread Ebola, but the origin of the virus in the tropical forest remains unknown. Developed countries, where outbreaks may occur in the future, will need to learn from the developing countries' experiences. To accomplish this and avoid potential global tragedies, more funds need to be provided for the study of infectious diseases in developing countries.

Toxic Heavy Metals

The major **heavy metals** (metals with relatively high atomic weight; see Chapter 5) that pose health hazards to people and ecosystems include mercury, lead, cadmium, nickel, gold, platinum, silver, bismuth, arsenic, selenium, vanadium, chromium, and thallium. Each of these elements may be found in soil or water not contaminated by humans. Each metal has uses in our modern industrial society, and each is also a by-product of the mining, refining, and use of other elements. Heavy metals often have direct physiological toxic effects. Some are stored or incorporated in living tissue, sometimes permanently. Heavy metals tend to be stored (accumulating with time) in fatty body tissue. A little arsenic each day may eventually result in a fatal dose—the subject of more than one murder mystery.

The content of heavy metals in our bodies is referred to as the *body burden*. The body burden of toxic heavy elements for an average human body (70 kg) is about 8 mg for antimony, 13 mg for mercury, 18 mg for arsenic, 30 mg for cadmium, and 150 mg for lead. Lead (for which we apparently have no biological need) has an average body burden of about twice that of the others combined, reflecting our heavy use of this potentially toxic metal.

Mercury, thallium, and lead are very toxic to humans. They have long been mined and used, and their toxic properties are well known. Mercury, for example, is the "Mad Hatter" element. At one time, mercury was used in making felt hats stiff; because mercury damages the brain, hatters in Victorian England were known to act peculiarly. Thus, the Mad Hatter in Lewis Carroll's *Alice in Wonderland* had real antecedents in history.

Toxic Pathways

Chemical elements released from rocks or human processes can become concentrated in humans (see Chapter 5) through many pathways (Figure 15.5). These pathways may involve what is known as **biomagnification**—the accumulation or increase in concentration of a substance in living tissue as it moves through a food web (also known as *bioaccumulation*). For example, cadmium, which influences the risk of heart disease, may enter the environment via ash from burning coal. The cadmium in coal exists in very low concentrations (less than 0.05 ppm). After coal is burned in a power plant, the ash is collected in a solid form and disposed of in a landfill. The landfill is covered with soil and revegetated. The low concentration of cadmium in the ash and soil is taken into the plants as they grow. But the concentration of cadmium in the plants is three to five times greater than the concentration in the ash. As the cadmium moves through the food chain, it becomes more and more concentrated. By the time it is incorporated in the tissue of people and other carnivores, the concentration is approximately 50 to 60 times the original concentration in the coal.

Mercury in aquatic ecosystems offers another example of biomagnification. Mercury is a potentially serious pollutant of aquatic ecosystems such as ponds, lakes, rivers, and the ocean. Natural sources of mercury in the environment include volcanic eruptions and erosion of natural mercury deposits, but we are most concerned with human input of mercury into the environment through processes such as burning coal in power plants, incinerating waste, and processing metals such as gold. Rates of input of mercury into the environment through human processes are poorly understood. However, it is believed that human activities have doubled or tripled the amount of mercury in the atmosphere, and it is increasing at about 1.5% per year.[11]

A major source of mercury in many aquatic ecosystems is deposition from the atmosphere through precipitation. Most of the deposition is of inorganic mercury (Hg^{++}, ionic mercury). Once this mercury is in surface water, it enters into complex biogeochemical cycles and a process known as *methylation* may occur. Methylation changes inorganic mercury to methyl mercury $[CH_3Hg]^+$ through bacterial activity.

Methyl mercury is much more harmful (toxic) than is inorganic mercury, and it is eliminated more slowly from

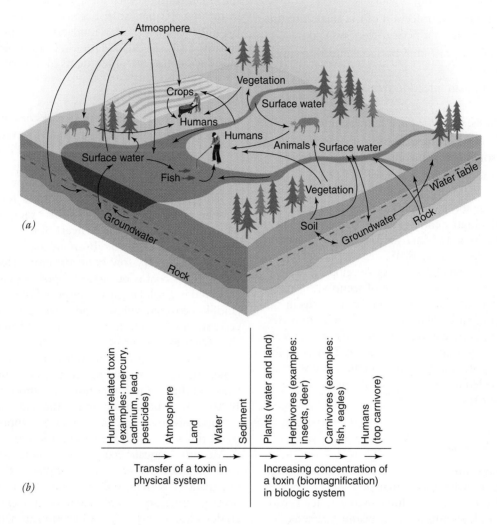

(a)

(b)

Human-related toxin (examples: mercury, cadmium, lead, pesticides)	Atmosphere	Land	Water	Sediment	Plants (water and land)	Herbivores (examples: insects, deer)	Carnivores (examples: fish, eagles)	Humans (top carnivore)

Transfer of a toxin in physical system

Increasing concentration of a toxin (biomagnification) in biologic system

Figure 15.5 ■ (*a*) Potential complex pathways for toxic materials through the living and nonliving environment. Note the many arrows into humans and other animals, sometimes in increasing concentrations, as they move through the food chain (*b*).

animals' systems. As the methyl mercury works its way through food chains, biomagnification occurs, so that higher concentrations of methyl mercury are found farther up the food chain. Thus, big fish in a pond that eat little fish contain a higher concentration of mercury than do smaller fish and the aquatic insects that the fish feed upon.

Selected aspects of the mercury cycle in aquatic ecosystems are shown in Figure 15.6. The figure emphasizes the input side of the cycle, from deposition of inorganic mercury through formation of methyl mercury, biomagnification, and sedimentation of mercury at the bottom of a pond. On the output side of the cycle, the mercury that enters fish may be taken up by animals that eat the fish; and sediment may release mercury by a variety of

processes, including resuspension in the water, where eventually the mercury enters the food chain or is released into the atmosphere through volatilization (conversion of liquid mercury to a vapor form).

Biomagnification also occurs in the ocean. Large fish such as tuna and swordfish have elevated mercury concentrations, and limiting consumption of these fish is recommended. Pregnant women are advised to not eat them at all.

The threat of mercury poisoning is widespread. Millions of young children in Europe, the United States, and other industrial countries are exposed to or have mercury levels that exceed health standards.[12] Even children in remote areas of the far north are exposed to mercury through their food chain.

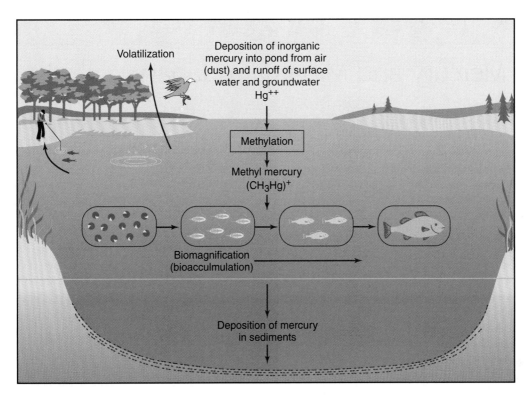

Figure 15.6 ■ Idealized diagram showing selected pathways for movement of mercury into and through an aquatic ecosystem. [*Source:* Modified from G. L. Waldbott, *Health Effects of Environmental Pollutants*, 2nd ed. (Saint Louis, MO: C. V. Mosby, 1978).]

During the twentieth century, several significant incidents of methyl mercury poisoning were recorded. One, in Minamata Bay, Japan, involved the industrial release of methyl mercury (see A Closer Look 15.2). Another, in Iran, involved a methyl mercury fungicide used to treat wheat seeds. In each of these cases, hundreds of people were killed and thousands were permanently damaged.[11]

Minamata Bay and Iran involved local-exposure mercury; what is being reported in the Arctic emphasizes mercury at the global level in a region far from emission sources of the toxic metal. The Inuit people in Quannea, Greenland, live above the Arctic Circle, far from any roads and 45 minutes by helicopter to the nearest outpost of modern society. Nevertheless, they are some of the most chemically contaminated people on Earth, with as much as 12 times the mercury in their blood as is recommended in U.S. guidelines. The mercury gets to the Inuit from the industrialized world by way of what they eat. The whale, seal, and fish they eat contain mercury that is further concentrated in the tissue and blood of the people. The process of increasing concentrations of mercury further up the food chain is an example of biomagnification.[12]

What needs to be done to stop mercury toxicity at the local to global level is straightforward. The answer is to reduce emissions of mercury by capturing it before emission or by using alternatives to mercury in industry. Success will require international cooperation and tech-nology transfer to countries such as China and India, who with their tremendous increases in manufacturing are the world's largest users of mercury today.[12]

Organic Compounds

Organic compounds are compounds of carbon produced naturally by living organisms or synthetically by human industrial processes. It is difficult to generalize about the environmental and health effects of artificially produced organic compounds because there are so many of them, they have so many uses, and they can produce so many different kinds of effects.

Synthetic organic compounds are used in industrial processes, pest control, pharmaceuticals, and food additives. We have produced over 20 million synthetic chemicals, and new ones are appearing at a rate of about 1 million per year! Most are not produced commercially, but up to 100,000 chemicals are now being, or in the past have been, used. Once used and dispersed in the environment, they may produce a hazard for decades or hundreds of years.

Persistent Organic Pollutants

Some synthetic compounds are called **persistent organic pollutants**, or **POPs**. Many were first produced decades ago when their harm to the environment was

Mercury and Minamata, Japan

In the Japanese coastal town of Minamata, on the island of Kyushu, a strange illness began to occur in the middle of the twentieth century. It was first recognized in birds that lost their coordination and fell to the ground or flew into buildings and in cats that went mad, running in circles and foaming at the mouth.[13] The affliction, known by local fishermen as the "disease of the dancing cats," subsequently affected people, particularly families of fishermen. The first symptoms were subtle: fatigue, irritability, headaches, numbness in arms and legs, and difficulty in swallowing. More severe symptoms involved the sensory organs; vision was blurred and the visual field was restricted. Afflicted people became hard of hearing and lost muscular coordination. Some complained of a metallic taste in their mouths; their gums became inflamed, and they suffered from diarrhea. Eventually, 43 people died and 111 were severely disabled; in addition, 19 babies were born with congenital defects. Those affected lived in a small area, and much of the protein in their diet came from fish from Minamata Bay.

A vinyl chloride factory on the bay used mercury in an inorganic form in its production processes. The mercury was released in waste that was discharged into the bay.

Mercury forms few organic compounds, and it was believed that the mercury, though poisonous, would not get into food chains. But the inorganic mercury released by the factory was converted by bacterial activity in the bay into methyl mercury, an organic compound that turned out to be much more harmful. Unlike inorganic mercury, methyl mercury readily passes through cell membranes. It is transported by the red blood cells throughout the body, and it enters and damages brain cells.[14] Fish absorb methyl mercury from water 100 times faster than they absorb inorganic mercury. (This was not known before the epidemic in Japan.) Once absorbed, methyl mercury is retained two to five times longer than is inorganic mercury.

Harmful effects of methyl mercury depend on a variety of factors, including the amount and route of intake, the duration of exposure, and the species affected. The effects of the mercury are delayed from three weeks to two months from the time of ingestion. If mercury intake ceases, some symptoms may gradually disappear, but others are difficult to reverse.[14]

The mercury episode at Minamata illustrates four major factors that must be considered in evaluating and treating toxic environmental pollutants.

1. *Individuals vary in their response to exposure to the same dose, or amount, of a pollutant.* Not everyone in Minamata responded in the same way; there were variations even among those most heavily exposed. Because we cannot predict exactly how any single individual will respond, we need to find a way to state an expected response of a particular percentage of individuals in a population.

2. *Pollutants may have a threshold*—that is, a level below which the effects are not observable and above which the effects become apparent. Symptoms appeared in individuals with concentrations of 500 ppb of mercury in their bodies; no measurable symptoms appeared in individuals with significantly lower concentrations.

3. *Some effects are reversible.* Some people recovered when the mercury-filled seafood was eliminated from their diet.

4. *The chemical form of a pollutant, its activity, and its potential to cause health problems may be changed markedly by ecological and biological processes.* In the case of mercury, its chemical form and concentration changed as the mercury moved through the food webs.

Table 15.1 • Selected Common Persistent Organic Pollutants (POPs)	
Chemical	**Example of Use**
Aldrin[a]	Insecticide
Atrazine[b]	Herbicide
DDT[a]	Insecticide
Dieldrin[a]	Insecticide
Endrin[c]	Insecticide
PCBs[a]	Liquid insulators in electric transformers
Dioxins	By-product of herbicide production

Source: Data in part from Anne Platt McGinn, "Phasing Out Persistent Organic Pollutants," in Lester R. Brown et al., *State of the World 2000* (New York: Norton, 2000).

[a] Banned in the United States and many other countries.

[b] Degrades in the environment. It is presistant when reapplied often.

[c] Restricted or banned in many countries.

not known, and they are now banned or restricted (see Table 15.1 and A Closer Look 15.3).

POPs have several properties that define them:[15]

- They have a carbon-based molecular structure, often containing highly reactive chlorine.
- Most are manufactured by humans; that is, they are synthetic chemicals.
- They are persistent in the environment; that is, they do not easily break down in the environment.
- They are polluting and toxic.
- They are soluble in fat and likely to accumulate in living tissue.
- They occur in forms that allow them to be transported by wind, water, and sediments for long distances.

For example, consider polychlorinated biphenyls (PCBs), which are heat-stable oils originally used as an insulator in electric transformers.[15]

■ A factory in Alabama manufactured PCBs in the 1940s, shipping them to a General Electric factory in Massachusetts. They were put in insulators and mounted on poles in thousands of locations.

■ The transformers deteriorated over time. Some were damaged by lightning, and others were damaged or destroyed during demolition. The PCBs leaked into the soil or were carried by surface runoff into streams and rivers. Others combined with dust and were transported by wind around the world.

■ The dust containing PCBs was deposited in ponds, lakes, or rivers, where it entered the food chain. First it entered algae along with nutrients it combined with. Insects ate the algae, which were eaten by shrimp and fish. In each stage up the food web, the concentration of PCBs increased.

■ Fish are caught by fishermen and eaten. The PCBs are then passed on to people where they are concentrated in fatty tissue and mother's milk.

Hormonally Active Agents (HAAs)

HAAs are also POPs. The opening case study discussed the feminization of frogs resulting from exposure to the herbicide atrazine, and you may wish to review that case study in context of the continued discussion here. There is an increasing body of scientific evidence that points to certain chemicals in the environment, known as **hormonally active agent (HAAs),** as having potential to cause developmental and reproductive abnormalities in animals, including humans. HAAs include a wide variety of chemicals, such as some herbicides, pesticides, phthalates (compounds found in many chlorine-based plastics), and PCBs. Evidence in support of the hypothesis that HAAs are interfering with the growth and development of organisms comes from studies of wildlife in the field and laboratory studies of human diseases such as breast, prostate, and ovarian cancer, as well as abnormal testicular development and thyroid-related abnormalities.[2]

Studies of wildlife, in addition to the previously discussed frogs, include evidence from alligator populations in Florida exposed to pesticides such as DDT that exhibit genital abnormalities, low egg production, and reduced penis size. Pesticides have also been linked to reproductive problems with several species of birds, including gulls, cormorants, brown pelicans, falcons, and eagles. Studies are also ongoing on Florida panthers, which apparently have abnormal ratios of sex hormones that may be affecting their reproductive capability. In summary, the major disorders that have been studied in wildlife have centered on abnormalities including thinning of eggshells of birds, decline in populations of various animals and birds, reduced viability of offspring, and changes in sexual behavior.[1]

With respect to human diseases, much research has been done on linkages between HAAs and breast cancer through exploring relationships between environmental estrogens and cancer. Other studies are ongoing to understand relationships between PCBs and neurological behavior that result in poor performance on standard intelligence tests. Finally, there is concern that exposure of people to phthalates that are found in chlorinated-based plastics is also causing problems. The consumption of phthalates in the United States is considerable, with the highest exposure in women of childbearing age. The products being tested as the source of contamination include perfumes and other cosmetics, such as nail polish and hairspray.[1]

In summary, there is good scientific evidence that some chemical agents in sufficient concentrations will affect human reproduction through endocrine and hormonal disruption. The endocrine system is of primary importance because it is one of the two main systems (along with the nervous system) that regulate and control growth, development, and reproduction. In humans the endocrine system is composed of a group of hormone-secreting glands, including the thyroid, pancreas, pituitary, ovaries (women), and testes (men). The hormones are transported in the bloodstream to virtually all parts of the body, where they act as chemical messengers to control growth and development of the body.[2]

The National Academy of Sciences completed a review of the available scientific evidence concerning HAAs and recommends that there should be continued monitoring of wildlife and human populations for abnormal development and reproduction. Furthermore, where wildlife species are known to be experiencing declines in population associated with abnormalities, experiments should be designed to study the phenomena with respect to chemical contamination. With respect to humans, the recommendation is for additional studies that will document the presence or absence of associations between HAAs and human cancers. When associations are discovered, the causality must also be investigated in terms of potential latency, relationships between exposure and disease, and indicators of susceptibility to disease of certain groups of people by age and sex.[1]

Radiation

Nuclear radiation is introduced here as a category of pollution. It is discussed in detail in Chapter 20 in conjunction with nuclear energy. We are concerned about nuclear radiation because excessive exposure is linked to serious health problems, including cancer. (See Chapter 25 for a discussion of radon gas as an indoor air pollutant.)

Dioxin: The Big Unknown

Dioxin, a persistant organic pollutant, or POP, may be one of the most toxic of the human-made chemicals in the environment. The history of the scientific study of dioxin and its regulation illustrate once again the interplay of science and values. Although science isn't entirely certain of the toxicity of dioxin to humans and ecosystems, society has made a number of value judgments involving regulation of the substance. Controversy has surrounded these judgments and will surely continue to do so.

Dioxin is a colorless crystal made up of oxygen, hydrogen, carbon, and chlorine. It is classified as an organic compound because it contains carbon. About 75 types of dioxin (and dioxin-like compounds) are known; they are distinguished from one another by the arrangement and number of chlorine atoms in the molecule.

Dioxin is not normally manufactured intentionally but is a by-product resulting from chemical reactions, including the combustion of compounds containing chlorine in the production of herbicides.[16] In the United States, there are a variety of sources for dioxin-like compounds (specifically, chlorinated dibenzo-p-dioxin, or CDD, and chlorinated dibenzofurans, or CDF). These compounds are emitted into the air through such processes as incineration of municipal waste (the major source), incineration of medical waste, burning of gasoline and diesel fuels in vehicles, burning of wood as a fuel, and refining of metals such as copper. The good news is that releases of CDDs and CDFs decreased about 75% from 1987 to 1995. However, we are only beginning to understand the many sources of emission of dioxins to the air, water, and land, and linkages and rates of transfer from dominant airborne transport to deposition in water, soil, and the biosphere. In too many cases the amounts of dioxins emitted are based more on expert opinion than on high-quality data or even on limited data.[17] As a result of scientific uncertainty, the controversy concerning dioxin is certain to continue.

Although dioxin is known to be extremely toxic to mammals, its actions in the human body are not well known. What is known is that sufficient exposure to dioxin (usually from meat or milk containing the chemical) produces a skin condition (a form of acne) that may be accompanied by loss of weight, liver disorders, and nerve damage.[18]

Studies of animals exposed to dioxin suggest that some fish, birds, and other animals are sensitive to even small amounts. As a result, it can cause widespread environmental damage to wildlife, including birth defects and death. However, the concentration necessary to cause human health hazards is still controversial. Studies suggest that workers exposed to high concentrations of dioxin for longer than a year have an increased risk of dying of cancer.[19]

The EPA has recently reclassified dioxin from a "probably" to a "known" human carcinogen. For most of the exposed people, such as those eating a diet high in animal fat, the EPA puts the risk of developing cancer between 1 in 1,000 and 1 in 100. This estimate represents the highest possible risk for the most exposed individuals. For most people the risk will likely be much lower, or near zero.[20]

The EPA has set an acceptable intake of dioxin at 0.006 pg per kilogram of body weight per day (1 pg = 10^{-12} g; see appendix for prefixes and multiplication factors). This level is deemed too low by some scientists, who argue that the acceptable intake ought to be 100 to 1,000 times higher, or approximately 1 to 10 pg/day.[19] The EPA believes that setting the level this much higher could result in health effects. However, some scientists assert that lack of data precludes establishment of a specific threshold concentration of dioxin at which health hazards begin.[21] As indicated by these uncertainties, toxicity of dioxin will remain unclear until additional studies better delineate the potential hazard.

Dioxin is a stable, long-lived chemical that is accumulating in the environment. Analysis of sediments taken from the bottom of Lake Superior suggests that the rate of deposition of dioxin increased eightfold from 1940 to 1970. However, since then, rates have slowly declined.[22] As yet, we have not been able to determine a safe, reliable, and economically feasible way to clean up areas contaminated by dioxin. Many old waste-disposal sites are contaminated by dioxin; it may also be found in soil and streams several kilometers around the sites.

The dioxin problem became well known in 1983 when Times Beach, Missouri, a river town just west of Saint Louis with a population of 2,400, was evacuated and purchased for $36 million by the government. The evacuation and purchase occurred after the discovery that oil sprayed on the town's roads to control dust contained dioxin and the entire area had been contaminated. Times Beach was labeled a dioxin ghost town (Figure 15.7). Today, the buildings have been bulldozed, and all that is left is a grassy and woody area enclosed by a barbed-wire-topped chain-link fence. The evacuation has since been viewed by some scientists (including the person who ordered the evacuation) as an overreaction by the government to a perceived dioxin hazard.

The EPA convened a panel of scientists to reevaluate the risk to the environment and to people from exposure to dioxin. Subsequent reports concluded that:

- Dioxin is a probable human carcinogen, and exposure is known to cause disruption of the endocrine, immune, and reproductive systems. However, it is not a widespread, significant cancer threat to people at ordinary (very low) levels of exposure.[23]

- The cancer risk to workers exposed to chemicals containing high concentrations of dioxin may be higher than previously thought.[23]

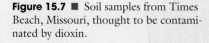 Very small levels of dioxin released into the environment can cause serious damage to wildlife sensitive to the chemical, potentially causing significant damage to ecosystems.[24]

As noted, the controversy concerning the toxicity of dioxin is not over.[25] Some environmental scientists argue that the regulation of dioxin must be tougher, whereas the industries producing the chemical argue that the dangers of exposure are overestimated.

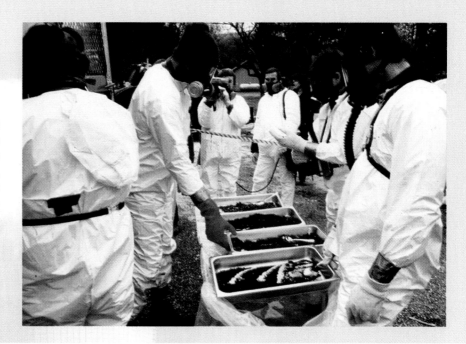

Figure 15.7 ■ Soil samples from Times Beach, Missouri, thought to be contaminated by dioxin.

Thermal Pollution

Thermal pollution, also called heat pollution, occurs when heat released into water or air produces undesirable effects. Heat pollution can occur as a sudden, acute event or as a long-term, chronic release. Sudden heat releases may result from natural events, such as brush or forest fires and volcanic eruptions, or from human-induced events, such as agricultural burning. The major sources of chronic heat pollution are electric power plants that produce electricity in steam generators.

The release of large amounts of heated water into a river changes the average water temperature and the concentration of dissolved oxygen (warm water holds less oxygen than cooler water), thereby changing the river's species composition (see the discussion of eutrophication in Chapter 22). Every species has a range of temperature within which it can survive and an optimal temperature for living. For some species of fish, the range is small, and even a small change in water temperature is a problem. Lake fish move away when the water temperature rises more than about 1.5°C above normal; river fish can withstand a rise of about 3°C.[26] Heating river water can change its natural conditions and disturb the ecosystem in several ways. Fish spawning cycles may be disrupted, and the fish may have a heightened susceptibility to disease; warmer water causes physical stress in some fish, and they may be easier for predators to catch; and warmer water may change the type and abundance of food available for fish at various times of the year.

There are several solutions to chronic thermal discharge into bodies of water. The heat can be released into the air by cooling towers (Figure 15.8), or the heated water can be temporarily stored in artificial lagoons until it is cooled to normal temperatures. Some attempts have been made to use the heated water to grow organisms of commercial value that require warmer water temperatures. Waste heat from a power plant can also be captured and used for a variety of purposes, such as warming buildings. (See Chapter 17 for a discussion of cogeneration.)

Particulates

Particulates are small particles of dust (including soot and asbestos fibers) released into the atmosphere by many natural processes and human activities. Modern farming and combustion of oil and coal add considerable amounts of particulates to the atmosphere, as do dust storms, fires (Figure 15.9), and volcanic eruptions. The 1991 eruptions of Mount Pinatubo in the Philippines were the largest volcanic eruptions of the twentieth century, explosively delivering huge amounts of volcanic ash, sulfur dioxide, and other volcanic material and gases into the atmosphere to elevations up to 30 km (18.6 mi). Eruptions can have a significant impact on the global environment and are linked to global climate change and stratospheric ozone depletion (see Chapters 23 and 26). In addition, many chemical toxins, such as heavy metals, enter the biosphere as particulates. Sometimes, nontoxic particulates link with toxic substances, creating a synergetic threat. (See discussion of particulates in Chapter 24.)

Asbestos

Asbestos is a term for several minerals that take the form of small, elongated particles, or fibers. Industrial use of asbestos has contributed to fire prevention and has provided protection from the overheating of materials. Asbestos is also used as insulation for a variety of purposes. Unfortunately, however, excessive contact with asbestos

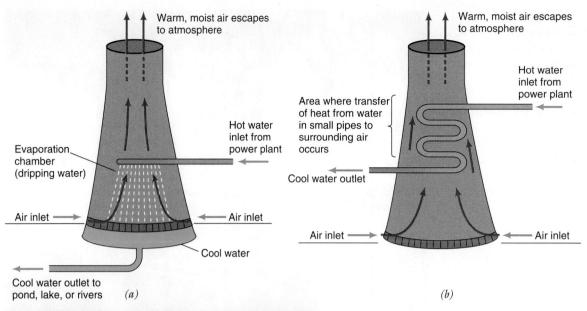

Warm, moist air escapes to atmosphere

Evaporation chamber (dripping water)

Hot water inlet from power plant

Air inlet → ← Air inlet

Cool water

Cool water outlet to pond, lake, or rivers

(a)

Warm, moist air escapes to atmosphere

Hot water inlet from power plant

Area where transfer of heat from water in small pipes to surrounding air occurs

Cool water outlet

Air inlet → ← Air inlet

(b)

(c)

Figure 15.8 ■ Two types of cooling towers. (*a*) Wet cooling tower. Air circulates through tower; hot water drips down and evaporates, cooling the water. (*b*) Dry cooling tower. Heat from the water is transferred directly to the air, which rises and escapes the tower. (*c*) Cooling towers emitting steam at Didcot power plant, Oxfordshire, England. Red and white lines are vehicle lights resulting from long exposure time (photograph taken at dusk).

Figure 15.9 ■ Fires in Indonesia in 1997 caused serious air pollution problems. People here are purchasing surgical masks in an attempt to breathe cleaner air.

has led to asbestosis (a lung disease caused by the inhalation of asbestos) and to cancer in some industrial workers. Experiments with animals have demonstrated that asbestos can cause tumors to develop if the fibers are embedded in lung tissue.[27] The hazard related to certain types of asbestos under certain conditions is thought to be so serious that extraordinary steps have been taken to reduce the presence of asbestos or to ban it outright. The expensive process of asbestos removal from old buildings (particularly schools) in the United States is one of those steps.

There are several types of asbestos, and they are not equally hazardous. The form most commonly utilized in the United States is white asbestos, which comes from the mineral chrysolite. It has been used as an insulation material around pipes, in floor and ceiling tiles, and for brake linings of automobiles and other vehicles. Approximately 95% of the asbestos that is now in place in the United States is of the chrysolite type. Most of this asbestos was mined in Canada, and environmental health studies of Canadian miners show that exposure to chrysolite asbestos is not particularly harmful. Studies involving another type of asbestos, known as crocidolite asbestos (blue asbestos), suggest that exposure to this mineral can be very hazardous and evidently does cause lung disease. Several other types of asbestos have also been shown to be harmful.[27]

A great deal of fear has been associated with nonoccupational exposure to chrysolite asbestos in the United States. Tremendous amounts of money have been expended to remove it from homes, schools, public buildings, and other sites in spite of the fact that there has been no asbestos-related disease recorded among those exposed to chrysolite in nonoccupational circumstances. It is now thought that much of the removal was unnecessary and that chrysolite asbestos doesn't pose a significant health hazard. Additional research into health risks from other varieties of asbestos is necessary to better understand the potential problem and to outline strategies to eliminate potential health problems.

Electromagnetic Fields

Electromagnetic fields (EMFs) are part of everyday urban life. Electric motors, electric transmission lines for utilities, and electrical appliances—such as toasters, electric blankets, and computers—all produce magnetic fields. There is currently a controversy over whether these fields produce a health risk.

Early on, investigators did not believe that magnetic fields were harmful, because fields drops off quickly with distance from the source and the strength of the fields that most people come into contact with are relatively weak. For example, the magnetic fields generated by power transmission lines or by a computer terminal are normally only about 1% of Earth's magnetic field; directly below power lines the electric field induced in the body is about what the body naturally produces within cells.

Several early studies, however, concluded that children exposed to EMFs from power lines have an increased risk of contracting leukemia, lymphomas, and nervous system cancers.[28] Investigators concluded that children so exposed are about one and a half to three times more likely to develop cancer than children with very low exposure to EMFs, but the results were questioned because of perceived problems with the research design (problems of sampling, tracking children, and estimating exposure to EMF).

A later study analyzed over 1,200 children, approximately half of whom suffered from acute leukemia. It was necessary to estimate residential exposure to magnetic fields generated by nearby power lines in the children's present and former homes. Results of that study, which is the largest such investigation to date, concluded that there is no association between childhood leukemia and measured exposure to magnetic fields.[28, 29]

Another study compared exposure of electric utility workers to magnetic fields with incidents of brain cancer and leukemia. That study revealed a weak association between exposure to magnetic fields and both brain cancer and leukemia. However, the associations are not strong and were not statistically significant.[30]

Saying that data are not statistically significant is another way of stating that the relationship between exposure and disease cannot be reasonably established, given the database that was analyzed. It does not mean that additional data in a future study will not find a statistically significant relationship. Statistics can predict the strength of the relationship between variables such as exposure to a toxin and the incidence of a disease, but statistics cannot prove a cause-and-effect relationship between them.

In summary, in spite of the many studies that have been completed to evaluate relationships between disease and exposure to magnetic fields in our modern urban environment, the jury is still out. There seems to be some indication that magnetic fields may cause problems, but so far the risks are relatively small and difficult to quantify.

Noise Pollution

Noise pollution is unwanted sound. Sound is a form of energy that travels as waves. We hear sound because our ears respond to sound waves through vibrations of the eardrum. The sensation of loudness is related to the intensity of the energy carried by the sound waves and is measured in units of decibels (dB). The threshold for human hearing is 0 dB; the average sound level in the interior of a home is about 45 dB; the sound of an automobile, about 70 dB; and the sound of a jet aircraft taking off, about 120 dB (see Table 15.2). A 10-fold increase in the strength of a particular sound adds 10 dB units on the scale. An increase of 100 times adds 20 units.[13] The decibel scale is

Table 15.2 • Examples of Sound Levels		
Sound Source	Intensity of Sound (dB)	Human Perception
Threshold of hearing	0	
Rustling of leaf	10	Very quiet
Faint whisper	20	Very quiet
Average home	45	Quiet
Light traffic (30 m away)	55	Quiet
Normal conversation	65	Quiet
Chain saw (15 m away)	80	Moderately loud
Jet aircraft flyover at 300 m	100	Very loud
Rock music concert	110	Very loud
Thunderclap (close)	120	Uncomfortably loud
Jet aircraft takeoff at 100 m	125	Uncomfortably loud
	140	Threshold of pain
Rocket engine (close)	180	Traumatic injury

logarithmic; it increases exponentially as a power of 10. For example, 50 dB is 10 times louder than 40 dB and 100 times louder than 30 dB.

Environmental effects of noise depend not only on the total energy but also on the sound's pitch, frequency, and time pattern and the length of exposure to the sound. Very loud noises (more than 140 dB) cause pain, and high levels can cause permanent hearing loss. Human ears can take sound up to about 60 dB without damage or hearing loss. Any sound above 80 dB is potentially dangerous. The noise of a lawn mower or motorcycle will begin to damage hearing after about eight hours of exposure. In recent years, there has been concern for teenagers (and older people, for that matter) who have suffered some permanent loss of hearing following extended exposure to amplified rock music (110 dB). At a noise level of 110 dB, damage to hearing can occur after an exposure time of only half an hour. Loud sounds at the workplace are another hazard. Levels of noise that are below the hearing-loss level may still interfere with human communication and may cause irritability. Noise in the range of 50–60 dB is sufficient to interfere with sleep, producing a feeling of fatigue upon awakening.

Voluntary Exposure

Voluntary exposure to toxins and potentially harmful chemicals is sometimes referred to as exposure to personal pollutants. The most common of these are tobacco, alcohol, and other drugs. Use and abuse of these substances have led to a variety of human ills, including death and chronic disease, criminal activity such as reckless driving and manslaughter, loss of careers, street crime, and the straining of human relations at all levels.

Scientific evidence has demonstrated that use of tobacco, in all of its forms, is both habit forming and dangerous to human health. Tobacco contains a variety of components that are toxic, carcinogenic, radioactive, and addictive. It has been estimated that 30% of all cancers in the United States are tied to smoking-related disorders. According to the American Cancer Society, cigarette smoking is responsible for approximately 80% of lung cancers; secondhand smoke is also a hazard, as are tobacco products such as chewing tobacco. Although the number of people who smoke in the United States, as a percentage of adults, has decreased in recent years, young people and people in developing countries are still becoming addicted to cigarettes and cigars.

Many people in our society use alcohol at social gatherings and celebrations. Approximately 70% of all American adults drink some alcohol, and moderate use of alcohol is legal and accepted in our society. However, when abused, alcohol causes very serious problems. Approximately one-half of all deaths in automobile accidents are related to alcohol use by drivers. Furthermore, significant numbers of violent crimes and other criminal activities are committed by people under the influence of alcohol. Some people believe that alcohol is the most abused drug in our society today. Young people

unfamiliar with the potential toxicity of alcohol have died from alcohol overdose (for example, by drinking 21 drinks on their 21st birthdays). Chronic alcoholism has many toxic consequences, including liver and heart disease.

A variety of illegal drugs are commonly used in the United States and other parts of the world. These drugs have various effects on their users, but the end result is often the degradation of the mind and/or body. Illegal drugs are particularly dangerous because their strength, composition, and other chemical characteristics are seldom subject to quality control. Of particular concern in recent years has been the development of synthetic (designer) drugs that are addictive and capable of causing significant health problems and even death—in some cases—in first-time users.

15.3 General Effects of Pollutants

Almost every part of the human body is affected by one pollutant or another, as shown in Figure 15.10*a*. For example, lead and mercury (remember the Mad Hatter) affect the brain; arsenic, the skin; carbon monoxide, the heart; and fluoride, the bones. Wildlife is affected as well. Sites of effects of major pollutants in wildlife are shown in Figure 15.10*b*; effects of pollutants on wildlife populations are listed in Table 15.3.

The lists of potential toxins and affected body sites for humans and other animals in Figure 15.10 may be somewhat misleading. For example, chlorinated hydrocarbons, such as dioxin, are stored in the fat cells of animals, but they cause damage not only to fat cells but to the entire organism through disease, damaged skin, and birth

Table 15.3 • Effects of Pollutants on Wildlife	
Effect on Population	**Examples of Pollutants**
Changes in abundance	Arsenic, asbestos, cadmium, fluoride, hydrogen sulfide, nitrogen oxides, particulates, sulfur oxides, vanadium, POPs[a]
Changes in distribution	Fluoride, particulates, sulfur oxides, POPs
Changes in birth rates	Arsenic, lead, POPs
Changes in death rates	Arsenic, asbestos, beryllium, boron, cadmium, fluoride, hydrogen sulfide, lead, particulates, selenium, sulfur oxides, POPs
Changes in growth rates	Boron, fluoride, hydrochloric acid, lead, nitrogen oxides, sulfur oxides, POPs

[a] Pesticides, PCBs, hormonally active agents, dioxin, and DDT are examples (see Table 15.1).
Source: J. R. Newman, *Effects of Air Emissions on Wildlife*, U.S. Fish and Wildlife Service, 1980. Biological Services Program, National Power Plant Team, FWS/OBS-80/40, U.S. Fish and Wildlife Service, Washington, D.C.

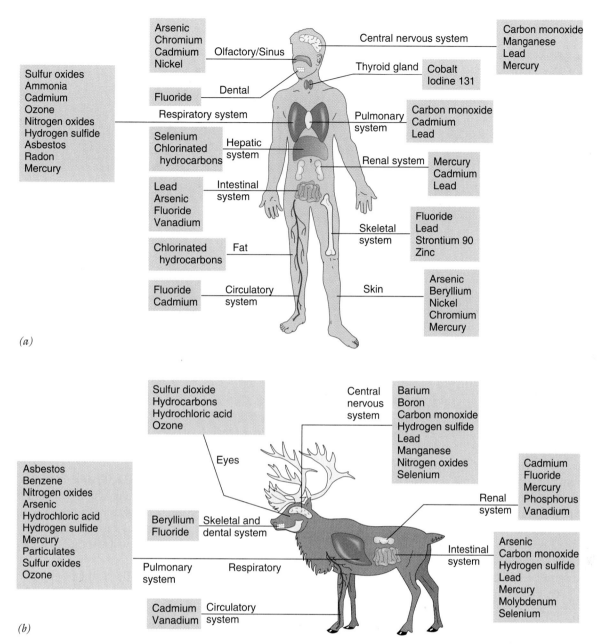

(a)

(b)

Figure 15.10 ■ (*a*) Effects of some major pollutants in human be-ings. (*b*) Known sites of effects of some major pollutants in wildlife. [*Source:* (*a*) G. L. Waldbott, *Health Effects of Environmental Pollutants*, 2nd ed. (St. Louis, MO: Mosby, 1978).

Copyright 1978 by C. V. Mosby. (*b*) J. R. Newman, *Effects of Air Emissions on Wildlife Resources*, U.S. Fish and Wildlife Services Program, National Power Plant Team, FWS/OBS-80/40 (Washington, D.C.: U.S. Fish and Wildlife Service, 1980).]

defects. Similarly, a toxin that affects the brain, such as mercury, causes a wide variety of problems and symptoms, as illustrated in the Minamata, Japan, example (discussed in A Closer Look 15.2). The value of Figure 15.10 is in helping us to understand in general the adverse effects of excess exposure to chemicals.

Concept of Dose and Response

Five centuries ago, the physician and alchemist Paracelsus wrote that "everything is poisonous, yet nothing is poiso-nous." By this, he meant essentially that a substance in too

great an amount can be dangerous, yet in an extremely small amount can be relatively harmless. Every chemical element has a spectrum of possible effects on a particular organism. For example, selenium is required in small amounts by living things but may be toxic or increase the probability of cancer in cattle and wildlife when it is pre-sent in high concentrations in the soil. Copper, chromium, and manganese are other chemical elements required in small amounts by animals but toxic in higher amounts.

It was recognized many years ago that the effect of a certain chemical on an individual depends on the dose. This concept is termed **dose response**. Dose dependency

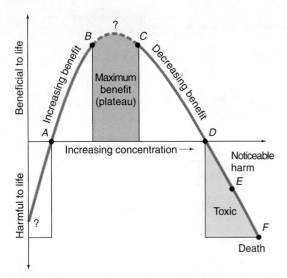

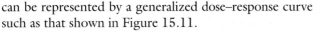

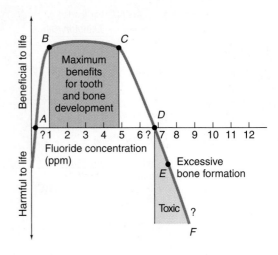

Figure 15.11 ■ Generalized dose–response curve. Low concentrations of a chemical may be harmful to life (below point *A*). As the concentration of the chemical increases from *A* to *B*, the benefit to life increases. The maximum concentration that is beneficial to life lies within the benefit plateau (*B–C*). Concentrations greater than this plateau provide less and less benefit (*C–D*) and will harm life (*D–F*) as toxic concentrations are reached. Increased concentrations above the toxic level may result in death.

Figure 15.12 ■ General dose–response curve for fluoride showing the relationship between fluoride concentration and physiological benefit.

can be represented by a generalized dose–response curve such as that shown in Figure 15.11.

When various concentrations of a chemical present in a biological system are plotted against the effects on the organism, two things are apparent. First, relatively large concentrations are toxic and even lethal (points *D*, *E*, and *F* in Figure 15.11). Second, trace concentrations may be beneficial for life (between points *A* and *D*); and the dose–response curve forms a plateau of optimal concentration and maximum benefit between two points (*B* and *C*). Points *A*, *B*, *C*, *D*, *E*, and *F* in Figure 15.11 are important thresholds in the dose–response curve. Unfortunately, the amounts at which points *E* and *F* occur are known only for a few substances, for a few organisms, including people; and the very important point *D* is all but unknown. Doses that are beneficial, harmful, or lethal may differ widely for different organisms and are difficult to characterize.

Fluorine provides a good example of the general dose–response concept. Fluorine forms fluoride compounds that prevent tooth decay and promote the development of a healthy bone structure.

Relationships between the concentration of fluoride (in a compound of fluorine, such as sodium fluoride, NaF) and health show a specific dose–response curve (Figure 15.12). The plateau for an optimal concentration of fluoride (point *B* to point *C*) to reduce dental caries (cavities) is from about 1 ppm to just less than 5 ppm. Levels greater than 1.5 ppm do not significantly decrease tooth decay but do increase the occurrence of discoloration of teeth. Concentrations of 4 to 6 ppm reduce the prevalence of osteoporosis, a dis-

ease characterized by loss of bone mass; and toxic effects are noticed between 6 and 7 ppm (point *D* in Figure 15.12).

Dose–Response Curve (LD-50, ED-50, and TD-50)

Individuals differ in their response to chemicals, and it is difficult to predict the dose that will cause a response in a particular individual. For this reason, it is practical to predict what percentage of a population will respond to a specific dose of a chemical.

For example, the dose at which 50% of the population dies is called the lethal dose 50, or **LD-50**. The **LD-50** is a crude approximation of a chemical's toxicity. It is a gruesome index that does not adequately convey the sophistication of modern toxicology and is of little use in setting a standard for toxicity. However, the LD-50 determination is required for new synthetic chemicals as a way of estimating their toxic potential. Table 15.4 lists, as examples, LD-50 values in rodents for selected chemicals.

The **ED-50** (effective dose 50%) is the dose that causes an effect in 50% of the population of observed subjects. For example, the ED-50 of aspirin would be the dose that relieves headaches in 50% of the people.[31]

The **TD-50** (toxic dose 50%) is defined as the dose that is toxic to 50% of the population. TD-50 is often used to indicate responses such as reduced enzyme activity, decreased reproductive success, or onset of specific symptoms, such as loss of hearing, nausea, or slurred speech.

For a particular chemical, there may be a whole family of dose–response curves, as illustrated in Figure 15.13. Which dose is of interest depends on what is being evaluated. For example, for insecticides we may wish to know the dose

Table 15.4 • Approximate LD-50 Values (for Rodents) for Selected Agents	
Agent	LD$_{50}$(mg/kg)[a]
Sodium chloride (table salt)	4,000
Ferrous sulfate (to treat anemia)	1,520
2,4-D (a weed killer)	368
DDT (an insecticide)	135
Caffeine (in coffee)	127
Nicotine (in tobacco)	24
Strychnine sulfate (used to kill certain pests)	3
Botulinum toxin (in spoiled food)	0.00001

[a] Milligrams per kilogram of body mass (termed mass weight, although it really isn't a weight) administered by mouth to rodents. Rodents are commonly used in such evaluations, in part because they are mammals (as we are), are small, have a short life expectancy, and their biology is well known.

Source: H. B. Schiefer, D. C. Irvine, and S. C. Buzik, *Understanding Toxicology* (New York: CRC Press, 1997).

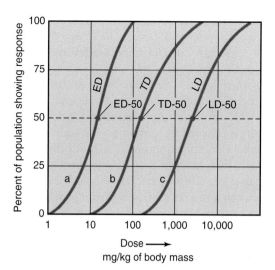

Figure 15.13 ■ Idealized diagram illustrating a family of dose–response curves for a specific drug: ED (effective dose), TD (toxic dose), and LD (lethal dose). Notice the overlap for some parts of the curves. For example, at ED-50, a few percent of the people exposed to that dose will suffer a toxic response, but none will die. At TD-50 dose, about 1% of the people exposed to that dose will die.

that will kill 100% of the insects exposed; therefore the LD-95 (the dose that kills 95% of the insects) may be the minimum acceptable level. However, when considering human health and the exposure to a particular toxin, we often want to know the LD-0—the maximum dose that does not cause any deaths.[31] For potentially toxic compounds such as insecticides, which may form a residue on food or food additives, we want to ensure that the expected levels of human exposure will have no known toxic effects. From an environmental perspective, this is important because of concerns about increased risk of cancer associated with exposure to toxic agents.[31]

For drugs used to treat a particular disease, the efficiency of the drug as a treatment is of paramount importance. In addition to knowing what the therapeutic value (ED-50) is, it is also important to know the relative safety of the drug. For example, there may be an overlap between the therapeutic dose (ED) and the toxic dose (TD) (see Figure 15.13). That is, the dose that causes a positive therapeutic response in some individuals might be toxic to others. A quantitative measure of the relative safety of a particular drug is the *therapeutic index*, defined as the ratio of the LD-50 to the ED-50. The greater the therapeutic index, the safer the drug is believed to be.[32] In other words, a drug with a large difference between the lethal and therapeutic dose is safer than one with a smaller difference.

Threshold Effects

Recall from A Closer Look 15.2 that a **threshold** is a level below which no effect occurs and above which effects begin to occur. If a threshold dose of a chemical exists, then a concentration of that chemical in the environment

below the threshold is safe. If there is no threshold dose, then even the smallest amount of the chemical has some negative toxic effect (Figure 15.14).

Whether or not there is a threshold effect for environmental toxins is an important environmental issue. For example, the U.S. Federal Clean Water Act originally stated as a goal to reduce to zero the discharge of pollutants into water. The goal implies there is no such thing as a threshold effect, since no level of toxin is to be legally

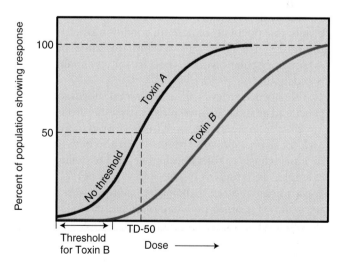

Figure 15.14 ■ In this hypothetical toxic dose–response curve, toxin *A* has no threshold; even the smallest amount has some measurable effect on the population. The TD-50 for toxin *A* is the dose required to produce a response in 50% of the population. Toxin *B* has a threshold (flat part of curve) where response is constant as dose increases. After the threshold dose is exceeded, the response increases.

permitted. However, it is unrealistic to believe zero discharge of a water pollutant can be achieved or to believe that we can reduce to zero the concentration of chemicals shown to be carcinogenic.

A problem in evaluating thresholds for toxic pollutants is that it is difficult to account for synergistic effects. Little is known about if or how thresholds might change if an organism is exposed to more than one toxin at the same time or to a combination of toxins and other chemicals, some of which are beneficial. Exposures of people to chemicals in the environment are complex, and we are only beginning to understand and conduct research on the possible interactions and consequences of multiple exposures.

Ecological Gradients

Dose–response effects differ among species. For example, the kinds of vegetation that can live nearest to a toxic source are often small plants with relatively short lifetimes (grasses, sedges, and weedy species usually regarded as pests) that are adapted to harsh and highly variable environments. Farther from the toxic source, trees may be able to survive. Changes in vegetation with distance from a toxic source define the **ecological gradient**.

Ecological gradients may be found around smelters and other industrial plants that discharge pollutants into the atmosphere from smokestacks. For example, ecological gradient patterns can be observed in the area around the smelters of Sudbury, Ontario, discussed earlier in this chapter (A Closer Look 15.1). Near the smelters, an area that was once forest is now a patchwork of bare rock and soil occupied by small plants.

Tolerance

The ability to resist or withstand stress resulting from exposure to a pollutant or harmful condition is referred to as **tolerance**. Tolerance can develop for some pollutants in some populations, but not for all pollutants in all populations.

Tolerance may result from behavioral, physiological, or genetic adaptation. *Behavioral tolerance* results from changes in behavior. For example, mice learn to avoid traps.

Physiological tolerance results when the body of an individual adjusts to tolerate a higher level of pollutant. For example, in studies at the University of California Environmental Stress Laboratory, students were exposed to ozone (O_3), an air pollutant often present in large cities (Chapter 24). The students at first experienced symptoms that included irritation of eyes and throat and shortness of breath. However, after a few days, their bodies adapted to the ozone, and they reported that they believed they were no longer breathing ozone-contaminated air, even though the concentration of O_3 stayed the same. This phenomenon explains why some people who regularly breathe polluted air report that they do not notice the pollution. Of course, it does not mean that the ozone is doing no dam-

age; it is, especially in people with existing respiratory problems. There are many mechanisms for physiological tolerance, including *detoxification*, in which the toxic chemical is converted to a nontoxic form, and for the internal transport of the toxin to a part of the body where it is not harmful, such as fat cells.

Genetic tolerance, or adaptation, results when some individuals in a population are naturally more resistant to a toxin than others. They are less damaged by exposure and more successful in breeding. Resistant individuals pass on the resistance to future generations, who are also more successful at breeding. Adaptation has been observed among some insect pests following exposure to some chemical pesticides. For example, certain strains of malaria-causing mosquitoes are now resistant to DDT (see the discussion in Chapter 12); and some organisms that cause deadly infectious diseases have become resistant to common antibiotic drugs, such as penicillin.

Acute and Chronic Effects

Pollutants can have acute and chronic effects. An *acute effect* is one that occurs soon after exposure, usually to large amounts of a pollutant. A *chronic effect* takes place over a long period, often as a result of exposure to low levels of a pollutant. For example, a person exposed all at once to a high dose of radiation may be killed by radiation sickness soon after exposure (an acute effect). However, that same total dose, received slowly in small amounts over an entire lifetime, may instead cause mutations and lead to disease or affect the person's DNA and offspring (a chronic effect).

15.4 Risk Assessment

Risk assessment can be defined as the process of determining potential adverse environmental health effects to people exposed to pollutants and potentially toxic materials (recall the discussion of measurements and methods of science in Chapter 2). Such an assessment generally includes four steps:[33]

1. *Identification of the hazard.* Identification consists of testing materials to determine whether exposure is likely to cause environmental health problems. One method used is to investigate populations of people who have been previously exposed. For example, to understand the toxicity of radiation produced from radon gas, researchers studied workers in uranium mines. Another method is to perform experiments to test effects on animals, such as mice, rats, or monkeys. This method has drawn increasing criticism from groups of people who believe such experiments are unethical. Another approach is to try to understand how a particular chemical works at the molecular level on cells. For example, research has been done to determine how dioxin interacts with living cells to produce an adverse response. After quantifying the response, scientists can develop math-

ematical models to predict or estimate dioxin's risk.[19] This relatively new approach might also be applicable to other potential toxins that work at the cellular level.

2. *Dose–response assessment.* The next step involves identifying relationships between the dose of a chemical (therapeutic drug, pollutant, or toxin) and the health effects to people. Some studies involve administering fairly high doses of a chemical to animals. The effects, such as illness or symptoms (rash, tumor development), are recorded for varying doses, and the results are used to predict the response in people. This is difficult, and the results are controversial for several reasons:

- The dose that results in a particular response may be very small and subject to measurement errors.
- There may be arguments over whether thresholds are present or absent.
- Experiments on animals such as rats, mice, or monkeys may not be directly applicable to humans.
- The assessment may rely on probability and statistical analysis. Although statistically significant results from experiments or observations are accepted as evidence to support an argument, statistics cannot establish that the substance tested *caused* the observed response.

3. *Exposure assessment.* Exposure assessment evaluates the intensity, duration, and frequency of human exposure to a particular chemical pollutant or toxin. The hazard to society is directly proportional to the total population exposed. The hazard to an individual is generally greater closer to the source of exposure. Like dose–response assessment, exposure assessment is difficult, and the results are often controversial, in part because of difficulties in measuring the concentration of a toxin present in doses as small as parts per million, billion, or even trillion. Some questions that exposure assessment attempts to answer are:

- How many people were exposed to concentrations of a toxin thought to be dangerous?
- How large an area was contaminated by the toxin?
- What are the ecological gradients for exposure to the toxin?
- How long were people exposed to a particular toxin?

4. *Risk characterization.* During this final step, the goal is to delineate health risk in terms of the magnitude of the potential environmental health problem that might result from exposure to a particular pollutant or toxin. To do this, it is necessary to identify the hazard, complete the dose-response assessment, and evaluate the exposure assessment as outlined above. This step involves all the uncertainties of the prior steps, and results are again likely to be controversial.

In summary, risk assessment is difficult, costly, and controversial. Each chemical is different, and there is no one method of determining responses of humans for specific EDs or TDs. Toxicologists use the scientific method of hypothesis testing with experiments (see Chapter 2) to generate predictions of how specific doses of a chemical may affect humans. Warning labels listing potential side effects of using a specific medication are required by law, and these warnings result from toxicology studies to determine a drug's safety.

Finally, risk assessment requires making scientific judgments and formulating actions to help minimize environmental health problems related to human exposure to pollutants and toxins. The process of *risk management* integrates the assessment of risk with technical, legal, political, social, and economic issues.[19] Scientific arguments concerning the toxicity of a particular material are often open to debate. For example, there is debate concerning whether the risk from dioxin is linear. That is, do effects start at minimum levels of exposure to dioxin and gradually increase, or is there a threshold exposure beyond which environmental health problems occur? (See A Closer Look 15.3).[19, 24] It is the task of people in appropriate government agencies assigned to manage risk to make judgments and decisions based on the risk assessment and then to take appropriate actions to minimize the hazard resulting from exposure to toxins. This might involve invoking the precautionary principle discussed in Chapter 1.

Summary

- Disease is an imbalance between an organism and the environment. Disease seldom has a one-cause to one-effect relationship, and there is often a gray zone between the state of health and the state of disease.

- Pollution produces an impure, dirty, or otherwise unclean state. Contamination means making something unfit for a particular use through the introduction of undesirable materials. Toxic materials are poisonous to people and other living things; toxicology is the study of toxic materials. A concept important in studying pollution problems is synergism, whereby actions of different substances produce a combined effect greater than the sum of the effects of the individual substances.

- How we measure the amount of a particular pollutant introduced into the environment or the concentration of that pollutant varies widely, depending on the substance. Common units for expressing the concentration of pollutants are parts per million (ppm) and parts per billion (ppb). Air pollutants are commonly measured in units such as micrograms of pollutant per cubic meter of air ($\mu g/m^3$).

- Categories of environmental pollutants include toxic chemical elements (particularly heavy metals), organic compounds, radiation, heat, particulates, electromagnetic fields, and noise.

Is Lead in the Urban Environment Contributing to Antisocial Behavior?

Lead is one of the most common toxic (harmful or poisonous) metals in our inner-city environments, and it may be linked to delinquent behavior in children. Lead is found in all parts of the urban environment (air, soil, older pipes, and some paint, for example) and in biological systems, including people (Figure 15.15). There is no apparent biological need for lead, but it is sufficiently concentrated in the blood and bones of children living in inner cities to cause health and behavior problems. In some populations, over 20% of the children have blood concentrations of lead that are higher than those believed safe.

Lead affects nearly every system of the body. Acute lead toxicity may be characterized by a variety of symptoms, including anemia, mental retardation, palsy, coma, seizures, apathy, uncoordination, subtle loss of recently acquired skills, and bizarre behavior. Lead toxicity is particularly a problem for young children, who apparently are more susceptible to lead poisoning than are adults. Following acute toxic response to lead, some children manifest aggressive, difficult-to-manage behavior.

The occurrence of lead toxicity or lead poisoning has cultural, political, and sociological implications. Over 2,000 years ago, the Roman Empire produced and used tremendous amounts of lead for a period of several hundred years. Production rates were as high as 55,000 metric tons per year. Romans had a wide variety of uses for lead. Lead was used in pots in which grapes were crushed and processed into a syrup for making wine, in cups and goblets from which wine was drunk, and as a base for cosmetics and medicines. In the homes of Romans wealthy enough to have running water, lead was used to make the pipes that carried the water. It has been argued that lead poisoning among the upper class in Rome was partly responsible for Rome's decline. Lead poisoning probably resulted in widespread stillbirths, deformities, and brain damage. Studies analyzing the lead content of bones of ancient Romans tend to support this hypothesis.

The occurrence of lead in glacial ice cores from Greenland has also been studied. Glaciers have an annual growth layer of ice. Older layers are buried by younger layers, allowing us to identify the age of each layer. Researchers drill glaciers, taking continuous samples of the layers that look like long solid rods of glacial ice; these are called cores. Measurements of the concentration of lead from cores show that during the Roman period, from approximately 500 B.C. to A.D. 300, lead concentrations in the glacial ice are about four times higher than before and after this period. This suggests that the mining and smelting of lead in the Roman Empire added small particles of lead to the atmosphere that eventually settled out in the glaciers of Greenland.

Lead toxicity, then, seems to have been a problem for a long time. Now, an emerging, interesting, and potentially significant hypothesis is that, in children, lead concentrations below the levels known to cause physical damage may be associated with an increased potential for antisocial, delinquent behavior. This is a testable hypothesis. (See Chapter 2 for a discussion of hypotheses.) If the hypothesis is correct, then some of our urban crime may be traced to environmental pollution!

A recent study in children aged 7 to 11 years old measured the amount of lead in bones and compared it with data concerning behavior over a four-year period. The study concluded that an above-average concentration of lead in children's bones was associated with an increased risk for attention-deficit disorder, aggressive behavior, and delinquency. The study took into account factors such as maternal intelligence, socioeconomic status, and quality of child rearing.

Figure 15.15 ■ The lead in urban soils (a legacy of our past use of lead in gasoline) is still concentrated where children are likely to play. Lead-based paint in older buildings, such as these in New York, also remains a hazard to young children, who sometimes ingest flakes of paint.

Critical Thinking Questions

1. What is the main point of the discussion about lead in the bones of children and behavior?

2. What are the main assumptions of the argument? Are they reasonable?

3. What other hypotheses might be proposed to explain the behavior?

- Organic compounds of carbon are produced by living organisms or synthetically by humans. Artificially produced organic compounds may have physiological, genetic, or ecological effects when introduced into the environment. Organic compounds vary with respect to their potential hazards: Some are more readily degraded in the environment than others; some are more likely to undergo biomagnification; and some are extremely toxic even at very low concentrations. Organic compounds that cause serious concern include persistant organic pollutants such as pesticides, dioxin, PCBs, and hormonally active agents.

- The effect of a chemical or toxic material on an individual depends on the dose. It is also important to determine tolerances of individuals as well as acute and chronic effects of pollutants and toxins.

- Risk assessment involves hazard identification, assessment of dose response, assessment of exposure, and risk characterization.

REEXAMINING THEMES AND ISSUES

Human Population

As the total population and population density increase, the probability that more people will be exposed to hazardous materials increases as well. Finding acceptable sites for the disposal of toxic chemicals also becomes more difficult as populations increase and people live closer to industrial areas and waste disposal sites.

Sustainability

Ensuring that future generations inherit a relatively unpolluted, toxin-free, healthy environment remains a difficult problem. Sustainable development requires that our use of chemicals and other materials not damage the environment.

Global Perspective

Release of toxins into the environment may result in global patterns of contamination or pollution. This is particularly true when a toxin or contaminant enters the atmosphere, surface water, or oceans and becomes widely dispersed. For example, pesticides, herbicides, or heavy metals emitted into the atmosphere in the midwestern United States may be transported by winds and deposited on glaciers in polar regions.

Urban World

Processes of industrial activity in urban areas concentrate potentially toxic materials that may be inadvertently, accidentally, or deliberately released into the environment. Human exposure to a variety of toxins—including lead, asbestos, particulates, organic chemicals, radiation, and noise—are often greater in urban areas.

People and Nature

The feminization of frogs and other animals resulting from exposure to human-produced hormonally active agents (HAAs) is an early warning or red flag that we are disrupting some basic aspects of nature. We are performing unplanned experiments on nature, and the consequences to us and other living organisms with which we share the environment are poorly understood. Control of HAAs seems to be an obvious candidate for the application of the precautionary principle, discussed in Chapter 1.

Science and Values

Because we value both human and nonhuman life, we are interested in expanding our knowledge concerning risk of exposure of living things to chemicals, pollutants, and toxins in the natural and human-made environment. Unfortunately, the state of our knowledge of risk assessment is often incomplete, and the dose–response for many chemicals is poorly understood.

What we decide to do about exposure to toxic chemicals reflects our values. Increased control of toxic materials in homes and the work environment is expensive. Do we sufficiently value the health of workers in countries where goods are manufactured to pay more for those goods in order to reduce environmental hazards for the workers at their work sites?

Key Terms

area sources **295**
asbestos **303**
biomagnification **297**
carcinogen **295**
contamination **294**
disease **293**
dose response **307**
ecological gradient **310**
ED-50 **308**

electromagnetic fields (EMFs) **305**
heavy metals **297**
hormonally active agent (HAA) **301**
LD-50 **308**
mobile sources **295**
noise pollution **305**
organic compounds **299**

particulates **303**
persistent organic pollutants (POPs) **299**
point sources **295**
pollution **294**
risk assessment **310**
synergism **295**
synthetic organic compounds **299**

TD-50 **308**
thermal pollution **303**
threshold **309**
tolerance **310**
toxic **295**
toxicology **295**

Study Questions

1. Do you think the hypothesis that some crime is caused in part by environmental pollution is valid? Why? Why not? How might the hypothesis be further tested? What are the social ramifications of the tests?

2. What kinds of life-forms would most likely survive in a highly polluted world? What would be their general ecological characteristics?

3. Some environmentalists argue that there is no such thing as a threshold for pollution effects. What is meant by this statement? How would you determine whether it was true for a specific chemical and a specific species?

4. What is biomagnification, and why is it important in toxicology?

5. You are lost in Transylvania while trying to locate Dracula's castle. Your only clue is that the soil around the castle has an unusually high concentration of the heavy metal arsenic. You wander in a dense fog, able to see only the ground a few meters in front of you. What changes in vegetation warn you that you are nearing the castle?

6. Distinguish between acute and chronic effects of pollutants.

7. Design an experiment to test whether tomatoes or cucumbers are more sensitive to lead pollution.

8. Why is it difficult to establish standards for acceptable levels of pollution? In giving your answer, consider physical, climatological, biological, social, and ethical reasons.

9. A new highway is built through a pine forest. Driving along the highway, you notice that the pines nearest the road have turned brown and are dying. You stop at a rest area and walk into the woods. One hundred meters away from the highway, the trees seem undamaged. How could you make a crude dose–response curve from direct observations of the pine forest? What else would be necessary to devise a dose–response curve from direct observation of the forest? What else would be necessary to devise a dose–response curve that could be used in planning the route of another highway?

10. Do you think that your personal behavior is placing you in the gray zone or in suboptimal health? If so, what can you do to avoid chronic disease in the future?

Further Reading

Amdur, M., J. Doull, and C. D. Klaasen, eds. 1991. *Casarett & Doull's Toxicology: The Basic Science of Poisons*, 4th ed. Tarrytown, N.Y.: Pergamon. A comprehensive and advanced work on toxicology.

Carson, R. 1962. *Silent Spring*. Boston: Houghton Mifflin. A classic book on problems associated with toxins in the environment.

Schiefer, H. B., D. G. Irvine, and S. C. Buzik. 1997. *Understanding Toxicology: Chemicals, Their Benefits and Risks*. Boca Raton, Fla.: CRC Press. A concise introduction to toxicology as it pertains to everyday life, including information about pesticides, industrial chemicals, hazardous waste, and air pollution.

Travis, C. C., and H. A. Hattemer-Frey. 1991. "Human Exposure to Dioxin," *The Science of the Total Environment* 104:97–127. An extensive technical review of dioxin accumulation and exposure.

Natural Disasters and Catastrophes

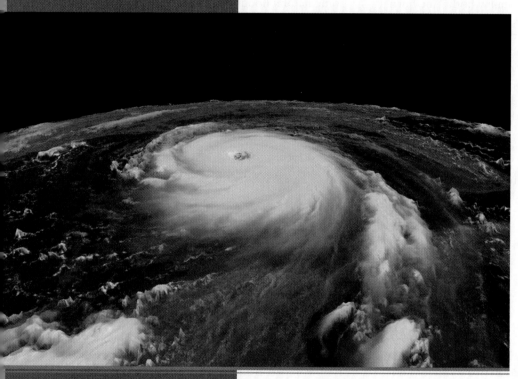

Hurricane Katrina approaching the Gulf Coast in late August 2005.

Learning Objectives

Population pressure along with poor land use decisions is creating catastrophes out of situations that more commonly used to be disasters. After studying this chapter, you should understand

■ What natural hazards, disasters, and catastrophes are.

■ That natural hazards are natural processes with service functions.

■ That hazardous events are predictable.

■ That linkages exist between hazards.

■ Why former disasters are now becoming catastrophes.

■ That the risk from hazards may be estimated.

■ That adverse effects of hazards can be anticipated and minimized.

■ The common adjustments to hazardous events.

CASE STUDY

Hurricane Katrina, Worst Natural Catastrophe in U.S. History

Hurricane Katrina was a catastrophe; it has been labeled an American tragedy. It was certainly both. The hurricane was a Category 5 in the Gulf of Mexico, weakening by the time it hit land to a Category 3 storm. Hurricanes are scaled from 1 to 5, and with each increase in category, wind speeds increase, as does storm surge—a mound of water pushed by the storm that moves ashore with the hurricane. Category 4 and 5 hurricanes produce catastrophes that bring great destruction (see A Closer Look: 16.1).

Hurricane Katrina made landfall in the early evening of August 29, 2005, about 45 km (30 mi) to the east of New Orleans. Katrina was a huge storm that caused serious damage up to 160 km (100 mi) from its center. The storm produced a storm surge of 3 to 6 m (9 to 20 ft). Much of the coastline of Louisiana and Mississippi was devastated as coastal barrier islands and beaches were eroded and homes destroyed. At first it was thought the city of New Orleans had dodged a bullet again, as the hurricane did not make a direct hit. However, the situation turned into catastrophe when water from Lake Pontchartrain north of the city and connected to the Gulf flooded the city. Levees capped with walls, constructed to keep the water in the lake and protect low-lying parts of the city, collapsed in two locations and water poured in. Another levee failed on the Gulf side of the city, contributing to the flooding. Approximately 80% of New Orleans was under water from knee deep to rooftop or greater depths. People who could have evacuated but didn't, and those who couldn't because they lacked transportation, took the brunt of the storm. New Orleans, with its population of about 1.3 million people, became a catastrophe of gigantic proportions. About the only part of the city that wasn't flooded was in the French Quarter (Old Town), the area of New Orleans famous for music and Mardi Gras.

The people who built New Orleans over 200 years ago realized that much of the area was low in elevation, and so they built on the higher places, which were on natural levees of the Mississippi River. The natural levees formed by periodic overbank deposition of sediment from the river over thousands of years. The natural levees are linear features parallel to the river channel.

Because they are higher than adjacent land, they provide some natural flood protection.

As marshes and swamps were drained, however, the city expanded into low areas that posed a much greater flood hazard. Much of the city is in a natural bowl, and parts are a meter or so (3 to 9 ft) below sea level (Figure 16.1). It has been known for a long time that if a large hurricane were to make a direct or near-direct hit on the city, extensive flooding and losses would result. Although the warnings were not completely ignored, sufficient funds were not provided to maintain the levee and system of floodwalls to protect low-lying areas of the city from a Category 3 hurricane.

Contributing to the problem is that the region is subsiding at highly variable rates from 1 to 4 m (3 to 12 ft) per 100 years. Over short periods, 50 to 75% of the subsidence in some areas is natural, resulting from geologic processes (movement along faults) that formed the Gulf of Mexico and the Mississippi River delta.[1] Subsidence has been as much as several meters (more than 10 ft) in the last 100 years, and during that period, sea level has risen about 20 cm (8 in.). The rise is due in part to global warming. As the gulf water and ocean water warm they expand, raising sea level. The subsidence results in part from a number of human processes, including extraction of groundwater and extraction of oil and gas, as well as loss of freshwater wetlands that compact and sink when they are denied sediment from the Mississippi River or are drained. Because the Mississippi River has human-constructed levees (embankments) and no longer delivers sediment to the wetlands, the wetlands have stopped building up from sediment accumulation. Before the levees were constructed in the Mississippi River delta, floodwater with its sediment spread across the delta, helping maintain wetland soils and plants. The freshwater wetlands near New Orleans were largely removed during past decades, replaced by saltwater ecosystems as sea level rose and the land continued to subside. The freshwater wetlands are a better buffer against winds and storm waves than are saltwater wetlands. The tall trees such the cyprus and other plants that grow in freshwater wetlands provide a roughness that slows down water from high waves or storm surges

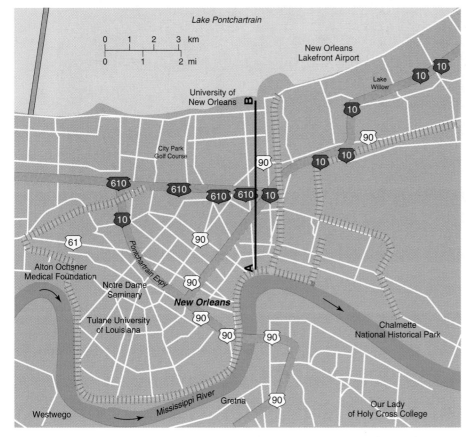

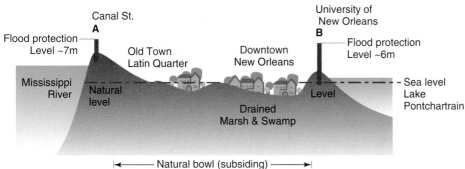

Figure 16.1 ■ New Orleans is located between Lake Pontchartrain and the Mississippi River, in an area that is subsiding. The city of New Orleans was built in a hazardous location. As a result of loss of wetlands and the extraction of oil and natural geologic processes, much of the area is subsiding and has become more vulnerable to flooding in recent years. Floodwalls and levees on the banks of the river and lake were designed for a Category 3 hurricane and its associated high water level. Unfortunately, Hurricane Katrina in 2005 was a large Category 3 storm with a Category 5 storm surge, and 80% of the city flooded.

moving inland. It is well known that one of the natural service functions of both saltwater and freshwater coastal wetlands is to provide protection to inland areas from storms.

Property damage from Hurricane Katrina and costs to rehabilitate or rebuild the area may exceed $100 billion, making it the most costly hurricane in U.S. history. The number of human deaths will never be known, for many bodies may have been washed out to sea or buried too deep to be found. The official number of deaths is 1,836. Initial loss of life and property from wind damage

and storm surge was immense. Entire coastal communities, together with their fishing industry, disappeared.

The hurricane and subsequent flooding set into motion a series of events that have caused significant environmental consequences. Following the flooding, thousands of homes and their contents were found to be a near-total loss in much of New Orleans (Figure 16.2). Think about what is in an average home or building as well as the automobiles that were left behind. A wide variety of chemicals, gasoline, and oil were released into the stagnant waters that filled the city. Figure 16.3 shows a person wading through

Figure 16.2 ■ A city flooded. Shown here is part of the city of New Orleans that was flooded in September 2005, following Hurricane Katrina. High water overtopped some flood protection structures, while others eroded from the front side.

polluted water. In addition, oil refineries and other facilities were damaged and oil spills were inevitable. After the water receded, some parts of New Orleans were left covered with a thick oily sludge. New Orleans was now a gigantic toxic soup consisting of a lot of organic material that included bodies of animals and people and everything that goes with humanity. Floodwaters were pumped into Lake Pontchartrain north of the city, which drains to the Gulf of Mexico. In ordinary times, it would have been illegal to dump toxic water into the lake and pollute the marine environment! As the oxygen-deprived toxic water entered relatively clean water, waters in the lake were polluted. When the water exited the lake, it flowed through valuable shallow saltwater areas where oyster beds have been harvested for decades. Fortunately, however, the volume of lake water is large relative to the volume of the polluted water, and there is active circulation to the gulf. As a result, contrary to what was first feared, the pollutants were apparently diluted and dispersed, and a large, water pollution hazard event did not materialize.

There is *little doubt* that New Orleans will be rebuilt, although by a year later, little rebuilding had started. The people of New Orleans are resilient, and most of them are expected to move back into this famous historic city. Hopefully, everyone has learned something from this event, and better measures will be taken to prevent this kind of catastrophe in the future. For example, in areas where floodwaters, even with future levee failure, will be less than one story high, reconstruction could build flood-proof buildings, placing the living areas on the second floor with the garage below.

The U.S. Army Corps of Engineers, which is largely responsible for the flood protection of New Orleans, presented a draft report in June 2006 concerning the hurricane protection system that had evolved piecemeal over a number of decades. In that report the Corps admitted that the flood protection system was a system in name only inasmuch as various parts were constructed at different times and did not comprise an overall system. The Corps is aware that errors were made in the design and construction of the levees in floodwalls over several decades. The failure of these structures, constructed to protect the city, was responsible for a majority of the flooding resulting from Hurricane Katrina in 2005.

A few of the major conclusions of the report are as follows.[2]

■ The hurricane protection for New Orleans did not function as a unified system. There were no fallback redundancies (second-tier protection) if the primary flood-control structures failed. Pumping stations deigned to remove floodwaters were the only example of a redundant system, and these were not designed to function in a major hurricane and extensive flooding.

■ Hurricane Katrina exceeded the designed criteria of the flood protection structures, but the structures did not perform as expected. Some levee and floodwall failures resulted from being overtopped by floodwaters; others failed through erosion from their front side. The engineers admitted that the soils used in building the flood

Figure 16.3 ■ New Orleans floodwaters were polluted. A man wades through polluted water in New Orleans. All sorts of toxic materials entered the floodwaters, which were ponded in the low parts of the city. The oil seen here probably came from automobiles or other vehicles.

defenses were sometimes inadequate, as were the hard floodwalls on top of the levees that allowed water to seep in and cause the structure to fail without being overtopped.

■ Southern Louisiana is sinking (subsiding) faster than was appreciated before Katrina, and the height of the flood protection structures was not adjusted for subsidence. Some of the floodwalls and levees were as much as 1 m (3 ft) below the elevation at which they were designed.

■ As the understanding of hurricanes and storm surges increased, the enhanced scientific information did not lead to updating the flood-control plan. The little adjustment that occurred was fragmented rather than consistent and uniform.

■ The consequences of the flooding were pervasive and greater than those of any previous disaster in New Orleans. Losses were so great that by themselves they created a challenge to recovery. Consequences were also concentrated. More than 75% of the people who died were over 60 years old and were concentrated in the areas with the greatest depth of flooding. A larger num-

ber of deaths occurred among the elderly because poor, elderly people and the disabled were the least able to evacuate without assistance.

■ The parts of the flood and hurricane protection system that have been repaired at a cost of about $800 million since Katrina are likely to be the strongest part of the flood protection system until the entire system can be updated. As of 2006, the protection level remains the same as before the hurricane struck.

For the Army Corps of Engineers to admit past errors is an encouraging sign in our efforts to better protect the city of New Orleans and the surrounding region from future storms. In fairness to the Corps, flood-control structures are often underfunded, and as with New Orleans, construction is spread over many years. Hopefully we will ultimately build a stronger, more effective hurricane protection system for New Orleans and other U.S. cities located in the paths of hurricanes.

The big question is this: Can flooding recur even if higher, stronger flood defenses are constructed? Of course it can—when a larger storm strikes, future damage is inevitable. If freshwater marshes are restored and the river waters of the Mississippi are allowed to flow through them again, a natural buffer from winds and waves will have been provided. As previously mentioned, the trees in freshwater marshes provide a roughness to the land that slows wind and retards the advance of waves from the gulf. Every 1.5 km (1 mi) of marsh land can reduce waves by about 25 cm (1 ft).

Tremendous sums of money are needed to make New Orleans more resistant to future storms. In light of the many billions of dollars that are spent following catastrophes, it would be prudent to spend the money in a proactive way to protect important resources, particularly those in our major cities.

Hurricanes can be observed and tracked by images from satellites for days to a week or more prior to their landfall. Scientists can also predict where a particular hurricane is most likely to strike land and the most likely areas to experience future hurricanes. As we have learned from Katrina, living in areas known to be hazardous carries a very high price. We have also learned that we need to be proactive with respect to natural hazards, particularly those that have a potential of producing catastrophes.

In the remainder of this chapter, we will discuss some of the principles of natural processes that are hazardous and how they produce disasters and catastrophes. Through case histories we will explore how poor land uses and changing land uses, when coupled with population increase, greatly increase the risk of some hazards.

16.1 Hazards, Disasters, and Catastrophes

Natural processes are physical, chemical, and biological changes that modify the landscape. Some processes are internal, such as earthquakes or volcanic eruptions, which are driven by changes deep in the Earth. Other processes act close to or at Earth's surface; these include landslides, flooding, coastal erosion, violent storms, flooding, and wildfires.

The science behind how most natural processes are produced is moderately well known. Earth and atmospheric scientists have developed the science necessary to be more proactive in how we adjust to natural hazards. Common denominators of vastly destructive natural events such as earthquake, volcano eruption, tsunami, landslide, hurricane, wildfire, and tornado are transport of material (water, air, and earth) and expenditure of energy. Heat wave and drought are two other natural hazards that are more related to weather links to atmospheric processes. The most devastating natural hazards are the following.

- **Earthquake:** Earthquakes result when the rocks that are under stress from the internal earth processes that produce our continents and ocean basins rupture (see plate tectonics in Chapter 5), mostly at depths of 10 to 15 km along faults (fracture, planes in rock with differential movement called displacement). Earthquakes release vast amounts of energy; they can release more energy than a large nuclear explosion.

- **Volcanic Eruption:** Volcanoes are the result of extrusion at the surfaces of molten rock (magma). Volcanic eruptions may be explosive and violent, or they may be less energetic lava flows. Volcanoes generally occur at boundaries between tectonic plates (see Chapter 5), where active geologic processes favor the melting of rocks and the upward movement of magma. Some volcanoes also occur in more central parts of tectonic plates where hot spots deep below heat the rocks above. Examples of hot spots are the volcanic activities at Yellowstone National Park and the Hawaiian Islands.

- **Landslides:** *Landslide* is a general term for the down slope movement of soil and rock. Landslides occur when the driving forces that tend to move soil, rock, vegetation, houses, and other materials down a slope exceed the resisting forces that hold the slope material in place. The resisting forces are produced by the strength of the material on slopes and result from interlocking grains of rock or soil, natural cementing material in rock and soil, or plant roots that bind the slope materials together and resist movement. Weak rocks on steep slopes provide for the combination of large driving forces and weak resisting forces that favors development of landslides. The dominant driving force on slopes is the weight of slope materials influenced by the force of gravity. The steeper the slope and the heavier the slope materials, the greater the driving forces. Human processes that add to or increase the slope angle (how steep it is) increase the drive forces. Resisting forces may be reduced by increasing the amount of water on or in a slope, or by removing vegetation that reduces the root strength of the soil or rock.

- **Hurricane:** A hurricane is a tropical storm with circulating winds in excess of 120 km (74 mi) that move across warm ocean waters of the tropics. Hurricanes gather and release huge quantities of energy as water is transformed from liquid in the ocean to vapor in the storm (see A Closer Look 16.1).

- **Tsunami:** A tsunami is a series of large ocean waves produced after the ocean water is suddenly disturbed vertically by processes such as earthquakes, volcanic eruptions, submarine landslides, or the impact of an asteroid or comet (see A Closer Look 16.2). Over 80% of all tsunamis are produced by earthquakes.

- **Wildfire:** A wildfire is a rapid, self-sustaining, biochemical oxidation process that releases light, heat, carbon dioxide, and other gases and particulates into the atmosphere. Fuel, plant material, is rapidly consumed during wildfires (see Chapter 5), helping maintain a balance between plant productivity and decomposition in ecosystems. The primary cause of periodic wildfire is vegetation. When microbes in the environment are not able to decompose plants fast enough to balance the carbon cycle, fire is necessary to achieve a long-term balance.

- **Tornado:** A tornado is a funnel-shaped cloud of violently rotating wind that extends downward from large cells of thunderstorms to the surface of Earth. Severe thunderstorms may occur when a cold air mass collides with a warmer one. Water vapor in the warmer part of the atmosphere is forced upward where it cools and produces precipitation. As more warm air is drawn in, the storm clouds grow higher and thunderstorm activity increases in intensity, forming lines of storm activity (squall lines hundreds of kilometers long or large cells of updraft called super cells). Tornadoes in the United States are concentrated in the Plains States between the Rocky and Appalachian mountains, where severe thunderstorms generally are more common. Some parts of this region have been called "tornado alley."

- **Flood:** A flood is the inundation of an area by water. Floods are produced by a variety of processes ranging from intense rainstorms to melting of snow, storm surge from a hurricane, tsunami, and rupture of flood protection structures, such as levees or dams. Flooding occurs as water is transported across the surface of land or inundates a particular location. River flooding, one of our most universally experienced hazards, shapes the landscapes through erosion and deposition. Erosion has produced features as small as gullies and as large as the Grand Canyon of the Colorado River.

Hurricane Form and Process

The word "hurricane" supposedly comes from a Caribbean Indian word for big wind or evil spirit. To be considered a hurricane, a storm must have sustained winds of at least 74 mph (119 km per hour). Hurricanes are a variation of the tropical cyclone, which is the general term for huge thunderstorm complexes that rotate around an area of low pressure forming over warm tropical ocean water.

Hurricanes start as tropical disturbances, which are large areas with unsettled weather spreading over a diameter of as much as 600 km (370 mi). Within this area is an organized mass of thunderstorms with a low pressure in which the movement of the storm and the rotation of Earth cause initial rotation. Once the storm has been classified as a tropical depression, it grows in size and strength as warm, moist air is drawn into the depression and begins to rotate counterclockwise in the Northern Hemisphere and clockwise in the Southern Hemisphere. As warm water evaporates from the sea and is drawn into the storm, the energy of the storm increases. Specifically, as warm seawater evaporates, it is transformed from liquid water in the sea to water vapor (gas)

in the air mass of the storm. When this happens, potential energy in the form of latent heat enters the storm. The latent heat of water is the amount of heat required to change liquid water to water vapor (gas). When this change occurs, a change in temperature of the atmosphere takes place (hence it is latent heat). Latent heat is one of the major sources of a hurricane's power (time rate of energy expenditure). As condensation (rain) occurs, the latent heat is released, warming the air and making it lighter. As the lighter air rises, more energy from the water is drawn in, and the hurricane increases in size, strength, and intensity. If wind speeds in the depression reach 63 km (39 m), the depression is called a tropical storm and receives a name.

Hurricanes are very high-energy storms. Their size can be incredible, and it's not unusual to see one moving toward the United States, which has an area about the same as the Gulf of Mexico stretching from Florida to Texas. A generalized diagram of a hurricane is shown on Figure 16.4. The hurricane has bands of spinning storms and thunder cells with an eye where air is subsiding in the center.

As the moist air rotates within the storm, intense rain bands are produced. When a hurricane moves over an island, the storm is deprived of some of its energy and often weakens. After a storm makes landfall on a continent and moves inland, it becomes less intense and eventually dies, but the storm and rainfall can cause river flooding and landslides well inland.

Hurricanes are classified on the basis of their size and intensity (see Table 16.1). There are five categories, with Category 1 being the smallest hurricane and 5 the largest and potentially the most damaging. However, even a Category 1 hurricane is a very dangerous storm.

Scientists know when a hurricane is coming because images from satellites show the tropical depressions and tropical storms. Changes can be watched to see if they will become hurricanes. To verify satellite data, special airplanes are flown through storms recording data on wind speed, air temperature, and air pressure. The paths of a few hurricanes of 2005 are shown in Figure 16.5.

As hurricanes near landfall, they often slow down in shallower water, but if they encounter warmer water they will actually

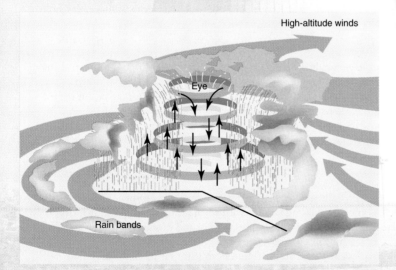

Figure 16.4 ■ Generalized diagram of a mature hurricane showing rain bands, eye, and patterns of wind flow. [*Source:* NUAA, modified after National Hurricane Center; R. W. Christopherson, *Geosystems,* 5th ed. (Upper Saddle River, N.J.: Prentice Hall, 2003.)

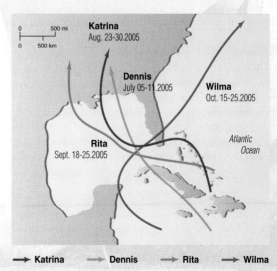

Figure 16.5 ■ Paths of four hurricanes in 2005. [*Source:* National Hurricane Center.]

increase in intensity. One of the most dangerous aspects of hurricanes is not the winds themselves, although they can be lethal, but the storm surge that causes coastal flooding. The storm surge, a local rise in sea level that results when hurricane winds push water toward the coast, may be several meters to more than 10 m (30 ft) high and may cause tremendous damage. Should a storm arrive at high tide, then the storm surge is even higher. Most deaths from hurricanes are caused by the storm surge as people are drowned or struck by solid objects within the surge. When the surge moves ashore, it is not a tall advancing line of water; rather, it is more like a continual increase in the rise of sea level as the hurricane approaches and makes landfall.

Table 16.1 • Saffir–Simpson Hurricane Scale

The Saffir–Simpson Hurricane Scale is a 1 to 5 rating based on the hurricane's present intensity. This rating is used to give an estimate of the potential property damage and flooding expected along the coast from a hurricane landfall. Wind speed is the determining factor in the scale, because storm-surge values are highly dependent on the slope of the continental shelf where the hurricane makes landfall.

CATEGORY 1 HURRICANE:

Winds 119 to 153 km (74 to 95 mi) per hour. Storm surge generally 1.2 to 1.5 m (4 to 5 ft) above normal. No real damage to building structures. Damage primarily to unanchored mobile homes, shrubbery, and trees. Some damage to poorly constructed signs. Also, some coastal road flooding and minor pier damage. Hurricanes Allison of 1995 and Danny of 1997 were Category 1 hurricanes at their peak intensity.

CATEGORY 2 HURRICANE:

Winds 154 to 177 km (96 to 110 mi) per hour. Storm surge generally 1.8 to 2.4 m (6 to 8 ft) above normal. Some roofing material, door, and window damage of buildings. Considerable damage to shrubbery and trees, with some trees blown down. Considerable damage to mobile homes, poorly constructed signs, and piers. Coastal and low-lying escape routes flood 2 to 4 hours before arrival of the hurricane center. Small craft in unprotected anchorages break moorings. Hurricane Bonnie of 1998 was a Category 2 hurricane when it hit the North Carolina coast, and Hurricane George of 1998 was a Category 2 hurricane when it hit the Florida Keys and later the Mississippi Gulf Coast.

CATEGORY 3 HURRICANE:

Winds 178 to 209 km (111 to 130 mi) per hour. Storm surge generally 2.7 to 3.7 m (9 to 12 ft) above normal. Some structural damage to small residences and utility buildings with a minor amount of wall failures. Damage to shrubbery and trees with foliage blown off trees and large trees blown down. Mobile homes and poorly constructed signs are destroyed. Low-lying escape routes are cut off by rising water 3 to 5 hours before arrival of the hurricane center. Flooding near the coast destroys smaller structures with larger structures damaged by battering of floating debris. Terrain continuously lower than 1.5 m (5 ft) above mean sea level may be flooded inland 13 km (8 mi) or more. Evacuation of low-lying residences within several blocks of the shoreline may be required. Hurricanes Roxanne of 1995, Fran of 1996, and Katrina in 2005 were Category 3 hurricanes at landfall on the Yucatan Peninsula of Mexico, North Carolina, and the Gulf Coast, respectively.

CATEGORY 4 HURRICANE:

Winds 210 to 249 km (131 to 155 mi) per hour. Storm surge generally 4 to 5.5 m (13 to 18 ft) above normal. More extensive wall failures with some complete roof structure failures on small residences. Shrubs, trees, and all signs are blown down. Complete destruction of mobile homes. Extensive damage to doors and windows. Low-lying escape routes may be cut off by rising water 3 to 5 hours before arrival of the hurricane center. Major damage to lower floors of structures near the shore. Terrain lower than 3.1 m (10 ft) above sea level may be flooded, requiring massive evacuation of residential areas as far inland a 10 km (6 mi.). Hurricane Luis of 1995 was a Category 4 hurricane while moving over the Leeward Islands; Hurricanes Felix and Opal of 1995 also reached Category 4 at peak intensity.

CATEGORY 5 HURRICANE:

Winds greater than 249 km (155 mi) per hour. Storm surge generally greater than 5.5 m (18 ft) above normal. Complete roof failure on many residences and industrial buildings. Some complete building failures with small utility buildings blown over or away. All shrubs, trees, and signs blown down. Complete destruction of mobile homes. Severe and extensive window and door damage. Low-lying escape routes are cut off by rising water 3-5 hours before arrival of the hurricane center. Major damage to lower floors of all structures located less than 4.6 m (15 ft) above sea level and within 458 m (500 yards) of the shoreline. Massive evacuation of residential areas on low ground within 8 to 16 km (5 to 10 mi) of the shoreline may be required. Hurricane Mitch of 1998 was a Category 5 hurricane at peak intensity over the western Caribbean. Hurricane Wilma in 2005 was a Category 5 hurricane at peak intensity and is the strongest Atlantic tropical cyclone of record.

- **Heat Wave:** A period of days or weeks of unusually hot weather is a recurring weather phenomenon related to heating of the atmosphere and the moving of air masses. It has been hypothesized that human-induced global warming has increased the number and intensity of heat waves in recent years.
- **Drought:** A period of months, or more commonly years, of unusually dry weather constitutes a drought. This phenomenon is related to natural cycles of wet years that alternate with series of dry years. The reasons for the cycles of drought are not well understood, but are thought to be related to the heating of ocean waters and the moving of major air masses. Droughts in California, for example, are thought to be due to the decadal shift in high-pressure zones that form in the central Pacific Ocean and to the jet stream that allows winter storms to extend south or remain further north. Dry years in southern California occur when storm tracks remain north of central California for several years. Prolonged droughts in the midwestern states, such as the Dust Bowl that developed in the 1930s in Kansas and other nearby regions, are associated with gigantic dust storms that commonly occur in desert regions. Droughts in central Africa have been devastating to human populations.

As the term implies, natural processes are "natural." They become hazards, disasters, or catastrophes when people interact with or live and work where they occur (see A Closer Look 16.3). We define a **natural hazard** as any natural process that is a potential threat to human life and property. The process and events themselves are not a hazard but become so because of human use of the land. A **disaster** is a hazardous event that occurs over a limited time span in a defined geographic area. When the loss of human life and property is significant, we say a disaster has occurred. Finally, a **catastrophe** is a massive disaster that requires significant expenditure of money and time for recovery to take place. Hurricane Katrina, which flooded the city of New Orleans and damaged much of the coastline of Mississippi in 2005, was the most damaging, most costly catastrophe in the history of the United States. Recovery from this enormous catastrophe will take years.

Figure 16.6 depicts the general occurrence of major hazards in the United States. Major hazards such as blizzards, ice storms, and droughts, as well as wildfire, are not shown. Droughts, wildfires, and heat waves also produce disasters and catastrophes. Droughts from the 1960s to the 1980s caused famine and killed about 1 million people in the Sahel region of Africa during those years.[3] Wildfires from Florida to the Pacific Coast and from Arizona–California

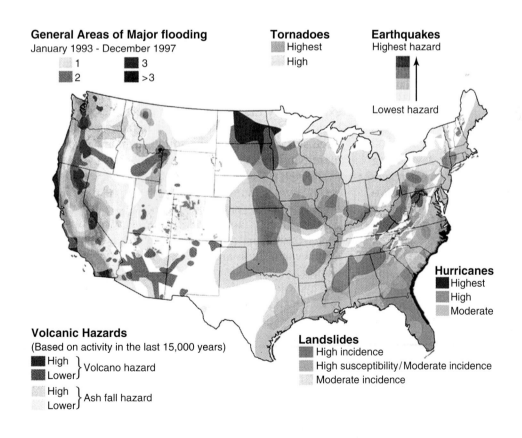

Figure 16.6 ■ Selected natural hazards in the United States. This simplified map of the United States shows areas at risk from flooding, hurricanes, earthquakes, landslides, tornadoes, and volcanic eruption. [*Source:* U.S. Geologic Survey.]

Figure 16.7 ■ Urban earthquake in the Los Angeles area. The Northridge earthquake in 1994 caused billions of dollars in property damage and claimed about 60 lives. The damage shown here is to the freeway system in the Los Angeles urban area.

to Washington State pose increased risk of catastrophes. Heat waves are particularly hazardous to the elderly and the young. A heat wave in Chicago in 1995 killed 465 people, but a heat wave in France during the summer of 2003 killed nearly 15,000 people. Heat waves are now considered the most deadly of all weather-related hazards. Global warming is thought to be contributing to the frequency and intensity of heat waves, wildfires, droughts, and other hazardous weather (see Chapter 23).

No area of the United States, or for that matter, the world is considered hazard-free. During the past few decades natural disasters such as hurricanes, floods, and earthquakes have taken the lives of several million people. Average annual loss of life from hazards around the globe has been around 150,000, with financial losses exceeding many billions of dollars. A single event such as a large earthquake or flood may exceed $100 billion. Not in-

cluded in these statistics are social losses such as employment, mental anguish, and reduced productivity, which are significant but more difficult to quantify.

The influence of human activity on hazards in the United States is summarized in Table 16.2. Hazards that produce the most extensive loss of property are not necessarily the same ones that cause the greatest loss of human life. Excluding heat waves and drought, the largest number of deaths per year from hazards is from tornado and windstorm, lightning, hurricane, and flooding. There is a fair amount of interaction between hazards. For example, hurricanes often cause both coastal and inland flooding. Loss of life from events such as earthquakes in the United States is difficult to calculate. When a large earthquake occurs, a number of people may be killed, and damages may be several tens of billions of dollars. For example, the 1994 Northridge earthquake in Los Angeles (Figure 16.7) killed

Table 16.2 • Selected U.S. Hazards: Occurrence Influenced by Human Activity and Potential to Produce a Catastrophe

Hazard	Occurrence Influenced by Human Activity	Catastrophe Potential
Flood	Yes	High
Earthquake	Yes[a]	High
Volcanic Eruption	No	High
Landslide	Yes	Medium
Hurricane	Probably	High[b]
Wildfire	Yes	High
Tornado	Probably[c]	High
Lightning	Probably[d]	Low

[a] Human activity can cause small earthquakes.
[b] Global warming with warmer sea temperatures increases the intensity of hurricanes.
[c] Global warming may increase the intensity of storms, including tornadoes.
[d] Global warming may increase the intensity of storms and the number of lightning strikes.

La Conchita Landslide, 2005

The small beachside community called La Conchita, meaning "Little Shell" in Spanish, located about 80 km northwest of Los Angeles, California, experienced a disaster on January 10, 2005. Ten people were killed and 30 homes were destroyed or damaged when a fast-moving debris flow (a type of landslide) roared through the upper part of the community (Figure 16.8). The debris flow was a partial reactivation of landslide that occurred in 1995, destroying several homes but inflicting no fatalities. The winter of 2004–2005 was a particularly wet one, and there was high-intensity rainfall at times. However, neither residents nor local officials recognized that another landslide was imminent. What differentiated the 2005 debris flow from that of 1995 was that the flow was faster at 45 km/hr (30 mph) and moved further into the community, trapping some people in their homes and causing those who could to run for their lives.

La Conchita in 2005, with a 200 m (600 ft) high slope directly behind the community, has a serious and continuing landslide hazard for humans living there. It should never have been constructed at the foot of the slope. Landslides have occurred in this area for about 100 years, or at least since people started keeping track of those events. The community is built on about 15 m (50 ft) of older landslide deposits. Landslides above La Conchita and to the east and west have been occurring for thousands of years, long before people decided to build beach homes at the base of the cliff.

Study of La Conchita and the surrounding area suggests that the debris flows and slides of 1995 and 2005, as well as older events are part of a much larger prehistoric landslide that had not been recognized when the more recent events occurred (Figure 16.9). Although there is no evidence that the entire large prehistoric slide is moving as a mass, parts are clearly active, especially at the western margin of the slide. The question is not if, but when, future landslides will occur on the slopes above La Conchita.

One way to reduce the hazard to people and property would be to transform La Conchita into a coastal park. Society as a whole could help relocate people through fair compensation for their valuable coastal property, transforming a hazardous site into a resource for future generations. We must also be diligent in future land-use planning and avoid unwise development on the large prehistoric slide above La Conchita. A similar but smaller prehistoric slide was reactivated in California's Malibu area in the mid-1980s, causing claimed damages in excess of $200 million.

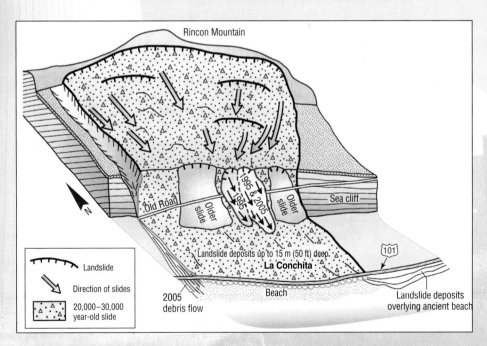

Figure 16.8 ■ Killer landslide in La Conchita, California. The landslide shown here was a rapid downslope movement of material mostly reactivated from a slide that occurred in 1995.

Figure 16.9 ■ This simple block diagram shows the slides of 1995 and 2005 at La Conchita, California. These two slides are part of a prehistoric slide younger than about 6,000 years old. The slides on the steep slope behind the community are a small part of a much larger, older prehistoric complex landslide, parts of which are active. Such large complex slides may be reactivated periodically over thousands of years.

about 60 people and caused as much as $30 billion in property damage. Some experts on losses in urban areas suggest that a large urban earthquake in southern California could inflict about $100 billion in damages while killing several thousand people.

The economic cost of natural disasters in the United States is increasing, primarily because the population is increasingly moving from the interior toward the coasts where hazards tend to occur. As a result, losses of life and property damage are likely to increase significantly in coming decades.

16.2 Disasters and Catastrophes: Taking a Historic Point of View

The well-known (but little heeded) lesson from history is that if we do not learn from past experiences, we will suffer the same consequences again. This is certainly true of natural hazards, which are repetitive events. As a result, study of their history provides basic information for any hazard reduction program. Recall the story of La Conchita, where a small town was built in an area that has experienced landslides for over a century and is built on top of prehistoric landslide deposits (see A Closer Look 16.2). Consider flooding, one of the most common of all hazards to humans. If we want to evaluate the flooding of a particular river, a good place to start is to evaluate the history of flooding for that river. This would involve examining existing flow records, aerial photos taken during times of floods, and the deposits and landforms produced by past flooding. In some parts of the world, people have been recording floods for centuries, if not thousands of years. In Egypt, people have kept careful records of floodwater heights for millennia, enabling them to predict when floods will recede and the likely height of the floods. Such knowledge was critical for predicting crop yields in the Nile Valley. In Great Britain, people used to mark the elevation of floods and the year they occurred on cathedral walls. High water marks, from floods of the distant past has helped extend the record of flooding to the modern era, when we regularly measure the elevation and amount of water from floods. By putting all the historic information together, we may be able to come up with more enlightened statements on the future of the flood hazard at a particular site. That is, we need to link the historic record with prehistoric records and modern measurements to gain greater insights into the flood hazard.

16.3 Fundamental Concepts Related to Natural Hazards

Some general concepts useful in understanding the nature and extent of natural processes and hazards, and how they might be reduced, minimized or eliminated, are as follows:

- Natural processes have service functions.
- Hazards are predictable.
- Linkages exist between hazards.

- Linkages exist between different hazards and between the physical and biological environment.
- Hazards that previously produced mostly disasters are now producing catastrophes.
- Risk from hazards can be estimated.
- Adverse effects of hazards can be minimized.

16.4 Natural Processes Have Natural Service Functions

Nature provides a number of natural service functions for people and the biosphere. For example, trees trap dust and other pollutants on the surface of leaves, helping to clean the air, and wetland plants extract nutrients that otherwise could cause problems in the environment. Coastal wetland plants, as we have learned, are also a natural buffer against winds and waves from storms that move inland.

Earthquakes, associated with mountain building, have produced much of the high topography and beautiful scenery on Earth. The high topography increases erosion and the forces that move sediments to lowlands and the ocean. Displacement on faults causes earthquakes and crushes the rocks along a fault plane. Beneath the surface, this crushed rock and its alteration to clay by weathering produce a barrier to migrating groundwater, often forcing the water to the surface in the form of seeps and springs. For example, along some parts of the San Andreas Fault in the very arid Coachella Valley, the fault produces desert oases with pools of clear water surrounded by palm trees and sometimes inhabited by rare fish (Figure 16.10). In Southern California much of the low flow water in mountain streams that provides habitat for the endangered southern steelhead trout is provided by water emerging from Earth as springs and seeps along faults and fractures in the side of the valley or banks of a stream.

Figure 16.10 ■ Earthquake faults dam groundwater, forcing it to the surface as springs or oases. This photograph shows palm oases along the San Andreas Fault in the Coachella Valley, California. These oases serve as a refuge for a variety of plants and animals, some of which are endangered.

A river and the flatland adjacent to it, known as the *floodplain*, together constitute a natural system. In most natural rivers, the water flows over the riverbanks and onto the floodplain every year or so. This natural process has many benefits for the environment.

▪ Water and nutrients are stored on the floodplain.
▪ Deposits on the floodplain contribute to the formation of nutrient-rich soils.
▪ Wetlands on the floodplain provide an important habitat for many birds, animals, plants, and other living things.

▪ The floodplain functions as a natural greenbelt that is distinctly different from adjacent environments and provides environmental diversity.

On one hand, volcanic eruptions can be catastrophic, but on the other hand, volcanoes emerging from the sea produce new land. The entire Hawaiian Island chain has been created by volcanic processes over several tens of millions of years (Figure 16.11). Volcanic ash may also produce young, fertile soils.

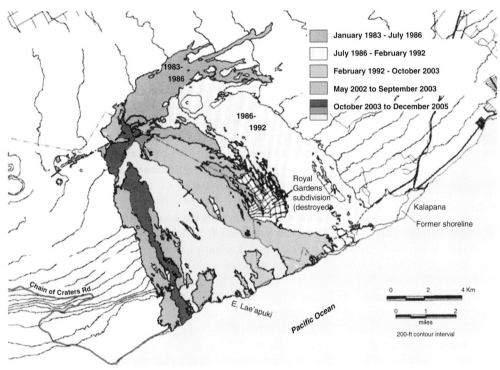

Figure 16.11 ▪ Volcanic eruption producing new land in Hawaii but devastated several communities. Volcanic activity on the Big Island of Hawaii has added considerable land to the island in recent years. In fact, the entire island has been built up of volcanic rocks over a relatively short period of geologic time. (*a*) The map shows flows from 1983 to 2005; (*b*) photo of lava delta at E. Lae Apunki on January 6, 2006. Cracks near the edge suggest that part or all of the delta will collapse into the sea. [*Source:* U.S. Geological Survey, 2006.]

(*a*)

(*b*)

Huge dust storms in Africa and Asia cause problems and are hazardous to the population. On the positive side, the dust may be transporting thousands of kilometers to enrich soils on far sides of the planet. Without this periodic nourishment, the soils would lose their fertility. This has been shown to be true in the Hawaiian Islands, particularly the older ones, which long ago would have had nutrients from original volcanic soils leached from the soil.

Landslides, though often damaging, also provide some service functions. For example, a landslide may block or dam a valley, forming a lake in mountains where otherwise lakes would be rare.

In summary, physical processes linked to the biological environment produce a varied landscape. Without periodic disturbance from natural processes, such as earthquakes, volcanic eruptions, and floods, soils would not be as fertile, water would not as available, the land would not be as diverse, and the diversity of life would be reduced. We also do not want to overlook the important aesthetic aspects of hazards that produce the topography of mountains and valleys and seascapes.

16.5 Hazards Are Predictable

Natural hazards are natural processes that are identified and studied through established scientific methods. Most hazardous events and processes can be mapped as to where they have occurred in the past and monitored in terms of present activity. On the basis of the location of past events, their frequency, patterns of occurrence, and precipitating events, some hazards may be predicted. Once a particular event such as a hurricane, tsunami, or flood wave has been identified, it is also possible to forecast when it might arrive at a specific location. For example, flooding of the Mississippi River in the spring in response to snowmelt, or a very large regional storm, is predictable downstream. It is possible to forecast when the river will reach flood stage and how fast the flood is likely to migrate down the valley toward the Gulf of Mexico. With regard to earthquakes, we know that one of the most likely locations for a large earthquake is where one has recently happened. This is because earthquakes often are clustered in time. Large earthquakes sometimes have precursor events such as foreshocks or patterns of smaller earthquakes prior to the main event. However, short-term earthquake prediction remains elusive.

Volcanic eruptions are often predicted over relatively short time periods because a dormant volcano that may have not erupted in a few hundred years often comes back to life with characteristic precursor events that suggest a coming eruption. These events include earthquake activity, release of gasses, heat that melts surface snow and ice, and swelling or growth of the mountain.

The role of statistics and probability is useful in evaluating the frequency of a particular event. For example,

although we are not able to predict when and how big the next flood will be, hydrologists study past floods to estimate the probability of a particular size (magnitude) flood occurring in a given year or over a given number of years.

A large (high-magnitude) flood of interest to planners is the "100-year flood." This flood is important because it is used to help determine flood insurance rates and the zoning of floodplains to restrict development on flood-prone areas and to design flood defenses. The 100-year flood is defined as the flood with an estimated 1 in 100 (1%) chance of occurring in a given year. By comparison, the 50-year flood has a 1 in 50 (2%) chance of occurring in a given year, and a 10-year flood has a 10% chance. However, once a 100-year flood occurs, the chances of another 100-year flood the next year remain at 1%. Therefore, two 100-year floods can occur in back-to-back years or even in a given year. This is like flipping a coin. There is a 50-50 chance of getting a heads or a tails with each flip. If you flip and get heads, the chance of flipping again and getting another head remains 50%.[4]

The probability of earthquakes can also be calculated. The probability is based on past events, their size or magnitude, and the time since the last earthquake of interest. Calculations for probabilities of earthquake and other events such as volcanic activity are more difficult, and different assumptions must be made than those for estimating the probability of a flood.

The probability of experiencing a hurricane or knowing where a hurricane will make landfall once it has passed over the ocean is also calculated from experience of past hurricanes, using mathematical and statistical calculations.

Another variable important in predicting and warning about hazardous events is location or geography. On a global scale, we know where most earthquakes and volcanoes are likely to occur. We regularly map landslide deposits, and on the basis of this data develop hazard maps showing where the risk to people and property is most likely to be. We can accurately predict where flooding is likely to occur based on mapping the landform known as the floodplain (which is flat land adjacent to the river) and on observing the extent of recent floods.

Sometimes it is possible to forecast an event and issue a warning, as can be done for tsunamis in the Pacific Ocean, where a tsunami warning system is in place (Figure 16.12). However, there was no warning system for tsunamis in the Indian Ocean in 2004. The Indonesian tsunami claimed an unprecedented loss of life—about 250,000 people (see A Closer Look 16.3). If there had been a tsunami warning system in the Indian Ocean similar to that of the Pacific, warnings would have been triggered automatically. Even after the earthquake, when it was known that a tsunami was headed toward Africa, communication lines were very poor, failing to provide that information to people in the path of the waves.

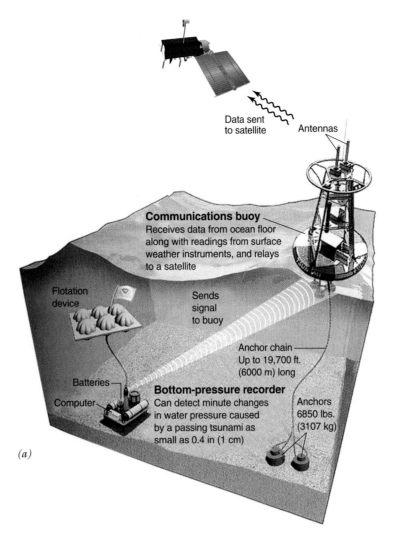

Data sent
to satellite

Antennas

Communications buoy
Receives data from ocean floor
along with readings from surface
weather instruments, and relays
to a satellite

Flotation
device

Sends
signal
to buoy

Anchor chain
Up to 19,700 ft.
(6000 m) long

Batteries

Computer

Bottom-pressure recorder
Can detect minute changes
in water pressure caused
by a passing tsunami as
small as 0.4 in (1 cm)

Anchors
6850 lbs.
(3107 kg)

(a)

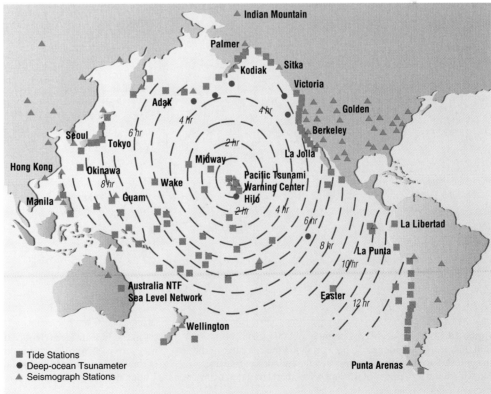

Indian Mountain

Palmer

Sitka

Kodiak

Victoria

Adak

Golden

4 hr

4 hr

6 hr

Berkeley

Seoul

Tokyo

La Jolla

2 hr

Hong Kong

Midway

Okinawa

Pacific Tsunami
Warning Center

Wake

Manila

Guam

8 hr

Hilo

2 hr

4 hr

La Libertad

6 hr

8 hr

La Punta

10 hr

Australia NTF
Sea Level Network

Easter

12 hr

Wellington

Punta Arenas

■ Tide Stations
● Deep-ocean Tsunameter
▲ Seismograph Stations

(b)

Figure 16.12 ■ *(a)* Tsunami
warning system includes a surface buoy and bottom sensor.
(b) Travel time (each band is
one hour) for a tsunami generated at Hawaii. The wave arrives
in Los Angeles in about 5 hours.
It takes about 12 hours for the
waves to reach South America.
Locations of existing six tsunami
instruments are shown. Others
are being planned for the
Atlantic and Caribbean. [Source:
NOAA.]

Indonesian Tsunami

The Indonesian tsunami of 2004 made us aware of these giant waves and their potential for destruction. Within a span of only a few hours about 250,000 people were killed, and millions were displaced as coastal area after coastal area around the Indian Ocean was struck by the series of tsunami waves produced by a magnitude 9 earthquake that occurred offshore of the Indonesian island of Sumatra. Tsunami waves come in as a series, and later waves may be higher than earlier ones. When the water from a wave retreats from the land flowing back to sea, the return flow can be as dangerous as an incoming wave.[3]

Over three-quarters of the deaths occurred in Indonesia, which suffered from both intense shaking from the earthquake that caused the tsunami and the tsunami itself. The site of the earthquake and the generation of the tsunami and its travel over several hours are shown in Figure 16.13. Notice that the time from generation of the tsunami to its arrival in Somalia was about 7 hours. In India, it was 2 hours, and in other places earlier or later depending on distance from the earthquake. The first tsunameter (Figure 16.12a) for the Indian Ocean was in place by December 2006, and sirens have been installed along some coastal areas

around the Indian Ocean to warn people of a tsunami.

Tsunamis (the Japanese word that is translated as large harbor wave) are produced by the sudden vertical displacement of ocean water.[3] They may be triggered by several types of events, such as large earthquakes that cause rapid uplift or subsidence of the seafloor; underwater landslides that may be triggered by an earthquake; collapse of part of a volcano that slides into the sea; submarine volcanic explosion; and impact of an extraterrestrial object, such as an asteroid or comet, in the ocean. Asteroid impact can produce a mega tsunami (a huge tsunami) about 100 times

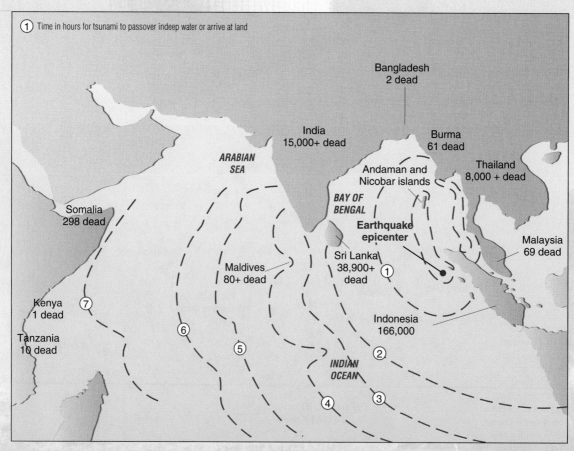

Figure 16.13 ■ A tsunami in December of 2004 killed about 250,000 people in Indonesia. This map shows the epicenter of the magnitude 9 earthquake that produced the Indonesian tsunami. Shown are the movement of tsunami waves that devastated many areas in the Indian Ocean. Notice that the waves took approximately 7 hours to reach Somalia, where almost 300 people were killed. Most of the deaths were in Indonesia, where the waves arrived only about 1 hour after the earthquake. [*Source:* NOAA.]

as high as the Indonesian tsunami and can put hundreds of millions of people at risk.[3] Fortunately, the probability of a large impact is very low. Of the potential causes listed earlier, earthquakes are by far the most commonly observed. The process is shown in Figure 16.14.

During the 2004 Indonesian earthquake, the floor of the ocean was suddenly uplifted a few meters along the entire fault line, which was hundreds of kilometers long. The fault movement vertically displaced the entire mass of water over the rupture. The sudden upward movement of the seawater also produced

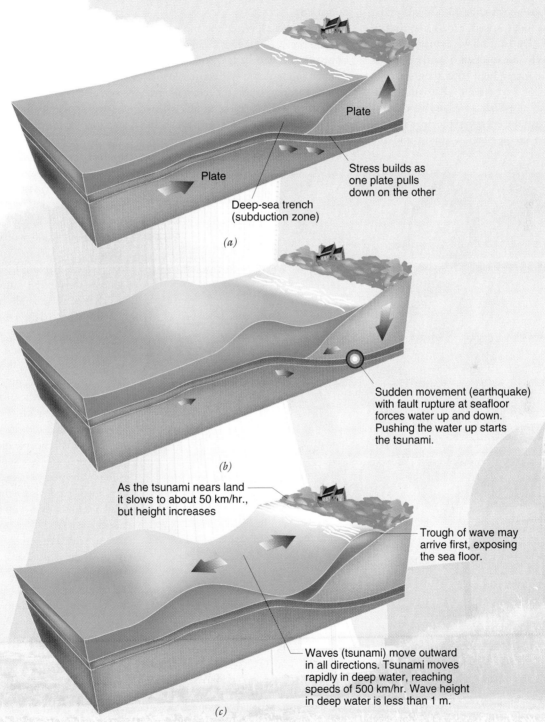

Plate

Plate

Deep-sea trench
(subduction zone)

Stress builds as
one plate pulls
down on the other

(a)

Sudden movement (earthquake)
with fault rupture at seafloor
forces water up and down.
Pushing the water up starts
the tsunami.

(b)

As the tsunami nears land
it slows to about 50 km/hr.,
but height increases

Trough of wave may
arrive first, exposing
the sea floor.

Waves (tsunami) move outward
in all directions. Tsunami moves
rapidly in deep water, reaching
speeds of 500 km/hr. Wave height
in deep water is less than 1 m.

(c)

Figure 16.14 ■ Idealized diagram of how a tsunami is generated by an earthquake. [*Source:* Modified after E. A. Keller and R.H. Blodgett, *Natural Hazards* (Upper Saddle River, N.J.: Prentice Hall, 2006).]

an adjacent mass of water that moves down. The effect was like throwing a giant boulder in your bathtub and watching the rings or waves of water spread out. In the bathtub analogy the boulder came down from the top, but the result was the same and waves radiated outward, moving at high speed across the Indian Ocean. The tsunami waves themselves are relatively low in open ocean, being less than a meter high, but they travel at speeds of a jet aircraft (750 km/hr, or 500 mi/hr). When a tsunami wave strikes the coast, the energy of the wave is compressed in the shallow waters and the height of the wave increases dramatically. Figure 16.15 shows the tsunami advancing on a tourist area on Phuket, Thailand, and Figure 16.16 shows before and after photographs of Banda Aceh, Indonesia. Notice that almost everything was destroyed.

Some people survived the Indonesian tsunami because of their education or because of tribal knowledge. Some schoolchildren and professionals recognized that water receding from the coastal area where they were vacationing was a danger sign that a wave was approaching. In some cases, people were warned by others and were able to evacuate to higher ground. Some native people in Indonesia have a collective memory of tsunamis, so that when the earthquake occurred, some people applied that knowledge and moved to high ground. This knowledge saved entire tribes on some islands. Remarkably, elephants saved a small group of about 12 tourists in Thailand from the 2004 tsunami.[5] At about the time the earthquake struck off the island of Sumatra in Indonesia, elephants started nervously trumpeting and they became agitated again about one hour later. Those elephants that were not taking tourists for rides broke loose from their strong chains and headed inland. Elephants that had tourists aboard for a ride didn't respond to their handlers and climbed a hill behind a beach resort where about 4,000 people would soon be killed by the tsunami. When handlers recognized the coming tsunami, additional tourists were lifted onto the elephants, which used their trunks to place the people on their backs (tourists generally mount elephants from a wooden platform) and moved inland. Tsunami waves surged about 1 km (0.5 mi) inland up from the beach, and the elephants stopped beyond where the waves ended their destructive path.

We may ask, did the elephants know something that people did not? As is well known, animals have sensory ability that differs from that of humans. It is possible that they heard the earthquake as it generated sound waves with low tones called infrasonic sound. Some people were also able to sense the sound waves but didn't perceive them as a hazard. Alternatively, the elephants could have sensed the actual movement of the ground as it vibrated from the earthquake, and so they fled. They went inland, which was the only way they could go. The linkage between the elephant's sensory ability and their behavior is speculative; nevertheless, the end result was the saving of a few lives.

Figure 16.15 ■ Tourists running for their lives. Man in foreground is looking back at the tsunami rushing toward him that is higher than the building. The location is Phuket, Thailand. Many of the people living there, as well as the tourists, did not initially think the wave would inundate the area where they were. When the waves arrived, they thought they would be able to outrun the rising water. In some cases people did escape, but more often people were drowned.

Figure 16.16 ■ Urban developments completely destroyed by Indonesian tsunami of 2004. The photographs shown here were taken before and after the tsunami that struck the Indonesian provincial capital of Banda Aceh on the northern end of the island of Sumatra. Essentially all of the development has been damaged or destroyed. Notice along the top of the photograph the beach with the extensive erosion, leaving what appear to be a number of small islands where a more continuous coast was formerly. This area subsided as a result of the earthquake. Photographs are satellite images from Digita Globe.

16.6 Linkages Exist between Hazards and between the Physical and Biological Environments

Understanding the links between hazards and between the physical and biological environment is an important part of understanding the consequences of natural hazards. First, hazards themselves may be linked. For example, volcanic eruptions often cause landslides. When a river valley is blocked with lava or when landslide deposits form a natural dam, the likelihood of flooding increases when the dam is overtopped or washed out. Volcanic eruptions can profoundly change the landscape and ecosystems. The eruption of Mount St. Helens in 1980, for example,

severely disrupted the landscape and rivers. However, recovery since 1980 has been dramatic (Figure 16.17).

Hurricanes are huge storms that produce high winds and coastal flooding. When hurricanes move inland, voluminous, intense rainfall may cause flooding and landslides in adjacent mountainous or hilly topography.

The occurrence of large submarine earthquakes (discussed earlier) is directly linked to tsunamis. The earthquakes can change the entire coastal environment by uplifting what was the ocean floor and submerging other areas of 100s to 1000s of square kilometers or more.

Natural hazards are linked to earth materials. For example, materials such as weak soils and rocks are prone to landsliding. Volcanic eruptions of hot ash on land, when

(a)

(b)

Figure 16.17 ■ Recovery following the 1980 eruption of Mount St. Helens in southwestern Washington. (*a*) After the eruption, much of the region was devastated as trees were blown down and the area was transformed into a desolate-looking landscape. In the several decades since the major eruption, considerable natural restoration has taken place and mountain has grown due to addition of new volcanic rock. (*b*), although the volcano still poses a threat for future eruptions.

they melt snow and ice on the flank of a volcano, are linked to mudflows and flooding.

Linkages to the biological environment from hazards are also common. Of particular importance are the disruption of ecosystems and the fragmentation of habitat from catastrophic events. On a global scale, the impact of a large extraterrestrial object may cause the extinction of many species. On a regional scale, storms, wildfires, floods, and landslides disrupt habitats and ecosystems over variable periods of time. Hurricanes erode beaches and tear out vegetation in coastal salt marshes and wetlands. These activities damage coastal ecosystems and may require years to recover. Storm waves may also erode near-shore coral reefs, disrupting habitat for marine organisms. Wildfire removes vegetation, increasing soil erosion and landslides. Eroded sediment enters streams, filling pools, and thereby damages stream habitat for fish. Floods, when severe, can cause extensive streambank erosion that widens the channel, removing streambank vegetation that is an important habitat for birds and mammals. Periodic wildfire has been sufficiently frequent that entire ecosystems have adapted to fire. Some species of plants regenerate (sprout new limbs) after fire, and others have seeds that need fire to make them viable.

16.7 Hazards That Previously Produced Disasters Are Now Producing Catastrophes

Human beings have changed in the past few tens of thousand years from a species of small numbers to over 6 billion people today. Early in our history, survival was a day-to-day struggle as we interacted with the natural environment. When our numbers were not large and we had little effect on Earth processes, our losses from hazardous processes were not as widespread as they are today. However, during the last 10,000 years of Earth history, our numbers experienced a dramatic increase. Of particular importance was the development of agricultural practices approximately 7,000 years ago. With a more stable food base, our population increased to about half a billion, with several hundred times the density of people living in the hunter–gatherer period that preceded agriculture. By approximately A.D. 1800, we were in the early industrial period and population had doubled to about one billion. With industrialization our cities grew larger, and eventually we learned more about sanitation, helping our numbers to rise until today we exceed 6.6 billion.

As human population has increased, 15 cities (urban regions) with populations exceeding 10 million people have been established on Earth. Table 16.3 lists population increases from 1950 projected to 2015 for these urban regions. Most of them are in areas vulnerable to several natural hazards. As population has grown, people (especially poor people) have been pushed into more hazardous areas because good building spots were already used up. Some more affluent people have deliberately moved to more hazardous areas for privacy and scenery. For example, the hills above Los Angeles, California, are steep, with numerous landslides. The most serious hazard is wildfire, which occurs every few decades. Nevertheless, people, not always conscious of hazards, have chosen to build luxury homes there.

Table 16.3 • Population Increase of Several Cities (Urban Regions) from 1950 Projected to 2015 (Populations are in millions of people.)			
City/Urban region	Population 1950	Population 2000 (estimated)	Population 2015 (projected)
Tokyo, Japan	6.2	27.7	28.7
Mumbai (Bombay), India	2.8	16.9	27.4
Lagos, Nigeria	1.0	12.2	24.4
Shanghai, China	4.3	13.9	23.4
Jakarta, Indonesia	2.8	9.5	21.2
São Paolo, Brazil	2.3	17.3	20.8
Karachi, Pakistan	1.1	11.0	20.6
Beijing, China	1.7	11.7	19.4
Mexico City, Mexico	3.5	17.6	19.0
Dhaka, China	4.0	18.0	19.0
New York City	12.0	16.5	17.6
Calcutta, India	4.5	12.5	17.3
Los Angeles	4.0	12.9	14.2
Cairo, Egypt	2.1	10.5	14.0
Buenos Aires, Brazil	5.3	12.2	13.9

[Source: Data from B. Boyle Torrey, *Forces of Change* (Smithsonian Institute; Washington, D. C.: National Geographic Society, 2000), 16s.]

As an example of how population trends can increase the severity of natural hazards from disaster to catastrophe, consider the volcano Nevado del Ruiz in Colombia (Figure 16.18a). When Nevado del Ruiz erupted in 1845, it produced a catastrophic mudflow that roared down the mountain, killing about 1,000 people in the Lagunilla River Valley. The deposits from that flow produced level ground in the valley, enticing people to move there. Thus an agricultural center was established. By 1985 the population of Armero, which was the center of the agricultural activity, had grown to about 23,000. On November 13, 1985, a catastrophe occurred at Armero when the volcano again erupted and produced a large mudflow that roared through the river valley. The flow killed not 1,000 people, as in 1845, but 21,000, essentially wiping out the town of Armero (Figure 16.18b).

In the 1845 event, there were no warnings of volcanic eruption; in 1985, however, there were a number of precursors, including increases in earthquake activity and hot spring activity in the year prior. As early as July 1985, volcanologists had begun to monitor the volcano and by October 1985 had completed a hazards map. That map and its accompanying report predicted the events of November 13. The risk assessment stated that there would be a potentially damaging mudflow should the ex-

pected eruption occur. Following the eruption, it took about two hours for the mudflows to reach the town, which was subsequently buried as buildings were swept off their foundations and people were killed in great numbers. The tragedy of the Nevado del Ruiz catastrophe is that the event had been predicted. The hazard maps that were circulated were largely ignored. If there had been better communication between civil defense headquarters and the local towns and a better appreciation of the hazard, then Armero could have been evacuated and thousands of lives saved. Today there is a permanent volcano observatory. Hopefully the lessons learned from the 1985 event will help minimize loss of life in the future.[7]

Land Use Transformation and Natural Hazards

How we choose to use the land has a direct effect on the magnitude or size as well as the frequency, or recurrence, of hazardous events. Landscape transformations such as changing forest land to agriculture or urban uses or large-scale logging may turn what were formerly disasters into catastrophes. This is demonstrated by two events in 1998: the flooding of the Yangtze River in China and Hurricane Mitch in Central America.

(a)

(b)

Figure 16.18 ■ Volcanic eruption triggers mudflow that kills 21,000 people. (a) Volcano Nevado del Ruiz in Columbia as viewed from the northeast on December 10, 1985, about a month following the distruction of Almero. The white plume is a minor eruption rising from the summit crater where large mudflows were generated. A previous eruption in 1845 also produced a disastrous mudflow

that killed about 1,000 people. (b) The town of Armero was nearly destroyed following the 1985 eruption. The rectangular pattern in the upper part of the photograph shows the outlines of the building foundations visible through thin, but destructive, volcanic mudflow deposits. This event shows how catastrophes may be repeated and may grow in size if warnings and lessons are not heeded.

The floods in the Yangtze River took approximately 4,000 lives and were probably magnified in intensity by land transformation. In the years prior to the flood, about 85% of the forest in the upper Yangtze River basin was lost as a result of both timber harvesting and shifting of the land to agricultural uses. As a result of these land-use changes, the amount and intensity of runoff increased, and flooding became much more common.[7] China eventually recognized the cause of the flood hazard and banned timber harvesting in the upper Yangtze River basin, allocating several billion dollars for reforestation. As we can learn from the Chinese, to reduce damages from natural hazards we need to practice good land conservation, with a goal of sustainable development to ensure that future generations do not live in a degraded environment that will subject them to more intense natural hazards.

With regard to Hurricane Mitch, the hurricane destroyed large areas in Central America, particularly Honduras, causing about 11,000 deaths. Prior to the hurricane, approximately one-half of the nation's forest had been removed and a large fire had occurred. The deforestation and fire reduced the strength of the materials on the hill slopes, causing them to wash away when the intense rains from the hurricane arrived. And with the hill slopes went the homes, roads, farms, bridges, and other infrastructure necessary for human existence.

As was learned in the flooding of the Yangtze River, care of the landscape and conservation of landforms and ecosystems is necessary if we are going to learn to live in harmony with natural processes. Doing otherwise is to invite catastrophe.[7]

16.8 Risk from Hazards Can Be Estimated

Before we decide what to do about a particular hazard, whether it be to protect homes, cities, nation, or region, we need to have a good idea about what the risks are. **Risk** for a particular event is defined as the product of the probability of that event occurring times the consequences should it occur.[8] Determining the probability of an event occurring is often the most difficult and controversial part of evaluating the risk.

Determining the consequences of a particular event is fairly straightforward and involves estimating property damage and loss of life on the basis of particular events. For example, prior to Hurricane Katrina, a number of studies had been published predicting what would happen to New Orleans should a large hurricane strike. Those studies were right on the mark. They predicted that the city, which is low-lying between Lake Pontchartrain and the Mississippi River, would probably flood and that the consequences of that flood would be catastrophic. Similarly, as we saw with the volcano Nevado del Ruiz in Colombia, the hazards maps and analysis stated that de-

bris and mudflows were almost a certainty and the consequences would be catastrophic for the river valleys affected. These also turned out to be on target. Finally, it is also straightforward to map development on floodplains, such as along the Mississippi River, and estimate what the losses would be if these areas were to be inundated.

More complicated than calculating a risk for a particular natural hazard is determining the **acceptable risk.** When we speak about acceptable risk, we mean the risks that individuals or society (institutions) are willing to take. For example, most of us drive cars and we know and accept the potential dangers of driving. Nevertheless, most of us accept that risk for the convenience of getting around our cities and traveling. The risk of being injured or dying in an automobile accident is high compared to being injured or dying from a nuclear power plant accident. However, the risk from the accident at the nuclear power plant is unacceptable, and as a result people in the United States have turned away from nuclear power. At a personal level, we frequently make choices as to the risks we are willing to take. For example, it is well known that earthquakes are much more likely to happen in California than in other parts of the country. Yet living in California is a risk millions of people accept, and the reasons are many, from the good climate to specific employment opportunities. The same could be said for the eastern coast from North Carolina to Florida and into the Gulf states, where hurricanes are common.

Institutions such as banks and the government approach the problem of defining acceptable risk from an economic point of view rather than a person's or a community's perception of the risk. The bank will ask how much risk it can tolerate with respect to flooding. The federal government may require that any property that receives financial assistance (loans) from it not have a flood hazard exceeding 1% per year that is protection up to and including the 100-year flood.[4]

In summary, risk analysis is a growing activity in evaluating natural disasters and catastrophes. We now know more about how to do the risk analysis, and are trying to implement particular responses to the perceived risk at the individual and societal levels.

16.9 Adverse Effects of Hazards Can Be Minimized

Active versus Reactive Response

Responses to natural hazards that produce disasters and catastrophes include search and rescue, firefighting, and provision of emergency food, water, and shelter following the event. These responses are all well and good and obviously necessary, and, as we saw from Hurricane Katrina, critical to providing support for people who need to be rescued or who are made homeless by such events. However, we need to move to a higher order of thinking about hazards and learn how to anticipate them and be more proactive. Among the proactive choices that anticipate hazardous events are (1) land-use planning to limit construction in hazardous locations; (2) construction of hazard-resistant structures, such as floodwalls and levees; (3) protection of ecosystems on coastal floodplains and wetlands that provide natural protection from hazards;[9] and (4) well-thought-out plans for evacuation and relief following a disaster. In the case of New Orleans, had the wetlands surrounding the city and along the coastline been maintained, the effects of the wind and storm surge would have been reduced. And had funds that had been requested for improving the system of floodwalls and dikes been allocated, the flooding might have been minimized.

The United States is not the only country to have failed to mount a quick response to disaster. In 1995 a large earthquake struck in Japan, and it was several days before the government responded. Governments, like people, may go into shock in the face of a catastrophe. Nonetheless, better planning and preparation are crucial in reducing the immediate and longer-term effects of catastrophe.

Impact and Recovery from Disasters and Catastrophes

Hazardous events have both direct or indirect effects on society. The **direct effects** involve the people killed, injured, dislocated, made homeless, or otherwise damaged by the event. The **indirect events** follow the disaster. They often include donations of money and goods, shelter for people, taxes to help finance recovery, and also emotional distress. Direct effects are experienced by relatively few individuals, whereas the indirect effects may affect society as a whole.[10]

Figure 16.19 presents a generalized view of recovery following disaster. Notice that the first few weeks after a disaster are often spent in a state of emergency, with normal activities ceasing or at least changing. In this period, power may be out; water may be pumped out of low-lying areas; and people are being rescued and sent to shelters. This period may last much longer than a few weeks, depending on the severity of the catastrophe, but it generally gives way to the restoration period. There is some patchy return of functions at first, and this will be followed by reconstruction, which may take several years.

It is during the planning period of restoration and reconstruction that we can be more proactive in trying to avoid future similar hazards. A good example of this can be seen in the 1972 floods of Rapid City, South Dakota. Flash flooding below an upstream flood-control reservoir killed more than 200 people and destroyed many homes on the floodplain. Unfortunately, the dam produced a false sense of security. Below the dam was a tributary to Rapid Creek, and that is where intense rain occurred that caused the catastrophic flood. Rather than quickly moving into the restoration and reconstruction phase, the people waited a number of weeks and thought out a solution that would minimize the problem in the future. As a result, today Rapid City uses land on the floodplain in an entirely different way than it did prior to the flood. This land now comprises greenbelts, golf courses, and other activities that are more appropriate for land that is likely to flood again. Rapid City, South Dakota, has therefore done what it can to minimize the flood hazard.[11, 12] One flood survivor, when asked if she was going to rebuild, replied, "Yes," pointing to an area high above the floodplain, far from future floods!

Figure 16.19 ■ Recovery following disasters may take years. Shown here is a highly generalized idea of what happens following a disaster, from the emergency to restoration and reconstruction. The period of emergency lasts a few days to a few weeks, and is followed by restoration activities, which can take a number of months or longer. The longest part is the reconstruction, which may take several years or more. [*Source:* Modified from R. W. Kates and D. Pijawka. *From Rubble to Monument: the Pace of Reconstruction* (Cambridge, Mass.: MIT Press, 1977).]

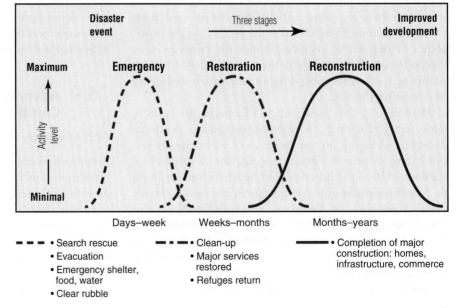

Perceiving, Avoiding, and Adjusting to Hazards

People are generally optimistic about natural hazards, choosing to believe their hill slope will never fail in a landslide or that their river will never flood. And how does one tell a family that has lived in a house for 75 years on a floodplain that is expected to be flooded by the 100-year flood that they live in a hazardous area, and that they have simply been lucky if they have not been flooded before? The same holds for landsliding. After the 1995 landslides in La Conchita, California, many people believed that once a small retaining wall was built there, they would be safe and could move back in. But geologists warned that future landslides were very likely and would occur again; they did, in 2005, and 10 people lost their lives.

Our cities and states often have laws and regulations designed to control what can be built in particular locations. In California, the Alquist–Priolo Earthquake Fault Zoning Act does not generally allow homes to be constructed within 17 m (50 ft) of active faults that may rupture during earthquake. In many places, floodplains are regulated so that people will not build homes on the most hazardous sites. Finally, landslides have been recognized in many areas, particularly in southern California where they are common; therefore, detailed geologic evaluation of slope stability is required prior to construction. In some cases homes cannot be built on slopes that are deemed unstable, while in other cases engineering solutions are required to stabilize a home site.

One of the best adjustments we can make to natural hazards is to avoid them. This means not building our homes on floodplains, active landslides, or directly over active faults. Land-use planning is one of the best tools we have for avoiding some hazards. For example, it is usually required that new construction be placed a minimum distance (called a setback) from the ocean. Minimum setback is usually 60 to 100 years of expected erosion (the expected life of a building). For example, if the erosion rate is 10 cm (4 in) per year, then the required setback is 6 to 10 m (20 to 30 ft). Of course, even with setbacks, erosion will continue, and structures that last long enough will eventually be threatened.

We commonly adjust to hazards through land-use planning, attempts to control natural processes, insurance, evacuation, disaster preparedness, and doing nothing. We are not going to dwell on the last option, but, unfortunately, it is one that is often taken with regard to many natural hazards. People know weather events are going to occur, are optimistic about their future, and assume that they will not be impacted. This can be dangerous thinking. Some people who rode out earlier hurricanes along the Mississippi Gulf Coast and declined to evacuate during Hurricane Katrina later acknowledged it was the worst decision they ever made.

Insurance represents an important adjustment to natural hazards. If you buy property in a flood area with a probability of flooding that exceeds 1% per year, you may be required to purchase flood insurance to get a loan for the purchase. If you live in an area with active earthquakes, you have the option of buying earthquake insurance. One problem with flood insurance programs has been that people will purchase it and then following floods will rebuild again in the same hazardous area. Steps are being taken to try to minimize and avoid this situation, although we still build far too many structures on floodplains that are subject to inundation. For example, a catastrophic flood occurred along the Mississippi River in 1993, causing billions of dollars of damage. Some towns were relocated following the flood; this certainly is a positive response. However, it wasn't the only response. The St. Louis area had floodplain regulations, but they are weak. Following the 1993 flood, over 20,000 new homes and buildings were built on the floodplain inundated by that flood. Residents are relying on new flood-control structures to protect their homes from flooding. This strategy cannot be completely successful inasmuch as floods larger than what the structures are designed to control will eventually happen. Also, after time, flood levees and walls may weaken and become more vulnerable to failure during floods. A number of communities have come to the conclusion that to minimize the hazard, regulation must be coupled with structural controls.[13, 14]

An important adjustment to catastrophes is evacuation. Since hurricanes can often be spotted days or weeks prior to their arrival, evacuation is a wise decision.

Prior to Hurricane Katrina, most of the population of the Gulf Coast and New Orleans did evacuate, but others who did not leave could not because they did not have an automobile, had no place to go, or had no funds for transportation. As a result of Katrina, plans to evacuate people in buses are now being formulated; these plans must consider those people who might not be able to evacuate without help. This includes people in hospitals, nursing homes, and other such facilities.

Disaster preparedness whereby supplies of water, food, and medical items are stockpiled and vehicles, including helicopters and trucks, are on standby, is an important part of minimizing the effects of hazards. This is particularly true if we know a large event, such as a hurricane or downstream movement of a flood wave, is likely to occur. Good disaster preparedness requires good communication at all levels, from individuals who are being affected by the hazard to city, state, and federal authorities. Of particular importance is a working chain of command so that everyone knows who is in control. Lack of central control was evidently a problem in the aftermath of Hurricane Katrina and greatly compounded rescue difficulties, particularly in the city of New Orleans. People outside the catastrophe could only stand by and watch

How Should New Orleans Be Rebuilt?

The case history of Hurricane Katrina that nearly destroyed the city of New Orleans in 2005 caused a catastrophe of gigantic proportions for the United States. The question is not if New Orleans should be rebuilt but how should it be rebuilt. After all the work done to shore up the flood defenses of the city, it remains just as vulnerable to flooding today as it was prior to Hurricane Katrina. In the future, the land will continue to subside and the site of the city will always be in a bowl shape depression. The French Quarter or "Old Town" is relatively safe because it's on the higher ground of a natural levee formed by the Mississippi River.

In the lower parts of town, debates are ongoing concerning how the city should be rebuilt. Should it include expensive high rise apartments and large luxury homes that would be for the more wealthy people? Or should it be rebuilt along the lines it was built prior to the hurricane, but with improved flood protection? There is fear that the city will never return to what it was and will likely become high priced real estate for wealthy people.

However, it's important to keep in mind that even if the flood protection program is improved an even-greater storm could still cause flooding in the future. The problem is that much of the city is below sea level and the big question is what to do about lowlying areas? Some say that homes should be flood-proofed by building garages to homes at street level and having the main structure or home above sea level on the second floor.

Critical Thinking Questions

1. What sorts of structures should be rebuilt in the areas flooded by Hurricane Katrina?
2. Do you expect there will be continued problems with flood defenses for the city? If so, what are problems likely to be and what can be done to minimize them?
3. Do you think that flood-proofing of homes and other buildings is sufficient to protect the city from future hurricanes? If not, why not?

the pain and suffering of people in New Orleans during the time it took for proper rescue and support to actually occur. Hopefully we have learned from this catastrophe and will never again experience such a miscommunication in disaster preparedness.

A final adjustment to hazards is to try to control them. Thus, we build dams and levees to hold water back and reduce flooding. The levees are often nothing more than earthen walls (embankments) that allow the flow in the river to be higher before it actually flows onto the floodplain. Many people have been led to believe they are

protected by levees, only to find out later that they are, in fact, in a hazardous area. This is particularly true for many of the areas along the Mississippi and Missouri rivers, but is also common throughout the world.

In California, just upstream of Sacramento, the state capital, the river is prone to periodic flooding. In recent years many homes have been constructed behind levees (Figure 16.20a), some having been built between the riverbank and levee (Figure 16.20b). Inevitably, many of these homes will one day be inundated by floodwaters. It is irresponsible for government to allow such widespread

(a)

(b)

Figure 16.20 ■ Homeowners lulled to a false sense of security behind levees of the Sacramento River, in California. (*a*) Homes built on the floodplain behind a large levee. The road is on the levee and the river is to the left. These homes are essentially sitting ducks subject to potential future flooding. (*b*) Home inside a levee with some floodproofing. The living area of the home is elevated above the floodplain. The river is to the left.

development on active floodplains, and yet it continues to do so almost everywhere. Downstream of Sacramento near Stockton is a large area of former wetlands known as the Sacramento–San Joaquin delta that is now largely drained for farming. (It is not a real delta, but that is another story.) As the wetlands were drained, the soils oxidized and the land sank, producing a bowl (similar to New Orleans) protected by levees. If the land were to be developed, new homes would be in a serious flood hazard area when the levees failed during flood or by earthquake. An earthquake near the delta could also damage the system that delivers water to agriculture and southern California's millions of people.

16.10 What Does the Future Hold with Respect to Disasters and Catastrophes?

Every year we seem to set new global records with respect to economic losses from disasters and catastrophes. If we examine the frequency of disasters by the decade, we see that the number of disasters has increased significantly in the last half-century (Figure 16.21). We have more disasters and catastrophes in part because the human population has increased and more people are living in hazardous areas. Also, as a result of population pressure we often make poor land-use choices, selecting areas that are prone to frequent hazards, such as flooding,

wildfires, and hurricanes. Finally, we may be directly affecting the severity of some hazardous events. For example, urbanization covers the land with buildings, streets, parking lots, and sidewalks. As a result, more water runs off the land more quickly. The increased runoff increases both the size and frequency of flooding of smaller streams and rivers.

As a second example, the oceans of the world are getting warmer in part in response to human-induced processes. As a result, more energy is fed into storms. Although the number of hurricanes has not increased in past decades, storms have intensified, and with more intense storms, the damage increases. With more catastrophes likely in the future, we must become better prepared for these events. It cannot be repeated too often: Anticipating hazards rather than simply reacting to them will help minimize economic losses and reduce pain and suffering.

Summary

■ Because of poor land-use planning combined with an ever-increasing population, the hazards and disasters of the past are more frequently becoming catastrophes. A disaster is a hazardous event that occurs over a relatively short time span in a defined geographic area with significant loss of human life and property. In contrast, a catastrophe is a massive disaster, requiring enormous expenditures of money and time for recovery to take place.

■ In studying natural processes and hazards, it is important to take a historic point of view. By examining when and where hazardous events have occurred in the past and what the consequences of those events were, we can better prepare for the future. This is particularly true for hazards that tend to occur repeatedly.

■ Some of the fundamental concepts related to minimizing consequences of natural hazards are as follows: (1) Hazards are predictable; (2) linkages exist between different hazards and between the physical and biological environments; (3) hazards that previously produced mostly disasters are now producing catastrophes; (4) the risk from hazards can be estimated; and (5) adverse effects of hazards can be minimized.

■ People are often optimistic about natural hazards and the likelihood that they will suffer damages; this is particularly true for events that occur rarely. The more people learn about hazards, however, the more likely they are to respond.

■ With the continuing increases in human population, more stress will be placed on the natural environment as people look for places to live and work. More people will therefore be in harm's way with respect to natural hazards, and so we must work harder to minimize loss of human life and property.

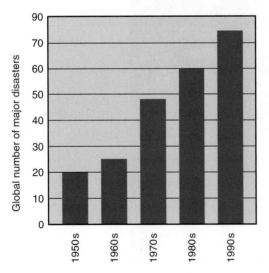

Figure 6.21 ■ Major disasters are increasing. The number of major disasters from 1950 to 1990 by decade is shown here. Notice the dramatic increase in the number of events. [*Source:* Data from J. M. Abramovitz, pp. 123–142, "Averting Unnatural Disasters," N. Brown, L. R. et al., *State of the World, 2001,* World Watch Institute (New York: W. W. Norton, 2001).]

REEXAMINING THEMES AND ISSUES

Human Population

As population increases and people move into more hazardous areas, damage and loss of life tend to increase. What were formerly disasters are becoming catastrophes.

Sustainability

Our use of living resources is not sustainable and is placing people more at risk from natural hazards such as flooding, landslide, and hurricane. When forests are clear-cut near human occupation, the risk of disasters increases as runoff increases and landslides become more frequent.

Global Perspective

Global change, especially human-induced global warming, is increasing the temperature of the oceans. As a result, we are changing to a hydrologic cycle and more energy is fed into the atmosphere. Some of the consequences are that hurricanes are apparently becoming more intense as are thunderstorms and other atmospheric hazards.

Urban World

More and more people are moving to urban centers in the United States and around the world. The increased concentration of people renders us more vulnerable to natural hazards that may impact a large number of people. An example is Hurricane Katrina, which flooded New Orleans.

People and Nature

More people are visiting natural places for activities such as hiking, rock climbing, and skiing. As more people engage in these activities, the deaths from rock fall and avalanche are increasing. On the other hand, when green belts and parks are incorporated in urban areas, that is nature moves to us, wind is reduced and flood producing runoff is decreased.

Science and Values

The science necessary to understand natural hazards is mature. In order to reduce impacts from hazards will require us to make a shift in our values. The major shift will be to provide funds necessary to reduce damages from natural hazards to property and to save lives.

Key Terms

acceptable risk **337**	earthquake **320**	landslides **320**	volcanic eruption **320**
catastrophe **323**	flood **320**	natural hazard **323**	wildfire **320**
direct effects **338**	heat wave **323**	risk **337**	
disaster **323**	hurricane **320**	tornado **320**	
drought **323**	indirect effects **338**	tsunami **320**	

Study Questions

1. Why are more disasters becoming catastrophes?
2. What are typical responses to natural hazards?
3. What is the difference between reacting to and anticipating hazards in hazard reduction programs?
4. Does global warming have any influence on hazards such as hurricanes?
5. What were the main social, environmental, and economic lessons learned from Hurricane Katrina, which devastated New Orleans in 2005?
6. What are some of the natural service functions of natural hazard events?
7. When do you think we should attempt to control natural processes, and when might it be preferable to learn to live with them? Examples of controlling natural processes: build a flood-control dam or restrict development on floodplains.
8. What are the trends in deaths and damage caused by natural hazards?

9. What is the role of history in trying to understand natural hazards?

10. What natural hazards might threaten your community and campus. What has been done to evaluate them and what could be done to minimize hazards?

11. You are attending a football game at your university's stadium and an announcement is made that a tornado or violent thunderstorm is moving your way. The storm is 20 miles away moving at 5 miles per hour. What action if any should you take? What factors would enter into your decision?

12. Design a research project to test the hypothesis that large U.S. cities such as New York, Chicago, and Los Angeles are more vulnerable to natural hazards than they were 50 years ago. What information is needed and how would you evaluate it?

13. Some islands in the Pacific are so low that rising sea levels are threating their very existance. How could you evaluate a particular island for its vulnerability to violent storms or tsunamis?

14. The response from government agencies following Hurricane Katrina has been criticized. What do you think could be done better to ensure that another Katrina catastrophe involving so many people is unlikely?

15. What have we learned from Hurricane Katrina that might be useful in evaluating of potential catastrophes from floods, earthquakes, and volcanic eruptions in other parts of the country?

Further Reading

Bolt, B. A. 2004. *Earthquakes*, 5th ed. San Francisco: W. H. Freeman.

Decker, R., and B. Decker. 1998. *Volcanoes*, 3rd ed. New York: W. H. Freeman.

Keller, E. A., and R. H. Blodgett. 2006. *Natural Hazards*. Upper Saddle River, N.J.: Prentice-Hall.

Payne, S. J., P. L. Andrews, and R. D. Laven. 1996. *Introduction to Wildland Fire*, 2nd ed. New York: John Wiley & Sons.

Pinter, N. 2005. "One Step Forward, Two Steps Back on U.S. Floodplains." *Science* 308: 207–208.

Yeats, R. S. 2001. *Living with Earthquakes in California: A Survivor's Guide*. Corvallis: Oregon State University Press.

CHAPTER 17

Energy: Some Basics

Manhattan Island from New Jersey during blackout, August 14, 2003. Millions of people during rush hour walked dark streets to go home.

Learning Objectives

Understanding the basics of what energy is, as well as the sources and uses of energy, is essential for effective energy planning. After reading this chapter, you should understand:

- That energy is neither created nor destroyed but is transformed from one kind to another.

- Why in all transformations energy tends to go from a more usable to a less usable form.

- What energy efficiency is and why it is always less than 100%.

- That people in industrialized countries consume a disproportionately large share of the world's total energy and how efficiency and conservation of energy can help make better use of global energy resources.

- Why some energy planners propose a hard-path approach to energy

provision and others a soft-path approach, and why both of these approaches have positive and negative points.

- Why moving toward global sustainable energy planning with integrated energy planning is an important goal.

- What elements are needed to develop integrated energy planning.

*Written with assistance from Mel S. Manalis.

C A S E   S T U D Y

National Energy Policy: From Coast-to-Coast Energy Crisis to Promoting Energy Independence

The most serious blackout (failure of electric power) in U.S. history occurred on August 14, 2003. New York City along with eight states and parts of Canada suddenly lost power at about 4:00 P.M. More than 50 million people were affected, some trapped in elevators or electric trains underground. People streamed into the streets of New York not certain whether or not the power failure was due to a terrorist attack. Power was restored within 24 hours to most places, but the event was an energy shock that demonstrates our dependence on aging power distribution systems and centralized electric power generation. Terrorists had nothing to do with the blackout, but the event caused harm, anxiety, and financial loss to millions of people.

In 2001, California was faced with "rolling blackouts"—electrical power outages that caused disruption in homes and industry. It was an "energy shock" that startled the entire country, from coast to coast, into thinking about energy policy in the United States.

A basic problem in California was that while the economic growth of the 1990s brought prosperity and more people to the state, energy demand rose and few new sources came online to meet the growing demand. Utility companies were forced to buy "emergency power" from other power suppliers at very high prices (as much as 900% higher). They also had to buy natural gas, which was increasing nationwide in price. By law, the utility companies could not pass on their increasing costs, and they claimed they were driven to near bankruptcy.

The California energy crisis sparked debate in Congress, resulting in a new national energy policy—something that has not happened since the 1970s. Lawmakers have recognized that we could face a similar, but nationwide, crisis again soon. Some congresspersons support the so-called hard path: Build more oil, natural gas, coal, and nuclear power plants. Others want more emphasis on energy conservation and alternative energy sources such as solar and wind. As of 2006, the price of energy in California and across the United States remains high, and the shock a wake-up call.

Seven presidents of the United States since the mid-1970s have attempted to address energy problems and becoming independent of foreign energy sources. Finally, in the summer of 2005, the Energy Policy Act of 2005 was passed by Congress and signed by the president, becoming law.

Today we are more dependent than ever on imported oil. In fact, since the 1970s there has been a 50% increase in the consumption of gasoline (where most oil is used) while domestic production of oil has dropped by nearly one-half. One of the reasons has been the dramatic reduction in oil production from Alaska, which has dropped by about one-half from the late 1980s and into the early twenty-first century. Short supply of gasoline in 2004–2006 drove up prices to as much as $3 per gallon in some states. Natural gas has followed a similar pattern with respect to production and consumption since the late 1980s. New power plants today use natural gas as the desired fuel because it is cleaner burning, it results in fewer pollutants, and the United States has abundant potential supplies of it. The problem with natural gas will be to bring the production in line to match consumption in the future. Energy planning at the national level in the first five years of the twenty-first century has been marked by an ongoing debate about future supplies of mostly fossil fuels, including coal, oil, and natural gas. Planning objectives have centered on providing a larger supply of coal and natural gas, and to a lesser extent, oil. Planners came to the conclusion that if the United States is to meet electricity demands by the year 2020, over 1,000 new power plants will have to be constructed. Working out the numbers, this means that about 60 per year will need to be constructed between now and 2020—more than one new facility per week!

The Energy Policy Act of 2005 emphasizes that we have made improvements in energy conservation and are continuing development of alternative energy sources, such as hydrogen, solar, and wind. Environmentalists criticize the plan as favoring large centralized energy production from fossil fuels and nuclear sources, with not enough emphasis on alternative energy sources. Alternative energy in the United States has been plagued by positive rhetoric followed by some funding that is then reduced in subsequent years. The key to real energy planning requires a diversity of energy sources with a better mix of the fossil fuels and alternative sources that must eventually replace them. What is apparent is that in the first decades of the twenty-first century we are going to be continually plagued with dramatic price changes in energy with accompanying shortages. This pattern will continue until we become much more energy-independent from foreign sources. Using our remaining fossil fuels, particularly the cleaner fuels such as natural gas, will represent a transitional phase to more sustainable sources. What is really necessary is a major program to develop sources such as wind and solar much more vigorously than has been done in the past or apparently will be done in the next few years. If we are unable to make the transition as world production of petroleum peaks and declines, then we will face an energy crisis that is unsurpassed in our history.

This case history demonstrates that the United States faces serious energy problems. With this in mind, we explore in this chapter some of the basic principles associated with what energy is, how much energy we consume, and how we might manage energy for the future.

17.1 Outlook for Energy

Energy crises are nothing new. People have encountered energy problems for thousands of years, as far back as the early Greek and Roman cultures.

Energy Crises in Ancient Greece and Rome

The climate in coastal areas of Greece 2,500 years ago was characterized by warm summers and cool winters, much as it is today. To warm their homes in winter, the Greeks used small, charcoal-burning heaters that were not very efficient. Since charcoal is produced from burning wood, wood was their primary source of energy, as it is today for half the world's people.

By the fifth century B.C., fuel shortages had become common, and much of the forest in many parts of Greece was depleted of firewood. As local supplies diminished, it became necessary to import wood from farther away. Olive groves became sources of fuel, which was made into charcoal for burning, reducing a valuable resource. By the fourth century B.C., the city of Athens had banned the use of olive wood for fuel.

At about this time, the Greeks began to build their houses facing the south, designing them so that the low winter sun entered the houses, providing heat, and the higher summer sun was partially blocked, cooling the houses. Recent excavations of ancient Greek cities suggest that large areas were planned so that individual homes could make maximum use of this solar energy. The Greeks' use of solar energy in heating homes was a logical answer to their energy problem.[1]

The use of wood in ancient Rome is somewhat analogous to the use of oil and gas in the United States today. Wealthy Roman citizens about 2,000 years ago had central heating in their large homes, burning as much as 125 kg (275 lb) of wood every hour. Not surprisingly, local wood supplies were exhausted quickly, and the Romans had to import wood from outlying areas. Eventually, wood had to be imported from as far away as 1,600 km (about 1,000 mi).[1]

The Romans turned to solar energy for the same reasons as the Greeks but with much greater application and success. The Romans used glass windows to increase the effectiveness of solar heating, developed greenhouses to raise food during the winter, and oriented large public bathhouses (some of which accommodated up to 2,000 people) to use passive solar energy (Figure 17.1). The Romans believed that sunlight in bathhouses was healthy, and it also saved greatly on fuel costs. The use of solar energy in ancient Rome was widespread and resulted in laws to protect a person's right to solar energy. In some areas, it was illegal for one person to construct a building that shaded another's.[1]

The ancient Greeks and Romans experienced an energy crisis in their urban environments. In turning to solar energy, they moved toward what today we call sustainability. We are on that same path as fossil fuels become scarce.

Figure 17.1 ■ Roman bathhouse (lower level) in the town of Bath, England. The orientation of the bathhouse and the placement of windows are designed to maximize the benefits of passive solar energy.

Energy Today and Tomorrow

The energy situation facing the United States and the world today is in some ways similar to that faced by the early Greeks and Romans. The use of wood in the United States peaked in the 1880s, when the use of coal became widespread. The use of coal in turn began to decline after 1920, when oil and gas started to become available. Today, we are facing the global peak of oil production, which is expected by about 2020. Fossil fuel resources, which took millions of years to form, may be essentially exhausted in just a few hundred years.

The decisions we make today will affect energy use for generations. Should we choose complex, centralized energy production methods, or use simpler and widely dispersed energy production methods, or use a combination of the two? Which sources of energy should be emphasized? Which uses of energy should be emphasized for increased efficiency? How can we rely on current energy sources and provide for developing a sustainable energy policy? There are no easy answers.

Use of fossil fuels, especially oil, has resulted in improvements in sanitation, medicine, and agriculture. These improvements have helped to make possible the global human population increase that we have discussed in other chapters. Many of us are living longer, with a higher standard of living than people before us. However, burning fossil fuels imposes growing environmental costs that are causing concerns ranging from urban pollution to a change in global climate.

One thing certain about the energy picture for tomorrow is that it will involve living with uncertainty when it comes to energy availability and cost. The sources and patterns of energy utilization will undoubtedly change. Problems with energy supply and cost can be expected to occur, as a result of growing demand and insufficient supply. Supplies will continue to be regulated, and there exists a growing potential for the disruption of supplies. Oil embargoes could cause significant economic impact in the United States and other countries; and a war or revolution in a petroleum-producing country would cause exports of petroleum to be reduced significantly.

It is clear that we need to rethink our entire energy policy in terms of sources, supply, consumption, and environmental concerns. We can begin by understanding basic facts about what energy is.

17.2 Energy Basics

The concept of **energy** is somewhat abstract; you cannot see it or feel it, even though you have to pay for it.[2] To understand energy, it is easiest to begin with the idea of a force. We all have had the experience of exerting force by pushing or pulling. The strength of a force can be measured by how much it accelerates an object.

What if your car stalls while you are going up a hill and you get out to push it uphill to the side of the road (Figure 17.2). You apply a force against gravity, which would otherwise cause the car to roll downhill. If the brake is on, the brakes, tires, and bearings might heat up from friction. The longer the distance over which you exert force in pushing the car, the greater the change in the car's position and the greater the amount of heat from friction in the brakes, tires, and bearings.

In physicists' terms, exerting the force over the distance moved is work. That is, **work** is the product of a force times a distance. Conversely, energy is the ability to do work. If you push hard but the car does not move, you have exerted a force but you have not done any work on the car (according to the definition), even if you feel very tired and sweaty.[2]

In pushing your stalled car, you have moved it against gravity and caused some of its parts (brakes, tires, bearings) to be heated. These effects have something in common: They are forms of energy. You have converted chemical energy in your body to the energy of motion of the car (kinetic energy). When the car is higher on the hill, the potential energy of the car has been increased, and friction produces heat energy.

Energy can be, and often is, converted or transformed from one kind to another, but the total energy is always conserved. The principle that energy cannot be created or destroyed but is always conserved is known as the **first law of thermodynamics**. (Recall the discussion of matter and

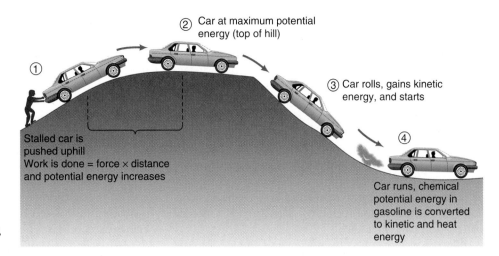

② Car at maximum potential
energy (top of hill)

①

③ Car rolls, gains kinetic
energy, and starts

④

Stalled car is
pushed uphill
Work is done = force × distance
and potential energy increases

Car runs, chemical
potential energy in
gasoline is converted
to kinetic and heat
energy

Figure 17.2 ■ Some basic energy concepts, including potential energy, kinetic energy, and heat energy.

energy in Chapter 5 and the detailed discussion of energy flow in the biosphere in Chapter 9.) Thermodynamics is the science that keeps track of energy as it undergoes various transformations from one type to another. We use the first law to keep track of the quantity of energy.[3]

To illustrate the conservation and conversion of energy, think about a tire swing over a creek (Figure 17.3). When the tire swing is held in its highest position, it is not moving.

① Energy is all potential.
② Energy is all kinetic.
③ Energy is potential and kinetic.

Figure 17.3 ■ Diagram of a tire swing, illustrating the relation between potential and kinetic energy.

It does, however, contain stored energy owing to its position. We refer to the stored energy as *potential energy*. Other examples of potential energy are the gravitational energy in water behind a dam; the chemical energy in coal, fuel oil, and gasoline, as well as in the fat in your body; and nuclear energy, which is related to the forces binding the nuclei of atoms.[2]

The tire swing, when released from its highest position, moves downward. At the bottom (straight down), the speed of the tire swing is greatest, and no potential energy remains. At this point, all the swing's energy is the energy of motion, called *kinetic energy*. As the tire swings back and forth, the energy continuously changes between the two forms, potential and kinetic. However, with each swing, the tire slows down a little more and goes a little less high because of friction created by the movement of the tire and rope through air and friction at the pivot where the rope is tied to the tree. The friction slows the swing, generating *heat energy*, which is energy from random motion of atoms and molecules. Eventually, all the energy is converted to heat and emitted to the environment, and the swing stops.[2]

The example of the swing illustrates the tendency of energy to dissipate and end up as heat. Indeed, physicists have found that it is possible to change all the gravitational energy in a tire swing (a type of pendulum) to heat. However, it is impossible to change all the heat energy thus generated back into potential energy.

Energy is conserved in the tire swing. All the initial gravitational potential energy has been transformed by way of friction to heat energy when the tire swing finally stops. If the same amount of energy, in the form of heat, were returned to the tire swing, would you expect the swing to start again? The answer is no! What, then, is used up? It is not energy, because energy is always conserved. What is used up is the energy *quality*—or the availability of the energy to perform work. The higher the quality of the energy, the more easily it can be converted to work; the lower the energy quality, the more difficult it is to convert to work.

This example illustrates another fundamental property of energy: Energy always tends to go from a more usable (higher-quality) form to a less usable (lower-quality) form. This is the **second law of thermodynamics**, and it means that, when you use energy, you lower its quality.

Let us return to the example of the stalled car, which you have now pushed to the side of the road. Having pushed the car a little way uphill, you have increased its potential energy. You can convert this to kinetic energy by letting it roll back downhill. You engage the gears to restart the car. As the car idles, the potential chemical energy (from the gasoline) is converted to waste heat energy and other energy forms, including electricity to charge the battery and play the radio.

Why can we not collect the wasted heat and use it to run the engine? Again, as the second law of thermodynamics tells us, once energy is degraded to low-quality heat, it can never regain its original availability or energy grade. When we refer to low-grade heat energy, we mean that relatively little of it is available to do useful work. High-grade energy, such as that of gasoline, coal, or natural gas, has high potential to do useful work. The biosphere continuously receives high-grade energy from the sun and radiates low-grade heat to the depths of space.[2, 3]

17.3 Energy Efficiency

Two fundamental types of energy efficiencies are derived from the first and second laws of thermodynamics: the first-law efficiency and the second-law efficiency. The **first-law efficiency** deals with the amount of energy without any consideration of the quality or availability of the energy. It is calculated as the ratio of the actual amount of energy delivered where it is needed to the amount of energy supplied to meet that need. Expressions for efficiencies are given as fractions; multiplying the fraction by 100 converts it to a percentage. As an example, consider a furnace system that keeps a home at a desired temperature of 18°C

(65°F) when the outside temperature is 0°C (32°F). The furnace, which burns natural gas, delivers 1 unit of heat energy to the house for every 1.5 units of energy extracted from burning the fuel. That means it has a first-law efficiency of 1 divided by 1.5, or 67% (see Table 17.1 for other examples).[3] The "unit" of energy for our furnace is arbitrary for the purpose of discussion; we could use British thermal units (Btu) or some other units (see A Closer Look 17.1).

First-law efficiencies are misleading because a high value suggests (often incorrectly) that little can be done to save energy through additional improvements in efficiency. This problem is addressed by the use of the **second-law efficiency**. Second-law efficiency refers to how well matched the energy end use is with the quality of the energy source. For our home-heating example, the second-law efficiency would compare the minimum energy necessary to heat the home with the energy actually used by the gas furnace. If we calculated the second-law efficiency (which is beyond the scope of this discussion), the result might be 5%—much lower than the first-law efficiency of 67%.[3] (We will see why later.) Table 17.1 also lists some second-law efficiencies for common uses of energy.

Values of second-law efficiency are important because low values indicate where improvements in energy technology and planning may save significant amounts of high-quality energy. Second-law efficiency tells us whether the energy quality is appropriate to the task. For example, you could use a welder's acetylene blowtorch to light a candle, but a match is much more efficient (and safer as well).

We are now in a position to understand why the second-law efficiency is so low (5%) for the house-heating example discussed earlier. This low efficiency implies that the furnace is consuming too much high-quality energy in carrying out the task of heating the house. In other words, the task of heating the house requires heat at a relatively low temperature, near 18°C (65°F), not heat with temperatures in excess of 1,000°C (1,832°F), such as is generated inside the gas furnace. Lower-quality energy, such

Table 17.1 • Examples of First- and Second-Law Efficiencies				
Energy (End Use)	First-Law Efficiency (%)	Waste Heat (%)	Second-Law Efficiency (%)	Potential for Savings
Incandescent lightbulb	5	95		
Fluorescent light	20	80		
Automobile	20–25	75–80	10	Moderate
Power plants (electric); fossil fuel and nuclear	30–40	60–70	30	Low to moderate
Burning fossil fuels (used directly for heat)	65	35		
Water heating			2	Very high
Space heating and cooling			6	Very high
All energy (U.S.)	50	50	10–15	High

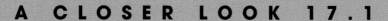

A CLOSER LOOK 17.1

Energy Units

When we buy electricity by the kilowatt-hour, what are we buying? We say we are buying energy, but what does that mean? Before we go deeper into the concepts of energy and its uses, we need to define some basic units.

The fundamental energy unit in the metric system is the *joule*; 1 joule is defined as a force of 1 newton[*] applied over a distance of 1 meter. To work with large quantities such as the amount of energy used in the United States in a given year, we use the unit *exajoule*, which is equivalent to 10^{18} (a billion billion) joules or roughly equivalent to 1 quadrillion, or 10^{15}, British thermal units (Btu), referred to as a *quad*. To put these big numbers in perspective, the United States today consumes approximately 100 exajoules (or quads) of energy per year, and world consumption is about 425 exajoules (quads) annually.

In many instances, we are particularly interested in the rate of energy use, or *power*, which is energy divided by time. In the metric system, power may be expressed as joules per second, or *watts*, W (1 joule per second is equal to 1 watt). When larger power units are required, we can use multipliers, such as *kilo* (thousand), *mega* (million), and *giga* (billion). For example, the rate of production of electrical energy in a modern nuclear power plant is 1,000 megawatts (MW) or 1 gigawatt (GW).

Sometimes it is useful to use a hybrid energy unit, such as the watt-hour, Wh (remember energy is power multiplied by time). Electrical energy is usually expressed and sold in *kilowatt-hours* (kWh, or 1,000 Wh). This unit of energy is 1,000 W applied for 1 hour (3,600 seconds), the equivalent energy of 3,600,000 J (3.6 MJ).

The average estimated electrical energy in kilowatt-hours used by various household appliances over a period of a year is shown in Table 17.2. The total annual energy used is the power rating of the appliance multiplied by the time the appliance was actually used. The appliances that use most of the electrical energy are water heaters, refrigerators, clothes dryers, and washing machines. A list of common household appliances and the amounts of energy they consume is useful in identifying those appliances that might help save energy through conservation or improved efficiency.

Table 17.2 • Average Estimated Electrical Energy Use per Year for Typical Household Appliances

Appliance	Power (W)	Average Hours Used per Year	Approximate Energy Used (kWh/yr)
Clock	2	8,760	17
Clothes dryer	4,600	228	1,049
Hair dryer	1,000	60	60
Lightbulb	100	1,080	108
Compact fluorescent[a]	18	1,080	19
Television	350	1,440	504
Water heater (150 L)	4,500	1,044	4,698
Energy-efficient model[a]	2,800	1,044	2,900
Toaster	1,150	48	552
Washing machine	700	144	1,008
Refrigerator	360	6,000	2,160
Energy-efficient model[a]	180	6,000	1,100

Source: Data from U.S. Department of Energy and D. G. Kaufman and C. M. Franz, *Biosphere 2000: Protecting Our Global Environment* (New York: HarperCollins, 1993).
[a] Newer, energy-efficient model.

[*] A newton (N) is the force necessary to produce an acceleration of 1 m per sec per sec (m/s^2) to a mass of 1 kg.

as solar energy, could do the task and yield a higher second-law efficiency, because there is a better match between the required energy quality and the house-heating end use. Through better energy planning, such as matching the quality of energy supplies to the end use, higher second-law efficiencies can be achieved, resulting in substantial savings of high-quality energy.

Examination of Table 17.1 indicates that electricity-generating plants have nearly the same first-law and second-law efficiencies. These generating plants are examples of *heat engines*. A heat engine produces work from heat. Most of the electricity generated in the world today comes from heat engines that use nuclear fuel, coal, gas, or other fuels. Our own bodies are examples of heat engines, operating with a capacity (power) of about 100 watts (W) and fueled indirectly by solar energy. (See A Closer Look 17.1 for an explanation of watts and other units of energy.) The internal combustion engine (used in automobiles) and the steam engine are additional examples of heat engines. A great deal of the world's energy is used in heat engines, with profound environmental effects such as thermal pollution, urban smog, acid rain, and global warming.

The maximum possible efficiency of a heat engine, known as *thermal efficiency*, was discovered by the French engineer Sadi Carnot in 1824, before the first law of thermodynamics was formulated.[4] Modern heat engines have thermal efficiencies that range between 60% and 80% of their ideal Carnot efficiencies. Modern 1,000-megawatt (MW) electrical generating plants have thermal efficiencies ranging between 30% and 40%; that means at least 60% to 70% of the energy input to the plant is rejected as waste heat. For example, assume that the electric power output from a large generating plant is 1 unit of power (typically 1,000 MW). Producing that 1 unit of power requires 3 units of input (such as burning coal) at the power plant, and the entire process produces 2 units of waste heat, for a thermal efficiency of 33%. The significant number here is the waste heat, 2 units, which amounts to twice the actual electric power produced.

Electricity may be produced by large power plants that burn coal or natural gas, by plants that use nuclear fuel, or by smaller producers, such as geothermal, solar, or wind sources (see Chapters 18, 19, and 20). Once produced, the electricity is fed into the grid, which is the network of power lines, or the distribution system. Eventually, it reaches homes, shops, farms, and factories, where it lights, heats, and drives motors and other machinery used by society. As electricity moves through the grid, there are losses.[5] The wires that transport electricity (power lines) have a natural resistance to electrical flow, known as *electrical resistivity*. This resistance converts some of the electric energy in the transmission lines to heat energy, which is radiated into the environment surrounding the lines.

17.4 Energy Sources and Consumption

People living in industrialized countries make up a relatively small percentage of the world's population, but they consume a disproportionate share of the total energy produced in the world. For example, the United States, with only 5% of the world's population, uses approximately 25% of the total energy consumed in the world. There is a direct relationship between a country's standard of living (as measured by gross national product) and energy consumption per capita. After the peak in oil production, expected in 2020–2050, supplies of oil and gasoline will be reduced and will be more expensive. Before then, use of these fuels may be curtailed to reduce potential global climate change. As a result, within the next 30 years, both developed and developing countries will need to find innovative ways to obtain energy. In the future, affluence may be as related to more efficient use of a wider variety of energy sources as to total energy consumption.

Fossil Fuels and Alternative Energy Sources

Today, approximately 90% of the energy consumed in the United States is produced by petroleum, natural gas, and coal. Because of their organic origin, these are called fossil fuels. They are produced from plant and animal material and are forms of stored solar energy that are part of our geologic resource base. They are essentially nonrenewable. Other sources of energy—which include geothermal, nuclear, hydropower, and solar, among others—are referred to as alternative energy sources. The term *alternative* designates these as sources that might replace fossil fuels in the future. Some of the alternative sources, such as solar and wind, are not depleted by consumption and are known as *renewable energy*.

The shift to alternative energy sources may be gradual, as fossil fuels continue to be utilized, or it could be accelerated as a result of concern over the potential environmental effects of burning fossil fuels. Regardless of which path we take, one thing is certain: Fossil fuel resources are finite. It has taken millions of years to form them, yet fossil fuels will be consumed in only a few hundred years of human history. Using even the most optimistic predictions, the fossil fuel epoch that started with the Industrial Revolution will represent only about 500 years of human history. Therefore, although fossil fuels have been extremely significant in the development of modern civilization, their use will be a short-lived event in the span of human history.

Energy Consumption in the United States Today

Energy consumption in the United States from 1950 to 2004 is shown in Figure 17.4. The figure dramatically illustrates our present dependence on the three major fossil fuels (coal; natural gas; and petroleum). From approximately 1950 through the late-1970s, energy consumption increased tremendously, from about 30 exajoules to 80 exajoules. (Energy units are defined in A Closer Look 17.1.) Since about 1980, energy consumption has increased by only about 20 exajoules. This situation is encouraging because it suggests that policies to improve energy conservation through efficiency improvements (such as requiring new automobiles to be more fuel efficient and buildings to be better insulated) have been at least partially successful.

What is not shown in the figure, however, is the tremendous energy loss. For example, energy consumption in the United States in 1965 was approximately 50 exajoules; of this total, about half was effectively used. Energy losses were about 50% (the number shown in Table 17.1 for all energy). In 2004, energy consumption in the United States was about 100 exajoules, and again about 50% was lost in conversion processes. Energy losses in 2004 were about equivalent to total energy consumption in 1965!

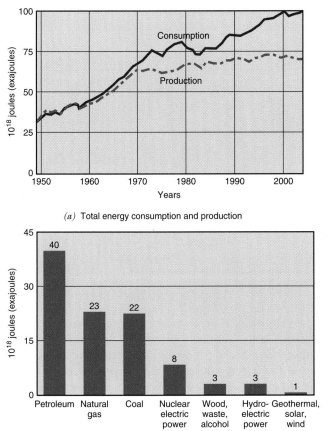

(a) Total energy consumption and production

(c) Energy consumption by source for 2004

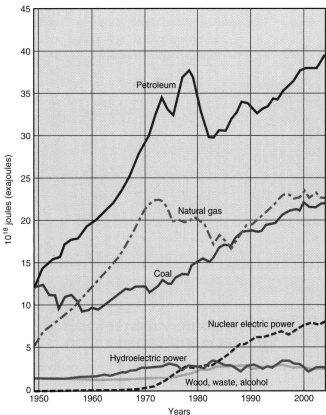

(b) Total energy consumption by source

Figure 17.4 ■ U. S. Energy from 1950–2004. *(a)* Total consumption and production; *(b)* consumption by source; *(c)* consumption by source in 2004. [*Source:* Modified after Energy Information Administration, *Annual Energy Review* 2004.]

The largest energy losses are associated with the production of electricity and with transportation.

We have already seen that energy is lost both in the generation of electricity at power plants and in the transmission of the electricity through the grid system. Major losses of energy in the transportation sector of our economy are related to the burning of fossil fuels in the engines of cars, buses, and trucks. Most of the losses related to the production of electricity and to transportation occur through the use of heat engines, which produce waste heat that is lost to the environment.

Another way to examine energy use is to look at the generalized energy flow of the United States for a particular year and by end use (Figure 17.5). The data on Table 17.3 show that for 2000 we imported considerably more oil than we produced and that consumption of energy is fairly evenly distributed in three sectors: residential/commercial, industrial, and transportation. It is clear that we remain dangerously vulnerable to changing world conditions affecting the production and delivery of crude oil. Evaluation of the entire spectrum of potential energy sources is necessary to ensure the availability of sufficient energy in the future while sustaining environmental quality. Energy consumption patterns have also come under scrutiny.

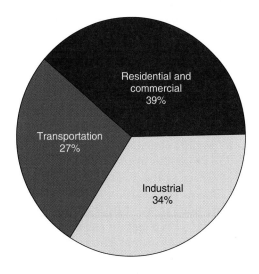

Figure 17.5 ■ Energy consumption in the United States by sector (approximate). [*Source:* Energy Information Administration, *Annual Energy Review* (Washington, D.C.: U.S. Department of Energy, 2000).]

Table 17.3 • Generalized (Approximate) Annual Energy Flow for the United States in 2000

Energy Source	Energy Production[a]	Net Imports + (Imports – Exports)	± Adjustments[b] =	Energy Consumed	Consumed by Sector
Coal	23.4	−0.8			Residential/commercial
Natural gas	19.9	3.7			38.0
Oil	12.4	22.8			Industrial
Nuclear	8.0	0			33.0
Hydropower	2.2	0			Transportation
Other	5.8	0.3			27.0
Total	71.7 +	26.0	± 0.3 =	98.0	98.0

Source: OECD, Energy Information Administration, *Annual Energy Review* (Washington, D.C.: U.S. Department of Energy, 2001).
[a]Exajoules (10^{18} J).
[b]Balancing to account for a variety of items, including unaccounted-for supply, blending components, and changes in stock.

17.5 Energy Conservation, Increased Efficiency, and Cogeneration

There is a movement to change patterns of energy consumption in the United States through measures such as conservation, increased energy efficiency, and cogeneration. **Conservation** of energy refers simply to getting by with less demand for energy. In a pragmatic sense, this has to do with adjusting our energy needs and uses to minimize the amount of high-quality energy necessary to accomplish a given task. Increased **energy efficiency** involves designing equipment to yield more energy output from a given amount of input energy (first-law efficiency) or better matches between energy source and end use (second-law efficiency).[6] **Cogeneration** includes a number of processes designed to capture and use waste heat rather than simply to release it into the atmosphere, water, or other parts of the environment as a thermal pollutant. An example of cogeneration is the *natural gas combined cycle power plant* that produces electricity in two ways: gas cycle and steam cycle. In the gas cycle, the natural gas fuel is burned in a gas turbine to produce electricity. In the steam cycle, hot exhaust from the gas turbine is used to create steam that is fed into a steam generator to produce additional electricity. The combined cycles capture waste heat from the gas cycle, nearly doubling the efficiency of the power plant from about 30% to 50–60%. Energy conservation is particularly attractive because it provides more than a one-to-one savings. Remember that it takes 3 units of fuel such as coal to produce 1 unit of power such as electricity (two-thirds is waste heat). Therefore, not using (conserving) 1 unit of power saves 3 units of fuel!

These three concepts—energy conservation, energy efficiency, and cogeneration—are all interlinked. For example,

when electricity is produced at large, coal-burning power stations, large amounts of heat may be emitted into the atmosphere. Cogeneration, by using that waste heat, can increase the overall efficiency of a typical power plant from 33% to as much as 75%, effectively reducing losses from 67% to 25%.[6] Cogeneration also involves generating electricity as a by-product from industrial processes that produce steam as part of their regular operations. Optimistic energy forecasters estimate that eventually we may meet approximately one-half the electrical power needs of industry through cogeneration.[6] Another source has estimated that more than 10% of the power capacity of the United States could be provided through cogeneration.

The average first-law efficiency of only 50% (Table 17.1) illustrates that large amounts of energy are currently lost in producing electricity and in transporting people and goods. Innovations in how we produce energy for a particular use can help prevent this loss, raising second-law efficiencies. Of particular importance will be energy uses with applications below 100°C (212°F), because a large portion of the total U.S. energy consumption (for uses below 300°C, or 572°F) is for space heating and water heating (Figure 17.6).

In considering where to focus our efforts to develop more energy efficiency, we need to look at the total energy-use picture. In the United States, space heating and cooling of homes and offices, water heating, industrial processes (to produce steam), and automobiles account for nearly 60% of the total energy use. By comparison, transportation by train, bus, and airplane account for only about 5%. Therefore, the areas we should target for development of more energy efficiency are building design, industrial energy use, and automobile design.[6, 7] We should note, however, that debate continues as to how much efficiency improvements and conservation can reduce future energy

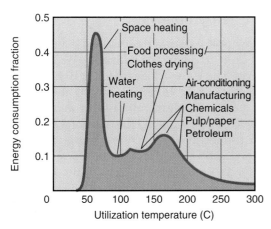

Figure 17.6 ■ Energy use below 300°C in the United States. [*Source:* Los Alamos Scientific Laboratory, *LASL 78–24*, 1978.]

demands and the need for increased production from traditional sources, such as fossil fuel.

Building Design

A spectrum of possibilities exists for increasing energy efficiency and conservation in residential buildings. For new homes, the answer is to design and construct homes that minimize the energy consumption necessary to ensure comfortable living.[8] For example, we can design buildings to take advantage of passive solar potential, as did the early Greeks and Romans and the Native American cliff dwellers. (Passive solar energy systems collect solar heat without using moving parts.) Windows and overhanging structures can be positioned so that the overhangs shade the windows from incoming summer solar energy, thereby keeping the house cool while allowing winter sun to penetrate the windows and warm the house.

The potential for energy savings through architectural design for older buildings is extremely limited. The position of the building on the site is already established, and reconstruction and modifications are often not cost-effective. The best approach to energy conservation for these buildings is insulation, caulking, weather-stripping, installation of window coverings and storm windows, and regular maintenance.

Buildings constructed to conserve energy are more likely to develop indoor air pollution problems, as pollutants emitted in buildings become concentrated due to reduced ventilation. Indoor air pollution is emerging as one of our most serious environmental problems. Potential problems can be reduced by better designs for air circulation systems that purify indoor air and bring in fresh, clean air (see Chapter 25). Construction that incorporates environmental principles is more expensive owing to higher

fees for architects and engineers as well as higher initial construction costs. Nevertheless, moving toward improved design of homes and residential buildings to conserve energy remains an important endeavor.

Industrial Energy

The graph of total energy consumption in the United States (Figure 17.4) shows that the rate of increase in energy utilization leveled off in the early 1970s. Nevertheless, industrial production of goods (automobiles, appliances, etc.) continued to grow significantly! Today, U.S. industry consumes about one-third of the energy produced. The reason we have had higher productivity with less energy use is that more industries are using cogeneration and more energy-efficient machinery, such as motors and pumps designed to use less energy.[6,9]

Automobile Design

Steady improvements have been made in the development of fuel-efficient automobiles during the last 30 years. In the early 1970s, the average U.S. automobile burned approximately 1 gallon of gas for every 14 miles traveled. By 1996, the miles per gallon (mpg) had risen to an average of 28 for highway driving and as high as 49 for some automobiles.[10] Fuel consumption rates did not improve much from 1996 to 1999. In 2004 many vehicles sold were SUVs and light trucks with fuel consumption of 10–20 mpg. A loophole in regulations permits these vehicles to have poorer fuel consumption than conventional automobiles. As a result of higher gasoline prices, sales of larger SUVs declined in 2006. Today the fuel consumption of some hybrid (gasoline-electric) vehicles exceeds 90 mpg on the highway and 60 mpg in the city. This improvement has several causes: increased efficiency and resulting conservation of fuel; cars that are smaller, with engines constructed of lighter materials[6]; and the combination of a fuel-burning engine with an electric motor.[11] Demand for hybrid vehicles is growing rapidly as gasoline prices continue to increase. Of course, there is a price to pay for this change. Smaller cars may be more prone to damage on impact, and as cars have gotten smaller, trucks have tended to stay the same or to have increased in size. As a result, the number of serious accidents between cars and trucks in the United States has increased.

Values, Choices, and Energy Conservation

A potentially effective method of conserving energy is to change our behavior by using less energy. This involves our values and the choices we make to act at a local level to address global environmental problems, such as human-induced warming caused by the burning of fossil fuels. For example, we make choices as to how far we commute to school or work and what method of transport we use to get there. Some people commute more than an hour by car to get to work, while others ride a bike, walk, or take a bus or train. Other ways of modifying behavior to conserve energy include the following:

- Using carpools to travel to and from work or school
- Purchasing a hybrid car (gasoline-electric)
- Turning off lights when leaving rooms
- Taking shorter showers (conserves hot water)
- Putting on a sweater and turning down the thermostat during winter
- Using energy-efficient compact florescent lightbulbs
- Purchasing energy-efficient appliances
- Sealing drafts in buildings with weather stripping and caulk
- Better insulating your home
- Washing clothes in cold water whenever possible
- Purchasing local foods to reduce energy in transport of food
- Reducing standby power for electronic devices and appliances by using powerstrips and turning them off when not in use

What other ways of modifying your behavior would help conserve energy?

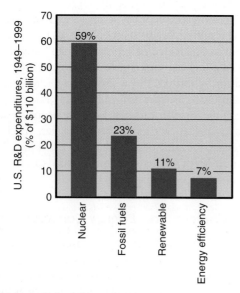

Figure 17.7 ■ Expenditures for U.S. energy research and development (R&D) from 1949 to 1999 (as a percent of $110 billion). [*Sources:* Energy Information Administration and Congressional Research Service; modified from Nona Yeats, *Los Angeles Times*, February 27, 2001.]

17.6 Energy Policy

U.S. energy policy during the past half-century has not moved us closer to energy self-sufficiency. We import more oil than ever. The United States spent $110 billion on energy research and development (R&D) from 1949 to 1999 (Figure 17.7). Nearly 60% of this went to nuclear energy, which supplies only 11% of our total energy. Over 20% went to fossil fuels. Renewable energy (wind, solar, biomass, and geothermal) and energy efficiency received only 11% and 7%, respectively. Investments in renewable energy are paying off slowly. Although renewable energy sources (water, solar and wind) provide only about 4% of the energy we use (Figure 17.7*b*), they are growing rapidly. The dollars spent on R&D for energy efficiency have yielded big returns. The average family in 1978 consumed about one-third more energy than in 2006. Savings came from such improvements as more efficient television sets, furnaces, refrigerators, and automobiles.

In the late 1990s, the United States spent about $2 billion per year on research and development for energy. By comparison, $45 billion per year went to research and development for the military. As we look to the future, we should attempt to separate politics and "pork barrel" expenditures (money politicians arrange for projects in their districts, regardless of their merit) from sound energy policy that will result in reaching sustainability with respect to energy production and use.

The Energy Policy Act of 2005 is the first national energy policy statement in over 10 years. Some of the provisions are as follows.

1. **Promotes conventional energy sources:** Recommends and supports using more coal, and natural gas with the objective of reducing our reliance on energy from foreign countries.
2. **Promotes nuclear power:** Recommends that the United States start building nuclear power plants again by 2010. Recognizes that nuclear power plants can generate large amounts of electricity without emitting air pollution or greenhouse gases.
3. **Encourages alternative energy:** Authorizes support or subsidies for wind energy and other alternative energy sources, such as geothermal, hydrogen, and biofuels (ethanol and biodiesel). The legislation also recognizes for the first time in U.S. energy policy wave power and tidal power as renewable energy technology. The act contains provisions to help make geothermal energy more competitive with fossil fuels in generation electricity. It increases the amount of biofuel (ethanol) that must be mixed with gasoline sold in the United States.
4. **Promotes conservation measures:** Sets higher efficiency standards for federal buildings and for household products. Directs federal attention to recommend fuel-efficiency standards for cars and trucks and SUVs.

Homeowners can claim new tax credits to install energy-efficient windows and appliances. The legislation also provides a tax credit on purchasing a fuel-efficient hybrid or clean-diesel vehicle.

5. **Promotes research:** Authorizes research into finding innovative ways to improve coal plants and to help construct cleaner coal plants; develop zero emission coal-burning power plants; determine how to tap into the vast amounts of oil trapped in oil shale and tar sands; and develop hydrogen-powered, pollution-free automobiles.

6. **Provides for energy infrastructure:** Provides incentives for oil refineries to expand their capacity. The act helps ensure that electricity is received over dependable modern infrastructure, and makes electric reliability standards mandatory. The legislation gives federal officials the authority to select sites for new power lines that are more independent of local pressure.

The act has been criticized for making most of the substantial incentives and subsidies to fossil fuels, especially coal and nuclear energy, at the expense of energy conservation and alternative energy. Regardless of perceived favoritism, politics, or lack of depth in understanding our energy crisis, the Energy Policy Act of 2005 initiates a new round of debate on the future of U.S. energy policy.

Hard Path versus Soft Path

Energy policy today is at a crossroads. One road leads to what is termed the **hard path**, which involves finding greater amounts of fossil fuels and building larger power plants. Taking the hard path means continuing the past emphasis on the quantity of energy we use. In this respect, the hard path is more comfortable. It requires no new thinking; no realignment of political, economic, or social conditions; and little anticipation of coming reductions in production of oil.

People heavily invested in the continued use of fossil fuels and nuclear energy favor the hard path. They argue that much environmental degradation around the world has been caused by people who have been forced to utilize local resources, such as wood, for energy. As a result, their lands have suffered from loss of plant and animal life and from soil erosion (as discussed in Chapters 10 and 11). They argue that the way to solve these environmental problems is to provide cheap, high-quality energy, such as fossil fuels or nuclear energy.

In countries like the United States, with sizable energy resources of coal, natural gas, and petroleum, people supporting the hard path argue that we should exploit those resources while finding ways to reduce the environmental impact of their use. According to these hard-path proponents, we should (1) let the energy industry develop the available energy resources and (2) let

industry, free from government regulations, provide a steady supply of energy with less total environmental damage.

The U.S. energy plan suggested by President George W. Bush is largely a hard-path proposal. The plan is to find and use more coal, oil, and natural gas; to use more nuclear power; and to build more than 1,000 new fossil fuel plants in the next 20 years. Energy conservation and development of alternative energy sources, while encouraged, are not considered primary in importance.

The second road of energy policy is called the **soft path**.[12] Amory Lovins, the scientist who has defined and championed this soft path, states that it involves energy alternatives that emphasize energy quality, are renewable, are flexible, and are environmentally more benign than those of the hard path. As defined by Lovins, these alternatives have several characteristics:

- They rely heavily on renewable energy resources, such as sunlight, wind, and biomass (wood and other plant material).
- They are diverse and are tailored for maximum effectiveness under specific circumstances.
- They are flexible, accessible, and understandable to many people.
- They are matched in energy quality, geographic distribution, and scale to end-use needs, increasing second-law efficiency.

Lovins points out that people are not particularly interested in having a certain amount of oil, gas, or electricity delivered to their homes; rather, they are interested in having comfortable homes, adequate lighting, food on the table, and energy for transportation.[12] According to Lovins, only about 5% of end uses require high-grade energy such as electricity. Nevertheless, a lot of electricity is used to heat homes and water. Lovins shows that there is an imbalance in using nuclear reactions at extremely high temperatures and in burning fossil fuels at high temperatures simply to meet needs where the necessary temperature increase may be only a few 10s of degrees. Such large discrepancies are thought wasteful and a misallocation of high-quality energy.

Energy for Tomorrow

The availability of energy supplies and the future demand for energy are difficult to predict because the technical, economic, political, and social assumptions underlying predictions are constantly changing. In addition, seasonal and regional variations in energy consumption must also be considered. For example, in areas with cold winters and hot, humid summers, energy consumption peaks during the winter months, with a secondary peak in the summer (the former resulting from heating and the latter from air-conditioning). Regional variations in energy sources and consumption are significant. For example, in the United States as a whole, the transportation sector uses about one-fourth of the energy consumed. However, in California, where people often commute long distances to work, about one-half of the energy is used for transportation, more than double the national average. Energy sources also vary by region. For example, in the eastern and southwestern United States, the fuel of choice for power plants is often coal, but on the West Coast, power plants are more likely to burn oil or natural gas or use hydropower from dams to produce electricity.

Future changes in population densities as well as intensive conservation measures will probably change existing patterns of energy use. This might involve a shift to more reliance on alternative (particularly renewable) energy sources. One prediction in 1991 was that energy consumption in the United States in the year 2030 may be as high as 120 exajoules or as low as 60 exajoules (Figure 17.8, scenarios A & C). The higher value assumes little change in energy policies, whereas the low value assumes the implementation of aggressive energy conservation policies. Since 1990 we have remained largely on the hard-path scenario. Barring new energy policy that mandates conservation, consumption will continue to increase in the future (scenario A in Figure 17.8). As of 2006, we are still on the scenario A growth path.

To bring about an environmental scenario that would also stabilize the climate in terms of global warming, use of energy from fossil fuels would need to be cut by about 50%. One reason such a large reduction in the use of fossil fuels may not come quickly is that some policy-makers are still not convinced that significant global warming will actually occur as a result of burning fossil fuels. (Global warming is discussed at length in Chapter 23.)

Over the past 30 years, energy scenarios have consistently overestimated the energy demands of the future. It seems unlikely that energy consumption in the United States in the year 2030 will be as low as 60 exajoules, and it is also unlikely that it will be as high as 120.

Low-energy scenarios for the future generally assume a moderate decrease in energy consumption accompanied by a shift from dependence on fossil fuels to the softer technologies (renewable, alternative energy resources). Reductions in energy use need not be associated with a lower quality of life. For example, Japan and some Western European nations with standards of living as high as or higher than that in the United States consume significantly less energy per person than the United States. What is needed is increased conservation of energy and more efficient use of energy, including the following:[7]

- More energy-efficient land-use planning that maximizes the accessibility of services and minimizes the need for transportation.

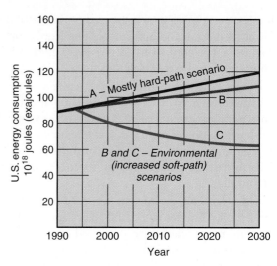

Figure 17.8 ■ Two possible future energy scenarios for the United States. [*Source:* Modified Union of Concerned Scientists, 2001.] Scenario C from 1991 prediction was overly optimistic. Scenario B is possible with intensive conservation.

- Agricultural practices and personal choices that emphasize (1) eating more locally grown foods and so reducing energy use for transporting crops and (2) eating more foods such as vegetables, beans, and grains. These foods require less total energy to produce than beef, chicken, and pork when crops are grown to feed these animals.

- Industrial guidelines for factories that promote energy conservation and minimize production of waste.

All projections of specific sources of energy use in the future must be considered speculative. Perhaps most speculative of all is the idea that we really can obtain most of our energy needs from alternative renewable energy sources in the next several decades. From an energy viewpoint, the next 20 to 30 years, as we move through the maximum production of petroleum, will be crucial to the United States and to the rest of the industrialized world.

The transformation of energy use from a hard to a softer path has a long history, as seen in the example of the early Greek and Roman cultures. In the latter part of the twentieth century, the United States experienced a shock in 1973 due to a shortage of oil. Long lines at the gas pumps caused anxiety concerning our energy supply and the lifestyle that depends on abundant oil. The 1973 oil shock spawned new research and development of alternative energy sources. It was also the impetus for the government to provide financial incentives for the utilization of sunlight, wind, and other alternative energy sources. However, with the return of abundant, inexpensive oil in the 1980s, there was much less government support of alternative energy. During this period, China, with one-fifth of the world's population, continued to develop its emerging big industry by burning huge quantities of coal.

Today, the industrialized countries of the world are even more dependent on imported oil than they were in the 1970s. Shortages and higher prices for oil are inevitable. One way to reduce oil consumption might be to establish new taxes on energy (see Critical Thinking Issue, Chapter 18).

The energy decisions we make in the very near future will greatly affect both our standard of living and our quality of life. From an optimistic point of view, we have the necessary information and technology to ensure a bright, warm, lighted, and moving future—but time may be running out, and we need action now. We can continue to take things as they come and live with the results of our present dependence on fossil fuels. Or we can choose to build a sustainable energy future based on careful planning, innovative thinking, and a willingness to move from our dependency on petroleum.

Integrated, Sustainable Energy Management

The concept of **integrated energy management** recognizes that no single energy source can provide all the energy required by the various countries of the world.[13] A range of options that vary from region to region will have to be employed. Furthermore, the mix of technologies and sources of energy will involve both fossil fuels and alternative, renewable sources.

A basic goal of integrated energy management is to move toward **sustainable energy development** that is implemented at the local level. Sustainable energy development would have the following characteristics:

■ It would provide reliable sources of energy.

■ It would not cause destruction or serious harm to our global, regional, or local environments.

■ It would help ensure that future generations inherit a quality environment with a fair share of the Earth's resources.

To implement sustainable energy development, leaders in various regions of the world will need to implement energy plans based on local and regional conditions. The plans will integrate the sources of energy most appropriate for a particular region with potential for conservation and efficiency and with desired end uses for energy. Such plans will recognize that the preservation of resources can be profitable and that degradation of the environment and poor economic conditions go hand in hand.[14] In other words, the degradation of air, water, and land resources results in a depletion of assets that ultimately will lower both the standard of living and the quality of life.

A good energy plan is part of an aggressive environmental policy with the goal of producing a quality environment for future generations. A good plan should do the following:[14]

■ Provide for sustainable energy development.

■ Provide for aggressive energy efficiency and conservation.

■ Provide for the diversity and integration of energy sources.

■ Provide for a balance between economic health and environmental quality.

■ Use second-law efficiencies as an energy policy tool (that is, strive to produce a good balance between quality of energy source and end uses for that energy).

Such a plan recognizes that energy demands can be met in environmentally preferred ways. An important element of the plan involves the energy used for automobiles. This builds on policies of the past 30 years to develop hybrid engines that are part electric and part internal combustion and improve fuel technology to reduce both fuel consumption and emission of air pollutants. Finally, the plan should factor in the marketplace through pricing that reflects the economic cost of using the fuel as well as its cost to the environment. In summary, the plan should be an integrated energy management statement that moves toward sustainable development. Those who develop such plans recognize that a diversity of energy supplies will be necessary and that the key components are (1) improvements in energy efficiency and conservation and (2) matching energy quality to end uses.[14]

The global pattern of ever-increasing energy consumption led by the United States and other nations cannot be sustained without a new energy paradigm that includes changes in human values rather than a breakthrough in technology. Choosing to own lighter, more fuel-efficient automobiles and living in more energy-efficient homes are consistent with a sustainable energy system that focuses on how to provide and use energy to improve human welfare. A sustainable energy paradigm establishes and maintains multiple linkages among energy production, energy consumption, human well-being, and environmental quality.[15] It might also involve using more distributed production of energy (see A Closer Look 17.2).

Summary

■ The first law of thermodynamics states that energy is neither created nor destroyed but is always conserved and is transformed from one kind to another. We use the first law to keep track of the quantity of energy.

■ The second law of thermodynamics tells us that as energy is used, it always goes from a more usable (higher-quality) form to a less usable (lower-quality) form.

■ Two fundamental types of energy efficiency are derived from the first and second laws of thermodynamics. In the United States today, first-law efficiencies average about 50%, which means that about 50% of the energy produced is returned to the environment as waste heat. Second-law efficiencies average 10 to 15%, so there is a high potential for saving energy through better match-

Micropower

It is likely that sustainable energy management will include the emerging concept of **micropower**—smaller, distributed systems for production of electricity. Such systems are not new. The inventor Thomas Edison evidently anticipated that electricity-generating systems would be dispersed; by the late 1890s, many small electrical companies were marketing and building power plants, often located in the basements of businesses and factories. These early plants evidently used cogeneration principles, since waste heat was reused for heating buildings.[16] Imagine if we had followed this early model: Homes would have their own power systems, electric power lines wouldn't snake through our neighborhoods, and we could replace older, less efficient systems as we do refrigerators.

Instead, in the twentieth century in the United States, power plants producing electricity grew larger. Industrializing countries, by the 1930s, had set up utility systems based on large-scale central power plants, as diagrammed in Figure 17.9a. Today, though, we are again evaluating the merits of distributive power systems, as shown in Figure 17.9b.

Large, centralized power systems are consistent with the hard path, while the distributive power system is more aligned with the soft path. Micropower devices rely heavily on renewable energy sources such as wind and sunlight, which feed into the electric grid system shown in Figure 17.9b. Use of micropower systems in the future is being encouraged because they are reliable and are associated with less environmental damage than are large fossil-fuel-burning power plants.[16]

Uses for micropower are emerging in both developed and developing countries. In countries that lack a centralized power-generating capacity, small-scale electrical power generation from solar and wind has become the most economical option. In nations with a high degree of industrialization, micropower may emerge as a potential replacement for aging electric power plants. For micropower to be a significant factor in energy production, a shift in policies and regulations to allow micropower devices to be more competitive with centralized generation of electrical power is required. Regardless of the obstacles that micropower devices face, distributive power systems will probably play an important role in our goal of achieving integrated, sustainable energy management for the future.

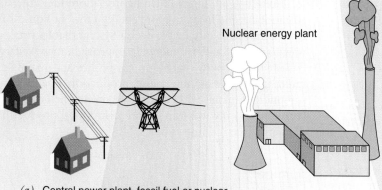

(a) Central power plant, fossil fuel or nuclear

Nuclear energy plant

Solar

Fuel cells in cars and homes

Biomass

(b) Distributive power systems, solar, fuel cell, wind, or biomass

Figure 17.9 ■ Idealized diagram comparing (a) a centralized power system such as those used in industrial developed countries today with (b) a distributive power system based on generating electricity from biomass, wind, solar, and other sources, all of which feed into the transmission and distribution system. [Source: Modified from S. Dunn, *Micropower, the Next Electrical Era*, Worldwatch Paper 151 (Washington, D.C.: Worldwatch Institute, 2000).]

Is There Enough Energy to Go Around?

The developing countries have most of the world's population (about 5 billion of 6 billion people) and are growing in population faster than developed countries. The average rate of energy use for individuals in developing countries is 1.0 kW, whereas that for persons in developed countries is 7.5 kW. If the current annual growth rate of 1.3% is maintained, the world's population will double in 54 years, to 12 billion people. More people will mean more energy use. In addition, people in developing countries will likely consume more energy per capita if they are to achieve a higher standard of living.

With a worldwide average energy-use rate of 2.6 kW per person, the 6 billion people on Earth use about 16 terrawatt (TW) years annually (a terrawatt is a trillion watts). A population of 12 billion with an average per capita energy use rate of 6.0 kW would use about five times as much as is presently used. Can the world support this much energy use?

John Holdren, an energy expert, believes that a realistic goal is for annual per capita energy use to reach 3 kW, with the world population peaking at 10 billion individuals by the year 2100. If this goal is to be achieved, developing nations will be able to increase their populations by no more than 60% and their energy use by no more than 100%; developed nations can increase their population by only 10% and will have to reduce their energy use by 2% each year.

Critical Thinking Questions

1. What would the energy use rate be if Holdren's goals were realized? How much total energy would be required for all people on Earth to have a standard of living supported by 7.5 kW per person? How do these totals compare with the present energy-use rate worldwide?

2. In what specific ways could energy be used more efficiently in the United States? Compare your list with those of your classmates, and compile a class list.

3. In addition to increasing efficiency, what other changes in energy consumption might be required to provide an average energy-use rate of 7.5 kW per person or 3.0 kW per person in the future?

4. Would you consider Holdren's vision of the energy future an example of a hard or a soft path? Explain.

---

ing of the quality of energy sources with their end uses.

■ Energy conservation and improvements in energy efficiency can have significant effects on energy consumption. It takes three units of a fuel such as oil to produce one unit of electricity. As a result, each electricity unit conserved or saved through improved efficiency saves three units of fuel.

■ There are arguments for both the hard and soft energy paths. The former has a long history of success and has produced the highest standard of living ever experienced. However, present sources of energy are causing serious environmental degradation and are not sustainable. Soft-path alternative energy sources are renewable, decentralized, diverse, and flexible; provide a better match between energy quality and end use; and emphasize second-law efficiencies.

■ Sustainable, integrated energy management is needed to make the transition from fossil fuels to other energy sources. The goal is to provide reliable sources of energy that do not cause serious harm to the environment and ensure that future generations inherit a quality environment.

Human Population

The industrialized and urbanized countries of the world produce and use most of the world's energy. As societies change from rural to urban, energy demands generally increase. Controlling the increase of human population is an important factor in reducing total demand for energy (total demand is the product of average demand per person and number of people).

Sustainability

It will be impossible to achieve sustainability in the United States if we continue with our present energy policies. Present use of fossil fuels is not sustainable. We need to rethink the sources, uses, and management of energy. Sustainability is the central issue in our decision to continue on the hard path or change to the soft path.

Global Perspective

Understanding trends in energy production and consumption on a global basis is important if we are to directly address the global impact of burning fossil fuels on problems of air pollution and global warming. Furthermore, the use of energy resources greatly influences global economics as these resources are transported and utilized around the world.

Urban World

A great deal of the total energy demand is in urban regions, such as Tokyo, Beijing, London, New York, and Los Angeles. How we choose to manage energy in our urban regions greatly affects the quality of urban environments. Burning cleaner fuels results in far less air pollution. This has been observed in several urban regions, such as London. Burning of coal in London once caused deadly air pollution events. Today, natural gas and electricity heat homes, and the air is cleaner. Burning coal in Beijing continues to cause significant air pollution and health problems for millions of people living there.

People and Nature

In our development and use of energy, we are changing nature in some significant ways. For example, burning of fossil fuels is changing the composition of the atmosphere, particularly through the addition of carbon dioxide. The carbon dioxide is contributing to the warming of the atmosphere, water, and land (see Chapter 23 for details). A warmer Earth is in turn changing the climates of regions of the Earth, weather patterns, and intensity of storms. The warming is causing some insects, such as mosquitoes, to move to formerly cooler environments, changing patterns of mosquito-borne diseases, such as malaria and other fevers, in people.

Science and Values

Public opinion polls consistently show that people value a quality environment. In response, energy planners are evaluating how to make more efficient use of our present energy resources, practice conservation of energy, and reduce adverse environmental effects of energy consumption. Science is providing options in terms of energy sources and uses; which choices we make will reflect our values.

Key Terms

cogeneration **354**
conservation **354**
energy **348**
energy efficiency **354**
first-law efficiency **350**

first law of
   thermodynamics **348**
hard path **357**
integrated energy
   management **359**

micropower **360**
second-law
   efficiency **350**
second law of
   thermodynamics **350**

soft path **357**
sustainable energy
   development **359**
work **348**

Study Questions

1. What evidence supports the notion that although present energy problems are not the first in human history, they are unique in other ways?

2. How do the terms *energy, work,* and *power* differ in meaning?

3. Compare and contrast potential advantages and disadvantages of a major shift from hard-path to soft-path energy development.

4. You have just purchased a 100-hectares wooded island in Puget Sound. Your house is uninsulated and built of raw timber. Although the island receives some wind, trees over 40 m tall block most of it. You have a diesel generator for electric power, and hot water is produced by an electric heater run by the generator. Oil and gas can be brought in by ship. What steps would you take in the next five years to reduce the cost of the energy you use with the least damage to the natural environment of the island?

5. How might better matching of end uses with potential sources yield improvements in energy efficiency?

6. Complete an energy audit of the building in which you live, and develop recommendations that might lead to lower utility bills.

7. How might plans using the concept of integrated energy management differ for the Los Angeles area and the New York City area? How might both of these plans differ from an energy plan for Mexico City, which is quickly becoming one of the largest urban areas in the world?

8. A recent energy scenario for the United States suggests that in the coming decades, energy sources might be natural gas (10%), solar power (30%), hydro-power (20%), wind power (20%), biomass (10%), and geothermal energy (10%). Do you think this is a likely scenario? What would be the major difficulties and points of resistance or controversy?

Further Reading

Berger, J. J. 2000. *Beating the Heat*. Berkeley, Calif.: Berkeley Hills Books. Excellent overview of all aspects of energy as they relate to global warming. Discusses how we can reduce or eliminate warming as an environmental threat.

Boyle, G., Everett, B. and Ramage, J. 2003. *Energy Systems and Sustainability*. Oxford, UK. Oxford University Press. An excellent summary of energy sources, uses, consumption, and sustainability.

Fay, J. A. and Golomb, D. S. 2002. *Energy and the Environment*. New York: Oxford University Press. A quantitative treatment of basic energy principles.

Lovins, A. B. 1979. *Soft Energy Path: Towards a Durable Peace*. New York: Harper & Row. An interesting book that presents the argument for the soft path. Its message is more important today than when it was written.

United Nations Development Program. 2000. *Energy and the Challenge of Sustainability*. New York: United Nations Development Program. Good presentation of energy sources and sustainable development.

CHAPTER 18

Fossil Fuels and the Environment

Oil is the most important fossil fuel today, and working in oil fields has never been easy.

Learning Objectives

We rely almost completely on fossil fuels (oil, natural gas, and coal) for our energy needs. However, these are nonrenewable resources, and their production and use have a variety of serious environmental impacts. After reading this chapter, you should understand:

■ Why we may have serious, unprecedented supply problems with oil and gasoline within the next 20 to 50 years.

■ How oil, natural gas, and coal form.

■ What the environmental effects are of producing and using oil, natural gas, and coal.

Peak Oil:
Myth or Reality

People in many countries have grown prosperous and lived longer during the past century as a result of abundant low-cost energy in the form of crude oil. The benefits of oil are undeniable, but so are potential problems, from air and water pollution to global warming. In any case, we are about to learn what life will be like with less, more expensive oil. The question is no longer if the peak will come in production, but when it will come and what the consequences to a society's economics and politics will be.[1] The peak, or **peak oil** is the time when one-half of Earth's oil has been exploited.

The global history of oil in terms of rate of discovery and consumption is shown in Figure 18.1. Notice that in 1940, five times as much oil was discovered as consumed; by 1980 the amount discovered equaled the amount consumed; and in the year 2000 the consumption of oil was three times what was discovered. Obviously, the trend is not sustainable; oil is being rapidly consumed relative to new resources being found.

The concept of peak oil production is shown in

Figure 18.2. We aren't sure what the peak production will be, but let's assume it to be about 50 billion bbl per year, and that the peak arrives some time during 2020 to 2050. In 2004 the growth rate for oil was 3.4%. Moving from the present production rate of about 30 billion bbl per year to 50 billion bbl in a few decades seems a reasonable estimate. When the peak production occurs, and if demand increases, then a gap between production and demand will result. If demand exceeds supply, cost will increase. The high cost of oil and gasoline in the first years of the 21st century reflects uncertainty in supplies related to wars and the delivery/refining processes.

We have time now to prepare for the eventual peak and to use the fossil fuels we have during the time of transition to other energy sources. If we have not prepared for the peak then disruption to society is likely. In the best scenario, the transition from oil will not occur until we have cost-competitive available alternatives in place.[2]

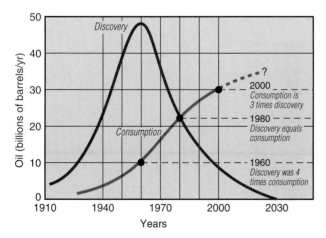

Figure 18.1 ■ Discovery of oil peaked in about 1960 and consumption exceeded discovery by 1980. Modified after K. Aleklett, "Oil: A Bumpy Road Ahead," *World Watch 19:1* (2006):10–12

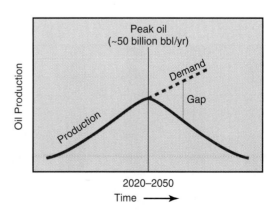

Figure 18.2 ■ Idealized diagram of world oil production and peak between 2020 to 2050. When production cannot meet demand, a gap (shortage) develops.

The approaching peak in oil production is a wake-up call to us that, although we will not run out of oil, it will become much more expensive. There will be problems supplying oil as demand increases by about 50% in the next 30 years. The peak in world oil production, when it arrives, will be unlike any problem we have faced in the past. Human population will increase several billion more in coming decades, and countries with growing economies like China and India will increase their oil consumption. China expects to double its import of oil in the next five years! As a result, the social, economic, and political ramifications of peak oil will be enormous. Planning now to conserve oil and plan for the transition to alternative energy sources will be critical in the coming decades. We cannot afford to depart the age of oil until alternatives are firmly in place. The remainder of this chapter will discuss the various fossil fuels and their uses.

18.1 Fossil Fuels

Fossil fuels are forms of stored solar energy. Plants are solar energy collectors because they can convert solar energy to chemical energy through photosynthesis (see Chapter 5). The main fossil fuels used today were created from incomplete biological decomposition of dead organic matter (mostly land and marine plants). This occurred when buried organic matter that was not completely oxidized was converted by chemical reactions over hundreds of millions of years to oil, natural gas, and coal. Biological and geologic processes in various parts of the geologic cycle produce the sedimentary rocks in which we find these fossil fuels.[3,4]

The major fossil fuels—crude oil, natural gas, and coal—are our primary energy sources; on a worldwide basis, they provide approximately 90% of the energy consumed (Figure 18.3). With the exception of Asia, which uses a lot of coal, oil and natural gas provide 70 to 80% of our primary energy. An exception is the Middle East where oil and gas provide nearly all of the energy. In this chapter, we focus primarily on these major fossil fuels. We also briefly discuss two other fossil fuels, oil shale and tar sands, that may become increasingly important as oil, gas, and coal reserves are depleted.

18.2 Crude Oil and Natural Gas

Most geologists accept the hypothesis that **crude oil** (petroleum) and **natural gas** are derived from organic materials (mostly plants) that were buried with marine or lake sediments in what are known as depositional basins. Oil and gas are found primarily along geologically young tectonic belts at plate boundaries, where large depositional basins are more likely to occur (see Chapter 5). However, there are exceptions, such as in Texas, the Gulf of Mexico, and the North Sea, where oil in depositional basins far from active plate boundaries has been discovered.

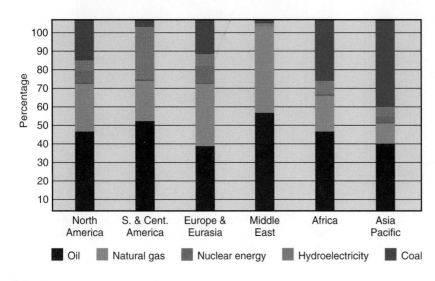

Figure 18.3 ■ World energy consumption by primary sources in 2004. [*Source*: Modified after British Petroleum Company, *BP Statistical Review of World Energy* (London: British Petroleum Company, 2005).]

The source material, or *source rock*, for oil and gas is fine-grained (less than 1/16 mm, or 0.0025 in., in diameter), organic-rich sediment buried to a depth of at least 500 m (1640 ft), where it is subjected to increased heat and pressure. The elevated temperature and pressure initiate the chemical transformation of the organic material in the sediment into oil and gas. The elevated pressure causes the sediment to be compressed; this, along with the elevated temperature in the source rock, initiates the upward migration of the oil and gas, which are relatively light, to a lower-pressure environment (known as the *reservoir rock*). The reservoir rock is coarser grained and relatively porous (has more and larger open spaces between the grains). Sandstone and porous limestone, which have a relatively high proportion (about 30%) of empty space in which to store oil and gas, are common reservoir rocks.

As mentioned, oil and gas are light; if their upward mobility is not blocked, they will escape to the atmosphere. This explains why oil and gas are not generally found in geologically old rocks. Oil and gas in rocks older than about 0.5 billion years have had ample time to migrate to the surface, where they have either vaporized or eroded away.[4]

The oil and gas fields from which we extract resources are places where the natural upward migration of the oil and gas to the surface is interrupted or blocked by what is known as a *trap* (Figure 18.4). The rock that helps form the trap, known as the *cap rock*, is usually a very fine-grained sedimentary rock, such as shale, which is composed of silt and clay-sized particles. A favorable rock structure, such as an anticline (arch-shaped fold) or a fault (fracture in the rock along which displacement has occurred), is necessary to form traps, as shown in Figure 18.4. The important concept is that the combination of favorable rock structure and the presence of a cap rock allow deposits of oil and gas to accumulate in the geologic environment, where they are then discovered and extracted.[4]

Petroleum Production

Production wells in an oil field recover oil through both primary and enhanced methods. *Primary production* involves simply pumping the oil from wells, but this method can recover only about 25% of the petroleum in the reservoir. To increase the amount of oil recovered to about 60%, enhanced methods are used. In *enhanced* recovery, steam, water, or chemicals, such as carbon dioxide or nitrogen gas, are injected into the oil reservoir to push the oil toward the wells, where it can be more easily recovered by pumping.

Next to water, oil is the most abundant fluid in the upper part of the Earth's crust. Most of the known, proven oil reserves, however, are located in a few fields. Proven oil reserves are the part of the total resource that has been identified and can be extracted now at a profit. Of the total reserves, 62% is located in 1% of the fields, the largest of which are in the Middle East (Figure 18.5a). The consumption of oil per person is shown in Figure 18.5b. Notice the domination of energy use in North America. Although new oil and gas fields have recently been and continue to be discovered in Alaska, Mexico, South America, and other areas of the world, the present known world reserves may be depleted in the next few decades.

The total resource always exceeds known reserves; it includes petroleum that cannot be extracted at a profit and petroleum that is suspected but not proved to be present. Several decades ago, the amount of oil that ultimately could be recovered was estimated to be about 1.6 trillion barrels. Today, that estimate is just over 3 trillion barrels.[5] The increases in proven reserves of oil in the last few decades have primarily been due to discoveries in the Middle East, Venezuela, Kazakhstan, and other areas.

Because so much of the world's oil is found in the Middle East, oil revenues have flowed into that area,

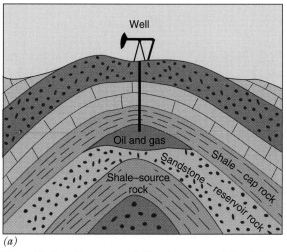

(a)

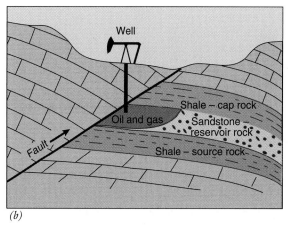

(b)

Figure 18.4 ■ Two types of oil and gas traps: *(a)* anticline and *(b)* fault.

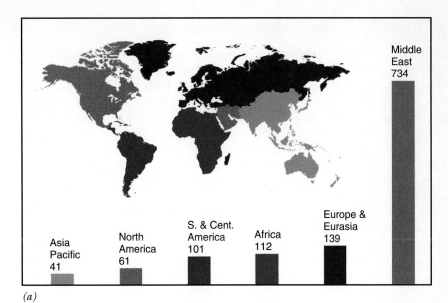

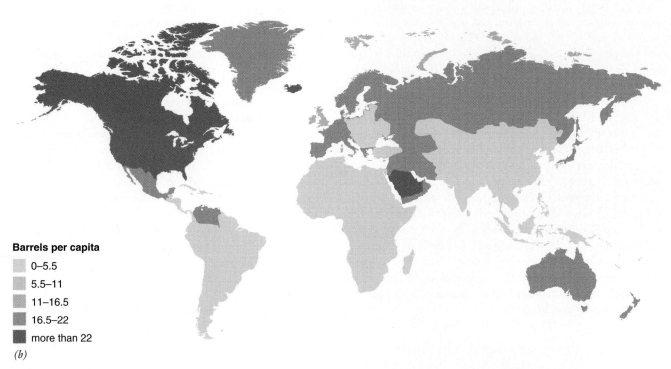

Figure 18.5 ■ (*a*) Proven world oil reserves (billions of barrels) in 2004. The Middle East dominates with 62% of total reserves. (*b*) Consumption of oil per person. [*Source:* Modified after British Petroleum Company, *BP Statistical Review of World Energy* (London: British Petroleum Company, 2005).]

Middle East 734

Europe & Eurasia 139

Africa 112

S. & Cent. America 101

North America 61

Asia Pacific 41

(*a*)

Barrels per capita

- 0–5.5
- 5.5–11
- 11–16.5
- 16.5–22
- more than 22

(*b*)

resulting in huge trade imbalances. Figure 18.6 shows the major routes of trade for oil. Interestingly, in recent years the United States has imported more oil from Venezuela than from the Middle East. Most of the Middle Eastern oil goes to Europe, Japan, and Southeast Asia.

Oil in the Twenty-First Century

Recent estimates of proven oil reserves suggest that, at present production rates, oil and natural gas will last only a few decades.[6, 7] The important question, however, is not how long oil is likely to last at present and future production rates but when we will we reach peak production. This is impor-

tant because following peak production, less oil will be available, leading to shortages and price shocks. World oil production, as mentioned in the opening case study on peak oil, is likely to peak between the years 2020 and 2050, within the lifetime of many people living today.[8] Even those who think peak oil production in the near future is a myth acknowledge that the peak is coming and that we need to be prepared.[2] Whichever projections are correct, there is finite time (a few decades or perharps a bit longer) left to adjust to potential changes in lifestyle and economies in a post-petroleum era.[8] We will never entirely run out of crude oil, but people of the world depend on oil for nearly 40% of their energy, and significant shortages will cause major problems.[9]

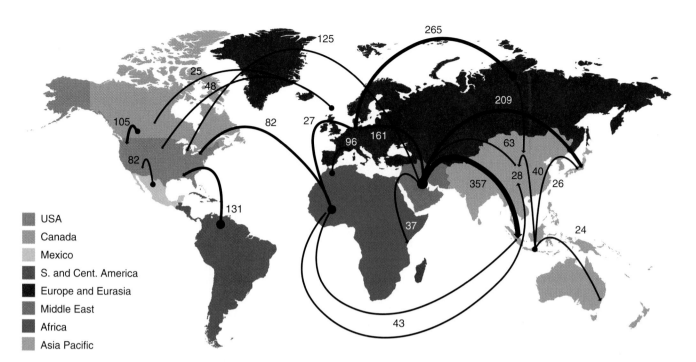

Figure 18.6 ■ Major trade routes for the world's oil in 2004, emphasizing the countries that use Middle Eastern oil. Units are millions of metric tons (1 metric ton = 2,205 pounds) [*Source:* Modified after British Petroleum Company, *BP Statistical Review of World Energy* (London: British Petroleum Company, 2005).]

Consider the following argument that we are heading toward a potential crisis in availability of crude oil:

■ We are approaching the time when approximately 50% of the total crude oil available from traditional oil fields will have been consumed.[8] Recent studies suggest that about 20% more oil awaits discovery than predicted a few years ago and that there is more oil in known fields than earlier thought. However, the volumes of new oil discovered and recovered in known fields will not significantly change the date when world production will peak and a decline in production will begin.[7] This point is controversial. Some experts believe that modern technology for exploration, drilling, and recovery of oil will ensure an adequate supply of oil for the distant future.[5]

■ Proven reserves are about 1 trillion barrels.[5] It is estimated that approximately 3 trillion barrels of crude oil may ultimately be recovered from remaining oil resources. World consumption today is about 30 billion barrels per year (82 million barrels per day). We are using what is left fast.

■ Today, for every three barrels of oil we consume, we are finding only one barrel.[9] In other words, output is four times higher than input. However, this could improve in the future.[5]

■ Forecasts that predict a decline in production of oil are based on the estimated amount of oil that may ultimately be recoverable (2 to 3 trillion barrels, two to three times today's proven reserves), along with projections of new discoveries and rates of future consumption. As mentioned, it has been estimated that the peak in world crude oil production, about 40 to 50 billion bbl/yr, will occur between the years 2020 and 2050.[1,7,9] The production of 50 billion bbl/yr is about a 50% increase over 2004. Whether you think this increase is optimistic or pessimistic depends on your view of past oil history, which has survived several predicted shortages or beliefs that the peak is inevitable, sooner than later.[1,2] Most oil experts believe peak oil is only a few decades away.

■ In the United States, it is expected that the production of oil as we know it now will end by about 2090. World production of oil will be nearly exhausted by 2100.[9]

What is an appropriate response to the likelihood that production rates of oil will likely fall in the mid–twenty-first century? First, we need an educational program to inform people and governments of the potential depletion of crude oil and the consequences of shortages. Presently, many people are operating in ignorance or denial in the face of a potentially serious situation. Planning and appropriate action are necessary to avoid military confrontation (we have already had one oil war), food shortages (oil is used to make the fertilizers modern agriculture depends on), and social disruption. Before significant shortages of oil occur, we need to develop alternative energy sources such as solar energy and wind power and perhaps rely more on nuclear energy. This is a proactive response to a potentially serious situation.

Natural Gas

We have only begun to seriously search for natural gas and to utilize this resource to its potential. One reason for this slow start is that natural gas is transported primarily by pipelines, and only in the last few decades have these been constructed in large numbers. In fact, until recently, natural gas found with petroleum was often simply burned off as waste; in some cases, this practice continues.[10]

The worldwide estimate of recoverable natural gas is about 165 trillion m^3, which at the current rate of world consumption will last approximately 70 years.[6] Considerable natural gas has recently been discovered in the United States, and at present U.S. consumption levels, this resource is expected to last about 30 years. Furthermore, new supplies are being found in surprisingly large amounts, particularly at depths that are deeper than those at which oil is found. Optimistic estimates of the total resource suggest that at current rates of consumption the supply may last approximately 120 years.[3]

This possibility has important implications. Natural gas is considered a clean fuel; burning it produces fewer pollutants than does burning oil or coal, so it causes fewer environmental problems than do the other fossil fuels. As a result, it is being considered as a possible transition fuel from other fossil fuels (oil and coal) to alternative energy sources, such as solar power, wind power, and hydropower.

In spite of the new discoveries and the construction of pipelines, long-term projections for a steady supply of natural gas are uncertain. The supply is finite, and at present rates of consumption, it is only a matter of time before the resources are depleted.

Coal-Bed Methane

The processes responsible for the formation of coal include the partial decomposition of plants buried by sediments that slowly convert the organic material to coal. This process also produces a lot of methane (natural gas) that is stored within the coal.[11] The methane is actually stored on the surfaces of the organic matter in the coal, and because there are many large internal surfaces in coal, the amount of methane for a given volume of rock is something like seven times more than could be stored in gas reservoirs associated with petroleum. The estimated amount of coal-bed methane in the United States is more than 20 trillion cubic meters, with about 3 trillion cubic meters that could be recovered economically today with existing technology. This is equivalent to about a five-year supply of methane at current rates of consumption in the United States.[12]

Two areas within the nation's coal fields that are producing methane are the Wasatch plateau in Utah and the Powder River basin in Wyoming. The Powder River basin is one of the world's largest coal basins, and presently there is an energy boom occurring in Wyoming, producing an "energy rush." The technology to recover coal-bed methane is a young one, but it is developing quickly. As of early 2003, there were approximately 10,000 shallow wells producing methane in the Powder River basin, and some say there will eventually be about 100,000 wells. The big advantage of the coal-bed methane wells is that they only need to be drilled to shallow depths (about 100 m, or a few hundred feet). Drilling can be done with conventional water-well technology, and the cost is about $100,000 per well compared to several million dollars for an oil well.[13]

Coal-bed methane is a promising energy source that comes at a time when the United States is importing vast amounts of energy and is attempting to evaluate a transition from fossil fuels to alternative fuels. However, there are several environmental concerns associated with coal-bed methane, including: (1) disposal of large volumes of water, which is produced when the methane is recovered; and (2) migration of methane, which may contaminate groundwater or migrate into residential areas.

A major environmental benefit from burning coal-bed methane, as with other sources of methane, is that combustion produces a lot less carbon dioxide than does the burning of coal or petroleum. Furthermore, production of methane gas prior to mining coal reduces the amount of methane that would be released into the atmosphere. Both methane and carbon dioxide are strong greenhouse gases that contribute to global warming. However, because methane produces a lot less carbon dioxide, it is considered one of the main transitional fuels from fossil fuels to alternative energy sources.

Of particular environmental concern in Wyoming is the safe disposal of salty water that is produced with the methane (the wells bring up a mixture of methane and water that contains dissolved salts from contact with subsurface rocks). Often the water is reinjected back into the subsurface, but in some instances the water flows into surface drainages or is placed into evaporation ponds.[12] Some of the environmental conflicts are between those producing the methane from wells and ranchers trying to raise cattle on the same land. Frequently, the ranchers do not own the mineral rights; and although energy companies may pay fees for the well, the funds are not sufficient to cover damage resulting from producing the gas. The problem results when the salty water produced is disposed of in nearby streams and increases the salinity of the stream flow. Ranchers then use the surface water to irrigate crops for cattle, and the soils become damaged from saltwater, reducing crop productivity. Although it has been argued that ranching is often a precarious economic venture and that ranchers have in fact been saved by the new money from coal-bed methane, many ranchers oppose coal-bed methane production without an assurance that salty waters will be safely disposed of. There is also concern for the sustainability of water resources as vast amounts of water are removed from the groundwater aquifers. In some instances, springs have been reported to have dried up following coal-bed methane extraction in the area.[13] In other words, the "mining" of groundwater for coal-bed methane extraction will remove water that has perhaps taken hundreds of years to accumulate in the subsurface environment.

There is also concern for the migration of methane away from the well sites, possibly to nearby, more urban areas. The problem is that methane in its natural state is odorless, as compared to the smelly variety in homes, and it is explosive. For example, in the 1970s an urban area near Gallette, Wyoming, was evacuated as a result of methane that was migrating into homes from nearby coal mines. Finally, coal-bed methane wells and related compressors and other needed equipment cause noise pollution. People living within a few hundred meters of coal-bed methane facilities have reported serious and distressing noise pollution.[13]

In summary, coal-bed methane is a tremendous source of energy. It is a relatively clean-burning fuel, but its extraction must be closely evaluated and studied to minimize environmental degradation.

Methane Hydrates

Beneath the seafloor, at depths of about 1,000 m, there exist deposits known as **methane hydrate**, a white, ice-like compound made up of molecules of methane gas (CH_4) molecular "cages" of frozen water. The methane has formed as a result of microbial digestion of organic matter in the sediments of the seafloor and has become trapped in these ice cages. Methane hydrates in the oceans were discovered over 30 years ago and are widespread in both the Pacific and Atlantic oceans. Methane hydrates are also found on land; the first ones discovered were in permafrost areas of Siberia and North America, where they are known as marsh gas.[14]

Methane hydrates in the ocean are found in areas where deep, cold seawater provides high pressure and low temperatures. They are not stable at lower pressure and warmer temperatures. At a water depth of less than about 500 m, methane hydrates decompose rapidly and methane gas is freed from the ice cages to move up as a flow of methane bubbles (like rising helium balloons) to the surface and the atmosphere.

Researchers from Russia in 1998 discovered the release of methane hydrates off the coast of Norway. During the release, scientists documented plumes of methane gas as tall as 500 m being emitted from methane hydrate deposits on the seafloor. It appears that there have been large emissions of methane from the sea. The physical evidence includes fields of depressions, looking something like bomb craters, that pock-mark the bottom of the sea near methane hydrate deposits. Some of the craters are as large as 30 m deep and 700 m in diameter, suggesting

that they were produced by rapid if not explosive eruptions of methane.

Methane hydrates in the marine environment are a potential energy resource with approximately twice as much energy as all the known natural gas, oil, and coal deposits on Earth.[14] Methane hydrates look particularly attractive to countries such as Japan that rely exclusively on foreign oil and coal for fossil fuel needs. Unfortunately, mining methane hydrates will be a difficult task, at least for the near future. The hydrates tend to be found along the lower parts of the continental slopes, where water depths are often greater than 1 km. The deposits themselves extend into the ocean floor sediments another few hundred meters. Most drilling rigs cannot operate safely at these depths, and development of a method to produce and transport the gas to land will be challenging.

Environmental Effects of Oil and Natural Gas

There is no escaping the fact that recovery, refining, and use of oil—and, to a lesser extent, natural gas—cause well-known, documented environmental problems, such as air and water pollution, acid rain, and global warming. Humans have gained many benefits from abundant, inexpensive energy, but at a price to the global environment and human health.

Recovery

Development of oil and gas fields involves drilling wells on land or beneath the seafloor (Figure 18.7). Possible environmental impacts on land include the following:

- Use of land to construct pads for wells, pipelines, and storage tanks and to build a network of roads and other production facilities.

- Pollution of surface waters and groundwater from: (1) leaks from broken pipes or tanks containing oil or other oil-field chemicals and (2) salty water (brine) that is brought to the surface in large volumes with the oil. The brine is toxic and may be disposed of by evaporation in lined pits, which may leak. Alternatively, brine may be disposed of by pumping it into the ground using deep disposal wells outside the oil fields. However, disposal wells may cause groundwater pollution (see Chapter 30).

- Accidental release of air pollutants, such as hydrocarbons and hydrogen sulfide (a toxic gas).

- Land subsidence (sinking) as oil and gas are withdrawn.

- Loss or disruption of and damage to fragile ecosystems, such as wetlands or other unique landscapes. This is the center of the controversy over the development of petroleum resources in pristine environments such as the Arctic National Wildlife Refuge in Alaska (see A Closer Look 18.1).

Figure 18.7 ■ Drilling for oil in (*a*) the Sahara Desert of Algeria and (*b*) the Cook Inlet of southern Alaska.

(*a*)

(*b*)

Environmental impacts associated with oil production in the marine environment include the following:

■ Oil seepage into the sea from normal operations or large spills from accidents, such as blowouts or pipe ruptures.

■ Release of drilling muds (heavy liquids injected into the bore hole during drilling to keep the hole open) containing heavy metals, such as barium, that may be toxic to marine life.

■ Aesthetic degradation from the presence of offshore oil-drilling platforms, which some people think are unsightly.

Refining

Refining crude oil and converting it to products also create environmental impacts. At refineries, crude oil is heated so that its components can be separated and collected (this process is called *fractional distillation*). Other industrial processes are then used to make products such as gasoline and heating oil.

Refineries may have accidental spills and slow leaks of gasoline and other products from storage tanks and pipes. Over years of operation, large amounts of liquid hydrocarbons may be released, polluting soil and groundwater resources below the site. Massive groundwater cleaning projects have been required at several West Coast refineries.

Crude oil and its distilled products are used to make fine oil, a wide variety of plastics, and organic chemicals used by society in huge amounts. The industrial processes involved in the production of these chemicals have the potential for releasing a variety of pollutants into the environment.

Delivery and Use

Some of the most extensive and significant environmental problems associated with oil and gas occur when the fuel is delivered and consumed. Crude oil is mostly transported on land in pipelines or across the ocean by tankers, and both methods present the danger of oil spills. For example, a bullet from a high-powered rifle punctured the trans-Alaskan pipeline in 2001, causing a small but damaging oil spill. In addition, strong earthquakes may pose a problem for pipelines in the future. However, proper engineering can minimize earthquake hazard. The large 2002 Alaskan earthquake ruptured the ground by several meters where it crossed the trans-Alaska pipeline. Because of its design, the pipeline allowed for the ground rupture and was not damaged, preventing environmental damage. Although most effects of oil spills are relatively short lived (days to years), marine spills have killed thousands of seabirds, temporarily spoiled beaches, and caused loss of tourist and fishing revenues (see Chapter 22).

Air pollution is perhaps the most familiar and serious environmental impact associated with the use (burning) of oil. Combustion of gasoline in automobiles produces pollutants that contribute to urban smog. The adverse effects of smog on vegetation and human health are well documented and are discussed in detail in Chapter 24.

The Arctic National Wildlife Refuge: To Drill or Not to Drill

The Arctic National Wildlife Refuge (ANWR) on the North Slope of Alaska is one of the few pristine wilderness areas remaining in the world (Figure 18.8). The U.S. Geological Survey estimates that the refuge contains about 3 billion bbl of recoverable oil. According to the oil industry, several times more oil than that can be recovered. The oil industry has long argued in favor of drilling for oil in the ANWR, but the idea was unpopular for decades among many members of the public and the U.S. government, and no drilling was permitted. When George W. Bush, who favored drilling in the ANWR, was elected president in 2000, however, controversy surrounding drilling emerged.

Argument in Favor of Drilling in the ANWR

Those in favor of drilling for oil in the refuge make the following arguments:

- The United States needs the oil, and it will help us to be more independent of imported oil.

- New oil facilities will bring jobs and dollars to Alaska.

- New exploration tools to evaluate the subsurface for oil pools require far fewer exploratory wells.

- New drilling practices have much less impact on the environment (Figure 18.9*a*). These include: (1) constructing roads of ice in the winter that melt in the summer instead of constructing permanent roads; (2) elevating pipelines to allow for animal migration (Figure 18.9*c*); (3) drilling in various directions from a central location, thus minimizing land needed for wells; and (4) disposing of fluid oil-field wastes by putting them back into the ground to minimize surface pollution.

- The land area affected will be small relative to the total area.

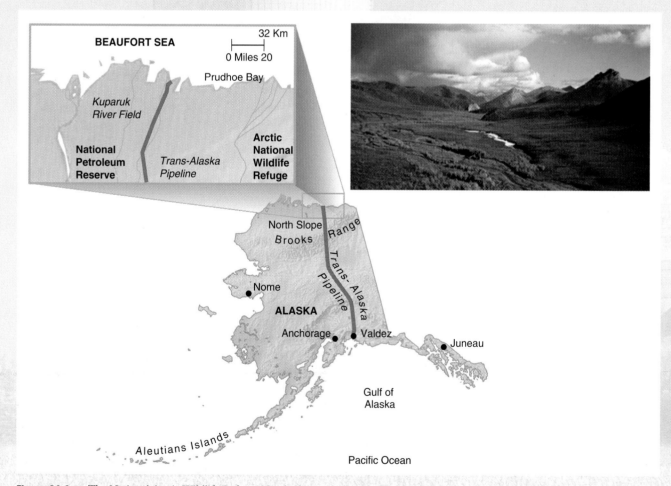

Figure 18.8 ■ The National Arctic Wildlife Refuge, North Slope, Alaska, is valued for its scenery, wildlife, and oil.

Argument against Drilling in the ANWR

Those who are opposed to drilling in the ANWR argue as follows:

■ Advances in technology are irrelevant to the question of whether or not the ANWR should be drilled. Some wilderness should remain wilderness!

Drilling will forever change the pristine environment of the North Slope.

■ Even with the best technology, oil exploration and development will impact the ANWR. Intensive activity, even in winter on roads constructed of ice, will probably disrupt wildlife.

■ Ice roads are constructed from water

from the tundra ponds. To build a road 1 km (0.63 mi) long requires about 3,640 m³ (1 million gallons) of water.

■ Heavy vehicles used in exploration permanently scar the ground—even if the ground is frozen hard when the vehicles travel across the surface of the open tundra.

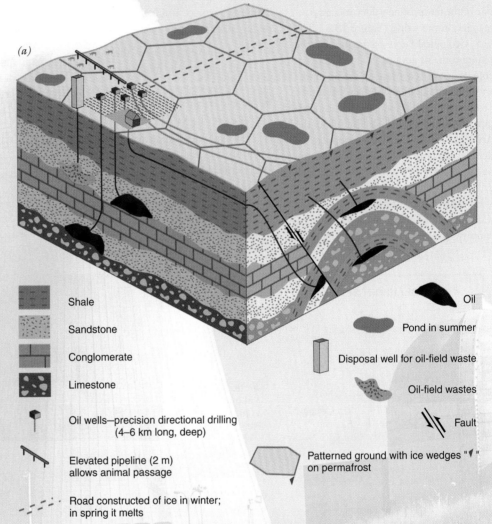

(a)

Figure 18.9 ■ (*a*) Those in favor of ANWR argue that new technology can reduce the impact of developing oil fields in the Arctic: wells are located in a central area and use directional drilling; roads are constructed of ice in the winter, melting to become invisible in the summer; pipelines are elevated to allow animals, in this case caribou, to pass through the area; and oil-field and drilling wastes are injected deep underground. See text for arguments against drilling. (*b*) Oil wells being drilled on frozen ground (tundra), North Slope of Alaska. (*c*) Caribou passing under a pipeline near the Arctic National Wildlife Refuge in Alaska.

Shale

Sandstone

Conglomerate

Limestone

Oil wells—precision directional drilling (4–6 km long, deep)

Elevated pipeline (2 m) allows animal passage

Road constructed of ice in winter; in spring it melts

Oil

Pond in summer

Disposal well for oil-field waste

Oil-field wastes

Fault

Patterned ground with ice wedges "▼" on permafrost

(b)

(c)

- Accidents may occur in even the best facilities.

- Oil development is inherently damaging because it involves a massive industrial complex of people, vehicles, equipment, pipelines, and support facilities.

The decision we make concerning the ANWR will reflect both science and values at a basic level. New technology will produce less environmental impact from drilling oil. How we value the need for energy from oil compared with how we value preservation of a pristine wilderness area will determine our action: to drill or not to drill. Although the president's energy plan calls for drilling in the ANWR, Congress in 2005 voted not to drill. The debate to drill or not to drill is not over yet.

18.3 Coal

Partially decomposed vegetation, when buried in a sedimentary environment, may be slowly transformed into the solid, brittle, carbonaceous rock we call **coal**. This process is shown in Figure 18.10. Coal is by far the world's most abundant fossil fuel, with a total recoverable resource of about 1,000 billion metric tons (Figure 18.11). The annual world consumption of coal is about 4 billion metric tons, sufficient for about 250 years at the current rate of use. However, if consumption of coal increases in the coming decades, the resource will not last nearly so long.[15]

Coal is classified according to its energy and sulfur content as anthracite, bituminous, subbituminous, or lignite (see Table 18.1). Energy content is greatest in anthracite coal and lowest in lignite coal. The distribution of coal in the contiguous United States is shown in Figure 18.12.

The sulfur content of coal is important because low-sulfur coal emits less sulfur dioxide (SO_2) and as a result is more desirable as a fuel for power plants. Most of the low-sulfur coal in the United States is relatively low-grade,

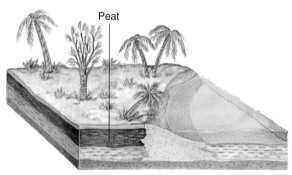

(a) Coal swamps form.

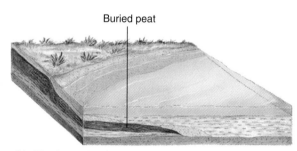

(b) Rise in sea level buries swamps in sediment.

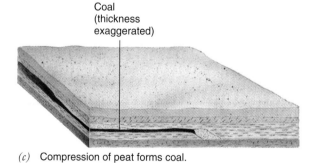

(c) Compression of peat forms coal.

Figure 18.10 ■ Processes by which buried plant debris (peat) is transformed into coal.

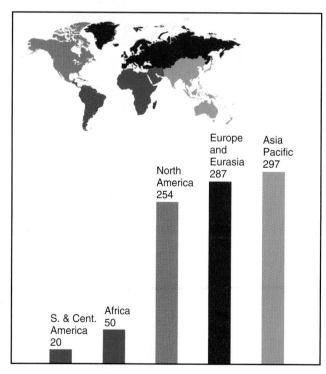

Figure 18.11 ■ World coal reserves (billions of tons) in 2004. [*Source:* British Petroleum Company, *BP Statistical Review of World Energy* (London: British Petroleum Company, 2005).]

Table 18.1 • U.S. Coal Resources

Type of Coal	Relative Rank	Energy Content (millions of joules/kg)	Sulfur Content (%)		
			Low (0–1)	Medium (1.1–3.0)	High (3 +)
Anthracite	1	30–34	97.1	2.9	—
Bituminous coal	2	23–34	29.8	26.8	43.4
Subbituminous coal	3	16–23	99.6	0.4	—
Lignite	4	13–16	90.7	9.3	—

Sources: U.S. Bureau of Mines Circular 8312, 1966; P. Averitt, "Coal," in D. A. Brobst and W. P. Pratt, eds., *United States Mineral Resources, U.S. Geological Survey, Professional Paper 820,* pp. 133–142.

low-energy lignite and subbituminous coal found west of the Mississippi River. Power plants on the East Coast treat the high-sulfur coal mined in their own region to lower its sulfur content before, during, or after combustion and thus avoid excessive air pollution. Although it is expensive, treating coal to reduce pollution may be more economical than transporting low-sulfur coal from the western states.

Coal Mining and the Environment

In the United States, thousands of square kilometers of land have been disturbed by coal mining, and only about half this land has been reclaimed. Reclamation is the process of restoring and improving disturbed land, often by re-forming the surface and replanting vegetation (see Chapter 10). Unreclaimed coal dumps from open-pit mines are numerous and continue to cause environmental problems. Because little reclamation occurred before about 1960, and mining started much earlier, abandoned mines are common in the United States. One surface mine in Wyoming, abandoned more than 40 years ago, caused a disturbance so intense that vegetation has still not been reestablished on the waste dumps. Such barren, ruined landscapes emphasize the need for reclamation.[15]

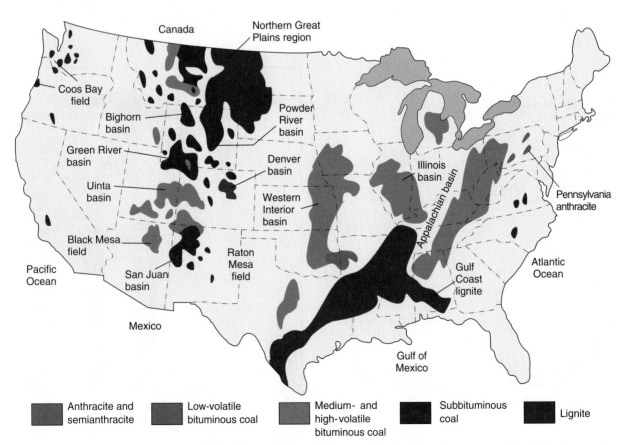

Figure 18.12 ■ Coal areas of the contiguous United States. This is a highly generalized map, and numerous relatively small occurrences of coal are not shown. [*Source:* S. Garbini and S. P. Schweinfurth, *U.S. Geological Survey Circular 979,* 1986.]

Figure 18.13 ■ Tar Creek near Miami, Oklahoma, runs orange in 2003 due to contamination of heavy metals from acid mine drainage.

Figure 18.14 ■ Strip coal mine in Wyoming. The land in the foreground is being mined, and the green land in the background has been reclaimed following mining.

Strip Mining

Over half of the coal mining in the United States is done by *strip mining*, a surface mining process in which the overlying layer of soil and rock is stripped off to reach the coal. The practice of strip mining started in the late nineteenth century and has steadily increased because it tends to be cheaper and easier than underground mining. More than 40 billion metric tons of coal reserves are now accessible to surface mining techniques. In addition, approximately another 90 billion metric tons of coal within 50 m (165 ft) of the surface are potentially available for strip mining. It is likely that more and larger strip mines will be developed as the demand for coal increases.

The impact of large strip mines varies from region to region, depending on topography, climate, and reclamation practices. One serious problem in areas of the eastern United States with abundant rainfall is *acid mine drainage*—the drainage of acidic water from mine sites (see Chapter 22). Acid mine drainage occurs when surface water (H_2O) infiltrates the spoil banks (rock debris left after the coal is removed). The water reacts chemically with sulfide minerals, such as pyrite (FeS_2), a natural component of some sedimentary rocks containing coal, to produce sulfuric acid (H_2SO_4). The acid then pollutes streams and groundwater resources (Figure 18.13). Acid water also drains from underground mines and roads cut in areas where coal and pyrite are abundant, but the problem of acid mine drainage is magnified when large areas of disturbed

material remain exposed to surface waters. Acid mine drainage can be minimized from active mines by channeling surface runoff or groundwater before it enters a mined area and diverting it around the potentially polluting materials.[16] However, diversion is not feasible in heavily mined regions where spoil banks from unreclaimed mines may cover hundreds of square kilometers. In these areas, acid mine drainage will remain a long-term problem.

In arid and semiarid regions, water problems associated with mining are not as pronounced as in wetter regions, but the land may be more sensitive to activities related to mining, such as exploration and road building. In some arid areas of the western and southwestern United States, the land is so sensitive that tire tracks can remain for years. (Indeed, wagon tracks from the early days of the westward migration reportedly have survived in some locations.) To complicate matters, soils are often thin, water is scarce, and reclamation work is difficult.

Strip mining has the potential to pollute or damage water, land, and biologic resources. However, good reclamation practices can minimize the damage (Figure 18.14). Reclamation practices required by law necessarily vary by site. Some of the principles of reclamation are illustrated in the case history of a modern coal mine in Colorado (see a Closer Look 18.2).

The Trapper Mine

The Trapper Mine on the western slope of the Rocky Mountains in northern Colorado is a good example of a new generation of large coal strip mines. The operation, in compliance with mining laws, is designed to minimize environmental degradation during mining and to reclaim the land for dryland farming and grazing of livestock and big game.

Over a 35-year period, the mine will produce 68 million metric tons of coal from the 20–24 km^3 (4.8–5.8 mi^3) site, to be delivered to a 1,300-MW power plant located adjacent to the mine. Today the mine produces about 2 million tons of coal per year, enough power for about one-half million homes. Four coal seams, varying from about 1 to 4 m (3.3–13.1 ft) thick, will be mined. The seams are separated by layers of rock, called overburden (rocks without coal), and there is additional overburden above the top seam of coal. The depth of the overburden varies from 0 to about 50 m (165 ft).

A number of steps are involved in the actual mining. First, bulldozers and scrapers remove the vegetation and topsoil from an area up to 1.6 km long and 53 m wide (1 mi by 175 ft), and the soil is stockpiled for reuse. Then the overburden is removed with a 23-m^3 (800-ft^3) dragline bucket. Next, the exposed coal beds are drilled and blasted to fracture the coal, which is removed with a backhoe and loaded onto trucks (Figure 18.15). Finally, the cut is filled, the topsoil replaced, and the land either planted with a crop or returned to rangeland.

At the Trapper Mine, the land is reclaimed without artificially applying water. Precipitation (mostly snow) is about 35 cm/year (about 14 in/year), which is sufficient to reestablish vegetation, provided there is adequate topsoil. The fact that reclamation is possible at this site emphasizes an important point about reclamation: It is site specific. What works at one location may not be applicable to other areas.

Water and air quality are closely monitored at the Trapper Mine. Surface water is diverted around mine pits, and groundwater is intercepted while pits are open. Settling basins, constructed downslope from the pit, allow suspended solids in the water to settle out before the water is discharged into local streams. Although air quality at the mine can be degraded by dust produced from the blasting, hauling, and grading of the coal, the dust is minimized by regular sprinkling of water on the dirt roads.

Reclamation at the Trapper Mine has been successful during the first years of operation. In fact, the U.S. Interior Department named it one of the best examples of mine reclamation. Although reclamation increases the cost of the coal by as much as 50%, it will pay off in the long-range productivity of the land as it is returned to farming and grazing uses. Wildlife also thrives; the local elk population has significantly increased, and the reclaimed land is home to sharp-tailed grouse, a threatened species. On the one hand, it might be argued that the Trapper Mine is unique in its combination of geology, hydrology, and topography, which has allowed for a successful reclamation. To some extent this is true, and perhaps the Trapper Mine presents an overly optimistic perspective on mine reclamation compared with other sites with less favorable conditions. On the other hand, the success of the mine operation demonstrates that with careful site selection and planning, the development of energy resources can be compatible with other land uses.

(a)

(b)

Figure 18.15 ■ (a) Mining an exposed coal bed at the Trapper Mine, Colorado; and (b) the land during restoration following mining. Topsoil (lower right) is spread prior to planting of vegetation.

Coal mining in the Appalachian Mountains of West Virginia is a major component of the state's economy. However, there is a growing environmental concern over the mining technique known as "mountaintop removal." This strip-mining method is very effective in obtaining coal as it levels the tops of mountains and fills valleys with coal mining waste rock. As mountains are destroyed, the flood hazard is increased as valleys are filled with mine waste and toxic wastewater is stored behind coal waste sludge dams. In October 2000 one of the worst environmental disasters in the history of mining in the Appalachian Mountains occurred in southeastern Kentucky. About 1 million cubic meters (250 million gallons) of toxic, thick black coal sludge, produced when coal is processed, was released into the environment. Part of the bottom of the impoundment (reservoir) where the sludge was being stored collapsed, allowing the sludge to enter an abandoned mine beneath the impoundment. The abandoned mine had openings to the surface, and sludge emerging from the mine flowed across people's yards and roads into a stream of the Big Sandy River drainage. About 100 km (65 mi) of stream was severely contaminated, killing several hundred thousand fish and other life in the stream. Mining also produces voluminous amounts of coal dust that settles on towns and fields, polluting the land and causing or facilitating lung diseases including asthma. Protests and complaints by communities in the path of mining that formerly were ignored are now receiving more attention by state mining boards. As people become better educated about mining laws, they are more effective in confronting mining companies to get them to reduce potential adverse consequences of mining. However, much more needs to be done. In May of 2002 a federal judge ordered the government to no longer allow mining companies to dump mining waste into streams and valleys. The decision upheld laws to protect our streams and rivers, but the ruling was overturned in January 2003.

Since the adoption of the Surface Mining Control and Reclamation Act of 1977, the U.S. government has required that mined land be restored to support its pre-mining use. The regulations also prohibit mining on prime agricultural land and give farmers and ranchers the opportunity to restrict or prohibit mining on their land, even if they do not own the mineral rights. Reclamation includes disposing of wastes, contouring the land, and replanting vegetation. Reclamation is often difficult, and it is unlikely that it will be completely successful. In fact, some environmentalists argue that success stories with reclamation are the exception and that strip mining should not be allowed in the semiarid southwestern states because reclamation is uncertain in that fragile environment.

Underground Mining

Underground mining accounts for approximately 40% of the coal mined in the United States. In addition, underground mines have been abandoned, particularly in the eastern U.S. coalfields of the Appalachian Mountains. Underground coal mining is a dangerous profession; there are always hazards of collapse, explosion, and fire. Respiratory illnesses are a risk, especially black lung disease, which is related to exposure to coal dust and has killed or disabled many miners.

Some of the environmental problems associated with underground mining include the following:

■ Acid mine drainage from the mines and waste piles has polluted thousands of kilometers of streams (see Chapter 22).

■ Land subsidence can occur over mines. Vertical subsidence occurs when the ground above coal mine tunnels collapses, often resulting in a crater-shaped pit at the surface (Figure 18.16). Coal mining areas in Pennsylvania and West Virginia, for example, are well known for serious subsidence problems. In recent years, a parking lot and crane collapsed into a hole over a coal mine in Scranton, Pennsylvania; and damage from subsidence caused condemnation of many buildings in Fairmont, West Virginia.

■ Coal fires in underground mines may be either naturally caused or deliberately set. The fires may belch smoke and hazardous fumes, causing people exposed to them to suffer from a variety of respiratory diseases. For example, in Centralia, Pennsylvania, a trash fire set in 1961 lit nearby underground coal seams on fire. They are still burning today and have turned Centralia into a ghost town.

Figure 18.16 ■ Subsidence below coal mines in the Appalachian coal belt.

Transport of Coal

Transporting coal from mining areas to large population centers where energy is needed is a significant environmental issue. Although coal can be converted at the production site to electricity, synthetic oil, or synthetic gas, these alternatives have their own problems. Power plants necessary to convert coal to electricity require water for cooling, and in semiarid coal regions of the western United States there may not be sufficient water. Furthermore, transmission of electricity over long distances is inefficient and expensive (see Chapter 17). The conversion of coal to synthetic oil or gas also requires a tremendous amount of water. Moreover, the conversion process is expensive.[17]

Freight trains and coal-slurry pipelines (designed to transport pulverized coal mixed with water) are options to transport the coal itself over long distances. Trains are typically used and will continue to be used because they provide relatively low-cost transportation compared with the cost of constructing pipelines. The economic advantages of slurry pipelines are tenuous, especially in the western United States, where large volumes of water to transport the slurry are difficult to obtain.[17]

The Future of Coal

The burning of coal produces nearly 60% of the electricity used and about 25% of the total energy consumed in the United States today.[18] Coal accounts for nearly 90% of the fossil fuel reserves in the United States, and we have enough coal to last at least several hundred years. However, serious concern has been raised about burning that coal. Giant power plants that burn coal as a fuel to produce electricity in the United States are responsible for about 70% of the total emissions of sulfur dioxide, 30% of the nitrogen oxides, and 35% of the carbon dioxide. (The effects of these pollutants are discussed in Chapter 24.)

Legislation as part of the Clean Air Amendments of 1990 mandated that sulfur dioxide emissions from coal-burning power plants be eventually cut by 70% to 90%, depending on the sulfur content of the coal, and that nitrogen oxide emissions be reduced by about 2 million metric tons per year. As a result of this legislation, utility companies are struggling with various new technologies designed to reduce emissions of sulfur dioxide and nitrogen oxides from burning coal. Options being used or developed include the following:[19]

- Chemical and/or physical cleaning of coal prior to combustion.
- New boiler designs that permit a lower temperature of combustion, which reduces emissions of nitrogen oxides.
- Injection of material rich in calcium carbonate (such as pulverized limestone or lime) into the gases following burning of the coal. This practice, known as **scrubbing**, removes sulfur dioxides. In the scrubber—a large, expensive component of a power plant—the carbonate reacts with sulfur dioxide, producing hydrated calcium sulfite as a sludge. The sludge has to be collected and disposed of, which is a major problem.

- Conversion of coal at power plants into a gas (syngas, a methane-like gas) before burning. This technology is being tested at the Polk Power Station in Florida. The syngas, though cleaner burning than coal, is still more polluting than natural gas.
- Consumer education about energy conservation and efficiency to reduce the demand for energy and thus the amount of coal burned and emissions released.
- Development of zero emission coal-burning electric power plants. Emissions of particulates, mercury, sulfur dioxdes, and other pollutants would be eliminated by physical and chemical processes. Carbon dioxide would be eliminated by injecting it deep into the earth or using a chemical process to sequester it (tie it up) with calcium or magnesium as a solid. The concept of zero emission is in the experimental stages of development.

The real shortages of oil and gas are still a few years away, but when they do come, they will put pressure on the coal industry to open more and larger mines in both the eastern and western coal beds of the United States. Increased use of coal will have significant environmental impacts for several reasons:

1. More and more land will be strip mined and thus will require careful and expensive restoration.
2. Unlike oil and gas, burning coal produces large amounts of air pollutants, as already mentioned. Also created is ash, which can be as much as 20% of the coal burned; boiler slag, a rocklike cinder produced in the furnace; and calcium sulfite sludge, produced from removing sulfur through scrubbing. Coal-burning power plants in the United States today produce about 90 million tons of these materials per year. Calcium sulfite from scrubbing can be used to make wallboard (by converting calcium sulfite to calcium sulfate, which is gypsum) and other products. Gypsum is being produced for wallboard in Japan and Germany. However, this practice is unlikely to be widely used in the United States, where wallboard can be made less expensively from abundant natural gypsum deposits. Another waste product, boiler slag, can be used for fill along railroad tracks and at construction projects. Nevertheless, about 75% of the combustion products of burning coal in the United States today end up on waste piles or in landfills.[20]
3. The handling of large quantities of coal through all stages (mining, processing, shipping, combustion, and final disposal of ash) will have potentially adverse environmental effects. These include aesthetic degradation, noise, dust, and—most significant from a health standpoint—release of harmful or toxic trace elements into the water, soil, and air.

It seems unlikely that coal will be abandoned in the near future in the United States, because we have so much of it and we have spent so much time and money developing coal resources. It has been suggested that we should now promote the use of natural gas in preference to coal because it burns so much cleaner. However, there is a concern that we might then become dependent on imports of natural gas. Regardless, it remains a fact that coal is the most polluting of all the fossil fuels.

Allowance Trading

An innovative approach to managing U.S. coal resources and reducing pollution is **allowance trading**. In this system, the Environmental Protection Agency grants utility companies tradable allowances for polluting. One allowance is good for 1 ton of sulfur dioxide emissions per year. In theory, some companies would not need all their allowances because they use low-sulfur coal or have installed equipment and methods that have reduced their emissions. Their extra allowances may then be traded and sold by brokers to other utility companies that are unable to stay within their allocated emission levels. The idea is to encourage competition in the utility industry and reduce overall pollution through economic market forces.[19]

Some environmentalists are not comfortable with the concept of allowance trading. They argue that, although buying and selling may be profitable to both parties in the transaction, it is less acceptable from an environmental viewpoint. They believe that companies should not be able to buy their way out of pollution problems (see Chapter 27 for more details).

18.4 Oil Shale and Tar Sands

Oil shale and tar sands play a minor role in today's mix of available fossil fuels. However, they may be more significant in the future when traditional oil from wells becomes scarce.

Oil Shale

Oil shale is a fine-grained sedimentary rock containing organic matter (kerogen). When heated to 500°C (900°F) in a process known as *destructive distillation*, oil shale yields up to nearly 60 L (14 gal) of oil per ton of shale. If not for the heating process, the oil would remain in the rock. The oil from shale is one of the so-called **synfuels** (from the words *synthetic* and *fuel*), which are liquid or gaseous fuels derived from solid fossil fuels. The best known sources of oil shale in the United States are found in the Green River formation, which underlies approximately 44,000 km^2 (17,000 mi^2) of Colorado, Utah, and Wyoming.

Total identified world oil shale resources are estimated to be equivalent to about 3 trillion bbl of oil. However, evaluation of the oil grade and the feasibility of economic recovery with today's technology is not complete. Oil shale resources in the United States amount to about 2 trillion bbl of oil, or two-thirds of the world total. Of this, 90%, or 1.8 trillion bbl, is located in the Green River oil shales.[21]

The environmental impact of developing oil shale varies with the recovery technique used. Both surface and subsurface mining techniques have been considered.

Surface mining is attractive to developers because nearly 90% of the shale oil can be recovered, compared with less than 60% by underground mining. However, waste disposal will be a major problem with either surface or subsurface mining. Both require that oil shale be processed, or *retorted* (crushed and heated), at the surface. The volume of waste will exceed the original volume of shale mined by 20% to 30%, owing to the fact that crushed rock, because of added pore spaces, has more volume than the solid rock from which it came. (For example, pour some concrete into a milk carton. When it hardens, remove the block of concrete from the carton and break it into small pieces with a hammer. Then try to place the pieces in the carton. You will find that not all of the pieces fit.) Thus the mines from which the shale is removed will not be able to accommodate all the waste, and its disposal will become a problem.[10]

In the 1970s, a tremendous rush to develop oil shale was expected. Interest in oil shale was heightened by the oil embargo in 1973 and by fear of continued shortages of crude oil. In the 1980s through the mid-1990s, however, plenty of cheap oil was available, and oil shale development was put on the back burner; it is much more expensive to extract a barrel of oil from oil shale than to pump it from a well. However, shortages of oil will occur in the future, and we are seeing signs and we will again turn to oil shale. This would result in significant environmental, social, and economic impacts in the oil shale areas resulting from rapid urbanization to house a large workforce, construction of industrial facilities, and an increased demand on water resources.

Tar Sands

Tar sands are sedimentary rocks or sands impregnated with tar oil, asphalt, or bitumen. Petroleum cannot be recovered from tar sands by pumping wells or other usual commercial methods because the oil is too viscous (thick) to flow easily. Oil in tar sands is recovered by first mining the sands (which are very difficult to remove) and then washing the oil out with hot water.

Some 75% of the world's known tar sand deposits are in the Athabasca Tar Sands near Alberta, Canada. The total Canadian resource represents about 2 trillion bbl, but it is not known how much of this will eventually be recovered. Today's production of the Athabasca Tar Sands is about

Should the Gasoline Tax Be Raised?

Petroleum, like other fossil fuels, is a limited and nonrenewable resource that must be conserved and used wisely during a period of transition to other energy sources. Because petroleum occurs naturally in liquid form and is easily portable, it is particularly well suited to transportation uses. The number of motor vehicles in the world, most of which use petroleum as fuel, has increased about nine times since 1950. There are about 750 million motor vehicles in the world, and about one-third of them are in the United States. Each day, the United States uses over 10 million bbl of oil for transportation—about 70% of the country's total oil consumption.

In addition to the drain on a limited resource, the use of petroleum in transportation contributes significantly to air pollution. Approximately 60% of the air pollutants annually emitted into the atmosphere in the United States are attributed to automobiles. Gasoline accounts for 25% of the carbon dioxide emissions from burning fossil fuels in the United States. Each gallon of gasoline burned adds 8.6 kg (19 lb) of carbon dioxide to the air. A car that runs for 160,000 km (100,000 mi) with an average fuel consumption of 11.7 km/L (27.5 mi/gal) emits 35 metric tons of carbon dioxide. Vehicles also emit carbon monoxide, nitrogen oxides, and organic chemicals called hydrocarbons. In the presence of sunlight, the hydrocarbons react with the nitrogen oxides to produce ozone. Carbon monoxide competes with oxygen in the blood, nitrogen oxides form nitric acid in the presence of water, some organic molecules are carcinogenic, and ozone is an irritant to the lungs.

Proposals for reducing the dependence on petroleum for transportation, and thereby slowing the drain on oil resources and decreasing air pollution, include designing more efficient vehicles; using alternative fuels; shifting to other modes of transportation, such as mass transit and bicycles; improving roads; and encouraging consumers to conserve gasoline. These proposals have to some extent been implemented.

Some energy policymakers have also suggested that increasing the tax on gasoline would encourage Americans to purchase more fuel-efficient vehicles and discourage unnecessary use of motor transportation, thereby lowering the demand for gasoline. One of the main arguments advanced for higher gasoline taxes is that U.S. gasoline taxes are only 12% to 18% of those of other industrialized countries (Figure 18.17). Opponents, however, argue that higher taxes at the pump might have a greater impact on poorer families and would also unfairly punish those living in western states, where driving distances are typically greater. As with many controversial issues, the facts themselves are in question. For example, do the poor spend proportionally more on transportation, and would they be disproportionately and adversely affected by a rise in gasoline taxes?

Not surprisingly, higher gas taxes have been advocated by many environmentalists. Automobile companies have also supported the proposal, hoping that it would eliminate pressure for more stringent fuel-efficiency standards. Americans are divided on the issue. In the past people have supported, by a slim majority, a modest tax increase of about 15 cents per gallon.

Critical Thinking Questions

1. What effect would increasing the price of gasoline have in terms of giving consumers an incentive to choose more fuel-efficient vehicles? What increase in the cost would motivate you to purchase a more fuel-efficient vehicle over other factors, such as size of car, comfort, and safety?

2. Make a table of the pros and cons of raising the price of gasoline through taxation in relation to environmental, economic, and social issues. What are the direct and indirect effects?

3. When you have completed the table, evaluate the arguments and decide where you stand on the issue. Write a paragraph explaining your position.

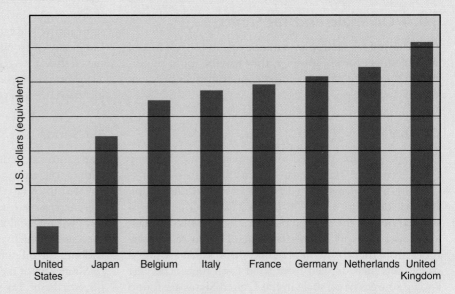

Figure 18.17 ■ Taxes for 1 gallon of gasoline in U.S. dollars equivalent for selected countries, 2002. [*Source:* Federal Highway Administration, U.S. Department of Transportation, *Monthly Motor Fuel Reports by State*, Washington, D.C., August 2003.]

1.5 million bbl of synthetic crude oil per day, about 15% of North America's production of oil.[22]

In Alberta, tar sand is mined in a large open-pit mine. The mining process is complicated by the fragile native vegetation, a water-saturated mat known as a muskeg swamp—a kind of wetland that is difficult to remove except when frozen. Restoration of this fragile, naturally frozen (permafrost) environment is difficult. In addition, there is a waste-disposal problem because the mined sand material (as discussed for oil shale) occupies a greater volume than unmined material. The land surface after mining can be up to 20 m (66 ft) higher than the original ground surface.

Summary

- The United States has an energy problem caused by dependence on fossil fuels, especially oil. Maximum global production (peak oil) is expected between 2020 and 2050, followed by a decline in production. The challenge is to plan now for the decline in oil supply and shift to alternative energy sources.

- Fossil fuels are forms of stored solar energy. Most are created from the incomplete biological decomposition of dead organic material that is buried and converted by complex chemical reactions in the geologic cycle.

- Because fossil fuels are nonrenewable, we will eventually have to develop other sources to meet our energy demands. We must decide when the transition to alternative fuels will occur and what the impacts of the transition will be.

- Environmental impacts related to oil and natural gas include those associated with exploration and development (damage to fragile ecosystems, water pollution, air pollution, and waste disposal); those associated with refining and processing (soil, water, and air pollution); and those associated with burning oil and gas for energy to power automobiles, produce electricity, run industrial machinery, heat homes, and so on (air pollution).

- Coal is a source of energy particularly damaging to the environment. The environmental impacts associated with mining, processing, transporting, and using coal are many. Problems associated with mining include fires, subsidence, acid mine drainage, and difficulties related to land reclamation. Burning coal can release air pollutants, including sulfur dioxide and carbon dioxide. Finally, burning coal produces a large volume of combustion products and by-products such as ash, slag, and calcium sulfite (from scrubbing). The environmental objective for coal is to develop a zero emission power plant.

REEXAMINING THEMES AND ISSUES

Human Population

As human population (particularly in developed countries such as the United States) has increased, so has the total impact from the use of fossil fuels. Total impact is the product of the impact per person times the total number of people. Reducing the impact will require that all countries adopt a new energy paradigm emphasizing use of the minimum energy needed to complete a task (end use) rather than the current prodigious overuse of energy.

Sustainability

It has been argued that we cannot achieve sustainable development and maintenance of a quality environment for future generations if we continue to increase our consumption of fossil fuels. Achieving sustainability will require wider use of a variety of alternative renewable energy sources and less dependence on fossil fuels.

Global Perspective

The global environment has been significantly affected by burning fossil fuels. This is particularly true for the atmosphere, where fast-moving processes operate (see Chapters 23 and 24). Solutions to global problems from burning fossil fuels are implemented at the local and regional levels, where the fuels are consumed.

Urban World

Burning of fossil fuels in urban areas has a long history of problems. Not too many years ago, black soot from burning coal covered the buildings of most major cities of the world, and historical pollution events killed thousands of people. Today, we are striving to improve our urban environments and reduce urban environmental degradation from burning fossil fuels.

People and Nature

Our exploration, extraction, and use of fossil fuels have changed nature in fundamental ways—from the composition of the atmosphere to the disturbance of coal mines, to the pollution of ground and surface waters. Some people buy giant SUVs supposedly to connect with nature, but using them causes more air pollution than automobiles, and if used off-road often degrades nature.

Science and Values

Scientific evidence for the adverse effects of burning fossil fuels is well documented. The controversy concerning their use is linked to our values. Do we value burning huge amounts of fossil fuels to increase economic growth more than living in a quality environment? Economic growth is possible without damaging the environment; developing a sustainable energy policy that doesn't harm the environment is possible with present technology. What is required are changes in values and lifestyle that are linked to energy production and use, human well-being, and environmental quality.

Key Terms

allowance trading **381**	fossil fuels **366**	oil shale **381**	synfuels **381**
coal **375**	methane hydrate **371**	peak oil **365**	tar sands **381**
crude oil **366**	natural gas **366**	scrubbing **380**	

Study Questions

1. Assuming that oil production will peak in about 2020 and then decline at about 3% per year, when will production be half of that in 2020? What could be the consequences? Why? How could these consequences be avoided?

2. Compare the potential environmental consequences of burning oil, burning natural gas, and burning coal.

3. What actions can you take at a personal level to reduce consumption of fossil fuels?

4. What environmental and economic problems could result from a rapid transition from fossil fuels to alternative sources?

5. What are some of the technical solutions to reducing air-pollutant emissions from burning coal? Which are best? Why?

6. What do you think about the idea of allowance trading as a potential solution to reducing pollution from burning coal?

7. Do you think we can develop a zero emission coal-burning power plant? What about for natural gas?

Further Reading

Boyl, G., Everett, B., and J. Ramage. 2003. *Energy Systems and Sustainability.* Oxford (UK): Oxford University Press. See excellent discussion of fossil fuel.

Fay, J. A., and D. S. Golomb. 2002. *Energy and the Environment.* New York: Oxford University Press. See Chapters 1–5 for fossil fuels.

Liu, P. I. 1993. *Introduction to Energy and the Environment.* New York: Van Nostrand Reinhold. A good summary of energy sources and issues, with discussions of the environmental effects of various energy sources.

Miller, E. W., and R. M. Miller. 1993. *Energy and American Society: A Reference Handbook.* Broomfield, Colo.: ABC-CLIO. Follows patterns of energy use from early American history through today and looks at future energy sources.

CHAPTER 19

Alternative Energy and the Environment

Wind power at Spirit Lake Elementary and Middle schools, Iowa. The wind turbine behind the playground is a symbol of the transition from oil and gas to alternative energy. Source: Spirit Lake Community School District.

Learning Objectives

Alternatives to fossil fuels include geo-thermal energy, solar energy, water power, wind power, and biofuels. Some of these alternatives are already being used, and efforts are under way to develop others. After reading this chapter, you should understand:

■ What passive, active and photovoltaic solar energy systems are, and the advantages, limitations, and environmental effects of developing and using them.

■ Why hydrogen may be an important fuel of the future.

■ The advantages, disadvantages, and environmental impacts of developing hydropower.

■ Why wind power has tremendous potential, and how its development and utilization could affect the environment.

■ What biofuels are and why they are an important energy source.

■ What geothermal energy is, and how developing and using it affects the environment.

■ What important policy issues will affect large-scale use of alternative energy sources.

Spirit Lake Community School District in Iowa, Going with the Wind

The transformation from our dependency on coal, oil, and natural gas (the fossil fuels) to other energy sources such as wind and solar power will be one of the iconic events of the twenty-first century.[1,2] Although we will not run out of oil, we will experience increasing supply problems when we pass the moment when one-half of Earth's total supply of oil has been used. That moment will likely occur in the next decade or so. To avoid economic and social problems resulting from a reduced supply of oil, we need to find alternatives sooner rather than later. In other words, we need to be proactive and find solutions to potential energy shortages rather than reacting after shortages are manifested. The transformation in many places at many levels is already happening. In many cases the environmental statement "think globally, act locally" is appropriate. The Spirit Lake Community School District in northwest Iowa made a decision in 1991 to act locally to reduce its dependency on fossil fuels. The goals were to help reduce environmental impacts (for example, air pollution) caused by the production of electricity from mainstream fossil fuel power plants; help reduce our nation's dependency on foreign oil and achieve energy independence; provide a

source of income at the local level; and provide students a hands-on learning experience.

To achieve these goals, the school district turned to wind power, one of the fastest-growing energy source in the world today. The district has two very visible wind turbines completed near the district's elementary school (see opening photograph). Online since 2001, the first turbine is turning a profit from selling excess power. With the second turbine online in 2007, the electric power demands of the district's schools and offices, including both the middle and high schools and football stadium, will be supplied by the wind that blows through district school lands. The wind turbine at the elementary school is already providing a valuable learning experience to students on the role of energy in society. Students learn that their windmill, visible from the playground, saves oil and coal that would otherwise be burned, helps reduce air pollution, and brings income to the school for needed maintenance and improvements. The Spirit Lake School District's energy transformation is one small step for Iowa and a larger symbolic step in the ongoing global transformation that is coming in the twenty-first century.

The growth of wind power, an alternative and renewable source of energy, is a positive step toward sustainable energy development. In this chapter, we explore alternative, renewable energy in terms of sources, advantages and disadvantages, environmental concerns, and policy issues.

19.1 Introduction to Alternative Energy Sources

As we have seen, the primary energy sources today are fossil fuels, which supply approximately 90% of the energy consumed by people. All other sources are considered **alternative energy** and are divided into **renewable energy**

and **nonrenewable energy**. Nonrenewable alternative energy sources include nuclear energy (discussed in Chapter 20) and geothermal energy. Nuclear energy is nonrenewable because it requires a mineral fuel mined from Earth. Geothermal energy is considered nonrenewable for the most part because heat can be extracted from Earth faster than it is naturally replenished (that is, output exceeds input; see Chapter 3). The renewable sources are solar energy, water (hydro) power, wind power, hydrogen produced from water, and energy derived from biomass (crops, wood, and so forth).

Renewable energy sources—sun, water, wind, and biomass—are often discussed as a group because they are all derived from the sun's energy. In other words, solar energy, broadly defined, comprises all the renewable energy sources, as shown in Figure 19.1. These energy sources are renewable because they are regenerated (renewed) by the sun within a time period useful to humans.

Water and biomass, though renewable, are fundamentally different from solar and wind energy, which are there as long as the sun shines and the wind blows. Water power and biomass energy are not always automatically renewed by nature. They may be depleted if the environment necessary for their renewal is not maintained. For biomass to be renewable for biofuels, both water and soil are necessary for plant growth. If either of these is depleted, then biomass production may decrease or even stop. Similarly, the water behind a dam depends on the climate to produce runoff. If the climate be-comes arid, runoff may be reduced and the dam will store less water and produce less electricity. If people divert river water for crops upstream of a dam, there will be less water in the reservoir and hydropower will be reduced.

The primary difference between renewable alternative energy sources and fossil fuels is analogous to the difference between a checking account that receives regular monthly deposits and a checking account that receives an initial large deposit but no further deposits. The account receiving regular deposits will, over the long term, not be depleted (similar to renewable energy sources). Even if you spend the present balance, more will be added later. The second account will eventually be depleted (similar to oil), although the timing will depend on the rate at which the funds are spent. Financial planning would differ for these two accounts. Similarly, energy planning based on oil differs fundamentally from planning based on renewable energy—for example, solar energy. Using solar energy requires that we live and plan within the constraints of the energy provided by the sun. Using oil requires planning based on a resource that is finite and is being depleted.[1]

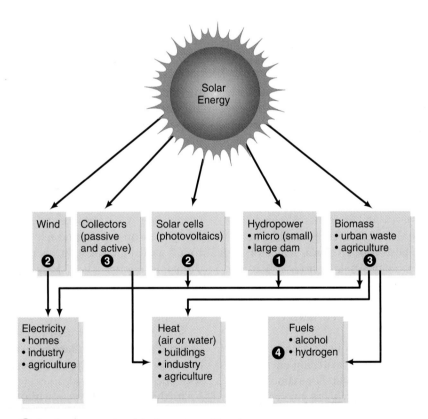

1 Produces most electricity from renewable solar energy

2 Rapidly growing, strong potential (Wind and solar are growing at 30% per year.)

3 Used today; important energy source

4 Potentially a very important fuel to transition from fossil fuels

Figure 19.1 ■ Routes of various types of renewable solar energy.

The total energy we may be able to extract from alternative energy sources is enormous (Table 19.1). For example, the estimated recoverable energy from solar energy is about 75 times the present annual human global energy consumption. The estimated recoverable energy from wind is comparable to current global energy consumption, as is the estimated recoverable energy from biomass.

It is true, however, that this energy may not necessarily always be available when we need it. Renewable energy sources, with the exception of biomass and water power (which can be stored), are intermittent, with daily and seasonal variations in supply. These variations will become a problem when renewable energy supplies make up about 40% of our total energy use. Beyond about 40%, the natural variation will periodically produce shortages in energy supply. As a result, the development of energy storage systems to smooth out the variations in supply will be necessary.[1] For example, suppose you have solar collectors on your roof to produce electricity. At night, no energy is produced, but if you store some of the energy produced during the day in batteries, you can use stored energy at night. For large-scale production of renewable energy, large energy storage systems are necessary.

In addition, renewable energy sources are not equally available in all locations. Therefore, matching energy production to appropriate sites is important. For example, in the United States, we might build solar energy power plants in the Southwest, where sunlight is often intense, and construct wind farms in the Great Plains, Texas, the Northwest, and California, where the wind is strong and steady. Likewise, we could build biofuel power plants in locations where forest and agriculture fuel resources are abundant.[3]

Renewable alternative energy sources are associated with minimal environmental degradation. In general, because no fuel is burned, these energy sources do not increase atmospheric carbon dioxide, which causes global warming. (An exception is the burning of biomass or its derivative, urban waste.) Renewable energy sources such as solar and wind power will not cause climate change or raise sea levels, increasing coastal erosion. Another advantage is that the construction lead time necessary to implement the technology for production of energy from renewable sources is often short compared with the construction of power plants that use fossil or nuclear fuels. Although in the United States most renewable energy sources are more expensive than energy from fossil fuels or nuclear energy, this is in part because in the past the federal subsidies they received were much smaller than those granted to domestic fossil fuel and nuclear energy production.[4]

Alternative energy sources, particularly solar and wind, are growing at tremendous rates. For the first time, it has become apparent that these energy sources may compete with fossil fuels. Alternative renewable energy sources offer our best chance to develop a truly sustainable energy policy that will not harm the planet.[5]

This section has supplied a brief introduction to alternative energy sources. Next, we discuss individual sources and, where appropriate, the environmental advantages and disadvantages of each.

19.2 Solar Energy

The total amount of solar energy reaching the Earth's surface is tremendous. For example, on a global scale, 10 weeks of solar energy is roughly equivalent to the energy stored in all known reserves of coal, oil, and natural gas on Earth. Solar energy is absorbed at Earth's surface at an average rate of 90,000 TW (1 TW is 10^{12} W), which is about 7,000 times the total global demand for energy.[1] In the United States, on average, 13% of the sun's original energy entering the atmosphere arrives at the surface (equivalent to approximately 177 W/m^2, or about 16 W/ft^2, on a continuous basis). The estimated year-round availability of solar energy in the United States is shown in Figure 19.2. However, solar energy is site specific, and detailed observation of a potential site is necessary to evaluate the daily and seasonal variability of its solar energy potential.[6]

Solar energy may be used through passive solar systems or active solar systems. **Passive solar energy systems** often involve architectural designs that enhance the absorption of solar energy by using and adjusting for natural changes that occur throughout the year without requiring mechanical power (Figure 19.3). Various societies during the past few thousand years have used passive solar energy (see Chapter 17). For example, Islamic architects have traditionally used passive solar energy in hot climates to cool buildings.[1]

One simple technique is to design overhangs on buildings that block summer (high-angle) sunlight but allow winter (low-angle) sunlight to penetrate and warm rooms. Another technique is to build a wall that absorbs the solar

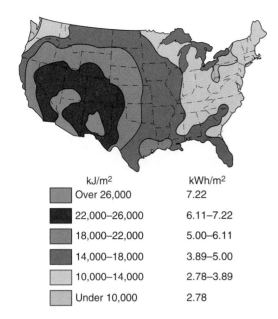

kJ/m²	kWh/m²
Over 26,000	7.22
22,000–26,000	6.11–7.22
18,000–22,000	5.00–6.11
14,000–18,000	3.89–5.00
10,000–14,000	2.78–3.89
Under 10,000	2.78

Figure 19.2 ■ Estimated solar energy for the contiguous United States. [*Source:* Modified from Solar Energy Research Institute, 1978.]

energy and then radiates heat that warms the room. Yet another technique makes use of trees that lose their leaves during the winter, which are common in the midwestern and eastern United States. When planted on the sunny side of a building, these trees provide summer shade that cools the building. In the winter, with the leaves gone, winter sunlight enters the house.

Many thousands of homes and other buildings in the southwestern United States, as well as other parts of the country, now use passive solar systems for at least part of their energy needs.[3] Passive solar energy also provides natural lighting to buildings through windows and skylights. A special glazing on the glass transmits light and provides insulation. Use of passive solar energy features is most cost effective when they are part of the design for a new building. However, in some cases, adding passive solar features to an existing building can also be cost-effective.

Active solar energy systems require mechanical power, usually electric pumps and other apparatuses, to circulate air, water, or other fluids from solar collectors to a location where the heat is stored until used. We discuss solar collectors next and then describe several specific types of active solar energy systems.

Solar Collectors

Solar collectors to provide space heating or, more commonly, hot water are usually flat panels consisting of a glass-covered plate over a black background where an absorbing fluid (water or a volatile liquid) is circulated through tubes (Figure 19.4). Solar radiation enters the glass and is absorbed by the black background. Heat is emitted from the black material, heating the fluid circulating in the tubes. If water is the absorbing fluid, it is heated to 38–93°C (100–200°F).[4] In cold climates, water is a poor absorbing fluid because it can freeze. As a result, a liquid with a low freezing temperature is used.

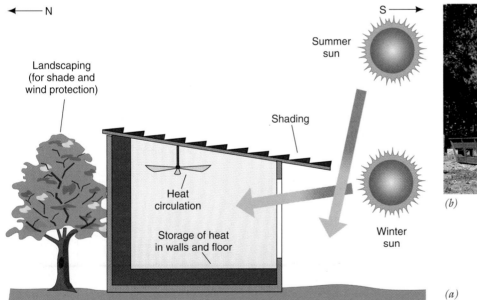

(b)

(a)

Figure 19.3 ■ (*a*) Essential elements of passive solar design. High summer sunlight is blocked by the overhang, but low winter sunlight enters the south-facing window. Other features are designed to facilitate the storage and circulation of passive solar heat. (*b*) Design of this home utilizes passive solar energy. Sunlight enters through the windows and strikes a specially designed masonry wall that is painted black. The masonry wall heats up and radiates this heat, warming the house during the day and into the evening. [*Source:* Moran, Morgan, and Wiersma, *Introduction to Environmental Science* (New York: Freeman, 1986). Copyright by W. H. Freeman & Company. Reprinted with permission.]

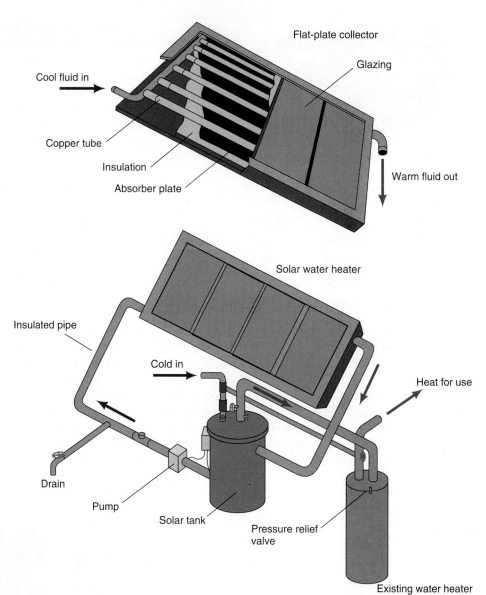

Figure 19.4 ■ Detail of a flat-plate solar collector and pumped solar water heater. [*Source:* Farallones Institute, *The Integral Urban House* (San Francisco: Sierra Club Books, 1979). Copyright 1979 by Sierra Club Books. Reprinted with permission.]

A second type of solar collector is called the evacuated tube collector. Its design is similar to that of the flat-plate collector. The difference is that each tube, along with its absorbing fluid, passes through a larger tube that helps reduce heat loss.[1] Solar collectors are experiencing very rapid growth. The global market grew about 50% from 2001 to 2004. Solar water heating systems in the United States are economically viable, paying for themselves in only four to eight years.[2]

Photovoltaics

Photovoltaics is a technology that converts sunlight directly into electricity (Figure 19.5). Photovoltaics are the world's fastest growing source of energy, with a growth rate of about 35% per year (it is doubling every two years). The photovoltaic industry is a $7 billion market and is expected to grow to $30 billion by 2010.[2] The systems use

solar cells, also called photovoltaic cells, made of thin layers of semiconductors (silicon or other materials) and solid-state electronic components with few or no moving

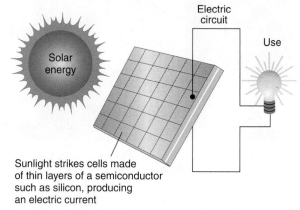

Figure 19.5 ■ Idealized diagram illustrating how photovoltaic solar cells work.

parts. Solar cell technology is advancing rapidly, and first-law efficiency is as high as 10%. The cells are constructed in standardized modules, encapsulated in plastic or glass, which can be combined to produce systems of various sizes so that power output can be matched to the intended use.

Electricity is produced when sunlight strikes the cell. The different electronic properties of the layers cause electrons to flow out of the cell through electrical wires. Photovoltaic cells can also work with light energy other than the sun. For example, using the light of your desk lamp, a photovoltaic cell can produce electricity to light a small bulb in a clock or to power your hand-held calculator.

An early and continuing use for photovoltaics is to supply power to satellites and space vehicles. Another commercial use is as a power source in remote areas for electric equipment such as water-level sensors, meteorological stations, and emergency telephones (Figure 19.6).

Photovoltaics is emerging as a significant contributor to developing countries that do not have the financial ability to build large central power plants that burn fossil fuels. Developing countries and their use of photovoltaics are demonstrating that solar technology can be simple, relatively inexpensive, and capable of meeting energy needs for people in many places in the world by matching the demand with an appropriate supply. For example, a solar company in the United States is providing people in a number of developing countries with photovoltaic systems that power lights and televisions at an installed cost of less than $400 per household.[7] About half a million homes, mostly in villages not linked to a countrywide electrical grid, now receive their electricity from photovoltaic cells.

Used in developing countries and other nations as well are solar roofing tiles that utilize the roof of a building as a platform for a power plant.[8] Panels of solar cells can also be placed on walls or window glass.[5] By 1997, Germany had installed photovoltaic cells on 10,000 roofs. The governments of India and the United States have both announced that a million roofs will be installed with photovoltaic systems by 2010.

Utility companies are interested in photovoltaics and are constructing power plants on a variety of scales, ranging from solar cell panels on roofs to larger power plants. A large photovoltaic power plant is located in Tucson, Arizona, at the Springerville Generating Station. This plant produces 24 MW with photovoltaics, and planned expansion will make it the largest solar power plant in the world.

In some remote locations today, the cost of using photovoltaics is comparable with the cost of using grid-connected power. These are places where the high cost of joining the grid exceeds the cost of photovoltaics.

In the future, the alternative energy source for many people and communities might well be lightweight photovoltaics in which the solar cells are mounted together to form modules. The growth and technology changes in photovoltaics suggest that in the twenty-first century, solar energy is likely to become a multigigawatt-per-year industry that will provide a significant portion of the energy we use.[9]

(a)

(b)

Figure 19.6 ■ (a) Panels of photovoltaic cells are being used here to power a small refrigerator to keep vaccines cool. The unit is designed to be carried by camels to remote areas in Chad.

(b) Photovoltaics are used to power emergency telephones along a highway on the island of Tenerife in the Canary Islands.

Power Towers

An interesting type of **solar energy system** is the solar power tower, shown in Figure 19.7. The system works by collecting heat from solar energy and delivering this energy in the form of steam to turbines that produce electric power. An experimental 10-MW power tower near Barstow, California, is approximately 100 m (330 ft) high and is surrounded by approximately 2,000 mirror modules, each with a reflective area of about 40 m² (430 ft²). The mirrors adjust continually to reflect as much sunlight into the tower as possible. In 1999, the power tower was able to produce electricity 24 hours per day by storing the heat produced during the day in a molten salt heat-storage system. The stored heat was used at night or on cloudy days to produce energy.

At the end of 1999, the power tower was shut down, in part because the plant was not economically competitive with other sources of electricity. Given the energy crisis in California in 2001, however, that decision was premature. It is expected that the cost of electricity from power towers will be reduced sufficiently by the year 2030 to be economically competitive with more traditional sources of electricity.[3]

Figure 19.7 ■ Solar power tower at Barstow, California. Sunlight is reflected and concentrated at the central collector, where the heat is used to produce steam to drive turbines and generate electric power.

Luz Solar Electric-Generating System

A system developed at Luz International is the most technically successful solar power experiment to date. The site, in the Mojave Desert, consists of nine solar farms. Each farm comprises a power plant surrounded by hundreds of solar collectors, and each produces 350 MW of electrical power, sufficient for about 550,000 people. The system uses solar collectors (curved mirrors) to heat a synthetic oil that flows through heat exchangers, which in turn drive steam turbine generators. The basic components of the system are shown in Figure 19.8. One advantage of this system is that it is modular, which allows for individual solar farms to be constructed relatively quickly.

The success of the system is based in part on the fact that it uses a natural gas burner backup system (75% solar and 25% natural gas), ensuring uninterrupted power generation both on cloudy days and during times of peak demand. Thus, the Luz system is really a combination of solar technology and conventional power generation.[10, 11]

Solar Energy and the Environment

The use of solar energy generally has a relatively low impact on the environment, but there are some environmental concerns. One disadvantage of solar energy is that it is relatively dispersed, arriving at Earth's surface like a fine mist, so that a large land area is required to generate a large amount of energy. This problem is negligible when solar collectors can be combined with existing structures, as with the addition of solar hot-water heaters on the roofs of existing houses. Highly centralized and high-technology solar energy units, such as solar power towers, have a greater impact on the land because they need considerable space. The impact of large centralized solar energy systems can be minimized by locating them in remote areas not used for other purposes and by making use of dispersed solar energy collectors on existing structures wherever possible.

Another concern involves the large variety of metals, glass, plastics, and fluids used in the manufacture and use of solar equipment. Some of these substances may cause environmental problems through production and by accidental release of toxic materials.

19.3 Hydrogen

Hydrogen, the fuel burned by our sun (producing solar energy), is the lightest, most abundant element in the universe. Hydrogen gas may be an important fuel of the future.[3]

Hydrogen is a high-quality fuel that can be easily used in any of the ways in which we normally use fossil fuels, such as to power automobile and truck engines and to heat water and buildings. Hydrogen is used and stored in **fuel cells** (see A Closer Look 19.1), which are similar to batteries in that electrons flow between negative and pos-

Power Block Assembly
1. Solar collector assembly
2. Natural gas boiler
3. Turbine generator
4. Steam generator and solar superheater
5. Control building
6. Cooling tower
7. Southern California Edison interconnect

(a)

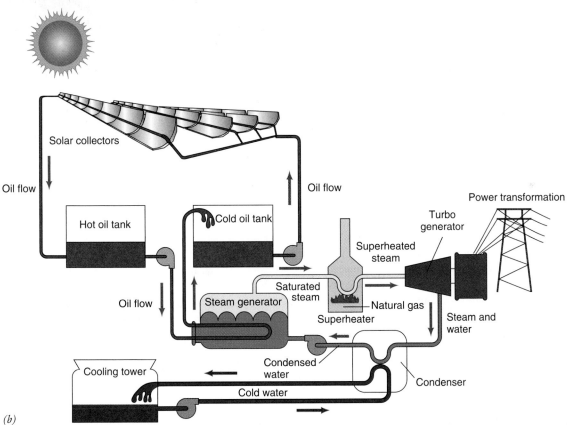

(b)

Figure 19.8 ■ (a) Luz International Solar Farm, showing the system of solar collectors, natural gas boiler, turbine generator, and other components. (b) Diagram illustrating how the Luz system works. [*Source:* Courtesy of Luz International.]

itive poles. However, a fuel cell generates electricity rather than just storing it as in a battery. Hydrogen, like natural gas, can be transported in pipelines and stored in tanks; and it can be produced using solar and other renewable energy sources. It is a clean fuel; the combustion product of burning hydrogen is water, so it does not contribute to global warming, air pollution, or acid rain (see Chapters 23 and 24). Technological improvements in producing hydrogen are certain, and the fuel price of hydrogen may be substantially reduced in the future.[12]

Fuel Cells—An Attractive Alternative

Power produced from burning fossil fuels, particularly coal and fuels used in internal combustion engines (cars, trucks, ships, and locomotives), is associated with serious environmental problems. As a result, we are searching for and developing environmentally benign technologies capable of generating power.[14] One promising technology uses fuel cells, which produce fewer pollutants, are relatively inexpensive, and have the potential to store and produce high-quality energy.

Fuel cells are highly efficient power-generating systems that produce electricity by combining fuel and oxygen in an electrochemical reaction. Hydrogen is the most common fuel type, although fuel cells that run on methanol, ethanol, and natural gas are available. Traditional generating technologies require combustion of fuel in order to convert the resultant heat into mechanical energy (to drive pistons or turbines), and this mechanical energy is then converted into electricity. With fuel cells, however, chemical energy is converted directly into electricity, thus increasing second-law efficiency (see Chapter 17) while reducing harmful emissions.

Basic components of a hydrogen-burning fuel cell are shown in Figure 19.9. Both hydrogen and oxygen are added to the fuel cell in an electrolyte solution. The reactants remain separated from one another, and a platinum membrane prevents electrons from flowing directly to the positive side of the fuel cell. The electrons are routed through an external circuit.[13, 14] The flow of electrons from the negative to the positive electrode is diverted along its path into an electrical motor, supplying current to keep the motor running. In order to maintain this reaction, hydrogen and oxygen are added as needed. When hydrogen is used in a fuel cell, the only waste product is water. Using natural gas (CH_4) in fuel cells produces some pollutants, but the amount is only about 1% of what would be produced by burning fossil fuels in an internal combustion engine or a conventional power plant.[14]

Fuel cells are efficient and clean, and they can be arranged in a series to produce the appropriate amount of energy for a particular task. In addition, the efficiency of a fuel cell is largely independent of its size and energy output. For these reasons, fuel cells are well suited to providing power for automobiles, homes, and large-scale power plants. They can also be used to store energy to be used as needed. Fuel cells are used in many locations. For example, they power buses at Los Angeles International Airport, in Vancouver, and provide heat and power at Vandenberg Air Force Base in California.[15]

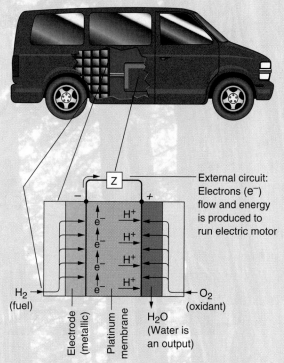

External circuit: Electrons (e^-) flow and energy is produced to run electric motor

H_2 (fuel)

O_2 (oxidant)

Electrode (metallic)

Platinum membrane

H_2O (Water is an output)

Figure 19.9 ■ Idealized diagram showing how a fuel cell works and its application to power a vehicle.

One way to produce hydrogen is to use an electric current to separate hydrogen (H) from water (H_2O), a process known as *electrolysis*. If the electricity for electrolysis is produced by solar or wind power, the hydrogen is produced in a clean, carbon-free process.[3] However, the most economical way to produce hydrogen is to use a thermal process in which steam is combined with natural gas (CH_4), removing the carbon (C) and leaving hydrogen (H). This process depends on a fossil fuel and produces some carbon dioxide (CO_2). Hydrogen can also be produced from gasification of biomass.

Hydrogen is an ideal medium for the storage of energy. It can help smooth out the natural variability of using renewable energy sources such as solar and wind power. As mentioned, hydrogen can be stored in fuel cells that are somewhat similar to batteries. Thus, hydrogen produced by electricity from solar or wind power and stored in fuel cells can be later used to produce electricity on cloudy days and at night or on days when the wind isn't blowing.[1]

The island nation of Iceland, with assistance from the European Union, is currently attempting to become the first hydrogen-based-energy economy. Although Iceland has enormous reserves of geothermal energy that can be used to produce hydrogen for fuel cells, it has no fossil fuels. The most important step will be to create the necessary infrastructure for storage, transport, and fueling stations for hydrogen, which is as flammable as gasoline.[13]

19.4 Water Power

Water power is a form of stored solar energy, as mentioned earlier, because climate, which dictates the flow of water on Earth, is driven in part by differential solar heating of the atmosphere. Water power has been successfully harnessed since the time of the Roman Empire. Waterwheels that convert water power to mechanical energy were turning in Western Europe in the seventeenth century; during the eighteenth and nineteenth centuries, large waterwheels provided energy to power grain mills, sawmills, and other machinery in the United States.

Today, hydroelectric power plants use the water stored behind dams. In the United States, hydroelectric plants generate about 80,000 MW of electricity—about 10% of the total electricity produced in the nation. In some countries, such as Norway and Canada, hydroelectric power plants produce most of the electricity used. Figure 19.10*a* shows the major components of a hydroelectric power station.

Hydropower can also be used to store energy through the process of pump storage (Figure 19.10*b* and *c*). During times when demand for power is low (for example, at night in the summer), electricity produced from oil, coal, or nuclear plants in excess of the demand is used to pump water uphill to a higher reservoir (high pool). Then, during times when demand for electricity is high (for example, on hot days in the summer), the stored water flows back down to a low pool through generators to help provide energy. The advantage of pump storage lies in the timing of energy production and use.

Small-Scale Systems

The total amount of electrical power produced by running water from large dams will probably not increase in the coming years in the United States, where most of the acceptable dam sites are already being utilized. However, small-scale hydropower systems, designed for individual homes, farms, or small industries, may be more common in the future. These small systems, known as *micro-hydropower* systems, have power output of less than 100 kW.[16]

Numerous sites in many areas have the potential for producing small-scale electrical power. This is particularly true in mountainous areas, where potential energy from stream water is often available. Micro-hydropower development is site specific, depending on local regulations, economic situations, and hydrologic limitations.

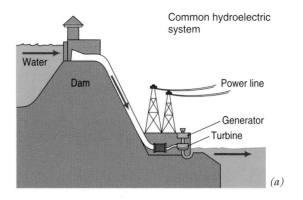

Common hydroelectric system

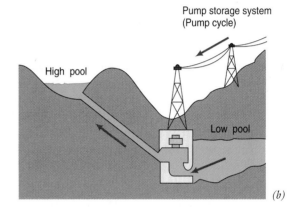

Pump storage system (Pump cycle)

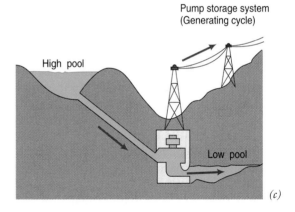

Pump storage system (Generating cycle)

Figure 19.10 ■ (*a*) Basic components of a hydroelectric power station. (*b*) A pump storage system. During light power load, water is pumped from low pool to high pool. (*c*) During peak power load, water flows from high pool to low pool through a generator. [*Source:* Modified from Council on Environmental Quality, *Energy Alternatives: A Comparative Analysis* (Norman: University of Oklahoma Science and Policy Program, 1975).]

Hydropower can be used to generate either electrical power or mechanical power to run machinery; its use may help reduce the high cost of importing energy. It may also help small operations become more independent of local utility providers.[16]

Water Power and the Environment

Water power is clean power; it requires no burning of fuel, does not pollute the atmosphere, produces no radioactive or other waste, and is efficient. However, there are environmental prices to pay (see Chapter 21):

■ Large dams and reservoirs flood large tracts of land that could have had other uses. For example, towns and agricultural lands may be lost.

■ Dams block the migration of some fish, such as salmon.

■ Water falling over high dams may pick up nitrogen gas; if the gas enters the blood of fish, it expands and kills them. This is similar to the condition known as the bends, which occurs if divers using a compressed-air source surface too rapidly from deep water. The fish respire nitrogen gas under pressure of the falling water; and when the pressure is reduced quickly, the gas bubbles in the blood expand and damage tissue. Nitrogen has killed migrating game fish in the Pacific Northwest in this manner.

■ Dams trap sediment that would otherwise reach the sea and eventually replenish the sand on beaches.

■ For a variety of reasons, many people do not want to turn wild rivers into a series of lakes.

■ Reservoirs with large surface areas increase evaporation of water compared to pre-dam conditions. In arid regions, evaporative loss of water from reservoirs is more significant than in more humid regions.

For all these reasons, and because many good sites for dams already have one, the likely growth of large-scale water power in the future (with the exception of a few areas, including Africa, South America, and China) appears limited. Indeed, in the United States, there is an emerging social movement to remove dams. Hundreds of dams, especially those with few useful functions, are being considered for removal; a few have already been removed (see Chapter 21).

As mentioned, there does seem to be continued interest in micro-hydropower to supply either electricity or mechanical energy. However, small dams and reservoirs tend to fill more quickly with sediment than large reservoirs, rendering their useful life much shorter. In fact, many dams likely to be removed are small ones filled with sediment. Because micro-hydropower development can adversely affect stream environments by blocking fish passage and changing downstream flow, careful consideration must be given to their construction. A few small dams cause little environmental degradation beyond the specific sites. However, if the number of dams in a region is large, the total impact may be appreciable. This principle applies to many forms of technology and development. The impact of a single development may be nearly negligible over a broad region; but as the number of such developments increases, the total impact may become significant.

19.5 Tidal Power

The use of **tidal power**, a type of water power derived from ocean tides, can be traced back to tenth-century Britain, where tides were used to power coastal mills.[1] However, only in a few places with favorable topography—such as the north coast of France, the Bay of Fundy in Canada, and the northeastern United States—are the tides sufficiently strong to produce commercial electricity. The tides in the Bay of Fundy have a maximum range of about 15 m (49 ft). A minimum range of about 8 m (26 ft) is necessary for development of tidal power to be considered.

To harness tidal power, a dam is built across the entrance to a bay or estuary, creating a reservoir. As the tide rises (flood tide), water is initially prevented from entering the bay landward of the dam. Then, when there is sufficient water (from the oceanside high tide) to run the turbines, the dam is opened, and water flows through it into the reservoir (the bay), turning the blades of the turbines and generating electricity. When the reservoir (the bay) is filled, the dam is closed, stopping the flow and holding the water in the reservoir. When the tide falls (ebb tide), the water level in the reservoir is higher than that in the ocean. The dam is then opened to run the turbines (which are reversible), and electric power is produced as the water is let out of the reservoir.

Figure 19.11 shows the La Rance tidal power plant on the north coast of France. La Rance, constructed in the 1960s, is the first and largest modern tidal power plant. The plant at capacity produces about 240,000 kW from 24 power units spread out across the dam. At the La Rance power plant, most electricity is produced from the ebb tide, which is easier to control.

Tidal power does have environmental impacts. The change in the hydrology of a bay or estuary caused by the dam can adversely affect the vegetation and wildlife. The dam restricts upstream and downstream passage of fish. Furthermore, the periodic rapid filling and emptying of the bay as the dam opens and closes with the tides rapidly changes habitats for birds and other organisms.

19.6 Wind Power

Wind power, like solar power, has evolved over a long period of time, from early Chinese and Persian civilizations to the present. Wind has propelled ships and has driven windmills to grind grain and pump water. In the past, thousands of windmills in the western United States were used to pump water for ranches. More recently, wind has been used to generate electricity.

Basics of Wind Power

Winds are produced when differential heating of Earth's surface creates air masses with differing heat contents and densities. The potential for energy from the wind is large,

Figure 19.11 ■ Tidal power station on the River Rance near Saint-Malo, France.

and yet there are problems with its use because wind tends to be highly variable in time, place, and intensity.[17]

Wind prospecting has become an important endeavor. On a national scale, regions with the greatest potential for wind energy are the Pacific Northwest coastal area, the coastal region of the northeastern United States, and a belt extending from northern Texas through the Rocky Mountain states and the Dakotas. Other good sites include mountain areas in North Carolina and the northern Coachella Valley in southern California. A site with sustained wind velocity of about 5 m/sec (16 ft/sec) or greater is considered a good prospect for wind energy development.[1]

Even in a particular area, the direction, velocity, and duration of wind may be quite variable, depending on local topography and temperature differences in the atmosphere.[17] For example, wind velocity often increases over hilltops, and wind may be funneled through a mountain pass (Figure 19.12). The increase in wind velocity over a mountain is due to a vertical convergence of wind, whereas in a pass, the increase is partly due to a horizontal convergence. Because the shape of a mountain or a pass is often related to the local or regional geology, prospecting for wind energy is a geologic as well as a geographic and meteorologic problem. The primary task in evaluating the wind energy potential of a region or site is to place instruments that measure and monitor over time the strength, direction, and duration of the wind.[11]

Significant improvements in the size of windmills and the amount of power they produce occurred from the late 1800s through approximately 1950, when many European countries and the United States became interested in large-scale generators driven by the wind. In the United States, thousands of small, wind-driven generators have been used on farms. Most small windmills generate approximately 1 kW of power, which is much too small to be considered for central power generation needs. Interest in wind power declined for several decades prior to the 1970s because of

the abundance of cheap fossil fuels; in recent years, interest in building windmills has been revived.

The cost of producing electricity from wind must be competitive with other sources to be economically viable. Today, electricity produced from wind often costs less than that from natural gas and is closing in on coal. Global

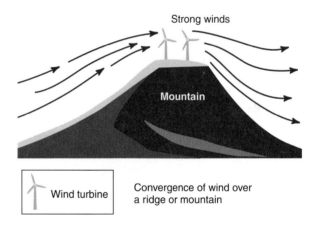

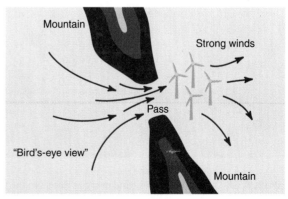

Figure 19.12 ■ Idealized diagram showing how wind energy is concentrated by topography.

wind energy was a $10 billion industry in 2004, employing over 100,000 people.[2] U.S. wind power capacity is about 7,000 MW at sites shown in Figure 19.13*a*. Individual, relatively small windmills produce from 60 to 75 kW of power and are arranged in wind farms, consisting of clusters of windmills located in mountain passes (Figure 19.13*b*). Electricity produced at these sites is connected to the general utility lines. In 1998, wind farms produced sufficient electricity to power, the city of San Francisco, making a significant contribution to the modern utility grid. Tax incentives have helped establish the wind power industry. Wind power will likely be California's second least expensive source of power by 2010, second only to hydropower. Today, wind farms in California produce about 1.5% of the state's electricity, but that percentage is increasing.

Wind energy is now used in many locations on Earth. Where land for wind farms is scarce, wind turbines can be sited offshore. Large utility companies are considering wind power in their long-range energy planning goals.

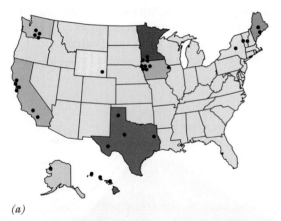

(a)

(b)

Figure 19.13 ■ (*a*) Location of major wind energy projects in the United States. [*Source:* National Renewable Energy Laboratory.] (*b*) Windmills on a wind farm near Altamont, California, a mountain pass region east of San Francisco.

European Union 71% (34,000 MW)
USA 15% (7,000 MW)
Asia 6% (3,000 MW)
Rest of world 8% (4,000 MW)

Figure 19.14 ■ Worldwide installed wind power totaled 18,000 MW at the end of 2004. [*Source:* Data from Worldwatch Institute, *Vital Signs 2005* (New York: W.W. Norton & Company, 2005).]

In 1996 the United States had the largest number of installed wind power systems, making it the world leader. However, by 2004, Germany, with over 16,000 MW installed, became the leader. The European Union produces 34,000 MW of wind power (Figure 19.14).[2] Total installed world wind power is 48,000 MW. To put this in perspective, one large fossil fuel or nuclear power plant produces about 1,000 MW.

Wind Power and the Environment

The use of wind power will not solve all our energy problems, but as one of the major alternative energy sources, wind power can be used at particular sites to reduce dependence on fossil fuel and help achieve sustainability. Wind energy does have a few disadvantages:

■ Windmills kill birds (birds of prey, such as hawks and falcons, are particularly vulnerable).

■ Large windmill farms use large areas of land for roads, pads for the windmills, and other equipment.

■ Windmills may degrade an area's scenic resources.

However, everything considered, wind energy has a relatively low environmental impact, and its continued use is certain.

Future of Wind Power

During recent years, wind energy has been growing at approximately 30% per year, nearly 10 times the growth rate of oil use. It's believed that there is sufficient wind energy in Texas, South Dakota, and North Dakota to satisfy the electricity needs of the entire United States. Wind power potential in Britain is more than twice the present demand for electricity.[2]

Consider the implications for nations such as China. China burns tremendous amounts of coal at a heavy environmental cost, including exposing millions of people to hazardous air pollution. In rural China, exposure to the smoke from burning coal in homes has increased the threat of lung cancer by a factor of nine or more. China could probably double its current capacity to generate electricity with wind alone![9]

As suggested earlier, power from wind is being taken very seriously. Although wind now provides less than 1% of the world's demand for electricity, its growth rate suggests that it could be a major supplier of power in the relatively near future. One scenario suggests that wind power could supply 10% of the world's electricity in the coming decades and, in the long run, could provide more power than hydropower, which today supplies approximately 20% of the electricity in the world.

The wind energy industry has created thousands of jobs in recent years; it is also becoming a major investment opportunity. Technology is producing more efficient wind turbines, thereby reducing the price of wind power. Today, a large state-of-the-art wind turbine is about 100 m in diameter (length of a football field) as tall as a 30 story building, and produces about 3 to 5 MW of electricity. One wind turbine provides enough power for several hundred homes. The world's largest wind turbine, at 5 MW, began operating in Germany by late 2004.[2]

In some regions, power generated from wind is price competitive with electricity produced from coal-fired and natural gas power plants. For example, one of the world's largest wind farms is on the Oregon–Washington border. The facility includes over 450 windmills sited on ridges above the Columbia River. Total power output at 300 MW is about one-third of that produced from a large fossil fuel or nuclear power plant. The cost of producing the electric energy is about 5 cents per kilowatt hour, which is competitive with electricity from burning natural gas.

19.7 Biofuels

Biofuel is energy recovered from *biomass*—organic matter, such as plant material and animal waste. Whereas the energy from most other renewable sources comes from physical processes, such as moving water and wind, the energy from biomass comes from chemical bonds formed through photosynthesis in living or once-living matter.[1] Biomass fuel is organic matter that can be burned directly or converted to

a more convenient form and then burned. For example, we can burn wood in a stove or convert it into charcoal to use as a fuel. Energy from biomass is the oldest fuel used by humans. Our Pleistocene ancestors burned wood in caves to keep warm and cook food.

Biomass continued to be a major source of energy for human beings throughout most of the history of civilization. When North America was first settled, there was more wood fuel than could be used. The forests often were cleared for agriculture by girdling trees (cutting through the bark all the way around the base of a tree) to kill them and then burning the forests.

Until the end of the nineteenth century, wood was the major fuel source in the United States. During the mid-twentieth century, when coal, oil, and gas were plentiful, burning wood became old-fashioned and quaint. Burning wood was something done for pleasure in an open fireplace that conducted more heat up the chimney than it provided for space heating. Now, with other fuels reaching a limit in abundance and production, there is renewed interest in the use of natural organic materials for fuel.

More than 1 billion people in the world today still use wood as their primary source of energy for heat and cooking. Today, in developing countries, biomass provides about 35% of the total energy supply.[1] Energy from biomass can take several routes: direct burning of biomass either to produce electricity or to heat water and air; heating of biomass to form a gaseous fuel (gasification); and distillation or processing of biomass to produce biofuels such as ethanol, biodiesel, or methane.[18]

Sources of Biofuel

Firewood is the best known and most widely used biomass fuel, but there are many others. In India and other countries, cattle dung is burned for cooking. Peat, a form of compressed dead vegetation, provides heating and cooking fuel in northern countries, such as Scotland, where it is abundant. The primary sources of biomass fuels in North America are forest products, agricultural residues, energy crops, animal manure, and urban waste (Table 19.2).

In some facilities, manure from livestock and other organic waste are converted (the technical term is *digested*) by microorganisms in specially designed digestion chambers to form methane (CH_4), which is burned to produce electricity, used in fuel cells, or used as fuel for tractors or other vehicles. Energy crops such as sugarcane are grown to be fermented to produce ethanol, a fuel that can be used in automobiles. For example, Brazil produces about 12 billion liters of ethanol per year from sugarcane. This is equivalent to about 60% of the fuel used for the nation's automobiles or about 15% of its total use of petroleum. The wastes from ethanol distillation were originally dumped in rivers, causing water pollution. The wastes are now treated to produce biogas and liquid fertilizers

Table 19.2 • Selected Examples of Biomass Energy Sources, Uses, and Products

Sources	Examples	Uses/Products	Comment
Forest products	Wood, chips	Direct burning,[a] charcoal[b]	Major source today in developing countries
Agriculture residues	Coconut husks, sugarcane waste, corncobs, peanut shells	Direct burning	Minor source
Energy crops	Sugarcane, corn, sorghum	Ethanol (alcohol),[c] gasification[d]	Ethanol is major source of fuel in Brazil for automobiles
Trees	Palm oil	Biodiesel	Fuel for vehicles
Animal residues	Manure	Methane[e]	Used to run farm machinery
Urban waste	Waste paper, organic household waste	Direct burning of methane from wastewater treatment or from landfills[f]	Minor source

[a] Principal biomass conversion.

[b] Secondary product from burning wood.

[c] Ethanol is an alcohol produced by fermentation, which uses yeast to convert carbohydrates into alcohol in fermentation chambers (distillery).

[d] Biogas from gasification is a mixture of methane and carbon dioxide produced by pyroclytic technology, which is a thermochemical process that breaks down solid biomass into an oil-like liquid and almost pure carbon char.

[e] Methane is produced by anaerobic fermentation in a digester.

[f] Naturally produced in landfills by anaerobic fermentation.

(recycled to sugarcane fields).[1] In Hawaii, sugarcane waste is burned to produce electricity.

Methane and biogas can also be produced from urban waste in landfills and sewage at wastewater treatment plants. In rural China, about 5 million small facilities are used to treat sewage. The original purpose was to reduce disease, but their potential as an energy source was soon recognized and used.[1]

Today, a number of facilities in the United States process urban waste to generate electricity or to be used as a fuel. If all such power plants were operating at full capacity, about 15% of the country's waste, or 35 million metric tons per year, could be burned to extract energy.[19] The United States has been slower than other countries to utilize urban waste as an energy source. For example, in Western Europe a number of countries utilize from one-third to one-half of their municipal waste for energy production. With the end of inexpensive, available fossil fuels, additional energy recovery systems utilizing urban waste will no doubt emerge.[20]

Biofuel and the Environment

The use of biofuels can pollute the air and degrade the land. For most of us, the smell of smoke from a single campfire is part of a pleasant outdoor experience; but under certain weather conditions, wood smoke from many campfires or chimneys in narrow valleys can lead to air pollution.

The use of biomass as fuel places pressure on an already heavily used resource. A worldwide shortage of firewood is adversely affecting natural areas and endangered species.

For example, the need for firewood has threatened the Gir Forest in India, the last remaining habitat of the Indian lion (not to be confused with the Indian tiger). The world's forests will also decrease if our need for forest products and forest biomass fuel exceeds the productivity of the forests.

If our forest and crop resources are managed properly (for sustainability), it may be possible to make biofuels more attractive. Current estimates for the United States suggest that many millions of hectares of land that are unsuitable for food production could be used for growing biomass crops (trees and other plants) using short rotation times (time between harvests). However, forest plantations would have to be managed for sustainability, because deforestation accelerates the process of soil erosion (soils without vegetation cover erode more quickly). When fine-grained (silt and clay) particles from soil erosion enter streams and rivers, the water becomes muddy, and its quality is degraded.

Advocates of using biomass for energy point out that, assuming areas harvested for bioenergy are replanted, the burning of biofuels contributes no net carbon dioxide emissions to the atmosphere. Replacing fossil fuels with sustainably produced biomass will actually reduce net CO_2 emissions, according to some estimates.[21] Finally, combustion of biomass-derived fuels generally releases fewer pollutants such as sulfur dioxide and nitrogen oxides than combustion of coal and gasoline. This is not always the case for burning urban waste, however. Although plastics, glass, and hazardous materials are removed before burning, some inevitably slip through the sorting process and are burned, releasing air pollutants, including heavy metals. Burning urban waste to recover energy is preferable to landfill disposal. However, incineration of waste paper competes with recycling, which is preferable to disposal or burning.

Future of Biofuel

Biomass energy in its various forms appears to have a future as an energy source. At the global scale, recoverable biomass energy is about equivalent to present energy consumption.[1] Biomass energy is also flexible in that solid biomass can be stored, as can biogas and biofuels. This allows biomass to be used as needed to meet demands.[1] However, questions remain about the amount of energy biomass can reasonably provide and the rate of depletion. Biomass energy is renewable, but it is limited and is dependent on favorable environmental conditions for renewal. Any use of biomass fuel must be part of the general planning for all uses of the land's resources.[22]

Biofuel could someday help free the United States from importing oil for gasoline. There is now sufficient idle farmland to produce enough ethanol to meet fuel needs for most, U.S. automobiles and light-duty trucks.[3] However, at present, this is not practical for economic reasons, and automobile engines would need modifications to run smoothly on pure ethanol.

It has been estimated that in the United States by 2030, biofuel crops could provide about 100,000 MW of electricity (12% of U.S. capacity), equivalent to 100 large coal-burning or nuclear power plants. Raising the necessary crops would require about 20 million hectares (50 million acres) of the 50 million hectares (128 million acres) that are expected to be idle in 2030. Today, biomass in the United States is a small but growing renewable source of electric power.[3] Sweden is a leader in biomass energy, with 17% of its primary energy consumption from biomass. The pulp and paper industry of Sweden provides much of the source for this fuel.[1]

19.8 Geothermal Energy

Geothermal energy is natural heat from the interior of the Earth that is converted to heat buildings and generate electricity. The idea of harnessing Earth's internal heat is not new. As early as 1904, geothermal power was used in Italy. Today, Earth's natural internal heat is being used to generate electricity in 21 countries, including Russia, Japan, New Zealand, Iceland, Mexico, Ethiopia, Guatemala, El Salvador, the Philippines, and the United States. Total worldwide production is approaching 9,000 MW (equivalent to nine large modern coal-burning or nuclear power plants)—double the amount in 1980. Some 40 million people today receive their electricity from geothermal energy at a cost competitive with that of other energy sources.[23] In El Salvador, geothermal energy is supplying 25% of the total electric energy used. However, at the global level, geothermal energy supplies less than 0.15% of the total energy supply.[1]

Geothermal energy may be considered a nonrenewable energy source when rates of extraction are greater than rates of natural replenishment. However, geothermal energy has its origin in the natural heat production within Earth, and only a small fraction of the vast total resource base is being utilized today. Although most geothermal energy production involves the tapping of high-heat sources, people are also using the low-temperature geothermal energy of groundwater in some applications.

Geothermal Systems

The average heat flow from the interior of the Earth is very low, about 0.06 W/m^2. This amount is trivial compared with the 177 W/m^2 from solar heat at the surface in the United States. However, in some areas, heat flow is sufficiently high to be useful for producing energy.[1] For the most part, areas of high-heat flow are associated with plate tectonic boundaries (see Chapter 5). Oceanic ridge systems (divergent plate boundaries) and areas where mountains are being uplifted and volcanic island arcs are forming (convergent plate boundaries) are areas where this natural heat flow is anomalously high. One such region is located in the western United States, where recent tectonic and volcanic activity has occurred.[23, 24]

On the basis of geologic criteria, several types of hot geothermal systems (with temperatures greater than about 80°C, or 176°F) have been defined, and the resource base is larger than that of fossil fuels and nuclear energy combined.

Lower temperature geothermal sources that cannot be used to produce electricity may be used for space heating of buildings, heating swimming pools, or heating soil to assist crop production in greenhouses. Such systems are extensively used in Iceland. In the United States, 18 communities, including Boise Idaho and Klamath Falls, Oregon, have geothermal heating systems.[25]

A common system for energy development is hydrothermal convection, characterized by the circulation of steam and/or hot water that transfers heat from depths to the surface. An example is the Geysers Geothermal Field, a geothermal system 145 km (90 miles) north of San Francisco, where about 1,000 MW of electrical energy is produced. The Geysers is the largest geothermal power operation in the world (Figure 19.15). At the Geysers the hot water is maintained in part by injecting treated wastewater from urban areas into hot rocks. The heated wastewater vaporizes and is extracted from geothermal production wells to produce electricity. The urban wastewater helps sustain the production of electricity at the Geysers geothermal power facility.[25]

Yellowstone National Park is an area with a significant potential for geothermal energy. The park contains about 10,000 hot springs and geysers (including the famous Old Faithful geyser). However, development in this area presents problems. Federal legislation has been passed to

protect the hot springs and geysers of Yellowstone. What is required to ensure protection is not clear. Certainly, though, there must be an adequate buffer zone to ensure that any geothermal energy development near the park will not damage Yellowstone's geothermal features. In other words, we want to make sure that geothermal energy use outside of the park will not "shut off" Old Faithful or other hot springs or geysers.

It may come as a surprise to learn that most groundwater can be considered a source of geothermal energy. It is geothermal because the normal internal heat flow from Earth keeps the temperature of groundwater at a depth of 100 m (320 ft) at about 13°C (55°F). Water at 13°C is cold for a shower, but compared with winter temperatures in much of the United States, it is warm and can help heat a house. Furthermore, compared with summer temperatures of 30–35°C, groundwater at 13°C is cool and can be used to provide air conditioning. In the summer, heat can be transferred from the warm air in a building to the cool groundwater. In the winter, when the outdoor temperature is below about 4°C (40°F), heat can be transferred from the groundwater to the air in the building, reducing the heating needed from other sources. Technology for such heat transfer is well known and available.

Geothermal Energy and the Environment

The environmental impact of geothermal energy is not as extensive as that of other sources of energy. Production of geothermal energy only releases about 12% of the carbon dioxide and sulfur dioxide released from burning coal to produce a given amount of electricity.[25] When geothermal energy is developed at a particular site, environmental problems include on-site noise, emissions of gas, and disturbance of the land at drilling sites, disposal sites, roads and pipelines, and power plants. Development of geothermal energy does not require large-scale transportation of raw materials or refining of chemicals, as development of fossil fuels does. Furthermore, geothermal energy does not produce the atmospheric pollutants associated with burning fossil fuels or the radioactive waste associated with nuclear energy. However, geothermal development often does produce considerable thermal pollution from hot wastewaters, which may be saline or highly corrosive.[24]

Geothermal power is not always popular. For instance, geothermal energy has been produced for years on the island of Hawaii, where active volcanic processes provide abundant near-surface heat. There is controversy, however, over further exploration and development. Native Hawaiians and others have argued that the exploration and development of geothermal energy degrade the tropical forest as developers construct roads, build facilities, and drill wells. In addition, religious and cultural issues in Hawaii relate to the use of geothermal energy. For example, some people are offended by the use of the "breath and water of Pele" (the volcano goddess) to make electricity. This issue points out the importance of being sensitive to the values and cultures of people where development is planned.

Future of Geothermal Energy

At present, the United States produces only 2,800 MW of geothermal energy.[26] However, if developed, known geothermal resources in the United States could produce about 20,000 MW, which is about 10% of the electricity needed for the western states.[23] Geohydrothermal resources not yet discovered could conservatively provide four times that amount (approximately 10% of total U.S. electric capacity), about equivalent to the electricity produced from water power today.[3]

19.9 Policy Issues

Renewable energy requires a broad energy policy that takes the long-term view. Policy issues and goals for alternative renewable energy include the following:[1]

- Encouraging and funding alternative energy research and development.
- Ensuring fair access to alternative energy sources.
- Developing a fiscal framework that takes into account the external costs of fossil fuels (air pollution and health costs) and the benefits of alternative energy sources (less environmental damage).
- Developing planning initiatives that promote environmentally preferred and socially acceptable alternative energy.
- Establishing industrial training programs to retrain employees in alternative energy industries, especially in rural agricultural regions and former coal mining communities.
- Encouraging funding institutions to support community investment in alternative energy.

These policy issues and goals are important in discussions concerning the future of alternative renewable energy sources. Before oil production declines, we need to rethink our entire energy planning strategy. Phasing in alternative renewable energy in time to meet our future energy demands is critical to avoiding a looming energy crisis in the years to come.

Summary

- The use of renewable alternative energy sources, such as wind and solar energy, is growing rapidly. These energy sources do not cause air pollution, create health problems, or cause climate changes. They offer our best chance to replace fossil fuels and develop a sustainable energy policy.
- Passive solar energy systems often involve architectural designs that enhance absorption of solar energy without requiring mechanical power or moving parts.
- Some active solar energy systems use solar collectors to heat water for homes. Systems to produce heat or electricity include power towers and solar farms.
- Photovoltaics is a technology that converts sunlight directly into electricity. Photovoltaic systems use solar cells for a variety of purposes, such as powering remote equipment. Experiments are ongoing to evaluate photovoltaic power plants. This emerging technology remains expensive.
- Hydrogen gas may be an important fuel of the future, especially when used in fuel cells.
- Water power today provides about 10% of the total electricity produced in the United States. Most good sites for large dams have already been utilized. Water power is clean, but there is an environmental price to pay in terms of disturbance of ecosystems, sediment trapped in reservoirs, loss of wild rivers, and loss of productive land.
- Wind power has tremendous potential as a source of electrical energy in many parts of the world. Many utility companies are presently experimenting with or using wind power as part of energy production or as part of long-term energy planning. Environmental impacts include loss of land to wind farms and killing of birds, as well as degradation of scenic resources.
- Biofuel is the oldest human fuel. More than 1 billion people still use wood as their primary source of energy. The environmental impacts of burning wood include deforestation, soil erosion, water pollution, and air pollution. Urban waste is another source of biofuel. However, burning organic urban waste can release toxins and cause air pollution. Two biofuels (ethanol and biodiesel) have the potential to substitute for a significant amount of oil and gas in coming decades to fuel vehicles. Conversion of crops to ethanol and biodiesel can also produce pollution but, considering total costs, may not be economically viable for some applications.
- Geothermal energy is natural heat from Earth's interior used as an energy source. The environmental effects of developing geothermal energy relate to specific site conditions and the type of heat utilized (steam, hot water, or warm water). Geothermal energy may involve disposal of saline or corrosive waters, as well as on-site noise, emission of gas, and industrial scars.
- Significant policy issues are related to the future of alternative renewable energy. These include financing of research and development, access to energy sources, education, and community involvement.

How Can We Evaluate Alternative Energy Sources?

The world is moving into a new era, one of transition from almost total dependence on fossil fuels to greater use of alternative renewable sources of energy. Although each of the alternatives offers a way out of the energy dilemma created by population growth and technological development, each has advantages and disadvantages. How can we evaluate the alternatives and select the right mix of energy sources for the coming decades? We can begin by comparing them on the basis of those characteristics most important to us: cost, jobs lost or gained, environmental impact, and potential for supplying energy.

Critical Thinking Questions

1. Using what you have learned about alternative energy in this chapter and elsewhere, evaluate the environmental impacts of the energy sources listed in the accompanying table. Complete the last column of the table. You may wish to subdivide the column into advantages and disadvantages.

2. Using the numbers 1–10, where 10 represents the best and 1 the worst, assign a rating to each value in the table. For example, in the column for carbon reduction, you might assign a rating of 10 to wind because it results in 100% reduction of carbon emissions. Solar

thermal energy would then receive a rating of 8.4. In rating environmental impact, you will have to use your judgment in assigning numerical values.

3. One way to evaluate the various alternatives would be to add up the rating scores for each energy source and see which ones received the highest score. However, you may feel that some of the characteristics are more important than others and therefore should be weighted more heavily. Assign a weight to each column of the table, taking into consideration the importance you believe each should have in decision making. For example, if you believe that costs are more important than land used, you will assign a higher value to costs. In order to be able to compare your evaluation with those of your classmates, use decimal fractions for the weights, such as 0.2. The total should add up to 1.0.

4. Now, for each energy source, multiply its rating in each column by the weight you have assigned to the column. What is the total weighted score for each energy source? What are the sources in order of score, from highest to lowest?

5. Based on this analysis, what policy and research recommendations would you make to the U.S. government concerning alternative sources of energy?

Energy Source	U.S. Recoverable Resource[a] (exajoule/yr)	Costs in 1998 Cents[b] (per KWh) 1988	2000	Land Use[c] (m^2/GWh for 30 years)	Carbon Reduction (%)	Carbon Avoidance Cost[d]($/ton)	Number of Jobs[e] (per thousand GWh/yr)	Environmental Impact
Wind	10–40	8	5	1,355	100	95	542	
Geothermal	Small	4	4	404	99	110	112	
Photovoltaic	35	30	10	3,237	100	819	—	
Solar thermal	65	8	6	3,561	84	180	248	
Biomass	13–26	5	NA	—	100[f]	125	—	
Combined-cycle coal	—	6[g]	—	3,642	10	954	116	
Nuclear	—	15[h]	—	—	86	535	100	

[a]*Recoverable resource* is a measure of how much of the energy can be captured or exploited. From M. Brower, *Cool Energy* (Washington, D.C.: Union of Concerned Scientists, 1990), p. 19.

[b]L. R. Brown, C. Flavin, and S. Postel, *Saving the Planet* (New York: Norton, 1991), p. 27.

[c]Ibid., p. 60.

[d]Based on comparison with existing coal-fired plants. From C. Flavin, "Slowing Global Warming," in *State of the World* (New York: W.W. Norton, 1990), p. 27.

[e]Brown et al., p. 62.

[f]Assumes that the amount of carbon dioxide released in combustion will be consumed by replanted vegetation.

[g]C. Flavin, "Building a Bridge to a Sustainable Future," in *State of the World* (New York: Norton, 1992), p. 35.

[h]A. K. Reddy and J. Goldenberg, "Energy for the Developing World," *Scientific American* 263 (3)(1990):116.

REEXAMINING THEMES AND ISSUES

Human Population

As the human population continues to increase, so does total world demand for and consumption of energy. Environmental problems related to increased consumption of fossil fuels could be minimized through controlling human population, increasing conservation efforts, and using those alternative renewable energy sources that do not harm the environment.

Sustainability

Use of fossil fuels is not sustainable. In order to plan for sustainability from an energy perspective, we need to rely more on alternative energy sources that are naturally renewable and do not pollute the atmosphere or otherwise damage the environment. To do otherwise is antithetical to the concept of sustainability.

Global Perspective

Evaluation of the potential of alternative energy sources requires us to understand global Earth systems and identify regions that are likely to produce high-quality alternative energy that could be utilized in urban regions of the world.

Urban World

Alternative renewable energy sources have a future in our urban environment. For example, the roofs of buildings can be used for solar collectors or photovoltaic systems. Patterns of energy consumption can be regulated through use of innovative systems such as pump storage to augment production of electrical energy when demands are high in urban areas.

People and Nature

Alternative energy sources such as solar and wind are perceived by many environmentalists as being linked more closely with nature than fossil fuels, nuclear energy, or even water power. This is because solar and wind energy development requires less human modification of the environment. Solar and wind energy allow us to live more in harmony with the environment, and thus we feel more connected to the natural world.

Science and Values

We are seriously considering alternative energy today because we value environmental quality. Recognizing that burning fossil fuels creates many serious environmental problems, we are trying to increase our scientific knowledge and improve our technology to produce energy supplies to meet our needs for the future while minimizing environmental damage. Our present science and technology can lead to a sustainable energy future, but we will need to change our values and our behavior to achieve it.

Key Terms

active solar energy
    systems **389**
alternative energy **386**
biofuel **399**
fuel cells **392**

geothermal energy **401**
nonrenewable
    energy **386**
passive solar energy
    systems **388**

photovoltaics **390**
renewable energy **386**
solar collectors **389**
solar energy system **392**
tidal power **396**

water power **395**
wind power **396**

Study Questions

1. What type of government incentives could be used to encourage use of alternative energy sources? Would their widespread use impact our economic and social environment?

2. Your town is near a large river that has a nearly constant water temperature of about 15°C (60°F). Could the water be used to cool buildings in the hot summers? How? What would be the environmental effects?

3. Which has greater future potential for energy production, wind or water power? Which causes more environmental problems? Why?

4. What are some of the problems associated with producing energy from biomass?

5. It is the year 2500, and natural oil and gas are rare curiosities that people see in museums. Given the technologies available today, what would be the most sensible fuel for airplanes? How would this fuel be produced to minimize adverse environmental effects?

6. When do you think the transition from fossil fuels to other energy sources will (or should) occur? Defend your answer.

Further Reading

Ahmed, K. 1994. "Renewable Energy Technologies: A Review of the Status and Costs of Selected Technologies." World Bank Technical Paper No. 240. Discussion of the relative costs of energy production with biomass, solar, and photovoltaic energy sources compared with traditional sources and how these technologies have and will continue to become more competitive with oil and coal.

Caleira, Ken, Atul Jain, and Martin Hoffert, 2003. "Climate Sensitivity Uncertainty and the Need for Energy without CO_2 Emission." *Science* 299: 2052–2054. A good in-depth discussion of climate change, energy needs, and possibilities for alternative energy sources.

Chartier, P., A. A. C. M. Beenackers, and G. Grassi, eds. 1995. *Biomass for Energy, Environment, Agriculture and Industry*. Proceedings of the 8th European Communities Conference, Vienna, Austria, October 3–5, 1994. London: Pergamon. A valuable, up-to-date summary of the status and promise of biomass use, including papers covering policy issues, assessments of the resource base, economic assessments, environmental impacts, and commercialization issues.

Gipe, P. 1995. *Wind Energy Comes of Age*. New York: John Wiley & Sons.

Hellemans, Alexander. 1999. "Solar Homes for the Masses." *Science* 285: 679. A look at the feasibility of solar homes with indications of the progress made to date.

Sheffield, J. 1997. "The Role of Energy Efficiency and Renewable Energies in the Future and World Energy Market." *Renewable Energy* 10(2): 315–318.

Turner, John. 1999. "A Realizable Renewable Energy Future." *Science* 285: 687–689. A good analytical treatment of the potential of alternative energy sources, particularly hydrogen.

U.S. Department of Energy. 1993. *Electricity from Biomass—Renewable Energy Today and Tomorrow*. Washington, D.C.: Solar Thermal Biomass Power Division, Office of Solar Energy Conversion. A brief introduction to the potential for biomass-generated electricity in the United States.

CHAPTER 20

Nuclear Energy and the Environment

Three Mile Island nuclear power plant in the mid 1990s, about 15 years following an accident at the plant.

Learning Objectives

As one of the alternatives to fossil fuels, nuclear energy generates a lot of controversy. After reading this chapter, you should understand:

- What nuclear fission is and what the basic components of a nuclear power plant are.

- What nuclear radiation is and what the three major types are.

- Why it is important to know the type of radiation and the half-life for a particular radioisotope.

- What the basic parts of the nuclear fuel cycle are, and how each is related to our environment.

- How radioisotopes affect the environment and the major pathways of radioactive materials in the environment.

- What the breeder reactor is, and why it is important for the future of nuclear energy.

- What the relationships between radiation doses and health are.

- What we have learned from accidents at nuclear power plants.

- How we might safely dispose of high-level radioactive materials.

- What the future of nuclear power is likely to be.

- That nuclear power produces electric energy without emitting air pollution or contributing to global warming.

CASE STUDY

Nuclear Energy and Public Opinion

In 1953, Dwight D. Eisenhower, a popular World War II general and U.S. president, described an optimistic vision of the future of atomic power, which he called "atoms for peace." In a 1954 address to science writers, the chairman of the Atomic Energy Commission, predicted that nuclear-powered generators would provide power so cheap, so nearly unlimited, and so clean that we would never need to meter it. More than 100 nuclear power plants were constructed in the next several decades, and orders had been placed for more than a hundred additional plants by 1978.

But the public mood changed in the early 1980s. No more nuclear reactors were ordered; today, there are still only 103 commercial nuclear power plants in the United States. Polls suggest that the majority of Americans have been opposed to construction of nuclear power plants since the early 1980s. The reversal of opinion occurred three years after the Three Mile Island accident and four years before the Chernobyl accident (discussed later in this chapter). People's anxiety concerning nuclear weapons and nuclear war peaked in the early 1980s at about the time attitudes changed toward constructing more nuclear power plants, suggesting that worry over nuclear weapons had spilled over onto commercial nuclear power.[1]

Public support for building new nuclear plants is increasing today, especially in the western United States. The California energy crisis caused many people to rethink the value of nuclear energy. In a May 2003 survey, about 50% of U.S. adults interviewed agreed that we should build more nuclear energy plants in the future.[2] It appears that nuclear power is at a crossroads. Polls like the one just mentioned suggest that many Americans believe nuclear power should and will be an important energy source in the future. However, many people still oppose construction of new nuclear power plants in the near future and remain uncomfortable with issues of safety. The Energy Policy Act of 2005 promotes nuclear energy and recommends that the United States start the process of building new nuclear power plants by 2010.

An important question is this: If nuclear energy were phased out, would it make any significant difference? The answer depends on whom you listen to. To some people, the answer is no. They argue that nuclear energy is too expensive and makes up a relatively small part of the U.S. energy package; only minor substitution of other sources would be required to compensate for its loss.[3] Others point out, however, that nuclear energy is supplying about 20% of our electricity and may be absolutely necessary if we are to sustain our environment.

Nuclear energy is one of several technologies that may eventually replace fossil fuels. Therefore, nuclear power, which does not contribute to global warming, may again be viewed as a major source of energy, particularly if issues of cost, availability of nuclear fuel, safety, and storage of waste can be resolved.[3]

The discussion of public opinion about nuclear power suggests a dilemma for the United States. Although many people remain uneasy about its use, nuclear power cannot be ignored as an alternative to fossil fuels. This chapter explores nuclear power reactors, radiation, accidents, waste management, and the future of nuclear power.

20.1 Nuclear Energy

Nuclear energy is the energy of the atomic nucleus. Two nuclear processes can be used to release that energy to do work: fission and fusion. Nuclear **fission** is the splitting of atomic nuclei, and nuclear **fusion** is the fusing, or combining, of atomic nuclei. A by-product of both fission and fusion reactions is the release of enormous amounts of energy. (You may wish to review the discussion of matter and energy in A Closer Look 5.1.)

Nuclear energy for commercial use is produced by the splitting of atoms in **nuclear reactors**, which are devices that produce controlled nuclear fission. In the United States, almost all of these reactors use a form of uranium oxide as fuel. Nuclear fusion is not yet used commercially, although it has been accomplished in experimental fusion reactors.

Fission Reactors

The first human-controlled nuclear fission, demonstrated in 1942 by Italian physicist Enrico Fermi at the University of Chicago, led to the development of nuclear energy to produce electricity. Today, in addition to power plants to supply electricity to homes and industry, nuclear reactors power submarines, aircraft carriers, and icebreaker ships. Russia is building ships that contain reactors to provide electricity to coastal towns, and the United States is designing reactors for deep-space missions.

Nuclear fission produces much more energy than other sources, such as burning fossil fuels. One kilogram (2.2 lb) of uranium oxide produces heat equivalent to approximately 16 metric tons of coal, making uranium an important source of energy in the United States and the world.

Three types, or isotopes, of uranium occur in nature: uranium-238, which accounts for approximately 99.3% of all natural uranium; uranium-235, which makes up about 0.7%; and uranium-234, which makes up about 0.005%. Uranium-235 and uranium-238 are two radioactive isotopes of uranium. However, uranium-235 is the only naturally occurring fissionable (or *fissile*) material; therefore, it is essential to the production of nuclear energy. Processing uranium (called *enrichment*) to increase the concentration of uranium-235 from 0.7% to about 3% produces enriched uranium, which is used as fuel for the fission reaction. Radiation is explained and related terms are defined in A Closer Look 20.1.

Fission reactors split uranium-235 by neutron bombardment (Figure 20.1). The reaction produces neutrons, fission fragments, and heat. The released neutrons strike other uranium-235 atoms, releasing more neutrons, fission products, and heat. The neutrons released are fast moving and must be slowed down slightly, or *moderated*, to increase the probability of fission. In *light*

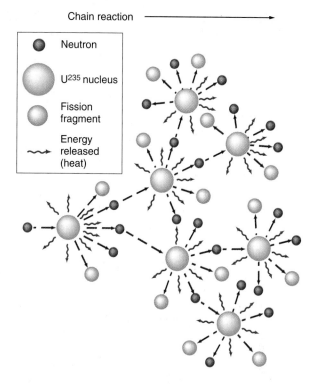

Chain reaction ⟶

Neutron

U235 nucleus

Fission fragment

Energy released (heat)

Figure 20.1 ■ Fission of uranium-235. A neutron strikes the U-235 nucleus, producing fission fragments and free neutrons and releasing heat. The released neutrons may then strike other U-235 atoms, releasing more neutrons, fission fragments, and energy. As the process continues, a chain reaction develops.

water reactors, the kind most commonly used in the United States, ordinary water is used as the moderator. As the process continues, a chain reaction develops as more and more uranium is split, releasing more neutrons and more heat.

Most reactors now in use consume more fissionable material than they produce and are known as **burner reactors**. The reactor is part of the nuclear steam supply system, which produces steam to run the turbine generators that produce the electricity.[4] Therefore, the reactor has the same function as the boiler that produces the heat in coal-burning or oil-burning power plants (Figure 20.4).

Figure 20.5 shows the main components of a reactor: the core (consisting of fuel and moderator), control rods, coolant, and reactor vessel. The core of the reactor is enclosed in the heavy, stainless steel reactor vessel; then, for extra safety and security, the entire reactor is contained in a reinforced concrete building.

In the reactor core, fuel pins, consisting of enriched uranium pellets in hollow tubes (3–4 m long and less than 1 cm, or 0.4 in., in diameter) are packed together (40,000 or more in a reactor) in fuel subassemblies. A minimum fuel concentration is necessary to keep the reactor *critical*—that is, to achieve a self-sustaining chain reaction.

Radioactive Decay

To many people, radiation is a subject shrouded in mystery. They feel uncomfortable with it, learning from an early age that nuclear energy may be dangerous because of radiation and that nuclear fallout from the detonation of atomic bombs can cause widespread human suffering. One thing that makes radiation scary is that we cannot see it, taste it, smell it, or feel it. In this feature, we try to demystify some aspects of radiation by discussing the natural process of radiation, or radioactivity.

First, we need to understand that radiation is a natural process that has been going on since the creation of the universe. Understanding the process of radiation involves understanding the *radioisotope*, which is a form of a chemical element that spontaneously undergoes **radioactive decay**. During the decay process, the radioisotope changes from one isotope to another and emits one or more forms of radiation.

You may recall from Chapter 5 that isotopes are atoms of an element that have the same atomic number (the number of protons in the nucleus) but that vary in atomic mass number (the number of protons plus neutrons in the nucleus). For example, two isotopes of uranium are $^{235}U_{92}$ and $^{238}U_{92}$. The atomic number for both isotopes of uranium is 92 (revisit Table 5.1); however, the atomic mass numbers are 235 and 238. The two different uranium isotopes may be written as uranium-235 and uranium-238 or ^{235}U and ^{238}U.

An important characteristic of a radioisotope is its *half-life*, the time required for one-half of a given amount of the isotope to decay to another form. For example, uranium-235 has a half-life of 700 million years, a very long time indeed! Radioactive carbon-14 has a half-life of 5,570 years, which is in the intermediate range, and radon-222 has a relatively short half-life of 3.8 days. Other radioactive isotopes have even shorter half-lives; for example, polonium-218 has a half-life of

about 3 minutes. Still other radioactive isotopes have half-lives as short as a fraction of a second.

The main point here is that each radioactive isotope has its own unique and unchanging half-life. Isotopes with very short half-lives are present for only a brief time, whereas those with long half-lives remain in the environment for long periods. Table 20.1 illustrates the general pattern for decay in terms of the elapsed half-lives and the fraction remaining. For example, suppose we start with 1 g polonium-218 with a half-life of approximately 3 minutes. After an elapsed time of 3 minutes, 50% of the polonium-218 remains. After 5 elapsed half-lives, or 15 minutes, only 3% is still present; and after 10 elapsed half-lives (30 minutes), 0.1% is still present. Where has the polonium gone? It has decayed to lead-214, another radioactive isotope, which has a half-life of about 27 minutes. The progression of changes associated with the decay process is often known as a *radioactive decay chain*. Now suppose we had started with 1 g uranium-235, with a half-life of 700 million years. Following

Table 20.1 • Generalized Pattern of Radioactive Decay		
Elapsed Half-life	Fraction Remaining	Percent Remaining
0	—	100
1	1/2	50
2	1/4	25
3	1/8	13
4	1/16	6
5	1/32	3
6	1/64	1.5
7	1/128	0.8
8	1/256	0.4
9	1/512	0.2
10	1/1024	0.1

10 elapsed half-lives, 0.1% of the uranium would be left—but this process would take 7 billion years.

Radioisotopes with short half-lives initially undergo a more rapid rate of change (nuclear transformation) than do radioisotopes with long half-lives. Conversely, radioisotopes with long half-lives have a less intense and slower initial rate of nuclear transformation but may be hazardous for a much longer time.[7]

There are three major kinds of nuclear radiation: *alpha particles*, *beta particles*, and *gamma rays*. Alpha particles consist of two protons and two neutrons (a helium nucleus) and have the greatest mass of the three types of radiation (Figure 20.2a). Because alpha particles have a relatively high mass, they do not travel far. In air, alpha particles can travel approximately 5–8 cm (about 2–3 in) before they stop. However, in living tissue, which is much denser than air, they can travel only about 0.005–0.008 cm (0.002–0.003 in). Because this is a very short distance, to cause damage to living cells, alpha particles must originate very close to the cells. Alpha particles can be stopped by a sheet or so of paper.

Beta particles are electrons and have a mass of 1/1,840 of a proton. Beta decay occurs when one of the protons or neutrons in the nucleus of an isotope spontaneously changes. What happens is that a proton turns into a neutron or a neutron is transformed into a proton (Figure 20.2b). As a result of this process, another particle, known as a neutrino, is also ejected. A neutrino is a particle with no rest mass (the mass when the particle is at rest with respect to an observer).[8] Beta particles travel farther through air than the more massive alpha particles but are blocked by even moderate shielding, such as a thin sheet of metal (aluminum foil) or a block of wood.

The third and most penetrating type of radiation comes from *gamma decay*. When gamma decay occurs, a gamma ray, a type

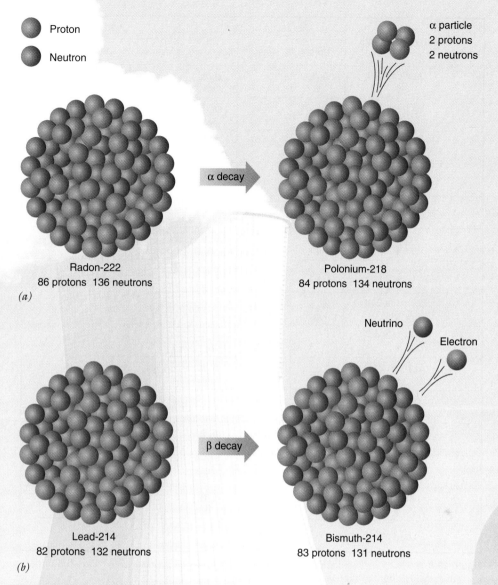

Proton

Neutron

α particle
2 protons
2 neutrons

Radon-222
86 protons 136 neutrons

(a)

α decay

Polonium-218
84 protons 134 neutrons

Neutrino

Electron

Lead-214
82 protons 132 neutrons

(b)

β decay

Bismuth-214
83 protons 131 neutrons

Figure 20.2 ■ Idealized diagrams showing (*a*) alpha and (*b*) beta decay processes. [*Source:* D. J. Brenner, *Radon: Risk and Remedy* (New York: Freeman, 1989). Copyright 1989 by W. H. Freeman and Company. Reprinted with permission.]

of electromagnetic radiation, is emitted from the isotope. Gamma rays are similar to X rays but are more energetic and penetrating; they travel the longest average distance of all types of radiation. Protection from gamma rays requires thick shielding, such as about a meter of concrete or several centimeters of lead.

Each radioisotope has its own characteristic emissions; some isotopes emit only one type of radiation, whereas others emit a mixture. In addition, the different types of radiation have different toxicities

(degrees or intensities of potential to harm or poison). In terms of human health and the health of other organisms, alpha radiation is most toxic or dangerous when inhaled or ingested. Because alpha radiation is stopped within a very short distance by living tissue, much of the damaging radiation is absorbed by the tissue. When alpha-emitting isotopes are stored in a container, however, they are relatively harmless. Beta radiation is intermediate in its toxicity, although most beta radiation is absorbed by the body when a beta

emitter is ingested. Gamma emitters are toxic and dangerous inside or outside the body; but when they are ingested, some of the radiation passes outside the body.

Some radioisotopes, particularly those of very heavy elements such as uranium, undergo a series of radioactive decay steps (a decay chain) before finally reaching a stable nonradioactive isotope. For example, uranium decays through a series of steps to the stable nonradioactive isotope of lead. A decay chain for uranium-238 (with a half-life of 4.5 billion years) to

Radioactive Elements	Radiation Emitted			Half-life		
	Alpha	Beta	Gamma	Minutes	Days	Years
Uranium-238 ↓	☢		☢			4.5 billion
Thorium-234 ↓		☢	☢		24.1	
Protactinium-234 ↓		☢	☢	1.2		
Uranium-234 ↓	☢		☢			247,000
Thorium-230 ↓	☢		☢			80,000
Radium-226 ↓	☢		☢			1,622
Radon-222 ↓	☢				3.8	
Polonium-218 ↓	☢	☢		3.0		
Lead-214 ↓		☢	☢	26.8		
Bismuth-214 ↓		☢	☢	19.7		
Polonium-214 ↓	☢			0.00016 (sec)		
Lead-210 ↓		☢	☢			22
Bismuth-210 ↓		☢			5.0	
Polonium-210 ↓	☢		☢		138.3	
Lead-206	None			Stable		

Figure 20.3 ■ Uranium-238 decay chain. [*Source:* F. Schroyer, ed., *Radioactive Waste*, 2nd printing (American Institute of Professional Geologists, 1985).]

stable lead-206 is shown in Figure 20.3. Also listed are the half-lives and types of radiation that occur during the transformations. Note that the simplified radioactive decay chain shown in Figure 20.3 involves 14 separate transformations and includes several environmentally important radioisotopes, including radon-222, polonium-218, and lead-210. The decay from one radioisotope to another is often stated in terms of parent and daughter products. For example, uranium-238, with a half-life of 4.5 billion years, is the parent of daughter product thorium-234.

To sum up, when considering radioactive decay, two important facts to remember are: (1) the half-life and (2) the type of radiation emitted.

A stable fission chain reaction in the core is maintained by controlling the number of neutrons that cause fission. The control rods, which contain materials that capture neutrons, are used to regulate the chain reaction. As the control rods are moved out of the core, the chain reaction increases; as they are moved into the core, the reaction slows. Full insertion of the control rods into the core stops the fission reaction.[5]

The function of the coolant is to remove the heat produced by the fission reaction. This is an important point—the rate of generation of heat in the fuel *must match* the rate at which heat is carried away by the coolant. All major nuclear accidents have occurred when something went wrong with the balance and heat built up in the reactor core.[6] A **meltdown** generally refers to a nuclear accident in which the nuclear fuel becomes so

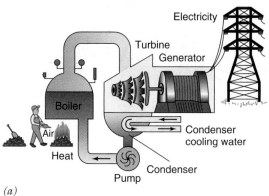

(a)

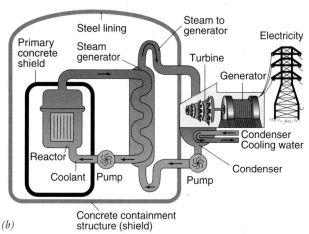

(b)

Figure 20.4 ■ Comparison of (*a*) a fossil fuel power plant and (*b*) a nuclear power plant with a boiling water reactor. Notice that the nuclear reactor has exactly the same function as the boiler in the fossil fuel power plant. The coal-burning plant (*a*) is Ratcliffe-on-Saw, located in Nottinghamshire, England, and the nuclear power station (*b*) is located in Leibstadt, Switzerland. [*Source:* American Nuclear Society, *Nuclear Power and the Environment,* 1973.]

hot that it forms a molten mass that breaches the containment of the reactor and contaminates the outside environment with radioactivity.

Other parts of the nuclear steam supply system are the primary coolant loops and pumps, which circulate the coolant through the reactor, extracting heat produced by fission, and heat exchangers or steam generators, which use the fission-heated coolant to make steam (Figure 20.4*a*). In light water reactors, water is used as the coolant as well as the moderator.

A design philosophy has emerged in the nuclear industry to build less complex, smaller reactors that are safer. Large nuclear power plants, which produce about 1,000 MW of electricity, require an extensive set of pumps and backup equipment to ensure that adequate cooling is available to the reactor. Smaller reactors can be designed with cooling systems that work by gravity and as a result are not as vulnerable to pump failure caused by power loss. Such cooling systems are said to have *passive stability,* and the reactors are said to be passively safe.[9] Another approach is the use of helium gas to cool reactors that have specially designed fuel capsules capable of withstanding temperatures as high as 1,800°C (about 3,300°F). The idea is to design the fuel assembly so that it cannot hold sufficient fuel to reach this temperature and thus cannot experience a core meltdown.

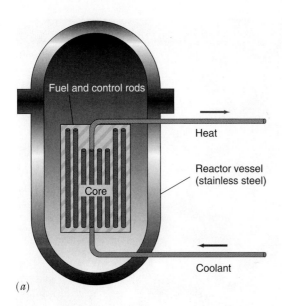

(a)

(b)

Figure 20.5 ■ (a) Main components of a nuclear reactor. (b) Glowing spent fuel elements being stored in water at a nuclear power plant.

Sustainability and Nuclear Power

Sustainability with respect to nuclear power has two aspects: (1) nuclear power's role in creating alternative fuel supplies and (2) the sustainability of nuclear fuel itself. In the first case, nuclear energy can be used to produce hydrogen from water or methane, to supply fuel cells in automobiles. Using hydrogen would help the United States transition from its dependence on oil to a less environmentally damaging energy source (hydrogen). This is a central theme of sustainability that has the objective of meeting our energy needs in the future without harming the environment. The second aspect of sustainability with respect to nuclear energy has to do with nuclear fuel. This is especially important because uranium-fueled nuclear power is a nonrenewable resource.

Nuclear power plants are becoming safer and more economical. Although we have not begun building any nuclear power plants in over 20 years, power plants provide an increasing amount of electrical energy. Since the early 1990s, America's nuclear plants have added over 23,000 MW, equivalent to 23 large fossil fuel–burning power plants. This increase is the result of more efficient use of existing nuclear power plants and has lowered the cost of energy production from nuclear plants.[10]

Our present light water reactors use uranium very inefficiently. Only about 1% of the uranium is used in the reactor; the other 99% ends up as waste. Therefore, our present reactors are part of the nuclear waste problem and not a long-term solution to the energy problem. One way for nuclear power to be sustainable for at least hundreds of years would be the use of a process known as *breeding*. **Breeder reactors** are designed to produce new

nuclear fuel. They do this through a process in which they transform waste or lower-grade uranium into fissionable material. Breeder reactors, if constructed in sufficient numbers (several thousand), could supply about half the energy presently produced by fossil fuels for about 2,000 years.[6]

Breeding is apparently the future of nuclear power, if sustainability in terms of nuclear fuel is the objective. Bringing breeder reactors "online" to produce safe nuclear power will take planning, research, and advanced reactor development. Also, fuel for the breeder reactors will have to be recycled, as reactor fuel must be replaced every few years. What is needed is a new type of breeder reactor comprising an entire system that includes reactor, fuel cycle (especially recycling and reprocessing of fuel), and less production of waste. Such a new reactor is possible but will require redefining our national energy policy and turning energy production in new directions. It remains to be seen whether this will happen.

Pebble-Bed Reactors

A gas-cooled reactor being developed, known as the pebble-bed reactor, may be available as early as 2006. The design of the reactor is centered around fuel elements called pebbles that are about the size of a billiard

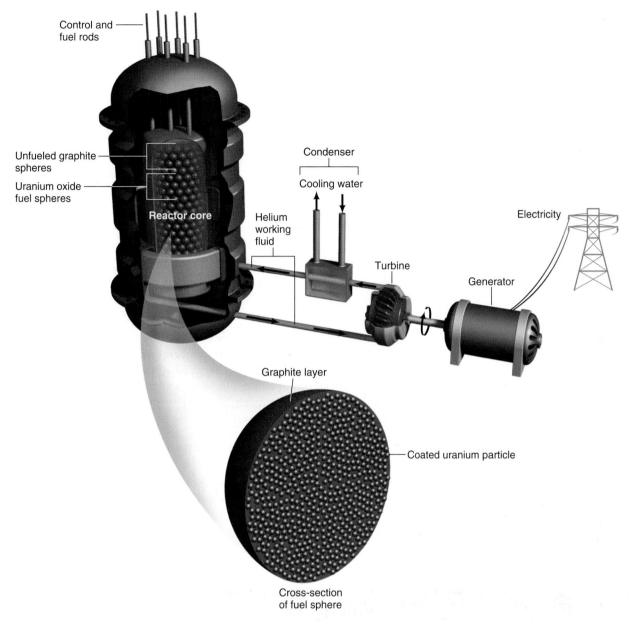

Control and fuel rods

Unfueled graphite spheres

Uranium oxide fuel spheres

Reactor core

Condenser

Cooling water

Helium working fluid

Turbine

Electricity

Generator

Graphite layer

Coated uranium particle

Cross-section of fuel sphere

Figure 20.6 ■ Idealized diagram of pebble-bed nuclear reactor used to provide electrical energy. [*Source:* Modified after J. A. Lake, R. G. Bennett, and J. F. Kotek, "Next-Generation Nuclear Power," *Scientific American* (January 2002): 73–81.]

ball (Figure 20.6). The pebbles have an outer graphite shell with an interior containing about 15,000 sand-grained particles of nuclear fuel (uranium oxide). Approximately 300,000 pebbles are loaded into a metal container shielded by a layer of graphite. Approximately 100,000 nonfuel graphite pebbles are interspersed with the fuel pebbles to assist in controlling production of heat from the reactor. Fuel pebbles are fed into the core, continuously refueling the nuclear reaction. As a spent fuel pebble leaves the core, another is added from the storage container.

The analogy is a gumball machine, where one gumball is removed and another takes it place. This is a safety feature of the reactor because the core at any time has just the right amount of fuel necessary for optimal production of energy. The pebble-bed reactors will probably be modular, with each unit producing about 120 MW of power, which is about one-tenth of that produced by a large centralized nuclear power plant. It is expected that pebble-bed reactors will compete economically with the new generation of natural gas power plants and be about 25% more efficient than present nuclear reactors.[10]

Fusion Reactors

In contrast to fission, which involves splitting heavy nuclei (such as uranium), fusion involves combining the nuclei of light elements (such as hydrogen) to form heavier ones

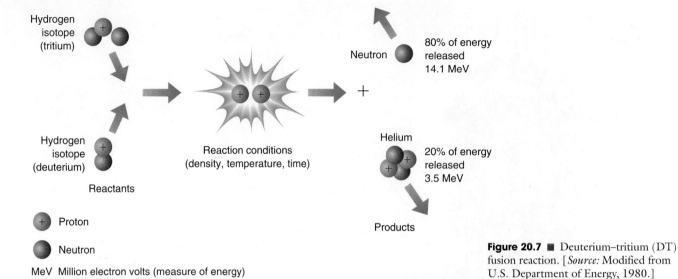

Hydrogen
isotope
(tritium)

Hydrogen
isotope
(deuterium)

Reactants

Reaction conditions
(density, temperature, time)

Neutron

80% of energy
released
14.1 MeV

Helium

20% of energy
released
3.5 MeV

Products

+ Proton

Neutron

MeV Million electron volts (measure of energy)

Figure 20.7 ■ Deuterium–tritium (DT) fusion reaction. [*Source:* Modified from U.S. Department of Energy, 1980.]

(such as helium). As fusion occurs, heat energy is released (Figure 20.7). Nuclear fusion is the source of energy in our sun and other stars.

In a hypothetical fusion reactor, two isotopes of hydrogen—deuterium and tritium—are injected into the reactor chamber, where the necessary conditions for fusion are maintained. Products of the deuterium–tritium (DT) fusion include helium, producing 20% of the energy released, and neutrons, producing 80% of the energy released (Figure 20.8).[11]

Several conditions are necessary for fusion to take place. First, the temperature must be extremely high (approximately 100 million degrees Celsius for DT fusion).

Second, the density of the fuel elements must be sufficiently high. At the temperature necessary for fusion, nearly all atoms are stripped of their electrons, forming a plasma. Plasma is an electrically neutral material consisting of positively charged nuclei, ions, and negatively charged electrons. Third, the plasma must be confined for a sufficient time to ensure that the energy released by the fusion reactions exceeds the energy supplied to maintain the plasma.[11, 12]

The potential energy available when and if fusion reactor power plants are developed is nearly inexhaustible. One gram of DT fuel (from a water and lithium fuel supply) has the energy equivalent of 45 barrels of oil. Deuterium

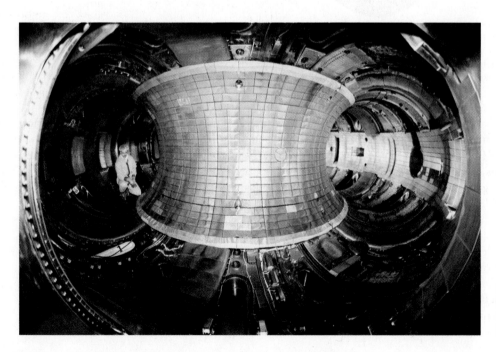

Figure 20.8 ■ Experimental fusion nuclear reactor that magnetically confines plasma at very high temperatures.

can be extracted economically from ocean water, and tritium can be produced in a reaction with lithium in a fusion reactor. Lithium can be extracted economically from abundant mineral supplies.

Many problems remain to be solved before nuclear fusion can be used on a large scale. Research is still in the first stage, which involves basic physics, testing of possible fuels (mostly DT), and magnetic confinement of plasma. Progress in fusion research has been steady in recent years, so there is optimism that useful power will eventually be produced from controlled fusion.[13]

20.2 Nuclear Energy and the Environment

The **nuclear fuel cycle** includes the processes involved in producing nuclear power from the mining and processing of uranium to controlled fission, the reprocessing of spent nuclear fuel, the decommissioning of power plants, and the disposal of radioactive waste. Throughout the cycle, radiation can enter and affect the environment (Figure 20.9). In order to understand the environmental effects of radiation, it is useful to be acquainted with the units used to measure radiation and the amount or dose of radiation that may cause a health problem. These are explained in A Closer Look 20.2.

Problems with Nuclear Power

Let's look a little closer at the environmental effects of the nuclear fuel cycle (Figure 20.9):

- Uranium mines and mills produce radioactive waste material that can pollute the environment. There have been instances in which radioactive mine tailings have been used for foundation and building materials and have contaminated dwellings. (Tailings are materials that are removed by mining activity but are not processed and generally remain at the site.)

- Uranium-235 enrichment and fabrication of fuel assemblies also produce waste materials that must be carefully handled and disposed of.

- Site selection and construction of nuclear power plants in the United States have been extremely controversial. The environmental review process is extensive and expensive, often centering on hazards related to the probability of such events as earthquakes.

- The power plant or reactor is the site most people are concerned about because it is the most visible part of the cycle. It is also the site of past accidents, including partial meltdowns that have released harmful radiation into the environment.

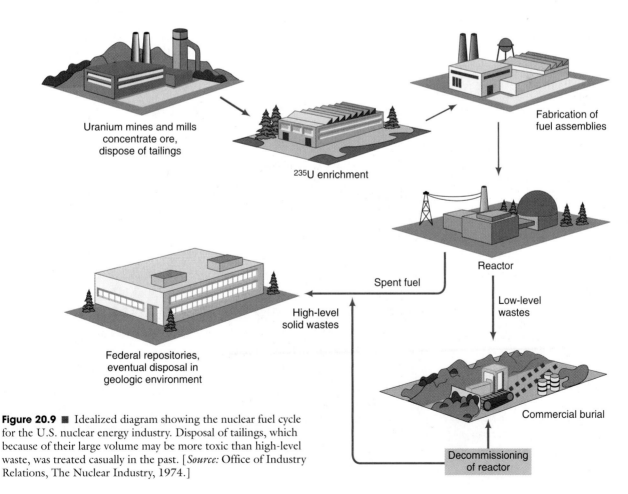

Figure 20.9 ■ Idealized diagram showing the nuclear fuel cycle for the U.S. nuclear energy industry. Disposal of tailings, which because of their large volume may be more toxic than high-level waste, was treated casually in the past. [*Source:* Office of Industry Relations, The Nuclear Industry, 1974.]

Radiation Units and Doses

The units used to measure radioactivity are complex and somewhat confusing. Nevertheless, a modest acquaintance with them is useful in understanding and talking about the effects of radiation on the environment.

A commonly used unit for radioactive decay is the *curie* (Ci), a unit of radioactivity equal to 37 billion nuclear transformations per second.[7]

The curie is named for Marie Curie and her husband, Pierre, who discovered radium in the 1890s. They also discovered polonium, which they named after Marie's homeland, Poland. The harmful effects of radiation were not known at that time, and both Marie Curie and her daughter died of radiation-induced cancer.[8] Her laboratory (Figure 20.10) is still contaminated today.

In the International System (SI) of measurement, the unit commonly used for radioactive decay is the *becquerel* (Bq), which is one radioactive decay per second. Units of measurement often used in discussions of radioactive isotopes, such as radon-222, are becquerels per cubic meter and picocuries per liter (pC/l). A picocurie is one-trillionth (10^{-12}) of a curie. Becquerels per cubic meter or picocuries per liter are therefore measures of the number of radioactive decays that occur each second in a cubic meter or liter of air.

When dealing with the environmental effects of radiation, we are most interested in the actual dose of radiation delivered by radioactivity. That dose is commonly measured in terms of *rads* (rd) and *rems*. In the International System, the corresponding units are *grays* (Gy) and *sieverts* (Sv). Rads and grays are the units of the absorbed dose of radiation; 1 gray is equivalent to 100 rads. Rems and sieverts are units of equivalent dose, or effective equivalent dose, where 1 sievert is 100 rems.[7] The energy retained by living tissue that has been exposed to radiation is called the *radiation absorbed dose*, which is where the term *rad* comes from. Because different types of ra-

diation have different penetrations and so result in different degrees of damage to living tissue, the rad is multiplied by a factor known as the *relative biological effectiveness* to produce the rem or sievert units. When very small doses of radioactivity are being considered, the millirem (mrem) or millisievert (mSv)—that is, one-thousandth (0.001) of a rem or sievert—is used.[7, 15, 16] For gamma rays, the unit commonly used is the roentgen (R), or, in SI units, coulombs per kilogram (C/kg).

People are exposed to a variety of sources of radiation from the sky, the air, and the food we eat (Figure 20.11). What is the background radiation received by people? This question is commonly asked by people concerned with radiation. The average American receives about 2 to 4 mSv/yr. Of this, about 1 to 3 mSv/yr, or 50% to 75%, is natural. The differences are primarily due to elevation and geology. More cosmic radiation from outer space (which delivers about 0.3 to 1.3 mSv/yr) is

received at higher elevations. Radiation from rocks and soils (such as granite and organic shales) containing radioactive minerals delivers about 0.3 to 1.2 mSv/yr. The amount of radiation delivered from rocks, soils, and water may be much larger in areas where radon gas (a naturally occurring radioactive gas) seeps into homes. As a result, mountain states that also have an abundance of granitic rocks, such as Colorado, have a greater background radiation than do states that have a lot of limestone bedrock and are low in elevation, such as Florida.[15] In spite of this general pattern, in locations in Florida where phosphate deposits occur, background radiation is above average because of a relatively high uranium concentration found in the phosphate rocks.[17]

The amount of radiation received by people from human sources is about 1.35 mSv/yr. Two sources include naturally occurring radioactive potassium-40 and carbon-14, which are present in our bodies and produce about 0.35 mSv/yr.

Figure 20.10 ■ Marie Curie in her laboratory.

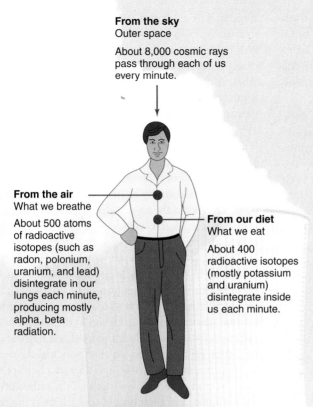

From the sky
Outer space

About 8,000 cosmic rays pass through each of us every minute.

From the air
What we breathe

About 500 atoms of radioactive isotopes (such as radon, polonium, uranium, and lead) disintegrate in our lungs each minute, producing mostly alpha, beta radiation.

From our diet
What we eat

About 400 radioactive isotopes (mostly potassium and uranium) disintegrate inside us each minute.

Figure 20.11 ■ The major sources of natural radiation are the sky, the air we breathe, and the food we eat. [*Source:* Modified from National Radiological Protection Board, *Living with Radiation*, 3rd ed. (Reading, England: National Radiological Protection Board, 1986).]

such as coal, oil, and natural gas, 0.03 mSv/yr; and nuclear power plants (under normal operating conditions), 0.002 mSv/yr.[7, 15, 16]

A person's occupation and lifestyle can also affect the annual dose of radiation received. If you fly at high altitudes in jet aircraft, you receive an additional small dose of radiation—about 0.05 mSv for each flight across the United States. If you work at a nuclear power plant, you can receive up to about 3 mSv/yr. Living next door to a nuclear power plant adds 0.01 mSv/year, and sitting on a bench watching a truck carrying nuclear waste pass by would add 0.001 mSv to your annual exposure. Sources of radiation are summarized in Figure 20.12a, assuming an annual total of 3 mSv/yr.[17, 18] The amount of radiation received at certain job sites, such as nuclear power plants and laboratories where X rays are produced, is closely monitored. At such locations, personnel wear badges that indicate the dose of radiation received.

Figure 20.12b lists some of the common sources of radiation to which we are exposed. Notice that exposure to radon gas can equal what people were exposed to as a result of the Chernobyl nuclear power accident, which occurred in the Soviet Union in 1986. In other words, in some homes, people are exposed to about the same radiation as that experienced by the people evacuated from the Chernobyl area. (Radon gas is discussed in detail in Chapter 25.)

Potassium is an important electrolyte in our blood, and one isotope of potassium (potassium-40) has a very long half-life. Although potassium-40 makes up only a very small percentage of the total potassium in our bodies, it is present in all of us, so we are all slightly radioactive. As a result, if you choose to share your life with another person, you are also exposed to a little bit more radiation.

Sources of low-level radiation from human processes include X rays for medical and dental purposes, which may deliver an average of 0.8 to 0.9 mSv/yr; nuclear weapons testing, approximately 0.04 mSv/yr; the burning of fossil fuels

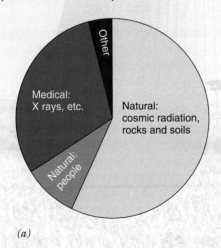

(a)

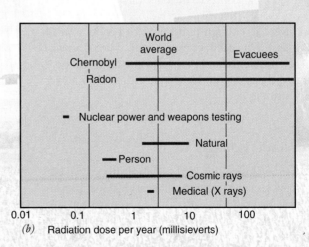

(b) Radiation dose per year (millisieverts)

Figure 20.12 ■ (a) Sources of radiation received by people; assumes annual dose of 3.0 mSv/yr, with 66% natural and 33% medical and other (occupational, nuclear weapons testing, television, air travel, smoke detectors, etc.). [*Sources:* U.S. Department of Energy, 1999; *New Encyclopedia Britannica*, 1997. Radiation. V26, p. 487.] (b) Range in annual radiation dose to people from major sources. [*Source:* Data in part from A. V. Nero, Jr., "Controlling Indoor Air Pollution," *Scientific American* 258(5) (1998): 42–48.]

- The United States does not reprocess spent fuel from reactors to recover uranium and plutonium at this time, so there is currently no environmental impact from them. However, many problems are associated with the handling and disposal of nuclear waste, as discussed later in this chapter.

- Waste disposal is a controversial part of the nuclear cycle because no one wants a nuclear waste-disposal facility nearby. There is general public concern that hazardous nuclear waste cannot be adequately isolated from the environment for the long period of time (millions of years) that it remains hazardous.

- Nuclear power plants have a limited lifetime of several decades. Decommissioning (removal from service) or modernization of a plant is a controversial part of the cycle with which we have little experience. Contaminated machinery needs to be disposed of or stored so that environmental damage will not occur. Decommissioning or refitting will be very expensive (perhaps several hundred million dollars) and is an important aspect of planning for the use of nuclear power. It is possible that the dismantling of decommissioned reactors may become one of the highest costs for the nuclear industry.[14]

In addition to the hazards of transporting and disposing of nuclear material, potential hazards are associated with supplying other nations with reactors. Terrorist activity and the possibility of irresponsible persons in governments add risks that are present in no other form of energy production. For example, Kazakhstan inherited a large nuclear weapons testing facility, covering hundreds of square kilometers, from the former Soviet Union. Several sites contain "hot spots" of plutonium in the soil that present a serious problem of toxic contamination. The facility also presents a security problem. There is some international concern that this plutonium could be collected and used by terrorists to produce dirty bombs (conventional explosives that disperse radioactive materials). There may even be enough plutonium to produce small nuclear bombs.[19] Nuclear energy may indeed be an answer to our energy problems, and perhaps someday it will provide unlimited cheap energy. However, with nuclear power comes responsibility.

Effects of Radioisotopes

As explained in A Closer Look 20.1, a **radioisotope** is an isotope of a chemical element that spontaneously undergoes radioactive decay. Radioisotopes affect the environment in two ways: by emitting radiation that affects other materials and by entering the normal pathways of mineral cycling and ecological food chains.

The explosion of a nuclear atomic weapon does damage in both ways. At the time of the explosion, intense radiation of many kinds and energies is sent out, killing organisms

directly. The explosion generates large amounts of radioactive isotopes, which are dispersed into the environment. Nuclear bombs exploded in the atmosphere produce a huge cloud that sends radioisotopes directly into the stratosphere, where the radioactive particles may be widely dispersed by winds. Atomic *fallout*—the deposit of these radioactive materials around the world—was an environmental problem in the 1950s and 1960s, when the United States, the former Soviet Union, China, France, and Great Britain were testing and exploding nuclear weapons in the atmosphere.

The pathways (Figure 20.13) of some of these isotopes illustrate the second way in which radioactive materials can be dangerous in the environment; they can enter ecological food chains. Let's consider an example. One of the radioisotopes emitted and sent into the stratosphere by atomic explosions was cesium-137. This radioisotope was deposited in relatively small concentrations but was widely dispersed in the Arctic region of North America. It fell on reindeer moss, a lichen that is a primary winter food of the caribou. A strong seasonal trend in the levels of cesium-137 in caribou was discovered; the level was highest in the winter, when reindeer moss was the principal food, and lowest in the summer. Eskimos who obtained a high percentage of their protein from caribou ingested the radioisotope by eating the meat, and their bodies concentrated the cesium. The more that members of a group depended on caribou as their primary source of food, the higher was the level of the isotope in their bodies (Figure 20.14).

It is possible to predict the environmental pathways that radioisotopes will follow because we know the normal pathways of nonradioactive isotopes with the same chemical characteristics. Our knowledge of biomagnification and of large-scale air and oceanic movements that transport radioisotopes throughout the biosphere will also help us to understand the effects of radioisotopes.

Radiation Doses and Health

The most important question in studying radiation exposure in people involves determining the point at which the exposure or dose becomes a hazard to health (see A Closer Look 20.2). Unfortunately, there are no simple answers to this seemingly simple question. We do know that a dose of about 5,000 mSv (5 sieverts) is considered lethal to 50% of people exposed to it. Exposure to 1,000–2,000 mSv is sufficient to cause health problems, including vomiting, fatigue, potential abortion of pregnancies of less than two months' duration, and temporary sterility in males. At 500 mSv, physiological damage is recorded. The maximum allowed dose of radiation per year for workers in industry is 50 mSv, which is approximately 30 times the average natural background radiation received by people.[7, 15] For the general public, the maximum permissible annual dose (for infrequent exposure) is set in the United States at 5 mSv, which is about three times the annual natural background.[7] For continuous or frequent exposure, the limit for the general public is 1 mSv.

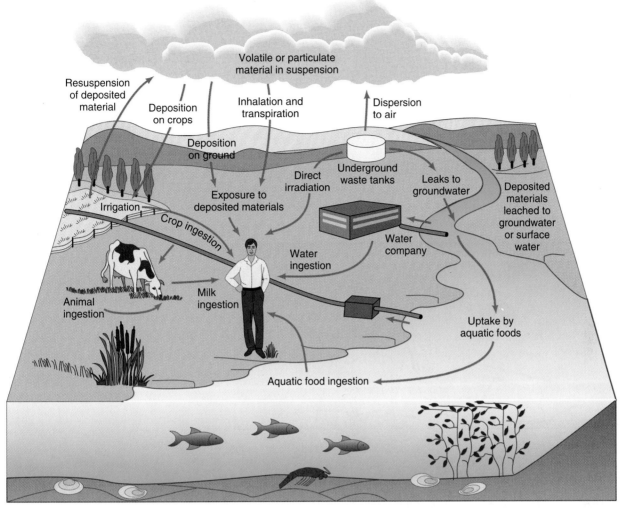

Figure 20.13 ■ How radioactive substances reach people. [*Source:* F. Schroyer, ed., *Radioactive Waste*, 2nd printing (American Institute of Professional Geologists, 1985).]

Most information concerning the effects of high doses of radiation on people comes from studies of the people who survived the atomic bomb detonations in Japan at the end of World War II. We also have information concerning people exposed to high levels of radiation in uranium mines, workers who painted watch dials with luminous paint containing radium, and people treated with radiation therapy for disease.[20] Workers in uranium mines who were exposed to high levels of radiation have been shown to suffer a significantly higher rate of lung cancer than the general population. Studies have shown that there is a delay of 10 to 25 years between the time of exposure and the onset of disease. Starting in about 1917 in New Jersey, approximately 2,000 young women were employed painting watch dials with luminous paint. To maintain a sharp point on their brushes, they licked them and as a result were swallowing radium, which was in the paint. By 1924, dentists in New Jersey were reporting cases of jaw rot; and within five years radium was

known to be the cause. Many of the women died of anemia or bone cancer.[8]

Although there is a vigorous and ongoing debate about the nature and extent of the relationship between radiation exposure and cancer mortality, most scientists agree that radiation can cause cancer. Some scientists believe that there is a linear relationship, such that any increase in radiation beyond the background level will produce an additional hazard. Others believe that the body is able to successfully handle and recover from low levels of exposure to radiation but that health effects (toxicity) become apparent beyond some threshold. The verdict is still out on this subject, but it seems prudent to take a conservative viewpoint and accept that there may be a linear relationship. Unfortunately, long-term chronic health problems related to low-level exposure to radiation are neither well known nor well understood.

Radiation has a long history in the field of medicine. Drinking waters that contain radioactive materials goes

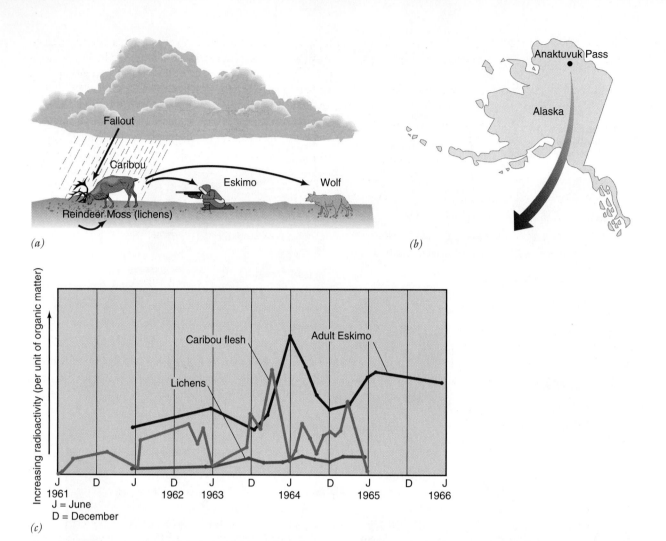

Figure 20.14 ■ Cesium-137 released into the atmosphere by atomic bomb tests was part of the fallout deposited onto soil and plants. (*a*) The cesium fell on lichens, which were eaten by caribou. The caribou were in turn eaten by the Eskimo. (*b*) Measurements of cesium were taken in the lichens, caribou, and Eskimo in the Anaktuvuk Pass of Alaska. (*c*) The cesium was concentrated by the food chain. Peaks in concentrations occurred first in the lichens, then in the caribou, and last in the Eskimo. [*Source:* (*c*) W. G. Hanson, "Cesium-137 in Alaskan Lichens, Caribou, and Eskimos," *Health Physics* 13 (1967): 383–389. Copyright 1967 by Pergamon Press. Reprinted with permission.]

back to Roman times. By 1899, the adverse effects of radiation had been studied and were well known; and in that year, the first lawsuit for malpractice in using X rays was filed. Because science had shown that radiation could destroy human cells, however, it was a logical step to conclude that drinking water containing radioactive material such as radon might help fight diseases such as stomach cancer. In the early 1900s, it became popular to drink water containing radon, and the practice was supported by doctors, who stated that there were no known toxic effects. Although we now know that statement to be incorrect, radiotherapy, which uses radiation to kill cancer cells in humans, has been widely and successfully used for a number of years.[8]

20.3 Nuclear Power Plant Accidents

Although the chance of a disastrous nuclear accident is estimated to be very low, the probability that an accident will occur increases with every reactor put into operation. For example, according to the U.S. Nuclear Regulatory Commission's performance goal for a single reactor, the probability of a large-scale core meltdown in any given year should be no greater than 0.01% (one chance in 10,000). However, if there were 1,500 nuclear reactors (about four times the present world total), a meltdown could, at the low annual probability of 0.01%, be expected every seven years. This is clearly an unacceptable risk.[6] Increasing safety by about 10 times would result in lower, more manageable risk; but the risk would still be appreciable because the potential consequences remain large.

Next, we discuss the two most well-known nuclear accidents, which occurred at the Three Mile Island and Chernobyl reactors. It is important to understand that these serious accidents resulted in part from human error.

Three Mile Island

One of the most dramatic events in the history of U.S. radiation pollution occurred on March 28, 1979, at the Three Mile Island nuclear power plant near Harrisburg, Pennsylvania. The malfunction of a valve, along with human errors (thought to be the major problem), resulted in a partial core meltdown. Intense radiation was released to the interior of the containment structure. Fortunately, the containment structure functioned as designed, and only a relatively small amount of radiation was released into the environment. Exposure from the radiation emitted into the atmosphere has been estimated at 1 mSv, which is low in terms of the amount of radiation required to cause acute toxic effects. Average exposure to radiation in the surrounding area is estimated to have been approximately 0.012 mSv, which is only about 1% of the natural background radiation received by people. However, radiation levels were much higher near the site. On the third day after the accident, 12 mSv/hour was measured at ground level near the site. By comparison, the average American receives about 2 mSv/year from natural radiation.

Because the long-term chronic effects of exposure to low levels of radiation are not well understood, the effects of Three Mile Island exposure, though apparently small, are difficult to estimate. However, the incident revealed many potential problems with the way in which U.S. society dealt with nuclear power. Historically, nuclear power had been relatively safe, and the state of Pennsylvania was unprepared to deal with the accident. For example, there was no state bureau for radiation help, and the state Department of Health did not have a single book on radiation medicine (the medical library had been dismantled two years earlier for budgetary reasons). One of the major impacts of the incident was fear; yet there was no state office of mental health, and no staff member from the Department of Health was allowed to sit in on important discussions following the accident.[21]

Chernobyl

Lack of preparedness to deal with a serious nuclear power plant accident was dramatically illustrated by events that began unfolding on the morning of Monday, April 28, 1986. Workers at a nuclear power plant in Sweden, frantically searching for the source of elevated levels of radiation near their plant, concluded that it was not their installation that was leaking radiation. Rather, the radioactivity was coming from the Soviet Union by way of prevailing winds. Confronted, the Soviets announced that an accident had occurred at a nuclear power plant at Chernobyl two days earlier, on April 26 (Figure 20.15). This was the first notice to the world of the worst accident in the history of nuclear power generation.

It is speculated that the system that supplied cooling waters for the Chernobyl reactor failed as a result of human error, causing the temperature of the reactor core to rise

Figure 20.15 ■ Guard halting entry of people into the forbidden zone evacuated in 1986 as a result of the Chernobyl nuclear accident.

to over 3,000°C (about 5,400°F), melting the uranium fuel. Explosions removed the top of the building over the reactor, and the graphite surrounding the fuel rods used to moderate the nuclear reactions in the core ignited. The fires produced a cloud of radioactive particles that rose high into the atmosphere. There were 237 confirmed cases of acute radiation sickness, and 31 people died of radiation sickness.[22]

In the days following the accident, nearly 3 billion people in the Northern Hemisphere received varying amounts of radiation from Chernobyl. With the exception of the 30-km (19-mi) zone surrounding Chernobyl, the world human exposure was relatively small. Even in Europe, where exposure was highest, it was considerably less than the natural radiation received during one year.[22]

In that 30-km zone, approximately 115,000 people were evacuated, and as many as 24,000 people were estimated to have received an average radiation dosage of 0.43 Sv (430 mSv). This group of people is being studied carefully.

It was expected, based on results from Japanese bomb survivors, that approximately 122 spontaneous leukemias would likely occur during the period from 1986 through 1998.[22] Surprisingly, as of late 1998, there was no significant increase in the incidence of leukemia, even among the most highly exposed people, but an increase of the incidences of leukemia could still be expected in the future.[23]

Studies have found that the number of childhood thyroid cancer cases per year has risen steadily in the three countries of Belarus, Ukraine, and the Russian Federation (those most affected by Chernobyl) since the accident. In 1994, a combined rate of 132 new thyroid cancer cases was identified. Since the accident, a total of 1,036 thyroid cancer cases have been diagnosed in children under 15. These cancer cases are believed to be linked to the released radiation from the accident, although other factors, such as environmental pollution, may also play a role. It is predicted that a few percent of the roughly 1 million children exposed to the radiation eventually will contract thyroid cancer.[24]

Outside the 30-km zone, the increased risk of contracting cancer is very small and not likely to be detected from an ecological evaluation.[24] Nevertheless, according to one estimate, Chernobyl will ultimately be responsible for approximately 16,000 deaths worldwide.[9]

Vegetation within 7 km of the power plant was either killed or severely damaged following the accident. Pine trees examined in 1990 around Chernobyl showed extensive tissue damage and still contained radioactivity. The distance between annual rings (a measure of tree growth) had decreased since 1986.[25]

Scientists returning to the evacuated zone in the mid-1990s found, to their surprise, thriving and expanding animal populations. Species such as wild boar, moose, otters, waterfowl, and rodents seemed to be enjoying a population boom in the absence of humans. The wild boar population had increased 10-fold since the evacuation. However, these animals may be paying a genetic price for living within the contaminated zone. So far the benefit of excluding humans apparently outweighs the negative factors associated with radioactive contamination.[26] The area now resembles a wildlife reserve.

In the areas surrounding Chernobyl, radioactive materials continue to contaminate soils, vegetation, surface water, and groundwater, presenting a hazard to plants and animals. The evacuation zone may be uninhabitable for a very long time unless some way is found to remove the radioactivity.[22] For example, the city of Prypyat, 5 km from Chernobyl, is a "ghost city." At the time of the accident, the population of the city was 48,000. Today, it is abandoned, with blocks of vacant apartment buildings and rusting vehicles. Roads are cracking and trees are growing as new vegetation transforms the urban land back to green fields. Cases of thyroid cancer are still increasing, and the number of cases is many times higher for people who lived as children in Prypyat at the time of the accident. The final story of the world's most serious nuclear accident is yet to completely unfold.[27] Estimates of the total cost of the Chernobyl accident vary widely, but the cost will probably exceed $200 billion.

Although the Soviets were accused of not giving attention to reactor safety and of using outdated equipment, people are still wondering if such an accident could happen again elsewhere. Because there are several hundred reactors producing power in the world today, the answer has to be yes. About 10 accidents have released radioactive particles during the past 34 years. Therefore, although Chernobyl is the most serious nuclear accident to date, it certainly was not the first, and it is not likely to be the last. Although the probability of a serious accident is very small at a particular site, the consequences may be great, perhaps resulting in an unacceptable risk to society. This is really not so much a scientific issue as a political one involving a question of values.

Advocates of nuclear power have argued that nuclear power is safer than other sources of energy. They say that the number of additional deaths caused by air pollution resulting from burning fossil fuels is much greater than the number of lives lost through nuclear accidents. For example, the 16,000 deaths that might eventually be attributed to Chernobyl are fewer than the number of deaths caused each year by air pollution from burning coal.[9] Those arguing against nuclear power state that as long as people build nuclear power plants and manage them, there will be the possibility of accidents. We can build nuclear reactors that are safer, but people will continue to make mistakes, and accidents will continue to occur.

20.4 Radioactive-Waste Management

Examination of the nuclear fuel cycle (Figure 20.9) illustrates some of the sources of waste that must be disposed of as a result of using nuclear energy to produce electricity. Radioactive wastes are by-products that must be expected when electricity is produced at nuclear reactors; they may be grouped into three general categories: low-level waste, transuranic waste, and high-level waste. In addition, the tailings from uranium mines and mills must also be considered hazardous. In the western United States more than 20 million metric tons of abandoned tailings will continue to produce radiation for at least 100,000 years.

Low-Level Radioactive Waste

Low-level radioactive waste contains sufficiently low concentrations or quantities of radioactivity that it does not present a significant environmental hazard if properly handled. Low-level waste includes a wide variety of items, such as residuals or solutions from chemical processing; solid or liquid plant waste, sludges, and acids; and slightly contaminated equipment, tools, plastic, glass, wood, and other materials.[28]

Low-level waste has been buried in near-surface burial areas in which the hydrologic and geologic conditions were thought to severely limit the migration of radioactivity.[28] However, monitoring has shown that several of the U.S. sites for disposal of low-level radiation have not provided adequate protection for the environment, and leaks of liquid waste have polluted groundwater. Of the original six burial sites, three had closed prematurely by 1979 due to unexpected leaks, financial problems, or loss of license. As of 1995, only two remaining government low-level nuclear waste repositories remained in operation in the United States, one in Washington and the other in South Carolina. In addition, there is a private facility in Utah run by Envirocare that accepts low-level waste. Construction of new burial sites, such as the Ward Valley site in southeastern California, has been met with strong public opposition, and controversy remains as to whether low-level radioactive waste can be disposed of safely.[29]

Transuranic Waste

Transuranic waste is composed of human-made radioactive elements heavier than uranium. It is produced in part by neutron bombardment of uranium in reactors and includes plutonium, americium, and einsteineum. Most transuranic waste is industrial trash, such as clothing, rags, tools, and equipment, that has been contaminated. The waste is low-level in terms of intensity of radioactivity, but plutonium has a long half-life and requires isolation from the environment for about 250,000 years. Most transuranic waste is generated from the production of nuclear weapons and, more recently, from cleanup of former nuclear weapons facilities.

Some transuranic wastes, as of 1999, are being transported to a disposal site near Carlsbad, New Mexico. The waste is isolated at a depth of 655 m (2,150 ft) in salt beds (rock salt) that are several hundred meters thick (Figure 20.16). Rock salt at the New Mexico site has several advantages:[30, 31]

■ The salt is about 225 million years old, and the area is geologically stable, with very little earthquake activity.

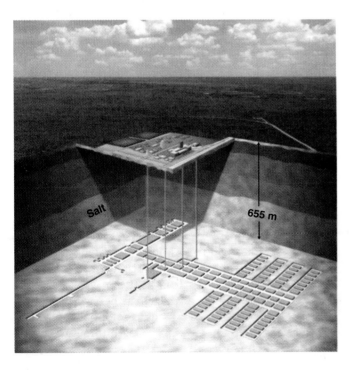

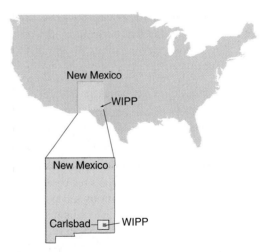

Figure 20.16 ■ Waste isolation pilot plant (WIPP) in New Mexico for disposal of transuranic waste. [*Source:* U.S. Department of Energy, 1999.]

- The salt has no flowing groundwater and is easy to mine. Excavated rooms in the salt about 10 m wide and 4 m high will be used for disposal.
- Rock salt flows slowly into mined openings. The waste-filled spaces in the storage facility will be naturally closed by the slow-flowing salt in 75 to 200 years, sealing the waste.

The New Mexico disposal site is important because it is the first geologic disposal site for radioactive waste in the United States. It is a pilot project that will be evaluated very carefully. Safety is the primary concern. Procedures to transport the waste to the disposal site as safely as possible and place it underground in the disposal facility have been established. Because the waste will be hazardous for many thousands of years and there are uncertainties concerning future cultures and languages, clear warnings above and below ground have been created. The site is clearly marked to help ensure that human intrusion in the future is avoided.[31]

High-Level Radioactive Waste

High-level radioactive waste consists of commercial and military spent nuclear fuel; uranium and plutonium derived from military reprocessing; and other radioactive nuclear weapons materials. It is extremely toxic, and a sense of urgency surrounds its disposal as the total volume of spent fuel accumulates. At present, in the United States, tens of thousands of metric tons of high-level waste are being stored at more than a hundred sites in 40 states. Seventy-two of the sites are commercial nuclear reactors.[32, 33]

Storage of high-level waste is at best a temporary solution, and serious problems with radioactive waste have occurred where it is being stored. Although improvements in storage tanks and other facilities will help, eventually some sort of disposal program must be initiated. Some scientists believe the geologic environment can best provide safe containment of high-level radioactive waste. Others disagree and have criticized proposals for long-term disposal of high-level radioactive waste underground. A comprehensive geologic disposal development program should have the following objectives:[34]

- Identification of sites that meet broad geologic criteria of ground stability and slow movement of groundwater with long flow paths to the surface.
- Intensive subsurface exploration of possible sites to positively determine geologic and hydrologic characteristics.
- Predictions of behavior of potential sites based on present geologic and hydrologic situations and assumptions for future changes in variables such as climate, groundwater flow, erosion, and ground movements.
- Evaluation of risk associated with various predictions.
- Political decision making based on risks acceptable to society.

The Nuclear Waste Policy Act of 1982 initiated a high-level nuclear waste-disposal program. The Department of Energy was given the responsibility to investigate several potential sites and make a recommendation. The 1982 act was amended in 1987; the amendment, along with the Energy Power Act of 1992, specified that high-level waste was to be disposed of underground in a deep, geologic waste repository. It was also specified that the Yucca Mountain site in Nevada was to be the only site evaluated. This does not mean that Yucca Mountain had been selected for actual disposal of nuclear waste, only that it is the only site being completely evaluated at this time. If the site is suitable, then it could accept high-level waste as early as 2010. Following are some of the key issues being addressed by the Department of Energy at the Yucca Mountain site:[33, 35]

- Assessment of the probability and consequences of volcanic eruptions.
- Evaluation of the earthquake hazard.
- Estimation of changes in the storage environment over long periods of time.
- Estimation of the time the waste may be contained and the types and rates of radiation that may escape from deteriorated waste containers.
- Evaluation of how heat generated by the waste may affect moisture in and around the repository and the design of the repository.
- Characterization of groundwater flow near the repository.
- Identification and understanding of major geochemical processes that control the transport of radioactive materials.

Extensive scientific evaluations of the Yucca Mountain site have been completed.[33] However, use of this site is controversial and is generating considerable resistance from the state and people of Nevada as well as those scientists not confident in the plan. Some of the scientific questions at Yucca Mountain have concerned natural processes and hazards that might allow radioactive materials to escape, such as surface erosion, groundwater movement, earthquakes, and volcanic eruptions. In 2002, Congress voted to submit a license of application for Yucca Mountain to the Nuclear Regulatory Commission.

One of the major questions concerning the disposal of high-level radioactive waste is this: How credible are long-range geologic predictions—those covering several thousand to a few million years?[34] Unfortunately, there is no easy answer to this question because geologic processes vary over both time and space. Climates change over long periods of time, as do areas of erosion, deposition, and groundwater activity. For example, large earthquakes even thousands of kilometers from a site may permanently change groundwater levels. The earthquake record for most of the United States extends back for only a few hundred years; therefore, estimates of future earthquake activity are tenuous at best.

The bottom line is that geologists can suggest sites that have been relatively stable in the geologic past but cannot absolutely guarantee future stability. This means that policymakers (not geologists) need to evaluate the uncertainty of predictions in light of pressing political, economic, and social concerns.[34] In the end, the geologic environment may be deemed suitable for safe containment of high-level radioactive waste, but care must be taken to ensure that the best possible decisions are made on this important and controversial issue.

20.5 The Future of Nuclear Energy

Nuclear energy as a power source for electricity is now being seriously evaluated. Advocates for nuclear power have argued that nuclear power is good for the environment for the following reasons:

- It does not produce potential global warming through release of carbon dioxide (see Chapter 23).

- It does not cause air pollution or emit precursors (sulfates and nitrates) that cause acid rain (see Chapter 24).

- If breeder reactors are developed for commercial use, the amount of fuel available will be greatly increased.

Those in favor of nuclear power argue that it is safer than other means of generating power and that we should build many more nuclear power plants in the future. This argument is predicated on the understanding that such power plants would be considerably safer than those being used today. That is, if we standardize nuclear reactors and make them safer and smaller, nuclear power could provide much of our electricity in the future.[9] The safety of nuclear power under normal operating procedures is not disputed, although the possibility of accidents and the disposal of spent fuel are concerns.

The argument against reviving nuclear power is based on political and economic considerations as well as scientific uncertainty concerning safety issues. Those opposed to expanding nuclear power argue that converting from coal-burning plants to nuclear power plants for the purpose of reducing carbon dioxide emissions would require an enormous investment in nuclear power to make a real impact. This is true. Furthermore, critics say, given the fact that safer nuclear reactors are only just being developed, there will be a time lag. As a result, nuclear power is not likely to have a real impact on environmental problems, such as air pollution, acid rain, and potential global warming, before at least the year 2050.[36] This point is debatable; the lag time could be shortened if research and development in advanced reactors become a higher priority and are better funded.

Another argument against nuclear power is that some countries may be interested in nuclear power as a path to nuclear weapons. Reprocessing used nuclear fuel from a power plant produces plutonium that can be used to make nuclear bombs. There is concern that rogue nations with nuclear power could divert plutonium to make weapons, or may sell plutonium to others, even terrorists, who would make nuclear weapons.[37]

Until 2001, the politics of nuclear energy was losing ground. Nearly all energy scenarios were based on the expectation that nuclear power would continue to grow slowly or perhaps even decline in the coming years. Since the Chernobyl accident, many countries in Europe have been reevaluating the use of nuclear power; and in most instances, the number of nuclear power plants being built has been significantly reduced. Indeed, in Germany, where about one-third of the country's electricity is produced by nuclear power, the decision has been made to shut down all nuclear power plants in the next 25 years as they become obsolete.

Nuclear power produces about 20% of electricity in the United States today. As mentioned earlier, there have been no new orders for nuclear power plants in the United States. Nevertheless, research and development into the smaller, safer nuclear power plants and breeder reactors are going forward. In addition, the Energy Policy Act of 2005 suggests that nuclear power use should be increased in the future. The nuclear option is again being evaluated in light of the environmental problems associated with fossil fuels. However, the benefits of nuclear power must be balanced with the safety and waste-disposal issues that have made nuclear energy an uncertain option for many people. The full impact of what began in 1942, when the atom was first split, is still to be determined.

Summary

- Nuclear fission is the process of splitting an atomic nucleus into smaller fragments. As fission occurs, energy is released. The major components of a fission reactor are the core, control rods, coolant, and reactor vessel.

- Nuclear radiation occurs when a radioisotope spontaneously undergoes radioactive decay and changes into another isotope.

- The three major types of nuclear radiation are alpha, beta, and gamma.

- Each radioisotope has its own characteristic emissions. Different types of radiation have different toxicities; and in terms of the health of humans and other organisms, it is important to know the type of radiation emitted and the half-life.

- The nuclear fuel cycle consists of the mining and processing of uranium, the generation of nuclear power through controlled fission, the reprocessing of spent fuel, the disposal of nuclear waste, and the decommissioning of power plants. Each part of the cycle is associated with characteristic processes, all with different potential environmental problems.

What Is the Future of Nuclear Energy?

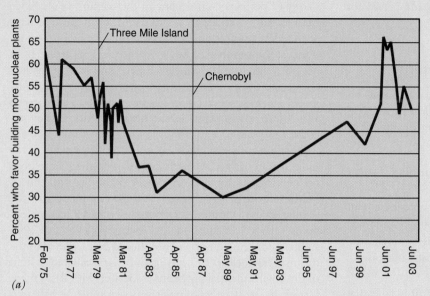

(a)

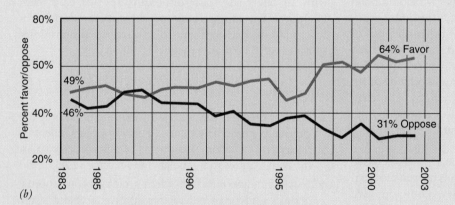

(b)

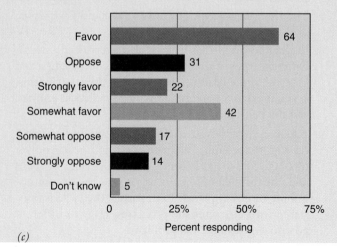

(c)

Nuclear energy in the United States and around the world is at a crossroads early in the twenty-first century. Although no new nuclear power plants had been ordered in the United States as of 2002, 31 new power plants were under construction worldwide (Table 20.2). Approximately 25% of these were in India. As of 2004 there were 440 operating power plants producing a total of about 362 GW of electricity. About 80% of the electricity produced in Lithuania and France is from nuclear energy. Figure 20.17a shows the result of public opinion polls from 1975 to 2003 in which people were asked to indicate whether they were favorably inclined to the idea of building new nuclear power plants in the United States. Figure 20.17b shows the trend from 1983 to 2003 of people in the United States who favor or oppose nuclear energy. Results from a greater breakdown, from those strongly favoring to those strongly opposing nuclear power, are shown on Figure 20.17c.

Figure 20.17 ■ (a) Public opinion concerning the construction of new nuclear power plants. [*Source:* E. A. Rosa and R. E. Dunlap, "Nuclear Power: Three Decades of Public Opinion," *Public Opinion Quarterly* 58 (1994): 295–325 (and www.nei.org).] (b) Percentages of people in the United States who favor and oppose nuclear power, 1938–2003. [*Source:* Nuclear Energy Institute, 2003.] (c) Results of 2003 survey showing how people in the United States feel about the use of nuclear power. [*Source:* Nuclear Energy Institute, 2003.]

Critical Thinking Questions

1. How might you interpret Figure 20.17a, which shows the percentage of people in the Unites States who favored building more nuclear power plants from 1975 to 2002? What do you think the effects of nuclear accidents and the California energy crisis of 2001 have had on the number of people who favor building new power plants?

2. According to Figure 20.17b and c, approximately two-thirds of people in the United States today favor nuclear energy, with the remaining one-third opposing it. Try to construct arguments for those who would strongly favor nuclear power and contrast them with the arguments of those who would strongly oppose it.

3. Why do you think there was a drop in the number of people who favor building more nuclear power plants from 2001 to 2002?

4. Putting together everything in this chapter and in Figure 20.17, what do you think the future of nuclear power will be in the United States and the world in the coming decades?

Table 20.2 • World Nuclear Power Reactors under Construction (2002)		
Country	**Units**	**Total MWe**
India	8	3,622
Ukraine	4	3,800
China	4	3,275
Russian Federation	3	2,825
Japan	3	3,696
Slovak Republic	2	776
Republic of Korea	2	1,920
Islamic Republic of Iran	2	2,111
Romania	1	655
Democratic People's Republic of Korea	1	1,040
Argentina	1	692
Total	**31**	**24,412**

Source: International Atomic Energy Agency. 2004. Power reactor information system.

- The present burner reactors (mostly light water reactors) use uranium-235 as a fuel. Uranium is a nonrenewable resource mined from the Earth. If many more burner reactors were constructed, we would face fuel shortages. Nuclear energy based on burning uranium-235 in light water reactors is thus not sustainable. For nuclear energy to be sustainable, safe, and economical, breeder reactors will need to be developed.

- Radioisotopes affect the environment in two major ways: by emitting radiation that affects other materials and by entering ecological food chains. Major environmental pathways by which radiation reaches people include uptake by fish ingested by people, uptake by crops ingested by people, inhalation from air, and exposure to nuclear waste and the natural environment.

- The dose–response for radiation is fairly well established. We know the dose–response for higher exposures, when illness or death occurs. However, there are vigorous debates concerning the health effects of low-level exposure to radiation and what relationships exist between exposure and cancer mortality. Most scientists believe that radiation can cause cancer. Ironically, radiation can be used to kill cancer cells, as in radiotherapy treatments.

- We have learned from accidents at nuclear power plants that it is difficult to plan for the human factor. People make mistakes. We have also learned that we are not as prepared for accidents as we would like to think. Some people believe that humans are not ready for the responsibility of nuclear power. Others believe that we can design much safer power plants where serious accidents are impossible.

- Transuranic nuclear waste is now being disposed of in salt beds—the first disposal of radioactive waste in the geologic environment in the United States.

- There is a consensus that high-level nuclear waste may be safely disposed of in the geologic environment. The problem has been to locate a site that is safe and not objectionable to the people who make the decisions and to those who live in the region.

- Nuclear power is again being seriously evaluated as an alternative to fossil fuels. On the one hand, it has advantages in that it emits no carbon dioxide, will not contribute to global warming or cause acid rain, and can be used to produce alternative fuels such as hydrogen. On the other hand, people are uncomfortable with nuclear power because of possible accidents and waste-disposal problems.

Human Population

As human population has increased, so has demand for electrical power; as a result, a number of countries have turned to nuclear energy. The California energy crisis has caused many people in the United States to rethink the value of nuclear energy. Though relatively rare, accidents at nuclear power plants such as Chernobyl have exposed people to increased radiation. There is considerable debate over potential adverse effects of that radiation. Nevertheless, the fact remains that as world population increases, and if the number of nuclear power plants increases, the total number of people exposed to a potential hazardous release of toxic radiation will increase as well.

Sustainability

It has been argued that sustainable energy development will require a return to nuclear energy, because nuclear energy does not contribute to a variety of environmental problems related to burning of fossil fuels. For nuclear energy to significantly contribute to sustainable energy development, however, we cannot depend on burner reactors that will quickly use Earth's uranium resources; rather, development of safer breeder reactors will be necessary.

Global Perspective

Use of nuclear energy fits into our global management of the entire spectrum of energy sources. In addition, testing of nuclear weapons has spread radioactive isotopes around the entire planet, as have nuclear accidents. Radioactive isotopes that enter rivers and other waterways may eventually enter the oceans of the world, where oceanic circulation may further disperse and spread them.

Urban World

Development of nuclear energy is a product of our technology and our urban world. In some respects, it is near the pinnacle of our accomplishments in terms of technology.

People and Nature

Nuclear reactions are the source of heat for our sun and are fundamental processes of the universe. Nuclear fusion has produced the heavier elements of the universe. Our use of nuclear reactions in reactors to produce useful energy is a connection to a basic form of energy in nature. Abuse of nuclear reactions in weapons carries the possibility of damaging, or even destroying, nature on Earth.

Science and Values

We have a good deal of knowledge concerning nuclear energy and nuclear processes. However, people remain suspicious and, in some cases, frightened by nuclear power—in part because of the value they place on a quality environment and their perception that nuclear radiation is toxic to that environment. As a result, the future of nuclear energy will be related to political decisions based in part on the risk acceptable to society. It will also depend on research and development to produce much safer nuclear reactors.

Key Terms

breeder reactors **414**
burner reactors **409**
fission **409**
fusion **409**

high-level radioactive
  waste **426**
low-level radioactive
  waste **425**

meltdown **412**
nuclear energy **409**
nuclear fuel cycle **417**
nuclear reactors **409**

radioactive decay **410**
radioisotope **420**
transuranic waste **425**

Study Questions

1. If exposure to radiation is a natural phenomenon, why are we worried about it?

2. What is a radioisotope, and why is knowing its half-life important?

3. What is the normal background radiation that people receive? Why is it variable?

4. What are the possible relationships between exposure to radiation and adverse health effects?

5. What processes in our environment may result in radioactive substances reaching people?

6. Suppose it is recommended that high-level nuclear waste be disposed of in the geologic environment of the region in which you live. How would you go about evaluating potential sites?

7. Are there good environmental reasons to develop and build new nuclear power plants? Discuss both sides of the issue.

Further Reading

Nuclear Energy Agency (NEA) and Organization for Economic Co-Operation and Development (OECD). 1994. *Power Generation Choices: Costs, Risks, and Externalities.* Proceedings of an international symposium, Washington, D.C., September 23–24, 1993. NEA, OECD. Includes discussion of the economics of nuclear power versus other energy sources.

Nuclear Energy Agency (NEA) and Organization for Economic Co-Operation and Development (OECD). 1995. *Environmental and Ethical Aspects of Long-Lived Radioactive Waste Disposal.* Proceedings of an international workshop, Paris, September 1–2, 1994. NEA, OECD. Essays covering topics of environmental policies, ethical and environmental considerations, cost-benefit analysis, and disposal issues of long-lived radioactive waste.

U.S. Department of Energy, Office of Environmental Management. 1995. *Closing the Circle on the Splitting of the Atom.* Washington, D.C.: U.S. Department of Energy. Descriptions of environmental, safety, and health problems associated with production of nuclear weapons and how the U.S. Department of Energy plans to deal with the problem.

Wald, M. 2003 (March). "Dismantling Nuclear Reactors." *Scientific American,* pp. 60–69. An in-depth discussion of steps in dismantling a nuclear power plant and some of the unforeseen difficulties.

World Health Organization. 1995. *Health Consequences of the Chernobyl Accident.* Geneva, Switzerland: World Health Organization. A short book covering the accident, response, health consequences, findings, and proposed future work.

Young, J. P., and R. S. Yalow, eds. 1995. *Radiation and Public Perception: Benefits and Risks.* Washington, D.C.: American Chemical Society. A comprehensive look at public perception of radiation risks and health effects of radiation through experimentation, occupational exposure, atomic detonation, and nuclear reactor accidents.

CHAPTER 21

Water Supply, Use, and Management

Catskill Mountains of upstate New York is an ecosystem and landscape that provides high-quality water to millions of people in New York City as a national service function.

Learning Objectives

Although water is one of the most abundant resources on Earth, many important issues and problems are involved in water management. After reading this chapter, you should understand:

■ Why water is one of the major resource issues of the twenty-first century.

■ What a water budget is, and why it is useful in analyzing water

supply problems and potential solutions.

■ What groundwater is, and what environmental problems are associated with its use.

■ How water can be conserved at home and in industrial and agricultural practice.

■ Why sustainable water management will become more difficult as the demand for water increases.

■ What the environmental impacts are of water projects such as dams, reservoirs, canals, and channelization.

■ What a wetland is, how wetlands function, and why they are important.

■ Why we are facing a growing global water shortage linked to our food supply.

What Is the Value of Clean Water to New York City?

The forest of the Catskill Mountains in upstate New York (see opening photograph) provides water to about 9 million people in New York City. The total contributing area in the forest is about 5,000 km^2 (2,000 square miles), of which the city of New York owns less than 8%. The water from the Catskills has historically been of high quality and in fact was once regarded as one of the largest municipal water supplies in the United States that did not require extensive filtering. Of course, what we are talking about here is industrial filtration plants where the water enters from reservoirs and groundwater and is then treated before being dispersed to users. In the past, the water from the Catskills has been filtered very effectively by natural processes. As water from rain or melting snow drips from trees or melts on slopes in the spring, some of it enters (infiltrates) the soil. The water then moves through the soil into the rocks below as groundwater. Some emerges to feed streams that flow into reservoirs. During its journey, the water enters into a number of physical and chemical processes that naturally treat and filter the water. These are natural service functions that the Catskill forest ecosystem provides to the people of New York.

These service functions were taken for granted until about the 1990s, when it became apparent that the water supply was becoming vulnerable to pollution from uncontrolled development in the watershed. A particular concern was runoff from buildings and streets, as well as seepage from septic systems that treat wastewater from homes and buildings, partly by allowing wastewater to seep through soil. At that time the Environmental Protection Agency warned the city that unless the water quality improved, New York City would have to construct a water treatment plant to filter the water. The cost of such a facility was estimated as between $6 and $8 billion, with an annual operating expense of several hundred million dollars. As an alternative, New York City chose to attempt to improve the water quality at the source. The city built a sewage treatment plant upstate in the Catskill Mountains at a cost of about $2 billion. This seems very expensive but is about one-third the cost of building the treatment plant to filter water. Thus, the city chose to invest in the "natural capital" of the forest, hoping that it will continue its natural service function of providing clean water. It will probably take several decades to tell if New York City's gamble will work in the long term.[1]

There have been unanticipated benefits to maintaining the Catskill Mountain Forest ecosystem. These benefits come from recreational activities, particularly trout fishing, which is a multibillion-dollar enterprise in upstate New York. In addition to the trout fishermen are people wanting to experience the Catskill Mountains through wildlife observation, bird watching, hiking, and winter sports.

You might wonder why the city has been successful in its initial attempt to maintain high-quality water when it only owns about 8% of the land the water comes from. The reason is that the city has provided financial incentives to farmers, homeowners, and other people living in the forest to maintain high-quality water resources. Although the amount of money is not large, it is sufficient to provide a sense of stewardship among the landowners, and they are attempting to abide by guidelines that help protect water quality. The real power in the case history of water from the Catskill Mountains to New York City is that of valuing natural ecosystems for the functions they perform. With a little help, many of our ecosystems can provide a variety of services, including improved water and air quality.[1]

The city of New York is not the only U.S. city that has chosen to protect watersheds to produce clean high-quality drinking water rather than constructing and maintaining expensive water treatment plants. Other cities using watershed protection to supply their water include Boston, Massachusetts; Seattle, Washington; and Portland, Oregon.

Water is a critical, limited, renewable resource in many regions on Earth. As a result, water is one of the major resource issues of the twenty-first century. This chapter discusses our water resources in terms of supply, use, management, and sustainability. It also addresses important environmental concerns related to water: wetlands, dams and reservoirs, channelization, and flooding. Near the end of this chapter we present a discussion of water use, management, and environmental concerns for the Colorado River and the system of dams that controls the flow of the river.

21.1 Water

To understand water as a necessity, as a resource, and as a factor in the pollution problem, we must understand its characteristics, its role in the biosphere, and its role in sustaining life. Water is a unique liquid; without it, life as we know it is impossible. Consider the following:

- Compared with most other common liquids, water has a high capacity to absorb and store heat. The capacity of water to hold heat has important climatic significance. Solar energy warms the oceans of the world, storing huge amounts of heat. The heat can be transferred to the atmosphere to develop hurricanes and other storms. The heat in warm oceanic currents such as the Gulf Stream warms Great Britain and Western Europe, making these areas much more hospitable for humans than would otherwise be possible at such high latitudes.

- Water is the universal solvent. Because many natural waters are slightly acidic, they can dissolve a great variety of compounds, from simple salts to minerals, including sodium chloride (common table salt) and calcium carbonate (calcite) in limestone rock. Water also reacts with complex organic compounds, including many amino acids found in the human body.

- Compared with other common liquids, water has a high surface tension, a property that is extremely important in many physical and biological processes that involve moving water through, or storing water in, small openings or pore spaces.

- Among the common compounds, water is the only one whose solid form is lighter than its liquid form. (It expands by about 8% when it freezes, becoming less dense.) That is why ice floats. If ice were heavier than liquid water, it would sink to the bottom of the oceans, lakes, and rivers. If water froze from the bottom up, shallow seas, lakes, and rivers would freeze solid. All life in the water would die, because cells of living organisms are mostly water, and as water freezes and expands, cell membranes and walls rupture. If ice were heavier than water, the biosphere would be vastly different from what it is, and life, if it existed at all, would be greatly altered.[2]

- Sunlight penetrates water to variable depths, permitting photosynthetic organisms to live below the surface.

A Brief Global Perspective

The water supply problem is as follows: We are facing a growing global water shortage that is linked to our food supply. We will return to this important concept at the end of the chapter, following a discussion of water use, supply, and management.

A review of the global hydrologic cycle, introduced in Chapter 5, is important here. The main process in the cycle is the global transfer of water from the atmosphere to the land and oceans and back to the atmosphere (Figure 21.1). Table 21.1 lists the relative amounts of water in the major storage compartments of the cycle. Notice that more than 97% of Earth's water is in the oceans; the next largest storage compartment, the ice caps and glaciers, accounts for another 2%. Together, these sources account for more than 99% of the total water, and both are generally unsuitable for human use because of salinity (seawater) and location (ice caps and glaciers). Only about 0.001% of the total water on Earth is in the atmosphere at any one time. However, this relatively small amount of water in the global water cycle, with an average atmosphere residence time of only about 9 days, produces all our freshwater resources through the process of precipitation.

Water can be found in either liquid, solid, or gaseous form at a number of locations at or near Earth's surface.

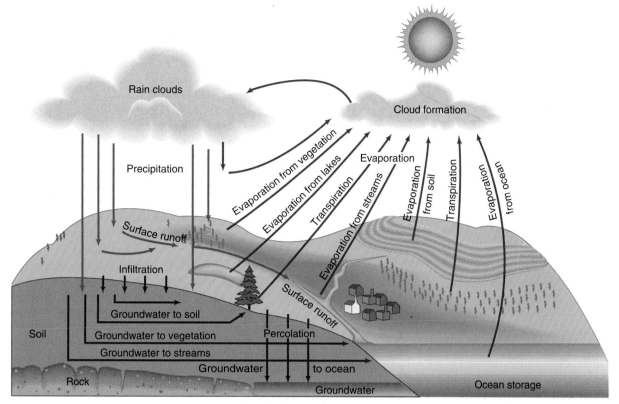

Figure 21.1 ■ The hydrologic cycle, showing important processes and transfer of water. [*Source:* Modified from Council on Environment Quality and Department of State, *The Global 2000 Report to the President*, vol. 2 (Washington, D.C.).]

Depending on the specific location, the residence time may vary from a few days to many thousands of years (see Table 21.1). However, as mentioned, more than 99% of Earth's water in its natural state is unavailable or unsuitable for beneficial human use. Thus, the amount of water for which all the people, plants, and animals on Earth compete is much less than 1% of the total.

As the world's population and industrial production of goods increase, the use of water will also accelerate. The world per capita use of water in 1975 was about 700 m^3/year, or 2,000 gal/day (185,000 gal/yr), and the total human use of water was about 3,850 km^3/year (about 10^{15} gal/yr). Today, world use of water is about 6,000 km^3/yr (about 1.58×10^{15} gal/yr), which is a significant fraction of the naturally available freshwater.

Compared with other resources, water is used in very large quantities. In recent years, the total mass (or weight) of water used on Earth per year has been approximately 1,000 times the world's total production of minerals, including petroleum, coal, metal ores, and nonmetals.[3] Because of its great abundance, water is generally a very inexpensive resource. However, in the southwestern United

Location	Surface Area (km^2)	Water Volume (km^3)	Percentage of Total Water	Estimated Average Residence Time of Water
Oceans	361,000,000	1,230,000,000	97.2	Thousands of years
Atmosphere	510,000,000	12,700	0.001	9 days
Rivers and streams	—	1,200	0.0001	2 weeks
Groundwater (shallow to depth of 0.8 km)	130,000,000	4,000,000	0.31	Hundreds to many thousands of years
Lakes (fresh water)	855,000	123,000	0.01	Tens of years
Ice caps and glaciers	28,200,000	28,600,000	2.15	Tens of thousands of years and longer

Table 21.1 • The World's Water Supply (Selected Examples)

Source: U.S. Geological Survey.

States, the cost of water has been kept artificially low as a result of government subsidies and programs.

Because the quantity and quality of water available at any particular time are highly variable, shortages of water have occurred and will probably continue to occur with increasing frequency. Such shortages can lead to serious economic disruption and human suffering.[4] In the Middle East and northern Africa, scarce water has resulted in harsh words and threats between countries. War over water is a possibility. The U.S. Water Resources Council estimates that water use in the United States by the year 2020 may exceed surface water resources by 13%.[4] Therefore, an important question is, how can we best manage our water resources, use, and treatment to maintain adequate supplies?

Groundwater and Streams

Next, before moving on to issues of water supply and management, we introduce groundwater and surface water and the terms used in discussing them. An acquaintance with this terminology is important in understanding many environmental issues, problems, and solutions.

The term **groundwater** usually refers to the water below the water table, where saturated conditions exist. The upper surface of the groundwater is called the *water table*.

Rain that falls on the land evaporates, runs off the surface, or moves below the surface and is transported underground. Locations where surface waters move into, or infiltrate, the ground are known as *recharge* zones. Places where groundwater flows or seeps out at the surface, such as springs, are known as *discharge zones or discharge points.*

Water that moves into the ground from the surface first seeps through pore spaces (empty spaces between soil particles or rock fractures) in the soil and rock known as the *vadose zone*. This area is seldom saturated (not all pore spaces are filled with water). The water then enters the groundwater system, which is saturated (all pore spaces are filled with water).

An *aquifer* is an underground zone or body of earth material from which groundwater can be obtained (from a well) at a useful rate. Loose gravel and sand with lots of pore space between grains and rocks or many open fractures generally make good aquifers. Groundwater in aquifers usually moves slowly at rates of centimeters or meters per day. When water is pumped from an aquifer, the water table is depressed around the well, forming a *cone of depression*. Figure 21.2 shows the major features of a groundwater and surface water system.

Streams may be classified as effluent or influent. In an **effluent stream**, the flow is maintained during the dry season by groundwater seepage into the stream channel from the subsurface. A stream that flows all year is called a perennial stream. Most perennial streams flow all year because they constantly receive groundwater to sustain flow. An **influent stream** is entirely above the water table and flows only in direct response to precipitation. Water from an influent stream seeps down into the subsurface. An influent stream is called an ephemeral stream because it doesn't flow all year.

A given stream may have reaches (unspecified lengths of stream) that are perennial and other reaches that are ephemeral. It may also have reaches, known as intermittent, that have a combination of influent and effluent flow varying with the time of year. For example, streams flowing from the mountains to the sea in Southern California often have reaches in the mountain that are perennial, supporting populations of trout or endangered southern steelhead, and lower intermittent reaches that transition

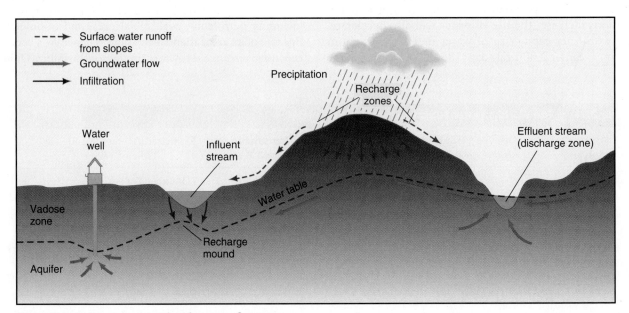

Figure 21.2 ■ Groundwater and surface water flow system.

to ephemeral reaches. At the coast, these streams may receive fresh or salty groundwater and tidal flow from the ocean to become a perennial lagoon.

Interactions between Surface Water and Groundwater

Surface water and groundwater interact in many ways and should be considered part of the same resource. Nearly all natural surface water environments, such as rivers and lakes, as well as human-constructed water environments, such as reservoirs, have strong linkages with groundwater. For example, withdrawal of groundwater by pumping from wells may reduce stream flow, lower lake levels, or change the quality of surface water. Reduction of effluent stream flow by lowering the groundwater level may change a perennial stream that flows all year to an intermittent influent stream. Similarly, withdrawal of sur-

face water by diversion from streams and rivers can deplete groundwater resources or change the quality of groundwater. Diversion of surface waters that recharge groundwaters may result in an increase in concentrations of dissolved chemicals in the groundwater. This happens because dissolved chemicals present in the groundwater are not diluted by mixing with infiltrated surface water. Finally, pollution of groundwater may result in pollution of surface water, and vice versa.[5]

Selected interactions between surface water and groundwater in a semiarid urban and agricultural environment are shown in Figure 21.3. Urban and agricultural runoff increase the volume of water in the reservoir. Pumping of groundwater for agricultural and urban uses lowers the groundwater level. Quality of surface water and groundwater is reduced by urban and agricultural runoff, which adds nutrients from fertilizers, oil from roads, and nutrients from treated wastewaters to streams and groundwater.

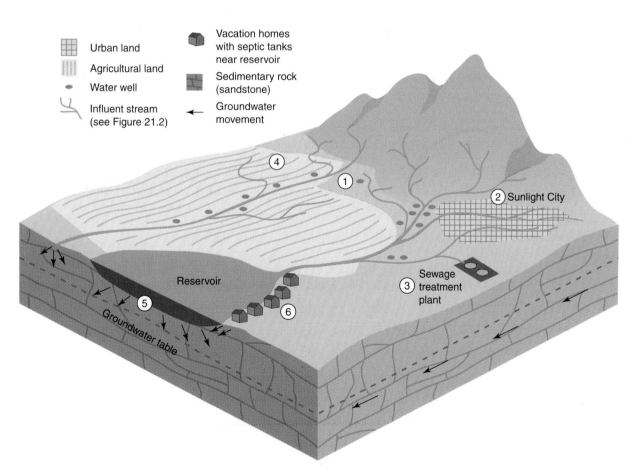

Figure 21.3 ■ Idealized diagram illustrating some interactions between surface water and groundwater for a city in a semiarid environment with adjacent agricultural land and reservoir. (1) Water pumped from wells lowers the groundwater level. (2) Urbanization increases runoff to streams. (3) Sewage treatment discharges nutrient-rich waters to stream, groundwater, and reservoir. (4) Agriculture uses irrigation waters from wells, and runoff to stream from fields contains nutrients from fertilizers. (5) Water from the reservoir is seeping down to the ground water. (6) Water from septic systems for homes is seeping down through the soil to the groundwater.

21.2 Water Supply: A U.S. Example

The water supply at any particular point on the land surface depends on several factors in the hydrologic cycle, including the rates of precipitation, evaporation, transpiration (water in vapor form that directly enters the atmosphere from plants through pores in leaves and stems), stream flow, and subsurface flow. A concept useful in understanding water supply is the **water budget**, which is a model that balances the inputs, outputs, and storage of water in a system. Simple annual water budgets (precipitation – evaporation = runoff) for North America and other continents are shown on Table 21.2. The total average annual water yield (runoff) from Earth's rivers is approximately 47,000 km^3 (1.2×10^{16} gal), but its distribution is far from uniform (see Table 21.2). Some runoff occurs in relatively uninhabited regions, such as Antarctica, which produces about 5% of Earth's total runoff. South America, which includes the relatively uninhabited Amazon basin, provides about one-fourth of Earth's total runoff. Total runoff in North America is about two-thirds that of South America. Unfortunately, much of the North American runoff occurs in sparsely settled or uninhabited regions, particularly in the northern parts of Canada and Alaska.

The daily water budget for the contiguous United States is shown in Figure 21.4. The amount of water vapor passing over the United States every day is approximately 152,000 million m^3 (40 trillion gal); and of this total, approximately 10% falls as precipitation in the form of rain, snow, hail, or sleet. Approximately 66% of the precipitation evaporates quickly or is transpired by vegetation. The remaining 34% enters the surface water or groundwater storage systems, flows to the oceans or across the nation's boundaries, is used by people, or evaporates from

reservoirs. Owing to natural variations in precipitation that cause either floods or droughts, only a portion of this water can be developed for intensive uses (only about 50% is considered available 95% of the time).[4]

Precipitation and Runoff Patterns

To put all this information in perspective, consider just the water in the Missouri River. In an average year, the water that flows down the Missouri River is enough to cover 25 million acres a foot deep—8.4 trillion gallons. The average water use in the United States is about 100 gallons a day per person—very high compared to the rest of the world. People in Europe use about one-half of that, and in some regions, such as sub-Saharan Africa, people make do with 5 gallons a day. At 100 gallons use a day, the Missouri's flow is enough to provide domestic water and public water use in the United States for about 230 million people. With a little water conservation and reduction in per capita use, the Missouri could provide enough water for all the people, so great is its flow. Not that people would actually use the Missouri's water that way, but you can stand on the shore of the Missouri, where the river flows under a major highway bridge, and get an idea of just how much water it would take to supply all those people.

In developing water budgets for water resources management, it is useful to consider annual precipitation and runoff patterns. Potential problems with water supply can be predicted in areas where average precipitation and runoff are relatively low, such as in the arid and semiarid parts of the southwestern and Great Plains regions of the United States. Surface water supply can never be as high as the average annual runoff because not all runoff can be successfully stored. Total storage of runoff is not possible because of evaporative losses from river channels, ponds,

Table 21.2 • Annual Water Budgets for the Continents[a]

Continental	Precipitation mm/yr	Precipitation km^3	Evaporation mm/yr	Evaporation km^3	Runoff km^3/yr
North America	756	18,300	418	10,000	8,180
South America	1,600	28,400	910	16,200	12,200
Europe	790	8,290	507	5,320	2,970
Asia	740	32,200	416	18,100	14,100
Africa	740	22,300	587	17,700	4,600
Australia and Oceania	791	7,080	511	4,570	2,510
Antarctica	165	2,310	0	0	2,310
Earth (entire land area)	800	119,000	485	72,000	47,000[b]

[a] Precipitation – evaporation = runoff.

[b] Surface runoff is 44,800; groundwater runoff is 2,200.

Source: I. A. Shiklomanov, "World Fresh Water Resources," in P. H. Gleick, ed., *Water in Crisis* (New York: Oxford University Press, 1993), pp. 3–12.

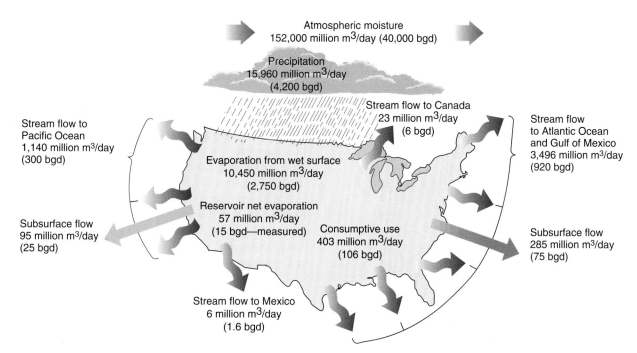

Figure 21.4 ■ Water budget for the United States (bgd = billion gallons per day). [*Source:* Water Resources Council, *The Nation's* *Water Resources 1975–2000* (Washington, D.C.: Water Resources Council, 1978).]

lakes, and reservoirs. As a result, shortages in the water supply are common in areas with natural low precipitation and runoff coupled with strong evaporation. In these areas, strong conservation practices are necessary to help ensure an adequate supply of water.[4]

Droughts

Because there are large annual and regional variations in stream flow, even areas with high precipitation and runoff may periodically suffer from droughts. For example, the dry years near the end of the twentieth century in the western United States produced serious water shortages. Fortunately for the more humid eastern United States, stream flow there tends to vary less than in other regions, and drought is less likely.[4] Nevertheless, droughts in the summers of some years in the southeastern United States will continue to cause hardships and billions of dollars of damage.

Groundwater Use and Problems

Nearly half the people in the United States use groundwater as a primary source for drinking water. It accounts for approximately 20% of all water used. Fortunately, the total amount of groundwater available in the United States is enormous. In the contiguous United States, the amount of shallow groundwater within 0.8 km (about 0.5 mi) of the surface of the land is estimated to be between 125,000 and 224,000 km^3 (3.3×10^{16} to 5.9×10^{16} gal). To put this in perspective, the lower estimate of the amount of shallow

groundwater is about equal to the total discharge of the Mississippi River during the last 200 years. However, the high cost of pumping limits the total amount of groundwater that can be economically recovered.[4]

In many parts of the country, groundwater withdrawal from wells exceeds natural inflow. In such cases of **overdraft**, we can think of water as a nonrenewable resource that is being *mined*. This can lead to a variety of problems, including damage to river ecosystems and land subsidence. Groundwater overdraft is a serious problem in the Texas–Oklahoma–High Plains area (which includes much of Kansas and Nebraska and parts of other states), as well as in California, Arizona, Nevada, New Mexico, and isolated areas of Louisiana, Mississippi, Arkansas, and the south Atlantic region.

In the Texas–Oklahoma–High Plains area, the overdraft amount per year is approximately equal to the natural flow of the Colorado River for the same period.[4] The Ogallala aquifer (also called the High Plains aquifer), which is composed of water-bearing sands and gravels that underlie an area of about 400,000 km^2 from South Dakota into Texas, is the main groundwater resource in this area. Although the aquifer holds a tremendous amount of groundwater, it is being used in some areas at a rate up to 20 times higher than the rate at which it is being naturally replaced. The water table in many parts of the aquifer has declined in recent years (Figure 21.5), causing yields from wells to decrease and energy costs for pumping the water to increase. The most severe water depletion problems in the Ogallala aquifer today are in locations where irrigation was first used in the 1940s. There

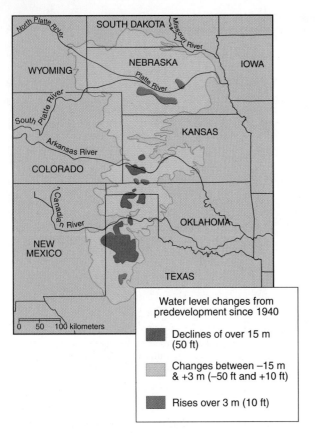

Water level changes from
predevelopment since 1940

- ■ Declines of over 15 m (50 ft)
- ■ Changes between −15 m & +3 m (−50 ft and +10 ft)
- ■ Rises over 3 m (10 ft)

Figure 21.5 ■ Groundwater level changes as a result of pumping in the Texas–Oklahoma–High Plains region. [*Source:* U.S. Geological Survey.]

is concern that eventually a significant portion of land now being irrigated will be returned to dryland farming as the resource is used up.

Some towns and cities in the High Plains are also starting to have water supply problems. Along the Platte River in northern Kansas there is still plenty of water and groundwater levels are high (Figure 21.5). Further south, in southwest Kansas and the panhandle in western Texas, where water levels have declined the most, supplies may only last another decade or so. In Ulysses, Kansas, (population 6,000) and Lubbock, Texas, (population 200,000) the situation is already getting serious. South of Ulysses, Lower Cimarron Springs, which was a famous water hole along a dry part of the Santa Fe Trail, dried up decades ago due to pumping groundwater. It was a symptom of what was coming. Both Ulysses and Lubbock are now facing water shortages and will need to spend millions of dollars to find alternative sources.

Desalination as a Water Source

Seawater is about 3.5% salt; that means each cubic meter of seawater contains about 40 kg (88 lb) of salt. **Desalination**, a technology to remove salt from water, is being used at several hundred production plants around the world to produce water with reduced salt. The salt content of the water must be reduced to about 0.05% for the water to be used as a freshwater resource. Large desalination plants produce 20,000–30,000 m^3 (about 5–8 million gal) of water per day. Today there are about 15,000 desalination plants in over 100 countries in operation. Improving technology is significantly reducing the cost of desalination.

The cost of desalinated water is several times that paid for traditional water supplies in the United States. Desalinated water has a *place value*, which means that the price increases quickly with the transport distance and the cost of moving water from the plant. Because the various processes that remove the salt require large amounts of energy, the cost of the water is also tied to ever-increasing energy costs. For these reasons, desalination will remain an expensive process used only when alternative water sources are not available.

Desalination also has environmental impacts. Discharge of very salty water from a desalination plant into another body of water, such as a bay, may locally increase salinity and kill some plants and animals intolerant to salt. The discharge from desalination plants may also cause wide fluctuations in salt content of local environments, which may damage ecosystems.

21.3 Water Use

In discussing water use, it is important to distinguish between off-stream and in-stream uses. **Off-stream use** refers to water removed from its source (such as a river or reservoir) for use. Much of this water is returned to the source after use; for example, the water used to cool industrial processes may go to cooling ponds and then be discharged to a river, lake, or reservoir. **Consumptive use** is an off-stream use in which water is consumed by plants and animals or used in industrial processes. The water enters human tissue or products or evaporates during use and is not returned to its source.[4]

In-stream use includes the use of rivers for navigation, hydroelectric power generation, fish and wildlife habitats, and recreation. These multiple uses usually create controversy, because each requires different conditions to prevent damage or detrimental effects. For example, fish and wildlife require certain water levels and flow rates for maximum biological productivity; these levels and rates will differ from those needed for hydroelectric power generation, which requires large fluctuations in discharges to match power needs. Similarly, in-stream uses of water for fish and wildlife will likely conflict with requirements for shipping and boating. Figure 21.6 demonstrates some of these conflicting demands on a graph that shows optimal discharge for various uses throughout the year. In-stream water use for navigation is optimal at a constant fairly high discharge. Some fish, however, prefer higher flows in the spring for spawning.

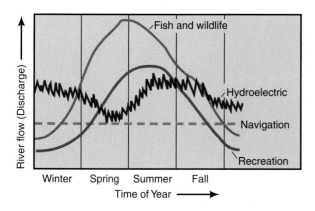

Figure 21.6 ■ In-stream water uses and optimal discharges (volume of water flowing per second) for each use. Discharge is the amount of water passing by a particular location and is measured in cubic meters per second. Obviously, all these needs cannot be met simultaneously.

Another problem for off-stream use is how much water can be removed from a stream or river without damaging the stream's ecosystem. This is an issue in the Pacific Northwest, where fish, such as steelhead trout and salmon, are on the decline partly because diversions (removal of water for agricultural, urban, and other uses) have reduced stream flow to the extent that fish habitats are damaged.

The Aral Sea in Kazakhstan and Uzbekistan provides a wake-up call regarding the environmental damage that can be caused by diverting water for agricultural purposes. Diversion of water from the two rivers that flow into the Aral Sea has transformed one of the largest bodies of inland water in the world from a vibrant ecosystem into a dying sea. The present shoreline is surrounded by thousands of square kilometers of salt flats that formed as the surface area of the sea was reduced by about 40% in the past 40 years (Figure 21.7). The volume of the sea was reduced by more than 50%. The salt content of the water increased fish kills, such as sturgeon that are an important component of the economy. Dust raised by winds from the dry salt flats is producing a regional air pollution problem, and the climate in the region has changed as the moderating effect of the sea has been reduced. Winters have grown colder and summers warmer. Fishing centers such as Muynak in the south and Aralsk to the north that were once on the shore of the sea are now many kilometers inland. Loss of fishing with decline of tourism has damaged the local economy.

A restoration of the small northern port of the Aral Sea is ongoing. A low, long dam was constructed across the lake bed just south of where the Syr Darya river enters the lake (see Figure 21.7). Conservation of water and the dam are producing dramatic improvement to the northern port of the lake and some fishing there is returning. The future of the lake has improved but great concern remains.

Transport of Water

In many parts of the world, demands are being made on rivers to supply water to agricultural and urban areas. This is not a new trend. Ancient civilizations, including the Romans and Native Americans, constructed canals and aqueducts to transport water from distant rivers to where it was needed. In our modern civilization, as in the past, water is often moved long distances from areas with abundant rainfall or snow to areas of high usage (usually agricultural areas). For instance, in California, two-thirds of the state's runoff occurs north of San Francisco, where there is a surplus of water. However, two-thirds of the water use in California occurs south of San Francisco, where there is a deficit. In recent years, canals of the California Water Project have moved tremendous amounts of water from the northern to the southern part of the state, mostly for agricultural uses but increasingly for urban uses as well.

On the opposite coast, New York City has imported water from nearby areas for more than 100 years. Water use and supply in New York City show a repeating pattern. Originally, local groundwater, streams, and the Hudson River itself were used. However, as population increased and the land was paved over, surface waters were diverted to the sea rather than percolating into the soil to replenish groundwater. Furthermore, what water did infiltrate the soil was polluted by urban runoff. Water needs in New York exceeded local supply, and in 1842 the first large dam was built.

As the city rapidly expanded from Manhattan to Long Island, water needs increased. The shallow aquifers of Long Island were at first a source of drinking water, but this water was used faster than the infiltration of rainfall could replenish it. At the same time, the groundwater became contaminated with urban and agricultural pollutants and from salt water seeping in underground from the ocean. (The pollution of Long Island groundwater is explored in more depth in the next chapter.) A larger dam was built at Croton in 1900. Further expansion of the population created the same pattern: initial use of groundwater; pollution, salinization, and overuse of the resource; and subsequent building of new, larger dams farther and farther upstate in forested areas.

In a broader perspective, the cost of obtaining water for large urban centers from long distances, along with competition for available water from other sources and users, will eventually place an upper limit on the water supply of the city. As shortages develop, stronger conservation measures are implemented, and the cost of water increases. As with other resources, as the water supply is reduced and demand for water increases, so does its price. If the price becomes high enough, more expensive sources may be developed—for example, pumping from deeper wells or using desalination.

Some Trends in Water Use

Trends in freshwater withdrawals and human population for the United States from 1950 to 1995 are shown in

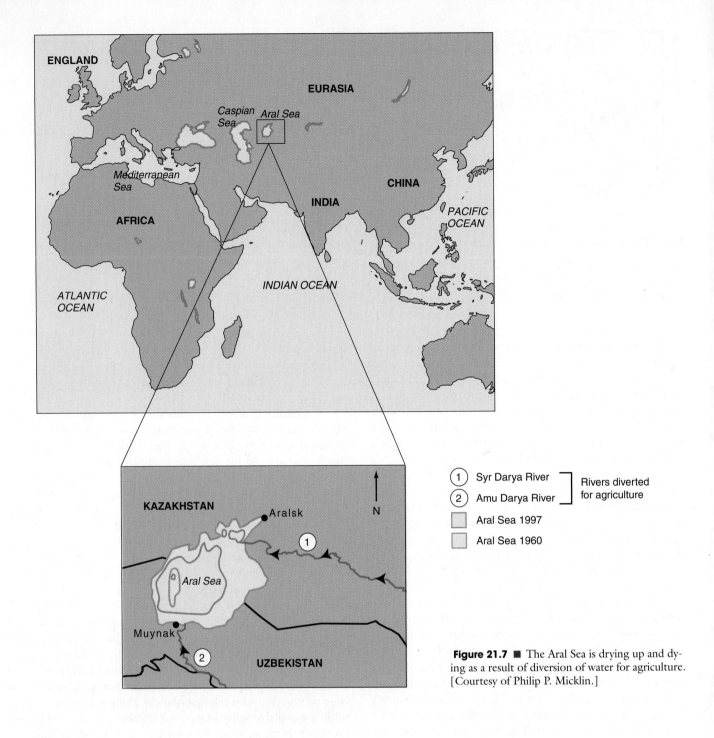

Figure 21.7 ■ The Aral Sea is drying up and dying as a result of diversion of water for agriculture. [Courtesy of Philip P. Micklin.]

Map legend:
① Syr Darya River ⎤ Rivers diverted
② Amu Darya River ⎦ for agriculture

Aral Sea 1997
Aral Sea 1960

Figure 21.8. You can see that withdrawal of surface water far exceeds withdrawal of groundwater. In addition, withdrawals of both surface water for human uses and groundwater increased between 1950 and 1980, reaching a total maximum of approximately 375 thousand million gal/day. However, since 1980, water withdrawals have decreased and leveled off. It is encouraging that water withdrawals have decreased since 1980 while the population of the United States has continued to increase. This suggests that

improvements have been made in water management and water conservation.[6, 7]

Trends in freshwater withdrawals by water-use categories for the United States from 1960 to 1995 are shown in Figure 21.9. Examination of this graph suggests that:

1. The major uses of water are for irrigation and the thermoelectric industry.

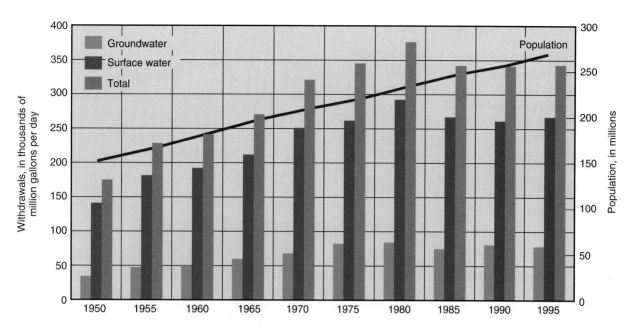

Figure 21.8 ■ Trends in U.S. fresh groundwater and surface water withdrawals and human population (1950–1995). [*Source:* W. B. Solley, R. P. Pierce, and H. A. Perlman. 1998. *Estimated Use of Water in the United States in 1995.* U.S. Geological Survey Circular 1200, 1998.]

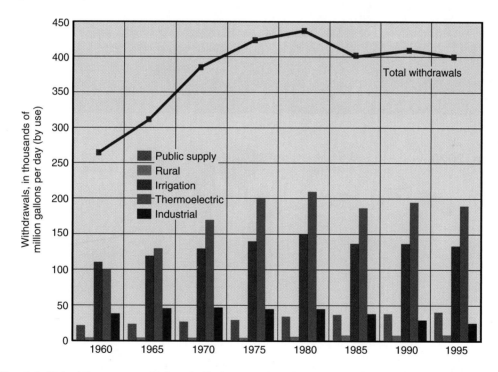

Figure 21.9 ■ Trends in United States water withdrawals (fresh and saline) by water-use category and total (fresh and saline) withdrawals (1960–1995). [*Source:* W. B. Solley, R. P. Pierce, and H. A. Perlman. 1998. *Estimated Use of Water in the United States in 1995.* U.S. Geological Survey Circular 1200, 1998.]

2. The use of water for irrigation by agriculture leveled off starting in about 1980.

3. Water use by the thermoelectric industry and other industries decreased slightly beginning in 1980.

4. Use of water for public and rural supplies continued to increase through the period from 1950 to 1995, presumably related to the increase in human population.[7]

21.4 Water Conservation

Water conservation is the careful use and protection of water resources. It involves both the quantity of water used and its quality. Conservation is an important component of sustainable water use. Because the field of water conservation is changing rapidly, it is expected that a number of innovations will reduce the total withdrawals of water for various purposes, even though consumption will continue to increase.[3]

Agricultural Use

Improved irrigation (Figure 21.10) could reduce agricultural withdrawals by 20 to 30%. Because agriculture is the biggest water user, this would be a tremendous savings. Suggestions for agricultural conservation include the following:

▨ Price agricultural water to encourage conservation (subsidizing water will encourage overuse).

Agriculture, 1990

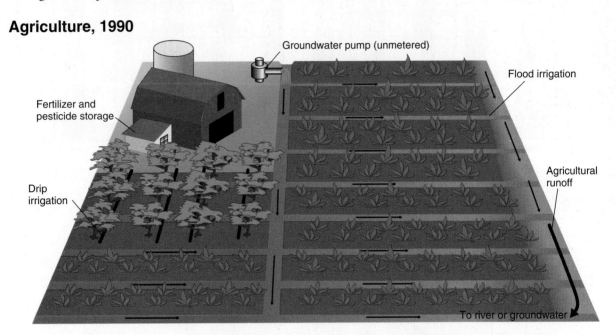

Agriculture, 2020

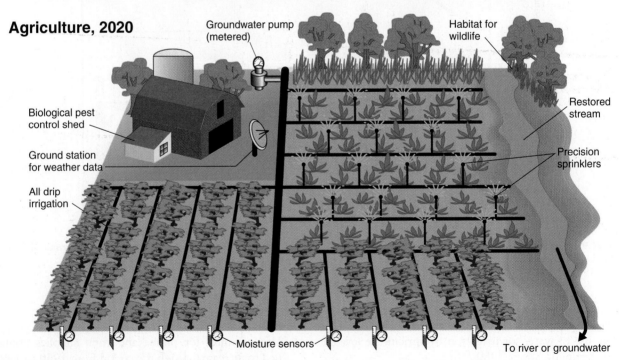

Figure 21.10 ■ Comparison of agricultural practices in 1990 with what they might be by 2020. The improvements call for a variety of agricultural procedures, from biological pest control to more efficient application of irrigation water to restoration of water resources and wildlife habitat. [*Source:* P. H. Gleick, P. Loh, S. V. Gomez, and J. Morrison, *California Water 2020, a Sustainable Vision* (Oakland, Calif.: Pacific Institute for Studies in Development, Environment, and Security, 1995).]

- Use lined or covered canals that reduce seepage and evaporation.
- Use computer monitoring and schedule release of water for maximum efficiency.
- Integrate the use of surface water and groundwater to more effectively use the total resource. That is, irrigate with surplus surface water when it is abundant, and also use surplus surface water to recharge groundwater aquifers by applying the surface water to specially designed infiltration ponds or injection wells. When surface water is in short supply, use more groundwater.
- Irrigate at times when evaporation is minimal, such as at night or in the early morning.
- Use improved irrigation systems, such as sprinklers or drip irrigation, that more effectively apply water to crops.
- Improve land preparation for water application; that is, improve the soil to increase infiltration and minimize runoff. Where applicable, use mulch to help retain water around plants.
- Encourage the development of crops that require less water or are more salt tolerant so that less periodic flooding of irrigated land is necessary to remove accumulated salts in the soil.

Domestic Use

Domestic use of water accounts for only about 10% of total national water withdrawals. However, because domestic water use is concentrated in urban areas, it may pose major local problems in areas where water is periodically or often in short supply. (See a Closer Look 21.1.) Most water in homes is used in the bathroom and for washing clothing and dishes. Water use for domestic purposes can be substantially reduced at a relatively small cost by implementing the following measures:

- In semiarid regions, replace lawns with decorative gravels and native plants.
- Use more efficient bathroom fixtures, such as low-flow toilets that use 1.6 gallons or less per flush rather than the standard 5 gallons and low-flow shower heads that deliver less but sufficient water.
- Turn off water when not absolutely needed for washing, brushing teeth, shaving, and so on.
- Flush the toilet only when really necessary.
- Fix all leaks quickly. Dripping pipes, faucets, toilets, or garden hoses waste water. A small drip can waste several liters per day; multiply this by millions of homes with a leak, and a large volume of water is lost.
- Purchase dishwashers and washing machines that minimize water consumption.

- Take a long bath rather than a long shower.
- Don't wash sidewalks and driveways with water (sweep them).
- Consider using gray water (from showers, bathtubs, sinks, and washing machines) to water vegetation. The gray water from washing machines is easiest to use, as it can be easily diverted before entering a drain.
- Water lawns and plants in the early morning, late afternoon, or at night to reduce evaporation.
- Use drip irrigation and place water-holding mulch around garden plants.
- Plant drought-resistant vegetation that requires less water.
- Learn how to read the water meter to monitor for unobserved leaks and record your conservation successes.

In addition, local water districts should encourage water pricing policies in which water is more expensive beyond some baseline amount determined by the number of people in a home and the size of the property.

Industry and Manufacturing Use

Water conservation measures taken by industry can be improved. For instance, water removal for steam generation of electricity could be reduced 25 to 30% by using cooling towers that use less or no water. Manufacturing and industry could curb water withdrawals by increasing in-plant treatment and recycling of water and by developing new equipment and processes that require less water.[4]

Perception and Water Use

How the water supply is perceived by people is important in determining how much water is used. Perception of water is based partly on its price and availability. If water is abundant and inexpensive, we don't think much about it. If water is scarce or expensive, it is another matter. For example, people in Tucson, Arizona, perceive the area as a desert (which it is) and use a lot of native plants (cactus and other desert plants) in yards and gardens around homes and buildings there. Tucson's water supply is mostly from groundwater, which is being mined (used faster than it is being naturally replenished). Tucson also receives some Colorado River water, used for irrigation and industrial purposes. Water use in Tucson is about 409 liters (108 gal) per person per day. Not far from Tucson, the people of Phoenix use about 60% more at 662 liters (175 gal) per person per day. In parts of Phoenix, as much as 3,780 liters (1,000 gal) of water per person per day is used to water mulberry trees and high hedges.

Phoenix has been accused of having an oasis mentality concerning water use, reflected in part by water rates. The people in Tucson often pay about 100% more for water

Water Supplies for Many Urban Areas in the United States Are in Trouble

The population of the United States continues to grow, and many urban areas in the United States are experiencing or will experience the impact of population growth on water supply. For example:

■ Southern California, in particular San Diego, is growing rapidly and its water needs are quickly becoming greater than local supplies. As a result, the city of San Diego has negotiated with farmers to the east in the Imperial Valley to purchase water for urban areas. The city is also building desalination plants and considering increasing the heights of dams so that more water can be stored for urban uses.

■ In Denver, city officials, fearing future water shortages, are proposing strict water conservation measures that include limits on water available for landscaping and the amount of grass that can be planted around new homes.

■ Chicago, the seventh-fastest-growing area in the United States from 1990 to 2000, reported significant groundwater depletion problems following a recent drought.

■ Tampa, Florida, fearing shortages of freshwater because of its continuing growth, began operating a desalination plant in 2003 that produces approximately 25 million gallons of water daily.

■ Atlanta, Georgia, the fourth-fastest-growing urban area in the United States from 1990 to 2000, is expecting increased demand on its water supplies as a result of population increase and is exploring ways to meet those demands.

■ New York City declared a drought emergency in 2002 and placed water restrictions on its more than 9 million citizens.

What is clear from these examples is that while there is no shortage of water in the United States or the world, there are local and regional shortages, particularly in large, growing urban areas in the semiarid western and southwestern United States.[8]

---

than the people in Phoenix, where the rates are among the lowest in the western United States. Water rates in Tucson are structured to encourage conservation, and some industries consider water conservation a cost control measure.[9] For example, for residential water use in Tucson (for a family of four), the price increases as water use per month increases. In Phoenix, the lower price of water encourages people to use more, and there is less incentive for water conservation. The message is that we could all do with a little of Tucson's desert mentality. This is particularly true for those in large urban areas, such as Los Angeles and San Diego, in southern California.

21.5 Sustainability and Water Management

Water is essential to sustain life and to maintain ecological systems necessary for the survival of humans. As a result, water plays important roles in ecosystem support, economic development, cultural values, and community well-being. Managing water use for sustainability is thus important in many ways.

Sustainable Water Use

From a water supply use and management perspective, **sustainable water use** can be defined as use of water

resources by people in a way that allows society to develop and flourish into an indefinite future without degrading the various components of the hydrologic cycle or the ecological systems that depend on it.[10] Some general criteria for water use sustainability are as follows.[10]

■ Develop water resources in sufficient volume to maintain human health and well-being.

■ Provide sufficient water resources to guarantee the health and maintenance of ecosystems.

■ Ensure minimum standards of water quality for the various users of water resources.

■ Ensure that actions of humans do not damage or reduce long-term renewability of water resources.

■ Promote the use of water-efficient technology and practice.

■ Gradually eliminate water pricing policies that subsidize the inefficient use of water.

Groundwater Sustainability

The concept of sustainability by its very nature involves a long-term perspective. With groundwater resources, the length of time for effective management for sustainability is even longer than for other renewable resources. Surface waters, for example, may be replaced over a relatively short time. In contrast, groundwater development may take place over many years, at relatively slow rates. Effects of pumping groundwater at rates greater than natural replenishment rates may take years to be recognized. Similarly, effects of withdrawal of groundwater, such as drying up of springs or reduction of stream flow, may not be recognized until years after pumping begins. The long-term approach to sustainability with respect to groundwater often involves balancing withdrawals of groundwater resources with recharge of those resources, which is an important component of water management.[11]

Water Management

Management of water resources for water supply is a complex issue that will become more difficult as demand for water increases in the coming years. This difficulty will be especially apparent in the southwestern United States and other semiarid and arid parts of the world where water is or soon will be in short supply. Options for minimizing potential water supply problems include locating alternative water supplies and managing existing supplies better. In some areas, location of new supplies is unlikely, and serious consideration is being given to ideas as original as towing icebergs to coastal regions where fresh water is needed. It seems apparent that water will become much

more expensive in the future; if the price is right, many innovative programs are possible.

A method of water management utilized by a number of municipalities is known as the *variable-water-source approach*. For example, the city of Santa Barbara, California, has developed a variable-water-source approach that uses several interrelated measures to meet present and future water demands. Details of the plan (shown in Figure 21.11) include importing state water, developing new sources, using reclaimed water, and instituting a permanent conservation program. In essence, this seaside community has developed a master water plan.

A Master Plan for Water Management

Luna Leopold, a famous U.S. hydrologist, suggests that a new philosophy of water management is needed, one based on geologic, geographic, and climatic factors as well as on the traditional economic, social, and political factors. He argues that the management of water resources cannot be successful as long as it is naively perceived from an economic and political standpoint.

The essence of Leopold's water management philosophy is that surface water and groundwater are both subject to natural flux with time. In wet years, there is plenty of surface water, and the near-surface groundwater resources are

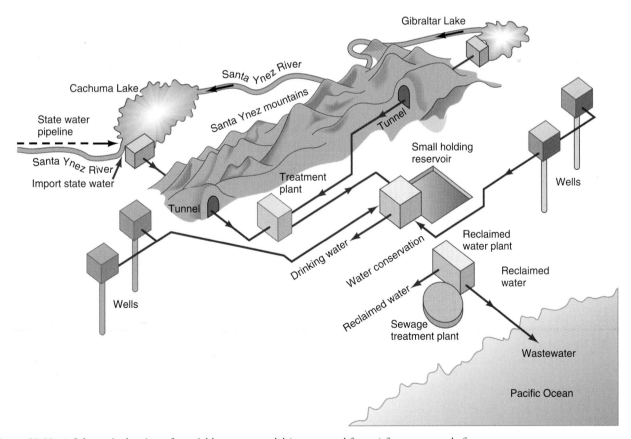

Figure 21.11 ■ Schematic drawing of a variable source model (present and future) for water supply for the city of Santa Barbara, California. [*Source:* Santa Barbara City Council, 1991.]

replenished. During dry years, which must be expected even though they may not be accurately predicted, specific plans to supply water on an emergency basis to minimize hardships must be in place and ready to use.

For example, subsurface waters in various locations in the western United States are too deep to be economically pumped from wells or have marginal water quality. These waters may be isolated from the present hydrologic cycle and therefore not subject to natural recharge. Such water might be used when the need is great. However, advance planning to drill the wells and connect them to existing water lines is necessary if they are to be ready when the need arises.

Another possible emergency plan might involve the treatment of wastewater. Reuse of such water on a regular basis might be too expensive, but advance planning to reuse treated water during emergencies could be a wise decision.

Finally, we should develop plans to use surface water when available, and we should not be afraid to use groundwater as needed in dry years. During wet years, natural recharge as well as artificial recharge (pumping excess surface water into the ground) will replenish the groundwater resources. This water management plan recognizes that excesses and deficiencies in water are natural and can be planned for.[12]

Water Management and the Environment

Many agricultural and urban areas require water to be delivered from nearby (and, in some cases, not-so-nearby) sources. To deliver the water, a system is needed for water storage and routing by way of canals and aqueducts from reservoirs. As a result, dams are built, wetlands may be modified, and rivers may be channelized to help control flooding. Often, a good deal of controversy surrounds water development.

The days of developing large projects in the United States without environmental and public review have passed. The resolution of development issues now involves input from a variety of government and public groups, which may have very different needs and concerns. These range from agricultural groups that see water development as critical for their livelihood to groups primarily concerned with wildlife and wilderness preservation. It is a positive sign that the various parties with interests in water issues are encouraged—and, in some cases, required—to meet and communicate their desires and concerns. Next, we address more directly the subjects of some of these concerns: wetlands, dams, channelization, and flooding.

21.6 Wetlands

Wetlands is a comprehensive term for landforms such as salt marshes, swamps, bogs, prairie potholes, and vernal pools (shallow depressions that seasonally hold water). Their common feature is that they are wet at least part of the year and as a result have a particular type of vegetation and soil. Figure 21.12 shows several types of wetlands.

(a)

(b)

(c)

Figure 21.12 ■ Several types of wetlands: (*a*) aerial view of part of the Florida Everglades at a coastal site; (*b*) cypress swamp, water surface covered with a floating mat of duckweed, northeast Texas; and (*c*) aerial view of farmlands encroaching on prairie potholes, North Dakota.

Wetlands may be defined as areas that are inundated by water or where the land is saturated to a depth of a few centimeters for at least a few days per year. Three major components used to determine the presence of wetlands are: hydrology, or wetness; type of vegetation; and type of soil. Of these, hydrology is often the most difficult to define, because some freshwater wetlands may be wet for only a few days a year. The duration of inundation or saturation must be sufficient for the development of wetland soils, which are characterized by poor drainage and lack of oxygen, and for the growth of specially adapted vegetation.[13]

Natural Service Functions of Wetlands

Wetland ecosystems may serve a variety of natural service functions for other ecosystems and for people, including the following:

- Freshwater wetlands are a natural sponge for water. During high river flow they store water, reducing downstream flooding. Following a flood, they slowly release the stored water, nourishing low flows.
- Many freshwater wetlands are important as areas of groundwater recharge (water seeps into the ground from a prairie pothole, for instance) or discharge (water seeps out of the ground in a marsh that is fed by springs).
- Wetlands are one of the primary nursery grounds for fish, shellfish, aquatic birds, and other animals. It has been estimated that as many as 45% of endangered animals and 26% of endangered plants either live in wetlands or depend on them for their continued existence.[13]
- Wetlands are natural filters that help purify water; plants in wetlands trap sediment and toxins.
- Wetlands are often highly productive and are places where many nutrients and chemicals are naturally cycled.
- Coastal wetlands provide a buffer for inland areas from storms and high waves.
- Wetlands are an important storage site for organic carbon; carbon is stored in living plants, animals, and rich organic soils.
- Wetlands are aesthetically pleasing to people.

Freshwater wetlands are threatened in many areas. One percent of the nation's total wetlands are lost every two years, and freshwater wetlands account for 95% of this loss. Wetlands such as prairie potholes in the midwestern United States and vernal pools in southern California are particularly vulnerable because their hydrology is poorly understood and establishing their wetland status is more difficult.[14] Over the past 200 years, over 50% of the wetlands in the United States have disappeared because they have been diked or drained for agricultural purposes or

filled for urban or industrial development. Perhaps as much as 90% of the freshwater wetlands have disappeared.

Although most coastal marshes are now protected in the United States, the extensive salt marshes at many of the nation's major estuaries, where rivers entering the ocean widen and are influenced by tides, have been modified or lost. These include deltas and estuaries of major rivers such as the Mississippi, Potomac, Susquehanna (Chesapeake Bay), Delaware, and Hudson.[15] The San Francisco Bay estuary, considered the estuary most modified by human activity in the United States today, has lost nearly all its marshlands to leveeing and filling (Figure 21.13).[15] Modifications result not only from filling and diking but also from loss of water. The freshwater inflow has been reduced by more than 50%, dramatically changing the hydrology of the bay in terms of flow characteristics and water quality. As a result of the modifications, the plants and animals in the bay have changed as habitats for fish and wildfowl have been eliminated.[15]

The delta of the Mississippi River includes some of the major coastal wetlands of the United States and the world. Historically, coastal wetlands of southern Louisiana were maintained by flooding of the Mississippi River, which delivered water, mineral sediments, and nutrients to the coastal

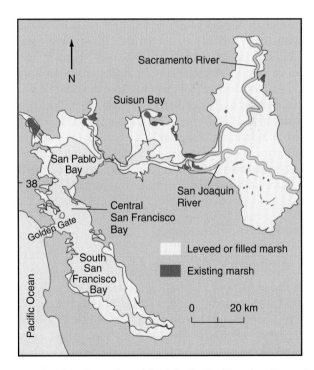

Figure 21.13 ■ Loss of marshlands in the San Francisco Bay and estuary from about 1850 to the present. [*Sources:* T. J. Conomos, ed., *San Francisco, the Urbanized Estuary* (San Francisco: American Association for the Advancement of Science, 1979); F. H. Nichols, J. E. Cloern, S. N. Luoma, and D. H. Peterson, "The Modification of an Estuary," *Science* 231 [1986]: 567–573. Copyright 1986 by the American Association for the Advancement of Science.]

environment. The mineral sediments contributed to the vertical accretion (building up) of wetlands. The nutrients enhanced growth of wetland plants, whose coarse organic components (leaves, stems, roots) also accreted. These accretion processes counter processes that naturally submerge the wetlands, including a slow rise in sea level and subsidence (sinking) due to compaction. If the rates of submergence of wetlands exceed the rates of accretion, then the area of open water increases, and the wetlands are reduced.

Today, human-constructed levees line the lower Mississippi River, confining the river and directing floodwaters, mineral sediments, and nutrients into the Gulf of Mexico rather than into the coastal wetlands. Deprived of water, sediments, and nutrients in a coastal environment where the sea level is rising, the coastal wetlands are being lost. Global sea level is rising 1 to 2 mm/yr as a result of human-induced and natural global warming. Regional and local subsidence in the Mississippi delta region combine with the global rise in sea level to produce a relative sea-level rise of about 12 mm/yr. To keep the coastal wetlands from declining, the rate of vertical accretion would thus need to be about 13 mm/yr. Currently, natural vertical accretion is only about 5 to 8 mm/yr.[16]

Most people agree that wetlands are valuable and productive lands for fish and wildlife. But wetlands are also valued as potential lands for agricultural activity, mineral exploitation, and building sites. Wetland management is drastically in need of new incentives for private landowners (who own the majority of several types of wetlands in the United States) to preserve wetlands rather than fill them in and develop the land.[14] Management strategies must also include careful planning to maintain the water quantity and quality necessary for wetlands to flourish or at least survive. Unfortunately, although laws govern the filling and draining of wetlands, no national wetland policy for the United States is in place. Debate continues as to what constitutes a wetland and how property owners should be compensated for preserving wetlands.[13, 17]

Restoration of Wetlands

A related management issue is restoration of wetlands. A number of projects have attempted to restore wetlands, with varied success. The most important factor to be considered in most freshwater marsh restoration projects is the availability of water. If water is present, wetland soils and vegetation will likely develop. The restoration of salt marshes is more difficult because of the complex interactions among the hydrology, sediment supply, and vegetation that allow salt marshes to develop. Careful studies of relationships between the movement of sediment and the flow of water in salt marshes is providing information crucial to restoration, which makes successful reestablishment of salt marsh vegetation more likely. The restoration of wetlands has become an important topic in the United States because of the mitigation requirement related to environmental impact analysis, as set forth in the National Environmental Policy Act of 1969 (see Chapter 30). According to this requirement, if wetlands are destroyed or damaged by a particular project, the developer must obtain or create additional wetlands at another site to compensate.[13] Unfortunately, the state of the art of restoration is not adequate to ensure that specific restoration projects will be successful.[18]

Constructing wetlands for the purpose of cleaning up agricultural runoff is an idea being attempted in areas with extensive agricultural runoff. Wetlands have the natural ability to remove excess nutrients, break down pollutants, and cleanse water. A series of wetlands are being created in Florida to remove nutrients (especially phosphorus) from agricultural runoff and thus help restore the Everglades to more natural functioning. The Everglades are a huge wetland ecosystem that functions as a wide, shallow river flowing south through southern Florida to the ocean. Fertilizers applied to farm fields north of the Everglades make their way directly into the Everglades by way of agricultural runoff, disrupting the ecosystem. (Phosphorus enrichment causes undesired changes in water quality and aquatic vegetation; see the discussion of eutrophication in the next chapter.) The human-made wetlands are designed to intercept and hold the nutrients so they do not enter and damage the Everglades.[19]

In southern Louisiana, restoration of coastal wetlands has recently included the application of treated wastewater, which adds nutrients, nitrogen, and phosphorous to accelerate plant growth. As plants grow, organic debris (stems, leaves, and so forth) build up on the bottom of the wetland and cause the wetland to grow vertically. This growth helps offset wetland submergence resulting from relative rise in sea level, maintaining and restoring the wetland.[16]

21.7 Dams and the Environment

Dams and their accompanying reservoirs generally are designed to be multifunctional structures. People who propose the construction of dams and reservoirs point out that reservoirs may be used for recreational activities and generating electricity as well as providing flood control and ensuring a more stable water supply. However, it is often difficult to reconcile these various uses at a given site. For example, water demands for agriculture might be high during the summer, resulting in a drawdown of the reservoir and the production of extensive mudflats or an exposed bank area subject to erosion (Figure 21.14). Recreational users find the low water level and the mudflats aesthetically displeasing. Also, high water demand may cause quick changes in lake levels, which may interfere with wildlife (particularly fish) by damaging or limiting spawning opportunities. Another consideration is that dams and

Figure 21.14 ■ Bank erosion along the shoreline of a reservoir in central California following release of water, exposing bare banks.

reservoirs tend to give a false sense of security to those living below these water retention structures. Dams may fail. Flooding may originate from tributary rivers that enter the main river below a dam; and dams cannot be guaranteed to protect people against floods larger than those for which they have been designed.

The environmental effects of dams are considerable and include the following:

■ Loss of land, cultural resources, and biological resources in the reservoir area.

■ Storage behind the dam of sediment that would otherwise move downstream to coastal areas, where it would supply sand to beaches. The trapped sediment also reduces water storage capacity, limiting the life of the reservoir.

■ Downstream changes in hydrology and in sediment transport that change the entire river environment and the organisms that live there.

■ Fragmentation of ecosystems above and below a dam.

For a variety of reasons that include displacement of people, loss of land, loss of wildlife, and permanent, adverse changes to river ecology and hydrology, many people today are vehemently against turning remaining rivers into a series of reservoirs with dams. In the United States, several dams have recently been removed, and others are being considered for removal as a result of the environmental damage they are causing. In contrast, China has the world's largest dam, as described in A Closer Look 21.2.

There is little doubt that if our present practices of water use continue we will need additional dams and reser-

voirs, and some existing dams will be heightened to increase water storage. However, there are few acceptable sites for new dams. Conflicts over the construction of additional dams and reservoirs are bound to occur. Water developers may view a canyon dam site as a resource for water storage, whereas others may view it as a wilderness area and recreation site for future generations. The conflict is common because good dam sites are often sites of high-quality scenic landscape.

There is also an economic aspect to dams: They are expensive to build and operate. They are often constructed with federal tax dollars in the western United States, where they provide inexpensive subsidized water for agriculture. This has been a point of concern to some taxpayers in the eastern United States, who do not have the benefit of federally subsidized water. Perhaps a different pricing structure for water would encourage conservation, and fewer new dams and reservoirs would be needed.

Canals

Water from upstream reservoirs may be routed downstream by way of natural watercourses or canals and aqueducts. Canals are not hydrologically the same as creeks or rivers. They often have smooth, steep banks; and the water moves deceptively fast. Canals are a hazard attracting children to swim in them and animals to swim across them. Where they flow, drownings of people and animals are an ever-present threat.

The construction of canal systems, especially in developing countries, has led to serious, unanticipated environmental problems. For example, when the High Dam on the Nile River at Aswan, Egypt, was completed in 1964, a system of canals was built to convey the water to agricultural sites. The canals became infested with snails that carry the disease schistosomiasis (snail fever). This disease has always been a problem in Egypt, but the swift currents of the Nile floodwaters flushed the snails out each year. The tremendous expanse of waters in irrigation canals now provides happy homes for these snails. The disease is debilitating and so prevalent in parts of Egypt that virtually the entire population of some areas may be affected by it.

Removal of Dams

In the United States, several dams, including the Edwards Dam near Augusta, Maine, have recently been removed. Removal of the Edwards Dam opened about 29 km (18 mi) of river habitat to migrating fish, including Atlantic salmon, striped bass, shad, alewives, and Atlantic sturgeon. Following removal, the Kennebec River came back to life as millions of fish migrated upstream for the first time in 160 years.[22]

Three Gorges Dam

In spite of the known adverse effects of large dams, the world's largest is now being constructed in China. Three Gorges Dam on the Yangtze River (Figure 21.15) has drowned cities, farm fields, important archeological sites, and highly scenic gorges while displacing approximately 2 million people from their homes. In the river, habitat for endangered dolphins will likely be damaged. On land, habitats will be fragmented and isolated as mountaintops become islands in the reservoir.

The dam, which is approximately 185 m high and more than 1.6 km wide, produces a reservoir nearly 600 km long. Raw sewage and industrial pollutants that are discharged into the river are discharged into the reservoir, and there is concern that the reservoir will become seriously polluted. In addition, the Yangtze River has a high sediment load, and it is feared that the upstream end of the reservoir, where sediments will likely be deposited, will fill with sediment, damaging deep-water shipping harbors.

The dam may produce a false sense of security in the people living in downstream cities. The presence of the dam may encourage further development in flood-prone areas, which will be damaged or lost if the dam and reservoir are unable to hold back floods in the future. If this happens, loss of property and life

Figure 21.15 ■ Three Gorges on the Yangtze River is a landscape of high scenic value. Shown here is the Wu Gorge, near Wushan, one of the gorges flooded by the water in the reservoir.

from flooding may be greater than if the dam had not been constructed. Contributing to this problem is the dam's location in a seismically active region where earthquakes and large landslides have been common in the past.

Should the dam fail, then downstream cities such as Wushan, with a population of several million people, might be submerged with catastrophic loss of life.[20]

A positive attribute of the giant dam and reservoir will be the capacity to produce about 18,000 MW of electricity, which is equivalent to about 18 large coal-burning power plants. As has been pointed out in earlier discussions, pollution from coal burning is a serious problem in China. Some opponents to the dam have pointed out, however, that a series of dams on tributaries to the Yangtze River could have produced similar electric power while not causing the anticipated environmental damage to the main river.[21]

A large number of U.S. dams (mostly small ones) have been removed or are in the planning stages for removal. The Elwha and Glines Canyon dams on the Elwha River (constructed in the early eighteenth century) in the Puget Sound of Washington State are scheduled to come down by 2010. The largest of the two is the Glines Canyon Dam, which is about 70 m (210 ft) high (the highest dam ever removed). The Elwha headwater is in Olympic National Park and prior to the dams's construction supported large salmon and steelhead runs. The dams denied the fish access to almost their entire spawning habitat. As a result, fish populations there have been greatly reduced. Large runs of fish also brought nutrients from the ocean to the river and landscape. Bears, birds, and other animals eat the salmon and transfer nutrients to the forest ecosystem. Without the salmon, both wildlife and forest suffer. The dams also store sediment, keeping it from reaching the sea. Denied sediment, beaches at the river's mouth have eroded, causing loss of clam beds. The dams will be removed in stages to minimize downstream impacts from the release of sediment. With the dams removed, the river will flow free for the first time in a century, and, hopefully, the ecosystem will recover and the fish will return in greater numbers.[23]

The Matilija Dam, completed by Ventura County in 1948, is about 190 m (620 ft) wide and 60 m (200 ft) high. The structure is in poor condition, with leaking, cracked concrete and a reservoir nearly filled with sediment. The dam serves no useful purpose and blocks endangered southern steelhead trout from their historic spawning grounds. The sediment trapped in the dam also reduces the natural nourishment of sand on beaches, increasing coastal erosion.

The trapped sediment presents a problem in removing the dam. If released quickly, it could damage the downstream river environment, filling pools and killing river organisms such as fish, frogs, and salamanders. If the sediment can be slowly, more naturally released, the downstream damage can be minimized.

The removal process began with much fanfare in October 2000, when a 27-m (90-ft) section was removed from the top of the dam. The entire removal process may take years, after scientists have determined how to safely remove the sediment stored behind the dam. The cost of the dam in 1948 was about $300,000. The cost to remove the dam and sediment will be more than 10 times that amount.[24]

The perception concerning dams as permanent structures similar to the pyramids of Egypt has clearly changed. What is learned from studying the removal of the Edwards Dam in Maine and the Matilija Dam in California will be useful in planning other dam removal projects. The studies will also provide important case histories to evaluate ecologic restoration of rivers following removal of dams.

21.8 Channelization and the Environment

Channelization of streams consists of straightening, deepening, widening, clearing, or lining existing stream channels. It is an engineering technique that has been used to control floods, improve drainage, control erosion, and improve navigation.[25] Flood control and drainage improvement are the most common objectives of so-called channel improvement projects.

Thousands of kilometers of streams in the United States have been modified by channelization. The practice all too often has produced adverse environmental effects, including the following:

- Degradation of the stream's hydrologic qualities, turning a meandering stream with pools (deep, slow flow) and riffles (faster, shallow flow) into straight channels that are nearly all riffle flow, resulting in loss of important fish habitats.

- Removal of vegetation along the watercourse, which removes wildlife habitats and shading of the water.

- Downstream flooding where the channelized flow ends, because the channelized section has a larger channel area and carries a greater amount of floodwater than the downstream natural channel can carry without overbank flow or flooding.

- Damage or loss of wetlands (because their source of water is removed by the channelization, which often lowers the groundwater table and thus drains the wetlands).

- Aesthetic degradation (channelized streams are much less attractive than natural streams).

Figure 21.16 compares selected environmental features of natural and channelized streams.

A case study in problems with channelization involves the Kissimmee River in Florida. Channelization of the river started in 1962. After nine years and $24 million of construction, the meandering river with many bends was converted into a straight ditch 83 km (about 50 miles) long (Figure 21.17). Unfortunately, not only did the channelization fail to provide the expected flood protection, but it also damaged a valuable wildlife habitat, contributed to water quality problems associated with the land drainage, and caused aesthetic degradation. Then, in the 1990s, efforts began to return the river to its original meandering channel. The restoration of the Kissimmee River may become the most ambitious restoration project ever attempted in the United States, and the cost may exceed the original cost of channelization. Work began on a 16-km (10-mi) stretch of river. By 2001, 12 km (7.5 miles) of nearly straight flood-control channel had been restored to a 24-km (15-miles) meandering channel with wetlands, returning ecosystems to a more natural state.

Not all channelization causes serious environmental degradation, and in many cases, drainage projects are beneficial. Moreover, channel design is being improved on the basis of past experience. Currently, more consideration is being given to the environmental aspects of channelization, and some projects are being designed with modified channels that behave more like natural streams than do straight ditches.

21.9 The Colorado River: Water Resources Management and the Environment

The history of the Colorado River emphasizes linkages among physical, biological, and social systems that are at the heart of environmental science.

The Colorado is the major river of the southwestern United States and extends into Mexico, where it ends in the Gulf of California (Figure 21.18). The river basin occupies approximately 632,000 km² (244,000 mi²). Considering its size, the river has only a modest flow. However, it is one of the most regulated and controversial bodies of water in the world. The total flow of water in the river was apportioned among various users, including seven U.S. states and Mexico, by the Colorado River Compact of 1922. The compact allocated no

Natural stream

Suitable water temperatures: adequate shading; good cover for fish life; minimal temperature variation; abundant leaf material output.

Low flow

Pool—riffle sequence

Pool—deep, slow flow, sand, and fine gravel on stream bed.

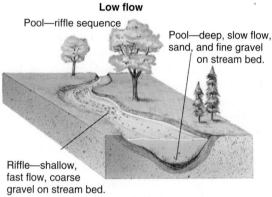

Riffle—shallow, fast flow, coarse gravel on stream bed.

Pools and riffles provide diversified habitats for many stream organisms.

Figure 21.16 ■ A natural stream compared with a channelized stream in terms of general characteristics and pool environments.

Channelized stream

Increased water temperatures: no shading; no cover for fish life; rapid daily and seasonal temperature fluctuations; reduced leaf material output.

Low flow

Mostly riffle

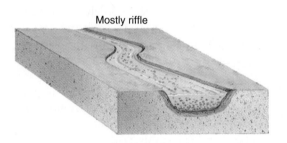

Lack of pools causes reduction in habitats; few organisms.

[*Source:* Modified from R. V. Corning, *Virginia Wildlife*, February 1975, pp. 6–8.]

water for environmental purposes, as the concept of sustainable water management was not considered.

Today, Colorado River water only occasionally flows into the Gulf of California—it's all stored in dams and used upstream. As a result, ecosystems of the lower river and delta, deprived of water and nutrients, have been damaged. The size of the delta has been reduced from about 7,500 km² (3,000 mi²) to less than half of that, damaging fish populations and forcing native people who depended on fishing to move away.

(a)

(b)

Figure 21.17 ■ (*a*) Channelized Kissimmee River and (*b*) restoration work in progress on the Kissimmee River near Cornwell, Florida.

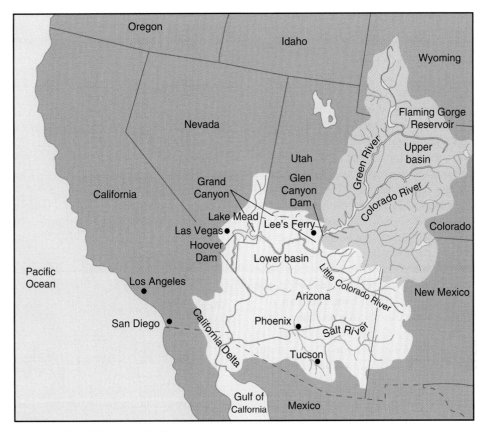

Figure 21.18 ■ The Colorado River basin.

The complex issues of water management for the Colorado River illustrate major problems that are likely to be faced by other semiarid regions of the world in the coming years: How are scarce water resources to be allocated? How can we best control water quality? How can we protect river ecosystems? There are no easy answers to these questions.

The Colorado River originates in the Wind River Mountains of Wyoming and, in its 2,300-km (1,400-mile) journey to the Gulf of California, flows through some of the most spectacular scenery in the world. Eight hundred years ago, Native Americans living in the Colorado River basin constructed a sophisticated water distribution system. In the 1860s, settlers cleared the debris from these early canals and used them once again for irrigation.[26]

Today, the water of the Colorado River basin is completely spoken for and distributed by canals and aqueducts to millions of people in urban areas as well as to agricultural areas. The basin itself remains sparsely populated; and the largest city on the river is Yuma, Arizona, with a population of 42,000. The historian Rod Nash, writing about his wilderness experience on the river, states that at the junction of the Green and Colorado rivers, in the heart of a national park, it is 80 km (50 mi) to the nearest video game.[27]

Managing the waters of the Colorado River has been frustrating in part because of inherent uncertainties concerning how much water can be expected in a given year.[26] Water resources in the basin include snowmelt, long-term winter precipitation, and short-term summer thunderstorms; as a result, the total water available on a year-to-year basis is tremendously variable. Table 21.3 compares the legal entitlements to water with the actual distribution to the users. Interestingly, the annual legal entitlements of 21.5 km^3 (17.5 million acre-feet) exceed the actual distribution of 17.8 km^3 (14.5 million acre-feet). (One acre-foot of water is equal to 1,233 m^3, or 325,829 gal. It is the volume of water necessary to cover one acre to a depth of one foot.)[28]

Table 21.3 shows that not all users take the full legal entitlement of water. If all did, serious problems would result, because the entitlements exceed the natural annual flow. The reason the distribution can be maintained is that the Colorado River is one of the most managed rivers in the world. Figure 21.19 shows the profile of the river and some of the major dams and reservoirs in the system, which store approximately 86.3 km^3 (70 million acre-feet) of water. The two largest reservoirs—Hoover Dam and Glen Canyon Dams—store about 80% of the total in the basin.

Total storage represents (with careful management) a buffer of several years' water supply. However, if there are several drought years in a row, maintaining a sufficient water supply for all users may not be possible.

The Glen Canyon Dam was completed in 1963 (Figure 21.20). From a hydrologic viewpoint, the Colorado River has been changed by the dam. The river has been tamed. The higher flows have been reduced, the average flow has increased, and the flow changes often because of fluctuating needs to generate electrical power. Changing the hydrology of the river has also changed other aspects, including the rapids; the distribution of sediments that form sandbars, called beaches by rafters; and the vegetation near the water's edge.[29] The sandbars, which are valuable wildlife habitats, shrank in size and number following construction of the dam because sediment that would have moved downstream to nourish them was trapped in the reservoir. All these changes affect the Grand Canyon, which is downstream from the dam.

A record snowmelt in the Rocky Mountains in June 1983 forced the release of about 2,500 m^3 (88,000 ft^3) of water per second from the Glen Canyon Dam—about three times the amount normally released and similar to the amount in a spring flood before the dam was built. Resulting floods scoured the river bed and banks, releasing stored sediment that replenished sandbars and breaking off some vegetation that had taken root.[30] This release of water was beneficial to the river environment and highlighted the importance of floods in maintaining the system in a more natural state. Natural disturbances are a necessary part of the river ecosystem if it is to function on a sustainable basis.

As an experiment, a flood with discharge about half the size of that in 1983 was deliberately released for a period of a week in 1996. Between March 26 and April 2, water was allowed to flow at full flood; then the flow was reduced for the last two days in order to redistribute the sand supply. The flood resulted in 55 new sandbars and increased the size of 75% of the existing sandbars. It also helped rejuvenate marshes and backwaters, which are critical habitats to the native fish and some endangered species.[30]

	Legal Entitlements (km^3/yr)	**Actual Distribution**[a] (km^3/yr)
Area		
Lower basin		
California	5.461[b]	5.461[c]
Arizona	4.674[b]	2.522[c]
Nevada	0.369[b]	0.369
Subtotal	10.455[d]	8.303
Upper basin		
Colorado	4.774[e]	2.959
Utah	2.122[e]	1.316
Wyoming	1.292[e]	0.801
New Mexico	1.038[e]	0.643
Subtotal	9.225[d]	5.720
Mexico	1.845[f]	1.845
Total	21.525	17.835

Table 21.3 • Legal and Actual Distribution of Colorado River Water

Source: W. L. Graf, *The Colorado River* (Washington, D. C.: Association of American Geographers, 1985).

[a] Includes losses to evaporation of 0.6 million acre-feet/year in the upper basin and 0.9 million acre-feet/year in the lower basin and 0.9 million acre-feet/year inflow to the lower basin from local streams.

[b] 1928 Boulder Canyon Project Act.

[c] Agreement at time of Central Arizona Project authorization by Congress. Upper basin amounts agreed to by states as percentages.

[d] 1922 Colorado River Compact.

[e] 1948 Upper Colorado River Basin Compact.

[f] 1944 Mexico–U.S. Treaty.

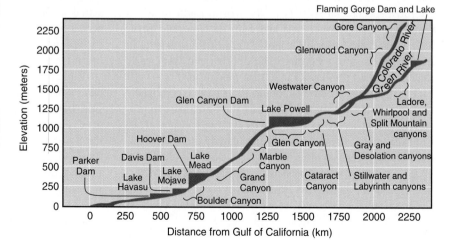

Figure 21.19 ■ Longitudinal profiles of the Colorado River and Green River, showing the major dams, reservoirs, and canyons. [*Source:* W. L. Graf, *The Colorado River*, Resource Publications in Geography (Washington, D.C.: Association of American Geographers, 1985).]

Figure 21.20 ■ Glen Canyon Dam on the Colorado River in Arizona.

This experimental release of high flows of water marked a turning point in river management—the first time the U.S. government had opened the floodgates of a dam to improve a river's ecosystem. Some scientists were concerned that floodwaters would not be of sufficient volume and that the duration of the flooding would be too short. The flood was hailed as a success, but it will take some time to see how long the results will last. It is hoped that what was learned will be used to restore river environments and improve ecosystems of other rivers impacted by dams. Today, the Colorado River in the Grand Canyon is more accessible to recreational river rafting, and recreation is better managed than ever before.

Before 1950, fewer than 100 explorers and river runners had made a trip through the Grand Canyon. Today, the number of people rafting through the canyon is limited to 15,000 per year; even with this limitation, long-range impacts on the canyon resources are appreciable. Even though all waste is removed by boat, increased traffic and camping are placing stress on canyon ecosystems.

We must concede that the Colorado River is changed and will remain changed as long as the Glen Canyon Dam is present. Although human-made floods have counteracted some of the changes, river restoration by itself cannot be expected to return the river to what it was before construction of the dam. However, better management of the flows and sediment transport will improve and better maintain river ecosystems.[27, 30, 31]

Conflicts over Colorado River water use and quality have spanned decades and extend beyond the river basin to urban centers and agricultural areas in California, Colorado, New Mexico, and Arizona. The crucial need for water in these semiarid regions has resulted in overuse of limited supplies of water as well as deterioration of water quality. This is a problem in the lower river particularly. By the time the water reaches Mexico, it is so salty that at times Mexican farmers are unable to use it. But Mexico is entitled to receive 1.8 billion m³ (1.5 million acre-feet) of Colorado River water per year according to a treaty with the United States.

The world's largest desalination plant went online in May 1992 on the Colorado River near Yuma, Arizona (Figure 21.21). The cost of the plant was about $260 million, and its purpose was to reduce the salinity of the water taken into the plant from about 1,200 ppm salt to 800 ppm. Plant capacity is about 97 million m³/yr (700 million gal/day). The upstream salt causing the problem comes mostly from agricultural runoff from the Welton-Mohawk Irrigation and Drainage District. The desalination plant had operated only a few months when floods destroyed canals that carried drainage water from the irrigation district to the plant. Since 1993, the plant has been on standby status (that is, it is closed but can be restarted). The annual cost of maintaining the standby status is about $6 million. There is some question whether the plant should resume operation. In some years, the Gila River has sufficient natural flows to

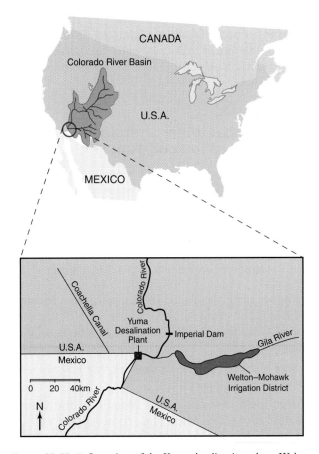

Figure 21.21 ■ Location of the Yuma desalination plant, Welton Mohawk Irrigation and Drainage District, and Coachella Canal.

dilute the salty agricultural waters, and desalination is not needed. In addition, salty waters from the irrigation district are diverted to the desert wetland of Cienga de Santa Clara, which has grown from about 200 to 20,000 hectares, becoming an important desert wetland. Using the desalination plant rather than diverting the salty water would cause the wetland to disappear. Finally, Mexico is receiving better-quality water from the Coachella Canal, which flows into the Colorado River. The canal was concrete lined to keep water from infiltrating the soil. The concrete lining saves more water than the desalination plant can produce. Thus, Mexico now receives its water as required by treaty without need for the desalination plant.

International and interstate agreements and court settlements have periodically intensified or eased tensions among water users of the Colorado River. These court decisions and laws, along with changes in water use, continue to significantly influence the lives of millions of people in the United States and Mexico.[32] For example, a political water war is heating up between Mexico and the United States over water that for decades has seeped from the All American Canal. The 36 km (23 mi) canal just inside the United States delivers Colorado River water west toward San Diego. The earth canal leaks billions of gallons to replenish groundwater used in Mexico by farmers. The canal may be concrete lined, which would damage farms in Mexico that now depend on the water.

Mexico has warned that if the farms dry up and fail, the people will be forced to illegally enter the United States for work. That issue aside, there are ethical reasons to honor past water use by Mexico.

21.10 Global Water Shortage Linked to Food Supply

As a capstone to this chapter, we present the hypothesis that we are facing a growing water shortage linked to our food supply. This is a potentially very serious problem. In the past few years, we have begun to realize that isolated water shortages are apparently indicators of a global pattern.[33] At numerous locations on Earth, both surface water and groundwater are being stressed and depleted:

- Groundwater in the United States, China, India, Pakistan, Mexico, and many other countries is being mined (used faster than it is being renewed) and so is being depleted.

- Large bodies of water—for example, the Aral Sea—are drying up (see Figure 21.7).

- Large rivers, including the Colorado in the United States and the Yellow in China, do not deliver any water to the ocean in some seasons or years. Others, such as the Nile in Africa, have had their flow to the ocean greatly reduced.

Water demand during the past half-century has tripled as human population has more than doubled. In the next half-century, the human population is expected to grow by another 2 to 3 billion. There is growing concern that there won't be sufficient water to grow the food to feed the 8–9 billion people expected to be inhabiting the planet by the year 2050. Therefore, a food shortage linked to water resources seems a real possibility. The problem is that our increasing use of groundwater and surface water resources for irrigation has allowed for increased food production—mostly crops such as rice, corn, and soybeans. These same water resources are being depleted, and as water shortages for an agricultural region occur, food shortages may follow.

The solution to avoid food shortages resulting from water resource depletion is clear. We need to control human population growth and conserve and sustain water resources. In this chapter, we have outlined a number of ways to conserve, manage, and sustain water. The good news is that a solution is possible—but it will take time, and we need to be proactive now before significant food shortages develop. For all the reasons discussed, one of the most important and potentially serious resource issues of the twenty-first century is water supply and management.

Summary

- Water is a liquid with unique characteristics that has made life on Earth possible.

- Although it is one of the most abundant and important renewable resources on Earth, more than 99% of Earth's water is unavailable or unsuitable for beneficial human use because of its salinity or location.

- The pattern of water supply and use on Earth at any particular point on the land surface involves interactions and linkages among the biological, hydrological, and rock cycles. To evaluate a region's water resources and use patterns, a water budget is developed to define the natural variability and availability of water.

- During the next several decades, it is expected that the total water withdrawn from streams and groundwater in the United States will decrease slightly, but the consumptive use will increase because of greater demands from a growing population and industry.

- Water withdrawn from streams competes with in-stream needs, such as maintaining fish and wildlife habitats and navigation, and may therefore cause conflicts.

- Groundwater use has led to a variety of environmental problems, including overdraft, loss of vegetation along watercourses, and land subsidence.

- Because agriculture is the biggest user of water, conservation of water in agriculture has the most significant effect on sustainable water use. However, it is also important to practice water conservation at the personal level in our homes and to price water to encourage conservation and sustainability.

How Wet Is a Wetland?

Areas where land meets water, whether fresh or salt, are places where *wetlands* are found. Characteristically, wetlands are covered by surface water or have saturated soils. Landscape features such as swamps, bogs, marshes, potholes, and sloughs are wetlands. For most of our history, wetlands were considered wastelands and were destroyed by filling, draining, or polluting. Only 230 million of the 530 million hectares (568.3 million of the 1,309.6 million acres) of wetlands that existed in the lower 48 states at the time of European colonization remain, and 40% of them suffer from some degree of pollution.

Today, however, it is generally recognized that wetlands have many values. They provide food, water, and cover for fish, shellfish, waterfowl, game animals, and many amphibians and reptiles. In addition, soil and plants in wetlands purify water by absorbing or destroying pollutants. Wetlands also help to recharge groundwater stores and, by holding water, control flooding and erosion. One-third of endangered and threatened species, two-thirds of the commercial saltwater fish and shellfish, one-third of the birds, and almost all the amphibians in our country depend on wetlands. Wetlands are among the most productive ecological communities in the world, many times more productive than a heavily fertilized cornfield.

Protection of these unique communities began with the goal of preserving wetlands used by wildlife, particularly ducks, but federal protection has since been broadened to include most of the remaining wetlands. Still, 81,000 to 162,000 hectares (200,151 to 400,302 acres) of wetlands are lost in the United States each year. Of particular concern are the freshwater wetlands, many of which are on private land. A recent policy of no net loss of wetlands has been praised by some environmental scientists but attacked by farmers and developers.

Particularly controversial are the small, seasonal potholes in the agricultural areas of the Midwest and northern Great Plains and other seasonal wet areas of the West that may not appear wet to the casual observer. They do, however, provide habitats for many species, including about half the 10 to 31 million waterfowl that nest in the lower 48 states. Critics of applying strict regulations to potholes say that if an area is not wet enough for a duck to land in and splash, it is not wet enough to qualify as a wetland.

Critical Thinking Questions

1. Results of a study comparing wildlife use of seasonal wetlands with wildlife use of wetlands that are permanently inundated or saturated in the San Francisco estuary are shown in the accompanying table. What conclusions can you draw from the data about the importance of seasonal wetlands to wildlife in the estuary? What additional data would you need to extend your conclusions to potholes in the Midwest?

2. Some people have proposed defining wetlands so as to exclude many of the seasonal ones, some 4.5 million hectares (11.1 million acres). What position would you take on such a proposal, and why? How would you reconcile the conflicting needs of the farmers and developers with the need to preserve wildlife habitats?

3. In what ways would you expect a substantial decrease in wetlands to affect populations of migratory birds?

Area and Wildlife Use of Wetlands in the San Francisco Estuary

Type of Wetland	Area ha	Area %	Wildlife Use Number of Species	Wildlife Use %[a]
Permanent	58,765	23		
Mudflat	25,949	10.2	49	11.9
Marsh	17,964	7.1		
Tidal salt			95	23.1
Brackish			192	46.6
Tidal fresh			174	42.2
Salt pond	14,852	5.8	82	19.9
Seasonal	195,709	5.8	82	19.9
Diked marsh and other	34,467	13.5		
Diked marsh			72	17.5
Other			—	
Farmed wetlands	156,176	61.3	92	22.3
Riparian woodlands	5,066	1.9	207	50.2
Total	**254,474**	**100**[b]	**412**	**100**

[a] Percent of total number of species found in all wetlands
[b] Permanent + seasonal = 100%

- There is a need for a new philosophy in water resource management that considers sustainability and uses creative alternatives and variable sources. Development of a master plan involves inclusion of normal sources of surface water and groundwater, conservation programs, and use of reclaimed water.

- Development of water supplies and facilities to more efficiently move water may cause considerable environmental degradation; construction of reservoirs and canals and channelization of rivers should be considered carefully in light of potential environmental impacts.

- Wetlands serve a variety of functions at the ecosystem level that benefit other ecosystems and people.

- The Colorado River in the southwestern United States and northern Mexico is one of the most regulated rivers in the world. Understanding links among physical, biological, and social systems of the Colorado River is necessary to manage its water resources and ecosystems.

- We are facing a growing global water shortage linked to the food supply.

- Water supply and management is one of the major resource issues of the twenty-first century.

REEXAMINING THEMES AND ISSUES

Human Population

As human population has increased, so has the demand for water resources. As a result, we must be more careful in managing Earth's water resources, particularly near urban centers.

Sustainability

The water resources of the planet are sustainable provided we manage them properly and they are not overused, polluted, or wasted. This requires good water management strategies. We believe that the move toward sustainable water use must be facilitated now to avoid conflicts in the future. Principles of water management presented in this chapter help delineate what needs to be done.

Global Perspective

The water cycle is one of the major global geochemical cycles. It is responsible for the transfer and storage of water on a global scale. Fortunately, on a global scale, the total abundance of water on Earth is not a problem. However, ensuring that it is available when and where it is needed in a sustainable way *is* a problem.

Urban World

Although urban areas consume only a small portion of the water resources used by people, it is urban areas where shortages are often most apparent. Thus, the concepts of water management and water conservation are critical in urban areas.

People and Nature

To many people, water is an icon of nature. Waves crashing on a beach, flowing in a river or over falls, or shimmering in lakes has inspired poets and countless generations of people to connect with nature.

Science and Values

Conflicts result from varying values related to water resources. We value natural areas such as wetlands and free-running rivers, but we also want water resources and protection from hazards such as flooding. As a result, we must learn to align more effectively with nature to minimize natural hazards, maintain a high-quality water resource, and provide the water necessary for the ecosystems of our planet. The test floods of the Colorado River discussed in this chapter are examples of new river management practices based on the science of understanding river processes linked to values that recognize the wish to sustain the Colorado as a vibrant, living river.

Key Terms

channelization **453**	groundwater **436**	off-stream use **440**	water budget **438**
consumptive use **440**	influent stream **436**	overdraft **439**	water conservation **444**
desalination **440**	in-stream use **440**	sustainable water use **446**	wetlands **449**
effluent stream **436**			

Study Questions

1. If water is one of our most abundant resources, why are we concerned about its availability in the future?

2. Which is more important from a national point of view, conservation of water use in agriculture or in urban areas? Why?

3. Distinguish between in-stream and off-stream uses of water. Why is in-stream use controversial?

4. What are some important environmental problems related to groundwater use?

5. How might your community better manage its water resources?

6. What are some of the major environmental impacts associated with the construction of dams and canals? How might these be minimized?

7. What are some of the environmental problems associated with channelizing rivers? How might the potential adverse effects of channelization be minimized?

8. How can we reduce or eliminate the growing global water shortage? Do you believe the shortage is related to our food supply? Why? Why not?

9. Why is water such an important resource issue?

Further Reading

Gleick, P. H. 2003. "Global Freshwater Resources: Soft-Path Solutions for the 21st Century," *Science* 302:1524–1528.

Gleick, P. H. 2000. *The World's Water 2000–2001.* Washington, D.C.: Island Press.

Graf, W. L. 1985. *The Colorado River.* Resource Publications in Geography. Washington, D.C.: Association of American Geographers. A good summary of the Colorado River water situation.

James, W., and J. Ncimczynowicz,eds. 1992. *Water, Development and the Environment.* Boca Raton, Fla.: CRC Press. Covers problems with water supplies imposed by a growing population, including urban runoff, pollution and water quality, and management of water resources.

La Riviere, J. W. M. 1989. "Threats to the World's Water," *Scientific American* 261(3): 80–84. Summary of supply and demand for water and threats to continued supply.

Spulber, N., and A. Sabbaghi. 1994. *Economics of Water Resources: From Regulation to Privatization.* London: Kluwer Academic. Discussions of water supply and demand, pollution and its ecological consequences, and water on the open market.

Twort, A. C., F. M. Law, F. W. Crowley, and D. D. Ratnayaka. 1994. *Water Supply*, 4th ed. Edward Arnold. Good coverage of water topics from basic hydrology to water chemistry, and water use, management, and treatment.

Wheeler, B. D., S. C. Shaw, W. J. Fojt, and R. A. Robertson. 1995. *Restoration of Temperate Wetlands.* New York: Wiley. Discussions of wetland restoration around the world.

CHAPTER 22

Water Pollution and Treatment

Pig stranded on a vehicle east of Burgaw, North Carolina, during a flood caused by Hurricane Floyd in 1999. Floodwaters inundated pig farms, killing many thousands of the animals. Their carcasses, feces, and urine flowed through homes, churches, and schools, causing a major pollution event.

Learning Objectives

Degradation of our surface water and groundwater resources is a serious problem, the effects of which are not fully known. There are a number of steps we can take to treat water and to minimize pollution. After reading this chapter, you should understand:

■ What constitutes water pollution and what the major categories of pollutants are.

■ Why the lack of disease-free drinking water is the primary water

pollution problem in many locations around the world.

■ How point and nonpoint sources of water pollution differ.

■ What biochemical oxygen demand is, and why it is important.

■ What eutrophication is, why it is an ecosystem effect, and how human activity can cause cultural eutrophication.

■ Why sediment pollution is a serious problem.

■ What acid mine drainage is, and why it is a problem.

■ How urban processes can cause shallow-aquifer pollution.

■ What the various methods of wastewater treatment are, and why some are more environmentally preferable than others.

■ What the environmental laws are that protect water resources and ecosystems.

North Carolina's Bay of Pigs

Hurricane Floyd struck the Piedmont area of North Carolina in September 1999. The killer storm took a number of lives while flooding many homes and forcing some 48,000 people into emergency shelters. The storm had another, more unusual effect as well. Floodwaters containing thousands of dead pigs along with their feces and urine flowed through schools, churches, homes, and businesses. The stench was reported to be overwhelming, and the count of pig carcasses may have been as high as 30,000. The storm waters had overlapped and washed out over 38 pig lagoons with as much as 950 million liters (250 million gal) of liquid pig waste, which ended up in flooded creeks, rivers, and wetlands. In all, something like 250 large commercial pig farms flooded out, drowning hogs whose floating carcasses had to be collected and disposed of (Figure 22.1).

Prior to the catastrophe brought on by Hurricane Floyd, the pig farm industry in North Carolina had been involved in a scandal reported by newspapers and television—and even *60 Minutes*. North Carolina has a long history of hog production, and the population of pigs swelled from about 2 million in 1990 to nearly 10 million in 1997. At that time, North Carolina became the second largest pig-farming state in the nation.[1] As the number of large commercial pig farms grew, the state allowed the hog farmers to build automated and very confining farms housing hundreds or thousands of pigs. There were no restrictions on farm location, and many farms were constructed on floodplains.

Each pig produces approximately 2 tons of waste per year. The North Carolina herd was producing approximately 20 million tons of waste a year, mostly manure and urine, which was flushed out of the pig barns and into open, unlined lagoons about the size of football fields. Favorable regulations, along with the availability of inexpensive waste-disposal systems (the lagoons), were responsible for the tremendous growth of the pig population in North Carolina during the 1990s.

(a)

(b)

Figure 22.1 ■ North Carolina's "Bay of Pigs." (*a*) Map of areas flooded by Hurricane Floyd in 1999 with relative abundance of pig farms. (*b*) Collecting dead pigs near Boulaville, North Carolina. The animals were drowned when floodwaters from the Cape Fear River inundated commercial pig farms.

Following the hurricane, mobile incinerators were moved into the hog region to burn the carcasses; but there were so many dead pigs that hog farmers had to bury some animals in shallow pits. The pits were supposed to be at least 1 m deep and dry, but there wasn't always time to find dry ground; and for the most part, the pits were dug and filled on floodplains. As these pig carcasses rot, bacteria will leak into the groundwater and surface water for some appreciable time.

The pig farmers blamed the hurricane for the environmental catastrophe. However, it was clearly a human-induced disaster that had been expected. An early warning occurred in 1995, when a pig-waste-holding lagoon failed and sent approximately 950 million liters (250 million gal) of concentrated pig feces down the New River past the city of Jacksonville and into the New River Estuary. Adverse environmental effects of that spill on marine life lasted for approximately three months.

The lesson to be learned from North Carolina's "Bay of Pigs" is that we are vulnerable to environmental catastrophes caused by large-scale industrial agriculture. Economic growth and production of livestock must be carefully planned to anticipate problems, and waste management facilities must be designed so as not to pollute local streams, rivers, and estuaries.

Was the lesson learned in North Carolina? The pig farmers had powerful friends in government and big money. Incredible as it may seem, following the hurricane, the farmers asked for $1 billion in grants to help repair and replace the pig facilities, including waste lagoons,

destroyed by the hurricane. Furthermore, they asked for exemptions from the Clean Water Act for a period of six months so that waste from the pig lagoons could be discharged directly into streams. This was not allowed.[2] With regard to future management, considering that North Carolina frequently is struck by hurricanes, not allowing pig operations to be located on floodplains seems obvious. However, this is only the initial step. The whole concept of waste lagoons needs to be rethought and alternative waste management practices put into effect if pollution of surface waters and groundwaters is to be avoided.

North Carolina's pig problem led to the formation of what is called the "Hog Roundtable," a coalition of civic, health, and environmental groups with the objective of controlling industrial-scale pig farming. As of 2004, its efforts, with others, resulted in a mandate to phase out pig waste lagoons and expand regulations to require buffers between pig farms and surface waters and water wells. The coalition also halted construction of a proposed slaughterhouse that would have allowed more pig farms to be established.

North Carolina's "Bay of Pigs" disaster produced a particularly visible and serious water pollution episode in a beautiful state with abundant natural resources. Other types of water pollution, such as waterborne disease in surface waters and pesticides in groundwater, are often much more difficult to identify without careful sampling and testing. This chapter discusses major categories of water pollution and traditional as well as innovative options for wastewater treatment.

22.1 Water Pollution

Water pollution refers to degradation of water quality. In defining pollution, we generally look at the intended use of the water, how far the water departs from the norm, its effects on public health, or its ecological impacts. From a public health or ecological view, a pollutant is any biological, physical, or chemical substance that, in an identifiable excess, is known to be harmful to other desirable living organisms. Water pollutants include heavy metals, sediment, certain radioactive isotopes, heat, fecal coliform bacteria, phosphorus, nitrogen, sodium, and other useful (even necessary) elements, as well as certain pathogenic bacteria and viruses. In some instances, a material may be considered a pollutant to a particular segment of the population although it is not harmful to other segments. For example,

excessive sodium as a salt is not generally harmful, but it may be harmful to people who must restrict salt intake for medical reasons.

Today, the primary water pollution problem in the world is the lack of clean, disease-free drinking water. In the past, epidemics (outbreaks) of waterborne diseases such as cholera have been responsible for the deaths of thousands of people in the United States. Fortunately, epidemics of such diseases have been largely eliminated in the United States as a result of treating drinking water prior to consumption. This certainly is not the case worldwide, however. Every year, several billion people are exposed to waterborne diseases. For example, an epidemic of cholera occurred in South America in the early 1990s, and outbreaks of waterborne diseases continue to be a threat even in developed countries.

It is a fundamental principle that the quality of water determines its potential uses. The major uses for water today are agriculture, industrial processes, and domestic (household) supply. Water for domestic supply must be free from constituents harmful to health, such as insecticides, pesticides, pathogens, and heavy-metal concentrations; it should taste good, should be odorless, and should not damage plumbing or household appliances. The quality of water required for industrial purposes varies widely depending on the process involved. Some processes may require distilled water; others need water that is not highly corrosive or that is free of particles that could clog or otherwise damage equipment. Because most vegetation is tolerant of a wide range of water quality, agricultural waters may vary widely in physical, chemical, and biological properties.[3]

Many different processes and materials may pollute surface water or groundwater. Some of these are listed in Table 22.1. All segments of society (urban, rural, industrial, agricultural, and military) may contribute to the problem of water pollution. Most of the sources result from runoff and leaks or seepage of pollutants into surface water or groundwater. Pollutants are also transported by air and deposited in water bodies.

Increasing population often results in the introduction of more pollutants into the environment as well as demands on finite water resources.[4] As a result, it can be expected that sources of drinking water in some locations will be degraded in the near future. More than one-quarter of drinking water systems in the United States have reported at least one violation of federal health standards.[5] Approximately 36 million people in the United States were recently supplied with water from systems that violated (at least once) federal drinking water standards.[6]

The U.S. Environmental Protection Agency has set thresholds, or limits, on water pollution levels for some (but not all) pollutants. As a result of difficulties in determining effects of exposure to low levels of pollutants, maximum concentration standards have been set for only a small fraction of the more than 700 identified drinking water contaminants. If the pollutant is present in a concentration greater than an established threshold, then the water is unsatisfactory for a particular use. A list of selected pollutants (contaminants) included in the national drinking water standards for the United States can be found in Table 22.2.

The following sections focus on several water pollutants to emphasize principles that apply to pollutants in general. (See Table 22.3 for categories and examples of water pollutants.) Other water pollutants are discussed elsewhere in this book (for example, heavy metals, organic chemicals, and thermal pollution in Chapter 15 and radioactive materials in Chapter 20). Before proceeding to our discussion of pollutants, however, we first consider biochemical oxygen demand and dissolved oxygen; dissolved oxygen is not a pollutant but rather is needed for healthy aquatic ecosystems.

22.2 Biochemical Oxygen Demand (BOD)

Dead organic matter in streams decays. Bacteria carrying out this decay use oxygen. If there is enough bacterial activity, the oxygen in the water available to fish and other organisms can be reduced to levels so low that they may die. A stream with low oxygen content is a poor environment for fish and most other organisms. A stream with an inadequate oxygen level is considered polluted for those organisms that require dissolved oxygen above the existing level.

Table 22.1 • Some Sources and Processes of Water Pollution	
Surface Water	**Groundwater**
Urban runoff (oil, chemicals, organic matter, etc.) (U, I, M)	Leaks from waste-disposal sites (chemicals, radioactive materials, etc.) (I, M)
Agricultural runoff (oil, metals, fertilizers, pesticides, etc.) (A)	Leaks from buried tanks and pipes (gasoline, oil, etc.) (I, A, M)
Accidental spills of chemicals including oil (U, R, I, A, M)	Seepage from agricultural activities (nitrates, heavy metals, pesticides, herbicides, etc.) (A)
Radioactive materials (often involving truck or train accidents) (I, M)	Saltwater intrusion into coastal aquifers (U, R, I, M)
Runoff (solvents, chemicals, etc.) from industrial sites (factories, refineries, mines, etc.) (I, M)	Seepage from cesspools and septic systems (R)
Leaks from surface storage tanks or pipelines (gasoline, oil, etc.) (I, A, M)	Seepage from acid-rich water from mines (I)
Sediment from a variety of sources, including agricultural lands and construction sites (U, R, I, A, M)	Seepage from mine waste piles (I)
Air fallout (particles, pesticides, metals, etc.) into rivers, lakes, oceans (U, R, I, A, M)	Seepage of pesticides, herbicide nutrients, and so on from urban areas (U)
	Seepage from accidental spills (e.g., train or truck accidents) (I, M)
	Inadvertent seepage of solvents and other chemicals including radioactive materials from industrial sites or small businesses (I, M)

Key: U = urban; R = rural; I = industrial; A = agricultural; M = military

Table 22.2 • National Drinking Water Standards

Contaminant	Maximum Contaminant Level (mg/l)
Inorganics	
Arsenic	0.05
Cadmium	0.01
Lead	0.015 action level[a]
Mercury	0.002
Selenium	0.01
Organic chemicals	
Pesticides	
Endrin	0.0002
Lindane	0.004
Methoxychlor	0.1
Herbicides	
2,4-D	0.1
2,4,S-TP	0.01
Silvex	0.01
Volatile organic chemicals	
Benzene	0.005
Carbon tetrachloride	0.005
Trichloroethylene	0.005
Vinyl chloride	0.002
Microbiological organisms	
Fecal coliform bacteria	1 cell/100 ml

[a] Action level is related to the treatment of water to reduce lead to a safe level. There is no maximum contaminant level for lead.
Source: U.S. Environmental Protection Agency.

The amount of oxygen required for biochemical decomposition processes is called the **biochemical oxygen demand (BOD)**. BOD is commonly used in water quality management (Figure 22.2). It measures the amount of oxygen consumed by microorganisms as they break down organic matter within small water samples, which are analyzed in a laboratory. The BOD is routinely measured as part of water quality testing, particularly at discharge points into surface water, such as at wastewater treatment plants. At treatment plants, the BOD of the incoming sewage water from sewer lines is measured, as is water from locations both upstream and downstream of the plant. This practice allows comparison of the upstream, or background BOD with the BOD of the water being discharged by the plant.

Dead organic matter—which produces BOD—is added to streams and rivers from natural sources (such as dead leaves from a forest) as well as from agricultural runoff and urban sewage. Approximately 33% of all BOD in streams results from agricultural activities. However, urban areas, particularly those with older, combined sewer systems (in which storm-water runoff and urban sewage share the same line), also considerably increase the BOD in streams. This results because during times of high flow, when sewage-treatment plants are unable to handle the total volume of water, raw sewage mixed with storm runoff overflows and is discharged untreated into streams and rivers.

When the BOD is high, as suggested earlier, the *dissolved oxygen content* of the water may become too low to support the life in the water. The U.S. Environmental Protection

Table 22.3 • Categories of Water Pollutants

Pollutant Category	Examples of Sources	Comments
Dead organic matter	Raw sewage, agricultural waste, urban garbage	Produces biochemical oxygen demand and diseases.
Pathogens	Human and animal excrement and urine	Examples: Recent cholera epidemics in South America and Africa; 1993 epidemic of cryptosporidiosis in Milwaukee, Wisconsin. See discussion of fecal coliform bacteria in Section 22.3.
Organic chemicals	Agricultural use of pesticides and herbicides (Chapter 12); industrial processes that produce dioxin (Chapter 15)	Potential to cause significant ecological damage and human health problems. Many of these chemicals pose hazardous-waste problems (Chapter 30).
Nutrients	Phosphorus and nitrogen from agricultural and urban land use (fertilizers) and wastewater from sewage treatment	Major cause of artificial eutrophication. Nitrates in groundwater and surface waters can cause pollution and damage to ecosystems and people.
Heavy metals	Agricultural, urban, and industrial use of mercury, lead, selenium, cadmium, and so on (Chapter 15)	Example: Mercury from industrial processes that is discharged into water (Chapter 15). Heavy metals can cause significant ecosystem damage and human health problems
Acids	Sulfuric acid (H_2SO_4) from coal and some metal mines; industrial processes that dispose of acids improperly	Acid mine drainage is a major water pollution problem in many coal mining areas, damaging ecosystems and spoiling water resources.
Sediment	Runoff from construction sites, agricultural runoff, and natural erosion	Reduces water quality and results in loss of soil resources.
Heat (thermal pollution)	Warm to hot water from power plants and other industrial facilities	Causes ecosystem disruption (Chapter 15).
Radioactivity	Contamination by nuclear power industry, military, and natural sources (Chapter 20)	Often related to storage of radioactive waste. Health effects vigorously debated (Chapters 15 and 20).

Figure 22.2 ■ Pollution control officer measuring oxygen content of the River Severn near Shrewsbury, England.

Agency defines the threshold for a water pollution alert as a dissolved oxygen content of less than 5 mg/l of water. Figure 22.3 illustrates the effect of high BOD on dissolved oxygen content in a stream when raw sewage is introduced as a result of an accidental spill. Three zones are identified:

1. *A pollution zone*, where a high BOD exists. As decomposition of the waste occurs, oxygen is used by microorganisms, and dissolved oxygen content of the water decreases.

2. *An active decomposition zone*, where the dissolved oxygen content reaches a minimum owing to rapid biochemical decomposition by microorganisms as the organic waste is transported downstream.

3. *A recovery zone*, where the dissolved oxygen increases and the BOD is reduced, because most oxygen-demanding organic waste from the input of sewage has decomposed and the natural stream processes are replenishing the water with dissolved oxygen. For example, in quickly moving water, the water at the surface mixes with air, and oxygen enters the water.

All streams have some capability to degrade organic waste. Problems result when the stream is overloaded with biochemical oxygen-demanding waste, overpowering the stream's natural cleansing function.

22.3 Waterborne Disease

As mentioned earlier, the primary water pollution problem in the world today is the lack of clean, disease-free drinking water. Each year, particularly in less developed countries, several billion people are exposed to waterborne diseases whose effects vary in severity from an upset stomach to death. As recently as the early 1990s, epidemics of cholera, a serious waterborne disease, caused widespread suffering and death in South America.

In the United States, we tend not to think much about waterborne illness. Although historically epidemics of waterborne disease killed thousands of people in U.S. cities, such as Chicago, public health programs have largely eliminated such epidemics by treating drinking water to remove disease-carrying microorganisms and not allowing sewage to contaminate drinking water supplies. As we will see, however, North America is not immune to **outbreaks**—or sudden occurrences—of waterborne disease.

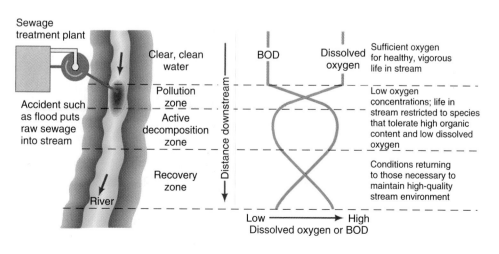

Figure 22.3 ■ Relationship between dissolved oxygen and biochemical oxygen demand (BOD) for a stream following the input of sewage.

Outbreak in Milwaukee, Wisconsin

The largest outbreak of waterborne disease in the history of the United States occurred in April 1993 in Milwaukee, Wisconsin. The disease, which causes flulike symptoms, is a gastrointestinal illness carried by a microorganism (a parasite) known as *Cryptosporidium* (the disease is known as *cryptosporidiosis*). Between March 11 and April 9, approximately 400,000 people of a total of 1.6 million people in a five-county area became ill following exposure to *Cryptosporidium* in the drinking water. Most people who contracted the illness were sick for approximately nine days; but the disease can be fatal to people with depressed immune systems, such as cancer patients and AIDS patients. Approximately 100 people died. The parasite is resistant to chlorination and evidently passed through one of the city's water treatment plants. The source of the parasite remains unknown, but possible sources include cattle grazing along rivers that flow into Milwaukee Harbor, slaughterhouses, and human sewage. It is possible that runoff from spring rains and snowmelt transported the parasites to Lake Michigan, where they entered the intake to water treatment plants.[7, 8]

The outbreak in Milwaukee was a wake-up call concerning the quality of U.S. drinking water. Many other cities in the United States that utilize surface water resources are just as vulnerable as Milwaukee.[7] In fact, recent tests suggest that *Cryptosporidium* is present in 65% to 97% of the surface waters in the United States.[9] In May 1994, another outbreak of the same disease, caused by the same parasite, killed 19 people in Las Vegas.[6]

The outbreak in Milwaukee happened even though the water treatment plants met all existing federal and state quality standards. Federal guidelines are even stricter now, but there still is concern as to whether waterborne parasites such as *Cryptosporidium* are being effectively removed. Although *Cryptosporidium* is very resistant to disinfectants, it may be removed by filtration. Upgrading water treatment plants is a cost-effective mechanism for reducing the threat of waterborne disease. The price of inaction is very high. Considering the high costs of illness and death associated with drinking contaminated water, future investments in technology and facilities to treat water are an important government service that should be considered a bargain.[6]

Fecal Coliform Bacteria

Because it is difficult to monitor disease-carrying organisms directly, we use the count of **fecal coliform bacteria** as a standard measure and indicator of disease potential. The presence of fecal coliform bacteria in water indicates that fecal material from mammals or birds is present, so organisms that cause waterborne diseases may be present as well. Fecal coliform bacteria are usually (but not always) harmless bacteria that are normal constituents of human and animal intestines. They are present in all human and animal waste. The threshold used by the U.S. Environmental Protection Agency for swimming water is not more than 200 cells of fecal coliform bacteria per 100 ml of water; if fecal coliform is above the threshold level, the water is considered unfit for swimming. Water with any fecal coliform bacteria is unsuitable for drinking.

One type of fecal coliform bacteria, *Escherichia coli*, or *E. coli*, has been responsible for causing human illness and death. Outbreaks have resulted from eating contaminated meat and drinking contaminated juices or water. There was an outbreak caused by contaminated meat at a popular fast-food chain in 1993. In 1998, 26 children became ill and one died after visiting a Georgia water park. As recently as July 1998, the community of Alpine, Wyoming, suffered a major outbreak of illness due to the presence of *E. coli* in the town's drinking water supply.[10] It is clear that *E. coli* bacteria can be a real threat to human health and must be carefully regulated.

Threat of waterborne diseases in nearshore coastal waters as well as lakes and rivers is responsible for thousands of warnings and beach closings per year in the United States. Advisories are often posted that swimming may be hazardous to health (Figure 22.4). In most cases, the identified pollutant is fecal coliform bacteria, which may indicate the presence of a specific disease-causing virus, such as hepatitis. Pollutants to coastal waters have a variety of sources, including urban storm runoff, leaking sewer lines, sewage-treatment plant overflow or failure, and leaking sewage treatment facilities from individual homes (septic tanks; see Section 22.10). Coastal communities in many areas are facing potential loss of income from tourism resulting from beach closures. As a result, increased research and testing of nearshore coastal waters is becoming more routine. As sources of pollutants are identified, management plans are being designed and implemented to reduce the threat of pollution and future beach closures.

Figure 22.4 ■ This beach in southern California is occasionally closed as a result of contamination by bacteria.

Outbreak in Walkerton, Ontario

One of the most serious outbreaks of *E. coli* bacterial infection in Canadian history developed in May of 2000 in Walkerton, Ontario, a town with about 5,000 inhabitants. The strain of *E. coli* involved was a dangerous one found in the digestive system of cows. The likely cause of the contamination was cow manure washed into the water supply during heavy rains and flooding on May 12, 2000. The local public utility commission knew as early as May 18 that water from wells serving the town was contaminated by the *E. coli*, but the commission did not report it immediately to health authorities. People were not advised to boil water (thus killing the bacteria) until it was too late to avoid the outbreak. By May 26, five people had died, more than 20 were in intensive care units, and over 500 were ill with severe symptoms that included cramps, vomiting, and diarrhea. The old and very young are most vulnerable to the ravages of the disease, which can damage the kidneys. Two of the first deaths were a 2-year-old child and an 82-year-old woman. Finally, officials took over management of the water supply, and bottled water was distributed.

The town of Walkerton experienced a full-blown outbreak of a waterborne disease. Doctors warned that additional deaths were likely and that modern medicine could do little to treat the disease. The best advice doctors could give was to consume large amounts of safe water and avoid dehydration as the infection ran its course. Soon, people were asking why the outbreak had occurred. An investigation was launched by authorities, and one of the questions asked was why there had been a delay between identifying the potential problem and issuing a warning. If people had been notified earlier, much of the sickness might have been avoided. The outbreak might also have been detected earlier if the government of Ontario had not cut back on the level of testing of the public water supply (earlier regulations had required more testing). A major lesson learned from Walkerton is that we should remain vigilant in testing our water supplies and raising the alarm quickly if potential problems arise.

22.4 Nutrients

Two important nutrients that cause water pollution problems are phosphorus and nitrogen, and both are released from sources related to land use. Forested land has the lowest concentrations of phosphorus and nitrogen in stream waters. In urban streams, concentrations of these nutrients are greater because of fertilizers, detergents, and products of sewage-treatment plants. Often, however, the highest concentrations of phosphorus and nitrogen are found in agricultural areas, where the sources are fertilized farm fields and feedlots (Figure 22.5). Over 90% of total nitrogen added to the environment by human activity comes from agriculture.

Medical Lake: An Example

An example will illustrate the effect nutrients can have on a body of water. During the summer of 1971, the waters of Medical Lake in Washington became clogged with algae and bacteria and turned a dark, turbid green. Fish, algae, and bacteria died, and the wind blew them into stinking masses onto the shore.[11] It was apparent that the Medical Lake ecosystem was dying.

Fish in the lake did not die from a direct effect of pollution, from disease organisms in the water, or from something in the water that was directly toxic to them. They died as the result of a complex effect produced by a community of interacting organisms linked to the local environment. (Recall the discussion of ecosystems and the community effect in Chapter 6.)

This is how the process worked. Phosphorus, a chemical element necessary for all living things and one that is often in short supply, became available in large concentrations in treated sewage water that entered the lake. Just as when phosphorus is added as a lawn fertilizer and causes increased growth of grass, the phosphorus added to the lake caused a population explosion of photosynthetic blue-green bacteria and algae. Mats of the algae and bacteria became so thick that those at the top shaded those at the bottom; not receiving enough light, those at the bottom died. The dead algae and bacteria became food for other bacteria, and as these bacteria increased,

Figure 22.5 ■ Cattle feedlot in Colorado. High numbers of cattle in small areas have the potential to create both surface water and groundwater pollution because of runoff and infiltration of urine.

they used more and more of the oxygen dissolved in the lake water. (Bacteria require oxygen for respiration, just as we do.) The supply of oxygen in the water was soon depleted. Algae, bacteria, and fish began to die. Because fish require a higher concentration of oxygen in water than do bacteria, most of the fish died, whereas some bacteria continued to live.

The fish did not die in Medical Lake from phosphorus poisoning. If you added phosphorus to water in an aquarium where there were only fish and no algae or bacteria, so that the phosphorus concentration was the same as that in Medical Lake, the phosphorus would not affect the fish. The fish died from a lack of oxygen resulting from a chain of events that started with the input of phosphorus and affected the whole ecosystem. The unpleasant effects resulted from the interactions among different species, the effects of the species on chemical elements in their environment, and the condition of the environment (the lake and the air above it). This is what we call an **ecosystem effect**.

Eutrophication

What happened at Medical Lake is called **eutrophication**—the process by which a body of water develops a high concentration of nutrients, such as nitrogen and phosphorus (in the forms of nitrates and phosphates). The nutrients cause an increase in the growth of aquatic plants in general, as well as production of photosynthetic blue-green bacteria and algae. Algae may form surface mats (Figure 22.6), shading the water and reducing light to algae below the surface, greatly reducing photosynthesis. The bacteria and algae die; as they decompose, the BOD increases, oxygen in the water is consumed, and oxygen content is re-

duced. If the oxygen content is sufficiently lowered, other organisms, such as fish, will die.

The process of eutrophication of a lake is shown in Figure 22.7. A lake that has a naturally high concentration of the chemical elements required for life is called a *eutrophic lake*. A lake with a relatively low concentration of chemical elements required by life is called an *oligotrophic lake*. Oligotrophic lakes have clear water that is pleasant for swimmers and boaters, with relatively low abundance of life. Eutrophic lakes have an abundance of life, often with mats of algae and bacteria and murky, unpleasant water.

When eutrophication is accelerated by human processes that add nutrients to a body of water, we say that **cultural eutrophication** is occurring. Problems associated with the artificial eutrophication of bodies of water are not restricted to lakes (see A Closer Look 22.1). In recent years concern has grown about the outflow of sewage from urban areas into tropical coastal waters and cultural eutrophication on coral reefs.[13, 14] For example, parts of the famous Great Barrier Reef of Australia, as well as some reefs fringing the Hawaiian Islands, are being damaged by eutrophication.[15, 16] The damage to corals occurs as nutrient input stimulates algal growth on the reef, which covers and smothers the coral.

The solution to artificial eutrophication is fairly straightforward and involves ensuring that high concentrations of nutrients from human sources do not enter lakes and other water bodies. This goal can be accomplished by reducing pollutants through use of phosphate-free detergents, controlling nitrogen-rich runoff from agricultural and urban lands, disposing of or reusing treated wastewater, and using more advanced water treatment methods, such as special filters and chemical treatments that remove more of the nutrients.

Figure 22.6 ■ Mats of dying green algae in a pond undergoing eutrophication.

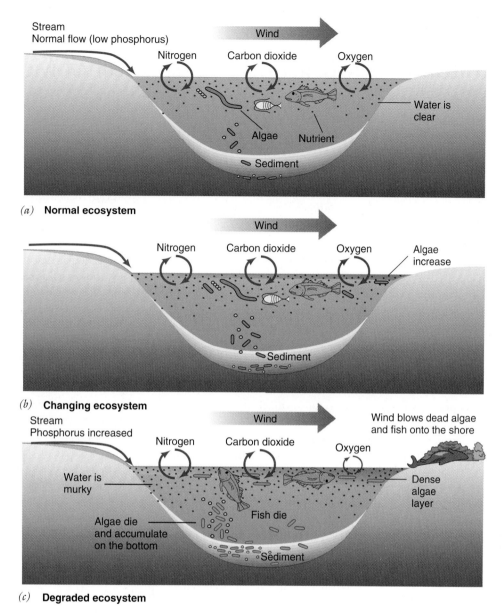

Stream
Normal flow (low phosphorus)

Wind

Nitrogen   Carbon dioxide   Oxygen

Water is
clear

Algae   Nutrient

Sediment

(a) **Normal ecosystem**

Wind

Nitrogen   Carbon dioxide   Oxygen   Algae
increase

Sediment

(b) **Changing ecosystem**

Stream
Phosphorus increased

Wind   Wind blows dead algae
and fish onto the shore

Nitrogen   Carbon dioxide   Oxygen

Water is
murky

Dense
algae
layer

Algae die
and accumulate
on the bottom

Fish die

Sediment

(c) **Degraded ecosystem**

Figure 22.7 ■ The eutrophication of a lake. (*a*) In an oligotrophic, or low-nutrient, lake, the abundance of green algae is low, the water clear. (*b*) Phosphorus is added to streams and enters the lake. Algae growth is stimulated, and a dense layer is formed. (*c*) The algae layer becomes so dense that the algae at the bottom die. Bacteria feed on the dead algae and use up the oxygen. Finally, fish die from lack of oxygen.

22.5 Oil

Oil discharged into surface water—usually in the ocean but also on land and in rivers—has caused major pollution problems. Several large oil spills from underwater oil drilling have occurred in recent years. However, although spills make headlines, normal shipping activities probably release more oil over a period of years than is released by the occasional spill. The cumulative impacts of these releases are not well known.

Exxon Valdez: Prince William Sound, Alaska

The best known oil spills are caused by tanker accidents. On March 24, 1989, the supertanker *Exxon Valdez* ran aground on Bligh Reef south of Valdez in Prince William Sound, Alaska. Alaskan crude oil that had been delivered to Valdez through the trans-Alaskan pipeline poured out of ruptured tanks of the vessel at about 20,000 barrels per hour. The tanker was loaded with about 1.2 million barrels of oil, and about 250,000 barrels (11 million gal)

Cultural Eutrophication in the Gulf of Mexico

Each summer, a so-called dead zone develops off the nearshore environment of the Gulf of Mexico, south of Louisiana. The zone varies in size from about 13,000 to 18,000 km^2 (5,000 to 7,000 mi^2), an area about the size of the small country of Kuwait or the state of New Jersey. Within the zone, bottom water generally has low concentrations of dissolved oxygen (less than 2 mg/l; a water pollution alert occurs if dissolved oxygen has a concentration of less than 5 mg/l). Shrimp and fish can swim away from the zone, but bottom dwellers such as shellfish, crabs, and snails are killed. Nitrogen is believed to be the most significant cause of the dead zone.

The low concentration of oxygen occurs because the nitrogen causes cultural eutrophication. Algae bloom, and as the algae die, sink, and decompose, the oxygen in the water is depleted. The source of nitrogen is believed to be in one of the richest, most productive agricultural regions of the world—the Mississippi River drainage basin.

The Mississippi River drains about 3 million km^2, which is about 40% of the land area of the lower 48 states. The use of nitrogen fertilizers in that area greatly increased beginning in the mid–twentieth century but leveled off in the 1980s and 1990s. The level of nitrogen in the river water has also leveled off, suggesting that the dead zone may have reached its maximum size. This provides the time necessary to study the cultural eutrophication problem carefully and make sound decisions to reduce or eliminate it.

A partial reduction of nitrogen (nitrates) reaching the Gulf of Mexico via the Mississippi River may be accomplished by the following actions:[12]

- Modify agricultural practices to reduce the nitrogen that enters the river from fertilizers by using fertilizers more effectively and efficiently.
- Restore and create river wetlands between farm fields and streams and rivers, particularly in areas known to contribute high amounts of nitrogen. The wetland plants use the nitrogen, lowering the amount that enters the river.
- Implement nitrogen reduction processes at wastewater treatment plants for towns, cities, and industrial facilities.
- Implement better flood control in the upper Mississippi River to retain floodwaters on floodplains, where nitrogen can be used by riparian vegetation.
- Divert floodwater from the Mississippi to backwaters and coastal wetlands of the Mississippi River Delta. At present, levees in the delta push river waters directly into the gulf. Plants in coastal wetlands will use the nitrogen, reducing the concentration of the nutrient that reaches the Gulf of Mexico.

Improving agricultural practices could result in a reduction of up to 20% in the nitrogen reaching the Mississippi. This would require a reduction of about 20% in fertilizers used, something farmers say would harm productivity. Restoration and creation of river wetlands and riparian forests hold the promise of reducing nitrogen input to the river by up to 40%. This would require some combination of wetlands and forest of about 10 million hectares (24 million acres), which is about 3.4% of the Mississippi River Basin.[11] That is a lot of land!

There is no easy solution to cultural eutrophication in the Gulf of Mexico. Clearly, however, a reduction in the amount of nitrogen entering the gulf is needed. Also needed is a better understanding of details of the nitrogen cycle within the Mississippi River basin and Delta. Developing this understanding will require monitoring of nitrogen and development of mathematical models of sources, sinks, and rates of transfer of nitrogen. With an improved understanding of the nitrogen cycle, a management strategy to reduce or eliminate the dead zone can be created and implemented.

The dead zone in the Gulf of Mexico is not unique in the world. Other dead zones include offshore of Europe, China, Australia, South America, and the northeastern United States. In all, about 150 others in the oceans of the world have been observed. Most are much smaller than the zone in the Gulf of Mexico. As with the Gulf, the other dead zones are due to oxygen depletion resulting from nitrogen, mostly from agricultural runoff. A few result from industrial pollution or runoff from urban areas, especially untreated sewage.

entered the sound. The spill could have been larger than it was, but fortunately some of the oil in the tanker was offloaded (pumped out) into another vessel (Figure 22.8). The *Exxon Valdez* spill produced an environmental shock that resulted in passage of the Oil Pollution Act of 1990 and in a renewed evaluation of cleanup technology.[17, 18]

An environmental shock resulted from the fact that the oil was spilled into what is considered one of the most pristine and ecologically rich marine environments of the world.[17] Many species of fish, birds, and marine mammals are present in the sound. The effects of the spill included the death of 13% of harbor seals, 28% of sea otters, and 100,000–645,000 seabirds.[18]

Within three days of the spill, winds began blowing the slick beyond any hope of containment. Of the 11 million gallons of spilled oil, about 20% evaporated and 50% was deposited on the shoreline. Only 14% was collected by skimming and waste recovery. The extent of oil sheens, tar

Figure 22.8 ■ *Exxon Valdez* tanker accident in Prince William Sound (1989). Oil is being offloaded from the leaking *Exxon Valdez* (left) to the smaller *Exxon Baton Rouge* (right).

balls (formed by sticky, less volatile components in the oil), and mousse (a thick, weathered patch of oil with the consistency of soft pudding) is shown in Figure 22.9.

Before the *Exxon Valdez* spill, it was generally believed that the oil industry was capable of dealing with oil spills. However, more than $3 billion has been spent to clean the

spill, and few people are satisfied with the results. Some scientists argue that the recovery might have been faster if some of the cleanup methods, such as spraying rocks and beaches with high-pressure hot water, had not been used. They argue that the coastal organisms that live under the rocks and that had survived the initial impact of the spill

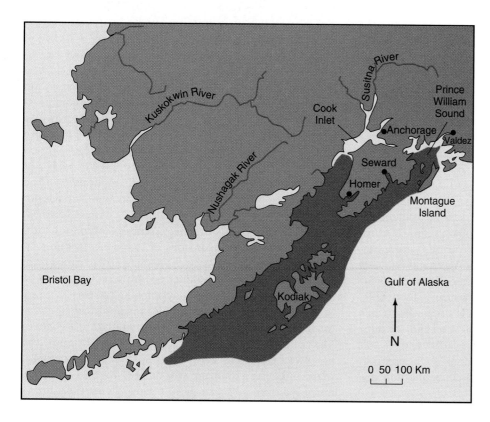

Figure 22.9 ■ Extent of Alaskan oil spill of 1989. [*Source:* Alaska Department of Fish and Game. 1989. *Alaska Fish and Game* 21(4), Special Issues.]

Figure 22.10 ■ Attempt to clean oil from the shoreline of Eleanor Island, Alaska, four months after the oil spill from the *Exxon Valdez*.

were killed by the high pressure and heat.[18] There is no doubt that the cleanup work posed enormous problems (Figure 22.10). Photographs and videotapes of workers attempting to clean individual pebbles on beaches are a vivid reminder of the difficulty and virtual futility of achieving an effective cleanup after an event of this magnitude. In addition, the oil spill disrupted the lives of the people who live and work in the vicinity of Prince William Sound.

The long-term effects of large oil spills are uncertain. We know that the effects can last several decades; toxic levels of oil have been identified in salt marshes 20 years following a spill.[18]

The *Exxon Valdez* spill demonstrated that the technology for dealing with oil spills is inadequate. The first and most important step is to avoid large spills; a primary method for doing so is to use supertankers with double hulls designed to minimize the release of oil on collision and rupture of tanks. The second most important step is to pump the oil out of a tanker as soon as an accident occurs, thereby preventing further spillage into the sea. Once a spill occurs, the collection of oil at sea using floating barriers and skimmers (oil is lighter than water and so floats on water) is a worthwhile endeavor; but if conditions include high winds and rough seas, it is nearly impossible. Cleaning oil from birds and mammals is also a worthwhile endeavor, but because the animals ingest oil and are difficult to clean, mortality is high. Oil on beaches may be collected by spreading absorbent material, such as straw, allowing the oil to soak in, and then collecting and disposing of the oily straw.

Jessica: Galápagos Islands

Another shock occurred on January 22, 2001, when a small tanker, the *Jessica*, made a navigational error off the coast of Ecuador near the Galápagos Islands and ran aground, spilling light diesel oil into the ocean. Although the spill (over 100,000 gallons) was small compared with the *Exxon Valdez* spill in Alaska, there was grave concern, and Ecuador declared a state of emergency. The Galápagos are environmental treasures and icons of the environment, where Charles Darwin worked in developing his theory of evolution of species. The United States responded quickly with a Coast Guard ship designed to pump oil from a damaged tanker. Some of the oil was washed ashore onto a small island, injuring birds, seals, and other marine life. The oil slick spread over 3,000 km^2 in the first week but was fortunately carried by currents and trade winds away from the nearby Galápagos. The spill was yet another warning regarding the use of better-built tankers with double hulls and, where possible, the routing of tankers away from ecological areas of particular concern.

22.6 Sediment

Sediment consisting of rock and mineral fragments ranging from gravel particles greater than 2 mm in diameter to finer sand, silt, clay, and even finer colloidal particles can produce a *sediment pollution* problem. In fact, by volume and mass, sediment is our greatest water pollutant. In many areas, it chokes streams; fills lakes, reservoirs, ponds, canals, drainage ditches, and harbors; buries vegetation; and generally creates a nuisance that is difficult to remove. Sediment pollution is a twofold problem: It results from erosion, which depletes a land resource (soil) at its site of origin (Figure 22.11), and it reduces the quality of the water resource it enters.[19]

Many human activities affect the pattern, amount, and intensity of surface water runoff, erosion, and sedimentation. Streams in naturally forested or wooded areas may be nearly stable; that is, there is relatively little excessive erosion or sedimentation. However, converting forested land to agriculture generally increases the runoff and sediment yield or erosion of the land. Application of soil conservation procedures to farmland can minimize but not eliminate soil loss. The change from agricultural, forested, or rural land to highly urbanized land has even more dramatic effects. Large quantities of sediment may be produced during the construction phase of urbanization. Fortunately, sediment production and soil erosion can be minimized by on-site erosion control measures.

Reduction of sediment pollution in an urbanizing area through control measures is demonstrated by a study in Maryland.[20] The suspended sediment transported by the northwest branch of the Anacostia River near Colesville, Maryland, with a drainage area of 54.6 km^2 (21 mi^2), was measured over a 10-year period. During that time, urban

Figure 22.11 ■ Erosion of a highway embankment in Georgia. The erosion produces gullies and removes vegetation. The sediment may be transported off-site to degrade water resources (streams, rivers, ponds, and lakes).

Figure 22.12 ■ Acid drainage from an abandoned mine site is entering a small stream channel and polluting the surface waters. This site is located in the mountains of southwestern Colorado.

construction within the basin involved about 3% of the area each year, and the total urban land area in the basin was about 20% at the end of the 10-year study. Sediment pollution was a problem because of the amount of rainfall and the type of soil, which was highly susceptible to erosion when not protected, during spring and summer rainstorms, by a vegetative cover.

A sediment control program reduced the sediment yield by an estimated 35%. The program's basic principles were to tailor the development to the natural topography, expose a minimum amount of land, provide temporary protection for exposed soil, minimize surface runoff from critical areas, and trap eroded sediment on the construction site.[20]

22.7 Acid Mine Drainage

The term **acid mine drainage** refers to water with a high concentration of sulfuric acid (H_2SO_4) that drains from mines—mostly coal mines but also metal mines (copper, lead, and zinc). Coal and the rocks containing coal are often associated with a mineral known as fool's gold or pyrite (FeS_2), which is iron sulfide. When the pyrite, which may be finely disseminated in the rock and coal, comes into contact with oxygen and water, it weathers. A product of the chemical weathering is sulfuric acid. In addition, pyrite is associated with metallic sulfide deposits, which, when weathered, also produce sulfuric acid. The acid is produced when surface water or shallow groundwater runs through or moves into and out of mines or tailings (Figure 22.12). If the acid-rich water runs off to a natural stream, pond, or lake, significant pollution and ecological damage may result. The acidic water is toxic to the plants and animals of

an aquatic ecosystem; it damages biological productivity, and fish and other aquatic life may die. Acid-rich water can also seep into and pollute groundwater.

Acid mine drainage is a significant water pollution problem in Wyoming, Indiana, Illinois, Kentucky, Tennessee, Missouri, Kansas, and Oklahoma and is probably the most significant water pollution problem in West Virginia, Maryland, Pennsylvania, Ohio, and Colorado. The total impact is significant because thousands of kilometers of streams have been damaged.

Even abandoned mines can cause serious problems. For example, subsurface mining in the tristate area of Kansas, Oklahoma, and Missouri for sulfide deposits containing lead and zinc began in the late nineteenth century and ended in some areas in the 1960s. When the mines were operating, they were kept dry by pumping out the groundwater that seeped in. However, since the termination of mining, some of them have flooded and overflowed into nearby creeks, polluting the creeks with acid-rich water. The problem was so severe in the Tar Creek area of Oklahoma that it was at one time designated by the U.S. Environmental Protection Agency as the nation's worst hazardous waste site.

22.8 Surface Water Pollution

Pollution of surface waters occurs when too much of an undesirable or harmful substance flows into a body of water, exceeding the natural ability of that water body to remove the undesirable material, dilute it to a harmless concentration, or convert it to a harmless form.

Water pollutants, like other pollutants, are categorized as being emitted from point or nonpoint sources (see Chapter 15). **Point sources** are distinct and confined,

Figure 22.13 ■ This pipe is a point source of chemical pollution from an industrial site entering a river in England.

such as pipes from industrial or municipal sites that empty into streams or rivers (Figure 22.13). In general, point source pollutants from industries are controlled through on-site treatment or disposal and are regulated by permit. Municipal point sources are also regulated by permit. In older cities in the northeastern and Great Lakes areas of the United States, most point sources are outflows from combined sewer systems. As mentioned earlier, such systems combine storm water flow with municipal wastewater. During heavy rains, urban storm runoff may exceed the capacity of the sewer system, causing it to overflow and deliver pollutants to nearby surface waters.

Nonpoint sources, such as runoff, are diffused and intermittent and are influenced by factors such as land use, climate, hydrology, topography, native vegetation, and geology. Common urban nonpoint sources include runoff from streets or fields; such runoff contains all sorts of pollutants, from heavy metals to chemicals and sediment. Rural sources of nonpoint pollution are generally associated with agriculture, mining, or forestry. Nonpoint sources are difficult to monitor and control.

From an environmental view, two approaches to dealing with surface water pollution are (1) to reduce the sources and (2) to treat the water to remove pollutants or convert them to forms that can be disposed of safely. Which of the options is used depends on the specific circumstances of the pollution problem. Reduction at the source is the most environmentally preferable way of dealing with pollutants. For example, air-cooling towers, rather than water-cooling towers, may be utilized to dispose of waste heat from power plants, thereby avoiding thermal pollution of water. The second method—water treatment—is used for a variety of pollution problems. Water treatments include chlorination to kill microorganisms such as harmful bacteria, and filtering to remove heavy metals.

There is a growing list of success stories in the treatment of water pollution. One of the most notable is the cleanup of the Thames River in Great Britain. For centuries, London's sewage had been dumped into that river, and there were few fish to be found downstream in the estuary. In recent decades, however, improvement in water treatment has led to the return of a number of species of fish, some not seen in the river in centuries.

Many large cities in the United States, such as Boston, Miami, Cleveland, Detroit, Chicago, Portland, and Los Angeles, grew on the banks of rivers; but the rivers were often nearly destroyed with pollution and concrete. Today, there are grassroots movements all around the country to restore urban rivers and adjacent lands as greenbelts, parks, or other environmentally sensitive developments. For example, the Cuyahoga River in Cleveland, Ohio, was so polluted by 1969 that sparks from a train ignited oil-soaked wood in the river, setting the surface of the river on fire! The burning of an American river became a symbol for a growing environmental consciousness. The Cuyahoga River today is cleaner and no longer flammable. From Cleveland to Akron, the river is a beautiful greenbelt (Figure 22.14). The greenbelt changed part of the river from a sewer into a valuable public resource and focal point for economic and environmental renewal.[21] However, in downtown Cleveland and Akron, the river remains an industrial stream, and parts remain polluted.

22.9 Groundwater Pollution

Approximately half of all people in the United States today depend on groundwater as their source of drinking water. (Water for domestic use in the United States is discussed in A Closer Look 22.2.) People have long believed that groundwater is in general pure and safe to drink. In fact, however, groundwater can be easily polluted by any one of several sources (see Table 22.1); and the pollutants, even though they are very toxic, may be difficult to recognize. (Groundwater processes were discussed in Section 21.1, and you may wish to review them.)

In the United States today, only a small portion of the groundwater is known to be seriously contaminated; but as mentioned earlier, the problem may become worse as human population pressure on water resources increases.

Principles of Groundwater Pollution: An Example

Some general principles of groundwater pollution are illustrated by an example. Pollution from leaking buried gasoline tanks belonging to automobile service stations is a widespread environmental problem that no one thought very much about until only a few years ago. Underground tanks are now strictly regulated. Many thousands of old, leaking tanks have been removed, and the surrounding soil and groundwater have been treated to remove the gasoline. Cleanup can be a very expensive process, involving removal and disposal of soil (as a hazardous waste) and treatment of the water using a process known as vapor extraction (Figure 22.15). Treatment may also be accomplished under the ground by micoorganisms that consume the gasoline. This is known as **bioremediation** and is much less expensive than removal, disposal, and vapor extraction.

Pollution from leaking buried gasoline tanks emphasizes some important points about groundwater pollutants:

- Some pollutants, such as gasoline, are lighter than water and thus float on the groundwater.

- Some pollutants have multiple phases: liquid, vapor, and dissolved. Dissolved phases chemically combine with the groundwater (e.g., salt dissolves into water).

- Some pollutants are heavier than water and sink or move downward through groundwater. Examples of sinkers include some particulates and cleaning solvents. Pollutants that sink may become concentrated deep in groundwater aquifers.

- The method used to treat or eliminate a water pollutant must take into account the physical and chemical properties of the pollutant and how these interact with surface water or groundwater. For example, the extraction well for removing gasoline from a groundwater resource (Figure 22.15) takes advantage of the fact that gasoline floats on water.

- Because cleanup or treatment of water pollutants in groundwater is very expensive, and undetected or untreated pollutants may cause environmental damage, the emphasis should be on preventing pollutants from entering groundwater in the first place.

Groundwater pollution differs in several ways from surface water pollution. Groundwater often lacks oxygen, a situation that kills aerobic types of microorganisms (which require oxygen-rich environments) but may provide a happy home for anaerobic varieties (which live in oxygen-deficient environments). The breakdown of pollutants that occurs in the soil and in material a meter or so below the surface does not occur readily in groundwater. Furthermore, the channels through which groundwater moves are often very small and variable. Thus, the rate of movement is low in most cases, and the opportunity for dispersion and dilution of pollutants is limited.

Figure 22.14 ■ The Cuyahoga River (lower left) flows toward Cleveland, Ohio, and Lake Erie. A trail along the Ohio and Erie Canal (lower right) is in the Cuyahoga National Park. The Skyline is that of industrial Cleveland.

Already, the extent of the problem is growing as the testing of groundwater becomes more common. For example, Atlantic City and Miami are two eastern cities threatened by polluted groundwater that is slowly migrating toward their wells.

It is estimated that 75% of the 175,000 known wastdisposal sites in the United States may be producing plumes of hazardous chemicals that are migrating into groundwater resources. Because many of the chemicals are toxic or are suspected carcinogens, it appears that we have inadvertently been conducting a large-scale experiment on the effects on people of chronic low-level exposure to potentially harmful chemicals. The final results of the experiment will not be known for many years.[22] Preliminary results suggest that we had better act now before a hidden time bomb of health problems explodes.

The hazard presented by a particular groundwater pollutant depends on several factors, including the concentration or toxicity of the pollutant in the environment and the degree of exposure of people or other organisms to the pollutants.[23] (See the section on risk assessment in Chapter 15.)

Water for Domestic Use: How Safe Is It?

Water for domestic use in the United States is drawn from surface waters and groundwater. Although some groundwater sources have high water quality and need little or no treatment, most sources are treated to conform to national drinking water standards (revisit Table 22.3).

Before treatment, water is usually stored in reservoirs or special ponds. Storage allows for solids, such as fine sediment and organic matter, to settle out, improving the clarity of water. The water is then run through a water plant, where it is filtered and chlorinated before it is distributed to individual homes. Once in people's homes, water may be further treated. For example, many people run their tap water through readily available charcoal filters before using it for drinking and cooking.

A growing number of people prefer not to drink tap water. Instead, they use bottled water for personal consumption; as a result, production of bottled water has become a multibillion-dollar industry.[6] Some people prefer not to drink water that contains chlorine or that runs through metal pipes. Furthermore, water supplies vary in clarity, hardness (concentration of calcium and magnesium), and taste; and the water available locally may not be to some people's liking. A common complaint about tap water is a chlorine taste, which may occur with chlorine concentrations as low as 0.2–0.4 mg/l. People may also fear contamination by minute concentrations of pollutants. The drinking water in the United States is some of the safest in the world; there is no doubt that treatment of water with chlorine has nearly eliminated waterborne diseases, such as typhoid and cholera, which previously caused widespread suffering and death in the developed world and still do in many parts of the world. However, we need to know much more about the long-term effects of exposure to low concentrations of toxins in our drinking water. How safe is the water in the United States? It's much safer than it was 100 years ago, but low-level contamination (below what is thought dangerous) of organic chemicals and heavy metals is a concern that requires continued research and evaluation.

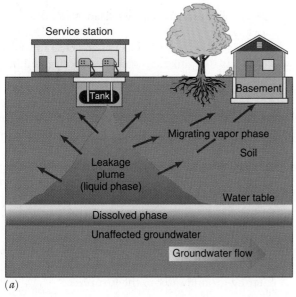

(a)

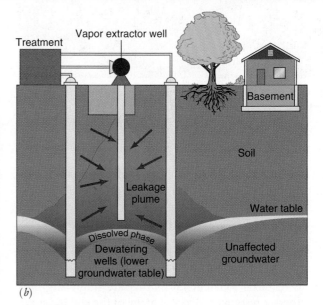

(b)

Figure 22.15 ■ Diagram illustrating (*a*) leak from a buried gasoline tank and (*b*) possible remediation using vapor extractor system. Notice that the liquid gasoline and the vapor from the gasoline are above the water table; a small amount dissolves into the water. All three phases of the pollutant (liquid, vapor, and dissolved) float on the denser groundwater. The extraction well takes advantage of this situation. The function of the dewatering wells is to pull the pollutants in where the extraction is most effective. [*Source:* Courtesy of the University of California Santa Barbara Vadose Zone Laboratory and David Springer.]

Long Island, New York

Another example—that of Long Island, New York—illustrates several groundwater pollution problems and how they affect people's water supply. Two counties on Long Island, New York (Nassau and Suffolk), with a population of several million people, depend entirely on groundwater. Two major problems associated with the groundwater in Nassau County are intrusion of salt water and shallow-aquifer contamination.[24] Saltwater intrusion is a problem in many coastal areas of the world (see Figure 22.16).

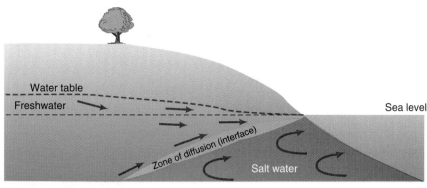

(a) **Before pumping**

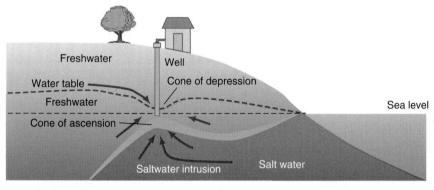

(b) **After pumping**

Figure 22.16 ■ How saltwater intrusion might occur. The upper drawing (*a*) shows the groundwater system near the coast under natural conditions, and the lower drawing (*b*) shows a well with both a cone of depression and a cone of ascension. If pumping is intensive, the cone of ascension may be drawn upward, delivering salt water to the well.

The general movement of groundwater under natural conditions for Nassau County is illustrated in Figure 22.17. Salty groundwater is restricted from migrating inland by the large wedge of freshwater moving beneath the island. Notice also that the aquifers are layered, with those closest to the surface being the most salty.

In spite of the huge quantities of water in Nassau County's groundwater system, intensive pumping in recent years has caused water levels to decline as much as 15 m (50 ft) in some areas. As groundwater is removed near coastal areas, the subsurface outflow to the ocean decreases, allowing saltwater to migrate inland. Saltwater intrusion has become a problem for south shore communities, which now must pump groundwater from a deeper aquifer, below and isolated from the shallow aquifers where saltwater intrusion problems exist.

The most serious groundwater problem on Long Island is shallow-aquifer pollution associated with urbanization. Sources of pollution in Nassau County include urban runoff, household sewage from cesspools and septic tanks, salt used to de-ice highways, and industrial and solid waste. These pollutants enter surface waters and then migrate downward, especially in areas of intensive pumping and declining groundwater levels.[24] Landfills for municipal solid waste have been a significant source of shallow-aquifer pollution on Long Island because pollutants (garbage) placed on sandy soil over shallow ground water quickly enter the water. For this reason, most Long Island landfills were closed in the last two decades.

22.10 Wastewater Treatment

Water used for industrial and municipal purposes is often degraded during use by the addition of suspended solids, salts, nutrients, bacteria, and oxygen-demanding material. In the United States, by law, these waters must be treated before being released back into the environment. **Wastewater treatment**, or sewage treatment, costs about $20 billion per year in the United States, and the cost continues to increase. Wastewater treatment will continue to be big business.

Conventional methods of wastewater treatment include septic-tank disposal systems in rural areas and centralized wastewater treatment plants in cities. Recent, innovative approaches include the application of wastewater to the land and waste water renovation and reuse. We discuss the conventional methods in this section and some newer methods in later sections.

Septic-Tank Disposal Systems

In many rural areas, no central sewage systems or waste water treatment facilities are available. As a result, individual septic-tank disposal systems, not connected to sewer systems, continue to be an important method of sewage disposal in rural areas as well as outlying areas of cities. Because not all land is suitable for the installation of a septic-tank disposal system, an evaluation of each site is required by law before a permit can be issued. An alert buyer should

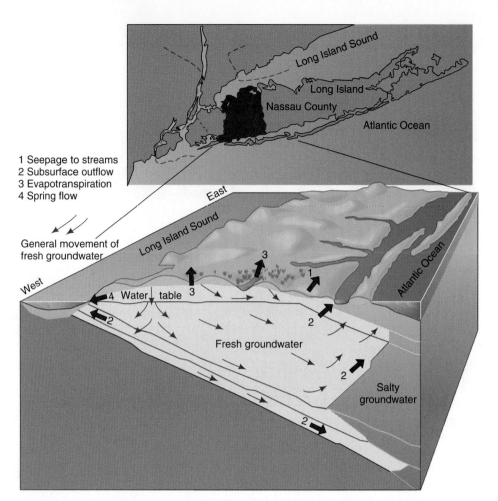

1 Seepage to streams
2 Subsurface outflow
3 Evapotranspiration
4 Spring flow

General movement of fresh groundwater

Figure 22.17 ■ The general movement of fresh groundwater for Nassau County, Long Island. [*Source:* G. L. Foxworth, Nassau County, Long Island, New York— Water Problems in Humid County, in G. D. Robinson and A. M. Spieke, eds., *Nature to Be Commanded*, U.S. Geological Survey Professional Paper 950, 1978, pp. 55–68.]

make sure that the site is satisfactory for septic-tank disposal before purchasing property in a rural setting or on the fringe of an urban area where such a system is necessary.

The basic parts of a septic-tank disposal system are shown in Figure 22.18. The sewer line from the house leads to an underground septic tank in the yard. The tank is designed to separate solids from liquid, digest (biochemically change) and store organic matter through a period of detention, and allow the clarified liquid to discharge into the drain field (absorption field) from a system of piping through which the treated sewage seeps into the surrounding soil. As the wastewater moves through the soil, it is further treated by the natural processes of oxidation and filtering. By the time the water reaches any fresh water supply, it should be safe for other uses.

Sewage absorption fields may fail for several reasons. The most common causes are failure to pump out the septic tank when it is full of solids, and poor soil drainage, which allows the effluent to rise to the surface in wet weather. When a septic-tank absorption field does fail, pollution of groundwater and surface water may result. Solutions to septic-tank system problems include siting septic tanks on well-drained soils, making sure systems are of adequate size, and practicing proper maintenance.

Wastewater Treatment Plants

In urban areas, wastewater treatment occurs at specially designed plants that accept municipal sewage from homes, businesses, and industrial sites. The raw sewage is delivered to the plant through a network of sewer pipes. Following treatment, the wastewater is discharged into the surface water environment (river, lake, or ocean) or, in some limited cases, used for another purpose, such as crop irrigation. The main purpose of standard treatment plants is to break down and reduce the BOD and kill bacteria with chlorine. A simplified diagram of a wastewater treatment plant is shown in Figure 22.19.

Wastewater treatment methods are usually divided into three categories: **primary treatment, secondary treatment,** and **advanced wastewater treatment**. Primary and secondary treatment are required by federal law for all municipal plants in the United States. However, treatment plants may qualify for a waiver to be exempt from secondary treatment if the installment of secondary treatment facilities poses an excessive financial burden. Where secondary treatment is not sufficient to protect the quality of the surface water into which the treated water is discharged—for example, a river with endangered fish species that must be protected—advanced treatment may be required.[25]

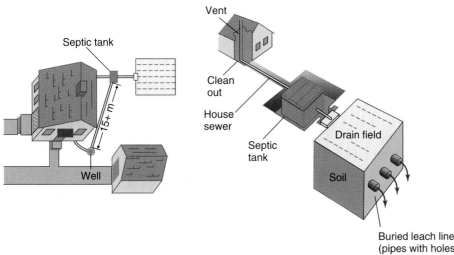

Figure 22.18 ■ Septic-tank sewage disposal system and location of the absorption field with respect to the house and well. [*Source:* Based on Indiana State Board of Health.]

Primary Treatment

Incoming raw sewage enters the plant from the municipal sewer line and is first passed through a series of screens, the purpose of which is to remove large floating organic material. The sewage next enters the grit chamber, where sand, small stones, and grit are removed and disposed of. The sewage then enters the primary sedimentation tank, where particulate matter settles out to form a sludge. Sometimes, chemicals are used to help the settling process. The sludge is removed and transported to the digester for further processing. Primary treatment removes approximately 30 to 40% of the BOD by volume from the wastewater, mainly in the form of suspended solids and organic matter.[25]

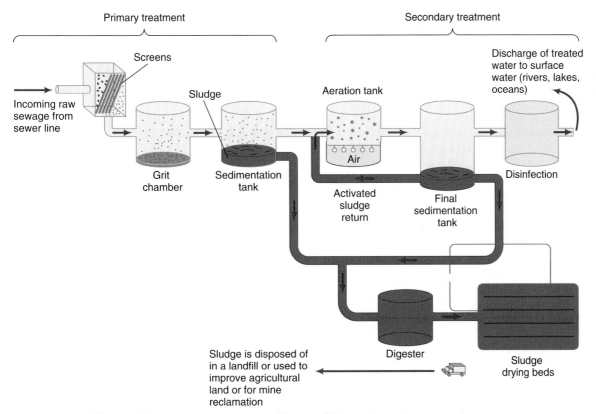

Figure 22.19 ■ Diagram of sewage-treatment processes. The use of digesters is relatively new, and many older treatment plants do not have them.

Secondary Treatment

There are several methods of secondary treatment. What is described here is known as *activated sludge*, the most common treatment. In this procedure, the waste water from the primary sedimentation tank enters the aeration tank (Figure 22.19), where the wastewater is mixed with air (which is pumped in) and some of the sludge from the final sedimentation tank. The sludge contains aerobic bacteria that consume organic material (BOD) in the waste. The wastewater then enters the final sedimentation tank, where sludge settles out. Some of this activated sludge, which is rich in bacteria, is recycled and mixed again in the aeration tank with air and new, incoming wastewater acting as a starter; the bacteria are used again and again. Most of the sludge from the final sedimentation tank, however, is transported to the sludge digester. There, along with sludge from the primary sedimentation tank, it is treated by anaerobic bacteria, which further degrade the sludge by microbial digestion.

Methane gas (CH_4) is a product of the anaerobic digestion and may be used at the plant as a fuel to run equipment or heat and cool buildings. In some cases it is burned off. Wastewater from the final sedimentation tank is then disinfected, usually by chlorination, to eliminate disease-causing organisms. The treated wastewater is then discharged into a river, lake, or ocean or, in some limited cases, used to irrigate farmland (see A Closer Look 22.3). Secondary treatment removes about 90% of the BOD that enters the plant in the sewage.[25]

The sludge from the digester is dried and disposed of in a landfill or applied to improve soil. In some instances, treatment plants in urban and industrial areas contain many pollutants, such as heavy metals, that are not removed in the treatment process. Sludge from these plants is too polluted to use to improve the soil, and the sludge must be disposed of. Some communities, however, require industries to pretreat sewage to remove heavy metals before the sewage is sent to the treatment plant; in these instances, the sludge can be more safely used for soil improvement.

Advanced Wastewater Treatment

Primary and secondary treatments do not remove all pollutants from incoming sewage. Some additional pollutants can be removed by adding more treatment steps. For example, nutrients such as phosphates and nitrates; organic chemicals; and heavy metals can be removed by specifically designed treatments such as sand filters, carbon filters, and chemicals applied to assist in the removal process.[25] Treated water is then discharged into surface water or may be used for irrigation of agricultural lands or municipal properties, such as golf courses, city parks, and grounds surrounding wastewater treatment plants.

Advanced wastewater treatment is used when it is particularly important to maintain good water quality. For example, if a treatment plant discharges treated waste water into a river and there is concern that nutrients remaining after secondary treatment may cause damage to the river ecosystem (eutrophication), advanced treatment may be used to reduce the nutrients.

Chlorine Treatment

As mentioned, chlorine is frequently used to disinfect water as part of wastewater treatment. Chlorine treatment is very effective in killing the pathogens that historically caused outbreaks of serious waterborne diseases, which killed many thousands of people. However, a recently discovered potential is that chlorine treatment produces minute quantities of chemical by-products, some of which have been identified as potentially hazardous to humans and other animals. For example, a recent study in Britain revealed that in some rivers, male fish sampled downstream from wastewater treatment plants had testes containing both eggs and sperm. This is likely related to the concentration of sewage effluent and the treatment method used.[26] Evidence also suggests that these by-products in the water may pose risk of cancer and other human health effects. The degree of risks is controversial and is currently being debated.[27]

22.11 Land Application of Wastewater

The practice of applying wastewater to the land results from the fundamental belief that waste is simply a resource out of place. Land application of untreated human waste was practiced for hundreds if not thousands of years before the development of wastewater treatment plants, which have sanitized the process through the reduction of BOD and the use of chlorination.

The Wastewater Renovation and Conservation Cycle

The ideal land application system is sometimes called the **wastewater renovation and conservation cycle** and is shown schematically in Figure 22.20. The major steps in the cycle are the following:

1. Return of treated wastewater (following primary treatment) to crops via a sprinkler or other irrigation system.
2. Renovation, or natural purification by slow percolation of the wastewater into the soil, to eventually recharge the groundwater resource with clean water (an advanced form of treatment).
3. Reuse of the treated water, which is pumped out of the ground for municipal, industrial, institutional, or agricultural purposes.

Recycling of wastewater is now being practiced at many sites around the United States. In a large-scale wastewater recycling program near Muskegon and Whitehall,

Boston Harbor: Cleaning Up a National Treasure

The city of Boston is steeped in early American history. The names of Samuel Adams and Paul Revere immediately come to mind when considering the late 1700s, when the colonies were struggling to obtain freedom from Britain. In 1773, Samuel Adams led a group of patriots aboard three British ships, dumping their cargo of tea into Boston Harbor. The issue that the patriots were emphasizing was what they believed to be an unfair tax on tea, and the event came to be known as the Boston Tea Party. The tea dumped into the harbor by the patriots did not pollute the harbor, but the growing city and the dumping of all sorts of waste eventually did. For over 200 years Boston Harbor had been a disposal site for the dumping of sewage, treated wastewater, and water contaminated from sewer overflows during storms into Massachusetts Bay. Late in the twentieth century, court orders demanded that measures be taken to clean up the bay.

After studies of Boston Harbor areas farther offshore in Massachusetts Bay, it was decided to relocate the areas of waste discharge (called outfalls) farther offshore from Boston Harbor. Pollution of the harbor resulted because the waste that was being placed there moved into a small, shallow part of Massachusetts Bay. Although there is vigorous tidal action between the harbor and the bay, the flushing time is about one week. The input of wastewater from the sewage outfalls was sufficient to cause water pollution. Study of Massachusetts Bay suggested that placing the outfalls farther offshore, where water is deeper and currents are stronger, would lower the pollution levels in Boston Harbor.

Moving the wastewater outfall offshore is definitely a step in the right direction, but the long-term solution to pollutants entering the marine ecosystem will require additional measures. Pollutants in the water, even when placed farther offshore in areas of greater circulation and greater water depth, will eventually accumulate and cause environmental damage. As a result, any long-term solution must include source reduction of pollutants. To this end the Boston Regional Sewage Treatment Plan included a new treatment plant designed to significantly reduce the levels of pollutants that are discharged into the bay. This acknowledges that dilution by itself cannot solve the urban waste management problem. Moving the sewage outfall offshore, when combined with source reduction of pollutants, is a positive example of what can be done to better manage our waste and reduce environmental problems.[35]

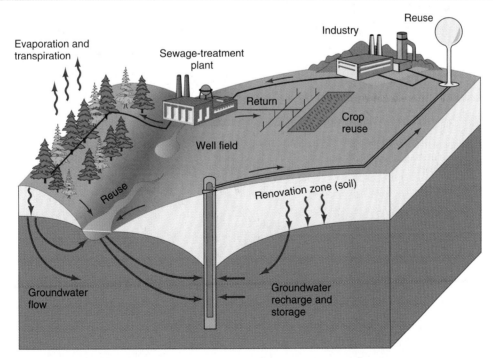

Figure 22.20 ■ The wastewater renovation and conservation cycle. [*Source:* R. R. Parizek, L. T. Kardos, W. E. Sopper, E. A. Myers, D. E. Davis, M. A. Farrel, and J. B. Nesbitt, "Pennsylvania State Studies: Waste Water Renovation and Conservation," *University Studies* 23. Copyright 1967 by the Pennsylvania State University. Reproduced by permission of the Pennsylvania State University Press.]

Michigan, raw sewage from homes and industry is transported by sewers to the treatment plant, where it receives primary and secondary treatment. The wastewater (over 32 million gals per day) is then chlorinated and pumped into a network that transports the effluent to a series of spray irrigation rigs, which apply the treated water to about 2,000 hectares (5,000 acres) of corn, soybeans, and alfalfa. After the wastewater trickles down through the soil, it is collected in a network of tile drains and transported to the Muskegon River for final disposal. This last step is an indirect advanced treatment that uses the natural physical and biological environment as a filter. This system removes most potential pollutants, meeting state water quality standards.

Technology for wastewater treatment is rapidly evolving. An important question being asked is: Can we develop environmentally preferred, economically viable wastewater treatment plants that are fundamentally different from those in use today? An idea for such a plant, called a resource recovery wastewater treatment plant, is shown in Figure 22.21. The term *resource recovery* here refers to the production of resources, including methane gas (which can be burned as a fuel), as well as ornamental plants and flowers that have commercial value.

The processes in the resource recovery treatment plant are as follows: First, the wastewater is run through filters to remove large objects. Second, the water undergoes anaerobic processing (this process produces the methane gas). Third, the nutrient-rich water flows over an incline surface containing plants (the plants use the nutrients and further purify the water). This process is thought to clean the water to the same standards obtained from secondary treatment in conventional wastewater treatment plants. If further purification is necessary, then the water may be processed by other living plants before being discharged into the environment.

Wastewater treatment that utilizes the resource recovery concept is in the experimental stage in small pilot plants. This technology must overcome several problems before it is likely to be used more widely. First, there has been a tremendous investment in traditional wastewater treatment plants, and engineers and other technicians are familiar with how to build and operate them. Second, economic incentives to provide for new technologies are not sufficient. Third (and perhaps most significant), there are not sufficient personnel trained to design and operate new types of wastewater treatment plants. This may be chang-

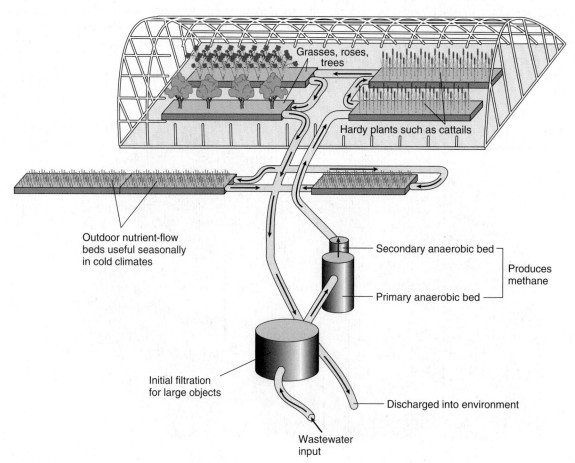

Figure 22.21 ■ Components of a resource recovery wastewater treatment plant. For this model, two resources are recovered: methane, which can be burned to produce energy from the anaero-bic beds, and ornamental plants, which can be sold. [*Source:* Based on W. J. Jewell, "Resource-Recovery Wastewater Treatment," *American Scientist* (1994) 82:366–375.]

(a)

(b)

(c)

Figure 22.22 ■ (*a*) Wetland Pointe au Chene Swamp, three miles south of Thibodaux, Louisiana, that receives wastewater; (*b*) one of the outfall pipes delivering wastewater; and (*c*) ecologists doing field work at the Pointe au Chene Swamp to evaluate the wetland.

ing, however, because more universities are developing environmental engineering programs that take a broader view of technological development and applications.[28]

Wastewater and Wetlands

Wastewater is being applied successfully to natural and constructed wetlands at a variety of locations.[29, 30, 31] Natural or human-constructed wetlands are potentially effective in treating the following water quality problems:

■ Municipal wastewater from primary or secondary treatment plants (BOD, pathogens, phosphorus, nitrate, suspended solids, metals).

■ Stormwater runoff (metals, nitrate, BOD, pesticides, oils).

■ Industrial wastewater (metals, acids, oils, solvents).

■ Agricultural wastewater and runoff (BOD, nitrate, pesticides, suspended solids).

■ Mining waters (metals, acidic water, sulfates).

■ Groundwater seeping from landfills (BOD, metals, oils, pesticides).

Treatment of wastewater through wetland systems is particularly attractive to communities that find it difficult to purchase traditional wastewater treatment plants. For example, the city of Arcata, in northern California, makes use of a wetland as part of its wastewater treatment system. The wastewater comes mostly from homes, with minor inputs from the numerous lumber and plywood plants in Arcata. It is treated by standard primary and secondary

methods, then chlorinated and dechlorinated before being discharged into Humboldt Bay.[29]

Louisiana Coastal Wetlands

The state of Louisiana, with its abundant coastal wetlands, is a leader in the development of advanced treatment using wetlands following secondary treatment (Figure 22.22). Nitrogen- and phosphorus-rich wastewater, when applied to coastal wetlands, increases the production of wetland plants, thereby improving water quality as these nutrients are used by the plants. When the plants die, their organic material (stems, leaves, roots) causes the wetland to grow vertically (or accrete), partially offsetting wetland loss due to sea level rise.[32] There are also significant economic savings in the application of treated wastewater to wetlands, because the financial investment needed is small compared with the cost of advanced treatment at conventional treatment plants. Over a 25-year period, a savings of about $40,000 per year is likely.[31, 33] In conclusion, the utilization of isolated wetlands, such as those in coastal Louisiana, is a practical solution to improving water quality in small, widely dispersed communities in the coastal zone. As water quality standards are tightened, wetland wastewater treatment will become a viable, effective alternative that is cost effective compared with traditional treatment.[32, 33]

Phoenix, Arizona: Constructed Wetlands

Wetlands can be constructed in arid regions to treat poor quality water. For example, at Avondale, Arizona, near Phoenix, a wetland treatment facility for agricultural wastewater is sited in a residential community (Figure 22.23).

The facility is designed to eventually treat about 17,000 m³/day (4.5 million gal/day) of water. The water entering the facility has nitrate (NO_3) concentrations as high as 20 mg/l. The artificial wetlands contain naturally occurring bacteria that reduce the nitrate to below the maximum contaminant level of 10 mg/l. Following treatment, the water flows by pipe to a recharge basin on the nearby Agua Fria River, where it seeps into the ground to become a groundwater resource. The wetland treatment facility cost about $11 million, about half the cost of a more traditional treatment facility.

22.12 Water Reuse

Water reuse can be inadvertent, indirect, or direct. *Inadvertent water reuse* results when water is withdrawn, treated, used, treated, and returned to the environment, followed by further withdrawals and use. Inadvertent water use is very common and a fact of life for millions of people who live along large rivers. Many sewage treatment plants are located along rivers and discharge treated water into the rivers. Downstream, other communities withdraw, treat, and consume the water.

Several risks are associated with inadvertent reuse:

1. Inadequate treatment facilities may deliver contaminated or poor-quality water to downstream users.
2. Because the fate of all disease-causing viruses during and after treatment is not completely known, the environmental health hazards of treated water remain uncertain.
3. Every year, new potentially hazardous chemicals are introduced into the environment. Harmful chemicals are often difficult to detect in the water; if they are ingested in low concentrations over many years, their effects on humans may be difficult to evaluate.[29]

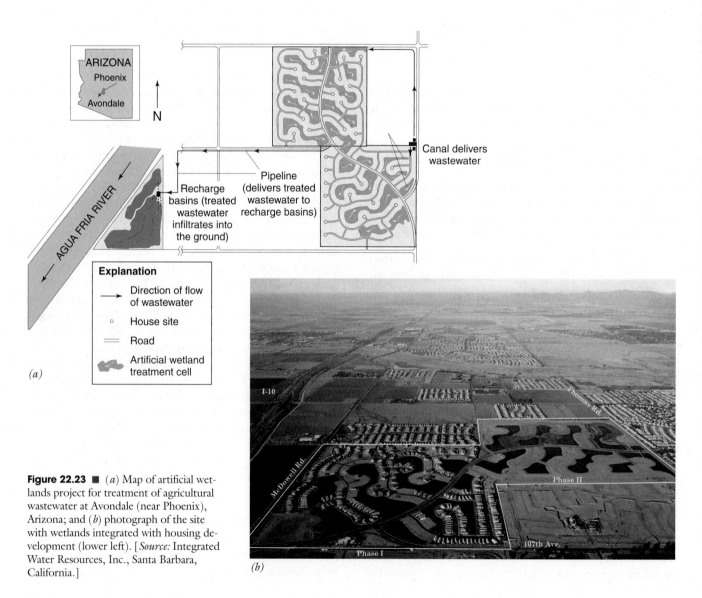

Figure 22.23 ■ (*a*) Map of artificial wetlands project for treatment of agricultural wastewater at Avondale (near Phoenix), Arizona; and (*b*) photograph of the site with wetlands integrated with housing development (lower left). [*Source:* Integrated Water Resources, Inc., Santa Barbara, California.]

Figure 22.24 ■ Water reuse at a Las Vegas, Nevada, resort hotel.

Indirect water reuse is a planned endeavor. An example is the wastewater renovation and conservation cycle previously discussed and illustrated in Figure 22.20. Similar plans have been used in many places in the southwestern United States, where several thousand cubic meters of treated wastewater per day have been applied to surface recharge areas. The treated water eventually enters groundwater storage to be reused for agricultural and municipal purposes.

Direct water reuse refers to use of treated wastewater that is piped directly from a treatment plant to the next user. In most cases, the water is used in industry, in agricultural activity, or for irrigation of golf courses, institutional grounds (such as university campuses), and parks. Direct water reuse is growing rapidly. Direct reuse of water by factories for industrial processes is the norm. In Las Vegas, Nevada, new resort hotels that use a great deal of water for fountains, rivers, canals, and lakes are required to treat wastewater and reuse it (Figure 22.24). Very little direct reuse of water is planned for human consumption (except in emergencies) because of perceived risks and negative cultural attitudes toward using treated wastewater.

22.13 Water Pollution and Environmental Law

Environmental law, the branch of law dealing with conservation and use of natural resources and control of pollution, is very important as we debate environmental issues and make decisions about how best to protect our environment. In the United States, laws at the federal, state, and local levels address these issues.

Federal laws to protect water resources go back to the Refuse Act of 1899, which was enacted to protect navigable streams, rivers, and lakes from pollution. Table 22.4 lists major federal laws that have a strong water resource/pollution component. Each of these major pieces of legislation has had a significant impact on water quality issues. Many federal laws have been passed with the purpose of cleaning up or treating pollution problems or treating wastewater. However, there has also been a focus on preventing pollutants from entering water. Prevention has the advantage of avoiding environmental damage and costly cleanup and treatment.

From the standpoint of water pollution, the mid-1990s in the United States was a time of debate and controversy. Congress in 1994 attempted to rewrite major environmental laws, including the Clean Water Act (1972, amended 1977). The purpose was to give industry greater flexibility in choosing how to comply with environmental regulations concerning water pollution. Industry interests favored proposed new regulations that, in their estimation, would be more cost-effective without causing an increase in environmental degradation. Environmentalists, on the other hand, viewed the attempts to rewrite the Clean Water Act as a giant step backward in the nation's fight to clean up our water resources. Apparently, Congress had incorrectly read the public's values on this issue. Survey after survey has established that there is strong support for a clean environment in the United States and that people are willing to pay to have clean air and clean water. Congress has continued to debate changes in environmental laws, but little has been resolved.[34]

In July 2000, the president of the United States, imposed new water pollution regulations aimed at protecting thousands of streams and lakes from nonpoint sources to agricultural, industrial, and urban pollution sources. The regulations are to be administered by the Environmental Protection Agency, which will work with local communities and states to develop detailed plans with the objective of reducing pollution to streams, rivers, lakes, and estuaries that do not now meet the minimum standards of water quality. The new regulations acknowledged that nonpoint sources of water pollution are a serious problem that is difficult to regulate. The plan, which will take at least 15 years to implement completely, has been opposed for years by Congress as well as some agricultural groups, the utility industry, and even the U.S. Chamber of Commerce. The main objections are that the requirements would be costly (billions of dollars) and that local and state governments are better suited to implement their own water pollution regulations. These regulations mark a new phase in water pollution control measures in the United States.

Summary

■ The primary water pollution problem in the world today is the lack of disease-free drinking water.

■ Water pollution is degradation of quality that renders water unusable for its intended purpose.

Table 22.4 • Federal Water Legislation

Date	Law	Overview
1899	Refuse Act	Protects navigable water from pollution.
1956	Federal Water and Pollution Control Act	Enhances the quality of water resources and prevents, controls, and abates water pollution.
1958	Fish and Wildlife Coordination Act	Mandates the coordination of water resources projects such as dams, power plants, and flood control must coordinate with U.S. Fish and Wildlife Service to enact wildlife conservation measures.
1969	National Environmental Policy Act	Requires environmental impact statement prior to federal actions (development) that significantly affect the quality of the environment. Included are dams and reservoirs, channelization, power plants, bridges, and so on.
1970	Water Quality Improvement Act	Expands power of 1956 act through control of oil pollution and hazardous pollutants and provides for research and development to eliminate pollution in Great Lakes and acid mine drainage.
1972 (amended in 1977)	Federal Water Pollution Control Act (Clean Water Act)	Seeks to clean up nation's water. Provides billions of dollars in federal grants for sewage treatment plants. Encourages innovative technology, including alternative water treatment methods and aquifer recharge of wastewater.
1974	Federal Safe Drinking Water Act	Aims to provide all Americans with safe drinking water. Sets contaminant levels for dangerous substances and pathogens.
1980	Comprehensive Environmental Response, Compensation, and Liability Act	Established revolving fund (Superfund) to clean up hazardous waste-disposal sites, reducing groundwater pollution.
1984	Hazardous and Solid Waste Amendments to the Resource Conservation and Recovery Act	Regulates underground gasoline storage tanks. Reduces potential for gasoline to pollute groundwater.
1987	Water Quality Act	Established national policy to control nonpoint sources of water pollution. Important in development of state management plants to control nonpoint water pollution sources.

- Major categories of water pollutants include disease-causing organisms, dead organic material, heavy metals, organic chemicals, acids, sediment, heat, and radioactivity.

- Sources of pollutants may be point sources, such as pipes that discharge into a body of water, or nonpoint sources, such as runoff, which are diffused and intermittent.

- Eutrophication is a natural or human-induced increase in the concentration in water of nutrients, such as phosphorus and nitrogen, required for living things. A high concentration of such nutrients may cause a population explosion of photosynthetic bacteria. As the bacteria die and decay, the concentration of dissolved oxygen in the water is lowered, leading to the death of fish.

- Sediment pollution is a twofold problem: soil is lost through erosion and water quality is reduced when sediment enters a body of water.

- Acid mine drainage is a serious water pollution problem that results when water and oxygen react with sulfide minerals, often associated with coal or metal sulfide deposits, forming sulfuric acid. Acidic water draining from mines or tailings pollutes streams and other bodies of water, damaging aquatic ecosystems and degrading water quality.

- Urban processes—for example, waste disposal in landfills, application of fertilizers, and dumping of chemicals such as motor oil and paint—can contribute to shallow-aquifer contamination. Overpumping of aquifers near the ocean may cause saltwater, found below the freshwater, to rise closer to the surface, thereby contaminating the water resource by a process called saltwater intrusion.

- Wastewater treatment at conventional treatment plants includes primary, secondary, and, occasionally, advanced treatment. In some locations, natural ecosystems, such as wetlands and soils, are being used as part of the treatment process.

- Water reuse is the norm for millions of people living along rivers where series of sewage treatment plants discharge treated wastewater back into the river water. People who withdraw river water downstream are reusing some of the treated wastewater.

- Industrial reuse of water is the norm for many factories.

- Deliberate use of treated wastewater for irrigating agricultural lands, parks, golf courses, and the like is growing rapidly as demand for water increases.

- Cleanup and treatment of water pollution for both surface water and groundwater resources are expensive and may not be completely successful. Furthermore, environmental damage may result before a pollution problem is identified and treated. Therefore, we should continue to focus on preventing pollutants from entering water, which is a goal of much water quality legislation.

How Can Polluted Waters Be Restored?

The Illinois River begins in the northeast part of the state and flows west and south, draining parts of Indiana and Wisconsin (see Figure 22.25). From Lake Michigan, which is connected to the river by a canal at Chicago, to the river's confluence with the Mississippi is a distance of 526 km (327 mi). The surrounding floodplains, once a mixture of prairie and oak–hickory forest, are now primarily used for raising crops. Formerly, the river was highly productive, especially in the lower 320 km (200 mi); it produced 10% of the U.S. freshwater fish catch in 1908 (11 million kg, or 24 million lb; 200 kg/ha, or 178 lb/acre). By the 1970s, the same stretch of river produced a mere 0.32% of the total freshwater fish harvest (4.5 kg/ha or 4 lb/acre). Two major factors are responsible for the change in the productivity of the Illinois River: diversion of Chicago's sewage from Lake Michigan to the river and agriculture. A brief history of events related to water quality in the Illinois River is given in Table 22.5.

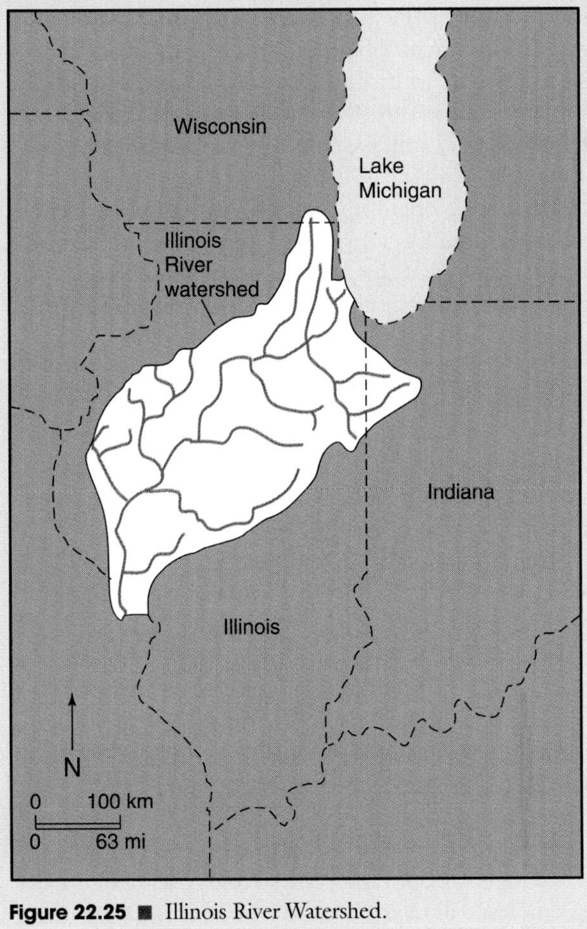

Figure 22.25 ■ Illinois River Watershed.

Critical Thinking Questions

1. Develop a hypothesis to explain why the fish population peaked in 1908, after the construction of the Chicago Sanitary and Ship Canal, and declined after that. Your hypothesis should also be able to explain the recovery of the fish in the 1920s and 1930s and the causes of the environmental problems of the 1940s and 1950s. Design a controlled experiment to test your hypothesis.

2. Why did water quality show some improvement by 1990, although the Tunnel and Reservoir Plan (TARP) was not yet completed? (*Hint:* See Table 22.5.)

3. The most important variables that affect the life of a river or stream are energy source (the amount of organic material entering the stream from sources outside it), water quality, habitat quality, water flow, and interactions among living things. In the case of the Illinois River, which variables are affected by human activities? For each variable, cite examples of specific activities, their environmental effects, and what could be done to further improve water quality in the river.

4. There is a conflict between managing the Illinois River for waterfowl and managing it for fish. Why is this so? How could the conflict be resolved?

Table 22.5 • Illinois River Water Quality History		
Year	**Critical Event**	**Environmental Impact**
1854–1855	After heavy rains, untreated sewage from Chicago entered Lake Michigan and then the city's drinking water	Cholera and typhoid epidemic in Chicago
1900	Chicago Sanitary and Ship Canal built to convey sewage away from Lake Michigan and into Illinois River	Waste entered Illinois River; commercial fish yield from river reached peak in 1908; by 1920, fish populations in river had declined
1920–1940	Most cities along river built sewage treatment plants	Some recovery in fish population
1940–1960	Rapid population growth in Chicago and other cities along the river; increase in agricultural acreage	Lower oxygen levels in river; further declines in fish populations; sport fish and ducks declined in backwaters and lakes of the river
1977	Construction of Chicago Tunnel and Reservoir Plan (TARP) to capture and treat sewage overflows initiated	Some improvement in water quality by 1990, but no change in turbidity or total phosphorus; sodium increased

Human Population

We state in this chapter that the number-one water pollution problem in the world today is the lack of disease-free drinking water. This problem is likely to get worse in the future as the number of people, particularly in developing countries, continues to increase. As population increases, so does the possibility of continued water pollution from a variety of sources relating to agricultural, industrial, and urban activities.

Sustainability

Any human activity that leads to water pollution—such as the building of pig farms and their waste facilities on floodplains, as discussed in the opening case—is antithetical to sustainability. Groundwater resources are fairly easy to pollute, and once degraded, these waters may remain polluted for long periods of time. Therefore, if we wish to leave a fair share of groundwater resources to future generations, we must ensure that these resources are not polluted, degraded, or made unacceptable for use by people and other living organisms on Earth.

Global Perspective

Several aspects of water pollution have global implications. For example, some pollutants may enter the atmosphere and be transported long distances around the globe, where they may be deposited and degrade water quality. Examples include radioactive fallout from nuclear reactor accidents or experimental detonation of nuclear devices. Waterborne pollutants from rivers and streams may enter the ocean and circulate with marine waters around the ocean basins of the world.

Urban World

Urban areas are centers of activities that may result in serious water pollution. A broad range of chemicals and disease-causing organisms are present in large urban areas, and these may enter surface waters and groundwaters to pollute them. An example is bacterial contamination of coastal waters, resulting in beach closures. Many large cities have grown along the banks of streams and rivers, and the water quality of those streams and rivers is often degraded as a result. There are positive signs that some U.S. cities are viewing their rivers as valuable resources, with a focus on environmental and economic renewal. Thus, rivers flowing through some cities are designated as greenbelts, with parks and trail systems along river corridors. Examples include New York City; Cleveland, Ohio; San Antonio, Texas; Corvallis, Oregon; and Sacramento and Los Angeles, California.

People and Nature

Polluting our water resources endangers people and ecosystems. When we connect with nature through dumping our waste in rivers, lakes, or the ocean, we are doing what other animals have done for millions of years—it is natural. For example, a large herd of hippopotamuses in a small pool may pollute the water with their waste, causing problems for other living things in the pond. The difference is that we understand that the consequences of dumping our waste is damaging the environment, and we know how to reduce our impact.

Science and Values

It is clear that the people of the United States place a high value on the environment and in particular on protection of critical resources such as water. Attempts to weaken water quality standards are viewed negatively by the public. There is also considerable concern for protection of water resources necessary for the variety of ecosystems found on Earth. This concern has led to research and development to find new technologies to reduce, control, and treat water pollution. Examples include development of new methods of wastewater treatment and support of laws and regulations that protect water resources.

Key Terms

acid mine drainage **475**
advanced wastewater treatment **480**
biochemical oxygen demand (BOD) **466**
bioremediation **477**

cultural eutrophication **470**
ecosystem effect **470**
environmental law **487**
eutrophication **470**
fecal coliform

bacteria **468**
nonpoint sources **476**
outbreaks **467**
point sources **475**
primary treatment **480**
secondary treatment **480**

wastewater renovation and conservation cycle **482**
wastewater treatment **479**
water reuse **486**

Study Questions

1. Do you think that outbreaks of waterborne diseases will be more common or less common in the future? Why? Where are outbreaks most likely to occur?

2. What was learned from the *Exxon Valdez* oil spill that might help reduce the number of future spills and their environmental impact?

3. What is meant by the term *water pollution*, and what are several major processes that contribute to water pollution?

4. Compare and contrast point and nonpoint sources of water pollution. Which is easier to treat, and why?

5. What is the twofold effect of sediment pollution?

6. In the summer, you buy a house with a septic system that appears to function properly. In the winter, effluent discharges at the surface. What could be the environmental cause of the problem? How could the problem be alleviated?

7. Describe the major steps in wastewater treatment (primary, secondary, advanced). Can natural ecosystems perform any of these functions? Which ones?

8. In a city along an ocean coast, rare waterbirds inhabit a pond that is part of a sewage treatment plant. How could this have happened? Is the water in the sewage pond polluted? Consider this question from the birds' and from your point of view.

9. How does water that drains from coal mines become contaminated with sulfuric acid? Why is this an important environmental problem?

10. What is eutrophication, and why is it an ecosystem effect?

11. How safe do you believe the drinking water is in your home? How did you reach your conclusion? Are you worried about low-level contamination by toxins in your water? What could be the sources of contamination be?

12. Do you think our water supply is vulnerable to terrorist attacks? Why? Why not? How could potential threats be minimized?

Further Reading

Borner, H., ed. 1994. *Pesticides in Ground and Surface Water.* Vol. 9 of *Chemistry of Plant Protection.* New York: Springer-Verlag. Essays on the fate and effects of pesticides in surface water and groundwater, including methods to minimize water pollution from pesticides.

Dunne, T., and L. B. Leopold. 1978. *Water and Environmental Planning.* San Francisco: W. H. Freeman. A great summary and detailed examination of water resources and problems.

Hester, R. E., and R. M. Harrison, eds. 1996. *Agricultural Chemicals and the Environment.* Cambridge: Royal Society of Chemistry, Information Services. A good source of information about the impact of agriculture on the environment, including eutrophication and the impact of chemicals on water quality.

Manahan, S. E. 1991. *Environmental Chemistry.* Chelsea, Mich.: Lewis. A detailed primer on the chemical processes pertinent to a broad array of environmental problems, including water pollution and treatment.

Newman, M. C. 1995. *Quantitative Methods in Aquatic Ecotoxicology.* Chelsea, Mich.: Lewis. Up-to-date text on fate, effects, and measurement of pollutants in aquatic ecosystems.

Nichols, C. 1989. "Trouble at the Waterworks," *The Progressive* 53:33–35. A concise report on the problem of tainted water supplies in the United States.

Rao, S. S., ed. 1993. *Particulate Matter and Aquatic Contaminants.* Chelsea, Mich.: Lewis. Coverage of the biological, microbiological, and ecotoxicological principles associated with interaction between suspended particulate matter and contaminants in aquatic environments.

CHAPTER 23

The Atmosphere, Climate, and Global Warming

Polar bear moving through thin ice while hunting for seals in Hudson Bay.

Learning Objectives

Earth's atmosphere is a dynamic system that is changing continuously while undergoing complex physical and chemical processes. After reading this chapter, you should understand:

■ What the basic composition and structure of the atmosphere are.

■ How the processes of atmospheric circulation, climate, and microclimate work.

■ What the four major processes that remove materials from the atmosphere are.

■ How the climate has changed during the last million years.

■ What the science behind human-induced global warming is.

■ How human activity has resulted in increased emissions of greenhouse gases.

■ How positive- and negative-feedback cycles in the atmosphere might affect global temperature change.

■ What effects global warming might have, and how we can adjust to those changes.

CASE STUDY

Global Warming and the Polar Bears of Hudson Bay

The mean surface temperature of Earth will likely increase by 1.5°C to 4.5°C during the period 1990–2100. Although scientific uncertainties remain, it is apparent that human-induced warming is now occurring. The questions have been reduced to how much, how fast, and where. Adverse effects of global warming will impact the goal of attaining sustainable development. The good news is that something can be done to minimize these effects if we act now.[1] The story of the plight of polar bears in the western Hudson Bay begins our exploration of climate change and global warming.

Polar bears are the largest carnivore in North America; they can reach 2.5 m in length and weigh over 700 kg (about as much as a three-quarter-ton truck). Polar bears are in trouble in the western Hudson Bay, and the early breaking up of sea ice each spring is thought to be the problem. During the past 40 years, sea ice, on a global basis, has thinned by as much as 40% and decreased in extent by about 10%, presumably in response to global warming. Some of the most significant impacts of global warming on wildlife are occurring in the Arctic, because temperature changes are more dramatic there than at lower latitudes.

For polar bears in the western Hudson Bay, sea ice is a critical habitat for hunting seals. In the spring,

polar bears prey on seals, especially on very young seals, providing the opportunity for the bears to fatten up before the annual melting of the sea ice. After the ice melts, the bears move to land, where they go on a fast that may last for months. In particular, pregnant female bears fast for up to eight months and need a significant reserve of fat to carry, care for, and feed their cubs until they can return to the ice to feed.

Since 1981, when studies on polar bears in the western Hudson Bay began, the bears there have weighed less than average and have given birth to fewer cubs. Biologists have established a link between the decline in polar bears and the earlier breakup of sea ice. If the trend continues, there will be a continued drop in the population of bears. Another consequence is that bears will be forced onto land earlier in the season, and dangerous contact with people is more likely to occur.[2, 3]

The situation in the eastern Hudson Bay at present is different, because sea ice is more permanent there, and some populations of polar bears live their entire lives on sea ice. These bears are less likely to be adversely affected by global warming. However, massive melting of sea ice would be a serious problem for all polar bears, including those of eastern Hudson Bay.[2]

Our story of polar bears in the Hudson Bay suggests that global change with warming can cause serious problems in the biosphere. This chapter addresses global issues related to weather, climate change, and global warming, with an emphasis on the role of humans in climate change.

Although global warming potentially has very serious consequences we do not need to despair. Science shows us that we can slow down and stop global warming. If we adopt sound policy and take appropriate action we will be able to have an atmosphere that is both cleaner and healthier for all life on earth.[4]

23.1 The Atmosphere

The **atmosphere** is the thin layer of gases that envelops Earth. We begin the chapter by examining basic features of atmospheric composition, structure, and processes.

Composition of the Atmosphere

The atmosphere is composed of gas molecules held close to Earth's surface by a balance between gravitation and thermal movement of air molecules (90% of the weight of the atmosphere is in the first 12 km above Earth's surface). Major gases in the atmosphere include nitrogen (78%), oxygen (21%), argon (0.9%), and carbon dioxide (0.03%). The atmosphere also contains trace amounts of numerous elements and compounds, including methane, ozone, hydrogen sulfide, carbon monoxide, oxides of nitrogen and sulfur, hydrocarbons, chlorofluorocarbons (CFCs), and various particulates or aerosols (small particles). Water vapor is also present in the lower few kilometers of atmosphere.

The atmosphere is a dynamic system, changing continuously. Physical movement of air masses, each with a different temperature, pressure, moisture, and aerosol content, produces weather and climate. A vast, chemically active system, the atmosphere is fueled by sunlight, high-energy compounds (for example, oxygen, methane, and carbon dioxide) emitted by living things, and human industrial and agricultural activities. Many complex chemical reactions take place in the atmosphere, changing from day to night and with the chemical elements available.

Structure of the Atmosphere

The atmosphere is made up of several layers. The structure of the atmosphere and the relationship between altitude and air temperature for the lower atmosphere are shown as an idealized diagram in Figure 23.1.

The lower part of the atmosphere (lower 10 to 12 km, or about 6 to 8 mi) is known as the *troposphere*, and it is here that weather occurs. In the troposphere, the temperature of the atmosphere decreases systematically with elevation (from about $17°C$ at the surface to $-60°C$ at 12 km above Earth's surface, the top of the troposphere) at a global average rate of decrease of approximately $6.5°C/km$ ($7.3°F/mi$). At the top of the troposphere, the tropopause (about 12 to 20 km above sea level), with a constant temperature of about $-60°C$, produces a lid, or cold trap, on the troposphere. The cold trap causes condensation of atmospheric water vapor. Therefore, there is very little water vapor in the *stratosphere*, which lies above the troposphere and which is the site of atmospheric warming that coincides with increases in altitude. Condensation of water in the troposphere produces clouds. The role of clouds, including how they develop and move, is an important area of research underway to understand the global processes that operate in the atmosphere.

Figure 23.1 also shows the *stratospheric ozone layer*, which extends from the tropopause to an elevation of approximately 40 km (25 mi), with a maximum concentration of ozone above the equator at about 25 to 30 km (16 to 19 mi). Stratospheric ozone (O_3) protects life in the lower atmosphere from receiving harmful doses of ultraviolet radiation and is discussed in detail in Chapter 26.

Atmospheric Processes: Temperature, Pressure, and Global Zones of High and Low Pressure

Two important measurable quantities in the atmosphere are pressure and temperature. *Pressure* is force per unit area. Atmospheric pressure is the weight of overlying atmosphere (air) per unit area; it decreases as altitude increases because there is less weight from overlying air. At sea level, atmospheric pressure is 10^5 N/m^2 (newtons per square meter), which is equivalent to 14.7 lb/in.^{2} *Temperature* refers to relative hotness or coldness of materials, such as air, water, soil, and living organisms, and is measured with a thermometer. In a quantitative sense, temperature is a measure of thermal energy, which is the kinetic energy of the motion of atoms and molecules in a substance. This book uses the Celsius ($°C$) scale.

Figure 23.1 ■ An idealized diagram of the structure of the atmosphere showing temperature profile and ozone layer of the atmosphere to an altitude of 110 km. Note that 99% of the atmosphere (by weight) is below 30 km, the ozone layer is thickest at about 25–30 km, and the weather occurs below about 11 km—about the elevation of the jet stream. [*Source:* A. C. Duxbury and A. B. Duxbury, *An Introduction to the World's Oceans,* © 1997. Wm. C. Brown Publishers, 5th ed.].

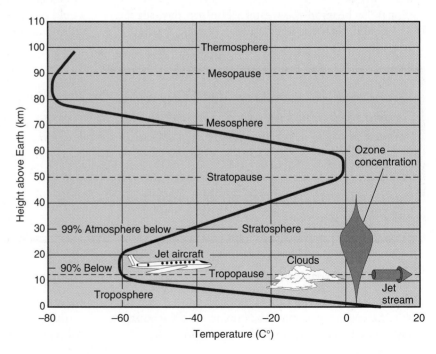

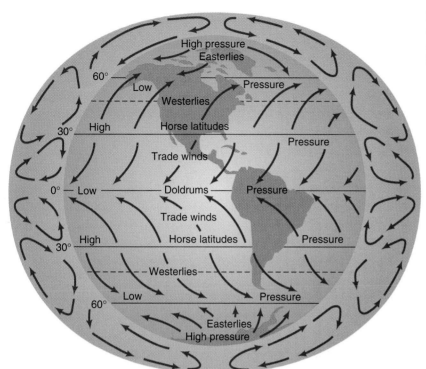

Figure 23.2 ■ Generalized circulation of the atmosphere. [*Source:* Williamson, *Fundamentals of Air Pollution*, ©1973, Figure 5.5. Reprinted with permission of Addison-Wesley, Reading, Mass.]

In addition to pressure and temperature, air is characterized by its water vapor content. In the lower atmosphere, water vapor content varies from approximately 1% to 4% by volume. The amount of water vapor present in the atmosphere at a particular location depends on many factors, including air temperature, air pressure, and availability of water vapor from processes such as evaporation from water bodies and soil and transpiration from vegetation (loss of water from plants to the air).

On a global scale, atmospheric circulation results primarily from Earth's rotation and the differential heating of Earth's surface and atmosphere. These processes produce global patterns that include prevailing winds and latitudinal belts of low and high air pressure from the equator to the poles (Figure 23.2). In general, belts of low air pressure develop at the equator and at 50° to 60° north and south latitude as a result of rising columns of air, producing precipitation. Belts of high pressure resulting from descending air develop at 25° to 30° north and south latitude, producing arid conditions.

The latitudinal belts have names, such as the "doldrums," regions at the equator with little air movement; "trade winds," northeast and southeast winds important in the early days of international trade, when clipper sailing ships moved the world's goods; and "horse latitudes," two belts centered about 30° north and south of the equator with descending air and high pressure. Earth's major deserts occur in the horse latitudes as a result of pervasive high pressure and low precipitation sandwiched between the equatorial and midlatitudinal zones of low pressure with higher precipitation.

Processes That Remove Materials from the Atmosphere

Understanding processes that remove materials from the atmosphere is important in solving atmospheric pollution problems. Four processes are responsible for removing human-induced particles and chemicals from the atmosphere:

■ *Sedimentation.* Particles heavier than air settle out as a result of gravitational attraction to Earth. For example, particulates from volcanic eruptions or burning coal will settle out over time as a dry deposition.

■ *Rain out.* Precipitation (rain, ice, or snow) can physically and chemically flush material from the atmosphere. For example, raindrops form by condensation of water on small particles in the atmosphere, bringing the particles to Earth with the raindrops. Carbon dioxide combines with water in the atmosphere to form weak carbonic acid by the following chemical reaction:

$$CO_2 + H_2O \rightarrow H_2CO_3$$

This process effectively removes some carbon dioxide from the atmosphere and explains why natural rainfall is slightly acidic (see Chapter 23).

■ *Oxidation.* Oxidation is a reaction in which oxygen is chemically combined with another substance. For example, sulfur dioxide in the atmosphere oxidizes easily to sulfur trioxide (SO^3), which may dissolve in water, forming sulfuric acid and producing acid rain (see Chapter 23).

■ *Photodissociation.* Solar radiation (light) can break down chemical bonds in a chemical process known as photodissociation. For example, ozone (O^3) in the atmosphere may break down to O^2 as a result of photodissociation.

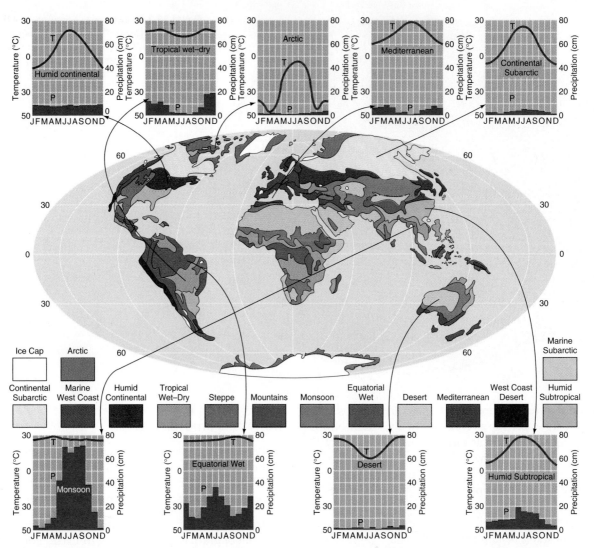

Figure 23.3 ■ The climates of the world and some of the major climate types in terms of characteristic precipitation and temperature conditions. [*Source:* Modified from W. M. Marsh and J. Dozier, *Landscape,* © 1981, John Wiley & Sons. Reprinted with permission of John Wiley & Sons, Inc.]

With our basic discussion of the composition of the atmosphere and some of the important atmospheric processes behind us, we next discuss climate and climatic change.

23.2 Climate

Climate refers to the representative or characteristic atmospheric conditions for a region on Earth. The term *climate* refers to these conditions over long time periods, such as seasons, years, or decades, whereas *weather* refers to shorter periods of time, such as hours, days, or weeks. When we say it's hot and humid in New York today or raining in Seattle, we are speaking of weather. When we say Los Angeles has cool, wet winters and warm, dry summers, we are referring to the Los Angeles climate. Climate depends in part on precipitation and temperature, both of which show tremendous variability on a global scale. Because the climate of a particular location may depend on extreme or infrequent conditions, however, climate is more than just the average temperature and precipitation of a region.

The simplest classification of climate is by latitude—tropical, subtropical, midaltitudinal (continental), subarctic (continental), and arctic. Several other categories are used as well, including humid continental, Mediterranean, monsoon, desert, and tropical wet–dry (Figure 23.3). Although detailed discussion of climatic types is beyond the scope of this book, it is important to recognize the significance of potential climatic variability in determining what kinds of organisms live where. Recall from the discussion of biogeography in Chapter 7 that similar climates produce similar kinds of ecosystems. This concept is important and useful for environmental science. Knowing the climate, we can predict a great deal about what kinds of life we will find in an area and what kinds could survive there if introduced.[5]

Climatic Change

An important aspect of climate is **climatic change**. The mean annual temperature of Earth has swung up and down by several degrees Celsius over the past million years

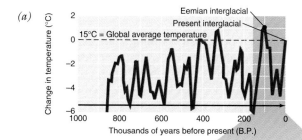

(a)

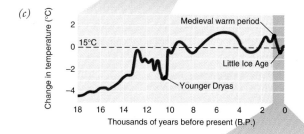

(b)

Figure 23.4 ■ Changes in Earth's temperature over varying time periods during the past million years. [*Sources:* UCAR/DIES, "Science Capsule, Changes in the Temperature of the Earth," *Earth Quest* 5, no. 1 (Spring 1991); Houghton, J. T., G. L. Jenkins, and J. J. Ephranns, eds. *Climate Change, the Science of Climate Change* (Cambridge: Cambridge University Press, 1996); U.K. Meteorological Office, *Climate Change and Its Impacts: A Global Perspective* 1997.]

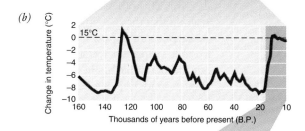

(c)

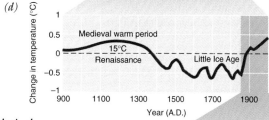

(d)

(Figure 23.4). Times of high temperature reflect relatively ice-free periods (interglacial periods) over much of the planet; times of low temperature reflect the glacial events (Figure 23.4a, b). It is not yet clear whether our current warm climate marks the end of the ice ages or is merely an interglacial period with another glacial age due.

Global climate also changes in shorter time frames. For example, continental glaciation ended about 12,500 years ago with very rapid warming, perhaps over a period as brief as a few decades.[6] This was followed by a short global cooling about 11,500 years ago (Figure 23.4c). Climatic change over the last 18,000 years reflects several warming and cooling trends that have greatly affected people. For example, during a major warming trend from 1100 to 1300 (medieval warm period, Figure 23.4c, d), the Vikings colonized Iceland, Greenland, and North America. When glaciers advanced during the cold period starting around 1400 (Little Ice Age), the Viking settlements in North America and parts of Greenland were abandoned.

(e)

Starting in approximately 1850, a warming trend became apparent. It lasted until the 1940s, when temperatures began to cool again. Figure 23.4e shows change over the last 140 years. On this scale, the 1940s warming event is clearer. As you can see, it was followed by a leveling off of temperature in the 1950s and then a further drop during the 1960s. After that time, temperature increased steadily through the 1990s. What is evident from the record of the last 100 years is that global mean annual temperature has increased by approximately 0.6° C. (Figure 23.4e). This period includes the warmest years of the twentieth century,[7, 8] known as the "late-twentieth-century increase in global temperature."[9] Global temperature data from the United States (NOAA) and Europe (WMO) help delineate this late twentieth-century rise in temperature that continues today.

- The first 5 years of the twenty-first century were some of the warmest in the 142 years since temperatures have been recorded, and in the last 1,000 years, according to geologic data.

- Warming since the mid-1970s has been approximately three times as rapid as in the preceding 100 years.

- The 10 warmest years have all occurred since 1990 and the five warmest since 1997.

- The warmest year on record was 2005, with 1998 second, and 2002 and 2003 tied for third.

- In the United States, 2003 was cooler and wetter than average in much of the eastern part of the country, and warmer and drier in much of the western part. Ten western states were much warmer than average; New Mexico had its warmest year on record. Alaska was warmer in all four seasons, and it was one of the five warmest years since Alaska began taking measurements in 1918.

- In 2003, Europe experienced summer heat waves, with the warmest seasonal temperatures ever recorded in Spain, France, Switzerland, and Germany. Approximately 15,000 people died in heat waves in Paris during the summer.

- Warm conditions with drought contributed to severe wildfires in Australia, southern California, and British Columbia.

Of course, a year or two of high temperatures with drought, heat waves, and wildfires is not by itself an indication of longer-term global warming. The persistent trend of increasing temperatures over three decades is more compelling evidence that global warming is real and happening.

The question that begs to be asked is: Why does climate change occur? Examination of Figure 23.4a suggests there are cycles of change about 100,000 years long separated by shorter cycles of 20,000 to 40,000 years in duration. These cycles were first identified by Milutin Milankovitch in the 1920s as a hypothesis to explain climate change. Milankovitch realized that the spinning Earth is like a wobbling top unable to keep a constant position in relation to the sun, and that position (in part) determines the amount of sunlight reaching and warming Earth. He discovered that variations in Earth's orbit around the sun follow a cycle of approximately 100,000 years, which correlates with the major glacial and interglacial periods of Figure 23.4a. Cycles of approximately 40,000 and 20,000 years are the result of changes in the tilt and wobble of Earth's axis. Milankovitch cycles are consistent with most of the long-term cycles we see in the climate. However, the cycles are not sufficient, by themselves, to produce the large-scale climatic variations in the geologic record. Therefore, the Milankovitch cycles can be looked at as natural mechanisms that, along with other processes, may produce climatic change.

Our climate system may be inherently unstable and capable of changing quickly from one state (cold) to another (hot) in as little as a few decades.[10] Part of what may drive the climate system and its changes is the "ocean conveyor belt"—a global circulation of ocean waters characterized by strong northward movement of upper warm waters of the Gulf Stream in the Atlantic Ocean. These waters are approximately 12–13°C when they arrive near Greenland and are cooled in the North Atlantic to a temperature of 2 to 4°C (Figure 23.5).[10] As the water is cooled, it becomes more salty and so

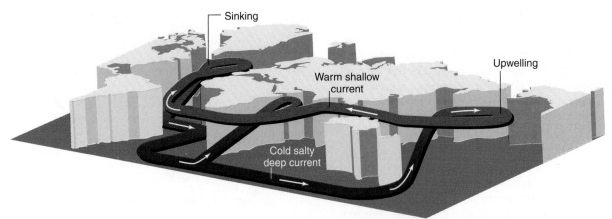

Figure 23.5 ■ Idealized diagram of the oceanic conveyor belt. The actual system is more complex; but in general, warm surface water (red) is transported westward and northward (increasing in salinity because of evaporation) to near Greenland, where it cools from contact with cold Canadian air. As the water increases in density, it sinks to the bottom and flows south, then east to the Pacific, then north where upwelling occurs in the north Pacific. The masses of sinking

and upwelling waters balance, and the total flow rate is about 20 million m^3/sec. The heat released to the atmosphere from the warm water keeps northern Europe 5°C to 10°C warmer than if the oceanic conveyor belt were not present. [*Source:* Modified from W. Broker, "Will Our Ride into the Greenhouse Future Be a Smooth One?" *Geology Today* 7, no. 5 (1997): 2–6.]

increases in density, causing it to sink to the bottom. The cold, deep current flows southward, then eastward, and finally northward in the Pacific Ocean. Upwelling in the north Pacific starts the warm shallow current again. The flow in this conveyor belt current is huge (20 million m³/sec), about equal to 100 Amazon Rivers. The amount of warm water and heat released into the atmosphere linked to the relatively warm winter moving east-northeast across the Atlantic is sufficient to keep northern Europe 5 to 10°C warmer than it would otherwise be. If the ocean conveyor belt was shut down by global warming there would be an effect on Europe's climate. However, the change would not be catastrophic or produce extreme cold or icebound conditions.[11] On the other hand, rapid global warming that reduces food production in the future when there are a few more billion people to feed could produce a global catastrophe.[10]

Some scientific uncertainties remain regarding the human role in the observed warming of Earth's climate. However, according to a 2001 report from the Intergovernmental Panel on Climate Change (IPCC), the bulk of scientific evidence suggests that the warming is due in part to human activities. As a consequence of projected emissions of carbon dioxide (CO_2), mean surface temperatures are expected to increase by 1.5 to 4.5°C by 2100.[1,4] The science of global warming is discussed in detail following consideration of the tools used to study global change.

23.3 Earth System Science and Global Change

Until very recently, it was generally thought that human activity could only cause local or, at most, regional environmental changes. We now know otherwise! The main goal of the emerging science known as **Earth system science** is to obtain a fundamental understanding of how our planet works as a system. From a pragmatic point of view, the research priorities of Earth system science and global change can be summarized as follows:[12]

- Establishment of worldwide measurement stations to better understand physical, hydrologic, chemical, and biological processes that are significant in the evolution of Earth on a variety of time scales.
- Documentation of global changes, especially those that occur in a time period of several decades, that are of particular interest to the human environment.
- Development of quantitative models useful in the prediction of future global change.
- Provide information needed by decision makers at regional and global levels.

The major tools for studying global change are:

- Evaluation of the geologic record
- Monitoring
- Mathematical models

Figure 23.6 ■ A scientist examines a glacial ice core stored in a freezer.

Geologic Record

The sediments deposited on floodplains, in bogs, and on lake and ocean bottoms can be read just as the pages of a history book can. Organic material, such as skeletal material, shells, pollen, and bits of wood, leaves, and other plant parts, is often deposited with sediments and can yield valuable information concerning Earth's history. In addition, organic material can be dated to provide the necessary chronology to establish past changes. Finally, both sediments and organic material can be used to evaluate the past climate—what lived where, what kinds of changes have occurred, and how extensive the changes were.

One interesting use of the geologic record has been the examination of glacial ice. The process of transformation of snow to glacial ice involves recrystallization and an increase in the density of the ice. The process also traps air bubbles, which can be analyzed to provide information concerning the concentration of carbon dioxide in the atmosphere at the time the ice formed. Thus, glacial ice can be thought of as a time capsule that stores information about the atmosphere in the past. To study the ice, scientists extract long cores of glacial ice (Figure 23.6) and carefully sample trapped air bubbles. This method has been used to analyze the carbon dioxide content of the atmosphere up to 160,000 years ago. Figure 23.7 shows the

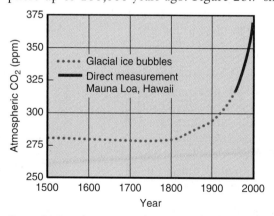

Figure 23.7 ■ Average concentration of atmospheric carbon dioxide, 1500–2000. [*Source:* Modified from Post W. M., T. Peng, W. R. Emanuel, A. W. King, V. H. Dale, and D. L. De Angelis, *American Scientist* 78, no. 4 (1990): 310–326. By permission of *American Scientist*, Journal of Sigma Xi, The Scientific Research Society.]

A CLOSER LOOK 23.1

Monitoring of Atmospheric Carbon Dioxide Concentrations

The activities of humans and other living things affect characteristics of Earth's surface, waters, and atmosphere, even in areas removed from such activities. For example, air pollutants and other artifacts of human society are found in glacial ice, and pesticides are found in lakes far from agricultural activity.

Air samples taken near the summit of Mauna Loa, Hawaii, have helped show another dimension of how life and human activity affect the atmosphere. Mauna Loa is one of the largest active volcanoes and the highest mountain in the world, based on elevation change from base to top. Air samples are taken there because it is located far away from direct effects of human life and other biological activity. One substance of interest in the Mauna Loa samples is carbon dioxide. Carbon dioxide is taken up by green plants during photosynthesis and released in respiration from all oxygen-breathing organisms; thus, a measure of carbon dioxide in the atmosphere is analogous to a measure of the breathing in and out of all life on Earth.

The Mauna Loa measurements of carbon dioxide are truly remarkable. Two important aspects of the record are shown in Figure 23.8: a strong upward trend over the years and an annual cycle, which is obvious and regular. Carbon dioxide concentration reaches a peak in winter and a trough in summer. The annual curve is a measure of life activities in the entire Northern Hemisphere. In summer, green plants are most active, and the total amount of photosynthesis exceeds the total amount of respiration. As a result, carbon dioxide is removed from the atmosphere in summer. In winter, photosynthesis decreases and be-

comes less than total respiration, so the carbon dioxide concentration in the atmosphere increases.

Similar observations have been made for the Southern Hemisphere in Antarctica. Because Antarctica is less accessible, measurements are more sporadic. Nevertheless, the same trends are observed: a strong upward trend in concentration of carbon dioxide in the atmosphere and an annual cycle. The annual cycle from Antarctica is smaller in amplitude than the Mauna Loa cycle because of the relatively smaller land area in the Southern Hemisphere and the smaller amount of vegetation.

The upward trend, or increase, in carbon dioxide concentration in the atmosphere shown in data from both Mauna Loa and Antarctica is believed to be due to the addition of carbon dioxide from burning of fossil fuels and other human activities, such as cutting of forests and burning of wood. (Burning releases the stored carbon in wood, which combines with oxygen, produc-

ing carbon dioxide). The Mauna Loa data provided some of the first evidence that directly indicated that life touches the entire Earth and that human activities have begun to affect the atmosphere of our planet.

A third, less obvious global trend is shown in Figure 23.8—an apparent increase in the amplitude of the annual cycle from 1960 to present. This evidently reflects the increase in photosynthetic biomass as a result of the increase in total atmospheric carbon dioxide.

The Mauna Loa data clearly demonstrate the benefits of long-term collection of information. Funding for long-term projects is often difficult to maintain because funding agencies may prefer to sponsor new projects rather than long-term monitoring. Nevertheless, understanding global change depends on collection and maintenance of supportive data. To that end, the Mauna Loa CO_2 measurement project is unique, and its effectiveness is a tribute to the people who initiated it and nurtured it for so many years.

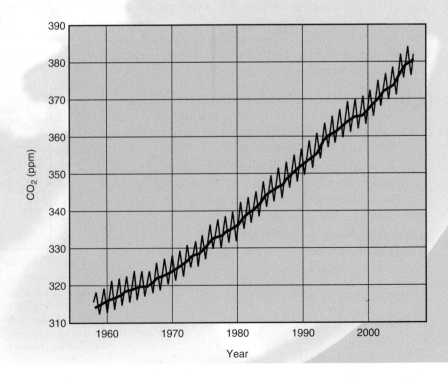

Figure 23.8 ■ Monthly average carbon dioxide concentration and long-term trend. The concentration in 1958 was about 315 ppm, and today it is about 380 ppm. [*Source:* Mauna Loa Observatory, http://mlo-noaa.gov. Accessed 12/02/06.

record from 1500 to 2000.[13] Ice cores have also yielded data on variability of solar radiation through measurement of the accumulation of cosmogenic isotopes in the ice.[8]

Real-Time Monitoring

Monitoring can be defined as the regular collection of data for specific purposes. For example, we monitor rainfall and the flow of water in rivers to evaluate water resources or flood hazards. We collect the data to provide baseline conditions from which to evaluate changes in the future. Similarly, samples of atmospheric gases, particulates, and chemicals are helping establish trends in the composition of the atmosphere (see A Closer Look 23.1). Finally, measurements of the temperature, composition, and chemistry of ocean waters can be used to help evaluate changes in the marine environment.

Mathematical Models

Mathematical models are attempts to represent, through numerical means, real-world phenomena, linkages, and interactions among physical, chemical, and biological processes. The models use equations to describe the phenomena and linkages being considered. Models have been developed to predict the flow of surface water and groundwater, ocean circulation, and atmospheric circulation. In the area of global change, the models that have gained the most attention are the **global circulation models (GCMs)**. The objective of GCMs is to predict atmospheric changes on a global scale.[14]

To organize the data necessary for the calculations in GCMs, the surface of Earth is divided into large cells measuring several degrees of latitude and longitude. The typical cell is about the size of Oregon or of Indiana and Ohio together. Several layers of data are necessary. Most GCMs use 6 to 20 levels of vertical data collected throughout the lower atmosphere (Figure 23.9). Mathematical relationships are used in the calculations to predict future atmospheric circulation. The GCMs are complex and require supercomputers for their operation.

Results from these models are relatively crude and so may not accurately predict future conditions.[14, 15] Furthermore, it is difficult to estimate interactions of other factors, such as cloud cover, which may significantly affect atmospheric energy relations. As a result, GCMs can only be considered a first approach to solving complex atmospheric problems. In spite of their limitations, mathematical models are providing information necessary for evaluating global change and Earth as a system. The GCMs are also helping to pinpoint what additional data are necessary to better predict future change. The models do predict, in a relative sense, areas or regions that are likely to be wetter or drier if certain changes in the atmosphere occur. Thus, predictions of future global change from mathematical models are being taken seriously, and

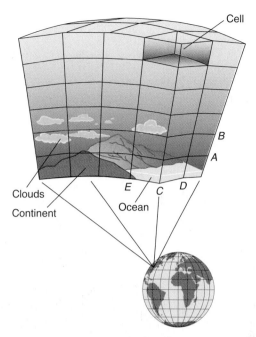

Figure 23.9 ■ Idealized diagram showing how cells are arranged for a global circulation model. A–B is a few kilometers in elevation; E–C is a few hundred kilometers in longitude; C–D is a few hundred kilometers in latitude.

their importance as a tool for studying global change will continue to increase.

Our discussion of the tools used to study global change leads us to consider human-induced climate change. The major question is: How much has human activity contributed to global warming?

23.4 Global Warming: Earth's Energy Balance and the Greenhouse Effect

Global warming is defined as a natural or human-induced increase in the average global temperature of the atmosphere near Earth's surface. The temperature at or near the surface of Earth is determined by four main factors:[14]

■ The amount of sunlight Earth receives.
■ The amount of sunlight Earth reflects.
■ Retention of heat by the atmosphere.
■ Evaporation and condensation of water vapor.

Electromagnetic Radiation and Earth's Energy Balance

In order to better understand global warming, it is necessary to have a modest acquaintance with electromagnetic radiation and Earth's energy balance. Our Earth is part of a planetary system receiving energy from the sun. This energy undergoes changes; affects life, oceans, atmosphere,

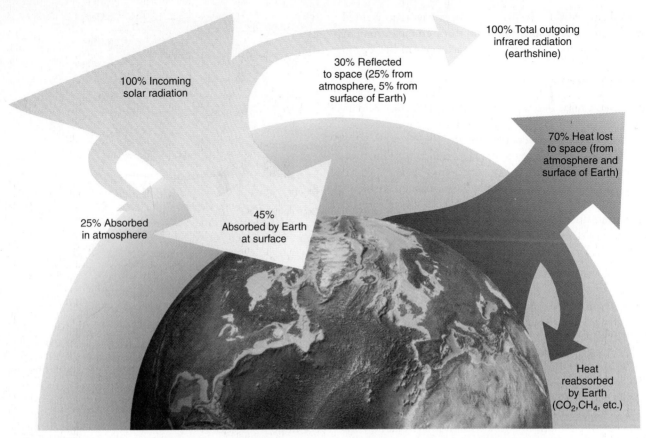

100% Incoming
solar radiation

30% Reflected
to space (25% from
atmosphere, 5% from
surface of Earth)

100% Total outgoing
infrared radiation
(earthshine)

70% Heat lost
to space (from
atmosphere and
surface of Earth)

25% Absorbed
in atmosphere

45%
Absorbed by Earth
at surface

Heat
reabsorbed
by Earth
(CO_2,CH_4, etc.)

Figure 23.10 ■ Idealized diagram showing Earth's energy balance. Incoming solar radiation amounts to approximately 5.5 million exajoules, and that is balanced by a similar amount leaving Earth. A very small component of heat (1,000 exajoules) entering the system is generated within Earth; it represents only about 0.002% of the budget and so is neglected in the values presented here.

Approximately 30% of the total incoming solar radiation (1.7 million exajoules) is reflected immediately in the atmosphere, mostly by clouds and at Earth's surface. Of the remaining 70% of incoming solar radiation, approximately 25% is absorbed in the upper atmosphere, and 45% reaches Earth and is absorbed at the surface. As the energy from the sun is absorbed by plants, soils, rocks, water, and other surface materials, it is transformed into heat, radiates as infrared radiation, and is eventually lost to space from Earth's upper atmosphere. Part of the infrared radiation thus released is reabsorbed by compounds, including carbon dioxide and methane, known as greenhouse gases (discussed later in the chapter). This energy, too, is ultimately lost to space as heat, completing the energy balance of Earth. About two-thirds of the incoming solar radiation is available to drive processes at or near Earth's surface, including the hydrologic cycle, wind, and so forth. [*Source:* Modified from N. L. Pruitt, L. S. Underwood, and W. Surver. *Bio Inquiry* (New York: Wiley, 2002).]

and climate; and is eventually emitted as heat back into the depths of space. In this system, Earth is an intermediate between the *source* (the sun) and the *sink* (space).

Nearly all the energy available at Earth's surface comes from the sun (Figure 23.10), with small additional amounts coming from human activities, geothermal energy (from the interior of Earth), and tides. Because energy is conserved, the amount emitted to space matches that received from the sun plus the other small contributions. The matching of input of energy from the sun with output from Earth defines Earth's energy balance. Although Earth intercepts only a very tiny fraction of the total energy emitted by the sun, solar energy sustains life on Earth and greatly influences climate and weather.

Energy is emitted from the sun in the form of electromagnetic radiation (EMR). Different forms of electromagnetic energy can be distinguished by their wavelengths. (See the special feature in Appendix A on EMR for a discussion of wavelength and two important laws concerning EMR.) The collection of all possible wavelengths of electromagnetic energy, considered a continuous range, is known as the **electromagnetic spectrum** (Figure 23.11). The electromagnetic spectrum is one of the most important phenomena in the physical sciences and is fundamental to understanding many environmental topics. Gamma rays, X rays, ultraviolet light, visible light, infrared radiation, television waves, radio waves, and radar are all different types of electromagnetic radiation. The relatively long wavelengths (greater than 1 m in the electromagnetic spectrum) include radio waves, and the shortest wavelengths are those of gamma rays and X rays. The electromagnetic radiation to which our eyes are sensitive, *visible electromagnetic radiation*, is a very small fraction of the total spectrum, as you can see

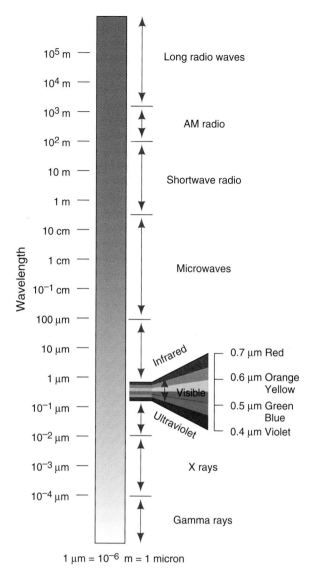

10^5 m —	Long radio waves	
10^4 m —		
10^3 m —	AM radio	
10^2 m —		
10 m —	Shortwave radio	
1 m —		
10 cm —		
1 cm —	Microwaves	
10^{-1} cm —		
100 μm —		
10 μm —	Infrared	0.7 μm Red
1 μm —	Visible	0.6 μm Orange Yellow
10^{-1} μm —	Ultraviolet	0.5 μm Green Blue
10^{-2} μm —		0.4 μm Violet
10^{-3} μm —	X rays	
10^{-4} μm —	Gamma rays	

Wavelength

1 μm = 10^{-6} m = 1 micron

Figure 23.11 ■ Types of electromagnetic radiation (EMR). Notice that the spectrum of wavelengths from radio waves to gamma rays is over nine orders of magnitude (1 μm = 10^{-6}, m = 1 micron).

in Figure 23.11. Other types of electromagnetic radiation with environmental significance include radar, microwaves, and infrared radiation.

The amount of energy per unit time radiated from a body such as the sun or Earth varies with the fourth power of the absolute temperature of the body. Thus, if a body's temperature doubles, the energy radiated increases by 2^4, or 16 times. This phenomenon explains why the sun, with a temperature of 5,800°C (10,500°F), radiates a tremendously greater amount of energy than does the surface of Earth, which radiates at an average temperature of 15°C (59°F). Figure 23.12 illustrates this and another important point: The sun emits strongly in the visible region (relatively short-wave radiation of

about 0.4–0.7 μm), whereas Earth emits in the infrared region (relatively long-wave radiation of about 10 μm). The hotter an object is, the more rapidly it radiates heat and the shorter the wavelength of its predominant radiation. Earth's surface and the surfaces of animals, plants, clouds, water, and rocks are cool enough to radiate heat predominantly in the infrared wavelength, which is invisible to us.

Energy from the sun travels to Earth at the speed of light through the vacuum of space. Less than half the solar energy that reaches our atmosphere is absorbed by the Earth and lower atmosphere. The rest is either absorbed by the upper atmosphere or reflected back into space. Much radiation that is harmful to living organisms, such as X rays and ultraviolet radiation, is filtered out by the upper atmosphere. The stratospheric ozone layer, extending from about 15 to 40 km (9 to 25 mi) above Earth's surface (Figure 23.1), is particularly important in absorbing ultraviolet radiation from the sun and protecting living organisms at Earth's surface. The depletion of the ozone layer is discussed in Chapter 26.

The Greenhouse Effect

Sunlight that reaches Earth warms both the atmosphere and the surface. Earth's surface and atmospheric system then re-radiate heat as infrared radiation.[14] Certain gases in Earth's atmosphere absorb and re-emit this radiation. Some of it returns to Earth's surface, making Earth warmer than it otherwise would be. In trapping heat, the gases act a little like the panes of glass in a greenhouse (although the process by which the heat is trapped is not the same as in a greenhouse). Accordingly, the effect is called the **greenhouse effect**, and the gases—which include water vapor, carbon dioxide, methane, and chlorofluorocarbons (CFCs)—are called **greenhouse gases**.

It is important to understand that the greenhouse effect is a natural phenomenon that has been occurring for millions of years on Earth as well as on other planets in our solar system. After learning how the greenhouse effect works, you will be on your way to understanding how the effect moderates the temperature of the lower atmosphere and helps keep Earth habitable for life. You will also understand how human activity is interacting with the greenhouse effect to cause human-induced global warming.

Most natural greenhouse warming is due to water in the atmosphere. On a global level, water vapor and small particles of water in the atmosphere produce about 85% and 12%, respectively, of our total greenhouse warming. The greenhouse gases we are concerned with here are those that result in part from *anthropogenic* processes—that is, from human activities. These include carbon dioxide, CFCs, methane, nitrous oxides, and ozone, all of which have increased significantly in the atmosphere in recent years. It has been hypothesized that Earth is warming because of the increases in these anthropogenic greenhouse gases.

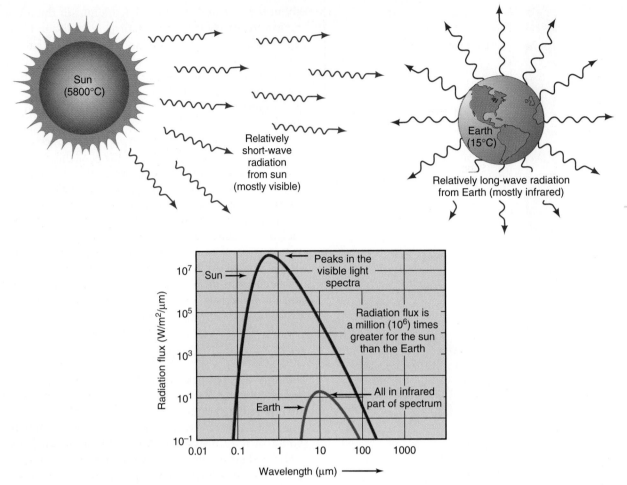

Figure 23.12 ■ Emission of energy from the sun compared with that from Earth. Notice that solar emissions have a relatively short wavelength, whereas those from Earth have a relatively long wavelength. [*Sources:* Modified from W. M. Marsh and J. Dozier, *Landscape* (Reading, Mass.: 1981); and L. R. Kump, J. F. Kasting, and R. G. Crane, *The Earth System* (Upper Saddle River, N.J.: Prentice Hall, 1999).]

The following discussion focuses on the anthropogenic greenhouse effect as it relates to three factors:

■ The burning of fossil fuels, which in recent years has added about 5.5 GtC (gigatons, or billions of metric tons, of carbon) per year to the atmosphere. The carbon combines with oxygen to form CO_2.

■ Deforestation, by burning trees, increases the concentration of atmospheric CO_2, adding 1.6 GtC per year. Burning trees releases carbon stored in the wood that combines with oxygen to form CO_2.

■ Human activities that emit other greenhouse gases, such as CFCs, ozone, methane, and nitrous oxides.

How the Greenhouse Effect Works

A highly idealized diagram showing some important aspects of the greenhouse effect is presented in Figure 23.13. The arrows labeled "energy input" represent the energy from the sun absorbed at or near the surface of Earth. The arrows labeled "energy output" represent energy emitted from the upper atmosphere and the surface of Earth, which balances the input, consistent with Earth's energy balance. The highly contorted lines near the surface of Earth represent the absorption of infrared radiation (IR) occurring there and producing the 15°C (59°F) near-surface temperature. Following many scatterings and absorptions and re-emissions, the infrared radiation emitted from levels near the top of the atmosphere (troposphere) corresponds to a temperature of approximately −18°C (0°F).

The one output arrow that goes directly through Earth's atmosphere represents radiation emitted through what is called the *atmospheric window* (Figure 23.14). The atmospheric window, centered on a wavelength of 10 μm, represents a region of wavelengths (8–12 μm) where outgoing radiation from Earth is not absorbed well by natural greenhouse gases (water vapor and carbon dioxide). Anthropogenic CFCs do absorb in this region, however;

Figure 23.13 ■ Idealized diagram showing the greenhouse effect. Incoming visible solar radiation is absorbed by Earth's surface, to be re-emitted in the infrared region of the electromagnetic spectrum. Most of this re-emitted infrared radiation is absorbed by the atmosphere, maintaining the greenhouse effect. [*Source:* Developed by M. S. Manalis and E. A. Keller, 1990.]

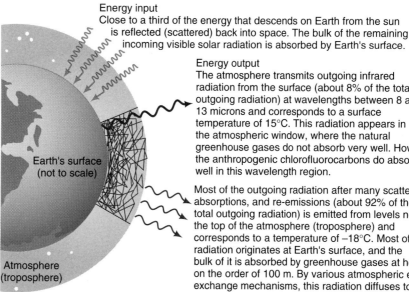

Energy input
Close to a third of the energy that descends on Earth from the sun is reflected (scattered) back into space. The bulk of the remaining incoming visible solar radiation is absorbed by Earth's surface.

Energy output
The atmosphere transmits outgoing infrared radiation from the surface (about 8% of the total outgoing radiation) at wavelengths between 8 and 13 microns and corresponds to a surface temperature of 15°C. This radiation appears in the atmospheric window, where the natural greenhouse gases do not absorb very well. However, the anthropogenic chlorofluorocarbons do absorb well in this wavelength region.

Most of the outgoing radiation after many scatterings, absorptions, and re-emissions (about 92% of the total outgoing radiation) is emitted from levels near the top of the atmosphere (troposphere) and corresponds to a temperature of −18°C. Most of this radiation originates at Earth's surface, and the bulk of it is absorbed by greenhouse gases at heights on the order of 100 m. By various atmospheric energy exchange mechanisms, this radiation diffuses to the top of the troposphere, where it is finally emitted to outer space.

Earth's surface (not to scale)

Atmosphere (troposphere)

and CFCs significantly contribute to the greenhouse effect in this way.

Let us look more closely at the relation of the greenhouse effect to Earth's energy balance, which was introduced in Figure 23.10 in a simplistic way. The figure showed that, of the incoming solar radiation, approximately 30% is reflected back to space from the atmosphere as short-wave solar radiation, while 70% is absorbed by Earth's surface and atmosphere. The 70% that is absorbed is eventually re-emitted as infrared radiation (IR) into space. Thus, the sum of the reflected solar radiation and the outgoing infrared radiation balances with the energy arriving from the sun.

This simple balance becomes much more complicated when we consider exchanges of IR within the atmospheric and Earth surface systems. In some instances, these internal radiation fluxes may have magnitudes greater than the amount of energy entering Earth's atmospheric system from the sun, as shown in Figure 23.15. A major contributor to the fluxes is the greenhouse effect.

At first glance, you might think it would be impossible to have internal radiation fluxes greater than the total amount of incoming solar radiation (shown as 100 units in Figure 23.15). It is possible because the infrared radiation bounces around many times in the atmosphere, resulting in high internal fluxes. For example, in terms of the figure, the amount of IR absorbed at the surface of Earth from the greenhouse effect is approximately 88 units, which is about twice the amount of short-wave solar radiation (45 units) absorbed by Earth's surface. In spite of the large internal fluxes, the overall energy balance remains the same. At the top of the atmosphere, the net downward solar radiation (70 units) balances the outgoing IR from the top of the atmosphere (70 units).

The important point in taking a more detailed look at Earth's energy balance is recognizing the strength of the greenhouse effect. For example, notice in the figure that, of the 104 units of IR emitted by the surface of Earth, only four go directly to the upper atmosphere and are emitted. The remainder are reabsorbed and re-emitted by greenhouse gases. Of these, 88 units are directed downward to Earth, and 66 units upward to the upper atmosphere.

All this may sound somewhat complicated, but if you read and study the points mentioned in Figure 23.15 carefully and work through the balances of the various parts of the energy fluxes, you will gain a deeper understanding of why the greenhouse effect is so important. The greenhouse effect keeps the lower atmosphere of Earth approximately 33°C warmer than it would otherwise be. In addition, the greenhouse effect provides other important

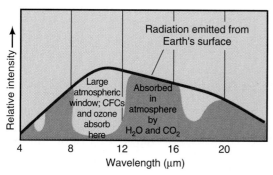

Figure 23.14 ■ Absorption spectra of greenhouse gases, water vapor, and carbon dioxide in the long-wave infrared radiation region. The atmospheric window is a region where neither water vapor nor carbon dioxide absorbs, but where CFCs do absorb. [*Source:* Modified from T. G. Spiro and W. M. Stigliani, *Environmental Science in Perspective* (Albany: State University of New York Press, 1980).]

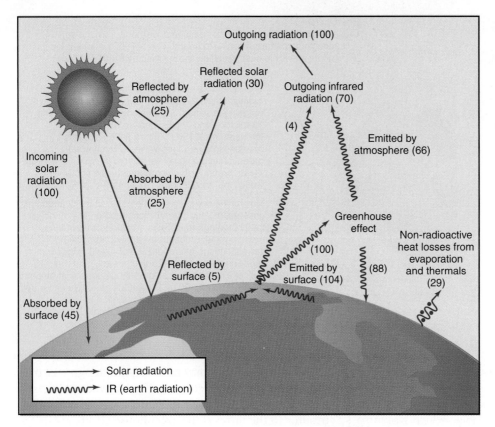

Figure 23.15 ■ Idealized diagram showing Earth's energy balance and the greenhouse effect. Incoming solar radiation is arbitrarily set at 100 units, and this is balanced by outgoing radiation of 100 units. Notice that some of the fluxes (rates of transfer) of infrared radiation (IR) are greater than 100, reflecting the role of the greenhouse effect. Some of these fluxes are explained in the diagram. [*Source:* Modified from D. L. Hartmann, *Global Physical Climatology*, International Geophysics Series, vol. 56 (New York: Academic Press, 1994); and S. Schneider, "Climate Modeling," *Scientific American* 256, no. 5 (1987): 72–80.]

•Total incoming solar radiation = 100 units

•Total absorbed by surface = 133 units
    45 from solar radiation (short wave)
    88 from greenhouse effect IR (infrared)

•Total emitted by surface = 133 units
    104 IR (of this, only 4 units pass directly to space without being
        absorbed or re-emitted in greenhouse effect)
    29 From evaporation and thermals (non-radioactive heat loss)

•Total IR emitted by upper atmosphere to space = 70 units
    66 units emitted by atmosphere
    4 units emitted by surface

•The 25 units of solar radiation absorbed by atmosphere are eventually
  emitted as IR (part of the 66 units)

•Total outgoing radiation = 100 units
    70 IR
    30 reflected solar radiation

service functions. For example, the strong downward emission of IR from the atmosphere, that results from the greenhouse effect, keeps variations in surface temperature from day to night relatively small. Without this effect, the land surface would cool much more rapidly at night and warm much more quickly during the day. Thus, the greenhouse effect not only maintains our relatively comfortable warm surface temperatures, but also helps limit temperature swings from day to night over the land.[3] It is, then, not the greenhouse effect itself that causes concern. Rather, it is the contribution of global warming from anthropogenic greenhouse gases.

Changes in Greenhouse Gases

The major anthropogenic greenhouse gases are listed in Table 23.1. The table also lists the recent rate of increase for each gas and its relative contribution to the anthropogenic greenhouse effect.

Carbon Dioxide

Approximately 200 billion metric tons of carbon in the form of carbon dioxide enter and leave Earth's atmosphere each year as a result of a number of biological and physical processes. Not surprisingly, carbon dioxide has received a lot of attention with respect to global

Table 23.1 • Relative Contribution of Trace Gases to the Anthropogenic Greenhouse Effect

Trace Gases	Relative Contribution (%)	Growth Rate (%/yr)
CFC	15^a–25^b	5
CH_4	12^a–20^b	0.4^c
O_3 (troposphere)	8^d	0.5
N_2O	5^d	0.2
Total	40–50	
Contribution of CO_2	50–60	0.3^e–$0.5^{d,f}$

[a] W. A. Nierenberg, "Atmospheric CO_2: Causes, Effects, and Options," *Chemical Engineering Progress* 85, no. 8 (August 1989): 27.

[b] J. Hansen, A. Lacis, and M. Prather, "Greenhouse Effect of Chlorofluorocarbons and Other Trace Gases," *Journal of Geophysical Research* 94 (November 20, 1989): 16, 417.

[c] Over the past 200 yrs.

[d] H. Rodhe, "A Comparison of the Contribution of Various Gases to the Greenhouse Effect," *Science* 248 (1990):1218, Table 2.

[e] W. W. Kellogg, "Economic and Political Implications of Climate Change," paper presented at Conference on Technology-based Confidence Building: Energy and Environment, University of California, Los Alamos National Laboratory, July 9–14, 1989.

[f] H. Abelson, "Uncertainties about Global Warming," *Science* 247 (March 30, 1990):1529.

warming; 50 to 60% of the anthropogenic greenhouse effect is attributed to this gas.

Measurements of carbon dioxide trapped in air bubbles in the Antarctic ice sheet suggest that during the 160,000 years prior to the Industrial Revolution, the atmospheric concentration of carbon dioxide varied from approximately 200 to 300 ppm.[13] The highest level or concentration of carbon dioxide in the atmosphere other than at present occurred during the major interglacial period about 125,000 years ago.

About 140 years ago, at the beginning of the Industrial Revolution, the atmospheric concentration of carbon dioxide was approximately 280 ppm, a level that had apparently remained constant over the previous 700 years.[14] Since the beginning of the Industrial Revolution, the concentration of carbon dioxide in the atmosphere has grown exponentially. Currently, the rate of increase is about 0.5% per year. If growth continues at this rate, we will see a doubling of the concentration before the end of the twenty-second century. Today, the concentration of carbon dioxide in the atmosphere is about 380 ppm, and it is predicted that the level may rise to approximately 450 ppm by the year 2050, more than 1.5 times the preindustrial level.[16]

Methane

The concentration of methane (CH_4) in the atmosphere more than doubled in the past 200 years, and it is thought to contribute approximately 12 to 20% of the anthropogenic greenhouse effect.[17] As with carbon dioxide, there

are important uncertainties in our understanding of the sources and sinks of methane in the atmosphere.[18, 19]

Natural environments release methane into the atmosphere. Major contributors are termites, which produce methane as they process wood; freshwater wetlands, where decomposing plants in oxygen-poor environments produce and release methane as a decay product; seepage from oil fields; and seepage from methane hydrates (see Chapter 18). The several anthropogenic sources of methane include emissions from landfills (the major source in the United States), burning of biomass, production of coal and natural gas, and agricultural activities, such as the cultivation of rice and the raising of cattle. (Methane is released by anaerobic activity in flooded lands where rice is grown, and cattle expel methane gas as part of their digestive processes.)

Chlorofluorocarbons

Chlorofluorocarbons, or CFCs, are inert, stable compounds that have been or are being used in spray cans as aerosol propellants and in refrigeration units. Use of CFCs as propellants was banned in the United States in 1978. The rate of increase of CFCs in the atmosphere in the recent past was about 5% per year. It has been estimated that approximately 15 to 25% of the anthropogenic greenhouse effect may be related to CFCs.[20]

In 1987, 24 countries signed a treaty, the Montreal Protocol, to reduce and eventually eliminate the production of CFCs and to accelerate the development of alternative chemicals. Because of the treaty, production of CFCs was nearly phased out by 2000. If CFCs had not been regulated by the Montreal Protocol, by the early 1990s they would have become the major contributor to the anthropogenic greenhouse effect.[14]

The potential global warming from CFCs is considerable, because they absorb in the atmospheric window, as explained earlier; and each CFC molecule may absorb hundreds or even thousands of times more infrared radiation emitted from Earth than is absorbed by a molecule of carbon dioxide. Furthermore, because CFCs are highly stable, their residence time in the atmosphere is long. Even though as production of these chemicals was drastically reduced, their concentrations in the atmosphere will remain significant (although reduced from today's concentrations) for many years, perhaps for as long as a century.[15, 20] (CFCs are discussed in more detail in Chapter 26, which examines stratospheric ozone depletion.)

Nitrous Oxide

Nitrous oxide (N_2O) is increasing in the atmosphere and is probably contributing as much as 5% of the anthropogenic greenhouse effect.[16] Anthropogenic sources of nitrous oxide include agricultural activities (application of fertilizers) and the burning of fossil fuels. Reductions in the use of fertilizers and the burning of fossil fuels would reduce emissions of nitrous oxide. However, this gas also has a long residence time; even if emissions were stabilized or reduced, elevated concentrations of nitrous oxide would persist for at least several decades.[17]

23.5 Science of Global Warming

As we suggested earlier, there is good reason to argue that increases in carbon dioxide and other greenhouse gases are related to an increase in the mean global temperature of Earth.[10, 20] Over the past 160,000 years, a strong correlation has existed between the concentration of atmospheric CO_2 and global temperature (Figure 23.16). When CO_2 has been high, temperature has also been high; conversely, low concentrations of CO_2 have correlated with a low global temperature.

Global models of the climate suggest that warming as a result of anthropogenic increases in greenhouse gases will occur (some scientists, however, find faults with the models). According to the models, the average global temperature will rise 1.5–4.5°C by 2100.[1,4] In the most optimistic case, if there are large reductions in emissions of greenhouse gases, the warming may be closer to 1°C.[1]

In order to better understand global warming, we need to consider both the positive and negative feedbacks that occur on Earth, on the sun, and in the atmospheric system. We also need to consider the major variables that may cause (force) a global change in temperature, including solar emissions, volcanic eruption, El Niño warming, and anthropogenic input of particulates and greenhouse gases. A process that changes global temperature is termed **forcing**, and forcing variables that cause or influence warming include solar emission, volcanic eruption and anthropogenic input of greenhouse gases. A particular forcing (say solar input) is not the cause of global warming by itself but is one of the causes of change.

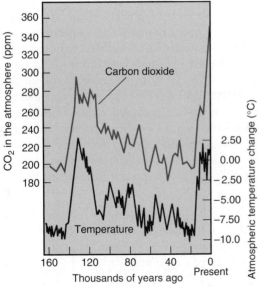

Figure 23.16 ■ Inferred concentration of atmospheric carbon dioxide and temperature change for the past 160,000 years. The relation is based on evidence from Antarctica and indicates a high correlation between temperature and CO_2. [*Source:* Adapted from S. H. Schneider "The Changing Climate," © September 1989, *Scientific American, Inc.* All rights reserved. vol. 261, p. 74.]

Negative and Positive Feedbacks

Greenhouse warming is very complex. The warming effect initiates both negative- and positive-feedback loops that can offset any temperature increase or enhance it. Negative-feedback loops are self-regulating and help stabilize global temperature change in response to a warming circumstance. Positive feedbacks are self-enhancing; thus, a process that causes an increase in global temperature leads to further increases in temperature. We first discussed positive and negative feedbacks with respect to Earth systems and changes in Chapter 3; you may wish to review those concepts.

Several of the potential negative and positive feedbacks concerning global warming are shown in Figure 23.17.[21] It is important to remember that if the negative feedbacks are strong and persistent, global warming may not occur. In contrast, if the negative-feedback systems are weak relative to positive feedback, warming is likely to occur more readily.

As shown in Figure 23.17a, negative feedbacks can be fairly complex. For example, negative-feedback cycle 1 is based on the hypothesis that as global warming occurs, there will be an increase in the algae populations in the warming ocean. Algae will absorb more carbon dioxide, reducing the concentration of CO_2 in the atmosphere and cooling it. Negative-feedback cycle 2 is related to terrestrial vegetation. Here, it is hypothesized that an increase in carbon dioxide concentration will stimulate plant growth (as it does in laboratory experiments), and the increased amount of vegetation will absorb more carbon dioxide from the atmosphere, facilitating cooling of the atmosphere. Negative-feedback cycle 3 may result if polar regions receive more precipitation from warmer air carrying more moisture. Increasing snowpack and ice buildup will cause solar energy to be reflected from Earth's surface, causing cooling. Finally, negative-feedback cycle 4 is related to cloud cover, which is poorly understood but extremely important. The idea is that as the global temperature increases, more water will evaporate from the ocean, leading to more water vapor in the atmosphere and thus more clouds. Because clouds tend to reflect incoming solar radiation, Earth will then be cooled by the increased cloud cover.

Positive-feedback processes are shown in Figure 23.17b. In positive-feedback cycle 5, the warming Earth causes an increase in the evaporation of water from the oceans, which adds water vapor to the atmosphere. But here, the water vapor causes additional warming (water vapor is an important greenhouse gas). The warming Earth causes increased melting of permafrost at high latitudes, which may result in additional release of the greenhouse gas methane (positive-feedback cycle 6). The methane is a by-product of decomposition of organic material in the melted permafrost layer. Another effect of the warming trend would be a reduction in the summer snowpack, which would reduce the amount of solar energy reflected from Earth (positive-feedback cycle 7). Collectively, these processes would result in additional warming and thus are part of a positive-feedback loop. In positive-feedback cycle 8, the warming of Earth is felt by people in urban areas, who use additional air conditioning,

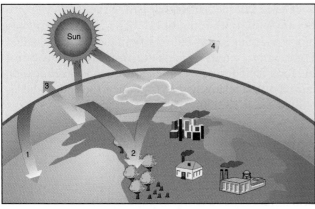

(a)

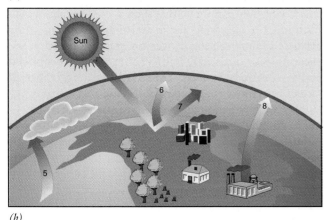

(b)

Figure 23.17 ■ (a) Negative-feedback cycles and (b) positive-feedback cycles associated with the greenhouse effect. See text for explanation. [*Source:* J. R. Luoma, "Gazing into Our Greenhouse Future," *Audubon* 93, no. 2 (1991): 57.]

thus increasing the burning of fossil fuels, which in turn releases additional carbon dioxide into the environment and results in additional global warming.

Both negative- and positive-feedback processes occur simultaneously in the atmosphere. Which are more important? The nod goes to positive feedbacks, because the mean global temperature is increasing. A great deal of research is currently being carried out to better understand negative-feedback processes associated with clouds and their water vapor. Many discussions on the greenhouse effect state that if Earth's atmosphere did not trap heat, our planet would be approximately 33°C (60°F) cooler at the surface; as a result, all water would be frozen. However, since water vapor is the major greenhouse gas in the atmosphere, no greenhouse effect implies no (or very little) water vapor in the atmosphere. This implies no clouds, which would lead to a substantial reduction in the atmospheric reflection of incoming sunlight, resulting in warmer surface temperatures on Earth. The dual role of atmospheric water vapor as both a negative and a positive feedback with respect to global

warming is extremely important to understanding possible climatic modifications created by an anthropogenic greenhouse effect.

Solar Forcing

The sun is responsible for heating Earth; therefore, in evaluating climatic change, we must consider solar variation a possible cause. As mentioned earlier, the Milankovitch cycles are consistent with most of the long-term variability of climate that has occurred during the past several hundred thousand years of Earth's history. However, the Milankovitch cycles by themselves are not sufficient to produce the magnitude of observed climate changes.

When we examine the history of climate during the past thousand years, the variability of solar energy plays a role. Evaluation of isotopes from the sun found in ice cores from glaciers reveals that during the medieval warm period, from approximately 1100 to 1300, the amount of solar energy reaching Earth was relatively high, comparable to the warming we see today. Evaluation of the data also suggests that minimum solar activity occurred during the fourteenth century, coincident with the beginning of the Little Ice Age (Figure 23.4). Thus, it appears that the variability of solar input of energy to Earth can indeed explain some of the climatic variability we have experienced in the past thousand years. The effect, however, is relatively small.[9]

Aerosols and Volcanic Forcing

An aerosol is a particle with a diameter less than 10 μm. Because the effects of collisions with air molecules dominate over gravity, these particles tend to remain in the atmosphere for a long time. Bigger particles (with diameters greater than 10 μm) drop out of the atmosphere faster because of gravity. Emissions of aerosols to the atmosphere by human processes have increased significantly since the Industrial Revolution, and volcanic eruptions have periodically released aerosols as well. Recent research has indicated that aerosols emitted from coal (sulfates) and volcanic eruptions may contribute to global cooling, because sulfates act as seeding for clouds. The aerosol particles provide surfaces for water to condense on; clouds form and reflect incoming solar energy. The aerosol particles also provide a "dust veil" that reflects a significant amount of sunlight.

In the early 1990s, net cooling owing to sulfate aerosols from burning fossil fuels offset some of the global warming expected from the anthropogenic greenhouse effect.[22, 23] However, emissions of the sulfate SO_2 (sulfur dioxide) from human activity are decreasing because of programs to reduce air pollution. The effect of reduction of SO_2 emissions will be a strengthening of the greenhouse effect from CO_2.

Reflection of solar radiation from air pollution particles has apparently reduced incoming solar radiation by as much as 10% in some regions, offsetting up to 50% of expected warming due to increases in greenhouse gases. The process of the reduction of incoming solar radiation by reflection from particles and their interaction with water vapor in the atmosphere (especially clouds) is known as **global dimming**.

Tremendous explosions from Mount Pinatubo in the Philippines in 1991 sent volcanic ash to elevations of 30 km (19 mi) into the stratosphere. The aerosol cloud of ash, with 20 million tons of sulfur dioxide, remained in the atmosphere circling Earth for several years. The particles of ash and sulfur dioxide scattered incoming solar radiation, resulting in a slight cooling of the global climate during 1991 and 1992. Calculations suggest that aerosol additions to the atmosphere from the Mount Pinatubo eruption counterbalanced the warming effects of greenhouse gas additions through 1992. However, by 1994, most aerosols from the eruption had fallen out of the atmosphere, and global temperatures had returned to previous higher levels. Volcanic climate forcing from pulses of volcanic eruptions is also believed to have significantly contributed to the cooling associated with the Little Ice Age, from about 1450 to 1850 (see Figure 23.4).[9]

El Niño

Another natural perturbation that affects global climate is the occurrence of El Niño events (see A Closer Look 23.2). During an El Niño, the normal conditions of equatorial upwelling of deep oceanic waters in the eastern Pacific are diminished or eliminated. Upwelling releases carbon dioxide to the atmosphere as carbon dioxide–rich deep water reaches the surface. El Niño events thus reduce the amount of oceanic carbon dioxide outgassing, influencing the global carbon dioxide cycle. Climatic models generally predict that as Earth warms, El Niño events will become more common.[1]

Methane Forcing

Methane (CH_4), or natural gas, is a strong greenhouse gas contributing as much as 20% of the anthropogenic greenhouse effect. There may also be a periodic natural methane forcing that causes rapid warming to end a glacial period. (See the discussion of methane hydrates in Chapter 18.)

One mechanism that releases large amounts of methane is linked to lowering of sea level. Remember from Chapter 18 that methane hydrates remain stable in deep (high-pressure) cold water. During glacial times, sea level drops because of the huge volume of water stored in glaciers. The lower sea level causes water pressure at the bottom of the sea to be reduced. Under these conditions, methane hydrates could become unstable and be released into the water. The gas, once released from the ocean bottom, would rise and enter the atmosphere. A second process favoring the release of methane from hydrates is the warming of ocean water. Warming of deep water occurs at the beginning of interglacial periods when sea level is low.[25]

If sufficient methane is released, either by lowering of the sea level or warming of the water, the gas may cause global warming, contributing to ending the glacial period. For example, release of methane may have been responsible for a rapid increase in the global temperature that occurred over a relatively short period at the end of the last ice age about 15,000 years ago.

Anthropogenic Forcing from Greenhouse Gases

Until recently, there was controversy concerning whether the anthropogenic component of global warming was significant or not. The question now seems to have been answered, and we can state that there is a significant human footprint on observed global warming. A recent study evaluated climate change over the past thousand years, allowing late–twentieth-century warming to be placed within a historical context. The study used a mathematical model to remove major natural forcing mechanisms, including solar radiation and volcanic eruptions, so that climatic forcing by greenhouse gases could be directly estimated. The research established that the natural variability in the climate system in the past thousand years is far less than what occurred at the end of the twentieth century. That is, present warming far exceeds natural variability and closely agrees with the warming predicted by global circulation models that take account of increasing greenhouse gases.[9] Results of the modeling are shown in Figure 23.18. Notice that the late–twentieth-century rise in temperature far exceeds the zone of maximum variability that could result from forcing mechanisms such as solar variability, volcanic activity, and global dimming.

It appears that the major scientific issues concerning global warming have been resolved. Significant global warming as a result of human activity is occurring. We now discuss some potential effects of and possible adjustments

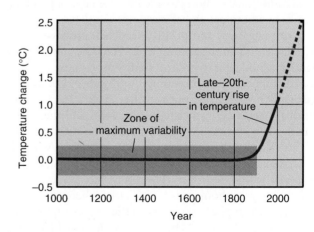

Figure 23.18 ■ Global temperature change during the past 1,000 years. Solar and volcanic forcing have been removed, leaving anthropogenic forcing from greenhouse gases. [*Source:* Modified from T. J. Crowley, 2000. "Causes of Climate Change over the Past 1000 Years," *Science* 289 (2000): 270–277]

El Niño

El Niño became a household word during the winter of 1997–1998, when it was blamed for everything from tornadoes and thunderstorms in Florida to catastrophic fires in Indonesia (see Chapter 24). The term *El Niño* means "little boy" in Spanish and refers in particular to the Christ Child, because the event often begins off the coast of South America near Christmas time.

El Niño events disrupt the ocean-atmosphere system in the tropical Pacific. They are in part responsible for weather phenomena that can cause billions of dollars in property damage and the loss of thousands of human lives. Occurring at intervals of two to seven years, El Niño events typically last for 12 to 18 months. They start with the weakening of east-to-west trade winds and the warming of eastern Pacific Ocean waters. As a result, tropical rainfall shifts from Indonesia to South America, as shown in Figure 23.19.

Let's look at this process in more detail. Under non–El Niño conditions, trade winds blow west across the tropical Pacific. The warm surface water in the western Pacific tends to pile up, so that the sea surface can be as much as 0.5 m higher at Indonesia than at Peru. In contrast, during El Niño, the trade winds weaken and may even reverse. As a result, the eastern equatorial Pacific Ocean becomes unusually warm, and the westward moving equatorial ocean current weakens or reverses. The rise in temperature of sea surface waters off the South American coast inhibits the upwelling of nutrient-rich cold water from deeper levels; the upwelling normally supports a diverse marine ecosystem and major fisheries. Because rainfall follows warm water eastward during El Niño years, there are high rates of precipitation and flooding in Peru, while droughts and fires are commonly observed

in Australia and Indonesia. Because warm ocean water provides an atmospheric heat source, El Niño changes global atmospheric circulation, which causes changes in weather in regions that are far removed from the tropical Pacific.

It is important to remember that El Niño events are natural phenomena and part of a dynamic system involving the coupling of Earth's atmosphere and ocean. El Niño events can alternate with contrasting phenomena called La Niña events. La Niña events are characterized by unusually cool ocean water temperatures, and the effects are opposite those of El Niño. For example, during a 1998–1999 La Niña, winter temperatures were cooler than normal in the U.S. Northwest. A La Niña event does not necessarily follow every El Niño event; but together, they constitute a natural cycle of ocean-atmospheric change with global consequences.[24]

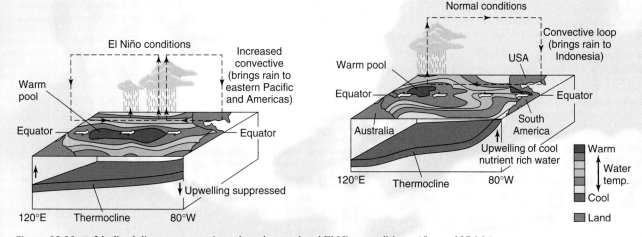

Figure 23.19 ■ Idealized diagram comparing selected normal and El Niño conditions. (*Source:* NOAA.)

to global warming. What we ultimately decide to do about global warming will reflect our values.

23.6 Potential Effects of Global Warming

If we continue emitting large amounts of carbon dioxide into the atmosphere, it is estimated that by 2030 the concentration of carbon dioxide in the atmosphere will have

doubled from pre–Industrial Revolution concentrations. The average global temperature (according to mathematical models) will have risen approximately 1° to 2°C (2° to 4°F), with significantly greater temperature change at the polar regions.[16] Global warming causes greater temperature increases at polar regions in part because of a positive-feedback mechanism; sea ice melts from warming, and water reflects much less light than white ice, resulting in enhanced warming. Solar energy that would have been reflected by sea ice is absorbed by the ice-free water. This mechanism is part of what is termed **polar amplification**.[3]

Specific effects of global warming are difficult to predict from global models. However, it is expected that warming will likely have the following consequences:[1]

- As a result of regional changes in climate, semi-arid land areas will become drier, while other regions will become wetter.
- World distribution of biomes will likely change. Some, such as alpine tundra, may be lost, while others, such as midlatitude deserts, may expand.
- Agricultural production in some regions will decrease.
- The incidence of diseases such as malaria and dengue fever will increase in tropical countries.
- Rising sea levels and accompanying coastal erosion will threaten lowlying islands and will displace tens of millions of people worldwide.
- The biosphere will change as a result of damage to ecosystems, including tropical coral reefs and birds and bears in the Arctic.

Next, we briefly discuss selected aspects of these consequences.

Changes in Climate

Various estimates have been made of what changes in annual temperature and precipitation are likely to occur as a result of global warming. Figure 23.20 shows changes expected to occur by 2050 assuming concentrations of greenhouse gases increasing at about 1% per year.[1, 26] In central North America (the grain-growing region), warming is expected to vary from approximately 2° to 4°C (4° to 7°F), with a small increase in precipitation. As a result, soil moisture may decrease in the summer by as much as 20%. Clearly, this could have a significant effect on the grain-growing areas of the United States.

Changes in relative runoff of water are linked to changes in temperature and precipitation. California, which depends on snow melt from the Sierra Nevada for water to irrigate one of the richest agriculture regions in the world, will have problems storing water in reservoirs. Rainfall will likely increase, but there will be less snow pack with warming. Runoff will be more rapid than if snow slowly melts. As a result, reservoirs will fill sooner and more water will escape to the Pacific Ocean. Lower runoff is projected for much of Mexico, South America, southern Europe, India, southern Africa, and Australia. It is important to keep in mind that these projections are based on global circulation models that are controversial and subject to variability. Nevertheless, most of the models predict changes in the directions indicated, and as a result are being taken seriously by both scientists and policymakers.

As already suggested, global rise in temperature is expected to significantly change patterns of rainfall, soil moisture, and other climatic factors related to agricultural productivity. Studies using global circulation models to predict patterns for the Northern Hemisphere suggest that some of the more northern areas, such as Canada and Russia, may become more productive. Although global warming might move North America's prime farming climate north from the midwestern United States to the region of Saskatchewan, Canada, the U.S. loss would not simply be translated into a gain for Canada. Saskatchewan would have the optimum climate for growing, but the Canadian soils are thinner and less fertile than the prairie-formed soils of the U. S. Midwest. Therefore, a climate shift could have serious negative effects on midlatitude food production. In addition, lands in the southern part of the Northern Hemisphere may become more arid; and as a result, soil moisture relationships will change.

People are anxious when uncertainty exists and they are particularly anxious when that uncertainty involves their food supply. There is real concern that hydrologic changes associated with climatic change resulting from global warming may seriously affect the global food supply. Today, 800 million people are malnourished; and as population increases, so will demand for food. During the twenty-first century, agricultural production in many subtropical and tropical regions (especially in South America and Africa) will likely decrease if the mean annual temperature increases by more than 2°C. Unfortunately, millions of the poorest people on Earth live in the tropics and subtropics, where risk of hunger is the greatest.[1]

There is also concern that global warming will alter normal weather and climatic patterns, including the frequency or intensity of violent storms. This possibility may be more important than changes in climate. The hypothesis is that warming ocean waters could feed more energy into high-magnitude storms, such as cyclones and hurricanes, causing a significant increase in their frequency or intensity. The research to date suggests that the number of hurricanes has not increased, but the average intensity has. Approximately half of Earth's human population live in coastal areas. Potential problems are exacerbated by the fact that many of these areas are low lying and are experiencing rapid population growth. In addition, greenhouse warming is expected to result in wetter winters, hotter and drier summers, an increased frequency of large storm events, and an increased possibility of droughts in the northern temperate latitudes.[27]

In summary, global warming could affect people and ecosystems on Earth in various ways, often for the worse. These changes may affect hydrology, crop production, forestry, and human health. Midlatitude climate zones could shift northward by as much as 550 km (330 mi) over the next century. At this rapid rate, some tree species may be nearly eliminated. Furthermore, the expected expan-

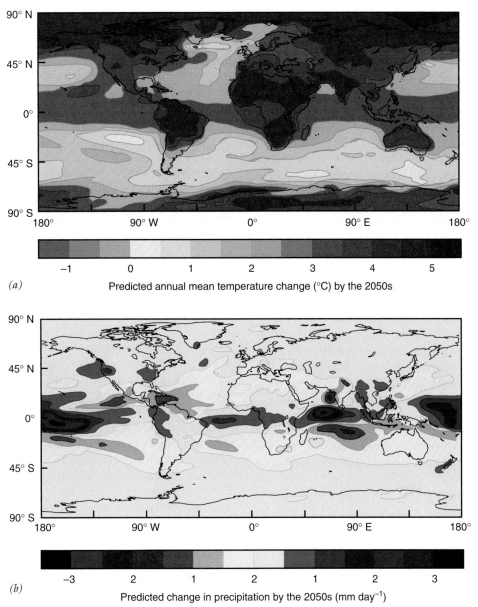

(a)

Predicted annual mean temperature change (°C) by the 2050s

(b)

Predicted change in precipitation by the 2050s (mm day^{-1})

Figure 23.20 ■ (*a*) Projected changes in annual temperatures from today to the 2050s. Notice that changes are greatest at the polar regions. (*b*) Projected changes in annual precipitation from today to the 2050s. Highest increases are near the equator. Modeling that predicted these changes assumes an increase in greenhouse gas of about 1% per year. [*Source:* Met Office, Hadley Center for Climate Prediction and Research, in R. T. Watson, presentation at the Sixth Conference of the Parties of the United Nations Framework Convention on Climate Change, Intergovernment Panel on Climate Change, November 13, 2000. Accessed December 1, 2000, www.ipcc.ch/press/sp-COPG.htm.]

sion of tropical climate zones will lead to an increase in tropical diseases such as malaria, dengue fever, yellow fever, and viral encephalitis.[28]

Rise in Sea Level

A rise in sea level is a potentially serious problem related to global warming. As mentioned, about half of the people on Earth live in the coastal zone, and about 50 million people per year experience flooding due to storm surges. As sea level rises and population increases, the number of people vulnerable to coastal flooding increases. For ex-

ample, two cyclones that hit highly populated Bangladesh in the last 25 years killed more than 400,000 people and caused over $1.6 billion in property damage (see Chapter 4). The double impact of rising sea level and more frequent and powerful cyclones and other tropical disturbances (owing to warmer oceans, as discussed earlier) would have devastating effects on people in developing countries.

Although a precise estimate of the total potential rise in sea level is not possible at this time, there is a consensus that the level of the sea will in fact rise. In fact, sea level along much of the U.S. coast is already rising at a rate of

1 to 2 mm/yr, or about 4 to 8 in/century.[29] The cause for the rise is thought to be twofold: thermal expansion of warming ocean water (the primary cause) and melting of glacial ice (a secondary cause).

Various models predict that sea level may rise anywhere from 20 cm to approximately 2 m (8–80 in) in the next century; the most likely rise is probably 20–40 cm (8–16 in). One estimate is that sea level will likely rise 15 cm by 2050 and 34 cm by 2100. When other factors—such as land subsidence and compaction, groundwater depletion, and natural climate variation—are considered, some coastal regions could experience a sea level rise of 45–55 cm by 2100.[29]

Such a change would have significant environmental impacts. It could cause increased coastal erosion on open beaches of 50–100 m (165–230 ft), making buildings and other structures in the coastal zone more vulnerable to damage from waves generated by high-magnitude storms. It could also cause a landward migration of estuaries and salt marshes, lead to loss of coastal wetlands (see Chapter 21), and put additional pressure on human structures in the coastal zone.[16] Finally, groundwater supplies for coastal communities may be threatened by saltwater intrusion should sea levels rise (see Chapter 22).

A rise in sea level of approximately 1 m (3.3 ft) would have even more serious consequences, threatening the existence of some lowlying islands. People would have to alter the coastal environment significantly to protect investments, and communities would be forced to choose either to make very heavy expenditures to control coastal erosion or to allow considerable loss of property.[16]

It seems inevitable that a rise in sea level will increase coastal erosion and lead to further investments to protect cities in the coastal zone. Construction of seawalls, dikes, and other erosion-controlling structures will become more common as coastal erosion threatens urban property. In more rural areas, where development is set well back from the coastal zone, the most likely response to rising sea level will be to adjust to the erosion that occurs. Coastal erosion is a difficult problem that is very expensive to deal with. In many cases, it is best to allow erosion to take place naturally where feasible and only defend against coastal erosion where absolutely necessary.

Glaciers and Antarctic Ice Cap

Many more glaciers in North America, Europe and other areas are retreating than advancing. In the Cascades of the Pacific Northwest and the Alps in Switzerland and Italy retreats are accelerating. For example, on Mt. Baker in the Northern Cascades of Washington, all eight glaciers on the mountain were advancing in 1976. Today all eight are retreating.[30] Glaciers in Glacier National Park, Montana may disappear by 2030. Already the number has been reduced from 150 in 1850 to 35 today. While many

glaciers are melting, the central ice cap on Antarctica is growing. Surprisingly, this is consistant with predictions of global climate models. As Earth warms, more snow falls on Antarctica. Satellite measurement from 1992–2003 suggest the East Antarctica ice sheet increased in mass by about 50 billion tons per year during the period of measurement.[31]

Changes in Biosphere

A growing body of evidence indicates that global warming is probably initiating a number of changes in the biosphere, sometimes threatening ecological systems and people. These changes include shifts in the ranges of plants and animals. Some such shifts are summarized in Figure 23.21.

Spring now arrives up to two weeks earlier than it did three decades ago, and this is having ecological effects.[32] An endangered species of woodpecker in North Carolina is laying eggs about a week earlier than it did two decades ago. Marmots in the mountains of Colorado are waking from winter hibernation about six weeks earlier than they did 17 years ago. Some cherry trees in Washington, D.C., are blooming about a month earlier than they did 50 years ago. Robins are arriving in Wisconsin a few days earlier than they did just 10 years ago.

Early arrival of spring can stress some species, changing communities of organisms. Rapid change may require adaptions that some species can't make. Earlier-arriving birds may compete for food with other birds that migrate later, when days are longer. Some plants may flower earlier, then be damaged by spring snowstorms.[33] Wild relatives of domesticated plants exist in remnants of their original habitats, some of which have become ecological islands. As the climate warms, these remaining habitats may no longer be suitable for these wild plants, but the wild strains are important for food production because they provide genetic diversity that is valuable in developing new hybrids to combat diseases and to adapt to new climate conditions.

A change mentioned in Figure 23.21 is a shift in the range of mosquitoes that carry diseases including malaria and dengue fever. Malaria, which causes chills and fever, is particularly worrisome, as it kills about 3,000 people per day, mostly children. Projected global warming will increase the land area where the disease can be transmitted. Today, that area contains 45% of the world's population. With global warming, malaria could threaten 60% of the world's population.[34, 35]

The possible relationship between global warming and the emergence of the West Nile virus in New York City in 1999 is an example of linkages between physical, biological, and social systems—the essence of environmental science. How the West Nile virus arrived in North America is not known, but its spread from mosquitoes to birds to people was enhanced by a warm winter followed by a dry spring

1. Black guillemots, sea birds living in the Arctic environment, have been declining in numbers on Cooper Island, Alaska. As sea ice receded earlier each spring during the 1990s, the birds' prey, the Arctic cod, moved with the ice, causing the decline in numbers of the sea birds.

2. Edith's checkerspot butterflies are moving north and to higher elevations. This migration has been going on over the last century and represents a shift that matches global warming in the butterflies' western North American habitat.

3. Sachem skipper butterflies have been expanding their range from northern California into Oregon and southeastern Washington. This has been going on over a 50-year period, coinciding with gradual global warming. Each move northward has occurred during an unusually warm summer.

4. Snails, sea stars, and other intertidal organisms have been shifting northward in Monterey Bay as a result of increases in shoreline water temperatures over the past 60 years.

5. Mexican jays are breeding earlier in the Chiricahua Mountains, according to a 30-year study. It is hypothesized that the birds have adjusted to earlier warm spring temperatures by breeding earlier to match emergence of food sources, including insects.

6. Mexican voles (small rodents with short tails and stout bodies) have expanded their range during the past 100 years into northeastern Arizona and Colorado from the south and west. Studies suggest that a number of other mammal species are also shifting northward to adjust to changes in climate and habitat.

7. Zone-tailed hawks and other birds have gradually moved northward over a 50-year period. The trend is observed for a number of bird species of the southwest and Mexico, including common blackhawks, whippoorwills, and brown-crested flycatchers.

8. Prairie grasses are on the decline in the central plains. In some cases, exotic weeds are replacing the native grasses. This phenomenon correlates with global warming since 1964.

9. Fire ants are expanding their range in all directions from the Deep South as far north as North Carolina. The ants' spread is limited by cold temperatures and correlates with warming trends in recent years.

10. Palm trees along Florida's west coast are dying as a result of exposure to salt water. This has resulted in part from a rise in global sea level associated with a trend over several thousand years. However, more recently, sea levels have been rising faster as a result of human-induced climate change, increasing the threat to low-lying palm trees.

11. Polar bears in western Hudson Bay are losing weight and bearing fewer cubs. Warmer temperatures are melting sea ice earlier in the spring. The sea ice is a critical habitat for the bears, who use the ice surface as a hunting platform for capturing seals.

12. Peary caribou populations are declining in the high Arctic environment of Canada. This may be related to a series of unusually warm, wet, and snowy winters, which have created vast ice cover. The animals are unable to forage below the solid crust of the ice.

13. Mosquitoes carrying dengue fever (a severe flu-like viral illness that may cause fatal internal bleeding) are moving to higher elevations in Mexico.

14. Coral reefs in the Florida Keys, Bermuda, and the Caribbean are declining as a result of bleaching thought to be caused by warming marine waters.

15. Subalpine forests are moving to higher elevations in the Olympic Mountains of Washington.

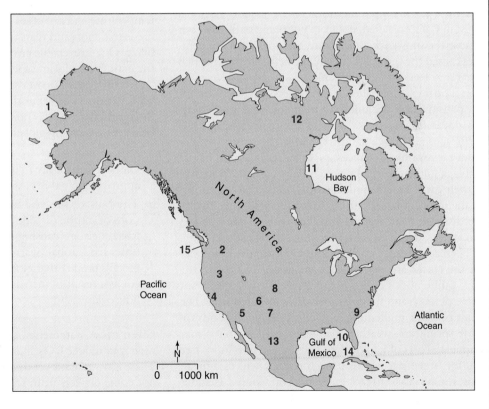

Figure 23.21 ■ Selected examples of how rising global temperatures are shifting the range of plants and animals in North America. (*Sources:* Modified from S. Levy, "Wildlife on the Hot Seat," *National Wildlife* 38, no. 5(2000):20–27; and N. Holmes, "Has Anyone Checked the Weather [Map]?" *Amicus Journal* 21, no. 4(2000):50–51.]

and a hot, wet summer; these conditions are symptoms of global warming. This is how it works:[35]

- During a mild winter, more mosquitoes survive in sewers as well as in still water in a variety of locations, including ponds, waste cans, and abandoned tires.
- During a dry spring, surface water sites, such as small ponds, decrease in size, concentrating both birds and mosquitoes at the water sites.
- Mosquitoes infected with the virus bite uninfected birds, passing the virus on to them.
- Infected birds are bit by uninfected mosquitoes, passing the virus to more mosquitoes.
- Hot summer months, with their warm air and heavy rains, cause the mosquito population to mature and grow rapidly. More mosquitoes become infected and pass on the virus both to birds and, eventually, to people.

Thus, the unusual weather, possibly from global warming, results in an increase in the population of mosquitoes. Once the virus appears, a strong positive-feedback cycle between birds and mosquitoes increases the number of infected mosquitoes, eventually producing a health problem for people.

What is happening to some butterfly species is also of concern. For example, as early as 1989, Edith's checkerspot butterflies in some high meadows of the Sequoia National Forest in California died, not from poisoning or disease, but because the snowpack melted early and the warm temperatures caused the butterflies to emerge before the nectar-rich plants they feed on were available.

As a final example of ecological effects of global warming, consider the declining numbers of black guillemots on Cooper Island, Alaska. Increases in temperature in the 1990s caused the sea ice to recede farther from Cooper Island each spring. The recession of sea ice occurred before the black guillemot chicks were mature enough to survive on their own. Parent birds feed on Arctic cod found under the sea ice, then return to the nest to feed their chicks. The distance from feeding grounds to nest must be less than about 30 km; but in recent years, the ice in the spring has been receding as much as 250 km from the island before the chicks are able to leave the nests. As a result, the black guillemots on the island have lost an important source of food, and their numbers have been reduced. The future of black guillemots on Cooper Island depends on future springtime weather. Too warm, and the birds may disappear. Too cold, and there may be too few snow-free days for breeding; in this case, too, they will disappear. Thus, the survival of black guillemots on Cooper Island is precarious, given our changing global climate.[2]

An increase in the risk of species extinction is a potential change to the biosphere resulting from global warming. This results because warming may cause a shifting of climate that impacts the distribution and abundance of a species. As climate changes, a species may attempt to move to a climatically suitable area. Some will be able to do so, but others will not. Furthermore, efforts of some species to relocate may be restricted by other environmental disruptions, such as habitat loss or fragmentation of habitat from land-use change.[36]

What is apparent from our examples is that the ecological consequences of global warming are not the concern of a distant future. Changes in the biosphere and ecological communities as a result of warming are happening now. As individual species in an ecosystem are affected, changes will occur in other species. This is the principle of environmental unity discussed in Chapter 3.

23.7 Adjustments to Potential Global Warming

Before we consider possible adjustments to global warming, we need to recognize that the problem is complex. We learned in Chapter 3 that solving complex environmental problems is often difficult due to exponential growth, lag times, and the possibility of irreversible change. We now return to that concept when considering global warming. Peak emissions of CO_2 will likely occur in the next 50 to 100 years (Figure 23.22). Response to the emission of carbon dioxide up to and beyond the peak will include exponential growth of the concentration of carbon dioxide in the atmosphere as well as exponential increase in global temperature and rise in sea level. Following lag time of hundreds to a thousand years or more, these will eventually stabilize. One irreversible environmental consequence may be the extinction of some species.

Debate continues over some details of climatic change related to human activities that release greenhouse gases into the atmosphere. At this point, two important statements can be made concerning global warming and climate change:

- Global temperature in the last few decades has increased approximately 0.5°C (0.9°F).
- The bulk of the evidence suggests that massive emissions of carbon dioxide into the atmosphere from use of fossil fuels are causing some significant portion of the observed global warming. Furthermore, computer models predict that global warming will continue to occur as a result of increased atmospheric levels of greenhouse gases.

Given these statements, should we spend large sums of money to reduce CO_2 and other greenhouse gas emissions? There is no easy answer to this question, and the answer will not come entirely from scientific evaluation. For example, the estimated cost to stabilize atmospheric concentration at 550 ppm (double the preindustrial level) is about $40 billion per year over a century.

In terms of science, the question of whether human-induced warming is occurring is solved: A significant part of global warming is human-induced. We now must eval-

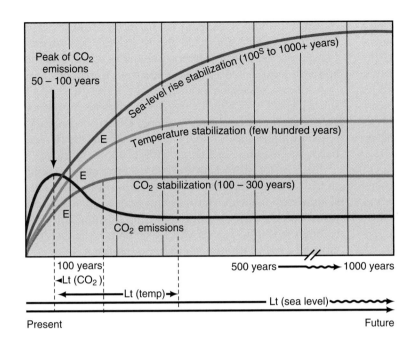

Figure 23.22 ■ Exponential growth (E) with lag times of atmospheric concentration (Lt) following peak in CO_2 emissions for stabolization of CO_2, global temperature and sealevel rise are hundreds to more than 1000 years. Modified after Intergovernmental Panel on Climate Change 2001 at http://www.ipcc.

uate potential consequences and risks to society and the biosphere. Policymakers must take action to address risks posed by global warming. They must do this even though significant scientific uncertainties remain concerning whether global warming and associated climate-induced changes can be reversed.[1]

What should we do about potential global warming? There are two basic adjustments:

■ Adapt; that is, do nothing to combat it, and live with future global climatic change.

■ Work to mitigate the situation (reduce its severity) by reducing emissions of greenhouse gases and enhancing biospheric sinks for CO_2.

Living with Global Change

Considering our present understanding of global climate and recent trends in energy use and deforestation, a likely adjustment will be learning to live with the changes. As discussed earlier, these include a warmer climate, more variability in weather patterns, changes in the biosphere, and a higher sea level. If the changes are relatively slow and total mean global warming is less than 2°C, then learning to live with new conditions may be feasible. In some cases, the changes will offer opportunities. Overall, however, this may not be the best adjustment. Surprises and unexpected problems will emerge. A 2°C change is at the low end of predicted global warming; at increases greater than 2°C, potential consequences to humans increase significantly. Investigation of sea surface temperature over the past several hundred thousand years from the geologic record suggests that warming will more likely be about 5°C,[37] consistent with results from modeling climate change that

warming will be as great as 4.5°C. Finally, although yields of crops will vary on a regional and local level, problems in food production will occur in tropical and subtropical regions, where many millions of people live.[1]

In summary, one possible adjustment to global warming will be to live with it. However, if we are prudent, we will plan instead to reduce emissions of carbon dioxide and other greenhouse gases. By doing so, we may be able to limit warming to less than 2°C.

Mitigating Global Warming

The attempt to control emissions of greenhouse gases, particularly carbon dioxide, was initiated when a major scientific conference on the issue of global warming was held in 1988 in Toronto, Canada. At that meeting, scientists recommended a 20% reduction in carbon dioxide emissions by 2005. The meeting was a catalyst for scientists and other concerned people to work with politicians to initiate international agreements for reductions in emissions of greenhouse gases. Although at that time many uncertainties remained concerning global warming, the prevailing attitude was that it was advisable to be conservative and reduce emissions before problems became apparent.

In 1992, at the Earth Summit in Rio de Janeiro, Brazil, a general blueprint for the reduction of global emissions was suggested. Some in the United States, however, objected that the reductions in CO_2 emissions would be too costly. Furthermore, agreements from the Earth Summit did not include legally binding limits. Following the meetings in Rio de Janeiro, governments worked to strengthen a climate-control treaty that included specific limits on the amounts of greenhouse gases that could be admitted into the atmosphere by each industrialized country.

Legally binding emission limits were discussed in Kyoto, Japan, in December 1997, but specific aspects of the agreement divided the delegates. The United States eventually agreed to cut emissions to about 7% below 1990 levels. However, that was far below the reductions suggested by scientists, who recommended reductions of 60–80% below 1990 levels. In fact, following the conference, it was realized that emissions of carbon dioxide in 2010 would likely be about 30% above the 1990 emissions.

Since that time, the science question has been answered, and the human component of global warming has increased, as evidenced by a series of very warm years and biosphere changes. In spite of this, some sectors of the fossil fuels industry continue to resist lowering emissions, arguing that the economic cost would be too great.

Every time an environmental solution to a problem has been suggested, there have been those who said it couldn't be done, yet it has been proved time and time again that it can be done. It appears that advances in alternative renewable energy are sufficient that society could in fact meet goals for reductions of carbon dioxide emissions without jeopardizing our economic future.

The United States, with 5% of the world's population, emits about 25% of the atmospheric carbon dioxide. It is encouraging to note that the U.S. government has acknowledged that reduction of emissions of carbon dioxide into the atmosphere is an important goal. Providing economic incentives to improve and design new energy-efficient technologies and alternative energy sources would be a positive step. However, the U.S. Congress has been slow to act or to fund necessary studies. It is discouraging that the United States, in the Hague meetings of late 2000, refused to honor reductions in emissions of CO_2 agreed to at Kyoto in 1997. A problem is that the fast growing economices of China and India are rapidly increasing their emissions of carbon dioxide, and are not bound by the Koyoto Protocol. In March 2001, the new administration in Washington disappointed European allies by announcing that the United States would not abide by the Kyoto Protocol. Nevertheless, the protocol was signed by 166 nations and became a formal international treaty in February of 2006. Perhaps the positive action of so many nations will spur the United States to rethink its position.

Although U.S. government rejected the Kyoto Protocol, some states are taking independent action. California, which by itself is twelfth in the world in emissions of carbon dioxide, passed legislation in 2006 to reduce emissions by 25% by 2020. Some have labeled the action as a "job killer," but environmentalists point out that the legislation will bring opportunity and new jobs to the state. California is often a leader, and others states are considering how to control greenhouse gases.

How can carbon dioxide emissions be reduced? Energy planning that relies heavily on energy conservation and efficiency, along with use of alternative energy sources, has the potential to reduce emissions of carbon dioxide. Increased use of nuclear power would also assist in reducing carbon dioxide emissions. Changing the balance of fossil fuels to burn more natural gas would also be helpful, because natural gas releases 28% less carbon per unit of energy than oil and 50% less than coal.[38] Other strategies to reduce emissions of carbon dioxide include the use of mass transit and subsequent decrease in the use of automobiles; providing greater economic incentives to energy-efficient technology; requiring higher fuel-economy standards for cars, trucks, and buses; and requiring higher standards of energy efficiency.

Burning forests to convert land for protecting agricultural uses accounts for about 20% of anthropogenic emissions of carbon dioxide into the atmosphere. Management plans with the objective of minimizing burning and protecting the world's forests would help reduce the threat of global warming. Planting more trees (reforestation) is also a potential strategy: Reforestation would increase biospheric sinks for carbon dioxide. Other natural sinks for carbon dioxide—such as soils, forests, and grasslands—might be enhanced and managed better to sequester more carbon dioxide than they currently do.[39]

Geologic (rock) sequestration of carbon is another possible mitigation measure to reduce the amount of carbon dioxide that would otherwise enter the atmosphere. The general principle of geologic sequestration of carbon is fairly straightforward. The idea is to capture carbon dioxide from power plants and industrial smokestacks and inject the carbon into deep subsurface geologic reservoirs. Geologic environments suitable for carbon sequestration are sedimentary rocks that contain salt water and sedimentary rocks that are the sites of depleted oil and gas fields. Sedimentary rock environments, which are widespread at numerous locations on Earth, have large capacity or potential to sequester as much as 1,000 gigatons of carbon. In order to significantly mitigate the adverse effects of carbon dioxide emissions that result in global warming, we need to sequester approximately 2 gigatons of carbon per year.[40]

The process of placing carbon dioxide in the geologic environment involves compressing the gas and changing it to a mixture of both liquid and gas and then injecting it deep underground. Individual injection projects can sequester approximately 1 million tons of carbon dioxide per year.

A carbon-sequestration project is under way in Norway beneath the North Sea. The carbon dioxide from a large natural gas production facility is injected approximately 1,000 meters deep into sedimentary rocks below a natural gas field. The project, started in 1996, injects about 1 million tons of carbon dioxide every year. It is estimated that the entire reservoir can hold up to about 600 billion tons of carbon. To put this number in perspective, it is about the amount of carbon dioxide that is likely to be produced from all of Europe's fossil fuel plants in the next several hundred years.[40] The cost of sequestering carbon beneath the North Sea is expensive, but it saves the company from paying carbon dioxide taxes for emissions into the atmosphere.

Pilot projects to demonstrate the potential of sequestering carbon in sedimentary rocks have been initiated in Texas beneath depleted oil fields. The potential to store carbon at sites in Texas and Louisiana is immense. One estimate is that between 200 and 250 billion tons of carbon dioxide could be sequestered in this region.[41]

Summary

- The atmosphere, a layer of gases that envelops Earth, is a dynamic system that is constantly changing. A great number of complex chemical reactions take place in the atmosphere, and atmospheric circulation takes place on a variety of scales, producing the world's weather and climates.

- Nearly all the compounds found in the atmosphere either are produced primarily by biological activity or are greatly affected by life.

- The four main processes that remove particles and pollutants from the atmosphere are sedimentation, rain out, oxidation, and photodissociation.

- Major climatic changes have occurred during the past 2 million years, with periodic appearances and retreats of glaciers. During the past 1,500 years, several warming and cooling trends have affected people. During the past 100 years, the mean global annual temperature has apparently increased by about 0.5°C.

- Water vapor and several other gases, including carbon dioxide, tend to warm Earth's atmosphere through the greenhouse effect. The vast majority of the greenhouse effect is produced by water vapor, which is a natural constituent of the atmosphere. Carbon dioxide and other greenhouse gases also occur naturally in the atmosphere. However, especially since the Industrial Revolution, human activity has added substantial amounts of carbon dioxide to the atmosphere, along with such greenhouse gases as methane and CFCs.

- The mean global temperature of Earth is rising. Climatic models suggest that when carbon dioxide has doubled from its preindustrial levels in the next few decades, the mean global temperature may rise by 1 to 2°C. Total warming during the twenty-first century may range from 1.5°C to 4.5°C.

- Many complex positive-feedback and negative-feedback cycles affect the atmosphere. Nevertheless, it appears that increases in the emission of greenhouse gases from human activity is a significant cause of late-twentieth-century and present global warming. Natural cycles, solar forcing, aerosols, volcanic eruptions, and El Niño events also affect the temperature of Earth.

- Major effects of global warming include: (1) changes in climatic patterns and frequency and intensity of storms, (2) a rise in sea level, (3) change in glaciers, and (4) changes in the biosphere.

- Changes in climatic patterns and storms are worrisome because they may adversely affect agriculture and thus food supply. A rise in sea level is a potentially serious problem because so many people live in or near coastal areas and coastal erosion is a difficult problem. As sea level rises and more heat energy is fed into the atmosphere, cyclones and tropical disturbances will pose a greater hazard for people living in vulnerable areas.

- Many more glaciers in North America and Europe are retreating than are advancing as the world warms. At the same time the central ice cap on Antarctica is growing as more snow falls there in a warmer world.

- Changes in the biosphere include shifts in where specific plants and animals live, bleaching of coral, and spreading of diseases such as malaria to higher elevations.

- Adjustments to global warming include learning to live with the changes and attempting to mitigate warming through reduction of emissions of greenhouse gases. It seems likely that the adjustment chosen will be to learn to live with change. A danger of this path is that climatic change may be rapid, resulting in problems that will be difficult to address.

- Because a significant component of global warming appears to result from anthropogenic processes, it is prudent to reduce emissions of greenhouse gases. With control of CO_2 emissions, global warming may be limited to less than 2°C, which would cause far less harm than would a larger increase. Carbon dioxide emissions can be reduced through energy conservation, sequestriation of carbon in the biosphere and lithosphere, and the use of alternative energy sources.

- The science of global warming is well understood, and human-induced warming is occurring. There is still time to take appropriate action to slow or stop the warming, but we need to act soon.

Should the Precautionary Principle Be Applied to Global Warming?

Global warming is emerging as one of the most controversial environmental issues of the twenty-first century. Many scientific studies have concluded that global warming is occurring and that human activities have contributed to it. The Intergovernmental Panel on Climate Change has predicted that by 2100 the surface temperature of Earth will likely increase by about 1.5° C to 4.5° C. If global warming over the next century is less than about 2.0°C, then it is highly likely we will be able to successfully adapt to the change. Potential adverse effects increase dramatically above a warming of about 2°C and approach very serious at levels above 4°C. Many countries around the world, including those in the European Union, had advocated that we apply the precautionary principle to global warming. Recall that we introduced the precautionary principle in Chapter 1 in our discussion of environmental health and toxicology. The idea behind the principle is that when there is evidence that environmental damage is resulting from a particular practice or process, scientific proof is not required to take cost-effective measures to protect the environment. The analogy in everyday life is "better safe than sorry." Applying the precautionary principle is a proactive decision to act and develop environmental policy as compared to a reactive management strategy of not acting until a high degree of scientific certainty has identified the problem and damage to the environment.

With respect to global warming and the precautionary principle, two important questions are: Is there sufficient scientific evidence to suggest that warming will result in environmental damage? And are cost-effective solutions to global warming available? Countries including the United Kingdom answer both these questions in the affirmative and are seeking international commitments to reduce emissions of carbon dioxide by approximately 60% from 1990 levels by the year 2050. The United Kingdom is taking a leadership role, working within the United National Framework Convention on Climate Change. The British government has determined that delaying action on global warming is not a viable alternative, and it has committed to reducing emissions consistent with the UN framework. The United States, on the other hand, while declaring support for the objectives of the UN framework, has refused to ratify the Kyoto Accord, which would result in emission reductions. In fact, the U.S. government has formally stated that additional scientific study is needed to prove that global warming has resulted from human activities and that serious harm is likely. It has been the position of the U.S. government not to take remedial action until such proof is available. The United States, which produces about 15% of the carbon dioxide emissions in the world, has determined that taking measures to reduce emissions would simply be too costly, given the state of scientific uncertainty about global warming. The European Union has taken preliminary steps to set up an emissions trading market, expected to be operating by 2005, to reduce emissions of carbon dioxide by 2050 to the level recommended by the UN framework.

So who is right? The United States or the European Union?

Critical Thinking Questions

1. Review the evidence for human-induced global warming presented in this chapter and decide whether there is a compelling case, a good case, or a poor case that warming is in fact occurring and will lead to environmental damage.

2. Is a cost-effective solution to global warming available? To help answer this question, consider the cost estimates for a target concentration of atmospheric CO_2 of 550 ppm. This is the level at which the concentration of carbon dioxide is approximately twice the preindustrial level and is likely to be reached by the end of the twenty-first century. The estimated total cost up until the year 2100 is about 4 trillion dollars, so you will have to consider the cost per year relative to potential benefits from controlling emissions. To help address this, consider the British experience in which overall emissions have been reduced by about 12% for the period 1990–2000. During this same period the British economy grew by 30% and employment increased by nearly 5%. The intergovernmental Panel on Climate Change has determined that stabilizing atmospheric carbon dioxide at 550 ppm would lead to an average gross domestic product loss for the developed countries of 1% by the year 2050. The panel further suggested that these costs would be more than offset by reduction of risk from hazards associated with global warming, including flooding and heat waves. Developing new technologies to reduce carbon emissions would also put money into research and development and eventually produce new employment opportunities. At any rate, consider all of this and decide whether in fact there is a cost-effective solution to global warming.

3. Should the United States, as a leader in science and technology, stand with much the rest of the world and work to both better understand global warming and its potential consequences and to move now to find ways to control emissions consistent with the UN framework and the Kyoto Accord?

4. Do you think the rest of the world is looking to the United

Human Population

Much of this chapter centers on human-induced climatic change. Burning of fossil fuels and trees, among other processes, has resulted in increased emissions of carbon dioxide into the atmosphere, forcing climate change. Continued increases in the world's human population will place further demands on resources, which will result in continued increases in emissions of greenhouse gases. Thus, as stated many times in this book, control of increases in human population is needed to maximize the chances of solving many environmental problems—such as global warming.

Sustainability

Through our emissions of greenhouse gases, we are conducting global experiments, the results of which are difficult to predict. Because of this, achieving sustainability in the future will be more difficult. If we do not know what the consequences or magnitude of human-induced climatic change will be, then it is difficult to predict how we might achieve sustainable development for future generations. Some of the uncertainties concerning global climatic change will be resolved in the coming years, and we will be better prepared to develop plans that will work in addressing variables such as providing a sustainable food supply for the world's population.

Global Perspective

One of the major topics of this chapter is global warming. In particular, we are interested in the potential effects of human activities. Our research on global warming centers on developing ways to look at what Earth is doing as a system with respect to climate; this is the essence of having a global perspective.

Urban World

In this chapter, we explore the effects of urbanization on the local climate. As large urban areas continue to expand, we can expect that urban microclimate effects will become more apparent. As a result, it is important in urban planning to incorporate the development of green zones and other such areas to partially offset potentially adverse urban microclimate effects.

People and Nature

Our ancestors adapted to natural climate change during the past million years. During that period, Earth experienced glacial and interglacial periods colder and warmer than today. However, only in the past century have human activities begun to modify climate. These human-induced changes to climate are significantly different in their scope from past changes, which resulted mostly from physical processes.

Science and Values

It is becoming increasingly clear that people in industrialized countries, as well as other areas of the globe, are placing more value on the quality of the environment. Numerous research programs have been established to try to understand what factors affect global climate change and, in particular, what impact human activities are having. It is expected that people will continue to place a high priority on understanding the global climate system and how it is likely to change in the future. The tough questions will be the choices we make. Will we choose to significantly reduce emissions of carbon dioxide by greatly reducing our use of the fossil fuels for the good of the environment, or will we continue our economically successful use of fossil fuels?

Key Terms

atmosphere **493**
climate **496**
climatic change **496**
Earth system science **499**

electromagnetic spectrum
502
forcing **508**
global circulation models

(GCMs) **501**
global dimming **510**
global warming **501**
greenhouse effect **503**

greenhouse gases **503**
polar amplification **512**

Study Questions

1. What is the composition of Earth's atmosphere and how has life affected the atmosphere during the past several billion years?

2. How do midlatitude cities differ from surrounding rural areas in climate?

3. What is the greenhouse effect? What is its importance to global climate?

4. What is an anthropogenic greenhouse gas? Discuss the various anthropogenic greenhouse gases in terms of their potential to cause global warming.

5. What are some of the major negative-feedback cycles and positive-feedback cycles that might increase or decrease global warming?

6. In terms of the effects of global warming, do you think that a change in climate patterns and storm frequency and intensity is likely to be more serious than a global rise in sea level? Illustrate your answer with specific problems and areas where the problems are likely to occur.

7. How would you refute or defend the statement that the best adjustment to global warming is to do little or nothing and learn to live with change?

Further Reading

Anthes, A. R. 1992. *Meteorology*, 6th ed. New York: Macmillan. A short text providing a good overview of basic meteorology and atmospheric processes.

Fay, J. A., and D. Golumb. 2002. *Energy and the Environment*. New York: Oxford University Press. See Chapter 10 on global warming.

Goodess, C. M., J. P. Palutikof, and T. D. Davies. 1992. *The Nature and Causes of Climate Change: Assessing the Long-Term Future*. Lewis. A good text discussing short- and long-term climate changes and anthropogenic versus natural forcing in global warming.

IPCC. 2001. *The Intergovernmental Panel on Climate Change Scientific Assessment*. New York: Oxford University Press. A detailed scientific review and assessment of global warming.

Mohnen, V. A., W. Goldstein, and C. W. Wang. 1991. "The Conflict over Global Warming," *Global and Environmental Change* 1(2):109–123. A paper providing detailed information on the global warming controversy that existed at the time.

Organisation for Economic Cooperation and Development, International Energy Agency. 1994. *The Economics of Climate Change*. Proceedings of an OECD/IEA Conference, OECD, IEA. These proceedings discuss the economics of climate change, differences of opinion on the matter, methods to link economic studies and climate change policy, and directions of international policy to deal with climate change.

Titus, J. G., and V. K. Narayanan. 1995. *The Probability of Sea Level Rise*. Washington, D.C.: U.S. Environmental Protection Agency. A close look at future sea-level rise due to global warming using mathematical models to predict changes.

CHAPTER 24

Air Pollution

A person guides a London bus at midday through thick fog on December 8, 1952.

Learning Objectives

The atmosphere has always been a sink—a place for deposition and storage—for gaseous and particulate wastes. When the amount of waste entering the atmosphere in an area exceeds the ability of the atmosphere to disperse or break down the pollutants, problems result. After reading this chapter, you should understand:

- Why human activities that pollute the air, combined with meteorological conditions, may exceed the natural abilities of the atmosphere to remove wastes.

- What the major categories and sources of air pollutants are.

- Why air pollution problems are different in different regions.

- What acid rain is, how it is produced, what its environmental impacts are, and how they might be minimized.

- What methods are useful in the collection, capture, and retention of pollutants before they enter the atmosphere.

- What air quality standards are, and why they are important.

- Why determining the economics of air pollution is controversial and difficult.

London Smog and Indonesian Fires

In December 1952, air in London became stagnant, and cloud cover blocked incoming solar radiation. The temperature dropped rapidly. At noontime it was about −1°C (30°F), and humidity climbed to 80%. Thick fog developed; cold and dampness increased demand for home heating. Because the primary fuel used in homes was coal, emissions of ash, sulfur oxides, and soot increased quickly. Stagnant air was filled with pollutants from home-heating fuels and automobile exhaust. At the height of the crisis, visibility was so reduced that automobiles used headlights at midday. Between December 4 and 10, about 4,000 Londoners died from the pollution. The siege of smog ended when the weather changed and the air pollution dispersed. Environment, not human activities, finally solved this problem. What went wrong?

During the London smog crisis, stagnant weather conditions, combined with emissions from homes burning coal and from cars burning gasoline, exceeded the atmosphere's ability to remove or transform pollutants. As a result, sulfur dioxide remained in the air and fog became acidic, adversely affecting people and other organisms. Human health effects were especially destructive because small acid droplets became fixed on larger particulates, which were easily drawn into the lungs.

The 1952 London smog crisis was a landmark event. Finally, the natural abilities of the atmosphere to serve as a sink for removal of wastes had been exceeded by human activities. The crisis was due, in part, to a reinforcing feedback situation. Burning fossil fuels added particulates to the air, increasing the formation of fog and decreasing visibility and light transmission. The dense, smoggy layer increased the dampness and cold and accelerated use of home-heating fuels. As the weather and pollution worsened, more people burned fuel, which further worsened weather and pollution. London had been known for its fogs long before 1952; what had been little known was the role of coal burning in exacerbating the fog conditions. The good news is that since 1952, London fogs associated with air pollution have been greatly reduced because coal has been replaced by cleaner natural gas as the primary home-heating fuel.

Forty-five years after the London smog crisis and following the most severe drought in 50 years, a serious air pollution event in Indonesia dwarfed the London event. As a result of a strong El Niño (see Chapter 23), huge fires ravaged Indonesia for months (Figure 24.1), producing a thick, toxic haze of smoke. What had gone wrong in this instance?

Slash-and-burn practices have been part of farming in tropical rain forests for centuries. Each family burns a few hectares of rain forest, plants it, and after harvest moves on to a new area. Fire is the preferred method of land clearing because clearing land with bulldozers is too expensive. It is generally thought that slash-and-burn farming is not particularly disturbing to the ecology of rain forests, as long as it is not too widespread and fires are carefully controlled.[1] More recently, large plantations and logging companies have been responsible for burning that they were unable to control. Indonesian countries have joined forces to adopt a zero-burn policy to control environmental damage from fires. The ban on burning has had mixed success; fires continue to occur annually in the dry season (some natural and some caused by human activity).

A combination of events in 1997 related to El Niño and out-of-control burns resulted in what some consider one of the world's greatest environmental disasters. Monsoon rains were late that year, and people took advantage of dry weather to burn more forest. Although much of the blame was placed on small farmers, agriculture on an industrial scale was also responsible for clear-cut logging and burning. During the 1997 dry season, at least 20,000 hectares of land burned. Vast amounts of smoke and particulate matter entered the atmosphere, causing a severe pollution problem. At one point, people were so desperate for relief from the dense, blinding haze that they attempted to use surgical masks as filters, but this provided little protection. Some 20 million Indonesians were treated for a variety of illnesses directly caused or aggravated by the fires. Smoke was so dense that a passenger airline crashed in Sumatra, killing 234 people. The Air Quality Index (AQI)—a measure of air quality—was at about 800 because of the fire. An AQI of 500 is considered very hazardous, and it is recommended that all people remain indoors. (The AQI is discussed later in this chapter.)

Near the end of summer, haze began to drift across the South China Sea to Malaysia and Singapore. Many of the fires were extinguished with the arrival of monsoons, but when the rains stopped in February 1998, more fires followed.

There is worry that annual cycles of fire have begun that will have cumulative adverse environmental impacts.

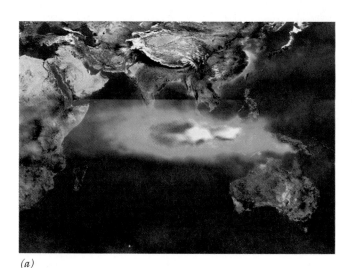

(a)

(b)

Figure 24.1 ■ (*a*) Image of air pollution (smoke from forest fires) over Indonesia and Indian Ocean on October 22, 1997. White is smoke near fires. Red, yellow, and green represent decreasing amounts of lower atmosphere pollution being transported westward. (*b*) Indonesian family members in Sumatra use air filters to help protect them from hazardous smoke on October 6, 1997.

Fires in 2002 caused significant air pollution that reduced visibility in some places to about 10 m, prompting schools to be closed. The fires were caused by annual drought, lightning, and timber harvesting; they threatened people and a major wildlife reserve for endangered orangutans.

The London smog of 1952 and the Indonesian fires of 1997–1998 are examples of serious, human-induced air pollution. This chapter discusses the major air pollutants, urban air, acid rain, and control of air pollution.

24.1 A Brief History of Air Pollution

As the fastest moving fluid medium in the environment, the atmosphere has always been one of the most convenient places to dispose of unwanted materials. Ever since we first used fire, the atmosphere has been a sink for waste disposal.

People have long recognized the existence of atmospheric pollutants, both natural and human-induced. Leonardo da Vinci wrote in 1550 that a blue haze formed from materials emitted into the atmosphere from trees. He had observed a natural photochemical smog resulting from hydrocarbons given off by living trees, the cause of which is still not completely understood. This haze gave rise to the name Smoky Mountains for the range in the southeastern United States. The phenomenon of acid rain was first described in the seventeenth century, and by the eighteenth century it was known that smog and acid rain damaged plants in London. Beginning with the Industrial Revolution in the eighteenth century, air pollution became more noticeable. The word *smog* was introduced by a physician at a public health conference in 1905 to denote poor air quality resulting from a mixture of smoke and fog.

Figure 24.2 ■ This steel mill in Beijing, China, is a major source of air pollution.

A major event in Donora, Pennsylvania, in 1948, was responsible for increasing research on air pollution in the United States. The Donora event remains the worst industrial air pollution incident in U.S. history, causing 20 deaths and 5,000 illnesses. What was called the "Donora fog" involved pollutants from the Donora Zinc Works metal-smelting plant and other sources. Pollutants, including sulfur dioxide, carbon monoxide, and heavy metals, were trapped by weather conditions in a narrow valley. The event lasted about three days until pollutants were washed out and dispersed by rainstorms.

The Donora event was followed in 1952 by the London smog crisis described in the opening case study. Following the Donora and London events, regulations to control air quality began. Today, in the United States and other countries, legislation to reduce emission of air pollutants has been successful, but more needs to be done. Chronic exposure to high levels of air pollutants continues to contribute to illnesses that kill people around the world.

What are the chances that another killing smog will occur somewhere in the world? Unfortunately, the chances are all too good, given the tremendous amount of air pollution in some large cities. For example, Beijing might be a candidate; the city uses an immense amount of coal, and coughing is so pervasive among its residents that they often refer to it as the "Beijing cough." Another likely candidate is Mexico City, which has one of the worst air pollution problems anywhere in the world today.

24.2 Stationary and Mobile Sources of Air Pollution

What are the sources of air pollution? The two major categories are stationary sources and mobile sources. **Stationary sources** are those that have a relatively fixed location. These include point sources, fugitive sources, and area sources.

Figure 24.3 ■ Burning sugarcane fields, Maui, Hawaii—an example of a fugitive source of air pollution.

- *Point sources*, as discussed in Chapter 15, emit pollutants from one or more controllable sites, such as power-plant smokestacks (Figure 24.2).
- *Fugitive sources* generate air pollutants from open areas exposed to wind processes. Examples include burning for agricultural purposes (Figure 24.3), as well as dirt roads, construction sites, farmlands, storage piles, surface mines, and other exposed areas from which particulates may be removed by wind.
- *Area sources*, also discussed in Chapter 15, are well-defined areas within which are several sources of air pollutants—for example, small urban communities, areas of intense industrialization within urban complexes, and agricultural areas sprayed with herbicides and pesticides.

Mobile sources of air pollutants move from place to place while emitting pollutants. These include automobiles, trucks, buses, aircraft, ships, and trains (Figure 24.4).[2]

24.3 General Effects of Air Pollution

Air pollution affects many aspects of our environment: its visual qualities, vegetation, animals, soils, water quality, natural and artificial structures, and human health. Air pollutants affect visual resources by discoloring the atmosphere and by reducing visual range and atmospheric clarity so that the visual contrast of distant objects is decreased. We cannot see as far in polluted air, and what we do see has less color contrast.

Figure 24.4 ■ Exhaust from older diesel buses such as this one in Thailand are a major source of air pollution.

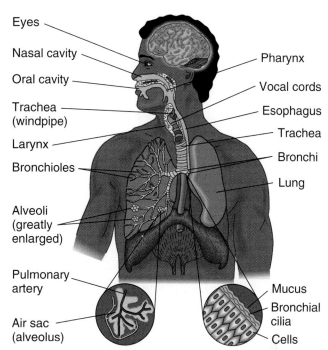

Figure 24.5 ■ Idealized diagram showing some of the parts of the human body (brain, cardiovascular system, and pulmonary system) that can be damaged by common air pollutants. The most severe health risks from normal exposures are related to particulates. Other substances of concern include carbon monoxide, photochemical oxidants, sulfur dioxide, and nitrogen oxides. Toxic chemicals and tobacco smoke also can cause chronic or acute health problems.

These effects were once limited to cities, but they now extend to some wide-open spaces of the United States. For example, near the area where the borders of New Mexico, Arizona, Colorado, and Utah meet, emissions from the Four Corners fossil fuel–burning power plant are altering the visibility in a region where in the past visibility was normally 80 km (50 mi) from a mountaintop on a clear day.[2]

The effects of air pollution on vegetation are numerous. They include damage to leaf tissue, needles, and fruit; reduction in growth rates or suppression of growth; increased susceptibility to a variety of diseases, pests, and adverse weather; and disruption of reproductive processes.[2]

Air pollution is a significant factor in the human death rate for many large cities. For example, it has been estimated that in Athens, Greece, the number of deaths is several times higher on days when the air is heavily polluted; in Hungary, where air pollution has been a serious problem in recent years, it may contribute to as many as 1 in 17 deaths. The United States is certainly not immune to health problems related to air pollution. The most polluted air in the country is found in the Los Angeles urban area, where millions of people are exposed to unhealthy air. It is estimated that as many as 150 million people live in areas of the United States where exposure to air pollution contributes to lung disease, which causes more than 300,000 deaths per year. Air pollution in the United States is directly responsible for annual health costs of about $50 billion.[3] In China, whose large cities have serious air pollution problems, mostly from burning coal, the health cost is now about $50 billion per year and may increase to about $100 billion per year by 2020.

Air pollutants can affect human health in several ways (Figure 24.5). The effects on an individual depend on the dose or concentration (see the discussion of dose–response in Chapter 15) and other factors, including individual susceptibility. Some of the primary effects of air pollutants include toxic poisoning, cancer, birth defects, eye irritation, and irritation of the respiratory system; increased susceptibility to

viral infections, causing pneumonia and bronchitis; increased susceptibility to heart disease; and aggravation of chronic diseases, such as asthma and emphysema. People suffering from respiratory diseases are the most likely to be affected by air pollutants. Healthy people tend to acclimate to pollutants in a relatively short period of time. However, this is a physiological tolerance; as explained in Chapter 15, it does not mean that the pollutants are doing no harm.

Many air pollutants have *synergistic effects* (in which the combined effects are greater than the sum of the separate effects). For example, sulfate and nitrate may attach to small particles in the air, facilitating their inhalation deep into lung tissue. There, they may do greater damage to the lungs than a combination of the two pollutants would be expected to do based on their separate effects. This phenomenon has obvious health consequences; consider joggers breathing deeply of particulates as they run along the streets of a city.

The effects of air pollutants on vertebrate animals in general include impairment of the respiratory system; damage to eyes, teeth, and bones; increased susceptibility to disease, parasites, and other stress-related environmental hazards; decreased availability of food sources (such as vegetation affected by air pollutants); and reduced ability for successful reproduction.[2]

Air pollution can also degrade soil and water resources when pollutants from the air are deposited. Soils and water may become toxic from the deposition of various pollutants. Soils may also be leached of nutrients by pollutants that form

Pollutant	Description	Effects on People [a,b]	Effects on Plants [a,c]	Effects on Materials [a,d]
Ozone (O₃)	Colorless gas with slightly sweet odor	Strong irritant, aggravates asthma; causes injury to cells in respiratory system, decreased elasticity of lung tissue, coughing, chest discomfort; eye irritation	Flecking, stippling, spotting, and/or bleaching of plant tissue (leaves, stems, etc.); oldest leaves are most sensitive; tips of needles of conifers become brown and die; reduction of yields and damage to crops including lettuce, grapes, and corn	Cracks rubber; reduces durability and appearance of paint, causes fabric dyes to fade
Sulfur dioxide (SO₂)	Colorless, odorless gas	Increase in chronic respiratory disease; shortness of breath; narrowing of airways for people with asthma	Bleaching of leaves; decay and death of tissue; younger leaves aremore sensitive than older ones; sensitive crops and trees include alfalfa, barley, cotton, spinach, beets, white pine, white birch, andtrembling aspen; if oxidized to sulfuric acid, causes damage associated with acid rain	If oxidized to sulfuric acid, damages buildings and monuments, corrodes metal; causes paper to become brittle; turns leather to red-brown dust; SO₂ fades dyes of fabrics, damages paint
Nitrogen oxides (NOₓ)[e]	Most are colorless and odorless; NO₂ with particles is reddish-brown	A mostly nonirritating gas; may aggravate respiratory infections and symptoms (some throat, cough, nasal congestion, fever) and increase risk of chest cold, bronchitis, and pneumonia in children	No perceptible effects on many plants, but may suppress plant growth for some, and may be beneficial at low concentrations; if oxidized to nitric acid, causes damage associated with acid rain	Causes fading of textile dyes; if oxidized to nitric acid, may damage buildings and monuments
Carbon monoxide (CO)	Colorless, odorless gas	Reduces the ability of the circulatory system to transport oxygen; causes headache, fatigue, nausea; impairs performance of tasks that require concentration; reduces endurance; may be lethal, causing asphyxiation	None perceptible	None perceptible
Particulate matter (PM 2.5, PM 10)	Very small particles; PM 2.5 less than 2.5 μm, PM 10 less than 10 μm in diameter[f]	Increased chronic and acute respiratory diseases; depending on chemical composition of particulates, may irritate tissue of throat, nose, lungs, and eyes	Depending on chemical composition of particles, may damage trees and crops; dry deposition of SO₂, when oxidized, is a form of acid rain	Contributes to and may accelerate corrosion of metal; may contaminate electrical contacts; damages paint appearance and durability; fades textile dyes
Lead (Pb)	Heavy metal	Children most at risk; brain damage, behavior problems, nerve disorders, digestive problems	May be toxic in soils; changes metabolism of plants	

[a] Effects depend on dose (concentration of pollutant and time of exposure) and susceptibility of people, plants, and materials to a particular pollutant. For example, older people, children, and those with chronic lung diseases are more susceptible to O₃, SO₂, and NOₓ.

[b] Annual U.S. losses exceed $50 billion.

[c] Annual U.S. losses to crops are $1 billion to $5 billion.

[d] Annual U.S. losses exceed $5 billion.

[e] In NOₓ, x refers to the number of oxygen atoms in the gas molecules, as in NO (nitric oxide) and NO₂ (nitrogen dioxide).

[f] Visible as soot, smoke, dust.

Sources: Modified from U.S. Environmental Protection Agency; R. W. Bunbel, D. L. Fox, D. B. Turner, and A. C. Stern. *Fundamentals of Air Pollution*, 3rd ed. (San Diego: Academic Press, 1994); and T. Godish. *Air Quality*, 3rd ed. (Boca Raton, Fla.: Lewis Publishers, 1997).

acids. The effects of air pollution on human-made structures include discoloration, erosion, and decomposition of building materials. These effects are described when we explore the topic of acid rain later in this chapter.

24.4 Air Pollutants

There are nearly 200 air pollutants recognized and assessed by the U.S. Environmental Protection Agency and listed in the Clean Air Act. Six of the most common are called **criteria pollutants** and are responsible for most of our air pollution problems. Table 24.1 lists these criteria pollutants and gives a short description of characteristics and effects. These six, along with other major air pollutants, are discussed following a brief introduction to the definition of primary and secondary pollutants and natural and human emissions. Most other air pollutants that cause problems are called **air toxics**. These are further divided into those that cause cancer or other serious health problems.

Primary and Secondary Pollutants, Natural and Human

The major air pollutants occur either in gaseous forms or as particulate matter (PM). Particulate matter pollutants are very small particles of solid or liquid substances less than 10 μm in diameter and may be organic or inorganic. The gaseous pollutants include sulfur dioxide (SO_2), nitrogen oxides (NO_x), carbon monoxide (CO), ozone (O_3); and volatile organic compounds (VOC) such as hydrocarbons, (compounds containing only carbon and hydrogen that include petroleum products), hydrogen sulfide (H_2S), and hydrogen fluoride (HF).

Air pollutants can be classified as primary or secondary. **Primary pollutants** are those emitted directly into the air. They include particulates, sulfur dioxide, carbon monoxide, nitrogen oxides, and hydrocarbons. **Secondary pollu-**tants are produced through reactions between primary pollutants and normal atmospheric compounds. For example, ozone forms over urban areas through reactions of primary pollutants, sunlight, and natural atmospheric gases. Thus, ozone is a secondary pollutant.

The primary pollutants that account for nearly all air pollution problems are carbon monoxide (58%), volatile organic compounds (11%), nitrogen oxides (15%), sulfur oxides (13%), and particulates (3%). In the United States today, about 140 million metric tons of these materials enter the atmosphere from human-related processes. If these pollutants were uniformly distributed in the atmosphere, the concentration would be only a few parts per million by weight. Unfortunately, pollutants are not uniformly distributed but tend to be released, produced, and concentrated locally or regionally—for example, in large cities.

In addition to pollutants from human sources, our atmosphere contains many pollutants of natural origin. Examples of natural emissions of air pollutants include the following:

- Release of sulfur dioxide from volcanic eruptions. For example, volcanic activity on the island of Hawaii emits SO_2 and other pollutants, which react in the atmosphere to produce volcanic smog called "vog." The smog can present a health hazard to people and cause local acid rain.
- Release of hydrogen sulfide from geysers and hot springs and from biological decay in bogs and marshes.
- Release of ozone in the lower atmosphere as a result of unstable meteorological conditions, such as violent thunderstorms.
- Emission of a variety of particles from wildfires and windstorms.[2]
- Natural hydrocarbon seeps—for example, the La Brea Tar Pits in Los Angeles.

The data in Table 24.2 suggest that, with the exception of sulfur and nitrogen oxides, natural emissions of air

Table 24.2 • Major Natural and Human-Produced Components of Selected Air Pollutants

Air Pollutants	Emissions (% of total)		Major Sources of Human-Produced Components	Percent
	Natural	Human-Produced		
Particulates	85	15	Fugitive (mostly dust)	85
			Industrial processes	7
			Combustion of fuels (stationary sources)	8
Sulfur oxides (SO_x)	50	50	Combustion of fuels (stationary sources, mostly coal)	84
			Industrial processes	9
Carbon monoxide (CO)	91	9	Transportation (automobiles)	54
Nitrogen dioxide (NO_2)		Nearly all	Transportation (mostly automobiles)	37
			Combustion of fuels (stationary sources, mostly natural gas and coal)	38
Ozone (O_3)	A secondary pollutant derived from reactions with sunlight, NO_2, and oxygen (O_2)		Concentration present depends on reactions in lower atmosphere involving hydrocarbons and thus automobile exhaust	
Hydrocarbons (HC)	84	16	Transportation (automobiles)	27
			Industrial processes	7

pollutants exceed human-produced emissions. Nevertheless, it is the human component that is most abundant in urban areas and that leads to the most severe problems for human health.

Criteria Pollutants

There are six criteria pollutants: sulfur dioxide; nitrogen oxides; carbon monoxide; ozone; particulates; and lead.

Sulfur Dioxide: Sulfur dioxide (SO_2) is a colorless and odorless gas normally present at Earth's surface at low concentrations. A significant feature of SO_2 is that once it is emitted into the atmosphere, it can be converted through complex oxidation reactions into fine particulate sulfate (SO_4) and removed from the atmosphere by wet or dry deposition. The major anthropogenic source of sulfur dioxide is the burning of fossil fuels, mostly coal in power plants (see Table 24.2). Another major source comprises a variety of industrial processes, ranging from petroleum refining to the production of paper, cement, and aluminum.[2, 4]

Adverse effects associated with sulfur dioxide depend on the dose or concentration present (see Chapter 15) and include injury or death to animals and plants, as well as corrosion of paint and metals. Crops such as alfalfa, cotton, and barley are especially susceptible. Sulfur dioxide is capable of causing severe damage to the lungs of human and other animals, particularly in the sulfate form. It is also an important precursor to **acid rain**.[2, 4] (See A Closer Look 24.1.)

U.S. emission rates of SO_2 for 1970 and 2002 are shown in Figure 24.6. Emissions peaked at about 32 million tons in the early 1970s and since then have been reduced to about 16 million tons, a reduction of about 50%, as a result of effective emission controls.

Nitrogen Oxides: Although nitrogen oxides (NO_x) occur in many forms in the atmosphere, they are emitted largely in two forms: nitric oxide (NO) and nitrogen dioxide (NO_2); only these two forms are subject to emission regulations. The more important of the two is NO_2, a yellow-brown to reddish-brown gas. A major concern with nitrogen dioxide is that it may be converted by complex reactions in the atmosphere to an ion, NO_3^{2-}, within small water particulates, impairing visibility. Both NO and NO_2 are major contributors to the development of smog, and NO_2 is a major contributor to acid rain (see A Closer Look 24.1). Nearly all NO_2 is emitted from anthropogenic sources. The two main sources are automobiles and power plants that burn fossil fuels.[2]

The environmental effects of nitrogen oxides on humans are variable but include irritation of eyes, nose, throat, and lungs and increased susceptibility to viral infections, including influenza (which can cause bronchitis and pneumonia).[2] Nitrogen oxides may suppress plant growth. When the oxides are converted to their nitrate form in the atmosphere, they impair visibility. However, when nitrate is deposited on the soil, it can promote plant growth through nitrogen fertilization.

U.S. emission rates of NO_x for 1970 and 2002 are shown in Figure 24.6. Emissions are primarily from combustion of fuels in power plants and vehicles. They have been reduced by about 17% since 1970.

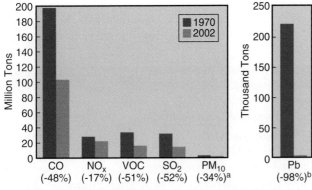

a Based on 1985 emission estimates. Emission estimates prior to 1985 are uncertain.

b Values for lead are based on 2001 data; 2002 data for lead are not yet available.

Figure 24.6 ■ U.S. emissions of six major air pollutants in 1970 compared to 2002 (carbon monoxide, nitrogen oxides, volatile organic compounds, sulfur dioxide, particulate matter, and lead). Notice that there have been significant reductions. [*Source:* After U.S. Environmental Protection Agency 2002 highlights, accessed March 24, 2004 at www.epa/gov.]

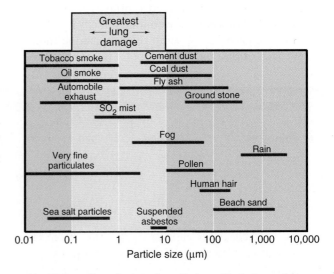

Figure 24.7 ■ Size of selected particulates. Shaded area shows size range that produces the greatest lung damage. [*Source:* Modified from Fig. 7–8, p. 244 in *Chemistry, Man and Environmental Change: An Integrated Approach*, by J. Calvin Giddings. Copyright © 1973 by J. Calvin Giddings. Reprinted by permission of HarperCollins Publishers, Inc.]

Carbon Monoxide: Carbon monoxide (CO) is a colorless, odorless gas that even at very low concentrations is extremely toxic to humans and other animals. The high toxicity results from a physiological effect—carbon monoxide and hemoglobin in blood have a strong natural attraction for one another. The hemoglobin in our blood will take up carbon monoxide nearly 250 times more rapidly than it will oxygen. Therefore, if there is any carbon monoxide in the vicinity, a person will take it in very readily, with potentially dire effects. Many people have been accidentally asphyxiated by carbon monoxide produced from incomplete combustion of fuels in campers, tents, and houses. The effects depend on the dose or concentration of exposure and range from dizziness and headaches to death. Carbon monoxide is particularly hazardous to people with known heart disease, anemia, or respiratory disease. In addition, it may cause birth defects, including mental retardation and impairment of growth of the fetus.[2] Finally, the effects of carbon monoxide tend to be worse at higher altitudes, where oxygen levels are naturally lower. Detectors (similar to smoke detectors) are now in common use to warn people if CO in a building becomes concentrated at a potentially harmful level.

Approximately 90% of the carbon monoxide in the atmosphere comes from natural sources. The other 10% comes mainly from fires, automobiles, and other sources of incomplete burning of organic compounds. Concentrations of carbon monoxide can build up and cause serious health effects in a localized area.

Emissions of CO in the United States for 1970 and 2002 are shown in Figure 24.6. Most emissions are through tailpipes of vehicles. Emissions peaked in the early 1970s at about 200 million metric tons. The rate in 2002 was about 100 million metric tons, a significant reduction of 50% considering the increase in vehicles. Reductions have resulted largely from cleaner-burning automobile engines.

Ozone and Other Photochemical Oxidants: Photochemical oxidants result from atmospheric interactions of nitrogen dioxide and sunlight. The most common photochemical oxidant is ozone (O_3), a colorless gas with a slightly sweet odor. In addition to ozone, a number of photochemical oxidants known as PANs (peroxyacyl nitrates) occur with photochemical smog (discussed later in the chapter).

Ozone is a form of oxygen in which three atoms of oxygen occur together rather than the normal two. Ozone is relatively unstable and releases its third oxygen atom readily, so it oxidizes or burns things more readily and at lower concentrations than does normal oxygen. Ozone is sometimes used to sterilize; for example, bubbling ozone gas through water is a method used to purify water. The ozone is toxic to and kills bacteria and other organisms in the water. When it is released into the air or produced in the air, ozone may injure living things.

Ozone is very active chemically, and it has a short average lifetime in the air. Because of the effect of sunlight on normal oxygen, ozone forms a natural layer high in the atmosphere (stratosphere). This ozone layer protects us from harmful ultraviolet radiation from the sun. Thus, although ozone is considered a pollutant in the lower atmosphere when concentrations are above the National Ambient Air Quality Standard threshold, it is beneficial in the stratosphere. The important topic of stratospheric ozone depletion is discussed in Chapter 26.

Ozone is a secondary pollutant produced on bright, sunny days in areas where there is much primary pollution. The major sources of the chemicals that produce ozone, as well as other oxidants, are automobiles, fossil fuel burning, and industrial processes that produce nitrogen dioxide. Because of the nature of its formation, ozone is difficult to regulate. It is the pollutant whose health standard is most frequently exceeded in urban areas of the United States.[5, 6] The adverse environmental effects of ozone and other oxidants, like those of other pollutants, depend in part on the dose or concentration of exposure and include damage to plants and animals as well as to materials such as rubber, paint, and textiles.

Effects of ozone on plants can be subtle. At very low concentrations, ozone can reduce growth rates while not producing any visible injury. At higher concentrations, ozone kills leaf tissue and, if pollutant levels remain high, whole plants. The death of white pine trees along highways in New England is believed to be due in part to ozone pollution. Ozone's effect on animals, including people, involves various kinds of damage, especially to the eyes and the respiratory system. Many millions of Americans are often exposed to ozone levels that damage cell walls in lungs and airways. This causes tissue to redden and swell, inducing cellular fluids to seep into the lungs. Eventually, the lungs decrease in elasticity and are more susceptible to bacterial infection; and scars and lesions may form in the airways. Even young, healthy people may not be able to breathe normally, and on especially polluted days, breathing may be shallow and painful.[6]

Particulate Matter (PM 10 and PM 2.5): Particulate matter (PM 10) is made up of particles less than 10 μm in diameter. The term is used for varying mixtures of particles suspended in the air we breathe. Particles are present everywhere, but high concentrations and/or specific types of particles have been found to present a serious danger to human health. Asbestos, for example, is particularly dangerous.

Farming adds considerable particulate matter to the atmosphere, as do windstorms in areas with little vegetation and volcanic eruptions. Nearly all industrial processes, as well as the burning of fossil fuels, release particulates into the atmosphere. Much particulate matter is easily visible as smoke, soot, or dust; other particulate matter is not easily visible. Particulates include materials such as airborne asbestos particles and small particles of heavy metals, such as arsenic, copper, lead, and zinc, which are usually emitted from industrial facilities such as smelters.

Of particular concern are very fine particle pollutants (PM 2.5) less than 2.5 μm in diameter (2.5 millionths of a meter; Figure 24.7). For comparison, the diameter of human hair is about 60 μm to 150 μm. Fine particles are easily inhaled into the lungs, where they can be absorbed into the bloodstream or remain embedded for a long period of time. Among the most significant fine particulate pollutants are sulfates and nitrates. These are mostly secondary pollutants produced in the atmosphere through chemical reactions between normal atmospheric constituents and sulfur dioxide and nitrogen oxides. These reactions are important in the formation of sulfuric and nitric acids in the atmosphere and are further discussed when we consider acid rain.[2]

When measured, particulate matter is often referred to as *total suspended particulates* (TSPs). Values for TSPs tend to be much higher in large cities in developing countries such as Mexico, China, and India than in developed countries such as Japan and the United States (Figure 24.8).

Particulates affect human health, ecosystems, and the biosphere. In the United States, particulate air pollution contributes to the death of 60,000 people annually.[7] Recent studies estimate that 2 to 9% of human mortality in cities is associated with particulate pollution; risk of mortality is about 15 to 25% higher in cities with the highest levels of fine particulate pollution.[8] As mentioned, particulates that enter lungs may lodge there, with chronic effects on respiration. Particulates are linked to both lung cancer and bronchitis. Particulate matter is especially hazardous to the elderly and to individuals with respiratory

problems, such as asthma. There is a direct relationship between particulate pollution and increased hospital admissions for respiratory distress.

Dust raised by road building and plowing not only makes breathing more difficult for animals (including humans) but also can be deposited on surfaces of green plants, where it may interfere with absorption of carbon dioxide and oxygen and release of water (transpiration). On a larger scale, particulates associated with large construction projects, such as housing developments, shopping centers, and industrial parks, may injure or kill plants and animals and damage surrounding areas, changing species composition, altering food chains, and thereby affecting ecosystems.

Modern industrial processes have greatly increased total suspended particulates in Earth's atmosphere. Particulates block sunlight and may cause changes in climate. Such changes have lasting effects on the biosphere. The process of gradual reduction in the solar energy that reaches the surface of Earth due to particulate air pollution is called **global dimming**. Global dimming cools the atmosphere and has lessened the global warming that has been predicted. The effects of global dimming are most apparent in the midlatitudes of the Northern Hemisphere, particularly over urban regions or where jet air traffic is more common. Jet plane exhaust emits particulate pollutants high in the atmosphere. This hypothesis was tested in 2001 when civil air traffic was shut down for two days following the September 11 attacks in New York. During those two days, the daily temperature range over the United States increased about 1° C. over the normal expected value.[9]

Anthropogenic emissions of PM 10 particles in the United States for 1970 and 2002 are shown on Figure 24.6. Emissions since 1970 have been reduced by about one-third (34%).

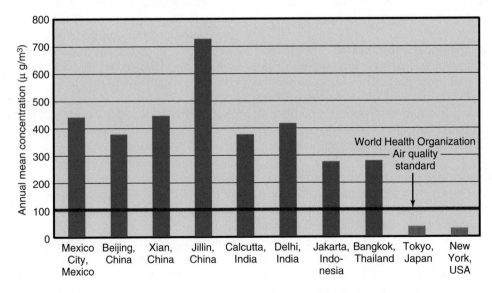

Figure 24.8 ■ Total suspended particulates (TSP) for several large cities in developing countries (blue) and developed countries (green). The value of 100 μg/m3 is the air quality standard set by the World Health Organization. [*Source:* Modified from R. T.

Watson, Intergovernment Panel on Climate Change, presentation at the Sixth Conference of Parties to the United Nations Framework Convention on Climate Change, November 13, 2000, Figure 20.]

Acid Rain

Acid rain is precipitation in which the pH is below 5.6. The pH of a solution is an expression of relative acidity and alkalinity. It is the negative logarithm of the concentration of the hydrogen ion (H^+). Many people are surprised to learn that all rainfall is slightly acidic; water reacts with atmospheric carbon dioxide to produce weak carbonic acid. Thus, pure rainfall has a pH of about 5.6, where 1 is highly acidic and 7 is neutral (see Figure 24.9). (Natural rainfall in tropical rain forests has been observed in some instances to have a pH of less than 5.6; this is probably related to acid precursors emitted by the trees.) Because the pH scale is logarithmic, a pH value of 3 is 10 times more acidic than a pH value of 4 and 100 times more acidic than a pH value of 5. Automobile battery acid has a pH value of 1.

Acid rain includes both wet (rain, snow, fog) and dry (particulate) acidic depositions. The depositions occur near and downwind of areas where the burning of fossil fuels creates major emissions of sulfur dioxide (SO_2) and nitrogen oxides (NO_x). Although these oxides are the primary contributors to acid rain, other acids are also involved. An example is hydrochloric acid, which is emitted from coal-fired power plants.

Acid rain has likely been a problem at least since the beginning of the Industrial Revolution. In recent decades, however, acid rain has gained more and more attention; today, it is a major global environmental problem affecting all industrial countries. In the United States, nearly all of the eastern states are affected, as well as West Coast urban centers such as Seattle, San Francisco, and Los Angeles. The problem is also of great concern in Canada, Germany, Scandinavia, and Great Britain. Developing countries that rely heavily on coal, such as China, are facing serious acid rain problems as well.

Causes of Acid Rain

As noted, sulfur dioxide (SO_2) and nitrogen oxides (NO_x) are the major contributors to acid rain. Amounts of these substances emitted into the environment in the United States are shown in Figure 24.6. Emissions of SO_2 peaked in the 1970s at about 30 million metric tons per year and had declined to about 16 million metric tons per year by 2002. Nitrogen oxides leveled off at about 23 million metric tons per year in the mid-1980s.

In the atmosphere, sulfur dioxide and nitrogen oxides are transformed by reactions with oxygen and water vapor to sulfuric and nitric acids. These acids may travel long distances with prevailing winds to be deposited as acid precipitation (Figure 24.10). As mentioned, such precipitation may take the form of rainfall, snow, or fog. Sulfate and nitrate particles may also be deposited directly on the surface of the land as dry deposition.

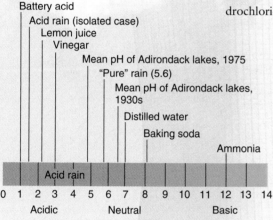

Figure 24.9 ■ The pH scale. [*Source:* Modified from U.S. Environmental Protection Agency, 1980.]

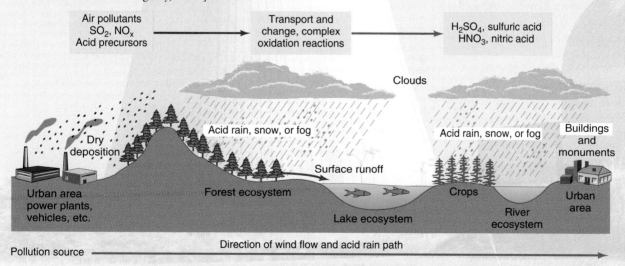

Figure 24.10 ■ Idealized diagram showing selected aspects of acid rain formation and paths.

These particles may later be activated by moisture to become sulfuric and nitric acids.

Sulfur dioxide is emitted primarily from stationary sources, such as power plants that burn fossil fuels, whereas nitrogen oxides are emitted from both stationary sources and transport-related sources, such as automobiles. Approximately 80% of sulfur dioxide and 65% of nitrogen oxides in the United States come from states east of the Mississippi River.

In some areas, stationary sources have attempted to reduce the local effects of emissions by constructing taller emission stacks. Taller stacks have reduced local concentrations of air pollutants but increased regional effects by spreading pollution more widely. Tall stacks increase average residence time of pollutants emitted into the atmosphere from 1–2 days to 10–14 days, because pollutants enter the atmosphere at a greater altitude, where mixing and transport by wind are more effective. Thus, this practice simply created more widespread problems. For example, problems associated with acid precipitation in Canada can be traced to emissions of sulfur dioxide and other pollutants in the Ohio Valley.

Analysis of the distances over which sulfur compounds can be transported before deposition suggests that approximately one-third of the total amount deposited over the eastern United States originates from sources farther away than 500 km (300 mi). Another one-third comes from sources between 200 and 500 km (about 125–300 mi) away, and the remainder comes from sources less than 200 km away.[10]

Sensitivity to Acid Rain

Geology and climatic patterns as well as types of vegetation and soil composition affect the potential impact of acid rain. Figure 24.11, showing areas of the United States and Canada sensitive to acid rain, is based on some of these factors. Sensitive areas are those in which bedrock or soil cannot buffer acid input. Materials (chemicals) that have the ability to neutralize acids are called *buffers*. Calcium carbonate ($CaCO_3$), the mineral calcite, which is present in many soils and rock (limestone), is an important natural

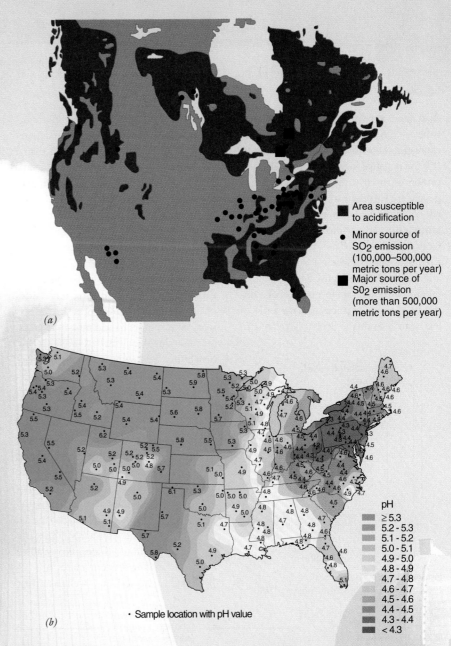

Figure 24.11 ■ (*a*) Areas in Canada and the United States that are sensitive to acid rain. [*Source:* "How Many More Lakes Have to Die?" *Canada Today* 12, no. 2 (1981).] (*b*) pH of precipitation over the United States in 2000. Notice the relationship between the numerous sources of SO_2 in the eastern United States in (*a*) and low (more acidic) pH values. [*Source:* National Atmospheric Deposition Program/National Trends Network, 2001.]

buffer to acid rain. Hydrogen in acid reacts with calcium carbonate, and the reaction neutralizes acid. Thus, areas less likely to suffer damage from acid rain are those in which bedrock contains limestone or other carbonate material or where soils contain calcium carbonate, which buffers the acid. In contrast, areas with abundant granitic rocks and areas in which soils have little buffering action are

sensitive to acid rain. Soils may lose their fertility when exposed to acid rain, either because nutrients are leached out by acid water moving through the soil or because the acid in the soil releases elements that are toxic to plants.

Forest Ecosystems

It has long been suspected that acid precipitation, whether snow, rain, fog, or dry deposition, adversely affects trees. Studies in Germany led scientists to cite acid rain and other air pollution as the cause of death for thousands of acres of evergreen trees in Bavaria. Similar studies in the Appalachian Mountains of Vermont (where many soils are naturally acidic) suggest that in some locations half the red spruce trees have died in recent years. Damage is attributed in part to acid rain and fog with pH levels of 4.1 and 3.1, respectively.

We have already seen that acid rain can cause loss of nutrients in the soil. This loss of nutrients results in weakening of trees, and they become more susceptible to disease, drought, and consumption by herbivores. As trees weaken and die, there is less habitat and food for birds and animals. Trees stripped of leaves and needles allow more light through to the forest floor, changing the temperature and water content of soil at the surface. This in turn affects what grows on the forest floor and what lives in the soil. Thus, the entire forest ecosystem is affected by acid rain. This is an example of the principle of environmental unity introduced in Chapter 3.

Lake Ecosystems

Records from Scandinavian lakes show an increase in acidity accompanied by a decrease in fish. The increased acidity has been traced to acid rain, the result of industrial processes in other countries, particularly Germany and Great Britain.

Acid rain affects lake ecosystems in two ways. First, it damages aquatic species (fish, amphibians, and crayfish) directly by disrupting their life processes in ways that limit growth or cause death. For example, crayfish produce fewer eggs in acidic water, and the eggs produced often grow into malformed larvae. Second, acid rain dissolves chemical elements necessary for life in the lake. Once in solution, the necessary elements leave the lake with water outflow. Thus, elements that once cycled in the lake are lost. Without these nutrients, algae do not grow, animals that feed on the algae have little to eat, and animals that feed on these animals also have less food.

To better study the effects of acidification on lakes, scientists in Canada added

sulfuric acid to a lake in northwest Ontario over a period of years and observed the effects. When the experiment started, the pH of the lake was 6.8. The following year, owing to addition of the acid, the pH dropped to 6.1. The initial drop in pH was not harmful to the lake; but as more and more acid was added, the pH dropped first to 5.8, then to 5.6, then to 5.4, and finally, five years after the project started, to 5.1. The problems started when the pH was lowered to 5.8. Some species disappeared, and others experienced reproductive failure. At a pH of 5.6, the death rate among lake trout embryos increased. When the pH was lowered to 5.4, lake trout reproduction failed.[11]

These experiments have proved valuable in pointing out what we can expect in thousands of other lakes that are now becoming increasingly acidified. The precise processes involved in the toxicity and damage to the lake are poorly understood. However, it is known that acid rain leaches metals, such as aluminum, lead, mercury, and calcium, from the soils and rocks in a drainage basin and discharges them into rivers and lakes. The elevated concentrations of aluminum are particularly damaging to fish, because the metal can clog the gills and cause suffocation. The heavy metals may pose health hazards to humans, because they may become concentrated in fish and then passed on to people, mammals, and birds when the fish are eaten. Drinking water taken from acidic lakes may also have high concentrations of toxic metals.

Not all lakes are as vulnerable to acidification as was the lake in the Ontario experiment. Acid is neutralized in waters with a high carbonate content (in the form of the ion HCO_3). Lakes on limestone or other rocks rich in calcium or magnesium carbonates can therefore readily buffer river and lake water against the addition of acids. Lakes with high concentrations of such elements are called hard-water lakes. Lakes on sand or igneous rocks, such as granite, tend to lack sufficient buffering to neutralize acids and are more susceptible to acidification.[12]

Thousands of rivers and lakes in the United States and Canada located in areas

sensitive to acid rain are currently in various stages of acidification. In Nova Scotia, at least a dozen rivers have water so acidic part of the year that they no longer support healthy populations of Atlantic salmon. In the northeastern United States, about 200 lakes in the Adirondacks are no longer able to support fish; thousands more are slowly losing the battle with acid rain.

Human Society

Acid rain damages not only forests and lakes but also many building materials, including steel, galvanized steel, paint, plastics, cement, masonry, and several types of rock, especially limestone, sandstone, and marble (Figure 24.12). Classical buildings on the Acropolis in Athens and in other cities show considerable decay (chemical weathering) that accelerated in the twentieth century as a result of air pollution. The problem has grown to such an extent that buildings require restoration and statues and other monuments must have protective coatings replaced quite frequently, resulting in costs of billions of dollars a year. Particularly important statues in Greece and other areas have been removed and placed in protective glass containers, with replicas standing in their former outdoor locations for tourists to view.[11]

In the United States, cities along the eastern seaboard are more susceptible to acid rain today because emissions of sulfur dioxide and nitrogen oxide are more abundant there. However, as noted, the problem is moving westward; acid precipitation has been recorded in California. Acid fog in Los Angeles may have a pH as low as 3—over 10 times as acidic as the average acid rain in the eastern United States. In contrast to acid rain, which may form relatively high in the atmosphere and travel long distances, acid fog forms near the ground. Water vapor mixes with pollutants and turns into an acid, which evidently condenses around very fine particles of smog; if the air is sufficiently humid, a fog may form. When the fog eventually dissipates, nearly pure drops of sulfuric acid may be left behind. Tiny particles containing the acid may be inhaled deeply into people's lungs—a considerable health hazard.

Figure 24.12 ■ Damage to a statue in Chicago resulting from an acid deposition (left) and the same statue following restoration (right).

Stone decay occurs about twice as rapidly in cities as it does in less urban areas. The damage comes mainly from acid rain and humidity in the atmosphere, as well as from corrosive groundwater.[13]

This implies that measuring rates of stone decay will tell us something about changes in the acidity of rain and groundwater in different regions and ages. It is now possible, where the ages of stone buildings and other structures are known, to determine if the acid rain problem has changed through time.

Control of Acid Rain

The cause of acid precipitation is known. It is the solution we are struggling with. One solution to lake acidification is rehabilitation by the periodic addition of lime, as has been done in New York State, Sweden, and Ontario. This solution is not satisfactory over a long period, however, because it is expensive and requires a continuing effort. The solution to the acid rain problem is to ensure that the production of acid-forming components in the atmosphere is minimized. The only long-term solution involves decreasing emissions of sulfur dioxide and nitrogen oxides. From an environmental point of view, the best strategy is increasing energy efficiency and conservation measures that result in burning less coal and using nonpolluting alternative energy sources. Another strategy is to utilize pollution abatement technology at power plants to lower emissions of air pollutants. Such technology is expensive and an additional cost in producing energy. Sulfer dioxide emissions in the United States have been reduced about 50% since 1970. This is a big improvement that will reduce acid rain.

Lead: Lead is an important constituent of automobile batteries and many other industrial products. When lead is added to gasolines, it helps protect engines and promotes more even fuel consumption. Lead in gasoline (which is still used in some countries) is emitted into the air with the exhaust. By this process, lead has been spread widely around the world and has reached high levels in soils and waters along roadways.

Once released, lead can be transported through the air as particulates to be taken up by plants through the soil or deposited directly on plant leaves. Thus, it enters terrestrial food chains. When lead is carried by streams and rivers, deposited in quiet waters, or transported to oceans or lakes, it is taken up by aquatic organisms and enters aquatic food chains.

Lead reaches Greenland as airborne particulates and via seawater and is stored in glacial ice. The concentration of lead in Greenland glaciers was essentially zero in A.D. 800 and reached measurable levels with the beginning of the Industrial Revolution in the mid–eighteenth century. The lead content of the glacial ice increased steadily from 1750 until about 1950, when the rate of lead accumulation began to increase rapidly. This sudden upsurge reflects rapid growth in the use of lead additives in gasoline. The accumulation of lead in the Greenland ice illustrates that our use of heavy metals in the twentieth century reached the point of affecting the entire biosphere.

Lead has now been removed from nearly all gasoline in the United States, Canada, and much of Europe. In the United States, lead emissions have been reduced by about 98% since the early 1970s (Figure 24.6). The reduction and eventual elimination of lead in gasoline is a good start in reducing levels of anthropogenic lead in the biosphere.

Air Toxics

Toxic air pollutants or air toxics are among those pollutants that are known or suspected to cause cancer or other serious health problems. Disease may be associated with both long-term and short-term exposure to these pollutants. The category of air toxics is used for air pollutants such as gases, metals, and organic chemicals that are emitted in relatively small volumes at a particular site. Similar to the toxins discussed in Chapter 15, the air toxics are known to cause respiratory, neurological, reproductive, or immune diseases. They are further catalogued in terms of whether or not they cause cancer. The degree to which a particular toxic air pollutant affects an individual's health is dependent on a number of factors, including duration and frequency of exposure, toxicity of the chemical, concentration of the pollutant the individual is exposed to, and method of exposure as well as an individual's general health.[14]

Examples of the more than 150 chemicals considered for analysis and identification as toxic air pollutants are hydrogen sulfide, hydrogen fluoride, chlorine gases, benzene, methanol, and ammonia. In 2006, the U.S. Environmental Protection Agency released an assessment of the national health risk from air toxics. The focus was on exposure to air toxins from breathing the pollutants and did not address other ways people are exposed.

The EPA assessment estimated that the average risk for cancer from exposure to air toxics is about 1 in 21,000. The most serious exposure to air toxics occurs in California and New York, with Oregon, Washington, D.C., and New Jersey making up the rest of the top five. States with the cleanest air include Montana, Wyoming, and South Dakota. The assessment concluded that benzene is the most significant air toxin posing a risk for cancer, accounting for 25% of the average individual cancer risk from all air toxics.[14]

The evaluation of exposure and risk from air toxics is a relatively new endeavor, and continued analysis is ongoing. Understanding air toxics as well as the more commonly known air pollutants increases our knowledge of risk from air pollution. New assessments are forthcoming based on more recent exposure to air toxics. These newer evaluations will better characterize potential public health risks, including cancer and other adverse health effects from the air we breath.

Standards have been set for more than 150 air toxics; when fully implemented, these standards and related regulations are expected to reduce annual emissions of air toxics from 1990 levels. It is projected that even as vehicle miles are expected to increase significantly by 2020, the emissions of gaseous air toxics (such as benzene) from vehicles on highways will decrease by about 80% from 1990 levels. We will now look at examples of several air toxics.

Hydrogen Sulfide: Hydrogen sulfide (H_2S) is a highly toxic corrosive gas easily identified by its rotten egg odor. Hydrogen sulfide is produced from natural sources such as geysers, swamps, and bogs and from human sources such as industrial plants that produce petroleum or that smelt metals. Potential effects of hydrogen sulfide include functional damage to plants and health problems ranging from toxicity to death for humans and other animals.[4]

Hydrogen Fluoride: Hydrogen fluoride (HF) is a gaseous pollutant released by some industrial activities such as production of aluminum, coal gasification, and burning of

coal in power plants. Hydrogen fluoride is extremely toxic. Even a small concentration (as low as 1 ppb) of HF may cause problems for plants and animals. HF is potentially dangerous to grazing animals because some forage plants can become toxic when exposed to this gas.[2]

Methyl Isocyanate: Some chemicals are so toxic that extreme care must be taken to ensure they do not enter the environment. This was demonstrated on December 3, 1984, when a toxic liquid from a pesticide plant leaked, vaporized, and formed a deadly cloud of gas that settled over a 64-km^2 area of Bhopal, India. The gas leak lasted less than an hour; yet more than 2,000 people were killed and more than 15,000 were injured. The colorless gas that resulted from the leak was methyl isocyanate, which causes severe irritation (burns on contact) to eyes, nose, throat, and lungs. Breathing the gas in concentrations of only a few parts per million (ppm) causes violent coughing, swelling of the lungs, bleeding, and death. Less exposure can cause a variety of problems, including loss of sight.

Methyl isocyanate is an ingredient of a common pesticide known in the United States as Sevin, as well as two other insecticides used in India. An industrial plant in West Virginia also makes the chemical. Small leaks not leading to major accidents occurred there both before and after the catastrophic accident in Bhopal.

Clearly, chemicals that can cause catastrophic injuries and death should not be stored close to large population centers. In addition, chemical plants need to have reliable accident-prevention equipment and personnel trained to control and prevent potential problems.

Volatile Organic Compounds: Volatile organic compounds (VOCs) include a variety of organic compounds used as solvents in industrial processes such as dry cleaning, degreasing, and graphic arts. Hydrocarbons—compounds composed of hydrogen and carbon—comprise one group of VOCs. Thousands of hydrocarbon compounds exist, including natural gas, or methane (CH_4); butane (C_4H_{10}); and propane (C_3H_8). Analysis of urban air has identified many hydrocarbons, some of which react with sunlight to produce photochemical smog. Potential adverse effects of hydrocarbons are numerous. Many are toxic to plants and animals, and some may be converted to harmful compounds through complex chemical changes that occur in the atmosphere.

On a global basis, only about 15% of hydrocarbon emissions (primary pollutants) are anthropogenic. In the United States, however, nearly half of hydrocarbons entering the atmosphere are emitted from anthropogenic sources. The largest human source of hydrocarbons in the United States is automobiles. Anthropogenic sources are particularly abundant in urban regions. However, in some southeastern U.S. cities, such as Atlanta, Georgia, natural emissions probably exceed those from automobiles and other human sources.[3]

Emission of VOCs for 1970 and 2002 are shown in Figure 24.6. Like sulfur dioxide and nitrogen oxide emissions, VOCs peaked in the early 1970s. As noted, a major source of hydrocarbons (VOC) is automobiles. Effective government-mandated emission controls for automobiles are thus responsible for the 50% lower emissions.

Benzene: Benzene is an additive in gasoline and is an important industrial solvent. Generally, it is produced when carbon-rich material such as oil and gasoline undergo incomplete combustion. It is also a component in cigarette smoke. Major environmental sources for benzene are both on- and off-road vehicles (automobiles, trucks, airplanes, trains, and farm machinery).[14]

Arcolein: Arcolein is a volatile hydrocarbon that is extremely irritating to the nose, eyes, and respiratory system in general. It is produced from manufacturing processes that involve combustion of petroleum fuels and is a component of cigarette smoke.[14]

24.5 Variability of Air Pollution

Pollution problems vary in different regions of the world. There is great variance even within the United States. For example, as we will see, in the Los Angeles basin and many U.S. cities, nitrogen oxides and hydrocarbons are particularly troublesome because they combine in the presence of sunlight to form photochemical smog. Most of the nitrogen oxides and hydrocarbons are emitted from automobiles, a collection of mobile sources. In other U.S. regions, such as in Ohio and the Great Lakes region, air quality problems also result from emissions of sulfur dioxide and particulates from industry and from coal-burning power plants, which are point sources.

Air pollution also varies with the time of year. For example, smog is usually a problem mostly in the summer months when there is a lot of sunlight; particulates are a problem in dry months when wildfires are likely and during months when the wind blows across the desert.

Las Vegas: Particulates

Pollution from particulates is a problem in arid regions where little vegetation is present and wind can easily pick up and transport fine dust. For example, the brown haze over Las Vegas, Nevada, is mostly due to naturally occurring particles (PM 10) from the desert environment. Las Vegas in the 1990s was the most rapidly growing urban area in the United States. Population in Clark County, which includes Las Vegas, increased from less than 300,000 in 1970 to over 1.5 million in 2005. Las Vegas also has some of the most polluted air in the southwestern United States (Figure 24.13). As mentioned, the main problem is the nearly 80,000 metric tons of PM 10 particles that enter the air in the Las Vegas region. About 60% of the dust comes from new construction sites, dirt roads, and vacant land. The remainder is natural windblown dust.

Figure 24.13 ■ Las Vegas haze resulting from particulate (PM 10) pollution. Sources of particulates include construction sites and dirt roads (60% of total particulates) and natural and other sources (40%).

Las Vegas also has a carbon monoxide pollution problem from vehicles; but it is the particulates that are causing concern, possibly leading to future Environmental Protection Agency sanctions and growth restrictions.

Haze from Afar

Air quality concerns are not restricted to urban areas. For example, the North Slope of Alaska is a vast strip of land approximately 200 km (125 mi) wide that is considered by many to be one of the last unspoiled wilderness areas left on Earth. It seems logical to assume that air in the Arctic environments of Alaska would be pristine in quality, except perhaps near areas where petroleum is being vigorously developed. However, ongoing studies suggest that the North Slope has an air pollution problem that originates from sources in Eastern Europe and Eurasia.

It is suspected that pollutants from the burning of fossil fuels in Eurasia are transported via the jet stream, moving at speeds that may exceed 400 km/hr (250 mi/hr), northeast from Eurasia over the North Pole and eventually to the North Slope of Alaska. There, the air mass slows, stagnates, and produces what is known as the Arctic haze. Concentrations of air pollutants, which include oxides of sulfur and nitrogen, are high enough that the air quality is being compared with that of some eastern U.S. cities, such as Boston. Air quality problems in remote areas such as Alaska have significance as we try to understand air pollution at the global level.[15]

Another global event occurred in the spring of 2001, when a white haze consisting of dust from Mongolia and industrial particulate pollutants arrived in North America. The haze affected one-fourth of the United States and could be seen from Canada to Mexico. The particulates were close enough to the ground to cause respiratory problems for people. In the United States, pollution levels from the haze alone were as high as two-thirds of federal health limits. The haze demonstrates what was formerly believed—that pollution from Asia is carried by winds across the Pacific Ocean.

24.6 Urban Air Pollution

Wherever many sources emit air pollutants over a wide area (whether automobile emissions in Los Angeles or smoke from wood-burning stoves in Vermont), air pollution can develop. Whether air pollution does develop depends on topography and meteorological conditions; it is these factors that determine the rate at which pollutants are transported away from their sources and converted to harmless compounds in the air. When the rate of production exceeds the rate of degradation and of transport, dangerous conditions can develop, as illustrated in the case study that opened this chapter.

Influences of Meteorology and Topography

Meteorological conditions can determine whether air pollution is a nuisance or a major health problem. The primary adverse effects of air pollution are damage to green plants and aggravation of chronic illnesses in people; most of these effects are due to relatively low-level concentrations of pollutants over a long period of time. Periods of pollution generally do not directly cause large numbers of deaths. However, as with the London and Pennsylvania cases described earlier, serious pollution events (disasters) can develop over a period of days and lead to increases in illnesses and deaths.

In the lower atmosphere, restricted circulation associated with inversion layers may lead to pollution events. An **atmospheric inversion** occurs when warmer air is found above cooler air, and it poses a particular problem when there is a stagnant air mass. Figure 24.14 shows two types of atmospheric inversion that may contribute to air pollution problems. In the upper diagram, which is somewhat analogous to the situation in the Los Angeles area, descending warm air forms a semipermanent inversion layer. Because the mountains act as a barrier to the pollution, polluted air moving in response to the sea breeze and other processes tends to move up canyons, where it is trapped. The air pollution that develops occurs primarily during the summer and fall.

The lower part of Figure 24.14 shows a valley with relatively cool air overlain by warm air. This type of inversion can occur when cloud cover associated with a stagnant air mass develops over an urban area. Incoming solar radiation is blocked by the clouds, which reflect and absorb some of the solar energy and are warmed. On the ground, or near Earth's surface, the air cools. If there is moisture in the air (humidity), then as the air cools, the dew point (the temperature at which water vapor condenses) is reached, and a fog may form. Because the air is cold, people burn more fuel to heat their homes and factories, and more pollutants are delivered into the atmosphere. As long as the stagnant conditions exist, the pollutants will build up. It was this mechanism that caused the deadly 1952 London smog.

Cities situated in a valley or topographic bowl surrounded by mountains are more susceptible to smog problems than are cities in open plains. Surrounding mountains and the occurrence of temperature inversions prevent the pollutants

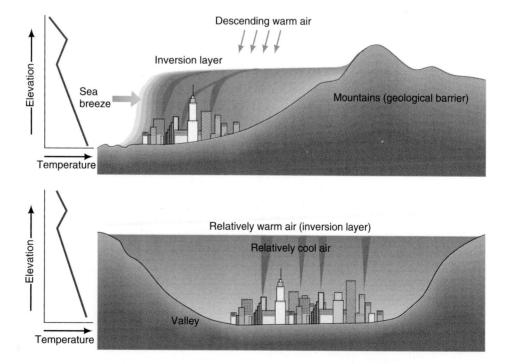

Figure 24.14 ■ Two causes for the development of atmospheric inversion, which may aggravate air pollution problems.

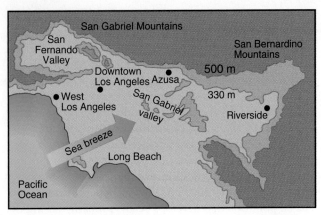

Figure 24.15 ■ Part of Southern California showing the Los Angeles basin (south coast air basin). [*Source*: Modified from S. J. Williamson, Fundamentals of Air Pollution, © 1973, by Addison-Wesley, Reading, Mass.]

from being transported by winds and weather systems. The production of air pollution is particularly well documented for Los Angeles, which has mountains surrounding part of the urban area and lies within a region where the air lingers, allowing pollutants to build up (Figure 24.15).

Potential for Urban Air Pollution

We have seen that topographical and meteorological conditions are important in the development of air pollution. More specifically, the potential for air pollution in urban areas is determined by the following factors:

1. The rate of emission of pollutants per unit area.
2. The downwind distance that a mass of air moves through an urban area.
3. The average speed of the wind.
4. The elevation to which potential pollutants can be thoroughly mixed by naturally moving air in the lower atmosphere (Figure 24.16).[16]

Figure 24.16 ■ The higher the wind velocity and the thicker the mixing layer (shown here as H), the less the air pollution. The greater the emission rate and the longer the downwind length of the city, the greater the air pollution. The chimney effect allows polluted air to move over a mountain and down into an adjacent valley.

The concentration of pollutants in the air is directly proportional to the first two factors. That is, as either the emission rate or downwind travel distance increases, so will the concentration of pollutants in the air. A good example is provided by the Los Angeles basin (see Figure 24.17). If there is a wind from the ocean, as is generally the case, coastal areas such as West Los Angeles will experience much less air pollution than inland areas such as Riverside.

Assuming a constant rate of emission of air pollutants, the air mass will collect more and more pollutants as it moves through the urban area. The inversion layer acts as a lid for the pollutants; however, near a geological barrier, such as a mountain, there may be a chimney effect, in which the pollutants spill over the top of the mountain (see Figures 24.15 and 24.16). This effect has been noticed in the Los Angeles basin, where pollutants may climb several thousand meters, damaging mountain pine trees and other vegetation and spoiling the air of mountain valleys.

City air pollution decreases with increases in the third and fourth factors, which are meteorological: the wind velocity and the height of mixing. The stronger the wind and the higher the mixing layer, the lower the pollution.

Smog

Smog, as noted earlier, is a general term first used in 1905 for a mixture of smoke and fog that produced unhealthy urban air. It is the most recognized term for urban air pollution. There are two major types of smog: **photochemical smog**, which is sometimes called L.A.-type smog or brown air; and **sulfurous smog**, which is sometimes referred to as London-type smog, gray air, or industrial smog.

Solar radiation is particularly important in the formation of photochemical smog (Figure 24.18). The reactions that occur in the development of photochemical smog are complex and involve both nitrogen oxides (NO_x) and organic compounds (hydrocarbons).

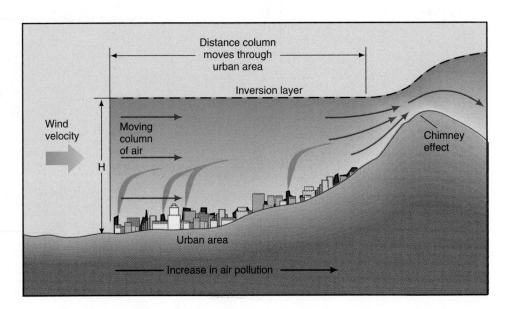

(a)

(b)

Figure 24.17 ■ The city of Los Angeles, California, on (*a*) a clear day and (*b*) a smoggy day.

The development of photochemical smog is directly related to automobile use. Figure 24.19 shows a characteristic pattern in terms of how the nitrogen oxides, hydrocarbons, and oxidants (mostly ozone) vary through a typically smoggy day in southern California. Early in the morning, when commuter traffic begins to build up, the concentrations of nitrogen oxide (NO) and hydrocarbons begin to increase. At the same time, the amount of nitrogen dioxide, NO_2 may decrease because sunlight breaks it down to produce NO plus atomic oxygen (NO + O). The atomic oxygen (O) is then free to combine with molecular oxygen (O_2) to form ozone (O_3). As a result, the concentration of ozone also increases after sunrise. Shortly thereafter, oxidized hydrocarbons react with NO to increase the concentration of NO_2 by midmorning. This reaction causes the NO concentration to decrease and allows ozone to build up, producing a midday peak in ozone and a minimum in NO. As the smog develops, visibility may be greatly reduced (Figure 24.17) as light is scattered by the pollutants.

Sulfurous smog is produced primarily by the burning of coal or oil at large power plants. Sulfur oxides and particulates combine under certain meteorological conditions to produce a concentrated sulfurous smog (Figure 24.20).

Future Trends for Urban Areas

What does the future hold for urban areas with respect to air pollution? The optimistic view is that urban air quality will continue to improve as it has in the past 35 years because we know so much about the sources of air pollution and have developed effective ways to reduce it. The pessimistic view is that in spite of this knowledge, population pressures and economics will dictate what happens in many parts of the world, and the result will be poorer air quality (more air pollution) in many locations.

The actual situation in the twenty-first century is likely to be a mixture of the optimistic and pessimistic points of view. Large urban areas in developing countries may

Figure 24.18 ■ How photochemical smog is produced.

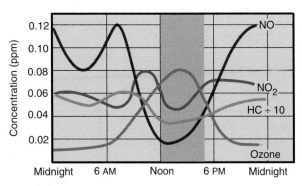

Figure 24.19 ■ Development of photochemical smog over the Los Angeles area on a typical warm day.

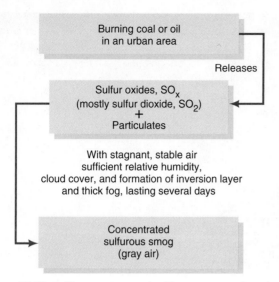

Figure 24.20 ■ How concentrated sulfurous smog and smoke might develop.

experience a reduction in air quality even as they attempt to improve the situation, because the population and economic factors will likely outweigh pollution abatement. Large urban areas in developed and more affluent countries (particularly in the United States) may well continue to experience improved air quality in coming years (see Figure 24.6).

The United States

As an example of trends in developed countries, consider the Los Angeles urban area. This region, which has the worst air quality in the United States, is coming to grips with the problem. The people studying air pollution in the Los Angeles region now understand that pollution abatement will take massive efforts much different from past strategies, which have taken a limited approach. A controversial, multifaceted air quality plan involving the entire Los Angeles urban region and includes the following features:[17]

■ Strategies to discourage automobile use and reduce the number of cars.

■ Stricter emission controls for automobiles.

■ A requirement for a certain number of zero-pollutant automobiles (electric cars) and hybrid cars with fuel cell and gasoline engines.

■ A requirement for more gasoline to be reformulated to burn cleaner.

■ Improvements in public transportation and incentives for people to use it.

■ Mandatory carpooling.

■ Increased controls on industrial and household activities known to contribute to air pollution.

At the household level, for example, common materials such as paints and solvents will be reformulated so that

their fumes will cause less air pollution. Eventually, certain equipment, such as gasoline-powered lawn mowers that contribute to air pollution, may be banned.

There are encouraging signs of improvement in Southern California's air quality. For example, from the 1950s to the present, the peak level of ozone (considered one of the best indicators of air pollution) has declined from about 0.68 to 0.3 ppm. The reduction has occurred in spite of the fact that during this period the population nearly tripled and the number of motor vehicles quadrupled.[18] Nevertheless, air quality in Southern California remains the nation's worst. Even if all the aforementioned controls in urban areas are implemented, air quality will continue to be a significant problem in coming decades, particularly if the urban population continues to increase.

We have focused on air pollution in Southern California because its air quality is especially poor. However, many large and not-so-large U.S. cities have poor air quality a significant part of the year. Based on the criteria of 30 days per year of unhealthy air resulting from ozone pollution, many millions of Americans live in cities where hazardous air pollution exists. The most polluted metropolitan areas in the United States include Riverside, California; Houston, Texas; Baltimore, Maryland; Charlotte, North Carolina; and Atlanta, Georgia. In contrast, some of the cities with the cleanest air in the United States include Bellingham, Washington; Cedar Rapids, Iowa; Colorado Springs, Colorado; and Des Moines, Iowa.[6] However, with the exception of the Pacific Northwest, no U.S. region is free from air pollution and associated adverse health effects.

Developing Countries

As mentioned earlier, cities in less developed countries with burgeoning populations are particularly susceptible to air pollution now and will be in the future (see Figure 24.8). They often do not have the financial base necessary to fight air pollution; they are more concerned with finding ways to house and feed their growing populations.

A good example is Mexico City. With a population of about 25 million, Mexico City is one of the four largest urban areas in the world. Cars, buses, industry, and power plants in the city emit hundreds of thousands of metric tons of pollutants into the atmosphere each year. The city is at an elevation of about 2,255 m (7,400 ft) in a natural basin surrounded by mountains, a perfect situation for a severe air pollution problem. It is becoming a rare day in Mexico City when the mountains can be seen, and physicians report a steady increase in respiratory diseases. Headaches, irritated eyes, and sore throats are common when the pollution settles in. Doctors advise parents to take their children out of the city permanently. The people in Mexico City do not need to be told they have an air pollution problem; it is all too apparent. However, developing a successful strategy to improve the quality of the air is difficult.

A major source of pollutants in Mexico City is motor vehicles. There are some 50,000 buses and taxis and several million automobiles in the city. Most are old and in poor running condition, and they pump immense amounts of pollutants into the atmosphere. Another major source of air pollution is leaks of liquefied petroleum gas (LPG, a hydrocarbon) used in homes for cooking and heating water. The leaking LPGs produce atmospheric precursors to the formation of ozone, a major component of urban photochemical smog, and LPG leaks in Mexico City may be responsible for a significant portion of the city's ozone pollution.[19]

In an attempt to reduce air pollution in the urban area, officials shut down a large oil refinery. For almost 60 years, the refinery had emitted nearly 90,000 metric tons of air pollutants into the atmosphere annually. Thousands of other industrial plants have been ordered to relocate. Although these measures will help the air quality of the urban area, industrial facilities are not the primary source of pollutants. Air pollution will continue to be a serious problem for many years if the city is unable to control increases in population; the use of buses, taxis, and automobiles; and leaks of liquefied petroleum gas. Indeed, Mexico City may eventually experience a pollution event of catastrophic proportions.

In summary, future trends in urban air problems and solutions will include a mixture of success stories and potential or actual tragedies. What is apparent is that urban air pollution is important to people, and ambitious air pollution control plans are being drawn up in many urban areas. Whether these plans are put into action will depend on numerous factors: global, regional, and local economies (reducing air pollution is expensive); population growth (more people means more air pollution); international cooperation (air pollutants travel across international borders); and the priority given to pollution abatement relative to other environmental concerns such as sanitation and clean water. With these thoughts in mind, we turn to a discussion of how to reduce air pollution.

24.7 Pollution Control

For both stationary and mobile sources of air pollutants, the most reasonable strategies for control have been to reduce, collect, capture, or retain the pollutants before they enter the atmosphere. From an environmental viewpoint, the reduction of emissions through energy efficiency and conservation measures (for example, burning less fuel) is the preferred strategy, with clear advantages over all other approaches (see Chapters 17–19). Here, we discuss pollution control for selected air pollutants.

Pollution Control: Particulates

Particulates emitted from fugitive, point, or area stationary sources are much easier to control than are the very small particulates of primary or secondary origin released from mobile sources, such as automobiles. As we learn more about these very small particles, we will have to devise new methods to control them.

A variety of settling chambers or collectors are used to control emissions of coarse particulates from power plants and industrial sites (point or area sources) by providing a mechanism that causes particles in gases to settle out in a location where they can be collected for disposal in landfills. In recent decades, tremendous gains have been made in the control of particulates, such as ash, from power plants and industry. Cities in the eastern United States, where buildings used to turn black from soot and ash, are now much cleaner. Particulate pollutants no longer pose serious health risks in these cities. In contrast, such risks have plagued parts of Eastern Europe in recent years.

Particulates from fugitive sources (such as waste piles) must be controlled on site so that the wind does not blow them into the atmosphere. Methods include protecting open areas, controlling dust, and reducing the effects of wind. For example, waste piles can be covered by plastic or other material, and soil piles can be vegetated to inhibit wind erosion; water or a combination of water and chemicals can be spread to hold dust down; structures or vegetation can be positioned to lessen wind velocity near the ground, thus retarding wind erosion of particles.

Pollution Control: Automobiles

Control of pollutants such as carbon monoxide, nitrogen oxides, and hydrocarbons in urban areas is best achieved through pollution control measures for automobiles. Control of these materials will also limit ozone formation in the lower atmosphere, since ozone forms through reactions with nitrogen oxides and hydrocarbons in the presence of sunlight.

Nitrogen oxides from automobile exhausts are controlled by recirculating exhaust gas, diluting the air-to-fuel mixture being burned in the engine. Dilution reduces the temperature of combustion and decreases the oxygen concentration in the burning mixture, resulting in the production of fewer nitrogen oxides. Unfortunately, the same process increases hydrocarbon emissions. Nevertheless, exhaust recirculation to reduce nitrogen oxide emissions has been common practice in the United States for more than 20 years.[20]

The most common device to reduce carbon monoxide and hydrocarbon emissions from automobiles is the exhaust system's catalytic converter. In the converter, oxygen from outside air is introduced and exhaust gases from the engine are passed over a catalyst, typically platinum or palladium. Two important chemical reactions occur: (1) carbon monoxide is converted to carbon dioxide; and (2) hydrocarbons are converted to carbon dioxide and water.

As government regulations controlling emissions became stronger, it became difficult to meet new standards without the aid of computer-controlled engine systems. Computer-controlled fuel injection began to replace carburetors in the 1980s and has resulted in lower fuel consumption and lower exhaust emissions.[20]

It has been argued that the automobile emission regulation plan in the United States has not been effective in reducing pollutants. Pollutants may be relatively low when a car is new, but many people do not take care of their automobiles well enough to ensure that the emission control devices continue to work. Some people even disconnect smog control devices. Evidence suggests that these devices tend to become less efficient with each year following purchase.

It has been suggested that effluent fees replace emission controls as the primary method of regulating air pollution from automobiles in the United States.[21] Under this scheme, vehicles would be tested each year for emission control, and fees would be assessed on the basis of test results. Fees would provide an incentive to purchase automobiles that pollute less, and annual inspections would ensure that pollution control devices were properly maintained. Although there is considerable controversy regarding enforced pollution inspections, such inspections are common in a number of areas and are expected to increase as air pollution abatement becomes essential.

Another approach to reducing urban air pollution from vehicles involves various measures aimed at reducing the number and type of cars on roads. Some of these methods were mentioned earlier and are being tried or discussed in Los Angeles and other areas. Other measures include developing cleaner automobile fuels through use of fuel additives and reformulation; requiring new cars to use less fuel; and encouraging the use of cars with electric engines and hybrid cars that have both an electric engine and an internal combustion engine.

Pollution Control: Sulfur Dioxide

Sulfur dioxide emissions can be reduced through abatement measures performed before, during, or after combustion. Technology to clean up coal so it will burn more cleanly is already available. Although the cost of removing sulfur makes fuel more expensive, the expense must be balanced against the long-term consequences of burning sulfur-rich coal.

Changing from high-sulfur coal to low-sulfur coal seems an obvious solution to reducing emissions of sulfur dioxide. In some regions, this change will work. Unfortunately, most low-sulfur coal in the United States is located in the western part of the country, whereas most coal is burned in the East. Thus, transportation is an issue; and use of low-sulfur coal is a solution only in cases where it is economically feasible.

Another possibility is cleaning up relatively high-sulfur coal by washing it to remove sulfur. In this process, finely ground coal is washed with water. Iron sulfide (mineral pyrite) settles out because of its relatively high density. Although the washing process is effective in removing nonorganic sulfur from minerals such as pyrite (FeS_2), it is ineffective for removing organic sulfur bound up with carbonaceous material. Cleanup by washing is therefore limited, and it is also expensive.

Another option is **coal gasification**, which converts coal that is relatively high in sulfur to a gas in order to remove the sulfur. The gas obtained from coal is quite clean and can be transported relatively easily, augmenting supplies of natural gas. The synthetic gas produced from coal is still fairly expensive compared with gas from other sources, but its price may become more competitive in the future.

Sulfur oxide emissions from stationary sources, such as power plants, can be reduced by removing the oxides from the gases in the stack before they reach the atmosphere. Perhaps the most highly developed technology for the cleaning of gases in tall stacks is flue gas desulfurization, or **scrubbing** (Figure 24.21). The technology to scrub sulfur dioxide and other pollutants at power plants was developed in the 1970s in the United States in response to passage of the Clean Air Act. However, the technology was not initially implemented in the United States because regulators chose to allow plants to dis-

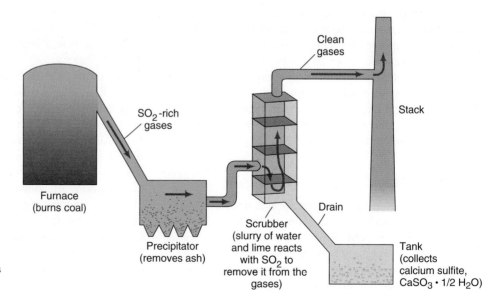

Figure 24.21 ■ Scrubber used to remove sulfur oxides from the gases emitted by tall stacks.

Clean gases

Stack

SO_2-rich gases

Furnace (burns coal)

Precipitator (removes ash)

Scrubber (slurry of water and lime reacts with SO_2 to remove it from the gases)

Drain

Tank (collects calcium sulfite, $CaSO_3 \cdot 1/2\ H_2O$)

perse pollutants using tall smokestacks rather than using scrubbing to remove them. This increased the regional acid rain problem.

Scrubbing occurs after coal is burned. The SO_2-rich gases are treated with a slurry (a watery mixture) of lime (calcium oxide, CaO) or limestone (calcium carbonate, $CaCO_3$). The sulfur oxides react with the calcium to form calcium sulfite, which is collected and then disposed of, usually in a landfill.

A German company in 1980 purchased coal-scrubbing technology and improved on it as part of efforts to reduce air pollution and acid rain. Rather than disposing of the calcium sulfite-rich sludge formed during the scrubbing process, the company further processes it to produce building materials (gypsum, $CaSO_4 \cdot 2H_2O$, which is sheet rock or wallboard) that are sold worldwide.

Another innovative approach to removing sulfur has been taken at a large coal-burning power plant near Mannheim, Germany. The smoke from combustion is cooled, then treated with liquid ammonia (NH_3), which reacts with the sulfur to produce ammonium sulfate. In this process, the sulfur-contaminated smoke is cooled in a heat-exchange process by outgoing clean smoke to a temperature that allows the chemical reaction between the sulfur-rich smoke and ammonia to take place. The cooled, cleaned, outgoing smoke is then heated by dirty smoke in the same sort of heat-exchange process to force it out the vent. Waste heat from the cooling towers is used to heat nearby buildings, and the plant sells the ammonium sulfate in a solid granular form to farmers to use as fertilizer.

Thus, Germany, in response to tough pollution control regulations, has substantially reduced its sulfur dioxide emissions (as well as many other pollutant emissions) and has boosted its economy in the process.[22]

24.8 Air Pollution Legislation and Standards

The killer smog events in Donora, Pennsylvania, in 1948 and London, England, in 1952 were the impetus for legislation to control air pollution in both England and the United States.

Clean Air Act Amendments of 1990

The **Clean Air Act Amendments of 1990** are comprehensive regulations enacted by the U.S. Congress that address acid rain, toxic emissions, ozone depletion, and automobile exhaust. In dealing with acid deposition (acid rain), the amendments establish limits on the maximum permissible emissions of sulfur dioxide from utility companies burning coal. The goal of the legislation—to reduce such emissions by about 50% to 10 million tons a year by 2000—was achieved (see Figure 24.6).

An innovative aspect of the legislation is to provide incentives to utility companies to reduce emissions of sulfur dioxide by providing marketable permits that allow companies to buy and sell the right to pollute.[23] The total amount of pollution allowed is divided into a given number of permits. Utilities with clean power plants that don't need their permits sell the permits to those that do need them. Environmentalists can also purchase these permits to keep them from being bought by utility companies, forcing the utility companies to use more vigorous pollution-abatement technology. Buying of permits by environmentalists, however, has not been a major factor. As the permits are bought and sold, they take on an economic value, and polluters begin to view polluting as an expensive way to do business.[23] One step back for air pollution legislation has been the 2003 decision of the president and the Environmental Protection Agency to allow utility companies to upgrade their systems without installing new pollution controls.

The 1990 amendments also call for a reduction in emissions of nitrogen dioxides by approximately 2 million tons from the 1980 level. Greater reductions would be difficult because large amounts of nitrogen oxide emissions are related to automobiles rather than to coal-burning power plants.[24]

Emissions of toxins into the atmosphere are targeted by the legislation to be reduced by as much as 90%. The toxins chosen are those thought to have the most potential for damaging human health, including causing cancer. Abatement will depend heavily on pollution control equipment, which will be required for large manufacturers and small businesses alike. Although this requirement will undoubtedly result in an increase in the cost of many goods and services, there should be a compensating improvement in the health of people.

The Clean Air Amendments also deal with ozone depletion in the stratosphere (see Chapter 26). The goal is to end the production of all chlorofluorocarbons (CFCs) and other chlorine chemicals in steps by 2030.[24]

As we have seen, air pollution in urban areas is commonly associated with automobile exhaust. Strategies outlined in the legislation include more stringent emission controls on automobiles and requirements for cleaner-burning fuels. The aim is to reduce the occurrence of urban smog. Expected impacts of the legislation include increases in the cost of automobile fuels and in the price of new automobiles.

Ambient Air Quality Standards

Air quality standards are important because they are tied to emission standards that attempt to control air pollution. Many countries have developed air quality standards, including France, Japan, Israel, Italy, Canada, Germany, Norway, and the United States. National Ambient Air Quality Standards (NAAQS) for the United States, defined to comply with the Clean Air Act, are shown in Table 24.3. Tougher standards for ozone and PM 2.5 in recent years were set to reduce adverse health effects for children and elderly people, who are most susceptible to air pollution. The

Table 24.3 • U.S. National Ambient Air Quality Standards (NAAQS)

Pollutant	Standard Value[a]		Standard Type
Carbon monoxide (CO)			
8-hour average	9 ppm	(10 mg/m³)	Primary [c]
1-hour average	35 ppm	(40 mg/m³)	Primary
Nitrogen dioxide (NO₂)			
Annual arithmetic mean	0.053 ppm	(100 µg/m³)	Primary and secondary [d]
Ozone (O₃)			
1-hour average	0.12 ppm	(235 µg/m³)	Primary and secondary
Lead (Pb)			
Quarterly average	1.5 µg/m³		Primary and secondary
Particulate (PM 10) *Particles with diameters of 10 micrometers or less*			
Annual arithmetic mean	50 µg/m³		Primary and secondary
24-hour average	150 µg/m³		Primary and secondary
Particulate (PM 2.5)[b] *Particles with diameters of 2.5 micrometers or less*			
Annual arithmetic mean	15 µg/m³		Primary and secondary
24-hour average	65 µg/m³		Primary and secondary
Sulfur Dioxide (SO₂)			
Annual arithmetic mean	0.03 ppm	(80 µg/m³)	Primary
24-hour average	0.14 ppm	(365 µg/m³)	Primary
3-hour average	0.50 ppm	(1300 µg/m³)	Secondary

[a] Parenthetical value is an approximately equivalent concentration.

[b] The ozone 8-hour standard and the PM 2.5 standards are included for information only. A 1999 federal court ruling blocked implementation of these standards, which the EPA proposed in 1997. EPA has asked the U.S. Supreme Court to reconsider that decision. (*Note:* In March 2001, the Court ruled in favor of the EPA, and the new standards are expected to take effect within a few years.)

[c] Primary standards set limits to protect public health, including the health of sensitive populations such as asthmatics, children, and the elderly.

[d] Secondary standards set limits to protect public welfare, including protection against decreased visibility and damage to animals, crops, vegetation, and buildings.

Source: U.S. Environmental Protection Agency.

new standards are saving the lives of thousands and improving the health of hundreds of thousands of children.

The new standards were opposed by U.S. business leaders, who argued that implementing them would cost hundreds of billions of dollars and up to a million jobs. In 1999, a lower federal court blocked the implementation of the new standards, and the EPA requested that the U.S. Supreme Court hear the case. In early March of 2001, the Supreme Court, in a unanimous decision, upheld the new, tougher standards. The justices held that the EPA's responsibility is to consider benefits to public health from pollution abatement—not financial costs.

Implementing new standards takes years; ways will have to be found to enforce them, and the EPA will have to defend the standards in the lower courts. Nevertheless, it is expected that the new standards will eventually be implemented. The decision of the Supreme Court is a turning point in the fight to reduce air pollution and its known adverse health effects. Standards from now on will hopefully be based on improving human health rather than on the economic costs of implementing standards.

Air Quality Index

In the United States, the Air Quality Index (AQI) (Table 24.4) is used to describe air pollution on a given day. For example, air quality in urban areas is often reported as good, moderate, unhealthy for sensitive groups, unhealthy, very unhealthy, or hazardous, corresponding to a color code of the Air Quality Index. The AQI is determined from measurements of the concentration of five major pollutants: particulate matter, sulfur dioxide, carbon monoxide, ozone, and nitrogen dioxide. An AQI value of greater than 100 is unhealthy. In most U.S. cities, values of AQI range between 0 and 100. AQI values greater than 100 are generally recorded for a particular city only a few times per year. However, some cities with serious air pollution problems may exceed an AQI of 100 many times per year. In a typical year, AQI values above 200 (for all U.S. sites) are rare, and those above 300 are very rare. In comparison, for large urban cities outside of the United States with dense human populations and numerous uncontrolled sources of pollution, AQIs greater than 200 are frequent.

Table 24.4 • Air Quality Index (AQI) and Health Conditions

Index Values	Descriptor	Cautionary Statement	General Adverse Health Effects	Action Level (AQI)[a]
0–50	Good	None	None	None
51–100	Moderate	Unusually sensitive people should consider limiting prolonged outdoor exertion.	Very few symptoms[b] for the most susceptible people[c]	None
101–150	Unhealthy for sensitive groups	Active children and adults, and people with respiratory disease, such as asthma, should limit prolonged outdoor exertion.	Mild aggravation of symptoms in susceptible people, few symptoms for healthy people	None
151–199	Unhealthy	Active children and adults, and people with respiratory disease, such as asthma, should avoid prolonged outdoor exertion; everyone else, especially children, should limit prolonged outdoor exertion.	Mild aggravation of symptoms in susceptible people, irritation symptoms for healthy people	None
200–300	Very unhealthy	Active children and adults, and people with respiratory disease, such as asthma, should avoid outdoor exertion; everyone else, especially children, should limit outdoor exertion.	Significant aggravation of symptoms in susceptible people, widespread symptoms in healthy people	Alert (200+)
Over 300	Hazardous	*Everyone* should avoid outdoor exertion.	300–400: Widespread symptoms in healthy people 400–500: Premature onset of some diseases Over 500: Premature death of ill and elderly people; healthy people experience symptoms that affect normal activity	Warning (300+) Emergency (400+)

[a] Triggers preventative action by state or local officials.

[b] Symptoms include eye, nose, and throat irritation; chest pain; breathing difficulty.

[c] Susceptible people are young, old, and ill people and people with lung or heart disease.

AQI 51–100	Health advisories for susceptible individuals.
AQI 101–150	Health advisories for all.
AQI 151–200	Health advisories for all.
AQI 200+	Health advisories for all; triggers an alert; activities that cause pollution might be restricted.
AQI 300	Health advisories to all; triggers a warning; probably would require power plant operations to be reduced and carpooling to be used.
AQI 400+	Health advisories for all; triggers an emergency; cessation of most industrial and commercial activities, including power plants; nearly all private use of vehicles prohibited.

Source: U.S. Environmental Protection Agency.

During a pollution episode, hourly ozone levels are reported; and a smog episode begins if the primary National Ambient Air Quality Standard (NAAQS) of 0.12 ppm (equivalent to 235 $\mu g/m^3$ in Table 24.2) is exceeded. This measure corresponds to unhealthy air with an AQI between 100 and 300 (Table 24.4). An air pollution alert is issued if the AQI exceeds 200. An air pollution warning is issued if the AQI exceeds 300, a point at which air quality is hazardous to all people. If the AQI exceeds 400, an air pollution emergency is declared, and people are requested to remain indoors and minimize physical exertion. Driving private automobiles may be prohibited, and industry may be required to reduce emissions to a minimum during the episode. Recall that the AQI was 800, twice as high as the level that constitutes an emergency, during the Indonesian fires of 1997–1998!

24.9 Cost of Air Pollution Control

The cost of air pollution control varies tremendously from one industry to another. For example, consider the incremental control costs (costs to remove an additional unit of pollution) for utilities burning fossil fuels and for an aluminum plant. The cost for incremental control in a fossil fuel–burning utility is a few hundred dollars per additional ton of particulates removed. For the aluminum plant, the cost to remove an additional ton of particulates may be as much as several thousand dollars. Some economists would argue that it is wise to increase the standards for utilities and relax or at least not increase them for aluminum plants. This practice would lead to more cost-efficient pollution control while maintaining good air quality. However, the geographic distribution of various facilities will obviously determine the trade-offs possible.[25]

Another economic consideration is that, as the degree of control of a pollutant increases, a point is reached at which the cost of incremental control is very high in relation to the additional benefits of the increased control. Because of this and other economic factors, it has been argued that enforcing fees or taxes for emitting pollutants might make more economic sense than attempting to evaluate the uncertain costs and benefits associated with enforcement of standards. Another approach is to issue vouchers to allow businesses to emit a certain total amount of pollution in a region. These vouchers are bought and sold on the open market. All these economic alternatives are controversial and may be objectionable to people who believe that polluters should not be allowed to buy their way out of doing what is socially responsible (that is, not polluting our atmosphere).

Economic analysis of air pollution is not simple. There are many variables, some of which are hard to quantify. We do know the following:

- With increasing air pollution controls, the capital cost for technology to control air pollution increases.
- As the controls for air pollution increase, the loss from pollution damages decreases.
- The total cost of air pollution is the cost of pollution control plus the environmental damages of the pollution.

Although the cost of pollution-abatement technology is fairly well known, it is difficult to adequately determine the loss from pollution damages, particularly when considering health problems and damage to vegetation, including food crops. For example, exposure to air pollution may cause or aggravate chronic respiratory diseases in human beings, with a very high cost. A recent study of the health benefits of cleaning up the air quality in the Los Angeles basin estimated that the annual cost associated with air pollution in the basin is 1,600 lives and about $10 billion.[26] Air pollution also leads to loss of revenue from people who choose not to visit areas such as Los Angeles and Mexico City because of known air pollution problems.

How do we determine the real and total benefits and costs of controlling or reducing air pollution? As we have seen, there are no easy answers to this question. In spite of our inability to determine all benefits and costs, it seems worthwhile to reduce the air pollution level below some

particular standard. Thus, in the United States, the National Ambient Air Quality Standards have been developed as a minimum acceptable air quality level. However, as discussed, it is also a good idea to consider alternatives, such as charging fees or taxes for emissions. If such charges are determined carefully and emissions are closely monitored, the charges should provide an incentive for the installation of control measures. The end result would be better air quality.[27, 28]

Summary

- Air pollutants are grouped into the "criteria pollutants" in the Clean Air Act (ozone, sulfur dioxide, nitrogen oxides, particulates, carbon monoxide, and lead), and air toxics that can cause serious health problems (for example, benzene, chlorine gas, and hydrogen fluoride).

- Every year, approximately 140 million metric tons of primary pollutants enter the atmosphere above the United States from processes related to human activity. Considering the enormous volume of the atmosphere, this is a relatively small amount of material. If it were distributed uniformly, there would be little problem with air pollution. Unfortunately, the pollutants generally are not evenly distributed but are concentrated in urban areas or in other areas where the air naturally lingers.

- The two main types of pollution sources are stationary and mobile. Stationary sources have a relatively fixed position and include point sources, area sources, and fugitive sources.

- There are two main groups of air pollutants: primary and secondary. Primary pollutants are those emitted directly into the air: particulates, sulfur dioxide, carbon monoxide, nitrogen oxides, and hydrocarbons. Secondary pollutants are those produced through reactions between primary pollutants and other atmospheric compounds. A good example of a secondary pollutant is ozone, which forms over urban areas through photochemical reactions between primary pollutants and natural atmospheric gases.

- The effects of the major air pollutants are considerable. They include effects on the visual quality of the environment, vegetation, animals, soil, water quality, natural and artificial structures, and human health.

- The combustion of large quantities of fossil fuels results in the emission of sulfur and nitrogen oxides into the atmosphere, creating acid rain. Environmental degradations associated with acid rain include loss of fish and other life in lakes, damage to trees and other plants, leaching of nutrients from soils, and damage to stone statues and buildings in urban areas.

- Meteorological conditions greatly affect whether polluted air is a problem in an urban area. In particular, restricted circulation in the lower atmosphere associated

Where Does Arctic Haze Come from, and How Does It Affect the Environment?

A dark gray haze, full of industrial pollutants, hovers over the ground and extends to an altitude of 8 km (5 mi). It is not Los Angeles but the frozen Arctic, thousands of miles from heavy industry. The polluted air mass includes all of the atmosphere above the Arctic Circle, as well as lobes extending into Eurasia and North America. The total area of polluted air mass, about as large as the African continent, is present during winter months and disappears in summer.

Scientists describe this haze as an *aerosol*—that is, microscopic particles dispersed in a gas, smoke, or fog. They have found dust from Mongolia and sea salt in the haze; but of greater concern are the pollutants, mainly sulfates, carbon soot (dark carbon), organic compounds, and toxic metals, including mercury, lead, and vanadium. The gaseous atmosphere itself contains elevated levels of carbon dioxide, methane, and carbon monoxide, as well as chemicals destructive of the ozone layer.

Like detectives searching for fingerprints, scientists used ratios of six elements (arsenic, antimony, zinc, indium, manganese, and vanadium) to a seventh, selenium, to track down sources of Arctic haze. Emissions from burning fossil fuels differ in amounts of these elements, depending on the type of fuel used. The ratios are so characteristic that scientists can tell whether pollution is from hard or soft coal. The ratios found in the haze were compared with ratios characteristic of types of fuel known to be used in various regions of the world. For example, manganese was present in greater amounts and vanadium in lesser amounts in the haze than in emissions typical of North America.

Based on the ratios and on knowledge of air circulation patterns, scientists identified Eastern Europe and Russia as major sources of Arctic air pollution, with a significant but lesser contribution from the United Kingdom and Western Europe. Two major routes (see Figure 24.22) are (1) from Eastern Europe and Russia across the Taimyr Peninsula and the North Pole to the Alaskan Arctic and (2) from the United Kingdom and Western Europe across Scandinavia to the Norwegian Arctic. (A different study argued that none of the pollution comes from Western Europe or from China. Rather, it comes from Eastern Europe and Russia, with some possibly originating within the Arctic Circle itself.)

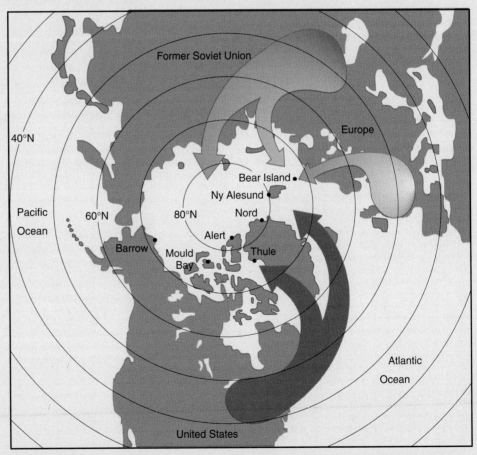

Figure 24.22 ■ Two airstreams out of the former Soviet Union (top arrows) carry most of the pollution that becomes Arctic haze. According to some findings, Europe is also an important source of smog (arrow at right). But the northeastern United States (bottom arrows) is not a big contributor to the haze, because pollutants are often washed out by storms before they reach the Arctic. The dots represent air-sampling stations.

Major atmospheric forces driving pollution along these routes begin with large temperature differences between the equator and the poles in winter, creating strong air currents from 0° to 90° latitude. The air currents are also propelled by seasonal lows in the North Atlantic and highs on the Eurasian continent. Once masses of pollutant-laden air reach the dry, stable air of Arctic winter, they form layers, which remain relatively intact. In spring, when the northward flow of air diminishes, the haze disperses and is carried to higher levels in the atmosphere and back to the midlatitudes.

Now that scientists have a better idea of the sources of Arctic haze, they are beginning to speculate on the effects it has on the ecology of the Arctic and on global climate. Furthermore, they are calling for research into environmental impacts of the haze and for cooperation among countries in the Northern Hemisphere to reduce the amount of haze. In the meantime, humans are producing enough toxic emissions to carry as far as 10,000 km (6,215 mi) and produce pollution levels at the North Pole as great as those found in medium-sized industrial cities.

Critical Thinking Questions

1. Although laden with pollutants, Arctic haze drops less of its aerosol on the ground than would happen in other areas of the world. What is it in the Arctic environment that could account for this?

2. Elements such as mercury and lead are present in small amounts in Arctic haze (the levels are often comparable to those near industrial sites and 10 to 20 times those in Antarctica). If the amounts are so low, why are scientists concerned about their effects? (You may wish to review Chapter 15.)

3. Sulfates in haze react with water to form acids. Scientists predict that the Arctic will be sensitive to additional acids. Why might this be so?

4. What are possible effects of dark carbon particles such as soot on the Arctic climate? Make a diagram. Put the term *dark carbon particles* in the center of a sheet of paper. Write as many direct consequences of dark carbon as you can around the term, and connect the consequences to the term with arrows. Now, what would be the secondary consequences of these direct effects? Place them around the edge of the diagram, connecting them by arrows to the relevant direct effect. If you can think of third- and fourth-order effects, place these on the diagram in a similar manner.

with temperature inversion layers may lead to pollution events.

- There are two major types of smog: photochemical and sulfurous. Each type of smog brings particular environmental problems that vary with geographic region, time of year, and local urban conditions.

- From an environmental viewpoint, the preferred method of reducing the emission of air pollutants produced from burning fossil fuels is to practice energy efficiency and conservation so that smaller amounts of fossil fuels are burned. Another option is to increase the use of alternative energy sources, such as solar and wind power, that do not emit air pollutants.

- Methods to control air pollution are tailored to specific sources and types of pollutants. These methods vary from settling chambers for particulates to scrubbers that use lime to remove sulfur before it enters the atmosphere.

- Efforts to reduce air pollution in urban regions center on automobiles, buses, and other vehicles, because they account for most of the pollutants that enter the urban atmosphere.

- Emissions of air pollutants in the United States are decreasing. In developing countries, air pollution in large urban centers is often a serious problem.

- Air quality in U.S. urban areas is usually reported in terms of whether the quality is good, moderate, unhealthy for sensitive groups, unhealthy, very unhealthy, or hazardous. These levels are defined in terms of the Air Quality Index (AQI).

- The relationships between emission control and environmental cost are complex. The minimum total cost is a compromise between capital costs to control pollutants and losses or damages resulting from pollution. If additional controls are necessary to lower the pollution to a more acceptable level, additional costs are incurred. Beyond a certain level of pollution abatement, these costs can increase rapidly.

Human Population

Increases in human population are expected to continue to have a significant adverse effect on air pollution problems. As the human population increases, so does the total use of resources, many of which are related to emissions of air pollutants. This may be partially offset in developed countries, where the per-capita emissions of air pollutants have been reduced in recent years.

Sustainability

Ensuring that future generations inherit a quality environment with minimal air pollution is an important objective of sustainability. As a result, it remains an important objective to find and develop technology that minimizes air pollution.

Global Perspective

Atmospheric processes and pollution of the atmosphere occur by their very nature on regional and global scales. Pollutants emitted into the atmosphere at a particular site may join the global circulation pattern and spread pollutants throughout the world. Air pollutants emitted from urban or agricultural areas may be dispersed to pristine areas far removed from human activities. Therefore, an understanding of global atmospheric processes is critical to finding solutions to many air pollution problems, including acid deposition.

Urban World

Cities and urban corridors are sites of intense human activity, and many of these activities are associated with emission of air pollutants. Some of the most significant adverse effects of air pollution are found in our urban areas. Some large cities have air pollution problems so severe in extent that the health and very lives of people are being affected.

People and Nature

We often think of nature as pure and unpolluted. In reality, though, nature can be toxic and polluted. This is especially true with regard to air pollution—for example, the vast majority of particulates and carbon monoxide such as volcanic eruption and wildfire. Even hydrocarbons have local sources such as seeps, which in areas such as offshore Goleta, California, emit a significant amount of hydrocarbons that contribute to the production of smog.

Science and Values

The science and technology necessary to reduce air pollution is well known; what we do with these tools involves a value judgment. It is clear that people value a high-quality environment, and clean air is at the top of the list. The developed countries have an obligation to take a leadership role in finding ways to utilize resources while minimizing air pollution. Of particular importance is finding methods and technologies that will allow for reduction of air pollution while stimulating economies. What is considered waste in one part of the urban-industrial complex may be used as resources in another part. This idea is at the heart of what is sometimes called industrial ecology.

Key Terms

acid rain **530**
air toxics **529**
air quality standards **545**
atmospheric
   inversion **539**

Clean Air Act Amendments
   of 1990 **545**
coal gasification **544**
criteria pollutants **529**
global dimming **532**

mobile sources **526**
photochemical smog **540**
primary pollutants **529**
scrubbing **544**
secondary pollutants **529**

smog **540**
stationary sources **526**
sulfurous smog **540**

1. Compare and contrast the London 1952 fog event with smog problems in the Los Angeles basin.

2. Why do we have air pollution problems when the amount of pollution emitted into the air is a very small fraction of the total material in the atmosphere?

3. What is the difference between point and nonpoint sources of air pollution? Which type is easier to manage?

4. What are the differences between primary and secondary pollutants?

5. Carefully examine Figure 24.16, which shows a column of air moving through an urban area, and Figure 24.19, which shows relative concentrations of pollutants that develop on a typical warm day in Los Angeles. What linkages between the information in these two figures might be important in trying to identify and learn more about potential air pollution in an area?

6. Why is acid deposition a major environmental problem, and how can it be minimized?

7. Why will air pollution-abatement strategies in developed countries probably be much different in terms of methods, process, and results from air pollution-abatement strategies in developing countries?

8. Why is it so difficult to establish national air quality standards?

9. In a highly technological society, is it possible to have 100% clean air? Is it feasible or likely?

10. How good are the air quality standards being used by the United States? How might their usefulness be evaluated? Do you think the standards will change in the future? If so, what are the likely changes?

11. What are air toxics and how are they classified?

12. What are the trends in emissions of the criteria pollutants in the Clean Air Act?

Further Reading

Boubel, R. W., D. L. Fox, D. B. Turner, and A. C. Stern. 1994. *Fundamentals of Air Pollution*, 3rd ed. New York: Academic. A thorough book covering the sources, mechanisms, effects, and control of air pollution.

Bryner, G. C. 1995. *Blue Skies, Green Politics: The Clean Air Act of 1990 and Its Implementation*, 2nd ed. Washington, D.C.: CQ Press. A good text on the Clean Air Act and air pollution in the United States.

Hewitt, D. N., W. T. Sturges, and NOAA, eds. 1995. *Global Atmospheric Chemical Change*. New York: Chapman & Hall. A book describing aspects of global air pollution, including chemical changes in the atmosphere, climate change, acid deposition, and other anthropogenic pollutants.

Rose, J., ed. 1994. *Acid Rain: Current Situation and Remedies*. Philadelphia: Gordon and Breach Science. Essays covering issues and consequences of acid rain in Europe and the United States.

Stone, R. 2002. "Air Pollution: Counting the Cost of London's Killer Smog," *Science* 298: 2106–2107. An in-depth look at events involved in the opening case study.

Wang, L. 2002 (February). "Paving out Pollution," *Scientific American*, p. 20. Discussion of an innovative approach to reducing air pollution.

CHAPTER 25

Indoor Air Pollution

This modern art museum, which opened in 1998 in Stockholm, Sweden, closed for repairs in 2002 due to building-related illness. Employees in the kitchen and bookstore complained of a variety of symptoms, including persistent coughs, headaches, and difficulty in breathing. The indoor air pollution is thought to be in part the result of toxic mold related to the problems with the ventilation system in the building. Similar problems have occurred in American buildings including the Massachusetts Registry of Motor Vehicles Building.

Learning Objectives

Indoor air pollution from human fires for cooking and heating has affected human health for thousands of years. Today, lack of adequate ventilation in many energy-efficient homes and offices has increased the risk from pollutants. After reading this chapter, you should understand:

■ Why indoor air pollutants cause some of our most serious environmental health problems.

■ What the major indoor air pollutants are, and where they come from.

■ Why concentrations of pollutants found in the indoor environment may be much greater than concentrations of the same pollutants generally found outdoors.

■ Why environmental tobacco smoke (ETS) is a serious indoor air pollutant.

■ What radon gas is, and why it can be considered one of our most serious environmental health problems.

■ How radon gas enters homes and other buildings, and how its indoor concentration may be minimized.

■ What a green building is and how it is linked to indoor air pollution.

■ What the major strategies are to control and minimize indoor air pollution.

Massachusetts Registry of Motor Vehicles Building: Sick Building Syndrome

Indoor air pollution is a serious environmental problem. Buildings such as the modern art museum in Stockholm (opening photograph) have been closed for repair or even abandoned due to pollution. Consider the Massachusetts Registry of Motor Vehicles Building.

Employees of the Massachusetts Registry of Motor Vehicles moved into their newly constructed building on April 19, 1994. First hints of problems were reported in June 1994, shortly after the building was fully occupied. Early reports concerned a variety of air quality problems, including unpleasant odors and a multitude of symptoms that were experienced by a large number of employees in the building. Some of the symptoms reported were respiratory problems; irritation of the eyes, nose, and throat; skin rashes; and effects on the central nervous system. The symptom most often reported was fatigue, followed closely by headaches and problems with mucous membranes. Some type of respiratory problem was reported by 52% of the staff in the building, compared with 17% having similar problems before moving to the new building.[1]

Studies by the Massachusetts Department of Public Health and private consultants suggested that the problem in the building was related to contaminated air caused by a poorly constructed ventilation system. The air circulation system in the building drew outside air into a space at the top of the structure (called the plenum). There, the air was cooled and pumped through ventilation ducts throughout the building. This produced a problem, because when the warmer outside air was cooled, water condensed onto ceiling tiles. A major component of the ceiling tiles was a starch that fermented when wet, producing an acid that smells like vomit. More serious was the discovery that the fireproofing material sprayed on all surfaces in the ventilation system was also wet and was falling apart. Mineral and other fibers from the fireproofing material were then released and spread throughout the building in the ventilation system. This exposed people in the building to potentially dangerous particulate matter. After occupying the building only 15 months, the staff of the Registry of Motor Vehicles moved out, and the building was closed.[2]

Figure 25.1 ■ Coal miners in Eastern Europe covered with lung-damaging coal dust.

The museum of art in Stockholm and the Massachusetts Registry of Motor Vehicles buildings are recent examples of indoor air pollution problems that have been with us for many thousands of years. Indoor air pollution has been present for as long as people have constructed buildings for protection from the elements and burned fuels to heat human-made environments. A detailed autopsy of a fourth-century Native American woman, frozen shortly after death, revealed that she suffered from **black lung disease** from breathing very polluted air over many years. The pollutants included hazardous particles from lamps that burned seal and whale blubber.[3] This same disease has long been recognized as a major health hazard for underground coal miners and has been called "coal miners' disease" (Figure 25.1). As recently as the mid-1970s, black lung disease was estimated to be responsible for about 4,000 deaths per year in the United States.[4]

People today spend between 70% and 90% of their time indoors or in other enclosed places (homes, workplaces, automobiles, restaurants, and so forth). But only recently have we begun to fully study the indoor environment and how pollution of that environment affects our health. The World Health Organization has estimated that as many as one in three workers may be working in a building that causes them to become sick. As many as 20% of public schools in the United States have problems related to indoor air quality. The U.S. Environmental Protection Agency considers indoor air pollution to be one of the most significant environmental health hazards people face in the modern workplace.[5]

25.1 Sources of Indoor Air Pollution

The potential sources of indoor air pollution are incredibly varied (Figure 25.2). Two common pollutants are shown in Figure 25.3. Indoor air pollutants can arise from both human activities and natural processes. In recent years the public has been made aware of several of these sources, described in the following list:

- Environmental tobacco smoke (secondhand smoke) is the most hazardous common indoor air pollutant, associated with over 40,000 deaths per year (mostly heart disease and lung cancer) in the United States.

- *Legionella pneumophila*, a bacterium that normally lives in pond water, causes a type of pneumonia called Legionnaires' disease when inhaled. About 20 species of the disease-causing bacteria have been identified. Most commonly, this disease is spread by way of air-conditioning equipment, which harbors the disease-causing bacteria in pools of stagnant water in air ducts and filters. Bacteria are transported through a building as a bacterial aerosol when heating or cooling units are in use. However, spread of the disease is not limited to this pathway. One epidemic occurred in a hospital as a result of contamination from an adjacent construction site. A recent outbreak of Legionnaires' disease caused by bacteria in the air-conditioning system of an aquarium killed four people.[5]

- Some molds (fungal growths) in buildings release toxic spores. When inhaled over a period of time, the spores can cause chronic inflammation and scarring of lungs, as well as hypersensitivity pneumonitis and pulmonary fibrosis. These medical conditions are painful, disabling, and can even cause death.[5] It is believed that molds may be responsible for up to half of all health complaints resulting from indoor air environments.

- Radon gas seeps up naturally from soils and rocks below buildings and is thought to be the second most common cause of lung cancer.

- Pesticides that are deliberately or inadvertently applied in buildings to control ants, flies, fleas, moths, and rodents are toxic to people as well.

1. Heating, ventilation, and air-conditioning systems of buildings may be sources of indoor air pollutants, including molds and bacteria, if filters and equipment are not maintained properly. Gas and oil furnaces release carbon monoxide, nitrogen dioxide, and particles.

2. Restrooms may be sources of a variety of indoor air pollutants, including secondhand smoke, molds, and fungi resulting from humid conditions.

3. Furniture and carpets in buildings often contain toxic chemicals (formaldehyde, organic solvents, asbestos), which may be released over time in buildings.

4. Coffee machines, fax machines, computers, and printers can release particles and chemicals, including ozone (O_3), which is highly oxidizing.

5. Pesticides can contaminate buildings with cancer-causing chemicals.

6. Fresh-air intake that is poorly located—as, for example, above a loading dock or first-floor restaurant exhaust fan—can bring in air pollutants.

7. People who smoke inside buildings, perhaps in restaurants, and people who smoke outside buildings, particularly near open or revolving doors, may cause pollution as the environmental tobacco smoke (secondhand smoke) is drawn into and up through the building by the chimney effect.

8. Remodeling, painting, and other such activities often bring a variety of chemicals and materials into a building. Fumes from such activities may enter the building's heating, ventilation, and air-conditioning system, causing widespread pollution.

9. A variety of cleaning products and solvents used in offices and other parts of buildings contain harmful chemicals whose fumes may circulate throughout a building.

10. People can increase carbon dioxide levels; they can emit bioeffluents and spread bacterial and viral contaminants.

11. Loading docks can be sources of organics from garbage containers, of particulates, and of carbon monoxide from vehicles.

12. Radon gas can seep into a building from soil; rising damp (water), which facilitates the growth of molds, can enter foundations and rise up walls.

13. Dust mites and molds can live in carpets and other indoor places.

14. Pollen can come from inside and outside sources.

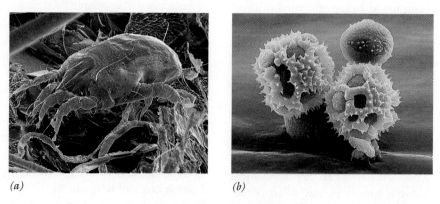

Figure 25.2 ■ Some potential sources of indoor air pollution.

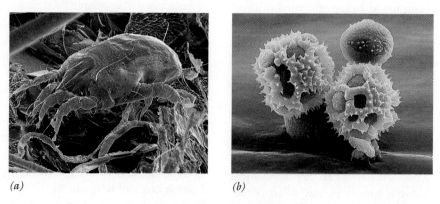

(a)   *(b)*

Figure 25.3 ■ (*a*) This dust mite (magnified about 140 times) is an eight-legged relative of spiders. It feeds on human skin in household dust. It lives in materials such as fabrics on furniture. Dead dust mites and their excrement can produce allergic reactions and asthma attacks in some people. (*b*) Microscopic pollen grains that in large amounts may be visible as a brown or yellow powder. The pollen here are dandelion and horse chestnut.

- Some varieties of asbestos, used as an insulating material and fireproofing material in homes, schools, and offices, are known to cause a particular type of lung cancer (see Chapter 15).
- Formaldehyde (a volatile organic compound, VOC, with a chemical formula of CH_2O) is used in some foam insulation materials, as a binder in particleboard and wood paneling, and in many other materials found in homes and offices. These materials can emit formaldehyde as a gas into buildings. Some mobile homes have been found to have high concentrations of formaldehyde because products containing the chemical are used in their construction (wood paneling, for example).
- Dust mites and pollen irritate the respiratory system, nose, eyes, and skin of people who are sensitive to them.

Common indoor air pollutants and guidelines for allowable exposure are listed in Table 25.1. Many of the

Table 25.1 • Sources, Concentrations, Occurrences, and Possible Health Effects of Indoor Air Pollutants

Pollutant	Source	Guidelines (Dose or Concentrations)	Possible Health Effects
Asbestos	Fireproofing; insulation, vinyl floor, and cement products; vehicle brake linings	0.2 fibers/mL for fibers larger than 5 µm	Skin irritation, lung cancer
Biological aerosols/ microorganisms	Infectious agents, bacteria in heating, ventilation, and air-conditioning systems; allergens	None available	Diseases, weakened immunity
Carbon dioxide	Motor vehicles, gas appliances, smoking	1,000 ppm	Dizziness, headaches, nausea
Carbon monoxide	Motor vehicles, kerosene and gas space heaters, gas and wood stoves, fireplaces; smoking	10,000 µg/m^3 for 8 hours; 40,000 µg/m^3 for 1 hour	Dizziness, headaches, nausea, death
Formaldehyde	Foam insulation; plywood, particleboard, ceiling tile, paneling, and other construction materials	120 µg/m^3	Skin irritant, carcinogen
Inhalable particulates	Smoking, fireplaces, dust, combustion sources (wildfires, burning trash, etc.)	55–110 µg/m^3 annual; 350 µg/m^3 for 1 hour	Respiratory and mucous irritant, carcinogen
Inorganic particulates			
Nitrates	Outdoor air	None available	
Sulfates	Outdoor air	4 µg/m^3 annual; 12 µg/m^3 for 24 hours	
Metal particulates			Toxic, carcinogen
Arsenic	Smoking, pesticides, rodent poisons	None available	
Cadmium	Smoking, fungicides	2 µg/m^3 for 24 hours	
Lead	Automobile exhaust	1.5 µg/m^3 for 3 months	
Mercury	Old fungicides; fossil fuel combustion	2 µg/m^3 for 24 hours	
Nitrogen dioxide	Gas and kerosene space heaters, gas stoves, vehicular exhaust	100 µg/m^3 annual	Respiratory and mucous irritant
Ozone	Photocopying machines, electrostatic air cleaners, outdoor air	235 µg/m^3 for 1 hour	Respiratory irritant, causes fatigue
Pesticides and other semivolatile organics	Sprays and strips, outdoor air	5 µg/m^3 for chlordane	Possible carcinogens
Radon	Soil gas that enters buildings, construction materials, groundwater	4 pCi/L	Lung cancer
Sulfur dioxide	Coal and oil combustion, kerosene space heaters, outside air	80 µg/m^3 annual; 365 µg/m^3 for 24 hours	Respiratory and mucous irritant
Volatile organics	Smoking, cooking, solvents, paints, varnishes, cleaning sprays, carpets, furniture, draperies, clothing	None available	Possible carcinogens

Sources: N. L. Nagda, H. E. Rector, and M. D. Koontz, 1987; M. C. Baechler et al., 1991; E. J. Bardana Jr. and A. Montaro (eds.), 1997; M. Meeker, 1996; D. W. Moffatt, 1997.

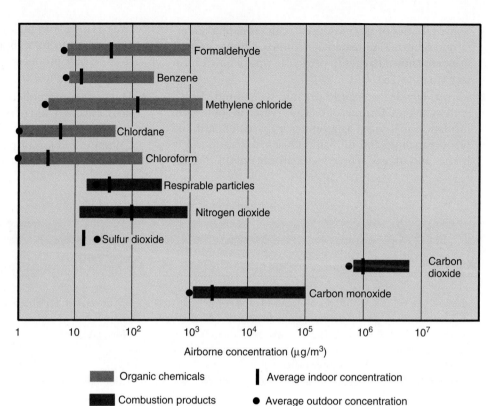

Figure 25.4 ■ Concentrations of common indoor air pollutants compared with outdoor concentrations plotted on a log scale, that is, 10^2=100; 10^3=1,000, 10^4=10,000, etc. [*Source:* A. V. Nero, Jr., "Controlling Indoor Air Pollution," *Scientific American,* 258, no. 5 (1998): 42–48.]

products and processes used in our homes and workplaces are sources of pollution. Furthermore, common indoor air pollutants are often highly concentrated compared with outdoor levels. For example, carbon monoxide, particulates, nitrogen dioxide, radon, and carbon dioxide are generally found in much higher concentrations indoors than outdoors. This important concept is shown in more detail in Figure 25.4, which provides a comparison of indoor with outdoor pollutants.

Why are concentrations of indoor air pollutants generally greater than those found outdoors? One obvious reason is that there are so many potential indoor sources of pollutants. Another reason is somewhat ironic: The effectiveness of the steps we have taken to conserve energy in homes and other buildings has led to the trapping of pollutants inside.

Two of the best ways to conserve energy in homes and other buildings are to increase the insulation and to decrease the infiltration of outside air. Constructing our buildings with windows that do not open and applying extensive caulking and weather-stripping does reduce energy consumption, but it also tends to affect the air quality of the building by reducing natural ventilation. An important function of ventilation is that it replaces the indoor air with outdoor air in which the concentrations of pollutants are generally much lower. With less natural ventilation, we must depend more on the ventilation systems that are part of heating and air-conditioning systems.

25.2 Heating, Ventilation, and Air-Conditioning Systems

Heating, ventilation, and air-conditioning systems are designed to provide a comfortable indoor environment for people. Design of these systems depends on a number of variables, including the activity of people in the building, air temperature and humidity, and air quality. The interaction among these factors determines whether people are comfortable indoors. If the heating, ventilation, and air-conditioning system is designed correctly and functions properly, it will provide thermal comfort for people inhabiting the building. It will also provide the necessary ventilation (utilizing outdoor air) and remove common air pollutants via exhaust fans and filters.[6]

Personal comfort levels in terms of temperature and humidity vary depending on age, physiology, and level of activity. Furthermore, different portions of buildings may have different temperatures and air quality because of their location in relation to heat sources, cold surfaces, and large windows. Humidity should be carefully controlled. High humidity may facilitate the growth of adverse mildews or molds, whereas low humidity may be a source of discomfort to some people.[6]

Regardless of the type of heating, ventilation, and air-conditioning system used in a home or other building, the effectiveness of that unit depends on the proper design of the equipment relative to the building, on proper installation, and on correct maintenance and operating procedures.[6] Indoor air pollution may result if any one of these factors concentrates pollutants from the many possible sources. If filters become plugged or contaminated with fungi, bacteria, or other potentially infectious agents, serious problems can result. In addition, as we see later in this chapter, ventilation systems are not generally designed to reduce some types of indoor pollution.[6, 7]

25.3 Pathways, Processes, and Driving Forces

Many air pollutants originate within buildings and may be concentrated there because of lack of proper ventilation with the outside atmosphere. Other air pollutants may enter a building by infiltration, either through cracks and other openings in the foundations and walls or by way of ventilation systems.

The driving forces that control or modify the flow of air in buildings result from a variety of processes related to both natural forces and human activity. Both natural and human processes in buildings create differential pressures that move air and contaminants from one area to another. Areas of high pressure may develop on the windward side of a building, whereas pressure is lower on the leeward, or protected, side. As a result, air is drawn into a building from the windward side. Opening and closing doors produces pressure differentials that induce air to move within buildings, and wind can affect the movement of air in a building, particularly if the structure is leaky.[6]

A **chimney effect** (or **stack effect**) occurs when there is a temperature differential between the indoor and outdoor environments. Warm air rises within a building. If the indoor air is warmer than that found outdoors, as the warmer air rises in the building to the upper levels, it is replaced in the lower portion of the building by outdoor air drawn in through a variety of openings, such as windows, doors, or cracks in the foundations and walls. Facilities such as elevator shafts and stairwells provide corridors through which air can move from one floor to another.[6] Environmental tobacco smoke, or secondhand smoke, may also be drawn into a building by the chimney effect if smokers move outside near revolving or open doors to smoke.[5]

Because air is such a fluid medium, the possible interactions between the driving forces and the building are complex and the distribution of potential air contaminants and pollutants is extensive. One outcome of this situation is that people in various parts of a building may complain about the air quality even if they are separated by considerable distances from each other and from potential sources of pollution.[6]

25.4 Building Occupants

Typically, the people living or working in particular indoor environments react to pollutants in different ways:

- Some groups of people are particularly susceptible to indoor air pollution problems.
- The symptoms reported by people in a particular environment vary.
- In some cases, the symptoms reported result from factors other than air pollution.

Particularly Susceptible People

People have varying sensitivity to air pollutants. One person may be adversely affected by a particular pollutant, whereas others in nearby areas may seem to be unaffected. Sometimes the problem is a matter of concentration rather than sensitivity; the person most affected by a particular indoor air pollutant might be experiencing the greatest exposure. A person's susceptibility to a particular air pollutant also depends on genetic factors, lifestyle, and age (see Chapter 15).

As a result of these individual differences, the response of one person bothered by a particular pollutant may differ from the response of another affected individual, making it difficult to evaluate indoor pollution problems.

Nevertheless, older people with impaired health and children (because of their activity level and developing lungs) are generally more sensitive to air pollutants.[7] People suffering from chronic lung or respiratory diseases, such as chronic bronchitis, allergies, or asthma, are especially likely to be affected adversely by poor indoor air quality. Another group more strongly affected comprises individuals who have suppressed immune systems owing to disease or medical treatment, such as chemotherapy or radiation therapy.[6] Some people, when exposed to chemicals, develop what is known as multiple chemical sensitivity (MCS), a controversial disease in which people are allergic to any material or product containing human-produced chemicals. Some sufferers of MCS are so sensitive that they effectively have to live in a "plastic bubble" that keeps out chemicals.[5]

Symptoms of Indoor Air Pollution

A great variety of symptoms can result from exposure to indoor air pollutants (Table 25.2). Some chemical pollutants can cause nosebleeds, chronic sinus infections, headaches, and irritation of the skin or eyes, nose, and throat. More serious problems include loss of balance and memory, chronic fatigue, difficulty in speaking, and allergic reactions, including asthma. For example, chlorine tablets, which are often used in swimming pools and hot tubs, are a tremendous irritant if dust from the tablets is inhaled; shortness of breath and coughing result. Other pollutants cause dizziness or nausea. Exposure to carbon monoxide results in shortness of breath at low concentrations. At high concentrations, extreme toxicity and death can result. Tissues sensitive to carbon monoxide include the brain, heart, and muscles.[8]

Table 25.2 • Some Symptoms of Indoor Air Pollution

Symptoms	ETS[a]	Combustion Products[b]	Biologic Pollutants[c]	VOCs[d]	Heavy Metals[e]	SBS[f]
Respiratory						
Inflammation of mucous membranes of the nose, nasal congestion	Yes	Yes	Yes	Yes	No	Yes
Nosebleed	No	No	No	Yes	No	Yes
Cough	Yes	Yes	Yes	Yes	No	Yes
Wheezing, worsening asthma	Yes	Yes	No	Yes	No	Yes
Labored breathing	Yes	No	Yes	No	No	Yes
Severe lung disease	Yes	Yes	Yes	No	No	Yes
Other						
Irritation of mucous membranes of eyes	Yes	Yes	Yes	Yes	No	Yes
Headache or dizziness	Yes	Yes	Yes	Yes	Yes	Yes
Lethargy, fatigue, malaise	No	Yes	Yes	Yes	Yes	Yes
Nausea, vomiting, anorexia	No	Yes	Yes	Yes	Yes	No
Cognitive impairment, personality change	No	Yes	No	Yes	Yes	Yes
Rashes	No	No	Yes	Yes	Yes	No
Fever, chills	No	No	Yes	No	Yes	No
Abnormal heartbeat	Yes	Yes	No	No	Yes	No
Retinal hemorrhage	No	Yes	No	No	No	No
Muscle pain, cramps	No	No	No	Yes	No	Yes
Hearing loss	No	No	No	Yes	No	No

[a] Environmental tobacco smoke.

[b] Combustion products include particles, NO_x, CO, and CO_2.

[c] Biologic pollutants include molds, dust mites, pollen, bacteria, and viruses.

[d] Volatile organic compounds, including formaldehyde and solvents.

[e] Heavy metals include lead and mercury.

[f] Sick building syndrome.

Source: Modified from American Lung Association, Environmental Protection Agency, and American Medical Association, "Indoor Air Pollution—An Introduction for Health Professionals," 523–217/81322 (Washington, D.C.: GPO, 1994).

The symptoms just described may have a quick onset after exposure. Other pollutants, including radon, asbestos, and chemicals such as benzene, may have long-term chronic health effects, including diseases such as cancer. Because of long lag times between exposure and disease, it may be difficult to establish relationships between a particular indoor air environment and disease in an individual.

Sick Buildings

As we saw in the opening case study, an entire building can be considered sick because of environmental problems. There are two types of sick buildings:

■ Buildings with identifiable problems, such as occurrences of toxic molds or bacteria known to cause disease. The diseases are known as *building-related illnesses* (BRI).

■ Buildings with **sick building syndrome (SBS)**, where the symptoms people report cannot be traced to any one known cause.

Sick building syndrome is a condition associated with an indoor environment that appears to be unhealthy. A number of people in such a building report adverse health effects that they believe are related to the amount of time they spend in the building. The range of complaints may vary from funny odors to more serious symptoms, such as headaches, dizziness, nausea, and so forth. In addition, a number of people in the building may be sick; or a group of people may have contracted a disease, such as cancer.

In many cases, it is difficult to establish what may be causing a particular sick building syndrome. Sometimes, the problem has been found to be related to poor management practices and worker morale rather than exposure to toxins in the building. When the occupants of a building report adverse health effects and a study follows, often the cause is not detected. A number of things may be happening:[6]

- The complaints result from the combined effects of a number of contaminants present in the building.
- Environmental stress from a source other than air quality—such as noise, high or low humidity, poor lighting, or overheating—is responsible.
- Employment-related stress—such as stress resulting from poor relations between labor and management, poor morale, or overcrowding—may be leading to the symptoms reported.
- Other unknown factors may be responsible. For example, pollutants or toxins may be present but not identified.

Of course, sick building syndrome may be the combined effect of various aspects of some or all of these factors. As noted, one common aspect of sick building syndrome is that often no one specific disease or cause is easily identified.[6]

We have reviewed some basics concerning sources, processes, and effects of indoor air pollution. Next, we consider two selected pollutants: environmental tobacco smoke and radon gas.

25.5 Environmental Tobacco Smoke

Environmental tobacco smoke (ETS), also known as *secondhand smoke*, comes from two sources: smoke exhaled by smokers and smoke emitted from burning tobacco in cigarettes, cigars, or pipes. People who are exposed to ETS are referred to as *passive smokers*.[9]

ETS is the best known of the hazardous indoor air pollutants. It is hazardous for the following reasons:[9, 10]

- Tobacco smoke contains several thousand chemicals, many of which are irritants. Examples include NOx, CO, hydrogen cyanide, and about 40 carcinogenic chemicals.
- Studies of nonsmoking workers exposed to ETS found that they have reduced airway functions comparable to what would be caused by smoking up to 10 cigarettes per day. They suffer more illnesses, such as coughs, eye irritation, and colds, and lose more work time than those not exposed to ETS.
- In the United States, about 3,000 deaths from lung cancer and 40,000 deaths from heart disease a year are thought to be associated with ETS.

ETS exposure depends on a variety of factors, including the number of people in a room smoking, the size of the room, and the rate of ventilation. Separating smokers from nonsmokers reduces but does not eliminate exposure to ETS. Thus, some U.S. state and local governments have banned smoking in restaurants, bars, and public buildings to protect citizens from ETS.

Smokers must realize that they are harming not only themselves but others, including their families, their friends, and the general public, by polluting the air that others breathe. Indeed, many people consider being exposed to ETS as a transgression of their right to breathe clean air in their homes, their workplaces, and public buildings.

The number of smokers in the United States has declined, but there are still about 40 million. The rate is higher in the developing world, where health warnings are few or nonexistent. Smoking is extremely addictive, because tobacco contains nicotine, a highly addictive substance. Nevertheless, education and social pressure have worked to some extent to influence some thoughtful people to quit smoking and encourage others to quit.

25.6 Radon Gas

It has become apparent within the past few decades that radon gas may constitute a significant environmental health problem in the United States (see A Closer Look 25.1).[11, 12] An interesting aspect of the radon gas hazard is that it comes from natural processes rather than human activities.

Radon is a naturally occurring radioactive gas that is colorless, odorless, and tasteless. It is a member of the naturally occurring radioactive decay chain from radiogenic uranium to stable lead (Figure 25.5). Radon-222, which

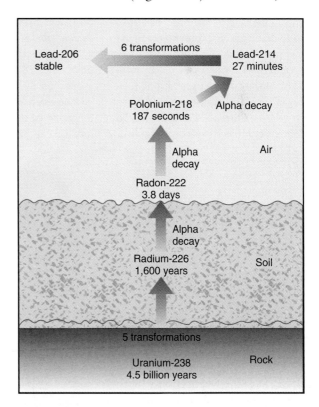

Figure 25.5 ■ Simplified diagram showing the radioactive decay chain for radon. Not all isotopes are shown. Half-lives and types of decay are shown for some. Radon is a gas and can move up from rock and soil into the air. The isotopes of polonium and lead are particles that may be suspended in air or attach to dust particles and move with air currents. They may also settle out of air.

Is Radon Gas Dangerous?

Many people today are anxious and worried about radon gas in homes because studies have shown that exposure to elevated concentrations of radon is associated with increased risk of lung cancer. It is believed that the risk increases with the level of exposure, the length of exposure, and certain habits, such as smoking.[12] The Environmental Protection Agency (EPA) estimates that 14,000 lung cancer deaths per year in the United States are related to exposure to radon gas and its daughter products (products that result from its radioactive decay), primarily polonium-218. (The estimate actually ranges from 7,000 to 30,000.) By comparison, there are approximately 140,000 total lung cancer deaths in the United States each year. If these estimates are correct—and they are controversial—approximately 10% of the lung cancer deaths in the United States can be attributed to radon gas. Exposure to radon gas has also been linked to other forms of cancer, such as melanoma (a deadly form of skin cancer) and leukemia, but such linkages are highly controversial.[18]

It is also believed that exposure to radon gas combined with smoking produces a synergistic effect that is particularly hazardous. One estimate is that the combination of exposure to radon gas and tobacco smoke is 10 to 20 times as hazardous as exposure to either pollutant by itself.[12]

Few comprehensive studies directly link radon gas exposure in houses to increased incidence of lung cancer. One study in Sweden did conclude that exposure to radon in homes is an important cause of lung cancer in the general population.[19] However, the linkage between radon and cancer is mostly based on studies of uranium miners, a group of people exposed to high concentrations of radon in mines.

It is believed that the health risk from radon gas is primarily related to its daughter products, such as polonium-218, which is a particle and thus adheres to dust. It is hypothesized that the dust is then inhaled into lungs, where cell-damaging alpha radiation can occur when polonium-218 decays (it has a half-life of approximately 3 minutes) (Figure 25.6).

The risk related to radon gas has been estimated by the EPA. Table 25.3 relates levels of exposure to radon gas to the estimated number of lung cancer deaths for smokers and nonsmokers.[12] The average concentration of radon gas in the outdoor environment is approximately 0.4 pCi/L; the average indoor level is approximately 1 pCi/L. The EPA has set the action level for radon at 4 pCi/L. This level is the concentration below which exposure is thought by the EPA to be an acceptable risk. In its risk charts (see Table 25.3), the EPA equates the risk associated with exposure to 4 pCi/L for a nonsmoker to about the risk of drowning. For a smoker, the risk increases to about 100 times the risk of dying in an airplane crash. These risks are calculated in terms of long-term (lifetime) exposure. Prediction of the number of deaths from lung cancer resulting from radon exposure is controversial. It has been argued that there is insufficient evidence to support direct correlation of smoking habits and exposure to radon gas with deaths from lung cancer. Furthermore, there are difficulties with measurements. Radon concentrations in one part of a home may be much different from those in another. If homes have basements, concentrations tend to be higher there than on upper level floors.

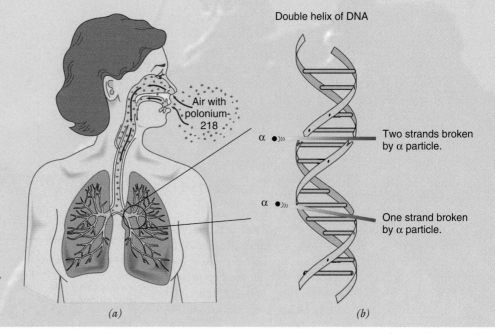

Figure 25.6 ■ (*a*) Deposition of polonium-218 in the lungs. (*b*) Alpha radiation (particle) breaking one or both DNA strands at the cell level. [*Source:* Modified from D. J. Brenner, *Radon: Risk and Remedy* (New York: Freeman, 1989). Reprinted with permission.]

Table 25.3 • Estimated Risk Associated with Radon

Radon Level (pCi/L)	If 1,000 People Were Exposed to This Level over a Lifetime	Risk of Cancer from Radon Exposure Compares to...	What to Do
Radon Risk If You Smoke[a]			Stop smoking and...
20	About 135 people could get lung cancer	100 times the risk of drowning	Fix your home
10	About 71 people could get lung cancer	100 times the risk of dying in a home fire	Fix your home
8	About 57 people could get lung cancer		Fix your home
4	About 29 people could get lung cancer	100 times the risk of dying in an airplane crash	Fix your home
2	About 15 people could get lung cancer	2 times the risk of dying in a car crash	Consider fixing between 2 and 4 pCi/L
1.3	About 9 people could get lung cancer	Average indoor radon level	
0.4	About 3 people could get lung cancer	Average outdoor radon level	Reducing radon levels below 2 pCi/L is difficult
Radon Risk If You Never Smoke[b]			
20	About 8 people could get lung cancer	The risk of being killed in a violent crime	Fix your home
10	About 4 people could get lung cancer		Fix your home
8	About 3 people could get lung cancer	10 times the risk of dying in an airplane crash	Fix your home
4	About 2 people could get lung cancer	The risk of drowning	Fix your home
2	About 1 person could get lung cancer	The risk of dying in a home fire	Consider fixing between 2 and 4 pCi/L
1.3	Less than 1 person could get lung cancer	Average indoor radon level	
0.4	Less than 1 person could get lung cancer	Average outdoor radon level	Reducing radon levels below 2 pCi/L is difficult

[a] If you are a former smoker, your risk may be lower.

[b] If you are a former smoker, your risk may be higher.

Source: U.S. Environmental Protection Agency, *A Citizen's Guide to Radon,* 2nd ed., ANR-464, 1992.

Nevertheless, all scientists agree that exposure to high levels of radon can cause cancer.

If the estimated risks from radon gas are anywhere close to the actual risk, then the hazard is a large one. The Surgeon General of the United States has stated that "indoor radon gas is a national health problem." The risks posed by exposure are thought to be hundreds of times higher than risks resulting from outdoor pollutants present in air and water. Such pollutants are generally regulated to reduce the risk of premature death and disease to less than 0.001%. Risks from some indoor pollutants, such as organic chemicals, may be as high as 0.1%.[20] These risks still are very small compared with the risk for radon. For example, people who live in homes for about 20 years with an average concentration of radon of about 25 pCi/L are estimated to have a 1% to 2% chance of contracting lung cancer.[12, 20]

has a half-life of 3.8 days, is the product of radioactive decay of radium-226. Radon decays with emission of an alpha particle to polonium-218, which has a half-life of approximately 3 minutes. (The discussion of radiation, radiation units, radiation doses, and health problems related to radiation in Chapter 20 will help you understand the following materials.)

Radon was discovered in 1900 by Ernest Dorn, a German chemist. The use and misuse of radon has an interesting history. In the early 1900s, bathing in radon water became a health fad. During this period, when radon was thought to be beneficial to health, many products containing radium, and thus radon, hit the market. These included chocolate candies, bread, and toothpaste. As recently as 1953, a contraceptive jelly containing radium was marketed in the United States.[11]

Geology and Radon Gas

The concentration of radon gas that reaches the surface of the Earth and thus can enter our dwellings is related to the concentration of radon in the rocks and soil, as well as the efficiency of the transfer processes from the rocks or soil to the surface. Some regions in the United States contain bedrock with an above-average natural concentration of uranium. A large area that includes parts of Pennsylvania, New Jersey, and New York is now famous for elevated concentrations of radon gas. This area, known as the Reading Prong, contains many homes with elevated radon gas concentrations.[11] Areas with elevated concentrations of radon have also been identified in a number of other states, including Florida, Illinois, New Mexico, South Dakota, North Dakota, Washington, and California.

How Does Radon Gas Enter Homes and Other Buildings?

A radon legend was born in Boyer Town, Pennsylvania, in 1984, when Stanley Watres, who had a job as a technical advisor to the Limerick Nuclear Power Station, set off radiation alarms on the way into the plant. The reactor in the power plant had not yet been turned on when the alarms went off, and extensive testing of Watres's clothing suggested that the contamination came not from where he worked but from where he lived. Investigators were astounded to find that the radiation level in his home was 3,200 pCi/L, 800 times higher than the action level of 4 pCi/L set by the Environmental Protection Agency! Until that time, scientists had not believed that radon could occur naturally in concentrations high enough to be hazardous to anyone.[11,13–15] The Watres home held the record until the latter part of the 1980s, when a home in Whispering Hills, New Jersey, was found to have a radiation level of 3,500 pCi/L.[14]

Radon gas enters homes and other buildings in three main ways (Figure 25.7):

1. It migrates up from soil and rock into basements and lower floors.

2. Dissolved in groundwater, it is pumped into wells and then into homes.

3. Radon-contaminated materials, such as building blocks, are used in construction.

It is very difficult to estimate the number of homes in the United States that may have elevated concentrations of radon gas. The EPA estimates that about 7% of U.S. homes have elevated radon levels and recommends that all homes and schools be tested.[12] The test is simple and inexpensive.

Radon-Resistant Techniques for Homes and Other Buildings

Protection of new homes from potential radon gas problems is straightforward and relatively inexpensive. It is also easy to upgrade a home to reduce radon. Some of the common radon-resistant construction techniques are shown in Figure 25.8. Although the techniques are variable depending on the type of foundation a particular home has, the basic strategy is to prevent radon from entering a home and to safely ensure that radon is removed from the home site. [16, 17]

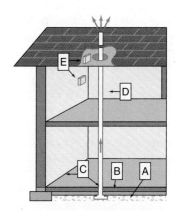

Figure 25.8 ■ Common methods to reduce radon exposure in homes and other buildings. *Source:* EPA.Radon-Resistant New Construction. Accessed 4/22/06 at www.epa.gov.
A. Gas Permeable Layer: A layer (often clean gravel) is placed beneath the slab or flooring system to allow the soil gas to move freely underneath the house.
B. Plastic Sheeting: Plasting sheeting is placed on top of the gas-permeable layer and under the slab to help prevent the soil gas from entering the home. In homes with a crawlspace, the sheeting is placed over the crawlspace floor.
C. Sealing and Caulking: Openings in the concrete foundation floor are sealed to reduce soil gas entry into the home.
D. Vent Pipe: A 3- or 4-inch PVC pipe (commonly used for plumbing) runs from the gas-permeable layer through the house to the roof to safely vent radon above the house.
E. Junction Box: An electrical junction box is installed if an electic venting fan is needed.

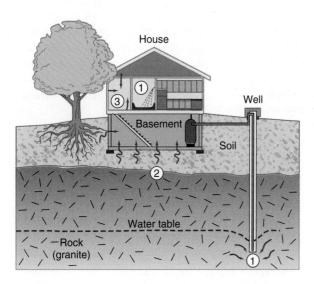

Figure 25.7 ■ How radon can enter homes. (1) Radon in groundwater enters a well and goes to a house, where it is used for water supply, dishwashing, showers, and other purposes. (2) Radon gas in rocks and soil migrates into a basement through cracks in the foundation and pores in construction. (3) Radon gas is emitted from construction materials used in building a house. [*Source:* Environmental Protection Agency.]

25.7 Indoor Air Pollution and Green Buildings

There is a movement underway in the United States and the world to create buildings that have a healthy environment for their occupants. The phrase for this objective is "making the building green." The processes involve using building designs that result in less pollution and better use of resources. When we are referring to the quality of the air and other aspects of the environment inside buildings, we are talking about the indoor environmental quality. That quality is based in part on the concentrations of pollutants and other conditions that may impact the health and comfort of occupants of a building. Providing for a good indoor environmental quality is a significant part of the **green building** concept. One of the basic objectives of green building design is to improve the indoor environmental quality through designing, constructing, and maintaining buildings that minimize indoor air pollutants, ensuring that fresh air is supplied and circulated, and managing moisture content to remove the threat of moisture-related problems such as mold.

25.8 Control of Indoor Air Pollution

In terms of the workplace, there are strong financial incentives to provide workers with a clean air environment. As much as $250 billion per year might be saved by decreasing illnesses and increasing productivity through improving the work environment.[5] A good starting point would be passing environmental legislation requiring minimum indoor air quality standards. This should include increasing the inflow of fresh air through ventilation. In Europe, systems of filters and pumps in many office buildings circulate air three times as frequently as is typical in buildings in the United States. Many building codes in Europe require that workers have access to fresh air (windows) and natural light. Unfortunately for U. S. workers,

no similar codes exist in the United States, and many buildings use central air-conditioning with windows permanently sealed.[5]

You might think that heating, ventilating, and air-conditioning systems, when operating properly and well maintained, will ensure good indoor air quality. Unfortunately, these systems are not designed to maintain all aspects of air quality. For example, commonly used ventilation systems do not generally reduce radon gas. Ventilation is one control strategy when faced with high concentrations of any indoor air pollutant, including radon. Other strategies, shown in Table 25.4, include source removal, source modification, and air cleaning.[8] These strategies do not constitute a complete list, and some combination of them may be the best approach.

One of the principal means for controlling the quality of indoor air is by dilution with fresh outdoor air via a ventilating air-conditioning system and windows that can be opened. Outside air is brought in and mixed with air in the return flow system from the building; the air is filtered, heated or cooled, and supplied to the building. Various types of air-cleaning systems for residential and nonresidential buildings are available to reduce potential pollutants, such as particles, vapors, and gases. These systems can be installed as part of the heating, ventilation, and air-conditioning system or as stand-alone appliances.[8]

Education also plays an important role in understanding and developing strategies for reducing indoor air pollution problems; it empowers people with knowledge necessary to make intelligent decisions. At one level, this may involve deciding not to install unvented or poorly vented appliances. A surprising (and tragic) number of people are killed each year by carbon monoxide poisoning resulting from poor ventilation in homes, campers, and tents. At other levels, educated people are more aware of their legal rights with respect to product liability and safety. Furthermore, education provides people with the information necessary to make decisions concerning exposure to chemicals, such as paints and solvents, and strategies to avoid potentially hazardous conditions in the home and workplace.[8]

Table 25.4 • Strategies to Control Indoor Air Pollution

Ventilation: General ventilation system; spot (zone or localized) ventilation (exhaust fans, etc.)

Source removal: Material or product substitution; restrictions on source use (e.g., establishment of smoking areas; restrictions on sale of particular items and on activities that cause indoor air pollution)

Source modification: Change in combustion design (e.g., maximize efficiency of a gas stove); material substitution (use materials that don't cause air pollution); reduction in emission rates by intervention of barriers (e.g., apply coatings over lead paint or asbestos)

Air cleaning (pollutant removal): Particle filtering; gas and vapor removal; passive scavenging or absorption

Education: Consumer information on products and materials; public information on health, productivity, and nuisance effects; resolution of legal rights and liabilities of consumer, tenant, manufacturer, and so on, related to indoor air quality

Source: Modified from Committee on Indoor Pollutants, *Indoor Pollutants* (Washington, D.C., 1981), p. 489.

Are Airplanes Adequately Ventilated?

The environment at 10,667 m (35,000 ft) above Earth, a typical cruising altitude for jet airplanes, is not fit for humans. The temperature is –54°C (–65°F), the humidity is close to zero, and the air pressure is low. To supply air for the passengers and crew of a jet airplane, outside air is compressed and heated by the jet engines. Because the air emerges from the engines at a higher pressure and temperature than would be comfortable, the air is cooled by mixture with outside air before entering the cabin, and the pressure is adjusted to that of an altitude of 1,524 m (5,000 ft). Humidity, between 5 and 20% at cruising altitudes, is supplied primarily by the breath and perspiration of passengers and crew.

Until the mid-1980s, commercial planes used 100% fresh air, which was recirculated every three minutes. But using fresh air taken in by the jet engines to provide air inside the plane reduces fuel efficiency (Figure 25.9). Providing fresh air in an airplane costs 22 to 37 times as much as supplying the same amount inside a building in Washington, D.C., in January. Airlines found they could save money by recirculating air already inside the cabin in combination with fresh air (usually in a 1:1 proportion). This practice reduced the flow rate, so that there was a complete change of air only every seven minutes or even less often.

Air flow is measured in cubic feet per minute (cfm); ventilation rate is the air flow per person (cfm/number of people). Ventilation rates vary in commercial planes from 150 cfm/person in the cockpit to 50 cfm/person in first class to 7 cfm/person in economy class. The latter is comparable to the rates found in trains and subways (5–7 cfm/person). Airline officials point out that the rate of air change in planes is much better than that found in office buildings (10–12 times an hour as opposed to 1–2 times).

Many flight attendants and passengers have complained about symptoms, such as headaches, dry eyes, fatigue, nausea, and upper respiratory problems, that could be associated with the air quality on planes. The carbon dioxide level, which is often used as an indicator of indoor air quality, averages 1,500 ppm in planes but has been measured as high as 2,000 ppm on 25% of flights studied. Even with ventilation rates as high as 35 cfm/person, carbon dioxide levels have been found at 1,200 ppm or more. The American Society of Heating, Refrigerating, and Air-Conditioning Engineers (ASHRAE) recommends a maximum of 1,000 ppm for indoor air, and the Occupational Safety and Health Administration (OSHA) has a standard of 5,000 ppm for industrial buildings. The Federal Aviation Administration (FAA) standard allows carbon dioxide levels up to 30,000 ppm.

Some experts are concerned that high carbon dioxide levels may be associated with other undesirable constituents, such as chemicals from cleaning fluids, pesticides, and fuels. Airborne contamination from sick passengers and animals below deck is another source of concern. Outbreaks of influenza have been tied to airplane air. Three passengers on one flight in 1992 became infected with tuberculosis, although it could not be clearly determined that infection occurred onboard. This occurrence and the fact that the incidence of tuberculosis has been on the rise worldwide since the 1980s sparked an inquiry by the U.S. Congress. Although medical experts concluded that long periods of close contact were required for transmission of tuberculosis, concern was great enough for the Centers for Disease Control (now called the Centers for Disease Control and Prevention) in Atlanta to undertake a study of the subject in 1993. The results of this study confirmed that although there is a possibility of tuberculosis transmission aboard airplanes, the possibility is extremely small. Thus, it is unlikely that airplane cabin air circulation standards will be affected.

Critical Thinking Questions

1. Airlines contend that higher ventilation rates are necessary in the cockpit. What reasons can you think of to support their argument? Flight attendants claim that their requirements for air are different from those of passengers. What do you think is the basis for this claim?

2. If air exchange in airliners is as good or better than that in trains and subways, is there any reason to question the air quality? Develop a rationale for a position on this question.

3. What criteria would you accept for establishing a standard for the level of carbon dioxide in airplanes?

4. Three options exist for reducing carbon dioxide levels in airliner cabins: reducing emissions, increasing ventilation rates, and absorbing carbon dioxide from the ambient air. What are the advantages and disadvantages of each option?

5. Very little research has been done on air quality and the health of employees and passengers. Why, even if research is conducted, might it be hard to prove a connection?

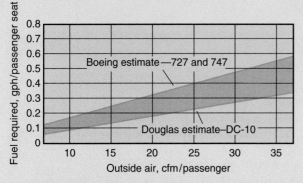

Figure 25.9 ■ Fuel required for ventilation with outside air on a jet airplane. [*Source:* Data from National Research Council, *The Airliner Cabin Environment* (Washington, D.C.: National Academy Press, 1989, p. 61).]

Summary

- Indoor air pollution has been with us for thousands of years, since people first built structures and burned fuel indoors. It is one of our most serious environmental health problems.

- Sources of indoor air pollution are extremely varied. They may be associated with the materials with which we build our buildings, the furnishings we put in them, and the types of equipment we use for heating and cooling, as well as natural processes that allow gases to seep into buildings.

- Concentrations of indoor air pollutants are generally greater than concentrations of the same pollutants found outdoors. Part of the reason is that in recent years we have attempted to conserve energy through better insulation of buildings. The common method for controlling indoor air pollution is ventilation. However, natural ventilation has been reduced through tighter construction of buildings, and commonly used ventilation systems are not generally designed to reduce certain types of indoor air pollutants. In addition, these systems require careful maintenance.

- A variety of pathways, processes, and driving forces affect the air quality of a building. The most common natural processes involve the differential pressure produced by wind and the chimney, or stack, effect, which occurs when there is a temperature differential between indoor and outdoor environments.

- Indoor air pollution has different effects on different people, and some groups of people are particularly susceptible to air pollution problems. Often, the symptoms reported by people working in a building vary. Some symptoms may result from factors other than air pollution.

- There are two basic types of sick buildings: those with an identifiable problem, such as mold or bacteria, and those with sick building syndrome, in which people's symptoms can't be traced to any one cause.

- Environmental tobacco smoke, or secondhand smoke, is the most hazardous indoor air pollutant.

- Radon gas that seeps into homes is thought to be a serious environmental health hazard in the United States today. Studies have suggested that exposure to elevated concentrations of radon is associated with an increased risk of lung cancer.

- Control of indoor air pollution involves several strategies, including ventilation, source removal, source modification, and installation of air-cleaning equipment, as well as education.

REEXAMINING THEMES AND ISSUES

Human Population

Indoor air pollution has been with us since we made the decision to build homes and move inside. As human populations grew and people constructed a broader variety of homes, they were exposed to additional pollutants. As the number of people on Earth continues to increase, people will use more and different resources to build homes, and more people are likely to be living in smaller spaces. As a result, effects of indoor air pollution are expected to increase in the future and to continue to present serious environmental health problems.

Sustainability

If we accept the premise that sustainability begins at home, then air quality in our homes and our workplaces is an important part of a sustainable future. Our health and that of future generations depend on breathing pollution-free air where we spend most of our time—indoors. Conflicts may arise, however, because we also wish to conserve resources and build energy-efficient homes as part of sustainability. As we have seen, that may lead to restricted circulation of air and resulting indoor air pollution problems. This conflict may be solved through technological advances that minimize air pollution while maximizing energy efficiency.

Global Perspective

Indoor air pollution occurs everywhere in the world where people live inside or work in buildings or other human-constructed enclosures such as mines. Strictly speaking, indoor air pollution is not a global problem; but it *is* a significant issue if we wish to address globally experienced human health concerns. Environmental tobacco smoke is a serious health problem, as is exposure to high levels of radon gas. These are examples of toxicity and health problems found almost everywhere people live.

Urban World

Indoor air pollution problems may occur anywhere people live inside. However, problems are more likely in urban areas that have significant outdoor air pollution problems. Therefore, in cities, if we wish to help eliminate indoor air pollution, we must also pay attention to outdoor air quality.

People and Nature

Indoor air pollutants have a variety of sources. Most are human-made chemicals. However, radon gas, a serious indoor air pollutant, is a natural product of Earth processes in soil and rocks. Thus, we see that nature beyond human activities can also pollute.

Science and Values

For too long, we have not paid sufficient attention to our indoor environment and the consequences of poor air quality there. This issue is now moving to the forefront, however. We believe that, in the future, technological advances will allow for better design of homes, buildings, and other structures to maximize air quality for the people who spend their time there. People value clean air, as can be seen in the attitude of many people toward secondhand smoke. Secondhand smoke presents a serious health risk to nonsmokers, and there is considerable social pressure to encourage people to quit smoking.

Key Terms

black lung disease **555**

chimney effect (stack effect) **559**

environmental tobacco smoke (ETS) **561**

green building **565**

radon **561**

sick building syndrome (SBS) **560**

Study Questions

1. What are some of the common sources of air pollutants where you live, work, or attend classes?
2. Develop a research plan to complete an audit of the indoor air quality in your local library. How might that research plan differ from a similar audit for the science buildings on your campus?
3. What do you think about the concept of sick building syndrome? If you were working for a large corporation and a number of your employees stated that they were becoming sick and listed a series of symptoms and problems, how would you react? What could you do? Play the role of the administrator, and develop a plan to look at the potential problem.
4. Some people argue that the potential hazard from radon gas in homes is much less than suggested by the Environmental Protection Agency. Do you agree or disagree? How might potential differences of opinion ultimately be answered?
5. Develop a plan to study the potential radon hazard in your community. Where would you start? How would you gather data, and so on? If your community has undergone extensive testing already, review the results and decide if further testing is necessary.
6. Do you think the concept of the green building is economically viable? Why? Why not?

Brenner, D. J. 1989. *Radon: Risk and Remedy*. New York: Freeman. A wonderful book concerning the hazard of radon gas. It covers everything from the history of the problem to what was happening in 1989, as well as solutions, and is highly recommended.

Brooks, B. O., and W. F. Davis. 1992. *Understanding Indoor Air Quality*. Ann Arbor, Mich.: CRC Press. A comprehensive evaluation of indoor air pollution. It discusses most of the sources of indoor air pollutants, as well as health effects and controls.

Kay, J. G., G. E. Keller, and J. F. Miller. 1991. *Indoor Air Pollution*. Chelsea, Mich.: Lewis. Essays on problems with biological and non-biological air pollution in the indoor environment.

Marconi, M., B. Seifert, and T. Lindvall. 1995. *Indoor Air Quality: A Comprehensive Reference Book*. New York: Elsevier. A thorough text covering all aspects of indoor air pollution and air quality.

U.S. Environmental Protection Agency. 1991. *Building Air Quality*. EPA/400/1-91/033, DHHS (NIOSH). Publication No. 91-114. A guide to air quality issues for people building structures and managing them. It has a good section on factors affecting indoor air quality and another on resolving problems.

U.S. Environmental Protection Agency. 1995. *The Inside Story: A Guide to Indoor Air Quality*. A brief introduction to the concepts of indoor air pollution.

U.S. Environmental Protection Agency. www.epa.gov. Very thorough Web site including information on most indoor air pollutants.

U.S. Environmental Protection Agency. 2003. *A Consumer's Guide to Radon Reduction*. EPA 402-K-03-002. A guide to radon remediation and the basics of radon gas exposure.

CHAPTER 26

Ozone Depletion

Sunscreen lotion being applied to a child's skin. Sunscreen reduces exposure to UVB, which is known to cause skin cancer.

Learning Objectives

Ozone depletion in the stratosphere is recognized as a major environmental problem with serious effects. After reading this chapter, you should understand:

- What ozone is and how ozone is naturally formed and destroyed in the stratosphere.

- What the so-called ozone shield is, and why it is important.

- How chemical and physical processes and reactions link emissions of chlorofluorocarbons (CFCs) to stratospheric ozone depletion.

- What role polar stratospheric clouds play in ozone depletion.

- Why ozone depletion is a long-term problem.

- What the environmental effects of ozone depletion are, and what options are available to minimize ozone depletion.

- Why international cooperation, including significant economic aid from wealthy to less wealthy nations, is necessary to encourage future reduction or elimination of emissions of ozone-depleting chemicals into the atmosphere.

Epidemic of Skin Cancer

People in the United States are experiencing an epidemic of skin cancer. A worldwide increase in skin cancer has occurred since early 1970. It is suspected that the increase in skin cancer rates is linked to ozone depletion.[1] Ozone depletion and, in particular, development of the "ozone hole" (discussed later in this chapter) have been most dramatic since about 1975, and this correlates with the increased incidence of skin cancers.

As ozone depletion occurs, ultraviolet B (UVB) radiation, a normal part of the solar radiation that reaches Earth's surface increases. UVB causes damage and mutation to DNA in skin cells, which may initiate cancer. It is believed that a 1% decrease in ozone causes an increase of UVB radiation of about 1 to 2%. For each 1% increase in UVB radiation, it is projected that skin cancer will increase about 2%.

The hightest rates of skin cancer in the world are in Australia. Most Australians of European descent are fair-skinned, rendering them more vulnerable to the clear sky and intense Australian sun. As a result of an increase in exposure to the UVB radiation, the average fair-skinned Australian is about 10 times more likely to develop skin cancer than fair-skinned people in Northern Europe. Australia is also closer to the Antarctic with its "ozone-hole" than are Europe and North America. As a result, southern Australian cities (including Melbourne and Sidney) experience an increase in exposure to UVB radiation if a region of depleted ozone air drifts overhead.

Since 1970, ozone depletion in the atmosphere above the United States has been about 10%. Under a worst-case scenario, this 10% decrease in ozone could cause an increase in skin cancer of 20 to 40% over the same period. However, the observed increase has been 90%! Thus, either we are incorrectly evaluating the effects of ozone depletion and increased UVB radiation on skin cancers, or diagnosis of skin cancer has improved, so more cases are found; or other factors are affecting the incidence of skin cancers. Indeed, because disease seldom involves a one cause–one effect relationship, the epidemic of skin cancers is probably due to multiple interrelated causes, including the following:

- People are living longer, and cancers are a disease of aging.
- Cancer seems to be more a problem of urbanized and industrial societies.
- People in the United States today are more affluent than in the past and spend more time outdoors exposed to cancer-inducing UVB radiation.

Older people who for years worked outside as fishermen, farmers, lifeguards, or in construction, as well as those who sunbathed regularly, are at greater risk of contracting skin cancer. Sunburns that produce blistering and severe sunburns in childhood are thought to be especially hazardous, increasing the risk of melanoma in later life.[1] The risk to an individual depends on two factors: the extent of exposure to UVB and the amount of protective melanin pigment in the skin. Melanin pigment absorbs UVB and provides a natural protection. Thus, fair-skinned people who have little melanin compared with darker-skinned people have a higher incidence of skin cancer.

Ozone depletion is a serious global environmental problem. We understand what causes ozone depletion and have implemented appropriate policies to solve the depletion problem.[2] The science of ozone depletion and the policies to reduce it are the two main subjects of this chapter.

26.1 Ozone

The air we breathe at sea level is composed of approximately 21% diatomic oxygen (O_2), which is two oxygen atoms bonded together. **Ozone** (O_3) is a triatomic form of oxygen in which three atoms of oxygen are bonded. Ozone is a strong oxidant and chemically reacts with many materials in the atmosphere. In the lower atmosphere, ozone is a pollutant produced by photochemical reactions involving sunlight, nitrogen oxides, hydrocarbons, and diatomic oxygen.

Figure 26.1 shows the structure of the atmosphere and concentrations of ozone. The highest concentrations are in the stratosphere, ranging from about 15 km to 40 km (9 to 25 mi) in altitude. Approximately 90% of the ozone in the atmosphere is found in the stratosphere, where peak concentrations are about 400 ppb. The altitude of peak concentration varies from about 30 km (19 mi) near the equator to about 15 km (9 mi) in polar regions.[3]

Ultraviolet Radiation and Ozone

The ozone layer in the stratosphere is often called the **ozone shield**, because it absorbs most of the potentially hazardous ultraviolet radiation that enters Earth's atmosphere from the sun. Figure 26.2 shows part of the electromagnetic spectrum, discussed in Chapter 23. Ultraviolet radiation consists of wavelengths between 0.1 and 0.4 μm and is subdivided into ultraviolet A (UVA), ultraviolet B (UVB), and ultraviolet C (UVC). Ultraviolet radiation with a wavelength of less than about 0.3 μm is potentially very hazardous to life. If much of this radiation reached Earth's surface, it would injure or kill most living things.

Ultraviolet C (UVC) has the shortest wavelength and is the most energetic of the types of ultraviolet radiation. It has sufficient energy to break down diatomic oxygen (O_2) in the stratosphere into two oxygen atoms. Each of these oxygen atoms may combine with an O_2 molecule to create ozone. Ultraviolet C is strongly absorbed in the stratosphere, and negligible amounts reach the surface of Earth.[3,4]

Ultraviolet A (UVA) radiation has the longest wavelength and the least energy of the three types of ultraviolet radiation. UVA can cause some damage to living cells, is not affected by stratospheric ozone, and is transmitted to the surface of Earth.[3]

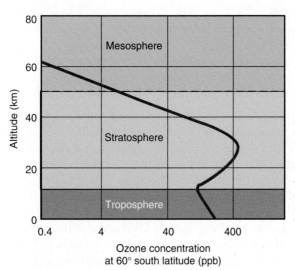

Stratosphere ozone (ozone layer):
Contains 90% of atmospheric ozone; it is the primary UV radiation screen.

Troposphere ozone:
Contains 10% of atmospheric ozone; it is smog ozone, toxic to humans, other animals, and vegetation.

(a)

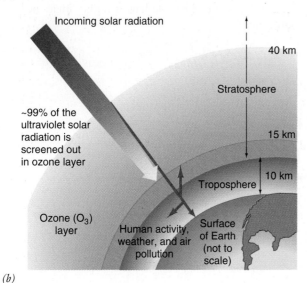

(b)

Figure 26.1 ■ (*a*) Structure of the atmosphere and ozone concentration. (*b*) Reduction of the most potentially biologically damaging ultraviolet radiation by ozone in the stratosphere. [*Source:* Ozone concentrations modified from R. T. Watson, "Atmospheric Ozone," in J. G. Titus, ed., *Effects of Change in Stratospheric Ozone and Global Climate*, vol. 1, *Overview*, p. 70 (U.S. Environmental Protection Agency).]

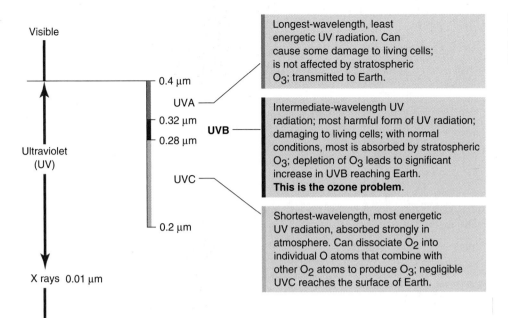

Longest-wavelength, least energetic UV radiation. Can cause some damage to living cells; is not affected by stratospheric O_3; transmitted to Earth.

Intermediate-wavelength UV radiation; most harmful form of UV radiation; damaging to living cells; with normal conditions, most is absorbed by stratospheric O_3; depletion of O_3 leads to significant increase in UVB reaching Earth. **This is the ozone problem.**

Shortest-wavelength, most energetic UV radiation, absorbed strongly in atmosphere. Can dissociate O_2 into individual O atoms that combine with other O_2 atoms to produce O_3; negligible UVC reaches the surface of Earth.

Ultraviolet B (UVB) radiation is energetic and is strongly absorbed by stratospheric ozone. Ozone is the only known gas that absorbs UVB. As a result, depletion of ozone in the stratosphere results in an increase in the UVB that reaches the surface of the Earth. Because UVB radiation is known to be hazardous to living things,[3, 4] this increase in UVB is the hazard we are talking about when we discuss the ozone problem.

Processes that produce ozone in the stratosphere are illustrated in Figure 26.3. The first process, or step, in ozone production occurs when intense ultraviolet radiation (UVC) breaks apart an oxygen molecule (O_2) through the process of photodissociation into two oxygen atoms. These atoms then react with another oxygen molecule to form two ozone molecules. Ozone, once produced, may absorb UVC radiation, which breaks the

ozone molecule into an oxygen molecule and an oxygen atom. This is followed by the recombination of the oxygen atom with another oxygen molecule to re-form into ozone. As part of this process, UVC radiation is converted to heat energy in the stratosphere.[5] Natural conditions that prevail in the stratosphere result in a dynamic balance between the creation and destruction of ozone.

In summary, approximately 99% of all ultraviolet solar radiation (all UVC and most UVB) is absorbed or screened out in the ozone layer. The absorption of ultraviolet radiation by ozone is a natural service function of the ozone shield and protects us from the potentially harmful effects of ultraviolet radiation.

Measurement of Stratospheric Ozone

Scientists first measured the concentration of atmospheric ozone in the 1920s, from the ground, using an instrument known as a Dobson ultraviolet spectrometer. The **Dobson unit (DU)** is still commonly used to measure the concentration of ozone; 1 DU is equivalent to a concentration of 1 ppb O_3. Today, we have a record of ozone concentrations from more than 30 locations around the world over about 30 years. Most of the measurement stations are in the midlatitudes, and the accuracy of the data varies with different levels of quality control.[3] Satellite measurements of concentrations of atmospheric ozone began in 1970 and continue today.

Ground-based measurements first identified ozone depletion over the Antarctic. Members of the British Antarctic Survey began to measure ozone in 1957 and in 1985 published the first data that suggested significant ozone depletion over Antarctica. The data are taken during October of each year—the Antarctic spring—and show that the concentration of ozone hovered around 300 DU from 1957 to

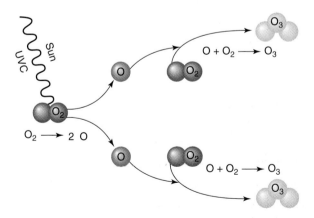

Figure 26.3 ■ Production of ozone (O_3) in the stratosphere. Photo dissociation of the oxygen molecule (O_2) yields two atoms of oxygen. Each combines with an oxygen molecule to form ozone (O_3). [*Source:* Modified from NASA-GSFC, "Stratospheric Ozone," accessed August 22, 2000 at http://see.gsfc.nasa.gov.]

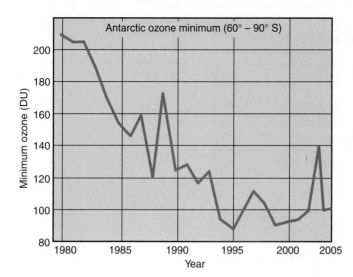

Figure 26.4 ■ Average Antarctic minimum ozone concentration, 1980 to 2005. Values in the 1970s were about 300 DU. [*Source:* Modified from NASA 2006. www.nasa.gov.]

about 1970, and then declined to about 200 DU by 1983. It then sharply dropped to approximately 150 DU by 1986. Since then, the variability of the minimum ozone concentration has been considerable, with a high of about 175 DU in 1989 and a low of about 90 DU in 1995 (Figure 26.4). By 2003, the value was about 140 DU, but dropped again to about 100 DU in 2005. In spite of the variations, the direction of change, with minor exceptions, is clear: Ozone concentrations in the stratosphere during the Antarctic spring have been decreasing since the mid-1970s.[6–9]

Satellite measurements of ozone recorded prior to 1985 had also indicated a significant reduction in ozone concentration; however, the values were so low that they were not believed. After the 1985 announcement of the decrease in ozone over Antarctica, the satellite measurements were reevaluated and found to confirm the observations reported by the British Antarctic Survey. This depletion in ozone was dubbed the *ozone hole*. However, there is no actual hole in the ozone shield where all the ozone is depleted; rather, the term describes a relative depletion in the concentration of ozone that occurs during the Antarctic spring.

26.2 Ozone Depletion and CFCs

The hypothesis that ozone in the stratosphere is being depleted by the presence of **chlorofluorocarbons (CFCs)** was first suggested in 1974 by Mario Molina and F. Sherwood Rowland.[10] This hypothesis, based for the most part on physical and chemical properties of CFCs and knowledge about atmospheric conditions, was immediately controversial. The idea received a tremendous amount of exposure both in newspapers and on television and was vigorously debated by scientists, companies producing CFCs, and other interested parties.

The public became concerned because everyday products, such as shaving cream, hair spray, deodorants, paints, and insecticides, were packaged in spray cans that carried CFCs as a propellant. The idea that these products could be responsible for threatening their health and the well-being of the environment captured the imagination of the American people, many of whom responded by writing to their senators and representatives and making individual decisions to purchase fewer products containing CFCs.[11]

The major features of the Molina and Rowland hypothesis are as follows:[3]

- The CFCs emitted in the lower atmosphere by human activity are extremely stable. They are unreactive in the lower atmosphere and therefore have a very long residence time (about 100 years). Another way of stating this is to say that no significant tropospheric sinks for CFCs are known. A possible exception is soils, which evidently do remove an unknown amount of CFCs from the atmosphere at Earth's surface.[12]

- Because CFCs have a long residence time in the lower atmosphere and because the lower atmosphere is very fluid, with abundant mixing, the CFCs eventually (by the process of dispersion) wander upward and enter the stratosphere. Once they have reached altitudes above most of the stratospheric ozone, they may be destroyed by the highly energetic solar ultraviolet radiation. This process releases chlorine, a highly reactive atom.

- The reactive chlorine released may then enter into reactions that deplete ozone in the stratosphere.

- The result of the depletion of ozone is an increase in the amount of UVB radiation that reaches Earth's surface. Ultraviolet B is a cause of human skin cancers and is also thought to be harmful to the human immune system.

Emissions and Uses of Ozone-Depleting Chemicals

Emissions of the commonly used chemicals related to ozone depletion are shown in Table 26.1. Emissions of chemicals thought to destroy stratospheric ozone amounted to approximately 1.5 million metric tons in 1989, with CFCs accounting for approximately 60% of the total emissions.

Table 26.1 also shows the approximate atmospheric lifetimes of these chemicals, which varies from about 5 years to more than 100 years. Because CFCs, in particular, have such long lifetimes in the atmosphere, they will be with us for many years.

Various uses of the chemicals are also shown in the table. CFCs have been used as aerosol propellants in spray cans, as a working gas in refrigeration and air-conditioning units, and in the foam-blowing process for the production of Styrofoam. A variety of cleaning solvents, such as carbon tetrachloride and methyl chloroform, contain chlorine and thus destroy ozone, as does halon, which contains bromine

Table 26.1 • Emissions of Some Chemicals Associated with Stratospheric Ozone Depletion in 1989, with Growth Rate of CFCs for 1998 and Substitutes for CFCs

Chemical	Emissions (thousands of tons)	Atmospheric Lifetime[a] (years)	Applications	Annual Growth Rate to 1989 (%)	Share of Contribution to Depletion (%)	Annual Growth Rate, 1998 (%)
CFC-12	454	139	Air-conditioning, refrigeration, aerosols, foams	5	45	0.5
CFC-11	262	76	Foams, aerosols, refrigeration	5	26	–0.1
CFC-113	152	92	Solvents	10	12	0.0
Carbon tetrachloride	73	35	Solvents	1	8	
Methyl chloroform	522	5	Solvents	7	5	
Halon 1301	3	65	Fire extinguishers	n.a.	4	
Halon 1211	79	11	Refrigeration, foams	11	0	
HCFC-22		12	Substitute for CFC			
HCFC-123		2	Substitute for CFC			
HCFC-124		6	Substitute for CFC			

[a]Sources for atmospheric lifetimes of substances include C. P. Shea, "Mending the Earth's Shield," *World Watch* (January–February 1989): 27–34; NASA-GSFC, "Stratospheric Ozone," accessed August 22, 2000 at http://see.gsfc.nasa.gov; and U.S. Environmental Protection Agency, "Ozone-Depleting Substances," accessed March 1, 2001 at http://www.epa.gov/ozone/ods/html.

(another chemical like chlorine) and is used in fire extinguishers.[3, 11]

One of the first restrictions on CFCs included its use as a propellant gas for spray cans. This practice was banned in the late 1970s in a number of countries, setting a trend that has continued, with the result that CFCs as aerosol propellants are no longer a problem.[3] In contrast, the use of CFCs as a refrigerant has increased dramatically in recent years, especially in developing countries, such as China.

Simplified Stratospheric Chlorine Chemistry

CFCs are considered responsible for most of the ozone depletion observed by scientists. Let us look more closely at how this effect occurs.

Earlier, we noted that there are no tropospheric sinks for CFCs. That is, the processes that remove most chemicals in the lower atmosphere—destruction by sunlight, rain-out, and oxidation—do not break down CFCs, because CFCs are transparent to sunlight, are essentially insoluble, and are nonreactive in the oxygen-rich lower atmosphere.[13] Indeed, the fact that CFCs are nonreactive in the lower atmosphere was one reason they were attractive for use as propellants.

When CFCs wander to the upper part of the stratosphere, however, reactions do occur. Highly energetic ultraviolet radiation (UVC) splits up the CFC releasing chlorine. When this happens, the following two reactions can take place:[13]

$$(1)\ Cl + O_3 \rightarrow ClO + O_2$$
$$(2)\ ClO + O \rightarrow Cl + O_2$$

These two equations define a chemical cycle that can deplete ozone (Figure 26.5). In the first reaction, chlorine combines with ozone to produce chlorine monoxide, which in the second reaction combines with monatomic oxygen to produce chlorine again. The chlorine can then enter another reaction with ozone and cause additional ozone depletion. This series of reactions is what is known as a *catalytic chain reaction*. Because the chlorine is not removed but reappears as a product of the second reaction, the process may be repeated over and over again. It has been estimated that each chlorine atom may destroy approximately 100,000 molecules of ozone over a period of one or two years before the chlorine is finally removed from the stratosphere through other chemical reactions and rain-out.[13] The significance of these reactions is apparent when we realize how many metric tons of CFCs have been emitted into the atmosphere (see Table 26.1).

It should be noted that what actually happens chemically in the stratosphere is considerably more complex than the two equations shown here. The atmosphere is essentially a chemical soup in which a variety of processes related to aerosols and clouds take place (some of these are addressed in the discussion of the ozone hole). Nevertheless, these equations show us the basic chemical chain reaction that occurs in the stratosphere to deplete ozone.

The catalytic chain reaction just described can be interrupted through storage of chlorine in other compounds in the stratosphere. Two possibilities are as follows:

1. Ultraviolet light breaks down CFCs to release chlorine, which combines with ozone to form chlorine monoxide (ClO), as already described. This is the first reaction

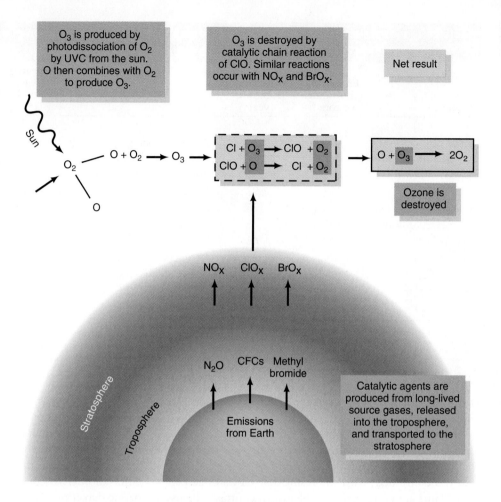

Figure 26.5 ■ Processes of natural formation of ozone and destruction by CFCs, N$_2$O, and methylbromide. [*Source:* Modified from NASA-GSFC, "Stratospheric Ozone," accessed August 22, 2000 at http://see.gsfc.nasa.gov.]

O$_3$ is produced by photodissociation of O$_2$ by UVC from the sun. O then combines with O$_2$ to produce O$_3$.

O$_3$ is destroyed by catalytic chain reaction of ClO. Similar reactions occur with NO$_X$ and BrO$_X$.

Net result

Sun

O$_2$ — O + O$_2$ → O$_3$

O

$$Cl + O_3 \longrightarrow ClO + O_2$$
$$ClO + O \longrightarrow Cl + O_2$$

$$O + O_3 \longrightarrow 2O_2$$

Ozone is destroyed

NO$_X$ ClO$_X$ BrO$_X$

N$_2$O   CFCs   Methyl bromide

Stratosphere

Troposphere

Emissions from Earth

Catalytic agents are produced from long-lived source gases, released into the troposphere, and transported to the stratosphere

discussed. The chlorine monoxide may then react with nitrogen dioxide (NO$_2$) to form a chlorine nitrate (ClONO$_2$). If this reaction occurs, ozone depletion is minimal. The chlorine nitrate, however, is only a temporary reservoir for chlorine. The compound may be destroyed, and the chlorine released again.

2. Chlorine released from CFCs may combine with methane (CH$_4$) to form hydrochloric acid (HCl). The hydrochloric acid may then diffuse downward. If it enters the troposphere, rain may remove it, thus removing the chlorine from the ozone-destroying chain reaction. This is the ultimate end for most chlorine atoms in the stratosphere. However, while the hydrochloric acid molecule is in the stratosphere, it may be destroyed by incoming solar radiation, releasing the chlorine for additional ozone depletion.

It has been estimated that the chlorine chain reaction that destroys ozone may be interrupted by the processes just described as many as 200 times while a chlorine atom is in the stratosphere.[3,14]

In part as a result of the ozone-depletion reactions, concentrations of ozone have declined in both northern and southern temperate latitudes. Figure 26.6 compares the stratospheric ozone from the 1970s to 2002 by latitude in the Southern Hemisphere. Notice that while ozone at the equator has been relatively constant, significant reduction has occurred in the Antarctic since the 1970s. Massive destruction of ozone identified in the Antarctic constitutes the ozone hole, which continues to be a source of concern.[3]

In discussing the global distribution of ozone, it is important to remember that, for the Southern Hemisphere, under natural conditions, the highest concentration of ozone is found in the polar regions (about 60° south latitude) and the lowest near the equator. (See Figure 26.6 for 1970, before massive ozone destruction occurred.) At first, this may seem strange, because ozone is produced in the stratosphere by solar energy, and more solar energy is found near the equator. Much of the world's ozone is produced near the equator, but the ozone in the stratosphere moves from the equator toward the poles with global air circulation patterns.[8]

26.3 The Antarctic Ozone Hole

Since the Antarctic ozone hole was first reported in 1985, it has captured the interest of many people around the world. Every year since then, ozone depletion has been observed in the Antarctic in October, the spring season there.

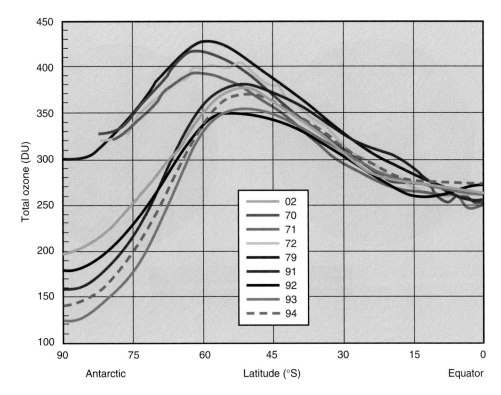

Figure 26.6 ■ Average change in concentration of stratospheric ozone, 1970–2002 (October mean total), by latitude. Notice that ozone reduction varies most in the Antarctic, whereas concentrations at the equator are nearly constant. [*Source:* Modified from NASA-GSFC, "Stratospheric Ozone," accessed October 2, 2003 at http://www.ccpp.edu. edu/SEES.]

The thickness of the ozone layer above the Antarctic during the springtime has been declining since the mid-1970s, and the geographic area covered by the ozone hole continues to increase.[15] The ozone hole has increased in size from a million or so square kilometers in the late 1970s and early 1980s to about 25 million square kilometers today, about the area of North America (Figure 26.7).

Development of the ozone hole from the 1970s to the early to middle 1990s is shown in Figure 26.8. Also shown are the percent differences between the two time periods, emphasizing the development of the massive ozone hole in the early to middle 1990s.[5]

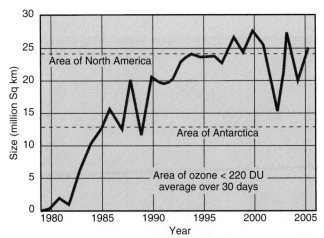

Figure 26.7 ■ Size of the Antarctic ozone hole, 1979–2006 (September 5–October 25 average). The size of the hole started increasing rapidly in the 1980s, and then grew slower and leveled off. [*Source:* Data from NASA, 2006. www.nasa.gov.]

Polar Stratospheric Clouds

The minimum concentration of ozone in the Antarctic since 1980 has varied from about 50 to 70% of that in the 1970s (300 DU) (refer to Figure 26.4). Lesser depletion in some years is thought to be related to fewer polar stratospheric clouds over the Antarctic. By contrast, in years when the ozone depletion was greater, the regions with the most ozone depletion were in the lower stratosphere at an altitude between 14 and 24 km (9 and 15 mi), where polar stratospheric clouds are present. What is the significance of these clouds?

Polar stratospheric clouds have been observed for at least the past hundred years at altitudes of approximately 20 km (about 12 mi) above the polar regions. The clouds are approximately 10–100 km (6–60 mi) in length and several kilometers thick.[14] They have an eerie beauty and an iridescent glow, with a color reminiscent of mother-of-pearl (Figure 26.9).[14]

Polar stratospheric clouds form during the polar winter (called the polar night because of the lack of sunlight, which results from the tilt of Earth's axis). During the polar winter, the Antarctic air mass is isolated from the rest of the atmosphere and circulates about the pole in what is known as the Antarctic **polar vortex**. The vortex, which rotates counterclockwise because of the rotation of Earth in the Southern Hemisphere, forms as the isolated air mass cools, condenses, and descends.[7] The cooling occurs because the isolated air mass continues to lose heat through radiation, and no more heat is supplied because of the lack of sunlight.

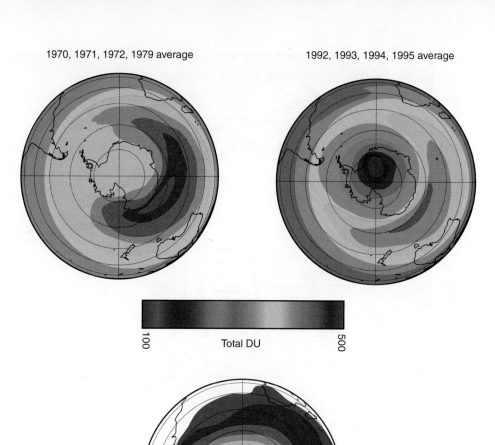

Figure 26.8 ■ Average development of the ozone hole in the 1970s compared with the early to middle 1990s, with percent differences between those two periods. [*Source:* Modified from NASA-GSFC, "Stratospheric Ozone," accessed August 22, 2000 at http://see.gsfc.nasa.gov.]

Figure 26.9 ■ Polar stratospheric clouds, February 12, 1989, photographed from an aircraft cruising at an altitude of approximately 12 km in the polar region north of Stavanger, Norway. The red haze and thin orange or brown layers at lower altitudes are Type I clouds. The red coloring is probably due to scattering from nitric acid particles. The higher white clouds are Type II polar stratospheric clouds, which consist mostly of water molecules frozen as ice.

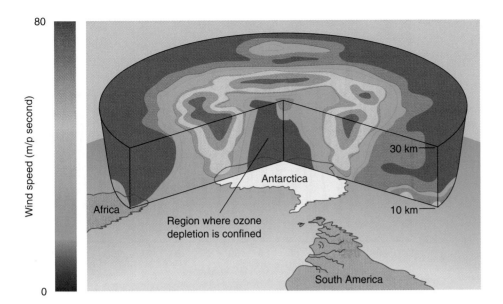

(a)

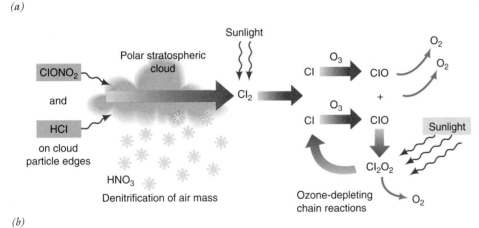

(b)

Figure 26.10 ■ (a) Idealized diagram of the Antarctic polar vortex and (b) the role of polar stratospheric clouds in the ozone depletion chain reaction. [*Source*: Based on O. B. Toon and R. P. Turco, "Polar Stratospheric Clouds and Ozone Depletion," *Scientific American*, 264, no. 6 (1991): 68–74.]

Clouds are formed in the vortex when the air mass reaches a temperature between 195 K and 190 K (–78° to –83°C; –108° to –117°F). At these very low temperatures, small sulfuric acid particles (approximately 0.1 μm) are frozen and serve as seed particles for nitric acid (HNO_3). These clouds are called Type I polar stratospheric clouds.

If temperatures drop below 190 K (–83°C; –117°F), water vapor condenses around some of the earlier-formed Type I cloud particles, forming Type II polar stratospheric clouds, which contain larger particles. Type II polar stratospheric clouds have a mother-of-pearl color visible in polar areas.

During the formation of polar stratospheric clouds, nearly all the nitrogen oxides in the air mass are held in the clouds as nitric acid. The nitric acid particles grow large enough to fall out by gravitational settling from the stratosphere. This phenomenon has the important result of leaving very little nitrogen oxide in the atmosphere in the vicinity of the clouds.[3,7,14] This process facilitates ozone-depleting reactions, which ultimately may reduce stratospheric ozone in the polar vortex by as much as 1 to 2% per day in the early spring, when sunlight returns to the polar region.

An idealized diagram of the polar vortex that forms over Antarctica is shown in Figure 26.10a. Ozone-depleting reactions that occur within the vortex are illustrated in Figure 26.10b. As this figure shows, in the dark Antarctic winter, almost all the available nitrogen oxides are tied up on the edges of particles in the polar stratospheric clouds or have settled out. Hydrochloric acid and chlorine nitrate (the two important sinks of chlorine) act on particles of polar stratospheric clouds to form dimolecular chlorine (Cl_2) and nitric acid through the following reaction:[16]

$$HCl + ClONO_2 \rightarrow Cl_2 + HNO_3$$

In the spring, when sunlight returns and breaks apart chlorine (Cl_2), the ozone-depleting reactions discussed earlier occur. Nitrogen oxides are absent from the Antarctic stratosphere in the spring; thus, the chlorine cannot be sequestered to form chlorine nitrate, one of its major sinks. Therefore chlorine is free to destroy ozone. In the early spring of the Antarctic, these ozone-depleting reactions can be rapid, producing the 70% reduction in ozone observed in 1995. Ozone depletion in the Antarctic vortex ceases

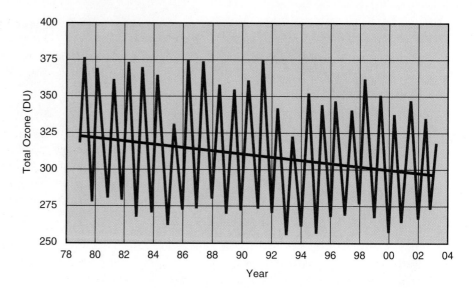

Figure 26.11 ■ Total ozone over the United States from 1979 to 2003. Although annual range is large, the average ozone concentration has slowly declined. [*Source:* NOAA (2006), www.noaa.gov.]

later in spring as the environment warms and the polar stratospheric clouds disappear, releasing nitrogen back into the atmosphere, where it can combine with chlorine and thus be removed from ozone-depleting reactions. Stratospheric ozone concentrations then increase as ozone-rich air masses again migrate to the polar region.

An Arctic Ozone Hole?

A polar vortex also forms over the North Pole area, but it is generally weaker than the Antarctic polar vortex and does not last as long. Nevertheless, ozone depletion occurs over the North Pole, and it is speculated that if the vortex persists for a month or more, ozone losses in the affected air mass may be as high as 30 to 40%.

A major concern regarding the northern polar vortex is that, as it breaks up, it sends ozone-deficient air masses southward, where they may drift over populated areas of Europe and North America. In January 1992, for example, satellite data indicated a large air mass containing high levels of chlorine monoxide (ClO) stretching from Great Britain eastward over Northern Europe.[17] (ClO is sometimes referred to as the "smoking gun" in the ozone problem because it plays a major part in ozone depletion—recall the earlier discussion of ozone reactions.) In contrast, the Antarctic polar vortex tends to remain more stationary, although in 1987 an ozone-depleted air mass that formed over Antarctica in October drifted northward over Australia and New Zealand in December, resulting in record low concentrations of stratospheric ozone in that region.[3]

In summary, although not as severe as the ozone depletion over the Antarctic, ozone depletion over the Arctic each winter is troublesome. Since the Arctic polar vortex is relatively weak, warmer air from the midlatitudes is usually able to dissipate the vortex before ozone depletion becomes severe. However, in 1995, ozone levels were as much as 40% below normal. Scientists studying the developing ozone hole speculated that the unusually cold Arctic winter of 1995 trig-

gered the record losses of ozone, leading to an ozone hole similar to the one that forms over the Antarctic.[18]

26.4 Tropical and Midlatitude Ozone Depletion

It has been firmly established that ozone depletion in polar regions occurs as a result of reactions that take place on particles in polar stratospheric clouds. Ice particles also occur in the stratosphere over the tropics; and at times, sulfuric acid aerosols are abundant in the stratosphere because of injection of sulfur by volcanic eruptions. That these particles may cause ozone depletion is only a hypothesis; there is no conclusive evidence.

At the South Pole, stratospheric ozone was lost at a rate of about 35% per year from the 1970s to the 1990s (see Figure 26.4).[19] However, evidence also suggests an increase in ozone depletion at midlatitudes over areas including the United States and Europe. Figure 26.11 shows the concentration of ozone above the United States (data are from four locations across the U.S.). Notice the large annual cycles, and the slow decline of average concentration of ozone from 1979 to 2003. In summary, although we know the most information about ozone depletion in polar regions (particularly Antarctica), depletion of ozone is a global concern, from the poles to the tropics.

26.5 The Future of Ozone Depletion

A troubling aspect of ozone depletion is that if the manufacture, use, and emission of all ozone-depleting chemicals were to stop today, the problem would not go away, because millions of metric tons of those chemicals are now in the lower atmosphere, working their way up to the stratosphere. As shown in Table 26.1, several CFCs have atmospheric lifetimes of 75 to 140 years. Thus, about 35% of the CFC-12

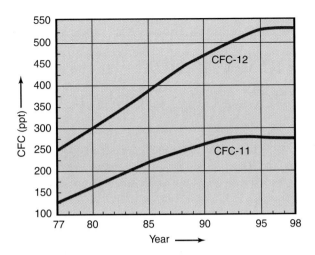

Figure 26.12 ■ Concentrations of CFC-11 and CFC-12 from 1977 to 1998. [*Source:* Data from Mauna Loa, Hawaii, modified from NOAA Climate Monitoring and Diagnostic Lab, accessed August 22, 2000 at http://www.cmdl.noaa.gov.]

molecules in the atmosphere are expected to still be there in 2100, and approximately 15% will be there in 2200.[3] In addition, approximately 10 to 15% of the CFC molecules manufactured in recent years have not yet been admitted to the atmosphere, because they remain tied up in foam insulation, air-conditioning units, and refrigerators.[3] Nevertheless, indicators suggest that growth in the concentrations of CFCs has been reduced and in some cases reversed.

Figure 26.12 shows concentrations of CFC-11 and CFC-12 from 1977 to 1998. CFC-11 concentrations reached a peak in about 1992 and then leveled off. CFC-12 concentrations peaked in 1995 and then leveled off. The growth in concentrations of CFC-12, which accounts for nearly 50% of ozone depletion (see Table 26.1), was about 5% per year from 1978 to 1995. It was about 0.5% per year by 1998.

Environmental Effects

Ozone depletion has several serious potential environmental effects, including damage to Earth's food chains on land and in the oceans and damage to human health, including increases in all types of skin cancers and cataracts and suppression of immune systems.[20, 21] As mentioned earlier, a 1% decrease in ozone can cause an increase of 1 to 2% in UVB radiation and an increase of 2% in the incidence of skin cancer.[21]

There has been speculation for some time that ozone depletion might lead to a reduction of primary productivity in the world's oceans. A loss of productivity of the phytoplankton (microscopic marine algae and photosynthetic bacteria that float near the surface of the ocean) would have a negative impact on a variety of other marine organisms because phytoplankton are at the base of the food chain. Because ozone over the Antarctic area has been depleted by as much as 70% in recent years, more UVB radiation is reaching the surface of the ocean there. A study

of the Antarctic waters beneath the mass of ozone-depleted air suggests that a reduction of primary productivity of at least 6 to 12% is associated with ozone depletion.[4] This disruption of the food chain might eventually affect people who consume marine resources, because fewer resources would be available for harvest. Also, because plankton are a sink for atmospheric carbon dioxide, their disruption might increase the concentration of CO_2 in the atmosphere, thereby increasing global warming.

If ozone depletion becomes more widespread and affects major food crops (such as beans, wheat, rice, and corn), serious social disruption could occur. Even a small decrease in food production might have large social and political consequences around the world. A loss of 10 to 15% production could be catastrophic.

The range of human health effects of ozone depletion is being vigorously researched and debated. There is general agreement that the effects will be negative and will result in increases in a variety of diseases, perhaps at epidemic levels compared with what would otherwise be expected. As mentioned, one of the most serious hazards anticipated is an increase in skin cancers of all types, including the often fatal melanoma.

In 1987, low ozone measurements over Antarctica in October were followed by low readings over Australia and New Zealand in December,[22] when a mass of ozone-depleted air separated from the Antarctic vortex as it broke up in late spring. There is fear that large ozone holes will increasingly threaten populated areas in midlatitude regions.

Already, around the world, the incidence of skin cancers has increased (see the case study at the beginning of this chapter). For many years, a suntan was considered a healthy look, and people deliberately exposed their bodies to sunlight. Today, health-conscious people are replacing tanning oils with sunblock lotions and hats. Newspapers in the United States are providing their readers with the **Ultraviolet (UV) Index** (Table 26.2), developed by the

Table 26.2 • Ultraviolet (UV) Index for Human Exposure		
Exposure Category	UV Index	Comment
Low	< 2	Sunblock recommended for all exposure
Moderate	3 to 5	Sunburn can occur quickly
High	6 to 7	Potentially hazardous
Very high	8 to 10	Potentially very hazardous
Extreme	11+	Potentially very hazardous

Note: At moderate exposure to UV, sunburn can occur quickly; at high exposure, fair-skinned people may burn in 10 minutes or less of exposure.

Source: Modified after U.S. Environmental Protection Agency 2004 (with the National Weather Service). Accessed June 16, 2004 at www.epa.gov.

Seasonal Changes in the UV Index: Implications for Antarctic Ozone Depletion

The link between the concentration of ozone in the stratosphere and levels of UV radiation arriving at different places on Earth's surface can be examined by determining the UV Index at different locations and days over the seasons of a year. This results because the index is a measure of the maximum level of sun burning UV radiation at a particular location on a given day. Figure 26.13 shows seasonal changes (from 1991 to 2001) in the UV Index at three locations: San Diego, California and Barrow, Alaska (both in the northern hemisphere), and Palmer, Antartica (in the Southern Hemisphere). San Diego, in Southern California, has higher values of the index for all four seasons than does Barrow, Alaska, as would be expected from San Diego's location closer to the equator. We would also expect the index, under normal conditions, to be higher than for any place in Antarctica. However, notice that the spring UV Index in Palmer, Antarctica, sometimes exceeds that of San Diego. The higher indices suggest higher levels of UV radiation at the surface. The dotted line for Palmer shows the UV Index from 1978 to 1983 before the occurrence of the ozone hole. Thus, we conclude that the differences of the index at Palmer (1978–1983 compared to 1991–2001) are due to ozone depletion in the Antarctic spring and development of the ozone hole that was observed, for this data, by 1991.[23]

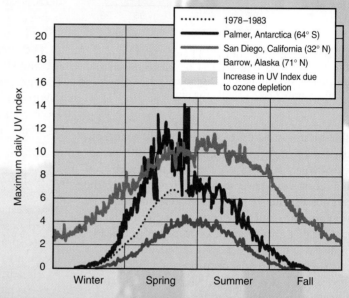

Figure 26.13 ■ Seasonal changes in the UV Index for two locations in the Northern Hemisphere and one in the Southern Hemisphere. Notice that San Diego generally has higher values on the index because it is closer to the equator than the other two sites. However, during the Antarctic spring the values are sometimes higher there than for San Diego. This is due to ozone depletion above the Antarctic. [*Source:* U.S. EPA, 2006.]

National Weather Service and the Environmental Protection Agency. The index predicts UV intensity levels on a scale from 1 to 11+ (see A Closer Look 26.1). Some news agencies also use the index to recommend the level of sunblock. It is speculated that the incidence of skin cancer caused by ozone depletion will increase until about 2060 and will then decline as the ozone shield recovers as a result of controls on CFC emissions.[19]

Individuals can reduce their risk of skin cancer and other skin damage from UV exposure by following a few simple precautions:

■ Limit exposure to the sun between the hours of 10 A.M. and 4 P.M. That is, avoid the hours of intense solar radiation.

■ When possible, remain in the shade.

■ Use a sunscreen with an SPF of at least 30. (Remember that protection from sunscreen with any SPF is diminished with time of exposure.)

■ Wear a wide-brimmed hat and, where possible, tightly woven full-length clothing (for example, loose-fitting light cotton).

■ Wear UV-protective sunglasses.

■ Avoid tanning salons and sunlamps.

■ Consult the UV Index before going out (see Table 26.2).

If your shadow is not as long as your height, you are in the part of the day with highest UV exposure. Take the precautions mentioned above. Following these precautions may help ensure that your skin will remain healthy, with fewer wrinkles into old age.

In addition to harming skin, ultraviolet radiation can damage eyes, causing cataracts, an eye disease in which the lens becomes opaque and vision is impaired. People are now more often choosing eyeglasses that block ultraviolet radiation. It is believed that an increased exposure to ultraviolet radiation may also damage or reduce the efficiency of the human immune system.[20] In turn, decreases in the effectiveness of the human immune system would result in higher numbers of a variety of diseases. Finally, a variety of environmental pollutants in the air and water could have synergistic effects, increasing potential health risks of exposure to ultraviolet radiation.

Management Issues

A key issue in management of ozone depletion is whether the depletion is natural or human-induced. If stratospheric ozone depletion were a natural process, then we would not expect to see the continuous, dramatic reductions since the mid-1970s, when the real impact of production of CFCs began. World production of CFCs increased significantly from 1970 to 1994. The most dramatic declines in ozone in Antarctica have occurred since 1980. Thus, it appears that chlorine from CFCs is the "smoking gun" mentioned earlier. Supporting this hypothesis, an investigation determined that the concentration of stratospheric chlorine (which is responsible for ozone destruction) is more than five times what could be expected from natural emissions from oceans or other natural processes. The authors concluded that, beyond a reasonable doubt, CFCs are responsible for ozone depletion in the stratosphere.[24] We turn next to a discussion of management issues and strategies to deal with CFC-induced ozone depletion.

The Montreal Protocol

A diplomatic achievement of monumental proportions was completed with the signing of the Montreal Protocol in September 1987; 27 nations signed the agreement originally, and 119 additional nations signed later. The protocol outlined a plan for the eventual reduction of global emissions of CFCs to 50% of 1986 emissions. The protocol originally called for elimination of the production of CFCs by 1999. Because of scientific evidence that stratospheric ozone depletion was occurring faster than predicted, the timetable for elimination of CFC production was shortened. Most industrialized countries, including the United States, had stopped production by the end of 1995; the deadline for developing countries was the end of 2005. An eventual phase-out of all CFC consumption is part of the Montreal Protocol.

Assessment of the effects of the protocol, along with the effects of other agreements and amendments made in London (1990) and Copenhagen (1992), suggests that stratospheric concentrations of ozone-depleting substances (CFCs) will return to pre-1980 levels by about 2050. Already, the rate of increase of CFC emissions has been reduced (see Table 26.1 and Figure 26.12). Nevertheless, as already mentioned, because of the long residence times of CFCs in the stratosphere, ozone depletion from CFCs is expected to continue for many years to come.[25, 26] That's the bad news. The good news is that ozone levels in the stratosphere will slowly increase in the next few decades.

Unfortunately, not all industrial nations are responding to the urgency of this issue, partly because of the economic gap between wealthy and poorer nations. For example, China and India chose not to participate in the protocol. Their refusal was probably related to the substantial investments they are making in refrigeration and the fact that replacement chemicals currently considered for CFCs are approximately six times as expensive. The poorer nations will probably not participate in the reduction of CFCs unless they are assisted by the wealthier countries.[27] The Montreal Protocol stipulates a transfer of both technology and funds from industrial to developing countries to speed up CFC phase-out. We return to this issue in the discussion of substitutes for CFCs.

A black-market trade in CFCs has hindered the attempt to eliminate their production and consumption. Illegal smuggling of CFCs into Europe from Russia in recent years has equaled as much as 10% of the total CFCs legally allowed in those countries. The United States has also received illegal imports of CFCs.[28]

Collection and Reuse of CFCs

One way to lower emissions of CFCs into the atmosphere is to develop ways to collect and reuse them. Every year, approximately 50 million refrigerators are discarded worldwide, and each one contains approximately 1.2 kg (2.6 lb) of CFCs, mostly tied up in the foam plastic insulation. Methods have been developed to liberate and collect these CFCs when refrigerators are recycled. The CFCs used as coolant gases can also be collected. One company in Germany recycles approximately 6,000 refrigerators a month.[29] The same techniques can be used to recover the CFCs in air conditioners used in automobiles and homes.

Substitutes for CFCs

Of primary importance in developing substitutes for CFCs is finding substitutes that are both safe and effective. Two substitutes for CFCs being experimented with today are **hydrofluorocarbons (HFCs)** and **hydrochlorofluorocarbons (HCFCs)**. These chemicals are controversial and more expensive than CFCs but do have advantages.

The advantage of HFCs is that they do not contain chlorine. They do contain fluorine, however, and when fluorine atoms are released into the stratosphere, they participate in reactions similar to those of chlorine. Thus, they can cause ozone depletion. However, ozone depletion is not thought to be a significant problem with fluorine, because it is approximately 1,000 times less efficient in those reactions.[3, 27] In addition, some blends of HFCs have no ozone-depleting potential.[30]

HCFCs contain an atom of hydrogen in place of a chlorine atom. They can be broken down in the lower atmosphere, in which case they do not inject chlorine into the stratosphere; however, they can cause ozone depletion if they do reach the stratosphere before being broken down. Although their atmospheric lifetime is much shorter than those of the CFCs (see Table 26.1), when HCFCs are used in tremendous quantities, they do cause ozone depletion.[3] HCFCs are at best a transitional chemical to be used until substitutes that do not cause ozone depletion are readily available. HCFCs will be phased out by 2030.

Human-Made Chemicals and the Ozone Hole: Why Was There Controversy?

By 1993, scientists had accumulated enough evidence to support earlier predictions that stratospheric ozone was being depleted over the Antarctic. Most of them blamed the damage on organic chlorine compounds (those containing both carbon and chlorine) manufactured by humans, such as CFCs. But consensus among most of the scientists in the field did not prevent a continuing storm of controversy over these findings. Critics included meteorologists, scientists from other fields, amateur scientists, journalists, talk-show hosts, and authors of nontechnical books on the environment. The critics charged that natural sources of chlorine, not those generated by humans, were responsible for ozone depletion and that the environmental and health threats of ozone depletion were greatly exaggerated.

Accusations leveled against scientists, NASA officials, and some industrialists included claims that the ozone scare was a sham, a scam, a hoax, or a conspiracy. Scientific uncertainties, such as those concerning the causes of ozone depletion in the Northern Hemisphere and whether depletion was allowing more ultraviolet radiation to reach Earth's surface, were emphasized.

In his 1993 presidential address to the American Association for the Advancement of Science, F. Sherwood Rowland, principal architect of the hypothesis that CFCs damage the ozone layer, cited poor communication between scientists and nonscientists as the cause of the controversy. His colleague, Mario Molina, stated that the arguments put forth by critics about natural causes of ozone depletion had been tested; no results had been found to support those arguments in the 20 years since the hypothesis that CFCs were responsible for ozone depletion had been proposed.

Others felt that the controversy fed on lack of understanding of the nature of science, particularly discomfort with uncertainty, on the part of most nonscientists. In making this point, science fiction writer Frederic Pohl quoted the late Nobel Prize physicist Richard Feynman: "Scientific knowledge is a body of statements of varying degrees of certainty—some most unsure, some nearly sure, but none absolutely certain."

Critical Thinking Questions

1. Table 26.3 summarizes the major points of argument about whether ozone depletion is primarily a result of natural or of human-made chemicals. Evaluate the arguments in the table as best you can in light of what you have learned about ozone in this chapter. Identify each statement as (a) a hypothesis, (b) an inference, (c) a fact confirmed by observation or experiment, (d) pseudoscience, or (e)

Table 26.3 • Major Arguments Regarding Sources of Ozone Depletion

Natural sources of chlorine are so large that CFCs are insignificant by comparison.

1. CFCs are heavier than air and would not reach the stratosphere in significant amounts.
2. No measurements have shown CFCs to be present in the stratosphere.
3. Volcanoes, which produce 20 times as much chlorine as do CFCs, caused the Antarctic ozone hole. Mount Erebus in Antarctica has been erupting since 1973 and emits more than 1,000 tons of chlorine a day.
4. Evaporation of sodium chloride from ocean water is a source of stratospheric chlorine.
5. Volcanoes release hydrogen chloride, which has increased in the stratosphere in the last 10 years.
6. Burning releases methyl chloride.

Humans are producing chemicals that significantly deplete stratospheric ozone in the Antarctic and possibly in the Northern Hemisphere.

1. Within the atmosphere, large masses of air are constantly mixing.
2. CFCs are found in stratospheric samples.
3. Volcanic activity has been around for a long time, and recent eruptions have not been unusual. Mount Erebus does not erupt forcefully enough to propel chlorine into the stratosphere in significant amounts. Furthermore, it produces only 15,000 tons of chorine a year.
4. Sodium chloride, unlike CFCs, is water soluble and is washed out of the atmosphere when it rains. No sodium is found in the lower stratosphere.
5. Both hydrogen chloride and hydrogen fluoride, also water soluble, are increasing, which is what would be predicted if CFCs were reaching the stratosphere. There are almost no natural sources of hydrogen fluoride, and increases in it would not be consistent with a volcanic source for chlorine. Sulfuric acid aerosols from volcanoes that do eject into the stratosphere promote destruction of ozone.

an absolute certainty. If you feel the information is incomplete, what additional observations or experiments would you suggest? Based on your analysis, what is your position on the question of whether natural or human-made chemicals are depleting the ozone layer?

2. Some critics have pointed out that ultraviolet radiation increases by 50 times (5000%) in going from pole to equator. The expected increase in ultraviolet radiation in the midlatitudes, they say, is like moving from New York City to Philadelphia or from sea level to an elevation of 1,500 ft. What is your reaction to this criticism?

3. In February 1992, NASA scientists reported unusually high levels of chlorine over the Northern Hemisphere and warned that this might lead to significant ozone loss over heavily populated areas. Congress reacted quickly to enact legislation to speed up discontinuing use of CFCs. When the ozone loss over the Northern Hemisphere was much less than the scientists had expected, they were attacked by some people as being unnecessarily alarmist and using scare tactics to obtain additional funding. Do you think these attacks were justified? Would it have been better if the scientists had presented a much more conservative scenario?

4. F. Sherwood Rowland identified two sources of public misunderstanding about scientific issues: poor communication about science and widespread scientific illiteracy. Do you agree or disagree with Rowland? Explain your position. If you agree, what remedies do you suggest?

Reduction or elimination of CFC use by replacement with HCFCs and HFCs in developing countries will require massive financial aid. They could do it on their own if an inexpensive, safe alternative were available. Research is being conducted to determine what chemicals may be satisfactory alternatives.[31]

Short-Term Adaptation to Ozone Depletion

As we have seen, there is good news in the ozone depletion story. Concentrations of CFCs in the upper atmosphere where ozone depletion occurs have apparently peaked. Depletion of stratospheric ozone will be a story of gradual recovery by the mid–twenty-first century.[20, 28] Recovery will take place as a result of restrictions in the production of ozone-depleting chemicals, especially CFCs.

Given the nature of the ozone-depletion problem and the atmospheric lifetimes of the chemicals that produce the depletion, the major short-term adaptation by people will be learning to live with higher levels of exposure to ultraviolet radiation. In the long term, achievement of sustainability with respect to stratospheric ozone will require management of human-produced ozone-depleting chemicals.

Summary

■ Concentration of atmospheric ozone has been measured for more than 70 years. In the last decade, measurements have been taken from instruments mounted on satellites. Evaluation of the available data shows a clear trend: Ozone concentrations in the stratosphere have been decreasing since the mid-1970s.

■ In 1974, Mario Molina and F. Sherwood Rowland advanced the hypothesis that stratospheric ozone might be depleted as a result of emissions of chlorofluorocarbons (CFCs) into the lower atmosphere. Major features of the hypothesis are as follows: CFCs are extremely stable and have a long residence time in the atmosphere; eventually, the CFCs reach the stratosphere, where they may be destroyed by highly energetic solar ultraviolet radiation, releasing chlorine; the chlorine may then enter into a catalytic chain reaction that depletes ozone in the stratosphere. An environmentally significant result of the depletion is that more ultraviolet radiation reaches the lower atmosphere, where it can damage living cells.

■ The Antarctic ozone hole was first reported in 1985 and since then has captured the imagination of people around the world. Of particular importance to understanding the ozone hole are the complex reactions that occur in the polar vortex and the development of polar stratospheric clouds. Reactions in the clouds tend to denitrify the air mass in the vortex; and during the polar spring, chlorine is released to react in the catalytic ozone depletion cycle. The reactions can be very rapid, producing the observed 70% reduction in stratospheric ozone in only a few weeks.

■ Tropical and midlatitude ozone depletion is also hypothesized. Polar stratospheric clouds may be present above the tropics; in addition, belts of sulfur dioxide–rich aerosol clouds may result from volcanic eruptions. These clouds may be related to processes that denitrify the atmosphere and facilitate ozone depletion.

■ Millions of tons of chemicals with the potential to deplete stratospheric ozone are now in the lower atmosphere and working their way to the stratosphere. As a result, if all production, use, and emissions of these chemicals were stopped today, the problem would continue. The good news is that the concentrations of CFCs in the atmosphere have apparently peaked and are now static or in slow decline.

■ Potential environmental effects related to ozone depletion include damage to Earth's food chain, both on land and in the ocean, and human health effects, including increases in skin cancers, cataracts, and suppression of the immune system.

■ Many nations around the world have agreed to the Montreal Protocol, which will reduce global emissions of CFCs to 50% of the 1986 levels. The agreement called for elimination of production of the chemicals by 1996 for industrialized nations and by 2006 for developing nations. A serious hurdle to compliance for some nations relates to the economic fact that most chemical replacements for CFCs are more expensive than CFCs. Therefore, it appears that financial aid will be required if the less wealthy nations are to eliminate CFC use.

■ Potential management strategies for the ozone depletion problem, along with eliminating production of CFCs, include: (1) collecting and reusing CFCs, and (2) using substitutes for CFCs.

■ Given the lifetimes of the ozone-depleting chemicals, people will have to continue to live with higher levels of exposure to ultraviolet radiation over the next few decades. Banning chemicals that can deplete stratospheric ozone is a step in the right direction and will ultimately result in the reduction of atmospheric ozone depletion.

REEXAMINING THEMES AND ISSUES

Human Population

As the number of humans on Earth has increased, so has our use of the chemicals that are currently causing ozone depletion. Of particular significance are developing countries wishing to increase their industrial output. Coupled with a rapid rise in human population, industrialization in these countries could dramatically increase emissions of the chemicals that are causing global changes in the atmosphere. Thus, control of human population is an important objective in addressing ozone depletion in the future.

Sustainability

Controlling emissions of chemicals that cause ozone depletion is at the heart of sustainability, which has the objective of providing a quality environment for future generations. Potential adverse environmental effects of ozone depletion related to biological productivity and disease are significant, and reduction of emissions of ozone-depleting chemicals has been a high priority. Results of reductions will become apparent during the first half of the twenty-first century, as stratospheric ozone concentrations recover to preindustrial levels.

Global Perspective

Stratospheric ozone depletion is a significant global environmental problem conclusively shown to result from human activity. Understanding the causes of and potential solutions to the problem has required careful evaluation of Earth and its atmosphere on a global scale. Interactions between emissions of chemicals and chemical reactions occurring in the atmosphere, as well as general circulation of the atmosphere, have all been involved in understanding the processes that have resulted in ozone depletion in the stratosphere. Recommended management strategies to solve this problem have come directly through our understanding of Earth system science and its applications.

Urban World

Significant quantities of CFCs have been used for air-conditioning and refrigeration in urban environments. Urban areas are the places where the chemicals are most often produced, used, and emitted into the environment.

People and Nature

Ozone depletion is an environmental problem, largely caused by the actions of people emitting chemicals into the atmosphere. The processes in the atmosphere that result in the depletion are part of how nature works—for example, the linkage of depletion with formation of the Antarctic polar vortex. Sometimes, as with ozone depletion, when we connect with nature we cause unanticipated problems.

Science and Values

The fact that human emission of chemicals is altering the stratosphere to the detriment of people and the environment in general has come as a surprise to many people. The science of ozone depletion is well understood, however, and we are moving through management decisions, based on our values, to ensure the problem will slowly go away. Solutions to ozone depletion are costly and will require sacrifices on the part of most people. Most of the cost will fall on the wealthiest countries. It is to be hoped that these nations will continue to have the moral and ethical courage to exercise leadership and pay the costs necessary for an improved global environment.

Key Terms

Study Questions

1. Do you think a gap ever existed between people's understanding of the ozone problem and their taking action to find a solution to the problem? Compare the situation with the global warming problem.

2. Given that primary productivity is reduced in the Antarctic by ozone depletion, how could you test the hypothesis that penguins might be adversely affected?

3. Consider the possibility of building a small chamber to test the hypothesis that polar stratospheric clouds facilitate ozone depletion. What might be some of the problems in attempting to do this? What are alternative approaches to studying the role of stratospheric clouds?

4. Suppose that next year all our understanding of ozone depletion is changed by the discovery that concentrations of stratospheric ozone have natural cycles and that the lower concentrations in recent years have resulted from natural rather than human-induced processes. How would you put all the information in this chapter into perspective? Would you think that science had let you down?

5. What types of economic and political changes will be necessary to encourage less developed countries to support plans to eliminate chemicals responsible for ozone depletion? Do you believe that richer countries have an obligation to help poorer ones?

6. Do you agree or disagree that most of the people in the world will adjust to ozone depletion by doing little to decrease their personal exposure to ultraviolet radiation?

7. How would you respond to the statement that "ozone depletion leads to increased exposure of humans to UV radiation that is increaing the incidence of skin cancer? Is this an environmental scare tactic?

Further Reading

Christie, M. 2000. *The Ozone Layer: A Philosophy of Science Perspective.* Cambridge: Cambridge University Press. A complete look at the history of the ozone hole from the first discovery of its existence to more recent studies of the hole over Antarctica.

Environmental Protection Agency. 1995. *Stratospheric Ozone Depletion: A Focus on EPA's Research.* Washington, D.C.: U.S. EPA Office of Research and Development. A brief but good overview of ozone depletion, current status of ozone holes, and consequences of ozone depletion.

Hamill, P., and O. B. Toon. 1991. "Polar Stratospheric Clouds and the Ozone Hole," *Physics Today* 44(12): 34–42. A good review of the ozone problem and important chemical and physical processes related to ozone depletion.

Jones, R. R., and T. Wigley. 1989. *Ozone Depletion: Health and Environmental Consequences.* New York: Wiley. Proceedings of the International Conference on the Health and Environmental Consequences of Stratospheric Ozone Depletion, held at the Royal Institute of British Architects, London, November 28–29, 1988.

Litfin, K. T. 1994. *Ozone Discourses: Science and Politics in Global Environmental Cooperation.* New York: Columbia University Press. Discussions on the importance of science and scientific discourse in shaping world politics and policy with regard to ozone depletion.

Makhijani, A., and K. R. Gurney. 1995. *Mending the Ozone Hole.* Cambridge, Mass.: MIT Press. A good overview of causes and consequences of stratospheric ozone depletion, sources of and alternatives to ozone-depleting chemicals, and policy recommendations to reverse ozone damage.

Reid, S. 2000. *Ozone and Climate Change: A Beginner's Guide.* Amsterdam: Gordon & Breach Science Publishers. A good look at the science behind the ozone hole and future predictions written for a general audience to make the science understandable.

Rowland, F. S. 1990. "Stratospheric Ozone Depletion by Chlorofluorocarbons." *AMBIO* 19(6–7): 281–292. An excellent summary of stratospheric ozone depletion that discusses some of the major issues.

Shea, C. P. 1989. "Mending the Earth's Shield," *World Watch* 2(1): 28–34. An article focusing on solutions to the ozone problem, primarily concerned with control strategies to stop ozone-depleting chemicals from being emitted into the atmosphere.

Toon, O. B., and R. P. Turco. 1991. "Polar Stratospheric Clouds and Ozone Depletion," *Scientific American* 246(6): 68–74. An article that provides valuable information concerning polar stratospheric clouds and their importance in ozone depletion. It offers a good explanation of the formation of polar stratospheric clouds and the chemistry that occurs there.

CHAPTER 27

Minerals and the Environment

This award-wining golf course in Golden, Colorado, was for a century an open-pit mine (quary) for clay to produce bricks. [Source: City of Golden, Colorado.]

Learning Objectives

Modern society depends on the availability of mineral resources, which can be considered a nonrenewable heritage from the geologic past. After reading this chapter, you should understand:

■ That the standard of living in modern society is related in part to the availability of natural resources.

■ What processes are responsible for the distribution of mineral deposits.

■ What the differences are between mineral resources and reserves.

■ What factors control the environmental impact of mineral exploitation.

■ How wastes generated from the use of mineral resources affect the environment.

■ What the social impacts are from mineral exploitation.

■ How sustainability may be linked to use of nonrenewable minerals.

C A S E   S T U D Y

Golden Colorado: Open-Pit Mine Becomes a Golf Course

The city of Golden, Colorado, has an award-winning golf course on land that for about 100 years was an open-pit mine (quarry) excavated in limestone rock.

The mine produced clay for making bricks from clay layers between limestone beds. Over the life of the mine, the clay was used as a building material at many sites, including prominent buildings in the Denver area, such as the Colorado Governor's Mansion. The site included unsightly mining pits with vertical limestone walls as well as a landfill for waste disposal. However, it had spectacular views of the foothills of the Rocky Mountains. Today the limestone cliffs with their exposed plant and dinosaur fossil have been trasformed into golf greens, fairways, and a driving range. The name of the course, Fossil Trace Golf Club, reflects its geologic heritage, and the course includes trails to fossil locations. It also includes channels, constructed wetlands, and three lakes that store floodwater runoff, helping protect Golden from flash floods. The reclamation project started with a grassroots movement by the people of Golden to have a public golf course. The reclamation is now a money-maker for the city and demonstrates that mining sites can not only be reclaimed, but also transformed into valuable property.

Fossil Trace Golf Club is a unique instance of mine reclamation. However, each renovation is somewhat unique, based on local physical, hydrological, and biological conditions. This chapter discusses the origin of mineral deposits as well as environmental consequences of mineral development.

27.1 The Importance of Minerals to Society

Modern society depends on the availability of mineral resources.[1,2] Many mineral products are found in a typical American home (see Table 27.1). Consider your breakfast this morning. You probably drank from a glass made primarily of sand, ate food from dishes made from clay, flavored your food with salt mined from Earth, ate fruit grown with the aid of fertilizers such as potassium carbonate (potash) and phosphorus, and used utensils made from stainless steel, which comes from processing iron ore and other minerals. While eating, you may have viewed the news on a television or computer screen, listened to music on your iPod, or made appointments using your cell phone. All these electronic items are made from metals and petroleum.

Minerals are so important to people that the standard of living increases with the availability of minerals in useful forms. The availability of mineral resources is one measure of the wealth of a society. Those who have been successful in locating and extracting or importing and using minerals have grown and prospered. Without mineral resources to grow food, construct buildings and roads, and manufacture everything from computers to televisions to automobiles, modern technological civilization as we know it would not be possible. In maintaining our standard of living in the United States, every person requires about 10 tons of nonfuel minerals per year.[3]

Minerals can be considered our nonrenewable heritage from the geologic past. Although new deposits are still forming from present Earth processes, these processes are producing new mineral deposits too slowly to be of use to us today. Because mineral deposits are generally located in small, hidden areas, they must be discovered. Unfortunately, most of the easy-to-find deposits have been exploited; if modern civilization were to vanish, our descendants would have a harder time discovering rich mineral deposits than we did. It is interesting to speculate that they might mine landfills for metals thrown away by

Table 27.1 • Mineral Products in a Typical U.S. Home

Electronic Equipment (cell phones, computers, iPods, e.g.): metals such as gold, silver, copper, and iron as well as petroleum for plastic

Building materials: sand, gravel, stone, brick (clay), cement, steel, aluminum, asphalt, glass

Plumbing and wiring materials: iron and steel, copper, brass, lead, cement, asbestos, glass, tile, plastic

Insulating materials: fiberglass, gypsum (plaster and wallboard)

Paint and wallpaper: mineral pigments (such as iron, zinc, and titanium) and fillers (such as talc and asbestos)

Plastic floor tiles, other plastics: mineral fillers and pigments, petroleum products

Appliances: iron, copper, and many rare metals

Furniture: synthetic fibers made from minerals (principally coal and petroleum products); steel springs

Clothing: natural fibers grown with mineral fertilizers; synthetic fibers made from minerals (principally coal and petroleum products)

Food: grown with mineral fertilizers; processed and packaged by machines made of metals

Drugs and cosmetics: mineral chemicals

Other items: windows, screens, lightbulbs, porcelain fixtures, china utensils, jewelry made from mineral products

our civilization. Unlike biological resources, minerals cannot be easily managed to produce a sustained yield; the supply is finite. Recycling and conservation will help, but eventually the supply will be exhausted.

27.2 How Mineral Deposits Are Formed

Metals in mineral form are generally extracted from naturally occurring, anomalously high concentrations of Earth materials. When metals are concentrated in anomalously high amounts by geologic processes, **ore deposits** are formed. The discovery of natural ore deposits allowed early peoples to exploit copper, tin, gold, silver, and other metals while slowly developing skills in working with metals.

The origin and distribution of mineral resources is intimately related to the history of the biosphere and to the entire geologic cycle (see Chapter 5). Nearly all aspects and processes of the geologic cycle are involved to some extent in producing local concentrations of useful materials. In this section, we first look at the distribution of mineral resources on Earth and then describe processes that form mineral deposits.

Distribution of Mineral Resources

Earth's outer layer, or crust, is silica rich, made up mostly of rock-forming minerals containing silica, oxygen, and a few other elements. The elements are not evenly distributed in the crust: Nine elements account for about 99% of the crust by weight (oxygen, 45.2%; silicon, 27.2%; aluminum, 8.0%; iron, 5.8%; calcium, 5.1%; magnesium, 2.8%; sodium, 2.3%; potassium, 1.7%; and titanium, 0.9%). In general, remaining elements are found in trace concentrations.

The ocean, covering nearly 71% of Earth, is another reservoir for many chemicals other than water. Most elements in the ocean have been weathered from crustal rocks on the land and transported to the oceans by rivers. Other elements are transported to the ocean by wind or glaciers. Ocean water contains about 3.5% dissolved solids, mostly chlorine (55.1% by weight). Each cubic kilometer of ocean water contains about 2.0 metric tons of zinc, 2.0 metric tons of copper, 0.8 metric ton of tin, 0.3 metric ton of silver, and 0.01 metric ton of gold. These concentrations are low compared with those in the crust, where corresponding values (in metric tons/km^3) are: zinc, 170,000; copper, 86,000; tin, 5,700; silver, 160; and gold, 5. After rich crustal ore deposits are depleted, we will be more likely to extract metals from lower-grade ore deposits or even from common rock than from ocean water; however, if mineral extraction technology becomes more efficient, this prognosis could change.

We have mentioned that minerals we mine occur in deposits—with anomalously high local concentrations. Why is this so? Planetary scientists now believe that Earth, like the other planets in the solar system, formed by condensation of matter surrounding the sun. Gravitational attraction brought together the matter dispersed around the forming sun. As the mass of the proto-Earth increased, the material condensed and was heated by the process. The heat was sufficient to produce a molten liquid core, consisting primarily of iron and other heavy metals, which sank toward the center. The crust formed from generally lighter elements and is a mixture of many different elements. The elements in the crust are not uniformly distributed because geologic processes and some biological processes selectively dissolve, transport, and deposit elements and minerals. We discuss these processes next.

Plate Boundaries

Plate tectonics is responsible for the formation of some mineral deposits. According to the theory of plate tectonics (see Chapter 5), the continents (which are crustal rocks and part of the lithosphere) are composed mostly of relatively light rocks. As the tectonic plates of the lithosphere slowly move across Earth's surface, so do the continents. Metallic ores are thought to be deposited in the crust both where the tectonic plates separate, or diverge, and where they come together, or converge.

At divergent plate boundaries, cold ocean water comes in contact with hot molten rock. The heated water is lighter and more active chemically. It rises through fractured rocks and leaches metals from them. The metals are carried in solution and deposited as metal sulfides when the water cools.

At convergent plate boundaries, rocks saturated with seawater are forced together, heated, and subjected to intense pressure, which causes partial melting. The combination of heat, pressure, and partial melting mobilizes metals in the molten rocks, or magma. Most major mercury deposits, for example, are associated with the volcanic regions that occur close to convergent plate boundaries. Geologists believe that the mercury is distilled out of the tectonic plate as the plate moves downward; as the plate cools, the mercury migrates upward and is deposited at shallower depths, where the temperature is lower.

Igneous Processes

Igneous processes are related to molten rock material known as magma. Ore deposits may form when magma cools. As the molten rock cools, heavier minerals that crystallize (solidify) early may slowly sink or settle toward the bottom of the magma, whereas lighter minerals that crystallize later are left at the top. Deposits of an ore of chromium, called chromite, are thought to be formed in this way. When magma containing small amounts of carbon is deeply buried and subjected to very high pressure during slow cooling (crystallization), diamonds (which are pure carbon) may be produced (Figure 27.1).[4]

Hot waters moving within the crust are perhaps the source of most ore deposits. It is speculated that circulating groundwater is heated and enriched with minerals on contact with deeply buried rocks. This water then moves up or laterally to other, cooler rocks, where the cooled water deposits the dissolved minerals.[5]

Sedimentary Processes

Sedimentary processes relate to the transport of sediments by wind, water, and glaciers. These processes often concentrate materials in amounts sufficient for extraction.

As sediments are transported, running water and wind help segregate the sediments by size, shape, and density. This sorting function is useful to people. The best sand or sand and gravel deposits for construction purposes, for example, are those in which the finer materials have been removed by water or wind. Sand dunes, beach deposits, and deposits in stream channels are good examples. The sand

Figure 27.1 ■ Diamond mine near Kimberley, South Africa. This is the largest hand-dug excavation in the world.

and gravel industry amounts to several billion dollars annually, and in terms of the total volume of materials mined, it is one of the largest nonfuel mineral industries in the United States.[3]

Stream processes transport and sort all types of materials according to size and density. Therefore, if the bedrock in a river basin contains heavy metals, such as gold, streams draining the basin may concentrate the metals in areas where there is less water turbulence or velocity. These concentrations, called placer deposits, are often found in open crevices or fractures at the bottoms of pools, on the inside curves of bends, or on riffles, where shallow water flows over rocks. Placer mining of gold (which was known as a poor man's method because a miner needed only a shovel, a pan, and a strong back to work the stream-side claim) played an important role in the settling of California, Alaska, and other areas of the United States.

Rivers and streams that empty into oceans and lakes carry tremendous quantities of dissolved material derived from the weathering of rocks. Over geologic time, a shallow marine basin may be isolated by tectonic activity that uplifts its boundaries. In other cases, climatic variations, such as the ice ages, produce large inland lakes with no outlets. These basins and lakes eventually dry up. As evaporation progresses, the dissolved materials precipitate (drop out of solution), forming a wide variety of compounds, minerals, and rocks that have important commercial value. Most of these *evaporates* (deposits originating by evaporation) can be grouped into one of three types:[6]

■ *Marine evaporates* (solids)—potassium and sodium salts, gypsum, and anhydrite.

■ *Nonmarine evaporates* (solids)—sodium and calcium carbonate, sulfate, borate, nitrate, and limited iodine and strontium compounds.

■ *Brines* (liquids derived from wells, thermal springs, inland salt lakes, and seawaters)—bromine, iodine, calcium chloride, and magnesium.

Heavy metals (such as copper, lead, and zinc) associated with brines and sediments in the Red Sea, Salton Sea, and other areas are important resources that may be exploited in the future. Evaporate minerals are widely used in industrial and agricultural activities, and their annual value is over $1 billion.[6] Evaporate and brine resources in the United States are substantial, ensuring an ample supply for many years.

Biological Processes

Some mineral deposits are formed by biological processes, and many are formed under conditions of the biosphere that have been greatly altered by life. Examples include phosphates (discussed in Chapter 5) and iron ore deposits.

The major iron ore deposits exist in sedimentary rocks that were formed more than 2 billion years ago.[7] There are several types of iron deposits. Gray beds, an important type, contain unoxidized iron. Red beds contain oxidized iron (the red color is the color of iron oxide). Gray beds formed when there was little oxygen in the atmosphere, and red beds formed when there was relatively more oxygen. Although the processes are not completely understood, it appears that major deposits of iron stopped forming when the atmospheric concentration of oxygen reached its present level.[8]

Organisms are able to form many kinds of minerals, such as the calcium minerals in shells and bones. Some of these minerals cannot be formed inorganically in the biosphere. Thirty-one different biologically produced minerals have been identified. Minerals of biological origin contribute significantly to sedimentary deposits.[9]

Weathering Processes

Weathering, the chemical and mechanical decomposition of rock, concentrates some minerals in the soil. When insoluble ore deposits, such as native gold, are weathered from rocks, they may accumulate in the soil unless removed by erosion. Accumulation occurs most readily when the parent rock is relatively soluble, as is limestone. Intensive weathering of certain soils derived from aluminum-rich igneous rocks may concentrate oxides of aluminum and iron. (The more soluble elements, such as silica, calcium, and sodium, are selectively removed by soil and biological processes.) If sufficiently concentrated, residual aluminum oxide forms an ore of aluminum known as bauxite. Important nickel and cobalt deposits are also found in soils developed from iron- and magnesium-rich igneous rocks.

Weathering produces sulfide ore deposits from low-grade primary ore through *secondary enrichment* processes.

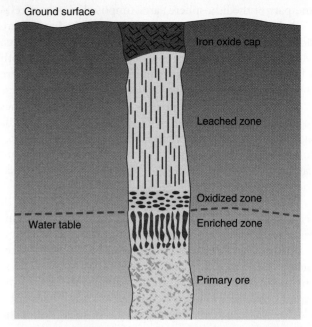

Figure 27.2 ■ Typical zones that form during secondary enrichment processes. Sulfide ore minerals in the primary ore vein are oxidized and altered and then are leached from the oxidized zone by descending groundwater and redeposited in the enriched zone. The iron oxide cap is generally a reddish color and may be helpful in locating ore deposits that have been enriched. [*Source:* R. J. Foster, *General Geology*, 4th ed. (Columbus, Ohio: Charles E. Merrill, 1983).]

Near the surface, primary ore containing minerals such as iron, copper, and silver sulfides is in contact with slightly acidic soil water in an oxygen-rich environment. As the sulfides are oxidized, they dissolve, forming solutions that are rich in sulfuric acid as well as silver and copper sulfate. These solutions migrate downward, producing a leached zone devoid of ore minerals (Figure 27.2). Below the leached zone and above the groundwater table, oxidation continues, and sulfate solutions continue their downward migration. Below the water table, if oxygen is no longer available, the solutions are deposited as sulfides, enriching the metal content of the primary ore by as much as 10 times. In this way, low-grade primary ore is rendered more valuable, and high-grade primary ore is made even more attractive.[10,11] For example, secondary enrichment of a disseminated copper deposit at Miami, Arizona, increased the grade of the ore from less than 1% copper in the primary ore to as much as 5% in some areas.[10]

27.3 Resources and Reserves

We can classify minerals as resources or reserves. **Mineral resources** are broadly defined as elements, chemical compounds, minerals, or rocks concentrated in a form that can be extracted to obtain a usable commodity—something that can be bought and sold. It is assumed that a resource can be extracted economically or at least has the potential

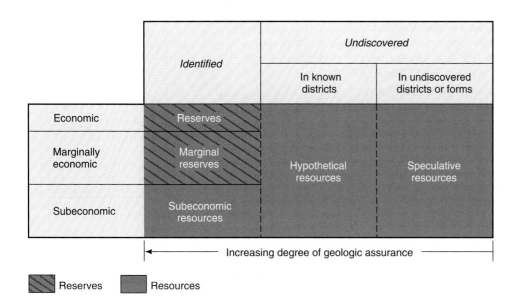

| | | Undiscovered | |
	Identified	In known districts	In undiscovered districts or forms
Economic	Reserves		
Marginally economic	Marginal reserves	Hypothetical resources	Speculative resources
Subeconomic	Subeconomic resources		

←——— Increasing degree of geologic assurance ———→

▨ Reserves   ▨ Resources

Figure 27.3 ■ Classification of mineral resources used by the U.S. Geological Survey and the U.S. Bureau of Mines. [*Source: Principles of a Resource Preserve Classification for Minerals*, U.S. Geological Survey Circular 831, 1980.]

for economic extraction. A **reserve** is that portion of a resource that is identified and from which usable materials can be legally and economically extracted *at the time of evaluation* (Figure 27.3).

Whether a mineral deposit is classified as part of the resource base or as a reserve may be a question of economics. For example, if an important metal becomes scarce, the price may rise, which would encourage exploration and extraction (mining). As a result of the price increase, previously uneconomic deposits (part of the resource base before the scarcity and price rise) may become profitable, and those deposits would be reclassified as reserves.

The main point here is that *resources are not reserves.* An analogy from a student's personal finances may help clarify this point. A student's reserves are liquid assets, such as money in the bank, whereas the student's resources include the total income the student can expect to earn during his or her lifetime. This distinction is often critical to the student in school, because resources that may become available in the future cannot be used to pay this month's bills.[12]

Regardless of potential problems, it is important for planning purposes to estimate future resources. This task requires continual reassessment of all components of a total resource through consideration of new technology, the probability of geologic discovery, and shifts in economic and political conditions.[2]

Silver provides an illustration of some important points about resources and reserves. Based on geochemical estimates of the concentration of silver in rocks, Earth's crust (to a depth of 1 km, or 0.6 mi) contains almost 2×10^{12} metric tons of silver, an amount much larger than annual world use, which is approximately 10,000 metric tons. If this silver existed as pure metal concentrated in one large mine, it would represent a supply sufficient for several hundred million years at current levels of use. Most of the silver, however, exists in extremely low concentrations, too low to be extracted economically with current technology.

The known reserves of silver, reflecting the amount we could obtain immediately with known techniques, are about 200,000 metric tons, or a 20-year supply at current use levels and no new reserves.

The problem with silver, as with all mineral resources, is not with its total abundance but with its concentration and relative ease of extraction. When an atom of silver is used, it is not destroyed but remains an atom of silver. It is simply dispersed and may become unavailable. In theory, given enough energy, all mineral resources could be recycled, but this is not possible in practice. Consider lead, which is mined from minerals in which it is concentrated. The lead that was used in gasoline for many years is now scattered along highways across the world and deposited in low concentrations in forests, fields, and salt marshes close to these highways. Recovery of this lead is, for all practical purposes, impossible.

27.4 Classification, Availability, and Use of Mineral Resources

Earth's mineral resources can be divided into several broad categories, depending on our use of them: elements for metal production and technology; building materials; minerals for the chemical industry; and minerals for agriculture. Metallic minerals can be further classified according to their abundance. The abundant metals include iron, aluminum, chromium, manganese, titanium, and magnesium. Scarce metals include copper, lead, zinc, tin, gold, silver, platinum, uranium, mercury, and molybdenum.

Some mineral resources, such as salt (sodium chloride), are necessary for life. Primitive peoples traveled long distances to obtain salt when it was not locally available. Other mineral resources are desired or considered necessary to maintain a particular level of technology.

When we think about mineral resources, we usually think of the metals; but, with the exception of iron, the predominant mineral resources are not metallic. Consider the annual world consumption of a few selected elements. Sodium and iron are used at a rate of approximately 100–1,000 million metric tons per year. Nitrogen, sulfur, potassium, and calcium are used at a rate of approximately 10–100 million metric tons per year, primarily as soil conditioners or fertilizers. Elements such as zinc, copper, aluminum, and lead have annual world consumption rates of about 3–10 million metric tons, and gold and silver have annual consumption rates of 10,000 metric tons or less. Of the metallic minerals, iron makes up 95% of all the metals consumed; and nickel, chromium, cobalt, and manganese are used mainly in alloys of iron (as in stainless steel). Thus, with the exception of iron, the nonmetallic minerals are consumed at much greater rates than are elements used for their metallic properties.

Availability of Mineral Resources

The basic issue with mineral resources is not actual exhaustion or extinction but the cost of maintaining an adequate stock within an economy through mining and recycling. At some point, the costs of mining exceed the worth of material. When the availability of a particular mineral becomes a limitation, there are four possible solutions:

1. Find more sources.

2. Recycle and reuse what has already been obtained.

3. Reduce consumption.

4. Find a substitute.

Which choice or combination of choices is made depends on social, economic, and environmental factors.

The availability of a mineral resource in a certain form, in a certain concentration, and in a certain amount is a geologic issue determined by Earth's history. What is considered a resource and at what point a resource becomes limited are ultimately social questions. Before metals were discovered, they could not be considered resources. Before smelting was invented, the only metal ores were those in which the metals appeared in their pure form. For example, originally gold was obtained as a pure, or native, metal. Now gold mines are deep beneath the surface, and the recovery process involves reducing tons of rock to ounces of gold.

Mineral resources are limited, which raises important questions. How long will a particular resource last? How much short-term or long-term environmental deterioration are we willing to accept to ensure that resources are developed in a particular area? How can we make the best use of available resources? These questions have no easy answers. We are now struggling with ways to better estimate the quality and quantity of resources.

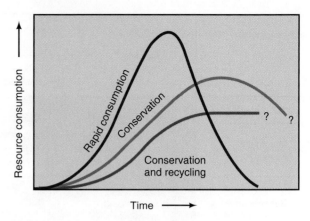

Figure 27.4 ■ Several hypothetical depletion curves.

Mineral Consumption

We can use a particular mineral resource in several ways: rapid consumption, consumption with conservation, or consumption and conservation with recycling. Which option is selected depends in part on economic, political, and social criteria. Figure 27.4 shows the hypothetical depletion curves corresponding to these three options. Historically, with the exception of precious metals, rapid consumption has dominated most resource utilization. However, as the supply of resources becomes short, increased conservation and recycling are expected. Certainly, the trend toward recycling is well established for metals such as copper, lead, and aluminum.

From a global viewpoint, limits on our mineral resources and reserves threaten our affluence. As the world population and the desire for a higher standard of living increase, the demand for mineral resources expands at a faster and faster rate. Today the more developed countries consume a disproportionate amount of the mineral resources extracted. For example, the United States, Western Europe, and Japan collectively use most of the aluminum, copper, and nickel that is extracted from the Earth.[5] Predicted increases in the world use of iron, copper, and lead, when linked with expected population increases, suggest that the rate of production of these metals will have to increase by several times if the world per-capita consumption rate is to rise to the level of consumption in developed countries today. Such an increase is very unlikely; affluent countries will thus have to find substitutes for some minerals or use a smaller proportion of the world's annual production. This situation parallels the Malthusian predictions discussed in Chapter 4: It is impossible in the long run to support an ever-increasing population on a finite resource base.

U.S. Supply of Mineral Resources

Domestic supplies of many mineral resources in the United States are insufficient for current use and must be supplemented by imports from other nations. For example, the

United States imports many of the minerals needed for its complex military and industrial system, called strategic minerals (examples include bauxite, manganese, graphite, cobalt, strontium, and asbestos). Of particular concern is the possibility that the supply of a much desired or much needed mineral may be interrupted by political, economic, or military instability in the supplying nation.

The fact that the United States—along with many other countries—depends on a steady supply of imports to meet the mineral demand of industries does not necessarily mean that the imported minerals do not exist within the country in quantities that could be mined. Rather, it suggests that there are economic, political, or environmental reasons that make it easier, more practical, or more desirable to import the material. This situation has resulted in political alliances that otherwise would be unlikely. Industrial countries often need minerals from countries with whose policies they do not necessarily agree. As a result, they make political concessions on human rights and other issues that they would not otherwise make.

27.5 Impacts of Mineral Development

The impact of mineral exploitation on the environment depends on such factors as ore quality, mining procedures, local hydrologic conditions, climate, rock types, size of operation, topography, and many more interrelated factors. The impact varies with the stage of development of the resource. For example, the exploration and testing stages involve considerably less impact than do the mining and processing stages. In addition, our use of mineral resources has a significant social impact.

Environmental Impacts

Exploration activities for mineral deposits vary from collection and analysis of remote-sensing data gathered from airplanes or satellites to fieldwork involving surface mapping and drilling. Generally, exploration has a minimal impact on the environment, provided that care is taken in sensitive areas, such as arid lands, marshes, and areas underlain by permafrost. Some arid lands are covered by a thin layer of pebbles over fine silt several centimeters thick. The layer of pebbles, called desert pavement, protects the finer material from wind erosion. When the desert pavement is disturbed by road building or other activities, the fine silts may be eroded, impairing physical, chemical, and biological properties of the soil and possibly scarring the land for many years. Similarly, marshes and other wetlands, such as the northern tundra, are very sensitive to even seemingly small disturbances such as vehicular traffic.

The mining and processing of mineral resources generally have a considerable impact on land, water, air, and biological resources. Furthermore, as it becomes necessary to use ores of lower and lower grades, negative effects on the environment tend to become greater problems. For example, there is concern about asbestos fibers in the drinking water (from Lake Superior) of Duluth, Minnesota, as a result of the disposal of iron mining waste (tailings) from mining low-grade iron ore.

A major practical issue is whether surface or subsurface mines should be developed in an area. There are several differences between surface (open-pit) and subsurface mining:[1]

- Subsurface mines are much smaller than open-pit mines.
- Mining activities at subsurface mines are less visible because less land at the surface is disturbed.
- Subsurface mining produces relatively little waste rock compared to open-pit mining. Open-pit mines often produce a tremendous volume of waste rock, with large visible impacts.
- Surface mining is cheaper but has more direct environmental effects.

The trend in recent years has been away from subsurface mining and toward large, open-pit (surface) mines, such as the Bingham Canyon copper mine in Utah (Figure 27.5). The Bingham Canyon mine is one of the world's largest human-made excavations, covering nearly 8 km^2 (3 mi^2) to a maximum depth of nearly 800 m (2,600 ft).

Surface mines and quarries today cover less than 0.5% of the total area of the United States. Even though the impact of these operations is a local phenomenon, numerous local occurrences will eventually constitute a larger problem. Environmental degradation tends to extend beyond the land actually being mined. Large mining operations disturb the land by directly removing material in some areas and dumping waste in others, thus changing topography. At the very least, these actions produce severe

Figure 27.5 ■ Aerial photograph of Bingham Canyon Copper Pit, Utah. It is one of the largest artificial excavations in the world.

Figure 27.6 ■ Tailings from a lead, zinc, and silver mine in Colorado. White streaks on the slope are mineral deposits apparently leached from the tailings.

aesthetic degradation. In addition, dust at mines may affect air resources, even though care is often taken to reduce dust production by sprinkling water on roads and on other sites that generate dust.

A potential problem associated with mineral resource development is the possible release of harmful trace elements to the environment. Water resources are particularly vulnerable to such degradation, even if drainage is controlled and sediment pollution is reduced. Surface drainage is often altered at mine sites, and runoff from precipitation may infiltrate waste material, leaching out trace elements and minerals. Trace elements (cadmium, cobalt, copper, lead, molybdenum, and others), when leached from mining wastes and concentrated in water, soil, or plants, may be toxic or may cause diseases in people and other animals who drink the water, eat the plants, or use the soil. Specially constructed ponds to collect such runoff help, but cannot be expected to eliminate all problems. The white streaks in Figure 27.6 are mineral deposits apparently leached from tailings from a zinc mine in Colorado. Similar-looking deposits may cover rocks in rivers for many kilometers downstream from some mining areas. Such obvious degradation of rivers is now a legacy of past mining in historical U.S. mining districts, where waste rock is still at the surface and reclamation has not occurred. In countries where mining regulations are weak or not enforced, degradation of air, soil, and water still occurs all too frequently.[1]

Groundwater may also be polluted by mining operations when waste comes into contact with slow-moving subsur-

face waters. Surface water infiltration or groundwater movement causes leaching of sulfide minerals that may pollute groundwater and eventually seep into streams to pollute surface water. Groundwater problems are particularly troublesome because reclamation of polluted groundwater is very difficult and expensive. (See Chapter 22 for a discussion of acid mine drainage, now called acid rock drainage.)

Physical changes in the land, soil, water, and air associated with mining directly and indirectly affect the biological environment. Plants and animals killed by mining activity or contact with toxic soil or water are examples of direct impacts. Indirect impacts include changes in nutrient cycling, total biomass, species diversity, and ecosystem stability owing to alterations in groundwater or surface water availability or quality. Periodic or accidental discharge of low-grade pollutants through failure of barriers, ponds, or water diversions or through breach of barriers during floods, earthquakes, or volcanic eruptions also may damage local ecological systems to some extent.

Social Impacts

Social impacts associated with large-scale mining result from the rapid influx of workers into areas unprepared for growth. Stress is placed on local services such as water supplies, sewage and solid-waste disposal systems, schools, and housing. Land use shifts from open range, forest, and agriculture to urban patterns. The additional people also increase the stress on nearby recreation and wilderness areas, some of which may be in a fragile ecological balance. Construction activity and urbanization affect local streams through sediment pollution, reduced water quality, and increased runoff. Air quality is reduced as a result of more vehicles, dust from construction, and generation of power.

Adverse social impacts also occur when mines are closed; towns surrounding large mines come to depend on the income of employed miners. Closures of mines produced the well-known ghost towns in the old American West. Today, the price of coal and other minerals directly affects the livelihood of many small towns. This relationship is especially evident in the Appalachian Mountain region of the United States, where closures of coal mines are taking their toll. These mine closings are partly the result of lower prices for coal and partly the result of rising mining costs.

One of the reasons mining costs are rising is the increased level of environmental regulation of the mining industry. Of course, regulations have also helped make mining safer and have facilitated land reclamation. Some miners, however, believe the regulations are not flexible enough, and there is some truth to their arguments. For example, some mined areas might be reclaimed for use as farmland now that the original hills have been leveled. Regulations, however, may require the restoration of the land to its original hilly state, even though hills make inferior farmland.

27.6 Minimizing Environmental Impact of Mineral Development

Minimization of environmental impacts of mineral development requires consideration of the entire cycle of mineral resources shown in Figure 27.7. Inspection of this diagram reveals that many components of the cycle are related to the generation of waste material. In fact, the major environmental impacts of mineral resource utilization are related to waste products. Waste produces pollution that may be toxic to humans, may harm natural ecosystems and the biosphere, and may be aesthetically undesirable. Waste may attack and degrade resources such as air, water, soil, and living things. Waste also depletes nonrenewable mineral resources and, when simply disposed of, provides no offsetting benefits for human society.

Environmental regulations at the federal, state, and local levels address topics such as sediment, air, and water

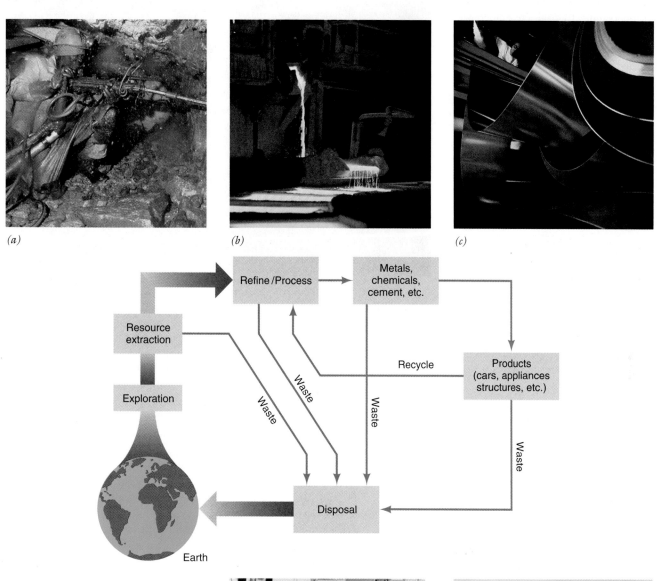

(a)   *(b)*   *(c)*

Figure 27.7 ■ Simplified flowchart of the resource cycle: (*a*) mining gold in South Africa; (*b*) copper smelter, Montana; (*c*) sheets of copper for industrial use; (*d*) appliances made in part from metals; and (*e*) disposal of mining waste from a Montana gold mine into a tailings pond.

(d)   *(e)*

Canada's Butchart Gardens: From Eyesore to Eden

It's often been stated that one person at the right time and in the right place has the opportunity to make a real difference. Jenny Butchart in the early 1900s had a vision to transform her husband's exhausted mine pit into a garden. Today, we would use the term *mine reclamation* for such a project. The story that follows was one of transformation. Gardens that began as a limestone quarry and an eyesore became Butchart Gardens, a tourist attraction on Vancouver Island, British Columbia, visited by about a million people each year. The Butchart Garden story offers an early illustration of the principles of environmental restoration and sequential land use.

The story is closely related to the cement industry. The process of producing Portland cement was developed in 1824 in England. The procedure includes fine grinding of an exact mixture of limestone and shale that has been heated in a rotary kiln to near-fusion temperature. This material is then finely ground again, placed in sacks, and sold to consumers as Portland cement. Adding aggregate such as gravel to the cement makes concrete, which is an integral part of construction in our urban environment.

Robert Butchart was educated in the process of making Portland cement while on honeymoon in England in 1884. In 1902, he learned of the availability of limestone located approximately 20 km north of the city of Victoria on Vancouver Island. He determined that the site was

Figura 27.8 ■ *Butchart Gardens, Vancouver Island, Canada. The site of the gardens was once a limestone quarry.*

ideal for the establishment of a cement plant and started the Cod Inlet Cement Plant there in 1904. The limestone used to produce the Portland cement was exhausted by 1908, and what was left behind was a large, unsightly open excavation about 20 m deep. The walls of the quarry were nearly vertical gray limestone.

That's when Jenny Butchart entered the picture. She had little experience with gardens but was fascinated with the site and the idea of transforming the quarry into a sunken garden. Her husband supported this endeavor, and with

the help of many workmen, she brought in massive amounts of topsoil by horse and cart to form garden beds on the floor of the quarry. The garden, which was completed in 1921, soon became a tourist attraction (Figure 27.8). Today, the garden has grown to include fountains, rose gardens, lakes, Japanese gardens, and other attractions in addition to the original sunken garden. The Butchart Gardens story illustrates the power of the vision of one person to transform the landscape from a degraded site to a beautiful and aestetically pleasing place.[20]

pollution resulting from all aspects of the mineral cycle. Regulations may also address reclamation of land used for mining. Today in the United States, approximately 50% of the land utilized by the mining industry has been reclaimed.

Minimization of environmental effects associated with mining takes several interrelated paths:[1]

■ *Reclaiming* areas where physical, hydrological, and biological disturbance has occurred. (See A Closer Look 27.1.)

■ *Stabilizing soils* that contain metals to minimize their release into the environment. Often this requires contaminated soils to be removed and placed in a waste facility.

- *Controlling air emissions* of metals and other materials from mining areas.

- *Preventing (by treatment) contaminated water from leaving or treating contaminated water that has left, a mining site.* The most common treatment for acid rock drainage (formerly called acid mine drainage) is to neutralize potential acidity with a material such as lime. This often requires construction of a waste treatment plant.

- *Treating waste on-site and off-site.* Minimizing on-site and off-site problems by controlling sediment, water, and air pollution through good engineering and conservation practices is an important goal. Of particular interest is the development of biotechnological processes such as biooxidation, bioleaching, and biosorption, as well as genetic engineering of microbes. These practices have enormous potential for both extracting metals and minimizing environmental degradation. For example, engineered (constructed) wetlands are being used at several sites; acid-tolerant plants in the wetlands remove metals from mine wastewaters and neutralize acids by biological activity.[13] At the Homestake Gold Mine in South Dakota, biooxidation is being used to convert contaminated water from the mining operation to substances that are environmentally safe; the process uses bacteria with a natural capacity to oxidize cyanide to harmless nitrates.[14]

- *Practicing the three R's of waste management.* That is, reduce the amount of waste produced; reuse materials in the waste stream as much as is feasible; and maximize recycling opportunities.

Let's look at the three R's in greater detail. Wastes from some parts of the mineral cycle may themselves be referred to as ores, because they contain materials that might be recycled and used again to provide energy or useful products.[15,16] The notion of reusing waste materials is not new. Such metals as iron, aluminum, copper, and lead have been recycled for many years and are still being recycled today. For example, the metal from almost all of the millions of automobiles discarded annually in the United States is recycled.[16,17]

The total value of recycled metals is about $50 billion. Of this, iron and steel are approximately 90% by weight and 40% by total value of recycled metals. Iron and steel are recycled in such large volumes for three reasons:[18] First, the market for iron and steel is huge, and as a result, there is a large scrap collection and processing industry. Second, an enormous economic burden would result from failure to recycle. Third, failing to recycle would create significant environmental impacts related to disposal of over 50 million tons of iron and steel.

It is estimated that each ton of recycled steel saves 1,136 kg (2,500 pounds) of iron ore, 455 kg (1,000 pounds) of coal, and 18 kg (40 pounds) of limestone. In addition, only one-third as much energy is required to produce steel from recycled scrap as from native ore.[17]

Other metals that are recycled in large quantities include lead (63%), aluminum (38%), and copper (36%).[3] Recycling aluminum reduces our need to import raw aluminum ore and saves approximately 95% of the energy required to produce new aluminum from bauxite.[17]

27.7 Minerals and Sustainability

Simultaneously considering sustainable development and mineral exploitation and use is problematic. This is because nonrenewable mineral resources are consumed over time and sustainability is a long-term concept that includes finding ways to provide future generations a fair share of Earth's resources. Recently, it has been argued that given human ingenuity and sufficient lead time, we can find solutions for sustainable development that incorporate nonrenewable mineral resources.

Human ingenuity is important because often it is not the mineral we need so much as what we use the mineral for. For example, we mine copper and use it to transmit electricity in wires or electronic pulses in telephone wires. It is not the copper itself we desire but the properties of copper that allow these transmissions. We can use fiberglass cables in telephone wires, eliminating the need for copper. Digital cameras have eliminated the need for film development that uses silver. The message is that it is possible to compensate for a nonrenewable mineral by finding new ways to do things. We are also learning that we can use raw mineral materials more efficiently. For example, when the Eiffel Tower was constructed in the late 1800s, 8,000 metric tons of steel were used. Today the tower could be constructed with a fourth of that amount of steel.[19]

Finding substitutes or ways to use nonrenewable resources more efficiently generally requires several decades of research and development. A measure of the time available for finding the solutions to depletion of nonrenewable reserves is the **R-to-C ratio**, where R is the known reserves (for example, hundreds of thousands of tons of a metal) and C is the rate of consumption (for example, thousands of tons per year used by people). The R-to-C ratio is often misinterpreted as the time a reserve will last at the present rate of consumption. During the past 50 years, the R-to-C ratios for metals such as zinc and copper have fluctuated around 30 years; during that time, consumption of the metals increased by about three times. This was possible because we discovered new deposits of the metals. The R-to-C ratio is a present analysis of a dynamic system in which both the amount of reserves and consumption may change over time. However, the ratio does provide a view of how scarce a particular mineral resource may be. Those metals with relatively small ratios can be viewed as being in short supply, and it is those resources for which we should find substitutes through technological innovation.[19]

Will Mining with Microbes Help the Environment?

Mining is an ancient technology first practiced at least 6,500 years ago. Modern mining methods are more technologically sophisticated but use the same basic processes (digging and smelting) to isolate valuable metals. To be economic, these methods have traditionally required high-grade ore and cheap sources of energy as well as an acceptance that mining to a lesser or greater extent would damage the environment. Although these conditions have prevailed for most of human history, they are changing. Earlier exploitation of mineral resources is pushing the mining industry toward lower-grade ores; nonrenewable energy sources are expensive and disappearing; and concern over degradation of the environment and health threats to humans and other species is growing. Demand for minerals, however, is increasing because of both population growth and technological development.

As an example, the average grade of copper ore has dropped from 6 to 0.6% over the last century, making copper mining more energy intensive and more wasteful. Five metric tons of coal are required to produce 1 metric ton of copper, and every kilogram of copper produced represents 89 kg (198 lb) of waste. Open-pit mining of copper causes acids and heavy metals, such as arsenic, to contaminate surface water and groundwater. Smelting produces sulfur dioxide and other gaseous compounds, as well as particles, which contribute to air pollution.

Microscopic organisms produced by biotechnology offer an entirely new approach to mining. By 1989, more than 30% of copper mined in the United States depended on a biochemical process that begins with a microbe, *Thiobacillus ferrooxidans.* Biological processes have also been used in mining uranium and gold. Research is under way to use microbes to remove sulfur from coal and cyanide from mining waste. The union of biological processes and mining is called *biohydrometallurgy.*

In the future, it may be possible to use microbes on ores without removing them from Earth. Metallurgists envision drilling wells into the ore and fracturing it, then injecting bacteria into the wells and fractures. The ore could be removed by flooding the wells with water, removing the ore, and recycling the water. Biotechnologists hope to use genetic engineering to develop bacteria to mine specific metals when no naturally occurring bacteria exist to extract them.

The disadvantages of biohydrometallurgy are that it is slow, requiring decades rather than years, and that methods for breaking ores into small enough particles for efficient extraction are not yet available. Already, however, biological methods are economically feasible for low-grade ores that elude conventional methods. Further technological innovations may make them competitive in more situations.

Critical Thinking Questions

1. What are the environmental advantages of biohydrometallurgy over conventional methods? What are the possible disadvantages?
2. How would you assess the possible dangers from genetically engineered organisms developed for mining compared with the dangers from organisms engineered for use in agriculture and medicine?
3. Some experts believe that without economic pressure (for example, the decline in high-grade ore, increased energy requirements for extracting metal, governmental regulations on clean air and water, economic recessions), the mining industry would not continue to explore biochemical methods. How could the government encourage the industry to devote more research and development efforts to expanding biochemical mining? Develop a proposal for a government policy.

In summary, we may approach sustainable development and use of nonrenewable mineral resources by finding ways to more wisely use resources, developing more efficient ways of mining resources, more efficiently using available resources, recycling more, and applying human ingenuity to find substitutes for a particular function that a nonrenewable mineral resource is used.

Summary

- Mineral resources are usually extracted from naturally occurring, anomalously high concentrations of Earth materials. Such natural deposits allowed early peoples to exploit minerals while slowly developing technological skills.

- The origin and distribution of mineral resources is intimately related to the history of the biosphere and the geologic cycle. Nearly all aspects and processes of the geologic cycle are involved to some extent in producing local concentrations of useful materials.

- Mineral resources are not mineral reserves. Unless discovered and developed, resources cannot be used to address present shortages.

- The availability of mineral resources is one measure of the wealth of a society. Modern technological civilization would not be possible without the exploitation of mineral resources. However, it is important to recognize that mineral deposits are not infinite and that we cannot maintain exponential population growth on a finite resource base.

- The United States and many other affluent nations rely on imports for their supplies of many mineral resources. As other nations industrialize and develop, such imports may be more difficult to obtain, and affluent countries may have to find substitutes for some minerals or use a smaller portion of the world's annual production.

- The environmental impact of mineral exploitation depends on many factors, including mining procedures, local hydrologic conditions, climate, rock types, size of operation, topography, and other factors.

- The mining and processing of mineral resources greatly affect the land, water, air, and biological resources and create social impacts as a result of the increased demand for housing and services in mining areas.

- Because the demand for mineral resources will increase in the future, we must strive to minimize both on-site and off-site problems by controlling sediment, water, and air pollution through good engineering and conservation practices.

- Sustainable development and use of nonrenewable resources are not necessarily incompatible. Reducing consumption, reusing, recycling, and finding substitutes are environmentally preferable ways to delay or alleviate possible crises caused by the convergence of a rapidly rising population and a limited resource base.

REEXAMINING THEMES AND ISSUES

Human Population

Increasing human population is a major issue in the availability and use of mineral resources. We would be experiencing a major mineral crisis today if the consumption rate for all the people of the world were anywhere near that of people in developed countries. Therefore, as human population increases and developing countries increase their own demands, we must find ways to reduce our per-capita mineral consumption. As we have emphasized before, satisfying the mineral needs of an ever-expanding population on a finite resource base at current levels is impossible in the long term.

Sustainability

Many sustainability issues arise from our use of mineral resources, because so many parts of the mineral cycle are related to waste management problems involving pollution of soil, air, and water. If we are to provide a quality environment for future generations, we must find ways to minimize adverse environmental effects of mineral exploitation, development, and use. Of particular importance will be conservation measures aimed at reducing our use of minerals, reusing materials whenever possible, and recycling as much as is feasible.

Global Perspective

Mineral resources are scattered over the Earth at various locations, where they were concentrated by geologic processes. Because the origin of many mineral deposits is directly related to global tectonics, understanding how our world works gives us a better perspective on our mineral resources.

Urban World

The use of mineral resources is concentrated in urban areas, where most people live. Therefore, urban areas are the places where recycling of minerals such as aluminum, iron, and copper is most economically viable.

People and Nature

Nature produced our nonrenewable resources in the distant geologic past. They are part of our heritage from Earth. We have used them throughout human history for our tools and built our civilization on their availability. Our task is to continue the use of minerals without damaging nature.

Science and Values

Because we value a quality environment and because mining and other activities related to the mineral cycle can cause environmental disruption, we have a great interest in developing the knowledge necessary to minimize adverse effects of mineral utilization. A variety of new technologies in this area have arisen in recent years, including biotechnology, which utilizes the biological environment to assist in solving environmental problems related to the mineral cycle, particularly as it relates to waste management.

Key Terms

mineral resources **592**   ore deposits **590**   reserve **593**   *R*-to-*C* ratio **599**

Study Questions

1. What is the difference between a resource and a reserve?
2. Under what circumstances might sewage sludge be considered a mineral resource?
3. If surface mines and quarries cover less than 0.5% of the land surface of the United States, why is there so much environmental concern about them?
4. When is recycling a mineral a viable option?
5. Which biological processes can influence mineral deposits?
6. A deep-sea diver claims that the oceans can provide all our mineral resources with no negative environmental effects. Do you agree or disagree?
7. What factors determine the availability of a mineral resource?
8. Utilizing a mineral resource involves four general phases: (a) exploration, (b) recovery, (c) consumption, and (d) disposal of waste. Which phase do you think has the greatest environmental effect?
9. How can use of nonrenewable mineral resources be compatible with sustainable development?

Further Reading

Brookins, D. G. 1990. *Mineral and Energy Resources.* Columbus, Ohio: Charles E. Merrill. A good summary of mineral resources.

Kesler, S. F. 1994. *Mineral Resources, Economics and the Environment.* Upper Saddle River, N.J.: Prentice Hall. A good book about mineral resources.

CHAPTER 28

Dollars and Environmental Sense: Economics of Environmental Issues

Then and now. Above: Photo by Ashed Curtis provided by the Washington State Historical Society shows Makah whalers landing a gray whale c. 1900. Inset: Photo by Theresa Parker (Makah) shows successful whale hunt May 17, 1999.

Learning Objectives

Why do people value environmental resources? To what extent are environmental decisions based on economics? Other chapters in this text have explained the causes of environmental problems and discussed technical solutions. The scientific solutions, however, are only part of the answer. This chapter introduces some basic concepts of environmental economics and shows how these concepts help us understand environmen-

tal issues. After reading this chapter, you should understand:

■ Why, when it comes to the environment, people sometimes seem to act against their own best interest.

■ When and how it is possible to put a dollar value on the environment.

■ What "the tragedy of the commons" is and how it leads to overexploitation of resources.

■ How the perceived future value of an environmental benefit affects our willingness to pay for it now.

■ What "externalities" are and why they matter.

■ How much risk we should be willing to accept for the environment and ourselves.

■ How we can place a value on environmental intangibles, such as landscape beauty.

CASE STUDY

Whale Burgers or Whale Conservation, or Both?

Attitudes about whales vary widely around the world. In June 2005 a fast-food chain in northern Japan began to offer a whale-meat burger (with lettuce and mayo). Eating whale meat is part of a historical cultural practice in Japan. That same month, at the annual meeting of the International Whaling Commission (IWC), Japan announced that it was going to double that nation's catch of minke whales, a small, highly abundant whale. The spokesman for New Zealand called this decision "shameful," and Australia's environment minister, Ian Campbell, called it an "outrage." Both New Zealand and Australia are making increasing amounts of money from whale-watching tourism.

The United States' official position is to oppose any whaling. But the International Whaling Commission has agreed that, for traditional cultural reasons, the Eskimo could take 67 bowhead whales a year.[22] And in the spring of 1999, eight young men of the Makan Indian tribe, who live near the mouth of the strait of Juan de Fuca on the Olympic Peninsula in the state of Washington, canoed out into the waters and within a week had killed a 40-ton gray whale, using a combination of traditional harpoons and a gun. They went out in a canoe last used in the 1920s. "Everybody felt like it was a part of making history," Micah L. McCarty, a tribal council member, said of the 1999 hunt. "It's inspired a cultural renaissance, so to speak. It inspired a lot of people to learn artwork and become more active in building canoes; the younger generation took a more keen interest in singing and dancing."[33]

The staunch supporters of the IWC want to end all whaling, while Norway, Japan, and several Native American groups continue to hunt whales as a cultural tradition. What is the "correct" way to think about whaling? Who is right? And to what extent does economics come into the picture? At one time, whaling on the high seas by Yankee fisherman was a big business (Figures 28.1 and 28.2). Is it still significant enough to argue about? And will it endanger whales?

Perhaps more than any other endangered-species issue, this one draws intense, formal, international debates. In this chapter we can gain some insight into why an issue like whaling matters so much, and how economics does and does not play a role.

Figure 28.1 ■ Medieval European drawing of a whale hunt.

Figure 28.2 ■ Bowhead whale baleen (their modified teeth) on a dock in San Francisco in the late 19th century when these flexible plates were important in women's corset stays and for other uses where strength and flexibility were important. These were soon replaced by new forms of steel, and the baleen market disappeared. Commercial bowhead whale hunting ended with the beginning of World War I.

> *The case study of whaling illustrates the importance of conflicts over values, including economic ones. In this chapter we will see that economic analysis can help us understand how to sustain renewable resources.*

28.1 The Economic Importance of the Environment

In the mid-1990s, the United States spent about $115 billion a year, about 2% of the nation's gross national product, to deal with pollution. The defense budget was only two and a half times larger. Present costs are closer to $170 billion, including amounts spent by consumers, corporations, and government.[2] This total is much greater than the $6 billion budget of the Environmental Protection Agency (EPA).[3]

Though costly, cleaning our environment has economic benefits. Populations subject to high levels of certain pollutants (people in inner cities, for example) have lower average life expectancies and higher incidences of certain diseases. Particulate air pollution in U.S. cities contributes to 60,000 deaths annually,[4] and 2% to 9% of total mortality in cities is associated with particulate air pollution.[5] By the year 2010, amendments to the Clean Air Act passed by Congress in 1990 would prevent 23,000 premature deaths in the United States, 1.7 million asthma attacks, and more than 60,000 hospital admissions due to respiratory problems. The value of the benefits from these amendments is estimated to be $110 billion in 2010, while the costs are estimated to be $27 billion. And these are from amendments alone.

Environmental decision-making often involves analysis of tangible and intangible factors. A mudslide that results from altering the slope of land is an example of a tangible factor; the beauty of the slope before the mudslide and its ugliness afterward is an example of an intangible factor. Of the two, the intangibles are obviously more difficult to deal with because they are hard to measure and to value economically. Nonetheless, evaluation of intangibles is becoming more important. One task of **environmental economics** is to develop methods for evaluating intangibles that provide good guidelines, are easy to understand, and are quantitatively credible. Not an easy goal!

28.2 The Environment as a Commons

Often people who use a natural resource do not act in a way that maintains that resource and its environment in a renewable state; that is, they do not seek sustainability. At first glance this seems puzzling. Why do people not act in their own best interest? Economic analysis suggests that the profit motive, by itself, will not always lead a person to act in the best interests of the environment. Here, we give two reasons why this may be so.

The first reason has to do with what the ecologist Garrett Hardin called "the tragedy of the commons."[6] When a resource is shared, an individual's personal share of profit from exploitation of the resource is usually greater than that individual's share of the resulting loss. The second has to do with the low growth rate, and therefore the low productivity, of a resource.

A **commons** is land (or another resource) owned publicly with public access for private uses. The term *commons* originated from land owned publicly and set aside in English and New England towns where all the farmers of the town could graze their cattle. The practice of sharing the grazing area worked as long as the number of cattle was low enough to prevent overgrazing. It would seem that people of goodwill would understand the limits of a commons. But take a dispassionate view and think about the benefits and costs to each farmer as if it were a game. Phrased simply, each farmer tries to maximize personal gain and must periodically consider whether to add more cattle to the herd on the commons. The addition of one cow has both a positive and a negative value. The positive value is the benefit when the herder sells that cow. The negative value is the additional grazing by the cow. The benefit to an individual of selling a cow for personal profit is greater than that individual's share of the loss in the degradation of the commons. The short-term successful game plan, therefore, is always to add another cow.

So an individual will act to increase use of the common resource. Eventually the common grazing land is so crowded with cattle that none can get adequate food and the pasture is destroyed. In the short run, everyone seems to gain, but in the long run, everyone loses. This applies generally: Complete freedom of action in a commons inevitably brings ruin to all. The implication seems clear: Without some management or control, all natural resources treated like a commons will inevitably be destroyed.

How can we deal with the tragedy of the commons? It is an unsolved puzzle. As several scientists wrote recently, "No single broad type of ownership—government, private or community—uniformly succeeds or fails to halt major resource deterioration." In trying to solve this puzzle, economic analysis can be helpful.

There are many examples of commons, both past and present. Of forests in the United States, 38% are on publicly owned lands. Resources in international regions, such as ocean fisheries away from coastlines and the deep-ocean seabed, where valuable mineral deposits lie, are international commons not controlled by any single nation (Figure 28.3).

Most of the continent of Antarctica is a commons, although there are some national territorial claims there, and international negotiations have continued for years about conserving Antarctica and possible use of its resources.

The atmosphere, too, is a commons, both nationally and internationally. Consider the possibility of global warming. Individuals, corporations, public utilities, motor vehicles, and nations add carbon dioxide to the air through the burning of fossil fuels. Just as Garrett Hardin suggested, people tend to

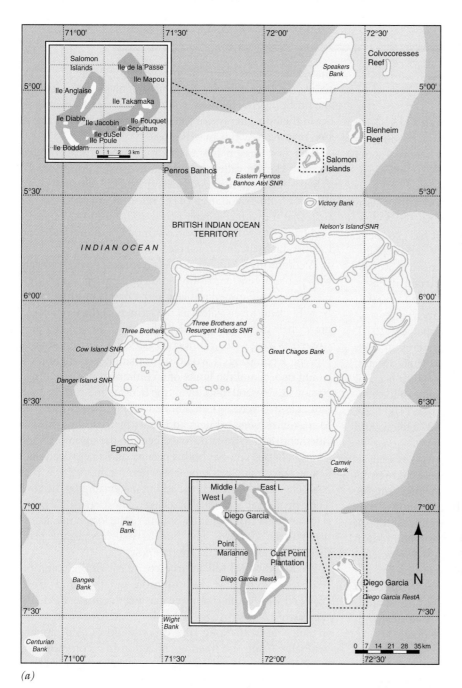

(a)

Fish in the waters of the Chagos coral reefs.
(b)

A red-footed booby on a palm tree in the Chagos atolls.
(c)

Figure 28.3 ■ British Indian Ocean Territory: a kind of commons. (*a*) This large region, called Chagos Archipelago, contains the Great Chagos Bank, the largest atoll structure in the world, covering 13,000 square kilometers. Among other uses, it functions as a major tuna fishery and is a global commons in this way. (*b*) It is also home to many rare species, such as beautiful water birds (*c*) and serves as a biodiversity commons as well.

respond by benefiting themselves (by burning more fossil fuel) rather than by benefiting the commons (burning less fossil fuel). The picture here is quite mixed, however, with much ongoing effort to bring cooperation to this common issue.

In the nineteenth century, burning wood in fireplaces was the major source of heating in the United States (and fuelwood is still the major source of heat in many nations). Until the 1980s, a wood fire in a fireplace or woodstove was considered a simple good, providing warmth and beauty; people enjoyed sitting around a fire and watching the flames. This activity must have a long history in human societies. But in the 1980s, with increases in populations and recreational houses in states such as Vermont and Colorado, home burning of wood began to pollute air locally. Especially in valley towns surrounded by mountains, the air became fouled, visibility declined, and there was a potential for effects on human health and environmental conditions. As a result, some communities restrict or prohibit the use of fireplaces and woodstoves. The local air is a commons, and its overuse required a societal change.

Recreation is a problem of the commons—overcrowding of national parks, wilderness areas, and other nature-recreation areas. An example is Voyageurs National Park in northern Minnesota. The park, located within the boreal forest biome (see Chapter 8) of North America, contains many lakes and islands and is an excellent place for fishing, hiking, canoeing, and viewing wildlife. Before the area became a national park, it was used for motorboating, snowmobiling, and hunting; a number of people in the region made their living from tourism based on these kinds of recreation. Some environmental groups argue that the Voyageurs National Park is ecologically fragile and needs to be legally designated a U.S. wilderness area in order to protect it from overuse and from the adverse effects of motorized vehicles. Others argue that the nearby million-acre Boundary Waters Canoe Area (see Chapter 13) provides ample wilderness, that Voyageurs can withstand a moderate level of hunting and motorized transportation, and that these uses should be allowed.

At the heart of this conflict is the problem of the commons, which in this case can be phrased as: What is the appropriate public use of public lands? Should all public lands be open to all public uses? Should some public lands be protected from people? At present, the United States has a policy of different uses for different lands. In general, national parks are open to the public for many kinds of recreation, whereas designated wildernesses have restricted visitorship and kinds of uses.

28.3 Low Growth Rate and Therefore Low Profit as a Factor in Exploitation

Recall that the second reason individuals tend to overexploit natural resources held in common is the low growth rate of the resource.[7] For example, one way to view whales economically is to consider them solely in terms of whale oil (see Chapter 14). Whale oil, a marketable product, can be thought of as the capital investment of the industry. How can whalers get the best return on their capital? (Here we need to remember that whale populations, like other populations, increase only if there are more births than deaths.) We will examine two approaches: resource sustainability and maximum profit. If whalers adopt a simple, one-factor resource sustainability policy, they will harvest only the net biological productivity each year and thus maintain the total abundance of whales at its current level; that is, they will stay in the whaling business indefinitely. In contrast, if they adopt a simple approach to maximizing immediate profit, they will harvest all the whales now, sell the oil, get out of the whaling business, and invest the profits.

Suppose they adopt the first policy. What is the maximum gain they can expect? Whales, like other large, long-lived creatures, reproduce slowly; typically, a calf is born every three or four years (Figure 28.4). The total net growth of a whale population is unlikely to be more than 5% per year. If all the oil in the whales in the oceans today represented a value of $100 million, then the most the whalers could expect to take in each year would be 5% of this amount, or $5 million. Meanwhile, they would have to pay the cost of upkeep on ships and other equipment, interest on loans, and salaries of employees—all of which would decrease profit. If whalers adopted the second policy and harvested all the whales, then they could invest the money from the oil. Although investment income varies, even a conservative investment of $100 million would very likely yield more than

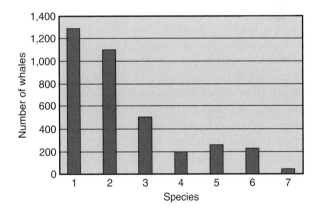

Figure 28.4 ■ Bowhead whales killed by Yankee whalers (caught and killed) from 1849 to 1914. The number killed is shown for each decade. The fact that the number killed declined rapidly, although hunting continued, indicates that the population rapidly decreased—it was unable to reproduce at a rate that could replace the large catches in the first two decades. [*Source:* Redrawn from J. R. Bockstoce and D. B. Botkin, *The Historical Status and Reduction of the Western Arctic Bowhead Whale* (Balaena mysticetus) *Population by the Pelagic Whaling Industry, 1849–1914.* Final report to the U.S. National Marine Fisheries Service by the Old Dartmouth Historical Society, 1980.]

5%, especially when this income could be received without the cost of paying a crew, maintaining the ships, buying fuel, marketing the oil, and so on.

It is quite reasonable and practical, if one considers only direct profit, to adopt the second policy: Harvest all the whales, invest the money, and relax. Whales simply are not a highly profitable long-term investment under the resource sustainability policy. Note that this discussion suggests that whaling on the open seas can be also viewed as a problem of a commons, complicated by low growth rate.

It is no wonder that there are fewer and fewer whaling companies and that companies left the whaling business when their ships became old and inefficient. Few nations support whaling; those that do have stayed with whaling for cultural reasons. For example, whaling is important to the Eskimo culture, and some harvest of bowheads takes place in Alaska; whale meat is a traditional Japanese food, and whale harvest is maintained for this reason.

Another factor to consider in resource use is the relative scarcity of a necessary resource, which affects its value and therefore its price. For example, if a whaler lived on an isolated island where whales were the only food and had no communication with other people, then his primary interest in whales would be in his remaining alive. He could not choose to sell off all whales to maximize profit, as he would have no one to sell them to. It would seem to make sense that he would harvest in a way that would maintain the population of whales. Or the whaler might estimate that his own life expectancy was only ten years and that to avoid starvation he had to consume the whales beyond their ability to reproduce. He might try to harvest the whales so that they would become extinct at the same time that he would die. "You can't take it with you" would be his attitude.

If ships began to land regularly at this island, he could trade and begin to benefit from some of the future value of whales. If ocean property rights existed so that he could "own" the whales that lived within a certain distance of his island, then he might consider the economic value of owning this right to the whales. He could sell rights to future whalers, or mortgage against them, and thus reap the benefits during his lifetime for whales that could be caught after his death. Causing the extinction of whales would not be necessary.

From harvesting whales, we see that we must think beyond the immediate, direct economic advantages of whaling. Policies that seem ethically good may not be the most profitable for an individual. Economic analysis clarifies how an environmental resource is used, what is perceived as its intrinsic value, and therefore its price—and this brings us to the question of externalities.

28.4 Externalities

One gap in our thinking about whales, an environmental economist would say, is that we must be concerned with externalities in whaling. An **externality**, also called an **indirect cost**, is an effect not normally accounted for in the cost–revenue analysis of producers and often not recognized by them as part of their costs and benefits.[7] Put simply, externalities are costs or benefits that don't show up in the price tag.[8] In the case of whaling, externalities include the loss of revenue to tourist boats used to view whales and the loss of an ecological role played by whales in marine ecosystems. Classically, economists agree that the only way for a consumer to make a rational decision is by comparing the *true* costs against the benefits the consumer seeks. If the true costs are not revealed, then the price will be wrong, and purchasers cannot act rationally.

Air and water pollution provide other good examples of externalities. Consider production of nickel from ore at the Sudbury, Ontario, smelters, which has serious environmental effects, as discussed in Chapter 15. Traditionally, the economic costs associated with the production of commercially usable nickel from an ore are the **direct costs**—that is, those borne by the producer and passed directly on to the user or purchaser. In this case, direct costs include purchasing the ore, buying energy to run the smelter, building the plant, and paying employees. Meanwhile, externalities include costs associated with degradation of the environment from the plant's emissions. For example, prior to implementation of pollution control, the Sudbury smelter destroyed vegetation over a wide area, which led to an increase in erosion. Although air emissions from smelters have been substantially reduced and restoration efforts have initiated a slow recovery of the area, pollution remains a problem, and total recovery of the local ecosystem may take a century or more.[9] There are costs associated with the value of trees and soil, with restoration of vegetation and land to a productive state.

Problem number one: What is the true cost of clean air over Sudbury? Economists say that there is plenty of disagreement about such a price but that everyone agrees that it is larger than zero. In spite of this, clean air and water are traded and dealt with in today's world as if their value were zero. How do we get the value of clean air and water and other environmental benefits to be recognized socially as greater than zero? In some cases, the dollar value can be determined. Water resources for power or other uses may be evaluated by the amount of flow of the rivers and the quantity of water storage in rivers and lakes; forest resources may be evaluated by the number, type, and sizes of trees and their subsequent yield of lumber; and mineral resources may be evaluated by the estimated number of metric tons of economically valuable mineral material at particular locations. Quantitative evaluation of the tangible natural resources—such as air, water, forests, and minerals—prior to development or management of a particular area is now standard procedure.

Problem number two: Who should bear the burden of these costs? Some suggest that environmental and ecological costs should be included in costs of production through taxation or fees. The expense would be borne by the corporation that benefits directly from the sale of the resource (nickel in the case of Sudbury) or would be passed on, in increased sales prices, to users (purchasers) of nickel. Others suggest that these costs should be shared by the entire society and paid for by general taxation (such as a sales tax or income tax). Stated simply, the question is whether it is better to finance pollution control using tax dollars or a "polluter pays" approach. Today, economists generally agree that the "polluter pays" approach provides much stronger incentives for cost-effective pollution reduction.

28.5 Natural Capital, Environmental Intangibles, and Ecosystem Services

Public Service Functions of Nature

A complicating factor in our perception of maintaining clean air and water is that ecosystems do some of this without our help—and have done so since before the Industrial Revolution. Forests absorb particulates, salt marshes convert toxic compounds to nontoxic forms, wetlands and organic soils treat sewage (see A Closer Look 21.4; see also Chapter 13). These are called the *public service functions* of nature. For example, it is estimated that bees pollinate $20 billion worth of crops in the United States. The cost of pollinating these crops by hand would be exorbitant, so a pollutant that eliminated bees would have large indirect economic consequences. We rarely think of this benefit of bees. Recently, however, an outbreak of bee parasites in the United States reduced the abundance of bees, bringing this once-intangible factor to public attention (Figure 28.5).

As another example, bacteria fix nitrogen in the oceans, lakes, rivers, and soils. The cost of replacing this function in terms of production and transport of artificially produced nitrogen fertilizers would be immense, but again we rarely think about this activity of bacteria. Bacteria also clean water in the soil by decomposing toxic chemicals.

The atmosphere performs a public service by acting as a large disposal site for toxic gases. For instance, carbon monoxide is eventually converted to nontoxic carbon dioxide either by inorganic chemical reactions or by bacteria.

Only when our environment loses a public service function do we usually begin to recognize its economic benefits. Then, what had been accepted as an economic externality (indirect cost) suddenly may become a direct cost.

Peolpe have attempted to estimate the dollar value of public service functions. At this time, we have to consider these estimates only rough approximations, as the value is difficult to measure. Public service functions of living things that benefit human beings and other forms of life have been estimated to provide between $3 tril-

Figure 28.5 ■ Public service functions of living things. Wild creatures and natural ecosystems perform public service functions for us—carrying out tasks important for our survival that would be extremely expensive for us to accomplish by ourselves. For example, bees pollinate millions of flowers important for food production, timber supply, and aesthetics.

lion and $33 trillion per year.[10,11] Economists refer to the ecological systems that provide these benefits as *natural capital*.

Valuing the Beauty of Nature

The beauty of nature—often referred to by the more general term *landscape aesthetics*—is another important environmental intangible. It has probably been important to people as long as our species has existed and certainly has been important since people have written, because the beauty of nature is a continuous theme in literature and art. Once again, as with forests cleaning the air, we face the difficult question: How do we arrive at a price for the beauty of nature? The problem is even more complicated because among the kinds of scenery we enjoy are many modified by people. For example, the open farm fields in Vermont improved the view of the mountains and forests in the distance, and as farming declined in the 1960s, the state began to provide tax incentives for farmers to keep their fields open and thereby help the tourism economy.

One of the perplexing problems associated with aesthetic evaluation is personal preference. One person may appreciate a high mountain meadow far removed from civilization, a second person may prefer visiting with others on a patio at a trailhead lodge, a third may prefer to visit a city park, and a fourth may prefer the austere beauty of a desert. If we are going to consider aesthetic factors in environmental analysis, we must develop a method of aesthetic evaluation that allows for individual differences—another yet unsolved topic.

Some philosophers suggest that there are specific characteristics of landscape beauty and that we can use these to help us set the value of intangibles. Some suggest that the three key elements of landscape beauty are coherence, complexity, and mystery—mystery in the form of something seen in part but not completely, or not completely explained. Other philosophers suggest that the primary aesthetic qualities are unity, vividness, and variety.[12] *Unity* refers to the quality or wholeness of the perceived landscape—not as an assemblage but as a single, harmonious unit. *Vividness* refers to that quality of landscape that makes a scene visually striking; it is related to intensity, novelty, and clarity. People differ in what they believe are the key qualities of landscape beauty, but, once again, almost everyone would agree that the value is greater than zero.

28.6 How Is the Future Valued?

The discussion about whaling—explaining why whalers may not find it valuable to conserve whales—reminds us of the old saying "A bird in the hand is worth two in the bush." In economic terms, a profit now is worth much more than a profit in the future. This brings up another economic concept important to environmental issues—the future value compared with the present value of anything.

As an example, suppose you are dying of thirst in a desert and meet two people; one offers to sell you a glass of water now, and the other offers to sell you a glass of water if you can be at the well tomorrow. How much is each glass worth? If you believe you will die today without water, the glass of water today is worth all your money, and the glass tomorrow is worth nothing. If you believe you can live another day without water, but will die in two days, you might place more value on tomorrow's glass than on today's.

In practice, things are rarely so simple and distinct. We know we are mortal, so we tend to value personal wealth and goods more if they are available now than if they are promised in the future. This evaluation is made more complex, however, because we are accustomed to thinking of the future—to planning a nest egg for retirement or for our children. Indeed, many people today argue that we have a debt to future generations and must leave the environment in at least as good a condition as we found it; these people would argue that the future environment is not to be valued less than the present one.

Since the future existence of whales and other endangered species has value to those interested in biological conservation, the question arises: Can we place an economic (quantitative dollar) value on the *future* existence of anything (Figure 28.6)? The future value depends on how long a period you are talking about. For example, the future times associated with some important global environmental topics, such as stratospheric ozone depletion and global warming, extend longer than a century. This is so because

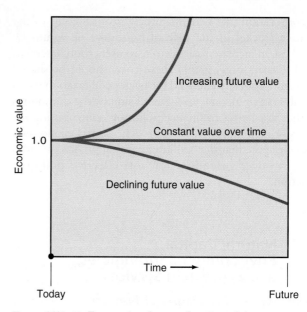

Figure 28.6 ■ Economic value as a function of time—a way of comparing the value of having something now with the value of having it in the future. A negative value means that there is more value attached to having something in the present than having it in the future. A positive value means that there is more value attached to having something in the future than having it today.

chlorofluorocarbons (CFCs) have such a long residence time in the atmosphere (see Chapter 26) and because of the time necessary to realize potential benefits from changing energy policy to offset global climate change.

Another aspect of future versus present value is that spending on the environment can be viewed as diverting resources from alternative forms of productive investment that will be of benefit to future generations. (This assumes that spending on the environment is not itself a productive investment.) For example, in the twentieth century, asbestos was used as an insulator in buildings, around hotwater pipes and similar devices. Then it was discovered that asbestos can cause cancer, and insulation containing asbestos was found in some schools. As long as the asbestos is well contained within an external covering, it does not pose an immediate threat. In some cases, people decided to spend the money now to remove the asbestos in order to reduce future risk to the students. Another approach is to leave the well-protected asbestos intact and spend the money that it would have cost to remove the asbestos on improving other educational facilities in a school, only acting to deal with the asbestos when it begins to enter the air in the school.

A further complicating issue is that as we get wealthier, the value we place on many environmental assets (like wilderness areas) increases dramatically. Thus, if society continues to grow in wealth over the next century as it has over the past century, the environment will be worth far more to our great-grandchildren than it was to our great-grandparents, at least in terms of willingness to pay to protect it. The implication—which complicates this topic even more—is that conserving resources and environment for the future is tantamount to taking from the poor today and giving to the rich in the future. If history is a guide, Americans in the twenty-first century will be far better off than Americans at the end of the twentieth century. To what extent should we ask the average American today to sacrifice now for much richer great-great-grandchildren? How can we know the future usefulness of today's sacrifices? Put another way, what would you have liked your ancestors in 1900 to have sacrificed for our benefit today? Should they have increased research and development on electric transportation? Should they have saved more tall-grass prairie or restricted whaling?

Economists observe that it is an open question whether something promised in the future will have *more* value then than it does today. Future economic value is difficult enough to predict because it is affected by how future consumers view consumption. But if, in addition, something has greater value in the future than it does today, then that leads to an impossible mathematical situation: In the very long run, the future value will become infinite, which of course is impossible. So in terms of the future, the basic issues are: (1) We are so much richer and better off than our ancestors that their sacrificing for us might have been inappropriate. (2) Even if they had wanted to sacrifice, how would they have known what sacrifices would be important to us?

As a general rule, one answer to these thorny questions about future value is: Do not throw away or destroy something that cannot be replaced if you are not sure of its future value. For example, if we do not fully understand the value of the wild relatives of potatoes that grow in Peru but do know that their genetic diversity might be helpful in developing future strains of potatoes, then we ought to preserve those wild strains.

28.7 Risk–Benefit Analysis

Death is the fate of all individuals, and every activity in life involves some risk of injury or death. How, then, do we place a value on saving a life by reducing the level of a pollutant? This question raises another important area of environmental economics: **risk–benefit analysis**, in which the riskiness of a present action in terms of its possible outcomes is weighed against the benefit, or value, of the action. Here, too, difficulties arise.

Acceptability of Risks and Costs

With some activities, the relative risk is clear. It is much more dangerous to stand in the middle of a busy highway than to stand on the sidewalk. Hang gliding has a much higher mortality rate than hiking. The effects of pollutants are often more subtle, so the risks are harder to pinpoint and quantify. Table 28.1 gives the risk associated with a variety of activities and some forms of pollution. The table shows the *lifetime* risk of death from each cause. In looking at the table, remember that since the ultimate fate of everyone is death, the total lifetime risk of death from all

Table 28.1 • Risk of Death from Various Causes

Cause	Result	Risk of Death (per lifetime)	Lifetime Risk of Death (%)	Comment
Cigarette smoking (pack a day)	Cancer, affect on heart, lungs, etc.	8 in 100	8.0%	
Breathing radon-containing air in the home	Cancer	1 in 100	1.0%	Naturally occurring
Automobile driving		1 in 100	1.0%	
Death from a fall		4 in 1,000	0.4%	
Drowning		3 in 1,000	0.3%	
Fire		3 in 1,000	0.3%	
Artificial chemicals in the home	Cancer	2 in 1,000	0.2%	Paints, cleaning agents, pesticides
Sunlight exposure	Melanoma	2 in 1,000	0.2%	Of those exposed to sunlight
Electrocution		4 in 10,000	0.04%	
Air outdoors in an industrial area		1 in 10,000	0.01%	
Artificial chemicals in water		1 in 100,000	0.001%	
Artificial chemicals in foods		less than 1 in 100,00	0.001%	
Airplane passenger (commercial airline)		less than 1 in 1,000,000	0.00010%	

Source: From *Guide to Environmental Risk* (1991), U.S. EPA Region 5 Publication Number 905/91/017.

causes must be 100%. So if you are going to die of something and you smoke a pack of cigarettes a day, you have 8 chances in 100 that your death will be a result of smoking. At the same time, your risk of death from driving an automobile is 1 in 100. Risk tells you the chance of an event but not its timing. So you might smoke all you want and die from the automobile risk first.

One of the striking things about Table 28.1 is that death from outdoor environmental pollution is comparatively low—even compared to the risks of drowning or of dying in a fire. This suggests that the primary reason we value lowering air pollution is an improvement in the quality of our lives, rather than an increase in the time we are alive. Considering the great interest people now show in air pollution, quality of life is much more important than is generally recognized. We are willing to spend money on improving that quality, not just the length of our lives.

Another striking observation in this table is that natural *indoor* air pollution is much more deadly than most outdoor air pollution—unless, of course, you live at a toxic waste facility (see Chapter 25).

Societies differ in socially, psychologically, and ethically acceptable levels of risk for any cause of death or injury. It is commonly believed that future discoveries will help to decrease various risks, perhaps eventually allowing us to approach a zero-risk environment. But complete elimination of risk is generally either technologically impossible or prohibitively expensive. We can make some generalizations about the acceptability of various risks. One factor is the number of people affected. Risks that affect a small population (such as employees at nuclear power plants) are usually more acceptable than those that involve all members of a society (such as risk from radioactive fallout).

In addition, novel risks appear to be less acceptable than long-established or natural risks, and society tends to be willing to pay more to reduce such risks. For example, France spends approximately $1 million to reduce the likelihood of one air-traffic death but only $30,000 for the same reduction in automobile deaths.[13] Some argue that the greater safety of commercial air travel compared with automobile travel is in part a function of the relatively novel fear of flying compared with the more ordinary fear of death from a road accident. That is, because the risk is newer to us and thus less acceptable, we are willing to spend more per life to reduce the risk from flying than to reduce the risk from driving.

People's willingness to pay for reducing a risk also varies with how essential and desirable the activity associated with the risk is. For example, many people accept much higher risks for athletic or recreational activities than they would for transportation- or employment-related activities (see Table 28.1). The risks associated with playing a sport or using transportation are assumed to be inherent in the activity. The risks to human health from pollution may be widespread and linked to a large number of deaths.

Although risks from pollution are often unavoidable and unseen, people want a lesser risk from pollution than from, say, driving a car or playing a sport.

In an ethical sense, it is impossible to put a value on a human life. However, it is possible to determine how much people are willing to pay for a certain reduction in risk or a certain probability of an increase in longevity. For example, a study by the Rand Corporation considered measures that would save the lives of heart-attack victims, including increasing ambulance services and initiating pretreatment screening programs. According to the study, which identified the likely cost per life saved and the willingness of people to pay, people were willing to pay approximately $32,000 per life saved, or $1,600 per year of longevity.[13] Although information is incomplete, it is possible to estimate the cost of extending lives in terms of the dollars per person per year for various actions (Figure 28.7 and Table 28.1). For example, on the basis of direct effects on human health, it costs more to increase longevity through a reduction in air pollution than to directly reduce deaths through the addition of a coronary ambulance system.

Such a comparison is useful as a basis for decision making. Clearly, though, when a society chooses to reduce air pollution, many factors beyond the direct, measurable health benefits are considered. Pollution directly affects more than just our health, and ecological and aesthetic damage can also indirectly affect human health (see Section 28.4). We might want to choose a slightly higher risk of death in a more pleasant environment (spend money to clean up the air instead of to increase ambulance services) rather than increase the chances of living longer in a poor environment (spend the money on reducing heart attacks).

Such comparisons may make you feel uncomfortable. But like it or not, we cannot avoid making choices of this kind. The issue boils down to whether we should improve the quality of life for the living or extend life expectancy regardless of the quality of life.[14]

The degree of risk is an important concept in our legal processes. For example, the U.S. Toxic Substances Control Act states that no one may manufacture a new chemical substance or process a chemical substance for a new use without obtaining a clearance from the EPA. The Act establishes procedures to estimate the hazard to the environment and to human health of any new chemical before it becomes widespread. The EPA examines the data provided and judges the degree of risk associated with all aspects of the production of the new chemical or process, including extraction of raw materials, manufacturing, distribution, processing, use, and disposal. The chemical can be banned or restricted in either manufacturing or use if the evidence suggests that it will pose an unreasonable risk to human health or to the environment.

But what is unreasonable?[15] This question brings us back to Table 28.1 and makes us realize that deciding what is "unreasonable" involves judgments about the quality of life as well as the risk of death. The level of acceptable pol-

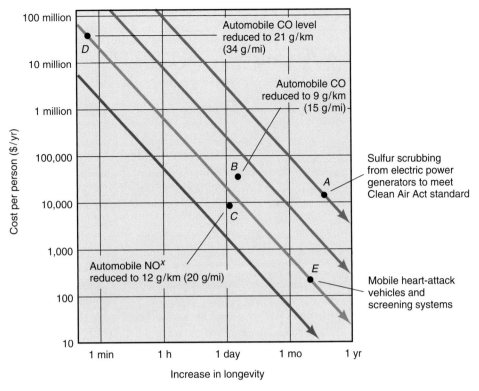

Figure 28.7 ■ One way to rank the effectiveness of various efforts to reduce pollutants is to estimate the cost of extending a life in dollars per year. This graph shows that reducing sulfur emissions from power plants to the Clean Air Act level (*A*) would extend a human life 1 year at a cost of about $10,000. Similar restrictions applied to automobile emissions (*B, C*) would increase lifetimes by 1 day. More stringent automobile controls would be much more expensive (*D*); mobile units and screening programs for heart problems would be much cheaper (*E*). This graph represents only one step in an environmental analysis. [*Source:* Based on R. Wilson, "Risk/Benefit Analysis for Toxic Chemicals," *Ecotoxicology and Environmental Safety* 4 (1980): 370–383.)]

lution (and thus risk) is a social–economic–environmental trade-off. Moreover, the level of acceptable risk changes over time in society, depending on changes in scientific knowledge, comparison with risks from other causes, the expense of decreasing the risk, and the social and psychological acceptability of the risk.

For example, when DDT was first used, no one understood the way that the chemical was transported within ecosystems, nor ecological effects of the chemical; scientific observations revealed these effects later (see A Closer Look 28.1). At that time, there was relatively little concern with environmental issues, and society was not yet willing to pay for many of these costs. Now, people widely agree that the environment is a major concern, and there is less willingness to accept indirect environmental effects. What had been considered externalities to the use of DDT have become internal cost factors.

As explained in Chapter 24, the total dollar cost of pollution is the sum of the costs to control pollution and the loss from pollution damages. In some cases, these two factors have opposite trends in terms of economic cost (as one goes down, the other goes up), and their intersection point is the minimum total cost, as shown by point *A* in Figure

28.7. In other cases, total costs may stabilize or even decline as companies using pollution control become more efficient and external costs in the form of environmental damage are minimized. If the minimum total cost involves a pollution level that is too high a risk, then additional control may add considerable expense.

The risk associated with a pollutant can be determined by the present levels of exposure and predicted future trends. These trends depend on the production and origin of the pollutant, pathways it follows through the environment, and changes it undergoes along these pathways. Dose–response curves establish risk to a population from a particular level of pollutant (see Chapter 15). Relative risks of different pollutants can be determined by comparing their current levels and their dose–response curves.

The above discussion suggests that when adequate data are available, it is possible to take scientific and technological steps to estimate the level of risk and from this to estimate the cost of reducing risk and compare the cost with the benefit. However, what constitutes an acceptable risk is more than a scientific or technical issue. Acceptability of a risk involves ethical and psychological attitudes of individuals and society. We must therefore ask

Risk–Benefit Analysis and DDT

A review of the history of the use of DDT illustrates the difficulty of completely eliminating a pollution risk. As noted in Chapter 12, DDT was first applied widely in the 1940s to control the spread of diseases such as malaria via insects and was subsequently used to control crop pests. At first, tests of the safety of DDT focused on human health effects, which were believed to be small. In the late 1950s, however, DDT was discovered in the livers of sharks. Because DDT had not been used in the ocean, at first it was believed that this observation was simply a measurement error or the

result of an unknown dumping of DDT directly into the ocean.

Gradually, and to their surprise, scientists began to understand that DDT had spread from farmland through surface water runoff and through the air into the ocean and that the chemical had become a worldwide contaminant. DDT was found in the tissues of penguins in the Antarctic and seals in the Pribilof Islands of the Bering Sea. By 1970, it had become clear that DDT was everywhere and was a global environmental problem.[16] Although DDT is banned for use in the United States, U.S. corporations continue

to be among the largest producers of this chemical, which is shipped to other nations. DDT is still used widely elsewhere, especially in developing nations of the tropics.

Complete elimination of DDT residues from all the environments of the world—and from all the sharks, penguins, seals, and birds—no longer seems feasible. Aside from the difficulty of eliminating pollutants, another lesson of this example is that what had seemed to be an economic externality of DDT (its indirect ecological effects on birds and ocean animals) became a major societal issue.

several questions: What risk from a particular pollutant is acceptable? How much is a given reduction in risk from that pollutant worth to us? How much will each of us, as individuals or collectively as a society, be willing to pay for a given reduction in that risk? The answers depend not only on facts but also on societal and personal values. What must also be factored into the equation is that costs associated with cleanup of pollutants and polluted areas and with restoration programs can be minimized or even eliminated if a recognized pollutant is controlled initially. The total cost of pollution control need not increase indefinitely.

Although pollution control may involve many dollars, the average cost per family in the United States is low, especially compared with other costs. It has been estimated that the per-family pollution-control cost is between $30 and $60 per year for a family with a median income. In addition to the low cost per family, pollution control has many benefits whose quantitative value can be estimated. For example, federal air quality standards are estimated to reduce the risk of asthma by 3% and the risk of chronic bronchitis and emphysema by 10% to 15% in locally exposed adults. Estimates of the total cost of the direct and indirect effects on human health from stationary sources of air pollution are $250 per family per year. Air pollution contributes to inflation by reducing the number of productive workdays, reducing work efficiency, adding to direct expenditures for health treatments, and necessitating repair of nonhuman environmental damage. On this basis, air pollution control appears to be cost-effective; in fact, it has economic benefits.[17]

28.8 Global Issues: Who Bears the Costs?

Global environmental problems make us more aware of the public service functions of the environment of our planet and of life around us, as well as raising new economic questions. An important case in point is the possibility of global warming. The problem is that our technological society is adding carbon dioxide and other greenhouse gases to the atmosphere that have the potential to warm the climate.[16, 18] (See the discussion of global warming in Chapter 22.) The direct solution is to decrease the release of these gases, but to do so would require a worldwide decrease in the burning of fossil fuels. Although most of the production of greenhouse gases today is from the industrial nations, in the future the developing nations, especially China and India, will contribute large quantities of these gases.

The economist Ralph d'Arge points out an economic problem arising from this global issue: The less developed countries did not share in the economic benefits of the burning of fossil fuels during the first two centuries of the Industrial Revolution, but they are sharing in the disadvantages of this activity.[19] Now the industrialized nations are suggesting that all nations, including the less developed ones, restrict their use of fossil fuels and therefore participate in future disadvantages without obtaining the benefits of cheap energy. Developing nations tend to think that industrial nations, which enjoyed the past benefits, should accept most of the future costs. At the same time, why shouldn't the developing nations proceed to develop and burn fossil fuels?

Table 28.2 • Approaches to Environmental Policy

Policy Instruments

1. Moral suasion (publicity, social pressure, etc.)
2. Direct controls
   a. Regulations limiting the permissible levels of emissions
   b. Specification of mandatory processes or equipment
3. Market processes[a]
   a. Taxation of environmental damage
      i. Tax rates based on evaluation of social damage
      ii. Tax rates designed to achieve preset standards of environmental quality
   b. Subsidies
      i. Specified payments per unit of reduction of waste emissions
      ii. Subsidies to defray costs of damage-control equipment
   c. Issue of limited quantities of pollution licenses
      i. Sale of licenses to the highest bidders
      ii. Equal distribution of licenses with legalized resale
   d. Refundable deposits against environmental damage
   e. Allocation of property rights to give individuals a proprietary interest in improved environmental quality
4. Government investment
   a. Damage-prevention facilities (e.g., municipal treatment plants)
   b. Regenerative activities (e.g., reforestation, slum clearance)
   c. Dissemination of information (e.g., pollution-control techniques, opportunities for profitable recycling)
   d. Research
   e. Education
      i. Of the general public
      ii. Of professional specialists (ecologists, urban planners, etc.)

Administrative Mechanisms

1. Administrative unit
   a. National agency
   b. Local agency
2. Financing
   a. Payment by those who cause the damage
   b. Payment by those who benefit from improvements
   c. General revenues
3. Enforcement mechanism
   a. Regulatory organization or police
   b. Citizen suits (with or without sharing of fines)

[a] Subsidies and taxes can also be distinguished by use of a property-rights framework. Per-unit subsidies implicitly confer ownership of the right to pollute on the polluter, and these rights are then purchased by the government via the subsidy. Taxes essentially say that there is public ownership of usage rights, which can be purchased from the public through its agent, the government, or by private parties on payment of the tax (price).

Source: W. J. Baumol and W. E. Oates, *Economics, Environmental Policy and the Quality of Life* (Englewood Cliffs, N.J.: Prentice Hall, 1979).

This perspective, limited to what benefits individual nations, may be too restricted in light of the global environmental effects of our technological civilization. It may be necessary to reduce the total production of greenhouse gases. If so, then the economic question is: Who pays, and how? At present, this is an unresolved issue in environmental economics. One suggestion is that the developed nations pay for the reduction in greenhouse emissions of the less developed nations. Another suggestion is that the developed countries share their technology with developing countries, thus helping the developing countries to reduce both local and global pollution. These issues were a major concern at the 1992 Earth Summit in Rio de Janeiro.[20]

28.9 How Do We Achieve a Goal? Environmental Policy Instruments

How does a society achieve an environmental goal, such as preservation and use of a resource or reduction of a pollutant? Any society has several methods to achieve such goals. Means to implement a society's policies are known among economists as **policy instruments** (see Tables 28.2 and 28.3). These include *moral suasion* (which politicians call "jawboning," i.e., persuading people by talk, publicity, and social pressure); *direct controls*, which include regulations; *market processes*, which affect the price of goods and include taxation of various kinds, subsidies, licenses, and deposits; and *government investments*, which include research and education. Society also has administrative mechanisms to ensure that the policy instruments chosen actually function.

Marginal Costs and the Control of Pollutants

How clean is clean? When have we done enough to think that the environment is "good" and that we have achieved a reasonable balance between benefits and costs? In deciding this, we have to look at the cost of each additional step we take, given the previous steps. This leads us to the concept of **marginal costs**. In controlling pollutants, marginal cost is the cost to reduce one additional unit of pollutant. By analogy, with conservation of an endangered species or a rare habitat or ecosystem, the marginal cost would be the cost of adding one more unit (however we might define a unit) to those already conserved.

With pollution control, the marginal cost often increases rapidly as the percentage of reduction increases. For example, the marginal cost of reducing the biological oxygen demand in wastewater from petroleum refining increases exponentially. When 20% of the pollutants have been removed, the cost of removing an additional kilogram is 5 cents. When 80% of the pollutants have been removed, it costs 49 cents to remove an additional kilogram. Extrapolating from these results, it would cost an infinite amount to remove all the pollution.

Tables 28.2 and 28.3 show three common methods of direct control of pollution: (1) setting maximum levels of pollution emission, (2) requiring specific procedures and processes that reduce pollution, and (3) charging fees for pollution emission. In the first case, a political body could

Table 28.3 • Performance of Various Policy Instruments

Policy Instrument	Reliability	Permanence	Adaptability to Growth	Resistance to Inflation	Incentive for Improved Effort	Economy	Feasibility without Metering	Noninterference in Private Decisions	Political Attraction Actual	Political Attraction Potential
Moral suasion	Good[a]	Poor	Good[b]	Good[b]	Fair	Poor[c]	Excellent	Excellent	Excellent	—
Direct controls										
By quota	Fair	Poor	Fair	Excellent	Poor	Poor	Poor	Poor	Excellent	—
By specification of technique	Fair	Poor	Good[b]	Good[b]	Poor	Poor	Excellent	Poor	Excellent	—
Fees	Excellent	Excellent	Fair	Fair	Excellent	Excellent	Poor	Excellent	Poor	Good
Sale of permits or licenses	Excellent	Excellent	Excellent	Excellent	Excellent	Excellent[d]	Poor	Excellent	Poor	Good
Subsidies										
Per-unit reduction	Fair[e]	Good	Fair	Fair	Excellent	Good	Poor	Excellent	Good	—
For equipment purchase	Fair	Good	Fair	Fair	Excellent	Good	Poor	Excellent	Good	—
Government investment	Good	?	?	?	—	?	Excellent	—	Good	—

[a] For short periods of time when urgency of appeal is made very clear.

[b] Baumol and Oates's judgment.

[c] Induces contributions from decision makers who are most cooperative, not necessarily from those able to do the job most effectively.

[d] Tends to allocate reduction quotas among firms in a cost-minimizing manner, but if the number of emissions permitted is too small it will force the community to devote an excessive quantity of resources to environmental protection.

[e] Tends to allocate reduction quotas among firms in cost-minimizing manner but introduces inefficiency into the environmental protection process by attracting more polluting firms into the subsidized industry, so that aggregate response is questionable.

Source: W. J. Baumol and W. E. Oates, *Economics, Environmental Policy and the Quality of Life* (Englewood Cliffs, N.J.: Prentice Hall, 1979).

Making Policy Work:
Fishing Resources and Policy Instruments

Ocean fishing illustrates different ways of making a policy work, referred to as policy instruments. The oceans outside of national territorial waters are commons, and thus the fish and mammals that live in them are common resources. What is a common resource may change over time, however. The move by many nations to define international waters as beginning 325 km (200 mi) from their coasts has turned some fisheries from completely open common resources to national resources open only to domestic fishermen.

In fisheries, there are four main management options:[22]

1. Establish total catch quotas for the entire fishery and allow anybody to fish until the total is reached.

2. Issue a restricted number of licenses but allow each licensed fisherman to catch many fish.

3. Tax the catch (the fish brought in) or the effort (the cost of ships, fuel, and other essential items).

4. Allocate fishing rights—that is, assign each fisherman a transferable and salable quota.

With total-catch quotas, the fishery is closed when the quota is reached. Whales, Pacific halibut, tropical tuna, and anchovies have been regulated in this way. When the practice was used in Alaska, all of the halibut were caught in a few days, with the result that restaurants no longer had halibut available for most of the year. This undesirable result led to a change in policy: The total-catch approach was replaced by the sale of licenses.

Although regulating the total catch can be done in a way that helps the fish, it tends to increase the number of fishermen and the capacity of vessels, and the end result is a hardship on fishermen. Recent economic analysis suggests that taxes that take into account the cost of externalities can work to the best advantage of fishermen and fish. Similar results are achieved by allocating a transferable and salable quota to each fisherman.

Determining which management method achieves the best use of a desirable environmental resource is not simple. The answer varies with the specific attributes of both the resource and the users. The tools of economics can be used to determine the methods that will work best within a given social framework.

---

set a maximum for the amount of sulfur emitted from the smokestack of an industry. In the second, it could restrict the kind of fuel the industry could use. Many areas have chosen the latter method by prohibiting the burning of high-sulfur coal.

The problem with the first approach—controlling emissions—is that careful monitoring is required indefinitely to make certain the allowable levels are not exceeded. Such monitoring may be costly and difficult to carry out. The disadvantages of the second approach—requiring specific procedures—are that the required methodology may impose a severe financial burden on the producer of the pollutant, restrict the kinds of production methods open to an industry, and become technologically obsolete. (See the discussion of air pollution and laws regulating it in Chapter 24.) However, after an initial expenditure on procedures to reduce pollution, production efficiency may be increased and other costs, such as monitoring and waste disposal, can be reduced. For example, Japan, whose economy is one of the most energy-efficient in the world, has reduced its air pollution more than any other industrial nation. In recent years, Japanese industry has used only 5 megajoules to produce $1 of gross domestic product (GDP), while U.S. industry required 12 megajoules. A large international chemical company, 3M, initiated a pollution-prevention program and was able to stop the release of a billion pounds of toxic chemicals and save the company $500 million in the process.

Although the United States has emphasized the use of direct regulation to control pollution, other countries have been successful in controlling pollution by charging effluent fees. For example, charges for effluents into the Ruhr River in Germany are assessed on the basis of both the concentration of pollutant and the total quantity of polluted water emitted into the river. In response, plants have introduced water recirculation and internal treatment to reduce emissions.[21]

Studies of the uses of different policy instruments for environmental matters have resulted in some ability to evaluate their relative success (see Table 28.3). For example, moral suasion is reliable but not very permanent in its effect. Sale of licenses or permits has been found to be among the more successful recourses.

In every environmental matter, there is a desire on the one hand to maintain individual freedom of choice and on the other to achieve a specific social goal. In ocean fishing, for example, how does a society allow every individual to choose whether or not to fish and yet prevent everyone from fishing at the same time and bringing fish species to extinction? This interplay between private good and public good is at the heart of environmental issues. Some argue that the market itself will provide the proper control. For example, it can be argued that people will stop fishing when there is no longer a profit to be made. We have already seen, however, that this may not be so. Two factors interfere with this

U.S. Fisheries: How Can They Be Made Sustainable?

Both overfishing and pollution have been blamed for the alarming decline in groundfish (cod, haddock, flounder, redfish, pollack, hake) off the northeastern coast of the United States. Most scientists and fisheries managers have focused on overfishing, but attempts to regulate fishing have generated bitter disputes with fishermen, many of whom contend that restrictions on fishing make them scapegoats for pollution problems. The controversy has become a classic battle between short-term economic interests and long-term environmental concerns.

The Massachusetts Division of Marine Fisheries conducted a study in 1992 to settle the question of overfishing versus pollution. It concluded that key characteristics in declining flounder populations were consistent with overfishing rather than pollution. Furthermore, if pollution had been the cause, the skate and dogfish populations would have been expected to decline as well. However, S. J. Correia, the author of the study, cautioned that the issue should not be seen as an either/or question and that "overfishing may currently mask any pollution-induced population growth constraints."

Issues related to the use of U.S. fisheries were hardly new in 1992. In 1977, in response to concerns about overfishing in U.S. waters by foreign factory ships, the U.S. government extended the nation's coastal waters from 12 to 200 mi (from 19 to 322 km). To encourage domestic fishermen, the National Marine Fisheries Service provided loan guarantees for replacing older vessels and equipment with newer boats with high-tech equipment for locating fish. During this same period, demand for fish increased as Americans became more concerned about cholesterol levels in red meat. Consequently, the number of fishing boats, the number of days at sea, and fishing efficiency increased sharply, and 50% to 60% of the populations of some species were landed each year.

A decision about Canadian and American fishing rights in the Georges Bank, the most prolific area in the North Atlantic, in favor of Canada in 1984 by the International Court of Justice in The Hague intensified competition for remaining fishing waters among U.S. fishermen. Overfishing knows no boundaries, however; in 1992, Canada was forced to suspend all cod fishing to save the stock from complete annihilation.

In 1982, the New England Fisheries Management Council attempted to enforce harvest quotas but rescinded the order under pressure from the commercial fishing industry. In the wake of a bitter controversy that resulted from this decision, as well as further declines in fish populations, the council later instituted a series of measures prohibiting fishing at certain times and in certain areas, mandating minimum net sizes, and enacting quotas on the catch.

In 1992, the council adopted a more ambitious plan intended to cut the fishing effort in half by 1997—but by means other than quotas and removal of current fishermen. The council decided to issue a limited number of fishing permits, limit the number of days at sea, use high-tech monitoring equipment to ensure compliance, and establish trip limits on some fish. Even more recently, portions of the Georges Bank were closed indefinitely to fishing for some fish species, including yellowtail, cod, and haddock. Landings of yellowtail in 1993 were 3,800 metric tons, the lowest on record and a mere 6% of the historical maximum in 1969.

The National Marine Fisheries Service instead advocates a system of individual transferable quotas (ITQs) by which permits are issued to boat owners to allow them to harvest a fixed amount of fish each year. They can lease, sell, or bequeath the permits to others. Although ITQs have been successful in several U.S. fisheries and in some other countries, some small operators in New England fear that large corporations will buy up permits and dominate the industry.

Critical Thinking Questions

1. Using the example of the New England fishery, what are some arguments for and against the proposition that all natural resources treated like commons will inevitably be destroyed unless controls are instituted?

2. Which measures described attempt to convert the fishing industry from a commons system to private ownership? How might these measures help prevent overfishing? Is it right to institute private ownership of public resources?

3. What approach to future value (approximately) does each of the following people assume for fish?

   Fisherman: If you don't get it now, someone else will.

   Fisheries manager: By sacrificing now, we can do something to protect fish stocks.

4. Develop a list of the environmental and economic advantages and disadvantages of ITQs. Would you support instituting ITQs in New England? Explain why or why not.

5. Do you think it possible to reconcile economic and environmental interests in the case of the New England fishing industry? If so, how? If not, why not?

argument: (1) By the time the reduced fish population results in no economic gain for fishermen, it may be too late to avoid eventual extinction. (2) Even when it is not possible to make a sustained annual profit, there may be an advantage in harvesting the entire resource and getting out of the business (see A Closer Look 28.2).

Summary

■ An economic analysis can help us understand why environmental resources have been poorly conserved in the past and how we might more effectively achieve conservation in the future.

■ Economic analysis is applied to two different kinds of environmental issues: the use of desirable resources (fish in the ocean, oil in the ground, forests on the land) and the minimization of pollution.

■ Resources may be common property or privately controlled. The kind of ownership affects the methods available to achieve an environmental goal. There is a

tendency to overexploit a common property resource and to harvest to extinction nonessential resources whose innate growth rate is low, as suggested in Hardin's "tragedy of the commons."

■ Future worth compared with present worth can be an important determinant of the level of exploitation.

■ The relation between risk and benefit affects our willingness to pay for an environmental good.

■ Evaluation of environmental intangibles, such as landscape aesthetics and scenic resources, is becoming more common in environmental analysis. When quantitative, such evaluation balances the more traditional economic evaluation and helps separate facts from emotion in complex environmental problems.

■ Societal methods to achieve an environmental goal include moral suasion, direct controls, market processes, and government investment. Many kinds of controls have been applied to the use of desirable resources and the control of pollution.

REEXAMINING THEMES AND ISSUES

Human Population

The tragedy of the commons will worsen as human population density increases, because there will be more and more individuals to seek gain at the expense of community values. For example, more and more individuals will try to make a living from harvesting natural resources. How people can use resources while at the same time conserving them requires an understanding of environmental economics.

Sustainability

From this chapter, we learn why people sometimes are not interested in sustaining an environmental resource from which they make a living. When the goal is simply to maximize profits, it is sometimes a rational decision to liquidate an environmental resource and put the money gained in a bank or another investment. To avoid such liquidation, we need to understand economic externalities and intangible values.

Global Perspective

Solutions to global environmental issues, such as global warming, require that we understand the different economic interests of developed and developing nations. These can lead to different economic policies and different valuation of global environmental issues.

Urban World

The tragedy of the commons began with grazing rights in small villages. As the world becomes increasingly urbanized, the pressure to use public lands for private economic gain is likely to increase. An understanding of environmental economics can help us find solutions to urban environmental problems.

People and Nature

This chapter brings us to the heart of the matter: How do we value the environment, and when can we attach a monetary value to the benefits and costs of environmental actions? People are intimately involved with nature. While we seek rational methods to put a value on nature, the values we choose often derive from intangible benefits, such as the appreciation of the beauty of nature.

Science and Values

One of the central questions of environmental economics concerns how to develop equivalent economic valuation for tangible and intangible factors. For example, how can we compare the value of timber that could be harvested with the beauty people attach to the scenery, trees intact? How can we compare the value of a dam that provides irrigation water and electrical power on the Columbia River with the scenery without the dam and the salmon that could inhabit that river?

Key Terms

commons **605**
direct costs **608**
environmental economics **605**

externality **608**
indirect costs **608**
marginal costs **615**

policy instruments **615**
risk–benefit analysis **611**

Study Questions

1. What is meant by the term *the tragedy of the commons?* Which of the following are the result of this tragedy?
   a. The fate of the California condor
   b. The fate of the gray whale
   c. The high price of walnut wood used in furniture

2. What is meant by risk–benefit analysis?

3. Cherry and walnut are valuable woods used to make fine furniture. Basing your decision on the information in the following table, which would you invest in? (*Hint:* Refer to the discussion of whales in this chapter.)
   a. A cherry plantation
   b. A walnut plantation
   c. A mixed stand of both species
   d. An unmanaged woodland where you see some cherry and walnut growing

Species	Longevity	Maximum Size	Maximum Value
Walnut	400 years	1 m	$15,000/tree
Cherry	100 years	1 m	$10,000/tree

4. Bird flu is spread in part by migrating wild birds. How would you put a value on (a) the continued existence of one species of these wild birds; (b) domestic chickens important for food but also a major source of the disease;

(c) control of the disease for human health? What relative value would you place on each (that is, which is most important and which least)? To what extent would an economic analysis enter into your valuation?

5. Which of the following are intangible resources? Which are tangible?
   a. The view of Mount Wilson in California
   b. A road to the top of Mount Wilson
   c. Porpoises in the ocean
   d. Tuna fish in the ocean
   e. Clean air

6. What kind of future value is implied by the statement "Extinction is forever"? Discuss how we might approach providing an economic analysis for extinction. (See Chapter 14 for additional information about extinction.)

7. Which of the following can be thought of as commons in the sense meant by Garrett Hardin? Explain your choice.
   a. Tuna fisheries in the open ocean
   b. Catfish in artificial freshwater ponds
   c. Grizzly bears in Yellowstone National Park
   d. A view of Central Park in New York City
   e. Air over Central Park in New York City

Further Reading

Daly, H. E., and J. Farley. 2003. *Ecological Economics: Principles and Applications.* Washington, D.C.: Island Press. Discusses an interdisciplinary approach to the economics of environment.

Goodstein, E. S. 2000. *Economics and the Environment*, 3rd ed. New York: Wiley.

Hardin, G. 1968. "Tragedy of the Commons," *Science* 162:1243–1248. One of the most cited papers in both science and social science, this classic work outlines the differences between individual interest and the common good.

Tietenberg, T. 2003. *Environmental Economics and Policy.* New York: Addison-Wesley.

CHAPTER 29

Urban Environments

Left: New Orleans before Katrina.
Right: New Orleans after Katrina.

Learning Objectives

Because the world is becoming increasingly urbanized, it is important to learn how to improve urban environments, to make cities more pleasant and healthier places in which to live, and to reduce undesirable effects on the environment. After reading this chapter, you should understand:

■ How to view a city from an ecosystem perspective: how location and site conditions determine the success, importance, and longevity of a city.

■ How cities have changed with changes in technology and with ideas about city planning.

■ How a city changes its own environment and affects the environment of the surrounding areas, and how we can plan cities to minimize some of these effects.

■ How trees and other vegetation not only beautify cities but also provide habitats for animals, and how we can alter the urban environment to encourage wildlife and to discourage pests.

■ How cities can be designed to promote biological conservation and become pleasant environments for people.

■ What fundamental choices we face in deciding what kind of future we want and what the role of cities will be in that future.

Should We Try to Restore New Orleans?

On August 29, 2005, Hurricane Katrina roared, slammed, and battered its way into New Orleans with 192 km/hr (120 mph) winds. Its massive storm surges breached the levees that had protected many of the city's residents from the Gulf Coast's waters, flooding 80% of the city and an estimated 40% of the houses (Figure 29.1). With so many people suddenly homeless and such major damage (Figure 29.2), New Orleans' mayor, Ray Nagin, ordered a first-time-ever complete evacuation of the city, an evacuation that became its own disaster. Some estimates claimed that 80% of the 1.3 million residents of the greater New Orleans metropolitan area evacuated.

By the time it was over, Katrina was the most costly hurricane in the history of the United States—between $75 billion and $100 billion, in addition to an estimated $200 billion in lost business revenue. And a year after the hurricane, much of the damage remains. An estimated 50,000 homes will have to be demolished.

Many former residents are still living elsewhere, scattered across the nation.[1] New Orleans has lost $1.5 million in tourist revenues every day since the levees broke. And this does not account for damage to the economy caused by potential interruption of the oil supply and exports of commodities such as grain. The storm also affected the casino and entertainment industry, as many of the Gulf Coast's casinos were destroyed or sustained considerable damage. New Orleans was home to roughly 115,000 small businesses. Some estimates suggest that half of these will never open again.

The problem with New Orleans is that it is built in the wetlands at the mouth of the Mississippi River, and much of it is below sea level (Figure 29.3). Although a port at the mouth of the Mississippi River has always been an important location for a city, there just wasn't a great place to build that city. The original development, the French Quarter, was above sea level (just barely), about the best that could be found.

(a)

(b)

Figure 29.1 ■ (*a*) Aerial photograph of New Orleans skyline before Hurricane Katrina struck. The Superdome is near the center leftr. (*b*) A similar view of New Orleans as in (*a*), but after Hurricane Katrina struck on August 31, 2005. The widespread flooding of the city from the hurricane is visible.

(a)

(b)

Figure 29.2 ■ (a) New Orleans residents struggle to escape Hurricane Katrina's flooding in the city.

(b) French Quarter: parked cars are crushed under bricks that have fallen from a building on Camp Street near Canal.

Figure 29.3 ■ Map of New Orleans showing how much of the city is below sea level. Originally the city was built in the French Quarter, which is above sea level, but as the population grew, levees were built to keep the water out and the city became an accident waiting to happen.

Hurricane Katrina was rated a Category 3 hurricane, but discussions about protecting the city from future storms focus on an even worse scenario, a Category 5 hurricane with winds up to 249 km/hr (155 mph). Protecting New Orleans from a Category 5 hurricane would cost an estimated $10 billion to $20 billion and take up to ten years. Restoring the coastline would cost $14 billion.[2] Even the most basic repairs will require about 3 million cubic yards of soil, the equivalent of a football field on which dirt is stacked 1,575 feet high.

Will New Orleans survive? More specifically, will it be restored to its former glory and importance? And will it continue in any fashion, even as a mere shadow of its former self? To know how to rebuild the city, and to decide whether this is worth doing, we have to understand the ecology of cities, how cities fit into the environment, the complex interplay between a city and its surroundings, and how a city acts as an environment for its residents.

29.1 City Life

In the past, the emphasis of environmental action has most often been on wilderness, wildlife, endangered species, and the impact of pollution on natural landscapes outside cities. Now it is time to turn more of our attention to city environments. In the development of the modern environmental movement in the 1960s and 1970s, it was fashionable to consider everything about cities bad and everything about wilderness good. Cities were thought of as polluted, dirty, lacking in wildlife and native plants, and artificial—and therefore bad. Wilderness was thought of as unpolluted, clean, full of wildlife and native plants, and natural—and therefore good.

Although it was fashionable to disdain cities, the majority of people live in urban environments and have suffered directly from their decline. Yet comparatively little public concern has focused on urban ecology. Many urban people see environmental issues as outside their realm, but the reality is just the opposite: City dwellers are at the center of some of the most important environmental issues. People are realizing that city and wilderness are inextricably connected. We cannot fiddle in the wilderness while our Romes burn from sulfur dioxide and nitrogen oxide pollution. Fortunately, we are experiencing a rebirth of interest in urban environments and in urban ecology. The National Science Foundation has added two urban areas, Baltimore and Phoenix, to its Long-Term Ecological Research Program, a program that supports long-term monitoring of, as well as research on, specific ecosystems and regions.

Worldwide we are becoming an increasingly urbanized species (Chapter 4). In the United States, about 75% of the population live in urban areas and about 25% live in rural areas. However, in the past decade more people have moved out of the largest cities in the United States than have moved into them, with the New York, Los Angeles, Chicago, and San Francisco/Oakland metropolitan areas each averaging a net loss of more than 60,000 a year. Chicago's Cook County lost a half million people between 2000 and 2004.[4, 5, 6, 7] Today approximately 45% of the world's population live in cities, and it is projected that 62% of the population, 6.5 billion people, will live in cities by the year 2025.[8] Economic development leads to urbanization; 75% of people in developed countries live in cities, but only 38% of people in the poorest developing countries are city dwellers.[7]

Megacities—huge metropolitan areas with more than 8 million residents—are cropping up more and more. In 1950, the world had only two: New York City and nearby urban New Jersey (12.2 million residents altogether) and greater London (12.4 million). By 1975, Mexico City, Los Angeles, Tokyo, Shanghai, and São Paulo, Brazil, had joined this list. By 2002, the most recent date for which data are available, 30 urban areas had more than 8 million people.[5]

In the future, most people will live in cities. In most nations, most urban residents will live in the country's single largest city. For most people, living in an environment of good quality will mean living in a city that is managed carefully to maintain that environmental quality.

29.2 The City as a System

We need to analyze a city as the ecological system that it is—but of a special kind. Like any other life-supporting system, a city must maintain a flow of energy, provide necessary material resources, and have ways of removing wastes. These ecosystem functions are maintained in a city by transportation and communication with outlying areas. A city is not a self-contained ecosystem; it depends on other cities and rural areas. A city takes in raw materials from the surrounding countryside: food, water, wood, energy, mineral ores, everything that a human society uses. In turn, the city produces and exports material goods and, if it is a truly great city, exports ideas, innovations, inventions, arts, and the spirit of civilization. A city cannot exist without a countryside to support it. As was said half a century ago, city and country, urban and rural, are one thing—one connected system of energy and material flows—not two things (Figure 29.4).[8]

As a consequence, if the environment of a city declines, almost certainly the environment of its surroundings will also decline. The reverse is also true: If the environment around a city declines, the city itself will be threatened. Some people suggest, for example, that the ancient Native American settlement in Chaco Canyon, Arizona, declined after the environment surrounding it either lost soil fertility from poor farming practices or suffered a decline in rainfall.

Cities also export waste products to the countryside, including polluted water, air, and solids. The average city resident in an industrial nation annually uses (directly or indirectly) about 208,000 kg (229 tons) of water, 660 kg (0.8 ton) of food, and 3,146 kg (3.5 tons) of fossil fuels and produces 1,660,000 kg (1,826 tons) of sewage, 660 kg (0.8 ton) of solid wastes, and 200 kg (440 lb) of air

pollutants. If these are exported without care, they pollute the countryside, reducing its ability to provide necessary resources for the city and making life in the surroundings less healthy and less pleasant.

With such dependencies and interactions between city and surroundings, it is no wonder that relationships between people in cities and in the countryside have often been strained. Why, country dwellers want to know, should they have to deal with the wastes of those in the city? The answer is that many of our serious environmental problems occur at the interface between urban and rural areas. People who live outside but near a city have a vested interest in maintaining a good environment for that city and maintaining a good system for managing the city's resources. The more concentrated the human population, the more land is available for other uses, including wilderness, recreation, conservation of biological diversity, and production of renewable resources. So cities benefit wilderness, rural areas, and so forth.

If people live in densely populated cities, ways must be found to make urban life healthy and pleasant and to keep the cities from polluting the very environment that their dense human population in theory frees for other uses. City planners have found many ways to make cities pleasing environments: developing parks and connecting cities in environmentally and aesthetically pleasing ways to rivers and nearby mountains. City planning has a long and surprising history, with the paired goals of defense and beauty. The long experience in city planning, combined with modern knowledge from environmental sciences, can make cities of the future healthier and more satisfying to people and better integrated within the environment. Beautiful cities are not only healthy but attract people, relieving pressures on the countryside.

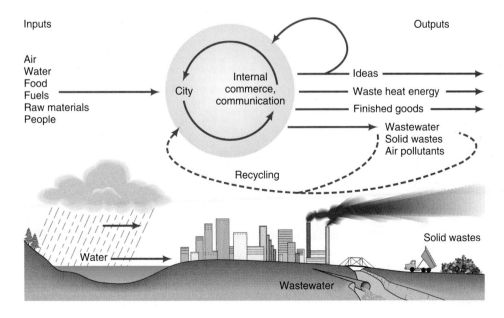

Figure 29.4 ■ The city as a system with flows of energy and materials. A city must function as part of a city–countryside ecosystem, with an input of energy and materials, internal cycling, and an output of waste heat energy and material wastes. As in any natural ecosystem, recycling of materials can reduce the need for input and the net output of wastes.

With the growing human population, we can imagine two futures. In one, cities are pleasing and livable, use resources from outside the city in a sustainable way, minimize pollution of the surrounding country, and allow room for wilderness, agriculture, and forestry. In the other future, cities continue to be seen as environmental negatives and are allowed to decay from the inside. People flee to grander and more expansive suburbs that occupy much land, and the poor who remain in the city live in an unhealthy and unpleasant environment. Without care for the city, its technological structure declines and it pollutes even more than in the past. Trends in both directions appear to be occurring.

In light of all these concerns, this chapter describes how a city can fit within, use, and avoid destroying the ecological systems on which it depends, and how the city itself can serve human needs and desires as well as environmental functions. With this information, you will have the foundations for making decisions, based on science and on what you value, about what kind of urban–rural landscape you believe will provide the most benefits for people and nature.

29.3 Site and Situation: The Location of Cities

Here is an idea that our modern life, with its rapid transportation and its many electronic tools, obscures: Cities are not located at random but develop mainly because of local conditions and regional benefits. In most cases, they grow up at crucial transportation locations (an aspect of what is called the city's situation) and can be readily defended, with good building locations, water supplies, and access to resources (qualities related to what is called site). The primary exceptions are cities that have been located primarily for political reasons. For example, Washington, D.C., was located to be near the geographic center of the area of the original 13 states; but the site was primarily swampland, and nearby Baltimore provided the major harbor of the region.

Importance of Site and Situation

As the case study of Venice, Italy, illustrates (see A Closer Look 29.1), the location of a city is influenced by the two factors just mentioned: **site**, which is the summation of all the environmental features of that location, and **situation**, which is the placement of the city with respect to other areas. A good site includes a geologic substrate suitable for buildings, such as a firm rock base and well-drained soils that are above the water table; nearby supplies of drinkable water; good nearby lands suitable for agriculture; and forests (Figure 29.5). It is also easier to build a city where the climate is benign—meaning that it does not suffer extremes of temperature and rainfall and is not subject to frequent storms. However, many important cities have been built in difficult climates. For example, Minneapolis–St. Paul has a cold winter and hot summer; Houston and Miami have hot, moist summers. In these cases, one negative aspect of site has been overcome with modern engineering technology.

The environmental situation strongly affects the development and importance of a city, particularly with regard to transportation and defense. Waterways are important for transportation. Before railroads, automobiles, and airplanes, cities depended on water for transportation. Most early cities were located on or near waterways. In the an-

(a)

(b)

Figure 29.5 ■ (*a*) Geologic, typographic, and hydrologic conditions greatly influence how successful the city can be. If these conditions, known collectively as the city's site, are poor, much time and effort is necessary to create a livable environment. As we learned in the opening case study, New Orleans has a poor site but an important situa-

tion. (*b*) In contrast, New York City's Manhattan is a bedrock island rising above the surrounding waters, providing a strong base for buildings and a soil that is sufficiently above the water table so that flooding and mosquitoes are much less of a problem.

Venice Sinking

The city of Venice, Italy, is slowly sinking, but for a long time no one knew the cause or a solution. Floods were becoming more and more common, especially during winter storms, when the winds drove waters from the Adriatic Sea into the city's streets[9] (Figure 29.6). At present Venice suffers 200 days of at least some flooding every year; a hundred years ago the average was seven days a year.[9]

Famous for its canals and architectural beauty, Venice is in danger of being destroyed by the very lagoon that had sustained its commerce for more than a thousand years. The city is sinking for three reasons: Sea level continues to rise as a consequence of effects of the last ice age; global warming is leading to an additional sea-level rise; and, as was discovered several decades ago, groundwater in the region was being pumped out for water supply.[10] Without the groundwater, the soil began to compress under the weight of the city. At least something could be done about this cause. Instead of obtaining water from directly below the city, Venice began to use water from the mainland. As a result, the city appeared to stop sinking for a while, and its rate of sinking has slowed.

Venice was founded by people escaping from the hordes that pillaged cities at the end of the Roman Empire. Its location in the marshes along the Adriatic Sea was easily defended, and the seaside was also a good location for transportation and trade. However, the site of Venice presented environmental problems that had to be solved before the city could become a major center. The shifting muds along the flat coast were a poor foundation for buildings. To improve the site, the early Venetians drove poles (made from saplings from forests on the mainland) into the mud and built on these. Venice still stands on this foundation, established more than 1,000 years ago.

Venice's success as a city depended on its situation—that is, its location relative to its surroundings. Nearby areas were a source of raw materials and resources for the city and provided a market for its prod-

Figure 29.6 ■ St. Mark's Square, Venice, under floodwaters, November 6, 2000. Floodwater covered more than 90% of the city in that incident.

ucts. Today, as in the past, Venice is affected by its surroundings in many ways, both positive and negative. It receives most raw goods and necessities from outside—food (except fish), fuel, and so forth. On the negative side, the highly decorated buildings and monuments for which the city is famous are deteriorating from acid rain generated by industries both nearby and far away. Its waters have high concentrations of fertilizers from farms and heavy metals from industry on the mainland.[11]

The problem is that the city continues to sink because of the rising sea level. The problem is exacerbated by global warming, which increases the rate of sea-level rise. Storm tides rush into the lagoon and flood the city. Floodgates that can temporarily block the flow of ocean water into the lagoon have been designed, but it is unclear how successful they will be, and they have remained controversial. In 2006 the Italian government decided to go ahead with the floodgates project, but environmentalists are concerned about its negative effects.[9] And if the floodgates are successful, they may stop the flushing of wastes from the lagoon, which has allowed Venice to persist despite antiquated sewage facilities.

Will Venice survive? At this time no one knows for certain. We have some ideas about what to do, but the real question is whether there is the political will in Italy to save this famous city. Whether Venice survives is therefore a question of science and values. The history of Venice also makes clear the connection, even in cities, between people and nature.

Venice is not the only city to experience problems with subsidence, or sinking. This can also result from the removal of oil, gas, minerals, and ore. Long Beach, California, has experienced subsidence as a result of the removal of oil and gas from wells, as well as the removal of groundwater.[12] In Mexico City, removal of groundwater has led to subsidence of as much as 7 m (30 ft).

As the people of Venice, Long Beach, and Mexico City have learned, cities influence and are influenced by their environment. The environment of a city affects its growth, success, and importance—and can also provide the seeds of its destruction. All cities are so influenced, and those who plan, manage, and live in cities must be aware of all aspects of the urban environment.

cient Roman Empire, for example, all important cities were near waterways. Waterways have continued to influence the locations of cities; most major cities of the eastern United States are situated either at major ocean harbors or at the **fall line** on major rivers. See A Closer Look 29.2 for the definition of the term *fall line* and more information that illustrates how much environment affects the location and success of cities (see Figure 29.7).

Cities are often founded at other kinds of crucial transportation points, growing up around a market, a river crossing, or a fort. Newcastle, England, and Budapest, Hungary, are located at the lowest bridging points on their rivers. Other cities, such as Geneva, are located where a river enters or leaves a major lake. Some well-known cities are located at the confluence of major rivers: Saint Louis lies at the confluence of the Missouri and Mississippi rivers; Manaus (Brazil), Pittsburgh (Pennsylvania), Koblenz (Germany), and Khartoum (Sudan) are at the confluence of several rivers. Many famous cities are located at crucial defensive locations, such as on or adjacent to easily defended rock outcrops. Examples include Edinburgh, Athens, and Salzburg. Other cities are situated on peninsulas—for example, Monaco and Istanbul.

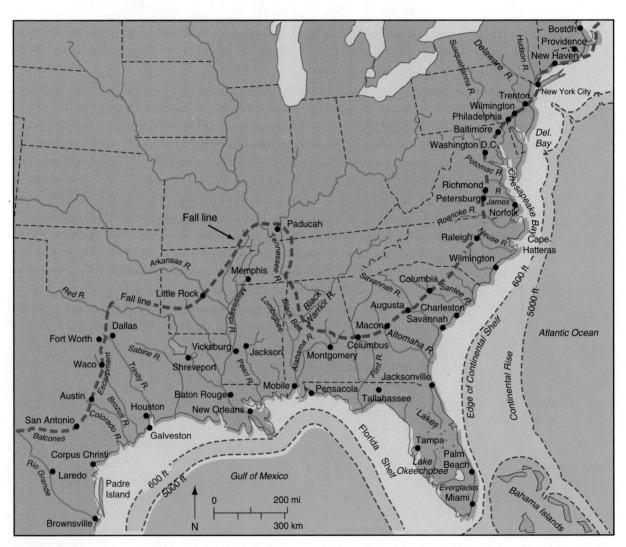

Figure 29.7 ■ Most major cities of the eastern and southern United States lie either at the sites of harbors or along a fall line (shown by the dashed line in the figure), which marks locations of waterfalls and rapids on major rivers. This is one way the location of cities is influenced by the characteristics of the environment. [*Source:* C. B. Hunt, *Natural Regions of the United States and Canada* (San Francisco: Freeman, 1974). Copyright 1974 by W. H. Freeman & Co.]

Cities and the Fall Line

A fall line on a river occurs where there is an abrupt drop in elevation of the land, creating waterfalls (Figure 29.7). A fall line typically occurs where streams pass from harder, more erosion-resistant rocks to softer rocks. In eastern North America the major fall line occurs at the transition from the granitic and metaphoric bedrock that forms the Appalachian Mountains and the softer, more easily eroded, and more recent sedimentary rocks. In general, the transition from major mountain range bedrock to another bedrock forms the primary fall line on continents.

Cities have frequently been established at fall lines, especially the major continental fall lines, for a number of reasons. Fall lines provide waterpower, an important source of energy in the eighteenth and nineteenth centuries, when the major eastern cities of the United States were established or rose to importance. At that time, the fall line was the farthest inland that larger ships could navigate; and just above the fall line was the farthest downstream that the river could be easily bridged. Not until the development of steel bridges in the late nineteenth century did it become practical to span the wider regions of a river below the fall line.[12] The proximity of a city to a river has another advantage: River valleys have rich, water-deposited soils that are good for agriculture. In early times, rivers also provided an important means of waste disposal, which today has become a serious problem.

Cities are often founded close to a mineral resource, such as salt (Salzburg, Austria), metals (Kalgoorlie, Australia), or medicated waters and thermal springs (Spa, Belgium; Bath, Great Britain; Vichy, France; and Saratoga Springs, New York). A successful city can grow and spread over surrounding terrain, so that its original purpose may be obscured to a resident. Its original market or fort may have evolved into a square or a historical curiosity. In most cases, though, cities originated where the situation provided a natural meeting point for people.

An ideal location for a city has both a good site and a good situation, but such a place is difficult to find. Paris is perhaps one of the best examples of a perfect location for a city—one with both a good site and a good situation. Paris began on an island more than 2,000 years ago, the situation providing a natural moat for defense and waterways for transportation. Surrounding countryside, a fertile lowland called the Paris basin, affords good local agricultural land and other natural resources.

Site Modification

Site is provided by the environment, but technology and environmental change can alter a site for better or worse. People can improve the site of a city and have done so when the situation of the city made it important and when its citizens could afford large projects. An excellent situation can sometimes compensate for a poor site. However, improvements are almost always required to the site so the city can persist. As the introductory case study made clear, New Orleans requires an expensive improvement in its site if it is to survive at all.

Changes in a site over time can have adverse effects on a city. For example, Bruges, Belgium, developed as an important center for commerce in the thirteenth century because its harbor on the English Channel permitted trade with England and other European nations. By the fifteenth century, however, the harbor had seriously silted in, and the limited technology of the time did not make dredging possible. This problem, combined with political events, led to a decline in the importance of Bruges—a decline from which it never recovered. Nevertheless, today, Bruges still lives, a beautiful city with many fine examples of medieval architecture. Ironically, the fact that these buildings were never replaced with modern ones makes Bruges a modern tourist destination. Ghent, Belgium, and Ravenna, Italy, are other examples of cities whose harbors silted in.[13] As human effects on the environment bring about global change, there may be rapid, serious changes in the sites of many cities. For instance, when global warming occurs and sea levels rise, many coastal cities will be subject to flooding.

29.4 City Planning and the Environment

A city can never be free of environmental constraints, even though its human constructions give us a false sense of security. Lewis Mumford, a historian of cities, wrote, "Cities give us the illusion of self-sufficiency and independence and of the possibility of physical continuity without conscious renewal."[8] But this security is only an illusion. (A brief history of the relation of cities to their environments is presented in A Closer Look 29.3.)

A danger in city planning is the tendency to transform a city center from natural to artificial features, to completely replace grass and soil with pavement, gravel, houses, and commercial buildings, creating an impres-

An Environmental History of Cities

The Rise of Towns

The first cities emerged on the landscape thousands of years ago during the New Stone Age with the development of agriculture, which provided food in sufficient quantity to maintain a city.[8] In this first stage, the number of people per square kilometer in cities was much higher than in the surrounding countryside, but the density was still too low to cause rapid, serious disturbance to the land. In fact, the waste from city dwellers and their animals was an important fertilizer for the surrounding farmlands. In this stage, the city's size was restricted by the primitive transportation methods for bringing food and necessary resources into the city and removing waste. Because of such limitations, no European medieval town served only by land transportation had a population greater than 15,000.[13]

The Urban Center

In the second stage, more efficient transportation made possible the development of much larger urban centers. Boats, barges, canals, and wharves, as well as roads, horses, carriages, and carts, made it possible for cities to be located farther from agricultural areas. Rome, originally dependent on local produce, became a city fed by granaries in Africa and the Near East.

The population of a city is limited by how far a person can travel in one day to and from work and by how many people can be packed into an area (density). In the second stage, the internal size of a city was limited by pedestrian travel. A worker had to be able to walk to work, do a day's work, and walk home the same day. The density of people per square kilometer was limited by architectural techniques and primitive waste disposal. These cities never exceeded a population of 1 million, and only a few approached this size, most notably Rome and some cities in China.

The Industrial Metropolis

The Industrial Revolution allowed greater modification of the environment than had been possible before. Two technological advances that had significant effects on the city environment were improved sanitation (Figure 29.8), which led to the control of many diseases, and improved transportation.

Modern transportation makes a larger city possible. Workers can live farther from their place of work and commerce, and communication can extend over larger areas. Air travel has freed cities even more from the traditional limitation of situation. We now have thriving urban areas where previously transportation was poor: in the Far North (Fairbanks, Alaska) and on islands (Honolulu). These changes increase urban dwellers' sense of separateness from their natural environment.

Subways and commuter trains have also allowed the development of suburbs. In some cities, however, the negative effects of urban sprawl have led many people back to the urban centers or to smaller, satellite cities surrounding the central city. The drawbacks of suburban commuting and the destruction of the landscape in suburbs have brought new appeal to the city center.

The Center of Civilization

We are at the beginning of a new stage in the development of cities. With modern telecommunications, people can work at home or at distant locations. Perhaps, as telecommunication frees us from the necessity for certain kinds of commercial travel and related activities, the city can become a cleaner, more pleasing center of civilization.

An optimistic future for cities requires a continued abundance of energy and material resources, which are certainly not guaranteed, and wise use of these resources. If energy resources are rapidly depleted, modern mass transit may fail, fewer people will be able to live in suburbs, and the cities will become more crowded. Reliance on coal and wood will increase air pollution. The continued destruction of the land within and near cities could compound transportation problems, making local production of food impossible. The future of our cities depends on our ability to plan and to conserve and use our resources wisely.

Figure 29.8 ■ The New York City aqueduct, built in the 1860s, brought clean, clear water from the Croton Reservoir, miles away, into the heart of the city. It was one of the most important factors in creating a livable environment in that city.

A Brief History of City Planning

Two dominant themes in formal city planning have been planning for defense and planning for beauty. Ancient Roman cities were typically designed along simple geometric patterns that had both practical and aesthetic benefits. The symmetry of the design was considered beautiful but also provided a useful layout for streets.

During the height of Islamic culture, in the first millennium, Islamic cities typically contained beautiful gardens, often within the grounds of royalty. Among the most famous urban gardens in the world are the gardens of the Alhambra, a palace in Granada, Spain (Figure 29.9). The gardens were created when this city was a Moorish capital, and they were maintained after Islamic control of Granada ended in 1492. Today, as a tourist attraction that receives 2 million visitors a year, the Alhambra gardens demonstrate the economic benefits of aesthetic considerations in city planning. They also illustrate that making a beautiful park a specific focus in a city benefits the city environment by providing relief from the city itself.

After the fall of the Roman Empire, the earliest planned towns and cities in Europe were walled fortress cities designed for defense. But even in these, city planners considered the aesthetics of the town. In the fifteenth century, one such planner, Leon Battista Alberti, argued that large and important towns should have broad and straight streets; smaller, less fortified towns should have winding streets to increase their beauty. He also advocated the inclusion of town squares and recreational areas, which continue to be important considerations in city planning.[14] One of the most successful of these walled cities is Carcasone, in southern France, now the third most visited tourist site in that country. Today, walled cities have become major tourist attractions, again illustrating the economic benefits of good aesthetic planning in urban development.

The usefulness of walled cities essentially ended with the invention of gunpowder.

The Renaissance sparked an interest in the ideal city, which in turn led to the development of the park city. A preference for gardens and parks, emphasizing recreation, developed in Western civilization in the seventeenth and eighteenth centuries. It characterized the plan of Versailles, France, with its famous formal parks of many sizes and tree-lined walks, and also the work of the Englishman Capability Brown, who designed parks in England and was one of the founders of the English school of landscape design, which emphasized naturalistic gardens.

Figure 29.9 ■ Planned beauty. The Alhambra gardens of Granada, Spain, illustrate how vegetation can be used to create beauty within a city.

sion that civilization has dominated the environment. Ironically, the artificial aspects of the city that make it seem so independent of the rest of the world actually make it more dependent on its rural surroundings for all resources. Although such a city appears to its inhabitants to grow stronger and more independent, it actually becomes more fragile.[8]

In this way, a city grows at the expense of surrounding countryside, destroying surrounding landscape on which it depends. As nearby areas are ruined for agriculture and as the transportation network extends, the use, misuse, and destruction of the environment increase. Many cities were built, prospered, and declined in this way.

City Planning for Defense and Beauty

Many cities in history grew without any conscious plan. **City planning**—formal, conscious planning for new cities—can be traced back as far as the fifteenth century. Sometimes cities have been designed for specific social purposes, with little consideration of the environment; in other cases, the environment and its effect on city residents have been major planning considerations.

Two dominant themes in formal city planning have been defense and beauty (see A Closer Look 29.4). We can think of these two types of cities as fortress cities and park cities. The ideas of the fortress city and the park city influenced the planning of cities in North America. The importance of aesthetic considerations is illustrated in the plan of Washington, D.C., designed by Pierre Charles L'Enfant. L'Enfant mixed a traditional rectangular grid pattern of streets (which can be traced back to the Romans) with broad avenues set at angles. The goal was to create a beautiful city, with many parks, including small ones at the intersections of avenues and streets. This design has made Washington, D.C., one of the most pleasant cities in the United States.

The City Park

Parks have become more and more important in cities. A significant advance for U.S. cities was the nineteenth-century planning and construction of Central Park in New York City, the first large public park in the United States. The park's designer, Frederick Law Olmsted, was one of the most important modern experts on city planning. He took site and situation into account and attempted to blend improvements to a site with the aesthetic qualities of the city.

For Olmsted, the goal of a city park was to provide psychological and physiological relief from city life through access to nature. A means to provide relief from city life was the creation of beauty in the park. Vegetation was one of the keys, and Olmsted carefully considered the opportunities and limitations of topography, geology, hydrology, and vegetation.

In contrast to the approach of a preservationist, who might simply have strived to return the area to its natural, wild state, Olmsted created a naturalistic environment, keeping the rugged, rocky terrain but putting ponds where he thought they were desirable. To add variety, he constructed "rambles" that were densely planted and followed circuitous patterns. He created a "sheep meadow" by using explosives. In the southern part of the park, where there were flat meadows, he created recreational areas. To help the park meet the needs of the city, he built transverse roads through the park. He created depressed roadways that allowed traffic to cross the park without detracting from the vistas seen by park visitors.

Central Park is an example of "design with nature," a term coined much later, and its design influenced other U.S. city parks. Olmsted remained a major figure in American city planning throughout the nineteenth century, and the firm he founded continued to be important in city planning into the twentieth century.[15, 16]

Olmsted's skill in creating designs that addressed both the physical and aesthetic needs of a city is further illustrated by his work in Boston. The original site of Boston had certain advantages: a narrow peninsula with several hills that could be easily defended, a good harbor, and a good water supply. As Boston grew, however, demand increased for more land for buildings, a larger area for docking ships, and a better water supply. The need to control ocean floods and to dispose of solid and liquid wastes grew as well. Much of the original tidal

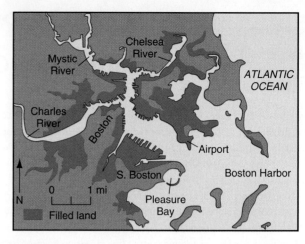

Figure 29.10 ■ Nature integrated into a city plan. Boston has been modified over time to improve the environment and provide more building locations. This map of Boston shows land filled in to provide new building sites as of 1982. Although such landfill allows for expansion of the city, it can also create environmental problems, which then must be solved. [*Source:* A. W. Spirn, *The Granite Garden: Urban Nature and Human Design* (New York: Basic Books, 1984).]

flats area, which had been too wet to build on and too shallow to navigate, had been filled in (Figure 29.10). Hills had been leveled and the marshes filled with soil. The largest project had been the filling of Back Bay, which began in 1858 and continued for decades. Once filled, however, the area had suffered from flooding and water pollution.[17]

Olmsted's solution to these problems was a water-control project called the "fens." His goal was to "abate existing nuisances" by keeping sewage out of the streams and ponds and building artificial banks for the streams to prevent flooding—and to do this in a natural-looking way. His solution included creating artificial watercourses by digging shallow depressions in the tidal flats, following meandering patterns like natural streams; setting aside other artificial depressions as holding ponds for tidal flooding; restoring a natural salt marsh planted with vegetation tolerant of brackish water; and planting the entire area to serve as a recreational park when not in flood. He put a tidal gate on the Charles River—Boston's major river—and had two major streams diverted directly through culverts into the Charles so that they flooded the fens only during flood periods. He reconstructed the Muddy River primarily to create new, accessible landscape.

The result of Olmsted's vision was that control of water became an aesthetic addition to the city. The blending of several goals made the development of the fens a landmark in city planning. Although it appears to the casual stroller to be simply a park for recreation, the area serves an important environmental function in flood control. Parks near rivers and the ocean are receiving more and more attention. For example, New York city is spending several hundred-million dollars to build two parks along Manhattan's Hudson River shoreline, where previously abandoned docks and warehouses littered the shoreline and separated the public from access to the river.

An extension of the park idea was the **garden city**, a term coined in 1902 by Ebenezer Howard. Howard believed that city and countryside should be planned together. A garden city was one that was surrounded by a **greenbelt**. The idea was to locate garden cities in a set connected by greenbelts, forming a system of countryside and urban landscapes. The idea caught on, and garden cities were planned and developed in Great Britain and the United States. Greenbelt, Maryland, just outside Washington, D.C., is one of these cities, as is Lecheworth, England. Howard's garden city concept, like Olmsted's use of the natural landscape in designing city parks, continues to influence city planning today.

29.5 The City as an Environment

A city changes the landscape, and because it does, it also changes the relationship between biological and physical aspects of the environment. Many of these changes were discussed in earlier chapters as aspects of pollution, water management, or climate. They are mentioned again as appropriate in the following sections, generally with a focus on how effective city planning can reduce the problems.

The Energy Budget of a City

Like any ecological and environmental system, a city has an "energy budget." The city exchanges energy with its environment in the following ways: (1) absorption and reflection of solar energy, (2) evaporation of water, (3) conduction of air, (4) winds (air convection), (5) transport of fuels into the city and burning of fuels by people within the city, and (6) convection of water (subsurface and surface stream flow). These in turn affect the climate within the city, and the city may affect the climate in the nearby surroundings, a possible landscape effect.

The Urban Atmosphere and Climate

Cities affect the local climate; as the city changes, so does its climate (see Chapter 23). Cities are generally less windy than nonurban areas because buildings and other structures obstruct the flow of air. But city buildings also channel the wind, sometimes creating local wind tunnels with high wind speeds. The flow of wind around one building is influenced by nearby buildings, and the total wind flow through a city is the result of the relationships among all the buildings. Thus, plans for a new building must take into account its location among other buildings as well as its shape. In some cases, when this has not been done, dangerous winds around tall buildings have resulted in blown-out windows, as happened to the John Hancock Building in Boston on January 20, 1973, a famous example of the problem.

A city typically receives less sunlight than the countryside because of the particulates in the atmosphere over cities. Often, urban areas have 10 or more times more particulates than surrounding areas.[18] In spite of the reduced sunlight, cities are warmer than surrounding areas (a city is a heat island), for two reasons. One is increased heat from the burning of fossil fuels and other industrial and residential activities. The other is a lower rate of heat loss, partly because buildings and paving materials act as solar collectors (Figure 29.11).[19]

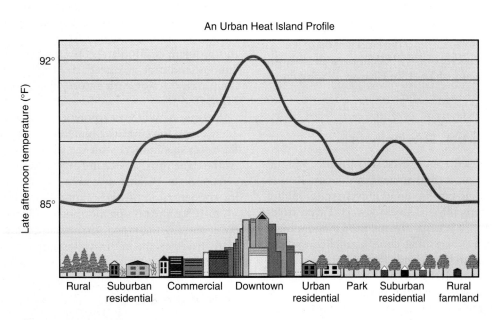

An Urban Heat Island Profile

Figure 29.11 ■ A typical urban heat island profile. The graph shows temperature changes correlated with the density of development and trees. [*Source:* Andrasko and Huang, in H. Akbari et al., *Cooling Our Communities: A Guidebook on Tree Planting and Light-Colored Surfacing* (Washington, D.C.: U.S. EPA Office of Policy Analysis, 1992).]

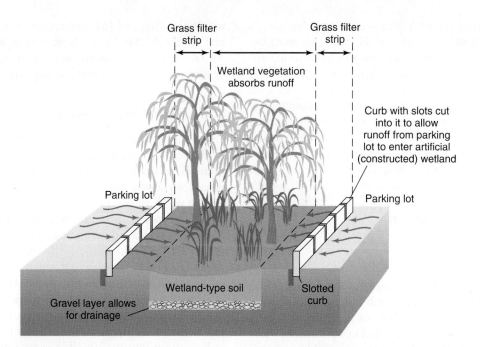

Grass filter strip

Grass filter strip

Wetland vegetation absorbs runoff

Curb with slots cut into it to allow runoff from parking lot to enter artificial (constructed) wetland

Parking lot

Parking lot

Wetland-type soil

Slotted curb

Gravel layer allows for drainage

Figure 29.12 ■ Planned for better drainage. A plan for the Alexandria, Virginia, central library parking lot includes wetland vegetation and soils, which temporarily absorb runoff from the parking lot (see arrows). The landscape architecture firm of Rhodeside and Harwell planned the project. [*Source:* Modified after Harwell, Rhodeside, & Rhodeside, Landscape Architects.]

Solar Energy in Cities

Until modern times, it was common to use solar energy to help heat city houses. Our century is a major exception to this approach, because cheap and easily accessible fossil fuels have led us to forget certain fundamental lessons. Cities in ancient Greece, Rome, and China were designed so that houses and patios faced south and passive solar energy applications were accessible to each household (Chapter 19).[19] Today, we are beginning to appreciate the importance of solar energy once again. Solar photovoltaic devices that convert sunlight to electricity are becoming a common sight in many cities, and some cities have enacted solar energy ordinances that make it illegal to shade another property owner's building in such a way that it loses solar heating capability. (See Chapter 19 for a discussion of solar energy.)

Water in the Urban Environment

Modern cities affect the water cycle, in turn affecting soils and consequently plants and animals in the city. Because paved city streets and city buildings prevent water infiltration, most rain runs off into storm sewers. Paved streets and sidewalks prevent water in the soil from evaporating to the atmosphere, a process that cools natural ecosystems; this adds to the urban heat island effect. Chances of flooding increase both within the city and downstream outside the city. New, ecological methods of managing storm water can alleviate these problems by controlling the speed and quality of water running off pavements and into streams. For example, a plan for the central library's parking lot in Alexandria, Virginia, includes wetland vegeta-

tion and soils that temporarily absorb runoff from the parking lot, remove some of the pollutants, and slow down water flow (Figure 29.12).

Most cities have a single underground sewage system. During times of no rain or light rain, this system handles only sewage. But during periods of heavy rain, the runoff is mixed with the sewage and can exceed the capacity of sewage-treatment plants, causing sewage to be emitted downstream without sufficient treatment. In most cities that already have such systems, the expense of building a completely new and separate runoff system is prohibitive, so other solutions must be found. One city that avoids this problem is Woodlands, Texas. It was designed by the famous landscape architect Ian McHarg, who originated the phrase "design with nature," the subject of A Closer Look 29.5.[20]

Because of reduced evaporation, midlatitude cities generally have a lower relative humidity (2% lower in winter to 8% lower in summer) than the surrounding countryside. At the same time, cities can have higher rainfall than their surroundings because dust above a city provides particles for condensation of water vapor. Some urban areas have 5% to 10% more precipitation and considerably more cloud cover and fog than their surrounding areas. Fog is particularly troublesome in the winter and may impede ground and air traffic.

Soils in the City

A modern city has a great impact on soils. Since most of a city's soil is covered by cement, asphalt, or stone, the soil no longer has its natural cover of vegetation, and the nat-

Design with Nature

In designing a new city, our knowledge of the problems of flooding and water runoff can be used to plan a better flow of water. The new town of Woodlands, a suburb of Houston, Texas, illustrates such planning. Woodlands was designed so that most houses and roads were on ridges and the lowlands were left as natural open space.

The lowlands provide areas for temporary storage of floodwater and, because the land is unpaved, allow the rain to penetrate the soil and recharge the aquifer for Houston. Preserving the natural lowlands has other environmental benefits as well. In this region of Texas, low-lying wetlands are habitats for native wildlife, such as deer. Large,

attractive trees, such as magnolias, grow there, providing food and a habitat for birds. The innovative city plan has economic as well as aesthetic and conservational benefits. It is estimated that a conventional drainage system would have cost $14 million more than the amount spent to develop and maintain the wetlands.[20]

---

ural exchange of gases between the soil and air is greatly reduced. Because they are no longer replenished by vegetation growth, these soils lose organic matter, and soil organisms die from lack of food and oxygen. In addition, the process of construction and the weight of the buildings compact the soil, which restricts water flow. City soils, then, are more likely to be compacted, waterlogged, impervious to water flow, and lacking in organic matter.

A kind of soil important in modern cities is the soil that occurs on **made lands**, which are lands created from fill, sometimes as waste dumps of all kinds, sometimes directly to create more land for construction. The soils of made lands are different from those of the original landscape. They may be made of all kinds of trash, from newspapers to bathtubs, and may contain some toxic materials. The fill material is unconsolidated, meaning that it is loose material without rock structure. Thus, it is not well suited to be a foundation for buildings. Fill material is particularly vulnerable to the shaking from earthquakes. In such events, the fill can act somewhat like a liquid and amplify the effects of the earthquake on buildings. However, some made lands have been turned into well-used parks. For example, a marina park in Berkeley, California, is built on a solid-waste landfill. It extends into San Francisco Bay, providing public access to beautiful scenery. It is a windy location, popular for kite flying and family strolls. (See Chapter 28 for more information about solid-waste disposal.)

Pollution in the City

In a city, everything is concentrated, including pollutants. City dwellers are exposed to more kinds of toxic chemicals in higher concentrations and to more human-produced noise, heat, and particulates than are their rural neighbors (see Chapter 15). This environment makes life riskier. Lives are shortened by an average of one to two years in the most polluted cities in the United States. The city with the greatest number of early deaths is Los Angeles, with an estimated 5,973 early deaths per year, followed by New

York with 4,024, Chicago with 3,479, Philadelphia with 2,590, and Detroit with 2,123.

Some urban pollution comes from motor vehicles, which have contributed lead where it is still used in gasoline, nitrogen oxides, ozone, carbon monoxide, and other air pollutants from exhaust. Stationary power sources also produce air pollutants. Home heating is a third source, contributing particulates, sulfur oxides, nitrogen oxides, and other toxic gases. Industries are a fourth source, contributing a wide variety of chemicals. The primary sources of particulate air pollution—which consists of smoke, soot, and tiny particles formed from emissions of sulfur dioxide and volatile organic compounds—are older, coal-burning power plants, industrial boilers, and gas- and diesel-powered vehicles.[21]

Although it is impossible to eliminate exposure to pollutants in a city, it is possible to reduce the exposure through careful design, planning, and development. For example, when lead was used in gasoline, exposure to lead was greater near roads. Exposure could be reduced by placing houses and recreational areas away from roads and by developing a buffer zone using trees that are resistant to the pollutant and that absorbed pollutants.

29.6 Bringing Nature to the City

A practical problem for planners and managers of cities is how to bring nature to the city—how to make plants and animals part of a city landscape. This has evolved into several specialized professions, including urban forestry (whose professionals are often called tree wardens), landscape architecture, city planning, and civil engineering specializing in urban development. Most cities have an urban forester on the payroll who determines the best sites for planting trees and the best species of trees to suit the environment. These professionals take into account climate, soils, and the general influences of the urban setting, such as the shade imposed by tall buildings and the pollution from motor vehicles.

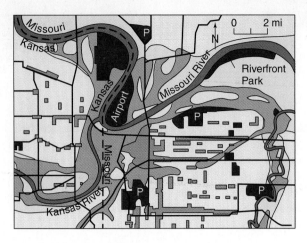

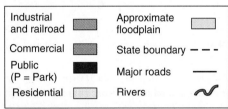

(b)

Figure 29.13 ■ A city and its (almost vanished) river. (*a*) Idealized general land use in Kansas City, Missouri. (*b*) A view of Kansas City's Missouri River waterfront.

Cities and Their Rivers

For cities on rivers, one way to bring nature to the city is to connect the city to its river. But, traditionally, rivers have been valued for their usefulness in transportation and waste disposal rather than their ability to make city life more pleasant or to help in the conservation of nature. Rivers have been viewed as places to dump wastes. The old story was that a river renewed and cleaned itself every mile or every three miles (depending on who said it). That might have been relatively correct when there was one person or

one family per linear river mile, but it is not true with the high density of cities or with modern chemicals.

Kansas City, Missouri, located at the confluence of the Kansas and Missouri rivers, illustrates the traditional treatment (Figure 29.13). The Missouri River's floodplain provided a convenient transportation corridor, so the south shore is dominated by railroads, while downtown the north shore forms the southern boundary of the city's airport. Except for a small riverfront park, the river has little place in this city as a source of recreation and relief for its citizens or in the conservation of nature.

Contrast Kansas City with Portland, Oregon. In Portland, Interstate Highway I-5 was originally built on the city's side of the Willamette River, separating Portland from its river. The city raised funds and moved the highway across the river, replacing the interstate corridor with Riverfront Park, a marina, housing, shops, and pleasant walkways and developing pocket parks that lead into the city's downtown.

Vegetation in Cities

Trees, shrubs, and flowers improve the beauty of a city. Plants fill different needs in different locations.[22] Trees provide shade, which reduces the need for air-conditioning and makes travel much more pleasant in hot weather. In parks, vegetation provides places for quiet contemplation; trees and shrubs can block some of the city sounds, and their complex shapes and structures create a sense of solitude. Plants also provide habitats for wildlife, such as birds and squirrels, which many urban residents consider pleasant additions to a city.

The use of trees in cities has expanded since the time of the Renaissance. In earlier times, trees and shrubs were set apart in gardens, where they were viewed as scenery but not experienced as part of ordinary activities. Street trees were first used in Europe in the eighteenth century; among the first cities to line streets with trees were Paris and London (Figure 29.14). In many cities, trees are now considered an essential element of the urban visual scene, and major cities have large tree-planting programs. For example, in New York City, 11,000 trees are planted each year; in Vancouver, Canada, 4,000 are planted per year.[22]

Trees are increasingly used to soften the effects of climate near houses. In colder climates, rows of conifers planted to the north of a house can protect it from winter winds. Deciduous trees to the south can provide shade in the summer, reducing requirements for air-conditioning, yet allowing sunlight to warm the house in the winter (Figure 29.15).[23]

Cities can even provide habitat for endangered plants. For example, Lakeland, Florida, uses endangered plants in local landscaping with considerable success.

However, vegetation in cities must be able to withstand special kinds of stress. Because trees along city streets are often surrounded by cement and because city soils tend to be

Figure 29.14 ■ Paris was one of the first modern cities to use trees along streets to provide beauty and shade, as shown in this picture of the famous Champs Elysées.

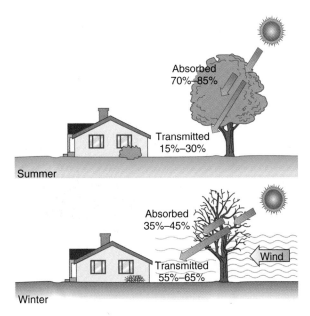

Figure 29.15 ■ Trees cool homes. Trees can improve the microclimate near a house, protecting the house from winter winds and providing shade in the summer while allowing sunlight through in the winter. [*Source:* J. Huang and S. Winnett, in H. Akbari et al., *Cooling Our Communities: A Guidebook on Tree Planting and Light-Colored Surfacing* (Washington, D.C.: U.S. EPA Office of Policy Analysis, 1992).]

compacted and to drain poorly, the root systems lack access to water and air and are more likely to suffer from extremes of drought on the one hand and soil saturation (immediately following or during a rainstorm) on the other. The solution is to specially prepare streets and sidewalks for tree growth. A tree-planting project was completed for the World Bank Building in Washington, D.C., in 1996. Special care was taken to provide good growing conditions for trees, including aeration, irrigation, and adequate drainage so that the soils did not become waterlogged. The trees continue to grow and remain healthy.[23]

Many species of trees and plants are very sensitive to air pollution. For example, the eastern white pine of North America is extremely sensitive to ozone pollution and does not do well in cities with heavy motor-vehicle traffic or along highways. Dust can interfere with the exchange of oxygen and carbon dioxide necessary for photosynthesis and respiration of the trees. City trees also suffer direct damage from the physical impact of bicycles, cars, and trucks and from vandalism. Trees subject to such stresses are more susceptible to attacks by fungus diseases and insects. The lifetime of trees in a city is generally shorter than in their natural woodland habitats unless they are given considerable care.

Some species of trees are more useful and successful in cities than are others. An ideal urban tree would be resistant to all forms of urban stress, have a beautiful form and foliage, and produce no messy fruit, flowers, or leaf litter that required cleaning. In most cities, in part because of these requirements, only a few species of trees are used for street planting. However, reliance on one or a few species results in ecologically fragile urban planting, as we learned when Dutch elm disease spread throughout the eastern United States, destroying urban elms. It is prudent to use a greater diversity of trees to avoid outbreaks of insect pests and tree diseases.[24]

Cities, of course, have many recently disturbed areas, including abandoned lots and corridors between lanes in boulevards and highways. Disturbed areas provide habitat for early-successional plants, including many that we call "weeds," which are often introduced (exotic) plants, such as European mustard. Therefore, wild plants that do particularly well in cities are those characteristic of disturbed areas and of early stages in ecological succession (see Chapter 10). City roadsides in Europe and North America have wild mustards, asters, and other early-successional plants.

Wildlife in Cities

With the exception of some birds and small, docile mammals such as squirrels, most forms of wildlife in cities are considered pests. But there is much more wildlife in cities, a great deal of it unnoticed. In addition, there is growing recognition that urban areas can be modified to provide more habitats for wildlife that people can enjoy. This can be an important method of biological conservation.[25, 26]

We can divide city wildlife into the following categories: (1) those species that cannot persist in an urban environment and disappear; (2) those that tolerate an urban environment but do better elsewhere; (3) those that have adapted to urban environments, are abundant there, and are either neutral or beneficial to human beings; and (4) those that are so successful they become pests.

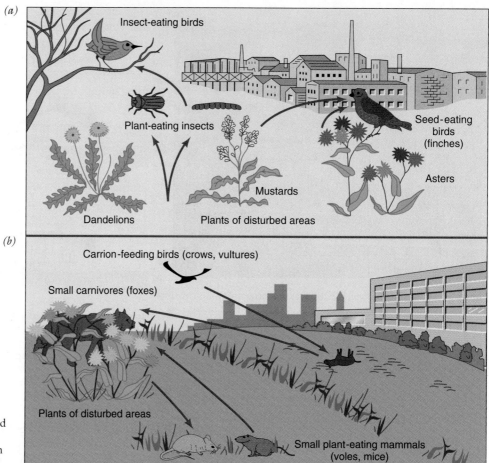

(a)

(b)

Insect-eating birds

Plant-eating insects

Seed-eating birds (finches)

Asters

Dandelions

Mustards

Plants of disturbed areas

Carrion-feeding birds (crows, vultures)

Small carnivores (foxes)

Plants of disturbed areas

Small plant-eating mammals (voles, mice)

Figure 29.16 ■ (*a*) An urban food chain based on plants of disturbed places and insect herbivores. (*b*) An urban food chain based on roadkill.

Urban "Wilds": The City as Habitat for Wildlife and Endangered Species

We do not associate wildlife with cities, but cities provide homes for many forms of wildlife. Consider, for example, how the fox has integrated itself into the urban landscape of London. Foxes in London feed on garbage and roadkill (animals run over by motor vehicles); shy and nocturnal, they are seen by few Londoners.[27] Their food chain includes pets and other animals, fruits and vegetables, earthworms and insects, and scavenged items. Automobile runovers account for about half of the deaths of urban foxes.

Cities are a habitat, albeit artificial. They can provide all the needs—physical structures and necessary resources such as food, minerals, and water—for many plants and animals. We can also identify ecological food chains in cities, as shown in Figure 29.16 for insect-eating birds and for a fox. For some species, cities' artificial structures are sufficiently like their original habitat to be home. For example, chimney swifts once lived in hollow trees but are now common in chimneys and other vertical shafts, where they glue their nests to the walls with saliva. A city can easily have more chimneys per square kilometer than a forest has hollow trees. Cities also include natural habitats in parks and preserves. Modern parks provide some of the

world's best wildlife habitats. In New York City's Central Park, approximately 260 species of birds have been observed—100 in a single day. Urban zoos play an important role in the conservation of endangered species, and the importance of parks and zoos will increase as truly wild areas shrink. Finally, cities that are seaports often have many species of marine wildlife at their doorsteps. New York City's waters include sharks, bluefish, mackerel, tuna, striped bass, and nearly 250 other species of fish.[27]

Cities can even be home to rare or endangered species. Peregrine falcons once hunted pigeons above the streets of Manhattan. Unknown to most New Yorkers, the falcons nested on the ledges of skyscrapers and dived on their prey in an impressive display of predation. The falcons disappeared when DDT and other organic pollutants caused a thinning of their eggshells and a failure in reproduction, but they have been reintroduced into the city. The first reintroduction into New York City took place in 1982; today there are 32 falcons living there (Figure 29.17).[28] The reintroduction of peregrine falcons illustrates an important recent trend: the growing understanding that city environments can assist in the conservation of nature, including the conservation of endangered species.

City environments can contribute to conservation of wildlife in a number of ways. Urban kitchen gardens, back-

Figure 29.17 ■ Falcon nest on top of a Manhattan building.

yard gardens that provide table vegetables and decorative plants, can be designed to provide habitats—for instance, flowers that provide nectar for threatened or endangered hummingbirds can be planted in such gardens. Rivers and their riparian zones, ocean shorelines, and wooded parks can provide habitat for endangered species and ecosystems. For example, prairie vegetation, which once occupied more land area than any other vegetation type in the United States, is rare today. One restored prairie exists within the city limits of Omaha, Nebraska. (Some urban nature preserves are not accessible to the public or offer only limited access, as is the case with the prairie preserve in Omaha.)

Urban drainage structures can also be designed as wildlife habitats. A typical urban runoff design depends on concrete-lined ditches that speed the flow of water from city streets to lakes, rivers, or the ocean. However, as with Boston's Back Bay design, discussed earlier, these features can be planned to maintain or create stream and marsh habitats with meandering waterways and storage areas that do not interfere with city processes. Such areas can become habitats for fish and mammals (Figure 29.18). Modified to promote wildlife, cities can provide urban

Stream

Naturalistic stream and marsh slow runoff and provide good wildlife and vegetation habitat

Marsh

Rapid runoff poor wildlife and vegetation habitat

Concrete-lined ditch

Figure 29.18 ■ How water drainage systems in a city can be modified to provide wildlife habitat. In the community on the right, concrete-lined ditches result in rapid runoff and have little value to fish and wildlife. In the community on the left, the natural stream and marsh were preserved; water is retained between rains and an excellent habitat is provided. [*Source:* D. L. Leedly and L. W. Adams, *A Guide to Urban Wildlife Management* (Columbia, MD.: National Institute for Urban Wildlife, 1984), pp. 20–21.]

How Can Urban Sprawl Be Controlled?

As the world becomes increasingly urbanized, individual cities are growing in area as well as population. Residential areas and shopping centers move into undeveloped land near cities, impinging on natural areas and creating a chaotic, unplanned human environment. Urban sprawl has become a serious concern in communities all across the United States. According to the U.S. EPA, in a recent six-month period approximately 5,000 people left Baltimore City to live in suburbs, with the result that nearly 10,000 acres of forests and farmlands were converted to housing. At this rate, the state of Maryland could use as much land for development in the next 25 years as it has used in the entire history of the state.[31] In the past ten years 22 states have enacted new laws to try to control urban sprawl.

The city of Boulder, Colorado, has been in the forefront of this effort since 1959, when it created the "blue line"—a line at an elevation of 1,761 m (the city itself is at 1,606 m) above which it would not extend city water or sewer services. Boulder's citizens felt, however, that the blue line was insufficient to control development and maintain the city's scenic beauty in the face of rapid population growth. (Boulder's population had grown in the decade before 1959 from 29,000 to 66,000 and reached 96,000 by 1998.) To prevent uncontrolled development in the area between the city and the blue line, in 1967 Boulder began to use a portion of the city sales tax to purchase land, creating a 10,800-hectares greenbelt around the city proper.

In 1976, Boulder went one step further and limited new residences to an increase of 2% a year. Two years later, recognizing that planned development requires a regional approach, the city and surrounding Boulder County adopted a coordinated development plan. By the early 1990s, it had become apparent that further growth control was needed for nonresidential building. The plan finally adopted by the city reduced the allowable density of many commercial and industrial properties, in effect limiting jobs rather than building space.

Boulder's methods to limit the size of its population have worked; in the most recent census, 2002, the population had barely increased by 2,000 people, numbering just a little more than 94,000.[32]

The benefits of Boulder's controlled-growth initiatives have been a defined urban–rural boundary; rational, planned development; protection of sensitive environmental areas and scenic vistas; and large areas of open space within and around the city for recreation. And in spite of its growth-control measures, Boulder's economy has remained strong. But with restraints on residential growth, many people who found jobs in Boulder were forced to seek affordable housing in adjoining communities, where populations ballooned. For example, the population of Superior, Colorado, grew from 225 in 1990 to 9,000 in 2000. Further, as commuting workers—40,000 a day—tried to get to and from their jobs in Boulder, traffic congestion and air pollution increased. In addition, because developers had not built stores in the outlying areas, shoppers flocked into Boulder's downtown mall. When plans for a competing mall in the suburbs were announced, however, Boulder officials worried about the loss of revenue if the new mall drew shoppers away from the city. At the same time, sprawl from Denver (only 48 km from Boulder), as well as its infamous "brown cloud" of polluted air, began to spill out along the highway connecting the two communities.

Critical Thinking Questions

1. Is a city an open or a closed system (see Chapter 3)? Use examples from the case of Boulder to support your answer.

2. As Boulder takes steps to limit growth, it becomes an even more desirable place to live, which subjects it to even greater growth pressures. What ways can you suggest to avoid such a positive-feedback loop?

3. Some people in Boulder think the next step is to increase residential density within the city. How do you think people living there will accept this plan? What are the advantages and disadvantages of increasing density?

4. To some, the story of Boulder is the saga of a heroic battle against commercial interests that would destroy environmental resources and a unique quality of life. To others, it is the story of an elite group building an island of prosperity and the good life for themselves while shifting the more unpleasant aspects of modern life elsewhere. How do you view Boulder's story, and why?

corridors that allow wildlife to migrate, following natural routes despite the imposition of the city.[29] Urban corridors also serve to prevent some of the effects of ecological islands (see Chapter 7) and are increasingly important to biological conservation.

Animal Pests

Pests are familiar to urban dwellers. The most common city pests are cockroaches, fleas, termites, rats, and pigeons, but there are many more, especially species of insects. In gardens and parks, pests include insects, birds, and mammals that feed on fruit and vegetables and destroy foliage of shade trees and plants. Pests compete with people for food and spread diseases. Indeed, before modern sanitation and medicine, such diseases played a major role in limiting human population density in cities. Bubonic plague is spread by fleas found on rodents; mice and rats in cities promoted the spread of the Black Death (see Chapter 1). Bubonic plague continues to be a health threat in cities. The World Health Organization reports several thousand cases a year.[30] Poor sanitation and high population densities of people and rodents set up a situation where the disease can strike.

An animal is a pest to people when it is in an undesired place at an undesirable time doing an unwanted thing. A termite in a woodland helps the natural regeneration of wood by hastening decay and speeding the return of chemical elements to the soil, where they are available to living plants. But termites in a house are pests because they threaten the physical structure of the house.

Animals that survive best in cities have certain characteristics in common. Animals that are urban pests usually are generalists in their food choice, so they can share the diet of people. They have a high reproductive rate and a short average lifetime.

Controlling Pests

We can best control pests by recognizing how they fit their natural ecosystem and identifying their natural controlling factors. People have often assumed that the only way to control animal pests is with poisons, but there are limitations to this approach. Early poisons used in pest control were generally toxic to people and pets (see Chapter 12). Another problem is that reliance on one toxic compound can cause a species to develop resistance, which can lead to rebound—that is, an increase in the pest once again. If a pesticide is used once and spread widely, it will greatly reduce the population of the pest. However, when the pesticide's effectiveness is lost, the population can increase rapidly as long as habitat is suitable and food plentiful. This situation occurred when an attempt was made to control Norway rats in Baltimore.

One of the keys to controlling pests is to eliminate their habitats. For example, the best way to control rats is to reduce the amount of open garbage and eliminate areas to hide and nest. Common access areas used by rats are the spaces between walls and openings between buildings where pipes and cables enter. Houses can be constructed to restrict access by rats. In older buildings, areas of access can be sealed.

Summary

▪ As an urban society, we must recognize the city's relation to the environment. A city influences and is influenced by its environment and is an environment itself.

▪ Like any other life-supporting system, a city must maintain a flow of energy, provide necessary material resources, and have ways of removing wastes. These functions are accomplished through transportation and communication with outlying areas.

▪ Because cities depend on outside resources, they developed only when human ingenuity resulted in modern agriculture and thus excess food production. The history of cities divides into four stages: (1) the rise of towns; (2) the era of classic urban centers; (3) the period of industrial metropolises; and (4) the age of mass telecommunication, computers, and new forms of travel.

▪ Locations of cities are strongly influenced by environment. It is clear that cities are not located at random but in places of particular importance and environmental advantage. A city's site and situation are both important.

▪ A city creates an environment that is different from surrounding areas. Cities change local climate; they are commonly cloudier, warmer, and rainier than surrounding areas. In general, life in a city is riskier because of higher concentrations of pollutants and pollutant-related diseases.

▪ Cities favor certain animals and plants. Natural habitats in city parks and preserves will become more important as wilderness shrinks.

▪ Trees are an important part of urban environments. Cities, however, create stresses on trees, and attention must be paid to the condition of urban soils and the supply of water for trees.

▪ Cities can help to conserve biological diversity, providing habitat for some rare and endangered species.

▪ As the human population continues to increase, we can envision two futures: one in which people are dispersed widely throughout the countryside and cities are abandoned except by the poor; and another in which cities attract most of the human population, freeing much landscape for conservation of nature, production of natural resources, and public service functions of ecosystems.

Human Population

As the world's human population increases, we are becoming an increasingly urbanized species. Present trends indicate that in the future, most citizens of most nations will live in their country's single largest city. Thus, a concern with urban environments will become increasingly important.

Sustainability

Cities contain the seeds of their own destruction: The very artificiality of a city gives its inhabitants the sense that they are independent of their surrounding environment. But the opposite is the case: The more artificial a city, the more it depends on its surrounding environment for resources and the more susceptible it becomes to major disasters, unless this susceptibility is recognized and planned for. The keys to sustainable cities are an ecosystem approach to urban planning and a concern with the aesthetics of urban environments.

Cities depend on the sustainability of all renewable resources. Cities greatly affect their surrounding environments. Urban pollution of rivers that flow into an ocean can affect the sustainability of fish and fisheries. Urban sprawl can have destructive effects on endangered habitats and ecosystems, including wetlands. At the same time, cities designed to support vegetation and some wildlife can contribute to the sustainability of nature.

Global Perspective

The great urban centers of the world produce global effects. As an example, because people are concentrated in cities and because many cities are located at the mouths of rivers, most major river estuaries of the world are severely polluted.

Urban World

The primary message of this chapter is that Earth is becoming urbanized and that environmental science must deal more and more with urban issues.

People and Nature

It has been a modern tendency to focus environmental conservation efforts on wilderness, large parks, and preserves outside of cities. Meanwhile, city environments have been allowed to decay. As the world becomes increasingly urbanized, a change in values is necessary. Conservation of biological diversity requires that we assign greater value to urban environments. The more pleasant city environments are, and the more recreation people can find in them, the less pressure there will be on the countryside.

Science and Values

Modern environmental sciences tell us much that we can do to improve the environments of cities and the effects of cities on the surrounding environments. What we choose to do with this knowledge depends on our values. Scientific information can suggest new options and choices, and we can select among these for the future of our cities, depending on our values.

Key Terms

city planning **631**
fall line **628**

garden city **633**
greenbelt **633**

made lands **635**
site **626**

situation **626**

Study Questions

1. Should we try to save New Orleans or just give up and move the port at the mouth of the Mississippi River elsewhere? Explain your answer in terms of environment and economics.

2. Which of the following cities are most likely to become ghost towns in the next 100 years? In answering this question, make use of your knowledge of changes in resources, transportation, and communications.
   a. Honolulu, Hawaii
   b. Fairbanks, Alaska
   c. Juneau, Alaska
   d. Savannah, Georgia
   e. Phoenix, Arizona

3. Some futurists picture a world that is one giant biospheric city. Is this possible? If so, under what conditions?

4. The ancient Greeks said that a city should have no more people than the number that can hear the sound of a single voice. Would you apply this rule today? If not, how would you decide to plan the size of a city?

5. You are the manager of Central Park in New York City and receive the following two offers. Which would you approve? Explain your reasons.
   a. A gift of $1 billion to plant trees from all the eastern states.
   b. A gift of $1 billion to set aside half the park to be forever untouched, thus producing an urban wilderness.

6. Your state asks you to locate and plan a new town. The purpose of the town is to house people who will work at a wind farm—a large area of many windmills, all linked to produce electricity. You must first locate the site for the wind farm and then plan the town. How would you proceed? What factors would you take into account?

7. Visit your town center. What changes, if any, would make better use of the environmental location? How could the area be made more livable?

8. In what ways does air travel alter the location of cities? The value of land within a city?

9. You are put in charge of ridding your city's parks of slugs, which eat up the vegetable gardens rented to residents. How would you approach controlling this pest?

10. It is popular to suggest that in the information age people can work at home and live in the suburbs and the countryside, so cities are no longer necessary. List five arguments for and five arguments against this point of view.

Further Reading

Beveridge, C. E., and P. Rocheleau. 1995. *Frederick Law Olmsted: Designing the American Landscape.* New York: Rizzoli International. The most important analysis of the work of the father of landscape architecture.

Howard, E. 1965. *Garden Cities of Tomorrow* (reprint). Cambridge, Mass.: MIT Press. A classic work of the nineteenth century that has influenced modern city design, as in Garden City, New Jersey, and Greenbelt, Maryland. It presents a methodology for designing cities with the inclusion of parks, parkways, and private gardens.

McHarg, I. L. 1995. *Design with Nature.* New York: Wiley. A classic book about cities and environment.

Ndubisi, F. 2002. *Ecological Planning: A Historical and Comparative Synthesis.* Baltimore: The Johns Hopkins University Press. An important discussion of an ecological approach to cities.

CHAPTER 30

Waste Management

New York City sanitation crew cleaning up after Yankees' victory parade in 2000.

Learning Objectives

The waste management concept of "dilute and disperse" (for example, dumping waste into a river) is a holdover from our frontier days, when we mistakenly believed land and water to be limitless resources. We next attempted to "concentrate and contain" waste in disposal sites—a practice that also proved to pollute land, air, and water resources. We are now focusing on managing materials to eliminate waste. Finally, we are getting it right! After reading this chapter, you should understand:

- What the advantages and disadvantages are of each of the major methods that constitute integrated waste management.

- How the physical and hydrologic conditions at a site affect its suitability as a landfill.

- What multiple barriers for landfills are, and how landfill sites can be monitored.

- Why management of hazardous chemical waste is one of our most serious environmental concerns.

- What the various methods are of managing hazardous chemical waste.

- What the major pathways are by which hazardous waste from a disposal site can enter the environment.

- What problems are related to ocean dumping, and why these problems are likely to persist for some time.

"e-waste": A Growing Environmental Problem

Hundreds of millions of computers and other electronic devices, such as cell phones, ipods, televisions, and computer games, are discarded every year. The average life of a computer is about three years, and it is not constructed with recycling in mind.

When we take **e-waste** to a location where computers are turned in, we assume that they will be handled properly and not cause environmental problems. It is becoming clear that this is too often far from what happens. The United States, which helped start the technology revolution, produces most of the e-waste. Where and how it is eventually disposed of may cause serious environmental problems. Electronic waste includes the plastic housing for computers that, when burned, may produce toxic material. Computer parts also include small amounts of heavy metals, including gold, tin, copper, cadmium, and mercury, that are harmful and toxic, or may cause cancer if breathed in, ingested, or absorbed through the skin. At present, many millions of computers are disposed by what is billed as recycling. However, in the United States the Environmental Protection Agency has no official process to ensure that this e-waste doesn't cause future problems. In fact, most of these computers are being exported under the label of recycling to countries such as Nigeria and China. China's largest e-waste facility is in Guiyu, which is located near Hong Kong. People in the Guiyu area process more than one million tons of e-waste each year with little thought to the workers' health or regulations concerning the potential toxicity of the material they are handling (Figure 30.1). In the United States, computers cannot be recycled profitably without charging the people who dump them a hefty fee. Even with that, many U.S. firms ship their e-waste out of the country where greater profits are possible. The revenue to the Guiyu area is about a million dollars per year, and so the central government is resistant to regulate the activity. The problem is that workers at the locations where computers are disassembled may not be aware of the toxic nature of some of the materials they are working with and thus have a hazardous occupation. Altogether, in the Guiyu area there are more than 5,000 family-run facilities that specialize in scavenging the e-waste for raw materials. While doing this, they are exposing themselves to a variety of toxins and potential health problems.

Figure 30.1 ■ e-waste being processed in China—a hazardous occupation.

To date, the United States has not taken a proactive stance to try to regulate the computer industry so that less waste is produced. In fact, the United States is the only major nation that did not ratify an international agreement that restricts and bans exports of hazardous e-waste.[1]

Present practices concerning our handling of e-waste are not sustainable. Our value of a quality environment should include the safe handling and recycling of e-waste. Hopefully, that is the path we will take in the future. There are positive signs as some companies are now processing e-waste to reclaim metals such as gold and silver. Others are designing computers that use less toxic materials and are easier to recycle .The European Union is taking a leadership role in requiring more responsible management of e-waste.

> *Our failure to manage e-waste reminds us that we have failed in the past 50 years to move from a throwaway, waste-oriented society to a society that sustains natural resources through improved materials management. In some cases we are moving in that direction by producing less waste and recycling more. With this in mind, we introduce in this chapter concepts of waste management applied to urban waste, hazardous chemical waste, and waste in the marine environment.*

30.1 Early Concepts of Waste Disposal

During the first century of the Industrial Revolution, the volume of waste produced in the United States was relatively small, and waste could be managed through the concept of "dilute and disperse." Factories were located near rivers because the water provided a number of benefits, including easy transport of materials by boat, sufficient water for processing and cooling, and easy disposal of waste into the river. With few factories and a sparse population, dilute and disperse was sufficient to remove the waste from the immediate environment.[2]

As industrial and urban areas expanded, the concept of dilute and disperse became inadequate, and a new concept, known as "concentrate and contain," came into use. It has become apparent, however, that containment was, and is, not always achieved. Containers, whether simple trenches excavated in the ground or metal drums and tanks, may leak or break and allow waste to escape. Health hazards resulting from past waste disposal practices have led to the present situation, in which many people have little confidence in government or industry to preserve and protect public health.[3]

In the United States, as well as in many other parts of the world, people are facing a serious solid-waste disposal problem. The problem exists because we are producing a great deal of waste and the acceptable space for permanent disposal is limited. It has been estimated that within the next few years approximately half the cities in the United States may run out of landfill space. Philadelphia, for example, is essentially out of landfill space now and is bargaining with other states on a monthly or yearly basis to dispose of its trash; the Los Angeles area has landfill space for only about 10 more years.

To say we are actually running out of space for landfills isn't altogether accurate. Land used for landfills is minute compared to the land area of the United States. Rather, existing sites are being filled, and it is difficult to site new landfills. After all, no one wants to live near a waste-disposal site, be it a sanitary landfill for municipal waste, an incinerator that burns urban waste, or a hazardous-waste disposal operation for chemical materials. This attitude is widely known as NIMBY ("not in my backyard").

Another major limiting factor is the cost of disposal. A decade or so ago, the cost of disposal of 1 metric ton of urban refuse was approximately $5 to $10. Today, the average cost is about $40; some cities, such as Philadelphia, pay as much as $75 per metric ton for waste disposal.[3,4] These costs are only a small part of waste management expenditures. Disposal or treatment of liquid and solid waste costs about $20 billion every year. It is one of our most costly environmental expenditures.[5]

30.2 Modern Trends

The environmentally correct concept with respect to waste management is to consider wastes as resources out of place. Although we may not soon be able to reuse and recycle all waste, it seems apparent that the increasing cost of raw materials, energy, transportation, and land will make it financially feasible to reuse and recycle more resources. Moving toward this objective is moving toward an environmental view that there is no such thing as waste. Under this concept, waste would not exist because it would not be produced or, if produced, would be a resource to be used again. This concept is referred to as the "zero waste" movement.

Zero waste is the essence of what is known as **industrial ecology**, the study of relationships among industrial systems and their links to natural systems. Under the principles of industrial ecology, our industrial society would function much as a natural ecosystem functions. Waste from one part of the system would be a resource for another part (see A Closer Look 30.1).[6]

The concept of zero waste production was until recently considered unreasonable in the waste management arena. The concept, however, is catching on. The city of Canberra, Australia, may be the first community to propose a plan to have zero waste, a goal it hopes to reach by 2010. Thousands

Industrial Ecology

As noted in the text, industrial ecology can be defined as the study of relationships among industrial systems and their linkages to natural systems. Industrial ecology and sustainable development are often linked. In part, this results from the analogy drawn between industrial ecology and natural ecology, with its focus on ecosystems. Just as sustaining life on Earth is a function of ecosystems, sustainability in our industrial society is a function of industrial ecology.[6]

To understand the concept of waste according to industrial ecology, consider waste in natural ecosystems. What might be thought of as waste in one part of an ecosystem is often a resource for another part. For example, the elephant eats and produces waste that becomes a resource for the dung beetle.

Now consider an example of industrial ecology. Suppose that there exists a coal-burning power plant that produces electricity for a town. Waste from producing the power includes ash from the coal, exhaust heat, and products from combustion that might go out a smokestack, including carbon dioxide, sulfur dioxide, and heat. For our hypothetical power plant and surrounding town, let's assume that the waste heat is used to warm homes and provide heat for industrial activities. Sulfur dioxide is removed from the system by scrubbing to produce gypsum, which is the major constituent of wallboard used in construction. Carbon dioxide is sequestered and used in local greenhouses along with waste heat to force and prolong the growing cycle of plants. We are left with

the ash from the coal, which is used for road surfacing.

This hypothetical scenario is being partly played out at several power plants today. It is expected that, as principles of industrial ecology are better understood and applied, we will see more applications in a variety of industries as well as in agriculture.

As we attempt to produce sustainable economic development on local, regional, and global scales, however, we need to acknowledge that industrial ecology based on application of science and technology alone is not adequate to achieve sustainability of global systems. We also need to clarify our values. Science can provide potential solutions to problems we face; however, which solutions we choose will reflect our values. That is one of the key themes of this book.

---

of kilometers away, in the Netherlands, a national waste reduction goal of 70 to 90% has been set. How this is to be accomplished is not entirely clear; but a large part of the planning involves taxation of waste in all its various forms, from emissions from smokestacks to solids delivered to landfills. Already in the Netherlands, discharges of heavy metals to waterways have been nearly eliminated by implementation of pollution taxes. At the household level, the government is considering programs—known as "pay as you throw"—in which people are charged by the volume of waste they produce. As disposal of waste, including household waste, is taxed, people produce less waste.[7]

Of particular importance to waste management is the growing awareness that many of our waste management programs involve moving waste from one site to another and not really managing it. For example, waste from urban areas may be placed in landfills; but eventually these landfills may cause new problems by producing methane gas or noxious liquids that leak from the site and contaminate the surrounding areas. Managed properly, however, methane produced from landfills is a resource that can be burned as a fuel (an example of industrial ecology).

Concepts involving our use of materials and the waste we produce are changing. Previous notions of waste disposal are no longer acceptable, and we are rethinking how we deal with

materials, with the objective of eliminating the concept of waste entirely. In this way, we can reduce the consumption of virgin materials, which depletes our environment, and can live within our environment more sustainably.[7]

30.3 Integrated Waste Management

The dominant concept today in managing waste is known as **integrated waste management (IWM)**, which is best defined as a set of management alternatives that includes *reuse, source reduction, recycling, composting, landfill, and incineration.*[3]

Reduce, Reuse, Recycle

The three R's of IWM are **reduce, reuse,** and **recycle**. The ultimate objective of the three R's is to reduce the amount of urban and other waste that must be disposed of in landfills, incinerators, and other waste management facilities. Study of the *waste stream* (the waste produced) in areas that utilize IWM technology suggests that the amount (by weight) of urban refuse disposed of in landfills or incinerated can be reduced by at least 50% and perhaps as much as 70%. A 50% reduction by weight of urban waste could be facilitated by:[3]

- Better design of packaging to reduce waste, an element of source reduction (10% reduction).
- Large-scale composting programs (10% reduction).
- Establishment of recycling programs (30% reduction).

As this list indicates, recycling is a major player in the reduction of the urban waste stream. Today in the United States we recycle about 30% of our total municipal solid waste, up from 10% 25 years ago. This amounts to 93% of automobile batteries, 60% of steel cans, 48% of paper and cardboard, 44% of alluminum cans, 36% of tires, 32% of plastic milk bottles, 25% of plastic soft drink containers, and 22% of glass containers.[8] This is encouraging news. Can recycling in fact reduce the waste stream by 50%? Recent work suggests that the 50% goal is reasonable. In fact, it has been reached in some parts of the United States. The potential upper limit for recycling is considerably higher. It is estimated that as much as 80 to 90% of the U.S. waste stream might be recovered through what is known as intensive recycling.[9] A pilot study involving 100 families in East Hampton, New York, achieved a level of 84%. More realistic for many communities is partial recycling, which targets a specified number of materials, such as glass, aluminum cans, plastic, organic material, and newsprint. Partial recycling can provide a significant reduction, and in many places it is approaching or even exceeding 50%.[10,11]

Public Support for Recycling

An encouraging sign associated with public support for the environment is an increase in the willingness of industry and business to support recycling on a variety of scales. For example, fast-food restaurants are using less packaging for their products and providing on-site bins for recycling used paper and plastic. Grocery stores are encouraging the recycling of plastic and paper bags by providing bins for their collection and recycling. Some food stores offer inexpensive canvas shopping bags to people who prefer them over disposable plastic and paper bags. Companies are redesigning products so that they can be more easily disassembled after use and the various parts recycled. As this concept catches on, small appliances, such as electric frying pans and toasters, may be recycled rather than ending up in landfills. The automobile industry is also responding by designing automobiles with coded parts so that they can be more easily disassembled (by professional recyclers) and recycled, rather than left to become rusting eyesores in junkyards.

On the consumer front, people are now more likely to purchase products that can be recycled or that come in containers that are more easily recycled or composted. Many consumers have purchased small home appliances that crush bottles and aluminum cans, reducing their volume and facilitating recycling. The entire arena is rapidly changing, and innovations and opportunities will undoubtedly continue.

Markets for Recycled Products

As with many other environmental solutions, implementing the IWM concept successfully can be a complex undertaking. In some communities where recycling has been successfully implemented, it has resulted in glutted markets for the recycled products, which has sometimes required temporary stockpiling or suspension of recycling of some items. It is apparent that if recycling is to be successful, markets and processing facilities will also have to be developed to ensure that recycling is a sound financial venture as well as an important part of IWM.

Recycling of Human Waste

The use of human waste or "night soil" on croplands is an ancient practice. In Asia, recycling of human waste has a long history. Chinese agriculture was sustained for thousands of years through collection of human waste, which was spread over agricultural fields. The practice spread, and by the early twentieth century, land application of sewage was one of the primary methods of disposal in many metropolitan areas in countries including Mexico, Australia, and the United States.[12] These early uses of human waste for agriculture occasionally spread infectious diseases through agents, including bacteria, viruses, and parasites, applied to crops along with the waste. Today, with the globalization of agriculture, we still see occasional warnings and outbreaks of disease from contaminated vegetables (see Chapter 22).

One of the major problems of recycling human waste is that, along with human waste, thousands of chemicals and metals flow through our modern waste stream. Even garden waste that is composted may contain harmful chemicals such as pesticides.

Heavy metals, petroleum products, industrial solvents, and pesticides may end up in our wastewater collection systems and sewage treatment plants. Because many toxic materials are likely to be present with the waste, we must be very skeptical of utilizing sewage sludge for land application. Of course, the contents of sewage sludge vary from place to place and even from day to day. Nevertheless, studies have shown that high levels of toxic chemicals may be present in the sludge of cities or towns with industries that use toxic materials.[12] The good news is that fewer toxic materials end up at sewage treatment plants than several decades ago because many industries are now pretreating their waste to remove materials that previously contaminated wastewaters.

Discussions involving various federal, local, and other government agencies, as well as industries, have addressed the question of how much toxic material in the waste stream constitutes a problem. This is really not the correct question to ask, however. The goal should be that sewage sludge contain no toxic materials at all. A problem is that the sewer lines from urban homes are the same ones utilized by industry. As a result, conventional waste-disposal technology is unlikely to produce a form of sludge that is safe for us and other living things. A possible solution is to separate urban waste from industrial waste. A second possibility is to pretreat waste from industrial sources to remove hazardous components before they enter the wastewater stream. Many industries today pretreat waste, as mentioned earlier. Finally, some communities are considering smaller wastewater treatment facilities intended to treat waste from homes; the recycled waste would be used locally by farms. In the future, as the cost of oil, which is necessary to produce fertilizers, continues to rise, the age-old practice of recycling human waste may again be financially viable and necessary in many more places than today.[12]

30.4 Materials Management

The recycling option of IWM has been seriously attempted for over two decades. Recycling has been responsible for generating entire systems of waste management that have produced tens of thousands of jobs in the United States and reduced the amount of urban waste sent from homes to landfills by about one-half. Many firms have combined waste reduction with recycling which significantly reduced the waste they deliver to landfills. However, in spite of this success, IWM has been criticized for overemphasizing recycling and failing to effectively advance policies to prevent waste production.

Futuristic waste management has the goal of zero production of waste, consistent with the ideals of industrial ecology. This visionary goal will require more sustainable use of materials combined with resource conservation in what is being termed **materials management**. It is believed that the goal could be pursued in the following ways:[13]

- Eliminate subsidies for extraction of virgin materials such as minerals, oil, and timber.
- Establish "green building" incentives that encourage the use of recycled-content materials and products in new construction.
- Assess financial penalties for production that uses poor materials management practices.
- Provide financial incentives for industrial practices and products that benefit the environment by enhancing sustainability (for example, reducing waste production and using recycled materials).
- Increase the number of new jobs in the technology of reuse and recycling of resources.

Materials management in the United States today is beginning to have effects on where industries are located. For example, approximately 50% of the steel produced in the country now comes from scrap. As a result, new steel mills are no longer located where resources such as coal and iron ore are close by. New steel mills are located in a variety of places, from California to North Carolina and Nebraska;

their resource is the local supply of scrap steel. Because they are starting with scrap metal, the new industrial facilities use far less energy and cause much less pollution than older steel mills that produce their product from virgin iron ore.[14] Similarly, recycling of paper is changing where new paper mills are constructed. In the past, mills were built near forested areas where timber necessary for paper production was being logged. As a result of the recycling of voluminous amounts of paper, mills are now being built near cities where supplies of recycled paper exist. For example, in New Jersey there are 13 paper mills utilizing recycled paper and eight steel "mini-mills" producing steel from scrap metal. What is remarkable is that New Jersey has little forested land and no iron mines. Resources for the paper and steel mills come from materials already in use and exemplify the power of materials management.[14]

Although materials management is providing alternatives to waste disposal, we still need to deal with the solid waste we produce in both urban and rural areas. We now discuss solid-waste management in more detail.

30.5 Solid-Waste Management

Solid-waste management continues to be a problem in the United States as well as other parts of the world. In many areas, particularly in developing countries, waste management practices are inadequate. These practices, which include poorly controlled open dumps and illegal roadside dumping, can spoil scenic resources, pollute soil and water resources, and produce potential health hazards.

Illegal dumping is a social problem as much as a physical one, because many people are simply disposing of their waste as inexpensively and as quickly as possible. They may not see dumping their garbage or trash as an environmental problem. If nothing else, this is a tremendous waste of resources; much of what is dumped could be recycled or reused. In areas where illegal dumping has been reduced, the keys have been awareness, education, and alternatives. Environmental problems of unsafe, unsanitary dumping of waste are made known to people through education programs, and funds are provided for cleanup and inexpensive collection and recycling of trash at sites of origin.

We look next at the composition of solid waste in the United States. Then, in the remainder of this section, we describe specific disposal methods: on-site disposal, composting, incineration, open dumps, and sanitary landfills.

Composition of Solid Waste

The general composition of solid waste before recycling likely to end up at a disposal site in the United States is shown in Figure 30.2. It is no surprise that paper is by far the most abundant of these solid wastes; however, this is only an average content, and considerable variation can be

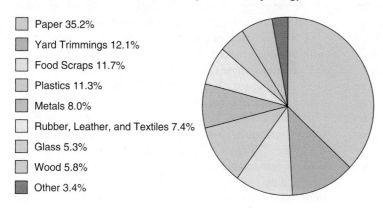

**2003 Total Waste Generation–
236 Million Tons (before recycling)**

Paper 35.2%

Yard Trimmings 12.1%

Food Scraps 11.7%

Plastics 11.3%

Metals 8.0%

Rubber, Leather, and Textiles 7.4%

Glass 5.3%

Wood 5.8%

Other 3.4%

Figure 30.2 ■ Composition of U.S. municipal solid waste (by weight) for 2003. [*Source:* U.S. Environmental Protection Agency, Municipal/solid waste. Accessed 4/21/06 at www.epa-gov.]

expected based on factors such as land use, economic base, industrial activity, climate, and season of the year.

Infectious wastes from hospitals and clinics sometimes make their way to disposal sites, where they can create potential health problems if they have not been properly sterilized before disposal. Some hospitals have facilities to incinerate such wastes, and incineration is probably the surest way to manage infectious medical waste. In addition, in urban areas, a large amount of toxic material may end up at disposal sites; as a result, many older urban landfills are now being considered hazardous-waste sites that will require costly cleanup.

People have many misconceptions about our waste stream.[15] There has been much publicity concerning urban waste associated with fast-food packaging, polystyrene foam, and disposable diapers. Therefore, many people assume that these products make up a large percentage of the total waste stream and are responsible for the rapid filling of landfills. However, excavations into modern landfills using archeological tools have cleared up some misconceptions concerning these items. We now know that fast-food packaging accounts for only about 0.25% of the average landfill; disposable diapers, approximately 0.8%; and polystyrene products, about 0.9%.[16] Paper, as mentioned, is the major constituent in landfills (perhaps as much as 50% by volume and 40% by weight). The largest single item is newsprint, which accounts for as much as 18% by volume.[15] Newsprint is one of the major items targeted for recycling because big environmental dividends can be expected. However (and this is a value judgment), the need to deal with the major contributors does not mean that we need not reduce our use of disposable diapers, polystyrene, and other paper products. In addition to creating a need for disposal, these products are made from resources that might be better managed.

On-Site Disposal

We turn now to specific methods of solid-waste disposal. A common on-site disposal method in urban areas is the mechanical grinding of kitchen food waste. Garbage-disposal

devices are installed in the wastewater pipe system at the kitchen sink, and the garbage is ground and flushed into the sewer system. This effectively reduces the amount of handling and quickly removes food waste. Final disposal is transferred to sewage-treatment plants, where solids remaining as sewage sludge still must be disposed of.[17]

Composting

Composting is a biochemical process in which organic materials such as lawn clippings and kitchen scraps decompose to a rich, soil-like material. The process involves rapid partial decomposition of moist solid organic waste by aerobic organisms. Although simple backyard compost piles may come to mind, as a waste management option, large-scale composting is generally carried out in the controlled environment of mechanical digesters.[17] This technique is popular in Europe and Asia, where intense farming creates a demand for the compost.[17] A major drawback of composting is the necessity of separating organic material from other waste. Therefore, it is probably economically advantageous only where organic material is collected separately from other waste. Composting plant debris previously treated with herbicides may produce a compost toxic to some plants. Nevertheless, composting is an important component of IWM, and its contribution continues to grow.

Incineration

In **incineration**, combustible waste is burned at temperatures high enough (900°–1,000°C, or 1,650°–1,830°F) to consume all combustible material, leaving only ash and noncombustibles to dispose of in a landfill. Under ideal conditions, incineration may reduce the volume of waste by 75 to 95%.[17] In practice, however, the actual decrease in volume is closer to 50% because of maintenance problems as well as waste supply problems. Besides reducing a large volume of combustible waste to a much smaller volume of ash, incineration has another advantage in that the process of incineration can be used to supplement other fuels and generate electrical power.

Incineration of urban waste is not necessarily a clean process. Incineration may produce air pollution and toxic ash. For example, incineration in the United States apparently is a significant source of environmental dioxin, a carcinogenic toxin (see Chapter 15).[18] Smokestacks from incinerators may emit oxides of nitrogen and sulfur, which lead to acid rain; heavy metals such as lead, cadmium, and mercury; and carbon dioxide, which is related to global warming.

In modern incineration facilities, smokestacks are fitted with special devices to trap pollutants; but the process of pollutant abatement is expensive. Furthermore, the plants themselves are expensive, and government subsidization may be needed to aid in their establishment. Recent evaluation of the urban waste stream suggests that with an investment of $8 billion, a sufficient number of incinerators could be constructed in the United States to burn approximately 25% of the solid waste that is generated. However, a similar investment in source reduction, recycling, and composting could result in diversion from landfills of as much as 75% of the nation's urban waste stream.[9]

The economic viability of incinerators depends on revenue from the sale of the energy produced by burning the waste. As recycling and composting are increased, they will compete with incineration for their portion of the waste stream, and sufficient waste (fuel) to generate a profit from incineration may not be available. The main conclusion that can be drawn based on IWM principles is that a combination of reusing, recycling, and composting could reduce the volume of waste requiring disposal at a landfill by at least as much as incineration.[9]

Open Dumps

In the past, solid waste was often disposed of in open dumps, where the refuse was piled up without being covered or otherwise protected. Thousands of open dumps have been closed in recent years, and new open dumps are banned in the United States and many other countries. Nevertheless, many are still being used worldwide (Figure 30.3).

Dumps have been located wherever land is available, without regard to safety, health hazards, or aesthetic degradation. Common sites are abandoned mines and quarries, where gravel and stone have been removed (sometimes by ancient civilizations); natural low areas, such as swamps or floodplains; and hillside areas above or below towns. The waste is often piled as high as equipment allows. In some instances, the refuse is ignited and allowed to burn. In others, the refuse is periodically leveled and compacted.[17]

As a general rule, open dumps create a nuisance by being unsightly, providing breeding grounds for pests, creating a health hazard, polluting the air, and sometimes polluting groundwater and surface water. Fortunately, open dumps are giving way to the better-planned and managed sanitary landfills.

Figure 30.3 ■ Urban garbage dump in Rio de Janeiro, Brazil. At this site, people are going through the waste and recycling materials that can be reused or resold.

Sanitary Landfills

A **sanitary landfill** (also called a municipal solid waste landfill) is designed to concentrate and contain refuse without creating a nuisance or hazard to public health or safety. The idea is to confine the waste to the smallest practical area, reduce it to the smallest practical volume, and cover it with a layer of compacted soil at the end of each day of operation, or more frequently if necessary. Covering the waste is what makes the landfill sanitary. The compacted layer restricts (but does not eliminate) continued access to the waste by insects, rodents, and other animals, such as seagulls. It also isolates the refuse, minimizing the amount of surface water seeping into and gas escaping from the waste.[19]

Leachate

The most significant hazard from a sanitary landfill is pollution of groundwater or surface water. If waste buried in a landfill comes into contact with water percolating down from the surface or with groundwater moving laterally through the refuse, **leachate**—noxious, mineralized liquid capable of transporting bacterial pollutants—is produced.[20] For example, two landfills dating from the 1930s and 1940s in Long Island, New York, have produced subsurface leachate trails (plumes) several hundred meters wide that have migrated kilometers from the disposal site. The nature and strength of the leachate produced at a disposal site depends on the composition of the waste, the amount of water that infiltrates or moves through the waste, and the length of time that infiltrated water is in contact with the refuse.[17]

Site Selection

The siting of a sanitary landfill is very important. A number of factors must be taken into consideration, including topography, location of the groundwater table,

Environmental Justice: Demographics of Hazardous Waste

An emerging field in the social sciences known as environmental justice focuses on investigating social issues related to the siting of facilities that many people find objectionable on environmental grounds. These facilities include sanitary landfills and other waste facilities, as well as chemical plants; people's concerns include risk to nearby residents in case of accidental spills, fires, explosions, or illegal discharge of waste or chemicals. In particular, environmental justice addresses the potential for inequitable impacts from the environmental hazards related to the waste facility and chemical industries.[24]

A study of 82 permitted hazardous-waste treatment, storage, and disposal fa-cilities in Los Angeles County, California, concluded that the communities most likely to contain such facilities are those located near industrial activity. These communities generally have a large residential concentration of working-class people of color. It appears that sites are chosen where local resistance is expected to be minimal or where it is perceived that the land has little value.

By and large, people living near hazardous-waste facilities in Los Angeles County are members of racial and ethnic minorities below the county average in terms of income, education, employment, and voting participation. These communities also generally lag behind the county mean in monetary values of residential homes and rental properties. In addition, industrial land use in these areas is often four to ten times the county average.

In summary, the study identified two dominant factors in location of waste facilities: the percentage of minority people living in the area and industrial land use. Of secondary importance was per-capita income. Race, then, appears to be a more important factor than income in locations of waste facilities. These findings in Los Angeles County suggest that people fighting for environmental justice have real cause for concern.[25]

---

amount of precipitation, type of soil and rock, and location of the disposal zone in the surface water and groundwater flow system. A favorable combination of climatic, hydrologic, and geologic conditions helps to ensure reasonable safety in containing the waste and its leachate.[21]

The best sites are in arid regions, where disposal conditions are relatively safe because little leachate is produced. In a humid environment, some leachate is always produced; therefore, an acceptable level of leachate production must be established to determine the most favorable sites in such environments. What is acceptable varies with local water use, regulations, and the ability of the natural hydrologic system to disperse, dilute, and otherwise degrade the leachate to harmless levels.

Elements of the most desirable site in a humid climate with moderate to abundant precipitation are shown in Figure 30.4. The waste is buried above the water table in relatively impermeable clay and silt soils, material through which water cannot easily move. Any leachate produced remains in the vicinity of the site and degrades by natural filtering action and chemical reactions between clay and leachate.[22,23]

The siting of waste-disposal facilities also involves important social considerations. Often, planners choose sites where they expect local resistance to be minimal or where they perceive land to have little value. Waste-disposal facilities are frequently located in areas where residents tend to have low socioeconomic status or belong to a particular racial or ethnic group. The study of social issues in siting waste facilities, chemical plants, and other such facilities is an emerging field known as **environmental justice**.[24] (See A Closer Look 30.2.)

Monitoring Pollution in Sanitary Landfills

Once a site is chosen for a sanitary landfill and before filling starts, monitoring the movement of groundwater should begin. **Monitoring** is accomplished by periodically taking samples of water and gas from specially designed monitoring wells. Monitoring the movement of leachate and gases should continue as long as there is any possibility of pollution. This procedure is particularly important after the site is completely filled and a final, permanent cover material is in place. Continued monitoring is neces-

Figure 30.4 ■ The most desirable landfill site in a humid environment. Waste is buried above the water table in a relatively impermeable environment. [*Source:* W. J. Schneider, *Hydraulic Implications of Solid-Waste Disposal*, U.S. Geological Survey Circular 601F, 1970.]

sary because a certain amount of settlement always occurs after a landfill is completed; and if small depressions form, surface water may collect, infiltrate, and produce leachate. Monitoring and proper maintenance of an abandoned landfill reduce its pollution potential.[19]

How Pollutants Can Enter the Environment from Sanitary Landfills

Pollutants from a solid-waste disposal site can enter the environment through as many as eight paths (Figure 30.5):[26]

1. Methane, ammonia, hydrogen sulfide, and nitrogen gases can be produced from compounds in the soil and the waste and can enter the atmosphere.

2. Heavy metals, such as lead, chromium, and iron, can be retained in the soil.

3. Soluble materials, such as chloride, nitrate, and sulfate, can readily pass through the waste and soil to the groundwater system.

4. Overland runoff can pick up leachate and transport it into streams and rivers.

5. Some plants (including crops) growing in the disposal area can selectively take up heavy metals and other toxic materials. These materials are passed up the food chain as people and animals eat the plants.

6. If plant residue from crops left in fields contains toxic substances, these substances return to the soil.

7. Streams and rivers may become contaminated by waste from groundwater seeping into the channel (3) or by surface runoff (4).

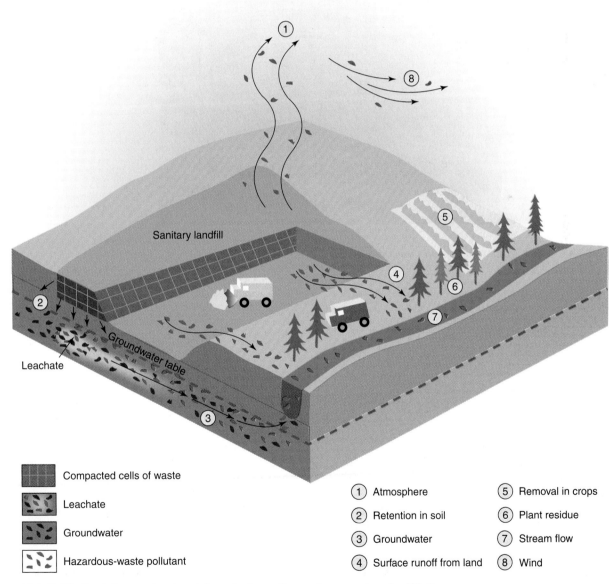

Figure 30.5 ■ Idealized diagram showing eight paths that pollutants from a sanitary landfill site may follow to enter the environment.

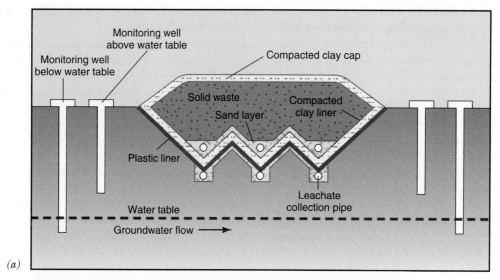

Figure 30.6 ■ (*a*) Idealized diagram of a solid-waste facility (sanitary landfill) illustrating multiple-barrier design, monitoring system, and leachate collection system. (*b*) Rock Creek landfill, Calaveras County, California, under construction. This municipal solid-waste landfill is underlain by a compacted clay liner (exposed light brown slope in the center left portion of the photograph). The black slopes, covered with gravel piles, overlie the compacted clay layer. These form a vapor barrier designed to keep moisture in the clay and help avoid cracking of the clay liner. The sinuous gray trench is lined with plastic and is part of the leachate collection system for the landfill. The excavated squared pond (upper part of photograph) is a leachate evaporation pond under construction. The landfill is also equipped with a system to monitor the vadose zone below the leachate collection system.

8. Toxic materials can be transported to other areas by the wind.

Modern sanitary landfills are engineered to include *multiple barriers*: clay and plastic liners to limit the movement of leachate; surface and subsurface drainage to collect leachate; systems to collect methane gas produced as waste decomposes; and groundwater monitoring to detect leaks of leachate below and adjacent to the landfill. A thorough monitoring program considers all eight possible paths by which pollutants enter the environment. In practice, however, seldom does monitoring include all pathways. It is particularly important to monitor the zone above the water table to identify potential pollution problems before they reach and contaminate groundwater resources, where correction would be very expensive. Figure 30.6 shows an idealized diagram of a landfill that utilizes the multiple-barrier approach and a photograph of a landfill site under construction.

Federal Legislation for Sanitary Landfills
New landfills opened in the United States after 1993 must comply with stricter requirements under the Resource Conservation and Recovery Act of 1980. The legislation is intended to strengthen and standardize design, operation, and monitoring of sanitary landfills. Landfills that cannot

Table 30.1 • Actions You Can Take to Reduce the Waste You Generate

Keep track of the waste you personally generate: Know how much waste you produce. This will make you conscious of how to reduce it.

Recycle as much as is possible and practical: Take your cans, glass, and paper to a recycling center or use curbside pickup. Take your hazardous materials such as batteries, cell phones, computers, paint, used oil, and solvents to a hazardous waste collection site.

Reduce packaging: Whenever possible buy your food items in bulk or concentrated form.

Use durable products: Choose automobiles, light bulbs, furniture, sports equipment, and tools that will last a longer time.

Reuse products: Some things may be used several times. For example, you can reuse boxes and shipping "bubble wrap" to ship packages.

Purchase products made from recycled material: Many bottles, cans, boxes, containers, cartons, carpets, clothing, floor tiles, and other products are made from recycled material. Select these whenever you can.

Purchase products designed for ease in recycling: Products as large as automobiles along with many other items are being designed with recycling in mind. Apply pressure to manufactures to produce items that can be easily recycled.

Source: Modified from U.S. Environmental Protection Agency. Accessed 4/21/06 at www.epa.gov.

comply with regulations face closure.

States may choose between two options:

1. Comply with federal standards.
2. Seek EPA approval of solid-waste management plans, which allows greater flexibility.

Provisions of federal standards include the following:

- Landfills may not be sited on floodplains, wetlands, earthquake zones, unstable land, or near airports (birds at sites are a hazard to aircraft).
- Landfills must have liners.
- Landfills must have a leachate collection system.
- Landfill operators must monitor groundwater for many specified toxic chemicals.
- Landfill operators must meet financial assurance criteria to ensure that monitoring continues for 30 years after the landfill is closed.

As mentioned, a state with EPA approval of its landfill program is allowed more flexibility:

- Groundwater monitoring may be suspended if the landfill operator can demonstrate that hazardous constituents are not migrating from the landfill.
- Alternative types of daily cover over the waste may be used.
- Alternative groundwater protection standards are allowed.
- Alternative schedules for documentation of groundwater monitoring are allowed.
- Under certain circumstances, landfills in wetlands and fault zones are allowed.
- Alternative financial assurance mechanisms are allowed.

Given the added flexibility, it appears advantageous for states to develop EPA-approved waste management plans.

Reducing the Waste You Produce

The average waste per person in the United States increased from about 1 kg (2.2 lb) in 1960 to 2 kg (4.5 lb) in 2003. This is an annual growth rate of about 1.5% per year and is not sustainable because the doubling time for waste production is only a few decades. The 236 million tons we produced in 2003 would be close to 500 million tons by 2050, and we are having disposal and management problems today. To reduce the waste you produce, you can start at your personal, household level. Table 30.1 lists some of the many ways you could reduce the waste you generate. What other ways can you think of?

30.6 Hazardous Waste

So far in this chapter, we have discussed integrated waste management and materials management for the everyday waste stream from homes and businesses. We now consider the important topic of hazardous waste.

Creation of new chemical compounds has proliferated in recent years. In the United States, approximately 1,000 new chemicals are marketed each year, and about 70,000 chemicals are currently on the market. Although many of these chemicals have been beneficial to people, approximately 35,000 chemicals used in the United States are classified as definitely or potentially hazardous to the health of people or ecosystems.

The United States currently produces about 700 million metric tons of hazardous chemical waste per year, referred to more commonly as **hazardous waste**. About 70% of the total is generated east of the Mississippi River, and about half of the total by weight is generated by chemical products industries, with the electronics industry and petroleum and coal products industries each contributing about 10%.[27, 28]

Another source of hazardous chemicals is buildings destroyed by events such as fires and hurricanes. Chemicals, such as paints, solvents, and pesticides that were stored in the buildings, may be released into the environment when debris from damaged buildings is burned or buried. As a result, collection of potentially hazardous chemicals following natural disasters is an important goal in managing hazardous materials.

In the mid 20th century, as much as half the total volume of hazardous waste produced in the United States was indiscriminately dumped.[28] Some hazardous waste was il-

Love Canal

The story of Love Canal is a well-known hazardous-waste horror story. In 1976, in a residential area near Niagara Falls, New York, trees and gardens began to die (Figure 30.7). Rubber on tennis shoes and bicycle tires disintegrated. Puddles of toxic substances began to ooze through the soil. A swimming pool popped from its foundation and floated in a bath of chemicals.

The story of Love Canal started in 1892 when a canal 8 km (5 mi) long was excavated by William Love as part of the development of an industrial park. The development didn't need the canal when inexpensive electricity arrived, and it was never completed. The canal remained unused for decades and became a dump for wastes. From 1920 to 1952, 20,000 tons of more than 80 chemicals were dumped in the canal. In 1953, the Hooker Chemical Company—which produced the insecticide DDT as well as an herbicide and chlorinated solvents and which had dumped chemicals into the canal—was pressured to donate the land to the city of Niagara Falls for $1. The city knew that chemical wastes were buried there, but no one expected any problems.[27] Eventually, several hundred homes and an elementary school were built on and near the site. For years, everything seemed fine. Then, in 1976–1977, heavy rains and snows triggered a number of events, making Love Canal a household word.

A study of the site identified many substances suspected of being carcinogens, including benzene, dioxin, dichlorethylene, and chloroform.

Although officials admitted that little was known about the impact of these chemicals, there was grave concern for people living in the area. Eventually, concern centered on alleged high rates of miscarriages, blood and liver abnormalities, birth defects, and chromosome damage. Although a study by New York health authorities suggested that no chemically caused health effects had been absolutely established,[29–31] the decision was made to clean up the site.

Cleanup of Love Canal demonstrates the technology for hazardous-waste treatment. The objective was to contain waste, stop migration of wastes through the groundwater flow system, and remove and treat contaminated soil and sediment in streambeds and sewers.[32] To minimize further contamination, the area was covered with 1 m (3.3 ft) of compacted clay and a polyethylene plastic cover to reduce infiltration of surface water. Water is inhibited from entering the site by specially designed barriers. These procedures greatly reduce subsurface seepage of water, and water that does seep out is collected and treated.[29–32]

By 1990, $275 million had been spent for cleanup and relocation projects at Love Canal, and in 1995 the chemical company agreed to pay an additional $129 million to reimborse cleanup costs.[33,34]

Homes in an adjacent area were purchased, and about 200 homes and a school had to be destroyed by the government. About 800 families were relocated and reimbursed for loss of their homes. The U.S. Environmental Protection Agency eventually declared the area clean, and about 280 remaining homes were sold.

Today the community around the canal is known as Black Creek Village, and many people live there.

Figure 30.7 ■ Aerial infrared photograph of the Love Canal area in New York State about 30 years ago. Healthy vegetation is bright red. The canal is a scar on the landscape where hazardous chemical waste rose to the surface, making "Love Canal" a household name for hazardous-waste problems.

legally dumped on public or private lands, a practice called midnight dumping. Buried drums of hazardous waste from illegal dumping have been discovered at hundreds of sites by contractors constructing buildings and roads. Cleanup of the waste has been costly and has delayed projects (see A Closer Look 30.3).[27]

In the United States, there are several tens of thousands of abandoned, waste disposal sites where past dumping was totally unregulated. Of these, probably up to 1,000 sites contain sufficient hazardous waste to be a threat to public health and the environment. For this reason, many scientists believe

management of hazardous chemical materials may be the most serious environmental problem in the United States.

Uncontrolled dumping of chemical waste has polluted soil and groundwater in several ways:

- Chemical waste may be stored in barrels, either stacked on the ground or buried. The barrels eventually corrode and leak, polluting surface water, soil, and groundwater.

- When liquid chemical waste is dumped into an unlined lagoon, contaminated water may percolate through soil and rock to the groundwater table.

Liquid chemical waste may be illegally dumped in deserted fields or even along roads.

Some sites pose particular dangers. The floodplain of a river, for example, is not an acceptable site for storing hazardous chemical waste. Yet that is exactly what occurred at a site on the floodplain of the River Severn near a village in one of the most scenic areas of England. Several fires at the site in 1999 were followed by a large fire of unknown origin on October 30, 2000. In that fire, approximately 200 tons of chemicals, including industrial solvents (xylene and toluene), cleaning solvents (methylenechloride), and various insecticides and pesticides, produced a fireball that rose into the night sky (Figure 30.8). The fire occurred during a rainstorm with wind gusts of hurricane strength. Toxic smoke and ash spread to nearby farmlands and villages, necessitating evacuation. People exposed to the smoke complained of a variety of symptoms, including headaches, stomachaches and vomiting, sore throats, coughs, and difficulty in breathing. Then, a few days later, on November 3, the site flooded (Figure 30.9). The floodwaters interfered with cleanup of the site after the fire and increased the risk of contamination of downstream areas by way of waterborne transport of hazardous chemical wastes. In one small village, contaminated floodwaters apparently inundated farm fields, gardens, and even homes.[35] Of course, the solution to this problem is to clean up the site and move waste storage facilities to a safer location.

30.7 Hazardous-Waste Legislation

The recognition in the 1970s that hazardous waste was a danger to people and the environment and that the waste was not being properly managed led to important federal legislation in the United States.

Resource Conservation and Recovery Act

Management of hazardous waste in the United States began in 1976 with passage of the Resource Conservation and Recovery Act (RCRA). At the heart of the act is identification of hazardous wastes and their life cycles. The idea was to issue guidelines and responsibilities to those who manufactured, transported, and disposed of hazardous waste. This is known as "cradle to grave" management. Regulations require stringent record keeping and reporting to verify that wastes do not present a public nuisance or a health problem.

RCRA applies to solid, semisolid, liquid, and gaseous hazardous wastes. A waste is hazardous if its concentration, volume, or infectious nature may contribute to serious disease or death or if it poses a significant hazard to people and the environment as a result of improper management (storage, transport, or disposal).[27] The act classifies hazardous wastes in several categories: materials highly toxic to people and other living things; wastes that may ignite when exposed to air; extremely corrosive wastes; and re-

Figure 30.8 ■ On October 30, 2000, fire ravaged a site on the floodplain of the River Severn in England where hazardous waste was being stored. Approximately 200 tons of chemicals burned.

Figure 30.9 ■ Flooding on November 3, 2000, followed a large fire at a hazardous-waste storage site on the floodplain of the River Severn in England (see Figure 30.8).

active unstable wastes that are explosive or generate toxic gases or fumes when mixed with water.

Comprehensive Environmental Response, Compensation, and Liability Act

In 1980, Congress passed the Comprehensive Environmental Response, Compensation, and Liability Act (CERCLA).

The act defined policies and procedures for release of hazardous substances into the environment (for example, landfill regulations). It also mandated development of a list of the sites where hazardous substances were likely to or already had produced the most serious environmental problems and established a revolving fund (*Superfund*) to clean up the worst abandoned hazardous-waste sites. In 1984 and 1986, CERCLA was strengthened by amendments that made the following changes:

- Improved and tightened standards for disposal and cleanup of hazardous waste (for example, requiring double liners, monitoring landfills).
- Banned land disposal of certain hazardous chemicals, including dioxins, polychlorinated biphenyls (PCBs), and most solvents.
- Initiated a timetable for phasing out disposal of all untreated liquid hazardous waste in landfills or surface impoundments.
- Increased the size of the Superfund. The fund was allocated about $8.5 billion in 1986. Congress approved another $5.1 billion for fiscal year 1998, which almost doubled the Superfund budget.[36]

Superfund has experienced significant management problems, and cleanup efforts are far behind schedule. Unfortunately, the funds available are not sufficient to pay for decontamination of all targeted sites. Furthermore, there is concern that present technology is not sufficient to treat all abandoned waste-disposal sites; it may be necessary to simply try to confine waste to those sites until better disposal methods are developed. It seems apparent that abandoned disposal sites are likely to persist as problems for some time to come.

Other Legislation

Federal legislation has changed the ways in which real estate business is conducted. For example, there are provisions by which property owners may be held liable for costly cleanup of hazardous waste present on their property even if they did not directly cause the problem. As a result, banks and other lending institutions might be held liable for release of hazardous materials by their tenants.

The Superfund Amendment and Reauthorization Act (SARA) of 1986 permits a possible defense against such liability, provided the property owner has completed an **environmental audit** prior to the purchase of the property. Such an audit involves studying past land use at the site, usually determined from analyzing old maps, aerial photographs, and reports. It may also involve drilling and sampling groundwater and soil to determine if hazardous materials are present. Environmental audits are now completed on a routine basis prior to purchase of property for development.

SARA legislation also required that certain industries report all releases of hazardous materials, and a list of compa-

nies releasing hazardous substances—known as the "Toxic 500"—became public. No property owner or industry wants to be on such a list, and the list is thought to have placed some pressure on industries identified as polluters to develop safer ways of handling hazardous materials.[37]

In 1990, the U.S. Congress reauthorized hazardous-waste control legislation. Priorities include:

- Establishing who is responsible (liable) for existing hazardous-waste problems.
- When necessary, assisting in or providing funding for cleanup at sites identified as having a hazardous-waste problem.
- Providing measures whereby people who suffer damages from the release of hazardous materials are compensated.
- Improving the required standards for disposal and cleanup of hazardous waste.

30.8 Hazardous-Waste Management: Land Disposal

Management of hazardous chemical waste involves several options, including recycling, on-site processing to recover by-products with commercial value, microbial breakdown, chemical stabilization, high-temperature decomposition, incineration, and disposal by secure landfill or deep-well injection. A number of technological advances have been made in toxic-waste management; as land disposal becomes more expensive, the recent trend toward on-site treatment is likely to continue. However, on-site treatment will not eliminate all hazardous chemical waste; disposal of some waste will remain necessary.

Table 30.2 compares hazardous-waste reduction technologies for treatment and disposal. Notice that all available technologies cause some environmental disruption. There is no simple solution for all waste management issues. In this section, we consider land disposal of hazardous waste; in the following section, we discuss alternative approaches.

Secure Landfill

A **secure landfill** for hazardous waste is designed to confine the waste to a particular location, control the leachate that drains from the waste, collect and treat the leachate, and detect possible leaks. This type of landfill is similar to the modern sanitary landfill; it is an extension of the sanitary landfill for urban waste. Because in recent years it has become apparent that urban waste contains much hazardous material, the design of sanitary landfills and that of secure landfills for hazardous waste have converged to some extent.

Design of a secure landfill is shown in Figure 30.10. A dike and liner (made of clay or other impervious material,

Table 30.2 • Comparison of Hazard Reduction Technologies

Parameter Compared	Disposal		Treatment			
	Landfills and Impoundments	Injection Wells	Incineration and Other Thermal Destruction	High-temperature Decomposition[a]	Chemical Stabilization	Microbial Breakdown
Effectiveness: how well it contains or destroys hazardous characteristics	Low for volatiles, high for unsoluble solids	High, for waste compatible with the disposal environment	High	High for many chemicals	High for many metals	High for many metals and some organic waste such as oil
Reliability issues	Siting, construction, and operation Uncertainties: long-term integrity and cover	Site history and geology, well depth, construction, and operation	Monitoring uncertainties with respect to high degree of DRE: surrogate measures, PICs, incinerability[b]	Mobile units; on-site treatment avoids hauling risks Operational simplicity	Some inorganics still soluble Uncertain leachate production	Monitoring uncertainties during construction and operation
Environment media most affected	Surface water and groundwater	Surface water and groundwater	Air	Air	Groundwater	Soil, groundwater
Least compatible wastes[c]	Highly toxic, persistent chemicals	Reactive; corrosive; highly toxic, mobile, and persistent	Highly toxic organics, high heavy-metal concentration	Some inorganics	Organics	Highly toxic persistent chemicals
Relative costs	Low to moderate	Low	Moderate to high	Moderate to high	Moderate	Moderate
Resource recovery potential	None	None	Energy and some acids	Energy and some metals	Possible building material	Some metals

[a] Molten salt, high-temperature fluid well, and plasma arc treatments.

[b] DRE = destruction and removal efficiency; PIC = product of incomplete combustion.

[c] Wastes for which this method may be less effective for reducing exposure, relative to other technologies. Wastes listed do not necessarily denote common usage.

Source: Modified after Council on Environmental Quality, 1983.

such as plastic) confine the waste, and a system of internal drains concentrates leachate in a collection basin, from which it is pumped and transported to a wastewater treatment plant. Designs of modern facilities include multiple barriers consisting of several impermeable layers and filters. The function of impervious liners is to ensure that leachate does not contaminate soil and groundwater resources. However, this type of waste-disposal procedure, like the sanitary landfill from which it evolved, must have several monitoring wells to alert personnel if leachates leak out of the system and threaten water resources.

It has recently been argued that there is no such thing as a really secure landfill, implying that they all leak to some extent. Impervious plastic liners, filters, and clay layers can fail, even with several backups; and drains can become clogged and cause overflow. Animals, such as gophers, ground squirrels, woodchucks, and muskrats, can chew through plastic liners, and some may burrow through clay liners, thus promoting or accelerating leaks. Yet careful siting and engineering can minimize problems. As with sanitary landfills, preferable sites are those with good natural barriers to migration of leachate: thick clay deposits, an arid climate, or a deep water table. Nevertheless, land disposal should be used only for specific chemicals compatible with and suitable for the method.

Land Application: Microbial Breakdown

Intentional application of waste materials to the surface soil is referred to as **land application**, land spreading, or land farming. We discussed land application of human waste earlier in the chapter. Land application of waste may be a desirable method of treatment for certain biodegradable industrial waste, such as oily petroleum waste and some or-ganic chemical-plant wastes. A good indicator of the use-fulness of land application for a particular waste is the *biopersistence* of the waste—how long the material remains in the biosphere. The greater the biopersistence, the less suitable the waste for land application procedures. Land application is not an effective treatment or disposal method for inorganic substances such as salts and heavy metals.[37]

Land application of biodegradable waste works because, when such materials are added to the soil, they are attacked by microflora (bacteria, molds, yeasts, and other organisms) that decompose the waste material in a process known as *microbial breakdown*. The soil thus can be thought of as a microbial farm that constantly recycles organic and inorganic matter by breaking it down into more fundamental forms useful to other living things in the soil. Because the upper soil zone contains the largest microbial populations, land application is restricted to the uppermost 15–20 cm (6–8 in) of the soil.[38]

Surface Impoundment

Both natural topographic depressions and human-made excavations have been used to hold hazardous liquid waste in a method known as **surface impoundment**. The depressions or excavations are formed primarily of soil or other surface materials but may be lined with manufactured materials such as plastic. Examples include aeration pits and lagoons at hazardous-waste facilities.

Surface impoundments have been criticized because they are especially prone to seepage, resulting in pollution of soil and groundwater. Evaporation from surface impoundments can also produce an air pollution problem. This type of storage or disposal system for hazardous waste is controversial, and many sites have been closed.

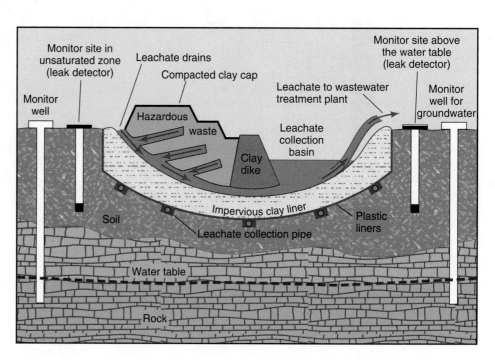

Figure 30.10 ■ A secure landfill for hazardous chemical waste. The impervious liners, systems of drains, and leak detectors are integral parts of the system to ensure that leachate does not escape from the disposal site. Monitoring in the unsaturated zone is important and involves periodic collection of soil water.

Deep-Well Disposal

Deep-well disposal, another controversial method of waste disposal, involves injection of waste into deep wells. A deep well must penetrate to a depth below and be completely isolated from all freshwater aquifers, thereby ensuring that injection of waste will not contaminate or pollute existing or potential water supplies. Typically, the waste is injected into a permeable rock layer several thousand meters below the surface in geologic basins capped by relatively impervious, fracture-resistant rock such as shale or salt deposits.[2]

Deep-well injection of oil-field brine (saltwater) has been important in the control of water pollution in oil fields for many years. Huge quantities of liquid waste (brine) pumped up with oil have been injected back into the rock.[39]

Deep-well disposal of industrial wastes should not be viewed as a quick and easy solution to industrial waste problems.[40] Even where geologic conditions are favorable for deep-well disposal, there are a limited number of suitable sites; within these sites, there is limited space for disposal of waste. Finally, disposal wells must be carefully monitored by additional wells, known as monitoring wells, to determine if the waste is remaining in the disposal site.

Summary of Land Disposal Methods

Direct land disposal of hazardous waste is often not the best initial alternative. There is consensus that even with extensive safeguards and state-of-the-art designs, land disposal alternatives cannot guarantee that the waste is contained and will not cause environmental disruption in the future. This concern holds true for all land disposal facilities, including landfills, surface impoundments, land application, and injection wells. Pollution of air, land, surface water, and groundwater may result from failure of a land disposal site to contain hazardous waste. Pollution of groundwater is perhaps the most significant risk because groundwater provides a convenient route for pollutants to reach humans and other living things. Some of the paths that pollutants may take from land disposal sites to contaminate the environment include leakage and runoff to surface water or groundwater from improperly designed or maintained landfills; seepage, runoff, or air emissions from unlined lagoons; percolation and seepage from failure of surface land application of waste to soils; leaks in pipes or other equipment associated with deep-well injection; and leaks from buried drums, tanks, or other containers.

30.9 Alternatives to Land Disposal of Hazardous Waste

Our methods for handling hazardous chemical waste should be multifaceted. In addition to the disposal methods just discussed, chemical waste management should include such processes as source reduction, recycling and resource recovery, treatment, and incineration.[41] Recently,

it has been argued that these alternatives to land disposal are not being utilized to their full potential; that is, the volume of waste could be reduced and the remaining waste could be recycled or treated in some form prior to land disposal of the residues of the treatment processes.[41] Advantages to source reduction, recycling, treatment, and incineration include the following:

- Useful chemicals can be reclaimed and reused.
- Treatment of wastes may make them less toxic and therefore less likely to cause problems in landfills.
- The actual waste that must eventually be disposed of is reduced to a much smaller volume.
- Because a reduced volume of waste is finally disposed of, there is less stress on the dwindling capacity of waste-disposal sites.

Although some of the following techniques have been discussed as part of IWM, the techniques have special implications and complications where hazardous wastes are concerned.

Source Reduction

The object of source reduction in hazardous-waste management is to reduce the amount of hazardous waste generated by manufacturing or other processes. For example, changes in the chemical processes involved, equipment used, raw materials used, or maintenance measures may successfully reduce the amount or toxicity of the hazardous waste produced.[41]

Recycling and Resource Recovery

Hazardous chemical waste may contain materials that can be recovered for future use. For example, acids and solvents collect contaminants when they are used in manufacturing processes. These acids and solvents can be processed to remove the contaminants and can then be reused in the same or other different manufacturing processes.[41]

Treatment

Hazardous chemical waste can be treated by a variety of processes to change the physical or chemical composition of the waste and so to reduce its toxic or hazardous characteristics. For example, acids can be neutralized, heavy metals can be separated from liquid waste, and hazardous chemical compounds can be broken up through oxidation.[41]

Incineration

Hazardous chemical waste can be successfully destroyed by high-temperature incineration. Incineration is considered a waste treatment rather than a disposal method because the process produces an ash residue, which must then be disposed of in a landfill operation. Hazardous waste has also been incinerated offshore on ships, creating potential air pollution and ash disposal problems for the marine environment.

30.10 Ocean Dumping

Waste management issues are not restricted to the land. Some waste ends up in the oceans of the world.

Oceans cover more than 70% of Earth. They play a part in maintaining our global environment and are of major importance in the cycling of carbon dioxide, which helps regulate the global climate. Oceans are also important in cycling many chemical elements important to life, such as nitrogen and phosphorus, and are a valuable resource to people because they provide necessities such as foods and minerals.

It seems reasonable that such an important resource would receive preferential treatment, and yet oceans have long been dumping grounds for many types of waste, including industrial waste, construction debris, urban sewage, and plastics (see A Closer Look 30.4). Ocean dumping contributes to the larger problem of ocean pollution, which has seriously damaged the marine environment and caused a health hazard. Locations in the oceans of the world that are accumulating pollution continuously,

that have intermittent pollution problems, or that have potential for pollution from ships in the major shipping lanes are shown on Figure 30.11. Notice that the areas with continual or intermittent pollution are located near shore. Unfortunately, these are also areas of high productivity with valuable fisheries. Shellfish have been found to contain organisms that produce diseases, such as polio and hepatitis. In the United States, at least 20% of the nation's commercial shellfish beds have been closed (mostly temporarily) because of pollution. Beaches and bays have been closed (again, mostly temporarily) to recreational uses. Lifeless zones in the marine environment have been created. Heavy kills of fish and other organisms have occurred, and profound changes in marine ecosystems have taken place (see Chapter 22).[42, 43]

Marine pollution has a variety of specific effects on oceanic life, including the following:

- Death or retarded growth, vitality, and reproductivity of marine organisms.
- Reduction in the dissolved oxygen content necessary for marine life because of increased biochemical oxygen demand.
- Eutrophication caused by nutrient-rich waste in shallow waters of estuaries, bays, and parts of the continental shelf, resulting in depletion of oxygen and

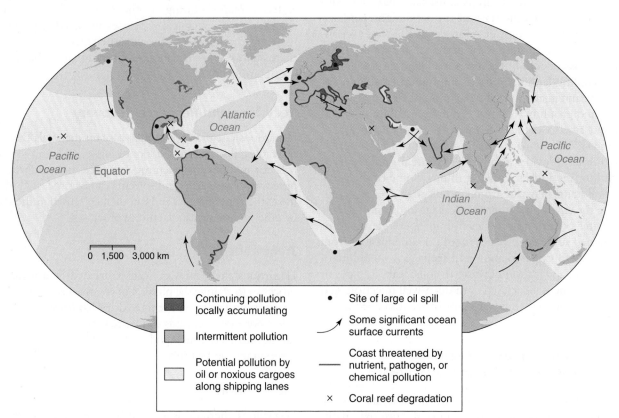

Figure 30.11 ■ Ocean pollution of the world. Notice that the areas of continuing and locally accumulating pollution, as well as the areas with intermittent pollution, are in nearshore environments. [*Source:* Modified from the Council on Environmental Quality, *Environmental Trends*, 1981, with additional data from A. P. McGinn, "Safeguarding the Health of the Oceans," WorldWatch Paper 145 (Washington, D.C.: WorldWatch Institute, 1999), pp. 22–23.]

A CLOSER LOOK 30.4

Plastics in the Ocean

Vast quantities of plastic are used for a variety of products ranging from beverage containers to cigarette lighters. For decades the oceans of the world have received plastics dumped into the waters by humans. Some are dumped from passing ships by passengers, while others are dropped as litter along beaches and moved into the water by tides. Once in the ocean, the plastics that float move with the ocean currents. They tend to accumulate in places of convergent currents that concentrate the debris. Convergent currents of the Pacific (Figure 30.12) have a whirlpool-like action that concentrates debris near the center of these zones. One such zone of convergence is located north of the equator where the northwestern Hawaiian Islands are located. These islands are very remote, with largely unspoiled land that most people would describe as being very remote and pristine. Yet there are literally hundreds of tons of plastics and other tpyes of human debris on these islands. Recently the National Oceanographic and Atmospheric Administration collected more than 80 tons of marine debris on Pearl and Hermes Atolls. The islands are home to ecosystems that include sea turtles, monk seals, and a variety of birds, including albatross. The marine scientist Jean-Michel Cousteau and colleagues have been studying the problem of plastics on the northwestern Hawaiian Islands, including Midway Island and Kure Atoll. They reported that the beaches of some of the islands and atolls look like a "recycling bin" of plastics. They found numerous cigarette lighters, some with fuel still in them, as well as caps from plastic bottles and all kinds of plastic toys and other debris. Birds on the islands pick up the plastic—attracted to it but not knowing what it is—and eat it. Figure 30.13 shows a dead albatross with debris in its stomach that caused its death. Plastic rings from a variety of products have been found around the snouts of seals, which ultimately then starve to death, and are also ingested by sea turtles. In some areas, the carcasses of albatrosses litter the shorelines. The solution to the problem of plastics in the ocean is to be more conscious of recycling plastic products to ensure they do not enter the marine environment. Collecting the plastic items on beaches where they accumulate is a step in the right direction but is a reactive response rather than being proactive and reducing the source of the pollution.

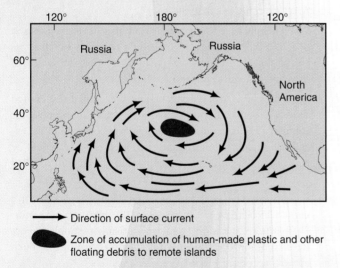

→ Direction of surface current

⬛ Zone of accumulation of human-made plastic and other floating debris to remote islands

Figure 30.12 ■ General circulation of the North Pacific Ocean. Arrows show direction of currents. Notice the tightening clockwise spiral pattern that carries floating debris to remote islands.

Figure 30.13 ■ Albatross killed on a remote Pacific island by ingesting a large volume of plastic and other debris delivered by ocean currents. Photograph is not staged! The bird actually ingested all the plastic shown.

subsequent killing of algae, which may wash up and pollute coastal areas. (See Chapter 22 for a discussion of eutrophication in the Gulf of Mexico.)

■ Habitat change caused by waste-disposal practices that subtly or drastically change entire marine ecosystems.[42]

Marine waters of Europe are in particular trouble, in part as a result of urban and agricultural pollutants that have raised concentrations of nutrients in seawater. Blooms (heavy, sudden growth) of toxic algae are becoming more common. In 1988 in the waterway connecting the North Sea to the Baltic Sea, for example, a bloom was responsible

CRITICAL THINKING ISSUE

Can We Make Recycling a Financially Viable Industry?

There is tremendous public support for recycling in the United States today. Many people understand that management of our waste has many advantages to society as a whole and the environment in particular. People like the notion of recycling because they correctly assume they are helping conserve resources, such as forests, that make up much of the nonurban environment of the planet. Large cities from New York to Los Angeles have initiated recycling programs, but there is continued concern that recycling is not yet "cost-effective." To be sure, there are success stories, such as a large urban paper mill on New York's Staten Island that recycles more than 1,000 tons of paper per day. It is claimed that this paper mill saves more than 10,000 trees a day and uses only about 10% of the electricity required to make paper from virgin wood processing. On the west coast, San Francisco has an innovative and ambitious recycling program that diverts nearly 50% of the urban waste from landfills to recycling programs. The city is even talking about the concept of zero waste, hoping to achieve total recycling of waste by 2020. In part, this is achieved by instigating a "pay-as-you-throw-away" approach; businesses and individuals are charged for disposal of garbage but not for materials that are recycled. Materials from the waste of the San Francisco urban area are shipped as far away as China and the Philippines to be recycled into usable products; organic waste is sent to agricultural areas; metals such as aluminum are sent around California and to other states, where they are recycled. New York, in contrast, recycles about 20% of its waste and in July of 2002 suspended all recycling of glass and plastic. By doing so, the city expected to save more than $40 million a year. This saving occurs because recycling metal, glass, and plastic costs approximately $100 more per ton than sending it to out-of-state landfills for disposal. The people of New York and their city officials are not against recycling, but in 2002 when the city was cash-poor, the economics simply did not add up. Having to decide to cut social programs or to cut recycling, they concluded that recycling of plastic and glass was simply not cost-effective. In an about-turn, New York then reinstated recycling of glass and plastic in 2003–2004.

To understand some of the issues concerning recycling and its cost, consider the following points:

- The average cost of disposal at a landfill is about $40/ton in the United States, and even at a higher price of about $80/ton, may be cheaper than the cost of recycling.

- Landfill fees in Europe range from $200 to $300/ton.
- Europe has been more successful in recycling, in part because countries such as Germany make manufacturers responsible for the disposal cost of packaging and the industrial goods they produce.
- In the United States, packaging accounts for approximately one-third of the entire waste produced by manufacturing.
- The cost to cities such as New York, which must export their waste out of state, is steadily rising and is expected to exceed the cost of recycling within about 10 years.
- Placing a 10-cent refundable deposit on all beverage containers except milk would greatly increase the number of such containers recycled. For example, states with a deposit system have an average recycling rate of about 70 to 95% of bottles and cans, whereas states that do not have a refundable deposit system average less than 30%.
- When people have to pay for the trash that has to be disposed of at a landfill, but are not charged for materials—such as paper, plastic, glass, and metals—that are recycled, the success of recycling is greatly enhanced.
- Companies that make beverages are not particularly excited about a proposal that would require a refundable deposit for containers. They claim that the additional costs would be several billion dollars but do agree that recovery rates would be higher and that this would help provide a steadier of recycled metal, such as aluminum, as well as plastic.
- Education is a big issue with recycling. Many people still don't know which items are in fact recyclable and which are not. Furthermore, they don't understand that commingling recyclable and nonrecyclable items in their waste results in much higher cost of separation at recycling centers.
- Global markets for recyclable materials such as paper and metals have potential for expansion, particularly for large urban areas on the seacoast where the shipping of materials is economically viable. Recycling in the United States today is a $14 billion industry, and if it is done right, it generates new jobs and revenue for participating communities.

Critical Thinking Questions

1. What can be done about the global problem of e-waste? Outline a plan.

2. What can be done to assist recycling industries to become more cost-effective?

3. What are some of the indirect benefits to society and the environment from recycling?

4. Defend or criticize the contention that if we really want to do something to improve the environment through reduction of our waste, we have to move beyond evaluating benefits of recycling based simply on the fact that it may cost more than dumping waste at a landfill.

5. What are the recycling efforts in your community and university, and how could improvements be made?

---

for killing nearly all marine life to a depth of about 15 m (50 ft). It is believed that urban waste and agricultural runoff contributed to the toxic bloom.

Although oceans are vast, they are basically giant sinks for materials from continents, and parts of the marine environment are extremely fragile.[43] One area of concern is the *microlayer*, the upper 3 mm of ocean water. The base of the marine food chain consists of planktonic life abundant in the microlayer. The young of certain fish and shellfish also reside in the upper few millimeters of water in the early stages of their life. Unfortunately, the upper few millimeters of the ocean also tend to concentrate pollutants, such as toxic chemicals and heavy metals. One study reported that the concentrations of heavy metals—including zinc, lead, and copper—in the microlayer are from 10 to 1,000 times higher than in the deeper waters. There is fear that disproportionate pollution of the microlayer will have especially serious effects on marine organisms.[43] There is also concern that some ecosystems in the oceans, such as coral reefs, estuaries and salt marshes, and mangrove swamps, are threatened by ocean pollution.

Marine pollution can have major impacts on people and society. Contaminated marine organisms may transmit toxic elements or diseases to people who eat them. When beaches and harbors become polluted by solid waste, oil, and other materials, there may be damage to marine life as well as a loss of visual appeal and other amenities (see A Closer Look 30.4). Economic loss is considerable. Loss of shellfish from pollution in the United States, for example, amounts to many millions of dollars per year. In addition, a great deal of money is spent cleaning up solid waste, liquid waste, and other pollutants in coastal areas.[41]

30.11 Pollution Prevention

As we saw at the beginning of the chapter, approaches to waste management are changing. During the first several decades of environmental concern and management (the 1970s and 1980s), the United States approached waste management through government regulations and control of waste. Control was accomplished through chemical, physical, or biological treatment and collection (for eventual disposal), transformation, or destruction of pollutants after they had been generated. This was thought to be the most cost-effective approach to waste management.

With the 1990s came a growing emphasis on **pollution prevention**, which involves identifying ways to prevent the generation of waste rather than finding ways to dispose of it. This approach, which is part of materials management, reduces the need for management of waste, because less waste is produced. Approaches to pollution prevention include the following:[44]

- Purchasing the proper amount of raw materials so that no excess remains to be disposed of.
- Exercising better control of materials used in manufacturing processes so that less waste is produced.
- Substituting nontoxic chemicals for hazardous or toxic materials currently in use.
- Improving engineering and design of manufacturing processes so that less waste is produced.

These approaches are often called P-2 approaches, for "pollution prevention." Probably the best way to illustrate the P-2 process is through a case history.[44]

A Wisconsin firm that produced cheese was faced with disposal of about 2,000 gallons a day of a salty solution generated as part of the cheese-making process. Initially, the firm spread the salty solution on nearby agricultural lands—common practice for firms that could not discharge wastewater into publicly owned treatment plants. This method of waste disposal, if the solution was applied incorrectly, caused the level of salts in the soil to rise so much that crops were damaged. As a result, the Department of Natural Resources in Wisconsin placed limitations on the discharge of salt to the land.

The cheese manufacturing firm decided to modify its cheese-making processes to recover salt from the solution and reuse it in production. This involved developing a recovery process using an evaporator. The recovery process reduced the salty waste by about 75% and at the same time reduced the amount of the salt the company needed to purchase by 50%. The operating and maintenance costs for recovery were approximately three cents per pound of salt recovered, and the time necessary to reclaim the extra cost of the new equipment was only two months. The firm saved thousands of dollars a year by purchasing less salt.

The case history of the cheese firm suggests that often rather minor changes can result in large reductions of waste produced. And this case history is not an isolated example. Thousands of similar cases exist today as we move from the era of recognizing environmental problems and regulating them at a national level to providing economic incentives and new technology to better manage materials.[44]

Summary

■ The history of waste-disposal practices since the Industrial Revolution has progressed from the practice of dilution and dispersion to the concept of integrated waste management (IWM), which emphasizes the three R's of reducing waste, reusing materials, and recycling.

■ The emerging concept of industrial ecology has as a goal a system in which the concept of waste doesn't exist, because waste from one part of the system would be a resource for another part.

■ The most common method for disposal of solid waste is the sanitary landfill. However, around many large cities, space for landfills is hard to find; few people wish to live near any waste-disposal operation.

■ Physical and hydrologic conditions of a site greatly affect its suitability as a landfill. These include landform, topography, rock and soil type, depth to groundwater, and amount of precipitation.

■ Hazardous chemical waste is one of the most serious environmental problems in the United States. Hundreds or even thousands of abandoned uncontrolled disposal sites could be time bombs that will eventually cause serious public health problems. We know that we will continue to produce some hazardous chemical waste. Therefore, it is imperative to develop and use safe disposal methods.

■ Management of hazardous chemical wastes involves several options, including on-site processing to recover by-products with commercial value, microbial breakdown, chemical stabilization, incineration, and disposal by secure landfill or deep-well injection.

■ Ocean dumping is a significant source of marine pollution. The most seriously affected areas are near shore, where valuable fisheries often exist.

■ Pollution prevention (P-2)—identifying and using ways to prevent the generation of waste—is an important emerging area of materials management.

REEXAMINING THEMES AND ISSUES

Human Population

Waste management strategies are inextricably linked to human population. As population increases, so does the waste generated because the total amount of waste is the product of per-capita waste generation times population. Furthermore, in developing countries, where population increase is the most dramatic, increases in industrial output, when linked to poor environmental control, produce or aggravate waste management problems.

Sustainability

The objective of providing for a quality environment for future generations is closely linked to waste management. Of particular importance here are the concepts of integrated waste management, materials management, and industrial ecology. Carried to their natural conclusion, the ideas behind these concepts would lead to a system in which the issue would no longer be waste management but instead would be resource management. Pollution prevention (P-2) is a step in this direction.

Global Perspective

Management of waste is becoming a global problem. Inappropriate management of waste contributes to air and water pollution, with the potential to cause environmental disruption on a regional or global scale. For example, waste generated by large inland cities and disposed of in river systems may eventually enter the oceans, where it may become dispersed by the global circulation patterns of ocean currents. Similarly, soils polluted by hazardous materials may erode, and the particles may enter the atmosphere or water system, to be dispersed widely.

Urban World

Because so much of our waste is generated in the urban environment, it is a focus of special attention for waste management. Where population densities are high, it is easier for the principles behind "reduce, reuse, and recycle" to be implemented. There are greater financial incentives for management of waste where waste is more concentrated.

People and Nature

Production of waste is a basic process of life. In nature, waste from one organism is a resource for another. Waste is recycled in ecosystems as energy flows and chemical cycles. As a result, the concept of waste in nature is much different than in the human waste stream. In the human system, waste may be stored in facilities such as landfills, where it may remain for long periods of time, far from natural cycling. Our activities to recycle waste or burn it for energy move us closer to transforming waste into resources. Converting waste to resources brings us closer to nature by causing urban systems to operate in parallel with natural ecosystems.

Science and Values

People today value a quality, pollution-free environment. The way in which waste has been managed has contributed to—and continues to contribute to—health and other environmental problems. An understanding of these problems has resulted in a considerable amount of work and research aimed at reducing or eliminating the impact of waste. How a society manages its waste is a sign of the maturity of the society and its ethical framework. Accordingly, we have become more conscious of environmental justice issues related to waste management.

Key Terms

composting **650**
deep-well disposal **661**
e-waste **645**
environmental audit **658**
environmental justice **652**
hazardous waste **655**

incineration **650**
industrial ecology **646**
integrated waste
  management
  (IWM) **647**
land application **660**

leachate **651**
materials management **649**
monitoring **652**
pollution prevention **665**
recycle **647**
reduce **647**

reuse **647**
sanitary landfill **651**
secure landfill **658**
surface impoundment **660**

Study Questions

1. Have you ever contributed to the hazardous-waste problem through disposal methods used in your home, school laboratory, or other location? How big a problem do you think such events are? For example, how bad is it to dump paint thinner down a drain?

2. Why is it so difficult to ensure safe land disposal of hazardous waste?

3. Would you approve the siting of a waste-disposal facility in your part of town? If not, why not, and where do you think such facilities should be sited?

4. Why might there be a trend toward on-site disposal rather than land disposal of hazardous waste? Consider physical, biological, social, legal, and economic aspects of the question.

5. Is government doing enough to clean up abandoned hazardous-waste dumps? Do private citizens have a role in choosing where cleanup funds should be allocated?

6. Considering how much waste has been dumped in the nearshore marine environment, how safe is it to swim in bays and estuaries near large cities?

7. Do you think we should collect household waste and burn it in special incinerators to make electrical energy? What problems and what advantages do you see for this method compared with other disposal options?

8. Many jobs will be available in the next few years in the field of hazardous-waste monitoring and disposal. Would you take such a job? If not, why not? If so, do you feel secure that your health would not be jeopardized?

9. Should companies that dumped hazardous waste years ago when the problem was not understood or recognized be held liable today for health-related problems to which their dumping may have contributed?

10. Suppose you found that the home you had been living in for 15 years was located over a buried waste-disposal site. What would you do? What kinds of studies could be done to evaluate the potential problem?

Further Reading

Allenby, B. R. 1999. *Industrial Ecology: Policy Framework and Implementation*. Upper Saddle River, N.J.: Prentice Hall. Primer on industrial ecology.

Ashley, S. 2002 (April). "It's Not Easy Being Green," *Scientific American*, pp. 32–34. A look at the economics of developing biodegradable products and a little of the chemistry involved.

Kreith, F. ed. 1994. *Handbook of Solid Waste Management*. New York: McGraw-Hill. Thorough coverage of municipal waste management, including waste characteristics, federal and state legislation, source reduction, recycling, and landfilling.

Rhyner, C. R., L. J. Schwartz, R. B. Wenger, and M. G. Kohrell. 1995. *Waste Management and Resource Recovery*. Boca Raton, Fla.: CRC, Lewis. Discussions of the archeology of waste, waste generation, source reduction and recycling, wastewater treatment, incineration and energy recovery, hazardous waste, and costs of waste systems and facilities.

Watts, R. J. 1998. *Hazardous Wastes*. New York: John Wiley. A to Z of hazardous wastes.

APPENDIX

A — Special Feature: Electromagnetic Radiation (EMR) Laws

Properties of Waves

- Direction of wave propagation

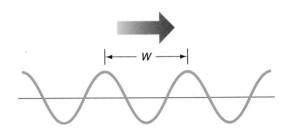

- W = wavelength (distance from one wave crest to the next)
- An EMR wave travels at the speed of light (C) in a vacuum, or about 300,000 km/s (3×10^8 m/s).
- The period T of a wave is the time it takes for a wave to travel a distance of one wavelength W. Then since distance is the product of speed (velocity) and time, W = CT.
- The frequency f of a wave is the number of cycles (each wavelength that passes a point is a cycle) of a wave that pass a particular point per unit time. Frequency f is measured in cycles per second (hertz). The frequency f is the inverse of T: f = 1/T and W = C/f. For example, the period T of a 6000 hertz EMF wave is: 6000 hertz = 1/T and T = 1/6000 S, or 1.7×10^{-4} S. The wavelength W = CT is 3×10^8 m/s times 1.7×10^{-4} S, or 5.1×10^4 m, which according to Figure 23.12 is a long radio wave.

Absolute Temperature Scale (kelvin, K)

- Zero is really zero; there are no negative values of K
- Temperature in K = temperature in °C + 273

$$K = °C + 273$$

- Example: water freezes at °C = 0 = 273 K

water boils at °C = 100 = 373 K

Stefan–Boltzmann Law

- All bodies with a temperature greater than absolute zero radiate EMR. These bodies are called thermal radiators. The amount of energy per second radiated from thermal radiators is called *intensity* and is given by the Stefan–Boltzmann law

$$E = aT^4$$

where E is the energy per second (intensity); T is the absolute temperature; and a is a constant (the nature of this constant involves physical ideas beyond the scope of this text).

- The Stefan–Boltzmann law states that the intensity of EMR coming from a thermal radiator is directly proportional to the fourth power of its absolute temperature.

Wien's Law

$$W_P = a/T$$

where W_P is the wavelength of the peak intensity of a thermal radiator; T is temperature in K; and a is a constant. For example, Figure 23.13 shows that W_P for the earth is about 10 μm. Wien's law states in a general way that the hotter a substance is, the shorter the wavelength of the emitted predominant electromagnetic radiation. That is, wavelength is inversely proportional to temperature.

B. Prefixes and Multiplication Factors

Number	$10\times$, Power of 10	Prefix	Symbol
1,000,000,000,000,000	10^{18}	exa	E
1,000,000,000,000,000	10^{15}	peta	P
1,000,000,000,000	10^{12}	tera	T
1,000,000,000	10^{9}	giga	G
1,000,000	10^{6}	mega	M
10,000	10^{4}	myria	
1,000	10^{3}	kilo	k
100	10^{2}	hecto	h
10	10^{1}	deca	da
0.1	10^{-1}	deci	d
0.01	10^{-2}	centi	c
0.001	10^{-3}	milli	m
0.000 001	10^{-6}	micro	μ
0.000 000 001	10^{-9}	nano	n
0.000 000 000 001	10^{-12}	pico	p
0.000 000 000 000 001	10^{-15}	femto	f
0.000 000 000 000 000 001	10^{-18}	atto	a

C. Common Conversion Factors

Length

1 yard = 3 ft, 1 fathom = 6 ft

	in	ft	mi	cm	m	km
1 inch (in) =	1	0.083	1.58×10^{-5}	2.54	0.0254	2.54×10^{-5}
1 foot (ft) =	12	1	1.89×10^{-4}	30.48	0.3048	—
1 mile (mi) =	63,360	5,280	1	160,934	1,609	1.609
1 centimeter (cm) =	0.394	0.0328	6.2×10^{-6}	1	0.01	1.0×10^{-5}
1 meter (m) =	39.37	3.281	6.2×10^{-4}	100	1	0.001
1 kilometer (km) =	39,370	3,281	0.6214	100,000	1,000	1

Area

1 square mi = 640 acres, 1 acre = 43,560 ft^2 = 4046.86 m^2 = 0.4047 ha
1 ha = 10,000 m^2 = 2.471 acres

	in^2	ft^2	mi^2	cm^2	m^2	km^2
1 in^2 =	1	—	—	6.4516	—	—
1 ft^2 =	144	1	—	929	0.0929	—
1 mi^2 =	—	27,878,400	1	—	—	2.590
1 cm^2 =	0.155	—	—	1	—	—
1 m^2 =	1,550	10.764	—	10,000	1	—
1 km^2 =	—	—	0.3861	—	1,000,000	1

Volume

	in³	ft³	yd³	m³	qt	liter	barrel	gal (U.S.)
1 in³ =	1	—	—	—	—	0.02	—	—
1 ft³ =	1,728	1	—	0.0283	—	28.3	—	7.480
1 yd³ =	—	27	1	0.76	—	—	—	—
1 m³ =	61,020	35.315	1.307	1	—	1,000	—	—
1 quart (qt) =	—	—	—	—	1	0.95	—	0.25
1 liter (l) =	61.02	—	—	—	1.06	1	—	0.2642
1 barrel (oil) =	—	—	—	—	168	159.6	1	42
1 gallon (U.S.) =	231	0.13	—	—	4	3.785	0.02	1

Mass and Weight

1 pound = 453.6 grams = 0.4536 kilogram = 16 ounces

1 gram = 0.0353 ounce = 0.0022 pound

1 short ton = 2000 pounds = 907.2 kilograms

1 long ton = 2240 pounds = 1008 kilograms

1 metric ton = 2205 pounds = 1000 kilograms

1 kilogram = 2.205 pounds

Energy and Power[a]

1 kilowatt-hour = 3413 Btus = 860,421 calories

2 Btu = 0.000293 kilowatt-hour = 252 calories = 1055 joules

1 watt = 3.413 Btu/hr = 14.34 calorie/min

1 calorie = the amount of heat necessary to raise the temperature of 1 gram (1 cm³) of water 1 degree Celsius

1 quadrillion Btu = (approximately) 1 exajoule

1 horsepower = 7.457×10^2 watts

1 joule = 9.481×10^{-4} Btu = 0.239 cal = 2.778×10^{-7} kilowatt-hour

[a]Values from Lange, N. A., 1967, *Handbook of Chemistry*, New York: McGraw-Hill.

Temperature

$F = \frac{9}{5}C + 32$

F is degrees Fahrenheit.
C is degrees Celsius (centigrade).

Fahrenheit		Celsius
32	Freezing of H₂0 (Atmospheric Pressure)	0
50		10
68		20
86		30
104		40
122		50
140		60
158		70
176		80
194		90
212	Boiling of H₂0 (Atmospheric Pressure)	100

Other Conversion Factors

1 ft³/sec = 0.0283 m³/sec = 7.48 gal/sec = 28.32 liter/sec

1 acre-foot = 43,560 ft³ = 1233 m³ = 325,829 gal

1 m³/sec = 35.32 ft³/sec

1 ft³/sec for one day = 1.98 acre-feet

1 m/sec = 3.6 km/hr = 2.24 mi/hr

1 ft/sec = 0.682 mi/hr = 1.097 km/hr

1 atmosphere = 14.7 lb(in.$^{-2}$) = 2116 lb(ft^{-2}) = 1.013×10^5 N(m^{-2})

Era	Approximate Age in Millions of Years Before Present	Period	Epoch	Life Form
	Less than 0.01		Recent (Holocene)	
	0.01–2	Quaternary	Pleistocene	Humans
	2			
Cenozoic	2–5		Pliocene	
	5–23		Miocene	
	23–35	Tertiary	Oligocene	
	35–56		Eocene	Mammals
	56–65		Paleocene	
	65			
Mesozoic	65–146	Cretaceous		
	146–208	Jurassic		Flying reptiles, birds
	208–245	Triassic		Dinosaurs
	245			
Paleozoic	245–290	Permian		Reptiles
	290–363	Carboniferous		Insects
	363–417	Devonian		Amphibians
	417–443	Silurian		Land plants
	443–495	Ordovician		Fish
	495–545	Cambrian		
	545			
	700			Multicelled organisms
	3,400			One-celled organisms
	4,000	Approximate age of oldest rocks discovered on Earth		
Precambrian				
	4,600	Approximate age of Earth and meteorites		

GLOSSARY

Abortion rate The estimated number of abortions per 1000 women aged 15 to 44 in a given year. (Ages 15 to 44 are taken to be the limits of ages during which women can have babies. This, of course, is an approximation, made for convenience.)

Abortion ratio The estimated number of abortions per 1000 live births in a given year.

Acceptable Risk The risk that individuals, society or institutions are willing to take.

Acid mine drainage Acidic water that drains from mining areas (mostly coal but also metal mines). The acidic water may enter surface water resources, causing environmental damage.

Acid rain Rain made acid by pollutants, particularly oxides of sulfur and nitrogen. (Natural rainwater is slightly acid owing to the effect of carbon dioxide dissolved in the water.)

Active solar energy systems Direct use of solar energy that requires mechanical power; usually consists of pumps and other machinery to circulate air, water, or other fluids from solar collectors to a heat sink where the heat may be stored.

Adaptive radiation The process that occurs when a species enters a new habitat that has unoccupied niches and evolves into a group of new species, each adapted to one of these niches.

Advanced wastewater treatment Treatment of wastewater beyond primary and secondary procedures. May include sand filters, carbon filters, or application of chemicals to assist in removing potential pollutants such as nutrients from the wastewater stream.

Aerobic Characterized by the presence of free oxygen.

Aesthetic justification for the conservation of nature An argument for the conservation of nature on the grounds that nature is beautiful and that beauty is important and valuable to people.

Age dependency ratio The ratio of dependent-age people (those unable to work) to working-age people. It is customary to define working-age people as those aged 15 to 65.

Age structure (of a population) A population divided into groups by age. Sometimes the groups represent the actual number of each age in the population; sometimes the groups represent the percentage or proportion of the population of each age.

Agroecosystem An ecosystem created by agriculture. Typically it has low genetic, species, and habitat diversity.

Air quality standards Levels of air pollutants that delineate acceptable levels of pollution over a particular time period. Valuable because they are often tied to emission standards that attempt to control air pollution.

Allowance trading Approach to managing coal resources and reducing pollution through buying, selling, and trading of allowances to emit pollutants from burning coal. The idea is to control pollution by controlling the number of allowances issued.

Alpha particles One of the major types of nuclear radiation, consisting of two protons and two neutrons (a helium nucleus).

Alternative energy Renewable and nonrenewable energy resources that are alternatives to the fossil fuels.

Anaerobic Characterized by the absence of free oxygen.

Aquaculture Production of food from aquatic habitats.

Aquifer An underground zone or body of earth material from which groundwater can be obtained from a well at a useful rate.

Area sources Sometimes also called nonpoint sources. These are diffused sources of pollution such as urban runoff or automobile exhaust. These sources include emissions that may be over a broad area or even over an entire region. They are often difficult to isolate and correct because of the widely dispersed nature of the emissions.

Asbestos A term for several minerals that have the form of small elongated particles. Some types of particles are believed to be carcinogenic or to carry with them carcinogenic materials.

Atmosphere Layer of gases surrounding Earth.

Atmospheric inversion A condition in which warmer air is found above cooler air, restricting air circulation; often associated with a pollution event in urban areas.

Autotroph An organism that produces its own food from inorganic compounds and a source of energy. There are photoautotrophs and chemical autotrophs.

Average residence time A measure of the time it takes for a given part of the total pool or reservoir of a particular material in a system to be cycled through the system. When the size of the pool and rate of throughput are constant, average residence time is the ratio of the total size of the pool or reservoir to the average rate of transfer through the pool.

Balance of nature An environmental myth that states that the natural environment, when not influenced by human activity, will reach a constant status, unchanging over time, referred to as an equilibrium state.

Barrier island An island separated from the mainland by a salt marsh. It generally consists of a multiple system of beach ridges and is separated from other barrier islands by inlets that allow the exchange of seawater with lagoon water.

Becquerel The unit commonly used for radioactive decay in the International System (SI) of measurement.

Beta particles One of the three major kinds of nuclear radiation; electrons that are emitted when one of the protons or neutrons in the nucleus of an isotope spontaneously changes.

Biochemical oxygen demand (BOD) A measure of the amount of oxygen necessary to decompose organic material in a unit volume of water. As the amount of organic waste in water increases, more oxygen is used, resulting in a higher BOD.

Biogeochemical cycle The cycling of a chemical element through the biosphere; its pathways, storage locations, and chemical forms in living things, the atmosphere, oceans, sediments, and lithosphere.

Biogeography The large-scale geographic pattern in the distribution of species, and the causes and history of this distribution.

Biohydrometallurgy Combining biological and mining processes, usually involving microbes to help extract valuable metals such as gold from the ground. May also be used to remove pollutants from mining waste.

Biological control A set of methods to control pest organisms by using natural ecological interactions, including predation, parasitism, and competition. Part of integrated pest management.

Biological diversity Used loosely to mean the variety of life on Earth, but scientifically typically used to consist of three components: (1) genetic diversity—the total number of genetic characteristics; (2) species diversity; and (3) habitat or ecosystem diversity—the number of kinds of habitats or ecosystems in a given unit area. Species diversity in turn includes three concepts: *species richness*, *evenness*, and *dominance*.

Biological evolution The change in inherited characteristics of a population from generation to generation, which can result in new species.

Biological production The capture of usable energy from the environment to produce organic compounds in which that energy is stored.

Biomagnification Also called *biological concentration*. The tendency for some substances to concentrate with each trophic level. Organisms preferentially store certain chemicals and excrete others. When this occurs consistently among organisms, the stored chemicals increase as a percentage of the body weight as the material is transferred along a food chain or trophic level. For example, the concentration of DDT is greater in herbivores than in plants and greater in plants than in the nonliving environment.

Biomass The amount of living material, or the amount of organic material contained in living organisms, both as live and dead material, as in the leaves (live) and stem wood (dead) of trees.

Biomass energy The energy that may be recovered from biomass, which is organic material such as plants and animal waste.

Biomass fuel A new name for the oldest fuel used by humans. Organic matter, such as plant material and animal waste, that can be used as a fuel.

Biome A kind of ecosystem. The rain forest is an example of a biome; rain forests occur in many parts of the world but are not all connected to each other.

Bioremediation A method of treating groundwater pollution problems that utilizes microorganisms in the ground to consume or break down pollutants.

Biosphere Has several meanings. One is that part of a planet where life exists. On Earth it extends from the depths of the oceans to the summit of mountains, but most life exists within a few meters of the surface. A second meaning is: the planetary system that includes and sustains life, and therefore is made up of the atmosphere, oceans, soils, upper bedrock, and all life.

Biota All the organisms of all species living in an area or region up to and including the biosphere, as in "the biota of the Mojave Desert" or "the biota in that aquarium."

Biotic province A geographical region inhabited by life forms (species, families, orders) of common ancestry, bounded by barriers that prevent the spread of the distinctive kinds of life to other regions and the immigration of foreign species into that region.

Birth rate The rate at which births occur in a population, measured either as the number of individuals born per unit of time or as the percentage of births per unit of time compared with the total population.

Black lung disease Often called coal miner disease because it is caused by years of inhaling coal dust, resulting in damage to the lungs.

Body burden The amount of concentration of a toxic chemical, especially radionuclides, in an individual.

Breeder reactor A type of nuclear reactor that utilizes between 40 and 70% of its nuclear fuel and converts fertile nuclei to fissile nuclei faster than the rate of fission. Thus breeder reactors actually produce nuclear fuels.

Brines With respect to mineral resources, refers to waters with a high salinity that contain useful materials such as bromine, iodine, calcium chloride, and magnesium.

Buffers Materials (chemicals) that have the ability to neutralize acids. Examples include the calcium carbonate that is present in many soils and rocks. These materials may lessen potential adverse effects of acid rain.

Burner reactors A type of nuclear reactor that consumes more fissionable material than it produces.

Capillary action The rise of water along narrow passages, facilitated and caused by surface tension.

Carbon cycle Biogeochemical cycle of carbon. Carbon combines with and is chemically and biologically linked with the cycles of oxygen and hydrogen that form the major compounds of life.

Carbon monoxide (CO) Colorless, odorless gas that at very low concentrations is extremely toxic to humans and animals.

Carbon-silicate cycle A complex biogeochemical cycle over time scales as long as one-half billion years. Included in this cycle are major geologic processes, such as weathering, transport by ground and surface waters, erosion, and deposition of crustal rocks. The carbonate-silicate cycle is believed to provide important negative feedback mechanisms that control the temperature of the atmosphere.

Carcinogen Any material that is known to produce cancer in humans or other animals.

Carnivores Organisms that feed on other live organisms; usually applied to animals that eat other animals.

Carrying capacity The maximum abundance of a population or species that can be maintained by a habitat or ecosystem without degrading the ability of that habitat or ecosystem to maintain that abundance in the future.

Cash crops Crops grown to be traded in a market.

Catastrophe A situation of event that causes sufficient damage to people, property or society from which recovery is a long and involved process. Also is defined as a very serious disaster.

Catch per unit effort The number of animals caught per unit of effort, such as the number of fish caught by a fishing ship per day. It is used to estimate the population abundance of a species.

Channelization An engineering technique that consists of straightening, deepening, widening, clearing, or lining existing stream channels. The purpose is to control floods, improve drainage, control erosion, or improve navigation. It is a very controversial practice that may have significant environmental impacts.

Chaparral A dense scrubland found in areas with Mediterranean climate (a long warm, dry season and a cooler rainy season).

Chemical reaction The process in which compounds and elements undergo a chemical change to become a new substance or substances.

Chemoautotrophs Autotrophic bacteria that can derive energy from chemical reactions of simple inorganic compounds.

Chemosynthesis Synthesis of organic compounds by energy derived from chemical reactions.

Chimney (or stack) effect Process whereby warmer air rises in buildings to upper levels and is replaced in the lower portion of the building by outdoor air drawn through a variety of openings, such as windows, doors, or cracks in the foundations and walls.

Chlorofluorocarbons (CFCs) Highly stable compounds that have been or are being used in spray cans as aerosol propellants and in refrigeration units (the gas that is compressed and expanded in a cooling unit). Emissions of chlorofluorocarbons have been associated with potential global warming and stratospheric ozone depletion.

Chronic disease A disease that is persistent in a population, typically occurring in a relatively small but constant percentage of the population.

Chronic hunger A condition in which there is enough food available per person to stay alive, but not enough to lead a satisfactory and productive life.

Chronic patchiness A situation where ecological succession does not occur. One species may replace another, or an individual of the first species may replace it, but no overall general temporal pattern is established. Characteristic of harsh environments such as deserts.

Clay May refer to a mineral family or to a very fine-grained sediment. It is associated with many environmental problems, such as shrinking and swelling of soils and sediment pollution.

Clean Air Act Amendments of 1990 Comprehensive regulations (federal statute) that address acid rain, toxic emissions, ozone depletion, and automobile exhaust.

Clear-cutting In timber harvesting, the practice of cutting all trees in a stand at the same time.

Climate The representative or characteristic conditions of the atmosphere at particular places on Earth. Climate refers to the average or expected conditions over long periods; weather refers to the particular conditions at one time in one place.

Climatic change Change in mean annual temperature and other aspects of climate over periods of time ranging from decades to hundreds of years to several million years.

Climax stage (or ecological succession) The final stage of ecological succession and therefore an ecological community that continues to reproduce itself over time.

Closed system A type of system in which there are definite boundaries to factors such as mass and energy such that exchange of these factors with other systems does not occur.

Closed-canopy forest Forests in which the leaves of adjacent trees overlap or touch, so that the trees form essentially continuous cover.

Coal Solid, brittle carbonaceous rock that is one of the world's most abundant fossil fuels. It is classified according to energy content as well as carbon and sulfur content.

Coal gasification Process that converts coal that is relatively high in sulfur to a gas in order to remove the sulfur.

Cogeneration The capture and use of waste heat; for example, using waste heat from a power plant to heat adjacent factories and other buildings.

Cohort All the individuals in a population born during the same time period. Thus all the people born during the year 2005 represent the world human cohort for that year.

Common law Law derived from custom, judgment, or decrees of courts rather than from legislation.

Commons Land that belongs to the public, not to individuals. Historically a part of old English and New England towns where all the farmers could graze their cattle.

Community, ecological A group of populations of different species living in the same local area and interacting with one another. A community is the living portion of an ecosystem.

Community effect (community-level effect) When the interaction between two species leads to changes in the presence or absence of other species or in a large change in abundance of other species, then a community effect is said to have occurred.

Competition The situation that exists when different individuals, populations, or species compete for the same resource(s) and the presence of one has a detrimental effect on the other. Sheep and cows eating grass in the same field are competitors.

Competitive exclusion principle The idea that two populations of different species with exactly the same requirements cannot persist indefinitely in the same habitat—one will always win out and the other will become extinct.

Composting Biochemical process in which organic materials, such as lawn clippings and kitchen scraps, are decomposed to a rich, soil-like material.

Comprehensive plan Official plan adopted by local government formally stating general and long-range policies concerning future development.

Cone of depression A cone-shaped depression in the water table around a well caused by withdrawal by pumping of water at rates greater than the rates at which the water can be replenished by natural groundwater flow.

Conservation With respect to resources such as energy, refers to changing our patterns of use or simply getting by with less. In a pragmatic sense the term means adjusting our needs to minimize the use of a particular resource, such as energy.

Consumptive use A type of off-stream water use. This water is consumed by plants and animals or in industrial processes or evaporates during use. It is not returned to its source.

Contamination Presence of undesirable material that makes something unfit for a particular use.

Continental drift The movement of continents in response to seafloor spreading. The most recent episode of continental drift started about 200 million years ago with the breakup of the supercontinent Pangaea.

Continental shelf Relatively shallow ocean area between the shoreline and the continental slope that extends to approximately a 600-foot (~200 m) water depth surrounding a continent.

Contour plowing Plowing land along topographic contours, as much in a horizontal plane as possible, thereby decreasing the erosion rate.

Controlled experiment A controlled experiment is designed to test the effects of independent variables on a dependent variable by changing only one independent variable at a time. For each variable tested, there are two setups (an experiment and a control) that are identical except for the independent variable being tested. Any difference in the outcome (dependent variable) between the experiment and the control can then be attributed to the effects of the independent variable tested.

Convection The transfer of heat involving the movement of particles; for example, the boiling water in which hot water rises to the surface and displaces cooler water, which moves toward the bottom.

Convergent evolution The process by which species evolve in different places or different times and, although they have different genetic heritages, develop similar external forms and structures as a result of adaptation to similar environments. The similarity in the shapes of sharks and porpoises is an example of convergent evolution.

Convergent plate boundary Boundary between two lithosphere plates in which one plate descends below the other (subduction).

Cosmopolitan species A species with a broad distribution, occurring wherever in the world the environment is appropriate.

Creative justification for the conservation of nature An argument for the conservation of nature on the grounds that people often find sources of artistic and scientific creativity in their contacts with the unspoiled natural world.

Crop rotation A series of different crops planted successively in the same field, with the field occasionally left fallow, or grown with a cover crop.

Crude oil Naturally occurring petroleum, normally pumped from wells in oil fields. Refinement of crude oil produces most of the petroleum products we use today.

Cultural eutrophication Human-induced eutrophication that involves nutrients such as nitrates or phosphates that cause a rapid increase in the rate of plant growth in ponds, lakes, rivers or the ocean.

Curie Commonly used unit to measure radioactive decay; the amount of radioactivity from 1 gram of radium 226 that undergoes about 37 billion nuclear transformations per second.

Death rate The rate at which deaths occur in a population, measured either as the number of individuals dying per unit time or as the percentage of a population dying per unit time.

Decomposers Organisms that feed on dead organic matter.

Deductive reasoning Drawing a conclusion from initial definitions and assumptions by means of logical reasoning.

Deep-well disposal Method of disposal of hazardous liquid waste that involves pumping the waste deep into the ground below and completely isolated from all freshwater aquifers. A controversial method of waste disposal that is being carefully evaluated.

Demand for food The amount of food that would be bought at a given price if it were available.

Demand-based agriculture Agriculture with production determined by economic demand and limited by that demand rather than by resources.

Demographic transition The pattern of change in birth and death rates as a country is transformed from undeveloped to developed. There are three stages: (1) in an undeveloped country, birth and death rates are high and the growth rate low; (2) the death rate decreases, but the birth rate remains high and the growth rate is high; (3) the birth rate drops toward the death rate and the growth rate therefore also decreases.

Demography The study of populations, especially their patterns in space and time.

Dependent variable See **Variable, dependent**

Denitrification The conversion of nitrate to molecular nitrogen by the action of bacteria—an important step in the nitrogen cycle.

Density-dependent population effects Factors whose effects on a population change with population density.

Density-independent population effects Changes in the size of a population due to factors that are independent of the population size. For example, a storm that knocks down all trees in a forest, no matter how many there are, is a density-independent population effect.

Desalination The removal of salts from seawater or brackish water so that the water can be used for purposes such as agriculture, industrial processes, or human consumption.

Desertification The process of creating a desert where there was not one before.

Dioxin An organic compound composed of oxygen, hydrogen, carbon, and chlorine. About 75 types are known. Dioxin is not normally manufactured intentionally but is a by-product resulting from chemical reactions in the production of other materials, such as herbicides. Known to be extremely toxic to mammals, its effects on the human body are being intensively studied and evaluated.

Direct effects With respect to natural hazards refers to the number of people killed, injured, dislocated, made homeless or otherwise damaged by a hazardous event.

Disaster A hazardous event that occurs over a limited span of time in a defined geographic area. Lose of human life and property damage is significant.

Disprovability The idea that a statement can be said to be scientific if someone can clearly state a method or test by which it might be disproved.

Divergent evolution Organisms with the same ancestral genetic heritage migrate to different habitats and evolve into species with different external forms and structures, but typically continue to use the same kind of habitats. The ostrich and the emu are believed to be examples of divergent evolution.

Divergent plate boundary Boundary between lithospheric plates characterized by the production of new lithosphere; found along oceanic ridges.

Diversity, genetic The total number of genetic characteristics, sometimes of a specific species, subspecies, or group of species.

Diversity, habitat The number of kinds of habitats in a given unit area.

Diversity, species Used loosely to mean the variety of species in an area or on Earth. Technically, it is composed of three components: species richness—the total number of species; species evenness—the relative abundance of species; and species dominance—the most abundant species.

Dobson unit Commonly used to measure the concentration of ozone. One Dobson unit is equivalent to a concentration of 1 ppb ozone.

Dominant species Generally, the species that are most abundant in an area, ecological community, or ecosystem.

Dominants In forestry, the tallest, most numerous, and most vigorous trees in a forest community.

Dose dependency Dependence on the dose or concentration of a substance for its effects on a particular organism.

Dose–response The principle that the effect of a certain chemical on an individual depends on the dose or concentration of that chemical.

Doubling time The time necessary for a quantity of whatever is being measured to double.

Drainage basin The area that contributes surface water to a particular stream network.

Drip irrigation Irrigation by the application of water to the soil from tubes that drip water slowly, greatly reducing the loss of water from direct evaporation and increasing yield.

Drought A period of months of more commonly years of unusually dry weather.

Early successional species Species that occur only or primarily during early stages of succession.

Earth system science The Science of Earth as a system. It includes understanding of processes and linkages between the lithosphere, hydrosphere, biosphere, and atmosphere.

Earthquake Generation of earthquake or seismic waves when rocks under stress fracture and break resulting in displacement along a fault.

Ecological community This term has two meanings. (1) A conceptual or functional meaning: a set of interacting species that occur in the same place (sometimes extended to mean a set that interacts in a way to sustain life). (2) An operational meaning: a set of species found in an area, whether or not they are interacting.

Ecological economics Study and evaluation of relations between humans and the economy with emphasis on long-term health of ecosystems and sustainability.

Ecological gradient A change in the relative abundance of a species or group of species along a line or over an area.

Ecological island An area that is biologically isolated so that a species occurring within the area cannot mix (or only rarely mixes) with any other population of the same species.

Ecological justification for the conservation of nature An argument for the conservation of nature on the grounds that a species, an ecological community, an ecosystem, or Earth's biosphere provides specific functions necessary to the persistence of our life or of benefit to life. The ability of trees in forests to remove carbon dioxide produced in burning fossil fuels is such a public benefit and an argument for maintaining large areas of forests.

Ecological niche The general concept is that the niche is a species' "profession"—what it does to make a living. The term is also used to refer to a set of environmental conditions within which a species is able to persist.

Ecological succession The process of the development of an ecological community or ecosystem, usually viewed as a series of stages—early, middle, late, mature (or climax), and sometimes postclimax. Primary succession is an original establishment; secondary succession is a reestablishment.

Ecology The science of the study of the relationships between living things and their environment.

Ecosystem An ecological community and its local, nonbiological community. An ecosystem is the minimum system that includes and sustains life. It must include at least an autotroph, a decomposer, a liquid medium, a source and sink of energy, and all the chemical elements required by the autotroph and the decomposer.

Ecosystem effect Effects that result from interactions among different species, effects of species on chemical elements in their environment, and conditions of the environment.

Ecosystem energy flow The flow of energy through an ecosystem—from the external environment through a series of organisms and back to the external environment.

Ecotopia A society based on sustainable development and sound environmental planning characterized by a stable human population within the carrying capacity of earth. Is an ideal state.

Ecotourism Tourism based on an interest in observation of nature.

ED-50 The effective dose, or dose that causes an effect in

50% of the population on exposure to a particular toxicant. It is related to the onset of specific symptoms, such as loss of hearing, nausea, or slurred speech.

Edge effect   An effect that occurs following the forming of an ecological island; in the early phases the species diversity along the edge is greater than in the interior. Species escape from the cut area and seek refuge in the border of the forest, where some may last only a short time.

Efficiency   The ratio of output to input. With machines, usually the ratio of work or power produced to the energy or power used to operate or fuel them. With living things, efficiency may be defined as either the useful work done or the energy stored in a useful form compared with the energy taken in.

Efficiency improvements   With respect to energy, refers to designing equipment that will yield more energy output from a given amount of energy input.

Effluent   Any material that flows outward from something. Examples include wastewater from hydroelectric plants and water discharged into streams from waste-disposal sites.

Effluent stream   Type of stream where flow is maintained during the dry season by groundwater seepage into the channel.

El Niño   Natural perturbation of the physical earth system that affects global climate. Characterized by development of warm oceanic waters in the eastern part of the tropical Pacific Ocean, a weakening or reversal of the trade winds, and a weakening or even reversal of the equatorial ocean currents. Reoccurs periodically and affects the atmosphere and global temperature by pumping heat into the atmosphere.

Electromagnetic fields (EMF)   Magnetic and electrical fields produced naturally by our planet and also by appliances such as toasters, electric blankets, and computers. There currently is controversy concerning potential adverse health effects related to exposure to EMF in the workplace and home from such artificial sources as power lines and appliances.

Electromagnetic spectrum   All the possible wavelengths of electromagnetic energy, considered as a continuous range. The spectrum includes long wavelength (used in radio transmission), infrared, visible, ultraviolet, X rays, and gamma rays.

Endangered species   A species that faces threats that might lead to its extinction in a short time.

Endemic species   A species that has evolved in, and lives only within, a specific location. E.g. the California condor is endemic to the pacific coast of North America.

Energy   An abstract concept referring to the ability or capacity to do work.

Energy efficiency   Refers to both first-law efficiency and second-law efficiency, where first-law efficiency is the ratio of the actual amount of energy delivered to the amount of energy supplied to meet a particular need, and the second-law efficiency is the ratio of the maximum available work needed to perform a particular task to the actual work used to perform that task.

Energy flow   The movement of energy through an ecosystem from the external environment through a series of organisms and back to the external environment. It is one of the fundamental processes common to all ecosystems.

Entropy   A measure in a system of the amount of energy that is unavailable for useful work. As the disorder of a system increases, the entropy in a system also increases.

Environment   All factors (living and nonliving) that actually affect an individual organism or population at any point in the life cycle. Environment is also sometimes used to denote a certain set of circumstances surrounding a particular occurrence (environments of deposition, for example).

Environmental audit   Process of determining the past history of a particular site, with special reference to the existence of toxic materials or waste.

Environmental economics   Economic effects of the environment and how economic processes affect that environment, including its living resources.

Environmental ethics   A school, or theory, in philosophy that deals with the ethical value of the environment.

Environmental geology   The application of geologic information to environmental problems.

Environmental impact   The effects of some action on the environment, particularly action by human beings.

Environmental impact report (EIR)   Similar to the environmental impact statement (EIS), a report describing potential environmental impacts resulting from a particular project, often at the state level.

Environmental impact statement   A written statement that assesses and explores possible impacts associated with a particular project that may affect the human environment. The statement is required in the United States by the National Environmental Policy Act of 1969.

Environmental justice   The principle of dealing with environmental problems in such a way as to not discriminate against people based upon socioeconomic status, race, or ethnic group.

Environmental law   A field of law concerning the conservation and use of natural resources and the control of pollution.

Environmental tobacco smoke   Commonly called second-hand smoke from people smoking tobacco.

Environmental unity   A principle of environmental sciences that states that everything affects everything else, meaning that a particular course of action leads to an entire potential string of events. Another way of stating this idea is that you can't only do one thing.

Environmentalism   A social, political, and ethical movement concerned with protecting the environment and using its resources wisely.

Epidemic disease   A disease that appears occasionally in the population, affects a large percentage of it, and declines or almost disappears for a while only to reappear later.

Equilibrium   A point of rest. At equilibrium, a system remains in a single, fixed condition and is said to be in equilibrium. Compare with **Steady state**.

Eukaryote   An organism whose cells have nuclei and

organelles. The eukaryotes include animals, fungi, vegetation and many single-cell organisms.

Eutrophic   Referring to bodies of water having an abundance of the chemical elements required for life.

Eutrophication   Increase in the concentration of chemical elements required for living things (for example, phosphorus). Increased nutrient loading may lead to a population explosion of photosynthetic algae and blue-green bacteria that become so thick that light cannot penetrate the water. Bacteria deprived of light beneath the surface die; as they decompose, dissolved oxygen in the lake is lowered and eventually a fish kill may result. Eutrophication of lakes caused by human-induced processes, such as nutrient-rich sewage water entering a body of water, is called cultural eutrophication.

Even-aged stands   A forest of trees that began growth in or about the same year.

Evolution, biological   The change in inherited characteristics of a population from generation to generation, sometimes resulting in a new species.

Evolution, nonbiological   Outside the realm of biology, the term *evolution* is used broadly to mean the history and development of something.

Exotic species   Species introduced into a new area, one in which it had not evolved.

Experimental errors   There are two kinds of experimental errors, random and systematic. Random errors are those due to chance events, such as air currents pushing on a scale and altering a measurement of weight. In contrast, a miscalibration of an instrument would lead to a systematic error. Human errors can be either random or systematic.

Exponential growth   Growth in which the rate of increase is a constant percentage of the current size; that is, the growth occurs at a constant rate per time period.

Externality   In economics, an effect not normally accounted for in the cost-revenue analysis.

Extinction   Disappearance of a life-form from existence; usually applied to a species.

Facilitation   During succession, one species prepares the way for the next (and may even be necessary for the occurrence of the next).

Fact   Something that is known based on actual experience and observation.

Fall line   The point on a river where there is an abrupt drop in elevation of the land and where numerous waterfalls occur. The line in the eastern United States is located where streams pass from harder to softer rocks.

Fallow   A field allowed to grow with a cover crop without harvesting for at least one season.

Fecal coliform bacteria   Standard measure of microbial pollution and an indicator of disease potential for a water source.

Feedback   A kind of system response that occurs when output of the system also serves as input leading to changes in the system.

First law of thermodynamics   The principle that energy may not be created or destroyed but is always conserved.

First-law efficiency   The ratio of the actual amount of energy delivered where it is needed to the amount of energy supplied in order to meet that need; expressed as a percentage.

Fission   The splitting of an atom into smaller fragments with the release of energy.

Flood   Inundation of an area by water. Our often produced by processes including intense rain storms, melting of snow, storms surge from a hurricane, tsunami, or failure of a flood protection structure such as a dam.

Flooding   The natural process whereby waters emerge from their stream channel to cover part of the floodplain. Natural flooding is not a problem until people choose to build homes and other structures on floodplains.

Floodplain   Flat topography adjacent to a stream in a river valley that has been produced by the combination of overbank flow and lateral migration of meander bends.

Fluidized-bed combustion   A process used during the combustion of coal to eliminate sulfur oxides. Involves mixing finely ground limestone with coal and burning it in suspension.

Food chain   The linkage of who feeds on whom.

Food-chain concentration   See **Biomagnification**.

Food web   A network of who feeds on whom or a diagram showing who feeds on whom. It is synonymous with food chain.

Force   A push or pull that affects motion. The product of mass and acceleration of a material.

Forcing   With respect to global change, processes capable of changing global temperature, such as changes in solar energy emitted from the sun, or volcanic activity.

Fossil fuels   Forms of stored solar energy created from incomplete biological decomposition of dead organic matter. Includes coal, crude oil, and natural gas.

Fuel cell   A device that produces electricity directly from a chemical reaction in a specially designed cell. In the simplest case the cell uses hydrogen as a fuel, to which an oxidant is supplied. The hydrogen is combined with oxygen as if the hydrogen were burned, but the reactants are separated by an electrolyte solution that facilitates the migration of ions and the release of electrons (which may be tapped as electricity, energy source).

Fugitive sources   Type of stationary air pollution sources that generate pollutants from open areas exposed to wind processes.

Fusion, nuclear   Combining of light elements to form heavier elements with the release of energy.

Gaia hypothesis   The Gaia hypothesis states that the surface environment of Earth, with respect to such factors as the atmospheric composition of reactive gases (for example, oxygen, carbon dioxide, and methane), the acidity-alkalinity of waters, and the surface temperature, are actively regulated by the sensing, growth, metabolism and other activities of the biota. Interaction between the physical and biological system on Earth's surface has led to a planetwide physiology that began more than 3 billion years ago and the evolution of which can be detected in the fossil record.

Gamma rays   One of the three major kinds of nuclear radiation. A type of electromagnetic radiation emitted from the isotope similar to X rays but more energetic and penetrating.

Garden city Land planning that considers a city and countryside together.

Gene A single unit of genetic information comprised of a complex segment of the four DNA base-pair-compounds.

Genetic drift Changes in the frequency of a gene in a population as a result of chance rather than of mutation, selection, or migration.

Genetic risk Used in discussions of endangered species to mean detrimental change in genetic characteristics not caused by external environmental changes. Genetic changes can occur in small populations from such causes as reduced genetic variation, genetic drift, and mutation.

Genetically modified crops Crop species modified by genetic engineering to produce higher crop yields and increase resistance to drought, cold, heat, toxins, plant pests, and disease.

Genetically modified organisms Also known as genetic engineering, refers to the altering of genes or genetic material with the objectives of producing new organisms or organisms with desired characteristics, or to eliminate undesirable characteristics in organisms.

Geochemical cycles The pathways of chemical elements in geologic processes, including the chemistry of the lithosphere, atmosphere, and hydrosphere.

Geographic Information System (GIS) Technology capable of storing, retrieving, transferring, and displaying environmental data.

Geologic cycle The formation and destruction of earth materials and the processes responsible for these events. The geologic cycle includes the following sub-cycles: hydrologic, tectonic, rock, and geochemical.

Geometric growth See **Exponential growth**.

Geopressurized systems Geothermal system that exists when the normal heat flow from the Earth is trapped by impermeable clay layers that act as an effective insulator.

Geothermal energy The useful conversion of natural heat from the interior of Earth.

Global circulation models (GCM) A type of mathematical model used to evaluate global change, particularly related to climatic change. GCMs are very complex and require supercomputers for their operation.

Global dimming The process of the reduction of incoming solar radiation by reflection from suspended particles in the atmosphere and their interaction with water vapor (especially clouds).

Global forecasting Process of predicting or forecasting future change in environmental areas such as world population, natural resource utilization, and environmental degradation.

Global warming Natural or human-induced increase in the average global temperature of the atmosphere near Earth's surface.

Gravel Unconsolidated, generally rounded fragment of rocks and minerals greater than 2 mm in diameter.

Green plans Long-term strategies for identifying and solving global and regional environmental problems. The philosophical heart of green plants is sustainability.

Green revolution Name attached to post-World War II agricultural programs that have led to the development of new strains of crops with higher yield, better resistance to disease, or better ability to grow under poor conditions.

Greenbelt A belt of recreational parts, farmland, or uncultivated land surrounding or connecting urban communities, forming a system of countryside and urban landscapes.

Greenhouse effect Process of trapping heat in the atmosphere. Water vapor and several other gases warm the Earth's atmosphere because they absorb and remit radiation; that is, they trap some of the heat radiating from the Earth's atmospheric system.

Greenhouse gases The suite of gases that have a greenhouse effect, such as carbon dioxide, methane, and water vapor.

Gross production (biology) Production before respiration losses are subtracted.

Groundwater Water found beneath the Earth's surface within the zone of saturation, below the water table.

Growth efficiency Gross production efficiency (P/C), or ratio of the material produced (P = net production) by an organism or population to the material ingested or consumed (C).

Growth rate The net increase in some factor per unit time. In ecology, the growth rate of a population, sometimes measured as the increase in numbers of individuals or biomass per unit time and sometimes as a percentage increase in numbers or biomass per unit time.

Habitat Where an individual, population, or species exists or can exist. For example, the habitat of the Joshua tree is the Mojave Desert of North America.

Half-life The time required for half the amount of a substance to disappear; the average time required for one-half of a radioisotope to be transformed to some other isotope; the time required for one-half of a toxic chemical to be converted to some other form.

Hard path Energy policy based on the emphasis of energy quantity generally produced from large, centralized power plants.

Hazardous waste Waste that is classified as definitely or potentially hazardous to the health of people. Examples include toxic or flammable liquids and a variety of heavy metals, pesticides, and solvents.

Heat energy Energy of the random motion of atoms and molecules.

Heat island Warmer air of a city than surrounding areas as a result of increased heat production and decreased rate of heat loss caused by the abundance of building and paving materials, which act as solar collectors.

Heat island effect Urban areas are several degrees warmer than their surrounding areas. During relatively calm periods there is an upward flow of air over heavily developed areas accompanied by a downward flow over nearby greenbelts. This produces an air-temperature profile that delineates the heat island.

Heat pumps Devices that transfer heat from one material to another, such as from groundwater to the air in a building.

Heat wave A period of days of weeks of unusually hot weather. A natural recurring weather phenomenon related to heating of the atmosphere and moving of air masses.

Heavy metals   Refers to a number of metals, including lead, mercury, arsenic, and silver (among others) that have a relatively high atomic number (the number of protons in the nucleus of an atom). They are often toxic at relatively low concentrations, causing a variety of environmental problems.

Herbivore   An organism that feeds on an autotroph.

Heterotrophs   Organisms that cannot make their own food from inorganic chemicals and a source of energy and therefore live by feeding on other organisms.

High-level radioactive waste   Extremely toxic nuclear waste, such as spent fuel elements from commercial reactors. A sense of urgency surrounds determining how we may eventually dispose of this waste material.

Historical range of variation   The known range of an environmental variable, such as the abundance of a species or the depth of a lake, over some past time interval.

Homeostasis   The ability of a cell or organism to maintain a constant environment. Results from negative feedback, resulting in a state of dynamic equilibrium.

Hormonally active agents   Chemicals in the environment able to cause reproductive and developmental abnormalities in animals, including humans.

Hot igneous systems   A geothermal system that involves hot, dry rocks with or without the presence of near-surface molten rock.

Human carrying capacity   Theoretical estimates of the number of humans who could inhabit Earth at the same time.

Human demography   The study of human population characteristics, such as age structure, demographic transition, total fertility, human population and environment relationships, death-rate factors, and standard of living.

Hurricane   A tropical storm with circulating winds in excess of 120 km (74 mi) per hour that moves across warm ocean waters of the tropics.

Hutchinsonian niche   The idea of a measured niche, the set of environmental conditions within which a species is able to persist.

Hydrocarbons   Compounds containing only carbon and hydrogen are a large group of organic compounds, including petroleum products, such as crude oil and natural gas.

Hydrochlorofluorocarbons   Also known as HCFCs, which are a group of chemicals containing hydrogen, chlorine, fluorine, and carbon, produced as a potential substitute for chlorofluorocarbons (CFCs).

Hydrofluorocarbons   Chemicals containing hydrogen, fluorine, and carbons, produced as potential substitutes for chlorofluorocarbons (CFCs).

Hydrologic cycle   Circulation of water from the oceans to the atmosphere and back to the oceans by way of evaporation, runoff from streams and rivers, and groundwater flow.

Hydrology   The study of surface and subsurface water.

Hydroponics   The practice of growing plants in a fertilized water solution on a completely artificial substrate in an artificial environment such as a greenhouse.

Hydrothermal convection systems   A type of geothermal energy characterized by circulation of steam and/or hot water that transfers to the surface.

Hypothesis   In science, an explanation set forth in a manner that can be tested and is capable of being disproved. A tested hypothesis is accepted until and unless it has been disproved.

Igneous rocks   Rocks formed from the solidification of magma. They are extrusive if they crystallize on the surface of Earth and intrusive if they crystallize beneath the surface.

Incineration   Combustion of waste at high temperature, consuming materials and leaving only ash and non-combustibles to dispose of in a landfill.

Independent variable   See **Variable, independent.**

Indirect costs   See **Costs, indirect.**

Indirect effects   Effects from a natural hazard or disaster that includes donations of money and goods as well as providing shelter for people and paying taxes that will help finance recovery and relief of emotional distress caused by natural hazardous events.

Inductive reasoning   Drawing a general conclusion from a limited set of specific observations.

Industrial ecology   Process of designing industrial systems to behave more like ecosystems where waste from one part of the system is a resource for another part.

Inference   (1) A conclusion derived by logical reasoning from premises and/or evidence (observations or facts), or (2) a conclusion, based on evidence, arrived at by insight or analogy, rather than derived solely by logical processes.

Influent stream   Type of stream that is everywhere above the groundwater table and flows in direct response to precipitation. Water from the channel moves down to the water table, forming a recharge mound.

Inspirational justification for the conservation of nature   An argument for the conservation of nature on the grounds that direct experience of nature is an aid to spiritual or mental well-being.

In-stream use   A type of water use that includes navigation, generation of hydroelectric power, fish and wildlife habitat, and recreation.

Integrated energy management   Use of a range of energy options that vary from region to region, including a mix of technology and sources of energy.

Integrated pest management   Control of agricultural pests using several methods together, including biological and chemical agents. A goal is to minimize the use of artificial chemicals; another goal is to prevent or slow the buildup of resistance by pests to chemical pesticides.

Integrated waste management (IWM)   Set of management alternatives including reuse, source reduction, recycling, composting, landfill, and incineration.

Interference   During succession, one species prevents the entrance of later successional species into an ecosystem. For example, some grasses produce such dense and thick mats that seeds of trees cannot reach the soil to germinate. As long as these grasses persist, the trees that characterize later stages of succession cannot enter the ecosystem.

Island arc   A curved group of volcanic islands associated with a deep oceanic trench and subduction zone (convergent plate boundary).

Isotope   Atoms of an element that have the same atomic number (the number of protons in the nucleus of the atom) but vary in atomic mass number (the number of protons plus neutrons in the nucleus of an atom).

Keystone species   A species, such as the sea otter, that has a large effect on its community or ecosystem so that its removal or addition to the community leads to major changes in the abundances of many or all other species.

Kinetic energy   The energy of motion. For example, the energy in a moving car that results from the mass of the car traveling at a particular velocity.

Kwashiorkor   Lack of sufficient protein in the diet, which leads to a failure of neural development in infants and therefore to learning disabilities.

Land application   Method of disposal of hazardous waste that involves intentional application of waste material to surface soil. Useful for certain biodegradable industrial waste, such as oil and petroleum, and some organic chemical waste.

Land ethic   A set of ethical principles that affirm the right of all resources, including plants, animals, and earth materials, to continued existence and, at least in some locations, to continued existence in a natural state.

Landscape perspective   The concept that effective management and conservation recognizes that ecosystems, populations, and species are interconnected across large geographic areas.

Landslide   Comprehensive term for earth materials moving down slope.

Land-use planning   Complex process involving development of a land-use plan to include a statement of land-use issues, goals, and objectives; a summary of data collection and analysis; a land-classification map; and a report that describes and indicates appropriate development in areas of special environmental concern.

Late successional species   Species that occur only or primarily in, or are dominant in, late stages in succession.

Law of the minimum (Liebig's law of the minimum)   The concept that the growth or survival of a population is directly related to the single life requirement that is in least supply (rather than due to a combination of factors).

LD-50   A crude approximation of a chemical toxicity defined as the dose at which 50% of the population dies on exposure.

Leachate   Noxious, mineralized liquid capable of transporting bacterial pollutants. Produced when water infiltrates through waste material and becomes contaminated and polluted.

Leaching   Water infiltration from the surface, dissolving soil materials as part of chemical weathering processes and transporting the dissolved materials laterally or downward.

Lead   A heavy metal that is an important constituent of automobile batteries and other industrial products. A toxic metal capable of causing environmental disruption and producing a health problem to people and other living organisms.

Liebig's law of the minimum   See **Law of the minimum**.

Life expectancy   The estimated average number of years (or other time period used as a measure) that an individual of a specific age can expect to live.

Limiting factor   The single requirement for growth available in the least supply in comparison to the need of an organism. Originally applied to crops but now often applied to any species.

Lithosphere   Outer layer of Earth, approximately 100 km thick, of which the plates that contain the ocean basins and the continents are composed.

Littoral drift   Movement caused by wave motion in nearshore and beach environment.

Local extinction   The disappearance of a species from part of its range, but continued persistence elsewhere.

Logistic carrying capacity   In terms of the logistic curve, the population size at which births equal deaths and there is no net change in the population.

Logistic equation   The equation that results in a logistic growth curve, that is, the growth rate $dN/dt = rN[(K-N)/N]$, where r is the intrinsic rate of increase, K is the carrying capacity, and N is the population size.

Logistic growth curve   The S-shaped growth curve that is generated by the logistic growth equation. In the logistic, a small population grows rapidly, but the growth rate slows down, and the population eventually reaches a constant size.

Low-level radioactive waste   Waste materials that contain sufficiently low concentrations or quantities of radioactivity so as not to present a significant environmental hazard if properly handled.

Luz solar electric generating system   Solar energy farms comprising a power plant surrounded by hundreds of solar collectors (curved mirrors) that heat a synthetic oil, which flows through heat exchangers to drive steam turbine generators.

Macronutrients   Elements required in large amounts by living things. These include the big six-carbon, hydrogen, oxygen, nitrogen, phosphorus, and sulfur.

Made lands   Man-made areas created artificially with fill, sometimes as waste dumps of all kinds and sometimes directly, to make more land available for construction.

Magma   A naturally occurring silica melt, a good deal of which is a liquid state.

Malnourishment   The lack of specific components of food, such as proteins, vitamins, or essential chemical elements.

Manipulated variable   See **Variable, independent**.

Marasmus   Progressive emaciation caused by a lack of protein and calories.

Marginal cost   In environmental economics, the cost to reduce one additional unit of a type of degradation, for example, pollution.

Marginal land   An area of Earth with minimal rainfall or otherwise limited severely by some necessary factor, so that it is a poor place for agriculture and easily degraded by agriculture. Typically, these lands are easily converted to deserts even when used for light grazing and crop production.

Mariculture   Production of food from marine habitats.

Marine evaporites   With respect to mineral resources,

refs to materials such as potassium and sodium salts resulting from the evaporation of marine waters.

Materials management Term related to waste management consistent with the ideals of industrial ecology that makes better use of materials, leading to more sustainable use of resources.

Matter Anything that occupies space and has mass. It is the substance of which physical objects are composed.

Maximum lifetime Genetically determined maximum possible age to which an individual of a species can live.

Maximum sustainable population The largest population size that can be sustained indefinitely.

Maximum sustainable yield (MSY) The maximum usable production of a biological resource that can be obtained in a specified time period without decreasing the ability of the population to sustain that level of production.

Mediation Negotiation process between adversaries, guided by a neutral facilitator.

Megacities Urban areas with at least 8 million inhabitants.

Meltdown Refers to a nuclear accident in which the nuclear fuel forms a molten mass that breaches the containment of the reactor, contaminating the outside environment with radioactivity.

Methane (CH_4) A molecule of carbon and form hydrogen atoms. It is a naturally occurring gas in the atmosphere, one of the so-called greenhouse gases.

Methane hydrate A white ice-like compound made up of molecules of methane gas trapped in "cages" of frozen water in the sediments of the deep seafloor.

Microclimate The climate of a very small local area. For example, the climate under a tree, near the ground within a forest, or near the surface of streets in a city.

Micronutrients Chemical elements required in very small amounts by at least some forms of life. Boron, copper, and molybdenum are examples of micronutrients.

Micropower The production of electricity using smaller distributed systems rather than relying on large central power plants.

Migration The movement of an individual, population, or species from one habitat to another or more simply from one geographic area to another.

Migration corridor Designated passageways among parks or preserves allowing migration of many life-forms among several of these areas.

Mineral Naturally occurring inorganic material with a definite internal structure and physical and chemical properties that vary within prescribed limits.

Mineral resources Elements, chemical compounds, minerals, or rocks concentrated in a form that can be extracted to obtain a usable commodity.

Minimum viable population The minimum number of individuals that have a reasonable chance of persisting for a specified time period.

Missing carbon sink Substantial amounts of carbon dioxide released into the atmosphere but apparently not reabsorbed and thus remaining unaccounted for.

Mitigated negative declaration Special type of negative declaration that suggests that the adverse environmental aspects of a particular action may be mitigated through modification of the project in such a way as to reduce the impacts to near insignificance.

Mitigation Process that identifies actions to avoid, lessen, or compensate for anticipated adverse environmental impacts.

Mobile sources Sources of air pollutants that move from place to place, for example, automobiles, trucks, buses, and trains.

Model A deliberately simplified explanation, often physical, mathematical, pictorial, or computer-simulated, of complex phenomena or processes.

Monitoring Process of collecting data on a regular basis at specific sites to provide a database from which to evaluate change. For example, collection of water samples from beneath a landfill to provide early warning should a pollution problem arise.

Monoculture (Agriculture) The planting of large areas with a single species or even a single strain or subspecies in farming.

Moral justification for the conservation of nature An argument for the conservation of nature on the grounds that aspects of the environment have a right to exist, independent of human desires, and that it is our moral obligation to allow them to continue or to help them persist.

Multiple use Literally, using the land for more than one purpose at the same time. For example, forestland can be used to produce commercial timber but at the same time serve as wildlife habitat and land for recreation. Usually multiple use requires compromises and trade-offs, such as striking a balance between cutting timber for the most efficient production of trees at a level that facilitates other uses.

Mutation Stated most simply, a chemical change in a DNA molecule. It means that the DNA carries a different message than it did before, and this change can affect the expressed characteristics when cells or individual organisms reproduce.

Mutualism See **Symbiosis**.

Natural catastrophe Sudden catastrophic change in the environment, not the result of human actions.

Natural gas Naturally occurring gaseous hydrocarbon (predominantly methane) generally produced in association with crude oil or from gas wells; an important efficient and clean-burning fuel commonly used in homes and industry.

Natural hazard Any natural process that is a potential threat to human life and property,

Natural selection A process by which organisms whose biological characteristics better fit them to the environment are represented by more descendants in future generations than those whose characteristics are less fit for the environment.

Nature preserve An area set aside with the primary purpose of conserving some biological resource.

Negative declaration Document that may be filed if an agency has determined that a particular project will not have a significant adverse effect on the environment.

Negative feedback A type of feedback that occurs when the system's response is in the opposite direction of the output. Thus negative feedback is self-regulating.

Net growth efficiency Net production efficiency (P/A), or ratio of the material produced (P) to the material assimilated (A) by an organism. The material assimilated is less than the material consumed, because some food taken is egested as waste (discharged) and never used by an organism.

Net production (biology) The production that remains after utilization. In a population, net production is sometimes measured as the net change in the numbers of individuals. It is also measured as the net change in biomass or in stored energy. In terms of energy, it is equal to the gross production minus the energy used in respiration.

New forestry The name for a new variety of timber harvesting practices to increase the likelihood of sustainability, including recognition of the dynamic characteristics of forests and of the need for management within an ecosystem context.

Niche (1) The "profession," or role, of an organism or species; or (2) all the environmental conditions under which the individual or species can persist. The fundamental niche is all the conditions under which a species can persist in the absence of competition; the realized niche is the set of conditions as they occur in the real world with competitors.

Nitrogen cycle A complex biogeochemical cycle responsible for moving important nitrogen components through the biosphere and other Earth systems. This is an extremely important cycle because nitrogen is required by all living things.

Nitrogen fixation The process by which atmospheric nitrogen is converted to ammonia, nitrate ion, or amino acids. Microorganisms perform most of the conversion, but a small amount is also converted by lightning.

Nitrogen oxides Occur in several forms (NO, NO_2, and NO_3). Most important as an air pollutant is nitrogen dioxide, which is a visible yellow brown to reddish brown gas. It is a precursor of acid rain and produced through the burning of fossil fuels.

Noise pollution A type of pollution characterized by unwanted or potentially damaging sound.

Nonmarine evaporites With respect to mineral resources, refers to useful deposits of materials such as sodium and calcium bicarbonate, sulfate, borate, or nitrate produced by evaporation of surficial waters on the land, as differentiated from marine waters in the oceans.

Nonpoint sources Sources of pollutants that are diffused and intermittent and are influenced by factors such as land use, climate, hydrology, topography, native vegetation, and geology.

Nonrenewable energy Energy sources, including nuclear and geothermal, that are dependent on fuels, or a resource that may be used up much faster than it is replenished by natural processes.

Nonrenewable resource A resource that is cycled so slowly by natural Earth processes that once used, it is essentially not going to be made available within any useful time framework.

No-till agriculture Combination of farming practices that includes not plowing the land and using herbicides to keep down weeds.

Nuclear cycle The series of processes that begins with the mining of uranium to be processed and used in nuclear reactors and ends with the disposal of radioactive waste.

Nuclear energy The energy of the atomic nucleus that, when released, may be used to do work. Controlled nuclear fission reactions take place within commercial nuclear reactors to produce energy.

Nuclear fuel cycle Processes involved with producing nuclear power from the mining and processing of uranium to control fission, reprocessing of spent nuclear fuel, decommissioning of power plants, and disposal of radioactive waste.

Nuclear reactors Devices that produce controlled nuclear fission, generally for the production of electric energy.

Obligate symbionts A symbiotic relationship between two organisms in which neither by themselves can exist without the other.

Observations Information obtained through one or more of the five senses or through instruments that extend the senses. For example, some remote sensing instruments measure infrared intensity, which we do not see, and convert the measurement into colors, which we do see.

Ocean thermal conversion Direct utilization of solar energy using part of a natural oceanic environment as a gigantic solar collector.

Off-site effect An environmental effect occurring away from the location of the causal factors.

Off-stream use Type of water use where water is removed from its source for a particular use.

Oil shale A fine-grained sedimentary rock containing organic material known as kerogen. On distillation, yields significant amounts of hydrocarbons including oil.

Old growth forest A nontechnical term often used to mean a virgin forest (one never cut), but also used to mean a forest that has been undisturbed for a long, but usually unspecified, time.

Oligotrophic Referring to bodies of waters having a low concentration of the chemical elements required for life.

Omnivores Organisms that eat both plants and animals.

On-site effect An environmental effect occurring at the location of the causal factors.

Open dumps Area where solid waste is disposed of by simply dumping it. Often causes severe environmental problems, such as water pollution, and creates a health hazard. Illegal in the United States and in many other countries around the world.

Open system A type of system in which exchanges of mass or energy occur with other systems.

Open woodlands Areas in which trees are a dominant vegetation form but the leaves of adjacent trees generally do not touch or overlap, so that there are gaps in the canopy. Typically, grasses or shrubs grow in the gaps among the trees.

Operational definitions Definitions that tell you what you need to look for or do in order to carry out an operation, such as measuring, constructing, or manipulating.

Optimal carrying capacity A term that has several meanings, but the major idea is the maximum abundance of a population or species that can persist in an ecosystem without degrading the ability of the ecosystem to maintain: (1) that population or species; (2) all necessary ecosystem processes; and (3) the other species found in that ecosystem.

Optimum sustainable population (OSP) The population that is in some way best for the population, its ecological community, its ecosystem, or the biosphere.

Optimum sustainable yield (OSY) The largest yield of a renewable resource achievable over a long time period without decreasing the ability of the population, its ecosystem or its environment to support the continuation of this level of yield. OSY differs from maximum sustainable yield (MSY) by taking the ecosystem of a population into account.

Ore deposits Earth materials in which metals are concentrated in high concentrations, sufficient to be mined.

Organic compound A compound of carbon; originally used to refer to the compounds found in and formed by living things.

Organic farming Farming that is more "natural" in the sense that it does not involve the use of artificial pesticides and, more recently, genetically modified crops. In recent years governments have begun to set up legal criteria for what constitutes organic farming.

Overdraft Groundwater withdrawal when the amount pumped from wells exceeds the natural rate of replenishment.

Overgrazing When the carrying capacity of land for an herbivore, such as cattle or deer, is exceeded.

Overshoot and collapse Occurs when growth in one part of a system over time exceeds carrying capacity, resulting in sudden decline in one or both parts of the system.

Ozone (O_3) Form of oxygen in which three atoms of oxygen occur together. Is chemically active and has a short average lifetime in the atmosphere. Forms a natural layer high in the atmosphere (stratosphere) that protects us from harmful ultraviolet radiation from the sun. Is an air pollutant when present in the lower atmosphere above the National Air Quality Standards.

Ozone shield Stratospheric ozone layer that absorbs ultraviolet radiation.

Particulate matter Small particles of solid or liquid substances that are released into the atmosphere by many activities, including farming, volcanic eruption, and burning fossil fuels. Particulates affect human health, ecosystems, and the biosphere.

Passive solar energy system Direct use of solar energy through architectural design to enhance or take advantage of natural changes in solar energy that occur throughout the year without requiring mechanical power.

Pasture Land plowed and planted to provide forage for domestic herbivorous animals.

Pebble A rock fragment between 4 and 64 mm in diameter.

Pedology The study of soils.

Pelagic whaling Practice of whalers taking to the open seas and searching for whales from ships that remained at sea for long periods.

Per-capita availability The amount of a resource available per person.

Per-capita demand The economic demand per person.

Per-capita food production The amount of food produced per person.

Permafrost Permanently frozen ground.

Persistent organic pollutants Synthetic carbon-based compounds, often containing chlorine, that do not easily break down in the environment. Many were introduced decades before their harmful effects were fully understood and are now banned or restricted.

Pesticides, broad spectrum Pesticides that kill a wide variety of organisms. E.g. arsenic, one of the first elements used as a pesticide, is toxic to many life-forms, including people.

Phosphorus cycle Major biogeochemical cycle involving the movement of phosphorus throughout the biosphere and lithosphere. This cycle is important because phosphorus is an essential element for life and often is a limiting nutrient for plant growth.

Photochemical oxidants Result from atmospheric interactions of nitrogen dioxide and sunlight. Most common is ozone (O_3).

Photochemical smog Sometimes called L.A.-type smog or brown air. Directly related to automobile use and solar radiation. Reactions that occur in the development of the smog are complex and involve both nitrogen oxides and hydrocarbons in the presence of sunlight.

Photosynthesis Synthesis of sugars from carbon dioxide and water by living organisms using light as energy. Oxygen is given off as a by-product.

Photovoltaics Technology that converts sunlight directly into electricity using a solid semiconductor material.

Physiographic province A region characterized by a particular assemblage of landforms, climate, and geomorphic history.

Pioneer species Species found in early stages of succession.

Placer deposit A type of ore deposit found in material transported and deposited by agents such as running water, ice, or wind. Examples include gold and diamonds found in stream deposits.

Plantations Managed forests, in which a single species is planted in straight rows and harvested at regular intervals.

Plate tectonics A model of global tectonics that suggests that the outer layer of Earth, known as the lithosphere, is composed of several large plates that move relative to one another. Continents and ocean basins are passive riders on these plates.

Point sources Sources of pollution such as smokestacks, pipes, or accidental spills that are readily identified and stationary. They are often thought to be easier to recognize and control than are area sources. This is true only in a general sense, as some very large point sources emit tremendous amounts of pollutants into the environment.

Polar amplification Processes in which global warming causes greater temperature increases at polar regions.

Polar stratospheric clouds Clouds that form in the stratosphere during the polar winter.

Polar vortex Arctic air masses that in the winter become

isolated from the rest of the atmosphere and circulate about the pole. The vortex rotates counterclockwise because of the rotation of Earth in the Southern Hemisphere.

Policy instruments   The means to implement a society's policies. Includes moral suasion (jawboning—persuading people by talk, publicity, and social pressure); direct controls, including regulations; and market processes affecting the price of goods and processes, such as subsidies, licenses, and deposits.

Pollutant   In general terms, any factor that has a harmful effect on living things or their environment.

Pollution   The process by which something becomes impure, defiled, dirty, or otherwise unclean.

Pollution prevention   Identifying ways to avoid the generation of waste rather than finding ways to dispose of it.

Pool (geology)   Common bed form produced by scour in meandering and straight channels.

Population   A group of individuals of the same species living in the same area or interbreeding and sharing genetic information.

Population age structure   The number of individuals or proportion of the population in each age class.

Population dynamics   The study of changes in population sizes and the causes of these changes.

Population momentum or lag effect   The continued growth of a population that occurs after replacement-level fertility is reached.

Population regulation   See **Density-dependent population effects and Density-independent population effects.**

Population risk   Term used in discussions of endangered species to mean random variation in population rates—birth rates and death rates—possibly causing species in low abundance to become extinct.

Positive feedback   A type of feedback that occurs when an increase in output leads to a further increase in output. This is sometimes known as a vicious cycle, since the more you have the more you get.

Potential energy   Energy that is stored. Examples include the gravitational energy of water behind a dam; chemical energy in coal, fuel oil, and gasoline; and nuclear energy (in the forces that hold atoms together).

Power   The time rate of doing work.

Precautionary principle   The idea that in spite of the fact that full scientific certainty is often not available to prove cause and effect, we should still take cost-effective precautions to solve environmental problems when there exists a threat of potential serious and/or irreversible environmental damage.

Predation-parasitism   Interaction between individuals of two species in which the outcome benefits one and is detrimental to the other.

Predator   An organism that feeds on other, live organisms, usually of other species. The term is usually applied to animals that feed on other animals.

Premises   In science, initial definitions and assumptions.

Primary pollutants   Air pollutants emitted directly into the atmosphere. Included are particulates, sulfur oxides, carbon monoxide, nitrogen oxides, and hydrocarbons.

Primary production   See **Production, primary.**

Primary succession   The initial establishment and development of an ecosystem.

Primary treatment (of wastewater)   Removal of large particles and organic materials from wastewater through screening.

Probability   The relative probability that an event will occur.

Production, ecological   The amount of increase in organic matter, usually measured per unit area of land surface or unit volume of water, as in grams per square meter (g/m^2). Production is divided into *primary* (that of autotrophs) and *secondary* (that of heterotrophs). It is also divided into *net* (that which remains stored after use) and *gross* (that added before any use).

Production, primary   The production by autotrophs.

Production, secondary   The production by heterotrophs.

Productivity, ecological   The *rate* of production; that is, the amount of increase in organic matter per unit time (for example, grams per meter squared per year).

Prokaryote   A kind of organism that lacks a true cell nucleus and has other cellular characteristics that distinguish it from the *eukaryotes*. Bacteria are prokaryotes.

Pseudoscientific   Ideas that are claimed to have scientific validity, but are inherently untestable and/or lack empirical support and/or were arrived at through faulty reasoning or poor scientific methodology.

Public service functions   Functions performed by ecosystems that improve other forms of life in other ecosystems. Examples include the cleansing of the air by trees and removal of pollutants from water by infiltration through the soil.

Public trust   Grants and limits the authority of government over certain natural areas of special character.

Qualitative data   Data that are distinguished by qualities or attributes that cannot or are not expressed as quantities. For example, blue and red are qualitative data about the electromagnetic spectrum.

Quantitative data   Data that are expressed as numbers or numerical measurements. For example, the wavelengths of specific colors of blue and red light (460 and 650 nanometers, respectively) are quantitative data about the electromagnetic spectrum.

R-to-C ratio   A measure of the time available for finding the solutions to depletion of nonrenewable reserves, where R is the known reserves (for example, hundreds of thousands of tons of a metal) and C is the rate of consumption (for example, thousands of tons per year used by people).

Radiation absorbed dose   Energy retained by living tissue that has been exposed to radiation.

Radioactive decay   A process of decay of radioisotopes that change from one isotope to another and emit one or more forms of radiation.

Radioactive waste   Type of waste produced in the nuclear fuel cycle; generally classified as high level or low level.

Radioisotope   A form of a chemical element that spontaneously undergoes radioactive decay.

Radon   Naturally occurring radioactive gas. Radon is colorless, odorless, and tasteless and must be identified through proper testing.

Rangeland   Land used for grazing.

Rare species   Species with a small total population size, or restricted to a small area, but not necessarily declining or in danger of extinction.

Realms (Ecological) Major biogeographic regions of Earth in which most animals have some common genetic heritage.

Record of decision Concise statement prepared by the agency planning a proposed project that outlines the alternatives considered and discusses which alternatives are environmentally preferable.

Recreational justification for the conservation of nature An argument for the conservation of nature on the grounds that direct experience of nature is inherently enjoyable and the benefits derived from it are important and valuable to people.

Recycle Integral part of waste management that attempts to identify resources in the waste stream that may be collected and reused.

Reduce With respect to waste management, refers to practices that will reduce the amount of waste we produce.

Reduce, reuse, and recycle The three Rs of integrated waste management.

Renewable energy Alternative energy sources, such as solar, water, wind, and biomass, that are more or less continuously made available in a time framework useful to people.

Renewable resource A resource, such as timber, water, or air, that is naturally recycled or recycled by artificial processes within a time framework useful for people.

Replacement level fertility Fertility rate required for the population to remain a constant size.

Representative natural areas Parks or preserves set aside to represent presettlement conditions of a specific ecosystem type.

Reserves Known and identified deposits of earth materials from which useful materials can be extracted profitably with existing technology and under present economic and legal conditions.

Resource-based agriculture Agricultural practices that rely on extensive use of resources, so that production is limited by the availability of resources.

Resources Reserves plus other deposits of useful earth materials that may eventually become available.

Respiration The complex series of chemical reactions in organisms that make energy available for use. Water, carbon dioxide, and energy are the products of respiration.

Responding variable See Variable, dependent.

Restoration ecology The field within the science of ecology with the goal to return damaged ecosystems to ones that are functional, sustainable, and more natural in some meaning of this word.

Reuse With respect to waste management, refers to finding ways to reuse production and materials so they need not be disposed of.

Riffle A section of stream channel characterized at low flow by fast, shallow flow. Generally contains relatively coarse bed-load particles.

Risk The product of the probability of an event occurring and the consequences should that event occur.

Risk assessment The process of determining potential adverse environmental health effects to people following exposure to pollutants and other toxic materials. Generally includes the four steps of identification of the hazard, dose-response assessment, exposure assessment, and risk characterization.

Risk-benefit analysis In environmental economics, the riskiness of the future that influences the value we place on things in the present.

Rock (engineering) Any earth material that has to be blasted in order to be removed.

Rock (geologic) An aggregate of a mineral or minerals.

Rock cycle A group of processes that produce igneous, metamorphic, and sedimentary rocks.

Rotation time Time between cuts of a stand or area of forest.

Rule of climatic similarity Similar environments lead to the evolution of organisms similar in form and function (but not necessarily in genetic heritage or internal makeup) and to similar ecosystems.

Ruminants Animals having a four-chambered stomach within which bacteria convert the woody tissue of plants to proteins and fats that, in turn, are digested by the animal. Cows, camels, and giraffes are ruminants; horses, pigs, and elephants are not.

Sand Grains of sediment between 1/16 and 2 mm in diameter; often sediment composed of quartz particles of this size.

Sand dune A ridge or hill of sand formed by wind action.

Sanitary landfill A method of disposal of solid waste without creating a nuisance or hazard to public health or safety. Sanitary landfills are highly engineered structures with multiple barriers and collection systems to minimize environmental problems.

Savanna An area with trees scattered widely among dense grasses.

Scientific method A set of systematic methods by which scientists investigate natural phenomena, including gathering data, formulating and testing hypotheses, and developing scientific theories and laws.

Scientific theory A grand scheme that relates and explains many observations and is supported by a great deal of evidence, in contrast to a guess, a hypothesis, a prediction, a notion, or a belief.

Scoping The process of early identification of important environmental issues that require detailed evaluation.

Scrubbing A process of removing sulfur from gases emitted from power plants burning coal. The gases are treated with a slurry of lime and limestone, and the sulfur oxides react with the calcium to form insoluble calcium sulfides and sulfates that are collected and disposed of.

Second growth A forest that has been logged and regrown.

Second law of thermodynamics A fundamental principle of energy that states that energy always tends to go from a more usable (higher quality) form to a less usable (lower quality) form. When we say that energy is converted to a less useful form we mean that entropy (a measure of the energy unavailable to do useful work) of the system has increased.

Secondary enrichment A weathering process of sulfide ore deposits that may concentrate the desired minerals.

Secondary pollutants Air pollutants produced through reactions between primary pollutants and normal atmospheric compounds. An example is ozone that forms over urban areas through reactions of primary pollutants, sunlight, and natural atmospheric gases.

Secondary production See **Production, secondary.**

Secondary succession The reestablishment of an ecosystem where there are remnants of a previous biological community.

Secondary treatment (of wastewater) Use of biological processes to degrade wastewater in a treatment facility.

Second-law efficiency The ratio of the minimum available work needed to perform a particular task to the actual work used to perform that task. Reported as a percentage.

Secure landfill A type of landfill designed specifically for hazardous waste. Similar to a modern sanitary landfill in that it includes multiple barriers and collection systems to ensure that leachate does not contaminate soil and other resources.

Sediment pollution By volume and mass, sediment is our greatest water pollutant. It may choke streams, fill reservoirs, bury vegetation, and generally create a nuisance that is difficult to remove.

Seed-tree cutting A logging method in which mature trees with good genetic characteristics and high seed production are preserved to promote regeneration of the forest. It is an alternative to clear-cutting.

Seismic Referring to vibrations in Earth produced by earthquakes.

Selective cutting In timber harvesting, the practice of cutting some, but not all, trees, leaving some on the site. There are many kinds of selective cutting. Sometimes the biggest trees with the largest market value are cut, and smaller trees are left to be cut later. Sometimes the best trees are left to provide seed for future generations. Sometimes trees are left for wildlife habitat and recreation.

Shelterwood cutting A logging method in which dead and less desirable trees are cut first; mature trees are cut later. This ensures that young, vigorous trees will always be left in the forest. It is an alternative to clear-cutting.

Sick building syndrome Condition associated with a particular indoor environment that appears to be unhealthy to the human occupants.

Silicate minerals The most important group of rock-forming minerals.

Silt Sediment between 1/16 and 1/256 mm in diameter.

Silviculture The practice of growing trees and managing forests, traditionally with an emphasis on the production of timber for commercial sale.

Sinkhole A surface depression formed by the solution of limestone or the collapse over a subterranean void such as a cave.

Site (in relation to cities) Environmental features of a location that influences the placement of a city. For example, New Orleans is built on low-lying muds, which form a poor site, while New York City's Manhattan is built on an island of strong bedrock, an excellent site.

Site quality Used by foresters to mean an estimator of the maximum timber crop the land can produce in a given time.

Situation (in relation to cities) The relative geographic location of a site that makes it a good location for a city. For example, New Orleans has a good situation because it is located at the mouth of the Mississippi River and is therefore a natural transportation junction.

Smog A term first used in 1905 for a mixture of smoke and fog that produced unhealthy urban air. There are several types of fog, including photochemical smog and sulfurous smog.

Soft path Energy policy that relies heavily on renewable energy resources as well as other sources that are diverse, flexible, and matched to the end-use needs.

Soil The top layer of a land surface where the rocks have been weathered to small particles. Soils are made up of inorganic particles of many sizes, from small clay particles to large sand grains. Many soils also include dead organic material.

Soil (in engineering) Earth material that can be removed without blasting.

Soil (in soil science) Earth material modified by biological, chemical, and physical processes such that the material will support rooted plants.

Soil fertility The capacity of a soil to supply the nutrients and physical properties necessary for plant growth.

Soil horizon A layer in soil (A,B,C) that differs from another layer in chemical, physical, and biological properties.

Solar cell (photovoltaic) Device that directly converts light into electricity.

Solar collector Device for collecting and storing solar energy. For example, home water heating is done by flat panels consisting of a glass cover plate over a black background on which water is circulated through tubes. Short-wave solar radiation enters the glass and is absorbed by the black background. As long-wave radiation is emitted from the black material, it cannot escape through the glass, so the water in the circulating tubes is heated, typically to temperatures of 38° to 93°C.

Solar energy Collecting and using energy from the sun directly.

Solar pond Shallow pond filled with water and used to generate relatively low-temperature water.

Solar power tower A system of collecting solar energy that delivers the energy to a central location where the energy is used to produce electric power.

Source reduction Process of waste management, the object of which is to reduce the amounts of materials that must be handled in the waste stream.

Species A group of individuals capable of interbreeding.

Stable equilibrium A condition in which a system will remain if undisturbed and to which it will return when displaced.

Stand An informal term used by foresters to refer to a group of trees.

Stationary sources Air pollution sources that have a relatively fixed location, including point sources, fugitive sources, and area sources.

Steady state When input equals output in a system, there is no net change and the system is said to be in a steady state. A bathtub with water flowing in and out at the same rate maintains the same water level and is in a steady state. Compare with **equilibrium**.

Stress   Force per unit area. May be compression, tension, or shear.

Strip cutting   In timber harvesting, the practice of cutting narrow rows of forest, leaving wooded corridors.

Strip mining   Surface mining in which the overlying layer of rock and soil is stripped off to reach the resource. Large strip mines are some of the largest excavations caused by people in the world.

Subduction   A process in which one lithospheric plate descends beneath another.

Subsidence   A sinking, settling, or otherwise lowering of parts of crust.

Subsistence crops   Crops used directly for food by a farmer or sold locally where the food is used directly.

Succession   The process of establishment and development of an ecosystem.

Sulfur dioxide (SD_2)   Colorless and odorless gas normally present at Earth's surface in low concentrations. An important precursor to acid rain. Major anthropogenic source is burning fossil fuels.

Sulfurous smog   Produced primarily by burning coal or oil at large power plants. Sulfur oxides and particulates combine under certain meteorological conditions to produce a concentrated form of this smog.

Suppressed   In forestry, tree species growing in the understory, beneath the dominant and intermediate species.

Surface impoundment   Method of disposal of some liquid hazardous waste. This method is controversial, and many sites have been closed.

Sustainability   Management of natural resources and the environment with the goals of allowing the harvest of resources to remain at or above some specified level, and the ecosystem to retain its functions and structure.

Sustainable development   The ability of a society to continue to develop its economy and social institutions and also maintain its environment for an indefinite time.

Sustainable ecosystem   An ecosystem that is subject to some human use, but at a level that leads to no loss of species or of necessary ecosystem functions.

Sustainable energy development   A type of energy management that provides for reliable sources of energy while not causing environmental degradation and ensuring that future generations will have a fair share of the Earth's resources.

Sustainable forestry   Effort to manage a forest so that a resource in it can be harvested at a rate that does not decrease the ability of the forest ecosystem to continue to provide that same rate of harvest indefinitely.

Sustainable harvest   An amount of a resource that can be harvested at regular intervals indefinitely.

Sustainable water use   Use of water resources that does not harm the environment and provides for the existence of high-quality water for future generations.

Symbiont   Each partner in symbiosis.

Symbiosis   An interaction between individuals of two different species that benefits both. For example, lichens contain an alga and a fungus that require each other

to persist. Sometimes this term is used broadly, so that domestic corn and people could be said to have a symbiotic relationship—domestic corn cannot reproduce without the aid of people, and some people survive because they have corn to eat.

Symbiotic   Relationships that exist between different organisms that are mutually beneficial.

Synergism   Cooperative action of different substances such that the combined effect is greater than the sum of the effects taken separately.

Synergistic effect   When the change in availability of one resource affects the response of an organism to some other resource.

Synfuels   Synthetic fuels, which may be liquid or gaseous, derived from solid fuels, such as oil from kerogen in oil shale or oil and gas from coal.

Synthetic organic compounds   Compounds of carbon produced synthetically by human industrial processes, as for example pesticides and herbicides.

System   A set of components that are linked and interact to produce a whole. For example, the river as a system is composed of sediment, water, bank, vegetation, fish, and other living things that all together produce the river.

Taiga   Forest of cold climates of high latitudes and high altitudes, also known as boreal forest.

Tar sands   Sedimentary rocks or sands impregnated with tar oil, asphalt, or bitumen.

Taxa   Categories that identify groups of living organisms based on evolutionary relationships or similarity of characters.

Taxon   A grouping of organisms according to evolutionary relationships.

TD-50   The toxic dose defined as the dose that is toxic to 50% of a population exposed to the toxin.

Tectonic cycle   The processes that change Earth's crust, producing external forms such as ocean basins, continents, and mountains.

Terminator gene   A genetically modified crop that has a gene to cause the plant to become sterile after the first year.

Tertiary treatment (of wastewater)   Advanced form of wastewater treatment involving chemical treatment or advanced filtration. An example is chlorination of water.

Theories   Scientific models that offer broad, fundamental explanations of related phenomena and are supported by consistent and extensive evidence.

Thermal (heat energy)   The energy of the random motion of atoms and molecules.

Thermal pollution   A type of pollution that occurs when heat is released into water or air and produces undesirable effects on the environment.

Thermodynamic system   Formed by an energy source, ecosystem, and energy sink, where the ecosystem is said to be an intermediate system between the energy source and the energy sink.

Thermodynamics, first law of   See **First law of Thermodynamics.**

Thermodynamics, second law of   See **Second law of Thermodynamics.**

Thinning   The timber-harvesting practice of selectively removing only smaller or poorly formed trees.

Threatened species   Species experiencing a decline in the number of individuals to the degree that a concern is raised about the possibility of extinction of that species.

Threshold   A point in the operation of a system at which a change occurs. With respect to toxicology, it is a level below which effects are not observable and above which effects become apparent.

Tidal power   Form of water utilizing ocean tides in places where favorable topography allows for construction of a power plant.

Time series   The set of estimates of some variable over a number of years.

Tolerance   The ability to withstand stress resulting from exposure to a pollutant or harmful condition.

Tornado   A funnel shaped cloud of violently rotating air that extends downwards from large thunder storms to contact the surface of earth.

Total fertility rate (TFR)   Average number of children expected to be born to a woman during her lifetime. (Usually defined as the number born to a woman between the ages of 15 and 44, taken conventionally as the lower and upper limit of reproductive ages for women.)

Toxic   Harmful, deadly, or poisonous.

Toxicology   The science concerned with study of poisons (or toxins) and their effects on living organisms. The subject also includes the clinical, industrial, economic, and legal problems associated with toxic materials.

Transuranic waste   Radioactive waste consisting of human-made radioactive elements heavier than uranium. Includes clothing, rags, tools, and equipment that has been contaminated.

Trophic level   In an ecological community, all the organisms that are the same number of food-chain steps from the primary source of energy. For example, in a grassland the green grasses are on the first trophic level, grasshoppers are on the second, birds that feed on grasshoppers are on the third, and so forth.

Trophic level efficiency   The ratio of the biological production of one trophic level to the biological production of the next lower trophic level.

Tundra   The treeless land area in alpine and arctic areas characterized by plants of low stature and including bare areas without any plants and areas covered with lichens, mosses, grasses, sedges, and small flowering plants, including low shrubs.

Ubiquitous species   Species that are found almost anywhere on Earth.

Ultraviolet A   The longest wavelength of ultraviolet radiation (0.32–0.4 micrometers), not affected by stratospheric ozone, and is transmitted to the surface of Earth.

Ultraviolet B   Intermediate-wavelength radiation which is the ozone problem. Wavelengths are from approximately 0.28–0.32 micrometers and is the most harmful of the ultraviolet radiation types. Most of this radiation is absorbed by stratospheric ozone, and depletion of ozone has led to increase in ultraviolet B radiation reaching Earth.

Ultraviolet C   Shortest wavelength of the ultraviolet radiation with wavelengths from approximately 0.2–0.28 micrometers. Is the most energetic of the ultraviolet radiation and is absorbed strongly in the atmosphere.

Only a negligible amount of Ultraviolet C reaches the surface of Earth.

Undernourishment   The lack of sufficient calories in available food, so that one has little or no ability to move or work.

Uneven-aged stands   A forest stand with at least three distinct age classes.

Unified soil classification system   A classification of soils, widely used in engineering practice, based on the amount of coarse particles, fine particles, or organic material.

Uniformitarianism   The principle stating that processes that operate today operated in the past. Therefore, observations of processes today can explain events that occurred in the past and leave evidence, for example, in the fossil record or in geologic formations.

Urban dust dome   Polluted urban air produced by the combination of lingering air and abundance of particulates and other pollutants in the urban air mass.

Urban forestry   The practice and profession of planting and maintaining trees in cities, including trees in parks and other public areas.

Utilitarian justification for the conservation of nature   An argument for the conservation of nature on the grounds that the environment, an ecosystem, habitat, or species, provides individuals with direct economic benefit or is directly necessary to their survival.

Utility   Value or worth in economic terms.

UVA   Least energetic form of ultraviolet radiation. It is capable of causing some damage to living cells, is not affected by stratospheric ozone, and is therefore transmitted to Earth.

UVB   Intermediate-wavelength ultraviolet radiation, damaging to living cells. Most is absorbed by stratospheric ozone, and therefore depletion of ozone leads to significant increase of this radiation. This is the ozone problem.

UVC   Shortest wavelength and most energetic of the ultraviolet radiation. It is strongly absorbed in the atmosphere, and negligible amounts reach the surface of Earth.

Vadose zone   Zone or layer above the water table where water may be stored as it moves laterally or down to the zone of saturation. Part of the vadose zone may be saturated part of the time.

Variable, dependent   A variable that changes in response to changes in an independent variable; a variable taken as the outcome of one or more other variables.

Variable, independent   In an experiment, the variable that is manipulated by the investigator. In an observational study, the variable that is believed by the investigator to affect an outcome, or dependent, variable.

Variable, manipulated   See **Variable, independent**.

Variable, responding   See **Variable, dependent**.

Virgin forest   A forest that has never been cut.

Volcanic eruption   Extrusion at the surface of earth of molten rock (magma). Maybe be explosive and violent or less energetic lava flows.

Vulnerable species   Another term for *threatened species*—species experiencing a decline in the number of individuals.

Waldsterben   German phenomenon of forest death as the result of acid rain, ozone, and other air pollutants.

Wallace's realms Six biotic provinces, or biogeographic regions, divided on the basis of fundamental inherited features of the animals found in those areas, suggested by A. R. Wallace (1876). His realms are Nearctic (North America), Neotropical (Central and South America), Palearctic (Europe, northern Asia, and northern Africa), Ethiopian (central and southern Africa), Oriental (the Indian subcontinent and Malaysia), and Australian.

Waste water renovation and conservation cycle Practice of applying wastewater to the land. In some systems, treated wastewater is applied to agricultural crops, and as the water infiltrates through the soil layer it is naturally purified. Reuse of the water is by pumping it out of the ground for municipal or agricultural uses.

Wastewater treatment Process of treating wastewater (primarily sewage) in specially designed plants that accept municipal wastewater. Generally divided into three categories: primary treatment, secondary treatment, and advanced wastewater treatment.

Water budget Inputs and outputs of water for a particular system (a drainage basin, region, continent, or the entire Earth).

Water conservation Practices designed to reduce the amount of water we use.

Water power Alternative energy source derived from flowing water. One of the world's oldest and most common energy sources. Sources vary in size from micro-hydropower systems to large reservoirs and dams.

Water reuse The use of wastewater following some sort of treatment. Water reuse may be inadvertent, indirect, or direct.

Water table The surface that divides the zone of aeration from the zone of saturation, the surface below which all the pore space in rocks is saturated with water.

Watershed An area of land that forms the drainage of a stream or river. If a drop of rain falls anywhere within a watershed to become surface runoff, it can flow out only through the same stream.

Weathering Changes that take place in rocks and minerals at or near the surface of Earth in response to physical, chemical, and biological changes; the physical, chemical, and biological breakdown of rocks and minerals.

Wetlands Comprehensive term for landforms such as salt marshes, swamps, bogs, prairie potholes, and vernal pools. Their common feature is that they are wet at least part of the year and as a result have a particular type of vegetation and soil. Wetlands form important habitats for many species of plants and animals, while serving a variety of natural service functions for other ecosystems and people.

Wilderness An area unaffected now or in the past by human activities and without a noticeable presence of human beings.

Wildfire Self-sustaining rapid oxidation that releases light, heat, carbon dioxide and other gases as it moves across the landscape. Also known as a fire in the natural environment that maybe initiated by natural processes such as lighting strike or deliberately set by humans.

Wind power Alternative energy source that has been used by people for centuries. More recently, thousands of windmills have been installed to produce electric energy.

Work (physics) Force times the distance through which it acts. When work is done we say energy is expended.

Zero population growth A population in which the number of births equals the number of deaths so that there is no net change in the size of the population.

Zone of aceration The zone or layer above the water table in which some water may be suspended or moving in a downward migration toward the water table or laterally toward a discharge point is often called the vadose zone.

Zone of saturation Zone or layer below the water table in which all the pore space of rock or soil is saturated.

Zooplankton Small aquatic invertebrates that live in the sunlit waters of streams, lakes, and oceans and feed on algae and other invertebrate animals.

NOTES

Chapter 1 Notes

1. Gordon, B. B. 1993. Pampering our coastlines. *Sea Frontiers* 39(2):5.

2. Mydans, S. 1996 (April 28). Thai shrimp farmers facing ecologist's fury. *New York Times.*

3. Pollution wiping out shrimp farms on main Indonesian Island of Java. 1996. *Quick Frozen Foods International* 37(3):50.

4. Quarto, A. 1994. Rainforests of the sea: Mangrove forests threatened by prawn aquaculture. *E* 5(1):16–19.

5. Wickramayanake, S. D. 1995. East Coast shrimp farms face trouble from both nature and protest groups. *Quick Frozen Foods International* 37(1):108–109.

6. Haub, C., and D. Cornelius. 2000. *Nine million world population by 2050.* Washington, D.C.: Population Reference Bureau.

7. Everett, G. D. 1961. One man's family. *Population Bulletin* 17:153–169.

8. Ehrlich, P. R., A. H. Ehrlich, and P. H. Holdren. 1977. *Ecoscience: Population, resources, environment,* 3d ed., San Francisco: Freeman.

9. Deevey, E. S. 1960. The human population. *Scientific American* 203:194–204.

10. Keyfitz, N. 1989. The growing human population. *Scientific American.* 261:118–126.

11. Gottfield, R. S. 1983. *The black death: Natural and human disaster in medieval Europe.* New York: Free Press.

12. Field, J. O., ed. 1983. *The challenge of famine: Recent experience, lessons learned.* Hartford, Conn.: Kumarian Press.

13. Glantz, M. H., ed. 1987. *Drought and hunger in Africa: Denying famine a future.* Cambridge: Cambridge University Press.

14. Seavoy, R. E. 1989. *Famine in East Africa: Food production and food politics.* Westport, Conn.: Greenwood.

15. United Nations Food and Agriculture Organization. 2002. *Global Information and Early Warning System (GIEWS),* Food supply situation and crop prospects in sub-Saharan Africa. New York.

16. Gower, B. S. 1992. What do we owe future generations? In D. E. Cooper and J. A. Palmers, eds., *The environment in question: Ethics and global issues,* pp. 1–12. New York: Routledge.

17. Haub, C., and D. Cornelius. 2000. *World population data sheet.* Washington, D.C.: Population Reference Bureau.

18. Bartlett, A. A. 1997–98. Reflections on sustainability, population growth, and the environment, revisited. *Renewable Resources Journal* 15(4):6–23.

19. Hawken, P., A. Lovins, and L. H. Lovins. 1999. *Natural capitalism.* Boston: Little, Brown.

20. Hubbard, B. M. 1998. *Conscious evolution.* Novato, Calif.: New World Library.

21. Cummins, K., D. B. Botkin, T. Dunne, H. Regier, M. J. Sobel, and L. M. Talbot. 1955. *Status and future of salmon of western Oregon and northern California: Findings and options.* Santa Barbara, Calif.: Center for the Study of the Environment.

22. Margulis, L., and J. E. Lovelock. 1989. *Urban growth.* Washington, D.C.: World Resources Institute.

23. Botkin, D. B., M. Caswell, J. E. Estes, and A. Orio, eds. 1989. *Changing the global environment: Perspectives on human involvement.* New York: Academic Press.

24. World Resources Institute. 1998. *Teacher's guide to world resources: Exploring sustainable communities.* Washington, D.C.: World Resources Institute. http://www.igc.org/wri/wr-98-99/citygrow.htm.

25. World Resources Institute. 1999. *Urban growth.* Washington, D.C.: World Resources Institute.

26. Foster, K. R., P. Vecchia, and M. H. Repacholi. 2000. Science and the precautionary principle. *Science* 288: 979–981.

27. Easton, T. A., and T. D. Goldfarb, eds. 2003. Taking sides, environmental issues, 10th ed. Issue 5. Is the Precautionary Principle a sound basis for international policy? pp. 76–101. Guilford, Conn.: McGraw-Hill/Dushkin.

28. Nash, R. F. 1988. *The rights of nature: A history of environmental ethics.* Madison: University of Wisconsin Press.

Chapter 1 Critical Thinking Issue Notes

Bryant, Dirk, Auretta Burke, John McManus, and Mark Spalding. 1998. *Reefs at risk: A map-based indicator of threats to the world's coral reefs.* Washington, D.C.: World Resources Institute.

Coles, S. L., and L. Ruddy. 1995. Comparison of water quality and reef coral mortality and growth in southeastern Kaneohe Bay, Oahu, Hawaii, 1990 to 1992, with conditions before sewage diversion, *Pacific Science* 49(3): 247–265.

Hinrichsen, D. 1997 (October). Coral Reefs in Crisis, *Bioscience* 47(9): 554–558.

Jameson, S. C., J. W. McManus, and M. D. Spalding. 1995 (May). State of the reefs: Regional and global perspectives. International Coral Reef Initiative Executive Secretariat (Background Paper).

Chapter 2 Notes

1. Wiens, J. A., D. T. Pattern, and D. B. Botkin. 1993. Assessing ecological impact assessment: Lessons from Mono Lake, California. *Ecological Applications* 3(4): 595–609; and Botkin, D. B., W. S. Broecker, L. G. Everett, J. Shapiro, and J. A. Wiens. 1988. *The future of Mono Lake* (Report No. 68). Riverside: California Water Resources Center, University of California.

2. Botkin, D. B. 2001. *No man's garden: Thoreau and a new vision for civilization and nature.* Washington, D.C.: Island Press.

3. Taylor, F. S. 1949. *Science and scientific thought.* New York: Norton.

4. Schmidt, W. E. 1991 (September 10). "Jovial con men" take credit (?) for crop circles. *New York Times,* p. B1.

5. Tuohy, W. 1991 (September 10). "Crop circles" their prank, 2 Britons say, *Los Angeles Times,* p. A14.

6. This information about crop circles is from *Crop Circle News,* http://cropcirclenews.com

7. Gibbs, A., and A. E. Lawson. 1992. The nature of scientific thinking as reflected by the work of biologists and by biology textbooks. *The American Biology Teacher* 54:137-152.

8. Pease, C. M., and J. J. Bull. 1992. Is science logical? *Bioscience* 42:293–298.

9. Lerner, L. S., and W. J. Bennetta. 1988 (April). The treatment of theory in textbooks. *The Science Teacher*, pp. 37–41.

10. Kuhn, T. S. 1970. *The structure of scientific revolutions.* Chicago: University of Chicago Press.

11. Trefil, J. S. 1978. A consumer's guide to pseudoscience. *Saturday Review* 4:16–21.

12. Hastings and Hastings. 1992. Telepathy. American Institute of Public Opinion poll from June 1990.

13. The Peregrine Fund, www.peregrinefund.org/cordorfactsheet.asp (3):March 7, 2006.

14. Feynman, R. P. 1998. *The meaning of it all: Thoughts of a citizen-scientist.* Reading, Mass.: Addison-Wesley, Perseus Books, p. 8. (From lectures at the University of Washington in 1963.)

15. Heinselman, H. M. (1973). Fire in the virgin forests of the boundary waters canoe area, Minnesota. *Journal of Quaternary Reasearch* 3:329–382.

Chapter 2 Critical Thinking Issue References

Ford, R. 1998 (March). "Critically Evaluating Scientific Claims in the Popular Press." *The American Biology Teacher* 60(3): 174–180.

Marshall, E. 1998 (May 15). "The Power of the Front Page of *The New York Times.*" *Science* 280: 996–997.

Chapter 3 Notes

1. Western, D., and C. Van Prat. 1973. Cyclical changes in habitat and climate of an East African ecosystem. *Nature* 241(549):104–106.

2. Dunne, T., and L. B. Leopold. 1978. *Water in environmental planning.* San Francisco: Freeman.

3. Altmann, J., S. C. Alberts, S. A. Altman, and S. B. Roy, 2002. Dramatic change in local climate patterns in the Amboseli basin, Kenya. *African Journal of Ecology*, 40:248-251.

4. Bartlett, A. A. 1980. Forgotten fundamentals of the energy crisis. *Journal of Geological Education* 28:4–35.

5. Meadows, D. H., D. L. Meadows, and J. Randers. 1992. *Beyond the limits: Confronting global collapse; envisioning a sustainable future.* Post Mills, Vt.: Chelsea Green Publishers.

6. Wootton, J. T., M. S. Parker, and M. E. Power. 1996. Effects of disturbances on river food webs. *Science* 273:1558–1561.

7. Leach, M. K., and T. J. Givnich. 1996. Ecological determinants of species loss in remnant prairies. *Science* 273:1555–1558.

8. Lovelock, J. 1995. *The ages of Gaia: A biography of our living earth,* rev. ed. New York: Norton.

9. Gardner, G. T., and P. C. Stern. 2002. *Environmental problems and human behavior,* 2d ed. Boston: Pearson Custom Publishing.

Chapter 3 Critical Thinking Issue References

Barlow, C. 1993. *From Gaia to selfish genes.* Cambridge, Mass.: MIT Press.

Kirchner, J. W. 1989. The Gaia hypothesis: Can it be tested? *Reviews of Geophysics* 27:223–235.

Lovelock, J. E. 1995. *Gaia: A new look at life on Earth.* New York: Oxford University Press.

Lyman, F. 1989. What hath Gaia wrought? *Technology Review* 92(5):55–61.

Resnik, D. B. 1992. Gaia: From fanciful notion to research program. *Perspectives in Biology and Medicine* 35(4):572–582.

Schneider, S. H. 1990. Debating Gaia. *Environment* 32(4):4–9, 29–32.

Chapter 3 Working It Out 3.1 Reference

Bartlett, A. A. 1993. The arithmetic of growth: Methods of calculation. *Population and Environment* 14(4):359–387.

Chapter 4 Notes

1. Population Reference Bureau, *2005 World Population Data Sheet.*

2. Keyfitz, N. 1992. Completing the worldwide demographic transition: The relevance of past experience. *AMBIO* 21:26–30.

3. Graunt, J. [1662] 1973. *Natural and political observations made upon the bill of mortality.* London, 662.

4. Dumond, D. E. 1975. The limitation of human population: A natural history. *Science* 187:713–721.

5. Zero Population Growth. 2000. *U.S. population.* Washington, D.C.: Zero Population Growth.

6. World Bank. 1984. *World development report 1984.* New York: Oxford University Press.

7. Erhlich, P. R. 1971. *The population bomb,* rev. ed. New York: Ballantine.

8. Malthus, T. R. [1803] 1992. *An essay on the principle of population.* Selected and introduced by Donald Winch. Cambridge, England: Cambridge University Press.

9. Bureau of the Census. 1990. *Statistical abstract of the United States 1990.* Washington, D.C.: U.S. Department of Commerce.

10. Xinhua News Agency, China's cross-border tourism prospers in 2002, December 31, 2002. From the Population Reference Bureau Web site available at http://www.prb.org/Template.cfm?Section=PRB&template=/ContentManagement/ContentDisplay.cfm&ContentID=8661.

11. U.S. Centers for Disease Control Web site available at http://www.cdc.gov/ncidod/sars/factsheet.htm and http://www.cdc.gov/ncidod/dvbid/westnile/qa/overview.htm

12. Population Reference Bureau Web site, available at http://www.prb.org/Template.cfm?Section=PRB&template=/ContentManagement/ContentDisplay.cfm&ContentID=8661.

13. Jamie Shreeve, *New York Times,* January 29, 2006. Why revive a deadly flu virus?

14. UN World Health Organization, http://www.who.int/mediacentre/news/statements/2006/s03/en/index.html February 12, 2006.

15. Central Intelligence Agency. 1999. *The world factbook.* Washington, D.C.: CIA.

16. World Bank. 1992. *World development report. The relevance of past experience.* Washington, D.C.: World Bank.

17. Guz, D., and J. Hobcraft. 1991. Breastfeeding and fertility: A comparative analysis. *Population Studies* 45:91–108.

18. Fathalla, M. F. 1992. Family planning: Future needs. *AMBIO* 21:84–87.

19. Alan Guttmacher Institute. *Sharing responsibility: Women, society and abortion worldwide.* New York: AGI, 1999.

20. Haupt, A., and T. T. Kane. 1978. *The Population Reference Bureau's population handbook*. Washington, D.C.: Population Reference Bureau.

21. Planned Parenthood Federation of America, Public Policy Division. 1997 (June). *International family planning: The need for services*. Planned Parenthood Federation of America. New York.

22. Xinhua News Agency March 13, 2002 untitled, available at http://www.16da.org.cn/english/archiveen/28691.htm.

Chapter 5 Notes

1. Lehman, J. T. 1986. Control of eutrophication in Lake Washington. In G. H. Orians, ed., *Ecological knowledge and environmental problem solving*, pp. 302–316. Washington, D.C.: National Academy of Science.

2. Henderson, L. J. [1913] 1966. *The fitness of the environment*. Boston: Beacon.

3. Van Koevering, T. E., and N. J. Sell. 1986. *Energy: A conceptual approach*. Englewood Cliffs, N.J.: Prentice Hall, p. 271.

4. Isacks, B., J. Oliver, and L. Sykes. 1968. Seismology and the new global tectonics. *Journal of Geophysical Research* 73:5855–5899.

5. Dewey, J. F. 1972. Plate tectonics. *Scientific American* 22:56–68.

6. Botkin, D. B. 1990. *Discordant harmonies: A new ecology for the 21st century*. New York: Oxford University Press.

7. Ehrlich, P. R., A. H. Ehrlich, and J. P. Holdren. 1970. *Ecoscience: Population, resources, environment*. San Francisco: W. H. Freeman, p. 1051.

8. Post, W. M., T. Peng, W. R. Emanuel, A. W. King, V. H. Dale, and D. L. De Angelis. 1990. The global carbon cycle. *American Scientist* 78:310–326.

9. Keeling, C. D., T. P. Whorf, M. Wahlen, and J. van der Plicht. 1995. Interannual extremes in the rate of rise of atmospheric carbon dioxide since 1980. *Nature* 375:666–670.

10. Hudson, R. J. M., S. A. Gherini, and R. A. Goldstein. 1994. Modeling the global carbon cycle: Nitrogen fertilization of the terrestrial biosphere and the "missing" CO_2 sink. *Global Biogeochemical Cycles* 8:307–333.

11. Woods Hole. 2004. The missing carbon sink. http://www.whrc.org/science/carbon/missingc.htm. Accessed February 15, 2006.

12. Houghton, R. 2003. Why are estimates of the global carbon balance so different? *Global Change Biology* 9:500–509.

13. Houghton, R. 2003. Revised estimates of the annual net flux of carbon to the atmosphere from changes in land use and land management 1850–2000. *Tellus* 55 B:378–390.

14. Herring, D., and R. Kannenberg. 2000. The mystery of the missing carbon. http://earthobservatory.nasa.gov/cgi-bin/printall?/study/BOREAS/missing_carbon.html. Accessed July 5, 2000.

15. Chameides, W. L., and E. M. Perdue. 1997. *Biogeochemical cycles*. New York: Oxford University Press.

16. Agren, G. I., and E. Bosatta. 1996. *Theoretical ecosystem ecology*. New York: Cambridge University Press.

17. Kasting, J. F., O. B. Toon, and J. B. Pollack. 1988. How climate evolved on the terrestrial planets. *Scientific American* 258:90–97.

18. Berner, R. A. 1999. A new look at the long-term carbon cycle. *GSA Today* 9(11):2–6.

19. Carter, L. J. 1980. Phosphate: Debate over an essential resource. *Science* 209:4454.

Chapter 5 Critical Thinking Issue References

Asner, G. P., T. R. Seastedt, and A. R. Townsend. 1997 (April). The decoupling of terrestrial carbon and nitrogen cycles. *Bioscience* 47 (4):226–234.

Hellemans, A. 1998 (February 13). Global nitrogen overload problem grows critical. *Science*, 279:988–989.

Smil, V. 1997 (July). Global populations and the nitrogen cycle. *Scientific American*, 76–81.

Vitousek, P. M., J. Aber, R. W. Howarth, G. E. Likens, P. A. Matson, D. W. Schindler, W. H. Schlesinger, and G. D. Tilman. 1997. Human alteration of the global nitrogen cycle: Causes and consequences. *Issues in Ecology*. http://esa.sdsc.edu/

Chapter 6 Notes

1. BBC news: http://news.bbc.co.uk/2/hi/science/nature/4547032.stm. Accessed March 14, 2006.

2. Ostfield, R. S., C. G. Jones, and J. O. Wolff. 1996 (May). Of mice and mast: Ecological connections in eastern deciduous forests. *BioScience* 46(5):323–330.

3. Morowitz, H. J. 1979. *Energy flow in biology*. Woodbridge: Conn.: Oxbow Press.

4. Brock, T. D. 1967. Life at high temperatures. *Science* 158: 1012–1019.

5. Brock, T. D. 1971. Life in the geysers basin. Washington, D.C.: National Park Service.

6. Lavigne, D. M., W. Barchard, S. Innes, and N. A. Oritsland. 1976. *Pinniped bioenergetics*. ACMRR/MM/SC/12. Rome: United Nations Food and Agriculture Organization.

7. Estes, J. A., and J. F. Palmisano. 1974. Sea otters: Their role in structuring nearshore communities. *Science* 185: 1058–1060.

8. The Marine Mammal Center, Sausalito, Calif.

9. Kenyon, K. W. 1969. The sea otter in the eastern Pacific Ocean. *North American Fauna*, no. 68. Washington, D.C.: Bureau of Sports Fisheries and Wildlife, U.S. Department of the Interior.

10. Duggins, D. O. 1980. Kelp beds and sea otters: An experimental approach. *Ecology* 61:447–453.

11. Kvitek, R. G., J. S. Oliver, A. R. DeGange, and B. S. Anderson. 1992. Changes in Alaskan soft-bottom prey communities along a gradient in sea otter predation. *Ecology* 73:413–428.

12. Paine, R. T. 1969. A note on trophic complexity and community stability. *American Naturalist* 100:65–75.

Chapter 7 Notes

1. Hughes, Catherine D., National Geographic Society, http://www.nationalgeographic.com/kids/creature_feature/0101/penguins2.html. Accessed. March 24, 2006.

2. Cicero, *The Nature of the Gods* (44 B.C.).

3. World Health Organization. 1998. *Malaria*. Fact Sheet No. 94. WHO.

4. United Nations World Health Organization. 2003. http://www.who.int/mediacentre/releases/2003/pr33/en/.

5. World Health Organization. 2000. *Overcoming antimicrobial resistance: World health report on infectious diseases 2000*. WHO.

6. James A. A. 1992. Mosquito molecular genetics: The hands that feed bite back. *Science* 257:37–38; Kolata, G. 1984. The search for a malaria vaccine. *Science* 226:679–682; Miller, L. H. 1992. The challenge of malaria. *Science* 257:36–37; World Health Organization. 1999. (June). Using malaria information. News Release No. 59. WHO; World Health Organization. 1999 (July). Sequencing the *Anopheles gambiae* genome. News Release No. 60. WHO.

7. Hutchinson, G. E. 1965. *The ecological theater and the evolutionary play.* New Haven, Conn.: Yale University Press.

8. Woese, C. R., O. Kandler, and M. L. Wheelis. 1990. Towards a natural system of organisms: Proposals for the domains Archaea, Bacteria, and Eucharya, *Proceedings of the National Academy of Sciences* (USA) 87:4576–4579.

9. Mather, J. R., and G. A. Yoskioka. 1968. The role of climate in the distribution of vegetation. *Annals of the Association of American Geography* 58:29–41.

10. Hardin, G. 1960. The competitive exclusion principle. *Science* 131:1292–1297.

11. Rogers, C. 1996. Red squirrel: *Sciurus vulgaris.* The Wild Screen Trust.

12. Schoener, T. W. 1983. Field experiments in interspecific competition. *American Naturalist* 1222:240–285.

13. Elton, C. S. 1927. *Animal ecology.* New York: Macmillan.

14. Hutchinson, G. E. 1958. Concluding remarks. *Cold Spring Harbor Symposium in Quantitative Biology* 22:415–427.

15. Miller, R. S. 1967. Pattern and process in competition. *Advances in Ecological Research* 4:1–74.

16. Botkin, D. B. 1985. The need for a science of the biosphere. *Interdisciplinary Science Reviews* 10:267–278.

17. Forman, Richard T. T. 1995. *Land mosaics: The ecology of landscapes and regions.* New York: Cambridge University Press.

18. Valentine, J. W. 1973. Plates and provinces, a theoretical history of environmental discontinuity. In N. F. Hughes, ed., *Organisms and continents through time,* pp. 79–92. Special Papers in Paleontology 12.

Organisms and continents through time, pp. 79–92. Special Papers in Paleontology 12.

10. Hallam, A. 1975. Alfred Wegener and the hypothesis of continental drift. *Scientific American* 232:88–97.

11. Hurley, P. M. 1968. The confirmation of continental drift. *Scientific American* 218:52–64.

12. Mather, J. R., and G. A. Yoshioka. 1968. The role of climate in the distribution of vegetation. *Ann. Ass. American Geography* 58:29–41.

13. Prentice, I. C., W. Cramer, S. P. Harrison, R. Leemans, R. A. Monserud, and A. M. Solomon. 1992. A global biome model based on plant physiology and dominance, soil properties and climate. *Journal of Biogeography* 19:117–134.

14. Botkin, D. B. 1977. The vegetation of the west. In H. R. Lamar, ed., *The reader's encyclopedia of the American west,* pp. 1216–1234. New York: Crowell.

15. Darwin, C. R. 1859. *The origin of species by means of natural selection or the preservation of favored races in the struggle for life.* London: Murray.

16. Grant, P. R. 1986. *Ecology and evolution of Darwin's finches.* Princeton, N.J.: Princeton University Press.

17. Cox, C. B., I. N. Healey, and P. D. Moore. 1973. *Biogeography.* New York: Halsted.

18. MacArthur, R. H., and E. O. Wilson. 1967. *The theory of island biogeography.* Princeton, N.J.: Princeton University Press.

19. Tallis, J. H. 1991. *Plant community history.* London: Chapman & Hall.

20. Missouri Botanic Garden, Botany in North America: In Honor of the XVI International Botanical Congress, August 1–7, 1999, St. Louis, Mo.

21. Waring, R. H., and J. F. Franklin. 1979. Evergreeen coniferous forests of the Pacific Northwest. *Science* 204:1380–1386.

22. North, M. P., J. F. Franklin, A. B. Carey, E. D. Forsman, and T. Hamer. 1999. Forest stand structure of the northern spotted owl's foraging habitat. *Forest Science* 45:520–527.

23. Botkin, D. B. 2001. *No man's garden: Thoreau and a new vision for civilization and nature.* Washington, D.C.: Island Press.

24. Botkin, D. B. 2004. *Our natural history: The lessons of Lewis and Clark.* New York: Oxford University Press.

Chapter 8 Notes

1. The National Zoo, http://nationalzoo.si.edu/Support/AdoptSpecies/AnimalInfo/BurmesePython/default.cfm. Accessed March 25, 2006.

2. CNN News http://www.cnn.com/2004/TECH/science/10/22/predators.in.paradise/index.html. Accessed March 25, 2006.

3. See also The National Park Service web page about exotic species in the Everglades http://www.nps.gov/ever/eco/exotics.htm.

4. Wallace, A. R. 1896. *The geographical distribution of animals.* Vol. 1. New York: Hafner. Reprint 1962.

5. Pielou, E. C. 1979. *Biogeography.* New York: Wiley.

6. Good, R. 1974. *The geography of the flowering plants,* 4th ed. London: Longman Group.

7. Takhtadzhian, A. L. 1986. *Floristic regions of the world.* Berkeley: University of California Press.

8. Udvardy, M. 1975. *A classification of the biogeographical provinces of the world.* IUCN Occasional Paper 18. Morges, Switzerland: IUCN.

9. Lentine, J. W. 1973. Plates and provinces, a theoretical history of environmental discontinuity. In N. F. Hughes, ed.,

Chapter 9 Notes

1. http://www.familytravelguides.com/articles/greatlakes/North_Central/hartwirk1.html. Accessed April 14, 2006.

2. Perlin, J. 1989. *A forest journey: The role of wood in the development of civilization.* New York: Norton.

3. Schrödinger, E. 1942. *What is life?* Cambridge: Cambridge University Press.

4. Morowitz, H. J. 1979. *Energy flow in biology.* Woodbridge, Conn.: Oxbow Press.

5. Slobodkin, L. B. 1960. Ecological energy relations at the population level. *American Naturalist* 95:213–236.

6. Peterson, R.O. 1995. *The wolves of Isle Royale: A broken balance.* Minocqua, Wis.: Willow Creek Press.

7. Jordan, J. D., D. B. Botkin, and M. I. Wolf, 1971. Biomass dynamics in a moose population. *Ecology* 52:147–152.

8. Kozlovsky, D. G. 1968. A critical evaluation of the trophic level concept: I. Ecological efficiencies. *Ecology* 49:147–160.

9. Schaefer, M., 1991. Secondary production and decomposition. In E. Rohrig and B. Ulrich, eds., *Temperate deciduous forests. Ecosystems of the world,* vol. 7. Amsterdam: Elsevier.

10. Golley, F. B. 1989. Energy dynamics of a food chain of an old-field community. *Ecol. Monographs* 30:187–291.

11. Bagley, P. B. 1989. Aquatic environments in the Amazon basin, with an analysis of carbon sources, fish production and yield. In D. P. Dodge, ed., *Proc. Int. Large Rivers Symp. Can. Spec. Publ. Fish. Aquat. Sci.* 106:385–398.

12. Fasham, M. J. R., ed. 1984. *Flows of energy and materials in marine ecosystems: Theory and practice.* New York and London: Plenum.

13. Gaill, F., B. Shillito, F. Menard, G. Goffinet, and J. Childress. 1997. Rate and process of tube production by the deep sea hydrothermal vent tubeworm *Riftia pachyptila. Marine Ecology Progress Series* 148:135–143.

14. Wills, J. 1996. Upwelling. *FMF Glossary.* First Millennial Foundation.

Chapter 10 Notes

1. CNN News, Tuesday, February 22, 2005 Posted: 2:28 PM EST (1928 GMT) http://www.cnn.com/2005/TECH/science/02/22/iraq.marshes/index.html. Accessed April 10, 2006.

2. UNEP project to help manage and restore the Iraqi Marshlands Iraqi Marshlands Observation System (IMOS)2006. http://imos.grid.unep.ch/

3. Hall, F. G., D. B. Botkin, D. E. Strebel, K. D. Woods, and S. J. Goetz, 1991. Large-scale patterns in forest succession as determined by remote sensing. *Ecology* 72:628–640.

4. Martin, P. S. 1963. *The last 10,000 years.* Tucson: University of Arizona Press.

5. Houseal, G., and D. Smith. 2000. Source-identified seed: The Iowa roadside experience. *Ecological Restoration* 18(3):173–183.

6. Gorham, E., P. M. Vitousek, and W. A. Reiners. 1979. The regulation of chemical budgets over the course of terrestrial ecosystem succession. *Annual Review Ecology and Systematics.* 10:53–84.

7. Vitousek, P. M., and L. R. Walker, 1987. Colonization, succession and resource availability: Ecosystem-level interactions. In A. J. Gray, M. J. Crawley, and P. J. Edwards, eds. *Colonization, succession and stability,* pp. 207–223. British Ecol. Soc. 26th Symp. Oxford: Blackwell Scientific Publications.

8. Connell, J. H., and R. O. Slatyer. 1977. Mechanism of succession in natural communities and their role in community stability and organization. *American Naturalist.* 111:1119–1144.

9. Pickett, S. T. A., S. L. Collins, and J. J. Armesto. 1987. Models, mechanisms and pathways of succession. *Botanical Review* 53:335–371.

10. Gomez-Pompa, A., and C. Vazquez-Yanes. 1981. Successional studies of a rain forest in Mexico. In D. C. West, H. H. Shugart, and D. B. Botkin, eds., *Forest succession: Concepts and application,* pp. 246–266. New York: Springer-Verlag.

11. MacMahon, J. A. 1981. Successional processes: Comparison among biomes with special reference to probable roles of and influences on animals. In D. C. West, H. H. Shugart, and D. B. Botkin, eds., *Forest succession: Concepts and application,* pp. 277–304. New York: Springer-Verlag.

12. Walthern, P. 1986. Restoring derelict lands in Great Britain. In G. Orians, ed., *Ecological knowledge and environmental problem-solving: Concepts and case studies,* pp. 248–274. Washington, D.C.: National Academy Press.

Chapter 10 Critical Thinking Issue References

Malakoff, D. 1998 (April 17). Restored wetlands flunk real-world test. *Science* 280: 371–372.

Pacific Estuarine Research Laboratory (PERL). 1997 (November). The status of constructed wetlands at Sweetwater Marsh National Wildlife Refuge. Annual Report to the California Department of Transportation.

Zedler, J. B. 1997. Adaptive management of coastal ecosystems designed to support endangered species. *Ecology Law Quarterly* 24:735–743.

Zedler, J. B., and A. Powell. 1993. Problems in managing coastal wetlands: Complexities, compromises, and concerns. *Oceanus* 36(2):19–28.

Chapter 11 Notes

1. Haub, Carl, and Diana Cornelius. 2000. *World population data sheet.* Washington, D.C.: Population Reference Bureau.

2. FAO statistics, April 28, 2006 http://faostat.fao.org/faostat/form?collection=Production.Crops.Primary&Domain=Production&servlet=1&hasbulk=&version=ext&language=EN.

3. China.cn Chinese Government Official Web Portal http://english.gov.cn/2005-08/08/content_27315.htm, April 28, 2006.

4. Brown, L. R. 1995. *Who will feed China? Wakeup call for a small planet.* New York: Norton.

5. World Resources Institute. 2000. *Facts and figures: Country environmental data.* Washington, D.C.: World Resources Institute.

6. A food crisis—or a blip? 1996 (February). *World Press Review* 43(2):34.

7. Cohen, J. E. 1995 (November/December). How many people can the earth support? *The Sciences,* pp. 18–23.

8. Holmes, B. 1993 (February 8). Feeding a world of 10 billion. *U.S. News & World Report* 114(5):55.

9. Livernash, R. 1995 (July/August). The future of populous economies: China and India shape their destinies. *Environment* 37(6):6–11, 25–34.

10. Ryan, M., and C. Flavin. 1995. *Facing China's limits. State of the world, 1995.* New York: Norton.

11. Tyler, P. E. 1996 (May 23). China's fickle rivers: Dry farms, needy industry bring a water crisis. *New York Times.*

12. Hawthorne, P. 1998 (April 3). Rebirth. *Time* 100/Africa 151(15), Time 100/Leaders and Revolutionaries.

13. United Nations Food and Agriculture Organization. 1998. FAOSTAT database. Rome: UNFAO.

14. Biological Resources Division USGS. 1998. *Historical interrelationships between population settlement and farmland in the conterminous United States, 1790–1992.* Land use history of North America. USGS.

15. Field, J. O., ed. 1993. *The challenge of famine: Recent experience, lessons learned.* Hartford, Conn.: Kumarian Press.

16. United Nations Food and Agriculture Organization. 2000 (April). Food emergencies persist in 34 countries throughout the world. *Food Outlook* 2, p. 4. Rome: UNFAO.

17. United Nations Food and Agriculture Organization. 1998 (September). Global information.

18. World Food Programme. 1998. *Tackling hunger in a world full of food: Tasks ahead for food aid.*

19. Raven, P. H., R. F. Evert, and S. E. Eichhorn. 1999. *Biology of plants.* New York: W. H. Freeman/Worth.

20. FAOSTAT 2003 Web site.

21. Bardach, J. E. 1968. Aquaculture. *Science* 161:1098–1106.
22. McConnell, Chai. 1995. Selecting new crops using strategic marketing management. The Sixth Conference of the Australasian Council on Tree and Nut Production. Available at http://www.newcrops.uq.edu.au/acotanc/papers/mcconnel.htm
23. U.S. Department of Agriculture, 2003, Alternative Farming Systems Information Center, available at http://www.nal.usda. gov/afsic/ofp/.
24. U.S. Department of Agriculture. *U.S organic farming emerges in the 1990s.* Available at http://www.ers.usda.gov/publications/aib770/aib770.pdf.
25. USDA Magriet Caswell communication to ACAB, July 2000.
26. Flannery, K. V. 1965. The ecology of early food production in Mesopotamia. *Science* 147:1247–1256.
27. Himnan, W. 1984. New crops for arid lands. *Science* 225:1445–1256.
28. California Agricultural Statistics Service. 1999. *California agricultural statistics.* Sacramento, Calif.: CASS.
29. Food and Agriculture Organization of the United Nations (FAO). 1996. *Food for all.* Rome: FAO.
30. PEW Biotechnology Initiative, Pew Charitable Trusts. Web site http://pewagbiotech.org/resources/factsheets/display.php3?FactsheetID=2.

Chapter 12 Notes
1. Haub, Carl, and Diana Cornelius. 2000. *World population data sheet.* Washington, D.C.: Population Reference Bureau.
2. FAO statistics, April 28, 2006 http://faostat.fao.org/faostat/form?collection=Production.Crops.Primary&Domain=Production&servlet=1&hasbulk=&version=ext&language=EN.
3. China.cn, Chinese Government Official Web Portal http://english.gov.cn/2005-08/08/content_27315.htm. Accessed April 28, 2006.
4. Encyclopedia Britannica online, 2003, http://www.britannicacom/eb/article?eu=61977&tocid=0&query=plow&ct=.
5. Lashof, J. C., ed. 1979. *Pest management strategies in crop protection.* Vol. 1. Washington, D.C.: Office of Technology Assessment, U.S. Congress.
6. Baldwin, F. L., and P. W. Santelmann. 1980. Weed science in integrated pest management. *BioScience* 30:675–678.
7. From UNEP Global Programme of Action for the Protection of Marine Environment from Lands-Based Activities, 2003, http://pops.gpa.unep.org/11aldi.htm.
8. Michigan State University Web site, 2003, http://www.msue.msu.edu/vanburen/ofm.htm.
9. Barfield, C. S., and J. L. Stimac, 1980. Pest management: An entomological perspective. *BioScience* 30:683–688.
10. EPA (2006). 2000–2001. *Pesticide market estimates: Usage,* EPA. http://www.epa.gov/oppbead1/pestsales/01pestsales/usage2001.html.
11. Jimmie D. Petty, J. N. H., Carl E. Orazio, Jon A Lebo, Barry C. Poulton, and Robert W. Gale (1995). Determination of Waterborne Bioavailable Organochlorine Pesticide Residues in the Lower Missouri River. Columbia, Missouri, U.S. Geological Survey Biological Resources Division, Columbia Environmental Research Center (formerly known as U.S. Department of the Interior, National Biological Service) http://www.cerc.usgs.gov/clearinghouse/data/usgs_brd_cerc_d_preflood.html.

12. Malaria Facts, U.S. Dept of Health and Human Services, 2004, National Center for Infectious Diseases, Division of Parasitic Diseases.
13. Cummins, 2004, GM Rice in Indian, www.i-sis.org.uk/GMRII.php.
14. United Nations. 1978. United Nations Conference on Desertification: Roundup plan of action and resolutions. New York: United Nations.
15. UN Food and Agriculture Organization. 1998. The United Nations convention to combat desertification: An Explanatory leaflet. Food and Agriculture Organization of the United Nations.
16. UN Food and Agriculture Organization. 1998. What is desertification? Food and Agriculture Organization of the United Nations.
17. Grainger, A. 1982. *Desertification: How people make deserts, how people can stop and why they don't,* 2d ed. London: Russell Press.
18. California Farm Bureau Federation. 2006. www.cfbf.com/info/agfacts.c/m
19. California State Dept. Of Water Resources, 2006, California land and water use, www.landwateruse.water.ca.gov.

Chapter 12 Critical Thinking Issue References
Martin, G. 1992 (June 29). Rice grower proud of his bird habitat. *San Francisco Chronicle,* pp. A1, A6.
Vogel, N. 1992 (December 6). Rice farmers change ways, reap good will. *Sacramento Bee,* pp. A1, A26.
Walker, S. L. 1992 (August 1). Rice growers sow good will. *San Diego Union-Tribune,* pp. A1, A15.
Wood, D. B. 1992 (September 3). California rice land does double duty. *Christian Science Monitor,* p. 10.
World Resources Institute. 1992. *The 1992 information please environmental almanac.* Boston: Houghton-Mifflin.
Mechanized irrigation of potato crops in Idaho illustrates the effect of modern agriculture on the environment.

Chapter 13 Notes
1. This and other statistics about U.S. wildfires are from the U.S. Federal Government, Wildland fire statistics, available at http://www.nifc.gov/stats/index.html. Accessed April 22, 2006.
2. CNN news for January 3, 2006. http://www.cnn.com/2006/US/01/02/wildfires/ index.html.
3. United Nations Food and Agriculture Organization. 2001. Rome: UN FAO. Available at ftp://ftp.fao.org/docrep/fao/003/y0900e/y0900e02.pdf
4. Botkin, D. B. 1990. *Discordant harmonies: A new ecology for the 21st century.* New York: Oxford University Press.
5. Likens, G. E., F. H. Borman, R. S. Pierce, J. S. Eaton, and N. M. Johnson. 1977. *The biogeochemistry of a forested ecosystem.* New York: Springer-Verlag.
6. The Hubbard Brook ecosystem continues to be one of the most active and long-term ecosystem studies in North America. An example of a recent publication is: Bailey, S. W., D. C. Buso, and G. E. Likens. 2003. Implications of sodium mass balance for interpreting the calcium cycle of a forested ecosystem. *Ecology* 84(2):471–484.

7. Swanson, F. J., and C. T. Dyrness. 1975. Impact of clearcutting and road construction on soil erosion by landslides in the western Cascade Range, Oregon. *Geology* 3:393–396.

8. Fredriksen, R. L. 1971. Comparative chemical water quality—natural and disturbed streams following logging and slash burning. In *Forest land use and stream environment*, pp. 125–137. Corvallis: Oregon State University.

9. Sedjo, R. A. 1983. The comparative economics of plantation forestry: A global assessment. Unpublished research paper, Johns Hopkins University, Baltimore.

10. Culbert Thesis, George Mason University.

11. Kimmins, H. 1995. Proceedings of the conference on certification of sustainable forestry practices. Malaysia.

12. Jenkins, Michael B. 1999. *The business of sustainable forestry*. Washington, D.C.: Island Press.

13. United Nations Food and Agriculture Organization. 2001. Rome: UN FAO. ftp://ftp.fao.org/docrep/fao/ 003/ y0900e/y0900e02.pdf.

14. Dombeck, M. 1997. Towards sustainable forest management. Speech to USDA Forest Service.

15. Busby, F. E., et al. 1994. Rangeland health: New methods to classify inventory and monitor rangelands. Washington, D.C.: National Academy Press.

16. World Resources Institute. *Disappearing land: Soil degradation*. Washington, D.C.: WRI.

17. World Resources Institute. 1993. *World resources 1992–93*. New York: Oxford University Press.

18. Manandhar, A. 1997. *Solar cookers as a means for reducing deforestation in Nepal*. Nepal: Center for Rural Technology.

19. Council on Environmental Quality and U.S. Department of State. 1981. *The global 2000 report to the president: Entering the twenty-first century*. Washington, D.C.: Council on Environmental Quality.

20. Botkin, D. B., and L. Simpson. 1990. The first statistically valid estimate of biomass for a large region. *Biogeochemistry* 9:161–174.

21. Perlin, J. 1989. *A forest journey: The role of wood in the development of civilization*. New York: Norton.

22. U.S. Department of State information service. http://usinfo.state.gov/gi/Archive/2005/Nov/15-850783.html, Accessed April 23, 2006.

23. Runte, A. 1997. *National parks: The American experience*. Lincoln: Bison Books of the University of Nebraska.

24. National Park Service Web site, http://www.wilderness.net/ index.cfm?fuse=NWPS&sec=fastfacts.

25. Quotations from Alfred Runte cited by the Wilderness Society on its Web site, available at http:// www.wilderness.org/NewsRoom/Statement/20031216.cfm.

26. Voyageurs National Park Web site, http://www.npsgov/voya/home.htm.

27. Costa Rica's TravelNet. National parks of Costa Rica. 1999. Costa Rica: Costa Rica's TravelNet.

28. Kenyaweb. 1998. *National parks and reserves*. Kenya: Kenyaweb.

29. Botkin, D. B. 1992. Global warming and forests of the Great Lakes states. In J. Schmandt, ed., *The regions and global warming: Impacts and response strategies*. New York: Oxford University Press.

30. Nash, R. 1978. International concepts of wilderness preservation. In J. C. Hendee, G. H. Stankey, and R. C. Lucas, eds., *Wilderness management*, pp. 43–59. United States Forest Service Misc. Pub. No. 1365.

31. National Park Service Web site, http:// www. wilderness.net/index.cfm?fuse=NWPS&sec=fastfacts.

32. The National Wilderness Preservation System. http://nationalatlas.gov/articles/boundaries/a_nwps.html 2006.

33. Hendee, J. C., G. H. Stankey, and R. C. Lucas. 1978. *Wilderness management*. United States Forest Service Misc. Pub. No. 1365.

Chapter 14 Notes

1. Jennifer A. Devine, and Richard L. Haedrich. 2006. Deep-sea fishes qualify as endangered. *Nature* 439(5): 29.

2. Botkin, D. B., and L. M. Talbot, 1992. Biological diversity and forests. In N. Sharma, ed., *Contemporary issues in forest management: Policy implications*. Washington, D.C.: World Bank.

3. Margulis, L., And K. V. Schwartz, *Five Kingdoms*, 3d ed. New York: W. H. Freeman.

4. http://www.coralreef.noaa.gov/outreach/protect/supp_medicines.html, April 10, 2006.

5. http://www.issues.org/18.3/p_bruckner.html.

6. Naess A. 1989. *Ecology, community, and lifestyle*. Cambridge, England: Cambridge University Press. Naess does admit to the need to eat. He writes, "It is against my intuition of unity to say 'I can kill you because I am more valuable' but not against the intuition to say 'I will kill you because I am hungry.' In the latter case, there would be an implicit regret: 'Sorry, I am now going to kill you because I am hungry.' In short, I find obviously right, but often difficult to justify, different sorts of behavior with different sorts of living beings. But this does not imply that we classify some as intrinsically more valuable than others" (p. 168).

7. www.nwf.org/wildlife/grizzlybear/. April 10, 2006.

8. Botkin, D. B. 2004. *Beyond the stony mountains: Nature in the American West from Lewis and Clark to today*. New York: Oxford University Press.

9. Mattson, D. J., and M. W. Reid. 1991. Conservation of the Yellowstone grizzly bear. *Conservation Biology* 5: 364–372.

10. The National Zoo. http://nationalzoo.si.edu/support/adoptspecies/Animalinfo/biosn/default.cfm. April 10, 2006.

11. www.bisoncentral.com. April 10, 2006.

12. Haines, F. 1970. *The buffalo*. New York: Thomas Y. Crowell.

13. Tom Stehn's whooping crane report, Aransas National Wildlife Refuge, December 10, 2003. Available at http://www.birdrockport.com/tom_stehn_whooping_crane_report.htm.

14. Whooping Crane Conservation Association. 2003. Available at http://whoopingcrane.com/wccaflockstatus.htm.

15. Friends of the Earth. 1979. *The whaling question: The inquiry by Sir Sidney Frost of Australia*. San Francisco: Friends of the Earth.

16. Bockstoce, J. R., and D. B. Botkin. 1980. The historical status and reduction of the western Arctic bowhead whale (*Balaena mysticetus*) population by the pelagic whaling industry, 1848–1914. New Bedford, Conn.: Old Dartmouth Historical Society.

17. United Nations Food and Agriculture Organization. 1978. *Mammals in the seas.* Report of the FAO Advisory Committee on Marine Resources Research, Working Party on Marine Mammals. FAO Fisheries Series 5, vol. 1. Rome: UNFAO.

18. NOAA Fisheries, Office of Protected Resources, 2006. Species under the Endangered Species Act (ESA). http://www.nmfs.noaa.gov/pr/species/esa.htm#delisted.

19. http://www.defenders.org/wildlife/dolphin/tundolph.html.

20. FAO Statistics 2006. http://www.fao.org/waicent/portal/statistics_en.asp.

21. NOAA. 2003. World fisheries. Available at http://www.st.nmfs.gov/st1/fus/current/04_world2002.pdf.

22. Species Survival Commission. 2006. Red list of threatened species. Geneva: IUCN.

23. U.S. Congress Office of Technology Assessment. 1987. *Technologies to maintain biological diversity.* Washington, D.C.: U.S. Government Printing Office, OTA-330, p. 45.

24. Martin, P. S. 1963. *The last 10,000 years.* Tucson: University of Arizona Press.

25. NOAA delisted species, http://www.nmfs.noaa.gov/pr/species/esa.htm#delisted and U.S. Fish and Wildlife Service Threatened and Endangered Species System (TESS) http://ecos.fws.gov/tess_public/StartTESS.do;jsessionid=B03B5D5596EF6E2A123A3A6F7F794C4B.

26. Lowry, Mark S. and Karin A. Forney. 2005. Abundance and distribution of California sea lions in central and northern California during 1998 and summer 1999. *Fishery Bulletin.*

27. California Department of Fish and Game, http://www.dfg.ca.gov/news/issues/lion/lion_faq.html, mountain lion abundances.

28. Information about the Kirtland's warbler and its habitat is from Byelich et al., p. 12; Mayfield, H. 1969. *The Kirtland's warbler.* Bloomfield Hills, Mich.: Cranbrook Institute of Science, pp. 24–25.

29. The discussion of the Kirtland's warbler is taken directly from Botkin, D. B. 1990. *Discordant Harmonies: A new ecology for the 21st century.* New York: Oxford University Press.

30. Michigan Department of Natural Resources http://www.michigan.gov/dnr/0,1607,7-153-10370_12145_12202-32591-,00.html. (The 2002 annual census counted over 1,000 Kirtland's warbler singing males for the second year in a row.)

Chapter 14 Critical Thinking Issue References

Hodgson, A. 1997, July. Wolf restoration in the Adirondacks? Wildlife Conservation Society, Working Paper No. 8.

Hopsack, D. A. 1996. Biological potential for eastern timber wolf re-establishment in the Adirondack Park. Wolves of America Conference Proceedings, November 14–16, 1996. Albany, N.Y., and Washington, D.C.: Defenders of Wildlife.

Stevens, W. K. 1997, March 3. Wolves may reintroduce themselves to East. *New York Times.*

Chapter 15 Notes

1. Committee on Hormonally Active Agents in the Environment, National Research Council, National Academy of Sciences. 1999. *Hormonally active agents in the environment.* Washington, D.C.: National Academy Press.

2. Krimsky, S. 2001. Hormone disrupters: A clue to understanding the environmental cause of disease. *Environment* 43(5): 22–31.

3. Royte, E. 2003. Transsexual frogs. Discover 24(2): 26–53.

4. Hayes, T., K. Haston, M. Tsui, A. Hong, C. Haeffele, and A. Vock. 2002. Feminization of male frogs in the wild. *Nature* 419:495–496.

5. Han, H. 2005. The grey zone, in C. Dawson and G. Gendreau. Healing our planet, healing ourselves. Santa Rosa, CA, Elite Books.

6. Warren, H. V., and R. E. DeLavault. 1967. A geologist looks at pollution: Mineral variety. *Western Mines* 40:23–32.

7. Evans, W. 1996. Lake Nyos. Knowledge of the fount and the cause of disaster. *Science* 379:21.

8. Krajick, K. 2003. Efforts to tame second African killer lake begin. *Science* 379:21.

9. Gunn, J., ed. 1995. *Restoration and recovery of an industrial region: Progress in restoring the smelter-damaged landscape near Sudbury, Canada.* New York: Springer-Verlag.

10. Blumenthal, D. S., and Ruttenber. 1995. *Introduction to environmental health,* 2d ed. New York: Springer.

11. U.S. Geological Survey. 1995. *Mercury contamination of aquatic ecosystems.* USGS FS 216-95.

12. Greer, L., M. Bender, P. Maxson, and D. Lennett, 2006. Curtailing mercury's global reach. In L. Starke (ed), *State of the world 2006,* pp. 96–114. New York: Norton.

13. Ehrlich, P. R., A. H. Ehrlich, and J. P. Holdren. 1970. *Ecoscience: Population, resources, environment.* San Francisco: Freeman.

14. Waldbott, G. L. 1978. Health effects of environmental pollutants, 2d ed. Saint Louis: Moseby.

15. McGinn, A. P. 2000 (April 1). POPs culture. *World Watch,* pp. 26–36.

16. Carlson, E. A. 1983. International symposium on herbicides in the Vietnam War: An appraisal. *BioScience* 33:507–512.

17. Cleverly, D., J. Schaum, D. Winters, and G. Schweer. 1999. Inventory of sources and releases of dioxin-like compounds in the United States. Paper presented at the 19th International Symposium on Halogenated Environmental Organic Pollutants and POPs, September 12–17, Venice, Italy. Short paper in *Organohalogen Compounds* 41:467–472.

18. Grady, D. 1983 (May). The dioxin dilemma. *Discover,* pp. 78–83.

19. Roberts, L. 1991. dioxin risks revisited. *Science* 251:624–626.

20. Kaiser, J. 2000. Just how bad is dioxin? *Science* 5473: 1941–1944.

21. Johnson, J. 1995. SAB advisory panel rejects dioxin risk characterization. *Environmental Science & Technology* 29:302A.

22. Thomas, V. M., and T. G. Spiro. 1996. The U.S. dioxin inventory: Are there missing sources? *Environmental Science & Technology* 30:82A–85A.

23. U.S. Environmental Protection Agency. 1994 (June). Estimating exposure to dioxin-like compounds. Review draft. Office of Research and Development, EPA/600/6-88/005 Ca-c.

24. U.S. Environmental Protection Agency. 1994. Health assessment document for 2,3,7,8-tetrachlorodibenzo-p-dioxin (TCDD) and related compounds. Vols. I–III. External review draft. Washington, D.C.: EPA.

25. Johnson, J. 1995. Dioxin risk assessment stalls: EPA to create new review panel. *Environmental Science & Technology* 29:492A.

26. Chanlett, E. T. 1979. *Environmental protection*, 2d ed. New York: McGraw-Hill.

27. Ross, M. 1990. Hazards associated with asbestos minerals. In B. R. Doe, ed., *Proceedings of a U.S. Geological Survey Workshop on Environmental Geochemistry*, pp. 175–176. U.S. Geological Survey Circular 1033.

28. Pool, R. 1990. Is there an EMF–cancer connection? *Science* 249:1096–1098.

29. Linet, M. S., E. E. Hatch, R. A. Kleinerman, L. L. Robison, W. T. Kaune, D. R. Friedman, R. K. Severson, C. M. Haines, C. T. Hartsock, S. Niwa, S. Wacholder, and R. E. Tarone. 1997. Residential exposure to magnetic fields and acute lymphoblastic leukemia in children. *New England Journal of Medicine* 337(1):1–7.

30. Kheifets, L. I., E. S. Gilbert, S. S. Sussman, P. Guaenel, S. D. Sahl, D. A. Savitz, and G. Thaeriault, G. 1999. Comparative analyses of the studies of magnetic fields and cancer in electric utility workers: Studies from France, Canada, and the United States. *Occupational and Environmental Medicine* 56(8):567–574.

31. Francis, B. M., 1994. *Toxic substances in the environment*. New York: John Wiley & Sons.

32. Poisons and poisoning. 1997. *Encyclopedia Britannica*. Vol. 25, p. 913. Chicago: Encyclopedia Britannica.

33. Air Risk Information Support Center (Air RISC), U.S. Environmental Protection Agency. 1989. *Glossary of terms related to health exposure and risk assessment*. EPA/450/3-88/016. Research Triangle Park, N.C.

Chapter 15 Critical Thinking Issue References

Needleman, H. L., J. A. Riess, M. J. Tobin, G. E. Biesecker, and J. B. Greenhouse. 1996. Bone lead levels and delinquent behavior. *Journal of the American Medical Association* 275:363–369.

Centers for Disease Control. 1991. *Preventing lead poisoning in young children*. Atlanta: Public Health Service, Centers for Disease Control.

Goyer, R. A. 1991. Toxic effects of metals. In M. O. Amdur, J. Doull, and C. D. Klaassen, eds., *Toxicology*, pp. 623–680. New York: Pergamon.

Bylinsky, G. 1972. Metallic nemesis. In B. Hafen, ed., *Man, health and environment*, pp. 174–185. Minneapolis: Burgess.

Hong, S., J. Candelone, C. C. Patterson, and C. F. Boutron. 1994. Greenland ice evidence of hemispheric lead pollution two millennia ago by Greek and Roman civilizations. *Science* 265:1841–1843.

Chapter 16 Notes

1. Dokka, R. K. 2006. Modern-day tectonic subsidence in coastal Louisiana. *Geology* 34:281–284.

2. U.S. Army Corps of Engineers 2006. Performance evaluation of the New Orleans and southeast Louisiana hurricane protection system. Vol. 1, Executive summary and overview. Washington, D.C.

3. Knauer, K., ed. 2006. *Nature's extremes*. Time Books. Des Moines, Iowa, p. 138.

4. Floodplain Management Association. 2006. Overview of flooding; frequency of flooding; and how much flood risk is acceptable? www.floodplain.org. Accessed on June 12, 2006.

5. Reuters, M. B. 2005. Elephants saved tourists from tsunami. www.savetheelephants.org. Accessed on June 13, 2006.

6. Herd, D. G. 1986. The 1985 Ruiz volcano disaster. *EOS, Transactions*, American Geophysical Union, 67(19):457–460.

7. Abramovitz, J. N., and S. Dunn, 1998. Record year for weather-related disasters. *Vital Signs Brief* 98-5. Washington, D.C.: World Watch Institute.

8. Crowe, B. W. 1986. Volcanic hazard assessment for disposal of high-level radioactive waste. In: Geophysics Study Committee, ed., *Active tectonics*, pp. 247–260. National Research Council. Washington, D.C.: National Academy Press.

9. Advisory committee on the International Decade for Natural Hazard Reduction. 1989. *Reducing disaster's toll*. National Research Council. Washington, D.C.: National Academy Press.

10. Kates, R. W., and D. Pijawka. 1977. From rubble to monument: The pace of reconstruction. In J. E. Haas, R. W. Kates, and M. J. Bowden, eds., *Reconstruction following disaster*, pp. 1–23. Cambridge, MA: MIT Press.

11. Costa, J. E., and V. R. Baker 1981. *Surficial geology: Building with the Earth*. New York: John Wiley & Sons.

12. Rahn, P. H. 1984. Flood-plain management program in Rapid City, South Dakota. *Geological Society of America Bulletin* 95:838–843.

13. Pinter, N. 2005. One step forward, two steps back on U.S. floodplains. *Science* 308:207–208.

14. Mount, J. F. 1997. *California rivers and streams*. Berkeley: University of California Press.

Chapter 17 Notes

1. Butti, K., and J. Perlin. 1980. *A golden thread: 2500 years of solar architecture and technology*. Palo Alto, Calif.: Cheshire Books.

2. Morowitz, H. J. 1979. *Energy flow in biology*. New Haven, Conn.: Oxbow Press.

3. Ehrlich, P. R., A. H. Ehrlich, and J. P. Holdren. 1970. *Ecoscience: Population, resources, environment*. San Francisco: W. H. Freeman.

4. Feynman, R. P., R. B. Leighton, and M. Sands. 1964. *The Feynman lectures on physics*. Reading, Mass.: Addison-Wesley.

5. Cuff, D. J., and W. J. Young. 1986. *The United States energy atlas*, 2d ed. New York: Macmillan.

6. Darmstadter, J., H. H. Landsberg, H. C. Morton, and M. J. Coda. 1983. *Energy today and tomorrow: Living with uncertainty*. Englewood Cliffs, N.J.: Prentice-Hall.

7. Steinhart, J. S., M. E. Hanson, R. W. Gates, C. C. Dewinkel, K. Briody, M. Thornsjo, and S. Kambala. 1978. A low energy scenario for the United States: 1975–2000. In L. C. Ruedisili and M. W. Firebaugh, eds., *Perspectives on energy*, 2d ed., pp. 553–588. New York: Oxford University Press.

8. Olkowski, H., B. Olkowski, and T. Javits. (Farallones Institute). 1979. *The integral urban house: Self reliant living in the city*. San Francisco: Sierra Club Books.

9. Flavin, C. 1984. *Electricity's future: The shift to efficiency and small-scale power*. Worldwatch Paper 61. Washington, D.C.: Worldwatch Institute.

10. Consumers' Research. 1995. Fuel economy rating: 1996 mileage estimates. *Consumers' Research* 78:22–26.

11. Berger, J. J. 2000. *Beating the heat*. Berkeley, Calif.: Berkeley Hills Books.

12. Lovins, A. B. 1979. *Soft energy paths: Towards a durable peace*. New York: Harper & Row.

13. Brown, L. R., C. Flavin, and S. Postel (Worldwatch Institute). 1991. *Saving the planet: How to shape an environmentally sustainable global economy*. New York: W. W. Norton.

14. California Energy Commission. 1991. *California's energy plan: Biennial report*. Sacramento, Calif.

15. Flavin, C., and S. Dunn. 1999. Reinventing the energy system. In L. R. Brown et al., eds., *State of the world 1999: A Worldwatch Institute report on progress toward a sustainable society*. New York: W. W. Norton.

16. Dunn, S. 2000. *Micropower, the next electrical era*. Worldwatch Paper 151. Washington, D.C.: Worldwatch Institute.

Chapter 17 Critical Thinking Issue References

Ehrlich, P. R., and A. H. Ehrlich, 1991. Healing the planet. Reading, Mass.: Addison-Wesley.

Fickett, A. P. 1990. Efficient use of electricity. *Scientific American* 263(3):65–74.

Holdren, J. P. 1990. Energy in transition. *Scientific American* 263(3):157–163.

Lean, G. 1990. *Atlas of the environment*. New York: Prentice-Hall.

U.S. Census Bureau. 1998. World population and growth rates. http://www.census.gov/ipc/www/world.html.

Chapter 18 Notes

1. Alekett, K. 2006. Oil: A bumpy road a head. *World Watch*. 19:1, 10–12.

2. Cavanay, R. 2006. Global oil about to peak? A recurring myth. *World Watch*. 19:1: 13–15.

3. Van Koevering, T. E., and N. J. Sell. 1986. *Energy: A conceptual approach*. Englewood Cliffs, N.J.: Prentice-Hall.

4. McCulloh, T. H. 1973. In D. A. Brobst and W. P. Pratt, eds., *Oil and gas in United States mineral resources*, pp. 477–496. U.S. Geological Survey Professional Paper 820.

5. Maugeri, L. 2004. Oil: Never cry wolf—when the petroleum age is far from over. *Science* 304:1114–1115.

6. British Petroleum Company. 1999. *B.P. statistical review of world energy*. London: British Petroleum Company.

7. Kerr, R. A. 2000. USGS optimistic on world oil prospects. *Science* 289:237.

8. Youngquist, W. 1998. Spending our great inheritance. Then what? *Geotimes* 43(7):24–27.

9. Edwards, J. D. 1997. Crude oil and alternative energy production forecast for the twenty-first century: The end of the hydrocarbon era. *American Association of Petroleum Geologists Bulletin* 81(8):1292–1305.

10. Darmstadter, J., H. H. Landsberg, H. C. Morton, and M. J. Coda. 1983. *Energy today and tomorrow: Living with uncertainty*. Englewood Cliffs, N.J.: Prentice-Hall.

11. Nuccio, V. 1997. *Coal-bed methane—an untapped energy resource and an environmental concern*. U.S. Geological Survey Fact Sheet. FS-019-97.

12. Nuccio, V. 2000. *Coal-bed methane: Potential environmental concerns*. US Geological Survey. USGS Fact Sheet. FS-123-00.

13. Wood, T. 2003. Prosperity's brutal price. *Los Angeles Times Magazine*. February 2, 2003.

14. Suess, E., G. Bohrmann, J. Greinert, and E. Lauch. 1999. Flammable ice. *Scientific American* 28(5):76–83.

15. Rahn, P. H. 1996. *Engineering geology: An environmental approach*, 2d ed. New York: Elsevier.

16. U.S. Environmental Protection Agency. 1973. *Processes, procedures and methods to control pollution from mining activities*. EPA-430/9-73-001. Washington, D.C.: U.S. Environmental Protection Agency.

17. Council on Environmental Quality. 1978. *Progress in environmental quality*. Washington, D.C.: Council on Environmental Quality.

18. Miller, E. W. 1993. *Energy and American society, a reference handbook*. Santa Barbara, Calif.: ABC-CLIO.

19. Corcoran, E. 1991. Cleaning up coal. *Scientific American* 264:106–116.

20. Energy Information Administration. 1995 (February). *Coal data: A reference*. Washington, D.C.: U.S. Department of Energy.

21. Knapp, D. H. 1995. Non-OPEC oil supply continues to grow. *Oil & Gas Journal* 93:35–45. Paris: International Energy Agency.

22. Peterson, G. 2003. New statute for Canadian Oil Sands. *Geotimes* 48(3) 7.

Chapter 18 Critical Thinking Issue References

Bleviss, D. L. 1988. *The new oil crisis and fuel economy technologies*. New York: Quorum Books.

Bleviss, D. L., and P. Walzer. 1990. Energy for motor vehicles. *Scientific American* 263(3):103–109.

Corson, W. H., ed. 1990. *The global ecology handbook*. Boston: Beacon.

Driving down the deficit. 1993. *U.S. News & World Report* 114(2):58–60.

Energy Information Administration. 1996 (January). *Monthly energy review*. Washington, D.C.: U.S. Department of Energy.

Greenwald, J. 1993. Why not a gas tax? *Time* 141(7):25–27.

Miller, E. W., and R. M. Miller. 1993. *Energy and American society. A reference handbook*. Santa Barbara, Calif.: ABC-CLIO.

Nadis, S., and J. J. MacKenzie. 1993. *Car trouble*. Boston: Beacon.

Chapter 19 Notes

1. Jackson, T., and R. Lofstedt. 1998. Royal commission on environmental pollution. Study on energy and the environment. Accessed November 29, 2000, at http://www.rcep.org.uk/studies/energy/98-6061/jackson.html.

2. Starke, L., ed. 2005. *Vital signs 2005*. New York: Norton.

3. Berger, J. J. 2000. *Beating the heat*. Berkeley, Calif.: Berkeley Hills Books.

4. Miller, E. W. 1993. *Energy and American society: A reference handbook*. Santa Barbara, Calif.: ABC-CLIO.

5. Flavin, C., and S. Dunn. 1999. Reinventing the energy system. In L. R. Browne et al., *State of the world 1999: A Worldwatch Institute report on progress toward a sustainable society*. New York: W. W. Norton.

6. Eaton, W. W. 1978. Solar energy. In L. C. Ruedisili and M. W. Firebaugh, eds., *Perspectives on energy*, 2d ed., pp. 418–436. New York: Oxford University Press.

7. Mayur, R., and B. Daviss. 1998 (October). The technology of hope. *The Futurist*, pp. 46–51.

8. Brown, L. R. 1999 (March–April). Crossing the threshold. *Worldwatch*, pp. 12–22.

9. Quinn, R. 1997 (March). Sunlight brightens our energy future. *The World and I*, pp. 156–163.

10. Johnson, J. T. 1990 (May). The hot path to solar electricity. *Popular Science*, pp. 82–85.

11. Demeo, E. M., and P. Steitz. 1990. The U.S. electric utility industry's activities in solar and wind energy. In K. W. Böer, ed., *Advances in solar energy*, Vol. 6, pp. 1–218. New York: American Solar Energy Society.

12. Schatz solar hydrogen project. N.D. Pamphlet. Arcata, Calif.: Humboldt State University.

13. Piore, A. 2002 (April 15). Hot springs eternal: hydrogen power. *Newsweek*, pp. 32H.

14. Kartha, S., and P. Grimes. 1994. Fuel cells: Energy conversion for the next century. *Physics Today* 47:54–61.

15. Haggin, J. 1995. Fuel-cell development reaches demonstration stage. *Chemical & Engineering News* 73:28–30.

16. Alward, R., S. Eisenbart, and J. Volkman. 1979. *Micro-hydro power: Reviewing an old concept*. Butte, Mont.: National Center for Appropriate Technology, U.S. Department of Energy.

17. Nova Scotia Department of Mines and Energy. 1981. *Wind power*.

18. Hunt, S. C., Sawn, J. L. and P. Stair, 2006. Cultivating renewable alternatives to oil. In L. Stark (ed), State of the World Ch. 4, pp. 61–77, New York: Norton.

19. U.S. Environmental Protection Agency, Office of Solid Waste and Emergency Response. 2002. *Municipal solid waste in the United States: 2000 facts and figures*. EPA530-R-02-001.

20. Council on Environmental Quality. 1979. *Environmental quality*.

21. Wihersaari, M. 1996. Energy consumption and greenhouse gas emissions from biomass production chains. *Energy Conversion and Management* 37:1217.

22. Sterzinger, G. 1995. Making biomass energy a contender. *Technology Review* 98:34–40.

23. Wright, P. 2000. Geothermal energy. *Geotimes* 45(7):16–18.

24. Duffield, W. A., J. H. Sass, and M. L. Sorey. 1994. *Tapping the Earth's natural heat*. U.S. Geological Survey Circular 1125.

25. Duffield, W. A. and J. H. Sass, 2003 Geothermal energy—clean power from Earth's heat. U. S. Geological Survey Circular 1249. p. 36.

26. Showstack, R. 2003. *Re-examining potential for geothermal energy in United States*. EOS 84(23):214.

Chapter 19 Critical Thinking Issue References

Brown, L. R., C. Flavin, and S. Postel. 1991. *Saving the planet*. New York: Norton.

Brower, M. 1990. *Cool energy*. Washington, D.C.: Union of Concerned Scientists.

Flavin, C. 1990. Slowing global warming. In L. R. Brown et al., eds., *State of the world*. New York: Norton.

Reddy, A. K. N., and J. Goldemberg. 1990. Energy for the developing world. *Scientific American* 263(3):110–113.

Chapter 20 Notes

1. Rosa, E. A., and R. E. Dunlap, 1994. Nuclear power: Three decades of public opinion. *Public Opinion Quarterly* 58(2):295–324.

2. Bisconti, A.S. 2003 (July). Two-thirds of Americans favor nuclear energy; public divided on building new nuclear plants. *Perspective on Public Opinion*.

3. Worldwatch Institute. 1994. Nuclear power levels peak. Worldwatch News Brief, March 5, 1994. Accessed February 9, 2001, at www.worldwatch.org/alerts/990304-html.V.16no6.

4. Churchill, A. A. 1993 (July). Review of WEC Commission: Energy for tomorrow's world. *World Energy Council Journal*, pp. 19–22.

5. Duderstadt, J. J. 1978. Nuclear power generation. In L. C. Ruedisili and M. W. Firebaugh, eds., *Perspectives on energy*, 2d ed., pp. 249–273. New York: Oxford University Press.

6. Till, C. E. 1989. Advanced reactor development. *Ann. Nucl. Energy* 16(6):301–305.

7. Ehrlich, P. R., A. H. Ehrlich, and J. P. Holdren. 1970. *Ecoscience: Population, resources, environment*. San Francisco: Freeman.

8. Brenner, D. J. 1989. *Radon: Risk and remedy*. New York: Freeman.

9. Cohen, B. L. 1990. *The nuclear energy option: An alternative for the 90s*. New York: Plenum.

10. Lake, J. A, R. G. Bennett, and J. F. Kotek. 2002 (January). Next-generation nuclear power. *Scientific American*, pp. 73–81.

11. U.S. Department of Energy. 1980. *Magnetic fusion energy*. DOE/ER-0059. Washington, D.C.: U.S. Department of Energy.

12. U.S. Department of Energy. 1979. *Environmental development plan, magnetic fusion*. DOE/EDP-0052. Washington, D.C.: U.S. Department of Energy.

13. Cordey, J. G., R. J. Goldston, and R. R. Parker. 1992. Progress toward a Tokamak fusion reactor. *Physics Today* 45:22–30.

14. Greenberg, P. A. 1993. Dreams die hard. *Sierra* 78:78.

15. Van Koevering, T. E., and N. J. Sell. 1986. Energy: *A conceptual approach*. Englewood Cliffs, N.J.: Prentice-Hall.

16. Waldbott, G. L. 1978. Health effects of environmental pollutants. 2d ed. Saint Louis: C. V. Moseby.

17. *New Encyclopedia Britannica*. 1997. Radiation, Vol. 26. p. 487.

18. U.S. Department of Energy. 1999. Radiation (in) waste isolation pilot plant. 1999. Carlsbad, New Mexico. Accessed at www.wipp.carlsbad.nm.us.

19. Stone, R. 2003. Plutonium fields forever. *Science* 300:1220–1224.

20. University of Maine and Maine Department of Human Services. 1983 (February). Radon in water and air. *Resource Highlights*.

21. MacLeod, G. K. 1981. Some public health lessons from Three Mile Island: A case study in chaos. *Ambio* 10:18–23.

22. Anspaugh, L. R., R. J. Catlin, and M. Goldman. 1988. The global impact of the Chernobyl reactor accident. *Science* 242:1513–1518.

23. Nuclear Energy Agency. 2002. Chernobyl Assessment of Radiological and Health Impacts: 2002 Update of Chernobyl: Ten Years On.

24. Balter, M. 1995. Chernobyl's thyroid cancer toll. *Science* 270:1758.

25. Skuterud, L., N. I. Goltsova, R. Naeumann, T. Sikkeland, and T. Lindmo. 1994. Histological changes in *Pinus sylvestris L.* in the proximal-zone around the Chernobyl power plant. *The Science of the Total Environment* 157:387–397.

26. Williams, N. 1995. Chernobyl: Life abounds without people. *Science* 269:304.

27. Fletcher, M. 2000 (November 14). The last days of Chernobyl. *Times* 2 (London), pp. 3–5.

28. Office of Industry Relations. 1974. *Development, growth and state of the nuclear industry.* Washington, D.C.: U.S. Congress, Joint Committee on Atomic Energy.

29. Weisman, J. 1996. Study inflames Ward Valley controversy. *Science* 271:1488.

30. Weart, W. D., M. T. Rempe, and D. W. Powers. 1998 (October). The waste isolation plant. *Geotimes.*

31. U.S. Department of Energy. 1999. Waste isolation pilot plant, Carlsbad, New Mexico. Accessed at www.wipp.carlsbad. nm.us.

32. Roush, W. 1995. Can nuclear waste keep Yucca Mountain dry—and safe? *Science* 270:1761.

33. Hanks, T. C., I. J. Winograd, R. E. Anderson, T. E. Reilly, and E. P. Weeks. 1999. *Yucca Mountain as a radioactive-waste repository.* U.S. Geological Survey Circular 1184.

34. Bredehoeft, J. D., A. W. England, D. B. Stewart, J. J. Trask, and I. J. Winograd. 1978. *Geologic disposal of high-level radioactive wastes—Earth science perspectives.* U.S. Geological Survey Circular 779. Arlington, Va.: U.S. Department of the Interior.

35. Nuclear Regulatory Commission. 2000. NRC's high level waste program. Accessed July 18, 2000, at http://www.nrc. gov/NMSS/DWM/hlw.htm.

36. Flavin, C. 1991. The case against reviving nuclear power. In L. R. Brown, ed., *The Worldwatch Reader,* pp. 205–220. New York: Norton.

37. Starke, L., ed. 2005. *Vital Signs 2005.* W. W. Norton & Company, New York: p. 139.

Chapter 20 Critical Thinking Issue References

Ahearne, J. F. 1993. The future of nuclear power. *American Scientist* 81(1):24–35.

Fox, M. R. 1987. Perspectives in risk: Compared to what? *Vital Speeches of the Day* 53(23):730–732.

Greenberg, P. A. 1993. Dreams die hard. *Sierra* 78(6):78.

Rosa, E. A., and R. E. Dunlap. 1994. Nuclear power: Three decades of public opinion. *Public Opinion Quarterly* 58:295–325.

Chapter 21 Notes

1. Morrison, J. 2005. How much is clean water worth? *National wildlife* 43(2):22–28

2. Henderson, L. J. 1913. *The fitness of the environment: An inquiry into the biological significance of the properties of matter.* New York: Macmillan.

3. Council on Environmental Quality and U.S. Department of State. 1980. *The global 2000 report to the president: Entering the twenty-first century,* Vol. 2. Washington, D.C.

4. Water Resources Council. 1978. *The nation's water resources,* 1975–2000, Vol. 1. Washington, D.C.

5. Winter, T. C., J. W. Harvey, O. L. Franke, and W. M. Alley. 1998. *Groundwater and surface water: A single resource.* U.S. Geological Survey Circular 1139.

6. Solley, W. B., R. R. Pierce, and H. A. Perlman. 1993. *Estimated use of water in the United States in 1990.* U.S. Geological Survey Circular 1081.

7. Solley, W. B., R. R. Pierce, and H. A. Perlman. 1998. *Estimated use of water in the United States in 1995.* U.S. Geological Survey Circular 1200.

8. U. S. General Accounting Office. 2003. *Freshwater supply: States' view of how federal agencies could help them meet the challenges of expected shortages.* Report GAO-03-514.

9. Alexander, G. 1984 (February/March). Making do with less. *National Wildlife,* special report, pp. 11–13.

10. Gleick, P. H., P. Loh, S. V. Gomez, and J. Morrison. 1995. *California water 2020, a sustainable vision.* Oakland, Calif.: Pacific Institute for Studies in Development, Environment and Security.

11. Alley, W. M., T. E. Reilly, and O. L. Franke. 1999. *Sustainability of ground-water resources.* U.S. Geological Survey Circular 1186.

12. Leopold, L. B. 1977. A reverence for rivers. *Geology* 5:429–430.

13. Holloway, M. 1991. High and dry. *Scientific American* 265:16–20.

14. Levinson, M. 1984 (February/March). Nurseries of life. *National Wildlife,* special report, pp. 18–21.

15. Nichols, F. H., J. E. Cloern, S. N. Luoma, and D. H. Peterson. 1986. The modification of an estuary. *Science* 231:567–573.

16. Day, J. W., Jr., J. M. Rybczyk, L. Carboch, W H. Conner, P. Delgado-Sanchez, R. I. Pratt, and A. Westphal. 1998. A review of recent studies of the ecology and economic aspects of the application of secondary treated municipal effluent to wetlands in southern Louisiana. From L. P. Rozas et al., eds., *Symposium on Recent Research in Coastal Louisiana,* February 3–5, 1998, Louisiana Sea Grant College Program, pp. 1–12.

17. Hileman, B. 1995. Rewrite of Clean Water Act draws praise, fire. *Chemical & Engineering News* 73:8.

18. Kaiser, J. 2001. Wetlands restoration: Recreated wetlands no match for original. *Science* 293:25a.

19. Gurardo, D., M. L. Fink, T. D. Fontaine, S. Newman, M. Chinmey, R. Bearzotti, and G. Goforth. 1995. Large-scale constructed wetlands for nutrient removal from stormwater runoff: An Everglades restoration project. *Environmental Management* 19:879–889.

20. Pearce, M. 1995 (January). The biggest dam in the world. *New Scientist,* pp. 25–29.

21. Zich, R. 1997. China's three gorges: Before the flood. *National Geographic* 192(3):2–33.

22. State of Maine. 2001. A brief history of the Edwards Dam. Accessed January 15, 2000 at http://janus.state.me.us/ spo/edwards/ timeline.htm.

23. American Rivers. Elwha River Restoration accessed March 1, 2006 at www.americanrivers.org.

24. Booth, W. 2000 (December 12). Restoring rivers—at a high price. *Washington Post,* p. A3.

25. U.S. Congress. 1973. *Stream channelization: What federally financed draglines and bulldozers do to our nation's streams.* House Committee Report No. 93-530. Washington, D.C.: U.S. Government Printing Office.

26. Graf, W. L. 1985. *The Colorado River: Instability and basin management.* Resource Publications in Geography. Washington, D.C.: Association of American Geographers.

27. Nash, R. 1986. Wilderness values and the Colorado River. In G. D. Weatherford and F. L. Brown, eds., *New courses for the Colorado River,* pp. 201–214. Albuquerque: University of New Mexico Press.

28. Hundley, N., Jr. 1986. The West against itself: The Colorado River—an institutional history. In G. D. Weatherford and F. L. Brown, eds., *New courses for the Colorado River*, pp. 9–49. Albuquerque: University of New Mexico Press.

29. Dolan, R., A. Howard, and A. Gallenson. 1974. Man's impact on the Colorado River and the Grand Canyon. *American Scientist* 62:392–401.

30. Hecht, J. 1996. Grand Canyon flood a roaring success. *New Scientist* 151 (2045):8.

31. Lucchitta, I., and L. B. Leopold. 1999. Floods and sand bars in the Grand Canyon. *Geology Today* 9:1–7.

32. Ballard, S. C., D. D. Michael, M. A. Chartook, M. R. Clines, C. E. Dunn, C. M. Hock, G. D. Miller, L. B. Parker, D. A. Penn, and G. W. Tauxe. 1982. *Water and western energy: Impacts, issues, and choices.* Boulder, Colo.: Westview Press.

33. Brown, L. R. 2003. *Plan B: Rescuing a planet under stress and a civilization in trouble.* New York: Norton.

Chapter 21 Critical Thinking Issues References

Baldwin, M. F. 1987. Wetlands: Fortifying federal and regional cooperation. *Environment* 29(7):16–20, 39.

Leidy, R. A., P. L. Fiedler, and E. R. Micheli. 1992. Is wetter better? *Bioscience* 40(9):58–61, 65.

National Institute for Urban Wildlife. *Wetlands conservation and use* (issue pak). Columbia, Md.: National Institute for Urban Wildlife.

Stevens, W. K. 1990 (March 13). Efforts to halt wetland loss are shifting to inland areas. *New York Times*, p. C1.

World Resources Institute. 1992. *The 1992 information please environmental almanac.* Boston: Houghton Mifflin.

Chapter 22 Notes

1. Mallin, M. A. 2000. Impacts of industrial animal production on rivers and estuaries. *American Scientist* 88(1): 26–37.

2. Bowie, P. 2000. No act of God. *The Amichs Journal* 21(4):16–21.

3. *Groundwater: Issues and answers.* 1985. Arvada, Colo.: American Institute of Professional Geologists.

4. Gleick, P. H. 1993. An introduction to global fresh water issues. In P. H. Gleick, ed., *Water in crisis.* New York. Oxford University Press, pp. 3–12.

5. Hileman, B. 1995. Pollution tracked in surface- and groundwater. *Chemical & Engineering News* 73:5.

6. Lewis, S. A. 1995. Trouble on tap. *Sierra* 80:54–58.

7. Smith, R. A. 1994. Water quality and health. *Geotimes* 39:19–21.

8. MacKenzie, W. R., et al. 1994. A massive outbreak in Milwaukee of *Cryptosporidium* infection transmitted through the public water supply. *The New England Journal of Medicine* 331:161–167.

9. Centers for Disease Control and Environmental Protection Agency. 1995. Assessing the public health threat associated with waterborne Cryptosporidiosis: Report of a workshop. *Journal of Environmental Health* 58:31.

10. Kluger, J. 1998. Anatomy of an outbreak. *Time* 152(5):56–62.

11. Maugh, T. H. 1979. Restoring damaged lakes. *Science* 203:425–427.

12. Mitch, W. J., J. W. Day, Jr., J. W. Gilliam, P. M. Groffman, D. L. Hey, G. W. Randall, and N. Wang. 2001. The Gulf of Mexico hypoxia—approaches to reducing nitrate in the Mississippi River or reducing a persistent large-scale ecological problem. *BioScience* (in press).

13. Hinga, K. R. 1989. Alteration of phosphorus dynamics during experimental eutrophication of enclosed marine ecosystems. *Marine Pollution Bulletin* 20:624–628.

14. Richmond, R. H. 1993. Coral reefs: Present problems and future concerns resulting from anthopogenic disturbance. *American Zoologist* 33:524–536.

15. Bell, P. R. 1991. Status of eutrophication in the Great Barrier Reef Lagoon. *Marine Pollution Bulletin* 23:89–93.

16. Hunter, C. L., and C. W. Evans. 1995. Coral reefs in Kaneohe Bay, Hawaii: Two centuries of Western influence and two decades of data. *Bulletin of Marine Science* 57:499.

17. Department of Alaska Fish and Game 1918. *Alaska Fish and Game* 21(4), Special Issue.

18. Holway, M. 1991. Soiled shores. *Scientific American* 265:102–106.

19. Robinson, A. R. 1973. Sediment, our greatest pollutant? In R. W. Tank, ed., *Focus on environmental geology*, pp. 186–192. New York: Oxford University Press.

20. Yorke, T. H. 1975. Effects of sediment control on sediment transport in the northwest branch, Anacostia River basin, Montgomery County, Maryland. *Journal of Research* 3:487–494.

21. Poole, W. 1996. Rivers run through them. *Land and People* 8:16–21.

22. Carey, J. 1984 (February/March). Is it safe to drink? *National Wildlife*, Special Report, pp. 19–21.

23. Pye, U. I., and R. Patrick. 1983. Groundwater contamination in the United States. *Science* 221:713–718.

24. Foxworthy, G. L. 1978. Nassau County, Long Island, New York—Water problems in humid county. In G. D. Robinson and A. M. Spieker, eds., *Nature to be commanded*, pp. 55–68. U.S. Geological Survey Professional Paper 950. Washington, D.C.: U.S. Government Printing Office.

25. Van der Leeden, F., F. L. Troise, and D. K. Todd. 1990. *The water encyclopedia*, 2d ed. Chelsea, Mich.: Lewis Publishers.

26. Jobling, S., M. Nolan, C. R. Tyler, G. Brighty, and J. P. Sumpter. 1998. Widespread sexual disturbance in wild fish. *Environmental Science and Technology* 32(17):2498–2506.

27. Environmental Protection Agency, Drinking Water Committee of the Science Advisory Board. 1995. *An SAB report: Safe drinking water. Future trends and challenges.* Washington, D.C.: Environmental Protection Agency.

28. Jewell, W. J. 1994. Resource—recovery wastewater treatment. *American Scientist* 82:366–375.

29. Task Force on Water Reuse. 1989. *Water reuse: Manual of practice SM-3.* Alexandria, Va.: Water Pollution Control Federation.

30. Kadlec, R. H., and R. L. Knight. 1996. *Treatment wetlands.* New York: Lewis Publishers.

31. Breaux, A. M., and J. W. Day, Jr. 1994. Policy considerations for wetland wastewater treatment in the coastal zone: A case study for Louisiana. *Coastal Management* (22):285–307.

32. Day, J. W., Jr., J. M. Rybczyk, L. Carboch, W. H. Conner, P. Delgado-Sanchez, R. I. Pratt, and W. Westphal. 1998. A review of recent studies of the ecology and economic aspects of the application of secondary treated municipal effluent to wet-

lands in southern Louisiana. From L. P. Rozas et al., eds., Symposium on Recent Research in Coastal Louisiana, February 1998, Louisiana Sea Grant College Program, pp. 1–12.

33. Breaux, A., S. Fuber, and J. Day, 1995. Using natural coastal wetland systems: An economic benefit analysis. *Journal of Environmental Management* (44):285–291.

34. Hileman, B. 1995. Rewrite of Clean Water Act draws praise, fire. *Chemical & Engineering News* 73:8.

35. U.S. Geologic Survey. 1997. Predicting the impact of relocating Boston's sewage outfall. UCGC fact sheet 185–97.

Chapter 22 Critical Thinking Issue References

Allan, J. D., and A. S. Flecker. 1993. Biodiversity conservation in running waters. *BioScience* 43(1):32–43.

Armour, C. L., D. A. Duff, and W. Elmore. 1991. The effects of livestock grazing on riparian and stream ecosystems. *Fisheries* 16(1):7 11.

Karr, J. R., L. A. Toth, and D. R. Dudley. 1985. Fish communities of midwestern rivers: A history of degradation. *BioScience* 35(2):90–95.

Sparks, R. 1992. The Illinois River floodplain ecosystem. In National Research Council, *Restoration of aquatic ecosystems,* pp. 412–432. Washington, D.C.: National Academy Press.

Stevens, W. K. 1993 (January 26). River life through U.S. broadly degraded. New York Times, pp. B5, B8.

Chapter 23 Notes

1. Intergovernmental Panel on Climate Change. 2001. Climate Change 2001. Series of 4 reports. Cambridge University Press.

2. Levy, S. 2000. Wildlife on the hot seat. *National Wildlife* 38(5):20–27.

3. Hartmann, D. L. 1994. *Global physical climatology.* International Geophysics Series, vol. 56. New York: Academic Press.

4. Hansen, J. 2003. Can we defuse the global warming time bomb? Natural Science. www.naturalscience.com.

5. Detwyler, T. R., and M. G. Marcus. 1972. *Urbanization and the environment: The physical geography of the city.* Belmont, Calif.: Duxbury Press.

6. Steig, E. J., E. J. Brook, J. W. C. White, C. M. Sucher, M. L. Bender, S. J. Lehman, D. L. Morse, E. D. Waddington, and G. D. Clow. 1998. Synchronous climate changes in Antarctica and the North Atlantic. *Science* 282:92–94.

7. Union of Concerned Scientists. 1989. *The greenhouse effect.* Cambridge, Mass.

8. Kerr, R. A. 1996. 1995 the warmest year? Yes and no. *Science* 271:137–138.

9. Crowley, T. J. 2000. Causes of climate change over the past 1000 years. *Science* 289:270–277.

10. Broecker, W. 1997. Will our ride into the greenhouse future be a smooth one? *GSA Today,* 7(5):2–6.

11. Steager, R. 2006. The source of Europe's mild climate. *American Scientist* 94: 334–341.

12. Earth System Sciences Committee. 1988. *Earth system science: A preview.* Boulder, Colo.: University Corporation for Atmospheric Research.

13. Post, W. M., T. Peng, W. R. Emanuel, A. W. King, V. H. Dale, and D. L. De Angelis. 1990. The global carbon cycle. *American Scientist* 78:310–326.

14. Moss, M. E., and H. F. Lins. 1989. *Water resources in the twenty-first century: A study of the implications of climate uncertainty.* U.S. Geological Survey Circular 1030. Washington, D.C.: U.S. Department of the Interior.

15. Hansen, J., A. Lacis, and M. Prather. 1989. Greenhouse effect of chlorofluorocarbons and other trace gases. *Journal of Geophysical Research* 94(D13):16417–16421.

16. Titus, J. G., S. P. Leatherman, C. H. Everts, Moffatt and Nichol Engineers, D. L. Kriebel, and R. G. Dean. 1985. *Potential impacts of sea level rise on the beach at Ocean City, Maryland.* Washington, D.C.: U.S. Environmental Protection Agency, Office of Policy Planning and Evaluation.

17. Rodhe, H. 1990. A comparison of the contribution of various gases to the greenhouse effect. *Science* 248:1217–1219.

18. Kerr, R. A. 1994. Methane increases put on pause. *Science* 263:751.

19. Dlugokencky, E. J., L. P. Steele, P. M. Lang, and K. A. Masarie. 1994. The growth rate and distribution of atmospheric methane. *Journal of Geophysical Research* 99(D8):17021–17043.

20. Council on Environmental Quality. 1990. *Environmental trends 1989.* Washington, D.C.

21. Luoma, J. R., and D. Hiser. 1991. Gazing into our greenhouse future. *Audubon* 93(2).

22. Charlson, R. J., S. E. Schwartz, J. M. Hales, R. D. Cess, J. A. J. Coakley, J. E. Hansen, and D. J. Hofmann. 1992. Climate forcing by anthropogenic aerosols. *Science* 255:423–430.

23. Kerr, R. A. 1995. Study unveils climate cooling caused by pollutant haze. *Science* 268:802.

24. NOAA. 1998. What is an El Niño? Accessed October 2, 1998 at http://www.elnino.noaa.gov.

25. Kennett, J. P., K. G. Cannariato, I. L. Hendy, and R. J. Behl. 2000. Carbon isotopic evidence for methane hydrate instability during Quaternary interstadials. *Science* 288:128–133.

26. Mohnen, V. A., W. Goldstein, and W. Wang. 1991. The conflict over global warming. *Global Environmental Change* 1:109–123.

27. Kerr, R. A. 1995. U.S. climate tilts toward the greenhouse. *Science* 268:363–364.

28. Kerr, R. A. 1995. Greenhouse report foresees growing global stress. *Science* 270:731.

29. Titus, J. G., and V. K. Narayanan. 1995. *The probability of sea level rise.* Washington, D.C.: U.S. Environmental Protection Agency.

30. Pelto, M.S. 1996. Recent changes in glacier and alpine runoff in the North Cascades, Washington. Hydrological Processes 10:1173–80.

31. Curt H. Davis, Yonghong Li, Joseph R. McConnell, Markus M. Frey, and Edward Hanna, 2005. Snowfall-Driven Growth in East Antarctic Ice Sheet Mitigates Recent Sea-Level Rise. Published online 19 May 2005 [DOI: 10.1126/science.1110662] in Science Express Reports.

32. Montaigne, F. 2004. Eco Signs, In The Heat is On. *National Geographic* 206(3):34–35.

33. Root, T.L., D.P. MacMynowsky, M.D. Mastrandrea, and S.H. Schneider, 2005. Human-modified temperature-induced species changes. Proceedings of National Academy of Sciences. 102(21):7465–7469.

34. Holmes, N. 2000. Has anyone checked the weather (map)? *Amicus Journal* 21(4):50–51.

35. Epstein, P. R. 2000. Is global warming harmful to health? *Scientific American* 283(2):50–57.

36. Thomas, C. D., et. al. 2004. Extinction risk of climate change. *Nature* 427:145–148.

37. Lea, D. W. 2004. The 100,000-yr cycle in tropical SST, greenhouse forcing, and climate sensitivity. *Journal of Climate* 17 (11) 2170–2179.

38. Dunn, S. 2001. Decarbonizing the energy economy. In *WorldWatch Institute state of the world 2001*. New York: W. W. Norton, pp. 83–102.

39. Rice, C. W. 2002. Storing carbon in soil: Why and how. *Geotimes* 47(1):14–17.

40. Friedman, S. J. 2003. Storing carbon in Earth. *Geotimes* 48(3):16–20.

41. Bartlett, K. 2003. Demonstrating carbon sequestration. *Geotimes* 48(3):22–23.

A Closer Look 23.1 References
Edahl, C. A., and C. D. Keeling. 1973. *Carbon and the biosphere*. Oak Ridge, Tenn.: Technical Information Service.

Keeling, C. D., T. P. Worf, and J. Van der Plicht. 1995. *Nature* 375:666–670.

Chapter 23 Critical Thinking Issue Reference
King, D. A. 2004. Climate change science: Adapt, mitigate, or ignore? *Science* 303:176–177.

Chapter 24 Notes
1. Simons, L. M. 1998. Plague of fire. *National Geographic* 194(2):100–119.

2. National Park Service. 1984. *Air resources management manual*.

3. American Lung Association. 1998. American Lung Association outdoor fact sheet. Accessed September 18, 1998 at http://www.lungusa.org.

4. Godish, T. 1991. *Air quality*, 2d ed. Chelsea, Mich.: Lewis Publishers.

5. Seitz, F., and C. Plepys. 1995. Monitoring air quality in healthy people 2000. *Healthy people 2000: Statistical notes no. 9*. Atlanta: Centers for Disease Control and Prevention, National Center for Health Statistics.

6. American Lung Association, 2001. *State of the Air 2000*.

7. Moore, C. 1995. Poisons in the air. *International Wildlife* 25:38–45.

8. Pope, C. A., III, D. V. Bates, and M. E. Raizenne. 1995. Health effects of particulate air pollution: Time for reassessment? *Environmental Health Perspectives* 103:472–480.

9. Travis, D. J., A. M. Carleton, and R. G. Lauritsen 2002. Contrails reduce daily temperature range. *Nature* 419:601.

10. Office of Technology Assessment. 1984. Balancing the risks. *Weatherwise* 37:241–249.

11. Canadian Department of the Environment. 1984. *The acid rain story*. Ottawa: Minister of Supply and Services.

12. Lippmann, M., and R. B. Schlesinger. 1979. *Chemical contamination in the human environment*. New York: Oxford University Press.

13. Winkler, E. M. 1998 (September). The complexity of urban stone decay. *Geotimes*, pp. 25–29.

14. U. S. Environmental Protection Agency 2006. National-scale air toxics assessment for 1999. Estimated emissions. Concentrations and risks. Technical Fact Sheet. Accessed April 10, 2006 at www.epa.gov.

15. Tyson, P. 1990. Hazing the Arctic. *Earthwatch* 10:23–29.

16. Pittock, A. B., L. A. Frakes, D. Jenssen, J. A. Peterson, and J. W. Zillman, eds. 1978. *Climatic change and variability: A southern perspective*. New York: Cambridge University Press.

17. Brown, L. R., ed. 1991. *The Worldwatch reader on global environmental issues*. New York: Norton.

18. Lents, J. M., and W. J. Kelly. 1993. Clearing the air in Los Angeles. *Scientific American* 269:32–39.

19. Blake, D. R., and F. S. Rowland. 1995. Urban leakage of liquefied petroleum gas and its impact on Mexico City air quality. *Science* 269:953.

20. Pountain, D. 1993 (May). Complexity on wheels. *Byte*, pp. 213–220.

21. Stern, A. C., R. T. Boubel, D. B. Turner, and D. L. Fox. 1984. *Fundamentals of air pollution*, 2d ed. Orlando, Fla.: Academic Press.

22. Moore, C. 1995. Green revolution in the making. *Sierra* 80:50.

23. Kolstad, C. D. 2000. *Environmental economics*. New York: Oxford University Press.

24. Molnia, B. F. 1991. Washington report. *GSA Today* 1:33.

25. Crandall, R. W. 1983. *Controlling industrial pollution: The economics and politics of clean air*. Washington, D.C.: Brookings Institution.

26. Hall, J. V., A. M. Winer, M. T. Kleinman, F. W. Lurmann, V. Brajer, and S. D. Colome. 1992. Valuing the health benefits of clean air. *Science* 255:812–816.

27. Krupnick, A. J., and P. R. Portney. 1991. Controlling urban air pollution: A benefits-cost assessment. *Science* 252:522–528.

28. Lipfert, F. W., S. C. Morris, R. M. Friedman, and J. M. Lents. 1991. Air pollution benefit-cost assessment. *Science* 253:606.

Chapter 24 Critical Thinking Issue References
Khalil, M., and R. A. Rasmussen 1993. Arctic haze—patterns and relationships to regional signatures of trace gases. *Global Biogeochemical Cycles* 7(1):27–36.

Rahn, K. A. 1984. Who's polluting the Arctic? *Natural History* 93(5):31–38.

Shaw, G. E. 1995. The Arctic haze phenomenon. *Bulletin of the American Meteorological Society* 76(12):2403–2413.

Soroos, M. S. 1992. The odyssey of Arctic haze. *Environment* 34(10):6–27.

Young, O. R. 1990. Global commons: The Arctic in world affairs. *Technology Review* 93:52–61.

Chapter 25 Notes
1. Massachusetts Department of Public Health, Bureau of Environmental Health Assessments. 1995. *Symptom prevalence survey related to indoor air concerns at the Registry of Motor Vehicles Building, Ruggles Station*.

2. Horton, W. G. B. 1995. *NOVA: Can buildings make you sick?* Video production. Boston: WGBH.

3. Zimmerman, M. R. 1985. Pathology in Alaskan mummies. *American Scientist* 73:20–25.

4. Ehrlich, P. R., A. H. Ehrlich, and J. P. Holdren. 1970. *Ecoscience: Population, resources, environment.* San Francisco: Freeman.

5. Conlin, M. 2000 (June 5). Is your office killing you? *Business Week,* pp. 114–124.

6. U.S. Environmental Protection Agency. 1991. *Building air quality: A guide for building owners and facility managers.* EPA/400/1-91/033, DHHS (NIOSH) Pub. No. 91–114. Washington, D.C.: Environmental Protection Agency.

7. Zummo, S. M., and M. H. Karol. 1996. Indoor air pollution: Acute adverse health effects and host susceptibility. *Environmental Health* 58:25–29.

8. Committee on Indoor Air Pollution. 1981. *Indoor pollutants.* Washington, D.C.: National Academy Press.

9. Godish, T. 1997. *Air quality,* 3d ed. Boca Raton, Fla.: Lewis Publishers.

10. O'Reilly, J. T., P. Hagan, R. Gots, and A. Hedge. 1998. *Keeping buildings healthy.* New York: Wiley.

11. Brenner, D. J. 1989. *Radon: Risk and remedy.* New York: W. H. Freeman.

12. U.S. Environmental Protection Agency. 1992. *A citizen's guide to radon: The guide to protecting yourself and your family from radon,* 2d ed. ANR-464. Washington, D.C.: Environmental Protection Agency.

13. Hurlburt, S. 1989 (June). Radon: A real killer or just an unsolved mystery? *Water Well Journal,* pp. 34–41.

14. Egginton, J. 1989. Menace of Whispering Hills. *Audubon* 91:28–35.

15. University of Maine and Maine Department of Human Services. 1983 (February). Radon in water and air. *Resource Highlights.*

16. U.S. Environmental Protection Agency. 1986. *Radon reduction techniques for detached houses: Technical guidance.* EPA 625/5-86-019. Research Triangle Park, N.C.: Air and Energy Engineering Research Laboratory, Office of Research and Development, U.S. Environmental Protection Agency.

17. U. S. Environmental Protection Agency. Radon-resistance new construction (RRNC). Accessed April 22, 2006 at www.epa.gov.

18. Henshaw, D. L., J. P. Eatough, and R. B. Richardson. 1990. Radon as a causative factor in induction of myeloid leukaemia and other cancers. *The Lancet* 335:1008–1012.

19. Pershagen, G., G. Akerblom, O. Axelson, B. Clavensjo, L. Damber, G. Desai, A. Enflo, F. Lagarde, H. Mellander, M. Svartengren, and G. A. Swedjemark. 1994. Residential radon exposure and lung cancer in Sweden. *New England Journal of Medicine* 330:159–164.

20. Nero, A. V., Jr. 1988. Controlling indoor air pollution. *Scientific American* 258:42–48.

Chapter 25 Critical Thinking Issue References

Castleman, M. 1995. Clean air up there. *Sierra* 80(3):16.

Douglass, W. C. 1992 (September). If you fly, don't breathe. *Second Opinion,* pp. 1–5.

Kenyon, T. A. 1996. Transmission of multidrug-resistant *Myobacterium tuberculosis* during a long airplane flight. *The New England Journal of Medicine* 334(15):933.

Manning, A. 1993 (June 22). Airborne ailments: Are diseases transmitted in flight? USA *Today,* pp. 1A, 2A.

Nagda, N. L. 1993 (July 29). Testimony before Committee on Science and Technology, U.S. House of Representatives Subcommittee on Technology, Environment and Aviation, Washington, D.C.

Nagda, N. L., M. D. Koontz, and A. G. Konheim. 1991 (August). Carbon dioxide levels in commercial airliner cabins. *ASHRAE Journal* 33:35–38.

National Research Council. 1986. *The airliner cabin environment: Air quality and safety.* Washington, D.C.: National Academy Press.

Tolchin, M. 1993 (June 21). Exposures to tuberculosis on planes are investigated. *New York Times,* p. A7.

Tolchin, M. 1993 (June 25). Inquiry will check air quality on airplanes. *New York Times,* p. A16.

Chapter 26 Notes

1. Kane, R. P. 1998. Ozone depletion, UV-B changes, and increased cancer incidence. *International Journal.*

2. Brown, L. R., C. Flavin, and S. Postel. 1991. *Saving the planet: How to shape an environmentally sustainable global economy.* New York: Norton.

3. Rowland, F. S. 1990. Stratospheric ozone depletion of chlorofluorocarbons. *AMBIO* 19:281–292.

4. Smith, R. C., B. B. Prezelin, K. S. Baker, R. R. Bidigare, N. P. Boucher, T. Coley, D. Karentz, S. Macintyre, H. A. Matlick, D. Menzies, M. Ondrusek, Z. Wan, and K. J. Waters. 1992. Ozone depletion: Ultraviolet saturation and phytoplankton biology in Antarctic waters. *Science* 255:952–959.

5. NASA-GSFC. 2000. Stratospheric ozone. Accessed August 22, 2000 at http://see.gsfc.nasa.gov.

6. Environmental Protection Agency. 1995. *Protection of the ozone layer.* EPA 230-N-95-00. Washington, D.C.: U.S. EPA Office of Policy, Planning, and Evaluation and Office of Air and Radiation.

7. NASA. 2004. Average Antarctic miminum ozone concentration. Accessed March 24, 2004 at http://jwocky.gsfc.nasa./gov/multi/min_ozone/gif.

8. Hamill, P., and O. B. Toon. 1991. Polar stratospheric clouds and the ozone hole. *Physics Today* 44:34–42.

9. Stolarski, R. S. 1988. The Antarctic ozone hole. *Scientific American* 258:30–36.

10. Molina, M. J., and F. S. Rowland. 1974. Stratospheric sink for chlorofluoromethanes: Chlorine-atom catalyzed distribution of ozone. *Nature* 249:810–812.

11. Brouder, P. 1986 (June). Annals of chemistry in the face of doubt. *New Yorker,* pp. 20–87.

12. Khalil, M. A. K., and R. A. Rasmussen. 1989. The potential of soils as a sink of chlorofluorocarbons and other manmade chlorocarbons. *Geophysical Research Letters* 16:679–682.

13. Rowland, F. S. 1989. Chlorofluorocarbons and the depletion of stratospheric ozone. *American Scientist* 77:36–45.

14. Toon, O. B., and R. P. Turco. 1991. Polar stratospheric clouds and ozone depletion. *Scientific American* 264:68–74.

15. Kerr, R. A. 1994. Antarctic ozone hole fails to recover. *Science* 266:217.

16. Webster, C. R., R. D. May, D. W. Toohey, L. M. Avallone, J. G. Anderson, P. Newman, L. Lait, M. Schoeberl, J. W.

Elkins, and K. R. Chay. 1993. Chlorine chemistry on polar stratospheric cloud particles in the Arctic winter. *Science* 261:1130–1134.

17. Kerr, R. A. 1992. New assaults seen on Earth's ozone shield. *Science* 255:797–798.

18. Zurer, P. 1995. Record low ozone levels observed over Arctic. *Chemical & Engineering News* 73:8.

19. Cutter Information Corp. 1996. Reports discuss present and future state of ozone layer. *Global Environmental Change Report* V, VIII, 21, no. 22, pp. 1–3. Dunster, B.C., Canada: Cutter Information Corp.

20. Shea, C. P. 1989. Mending the Earth's shield. *World Watch* 2:28–34.

21. Kerr, J. B., and C. T. McElroy. 1993. Evidence for large upward trends of ultraviolet-B radiation linked to ozone depletion. *Science* 262:1032–1034.

22. Atkinson, R. J., W. A. Matthews, P. A. Newman, and R. A. Plumb. 1989. Evidence of the mid-latitude impact of Antarctic ozone depletion. *Nature* 340:290–294.

23. U.S. Environmental Protection Agency. 2006. Ozone depletion. Accessed April 16, 2006 at www.epa.gov.

24. Russell, J. M., M. Luo, R. J. Cicerone, and L. E. Deaver. 1996. Satellite confirmation of dominance of chlorofluorocarbons in the global stratospheric chlorine budget. *Nature* 379:526.

25. Showstack, R. 1998. Ozone layer is on slow road to recovery, new science assessment indicates. *Eos* 79(27):317–318.

26. Spurgeon, D. 1998. Surprising success of the Montreal protocol. *Nature* 389(6648):219.

27. Makhijani, A., and A. Bickel, 1990. Still working on the ozone hole. *Technology Review* 93:52–59.

28. Brown, L. R., N. Lenssen, and H. Kane. 1995. CFC production plummeting. In Worldwatch Institute, *Vital Signs* 1995. New York: Norton.

29. Shea, C. P. 1991. Disarming refrigerators. *World Watch* 4:36.

30. U.S. Environmental Protection Agency. 2003. Ozone depletion: accessed October 8, 2003 at http://www.epa.gov/ozone/index.html.

31. MacKenzie, D. 1990. Cheaper alternatives for CFCs. *New Scientist* 126:39–40.

Chapter 26 Critical Thinking Issue References

Brasseur, G., and C. Branier. 1992. Mount Pinatubo: Aerosols, chlorofluorocarbons, and ozone depletion. *Science* 257:1239–1241.

Kerr, R. A. 1993. Ozone takes a nose dive after the eruption of Mt. Pinatubo. *Science* 260:490–491.

Pohl, F., and J. P. Hogan. 1993. Ozone politics: They call this science? *Omni* 15(8):34–42, 91.

Rowland, F. S. 1993. President's lecture: The need for scientific communication with the public. *Science* 260:1571–1576.

Russel, J. M., M. Luo, R. J. Cicerone, and L. E. Deaver. 1996. Satellite confirmation of dominance of chlorofluorocarbons in the global stratospheric chlorine budget. *Nature* 379(6565):526.

Tabazadeh, A., and R. P. Turco. 1993. Stratospheric chlorine injection by volcanic eruptions: HCl scavenging and implications for ozone. *Science* 260:1082–1085.

Taubes, G. 1993 (June 11). The ozone backlash. *Science* 260, pp. 1580–1583.

Chapter 27 Notes

1. Hudson, T. L., F. D. Fox, and G. S. Plumlee. 1999. *Metal mining and the environment.* Alexandria, Va.: American Geological Institute.

2. McKelvey, V. E. 1973. Mineral resource estimates and public policy. In D. A. Brobst and W. P. Pratt, eds., *United States mineral resources,* pp. 9–19. U.S. Geological Survey Professional Paper 820.

3. U.S. Department of the Interior, Bureau of Mines. 1993. *Mineral commodity summaries, 1993.* I 28.149:993. Washington, D.C.: U.S. Department of the Interior.

4. Meyer, H. O. A. 1985. Genesis of diamond: A mantle saga. *American Mineralogist* 70:344–355.

5. Kesler, S. F. 1994. *Mineral resources, economics, and the environment.* New York: Macmillan.

6. Smith, G. I., C. L. Jones, W. C. Culbertson, G. E. Erickson, and J. R. Dyni. 1973. Evaporites and brines. In D. A. Brobst and W. P. Pratt, eds., *United States mineral resources,* pp. 197–216. U.S. Geological Survey Professional Paper 820.

7. Awramik, S. A. 1981. The pre-Phanerozoic biosphere— three billion years of crises and opportunities. In M. H. Nitecki, ed., *Biotic crises in ecological and evolutionary time,* pp. 83–102. Spring Systematics Symposium. New York: Academic Press.

8. Margulis, L., and J. E. Lovelock. 1974. Biological modulation of the Earth's atmosphere. *Icarus* 21:471–489.

9. Lowenstam, H. A. 1981. Minerals formed by organisms. *Science* 211:1126–1130.

10. Bateman, A. M. 1950. *Economic mineral deposits,* 2d ed. New York: Wiley.

11. Park, C. F., Jr., and R. A. MacDiarmid. 1970. *Ore deposits,* 2d ed. San Francisco: Freeman.

12. Brobst, D. A., W. P. Pratt, and V. E. McKelvey. 1973. *Summary of United States mineral resources.* U.S. Geological Survey Circular 682.

13. Jeffers, T. H. 1991 (June). Using microorganisms to recover metals. *Minerals Today.* Washington, D.C.: U.S. Department of Interior, Bureau of Mines, pp. 14–18.

14. Haynes, B. W. 1990 (May). Environmental technology research. *Minerals Today.* Washington, D.C.: U.S. Bureau of Mines, pp. 13–17.

15. Sullivan, P. M., M. H. Stanczyk, and M. J. Spendbue. 1973. *Resource recovery from raw urban refuse.* U.S. Bureau of Mines Report of Investigations 7760.

16. Davis, F. F. 1972 (May). Urban ore. *California Geology,* pp. 99–112.

17. U.S. Geological Survey. 2000. *Minerals yearbook 1998— Recycling metals.* Accessed August 21, 2000 at http://minerals.usgs.gov.

18. Brown, L., N. Lenssen, and H. Kane. 1995. Steel recycling rising. In *Vital Signs* 1995. Worldwatch Institute.

19. Wellmar, F. W., and M. Kosinowoski. 2003. Sustainable development and the use of non renewable sources. *Geotimes* 48(12): 14–17.

20. The Butchart Gardens. 1998. Victoria, B.C.: The Butchart Gardens.

Chapter 27 Critical Thinking Issue Reference

Debus, K. H. 1990 (August/September). Mining with microbes. *Technology Review* 93:50.

Chapter 28 Notes

1. *New York Times* http://select.nytimes.com/search/restrict-ed/article?res=FA0716F734550C7A8DDDA00894DD404482. March 13, 2006. Anger over bid to hike whale catch June 20, 2005. CNN, http://www.cnn.com/2005/WORLD/asiapcf/06/20/whaling.meeting/index.html.

2. Roberts, L. 1991. Costs of a clean environment. *Science* 251:1182.

3. Fairley, P. 1995. Compromise limits EPA budget cut, removes House riders. *Chemical Week* 157:17.

4. Moore, C. E. 1995. Poisons in the air. *International Wildlife* 25:38–45.

5. Pope, C. A., III, D. V. Bates, and M. E. Raizenne. 1995. Health effects of particulate air pollution: Time for reassessment. *Environmental Health Perspectives* 103:472–480.

6. Hardin, G. 1968. The tragedy of the commons. *Science* 162:1243–1248.

7. Clark, C. W. 1973. The economics of overexploitation. *Science* 181:630–634.

8. Freudenburg, W. R. 2004. Personal communication.

9. Gunn, J. M., ed. 1995. *Restoration and recovery of an industrial region: Progress in restoring the smelter-damaged landscape near Sudbury, Canada.* New York: Springer-Verlag.

10. Costanza, R., et al. 1997. The value of the world's ecosystem services and natural capital. *Nature* 387:253–260.

11. James, A., K. T. Gaston, and A. Blamford. 2001. Can we afford to conserve biodiversity? *BioScience* 51(1):43–52.

12. Litton, R. B. 1972. Aesthetic dimensions of the landscape. In J. V. Krutilla, ed. *Natural environments.* Baltimore: Johns Hopkins University Press.

13. Schwing, R. C. 1979. Longevity and benefits and costs of reducing various risks: *Technological Forecasting and Social Change* 13:333–345.

14. Gori, G. B. 1980. The regulation of carcinogenic hazards. *Science* 208:256–261.

15. Cairns, J., Jr., 1980. Estimating hazard. *BioScience* 20:101–107.

16. James, A., K. T. Gaston, and A. Blamford. 2001. Can we afford to conserve biodiversity? *BioScience* 51(1):43–52.

17. Ostro, B. D. 1980. Air pollution, public health, and inflation. *Environmental Health Perspectives* 345:185–189.

18. Office of Technology Assessment. 1991. *Changing by degrees: Steps to reduce greenhouse gases.* Washington, D.C.: U.S. Superintendent of Documents.

19. D'Arge, R. 1989. Ethical and economic systems for managing the global commons. In D. B. Botkin, M. Caswell, J. E. Estes, and A. Orio, eds. *Changing the global environment: Perspectives on human involvement*, pp. 327–337. New York: Academic Press.

20. Rogers, A. 1993. *The Earth summit: A planetary reckoning.* Los Angeles: Global View Press

21. Baumol, W. J., and W. E. Oates, 1979. *Economics, environmental policy, and the quality of life.* Englewood Cliffs, N.J.: Prentice-Hall.

22. Clark, C. W. 1981. Economics of fishery management. In T. L. Vincent and J. M. Skowronski, eds., *Renewable resource management: Lecture notes in biomathematics*, pp. 9–111. New York: Springer-Verlag.

Chapter 28 Critical Thinking Issue References

Anthony, V. C. 1993. The state of groundfish resources off the north-eastern United States. *Fisheries* 18 (3):12–17.

Correia, S. J. 1992 (3d quarter). Flounder population declines: Overfishing or pollution! *Division of Marine Fisheries News.* Boston: Massachusetts Division of Marine Fisheries.

How to fish. 1988 (December 10). *The Economist* 309 (7580):93–96.

Keen, E. A. 1991. Ownership and productivity of marine fisheries resources. *Fisheries* 16:18–22.

Lawren, B. 1992. Net loss. *National Wildlife* 30(6):47–52.

Leal, D. R. 1992 (July 30). Using property rights to regulate fish harvest. *Christian Science Monitor* 84:18.

National Marine Fisheries Service. 1995. *Status of the fishery resources off the northeastern United States for 1994.* Woods Hole, Mass.: U.S. Department of Commerce, NOAA, NMFS Northeast Fisheries Science Center.

Pierce, D. 1992 (2d quarter). New England council to cut fishing effort in half over next 5 years. *Division of Marine Fisheries News.* Boston: Massachusetts Division of Marine Fisheries.

Satchell, M. 1992. The rape of the oceans. *U.S. News and World Report* 112(24):64–75.

Chapter 29 Notes

1. Wikipedia, the free encyclopedia, http://en.wikipedia.org/wiki/Hurricane_katrina.

2. Steinhauer, November 8, J. 2005. New Orleans is still grappling with the basics of rebuilding. *The New York Times*, p.A1.

3. Lalasz, R. 2006 (may). Americans flocking to outer suburbs in record numbers. Population Reference Bureau, http://www.prb.org.

4. EPA. May 11, 2006. What is Urban Sprawl? http://www.epa.gov/maia/html/sprawl.html.

5. Butler, Rhett, 2003, *World's Largest Urban Areas [Ranked by Urban Area Population], The World Gazetteer*, http://www.mongabay.com/citiesurban01.htm.

6. Neubauer, D. 2004 (September 24). Mixed blessings of the megacities, *Yale Global Online Magazine*, http://yaleglobal.yale.edu.

7. Haub, C., and D. Cornelius. 2000. *World population data sheet.* Washington, D.C.: Population Reference Bureau.

8. Mumford, L. 1972. The natural history of urbanization. In R. L. Smith, ed., *The ecology of man: An ecosystem approach*, pp. 140–152. New York: Harper & Row.

9. BBC News. http://news.bbc.co.uk/go/pr/fr/-/2/hi/europe/4292374.stm. September 29, 2005.

10. Fein, Melissa. Venice is sinking and Italy is watching, so why should the U.S. help? Accessed at http://www.law.harvard.edu/faculty/martin/art_law/fien_venice.htm.

11. Pavoni, B., A. Sfriso, D. Donazzolo, and A. A. Orio. 1990. Influence of wastewaters on the city of Venice and the hinterland on the eutrophication of the lagoon. *The Science of the Total Environment* 96:325–252.

12. Hunt, C. B. 1974. *Natural regions of the United States and Canada.* San Francisco: Freeman.

13. Leibbrand, K. 1970. *Transportation and town planning.* Translated by N. Seymer. Cambridge, Mass.: MIT Press.

14. Reps, J. W. 1965. *The making of urban America: A history of city planning in the United States*, 2d ed. Princeton, N. J.: Princeton University Press.

15. McLaughlin, C. C., ed. 1977. The formative years: 1822–1852. Volume 1 of *The Papers of Frederick Law Olmsted.* Baltimore: Johns Hopkins University Press.

16. Miller, L. B. 1987. Miracle on 104th Street. *American Horticulturalist* 66:14–17.

17. Spirn, A. W. 1984. *The granite garden: Urban nature and human design.* New York: Basic Books.

18. Detwyler, T. R., and M. G. Marcus, eds. 1972. *Urbanization and the environment: The physical geography of the city.* North Scituate, Mass.: Duxbury Press.

19. Butti, K., and J. Perlin. 1980. *A golden thread: 2500 years of solar architecture and technology.* New York: Cheshire.

20. McHarg, I. L. 1971. *Design with nature.* Garden City, N.Y.: Doubleday.

21. Ford, A. B., and O. Bialik. 1980. Air pollution and urban factors in relation to cancer mortality. *Archives of Environmental Health* 35:350–359.

22. Nadel, I. B., C. H. Oberlander, and L. R. Bohm. 1977. *Trees in the city.* New York: Pergamon.

23. Moll, G., P. Rodbell, B. Skiera, J. Urban, G. Mann, and R. Harris. 1991 (April/May). Planting new life in the city. *Urban Forests*, pp. 10–20. Washington, D.C.: American Forestry Association.

24. Dreistadt, S. H., D. L. Dahlsten, and G. W. Frankie. 1990. Urban forests and insect ecology. *BioScience* 40:192–198.

25. Leedly, D. L., and L. W. Adams. 1984. *A guide to urban wildlife management.* Columbia, Md.: National Institute for Urban Wildlife.

26. Tylka, D. 1987. Critters in the city. *American Forests* 93:61–64.

27. Burton, J. A. 1977. *Worlds apart: Nature in the city.* Garden City, N.Y.: Doubleday.

28. Department of Environmental Protection, New York. 2006. Peregrine falcons in New York City. http://www.nyc.gov/html/dep/html/news/falcon.html.

29. Adams, L. W., and L. E. Dove. 1989. Wildlife reserves and corridors in the urban environment. Columbia, Md.: National Institute for Urban Wildlife.

30. http://www.responsiblewildlifemanagement.org/bubonic_plague.htm

31. U.S. Environmental Protection Agency 2006. http://www.epa.gov/maia/html/sprawl. html. May 11, 2006.

32. www.epodunk.com/egi-bin/PopInfo.php?locIndex=936, May 11, 2006.

Chapter 30 Notes

1. Harder, B. 2005 (November 8). Toxic e-waste is conched in poor nations. *National Geographic News.*

2. Galley, J. E. 1968. Economic and industrial potential of geologic basins and reservoir strata. In J. E. Galley, ed., *Subsurface disposal in geologic basins: A study of reservoir strata*, pp. 1–19. American Association of Petroleum Geologists Memoir 10. Tulsa, Okla.: American Association of Petroleum Geologists.

3. Relis, P., and A. Dominski. 1987. *Beyond the crisis: Integrated waste management.* Santa Barbara, Calif.: Community Environmental Council.

4. Repa, D. W., and A. Blakey. 1996. Municipal solid waste disposal trends: 1996 update. *Waste Age* 27:42–54.

5. Council on Environmental Quality. 1973. *Environmental quality—1973.* Washington, D.C.: U.S. Government Printing Office.

6. Allenby, B. R. 1999. *Industrial ecology: Policy framework and implementation.* Upper Saddle River, N.J.: Prentice-Hall.

7. Garner, G., and P. Sampat. 1999 (May). Making things last: Reinventing of material culture. *The Futurist*, pp. 24–28.

8. U.S. Environmental Protection Agency. Municipal solid waste. Accessed April 21, 2006 at www.epa.gov.

9. Relis, P., and H. Levenson. 1998. *Discarding solid waste as we know it: Managing materials in the 21st century.* Santa Barbara, Calif.: Community Environmental Council.

10. Young, J. E. 1991. Reducing waste-saving materials. In L. R. Brown, ed., *State of the world, 1991*, pp. 39–55. New York: Norton.

11. Steuteville, R. 1995. The state of garbage in America: Part I. *BioCycle* 36:54.

12. Gardner, G. 1998 (January/February). Fertile ground or toxic legacy? *World Watch*, pp. 28–34.

13. McGreery, P. 1995. Going for the goals: Will states hit the wall? *Waste Age* 26:68–76.

14. Brown, L. R. 1999 (March/April). Crossing the threshold. *World Watch*, pp. 12–22.

15. Rathje, W. L., and C. Murphy. 1992. Five major myths about garbage, and why they're wrong. *Smithsonian* 23:113–122.

16. Rathje, W. L. 1991. Once and future landfills. *National Geographic* 179(5):116–134.

17. Schneider, W. J. 1970. Hydrologic implications of solid-waste disposal. 135(22). U.S. Geological Survey Circular 601F. Washington, D.C.: U.S. Geological Survey.

18. Thomas, V. M., and T. G. Spiro. 1996. The U.S. dioxin inventory: Are there missing sources? Environmental Science & Technology 30:82A–85A.

19. Turk, L. J. 1970. Disposal of solid wastes—acceptable practice or geological nightmare? In Environmental Geology, pp. 1–42. Washington, D.C.: American Geological Institute Short Course, American Geological Institute.

20. Hughes, G. M. 1972. Hydrologic considerations in the siting and design of landfills. Environmental Geology Notes, no. 51. Urbana: Illinois State Geological Survey.

21. Bergstrom, R. E. 1968. Disposal of wastes: Scientific and administrative considerations. Environmental Geology Notes, no. 20. Urbana: Illinois State Geological Survey.

22. Cartwright, K., and Sherman, F. B. 1969. Evaluating sanitary landfill sites in Illinois. Environmental Geology Notes, no. 27. Urbana: Illinois State Geological Survey.

23. Rahn, P. H. 1996. Engineering geology, 2nd ed. Upper Saddle River, N.J.: Prentice-Hall.

24. Bullard, R. D. 1990. Dumping in Dixie: Race, class and environmental quality. Boulder, CO: Westview Press.

25. Sadd, J. L., J. T. Boer, M. Foster, Jr., and L. D. Snyder. 1997. Addressing environmental justice: Demographics of hazardous waste in Los Angeles County. Geology Today 7(8):18–19.

26. Walker, W. H. 1974 Monitoring toxic chamical pollution from land disposal sites in humid regions. Ground Water 12: 213–218.

27. Watts, R. J. 1998. Hazardous wastes. New York: John Wiley & Sons.

28. Wilkes, A. S. 1980. Everybody's problem: Hazardous waste. SW-826. Washington, D.C.: U.S. Environmental Protection Agency, Office of Water and Waste Management.

29. Elliot, J. 1980. Lessons from Love Canal. Journal of the American Medical Association 240:2033–2034, 2040.

30. Kufs, C., and C. Twedwell. 1980. Cleaning up hazardous landfills. Geotimes 25:18–19.

31. Albeson, P. H. 1983. Waste management. Science 220:1003.

32. New York State Department of Environmental Conservation. 1994. Remedial chronology: The Love Canal hazardous waste site. New York State.

33. Kirschner, E. 1994. Love Canal settlement: OxyChem to pay New York state $98 million. Chemical & Engineering News 72:4–5.

34. Westervelt, R. 1996. Love Canal: OxyChem settles federal claims. Chemical Week 158:9.

35. Whittell, G. 2000 (November 29). Poison in paradise. (London) Times 2, p. 4.

36. U.S. Environmental Protection Agency. 2003. Key Dates in Superfund. Accessed November 6, 2003 at http://www.epa.gove/superfund/action/law/keydates.htm.

37. Bedient, P. B., H. S. Rifai, and C. J. Newell. 1994. Ground water contamination. Englewood Cliffs, N.J.: Prentice-Hall.

38. Huddleston, R. L. 1979. Solid-waste disposal: Land farming. Chemical Engineering 86:119–124.

39. McKenzie, G. D., and W. A. Pettyjohn. 1975. Subsurface waste management. In G. D. McKenzie and R. O. Utgard, eds., Man and his physical environment: Readings in environmental geology, 2nd ed., pp. 150–156. Minneapolis: Burgess Publishing.

40. National Research Council, Committee on Geological Sciences. 1972. The earth and human affairs. San Francisco: Canfield Press.

41. Cox, C. 1985. The buried threat: Getting away from land disposal of hazardous waste. No. 115-5. California Senate Office of Research.

42. Council on Environmental Quality. 1970. Ocean dumping: A national policy: A report to the president. Washington, D.C.: U.S. Government Printing Office.

43. Lenssen, N. 1989 (July–August). The ocean blues. World Watch, pp. 26–35.

44. U.S. Environmental Protection Agency. 2000. Forward pollution protection: The future look of environmental protection. Accessed August 12, 2000, at http://www.epa.gov/p2/p2case. htm#num4.

Chapter 30 Critical Thinking Issue References
Schueller, G. H. 2002. Is recycling on the skids? *Wasting Away on Earth* 24(3):21–23.

PHOTO CREDITS

Chapter 1 Ch Op 1: Jeff Rotman/Stone/Getty Images. Fig. 1.1a: Frans Lanting/Minden Pictures, Inc. Fig. 1.1b: Tom Bean. Fig. 1.4b: Corbis-Bettman. Fig. 1.5a: Peter Turnley/Corbis Images. Fig. 1.5b: Viviane Moos/SABA. Fig. 1.7: National Snow and Ice Data Center/Photo Researchers. Fig. 1.8: Bill Ross/Corbis Images. Fig. 1.9a: Prof. Randall Schaetzl. Fig. 1.9b: J.P. Ferrero/Jacana/Photo Researchers. Fig. 1.1: Courtesy Dan Botkin. Fig. 1.11: Natalie Fobes/Corbis Images. Fig. 1.12: Phil Dyer/iStockphoto.

Chapter 2 Ch Op 2: Tim Laman/NGS/Getty Images, Inc. Fig. 2.1: Phil Schermeister/Corbis Images. Fig. 2.4a: Gideon Mendel/Magnum Photos, Inc. Fig. 2.4b: D. Hudson/Corbis Sygma. Fig. 2.5: Alan Root/Photo Researchers. Fig. 2.6: Dietmar Nill/Nature Picture Library. Fig. 2.11a: AP/Wide World Photos. Fig. 2.11b: U.S. Fish and Wildlife Service. Fig. 2.12: Visuals Unlimited. Fig. 2.13a: Raymond Gehman/Corbis Images. Fig. 2.13b: Breck Kent.

Chapter 3 Ch Op 3: Steve Bloom Images/Alamy Limited. Fig. 3.2: M. Renaudeau/HOA-QUI. Fig. 3.3: Chris Johns. Fig. 3.4: Mark Segal/Stone/Getty Images. Fig. 3.5: Courtesy Ed Keller. Fig. 3.6a: Tim Tadder/AP/Wide World Photos. Fig. 3.7: Ed Keller. Fig. 3.1: Charles A. Mauzy/Stone/Getty Images. Fig. 3.11: Tom Till/DRK Photo.

Chapter 4 Ch Op 4: Dima Korotayev/Pressphotos/Getty Images, Inc. Fig. 4.1: Greg Baker/AP/Wide World Photos. Fig. 4.11: Jacobs Stock Photography/Getty Images, Inc.

Chapter 5 Ch Op 5: Corbis Images. Fig. 5.1a: TerraNova International/Photo Researchers. Fig. 5.1b: Ed Keller. Fig. 5.1: Kim Heacox/Peter Arnold, Inc. Fig. 5.11: Manfred Gottschalk/Tom Stack & Associates. Fig. 5.21: William Felger/Grant Heilman Photography.

Chapter 6 Ch Op 6: Courtesy Dan Botkin. Fig. 6.1f: Alvin E. Staffan/Photo Researchers. Fig. 6.1b: Bill Ivy/Ivy Images. Fig. 6.1a: O. Spielman/CNRI/Phototake. Fig. 6.1c: Jeanne Drake. Fig. 6.1e: Runk/Schoenberger/Grant Heilman Photography. Fig. 6.1d: Scott Nielsen/Bruce Coleman, Inc.. Fig. 6.2a: Courtesy Dan Botkin. Fig. 6.2b: Courtesy Dan Botkin. Fig. 6.3: Farrell Grehan/Photo Researchers. Fig. 6.9: Photographer's Choice/Getty Images. Fig. 6.10: 2000 Airphoto, Jim Wark.

Chapter 7 Ch Op 7: Bonne Pioche/Buena Vista/The Picture Desk. Fig. 7.1: Daniel J. Cox/Photographer's Choice/Getty Images, Inc. Fig. 7.2: The Metropolitan Museum of Art, Gift of John D. Rockefeller, Jr., 1937 (37.80.6) Photograph 1993 The Metropolitan Museum of Art. Fig. 7.2a: Oliver Meckes/Photo Researchers. Fig. 7.2b: Oliver Meckas/Photo Researchers. Fig. 7.3c: Adam Jones/Photo Researchers. Fig. 7.7a: John Weinstein/The Field Museum. Fig. 7.7b: Mark A. Klinger/CMNH. Fig. 7.8a: Dr. Patricia Schultz/Photo Researchers. Fig. 7.8b: Kari Lounatmaa/Photo Researchers.

Chapter 8 Ch Op 8: Bob DeGross/NPS. Fig. 8.1a: Jeff Gynane/iStockphoto. Fig. 8.1b: Courtesy Dan Botkin. Fig. 8.5a: Stephen J. Krasemann/DRK Photo. Fig. 8.5b: Kim Heacox/DRK Photo. Fig. 8.5c: David Hosking/Alamy Images. Fig. 8.7a: Ferrero/Labat/Auscape International Pty. Ltd. Fig. 8.7b: Toni Angermayer/Photo Researchers. Fig. 8.7c: Gary Unwin/iStockphoto. Fig. 8.16: John Shaw/Bruce Coleman, Inc. Fig. 8.17: Charles Glatzer/The Image Bank/Getty Images. Fig. 8.18: Tom & Pat Leeson. Fig. 8.19: Tom & Pat Leeson/Photo Researchers. Fig. 8.20: Frans Lanting Minden Pictures, Inc. Fig. 8.21: Steve Kaufman/Peter Arnold, Inc. Fig. 8.22: Bill Ivy/Ivy Images. Fig. 8.23: Richard A. Cooke/Corbis Images. Fig. 8.24: William Johnson/Stock Boston. Fig. 8.25: F. Denhz/BIOS/Peter Arnold, Inc.

Chapter 9 Ch Op 9: Gideon Mendel/Corbis Images. Fig. 9.1: Antony Njuguna/Reuters/Landov LLC. Fig. 9.3a: Darrell Gulin/DRK Photo. Fig. 9.3b: Darrell Gulin/DRK Photo. Fig. 9.4: Mona Lisa Production/Science Photo Library. Fig. 9.6a: Daniel Botkin. Fig. 9.6b: Daniel Botkin. Fig. 9.6c: Daniel Botkin. Fig. 9.6d: Daniel Botkin. Fig. 9.9a: Theo Allofs/Getty Images, Inc. Fig. 9.9b: Tom and Pat Lesson. Fig. 9.10: Peter Batson/Image Quest Marine.

Chapter 10 Ch Op 10: Albert Bierstadt, Sunset in the Yosemite Valley, 1868, The Haggin Museum. Fig. 10.1a: Nik Wheeler/Corbis Images. Fig. 10.1b: Norman Einstein. Fig. 10.2: Courtesy NASA Goddard Space Flight Center and the authors of Hall, F.G., D.B. Botkin, D.E. Strbel, K.D. Woods and S.J. Goetz, 1991, Large Scale Patterns in Forest Succession As Determined by Remote Sensing, Ecology, 72: 628-640. Fig. 10.4a: Masha Nordbye/Bruce Coleman, Inc. Fig. 10.4b: © Grant Heilman Photography. Fig. 10.5a: Breck Kent/Animals Animals NYC. Fig. 10.5b: Michael P. Gadomski/Animals Animals NYC. Fig. 10.5c: Michael P. Gadomski/Animals Animals NYC. Fig. 10.6: Terry Donnelly Tom Stack & Associates. Fig. 10.7: Daniel Botkin. Fig. 10.12: David Tomlinsin/Windrush Photos.

Chapter 11 Ch Op 11: Peter Dean Agripicture Images/Alamy Images. Fig. 11.1b: Stone/Getty Images. Fig. 11.2: John Vink/Magnum Photos, Inc. Fig. 11.5: Mark A. Ernste/UNEP GRID, Sioux Falls (http:/na.unep.net). Fig. 11.7: AP/Wide World Photos. Fig. 11.8a: Kevin Morris/Stone/Getty Images. Fig. 11.8b: JC Carton/Bruce Coleman, Inc.. Fig. 11.8c: Thomas Hovland/Grant Heilman Photography. Fig. 11.13: Doug Plummer/Photo Researchers. Fig. 11.15: J. Victolero/International Rice Research Institute. Fig. 11.19: EPA/Pedro Armestre/Corbis Images. Fig. 11.2: AJAY VERMA/Reuters/Landov LLC.

Chapter 12 Ch Op 12: Courtesy of Conor Watkins. Fig. 12.1: Courtesy Steve Burr. Fig. 12.2a: Lynn Betts. Fig. 12.2b: Courtesy Dr. Paul McDaniel, University of Idaho. Fig. 12.3a: Corbis-Bettmann. Fig. 12.3b: U.S. Dept. of Agriculture Photography Ctr. Fig. 12.4: Bettman/Corbis Images. Fig. 12.5a: Science Vu/Visuals Unlimited. Fig. 12.5b: K-State Research and Extension. Fig. 12.8a: David Epstein, Integrated Pest Management Program, Michigan State University. Fig. 12.8b: OPIE/Remi Coutin. Fig. 12.12a: Simko/Visuals Unlimited. Fig. 12.12b: Arthur C. Smith III/Grant Heilman Photography. Fig. 12.12c: Bob Gurr/DRK Photo. Fig. 12.13: FAO/17463/A. Odoul. Fig. 12.14: R. de la Harpe/Biological Photo Service/PO. Fig. 12.16a: Bill Bachman/Photo Researchers. Fig. 12.16b: Grant Heilman Photography. Fig. 12.18: Mark Gibson/Alamy Images.

Chapter 13 Ch Op 13: Dave Welling. Fig. 13.1a: NASA Earth Observatory/MODIS Rapid Response team. Fig. 13.1b: AP/Wide World Photos. Fig. 13.2a: Dave G. Houser/Post-Houserstock/Corbis Images. Fig. 13.3: AP/Wide World Photos. Fig. 13.4: David Muench Photography. Fig. 13.8: Tom Bean/DRK Photo. Fig. 13.10: Tony Arruza/Corbis Images. Fig. 13.15a: Steve McCurry/Magnum Photos, Inc. Fig. 13.15b: Robert Frerck/Odyssey Productions. Fig. 13.16a: NASA/Science Source/Photo Researchers. Fig. 13.16b: Gregory G. Dimijian, M.D./Photo Researchers. Fig. 13.17: Chee-Onn Leong/iStockphoto. Fig. 13.18: Stephen Krasemann/Nature Conservatory/Photo Researchers. Fig. 13.19: U.S. Department of Interior, Fish and Wildlife Services.

Chapter 14 Ch Op 14a. Ch Op 14b: age fotostock/SUPER-STOCK. Ch Op 14c: Courtesy NOAA. Fig. 14.2a: Florida Keys National Marine Sanctuary Staff/Courtesy NOAA. Fig. 14.2b: Courtesy Lance Horn, National Undersea Research Center/University of North Carolina at Wilmington. Fig. 14.5: PhotoDisc, Inc. Fig. 14.5a: Tom Bledsoe/DRK Photo. Fig. 14.5b: National Mueum of American Art, Washington, D.C./Art Resource, N.Y. 1985.66.415: George Catlin, Buffalo Chase, Mouth of the Yellowstone, 1832-33. Fig. 14.6a: Ken Lucas/Visuals Unlimited. Fig. 14.7a: Richard Ellis/Photo Researchers. Fig. 14.7b: Marc Epstein/DRK Photo. Fig. 14.8a: Francois Gohier/Photo Researchers. Fig. 14.8b: Seapics.com. Fig. 14.10a: Al Grillo/Alaska Stock Images/PictureQuest. Fig. 14.10b: Jeff Rotman/www.jeffrotman.com. Fig. 14.10c: Steven Kazlowsk/AlaskaStock. Fig. 14.12a: Courtesy NOAA. Fig. 14.15: John Cunningham/Visuals Unlimited. Fig. 14.16a: © AP/Wide World Photos. Fig. 14.16b: Ross Frid/Visuals Unlimited. Fig. 14.17a: Richard Elliott/Stone/Getty Images. Fig. 14.17b: Tom & Pat Leeson/DRK Photo. Fig. 14.17c: Dianne Blell/Peter Arnold, Inc. Fig. 14.18a: Gilbert Grant/Photo Researchers. Fig. 14.18b: Photo Researchers.

Chapter 15 Ch Op 15: Stephen Dalton/Photo Researchers. Fig. 15.2a: T. Orban/Corbis Sygma. Fig. 15.2b: David & Peter Turnley/Corbis Images. Fig. 15.3: Prof. Ed Keller. Fig. 15.4a: Bill Brooks/Masterfile. Fig. 15.4b: Mike Grandmaison Photography. Fig. 15.7: O. Franken/Corbis Sygma. Fig. 15.8c: Martin Bond/Photo Researchers. Fig. 15.9: Michael Yamashita. Fig. 15.15: © AP/Wide World Photos.

Chapter 16 Ch Op 15: Courtesy NOAA. Fig. 16.2: REUTERS/Allen Fredrickson/Landov LLC. Fig. 16.3: Bill Haber/AP/Wide World Photos. Fig. 16.7: Douglas C. Pizac/AP/Wide World Photos. Fig. 16.8: AP/Wide World Photos. Fig. 16.1: Frank Balthis/Courier. Fig. 16.11b: USGS/Earth Surface Processes Team. Fig. 16.15: John Russell/Zuma Press. Fig. 16.16a: Digital Globe/Getty Images, Inc. Fig. 16.16b: Digital Globe/Getty Images, Inc. Fig. 16.17a: Roger Ressmeyer/Corbis Images. Fig. 16.17b: Layne Kennedy/Corbis Images. Fig. 16.18a: N. Banks/USGS/Earth Surface Processes Team. Fig. 16.18b: Jacques Langevin/Corbis Sygma. Fig. 16.20a: Courtesy Ed Keller. Fig. 16.20b: Courtesy Ed Keller.

Chapter 17 Ch Op 17: AP/Wide World Photos. Fig. 17.1: Taxi/Getty Images.

Chapter 18 Ch Op 18: REUTERS/Ceerwan Aziz/Landov LLC. Fig. 18.7a: George Hunter/Stone/Getty Images. Fig. 18.7b: Ken Graham/Bruce Coleman, Inc. Fig. 18.8: U.S. Fish and Wildlife Service/Liaison Agency, Inc./Getty Images. Fig. 18.9b: © AP/Wide World Photos. Fig. 18.9c: U.S. Fish and Wildlife Service/Liaison Agency, Inc./Getty Images. Fig. 18.13: AP/Wide World Photos. Fig. 18.14: © Grant Heilman Photography. Fig. 18.15a: Prof. Ed Keller. Fig. 18.15b: Prof. Ed Keller. Fig. 18.16: William P. Hines/The Scranton Times Tribune Library.

Chapter 19 Ch Op 19: Courtesy Spirit Lake Community School District. Fig. 19.3a: Tom Bean. Fig. 19.6a: H. Gruyaeart/Magnum Photos, Inc. Fig. 19.6b: Prof. Ed Keller. Fig. 19.7: T. J. Florian/Rainbow. Fig. 19.8a: Courtesy Luz International. Fig. 19.11: C. Delis/Explorer. Fig. 19.13: Glen Allison/Stone/Getty Images. Fig. 19.15: Courtesy Pacific Gas & Electric Company.

Chapter 20 Ch Op 20: George D. Lepp/Corbis Images. Fig. 20.4a: Graham Finlayson/Stone/Getty Images. Fig. 20.4b: Martin Bond/Science Photo Library. Fig. 20.5b: Roger Ressmeyer/Starlight/Corbis Images. Fig. 20.8: Courtesy the Princeton Plasma Physics Laboratory. Fig. 20.1: Corbis-Bettmann. Fig. 20.15: Igor Kostin/Corbis Sygma.

Chapter 21 Ch Op 21: oote boe/Alamy Images. Fig. 21.12a: Stephen Krasemann/Stone/Getty Images. Fig. 21.12b: Gregory G. Dimijian, M.D./Photo Researchers. Fig. 21.12c: Jim Brandenburg/Minden Pictures, Inc. Fig. 21.14: Courtesy Ed Kller. Fig. 21.15: Liu Liqun/Corbis Images. Fig. 21.17a: Sterling Dimmitt. Fig. 21.17b: Courtesy Jacksonville Corp. of Engineers. Fig. 21.2: Rich Buzzelli/Tom Stack & Associates.

Chapter 22 Ch Op 22: John Althouse/Liaison Agency, Inc./Getty Images. Fig. 22.1b: AP/Wide World Photos. Fig. 22.2: Ben Osborne/Stone/Getty Images. Fig. 22.4: Courtesy Ed Keller. Fig. 22.5: McAllister/Liaison Agency, Inc./Getty Images. Fig. 22.6: William E. Ferguson. Fig. 22.8: Michelle Barnes/Liaison Agency, Inc./Getty Images. Fig. 22.1: Charles Mason/Black Star. Fig. 22.11: Jim Strawser/Grant Heilman Photography. Fig. 22.12: John Cancalosi/DRK Photo. Fig. 22.13: David Woodfall/DRK Photo. Fig. 22.14: © AP/Wide World Photos. Fig. 22.22a: Courtesy John Day, Louisiana State University. Fig. 22.22b: Courtesy John Day, Louisiana State University. Fig. 22.22c: Courtesy John Day, Louisiana Stae University. Fig. 22.23b: © Integrated Water Systmen, Inc. Fig. 22.24: Prof. Ed Keller.

Chapter 23 Ch Op 23: Johnny Johnson/DRK Photo. Fig. 23.2: Roger Ressmeyer/Corbis Images.

Chapter 24 Ch Op 24: Hulton/Archive by Getty Images. Fig. 24.1a: Courtesy NASA. Fig. 24.1b: Jason Reed/REUTERS/Liaison Agency, Inc. Fig. 24.2: F. Hoffman/The Image Works. Fig. 24.3: Courtesy Ed Keller. Fig. 24.4: Jules Bucher/Photo Researchers. Fig. 24.12a: Don & Pat Valenti/Stone/Getty Images. Fig. 24.12b: Don & Pat Valenti/Stone/Getty Images. Fig. 24.13: Bob Carey/Los Angeles Times. Fig. 24.19a: Jim Mendenhall. Fig. 24.19b: Jim Mendenhall.

Chapter 25 Ch Op 25: © AP/Wide World Photos. Fig. 25.1: Blanche/Liaison Agency, Inc./Getty Images. Fig. 25.3a: Andrew Syred/Photo Researchers. Fig. 25.3b: Oliver Meckes/Photo Researchers.

Chapter 26 Ch Op 26: Peter Cade/Stone/Getty Images. Fig. 26.1: Courtesy NASA.

Chapter 27 Ch Op 27: Courtesy of Fossil Trace Golf Course. Fig. 27.1: Helen Thompson/Animals Animals Earth Scenes. Fig. 27.5: Royce Bair/ProFiles West, Inc. Fig. 27.6: Ed Keller. Fig. 27.7a: Stone/Getty Images. Fig. 27.7b: Craig Aurness/Corbis Images. Fig. 27.7c: R. Maissonneuve/Publiphoto/Photo Researchers. Fig. 27.7d: Stone/Getty Images. Fig. 27.7e: Martin Miller. Fig. 27.8: Courtesy Ed Keller.

Chapter 28 Ch Op 28a: Theresa Parker. Ch Op 28b: Washington State Historical Society, Curtis, 1220. Fig. 28.1: New Bedford Whaling Museum. Fig. 28.2: The New Bedford Whaling Museum. Fig. 28.3b: Mark Spalding, PhD. Fig. 28.3c: Mark Spalding, PhD. Fig. 28.5: C. Bradley Simmons/Bruce Coleman, Inc.

Chapter 29 Ch Op 29a: Andre Jenny/Alamy Images. Ch Op 29b: AP/Wide World Photos. Fig. 29.1a: Alamy Images. Fig. 29.1b: Marc Serota/Reuters/Landov LLC. Fig. 29.2a: AP/Wide World Photos. Fig. 29.2b: Robert Kaufmann/FEMA Press. Fig. 29.3: Tim Vasquez/Weather Graphics. Fig. 29.5a: U. S. Army Corps of Engineers. Fig. 29.5b: Thinkstock/Getty Images. Fig. 29.6: Pizzoli Alberto/Corbis Sygma. Fig. 29.8: Collection of ESL Information Services, Engineering Societies Library, New York City. Fig. 29.9: Patrick Ward/Corbis Images. Fig. 29.13b: Jeri Gleiter/Peter Arnold, Inc. Fig. 29.14: Robert Holmes/Corbis Images. Fig. 29.17: Ralph Ginzburg/Peter Arnold, Inc.

Chapter 30 Ch Op 30: Newsmarkers/Getty Images. Fig. 30.1: Norman Ng. Fig. 30.3: Alex Quesada/Matrix International, Inc. Fig. 30.6b: Courtesy John H. Kramer. Fig. 30.7: Courtesy NY State Dept. of Environmental Conservation. Fig. 30.8: Courtesy of Gloucester Fire Service & Sandhurst Area Action Group. Fig. 30.9: Courtesy of Gloucester Fire Service & Sandhurst Area Action Group. Fig. 30.15: Courtesy of Cynthia Vanderlip/Algalita Marine Research Foundation.

INDEX